IC SCHEMATIC SOURCEMASTER

IC SCHEMATIC SOURCEMASTER
Kendall Webster Sessions

John Wiley & Sons

New York　　Chichester　　Brisbane　　Toronto

Copyright © 1978, by John Wiley & Sons, Inc.

All rights reserved. Published simultaneously in Canada.

Reproduction or translation of any part of this work beyond that permitted by Sections 107 or 108 of the 1976 United States Copyright Act without the permission of the copyright owner is unlawful. Requests for permission or further information should be addressed to the Permissions Department, John Wiley & Sons, Inc.

Library of Congress Cataloging in Publication Data:

Sessions, Ken W
   IC schematic sourcemaster.

  1. Integrated circuits. I. Title.
TK7874.S444   621.381'73   77-13404
ISBN 0-471-02623-9

Printed in the United States of America

10 9 8 7 6 5 4 3 2

PREFACE

When the integrated circuit first appeared, many of us then involved in electronics were naive enough to think that the day of the design engineer was over, that developmental work would become the exclusive province of engineering specialists whose function it was to integrate those circuits traditionally constructed with discrete devices. We were wrong. The integrated circuit (IC) has become a module of sorts, to be used in much the same fashion as the transistor; the chief difference is simply that the IC allows more functions to be performed within a single modular block of an overall functioning electronic "system."

Perhaps the most rewarding fallout advantage of the IC is that circuit economy is no longer a design constraint. When we work with ordinary discrete devices we try to accomplish as many needed functions as possible using as few active components as possible, since adding a stage means adding the cost of the transistors required to build the stage as well as the cost of the passive components that have to be used with it. But an IC of a given class costs about the same as other ICs in that class, whether it contains the equivalent of a hundred transistors or a thousand. And that cost is not the least bit expensive by any standards. A 2000-transistor IC does not cost 2000 times as much as a single transistor; it does not even cost 200 times that—or even 20. We simply do not have to worry about the number of transistors in an IC—and that means we do not have to worry about circuit complexity at all as long as our circuit design stays within the bounds of a single IC.

This book contains some 1500 schematic diagrams for electronic circuits built around integrated packages. So versatile are the IC devices presented here that in many cases a single package is used in a wide variety of differing applications—some of these applications that were not even conceived at the time the IC itself was developed. It is this "black box" circuit design concept that keeps electronics interesting indeed; it is the very element that has opened up an entirely new vista in design engineering.

Today it is accepted practice to employ *digital* ICs to perform linear functions. And it is perfectly correct to choose an IC for a job, even though perhaps only half the active devices within the IC are actively functional in the final circuit design.

Integrated circuits are sometimes interchangeable, particularly when designed for applications using the devices as originally intended. But many of the schematics presented here are representative of functions for which the devices employed were *not* specifically designed. A specific Motorola dual JK flip-flop, for example, is an IC designed to accomplish the functions for which it was named within an overall digital framework. But in this book the device is also used as a code practice oscillator (a linear function). However, it is not certain whether another manufacturer's dual JK flip-flop may be employed in a similar linear application using the given schematic. Therefore, every schematic in this book contains the part numbers assigned by the contributing manufacturer. I strongly recommend that you use devices bearing the numbers shown rather than attempt to treat these circuits generically.

In the main, the circuits themselves have been contributed by those who were responsible for their successful design and manufacture. Companies like National Semiconductors, RCA, and Motorola employ design specialists to create IC packages performing specific functions. Using discrete devices, these would require many individual stages and an even larger number of separate transistors, diodes, resistors, and the like. But once the ICs begin to come off the production line, other engineers in the same companies take on the task of making these devices do several other jobs. As you will see when you begin to put some of these remarkable devices to work, the effort put forth by these firms has really paid off for all of us!

To make this book easier to use, the schematics have been categorized into broad circuit classes. But that does not mean that you will not be able to adapt a circuit of one class into an application of another. As an example, the section containing audio schematics includes stereo preamplifiers, amplifiers, and related signal-conditioning circuits such as bass and treble controls and equalizers. But you will find other tone-control circuits and equalization networks in the section on filters. If you do not find what you need in the section where you think it is most apt to appear, spend a few minutes studying other sections that seem reasonably appropriate.

Wherever it could be done without compromise, schematics have been included with no accompanying text other than the briefest descriptive caption. However, since many of the devices employed in the individual circuits featured in this book are adapted to functions other than the IC designer intended, some circuit explanations seemed in order. But textual details have been minimized to keep you from having to plow through material you do not need and to leave room for more circuits.

The more astute reader will recognize certain style irregularities and inconsistencies. A digital-to-analog converter might be called a D/A converter on one page and a DAC on another. An IC amplifier might be shown as a triangular symbol on one diagram and as a box or circle with pin numbers on another. In the interest of accuracy, we have retained the terminology and conventions used by the original contributors of the schematics. Photographically reproducing the original serves to eliminate the error factor but does introduce some inconsistencies, as noted.

Experienced electronic circuit constructionists may recognize the IC makers by the part numbers—the letter L, for example, is a typical National prefix; M identifies Motorola; CA is RCA's number prefix; μA belongs to Fairchild Camera and Instrument Corporation; SN to Texas Instruments; TA to Amperex; IR to Workman Electronics (International Rectifier), etc. Lesser known manufacturers are identified on the appropriate diagrams as to IC source.

I am extremely grateful to the various manufacturers listed below who cooperated with me in sending publications and data sheets to prepare this circuit collection. It is they who have gone to the effort and expense of exploiting IC packages that might otherwise be destined to single purpose functions. And they have given us the basis of what I believe is the most valuable electronics "recipe" book ever published. In alphabetical order they are: Amperex, Solid State and Active Devices Division; Fairchild, Semiconductor Division; General Instrument, Semiconductor Components Division; GTE/Sylvania; Motorola Semiconductor Products Inc.; National Semiconductors, Inc.; RCA, Solid State Division; Signetics, Incorporated; Siliconix, Incorporated; Texas Instruments, Incorporated; Workman Electronics (source of devices preceded by IR designator)

Kendall Webster Sessions

CONTENTS

SECTION 1
VOLTAGE, CURRENT, AND POWER SOURCES   1

SECTION 2
FOLLOWERS, BUFFERS, AND SPECIAL-PURPOSE AMPLIFIERS   65

SECTION 3
INSTRUMENTATION AND TEST EQUIPMENT CIRCUITS   105

SECTION 4
WAVEFORM GENERATORS AND SHAPERS   145

SECTION 5
TELEVISION AND VIDEO CIRCUITS   189

SECTION 6
AUDIO CIRCUITS   223

SECTION 7
COMMUNICATIONS CIRCUITS   281

SECTION 8
SENSING, MONITORING, CONTROLLERS, AND POWER DRIVERS   339

SECTION 9
TIMERS   373

SECTION 10
AUTOMOTIVE CIRCUITS   383

SECTION 11
ACTIVE FILTERS   399

SECTION 12
OPTOELECTRONIC CIRCUITS   421

SECTION 13
ANALOG AND DIGITAL CONVERTERS   429

SECTION 14
DATA ACQUISITION, MULTIPLEXING, AND TRANSMISSION   445

SECTION 15
LOGIC AND FUNCTIONALLY RELATED CIRCUITS   485

SECTION 16
HOBBY, FUN, AND EXPERIMENTER CIRCUITS   519

INDEX   543

IC SCHEMATIC SOURCEMASTER

SECTION 1
VOLTAGE, CURRENT, AND POWER SOURCES

Every electronic circuit requires a power supply; thus, this section begins with an examination of this common circuit function. Bear in mind that the circuits that show positive voltages referred to ground can be reversed in polarity by simply turning all polarized devices (electrolytics, diodes) around. In some cases part values have been omitted from circuits because of the noncritical nature of the value in the application shown. For example, the diagrams of basic power supplies give no part numbers for the diodes, no capacitance values for the electrolytics, no inductance values for the chokes, and no winding information for transformer primaries and secondaries. Only the circuits themselves are shown; you may use large capacitance and inductance values for heavy filtering or low values for light filtering. In those circuits where part value is important to maintain a specific time constant or to achieve a specific amount of filtering with a stipulated input voltage, such values are given.

This section also includes circuits for voltage and current references, current and voltage regulators, inverters and switching regulators, current sinks, battery chargers, and other circuits of comparable functions. The 150 or so circuits in this section are not the only power sources and regulators presented in this book; many of the subsequent sections include integral power supplies, overload protection circuitry, regulation systems, and the like. If you're looking for a power source to fulfill a specific need, be sure to check those circuits included in the section that carries the schematic for the device you intend to power.

Basic power supplies.

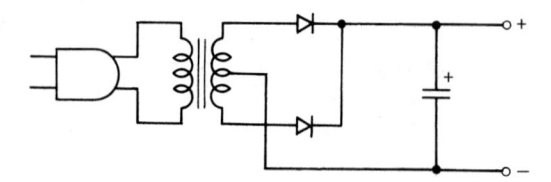

Full-wave, capacitor input

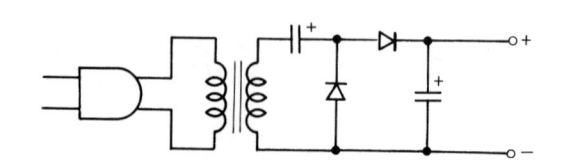

Half-wave doubler, capacitor input

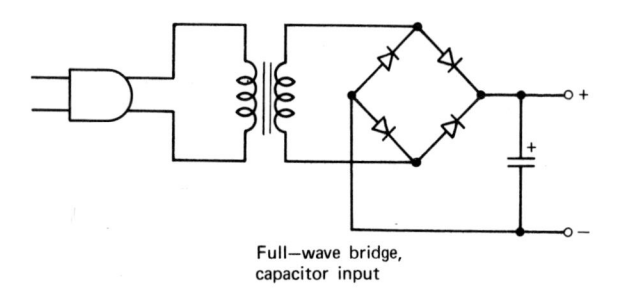

Full-wave bridge, capacitor input

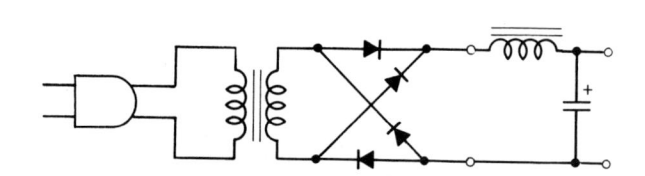

Full-wave bridge, choke input

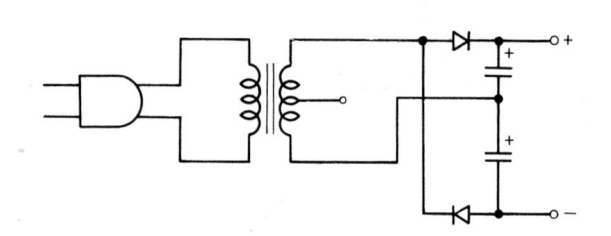

Full-wave doubler, capacitor input

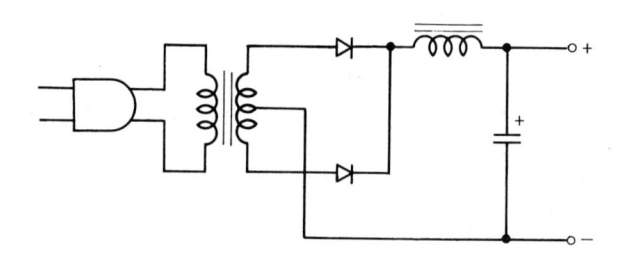

Full-wave, choke input

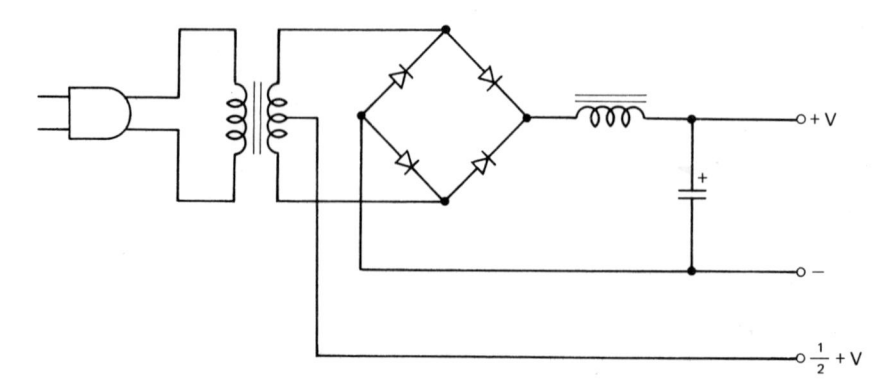

Two-output bridge, choke input

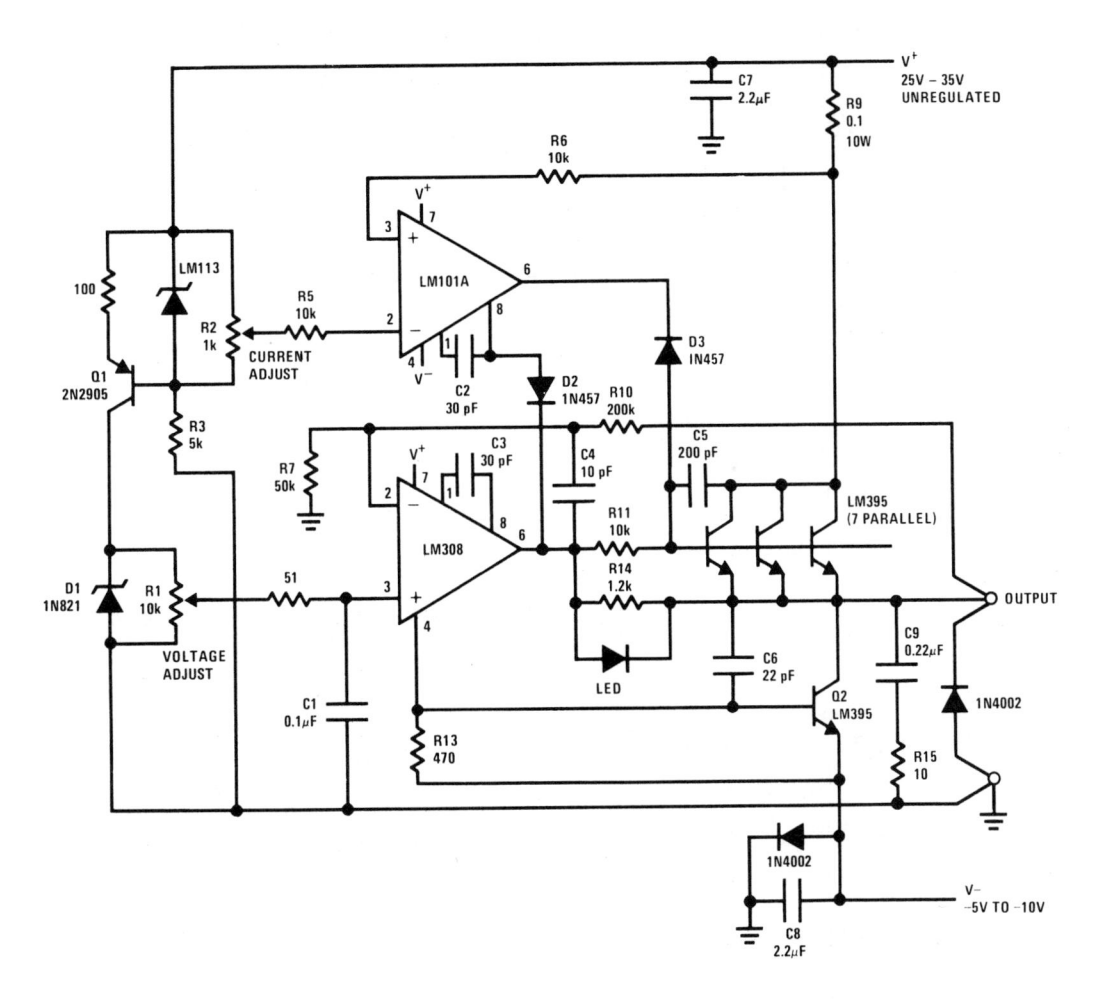

A general-purpose lab-type constant-voltage, constant-current power supply is easily made using standard integrated circuits. The circuit shown will provide up to 25V at up to 10A output, with both the output voltage and current adjustable to zero. Although relatively simple, very high performance is obtained. The LM395 acts as a 2A thermally limited high-gain power transistor. Since only a maximum of 10 μA is needed to drive the pass elements and complete overload protection is included on the chip, the external biasing and protection circuit is minimized. Only two control op-amps are needed— one for voltage control and one for current control.

In constant-voltage operation, a reference voltage is fed from pot R1 through a high-frequency filter into the noninverting input of an LM308 op-amp. The output of the LM308 drives seven paralleled LM395s as emitter followers to obtain a 10A capability.

When the circuit is used in the constant-current mode, the LM101A overcomes the constant-voltage loop to control the output. Output current is sensed in R9 and compared with the voltage between V+ and the arm of R2. R2 is connected across an LM113 low voltage reference diode to provide a 0V to 1.2V reference for 0 to 12A output. When the current increases to the control point, the output of the LM101A swings negative and decreases the drive to the output pass devices through D3, limiting the current. No separate positive supply is needed since the common-mode operative range of the LM101A is equal to the positive supply. Diode D2 clamps the output of the LM101A when it is not regulating, decreasing the switchover time from voltage to current mode operation. The LED lights up during current limit. Capacitors C7 and C8 should be solid tantalum.

SECTION 1 *Voltage, Current and Power Sources*

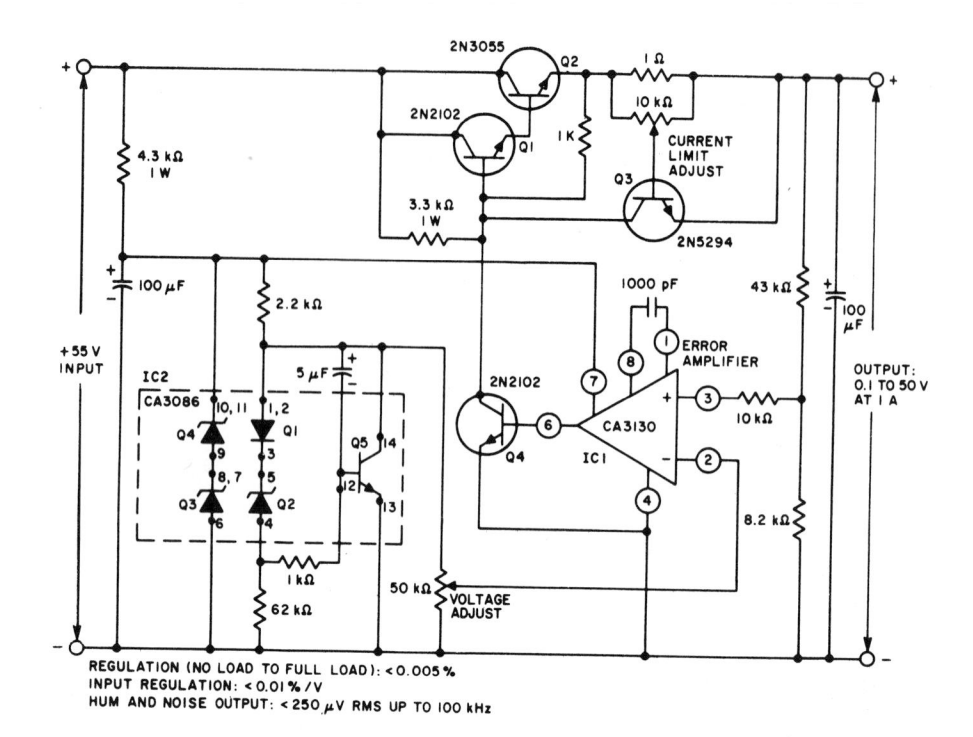

Regulated power supply capable of providing regulated output voltage by continuous adjustment over the range from 0.1 to 50V and currents up to 1A. The error amplifier (IC1) and circuitry associated with IC2 function normally, although the output of IC1 is boosted by a discrete transistor (Q4) to provide adequate base drive for Darlington-connected series-pass transistors Q1 and Q2.

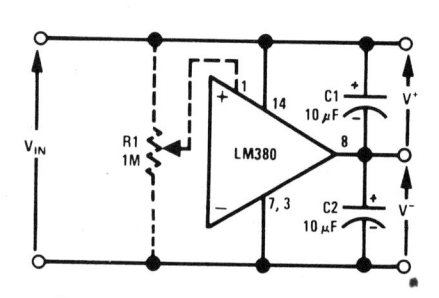

Low-cost dual supply. This circuit demonstrates a minimum-parts-count method of symmetrically splitting a supply voltage. Unlike the normal R, C, and power zener diode technique, the LM380 circuit does not require a high standby current and power dissipation to maintain regulation. With a 20V input voltage (±10V output), the circuit exhibits a change in output voltage of approximately 2% per 100 mA of unbalanced load change. Any balanced load change will reflect only the regulation of the source voltage V_{IN}.

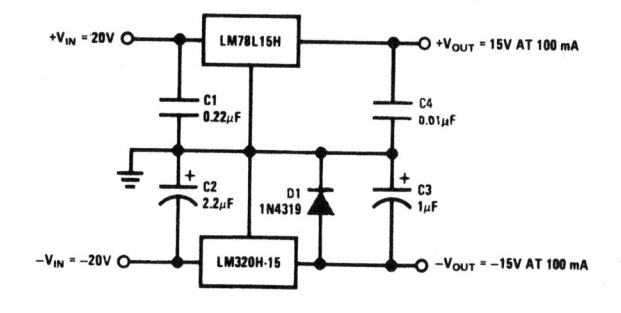

±15V, 100 mA dual power supply. Capacitors C2 and C3 should be solid tantalum.

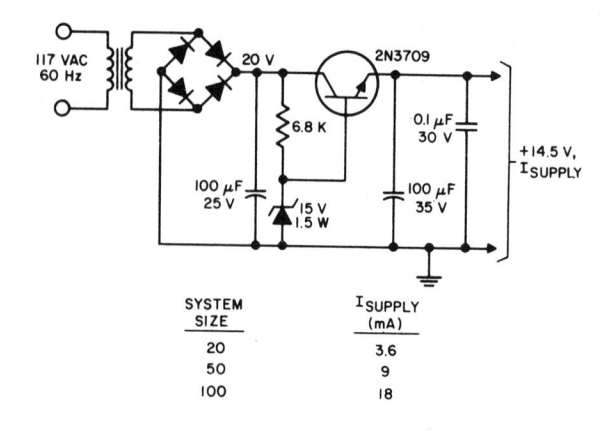

SYSTEM SIZE	I_{SUPPLY} (mA)
20	3.6
50	9
100	18

Power supply for RCA CMOS yielding highest obtainable noise immunity.

IC SCHEMATIC SOURCEMASTER

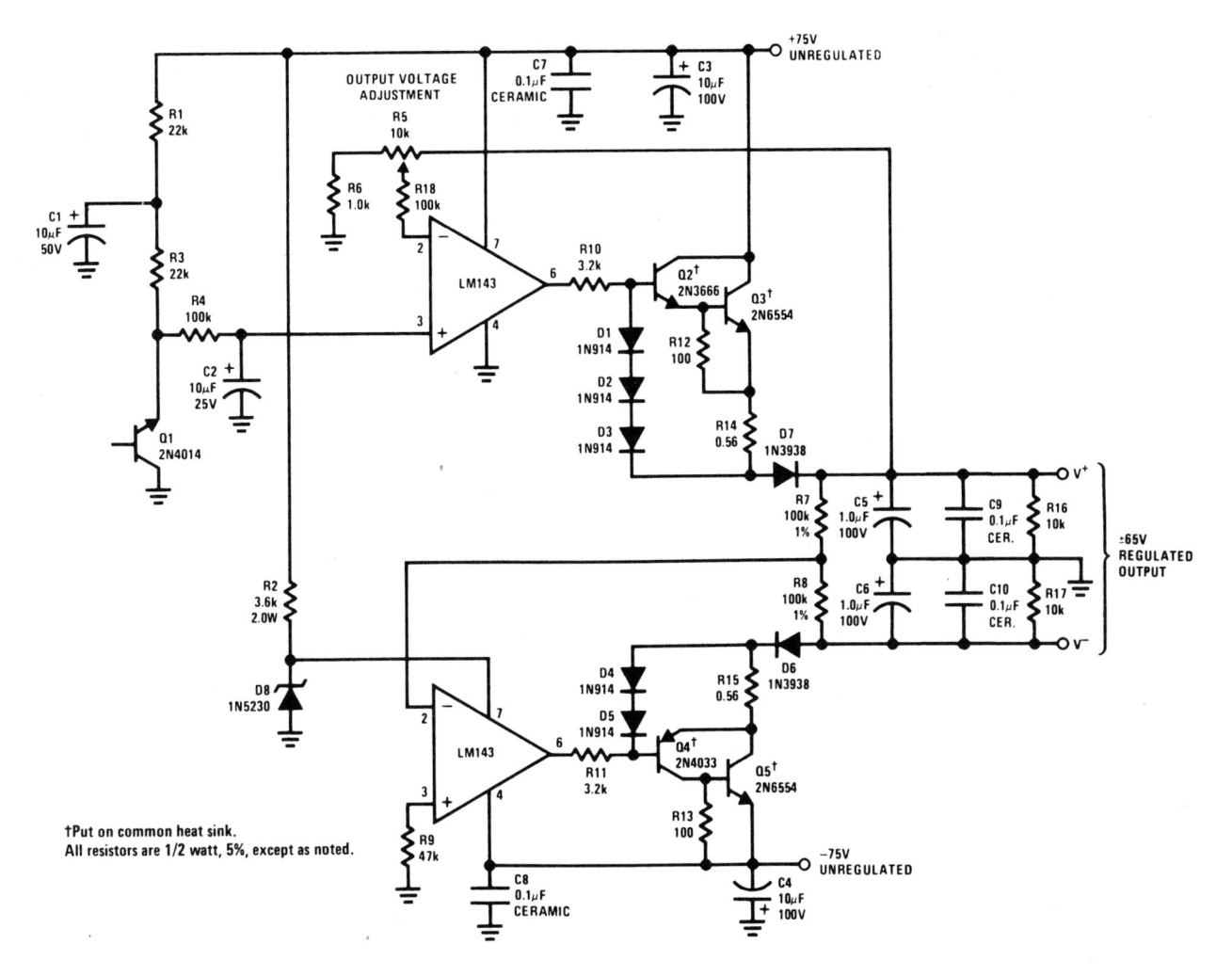

Tracking ±65V, 1A power supply with short-circuit protection. Units can be destroyed by any combination of high ambient temperature, high supply voltages, and high power dissipation that results in excessive die temperature. This is also true when driving low-impedance or reactive loads or loads that can revert to low impedance; for example, the LM143 can drive most general-purpose operational amplifiers (op-amps) outside the maximum input voltage range, causing heavy current to flow and possibly destroying both devices.

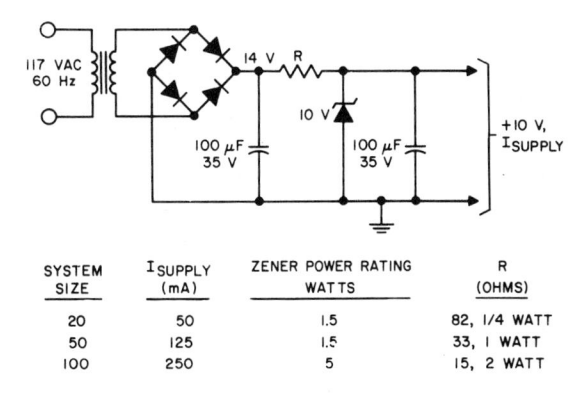

SYSTEM SIZE	I_{SUPPLY} (mA)	ZENER POWER RATING WATTS	R (OHMS)
20	50	1.5	82, 1/4 WATT
50	125	1.5	33, 1 WATT
100	250	5	15, 2 WATT

Power supply for mixed system consisting of 50% at 2 MHz and 50% at 50 kHz (CMOS circuits).

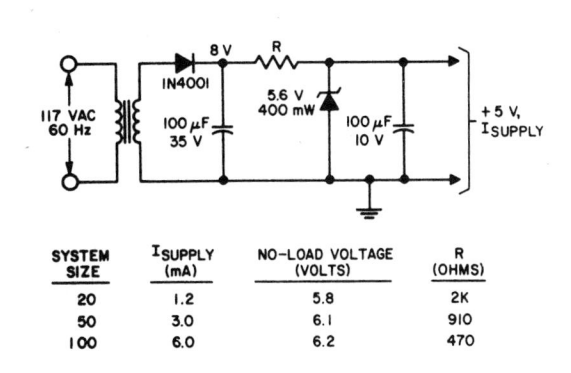

SYSTEM SIZE	I_{SUPPLY} (mA)	NO-LOAD VOLTAGE (VOLTS)	R (OHMS)
20	1.2	5.8	2K
50	3.0	6.1	910
100	6.0	6.2	470

Power supply for a low-frequency RCA CMOS system.

SECTION 1 *Voltage, Current and Power Sources*

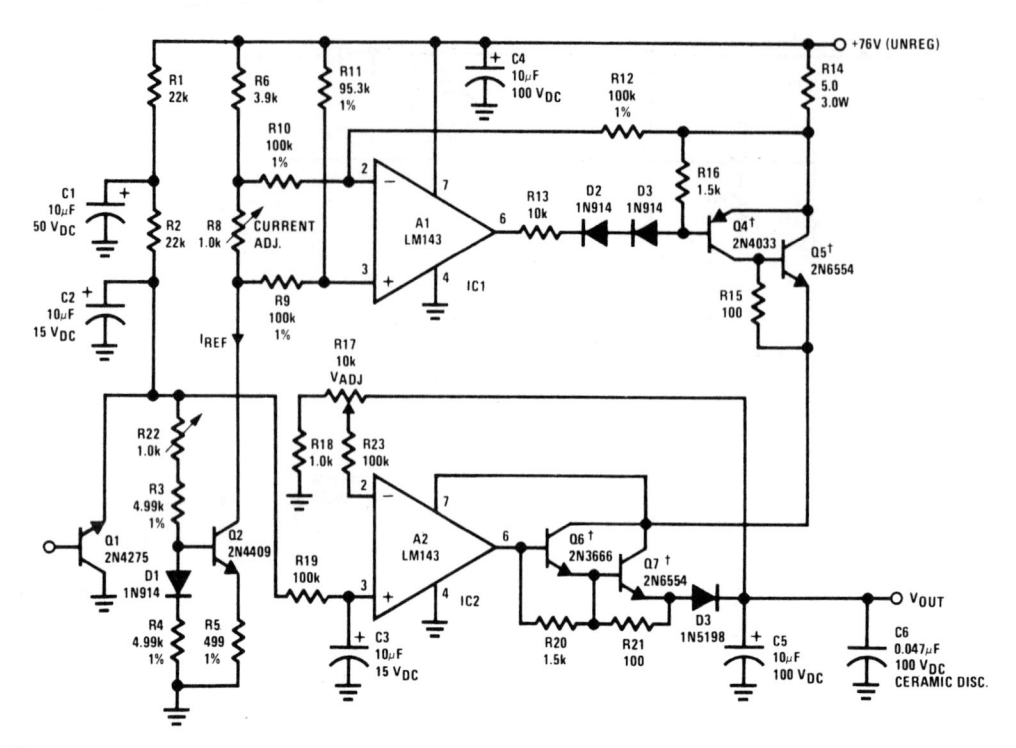

†Put on common heat sink, Thermalloy 6006B or equivalent.
All resistors 1/2W, 10% unless otherwise noted.
All capacitors 20%.

A +65V, 1.0A power supply with continuously variable output current and voltage. The power supply circuit is a nonideal voltage source in series with a nonideal current source. A reference voltage of approximately 6.5V is obtained by "zenering" the base-emitter junction of the 2N4275. The positive temperature coefficient of the zener voltage is compensated by the negative temperature coefficient of the forward-biased base-collector junction. The output of the voltage reference goes to the variable-gain power amplifier (IC2, Q6, Q7) and to a reference current source made up of Q2 and D1. The variable-gain power amplifier multiplies the reference voltage from one to ten times as a result of the variable-feedback resistor, R17. Since the maximum current output of IC2 is at most 20 mA, the Darlington-connected Q6 and Q7 devices are used to boost the available output current to 500 mA.

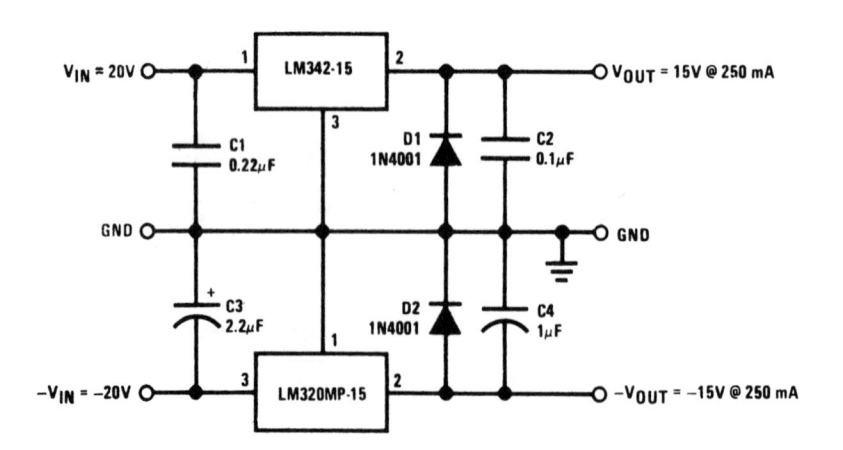

±15V, 250 mA dual power supply. Capacitors C3 and C4 should be solid tantalum.

IC SCHEMATIC SOURCEMASTER

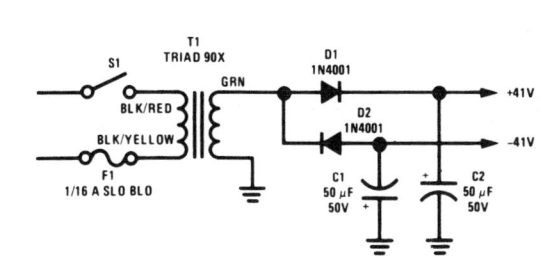

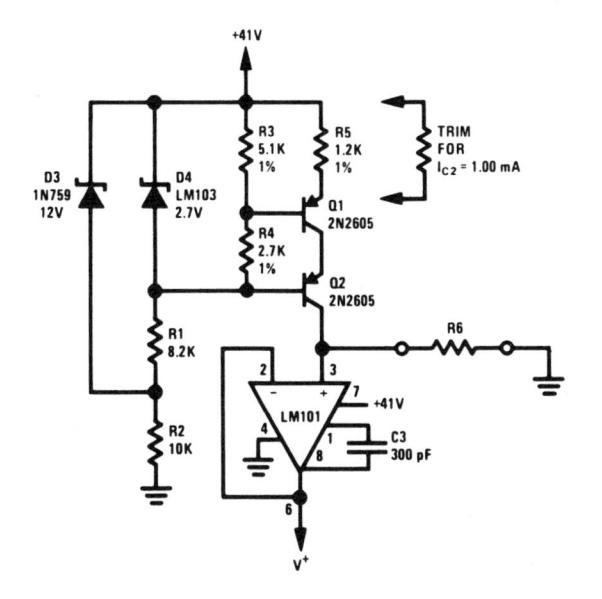

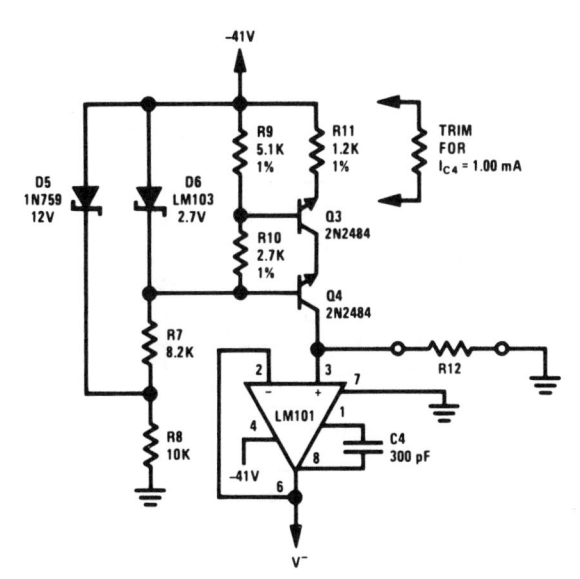

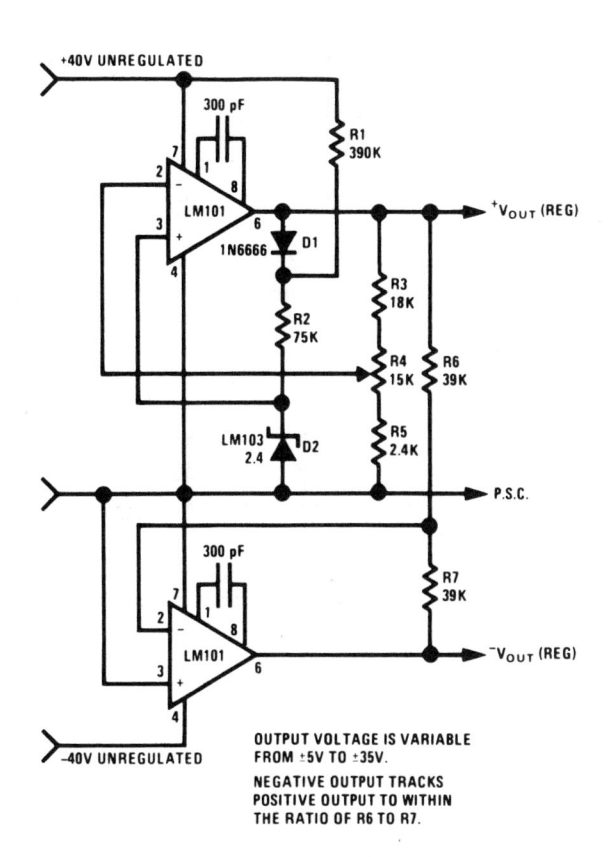

A tracking regulated power supply is very suitable for powering an operational amplifier system, since positive and negative voltages track, eliminating common-mode signals originating in the supply voltage. In addition, only one voltage reference and a minimum number of passive components are required.

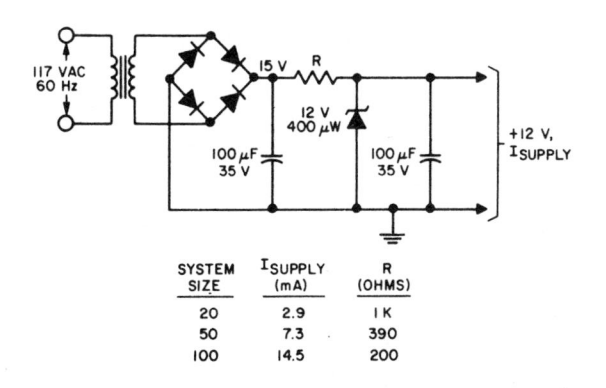

Programmable positive and negative power supply. The regulator section of the supply consists of two voltage followers whose input is provided by the voltage drop across a reference resistor of a precision current source.

Power supply yielding relatively high noise immunity for RCA CMOS.

SECTION **1** *Voltage, Current and Power Sources* **7**

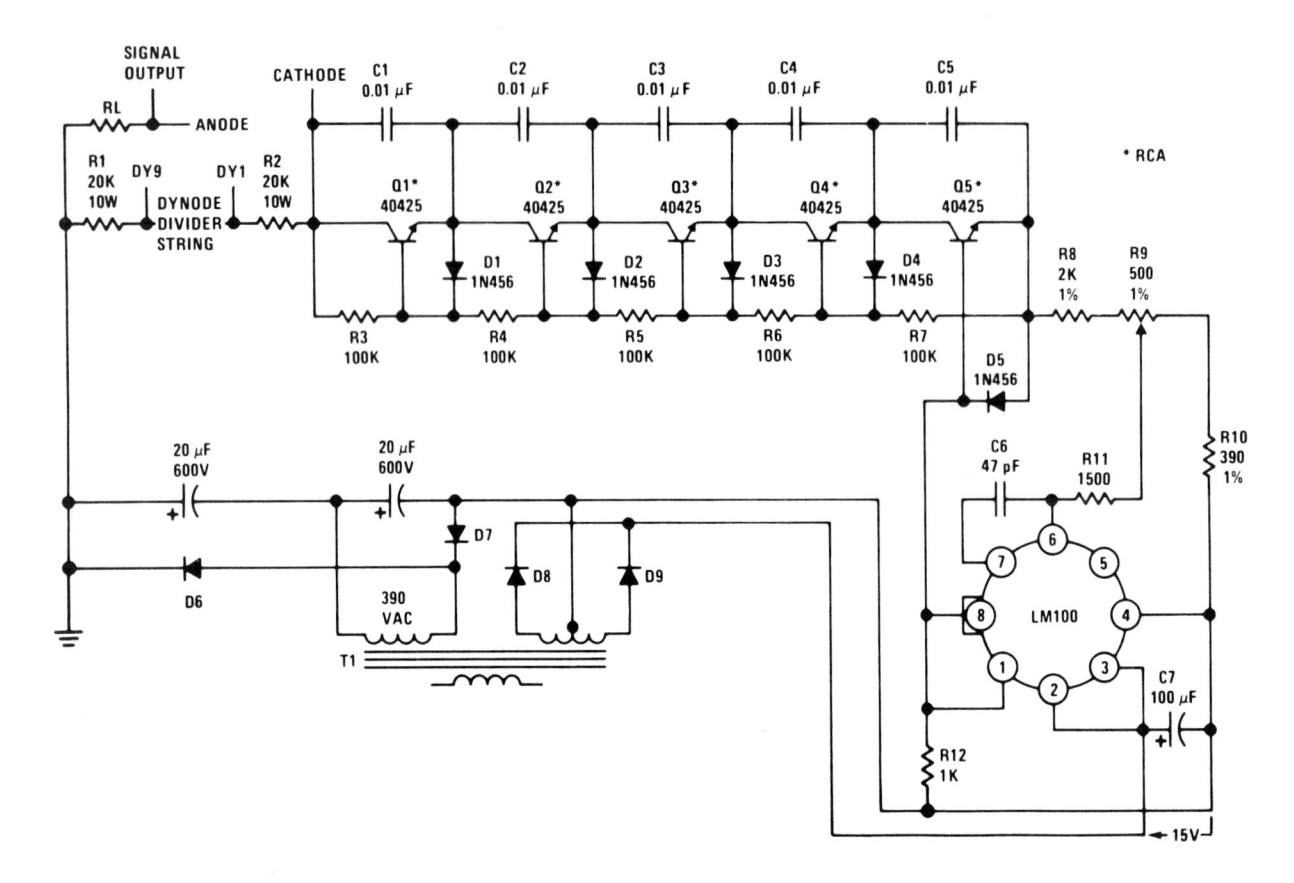

High-voltage supply for a 9-dynode photomultiplier tube. A full-wave rectifier operating off one winding of a power transformer provides a 15V bias voltage for the LM100. The high voltage is produced from a voltage doubler, which operates from a second winding. The circuit actually functions as a current regulator; the output current is passed through a resistive divider, which develops the operating voltages for the cathode and dynodes of the photomultiplier tube.

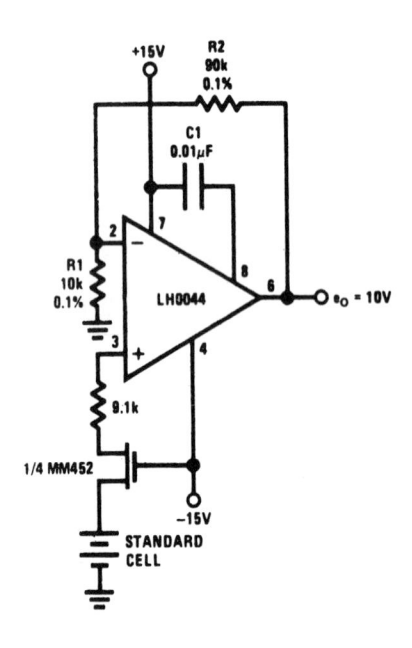

10V reference supply.

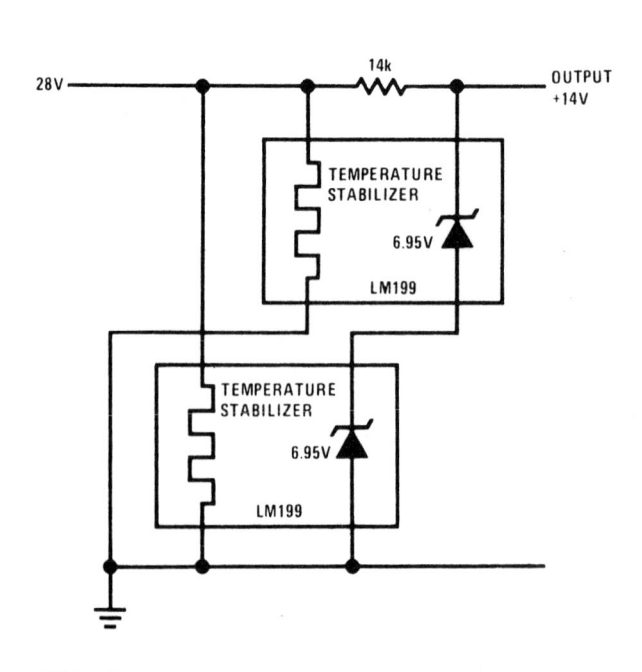

14V reference.

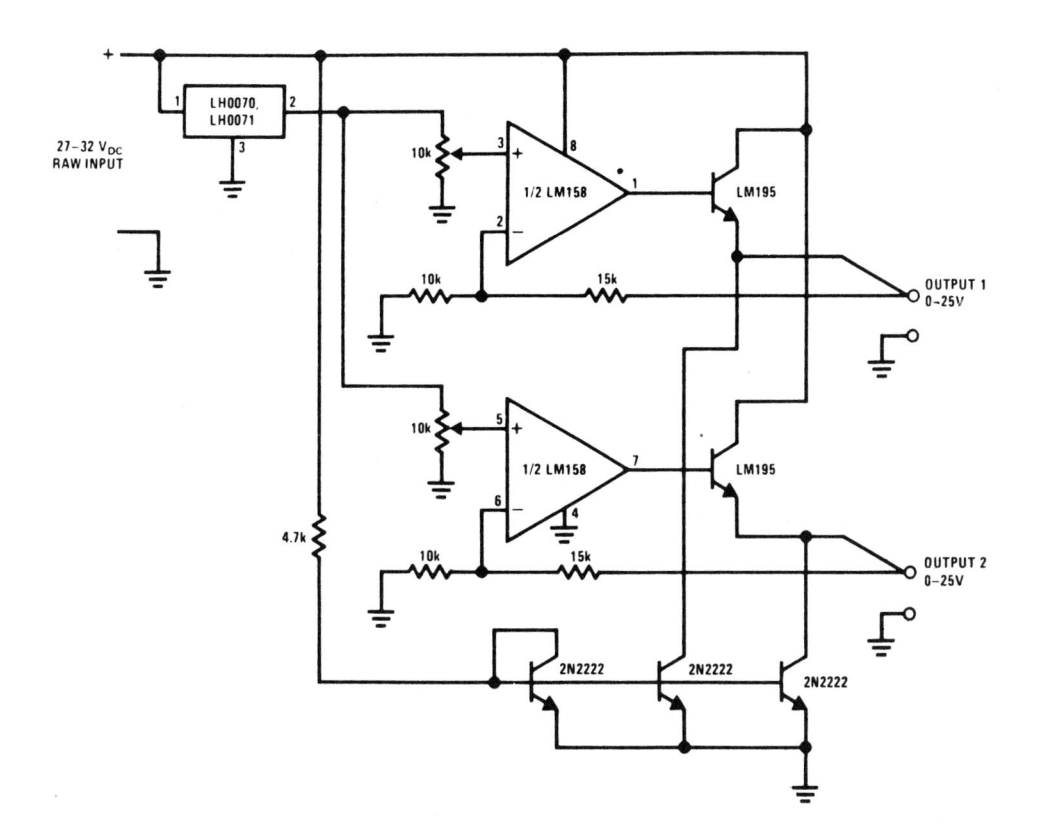

Dual-output bench power supply.

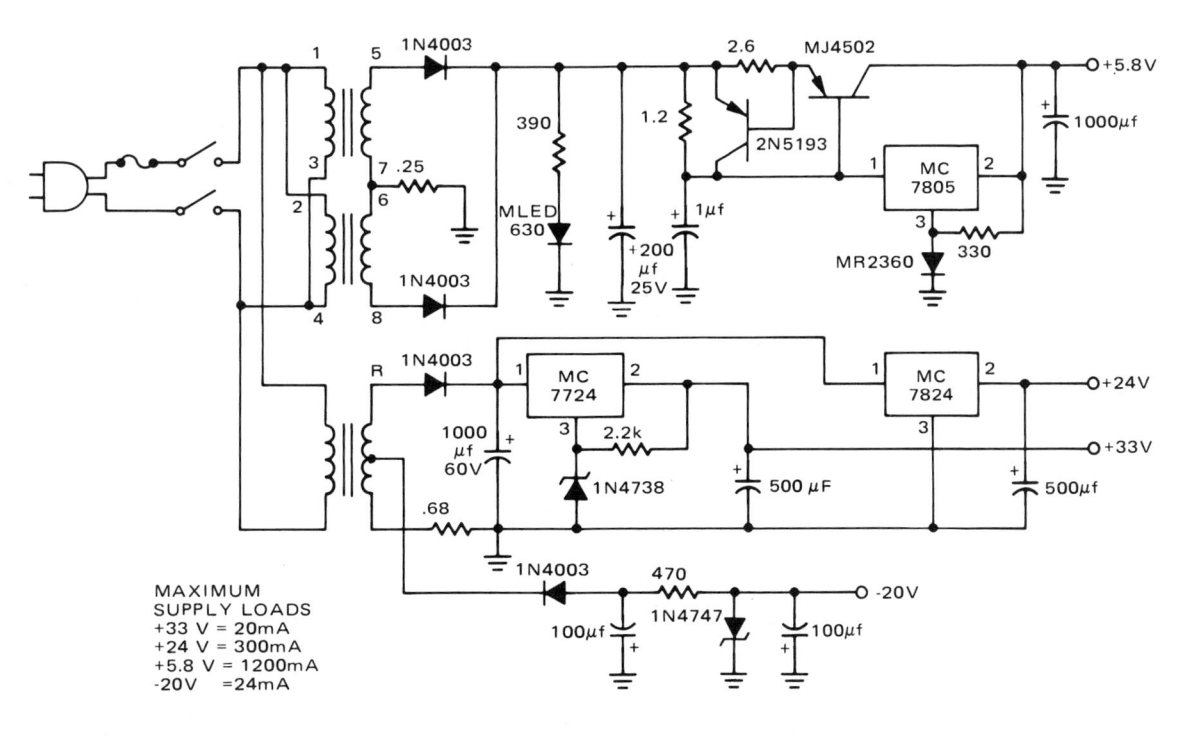

Power supply.

SECTION 1 *Voltage, Current and Power Sources*

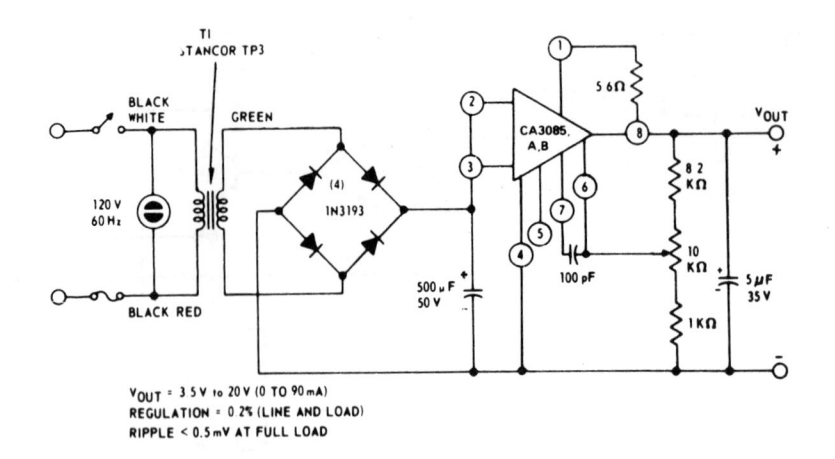

Application of the CA3085 in a **typical power supply**.

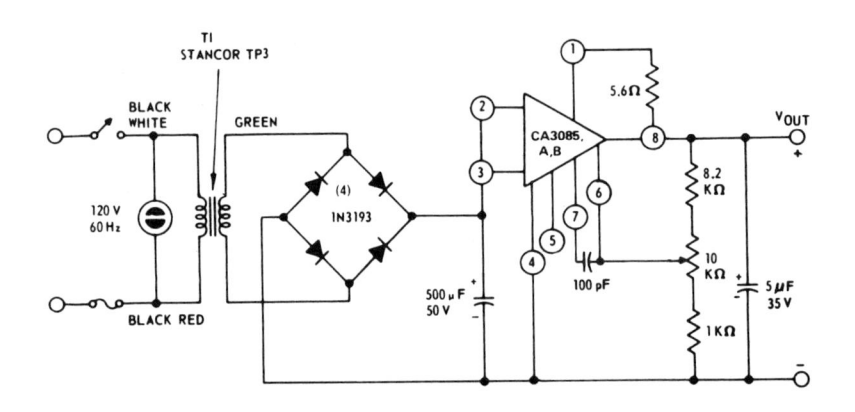

Straightforward low-voltage power supply using RCA's CA3085 as regulator. Output is 3.5 to 20V, 0 to 90 mA. Full load regulation 0.2% with less than 0.5 mV of ripple.

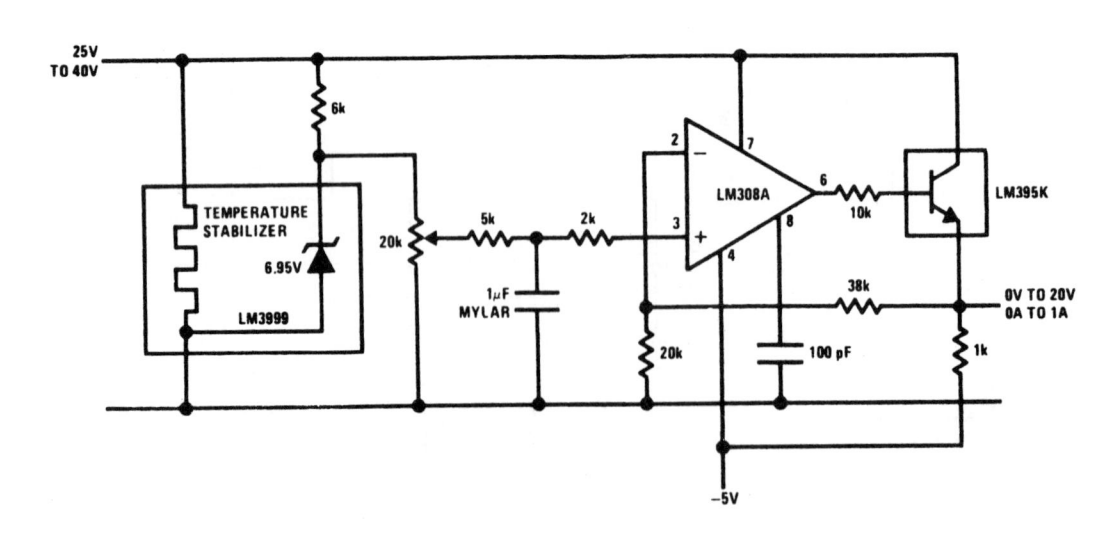

0 to 20V power reference.

10 IC SCHEMATIC SOURCEMASTER

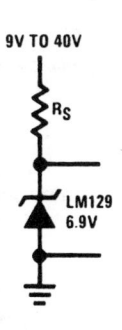

Simple reference. A new subsurface breakdown zener used in the LM129 gives lower noise and better long-term stability than conventional IC zeners. Virtually no voltage shifts in zener voltage as a result of temperature cycling, and the device is insensitive to stress on the leads.

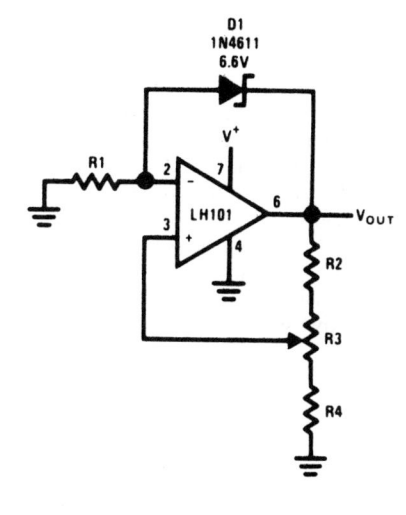

Positive Voltage Reference

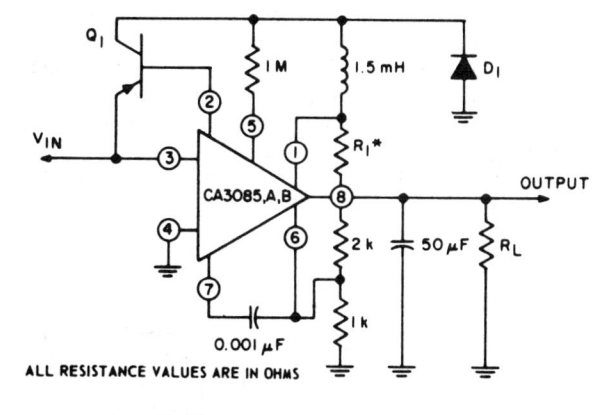

Switching regulator circuit. D1: RCA IN1763A or equivalent; Q1: RCA 2N5322 or equivalent; R1 = $0.7/I_L$ max.

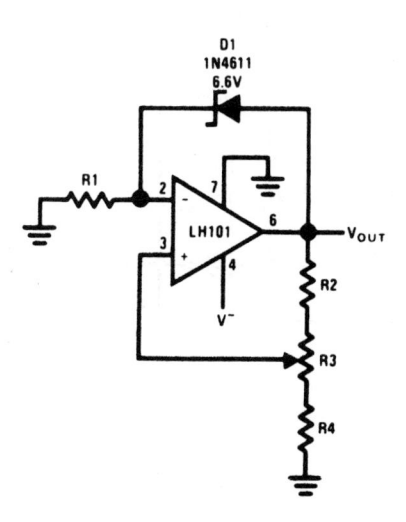

Negative Voltage Reference

Adjustable voltage references. The two circuits shown have different areas of applicability. The basic difference is that the V+ version illustrates a voltage source that provides a voltage greater than the reference diode, while the V− version illustrates a voltage source that provides a voltage lower than the reference diode.

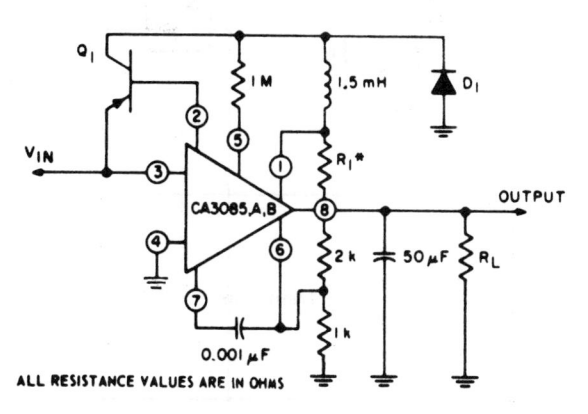

D1: RCA-IN1763A OR EQUIVALENT
Q1: RCA-2N5322 OR EQUIVALENT
*R1 = 0.7 I_L (MAX.)

Switching regulator circuit.

SECTION **1** Voltage, Current and Power Sources **11**

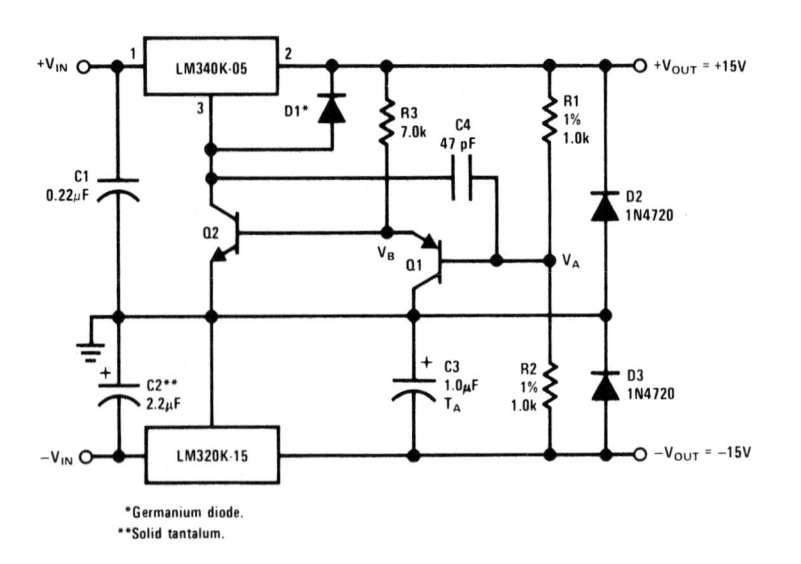

Tracking dual ±15V supply in which the positive regulator tracks the negative. Under steady-state conditions V_A is at a virtual ground and V_B at V_{BE} is above ground. Q2 then conducts the quiescent current of the LM340. If $-V_{OUT}$ becomes more negative, the collector-base junction of Q1 is forward biased, thus lowering V_B and raising the collector voltage of Q2. As a result, $+V_{OUT}$ rises, and the voltage V_A again reaches ground potential.

Assuming Q1 and Q2 to be perfectly matched, the tracking action remains unchanged over the full operating temperature range.

With R1 and R2 matched to 1%, the positive regulator tracks the negative within 100 mV (less than 1%). The capacitor C4 has been added to improve stability. Typical load regulations for the positive and negative sides from a 0 to 1.0A pulsed load (t_{ON} = 1.0 msec, t_{OFF} = 200 msec) are 10 mV and 45 mV, respectively.

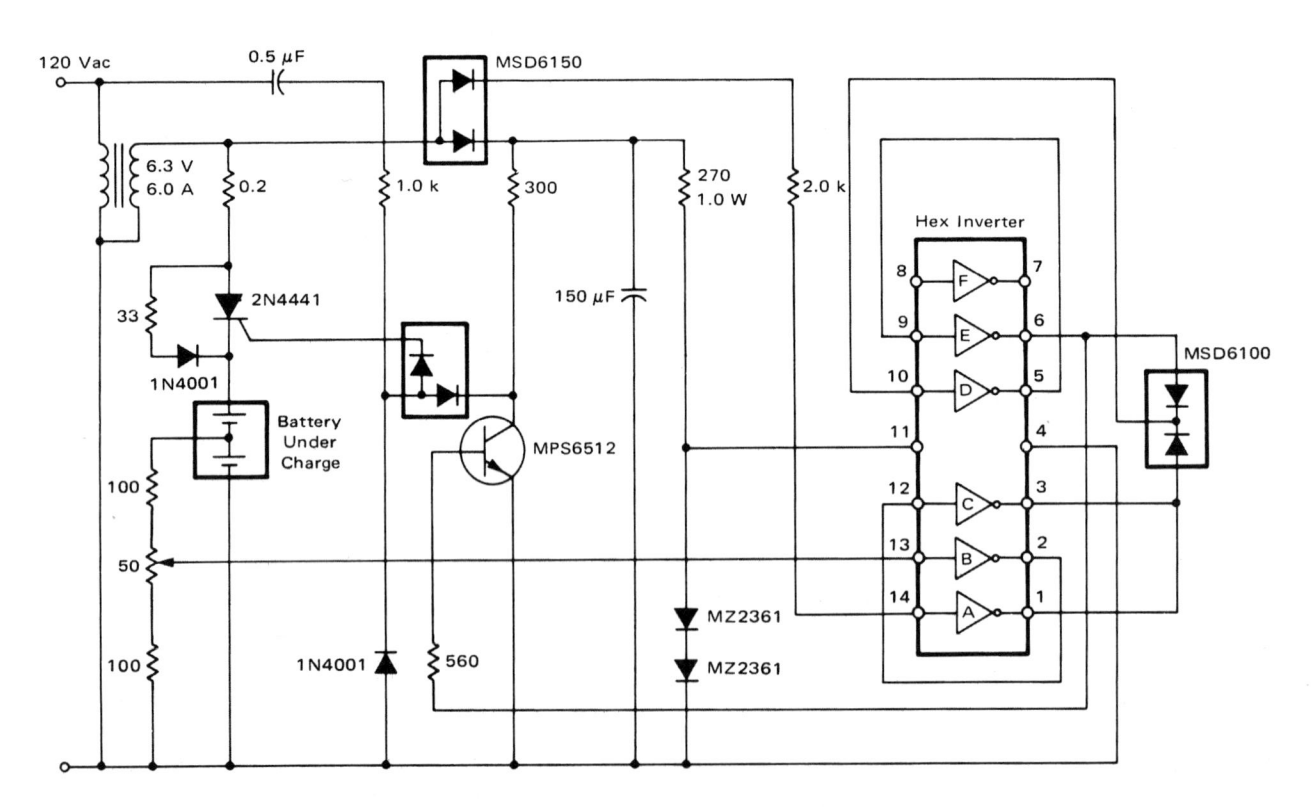

A 60 Hz battery charger with voltage-sensing circuit offers fast charging for ni-cads by pulsing, which is more efficient than dc charging. For a typical battery of sub-C size cells that have a capacity of 1.2 A-hr, the average charge currents required would be 1.2A for a 1-hour period (C-size), 2.4A for ½ hour (2C), and 4.8A for 15 minutes (4C).

12 IC SCHEMATIC SOURCEMASTER

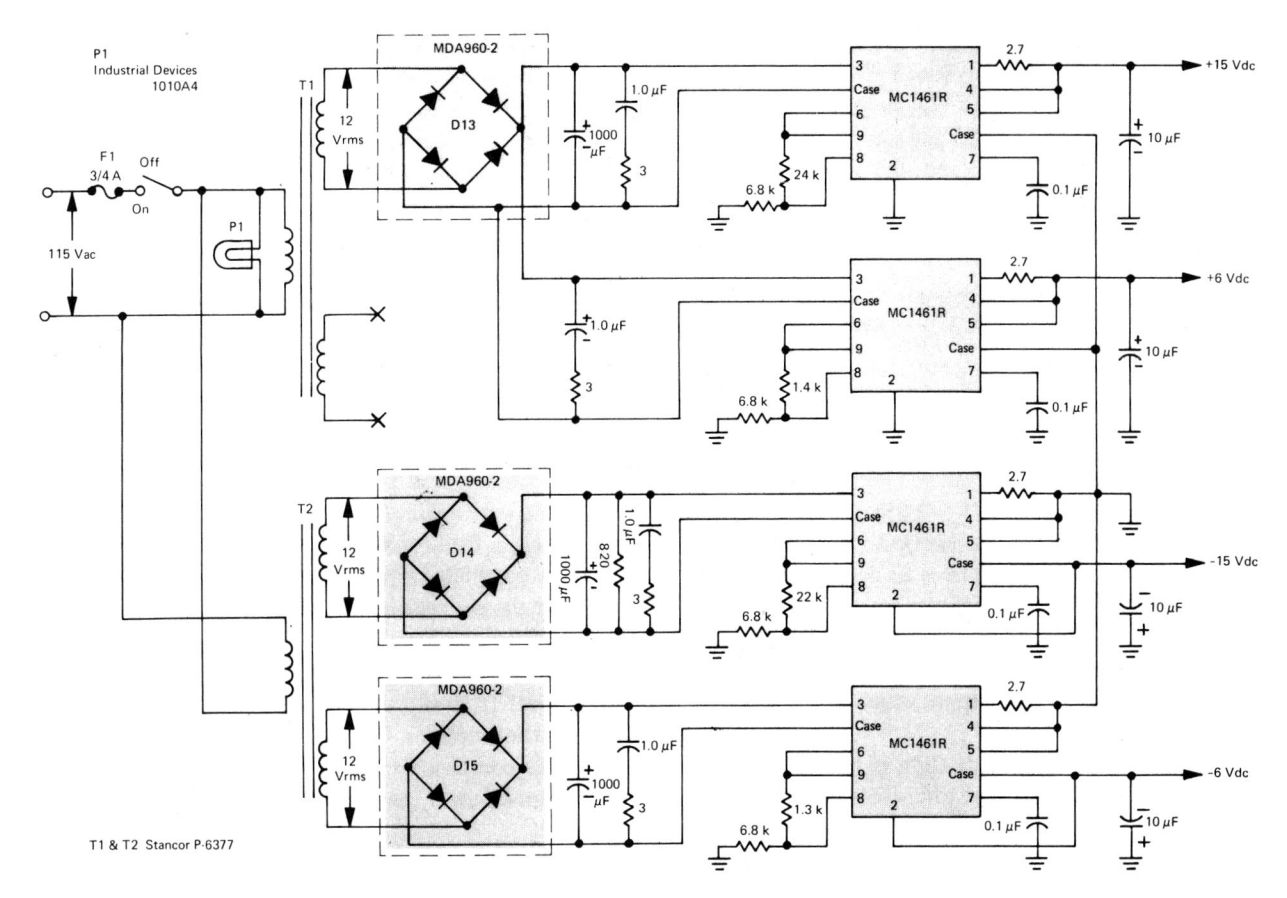

Power supply has four separate output lines and can deliver two common split-supply voltages: ±6 and ±15V. This circuit is not recommended for new designs; the chip is no longer being manufactured, but if you have a few of them around, it will prove a worthy application. This circuit was originally designed by Motorola engineers to power a proprietary function generator.

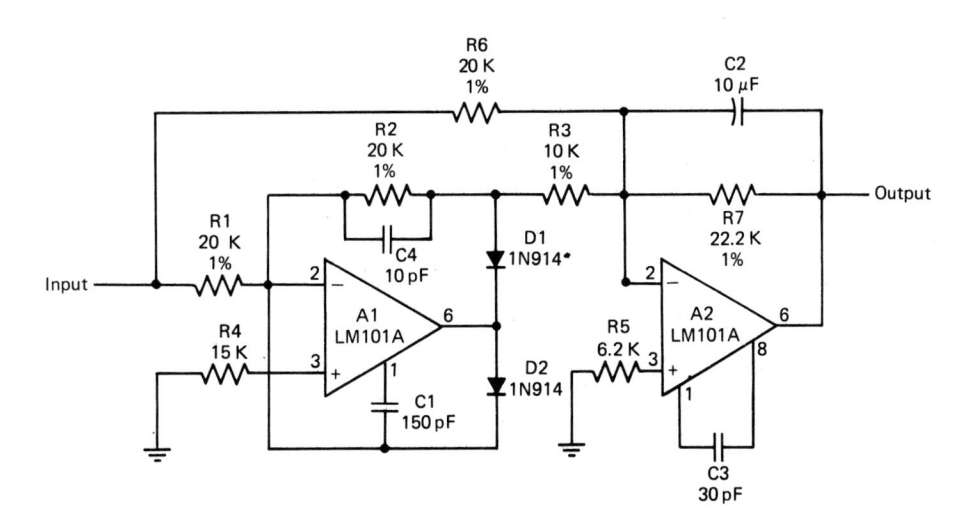

Fast ac/dc converter. Feedforward compensation can be used to make a fast full-wave rectifier without a filter.

SECTION 1 *Voltage, Current and Power Sources* 13

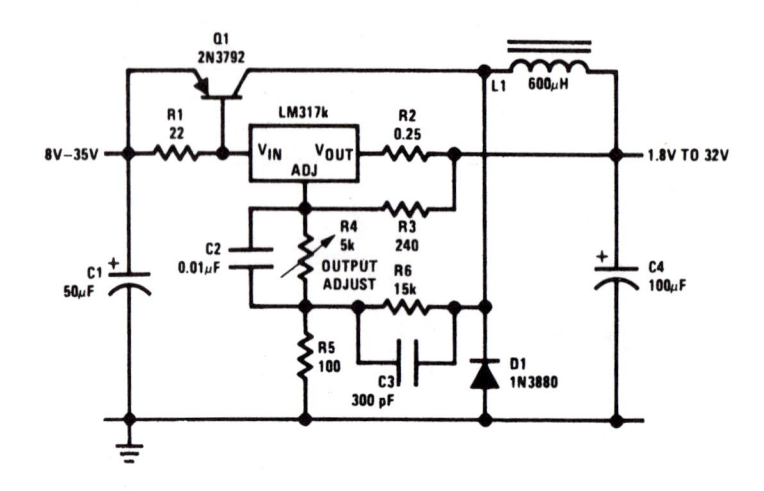

Low-Cost 3A switching regulator. Capacitors C1 and C4 are solid tantalum; coil core is Arnold A-254168-2, 60 turns.

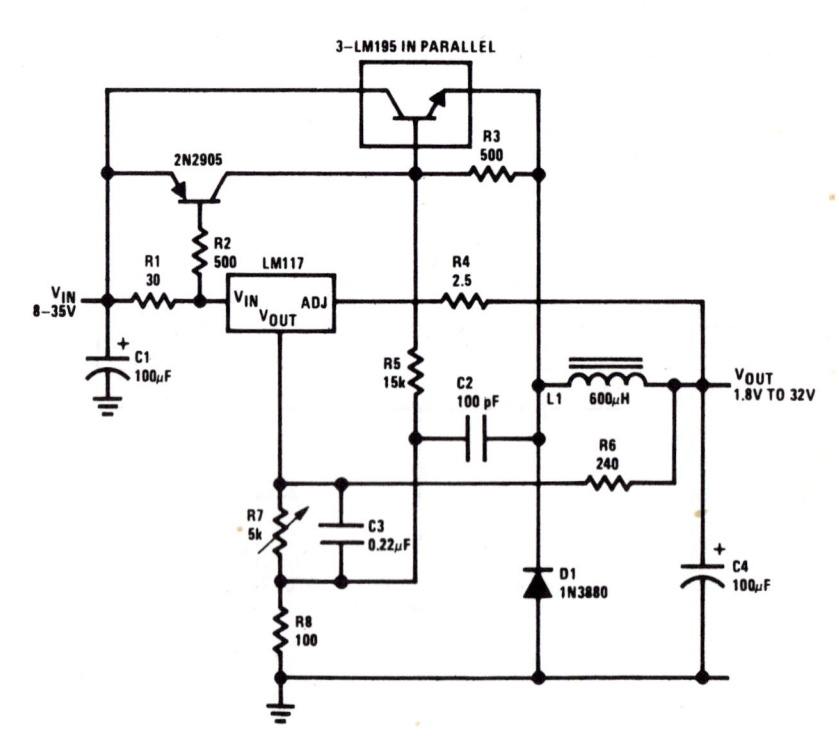

A 4A switching regulator with overload protection. Capacitors C1 and C4 are solid tantalum; coil core is Arnold A-254168-2, 60 turns.

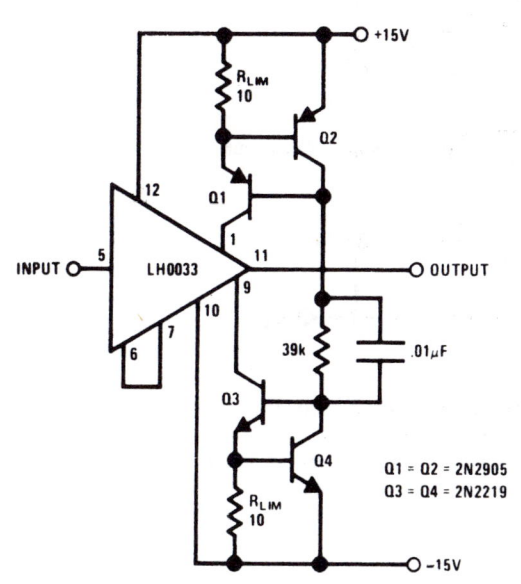

LH0033 current limiting

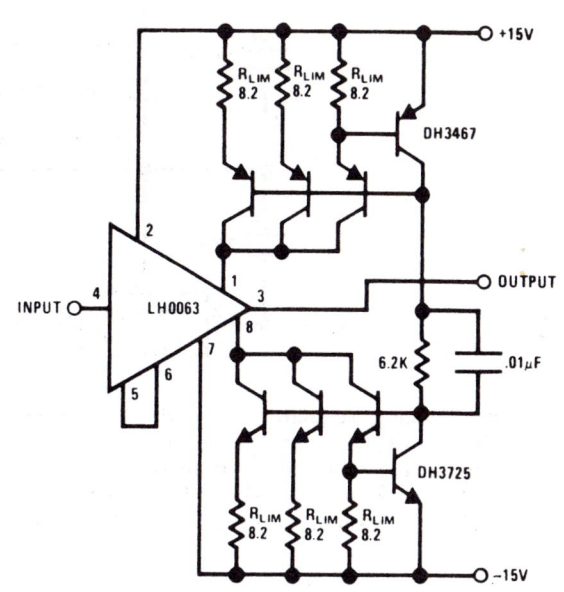

LH0063 current limiting

LH0033 and LH0063 current limiting using current sources. Both the LH0033 and LH0063 are designed to drive capacitive loads, such as coaxial cables, in excess of several thousand picofarads without susceptibility to oscillation. However, peak current resulting from $C(d_V/d_t)$ should be limited below absolute maximum peak current ratings for the devices. Thus, for the LH0033: $(\Delta V_{IN}/\Delta t)C_L \leq I_{OUT} \leq \pm 250$ mA; and for the LH0063, $(\Delta V_{IN}/\Delta t)C_L \leq I_{OUT} \leq \pm 500$ mA.

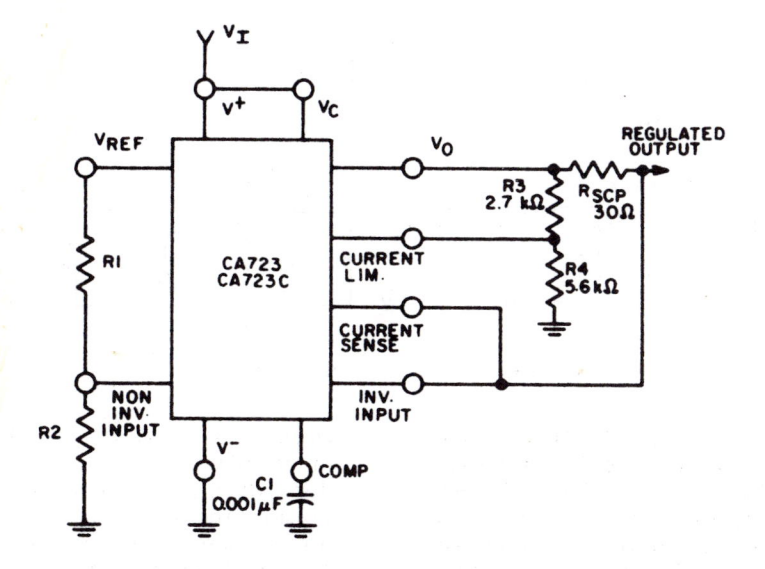

Foldback current-limiting circuit. Regulated output voltage, 5V; line regulation ($\Delta V_I = 3V$), 0.15 mV; load regulation ($\Delta I_L = 10$ mA), 1 mV; short-circuit current, 20 mA.

SECTION 1 *Voltage, Current and Power Sources* **15**

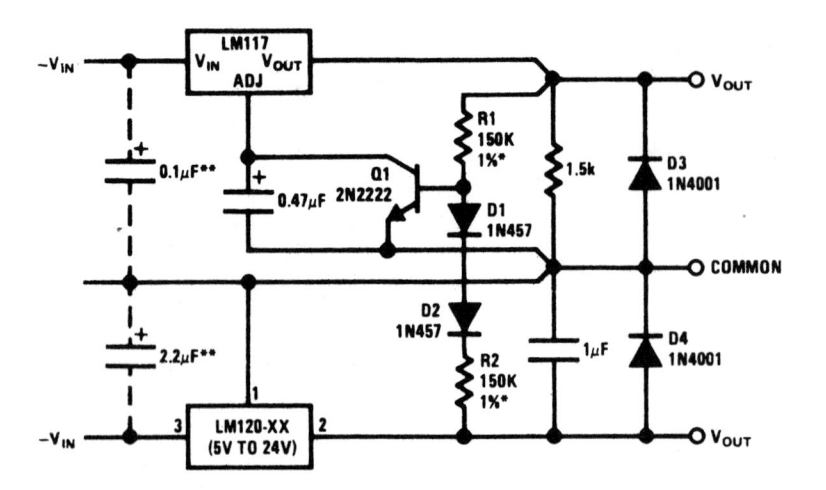

Wide-range tracking regulator. An LM3086N array may substitute for Q1, D1, and D2 for better stability and tracking. In the array diode, transistors Q5 and Q4 (in parallel) make up D2; similarly, Q1 and Q2 become D1, and Q3 replaces the 2N2222. The tolerance of R1 and R2 determine matching of positive and negative inputs. The 0.1 μF capacitor is only necessary if raw supply capacitors are more than 3 inches from regulators.

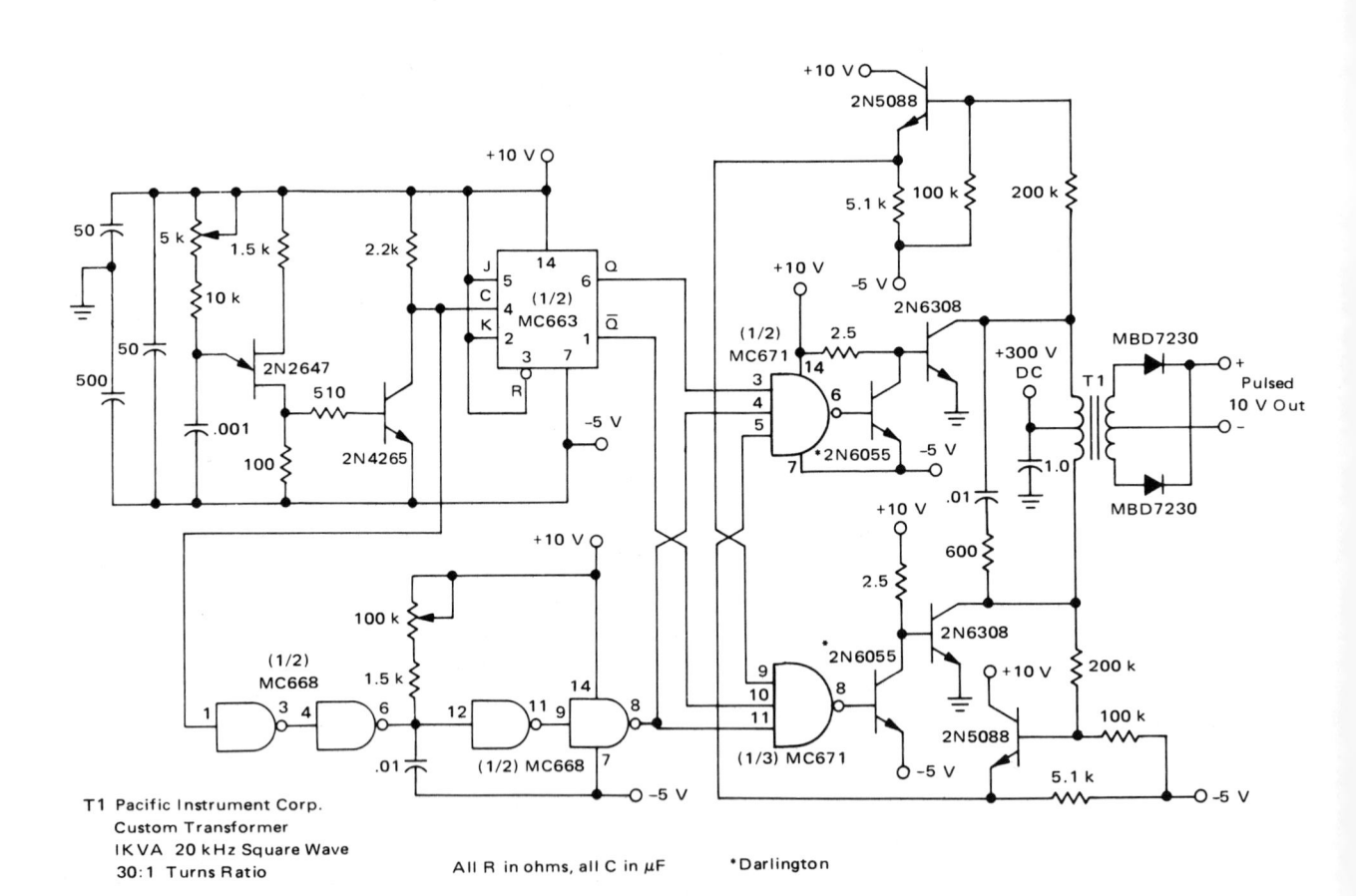

Line-operated inverter operates at 20 kHz and delivers a full kilowatt. The inverter requires three supplies to operate: +10V, −5V, and +300V. The +10/−5V supplies provide forward and reverse base drive to the output transistors and drive the discrete and IC logic circuits that are floating between these supplies. The 300V supply is for main power to the load and would normally be obtained from full-wave rectifiers on a 3-phase, 208V power line.

The relaxation oscillator uses a UJT to generate clock pulses at a 40 kHz rate. These pulses are shaped in the oscillator section and fed to an MHTL JK flip-flop, which performs phase splitting and frequency division. The outputs of the JK are two 20 kHz square waves phase-shifted by 180°. The next stages now have duplicate internal circuits in order to drive the final pair of inverter transistors. NAND gates are used as a comparator circuit to control the output stage.

16 IC SCHEMATIC SOURCEMASTER

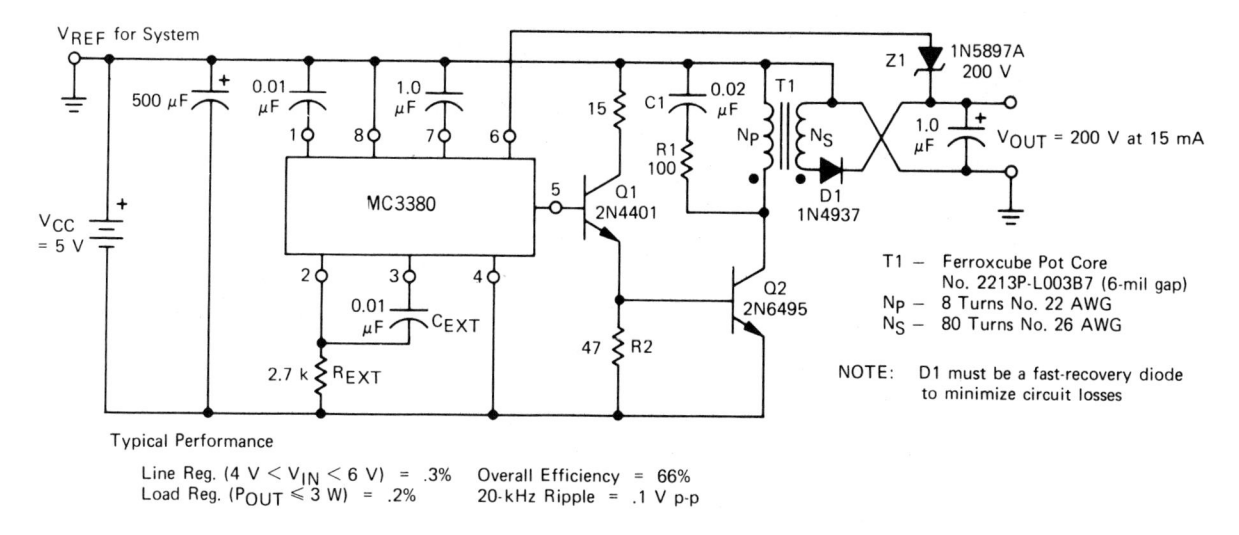

200V switching regulator. In this circuit, the MC3380 is used as the control element. The regulator converts 5V into 200V and provides up to 15 mA of output current, sufficient to drive 10 to 15 digits of most gas-discharge displays.

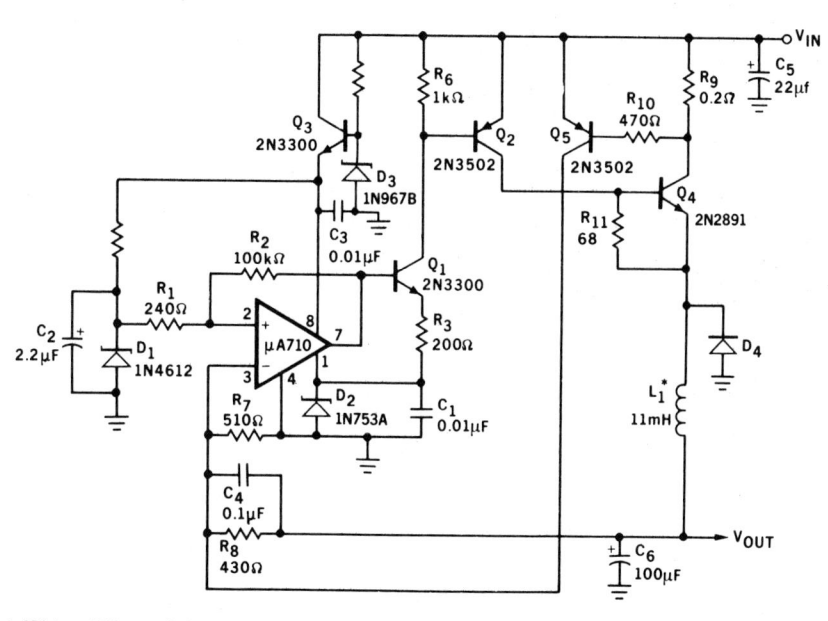

High-power switching regulator. A preregulator (Q_3, R_5, and D_3) is used to supply current for the temperature-compensated voltage reference diode (D_1). The comparator is operated from supply voltages of 0V, +6V, and +18V instead of the usual positive-negative voltage combination. The +6V is provided by zener D_2. Emitter follower Q_4 is added at the output to increase the output current capability; Q_3 supplies current directly to the inverting input of the comparator. Therefore, the power transistor still switches and does not have to handle anywhere near the full short-circuit current at the maximum supply voltage.

SECTION **1** *Voltage, Current and Power Sources*

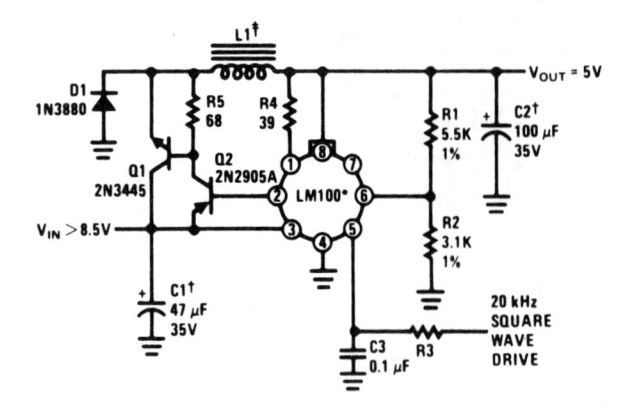

Circuit for synchronizing a switching regulator with a square-wave drive signal. Positive feedback is not used; instead, the square-wave drive signal is integrated, and the resulting triangular wave (about 40 mV peak-to-peak) is applied to the reference bypass terminal of the LM100. This triangular wave will cause the regulator to switch, since its gain is so high that the waveform overdrives it. The duty cycle of the switched waveform is controlled by the voltage on feedback terminal 6. If this voltage goes up, the duty cycle will decrease, since it is picking off a smaller portion of the triangular wave on pin 5. Similarly, the duty cycle will decrease if the voltage on pin 6 drops.

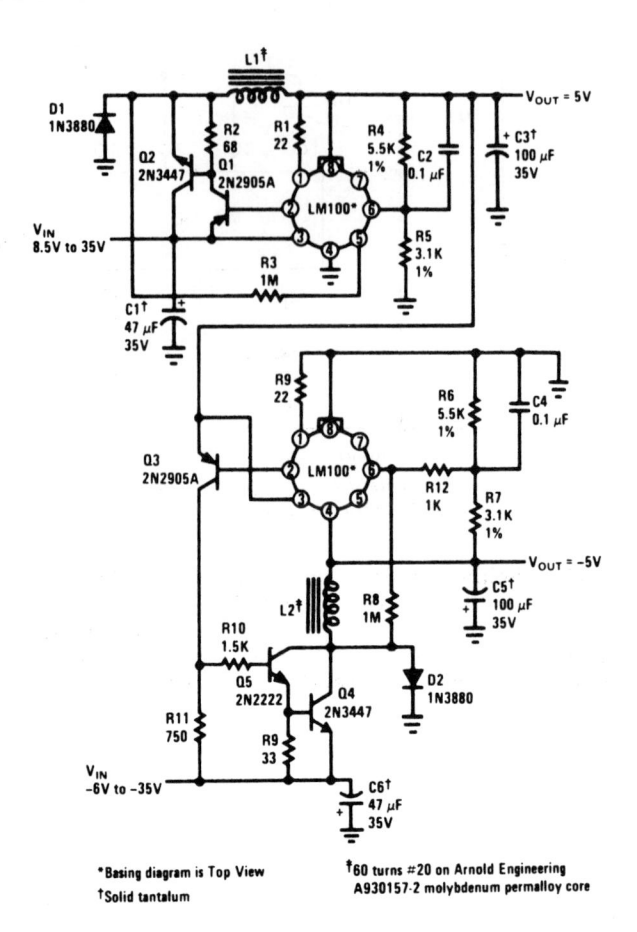

A negative switching regulator where the unregulated input and regulated output have a common ground. The only limitation of the circuit is that a positive voltage greater than 3V must be available in order to properly bias the negative regulator.

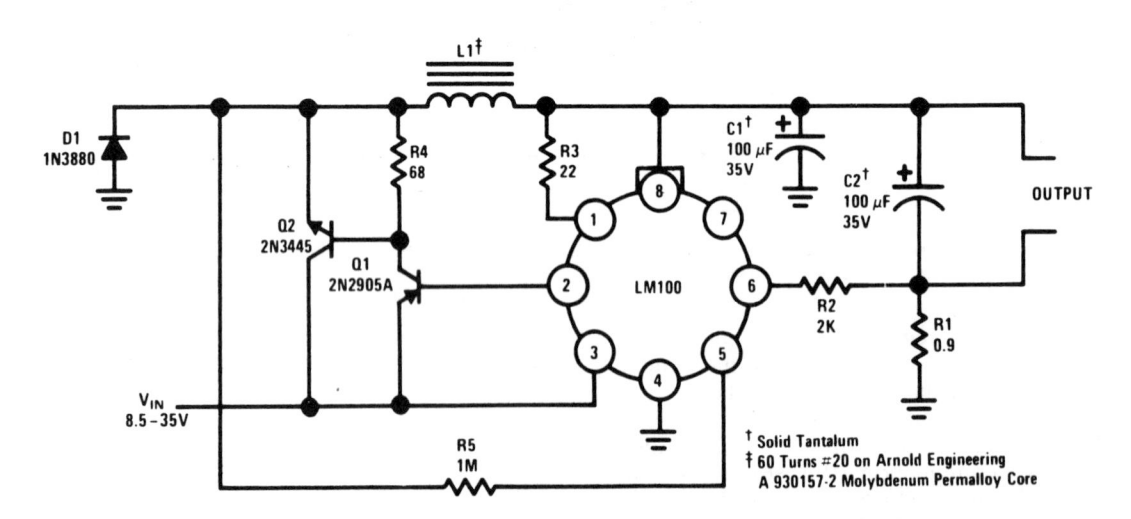

A switching current regulator where the input power for a fixed load current is roughly proportional to the voltage across the load. A standard switching regulator is used, except that the load is connected from the output to the feedback terminal of the LM100. A current-sense resistor, R1, is connected from the feedback terminal to ground to set the output current. If desired, an adjustment potentiometer can be connected across the current-sense resistor.

18 IC SCHEMATIC SOURCEMASTER

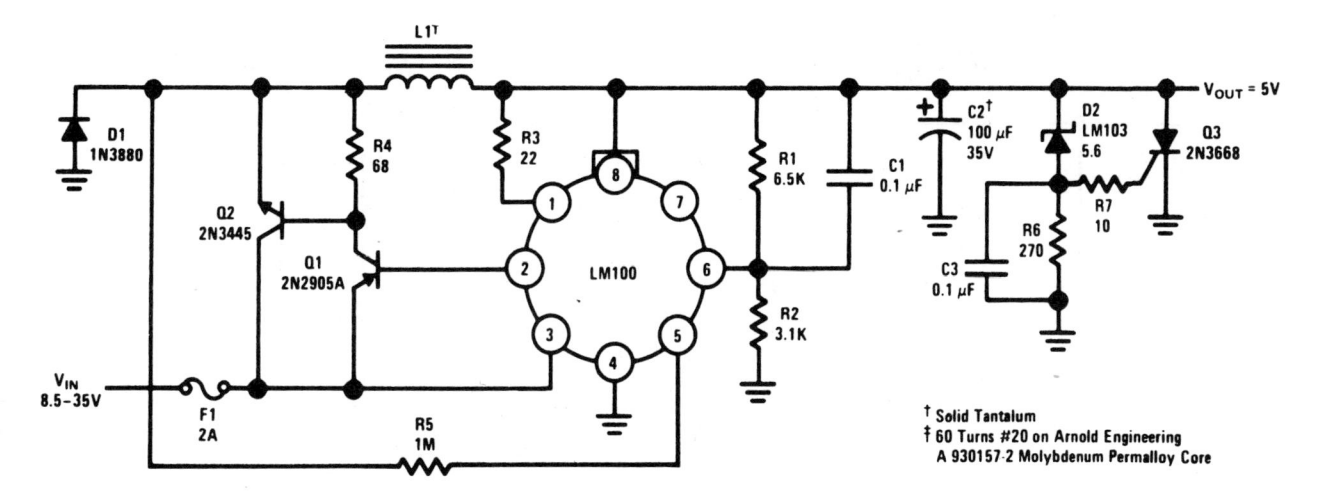

A switching regulator with crowbar overvoltage protection can be used in place of a power converter to reduce high input voltages to a considerably lower output voltage with good efficiency. In addition, it simultaneously regulates the output voltage. As a result, a switching regulator is simpler and more efficient than a power converter/regulator combination. This circuit overcomes the one objection to switching regulators: that they fail with the output voltage going up to the unregulated input voltage.

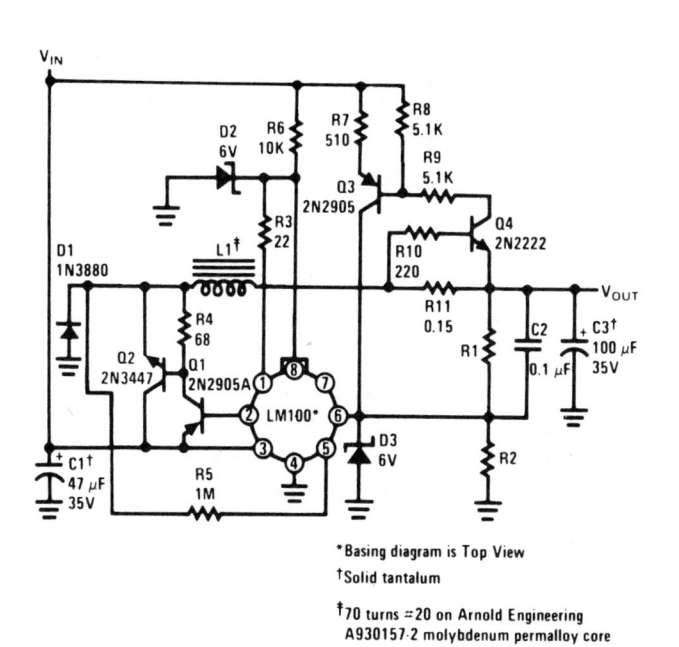

A switching regulator with continuous short-circuit protection under worst-case conditions. The current-sensing resistor is located in series with the inductor; therefore, the peak limiting current can be more precisely determined, since the current spike generated by pulling the stored charge out of the catch diode does not flow through the sense resistor.

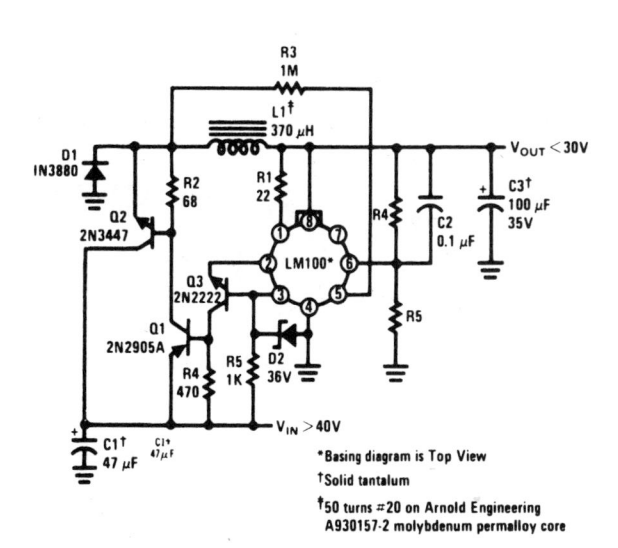

Switching regulator for high-voltage inputs. The voltage seen by the LM100 is maintained at a fixed level within ratings by zener D2. The zener voltage must be at least 3V greater than the output voltage. The output of the LM100 is level shifted up to the input voltage by an additional NPN transistor, Q3, which is operated common base. This drives the PNP switch driver.

SECTION **1** *Voltage, Current and Power Sources* **19**

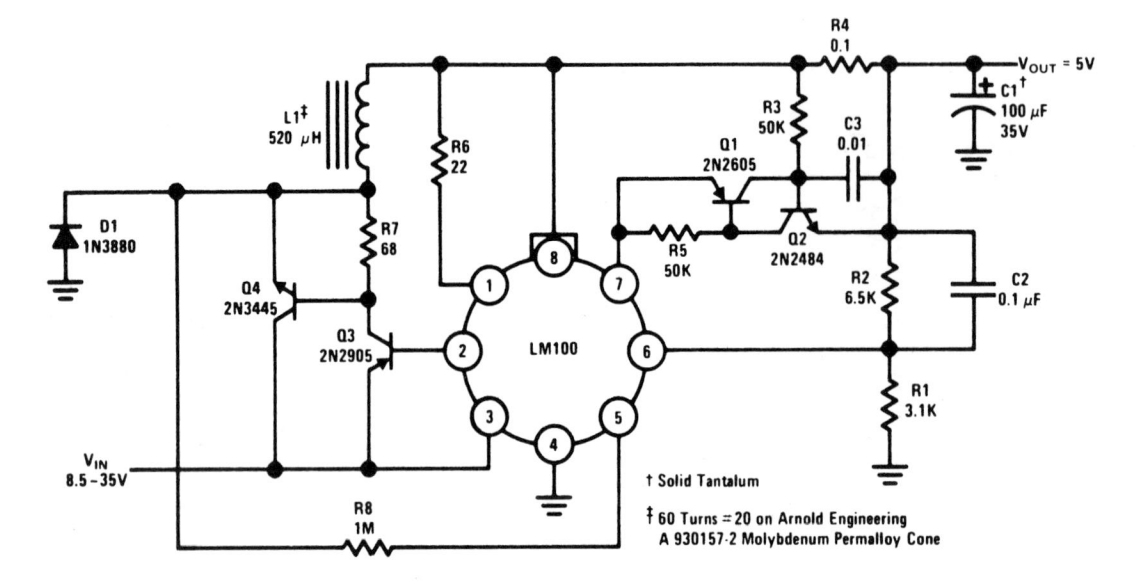

A 3A switching regulator with overload shutoff. When the output current becomes excessive, the voltage drop across a current-sense resistor fires an SCR, which shuts off the regulator. The regulator remains off, dissipating practically no power until it is reset by removing the input voltage.

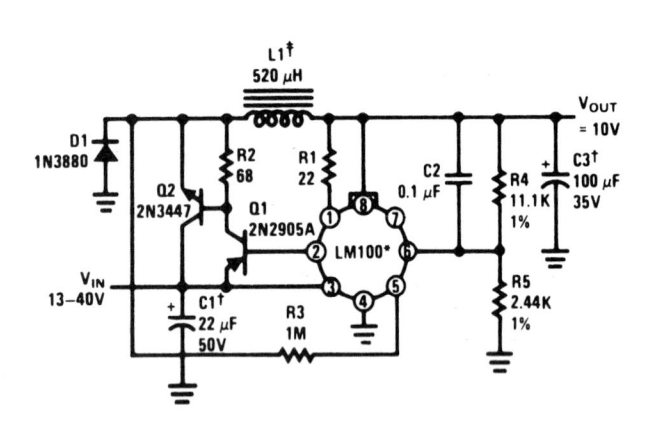

A switching regulator capable of delivering output currents of 3A continuously with only a small heatsink. Output currents to 5A can be obtained at reduced efficiency; however, the case temperature of the power switch and catch diode approaches 100°C under this condition, so continuous operation is not recommended unless more heatsink area is provided.

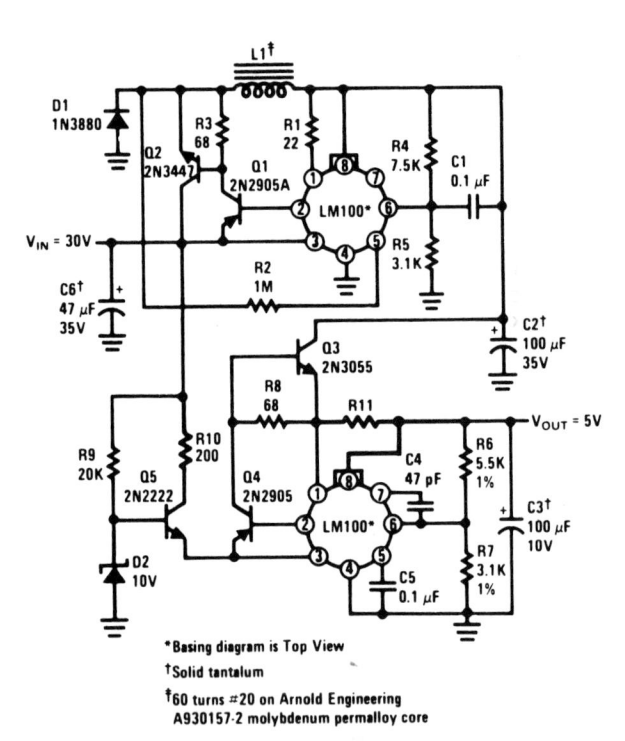

Switching and linear regulator combination. The switching regulator not only reduces the input voltage with high efficiency but also regulates it. The linear regulator operates with a fixed input/output voltage differential, which holds dissipation to a minimum.

Here, the linear regulator is biased by a zener preregulator (R9, D2, and Q5) to isolate it from noise on the unregulated supply. This separate bias supply permits the linear pass transistor, Q3, to operate right down into saturation. The collector of Q3 is supplied by the output of a switching regulator, which is made sufficiently higher than the linear regulator output to allow for the maximum overshoot of the switching regulator plus the saturation of Q3.

20 IC SCHEMATIC SOURCEMASTER

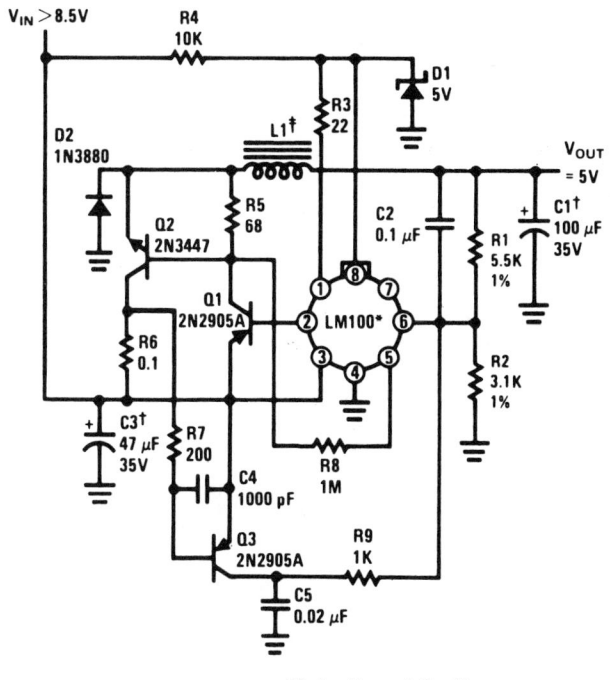

Switching regulator with current limiting. The peak current through the switch transistor is sensed by R6. When the voltage drop across this resistor becomes large enough to turn on Q3, the output voltage begins to fall, since current is being supplied to the feedback terminal of the regulator from the collector of Q3, thus requiring less current from the output through R1. The circuit will continue to oscillate even with a shorted output because of positive feedback through R6 and the relatively long discharge time constant of C2.

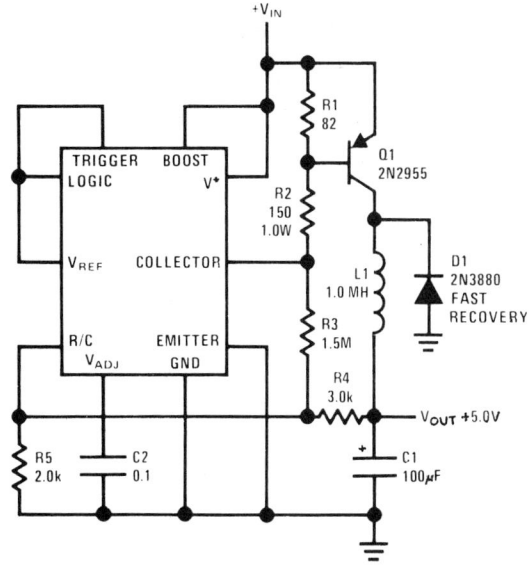

This switching regulator is made using the internal reference and comparator of the LM122 timer to drive a PNP switch transistor. Features include a 5.5V minimum input voltage at 1A output current, low parts count, and good efficiency (greater than 75%) for input voltages to 10V. Line and load regulation are less than 0.5%, and output ripple at the switching frequency is only 30 mV. Q1 is an inexpensive plastic device that does not need a heatsink up to 50°C. D1 should be a fast-switching diode. Output voltage can be adjusted between 1V and 30V by choosing proper values for R2, R3, R4, and R5. For outputs less than 2V, a divider with 250Ω resistance must be connected between V_{REF} and ground, with its tap point tied to V_{ADJ}.

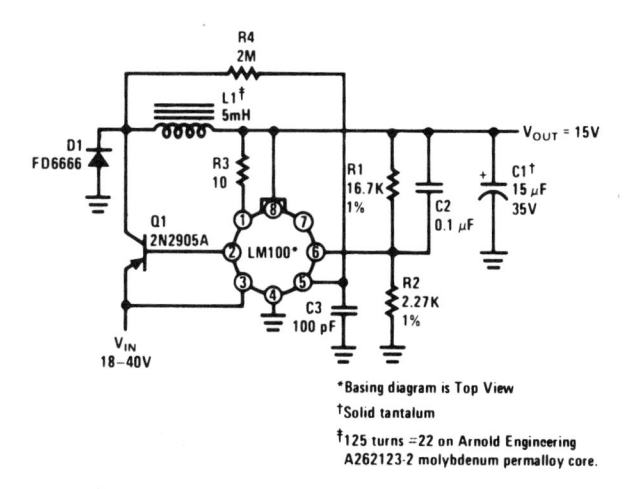

A switching regulator suitable for output currents as high as 500 mA. This limit is set by the output current available from the LM100 to saturate the switch transistor, Q1. For lower currents, the value of R3 should be increased so that the base of Q1 is not driven unnecessarily hard.

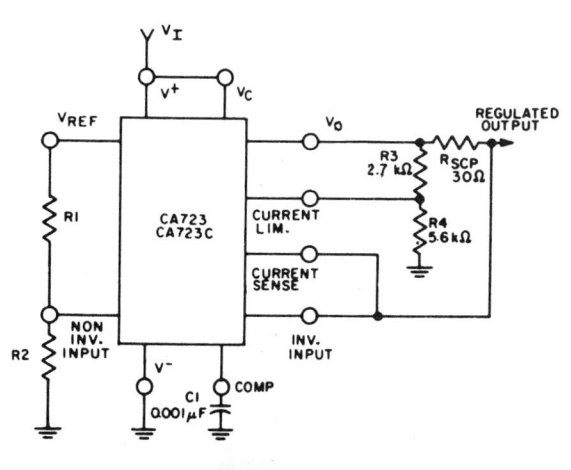

CIRCUIT PERFORMANCE DATA:
REGULATED OUTPUT VOLTAGE . . . . 5 V
LINE REGULATION (ΔV_I = 3 V) . . . . 0.5 mV
LOAD REGULATION (ΔI_L = 10 mA) . . 1 mV
SHORT-CIRCUIT CURRENT . . . . . . 20 mA

Voltage regulator (5V) with foldback current limiting.

SECTION **1** *Voltage, Current and Power Sources* **21**

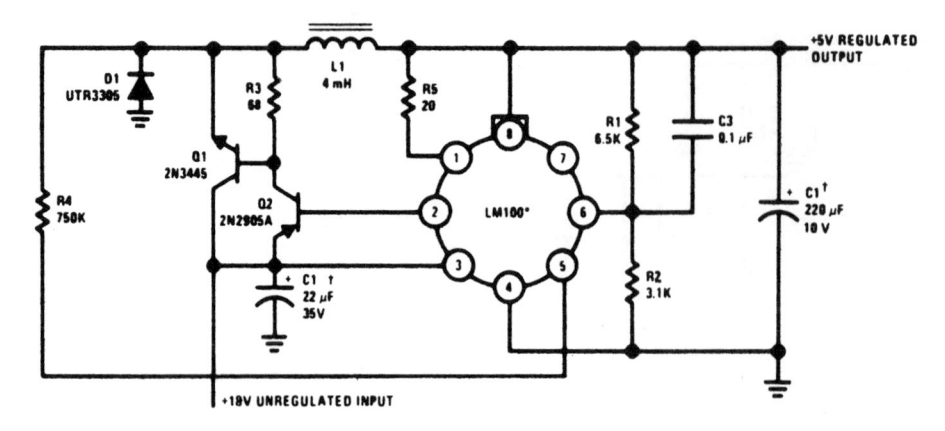

High-current switching regulator. Two external transistors, an NPN and a PNP, are connected in cascade to handle the output current. The regulated output is fed back through a resistive divider that determines the output voltage in the normal manner. The regulator is made to oscillate by applying positive feedback to the reference terminal through R4.

The switching transistors, Q1 and Q2, turn on when the voltage on the feedback terminal is less than that on the reference terminal. This action raises the reference voltage, since current is fed into this point from the switch output through R4. The switching transistors remain on until the voltage on the feedback terminal increases to the higher reference voltage. The regulator then switches off, lowering the reference voltage. It remains off until the voltage on the feedback terminal falls to the lower reference voltage.

When the switch transistors are on, power is delivered from the power source to the load through L1. When the transistors turn off, the inductor continues to deliver current to the load with D1 supplying a return path. Since fairly fast rise and fall times are involved, D1 cannot be an ordinary silicon rectifier. A fast-switching diode must be used to prevent excessive switching transients and large power losses. Capacitor C1 should be solid tantalum; L1 is 160 turns of No. 20 wire on Arnold Engineering A-548127-2 molybdenum Permalloy core.

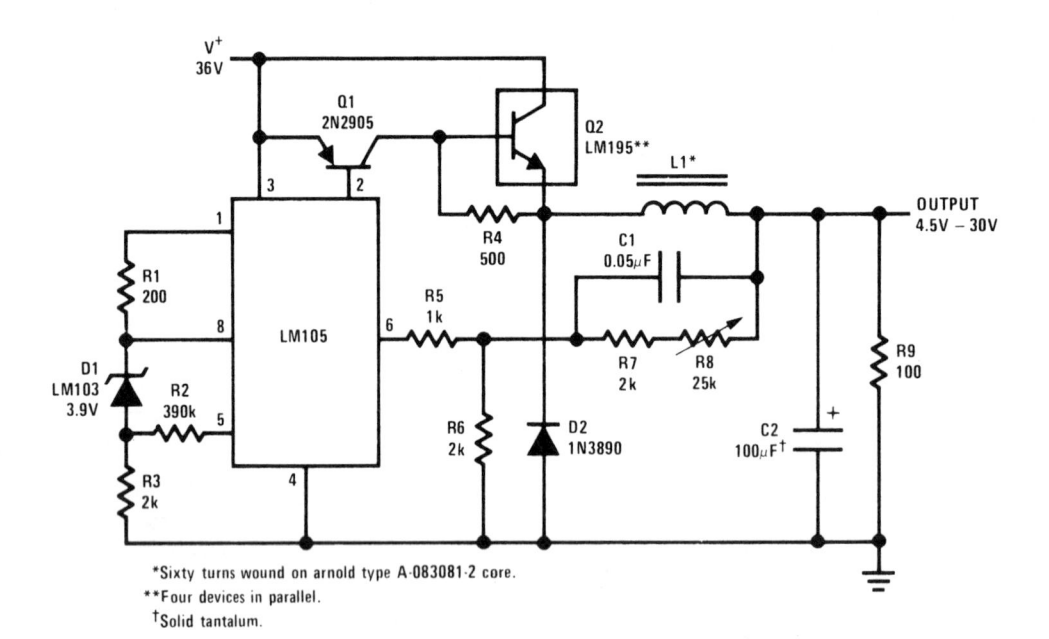

A 6A, variable-output switching regulator for general-purpose applications. An LM105 positive regulator is used as the amplifier reference for the switching regulator. Positive feedback to induce switching is obtained from the LM105 at pin 1 through an LM103 diode. Positive feedback is applied to the internal amplifier at pin 5 and is independent of supply voltage. This forces the LM105 to drive the pass devices either on or off, rather than linearly controlling their conduction. Negative feedback, delayed by L1 and output capacitor C2, causes the regulator to switch, with the duty cycle adjusting automatically to provide a constant output. Four LM195s are used in parallel to obtain a 6A output, since each device can only supply about 2A. Note that no ballasting resistors are needed for current sharing. When Q1 turns on, all bases are pulled up to V+, and no base current flows in the LM195 transistors.

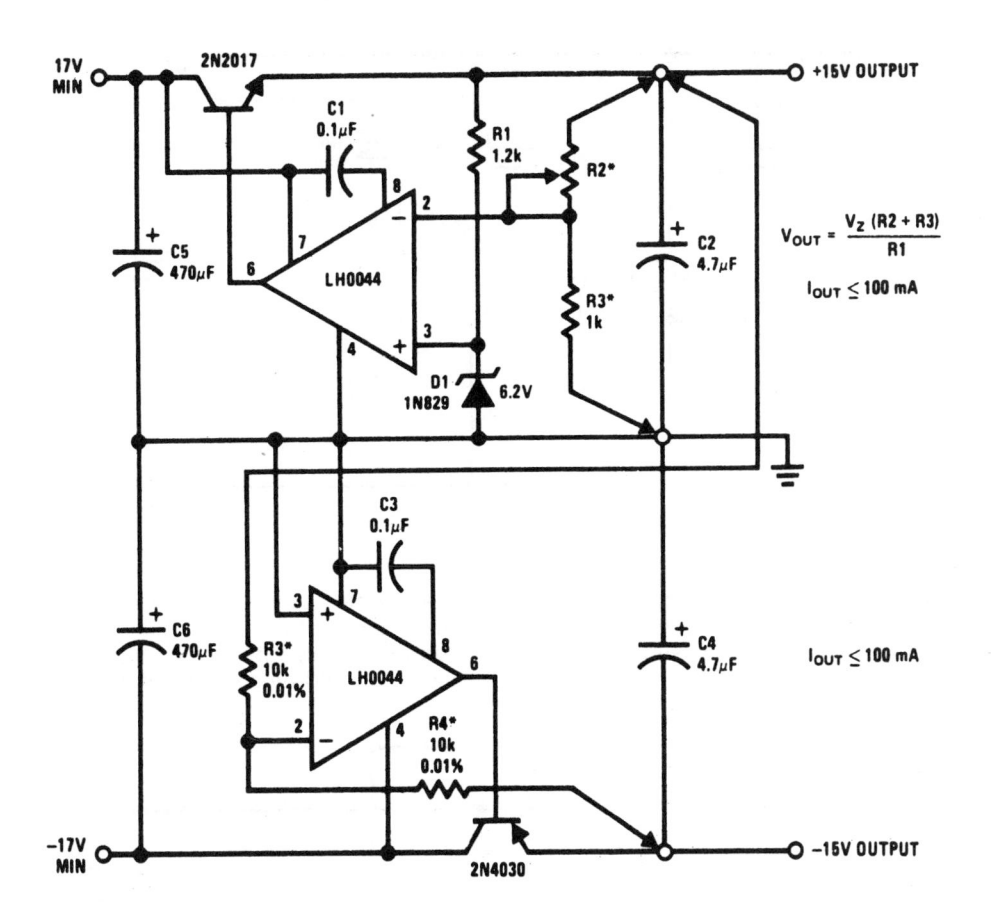

Precision dual tracking regulator. R2 through R4 are wire-wound for minimum drift. Line and load regulation ≤ 0.005%.

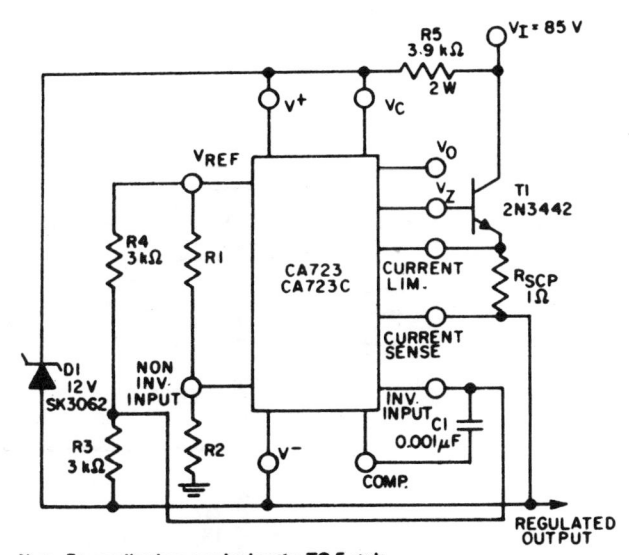

$$V_{OUT} = \frac{V_Z (R2 + R3)}{R1}$$

$I_{OUT} \leq 100$ mA

CIRCUIT PERFORMANCE DATA:
REGULATED OUTPUT VOLTAGE . . . 50 V
LINE REGULATION ($\Delta V_I = 20$ V) . . . 15 mV
LOAD REGULATION ($\Delta I_L = 50$ mA) . . . 20 mV

Note: For applications employing the TO-5 style package and where V_Z is required, an external 6.2-volt zener diode should be connected in series with V_O (Terminal 6).

Positive floating regulator circuit.

SECTION 1 *Voltage, Current and Power Sources* 23

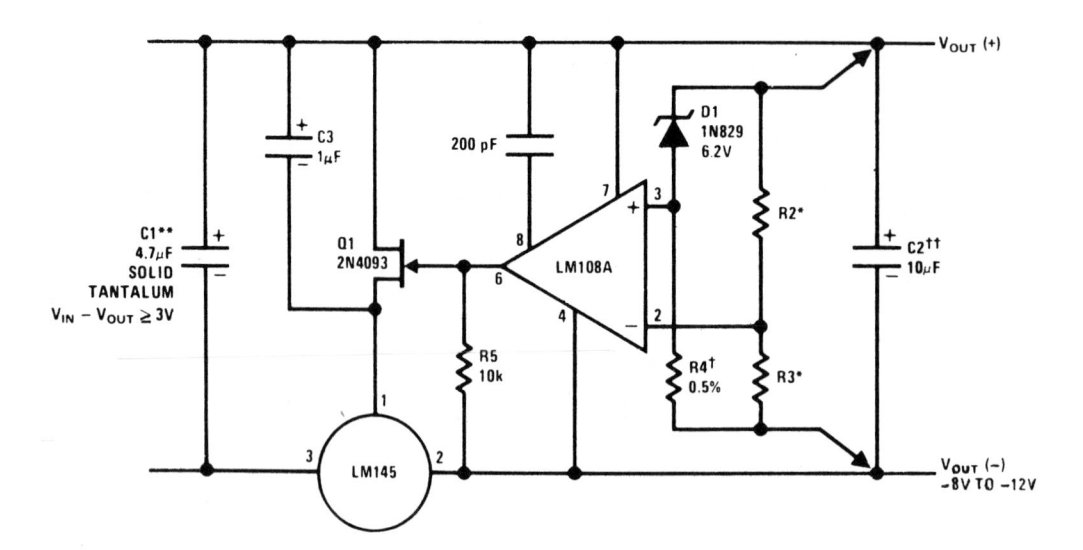

High-stability regulator. Load and line regulation is less than 0.01%, and temperature drift is less than 0.001%/°C. C1 is not needed if power supply filter capacitor is within 3 inches of regulator. R4 determines zener current and may be adjusted to minimize temperature drift. Select R2 and R3 to set output voltage; 1 ppm/°C is suggested. Capacitors C1 and C2 should be solid tantalum.

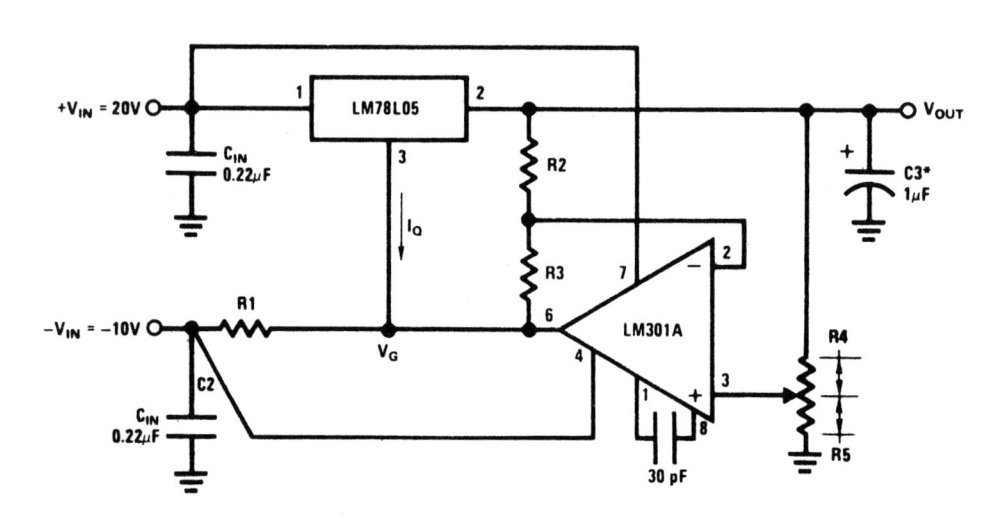

Variable-output regulator 0.5–18V. $V_{OUT} = V_G + 5V$, R1 = $-V_{IN}/I_Q$ of LM78L05; $V_{OUT} = 5V$ (R2/R4) for (R2 + R3) = (R4 + R5). A 0.5V output will correspond to (R2/R4) = 0.1, (R3/R4) = 0.9. Capacitor C3 should be solid tantalum.

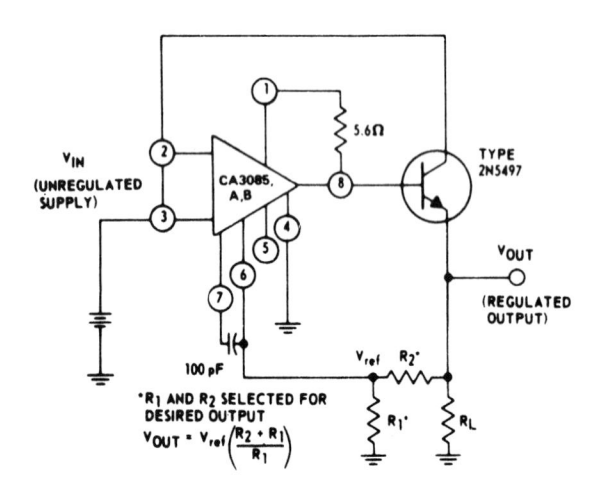

Typical high-current voltage regulator circuit.

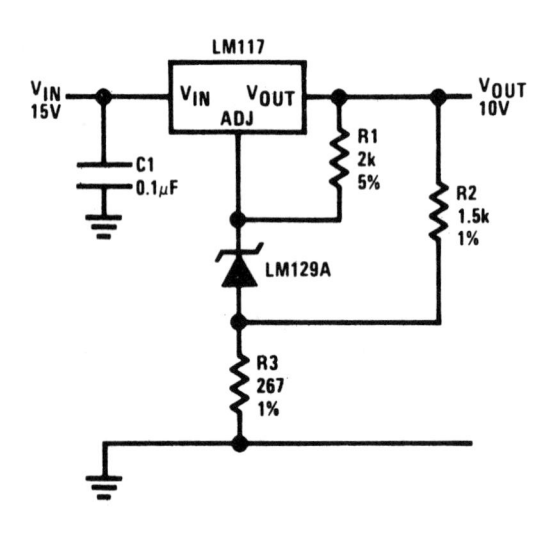

High-stability 10V regulator.

24 IC SCHEMATIC SOURCEMASTER

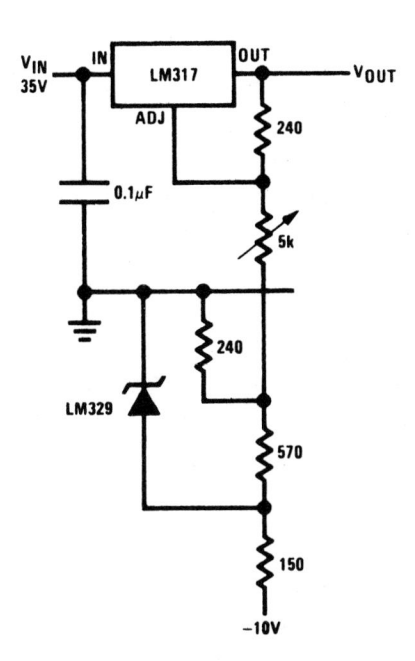

This low-cost 0–25V regulator uses precision multi-current temperature-compensated 7.9V reference, with dynamic impedance a factor of 10 to 100 less than discrete diodes.

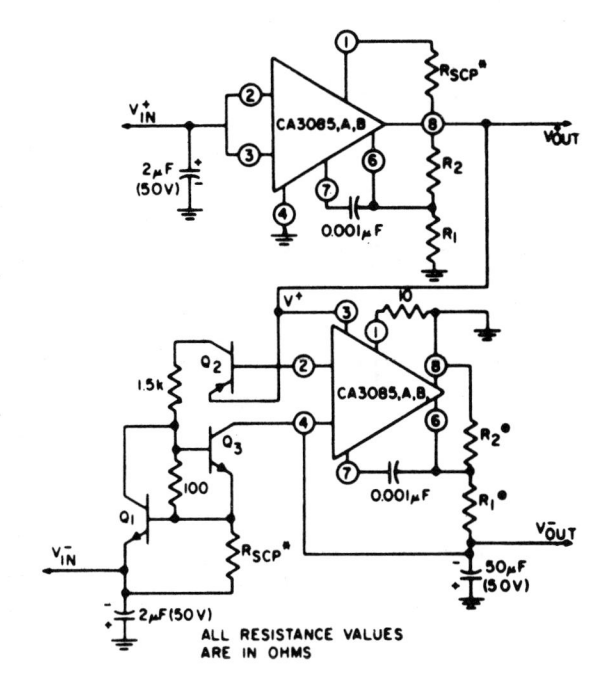

ALL RESISTANCE VALUES ARE IN OHMS

Q1 RCA-2N2102 OR EQUIVALENT
Q2 ANY P-N-P SILICON TRANSISTOR (RCA-2N5322 OR EQUIVALENT)
Q3 ANY N-P-N SILICON TRANSISTOR THAT CAN HANDLE THE DESIRED LOAD CURRENT (RCA-2N3772 OR EQUIVALENT)

*$V_{OUT} = \left(\frac{R_1 + R_2}{R_1}\right)$

*R_{SCP} SHORT-CIRCUIT PROTECTION RESISTANCE

Combination positive- and negative-voltage regulator circuit.

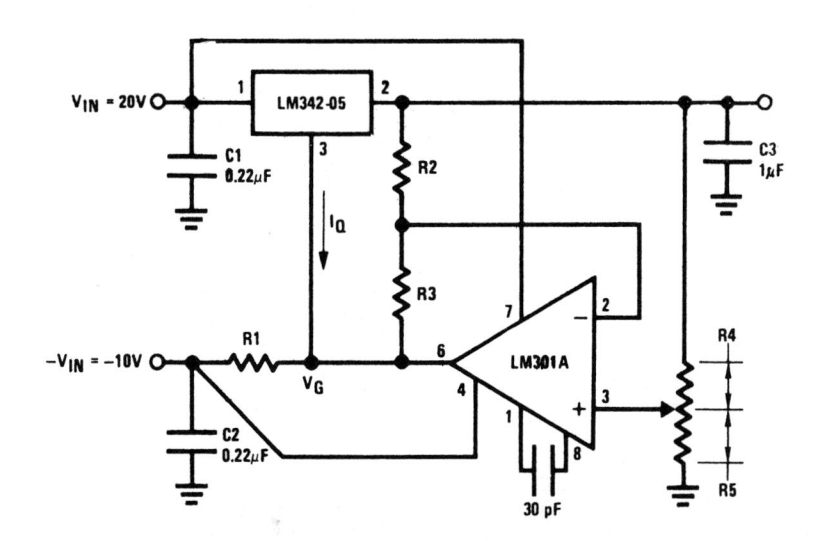

$V_{OUT} = V_G + 5V$, $R1 = (-V_{IN}/I_Q \text{ LM342})$
$V_{OUT} = 5V(R2/R4)$ for $(R2 + R3) = (R4 + R5)$

Variable-output regulator, 0.5–18V. A 0.5V output will correspond to $(R2/R4) = 0.1$, $(R3/R4) = 0.9$. Capacitor C3 should be solid tantalum.

SECTION 1 *Voltage, Current and Power Sources* 25

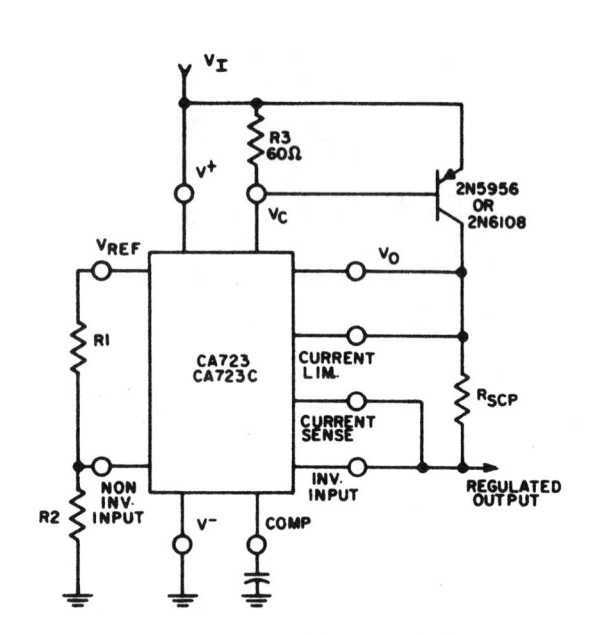

A positive-voltage regulator circuit with external PNP pass transistor. Regulated output voltage, 5V; line regulation ($\Delta V_I = 3V$), 0.5 mV; load regulation ($\Delta I_L = 1A$), 5 mV.

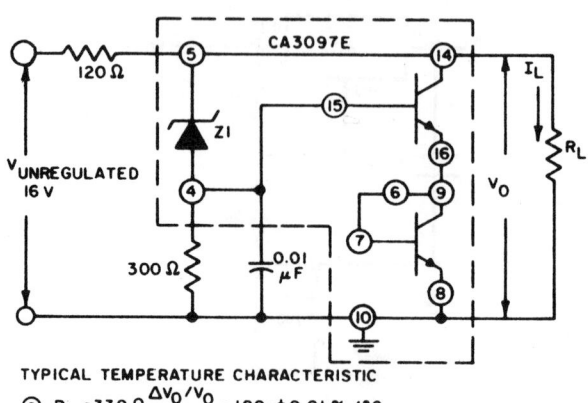

Temperature-compensated shunt regulator.

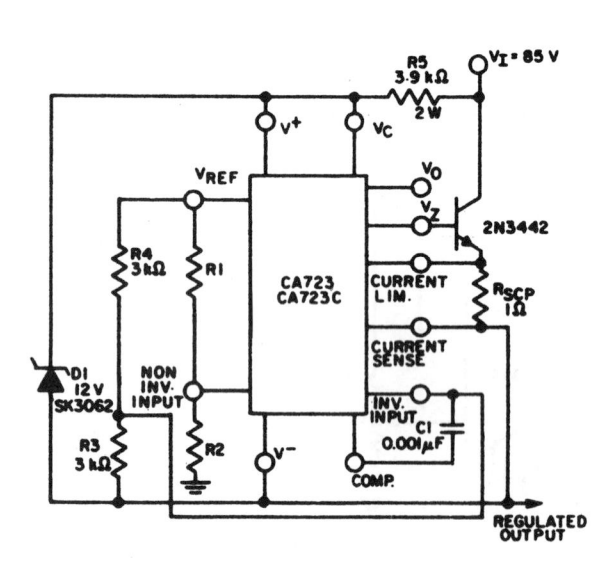

TYPICAL TEMPERATURE CHARACTERISTIC
@ $R_L = 330\,\Omega\ \dfrac{\Delta V_O/V_O}{\Delta T} \times 100 = \pm 0.01\%/°C$

TYP. LOAD REGULATION @ $I_L = 0$ TO 40 mA, ($\Delta V_O/V_O) \times 100 = -3\%$ (NO LOAD TO FULL LOAD)

TYP. LINE REGULATION @ $R_L = 330\,\Omega,\ \dfrac{\Delta V_O/V_O}{\Delta V_{UNREG}} \times 100 = \pm 0.55\%/V$

Positive floating regulator circuit. For applications employing the TO-5 style package and where V_Z is required, an external 6.2V zener diode should be connected in series with V_0 (terminal 6). Regulated output voltage, 50V; line regulation ($\Delta V_I = 20V$), 15 mV; load regulation ($\Delta I_L = 50$ mA), 20 mV.

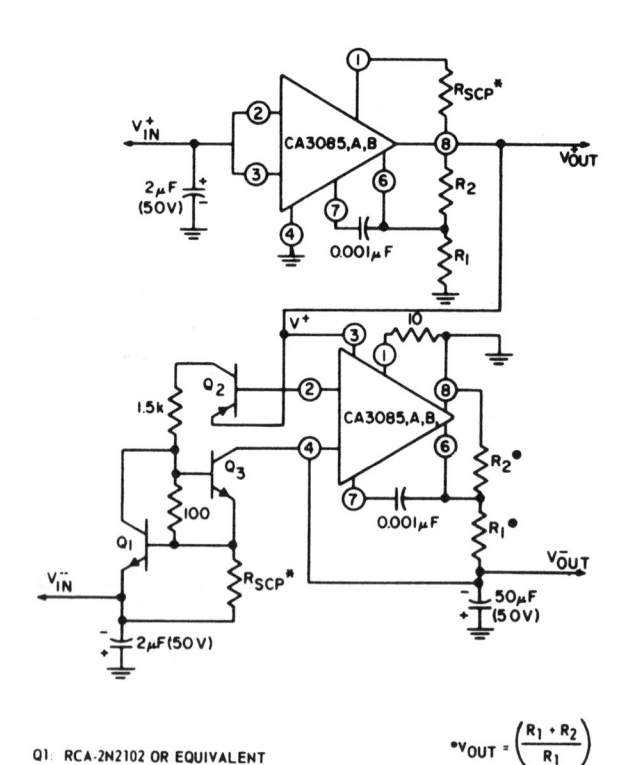

Q1: RCA-2N2102 OR EQUIVALENT
Q2: ANY P-N-P SILICON TRANSISTOR (RCA-2N5322 OR EQUIVALENT)
Q3: ANY N-P-N SILICON TRANSISTOR THAT CAN HANDLE THE DESIRED LOAD CURRENT (RCA-2N3772 OR EQUIVALENT)

*$V_{OUT} = \left(\dfrac{R_1 + R_2}{R_1}\right)$

*R_{SCP}: SHORT-CIRCUIT PROTECTION RESISTANCE

Combination positive- and negative-voltage regulator circuit.

26 IC SCHEMATIC SOURCEMASTER

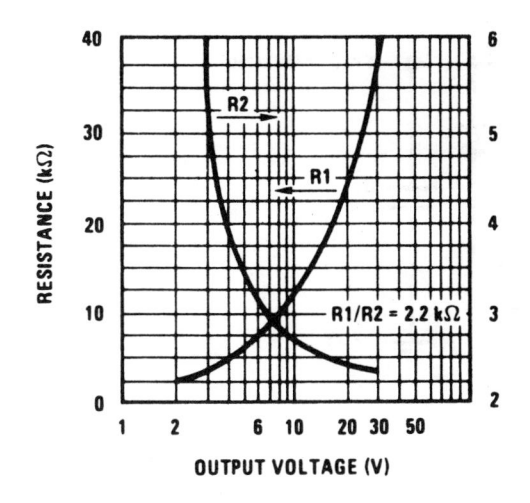

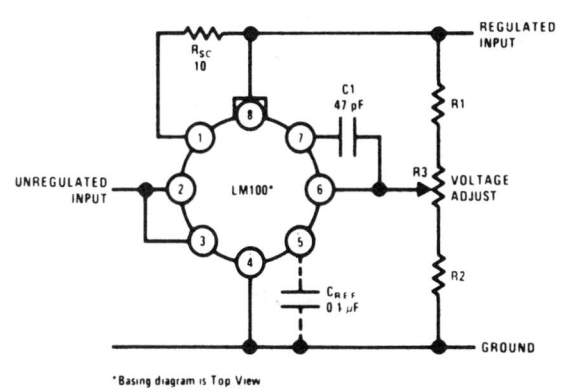

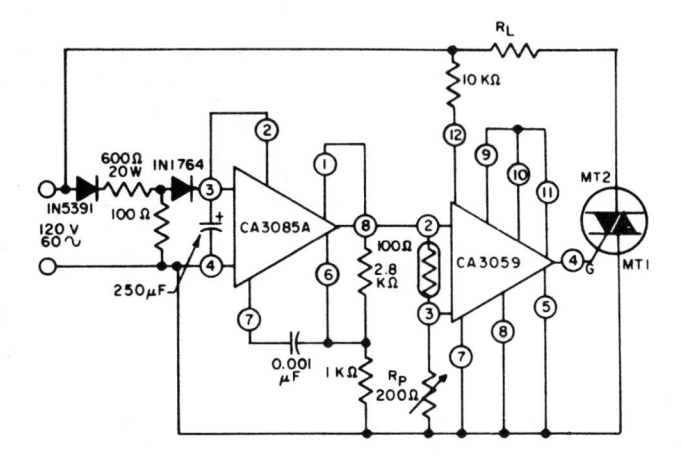

Voltage regulator for sensor and zero-voltage switch. Terminal 12 on the CA3059 provides the ac trigger signal that actuates the zero-voltage switch synchronously with the power line to control the load-switching triac.

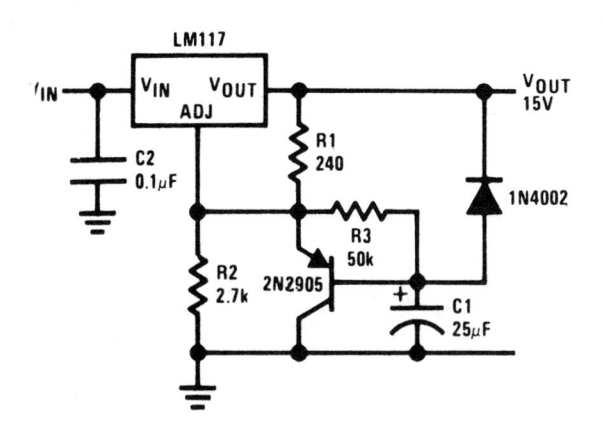

Basic regulator circuit. The output voltage is set by R1 and R2, with a fine adjustment provided by potentiometer R3. The resistance seen by the feedback terminal should be approximately 2.2k to minimize drift caused by the bias current on this terminal. The chart is based on this and gives the optimum values for R1 and R2 as a function of design-center output voltage. The potentiometer should be at ¼ of R2 to ensure that the output can be set to the desired voltage.

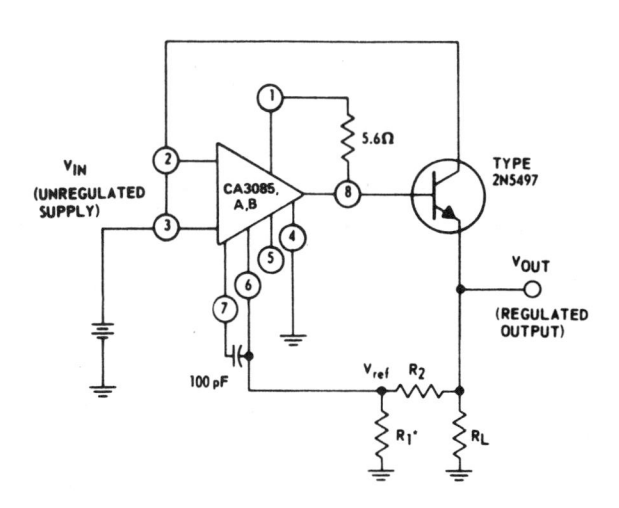

Slow turn-on 15V regulator.

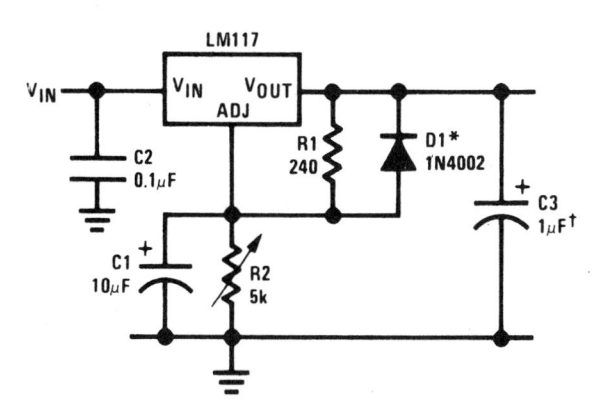

High-current voltage regulator circuit. Select R1 and R2 for desired output. $V_{OUT} = V_{ref}[(R2 + R1)/R1]$.

Adjustable regulator with improved ripple rejection. C1 and C3 should be solid tantalum. D1 discharges C1 if output is shorted to ground.

SECTION 1 *Voltage, Current and Power Sources* **27**

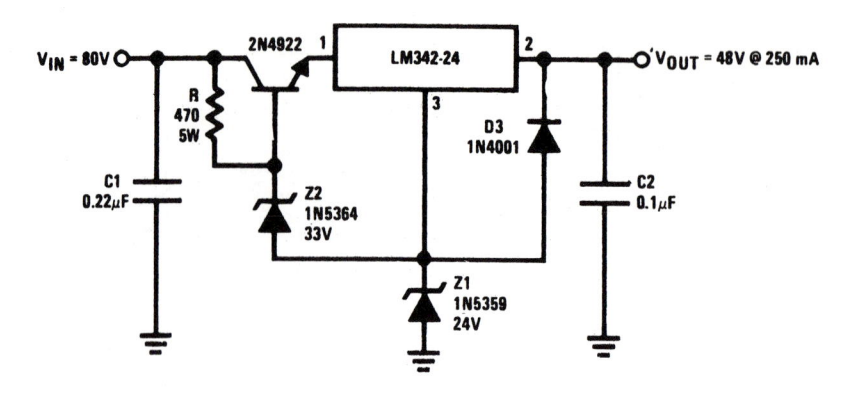

High-output voltage regulator. C1 is only necessary if regulator is located far from the power supply filter. D3 aids in full load startup and protects the regulator during short circuits from high input to output voltage differentials.

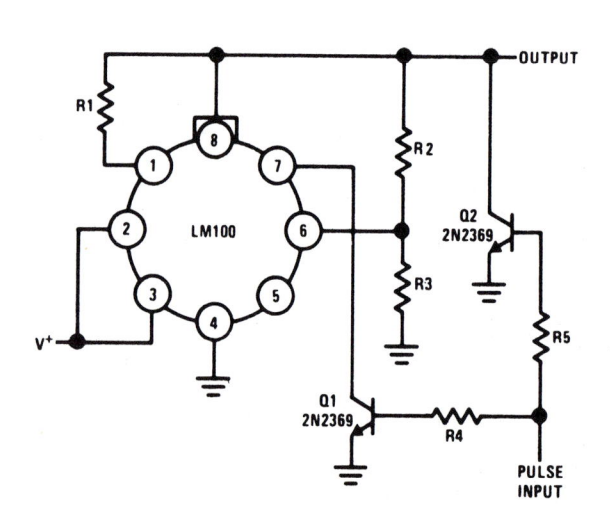

Pulse regulator. R2 and R3 are set up to give the desired pulse height (dc) from the LM100. A positive-pulse input turns on Q1, which disables the LM100 by grounding the base of the NPN emitter followers on the output of the integrated circuit. At the same time, Q2 grounds the regulator output, providing current-sinking capability.

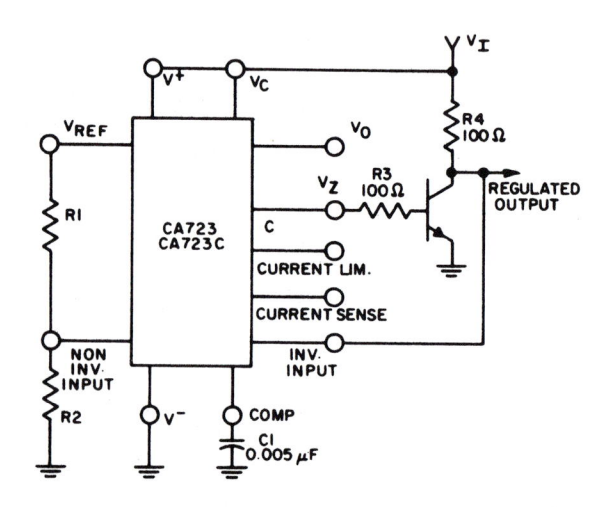

Shunt regulator circuit. For applications employing the TO-5 style package and where V_Z is required, an external 6.2V zener diode should be connected in series with V_O (terminal 6). Regulated output voltage, 5V; line regulation ($\Delta V_I = 10V$), 0.5 mV; load regulation ($\Delta I_L = 100$ mA), 1.5 mV.

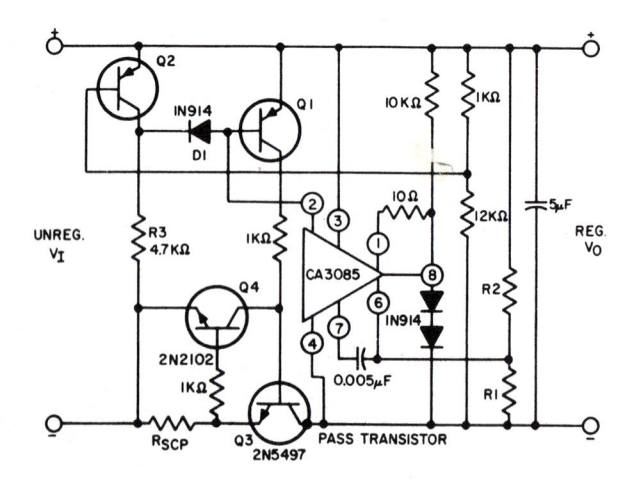

A negative-voltage regulator with constant-current-limiting circuit.

28 IC SCHEMATIC SOURCEMASTER

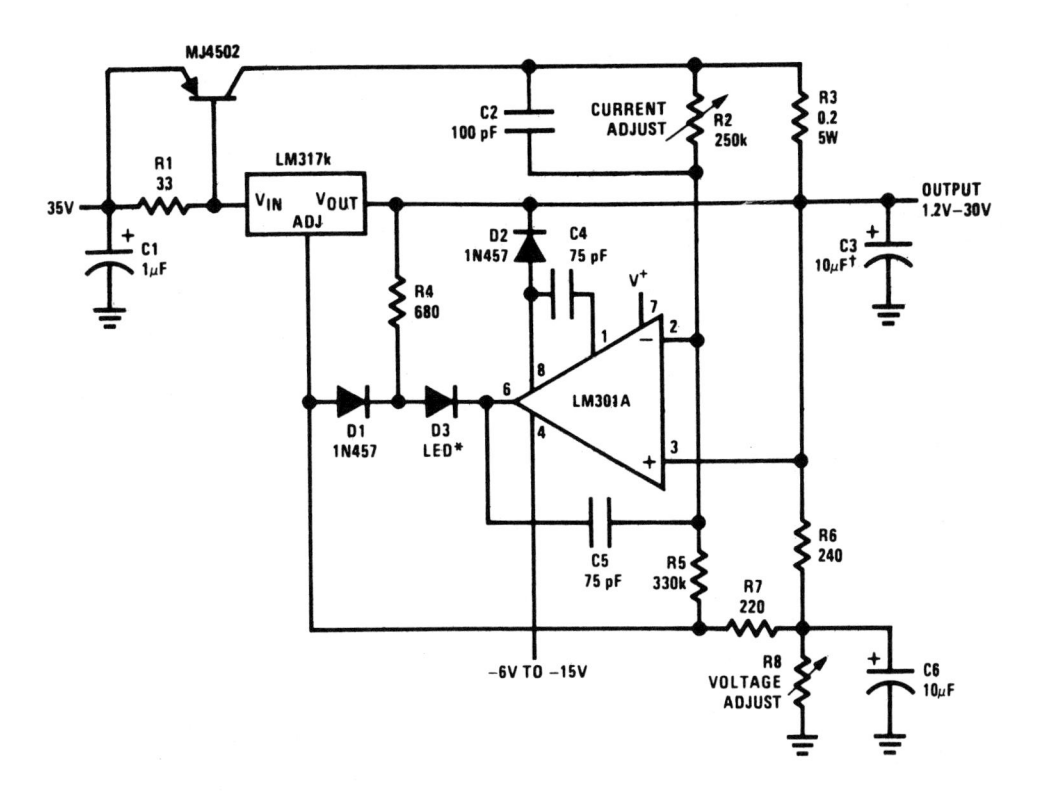

5A constant-voltage/constant-current regulator. C1 should be solid tantalum. D3 lights in constant-current mode.

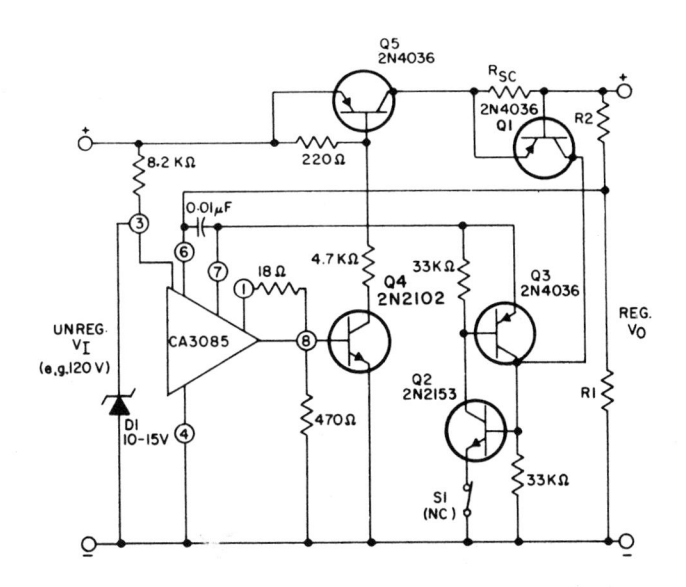

A high-voltage regulator incorporating current "snap-back" protection. When a sufficient voltage drop is developed across R_{SC}, transistor Q1 becomes conductive and current flows into the base of Q2 so that it also becomes conductive. Transistor Q3, in turn, is driven into conduction, thereby latching the Q2-Q3 combination (basic SCR action) so that it diverts (through terminal 7) base drive from the output stage (Q13, Q14) in the CA3085. By this means, base drive is diverted from Q4 and the pass transistor, Q5. To restore regulator operation, normally closed switch S1 is momentarily opened and unlatches Q2-Q3.

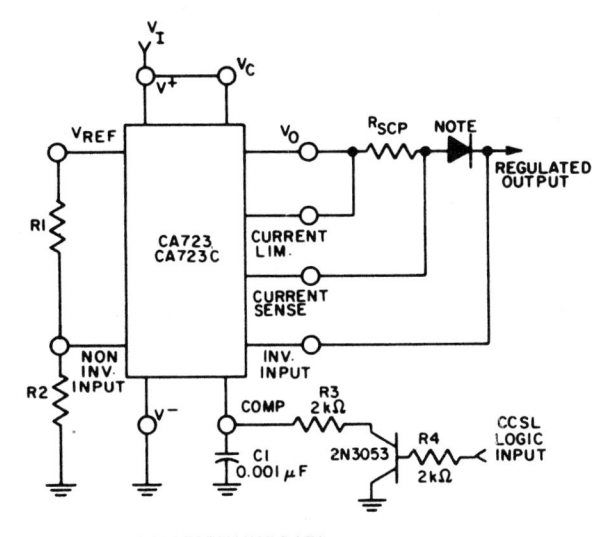

CIRCUIT PERFORMANCE DATA:
REGULATED OUTPUT VOLTAGE . . . . 5 V
LINE REGULATION (ΔV_I = 3 V) . . . . 0.5 mV
LOAD REGULATION (ΔI_L = 50 mA) . . 1.5 mV

Note : Add a diode if V_O > 10 V.

A remote shutdown regulator circuit with current limiting. A current-limiting transistor may be used for shutdown if current limiting is not required. Regulated output voltage, 5V; line regulation (ΔV_I = 3V), 0.5 mV; load regulation (ΔI_L = 50 mA), 1.5 mV.

SECTION **1** *Voltage, Current and Power Sources* **29**

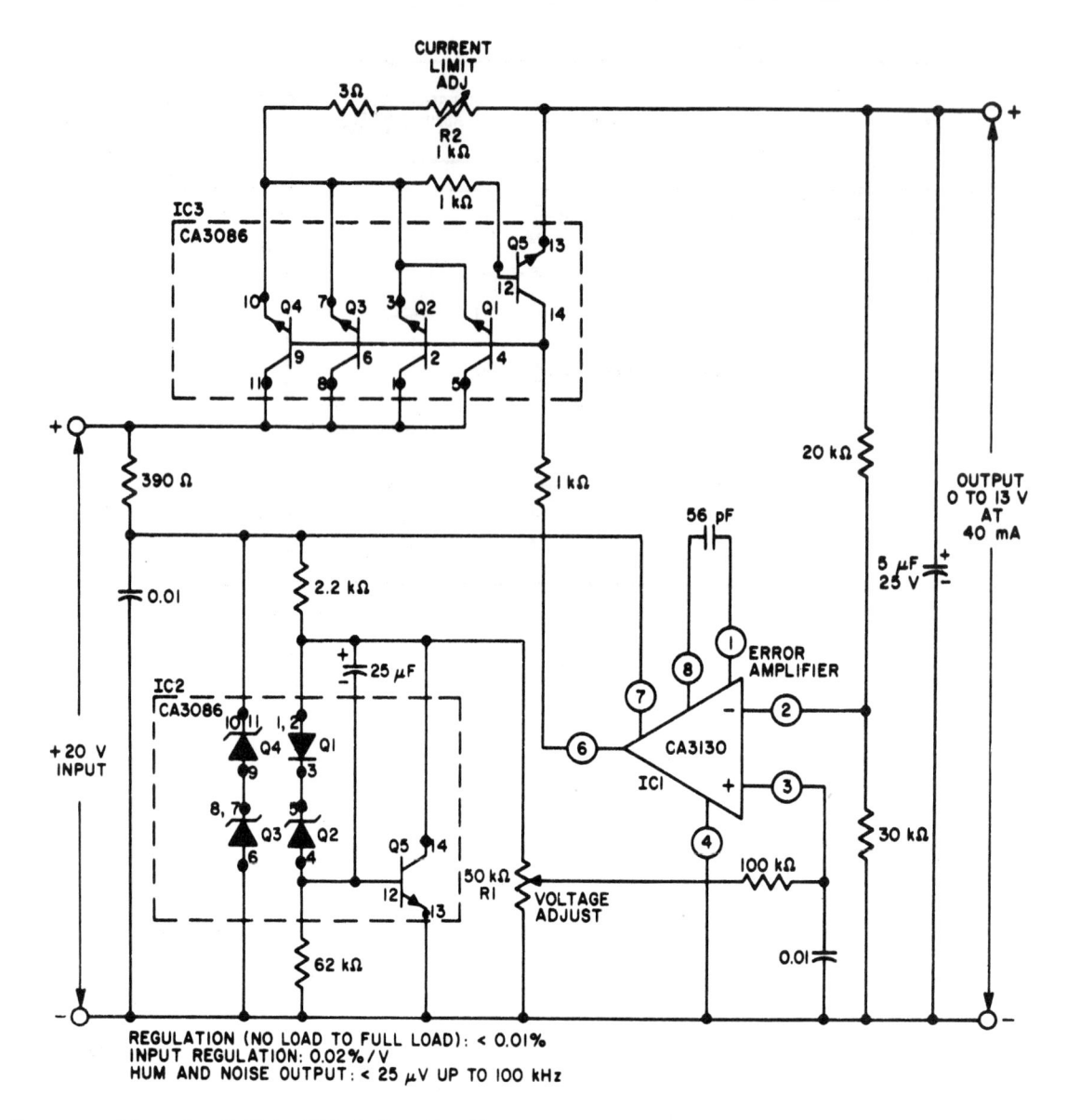

A **voltage regulator circuit** (0 to 13V at 40 mA) with error amplifier that functions when the regulated output voltage is required to approach zero. Shown is a 40 mA power supply capable of providing regulated output voltage by continuous adjustment over the range from 0 to 13V. Q3 and Q4 in IC2 (a CA3086 transistor-array IC) function as zeners to provide supply voltage for the CA3130 comparator (IC1). Q1, Q2, and Q5 in IC2 are configured as a low-impedance temperature-compensated source of adjustable reference voltage for the error amplifier. Transistors Q1, Q2, Q3, and Q4 in IC3 (another CA3086 transistor-array IC) are connected in parallel as the series-pass element. Transistor Q5 in IC3 functions as a current-limiting device by diverting base drive from the series-pass transistors in accordance with the adjustment of resistor R2.

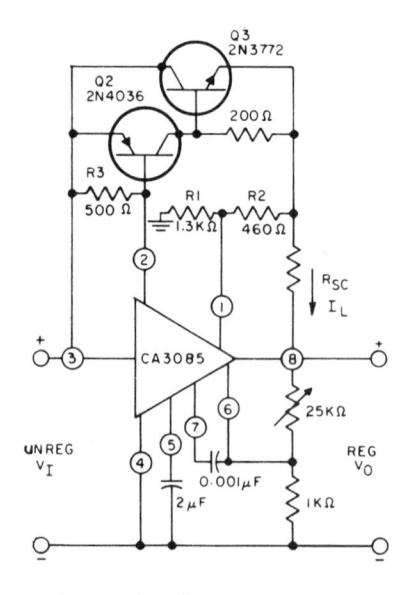

A **high-output-current voltage regulator** with foldback current limiting.

30   IC SCHEMATIC SOURCEMASTER

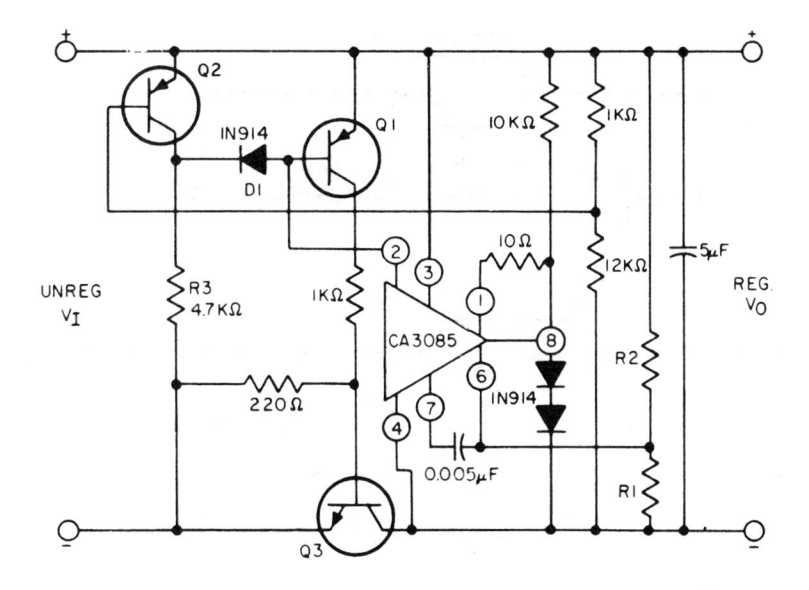

Negative-voltage regulator. Diode D1 is used initially in a circuit-starter function; transistor Q2 latches D1 out of its starter-circuit function so that the CA3085 can assume its role in controlling pass transistor Q3 by means of Q1.

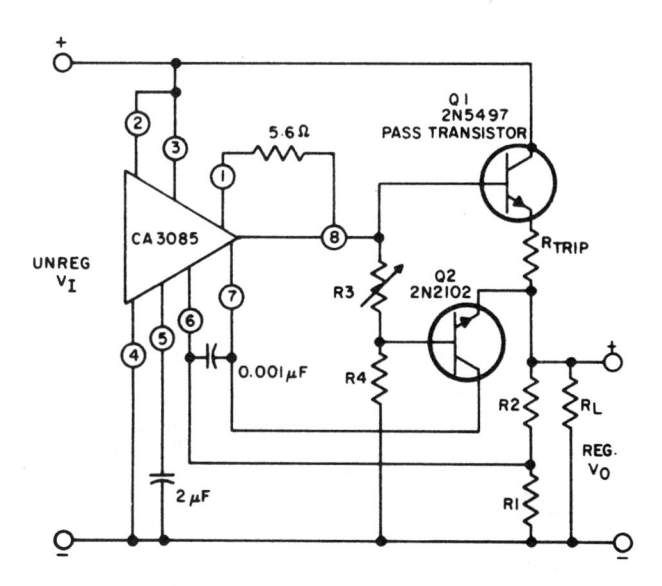

A high-output-current voltage regulator using auxiliary transistor to provide foldback current limiting.

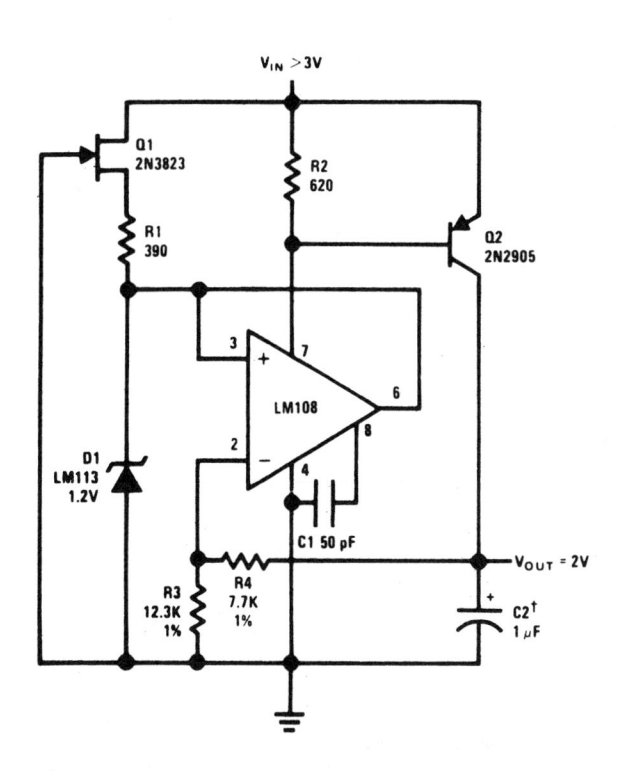

Low-voltage regulator. Capacitor C2 should be solid tantalum.

SECTION 1 *Voltage, Current and Power Sources* 31

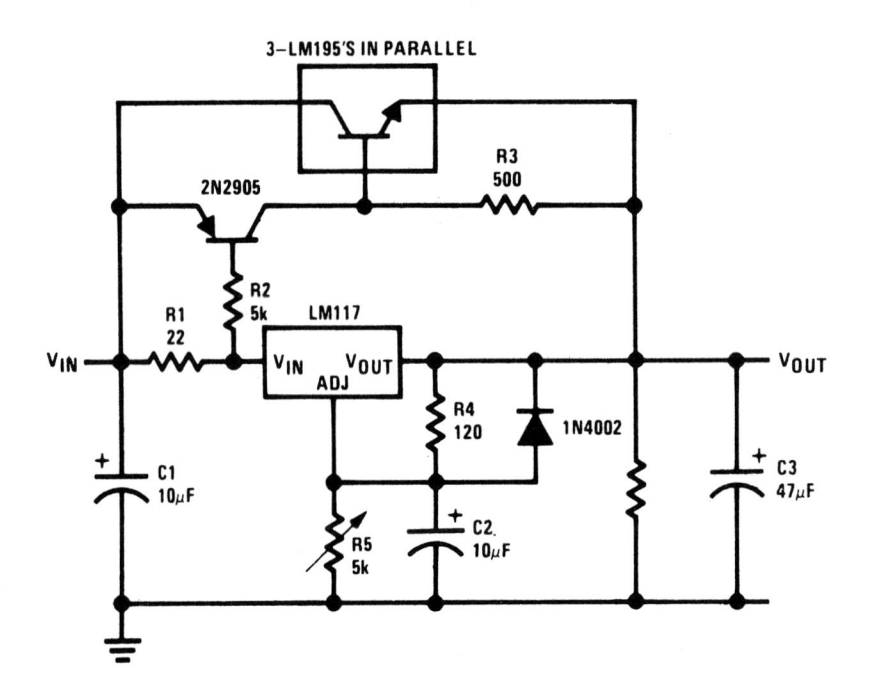

High-current adjustable regulator. Capacitors should be solid tantalum. C2 is optional for improved ripple rejection. Minimum load current, 30 mA.

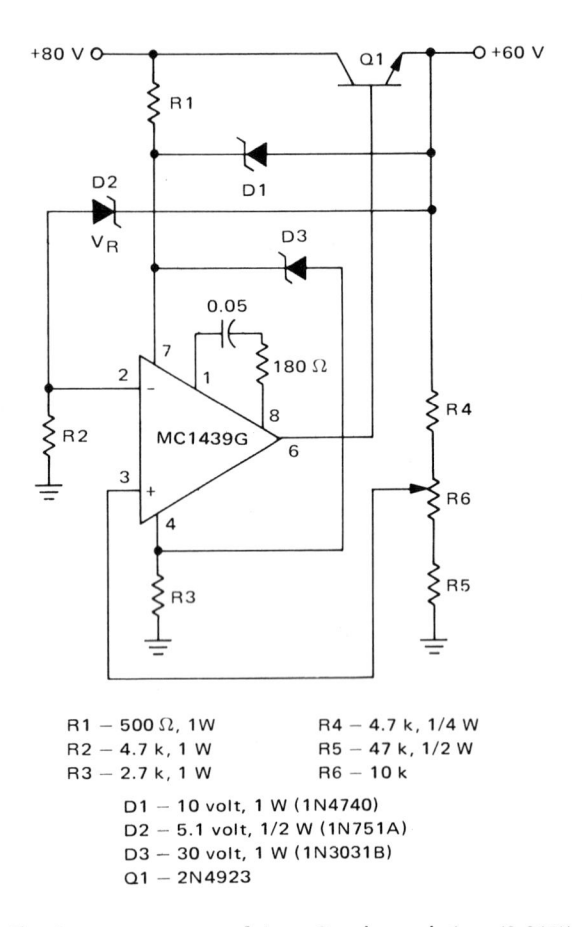

R1 — 500 Ω, 1 W R4 — 4.7 k, 1/4 W
R2 — 4.7 k, 1 W R5 — 47 k, 1/2 W
R3 — 2.7 k, 1 W R6 — 10 k
D1 — 10 volt, 1 W (1N4740)
D2 — 5.1 volt, 1/2 W (1N751A)
D3 — 30 volt, 1 W (1N3031B)
Q1 — 2N4923

Floating op-amp regulator. Good regulation (0.01%) can be achieved in this way, but loop gain decreases with higher output voltages and the result is not entirely satisfactory when $V_O \geq 10 \times V_R$.

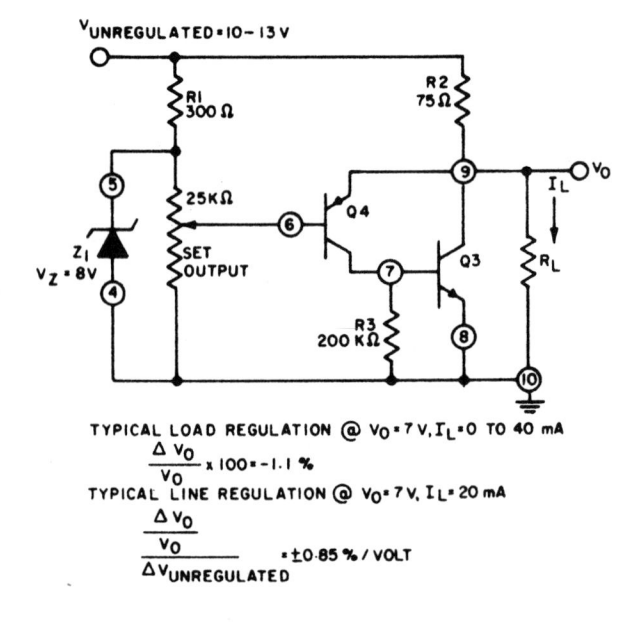

5 to 7.5V shunt regulator using CA3097E.

32 IC SCHEMATIC SOURCEMASTER

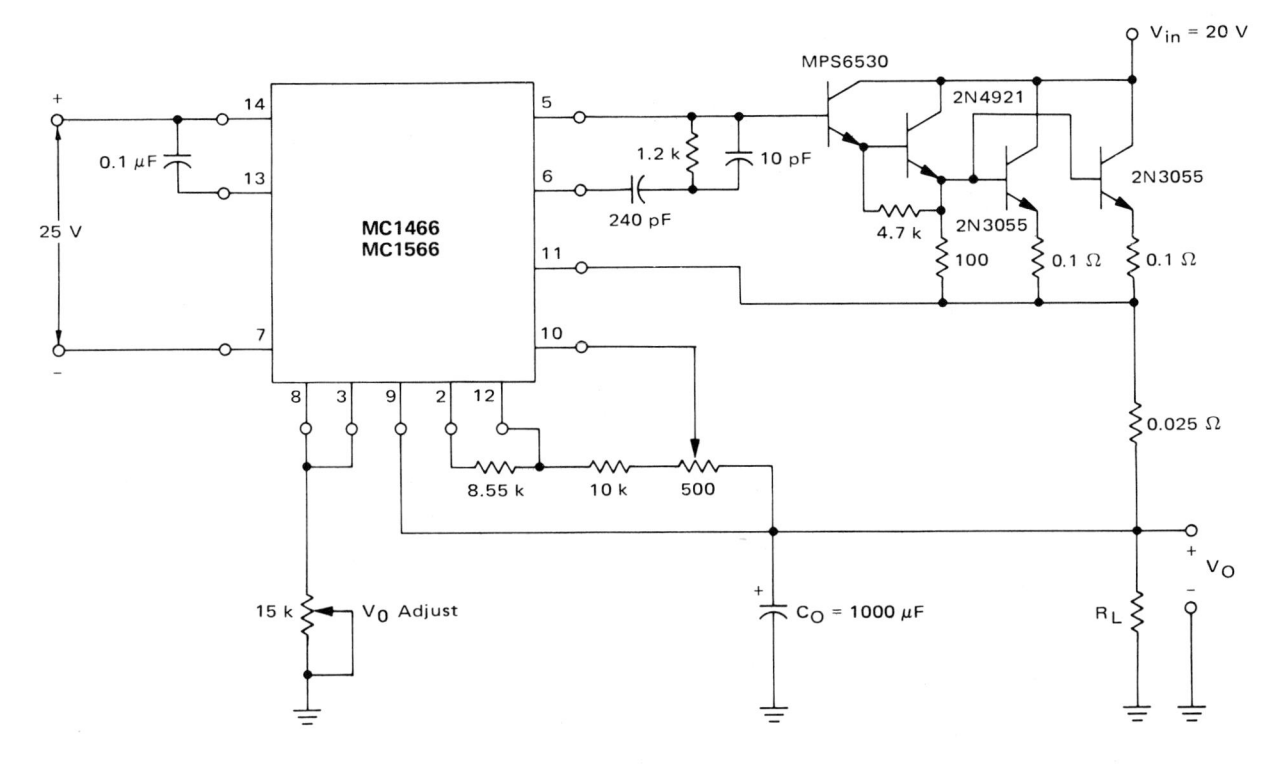

0 to 15V, 10A regulator with pass transistor to boost the load current to 10A.

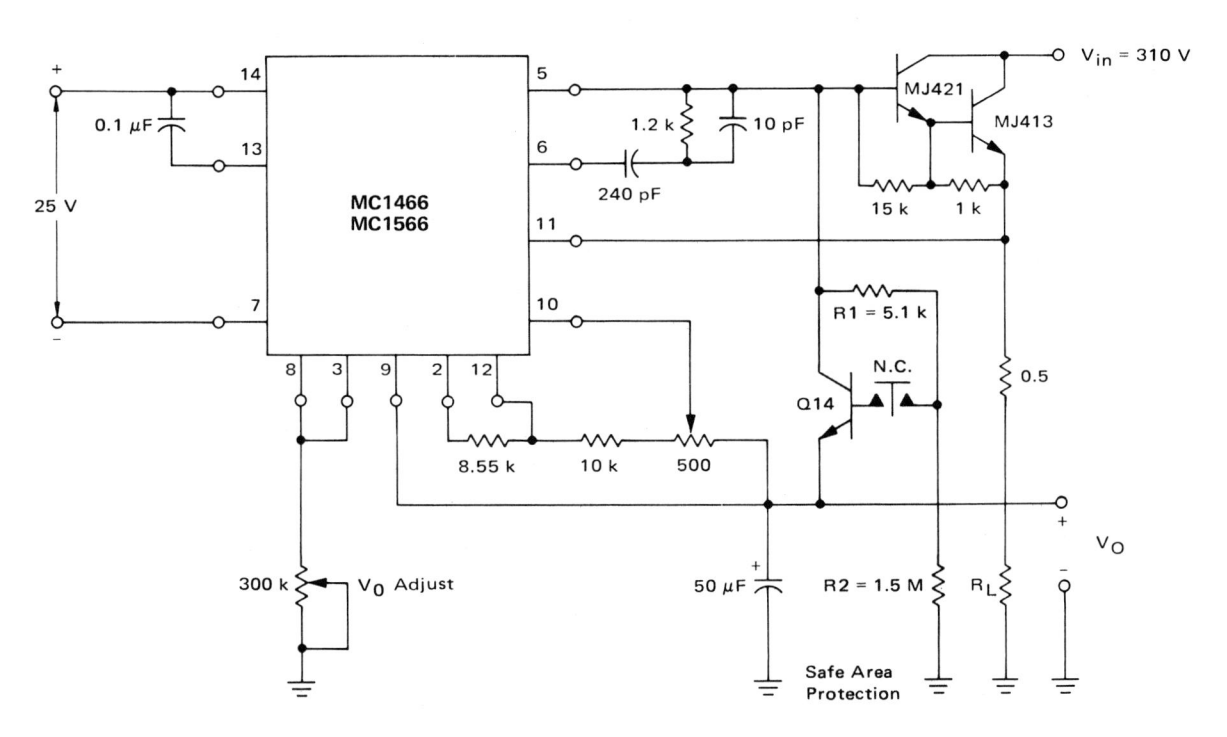

300V, 0.5A regulator with safe-area protection employs an automatic shutdown technique to protect the pass transistors under short-circuit conditions. Q14 provides a positive latching shutdown when output voltage drops to a value established by R1 and R2. Choose R1 according to shutdown voltage desired:

$$R1 \approx \frac{0.8R2}{V_{shutdown}}$$

SECTION 1 *Voltage, Current and Power Sources*

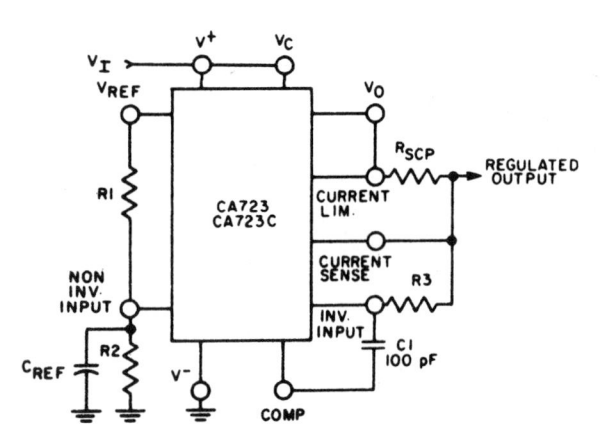

Low-voltage regulator (V_0 = 2–7V).

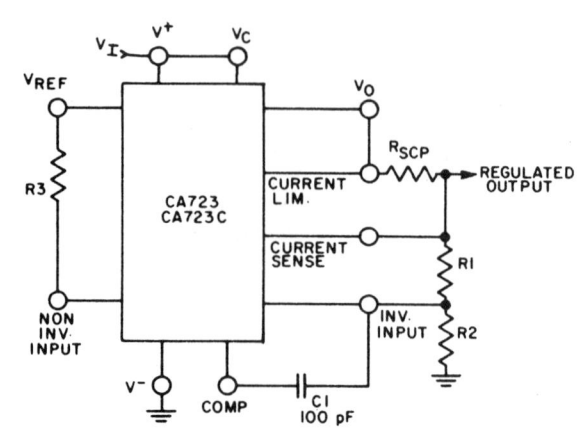

CIRCUIT PERFORMANCE DATA:
REGULATED OUTPUT VOLTAGE . . . 5 V
LINE REGULATION (ΔV_I = 3 V) . . . . 0.5 mV
LOAD REGULATION (ΔI_L = 50 mA) . . . 1.5 mV

Note: $R3 = \frac{R1\,R2}{R1+R2}$ for minimum temperature drift

High-voltage regulator (V_0 = 7–37V).

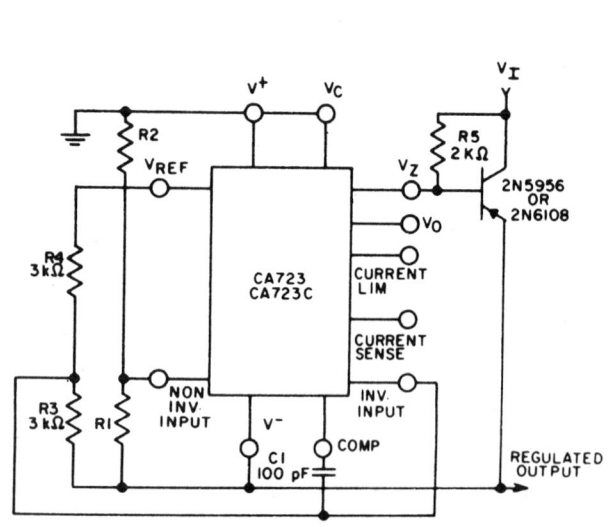

CIRCUIT PERFORMANCE DATA:
REGULATED OUTPUT VOLTAGE . . . 15 V
LINE REGULATION (ΔV_I = 3 V) . . . . 1.5 mV
LOAD REGULATION (ΔI_L = 50 mA) . . 4.5 mV

Note: $R3 = \frac{R1\,R2}{R1+R2}$ for minimum temperature drift

R3 may be eliminated for minimum component count.

Negative-voltage regulator.

CIRCUIT PERFORMANCE DATA:
REGULATED OUTPUT VOLTAGE . . . −15 V
LINE REGULATION (ΔV_I = 3 V) . . . . 1 mV
LOAD REGULATION (ΔI_L = 100 mA) . . 2 mV

Note: For applications employing the TO-5 style package and where V_Z is required, an external 6.2-volt zener diode should be connected in series with V_O (Terminal 6).

Positive-voltage regulator with external PNP pass transistor.

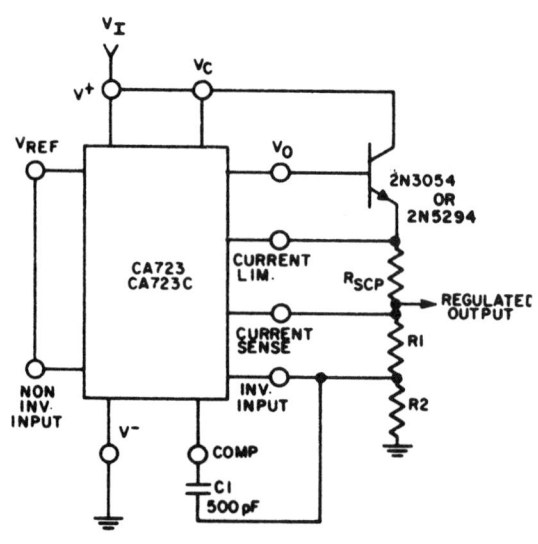

CIRCUIT PERFORMANCE DATA:
REGULATED OUTPUT VOLTAGE . . . 15 V
LINE REGULATION (ΔV_I = 3 V) . . . . 1.5 mV
LOAD REGULATION (ΔI_L = 1 A) . . . . 15 mV

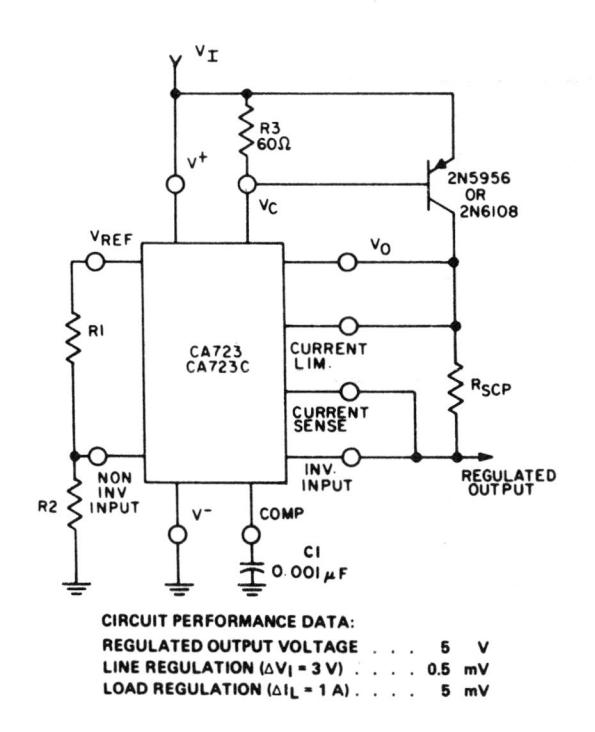

Positive-voltage regulator with external NPN pass transistor.

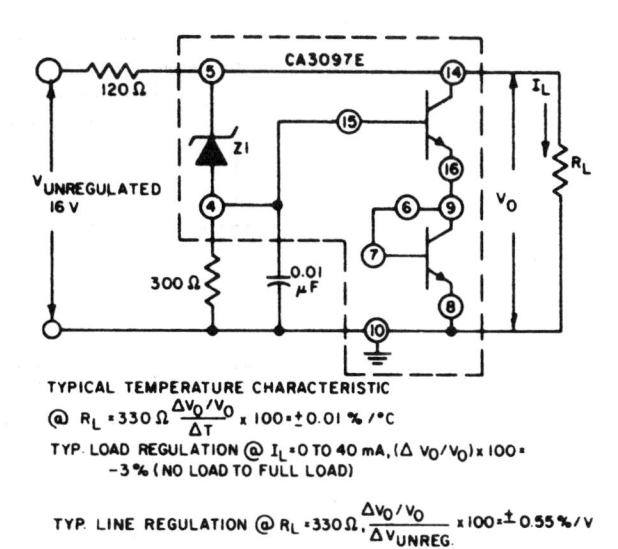

CIRCUIT PERFORMANCE DATA:
REGULATED OUTPUT VOLTAGE ... 5 V
LINE REGULATION ($\Delta V_I = 3$ V) ... 0.5 mV
LOAD REGULATION ($\Delta I_L = 1$ A) ... 5 mV

Temperature-compensated shunt regulator.

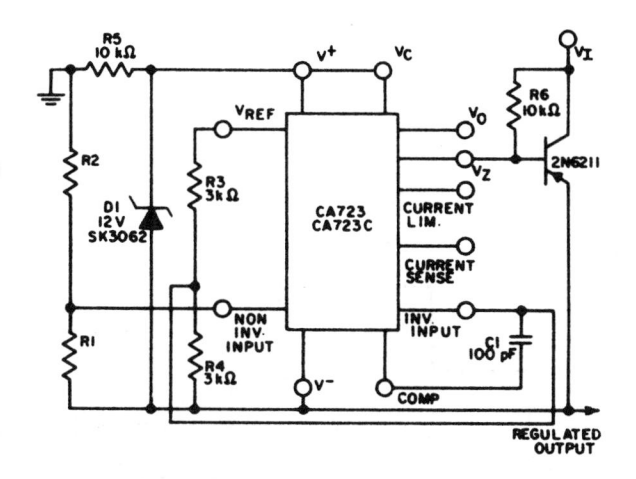

TYPICAL TEMPERATURE CHARACTERISTIC
@ $R_L = 330\,\Omega$, $\frac{\Delta V_O / V_O}{\Delta T} \times 100 = \pm 0.01\,\%/°C$

TYP. LOAD REGULATION @ $I_L = 0$ TO 40 mA, $(\Delta V_O/V_O) \times 100 = -3\%$ (NO LOAD TO FULL LOAD)

TYP. LINE REGULATION @ $R_L = 330\,\Omega$, $\frac{\Delta V_O / V_O}{\Delta V_{UNREG}} \times 100 = \pm 0.55\,\%/V$

Negative-floating regulator circuit. For applications employing the TO-5 style package and where V_Z is required, an external 6.2V zener diode should be connected in series with V_O (terminal 6). Regulated output voltage, −100V; line regulation ($\Delta V_I = 20$V), 30 mV; load regulation ($\Delta I_L = 100$ mA), 20 mV.

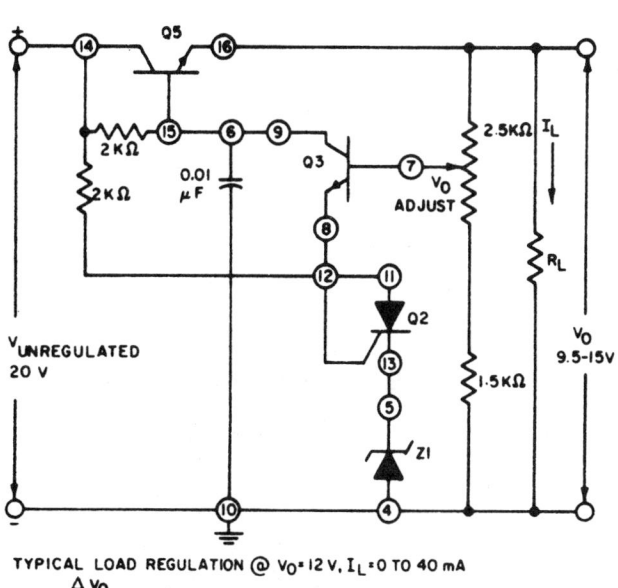

TYPICAL LOAD REGULATION @ $V_O = 12$ V, $I_L = 0$ TO 40 mA
$\frac{\Delta V_O}{V_O} \times 100 = \pm 0.4\%$ (NO LOAD TO FULL LOAD)

TYPICAL LINE REGULATION @ $V_O = 12$ V
$\frac{\Delta V_O / V_O}{\Delta V_{UNREG}} \times 100 = \pm 0.45\,\%/V$

Series voltage regulator. Numbers shown are for CA3097E.

SECTION **1** *Voltage, Current and Power Sources* **35**

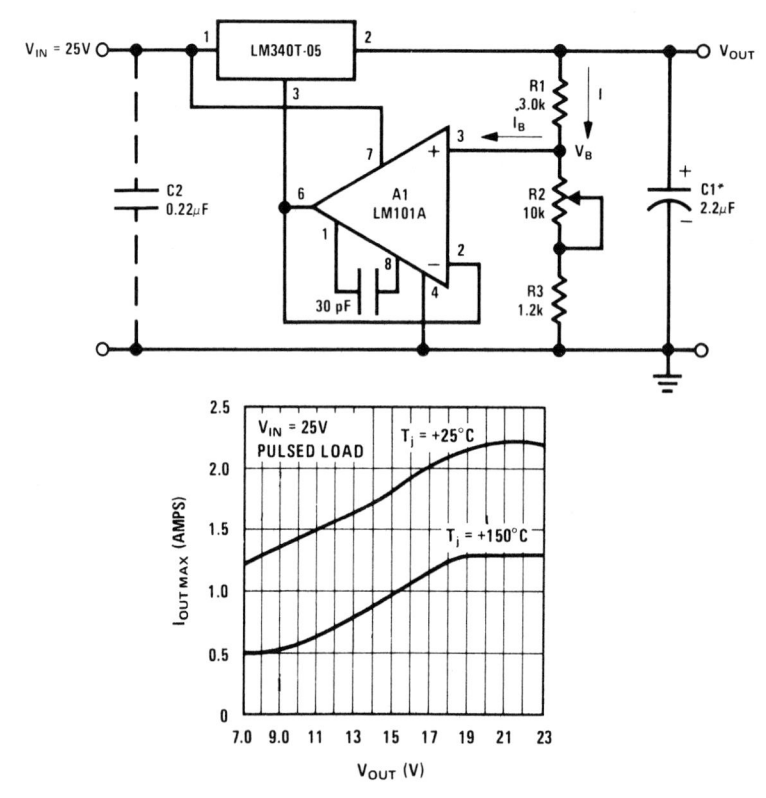

Maximum Output Current

In this **variable-output regulator**, the ground terminal is "lifted" by an amount equal to the voltage applied to the noninverting input of the operational amplifier (LM101A). The output voltage of the regulator is therefore raised to a level set by the value of the resistive divider R1, R2, and R3 and is limited by the input voltage. With the resistor values shown, the output voltage is variable from 7.0 to 23V, and the maximum output current (pulsed load) varies from 1.2 to 2.0A ($T_j = 25°C$) as shown in the graph. C2 is necessary if the regulator is far from the power supply filter: C1 should be solid tantalum.

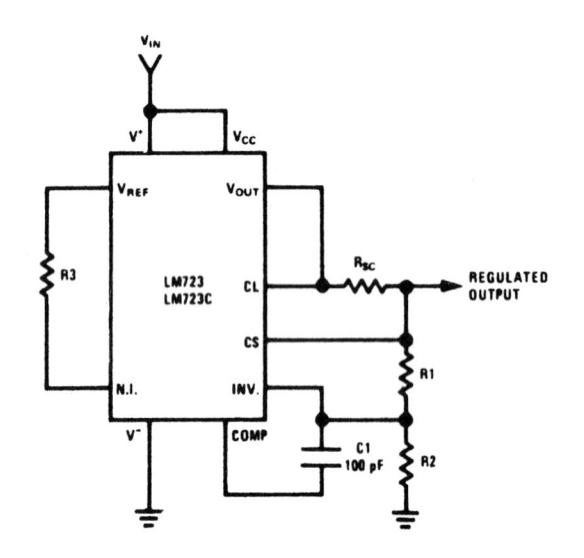

Basic high-voltage regulator. V_{OUT} = 7 to 37V, with 15V typical. Line regulation (ΔV_{IN} = 3V): 1.5 mV; load regulation (ΔI_L = 50 mA): 4.5 mV. R3 = R1R2/(R1 + R2) for minimum temperature drift. R3 may be eliminated for minimum component count.

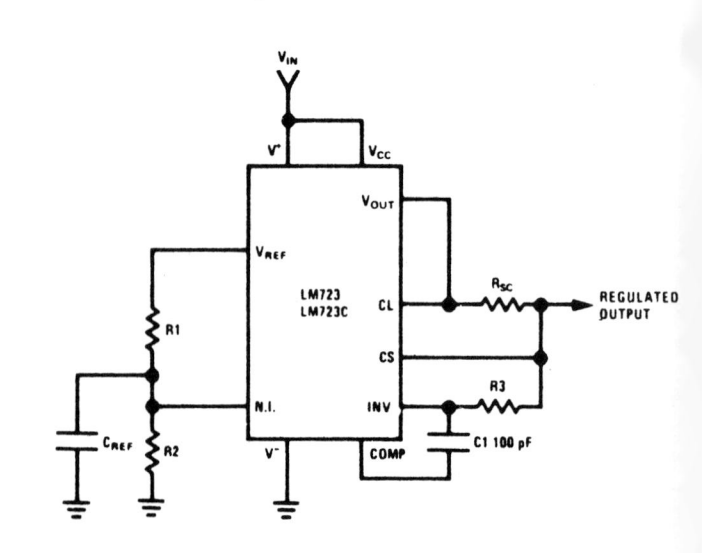

Basic low-voltage regulator. V_{OUT} = 2 to 7V, with 5V typical. Line regulation (ΔV_{IN} = 3V): 0.5 mV; load regulation (ΔI_L = 50 mA): 1.5 mV. R3 = R1R2/(R1 + R2) for minimum temperature drift.

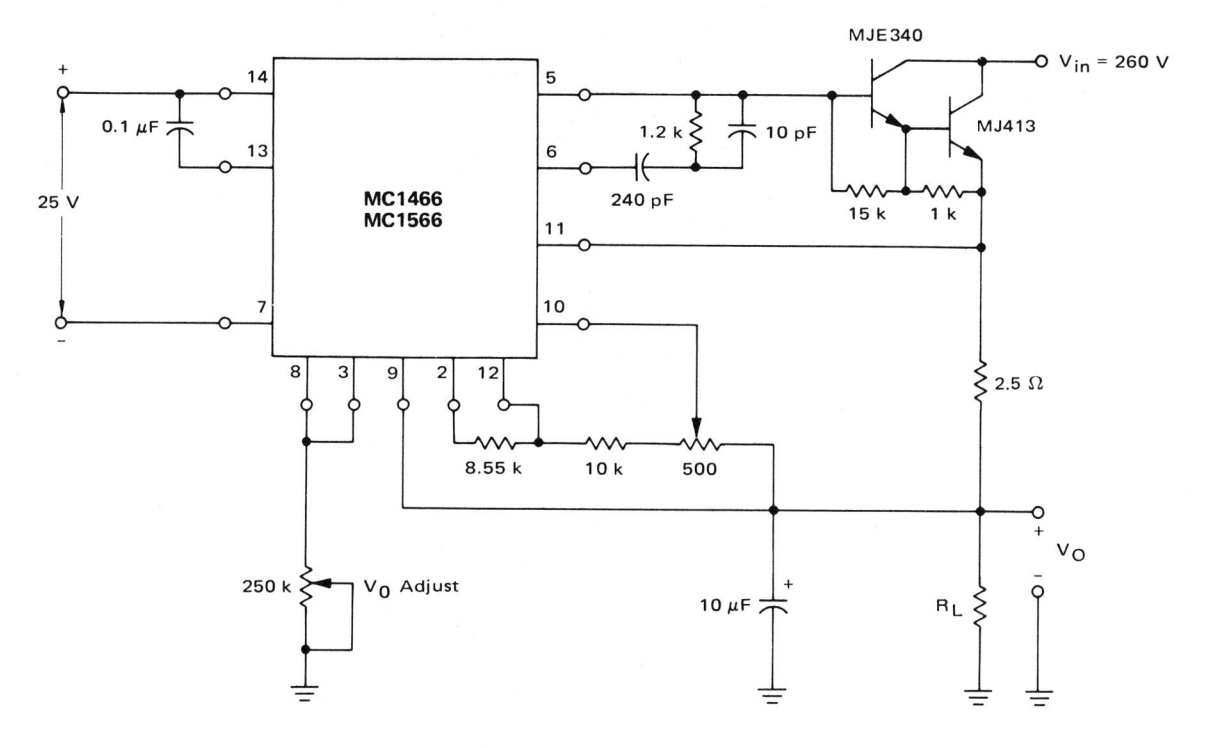

0 to 250V, 100 mA regulator is limited only by the breakdown and safe areas of the output pass transistors.

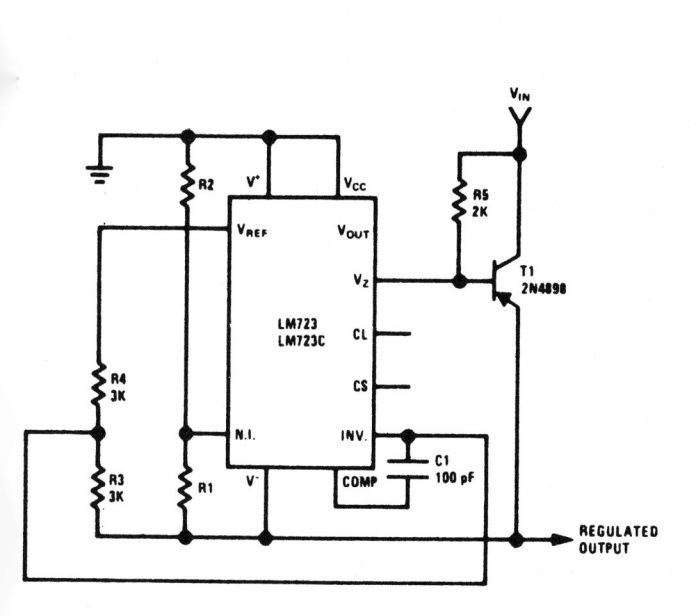

Negative-voltage regulator. $V_{OUT} = -15V$. Line regulation ($\Delta V_{IN} = 3V$): 1 mV; load regulation ($\Delta I_L = 100$ mA): 2 mV.

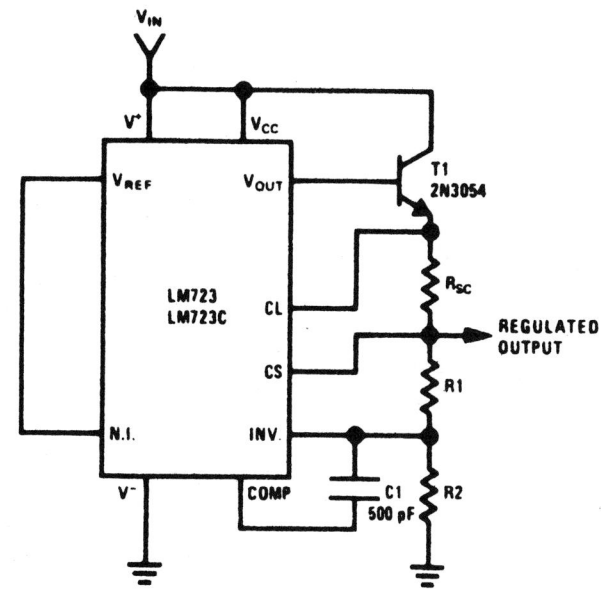

A positive-voltage regulator with external NPN pass transistor. $V_{OUT} = +15V$. Line regulation ($\Delta V_{IN} = 3V$): 1.5 mV; load regulation ($\Delta I_L = 1A$): 15 mV.

SECTION **1** Voltage, Current and Power Sources

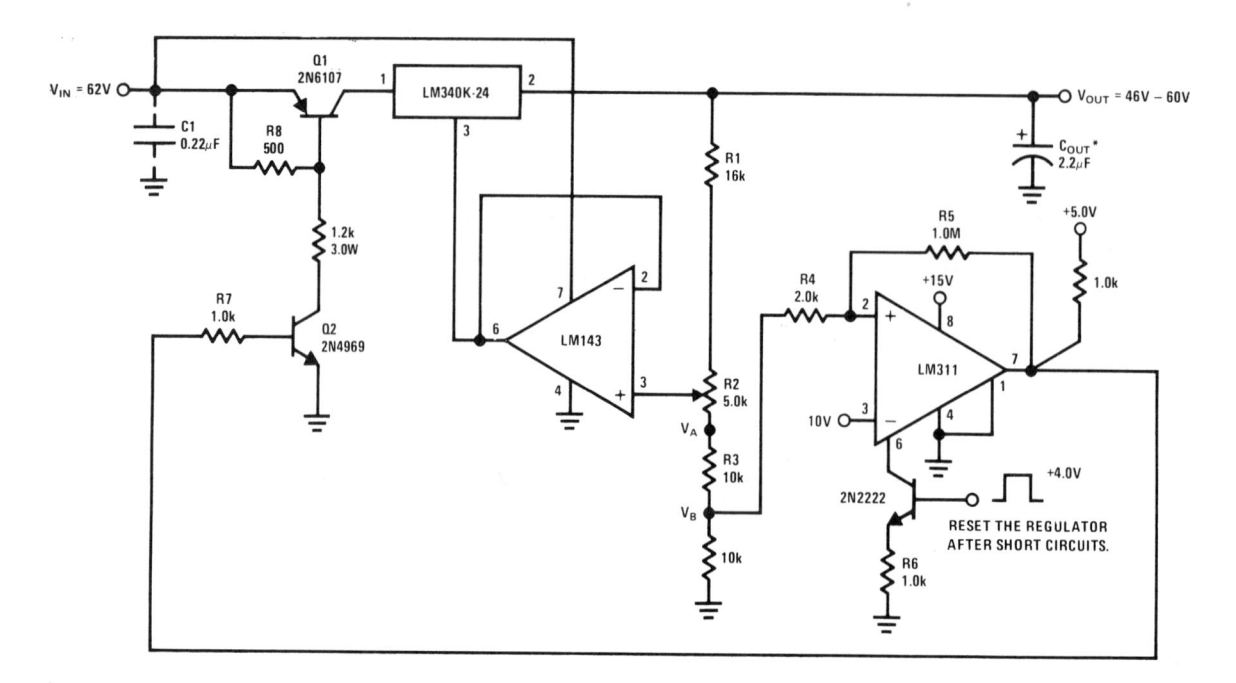

A variable high-voltage regulator with short-circuit and overvoltage protection. The principal inconvenience is that the voltage across the regulator must be limited to the maximum rating of the device; the higher the applied input voltage, the higher the ground pin of the LM340 must be lifted. The range of the variable output is limited by the supply voltage limit of the operational amplifier and the maximum voltage allowed across the regulator:

$$V_{OUT}(max) - V_{OUT}(min) = V_{supply}(max) - V(nom)$$

C_{OUT} should be solid tantalum.

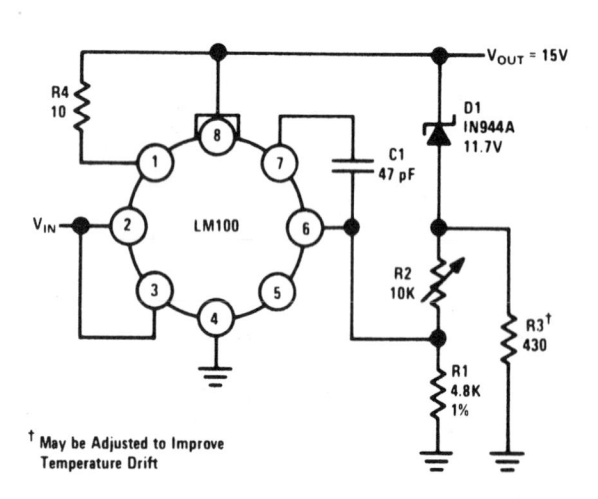

† May be Adjusted to Improve Temperature Drift

This high-stability regulator uses a reference diode in the feedback divider, which permits a lower divider ratio to be used and therefore improves regulation and drift characteristics.

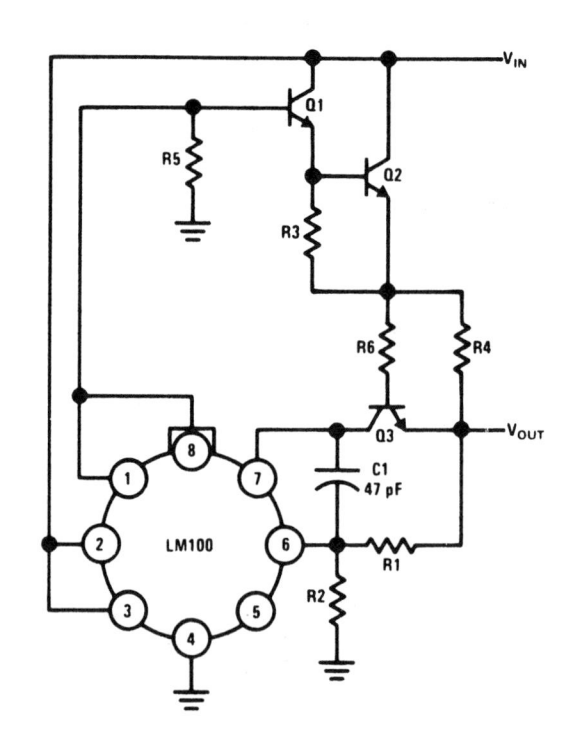

A regulator circuit for using the LM100 with all NPN pass elements. With this configuration, it is not possible to use the internal current limiting of the LM100, so an external transistor, Q3, must be added to provide this function. Limiting occurs when the voltage drop across R4 is equal to the emitter-base voltage of Q3. R5 is also required to make sure that the LM100 is operated above its minimum load current.

The main advantage of using all NPN pass transistors is that the circuit can be operated with less capacitance on the output of the regulator. When NPN and PNP transistors are used, relatively large (1–10 pF) bypass capacitors must be connected on both the input and output of the regulator. Without these, the circuit is susceptible to oscillations.

38 IC SCHEMATIC SOURCEMASTER

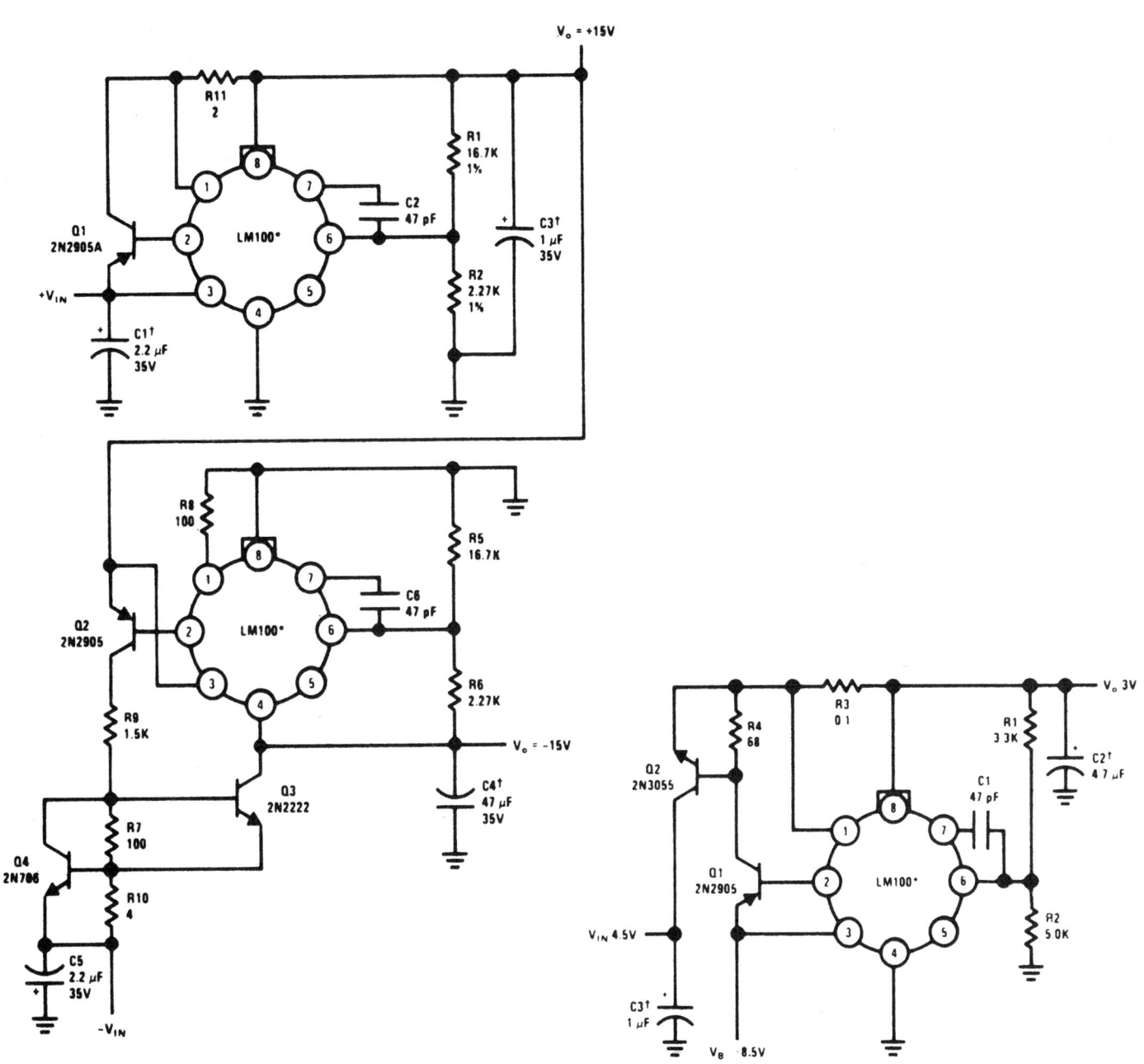

Positive- and negative-voltage regulator. The inputs and outputs of both regulators have a common ground. For the negative regulator, the normal output terminal (pin 8) of the LM100 is grounded, and the ground terminal (pin 4) is connected to the regulated negative output. Hence, as in the usual mode of operation, it regulates the voltage between the output and ground terminals. A PNP booster transistor, Q2, is connected in the normal manner and drives an NPN series-pass transistor, Q3. The additional components (R7, R8, R9, R10, and Q4) are included to provide current limiting. Capacitors C3 and C4 should be solid tantalum. Basing diagram is top view.

Regulator circuit for obtaining higher efficiency operation with low-output voltages. Here, the series-pass transistor, Q2, and the regulator are operated from separate supplies. The series-pass transistor is run from a low-voltage main supply that minimizes the input/output differential for increased efficiency. The regulator, on the other hand, operates from a low-power bias supply with an output greater than 8.5V. Care should be taken that Q2 never saturates. Otherwise, Q1 will try to supply the entire load current and destroy itself unless the bias supply is current limited. Capacitors C2 and C3 should be solid tantalum. Basing diagram is top view.

SECTION 1 *Voltage, Current and Power Sources*

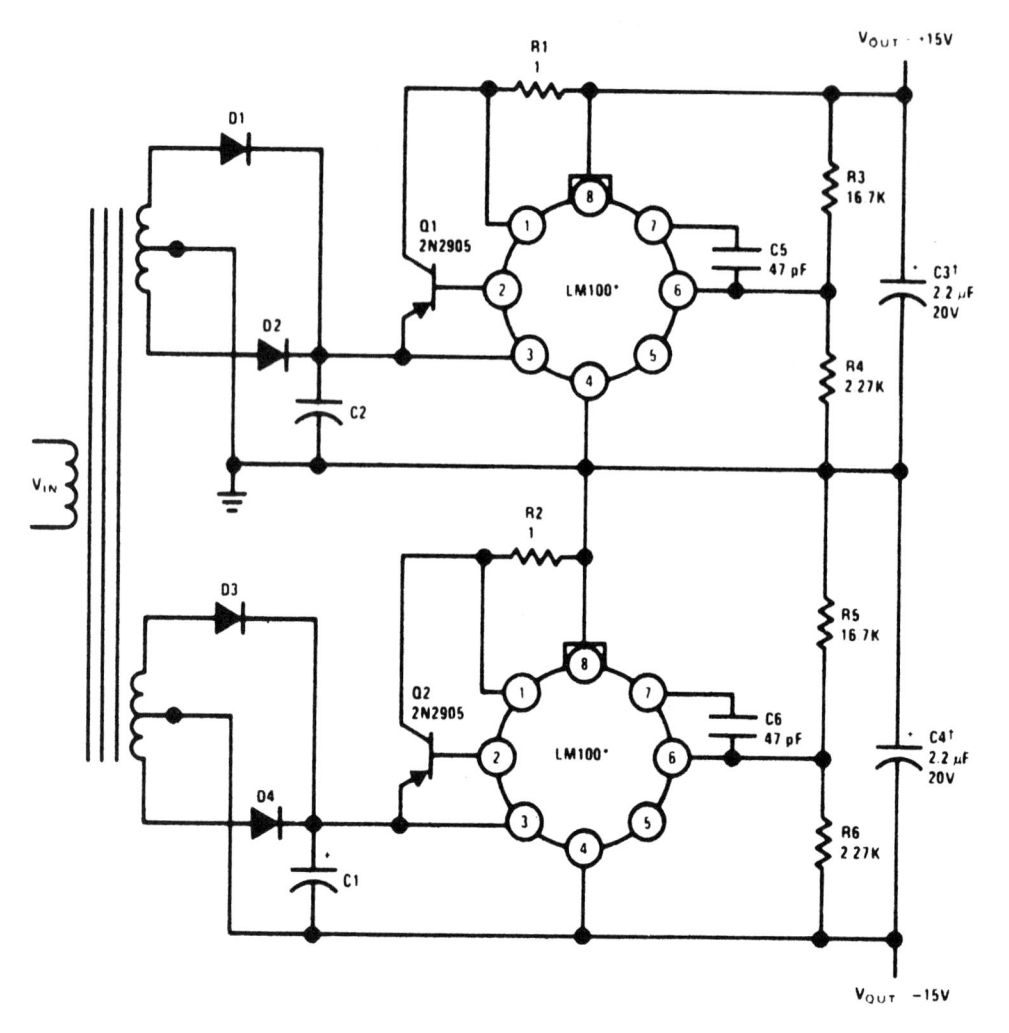

Circuit for using the LM100 as both a positive and a negative regulator. Split secondaries are used on a power transformer to create a floating voltage source for the negative regulator. With this floating source, the conventional regulator is used, except that the output is grounded. Capacitors C3 and C4 should be solid tantalum. Basing diagram is top view.

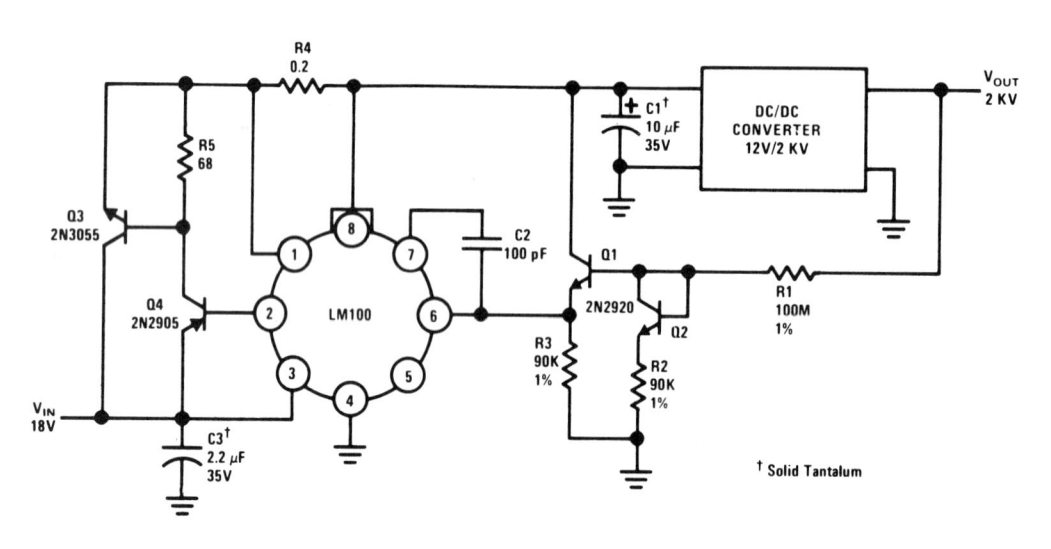

High-voltage regulator. The LM100 senses the output of the high-voltage supply through a resistive divider and varies the input to a dc/dc converter, which generates the high voltage. Hence, the circuit regulates without having any high voltages impressed across it.

40 IC SCHEMATIC SOURCEMASTER

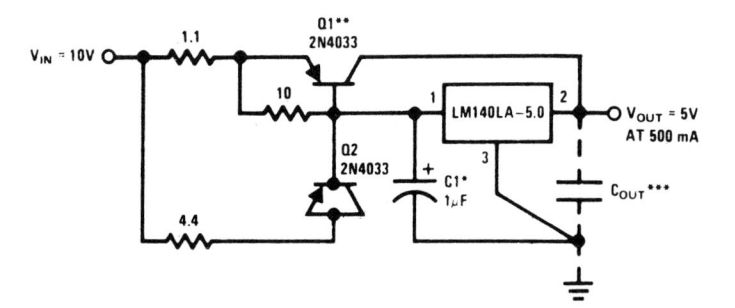

A 5V 500 mA regulator with short-circuit protection. Capacitor C1 is a solid tantalum type. The inclusion of C_{OUT} is optional for improved ripple rejection and transient response. Q1 requires a heatsink. Load regulation is 0.6% $0 \leq I_L \leq 250$ mA pulsed with t_{ON} = 50 m_{sec}.

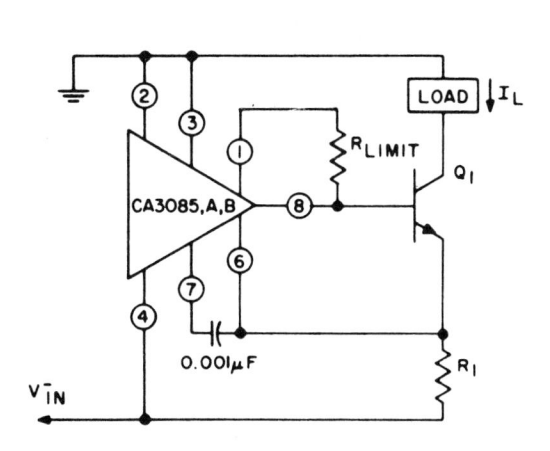

Current-regulator circuit. The NPN can be any general-purpose silicon device that can handle 2A current, such as 2N3772. Load current of regulator: 1.6/R1.

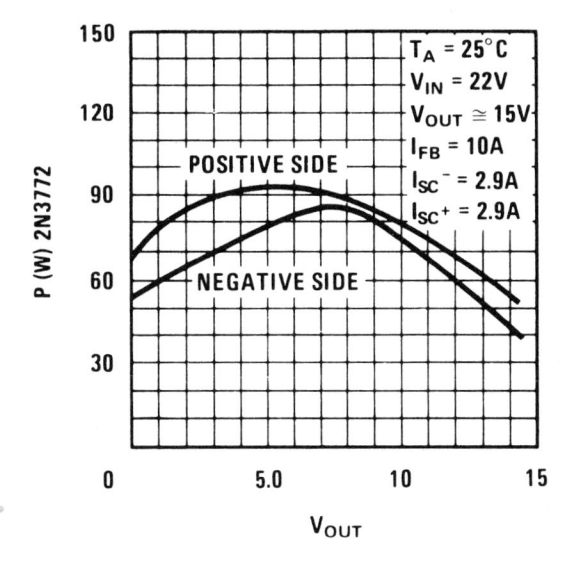

Power dissipation in the external pass transistors (Q5, Q7).

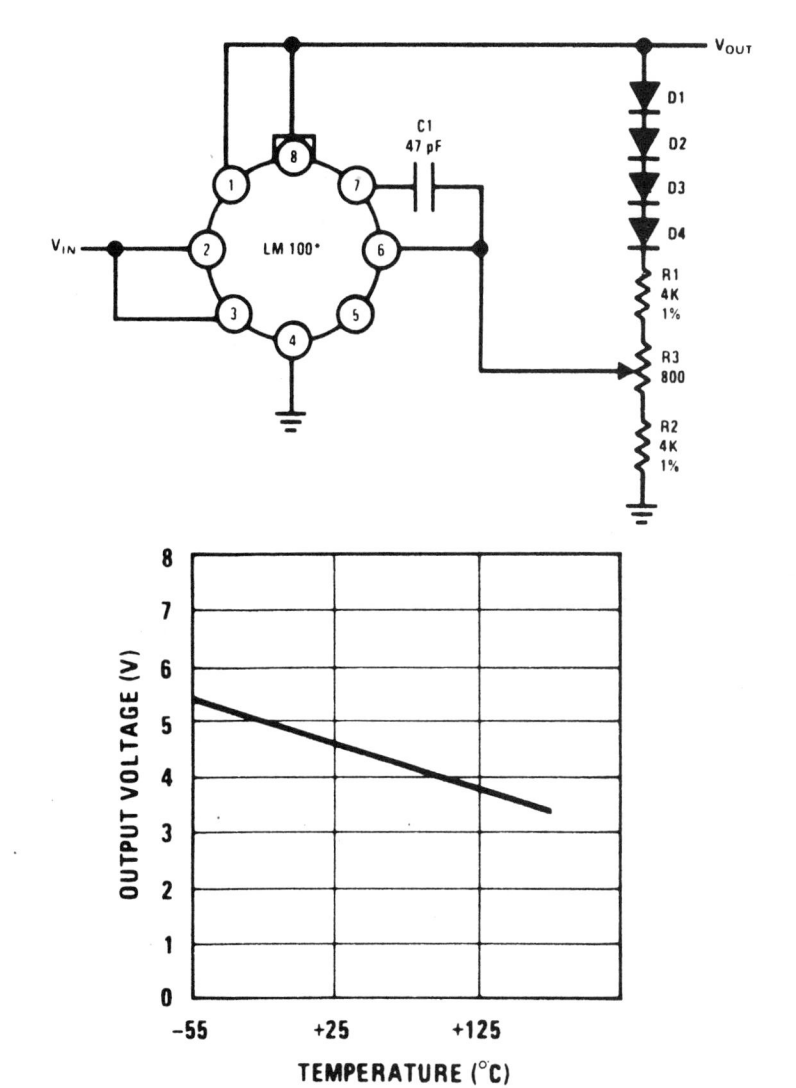

Temperature-compensating voltage regulator with negative temperature coefficient. Silicon diodes in the feedback divider give the required negative temperature coefficient. The advantage of using diodes rather than thermistors or other temperature-sensitive resistors is that their temperature coefficient is quite predictable, so it is not necessary to make trial-and-error adjustments in temperature testing. Basing diagram is top view.

SECTION **1** *Voltage, Current and Power Sources* **41**

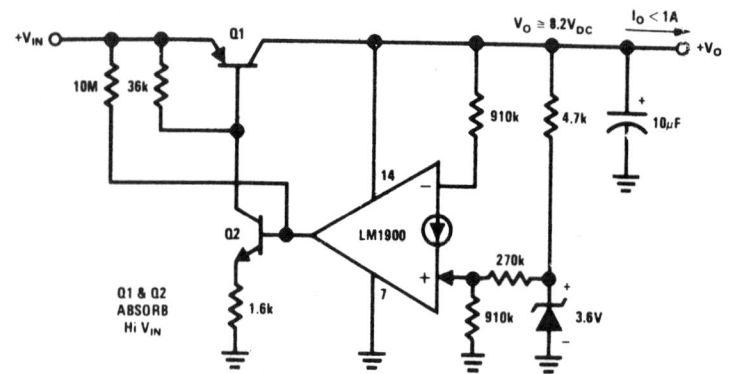

High V_{IN}, low($V_{IN} - V_O$) self-regulator

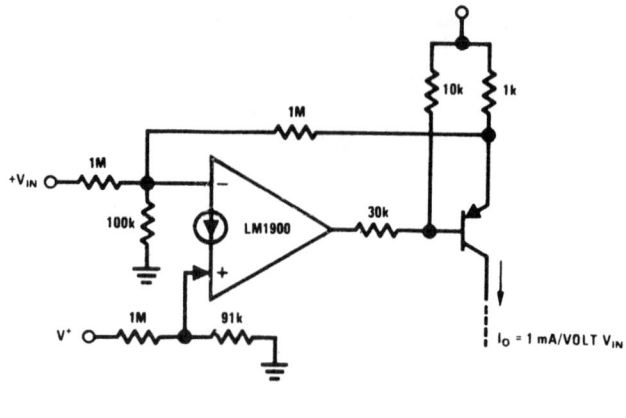

Voltage-controlled current source (transconductance amplifier)

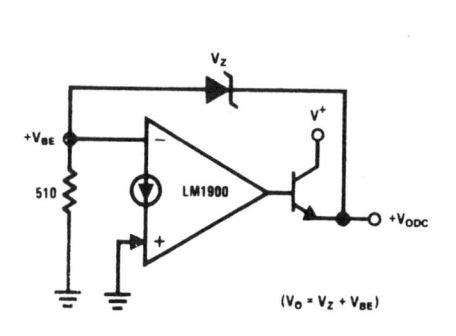

Voltage regulator

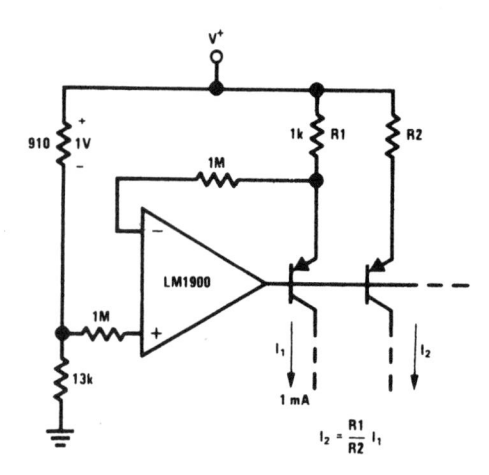

$(V_O = V_Z + V_{BE})$

Fixed current sources

$I_2 = \frac{R1}{R2} I_1$

Current sources and regulators.
Output short circuits either to ground or to the positive power supply should be of short duration. Units can be destroyed, not as a result of the short-circuit current causing metal fusing, but rather as a result of the large increase in IC chip dissipation, which will cause eventual failure due to excessive junction temperatures. For example, when operating from a well-regulated +5V dc power supply at $T_A = 25°C$ with a 100K shunt-feedback resistor (from the output to the inverting input), a short directly to the power supply will not cause catastrophic failure, but the current magnitude will be approximately 50 mA, and the junction temperature will be above T_J max.

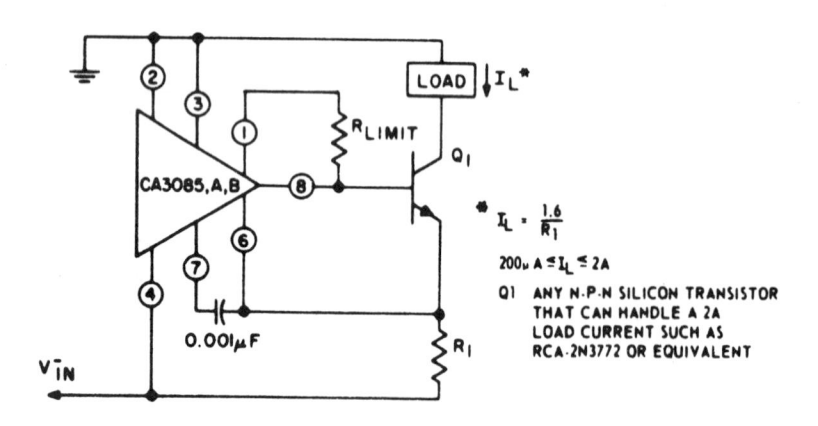

Typical **current-regulator circuit**.

$I_L = \frac{1.6}{R_1}$

$200\mu A \leq I_L \leq 2A$

Q1 ANY N·P·N SILICON TRANSISTOR THAT CAN HANDLE A 2A LOAD CURRENT SUCH AS RCA-2N3772 OR EQUIVALENT

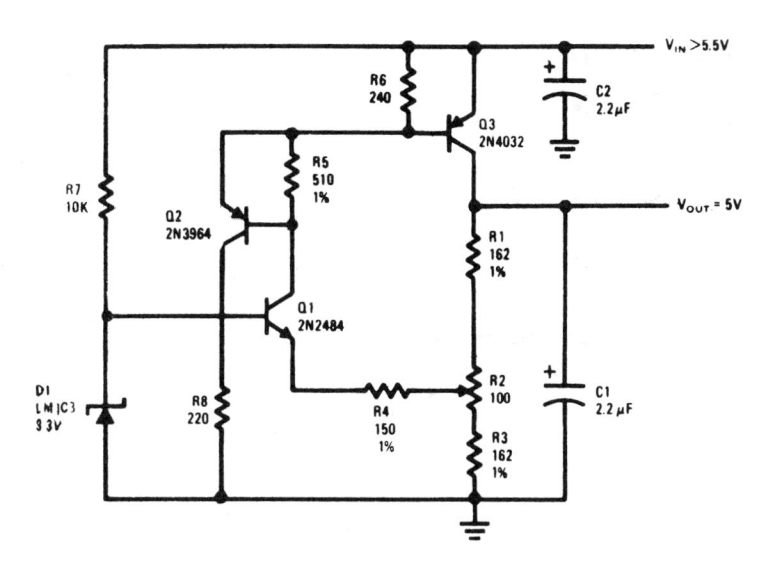

200 mA positive regulator. Capacitors should be solid tantalum. If necessary, select a value for R4 that will ensure minimum temperature drift.

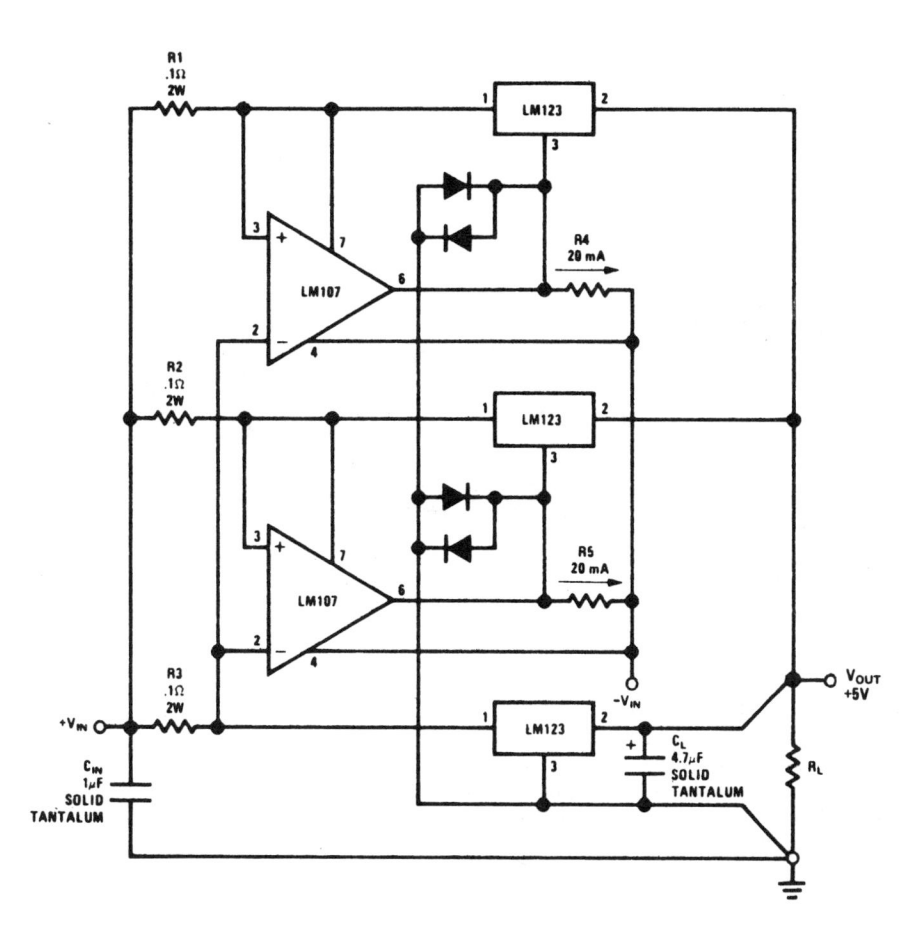

A 10A regulator with complete overload protection. Values for R4 and R5 should be selected for 20 mA current from unregulated negative supply.

SECTION 1 *Voltage, Current and Power Sources*

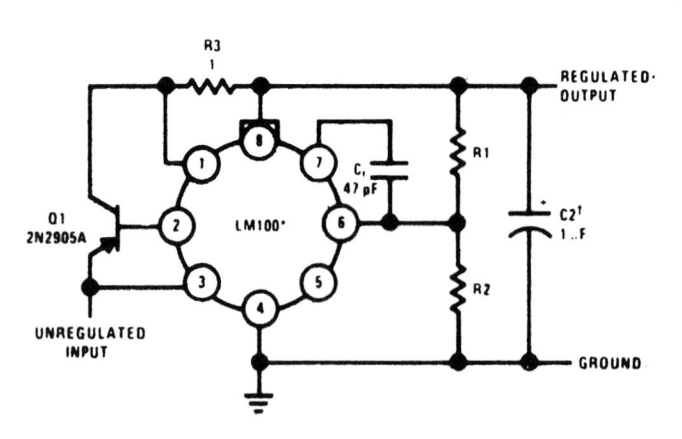

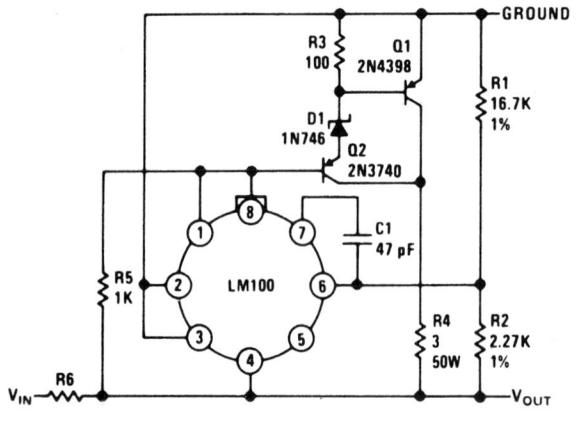

Regulator connected for 200 mA output current. When external transistors are used, it is necessary to bypass the output terminal close to the integrated circuit. This is required to suppress oscillations in the minor feedback loop around the external transistor and the output transistor of the integrated circuit. Since the instability usually occurs at high frequencies, a low-inductance (solid tantalum) capacitor must be used. Electrolytic capacitors that have a high equivalent series resistance are not effective.

3A negative shunt regulator. A zener diode provides a level shift so that the output transistors within the LM100 are properly biased. R5 supplies the base drive for Q2 and also the minimum load current for the LM100. R4 is included to minimize dissipation in the power transistors when the regulator is lightly loaded. The output voltage is determined in the normal fashion by R1 and R2. Although no output capacitor is used, it may be advisable to include one to reduce the output impedance at high frequencies.

Because a shunt regulator is a two-terminal device, this design can be used as either a positive or a negative regulator.

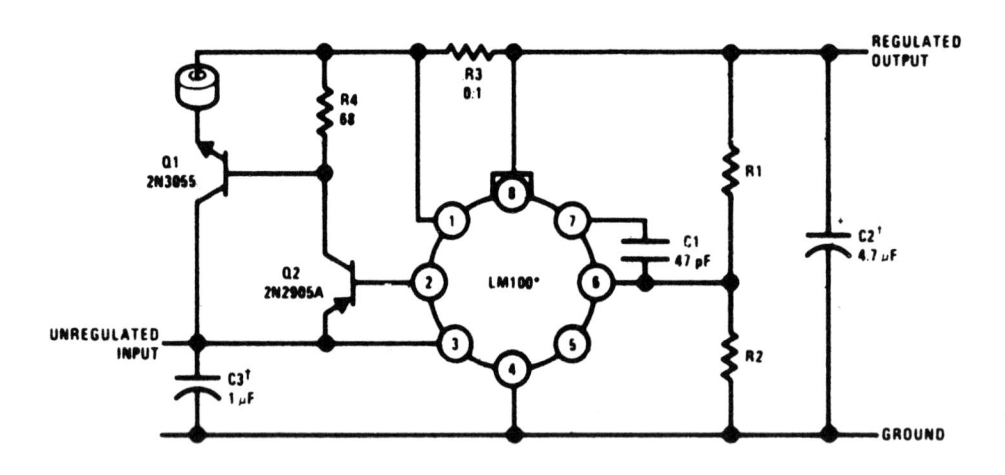

Ferroxcube K5-001-00/3B, if required

Regulator connected for 2A output current. The PNP transistor, Q2, is used to drive an NPN power transistor, Q1. With this circuit it is necessary to bypass both the input and output terminals of the regulator with low-inductance capacitors to prevent oscillation in the minor feedback loop through Q1, Q2, and the output transistor of the integrated circuit. In addition, with certain types of NPN power transistors, it may be necessary to install a ferrite bead in the emitter lead of the device to suppress parasitic oscillations in the power transistor. C2 and C3 should be solid tantalum. Basing diagram is top view.

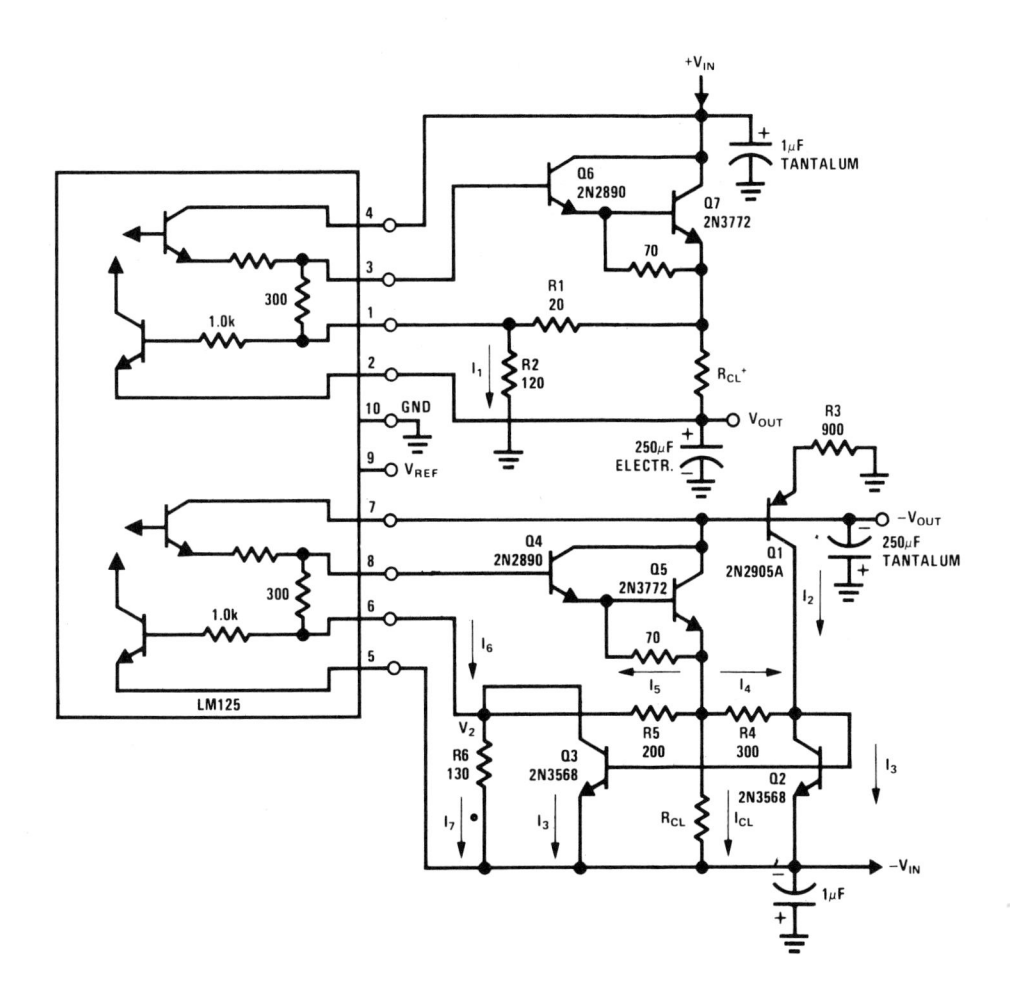

This 10A regulator features foldback current limiting. In this design, the current contribution from the internal 300Ω resistor is greater than comparable circuits because of the 2V$_{BE}$ drop across the Darlington pair.

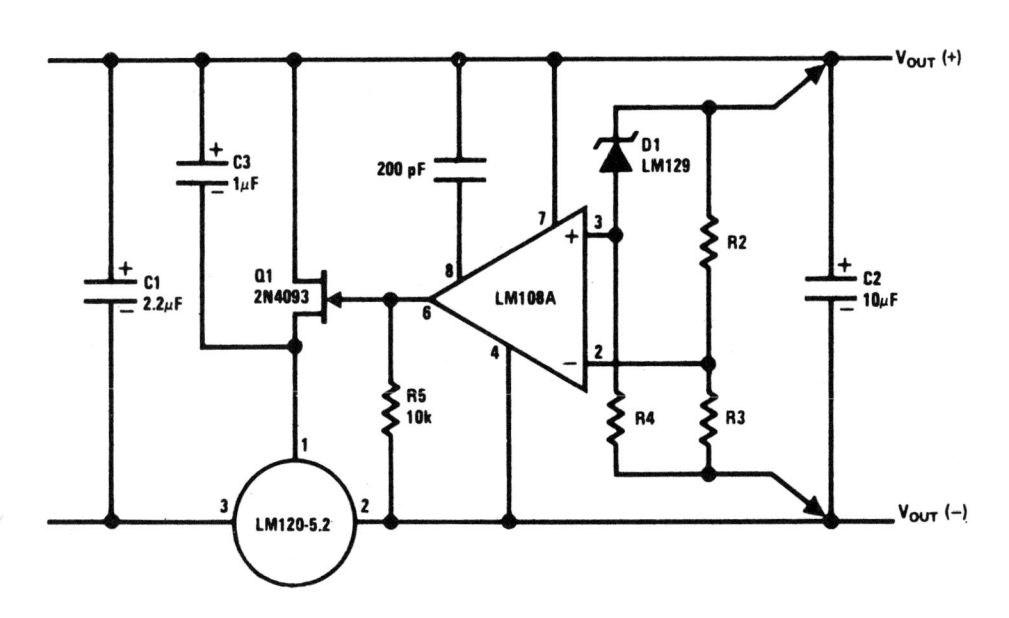

High-stability 1A regulator. An LM120-12 or LM120-15 may be used to permit higher input voltages, but the regulated output voltage must be at least −15V when using the LM120-12 and −18V for the LM120-15. Load and line regulation should be less than 0.01%; temperature stability should be less than 0.2%. R4 determines the zener current. R2 and R3 sets the output voltage; 2 ppm/°C are suggested. C1 and C2 should be solid tantalum.

SECTION 1 *Voltage, Current and Power Sources* 45

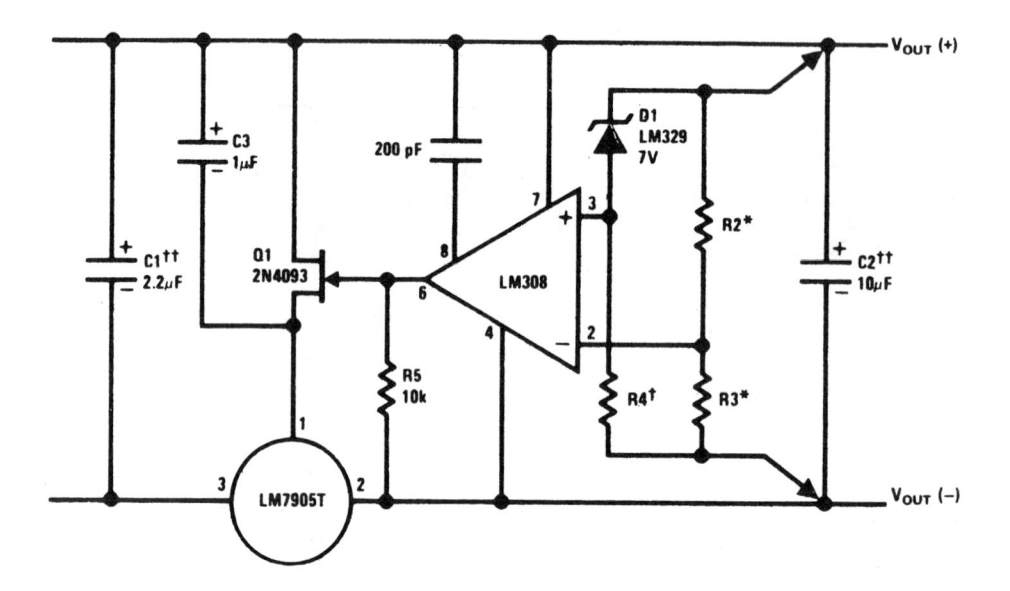

This high-stability 1A regulator offers load and line regulation of less than 0.01% and temperature stability of ≤0.2%. Select resistors marked with asterisk to set output voltage; 2 ppm/°C are suggested. C1 should be solid tantalum.

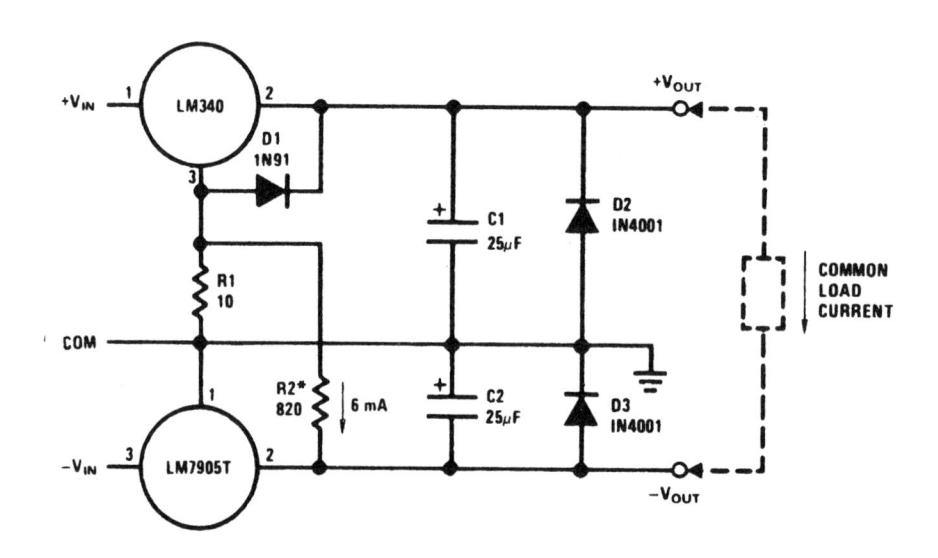

Positive regulator latch-up prevented. R1 and D1 allow the positive regulator to start up when $+V_{IN}$ is delayed relative to $-V_{IN}$ and a heavy load is drawn between the outputs. Without R1 and D1, most three-terminal regulators will not start with heavy (0.1–1A) load current flowing to the negative regulator, even though the positive output is clamped by D2. R2 is optional. Ground pin current from the positive regulator flowing through R1 will increase $+V_{OUT} \approx 60$ mV if R2 is omitted.

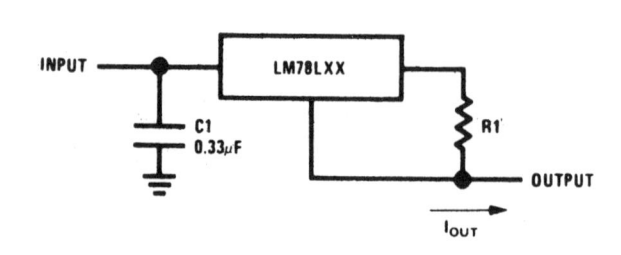

Current regulator. $I_{OUT} = (V_{23}/R1) + I_Q$; $\Delta I_Q = 1.5$ mA over line and load changes.

46 IC SCHEMATIC SOURCEMASTER

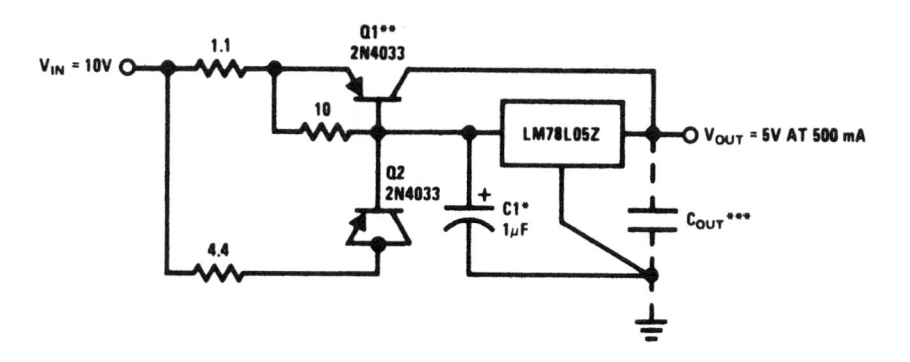

A 5V, 500 mA regulator with short-circuit protection. Capacitor C1 should be solid tantalum. Q1 requires a heatsink. C_{OUT} is optional for improved ripple rejection and transient response. Load regulation is 0.6% $0 \leq I_L \leq 250$ mA pulsed with $t_{ON} = 50$ msec.

LIMITING CHARACTERISTICS

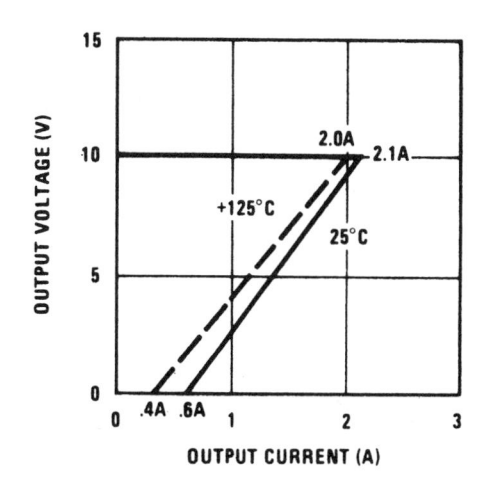

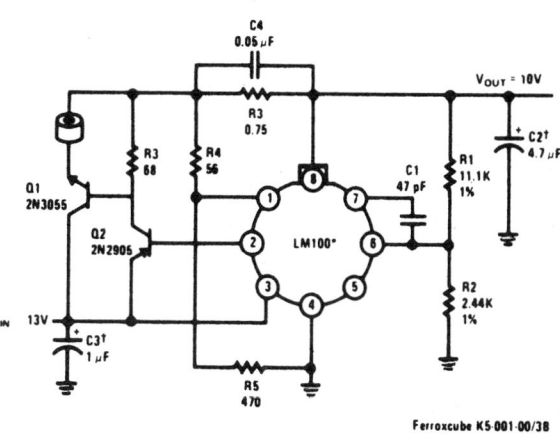

Circuit for obtaining switchback current limiting. The voltage drop across the current-sense resistor at full load is 1.5V. This does not increase the minimum input/output voltage differential, since the output of the LM100 does not see this increased voltage. With a 10V output and a 2A load, the circuit will still work with input voltages down to 13V, worst case.

In addition to providing the switchback characteristics, R4 and R5 also give a 20 mA preload on the regulator so that it can be operated without a load. C2 and C3 should be solid tantalum.

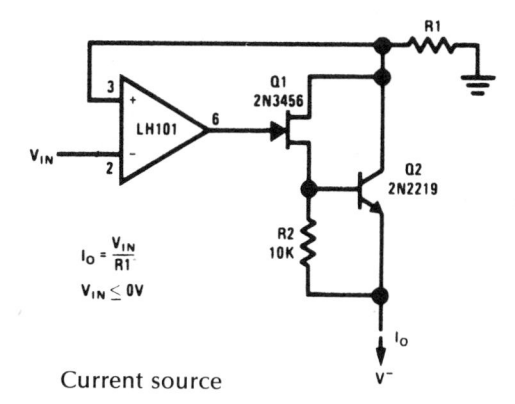

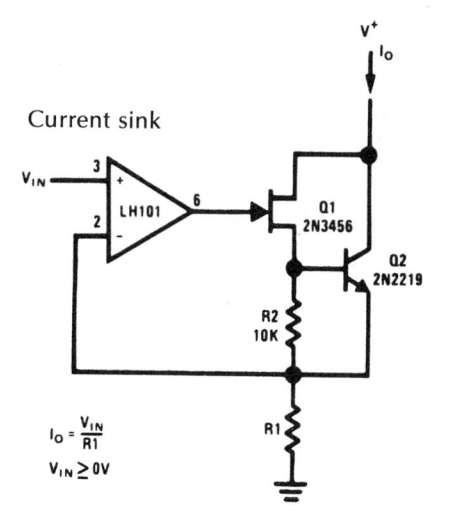

Current source

$I_O = \dfrac{V_{IN}}{R1}$

$V_{IN} \leq 0V$

Current sink

$I_O = \dfrac{V_{IN}}{R1}$

$V_{IN} \geq 0V$

Precision current source/sink. Caution must be exercised in applying these circuits. The voltage compliance of the source extends from BV_{CER} of the external transistor to approximately a more negative direction than V_{IN}. The compliance of the current sink is the same in the positive direction.

The impedance of these current generators is essentially infinite for small currents, and they are accurate as long as V_{IN} is much greater than V_{OS} and I_O is much greater than I_{bias}.

SECTION 1 *Voltage, Current and Power Sources*

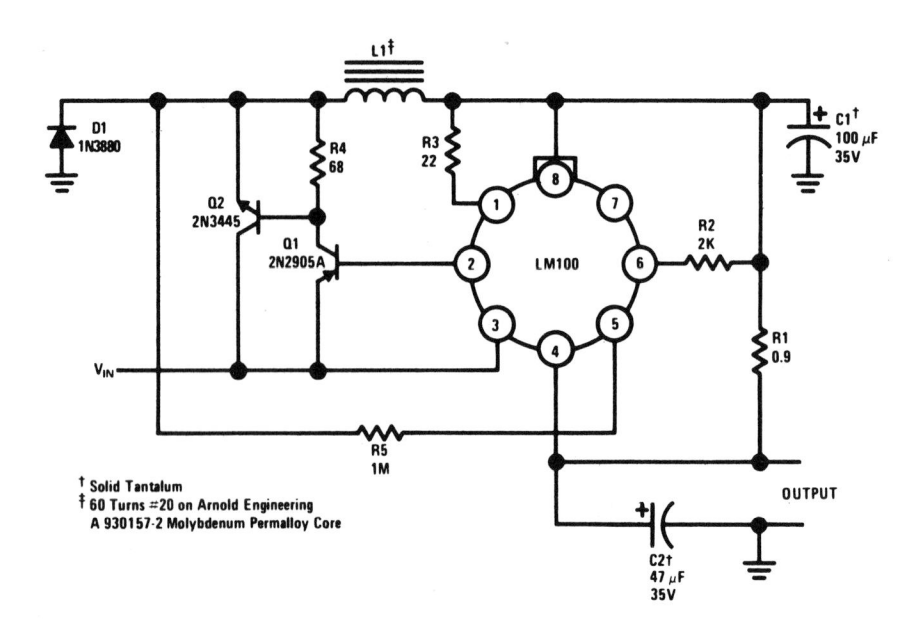

Current source with grounded load. Here, the load is inserted in the ground line. The quiescent current of the regulator, flowing out of pin 4, introduces an error term. However, since this current is only about 2 mA and is reasonably independent of changes in the input or load voltages, the error is usually not significant.

With this circuit, the difference between the input voltage and the load voltage cannot drop below 8.5V or the circuit will drop out of regulation because the voltage across the LM100 is insufficient to bias the reference circuitry.

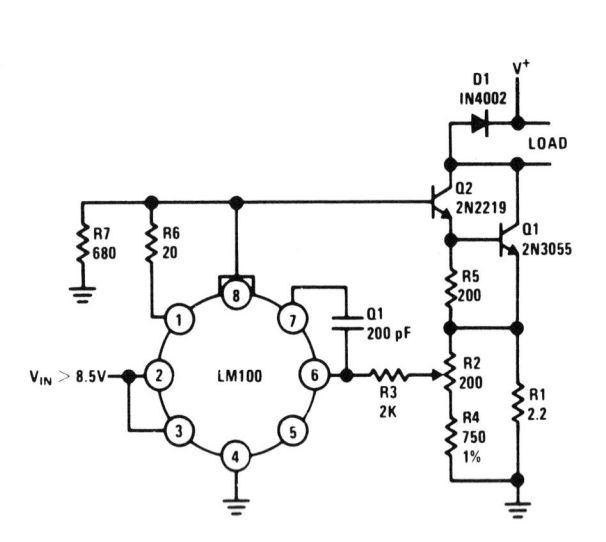

Current source. Here the LM100 regulates the emitter current of a Darlington-connected transistor, and the output current is taken from the collectors. The use of a Darlington connection for Q1 and Q2 improves the accuracy of the circuit by minimizing the base-current error between the emitter and collector current.

The output of the LM100, which drives the control transistors, must be short-circuit protected with R6 to limit the current when Q2 saturates. R7 is required to provide the minimum load current for the integrated circuit. D1 is included to absorb the kickback of inductive loads when power is shut off. The output current of the circuit is adjusted with R2.

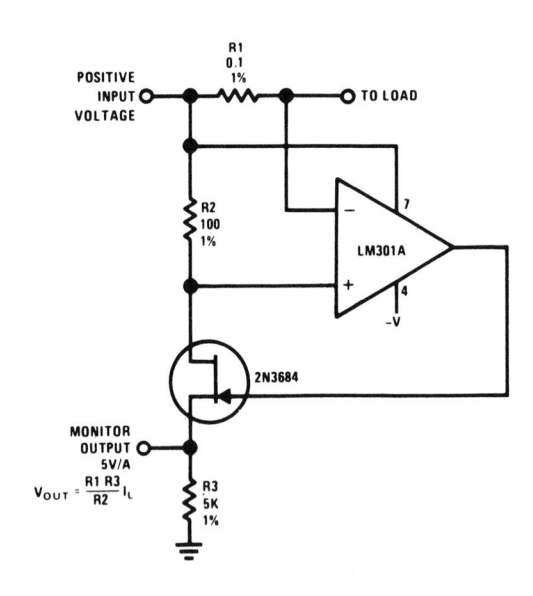

In this current monitor, R1 senses the current flow of a power supply. The JFET is used as a buffer because $I_D = I_S$; therefore, the output monitor voltage accurately reflects the power supply current flow.

48 IC SCHEMATIC SOURCEMASTER

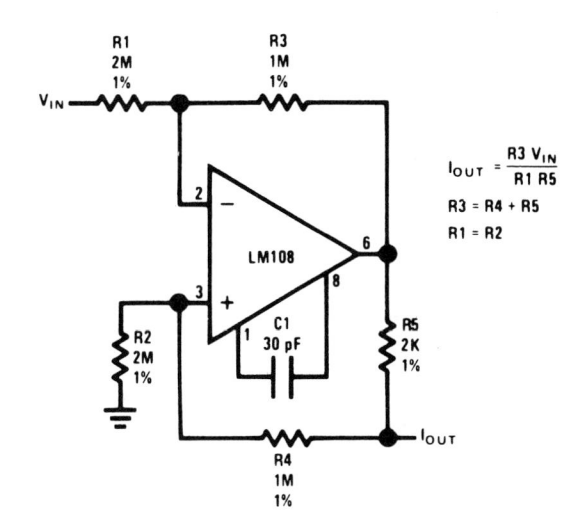

This bilateral current source is fairly flexible and has few restrictions as far as its use is concerned. It supplies a current that is proportional to the input voltage and drives a load referenced to ground or any voltage within the output-swing capability of the amplifier. With the output grounded, the output current will be determined by R5 and the gain setting of the op-amp, yielding

$$I_{OUT} = -\frac{R3 V_{IN}}{R1 R5}$$

When the output is not at zero, the output current will be independent of the output voltage. For R1 + R3 ≫ R5, the output resistance of the circuit is given by

$$R_{OUT} \cong R5 \left(\frac{R}{\Delta R}\right)$$

where R is any one of the feedback resistors (R1, R2, R3, or R4) and ΔR is the incremental change in the resistor value from design center.

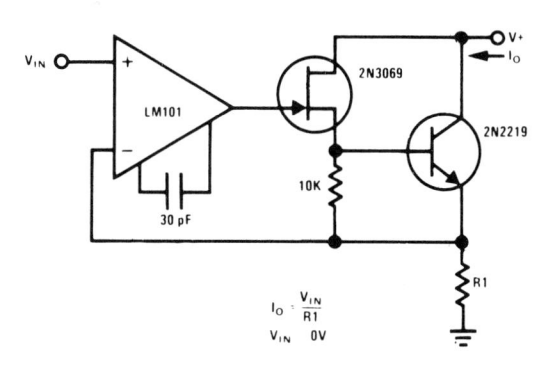

Precision current sink. The 2N3069 JFET and 2N2219 bipolar transistor have inherently high output impedance. Using R1 as a current-sensing resistor to provide feedback to the LM101 op-amp provides a large amount of loop gain for negative feedback to enhance the true "current sink" nature of this circuit. For small current values, the 10K resistor and 2N2219 may be eliminated if the source of the JFET is connected to R1.

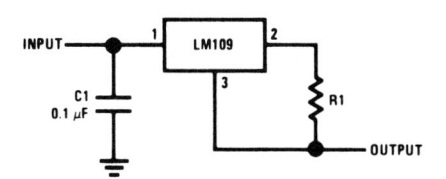

Current regulator. The regulated output voltage is impressed across R1, which determines the output current. The quiescent current is added to the current through R1, and this puts a lower limit of about 10 mA on the available output current.

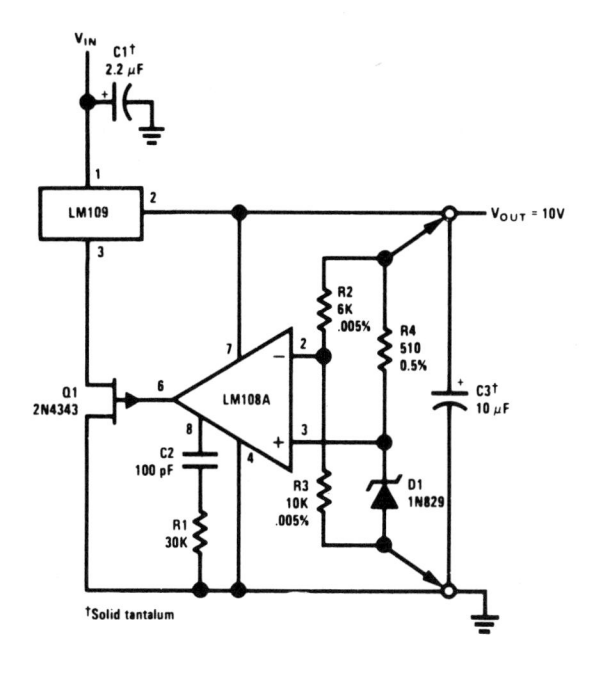

High-stability regulator. An operational amplifier compares the output voltage with the output voltage of a reference zener. The op-amp controls the LM109 by driving the ground terminal through a JFET. The load and line regulation of this circuit is better than 0.001%. Noise, drift, and long-term stability are determined by reference zener D1. Noise can be reduced by inserting 100K, 1% resistors in series with both inputs of the op-amp and bypassing the noninverting input to ground. A 100 pF capacitor should also be included between the output and the inverting input to prevent frequency instability. Temperature drift can be reduced by adjusting R4, which determines the zener current, for minimum drift. For best performance, remote sensing directly to the load terminals should be used.

SECTION 1 *Voltage, Current and Power Sources*

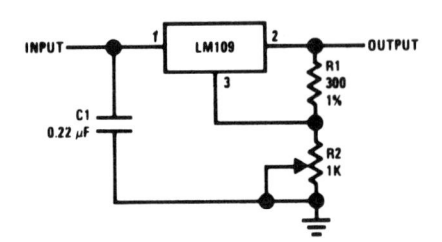

Adjustable-output regulator. The regulated output voltage is impressed across R1, developing a reference current. The quiescent current of the regulator, coming from the ground terminal, is added to this. These combined currents produce a voltage drop across R2 that raises the output voltage. Hence, any voltage above 5V can be obtained as long as the voltage across the integrated circuit is kept within ratings.

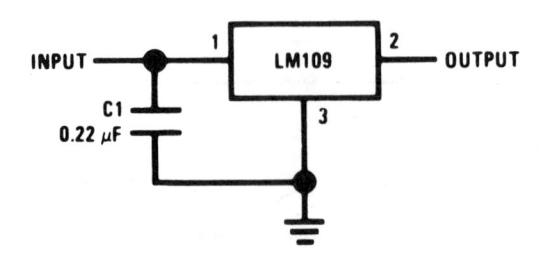

Fixed 5V regulator. It is necessary to bypass the unregulated supply with a 0.22 µF capacitor to prevent oscillations that can cause erratic operation. This, of course, is only necessary if the regulator is located an appreciable distance from the filter capacitors on the output of the dc supply.

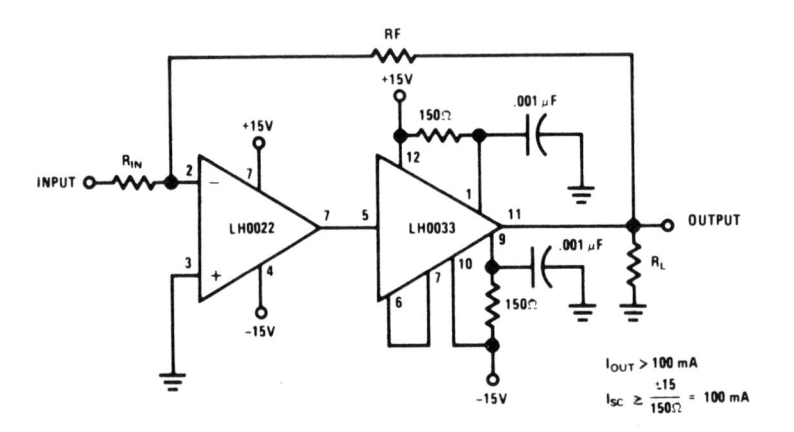

Current booster. Output currents in excess of 100 mA may be obtained. Inclusion of 150Ω resistors between pins 1 and 12 and pins 9 and 10 provide short-circuit protection, while decoupling pins 1 and 9 with 1000 pF capacitors allows nearly full output swing.

The value for the short-circuit current is given by:

$$I_{SC} \cong \frac{V^+}{R_{LIMIT}} = \frac{V^-}{R_{LIMIT}}$$

where $I_{SC} \leq 100$ mA.

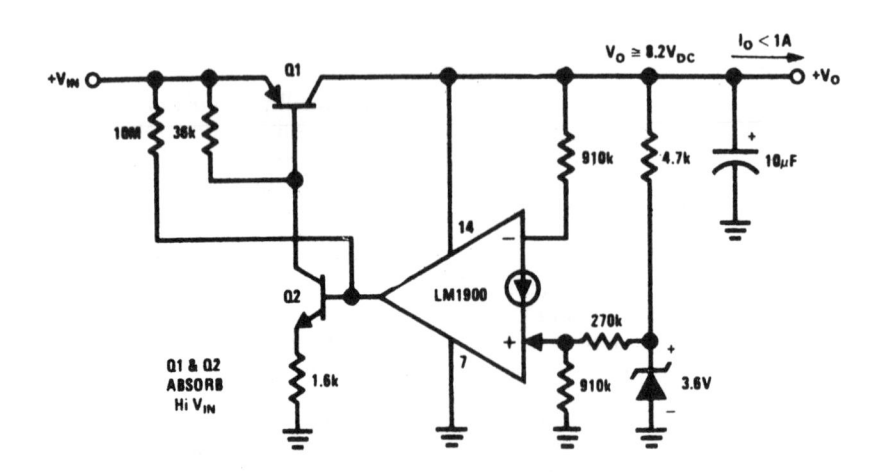

High V_{IN}, low ($V_{IN} - V_O$) **self-regulator.**

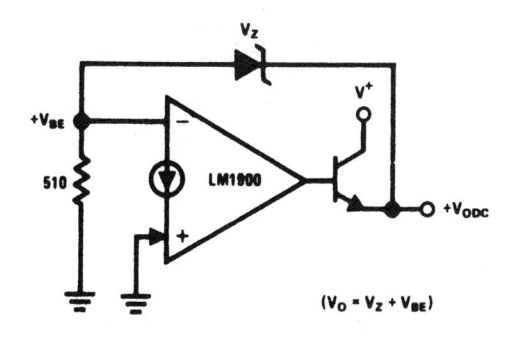

Voltage regulator.

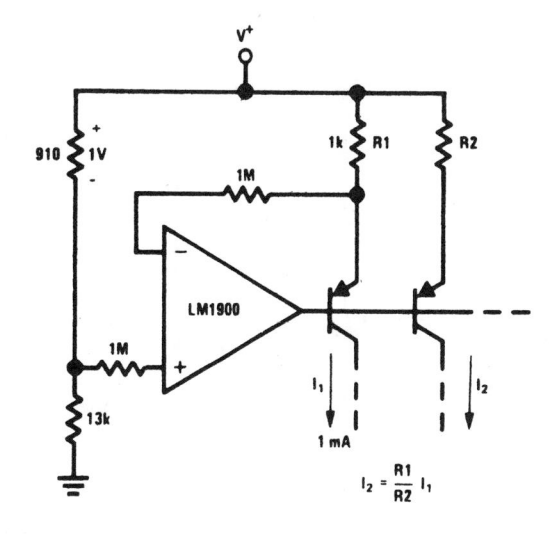

($V_O = V_Z + V_{BE}$)

Fixed current sources.

$I_2 = \dfrac{R1}{R2} I_1$

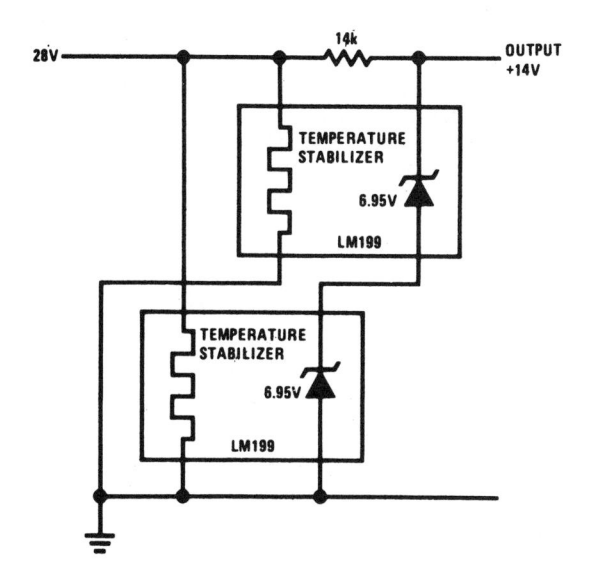

14V reference.

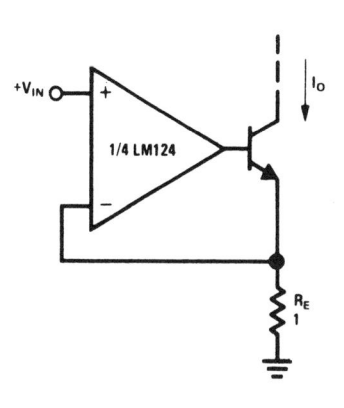

High-compliance current sink. $I_O = 1A/V\ V_{IN}$; increase R_E for I_O small.

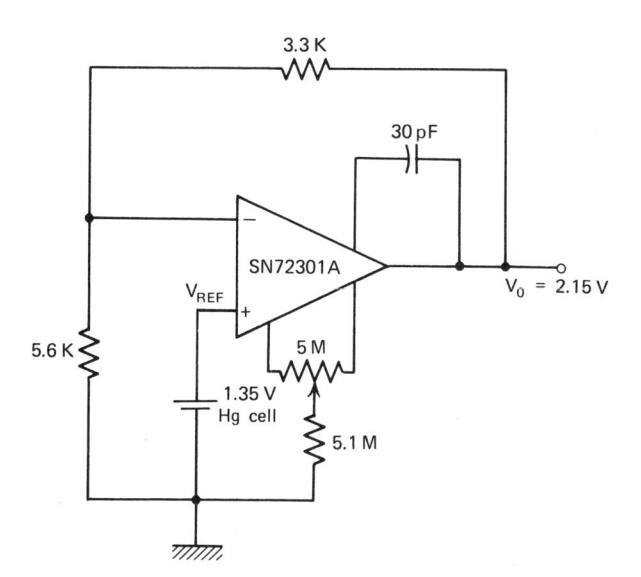

Stable voltage source. Use metal film resistors throughout.

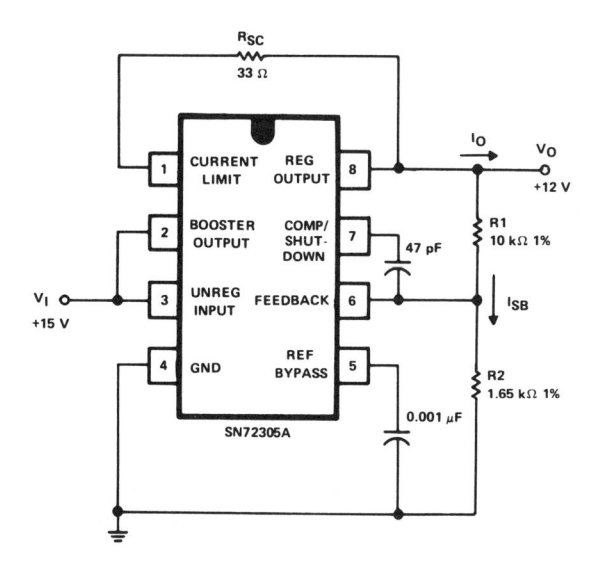

Basic positive-voltage regulator with short-circuit protection.

SECTION 1 *Voltage, Current and Power Sources* **51**

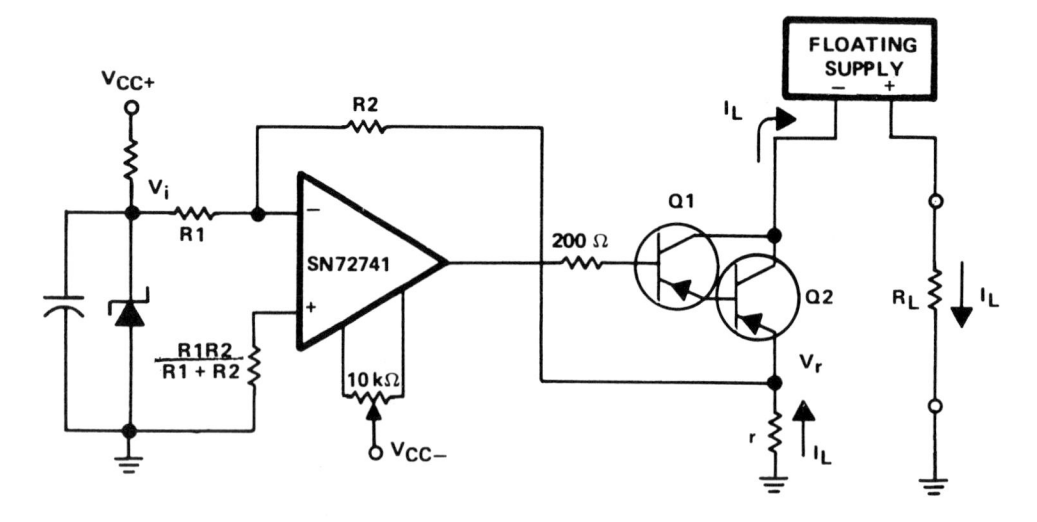

High-current generator. Improved output current-handling capability may be achieved by adding an FET or bipolar transistor to the output of the operational amplifier. A Darlington transistor pair achieves a high-level current source with only limited drive current required of the operational amplifier. The SN72741 furnishes the base drive required to maintain V_r at a desired level across r.

A 200Ω resistor is used in series with the op-amp output to damp capacitive loading effects of the transistor inputs and prevent oscillations. Depending on the transistors used, this circuit could furnish constant-current levels of 1A or more.

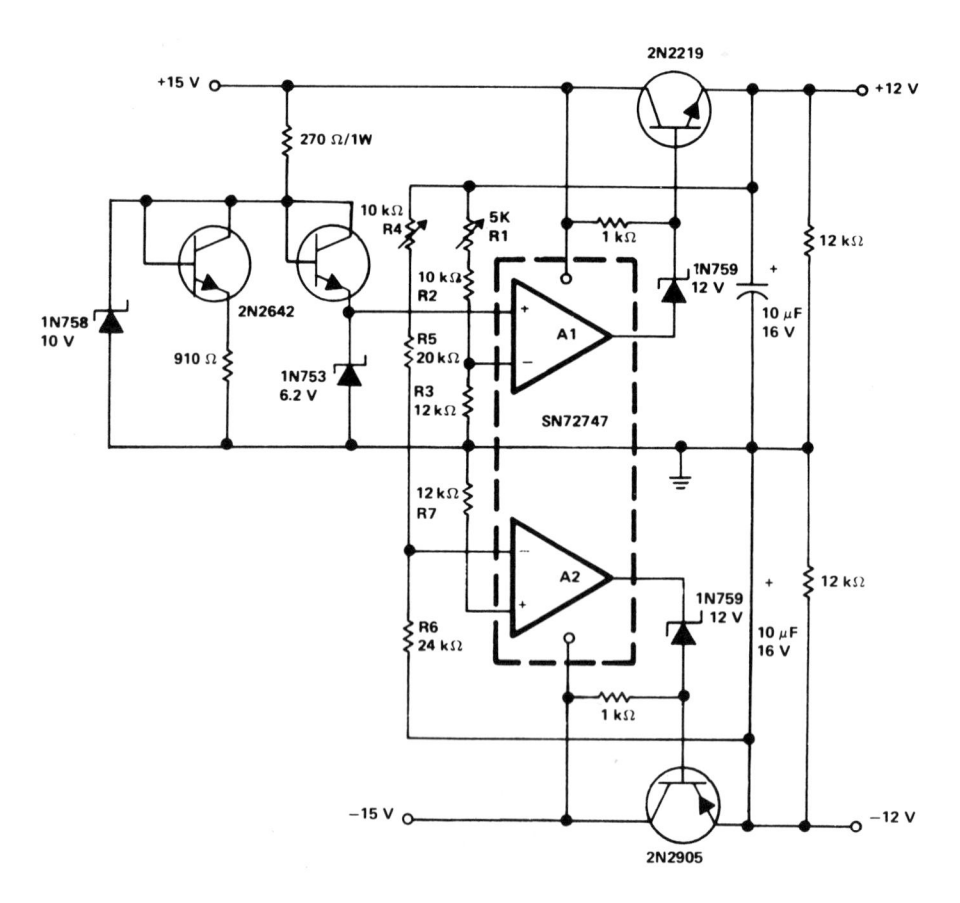

Dual tracking voltage regulator. In an extension of these principles, op-amps may be used to compare a reference voltage with a supply output voltage and to provide an error signal to the series-pass transistor for maintaining a stable output voltage under varying load conditions. In this application, voltage from a resistor divider (R1, R2, and R3) across the positive output is compared, in A1, with a stable 6.2V reference. Any difference voltage results in a correcting signal, which is fed to the 2N2219 pass transistor for adjusting the output voltage.

52 IC SCHEMATIC SOURCEMASTER

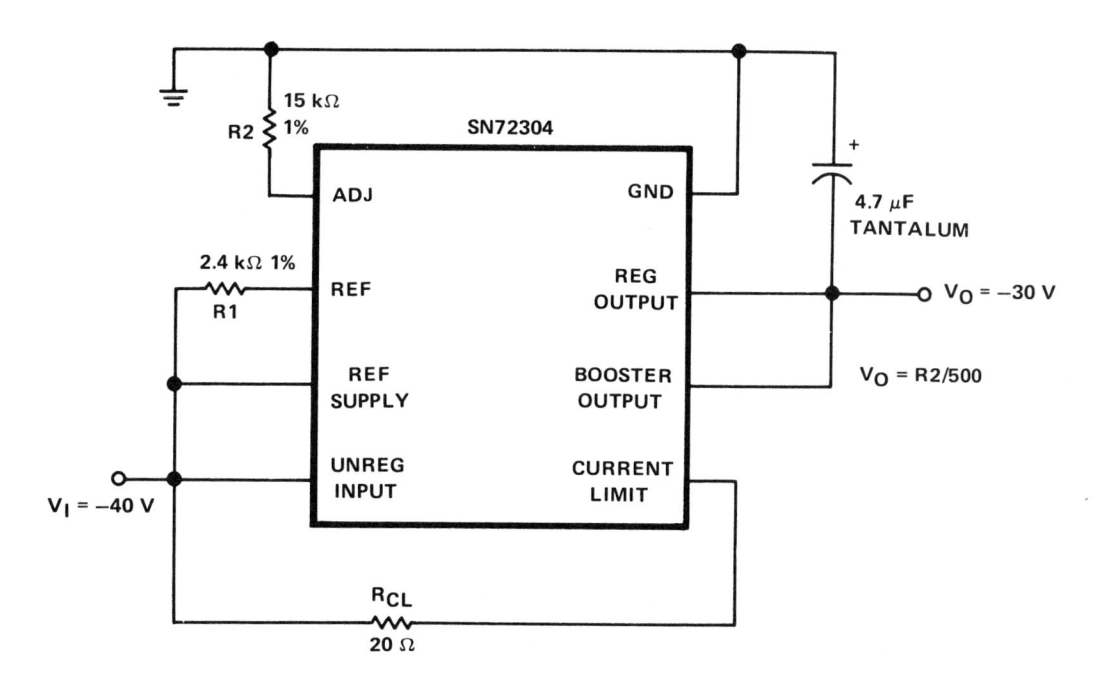

Basic −30V regulator.

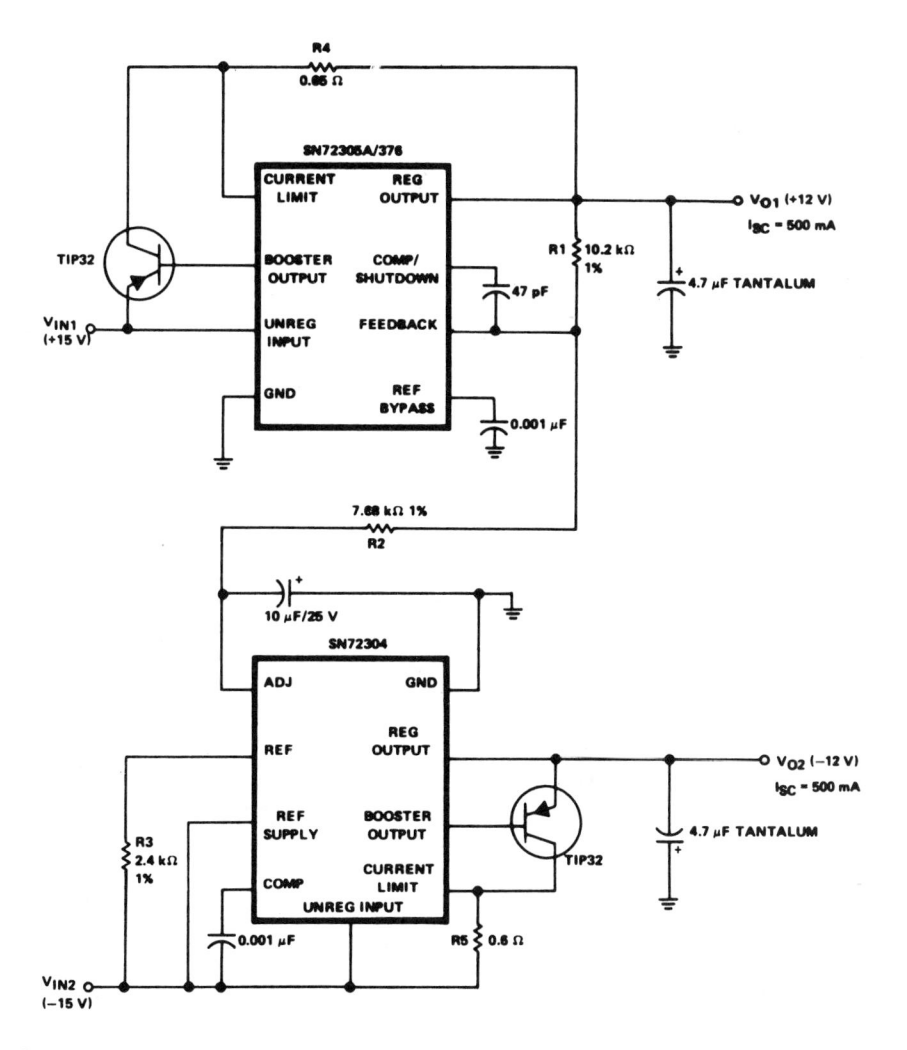

Dual tracking regulator. Tracking is accomplished by connecting the feedback control point of the SN72305A to the adjust terminal of the SN72304 with resistor R2. R1 and R2 form a sampling network affecting both supplies. Variation in either output voltage will result in a change in the other.

SECTION **1** *Voltage, Current and Power Sources*

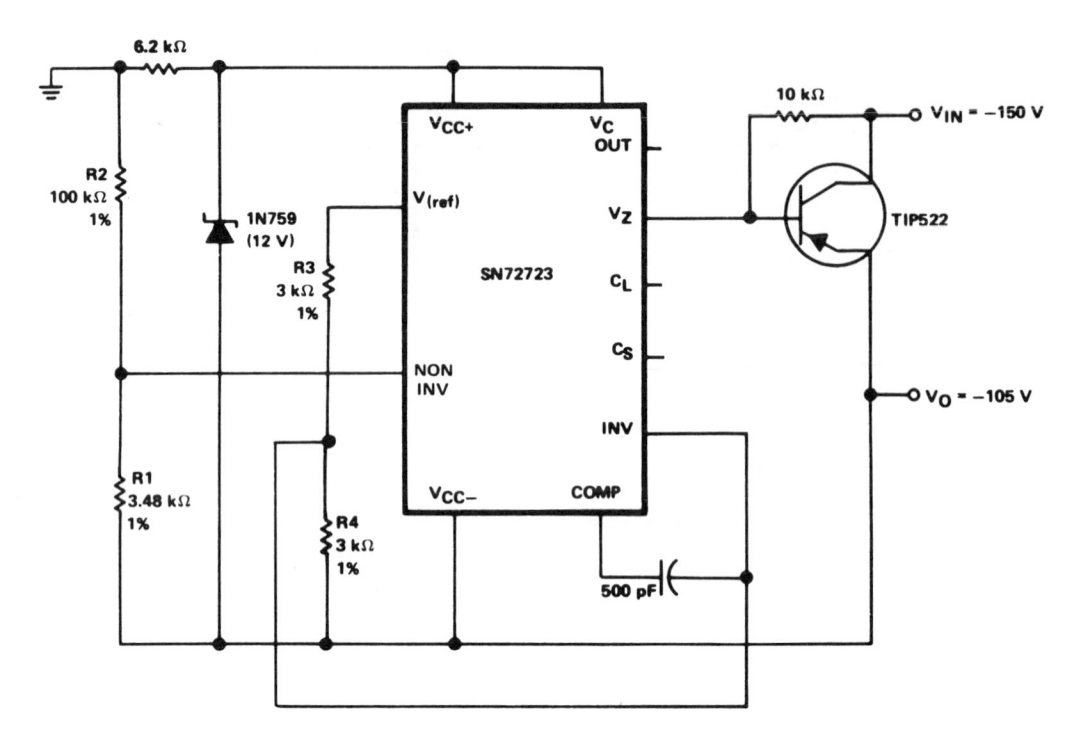

Negative 105V voltage regulator. In this example, a 1N759 12V zener is employed to clamp the voltage appearing across the floating regulator.

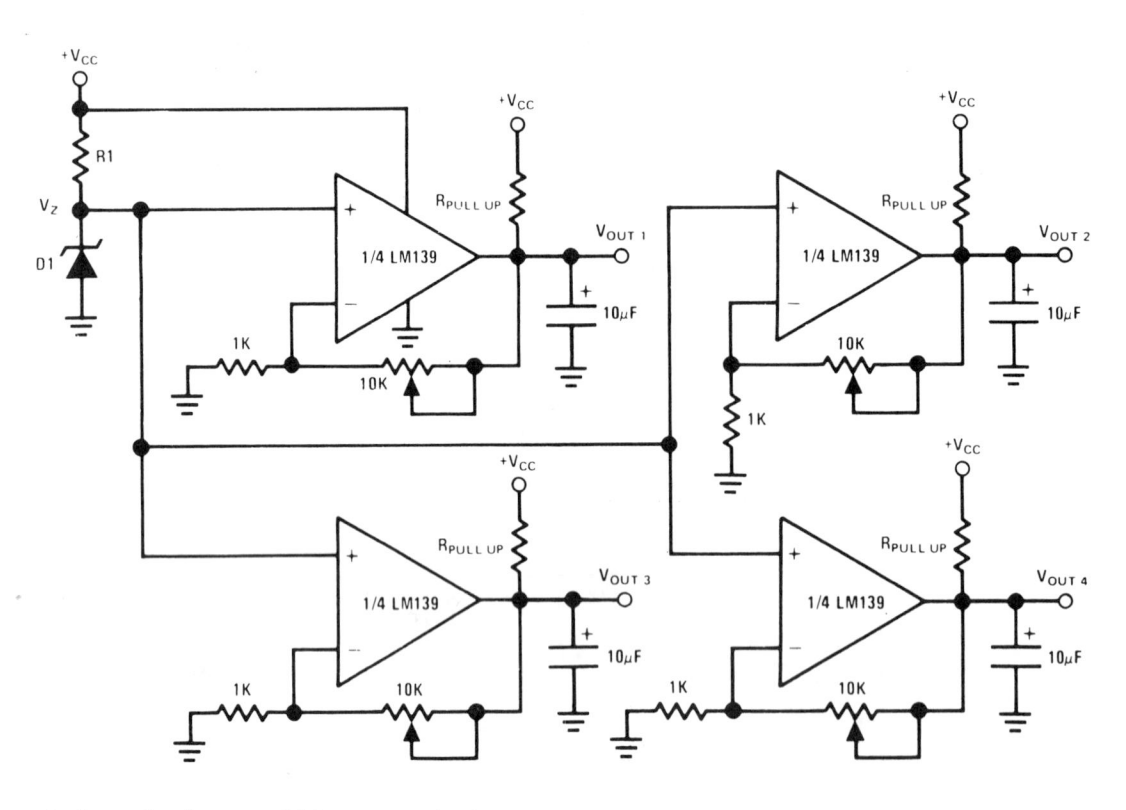

Four independently variable temperature-compensated reference supplies. If the proper zener is chosen, these four voltages will have a near-zero temperature coefficient. For industry-standard zeners, this will be somewhere between 5.0 and 5.4V at a zener current of approximately 10 mA.

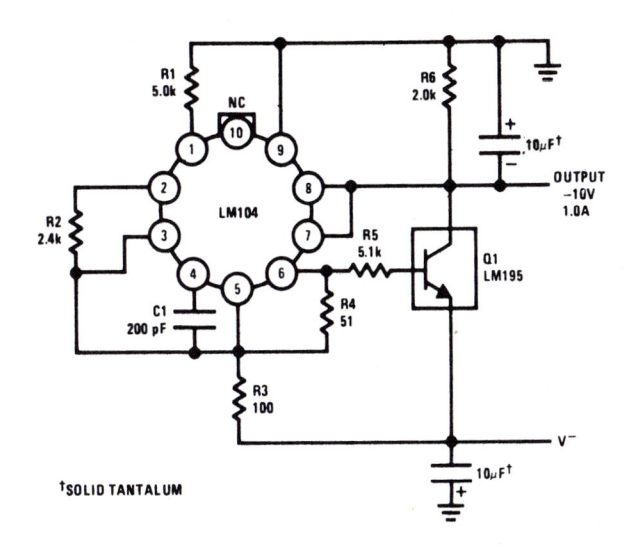

1.0A negative regulator.

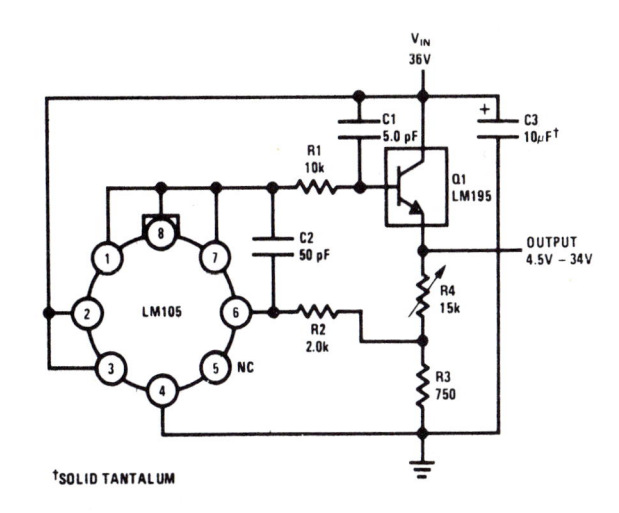

†SOLID TANTALUM

1.0A positive regulator.

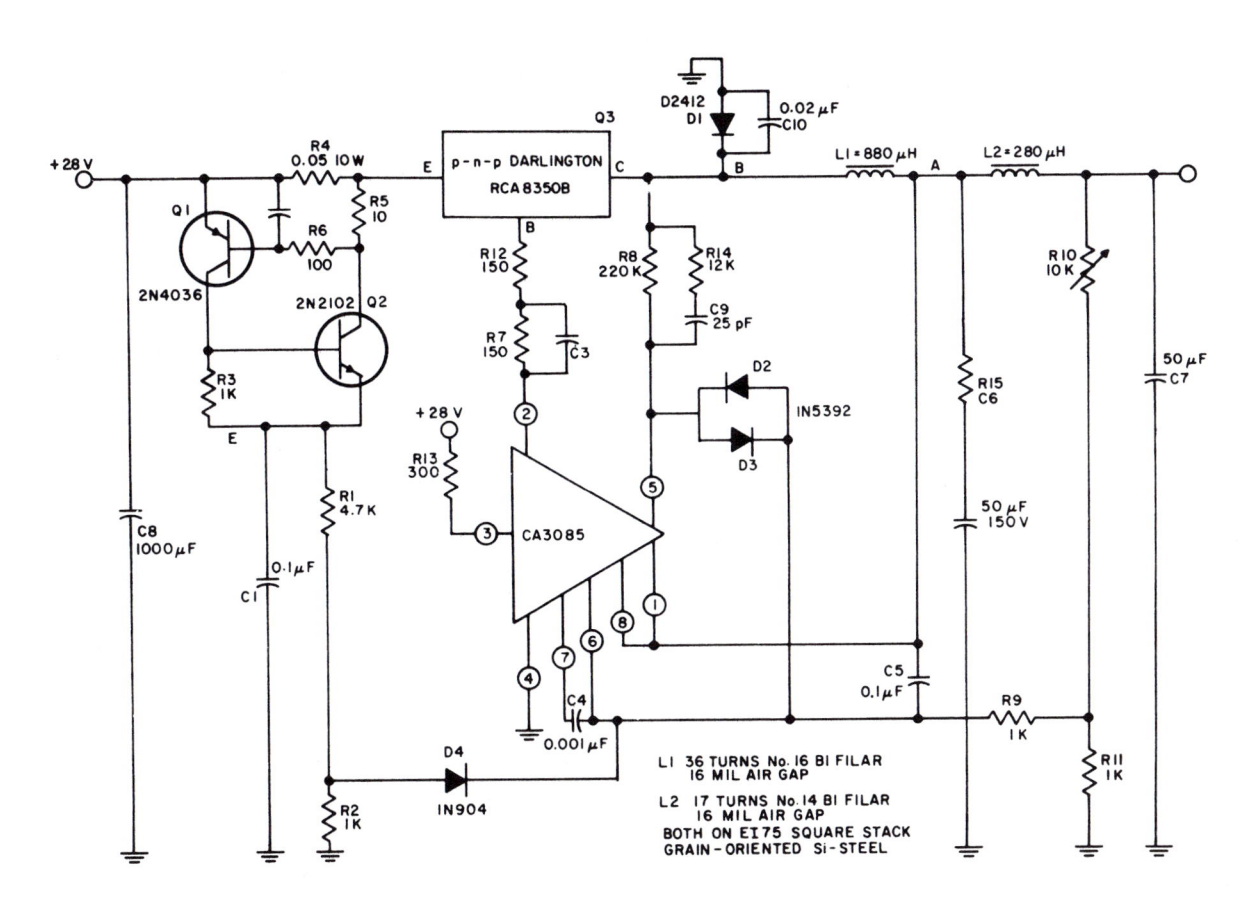

†SOLID TANTALUM

Stepdown switching regulator. When the pass unit, Q3 (a PNP Darlington, RCA8350B), is switched on, current is charged into L1; when Q3 switches off, the current through L1 continues to flow via the commutating diode, D1.

The dc output voltage is determined by the ratio of R10 to R11, just as in a linear series regulator. Switching action is accomplished by comparing a ripple voltage to a hysteresis voltage. The circuit switches on and off, triggered by the ripple of the output voltage. The voltage at pin 6 of the CA3085 is determined by R10 and R11 and is proportional to the output voltage plus the ripple voltage at point A, V_A, fed in by capacitor C5. This voltage is compared with the voltage at pin 5. The voltage at pin 5 consists of the built-in reference voltage of the CA3085 plus a variable component proportional to the voltage at point B, V_B, fed through R8.

SECTION 1 *Voltage, Current and Power Sources* 55

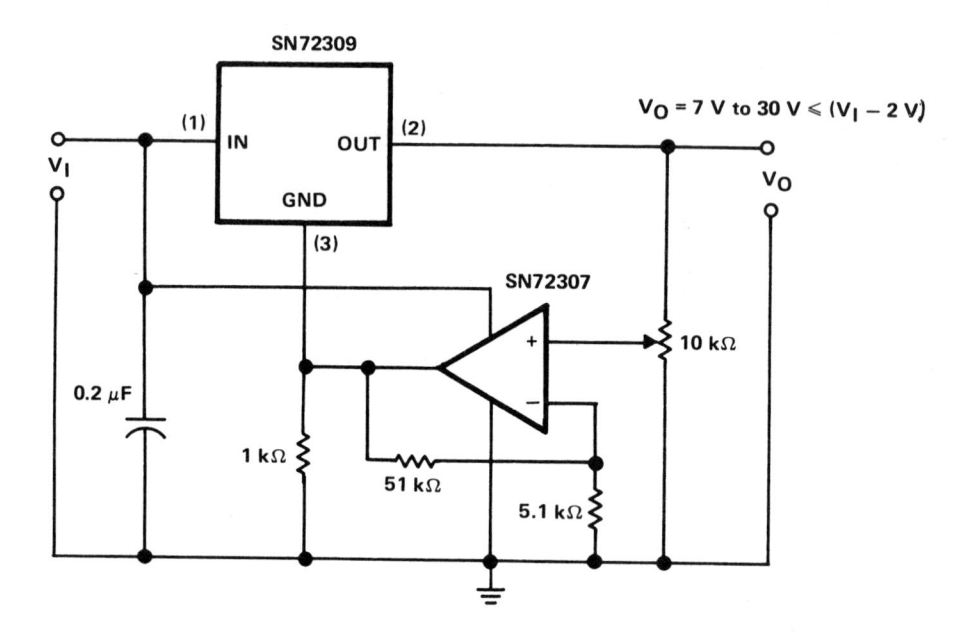

Highly stable adjustable-output voltage regulator.
The load regulation is within 50 mV for output loads of 5 to 50 mA. The loop gain of the feedback circuit can be increased as shown to obtain even better load regulation and an adjustable output voltage. Regulation to within 10 mV for a 10V output over an operating range of 5 to 50 mA is possible.

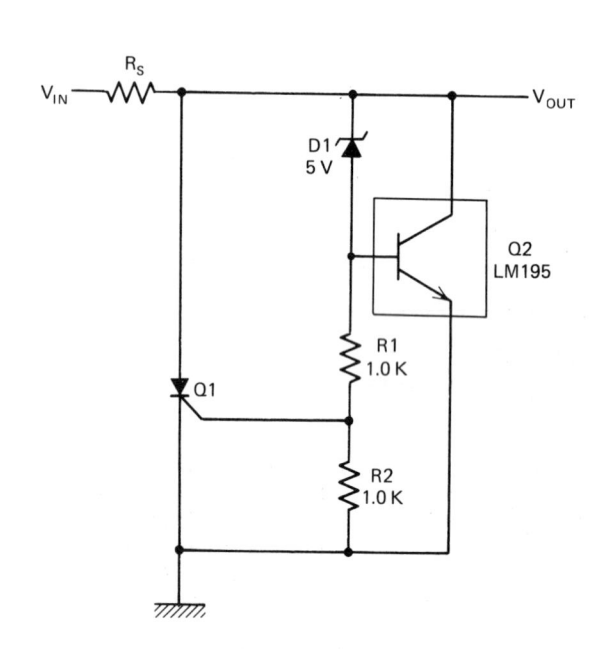

6.0V shunt regulator with crowbar.

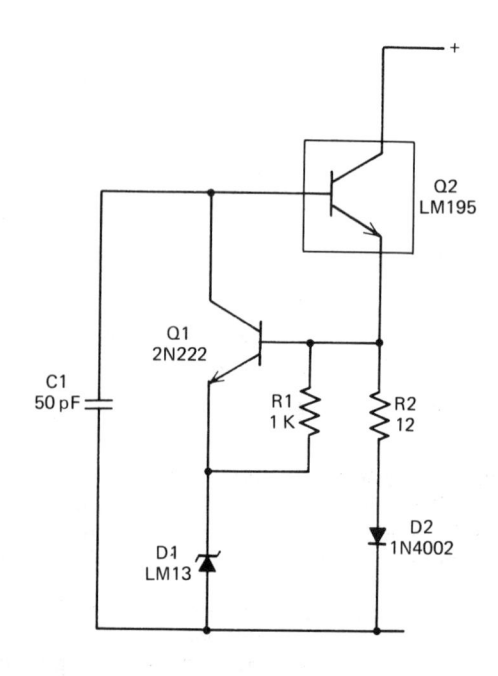

Two-terminal 100 mA current regulator.

56 IC SCHEMATIC SOURCEMASTER

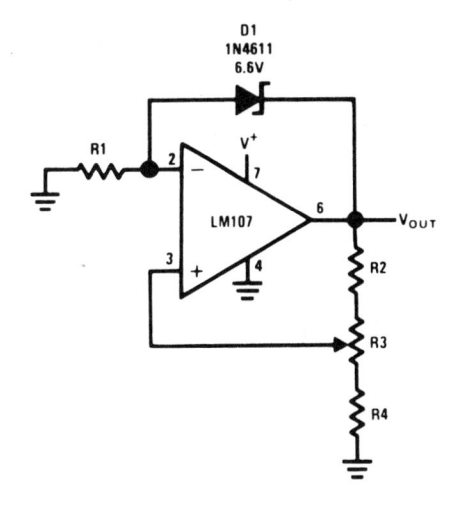

Positive-voltage references.

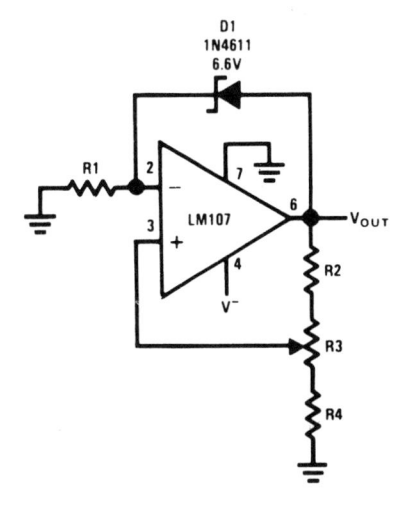

Negative-voltage references.

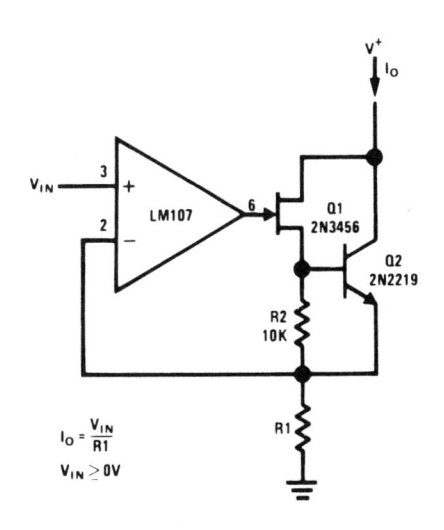

Precision current sink.

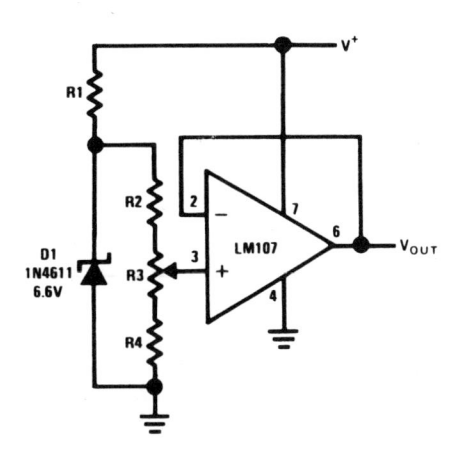

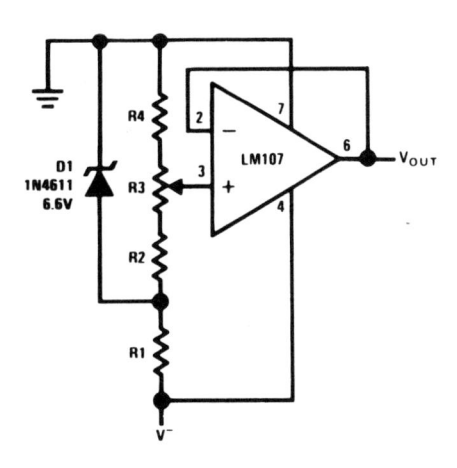

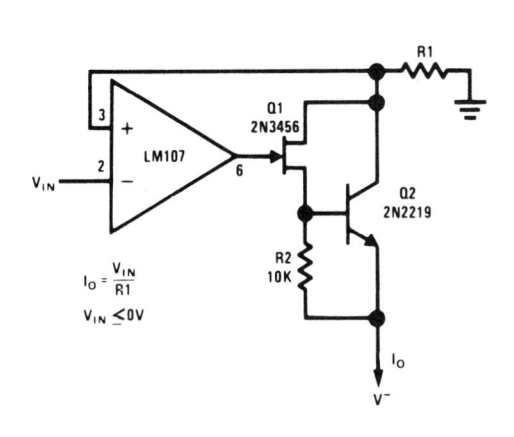

$I_O = \dfrac{V_{IN}}{R1}$
$V_{IN} \geq 0V$

Precision current source.

$I_O = \dfrac{V_{IN}}{R1}$
$V_{IN} \leq 0V$

SECTION 1 *Voltage, Current and Power Sources*

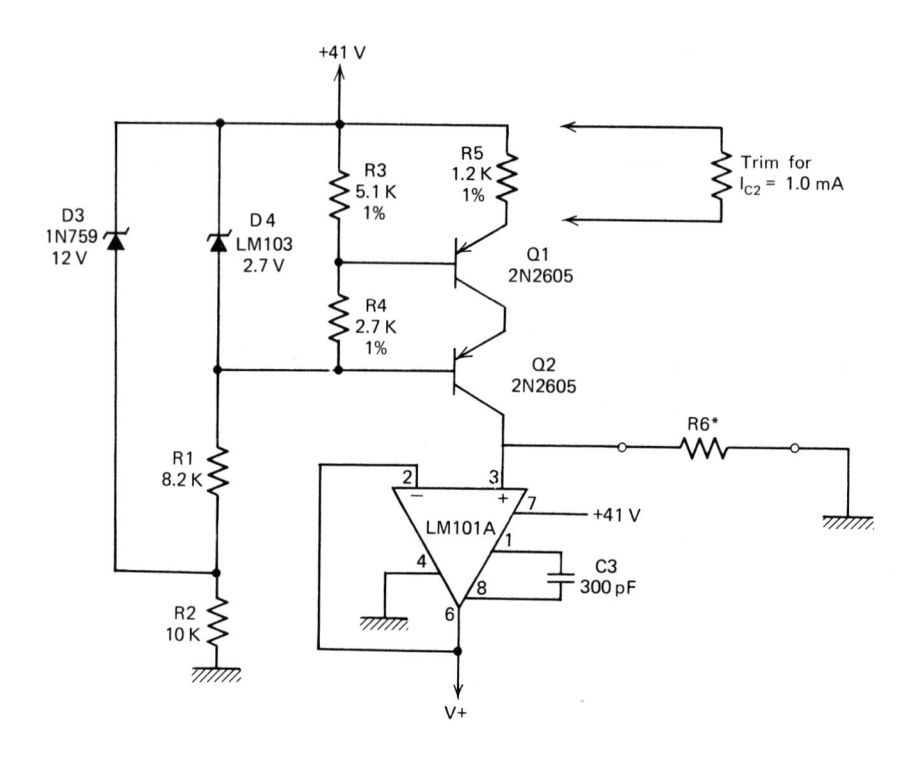

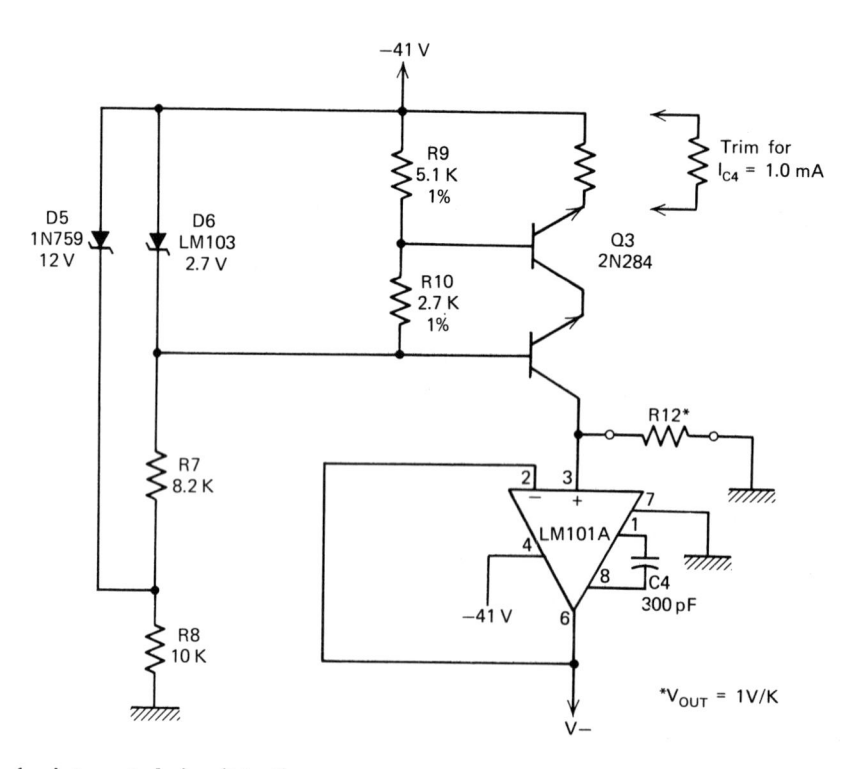

Low-power supply for integrated circuit testing.

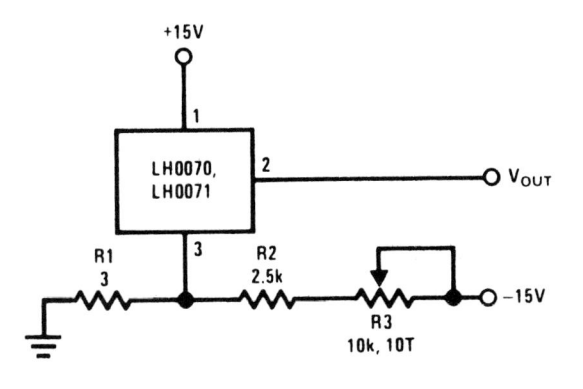

Output-voltage "fine" adjustment. The output of the LH0070 and LH0071 may be adjusted to a precise voltage by using this circuit, since the supply current of the devices is relatively small and constant with temperature and input voltage. Supply sensitivities are degraded slightly to 0.01%/V change in V_{OUT} for changes in V_{IN} and V^-.

An additional temperature drift of 0.0001%/°C is added as a result of the variation of supply current with temperature of the LH0070 and LH0071. Sensitivity to the value of R1, R2, and R3 is less than 0.001%/%.

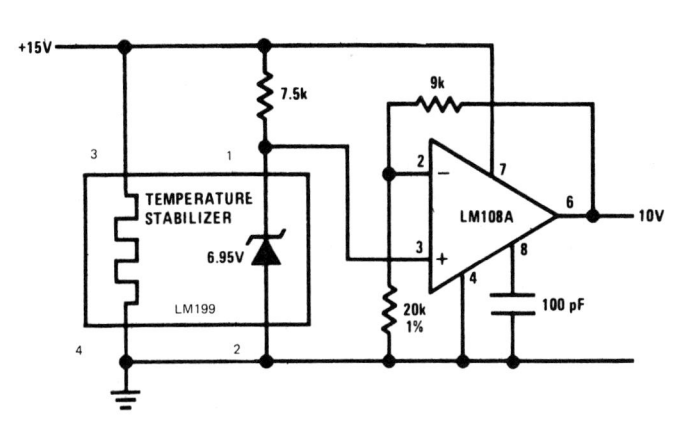

Buffered reference with single supply.

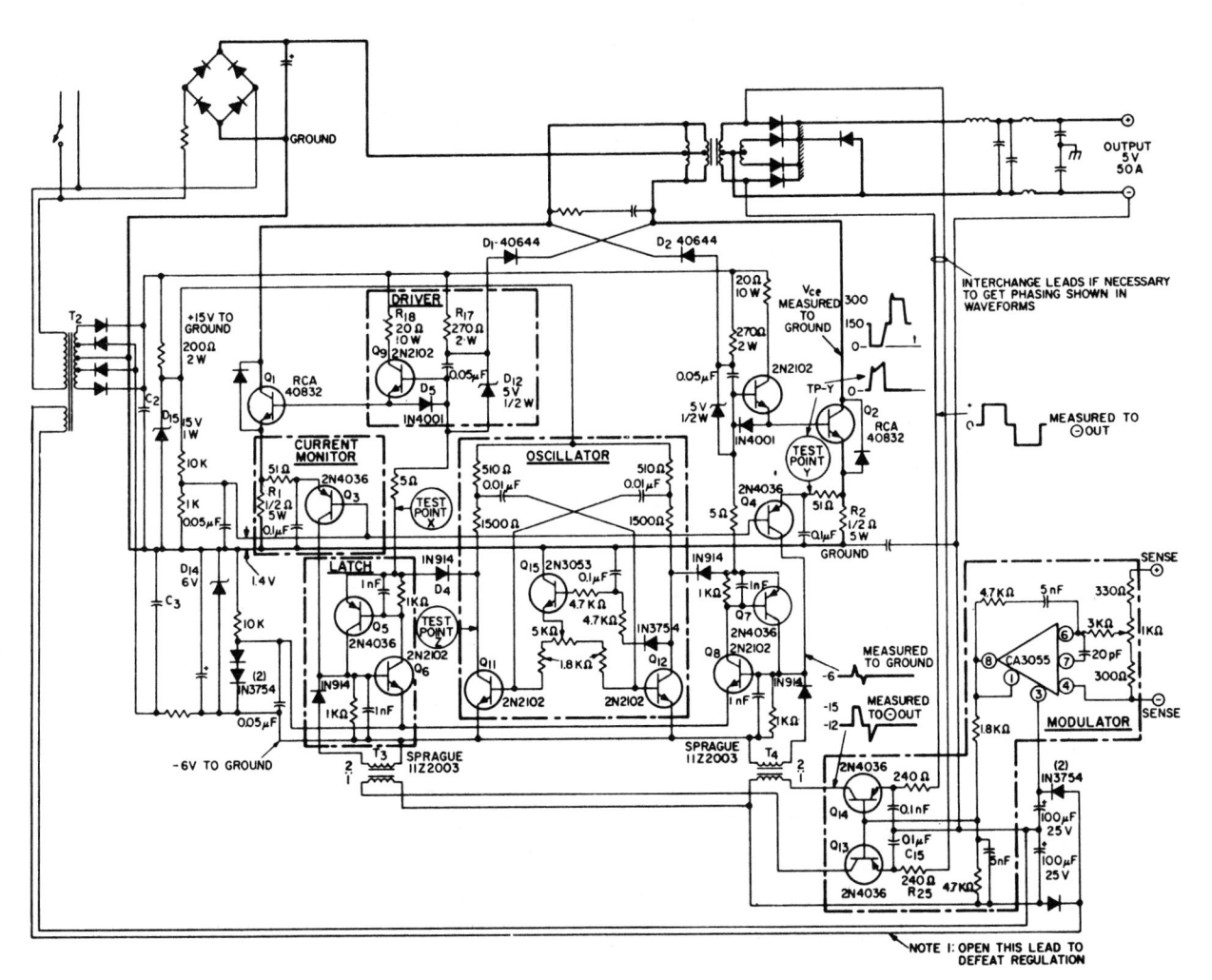

A switching-regulator power supply, with modulator circuits emphasized.

SECTION 1 *Voltage, Current and Power Sources* **59**

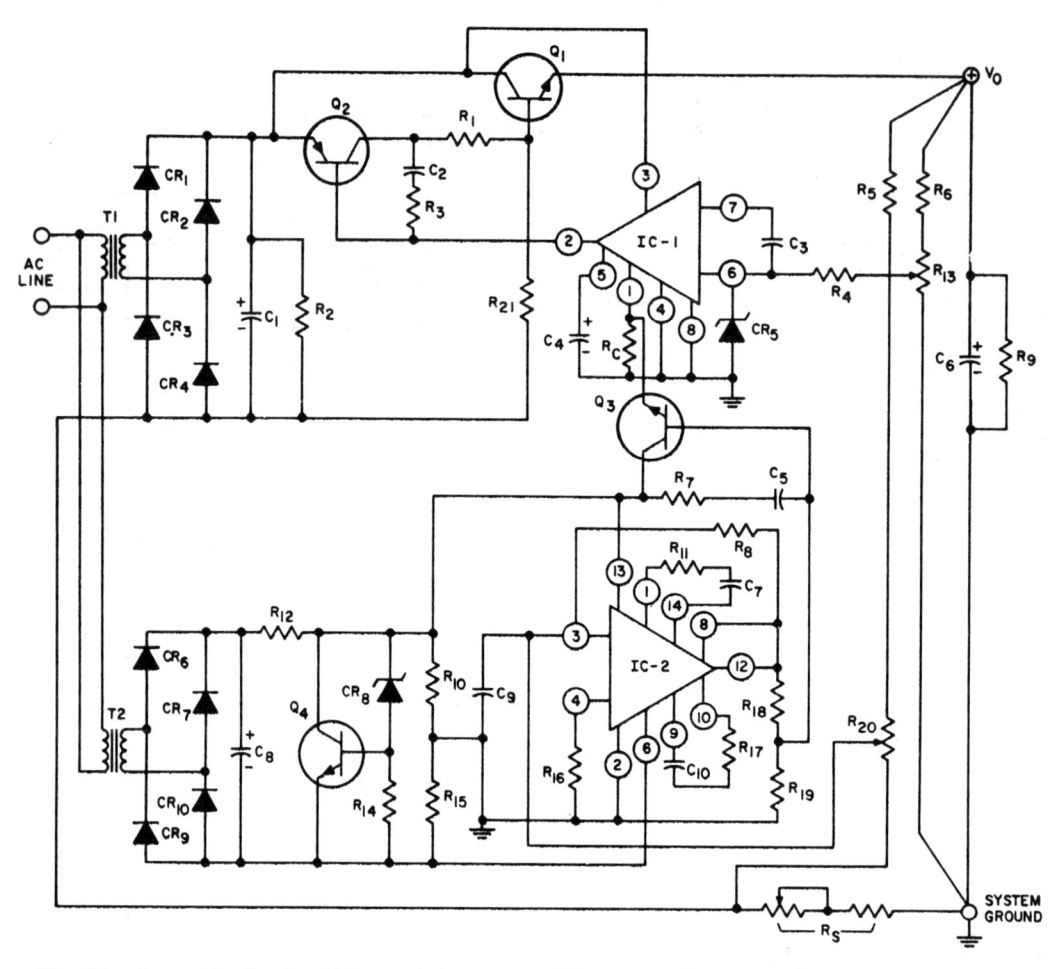

T1	Signal Transformer Co., Part No. 24-4 or equivalent	R4	100 ohms, 1/2 watt, carbon, IRC Type RC 1/2 or equivalent
T2	Signal Transformer Co., Part No. 12.8-0.25 or equivalent	R5	430 ohms, 2 watts, wire wound, IRC Type BWH or equivalent
CR1-CR4	RCA-1N1614		
CR5	Zener Diode, 1N5225 (3.3 V)	R6	9100 ohms, 2 watts, wire wound, IRC Type BWH or equivalent
CR6, CR7, CR9, CR10	Power Rectifier, RCA-1N3193	R7	470 ohms, 1/2 watt, carbon, IRC type RC 1/2 or equivalent
CR8	Zener Diode, 1N5242 (12 V)	R8	5100 ohms, 1/2 watt, carbon, IRC type RC 1/2 or equivalent
C1	5900 μF, 75 V, Sprague Type 36D592F075BC or equivalent		
C2	0.005 μF, ceramic disc, Sprague TGD50 or equivalent	R9, R14	1000 ohms, 2 watts, wire wound, IRC type BWH or equivalent
C3, C7, C10	50pF, ceramic disc, Sprague 30GA-Q50 or equivalent	R10, R15	250 ohms, 2 watts, 1% wire wound, IRC type AS-2 or equivalent
C4	2μF, 25 V, electrolytic, Sprague 500D G025BA7 or equivalent	R11, R17	1000 ohms, 1/2 watt, carbon, IRC type RC 1/2 or equivalent
C5	0.01 μF, ceramic disc, Sprague TG510 or equivalent	R12	82 ohms, 2 watts, IRC type BWH or equivalent
C6	500 μF, 50 V, Cornell-Dubilier No. BR500-50 or equivalent	R13	1000 ohms, potentiometer, Clarostat Series U39 or equivalent
C8	250 μF, 25 V, Cornell-Dubilier BR 250-25 or equivalent	R16	1200 ohms, 2 watts, wire wound, IRC type BWH or equivalent
C9	0.47 μF, film type, Sprague Type 220P or equivalent		
R1	5 ohms, 1 watt, IRC type BWH or equivalent	R18	510 ohms, 1/2 watt, carbon, IRC type RC 1/2 or equivalent
R2	1000 ohms, 5 watts, Ohmite type 200-5 1/4 or equivalent	R19	10,000 ohms, 1/2 watt, carbon, IRC type RC 1/2 or equivalent
R3	1200 ohms, 1/2 watt, carbon, IRC type RC 1/2 or equivalent		

A 60W, 20V regulated power supply with foldback current limiting.

60 IC SCHEMATIC SOURCEMASTER

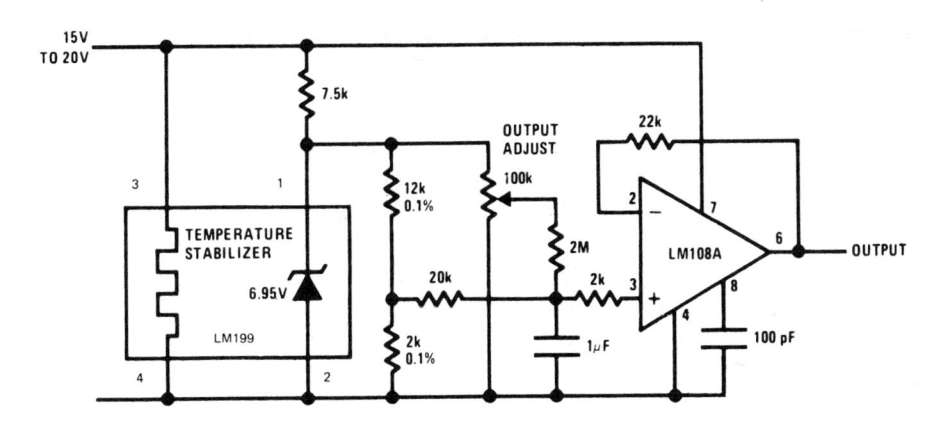

Standard cell replacement.

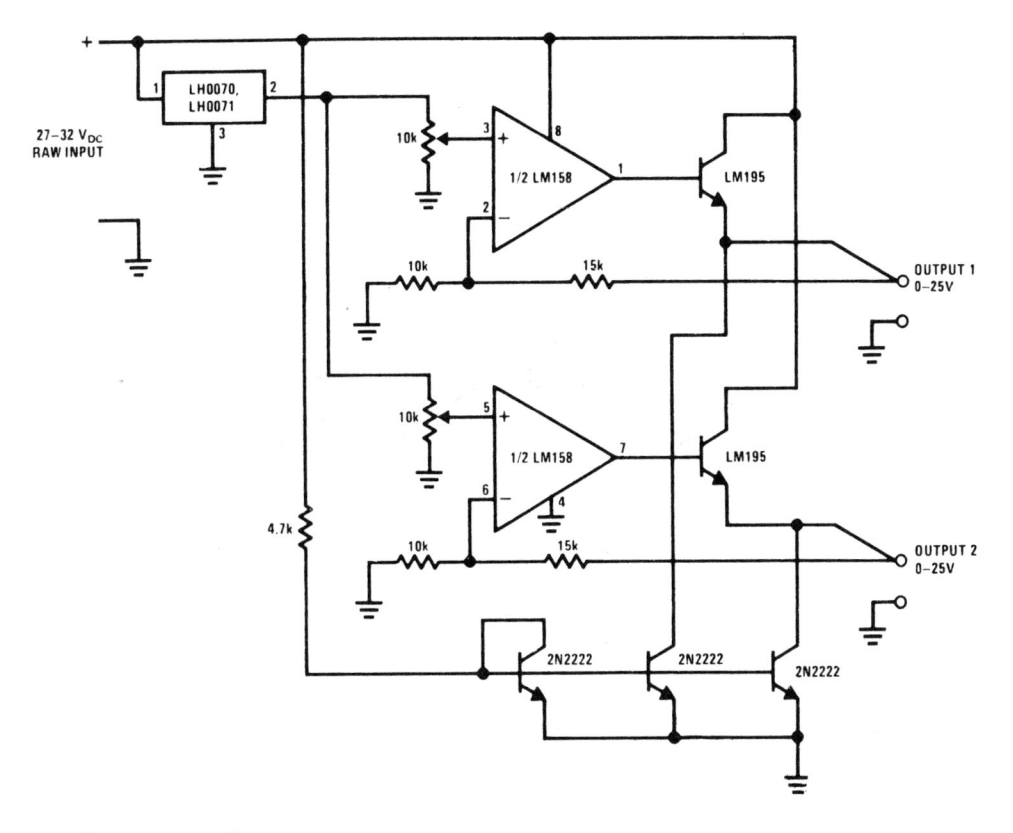

Dual-output bench power supply.

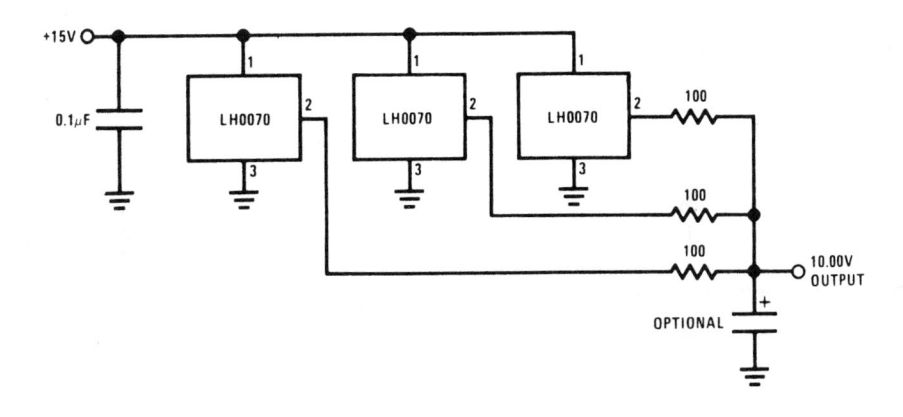

Statistical voltage standard.

SECTION 1 *Voltage, Current and Power Sources* 61

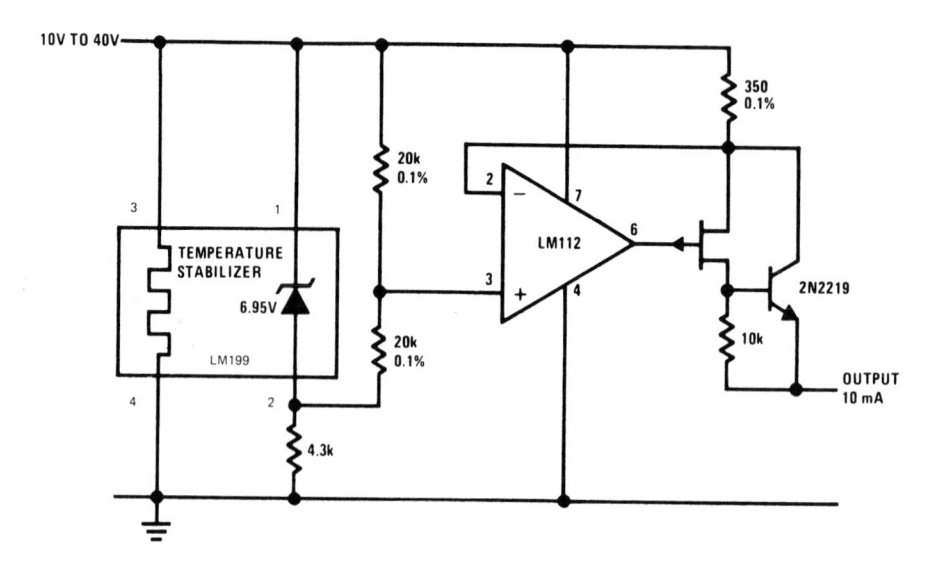

Positive-current source.

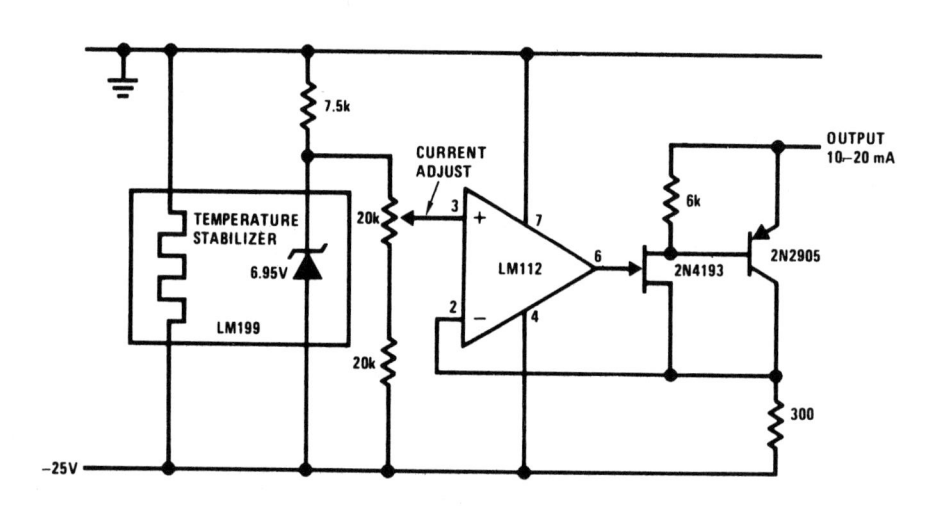

Negative-current source.

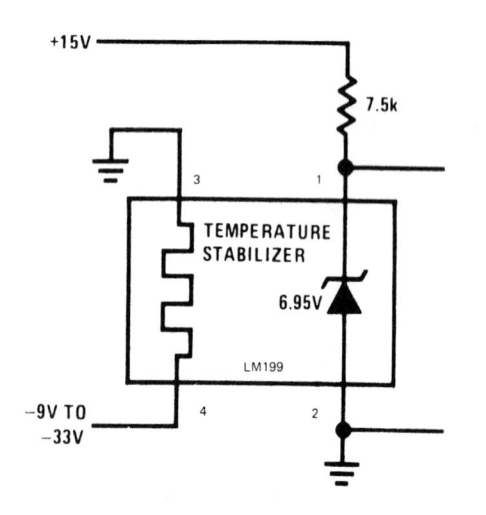

Negative heater supply with positive reference.

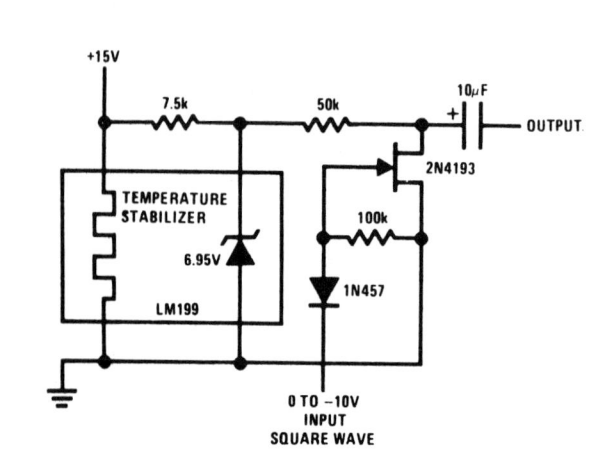

Square-wave voltage reference.

62 IC-SCHEMATIC SOURCEMASTER

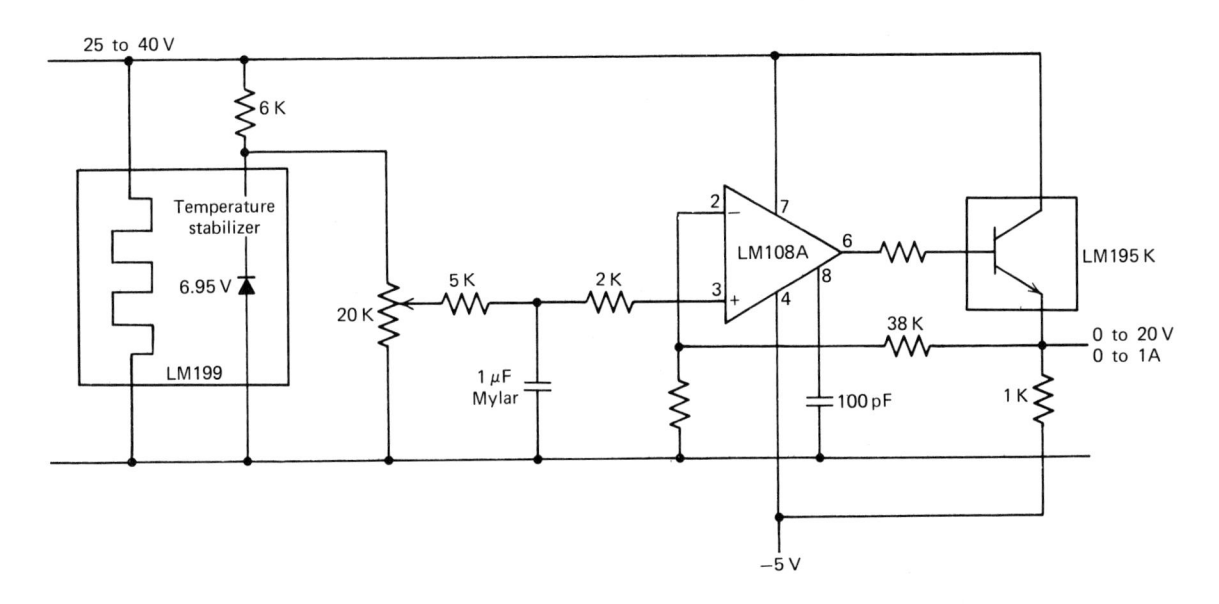

0–20V power reference.

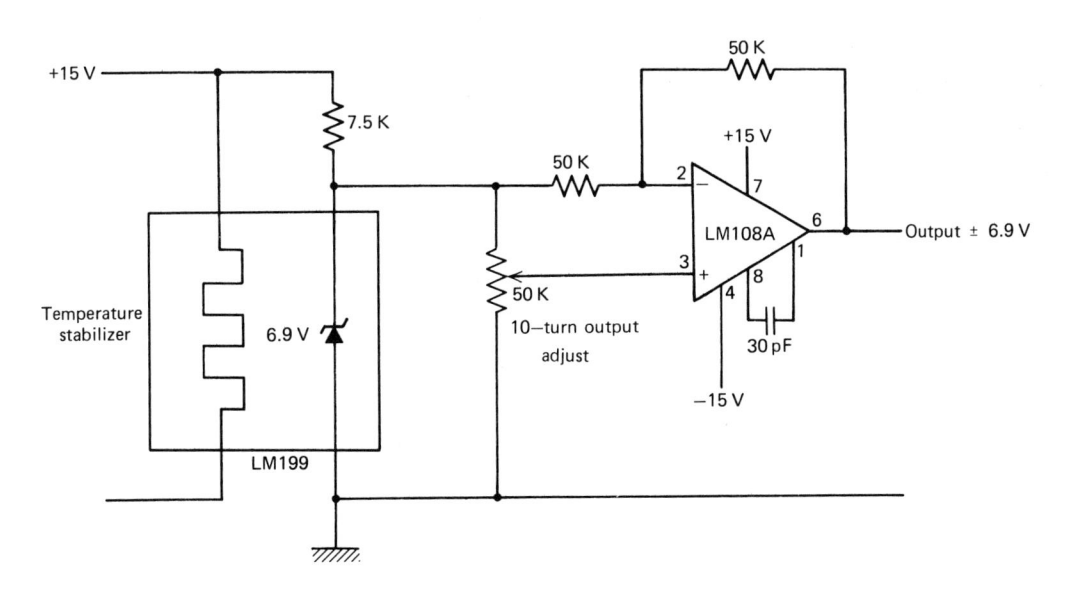

Bipolar output reference.

SECTION 1 *Voltage, Current and Power Sources*

SECTION 2
FOLLOWERS, BUFFERS, AND SPECIAL-PURPOSE AMPLIFIERS

Virtually every electronic circuit designed for a linear application involves amplification. This section includes circuits for those amplifiers that might better be classified under headings other than audio, instrumentation, radio, etc. Followers, of course, are not voltage amplifiers, since they have unity gain, but they are the conventional "first stage" of most amplification systems. Buffers may or may not amplify, depending on the function—but they are nonetheless universally classified as amplifiers. The follower serves primarily as an impedance-matching device and, like the buffer, is used extensively to isolate a stage or element with a "loading" characteristic from a subsequent stage or element that cannot tolerate such loading without signal degradation.

The integrated circuits featured in this section include both bipolar and complementary MOS types. RCA calls its integrated FET circuitry COS/MOS, while other makers refer to it as CMOS. Despite the nomenclature that appears in the body of the schematics, this text uses the more common CMOS designation for complementary FET ICs.

If you have trouble locating a circuit for your existing amplifying application, be sure to check other appropriate sections in this book.

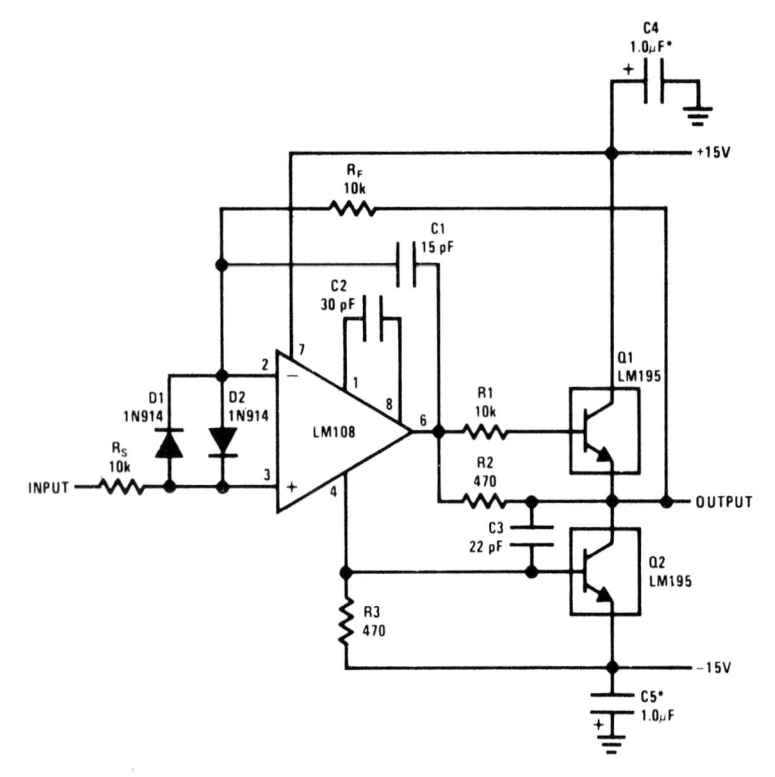

1.0A voltage follower. C4 and C5 should be solid tantalum.

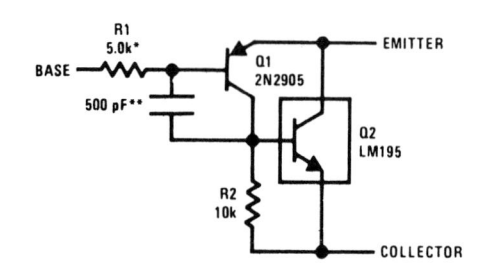

Power PNP. R1 protects against excessive base drive; the 500 pF capacitor is needed for stability.

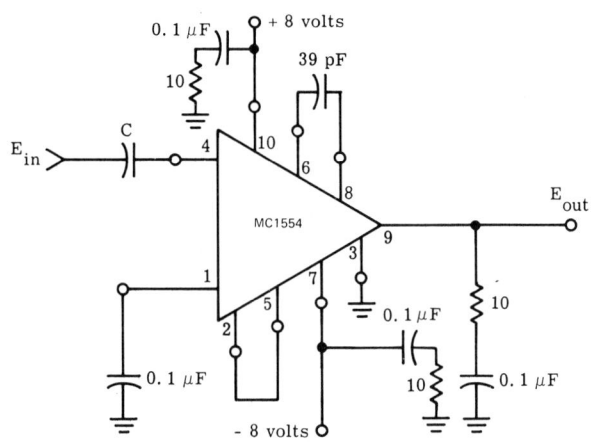

NOTE: Voltage Gain ≈ 35 For Connection Shown

1W inverting amplifier. Input impedance looking into pin 4 is 250Ω in the particular gain configuration shown; consequently, a large value of C will be required to couple to low frequencies. If direct coupling to the input is employed, a sacrifice in output offset will result. This offset can be compensated for by properly biasing pin 1 or terminating pin 1 with approximately 250Ω.

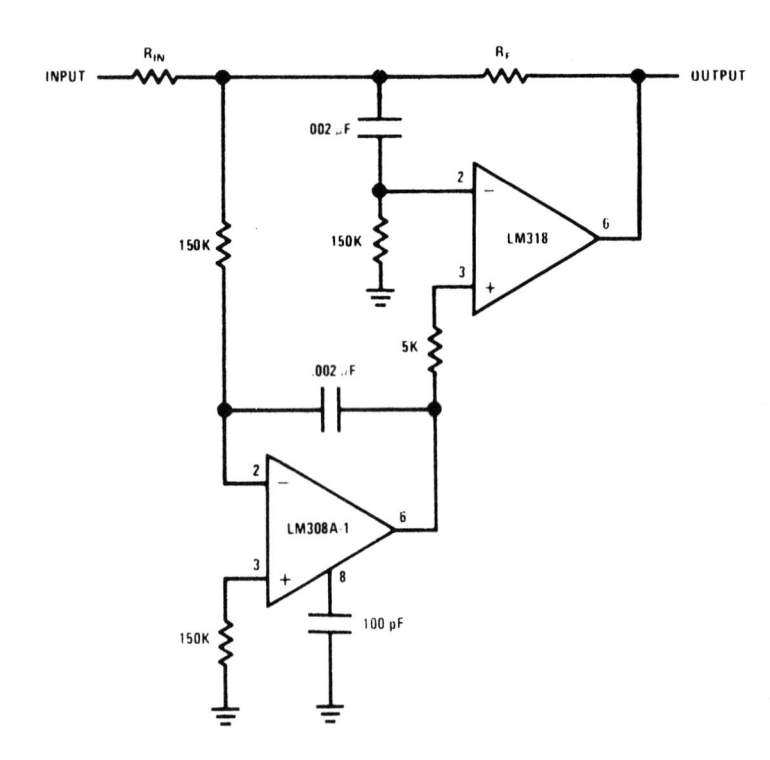

High-speed amplifier features low-drift and low-input current.

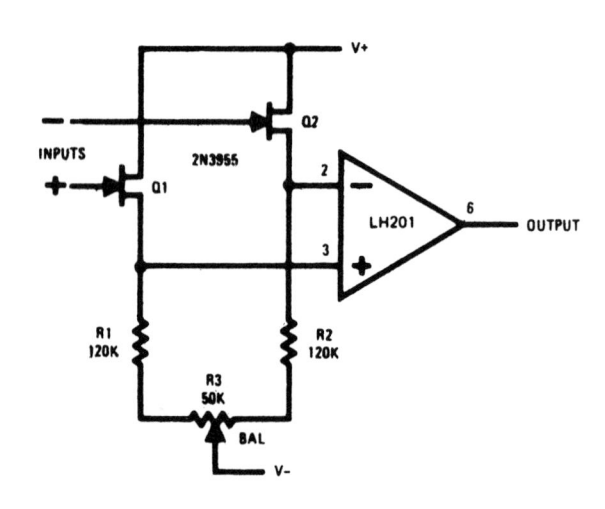

FET operatioal amplifier.

66 IC SCHEMATIC SOURCEMASTER

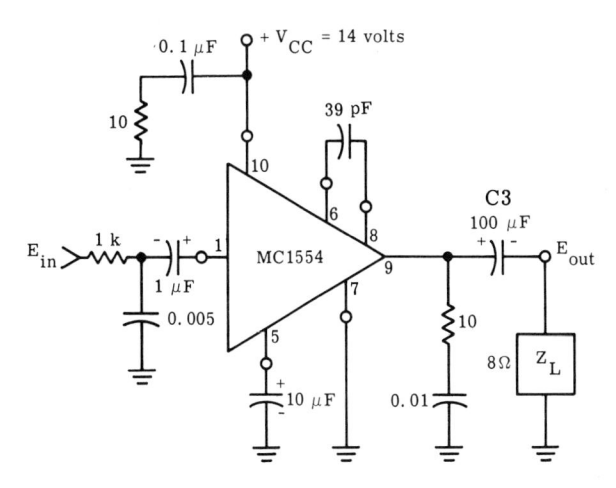

NOTE: Voltage Gain = 9 For Connection Shown

A 1W noninverting power amplifier operating with a single power supply and capacitively coupled to both the load and the source.

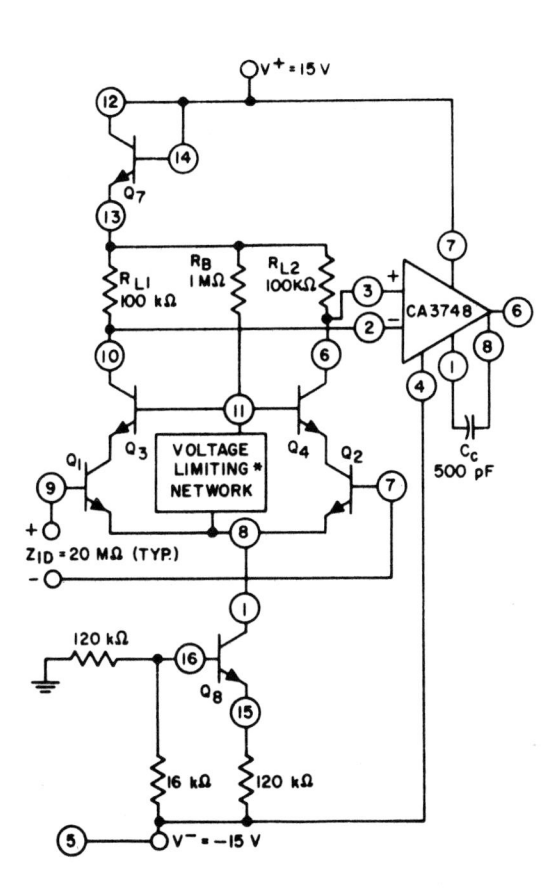

NOTE: Voltage Gain = 9 For Connection Shown

A 1W noninverting amplifier operating from split supply. Voltage gain of this circuit configuration is 9. An external heatsink is required, since internal power dissipation is nearly 1.5W. Frequency response is 40 Hz to 22 kHz (+1.5 dB, −3 dB).

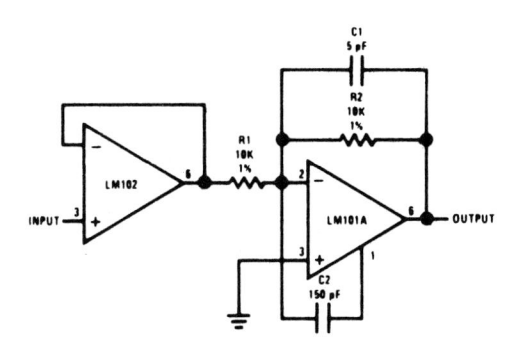

A superbeta op-amp with resistor drive network. Voltage-limiting network is built into the superbeta IC. To use it, just connect a resistor from the positive supply of the differential amplifier to pin 11; the value should be such that the current through the resistor is about half that measured at pin 8 under quiescent conditions.

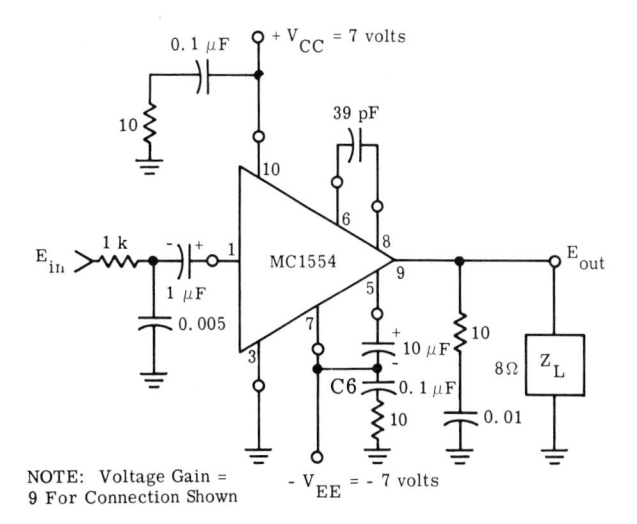

Fast inverting amplifier with high input impedance.

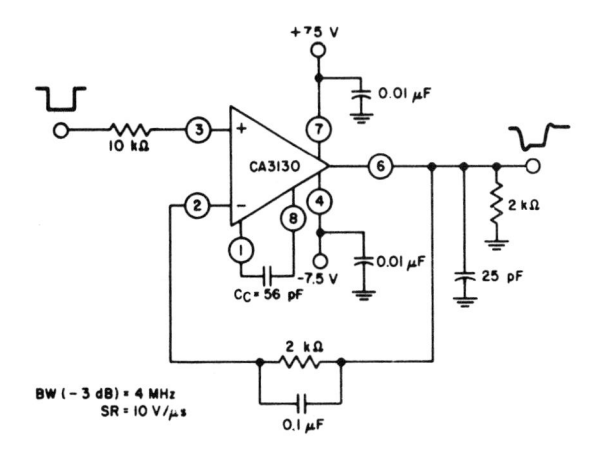

A split-supply voltage follower with associated waveforms.

SECTION **2** Followers, Buffers, and Special-Purpose Amplifiers **67**

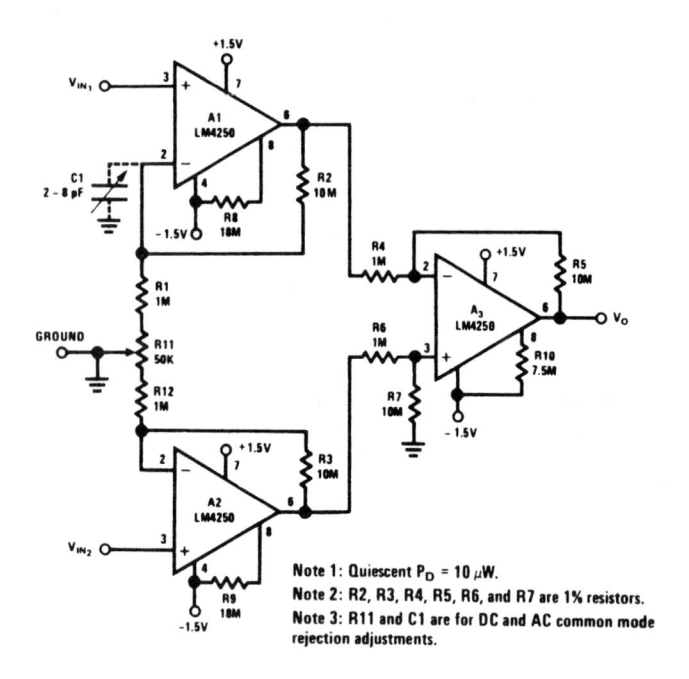

X100 instrumentation amplifier Quiescent P_D = 10 μW. R2, R2, R4, R5, R6, and R7 are 1%. R11 and C1 are for dc and ac common-mode rejection adjustments.

Note 1: Quiescent P_D = 10 μW.
Note 2: R2, R3, R4, R5, R6, and R7 are 1% resistors.
Note 3: R11 and C1 are for DC and AC common mode rejection adjustments.

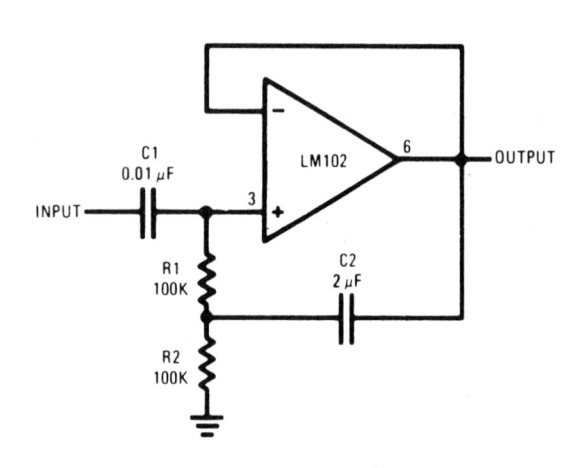

Ac amplifier. The LM102 has a minimum input resistance of 10,000M, so for dc amplifier applications this can be completely neglected. However, with an ac-coupled amplifier, a biasing resistor must be used to supply the input current. This drastically reduces the input resistance. Here is a method of bootstrapping the bias resistor to get higher input resistance. Even though a 200k bias resistor is used for good dc stability, the input resistance is about 12M at 100 Hz, increasing to 100M at 1 kHz.

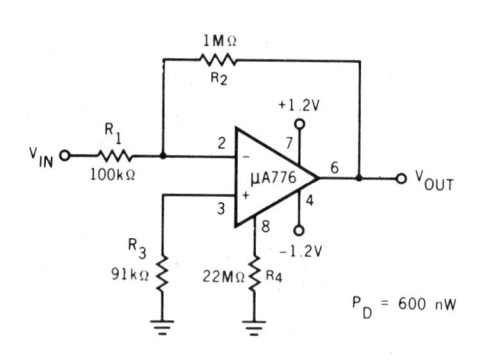

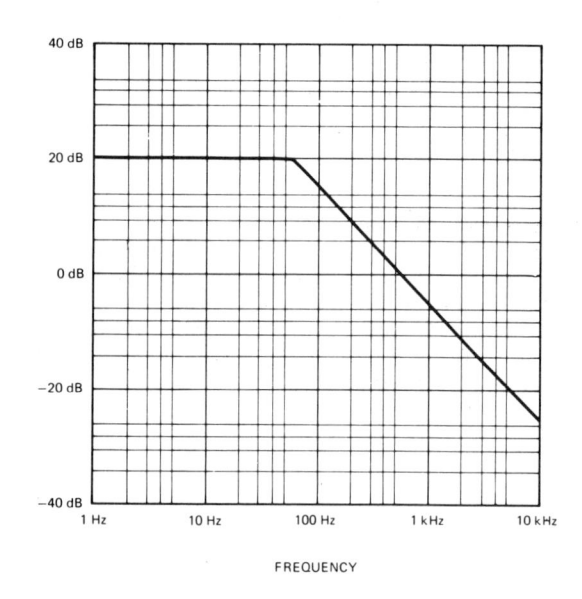

A 600 nW amplifier can be driven with pair of mercury cells for time equal to battery shelf life. The small-signal f_2 or turnover frequency shown on the response curve is 55 Hz. The maximum peak-to-peak output signal amplitude is 1.2V. The 30 nA set current yields a maximum slew rate of 0.7 V/msec. Combining this slew rate with the 1.2/2 or 0.6V peak output amplitude gives a large-signal f_2 of 185 Hz. Therefore, for this application, the response is limited only by the small-signal f_2 and is independent of output amplitude.

68   IC SCHEMATIC SOURCEMASTER

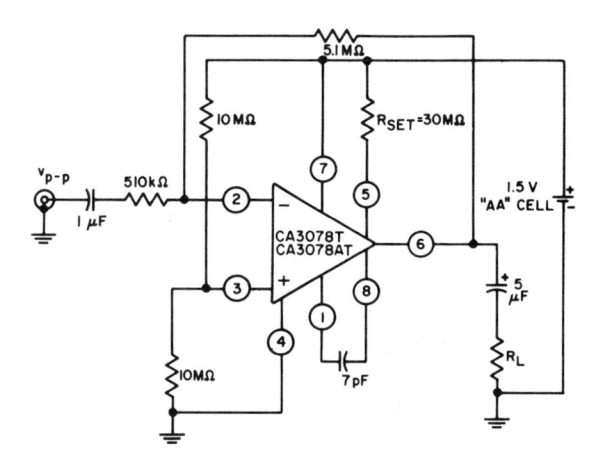

Inverting amplifier

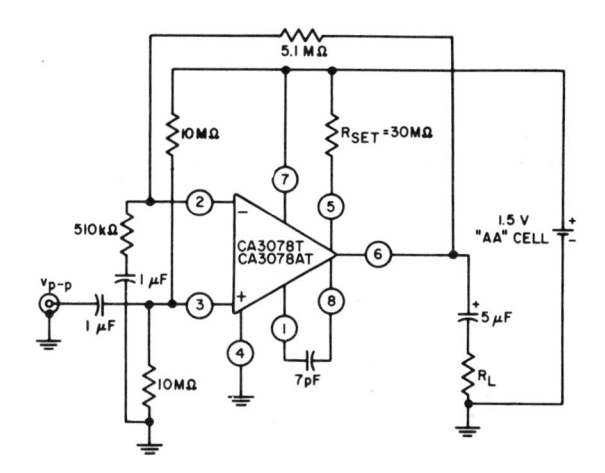

Noninverting amplifier

Amplifiers offer a 20 dB gain when driven with single 1.5V AA cell. Each circuit draws about 675 nW. Output voltage swing is 300 mV p-p with 20k load.

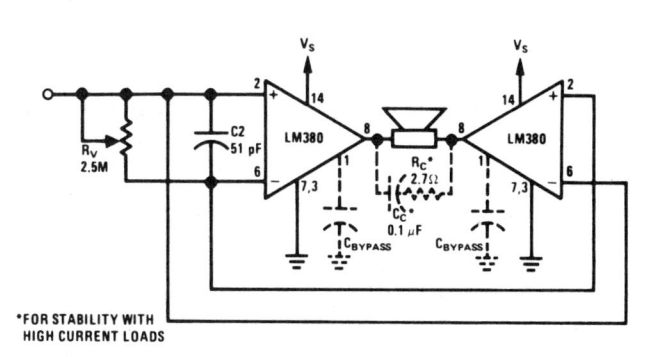

*FOR STABILITY WITH HIGH CURRENT LOADS

This bridge amplifier provides twice the voltage swing across the load for a given supply, thereby increasing the power capability by a factor of four over a single amplifier.

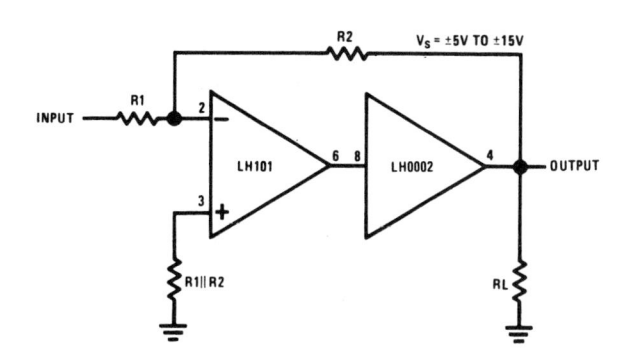

An LH0002 integrated with an LH101 in a **booster follower**. The configuration is stable without the requirement for any external compensation; however, it would behoove the designer to be conservative and bypass both the negative and positive power supplies with at least a 0.01 µF capacitor to cancel any power supply lead inductance. A 100Ω damping resistor located right at the input of the LH0002 might also be required between the operational amplifier and the booster amplifier. The physical layout will determine the requirement for this type of oscillation suppression. Current limiting can be added by incorporating series resistors from pins 2 and 6 to their respective power supplies. The exact value would be a function of power supply voltage and required operating temperature.

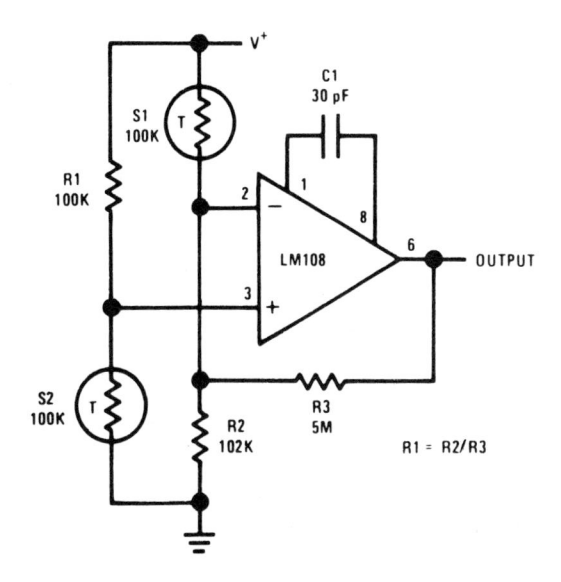

Amplifier for bridge transducers. When the bridge goes off balance, the op-amp maintains the voltage between its input terminals at zero, with the current fed back from the output through R3. This circuit does not act like a true differential amplifier for large imbalances in the bridge. The voltage drops across the two sensor resistors, S1 and S2, become unequal as the bridge goes off balance, causing some nonlinearity in the transfer function. However, this is not usually objectionable for small signal swings.

SECTION **2** Followers, Buffers, and Special-Purpose Amplifiers **69**

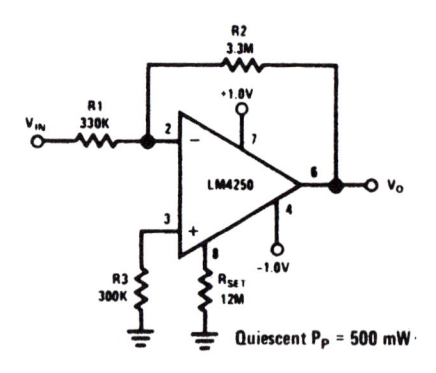

500 nW X10 amplifier

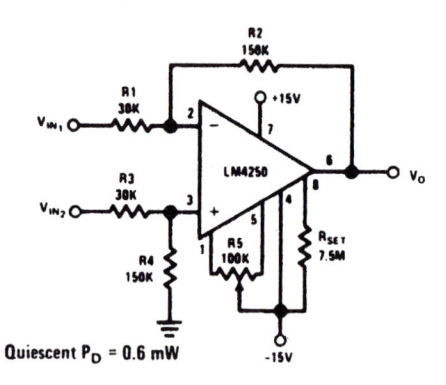

X5 difference amplifier

A single external master bias-current-setting resistor programs the input bias current, input offset current, quiescent power consumption, slew rate, input noise, and gain-bandwidth product. Standby power consumption as low as 500 nW is achievable, and no frequency compensation is required. Circuit can be powered by two flashlight batteries.

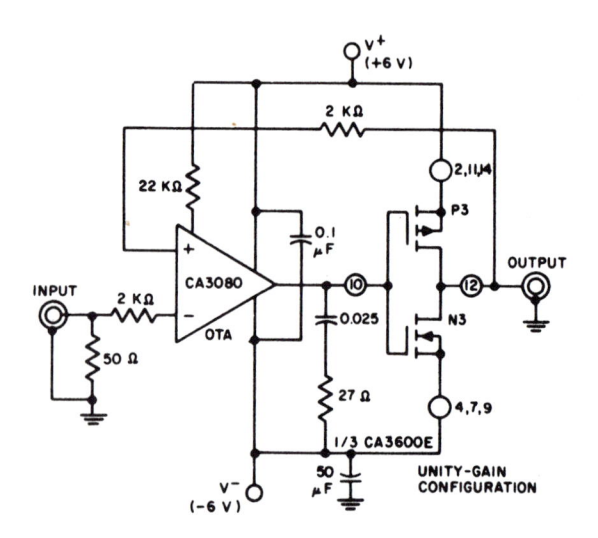

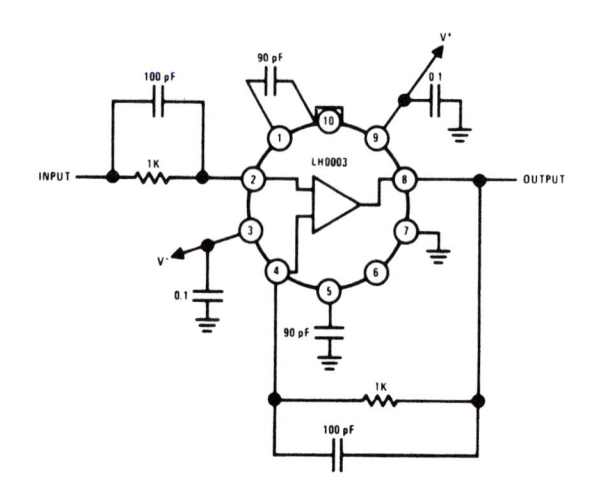

A unity-gain follower for application requiring wide input-impedance variations, low distortion.

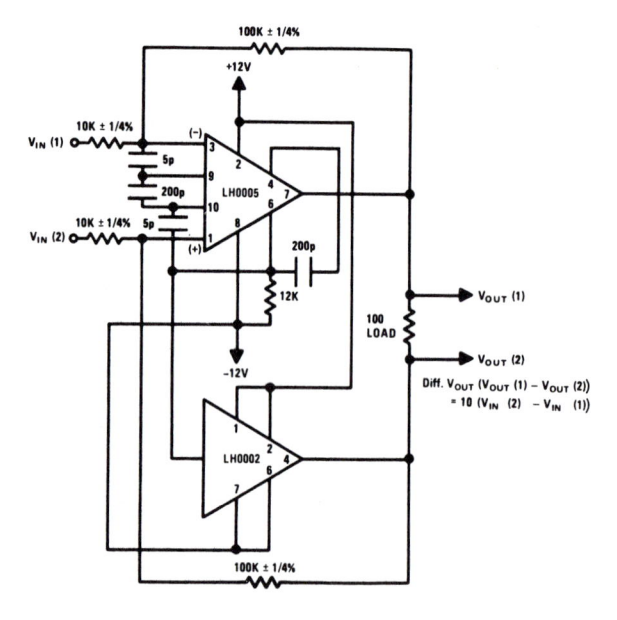

Differential input/output operational amplifier integration. For this circuit to function properly, a load must be floated between the outputs of the two devices to provide a complete loop of feedback. A differential head on a scope across the load presents a true waveform of the actual signal being applied to it. If only one end of the load is displayed, it will appear distorted because this information is being fed back negatively to the input to cancel the loop distortion of the overall amplifier. With the compensation shown, a 20V (peak-to-peak) signal can be applied to a 100Ω load to 80 kHz. The overall circuit is approximately 33% efficient under these conditions. A heatsink must be used at higher temperatures.

CMOS transistor pair used as **postamplifier to op-amp in unity-gain circuit.** When compensated for the unity-gain voltage-follower mode shown here, the slew rate is about 1V/μsec.

70   IC SCHEMATIC SOURCEMASTER

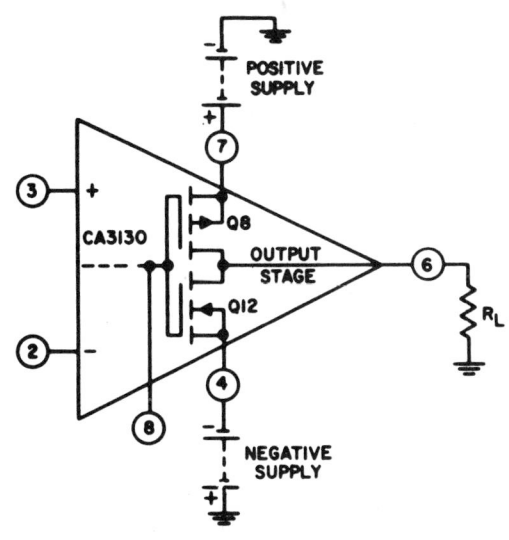

DUAL POWER-SUPPLY OPERATION

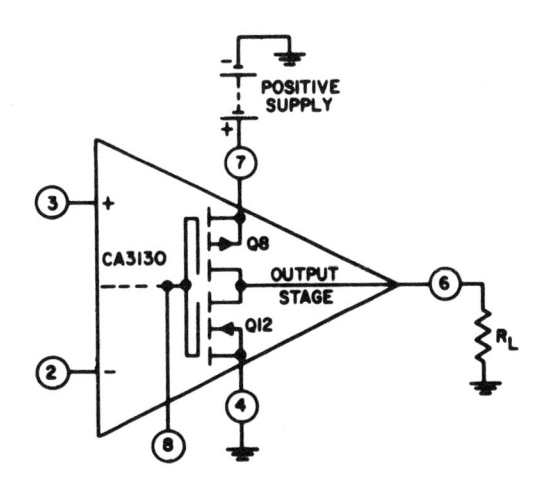

SINGLE POWER-SUPPLY OPERATION

CA3130 output stage in dual and single power-supply operation.

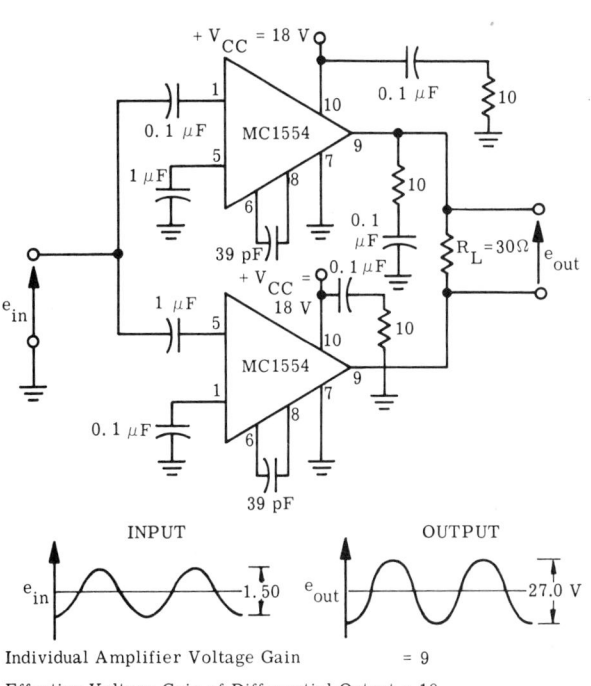

INPUT e_{in} — 1.50

OUTPUT e_{out} — 27.0 V

Individual Amplifier Voltage Gain = 9
Effective Voltage Gain of Differential Output = 18
NOTE: Supply Voltage = 18 volts
 Yet Peak-To-Peak Output Swing = 27.0 volts
Power Out ≈ 3 Watts

Differential output power amplifier with 3W capability. The top amplifier is connected in noninverting gain-of-9 configuration, while the lower amplifier is connected in an inverting (gain-of-9) mode, resulting in an effective overall voltage gain of twice that of either individual amplifier operating alone. The input impedance of the upper amplifier is 10k, while that of the lower amplifier is 1k. Because of the differential output connection, the peak-to-peak output-voltage swing capabilities of the amplifier can and do exceed the supply voltage. Effective use of this amplifier requires minimum lead lengths and proper power supply decoupling.

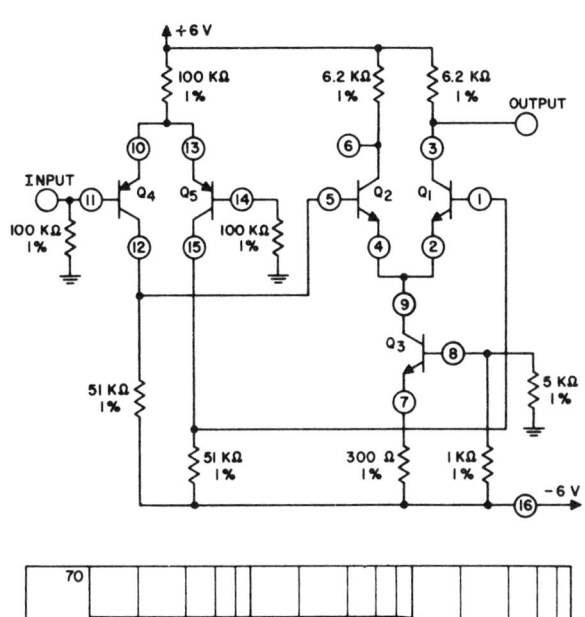

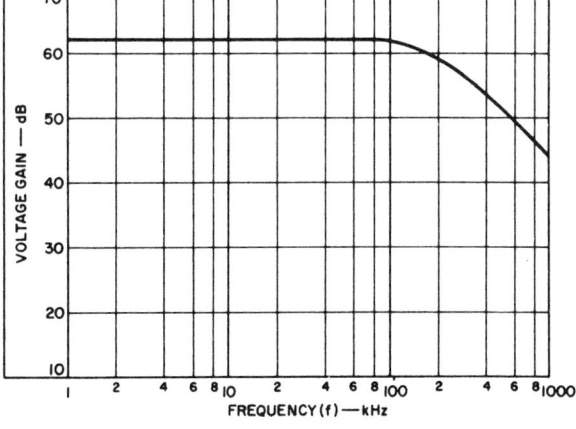

A cascade of differential amplifiers using CA3096AE features operation with either dual or single supply, wide input common-mode range of +5 to −5V, and bias current less than 1 μA. Chart shows gain/frequency characteristics.

SECTION 2 *Followers, Buffers, and Special-Purpose Amplifiers* 71

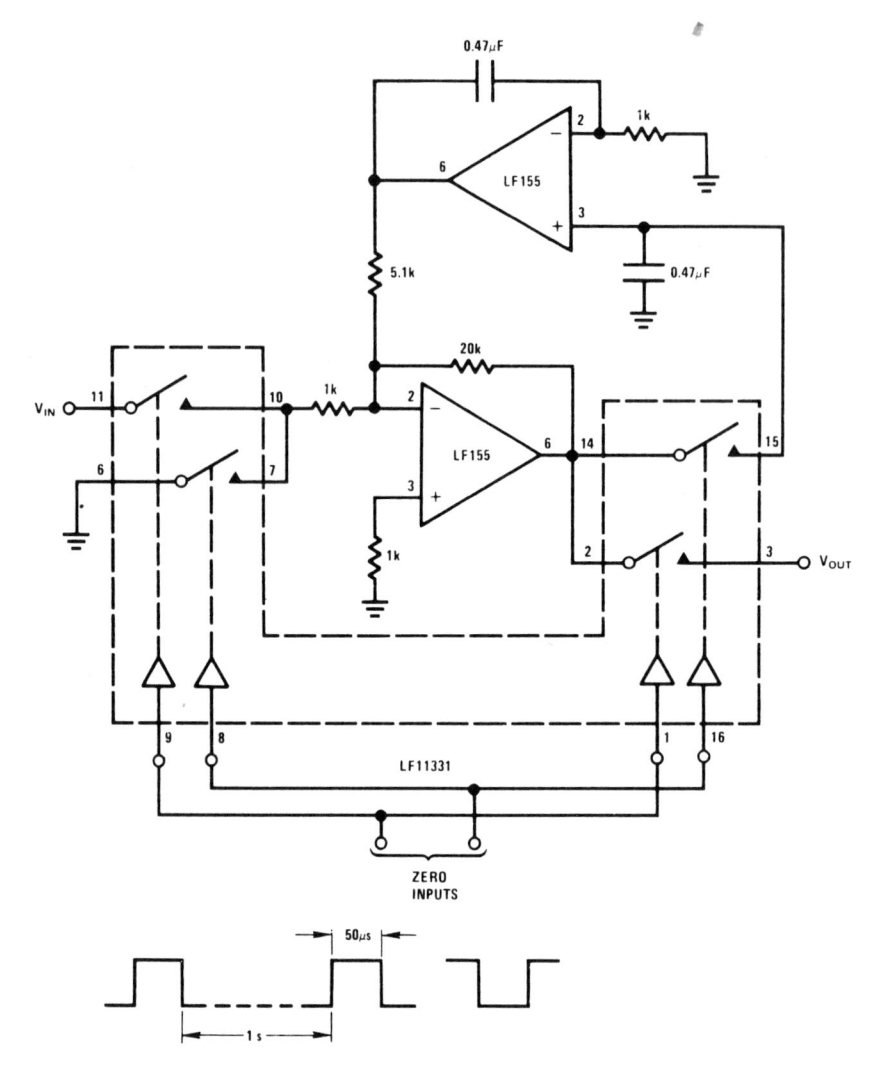

Self-zeroing operational amplifier.
±300 mA peak current to the load. The input impedance of the booster (typically 400k) alleviates the loading on the MC1556 and, hence, additional drifts that may occur from resulting thermal effects.

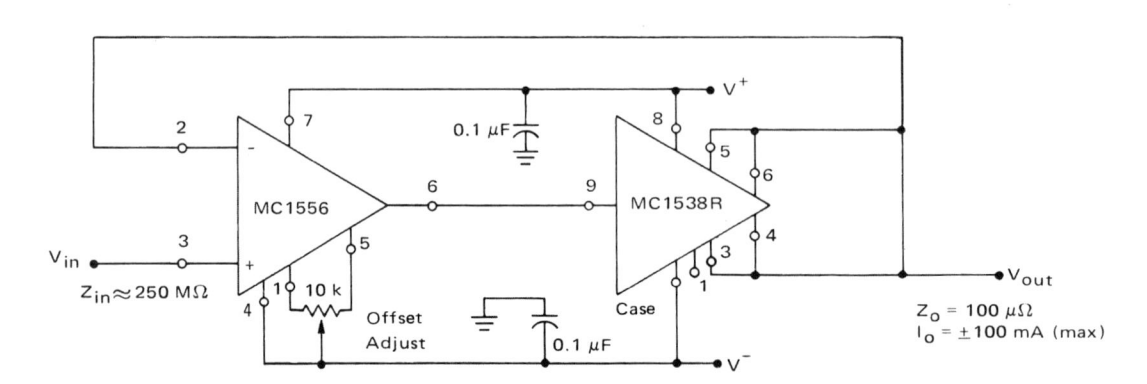

Current-boosted voltage follower. The booster's output impedance is typically 10Ω, which results in a factor-of-100 reduction in the closed-loop output impedance. This permits greater stability when driving a capacitive load. The MC1538R can supply up to ±300 mA peak current to the load. The input impedance of the booster (typically 400K) alleviates the loading on the MC1556 and, hence, additional drifts that may occur from resulting thermal effects.

72 IC SCHEMATIC SOURCEMASTER

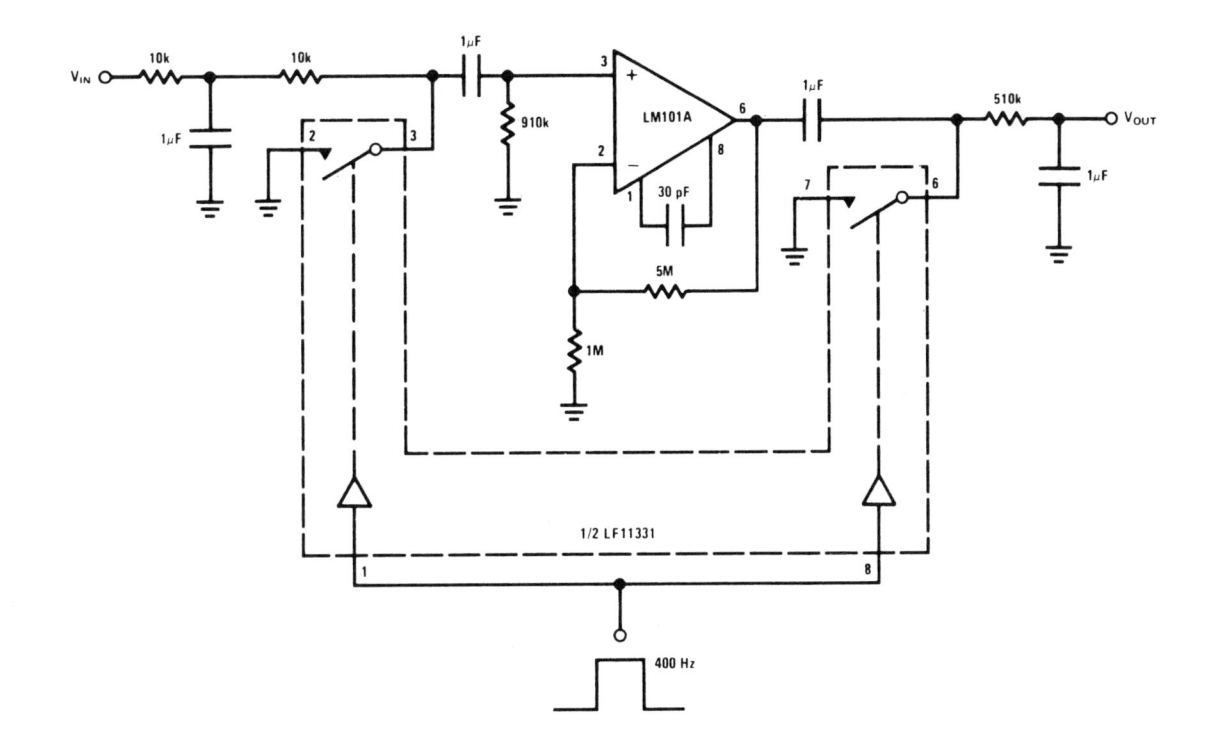

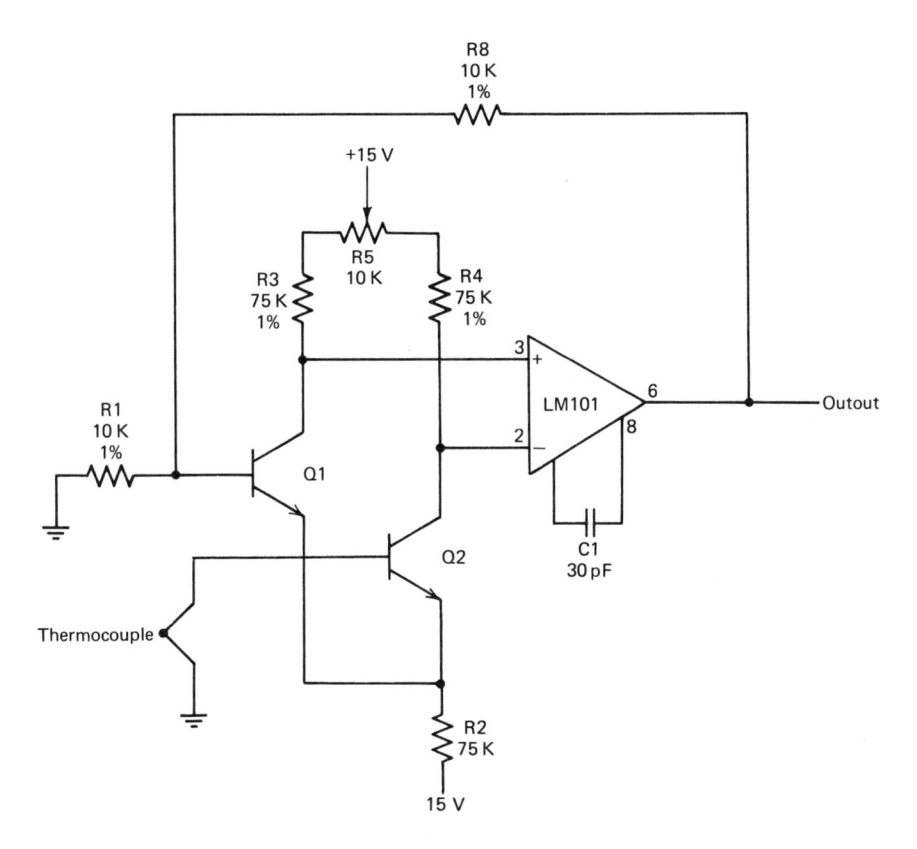

Dc amplifier using drift-compensation technique. A monolithic transistor pair is used as a preamplifier for a conventional operational amplifier. A null potentiometer, which is set for zero output for zero input, unbalances the collector load resistors of the transistor pair such that the collector currents are unbalanced for zero offset. This gives minimum drift. An interesting feature of the circuit is that the performance is relatively unaffected by supply-voltage variations; a 1V change in either supply causes an offset voltage change of about 10 μV.

SECTION **2** *Followers, Buffers, and Special-Purpose Amplifiers*

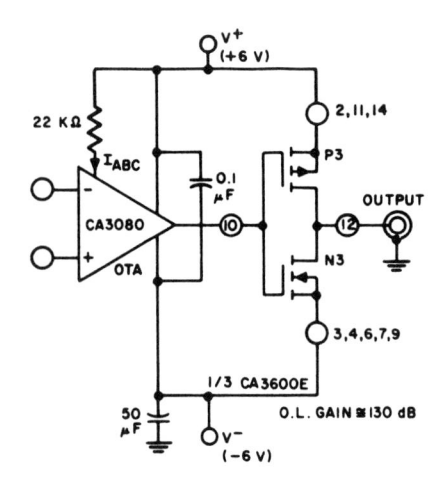

CMOS transistor pair used as postamplifier to op-amp in open-loop circuit. The 30 dB gain in a single CMOS transistor pair is an added increment to the 100 dB gain in the CA3080, yielding a total forward gain of about 130 dB.

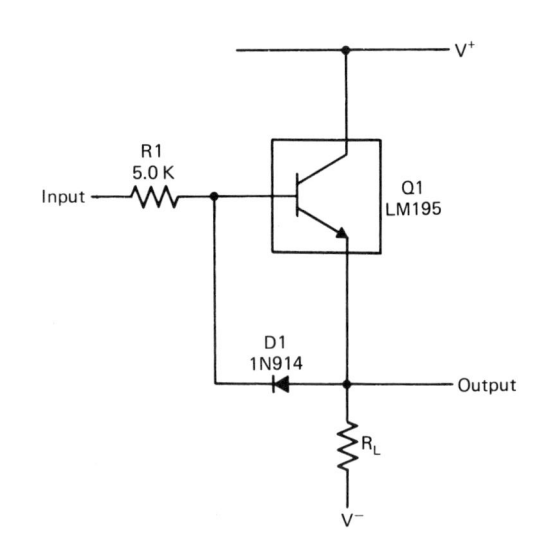

Fast follower. D1 prevents storage with fast-fall-time square-wave drive.

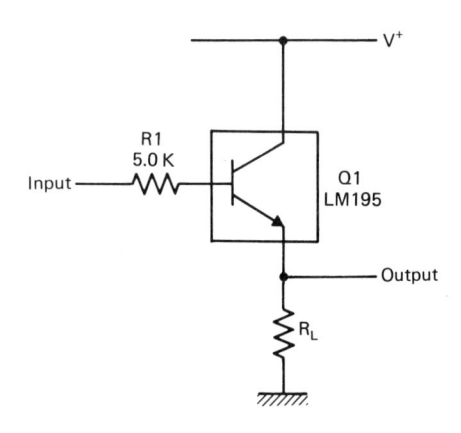

Emitter follower. R1 is needed for stability.

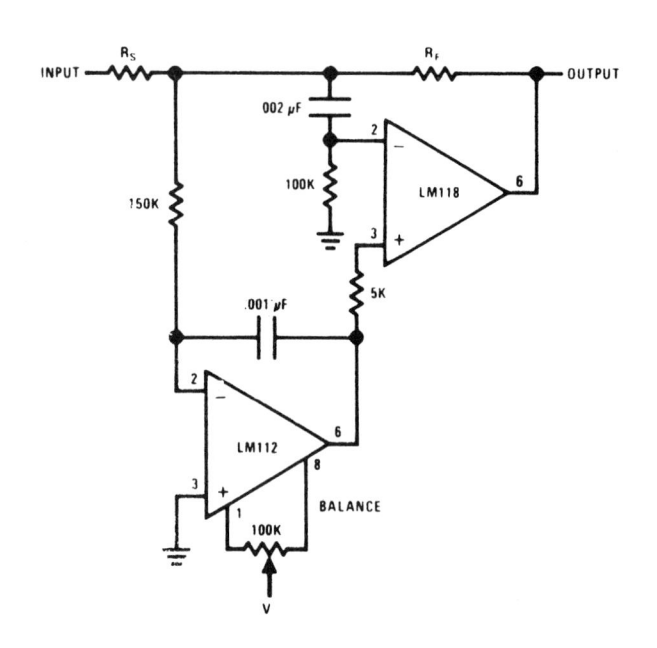

A fast-summing amplifer with low input current.

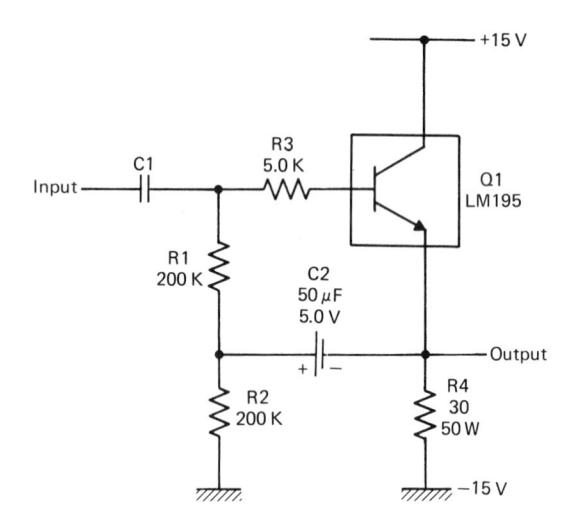

High-input-impedance ac emitter follower.

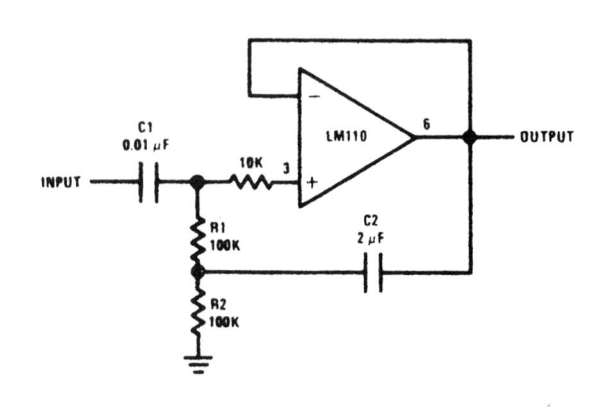

High-input-impedance ac amplifier.

74   IC SCHEMATIC SOURCEMASTER

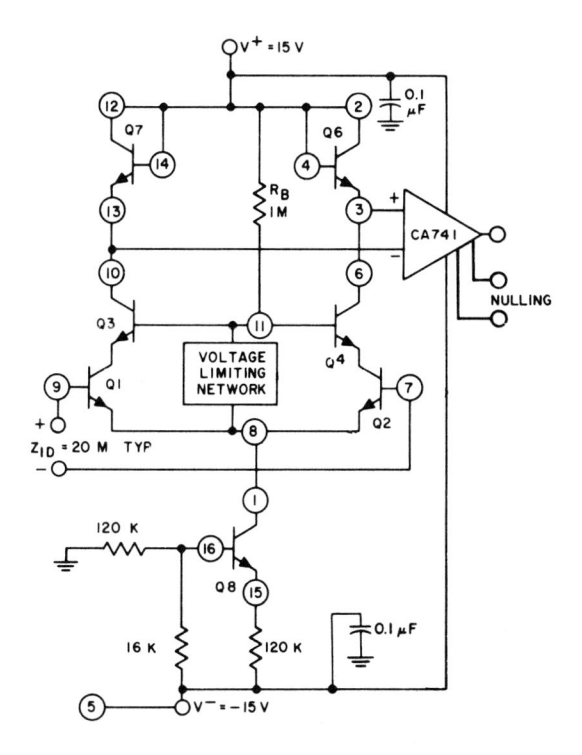

Op-amp with unity-gain preamplifier. This circuit boosts the input impedance, Z_{id}, of the op-amp to the order of 20M, typical. Transistors Q1–Q3 and Q2–Q4 operate as a differential-cascode amplifier, with transistor Q8 as their constant-current source. Transistors Q6 and Q7 are diode-connected to establish dc levels that are appropriate for direct connection to the CA741 input terminals. No additional external compensation is required with this circuit because the unity voltage gain provided by the preamplifier precedes the internally compensated CA741 op-amp. The resultant offset voltage of the combination circuit is the algebraic sum of the offsets due to Q1, Q7 vs Q2, Q6, and offset due to the CA741. The resultant offset can be nulled at the normal nulling terminals on the CA741. This circuit is ideal for amplification of signals emanating from sources with very high impedances. A voltage-limiting network is built into the superbeta IC. To use it just connect a resistor from the positive supply of the differential amplifier to pin 11; the value should be such that current through the resistor is about half that measured at pin 8 under quiescent conditions.

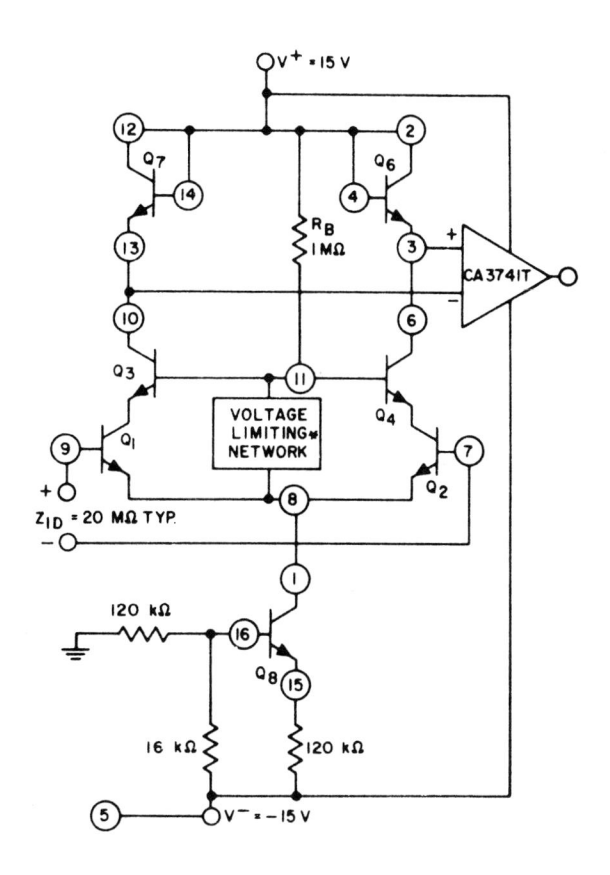

A superbeta op-amp with diode drive network. Voltage-limiting network is built into superbeta IC. To use it just connect a resistor from the positive supply of the differential amplifier to pin 11; the value should be such that the current through the resistor is about half that measured at pin 8 under quiescent conditions.

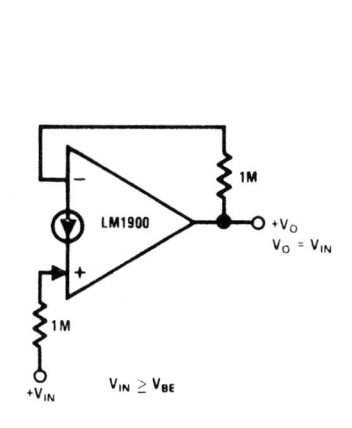

Buffer amplifier

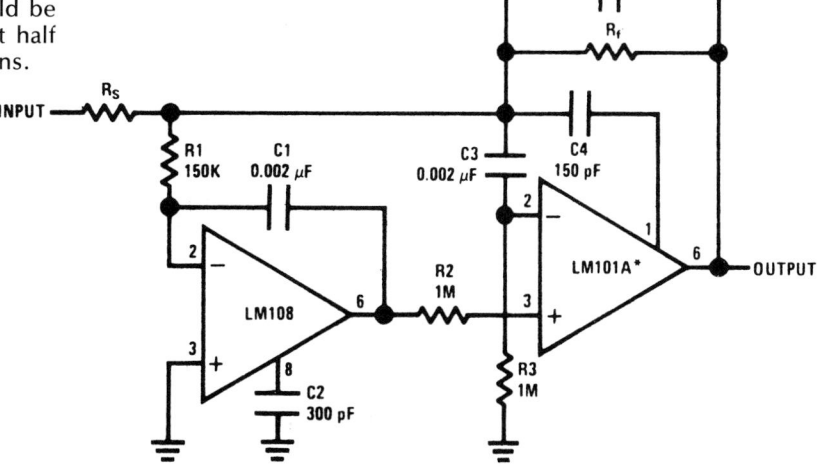

Fast-summing amplifier. Power bandwidth, 250 kHz; small signal bandwidth, 3.5 MHz; slew rate, 10V/μsec. $C5 = 6 \times 10^{-8}/R_f$.

SECTION 2 *Followers, Buffers, and Special-Purpose Amplifiers*

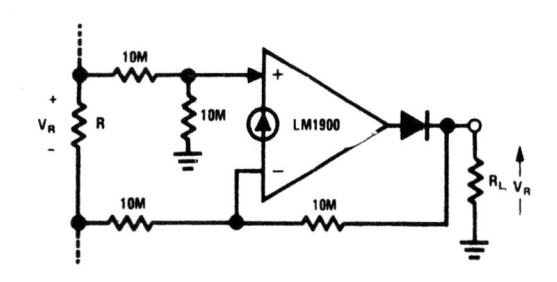

Ground-referencing a differential input signal

Operational amplifier applications. Use care in the layout of such circuits, as unintentional signal coupling from the output to the noninverting input can cause oscillations. This is likely only in breadboard hookups with long component leads and can be prevented by a more careful lead dress or by locating the noninverting-input biasing resistor close to the IC. A quick check of this condition is to bypass the noninverting input to ground with a capacitor. High-impedance biasing resistors used in the noninverting input circuit make this input lead highly susceptible to unintentional ac signal pickup.

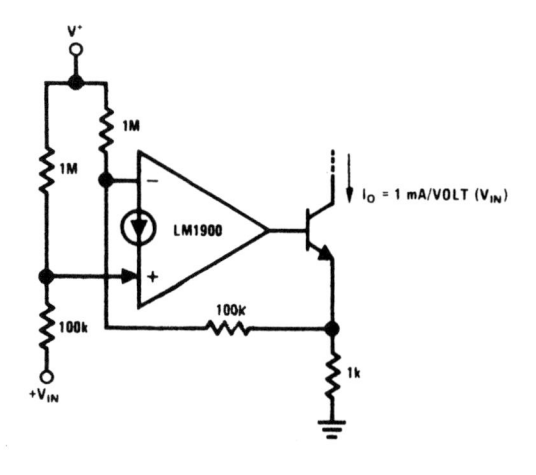

Voltage-controlled current sink (transconductance amplifier)

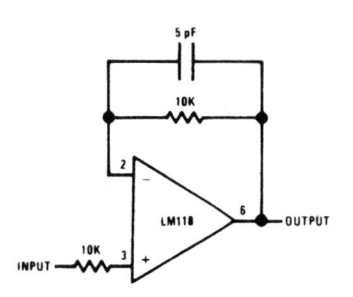

Fast voltage follower.

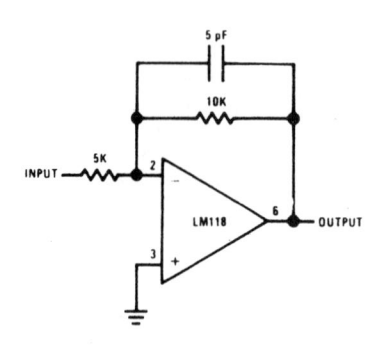

Fast-summing amplifier.

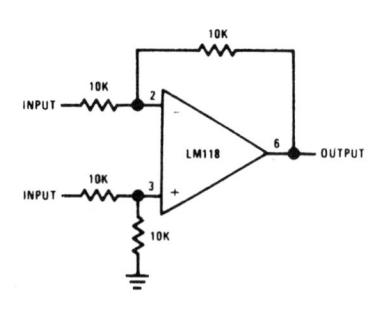

Differential amplifier.

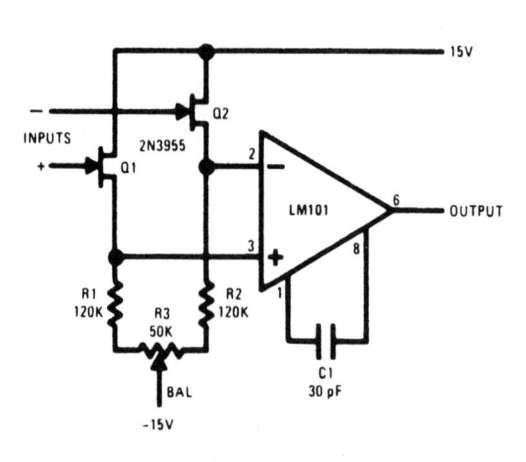

FET operational amplifier. The composite circuit has roughly the same gain as the integrated circuit by itself and is compensated for unity gain with a 30 pF capacitor as shown. Although it works well as a summing amplifier, the circuit is less than satisfactory in applications requiring high common-mode rejection. This happens because resistors are used for current sources and because the FETs by themselves do not have good common-mode rejection.

76 IC SCHEMATIC SOURCEMASTER

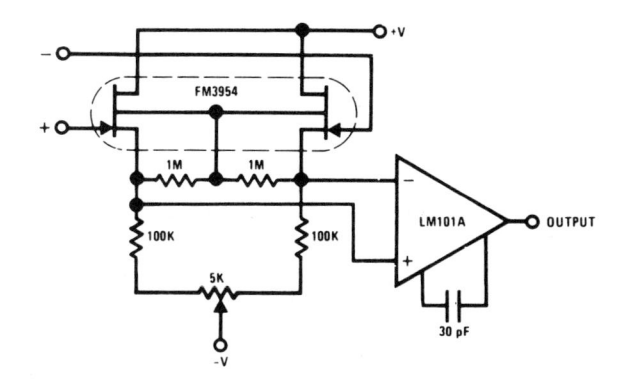

FET op-amp. The FM3954 monolithic dual FET provides an ideal low-offset, low-drift buffer function for the LM101A op-amp. The excellent matching characteristics of the FM3954 track well over its bias current range, thus improving common-mode rejection.

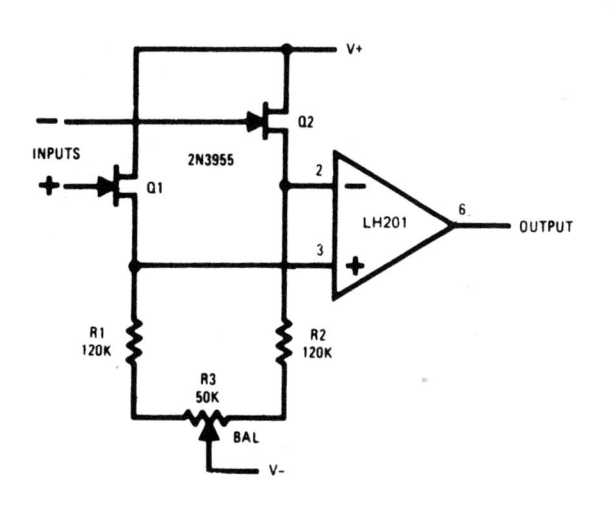

FET operational amplifier.

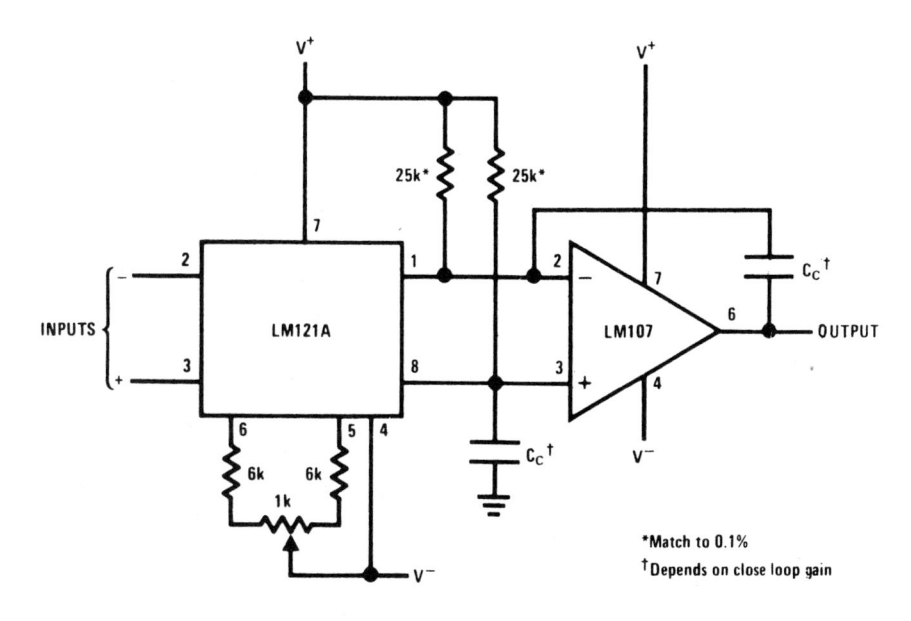

*Match to 0.1%
†Depends on close loop gain

Increased common-mode range at high operating currents.

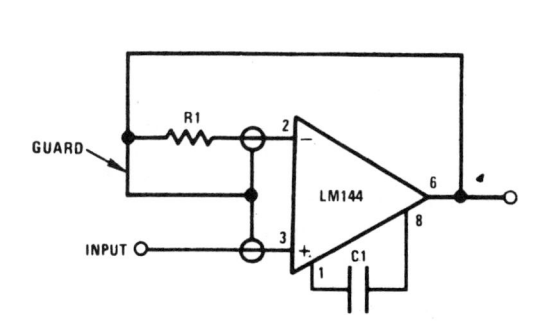

Guarded voltage follower. R1 = R_{SOURCE}.

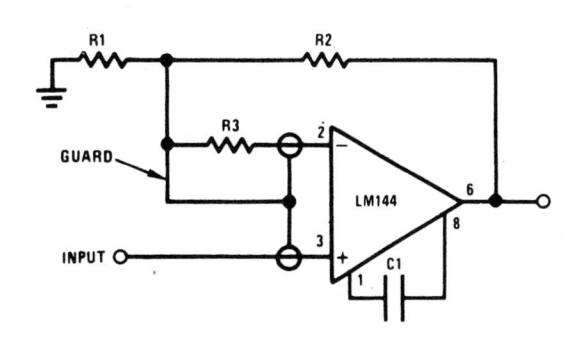

Guarded noninverting amplifier. R3 + (R1)(R2)/R1 + R2 = R_{SOURCE}.

SECTION **2** *Followers, Buffers, and Special-Purpose Amplifiers* **77**

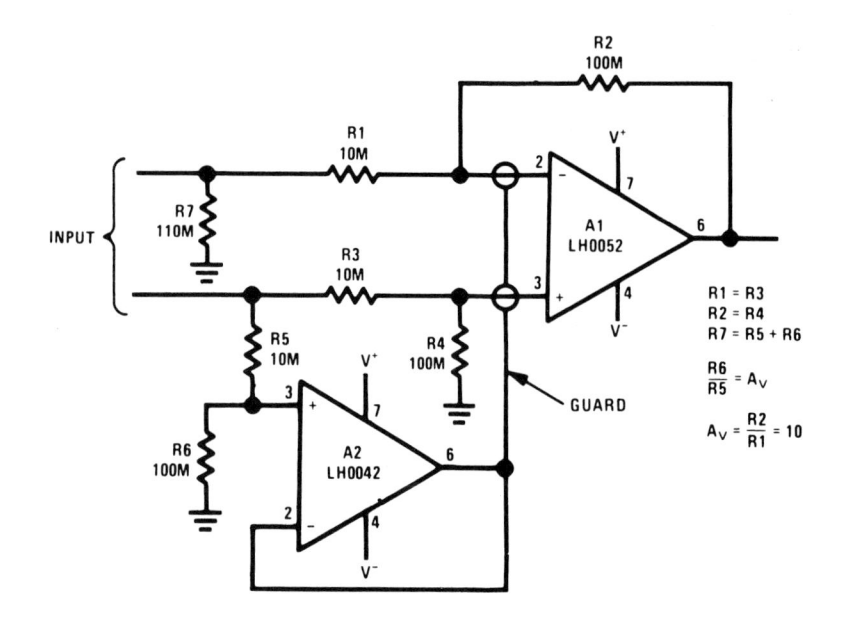

Guarded full-differential amplifier. Resistors R5 and R6 develop the proper voltage for the guard at their junction, but it will normally be impractical to make them low enough in resistance because of source loading. R7 is included to balance the effect of R5 plus R6 and thus not degrade the closed-loop common-mode rejection.

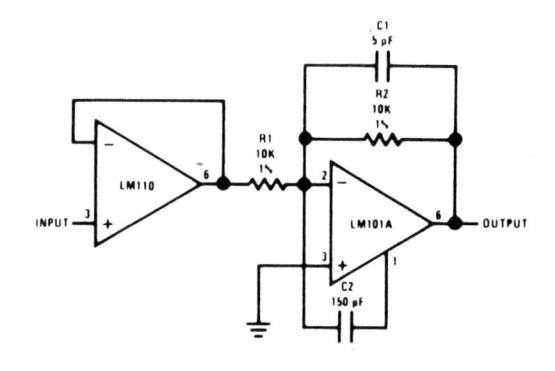

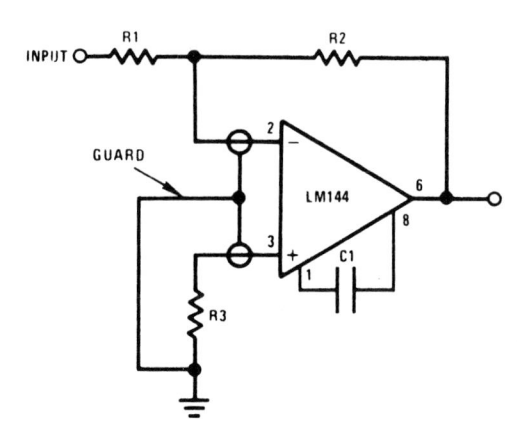

Guarded inverting amplifier. R3 = R1R2/R1 + R2.

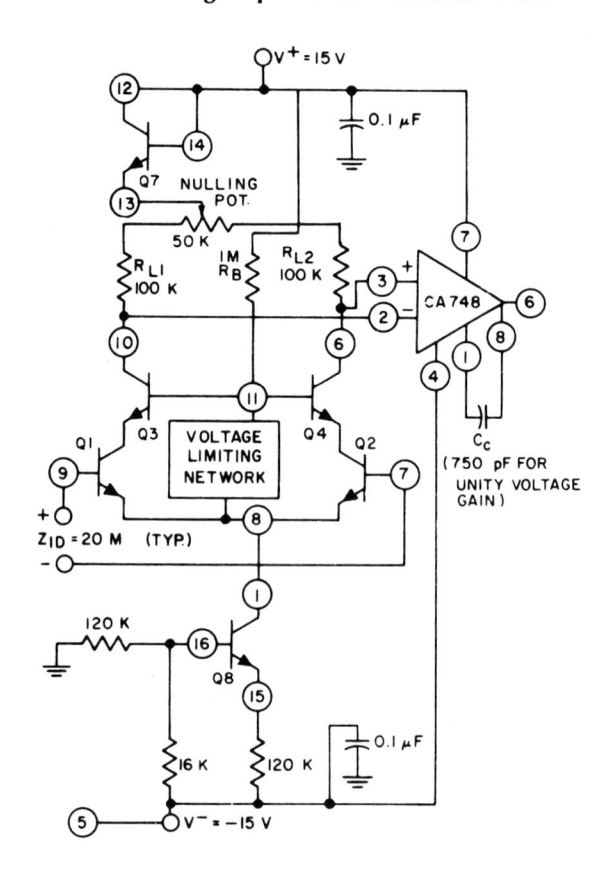

A fast-inverting amplifier with high input impedance.

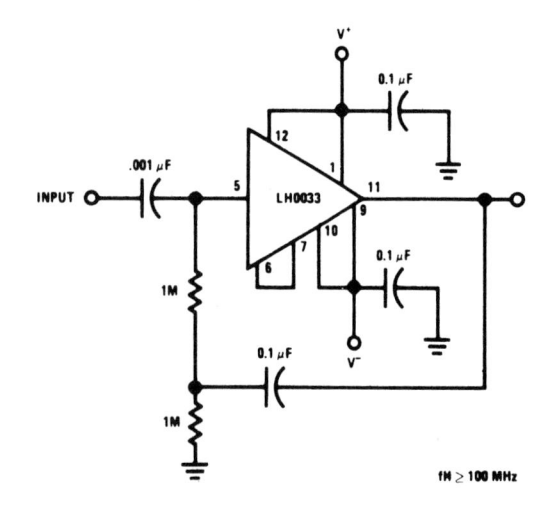

This high-input-impedance ac-coupled amplifier uses bootstrapping to achieve 10M input impedance.

A high-gain preamplifier using the transistors in the CA3095E to drive a CA748 op-amp. Transistors Q1–Q3 and Q2–Q4 operate as a differential cascode amplifier with transistor Q8 as their constant-current source. Transistor Q7 is diode-connected to drop the dc common-mode voltage at the input of the CA741 to within its linear operating range. A voltage-limiting network is built into the superbeta IC. To use it just connect a resistor from the positive supply of the differential amplifier to pin 11; the value should be such that current through the resistor is about half that measured at pin 8 under quiescent conditions.

78 IC SCHEMATIC SOURCEMASTER

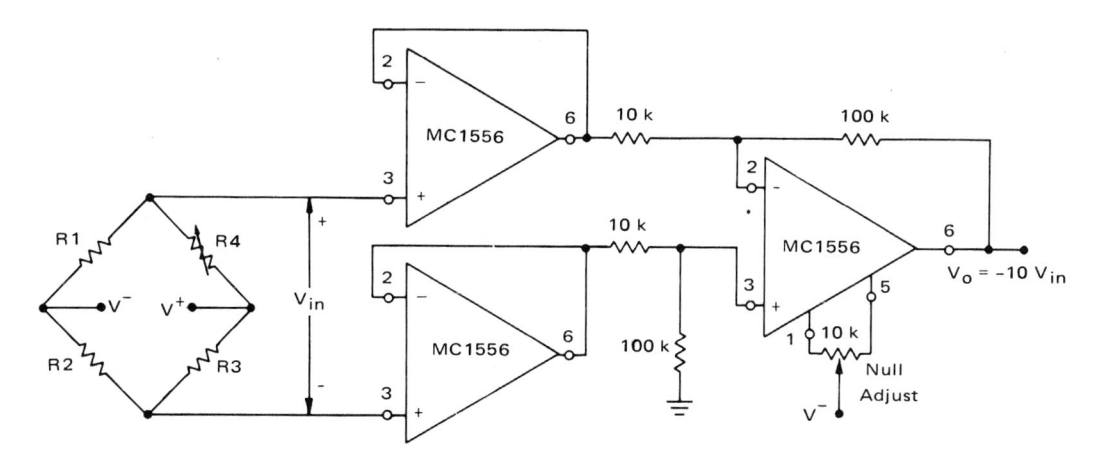

High-impedance bridge amplifier. The high input impedance avoids loading effects on the bridge and transforms the impedance down to a level where a third amplifier used in a differential mode can provide voltage gain. The third amplifier employs the standard offset adjust circuit to provide nulling capability for the configuration.

Although the circuit is shown for complementary supply voltages, it lends itself well to operation from a single supply since the bridge can be operated just as well from the single supply. The advantage of the MC1556 in this circuit is its low value of input bias current, which permits operation down to 1 mV input voltage at 1.5% accuracy without bias current compensation.

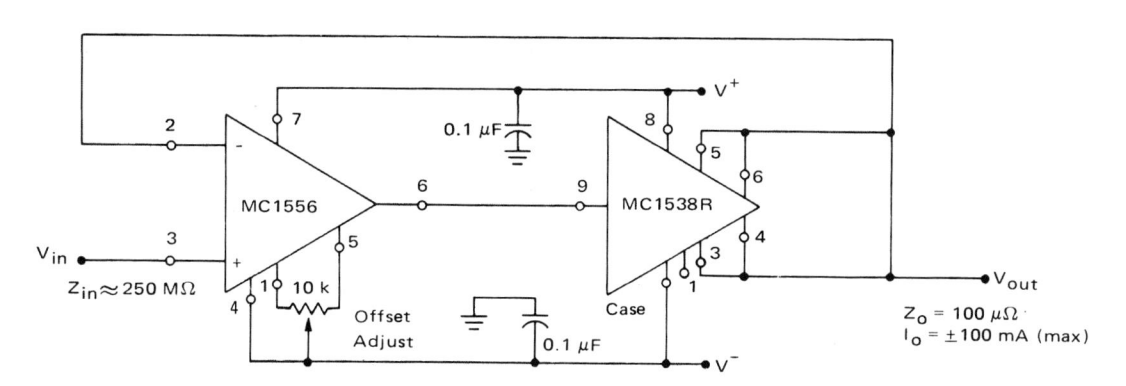

High-input-impedance, high-output-current voltage follower. The MC1538R booster's output impedance is typically 10Ω, which results in a factor-of-100 reduction in the closed-loop output impedance as well. This permits greater stability when driving a capacitive load. Additionally, the MC1538R can supply up to ±300 mA peak current to the load. The input impedance of the booster (typically 400k) alleviates the loading on the MC1556 and, hence, additional drifts that may occur from resulting thermal effects.

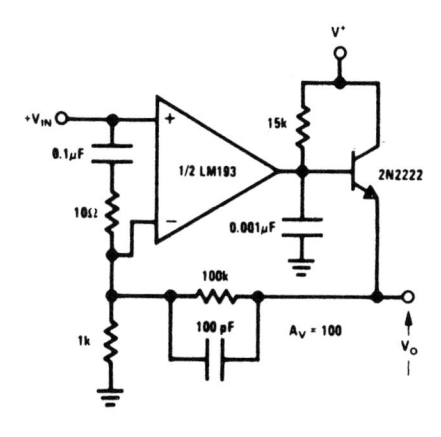

Improved op-amp. $A_v = 100$.

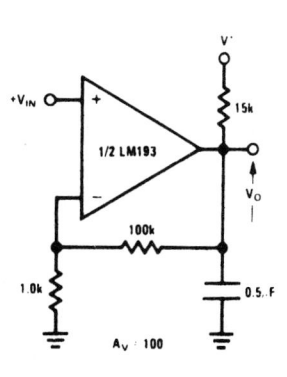

Low-frequency op-amp. $A_v = 100$.

SECTION **2** *Followers, Buffers, and Special-Purpose Amplifiers*

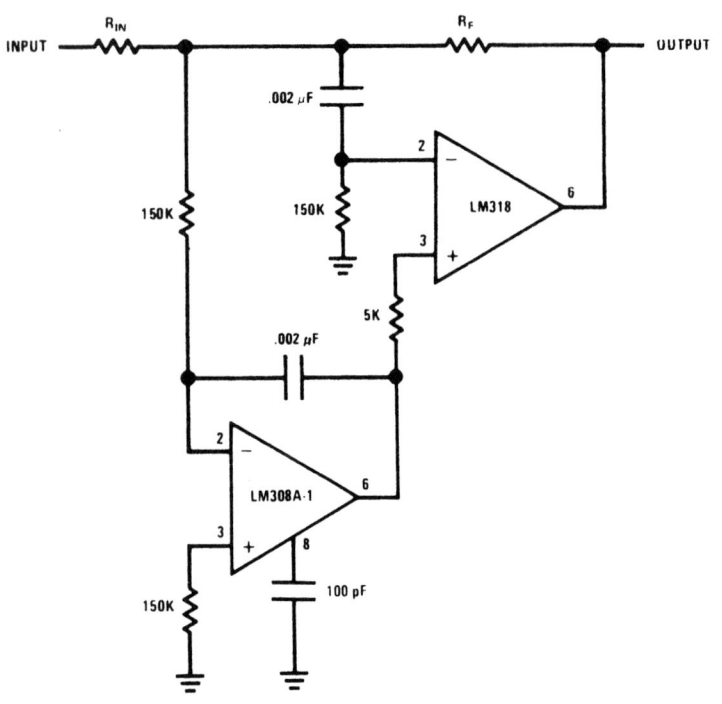

A high-speed amplifier with low drift and low input current.

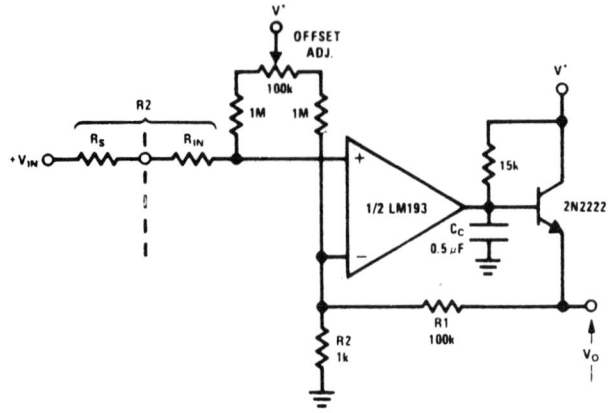

Low-frequency op-amp with offset adjust.

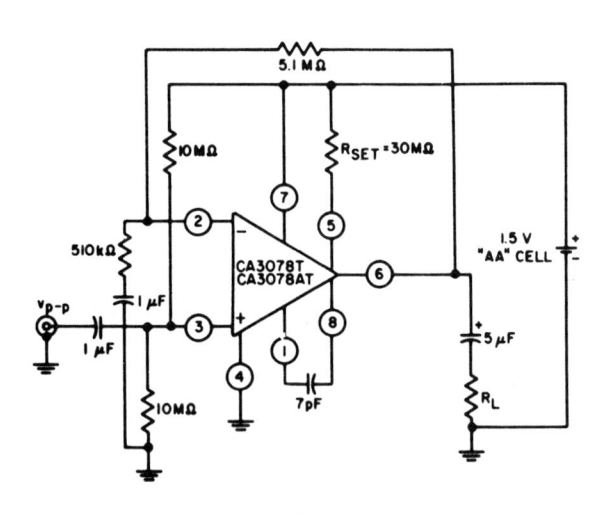

Inverting 20 dB amplifier.

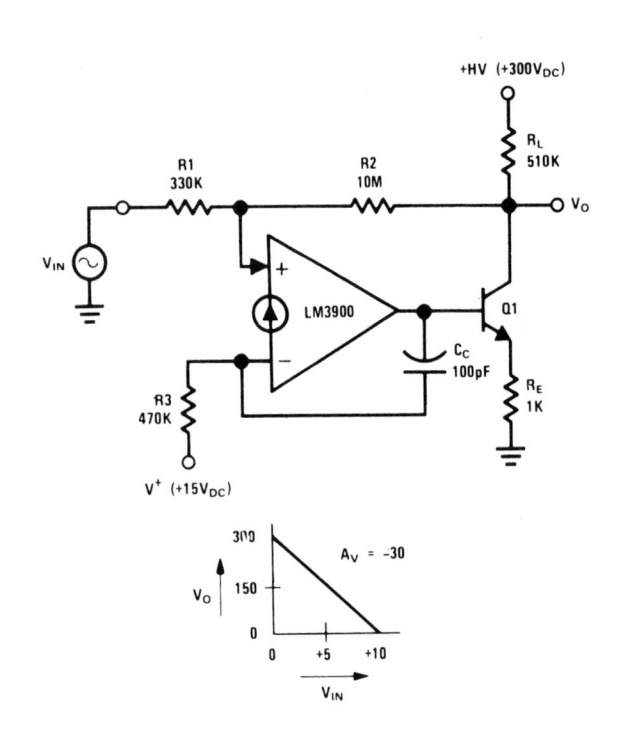

A high-voltage inverting amplifier with an output voltage swing from essentially 0 to +300V. Transistor Q1 must be a high-breakdown device, as it will have the full supply voltage across it. Biasing resistor R3 centers the transfer characteristic, and the gain is the ratio of R2 to R1. Load resistor R_L can be increased to reduce the current drain.

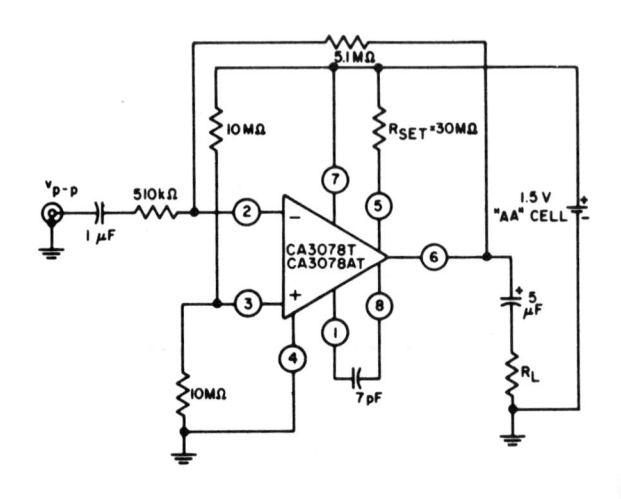

Noninverting 20 dB amplifier.

80 IC SCHEMATIC SOURCEMASTER

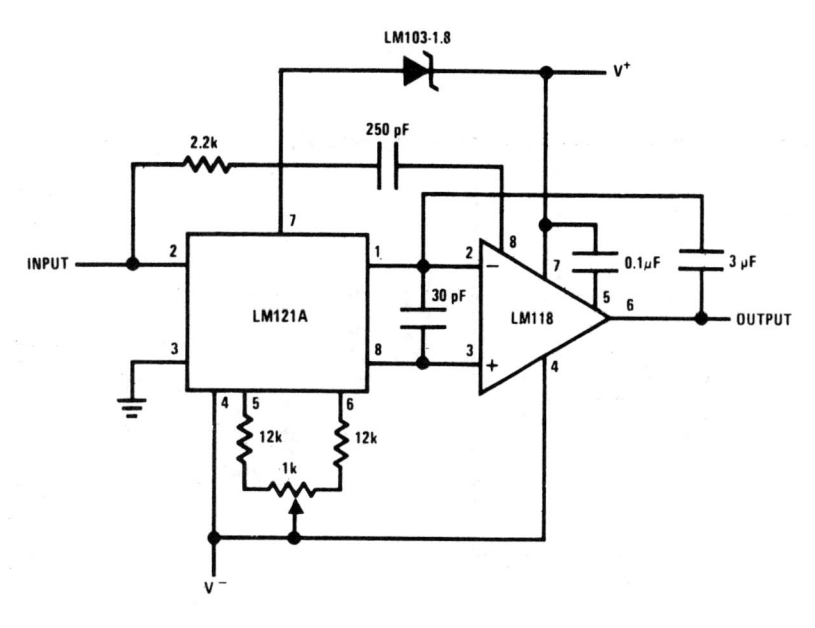

High-speed inverting amplifier with low drift. Bandwidth, 10 MHz; slew rate, 40V/μsec.

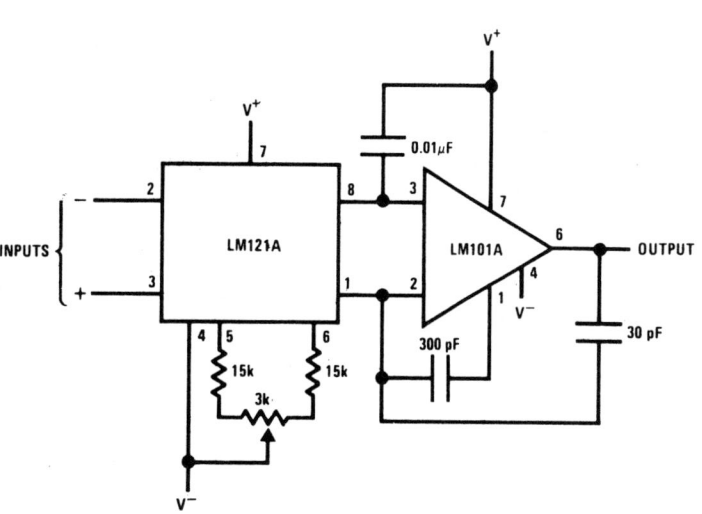

Medium-speed general-purpose amplifier. Bandwidth, 3.5 MHz; slew rate, 1.1V/μsec.

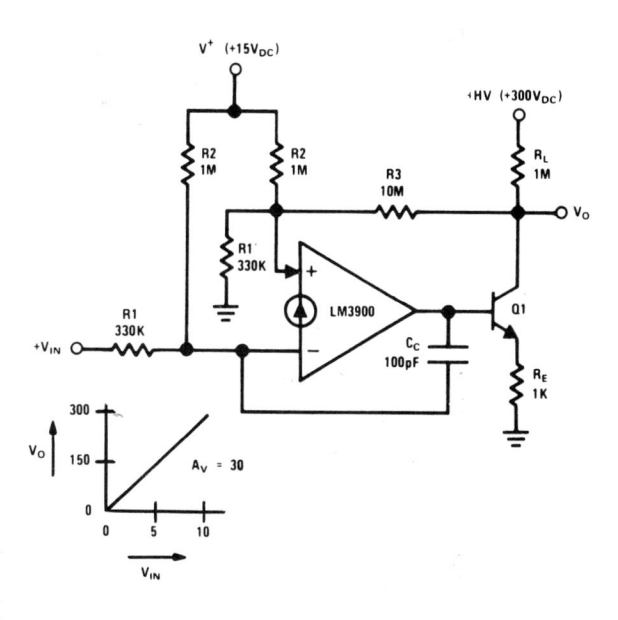

High-voltage noninverting amplifier. Common-mode biasing resistors (R2) are used to allow the input voltage to go to zero. The output voltage, V$_0$, will not actually go to zero because of R$_E$ but should go to approximately +0.3V. Again, the gain is 30, and a range of the input voltage of from 0 to +10V will cause the output voltage to range from approximately 0 to +300V.

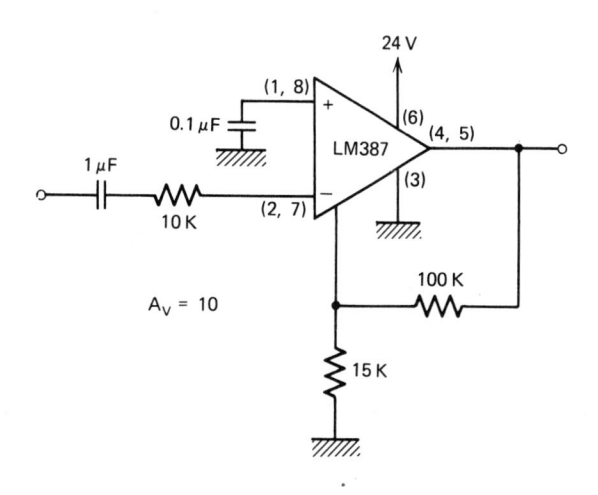

This inverting amplifier features ultralow distortion.

SECTION **2** *Followers, Buffers, and Special-Purpose Amplifiers* 81

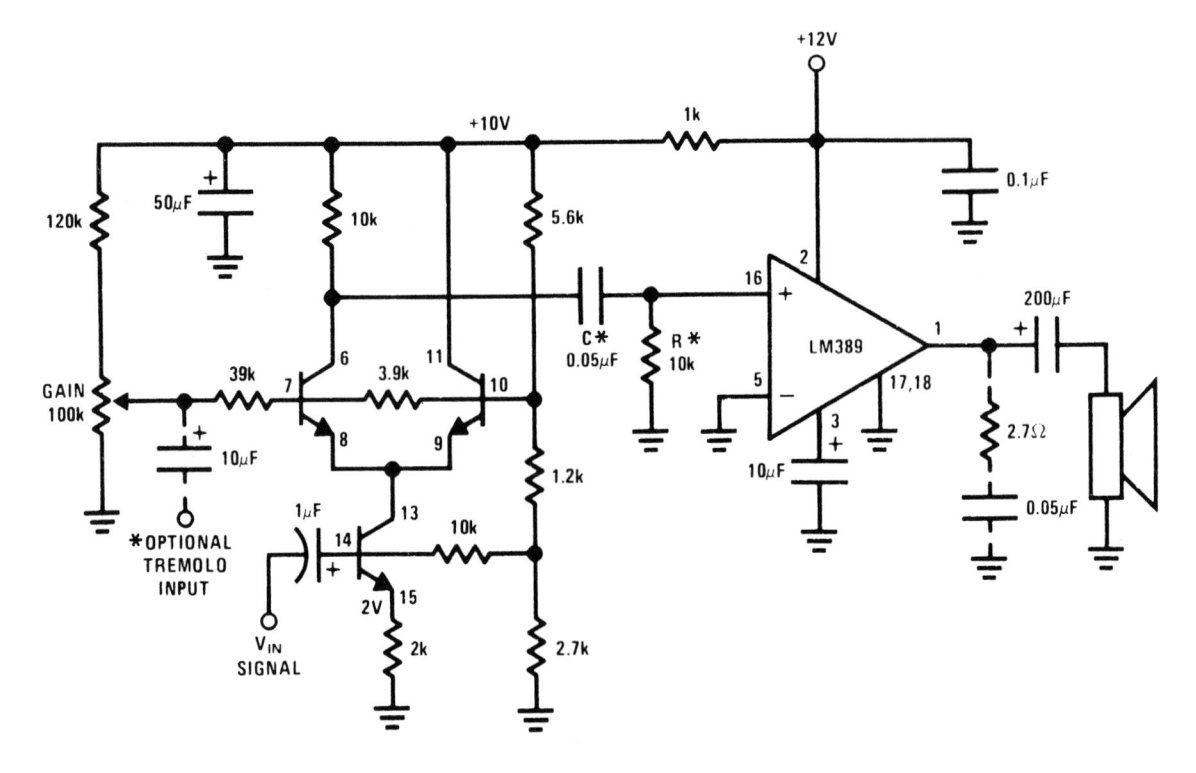

Voltage-controlled amplifier or **tremolo circuit.** Transistors are part of LM389. Tremolo frequency is about $1/[2\pi(R + 10k)C]$.

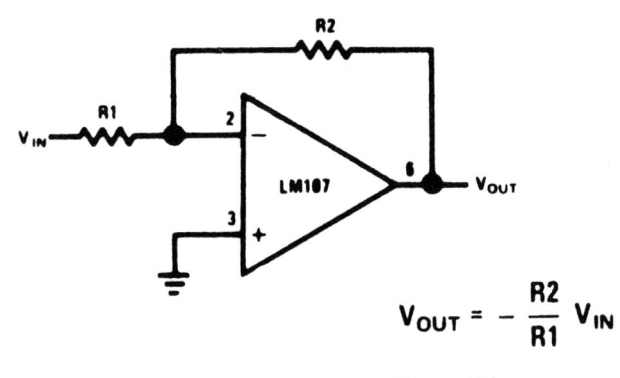

Inverting amplifier.

$$V_{OUT} = -\frac{R2}{R1} V_{IN}$$

$$R_{IN} = R1$$

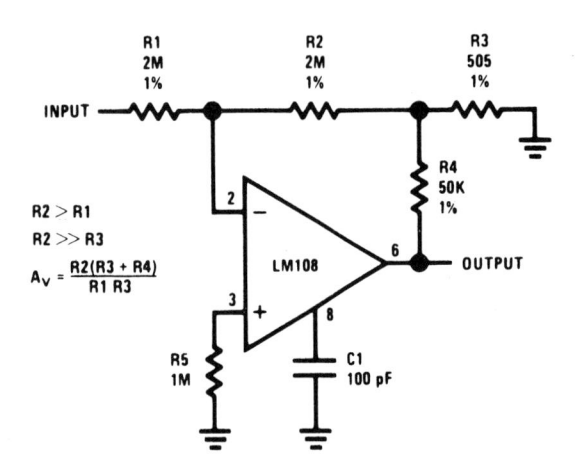

Inverting amplifier with high input resistance. This circuit does increase the offset voltage somewhat. The output offset voltage is given by

$$V_{OUT} = \left(\frac{R1 + R2}{R2}\right) A_V V_{OS}$$

The offset voltage is only multiplied by $A_V + 1$ in a conventional inverter. This circuit multiplies the offset by 200 instead of by 101. This multiplication factor can be reduced to 110 by increasing R2 to 20M and R3 to 5.55k.

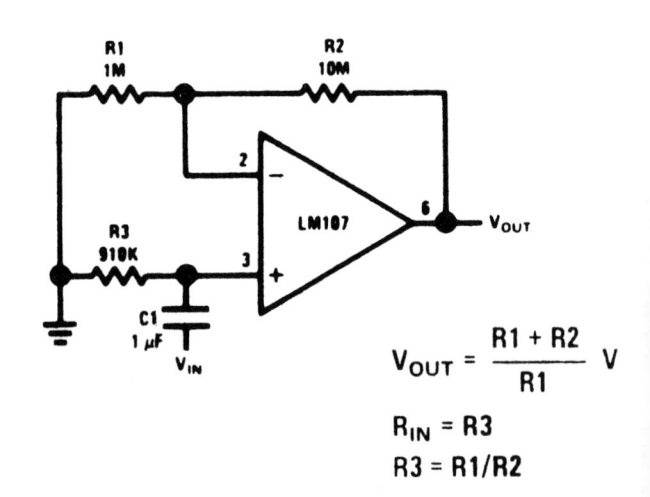

Noninverting ac amplifier.

$$V_{OUT} = \frac{R1 + R2}{R1} V$$

$$R_{IN} = R3$$

$$R3 = R1/R2$$

82 IC SCHEMATIC SOURCEMASTER

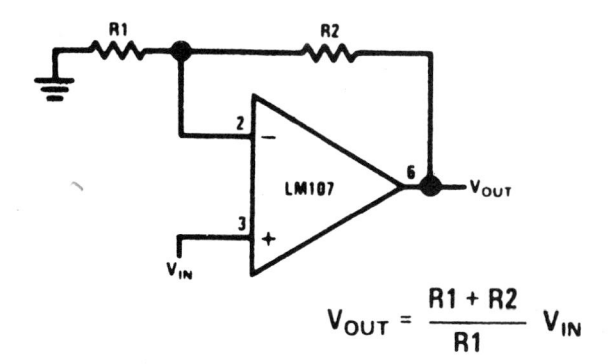

$$V_{OUT} = \frac{R1 + R2}{R1} V_{IN}$$

Noninverting amplifier.

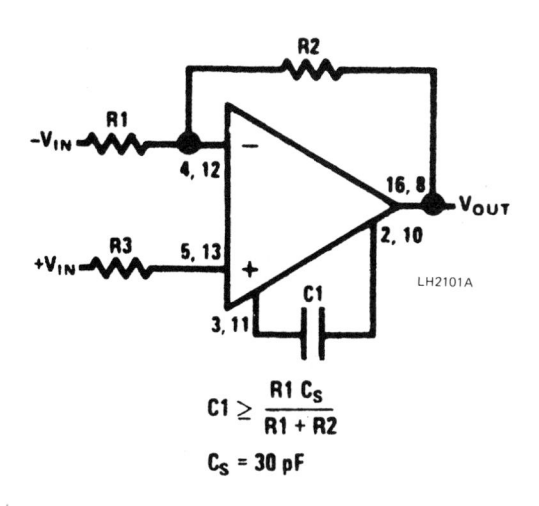

$$C1 \geq \frac{R1\, C_S}{R1 + R2}$$

$$C_S = 30\text{ pF}$$

Single-pole compensation.

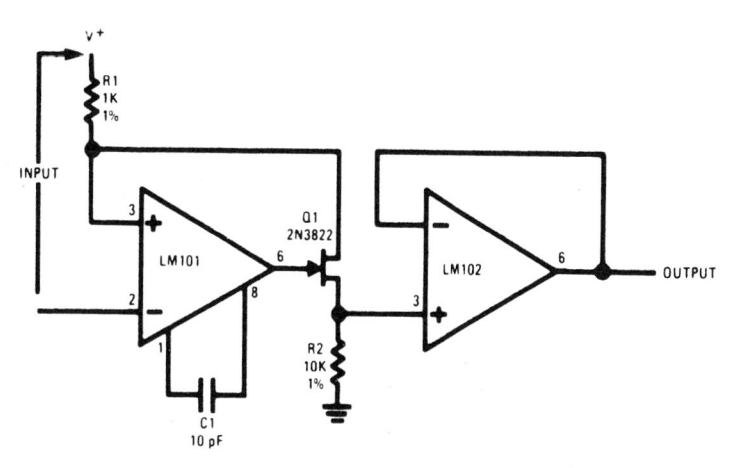

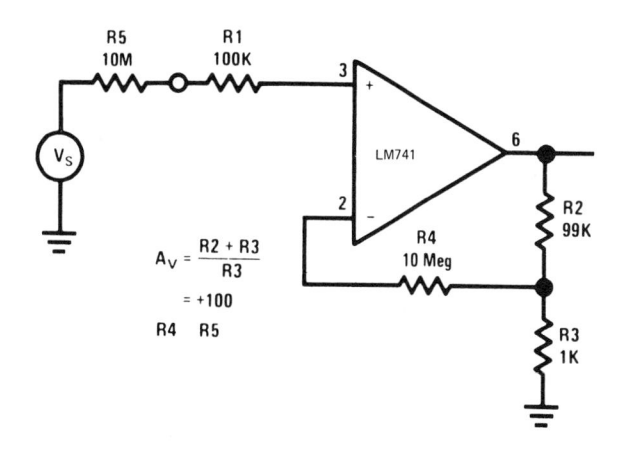

Level-isolation amplifier. Here, the FET on the output of the operational amplifier produces a voltage drop across feedback resistor R1 which is equal to the input voltage. The voltage across R2 will then be equal to the input voltage multiplied by the ratio R2:R1, and the common-mode rejection will be as good as the basic rejection of the amplifier independent of the resistor tolerances. This voltage is buffered by an LM102 voltage follower to give a low-impedance output.

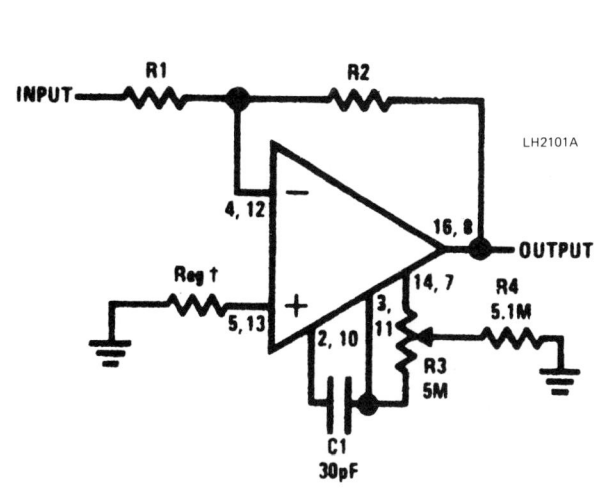

†May be zero or equal to parallel combination of R1 and R2 for minimum offset.

Inverting amplifier with balancing circuit.

$$A_V = \frac{R2 + R3}{R3} = +100$$

R4 ≈ R5

Noninverting amplifier. The voltage gain of this circuit is approximately 1 + R2/R3 (neglecting amplifier open-loop gain). R4 is included as a convenient variable to equalize resistances in the two amplifier inputs; R4 in series with the parallel combination of R2 and R3 should be set equal to the source resistance plus R1. All of these resistors may not be necessary, depending on the required voltage gain, source impedance, accuracy requirement, temperature range, and amplifier selected.

SECTION **2** *Followers, Buffers, and Special-Purpose Amplifiers*

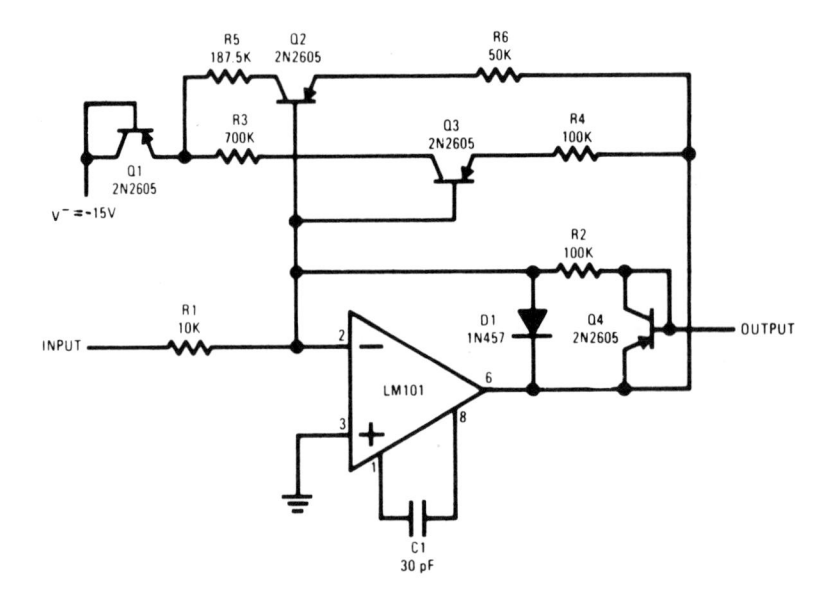

A nonlinear operational amplifier with temperature-compensated breakpoints. For small input signals, the gain is determined by R1 and R. Both Q2 and Q3 are conducting to some degree, but they do not affect the gain because their current gain is high and they do not feed any appreciable current back into the summing mode. When the output voltage rises to 2V (determined by R3, R4, and V$^-$), Q3 draws enough current to saturate, connecting R4 in parallel with R2. This cuts the gain in half. Similarly, when the output voltage rises to 4V, Q2 will saturate, again halving the gain.

Temperature compensation is achieved by including Q1 and Q4. Q4 compensates the emitter-base voltage of Q2 and Q3 to keep the voltage across the feedback resistors, R4 and R6, very nearly equal to the output voltage, while Q1 compensates for the emitter-base voltage of these transistors as they go into saturation, making the voltage across R3 and R5 equal to the negative supply voltage. A detrimental effect of Q4 is that it causes the output resistance of the amplifier to increase at high output levels. It may therefore be necessary to use an output buffer if the circuit must drive an appreciable load.

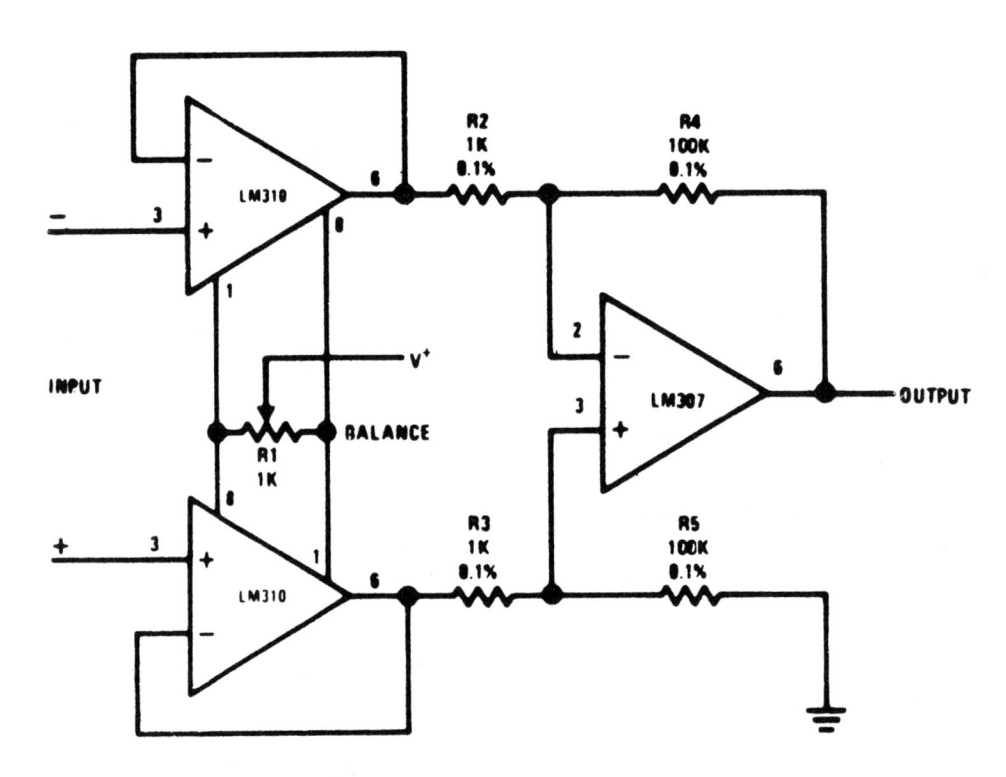

Differential-input instrumentation amplifier.

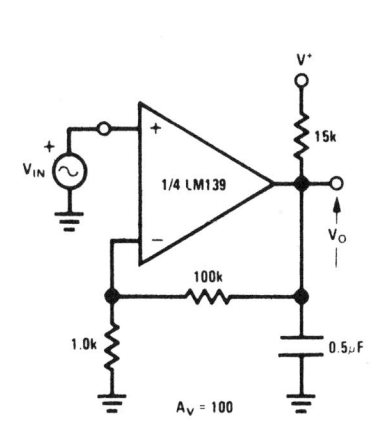

Low-frequency op-amp. $A_V = 100$.

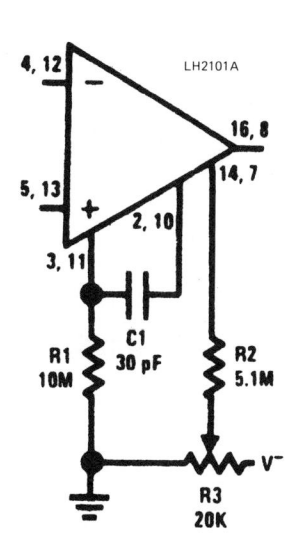

Alternate balancing circuit.

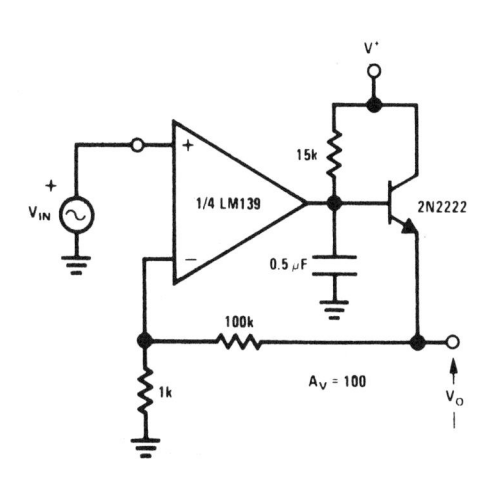

Low-frequency op-amp. $V_0 = 0V$ for $V_{IN} = 0V$; $A_V = 100$.

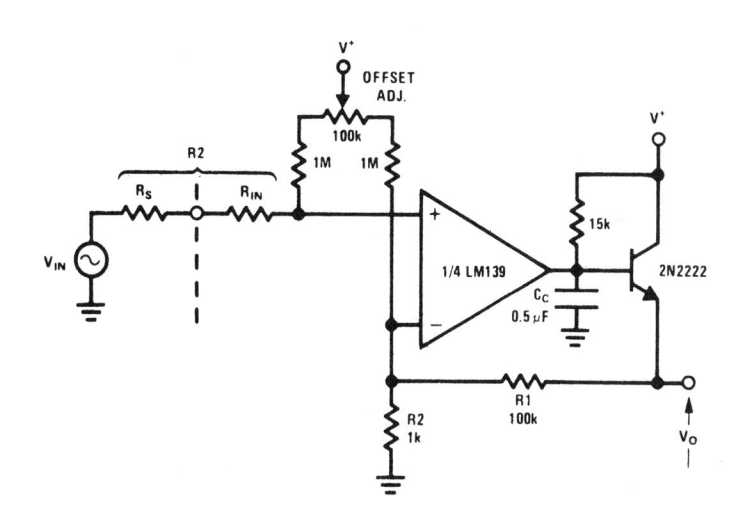

Low-frequency op-amp with offset adjust.

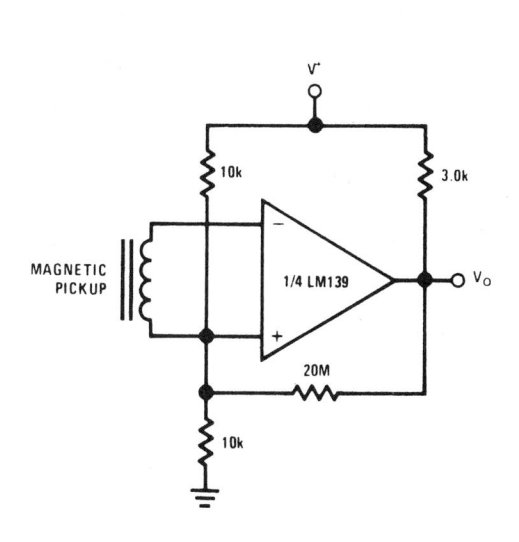

Transducer amplifier.

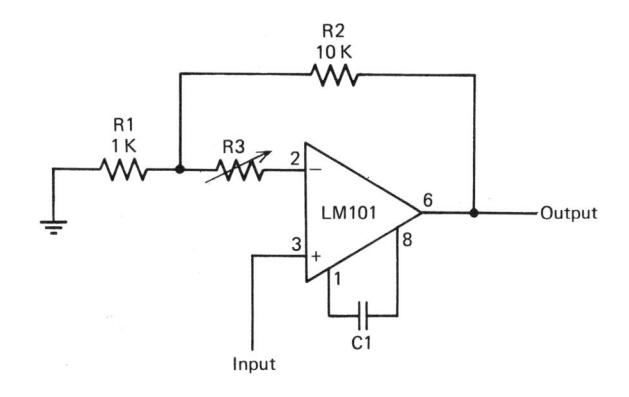

Noninverting amplifier with bias-current compensation for fixed source resistances. For proper adjustment range, R3 should have a maximum value of about three times that of the source resistance, and the equivalent parallel resistance of R1 and R2 should be less than one-third that of the input source resistance.

SECTION **2** *Followers, Buffers, and Special-Purpose Amplifiers*

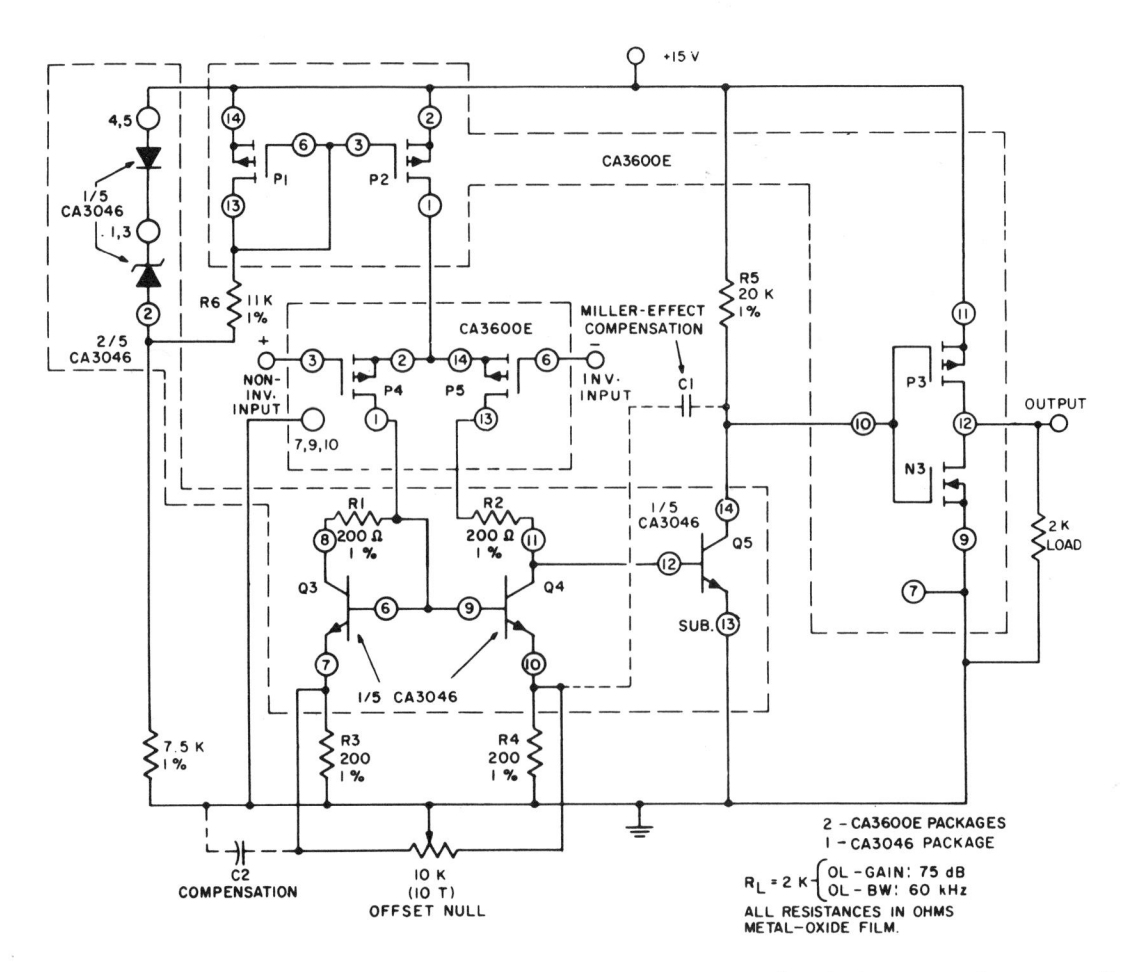

An operational amplifier using CMOS transistor pairs is particularly suited for single-supply operation (for example, in mobile and aircraft service). The op-amp is unique in that it is responsive to small-signal ground-reference inputs, and the output stage can easily be driven within 1 mV of ground potential.

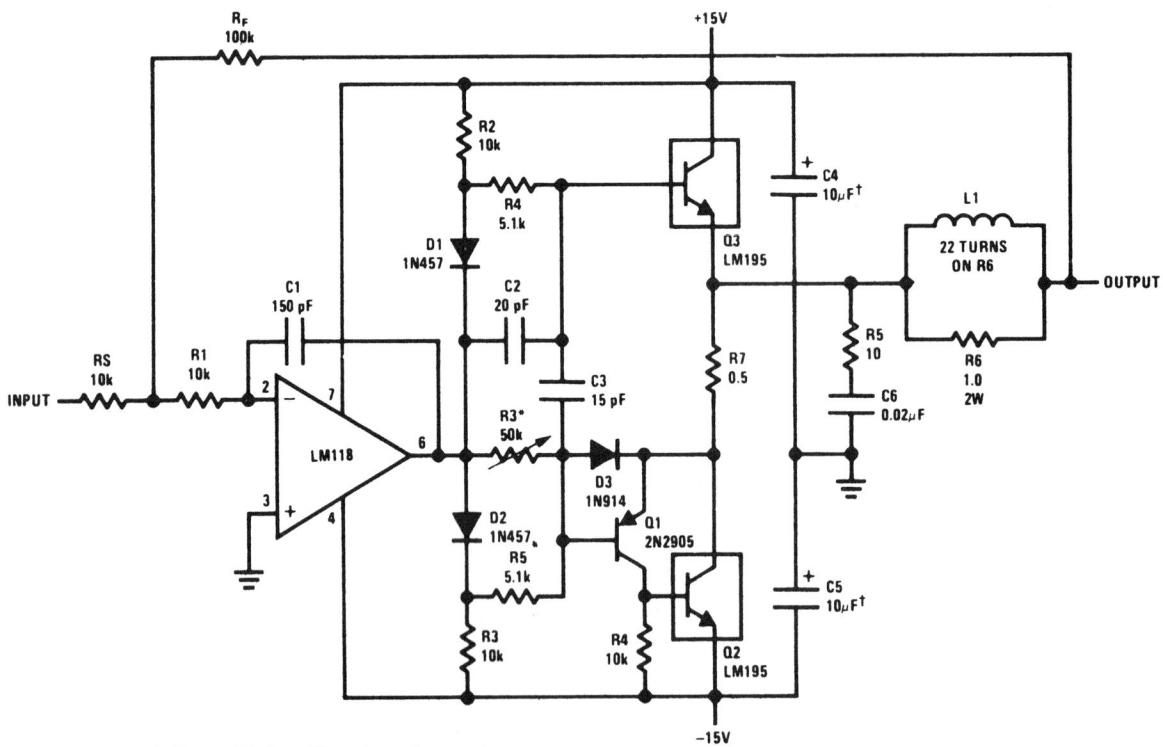

Power op-amp. Adjust R3 for 50 mA quiescent current. C5 should be solid tantalum.

86 IC SCHEMATIC SOURCEMASTER

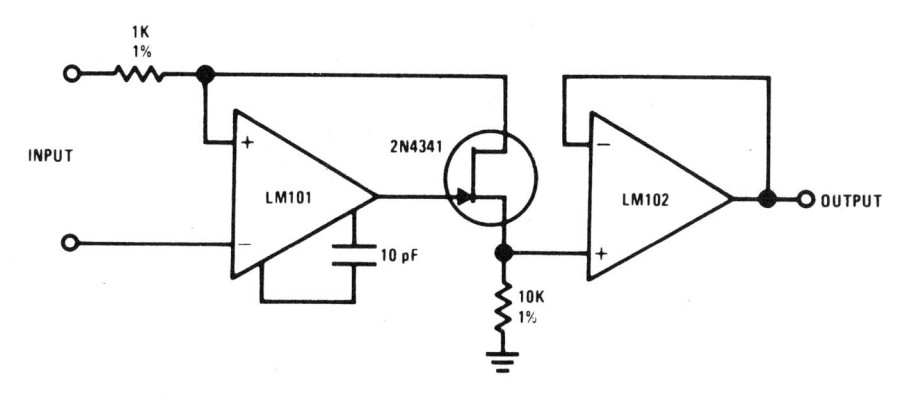

Level-shifting isolation amplifier. Here, the 2N4341 JFET is used as a level shifter between two op-amps operated at different power supply voltages.

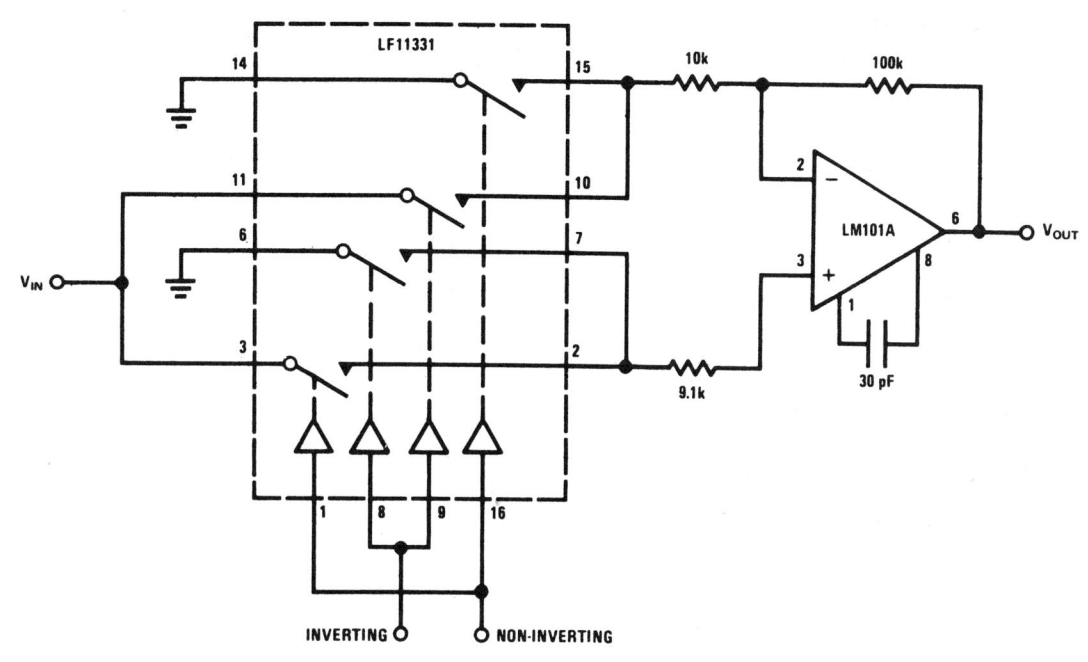

Programmable inverting/noninverting operational amplifier.

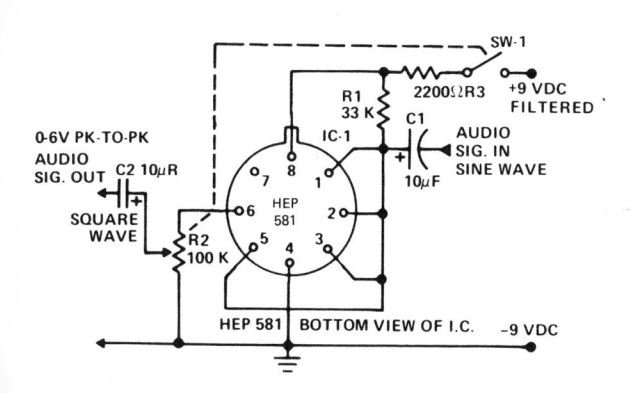

This squaring amplifier (sine- to square-wave converter) uses 4-input OR/NOR gate. A 2.0 to 10.0V sine-wave input produces symmetrical square-wave output that is $2V_{p-p}$ maximum. Current drain is 3.5 mA at 9V.

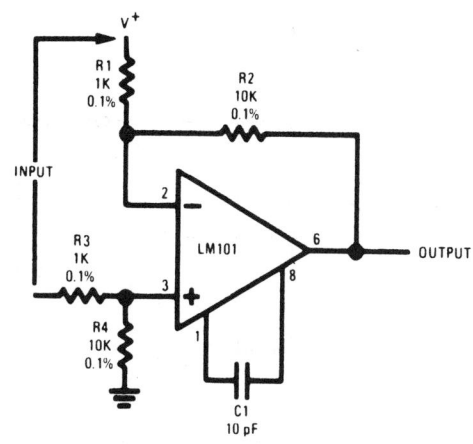

A standard differential amplifier, in which we take a voltage referred to some dc level and produce an amplified output referred to ground. However, this circuit has the disadvantages that the signal source is loaded by current from the input divider, R3 and R4, and the feedback resistors must be very well matched to prevent erroneous outputs from the common-mode input signal.

SECTION **2** *Followers, Buffers, and Special-Purpose Amplifiers*

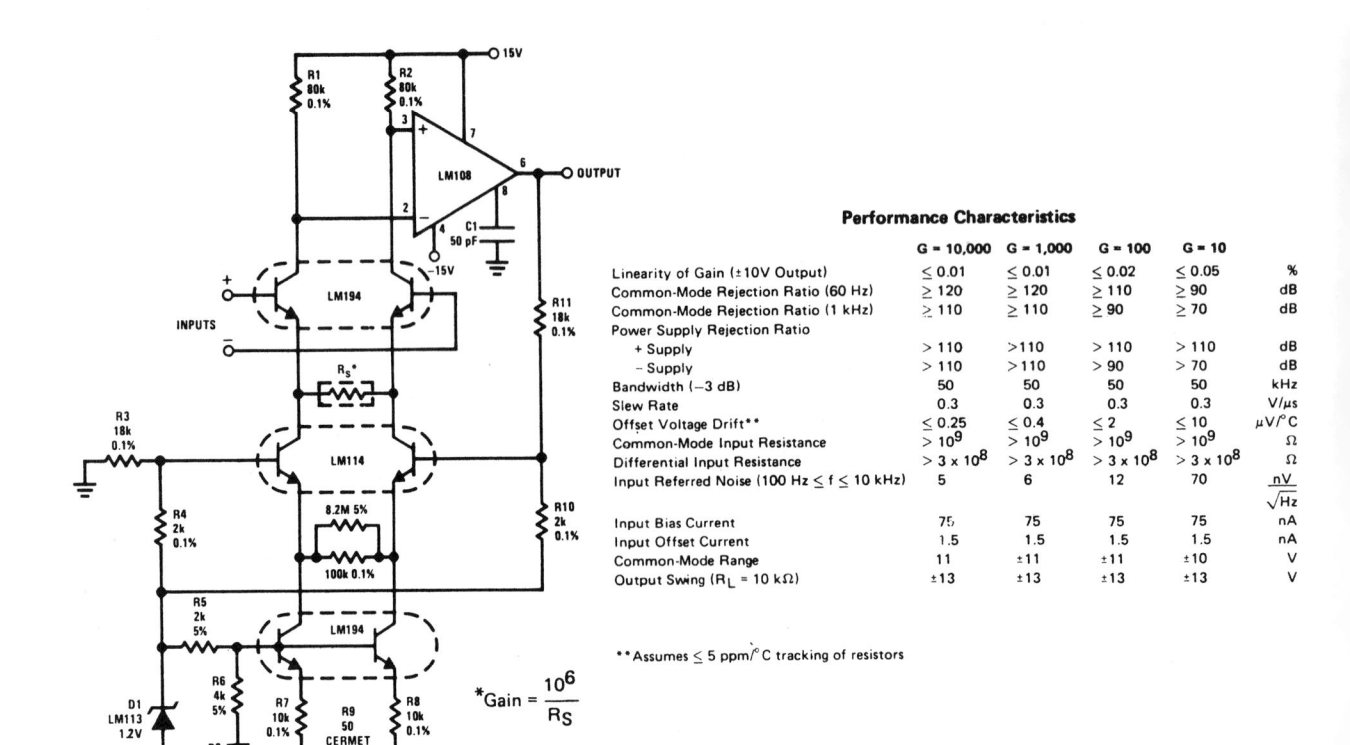

High-performance instrumentation amplifier.

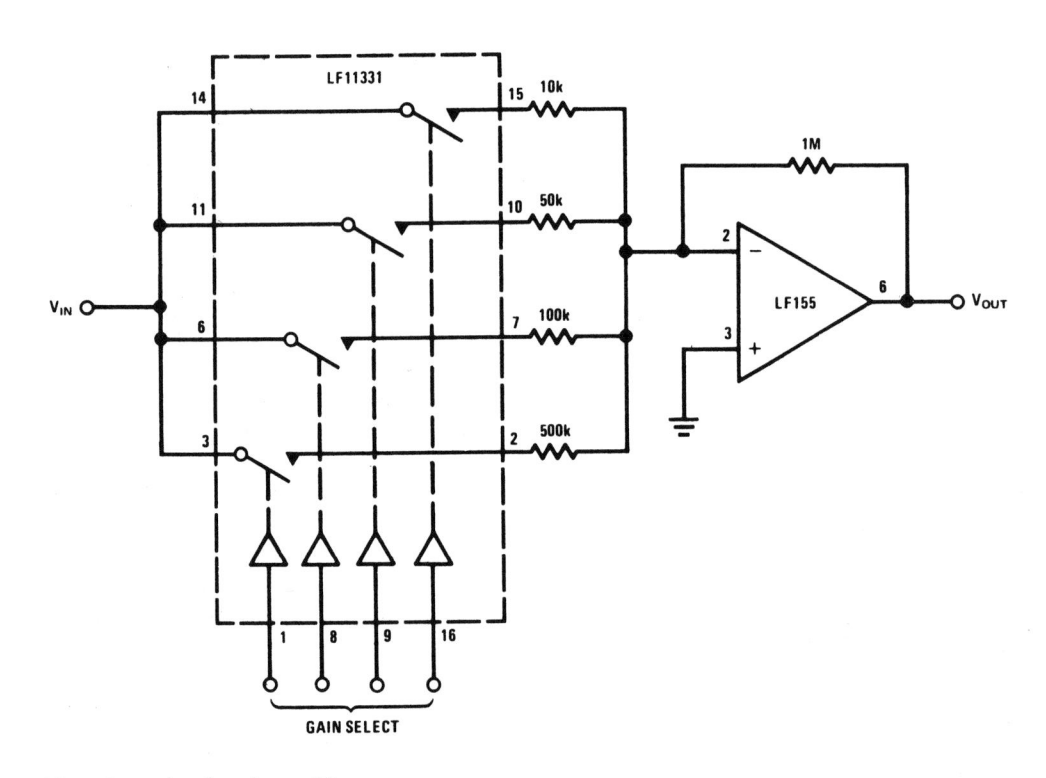

Performance Characteristics

	G = 10,000	G = 1,000	G = 100	G = 10	
Linearity of Gain (±10V Output)	≤ 0.01	≤ 0.01	≤ 0.02	≤ 0.05	%
Common-Mode Rejection Ratio (60 Hz)	≥ 120	≥ 120	≥ 110	≥ 90	dB
Common-Mode Rejection Ratio (1 kHz)	≥ 110	≥ 110	≥ 90	≥ 70	dB
Power Supply Rejection Ratio					
+ Supply	> 110	>110	> 110	> 110	dB
− Supply	> 110	>110	> 90	> 70	dB
Bandwidth (−3 dB)	50	50	50	50	kHz
Slew Rate	0.3	0.3	0.3	0.3	V/µs
Offset Voltage Drift**	≤ 0.25	≤ 0.4	≤ 2	≤ 10	µV/°C
Common-Mode Input Resistance	> 10^9	> 10^9	> 10^9	> 10^9	Ω
Differential Input Resistance	> 3×10^8	> 3×10^8	> 3×10^8	> 3×10^8	Ω
Input Referred Noise (100 Hz ≤ f ≤ 10 kHz)	5	6	12	70	$\frac{nV}{\sqrt{Hz}}$
Input Bias Current	75	75	75	75	nA
Input Offset Current	1.5	1.5	1.5	1.5	nA
Common-Mode Range	±11	±11	±11	±10	V
Output Swing (R_L = 10 kΩ)	±13	±13	±13	±13	V

**Assumes ≤ 5 ppm/°C tracking of resistors

$$*\text{Gain} = \frac{10^6}{R_S}$$

Programmable-gain operational amplifier.

88 IC SCHEMATIC SOURCEMASTER

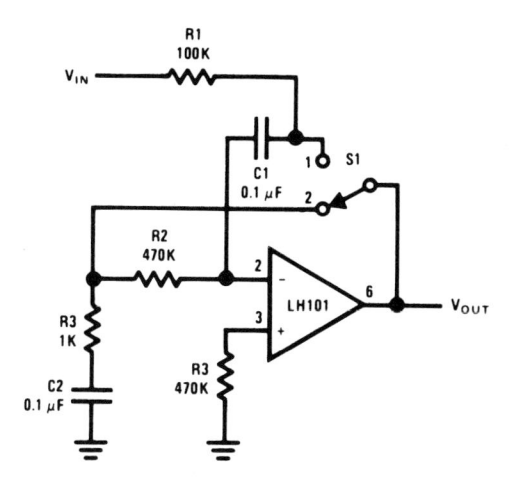

This reset-stabilized amplifier, a form of chopper-stabilized amplifier, is operated closed-loop with a gain of 1. The output of this circuit is a pulse whose amplitude is equal to V_{IN}.

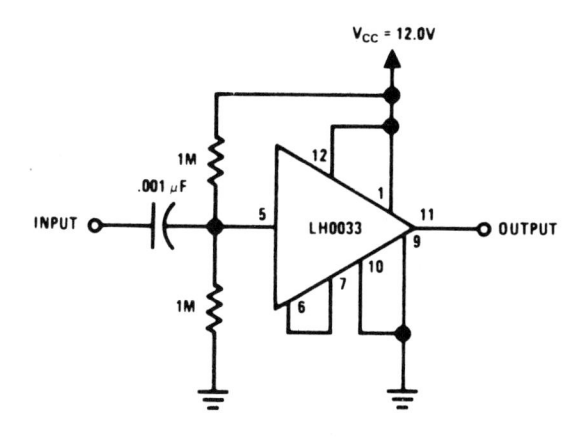

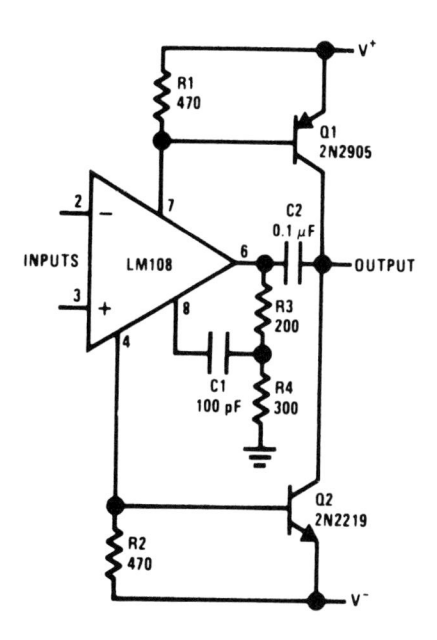

This simple power booster swings the output up to the supplies within a fraction of a volt. The increased voltage swing is particularly helpful in low-voltage circuits. The output transistors are driven from the supply leads of the op-amp. It is important that R1 and R2 be made low enough so Q1 and Q2 are not turned on by the worst-case quiescent current of the amplifier. The output of the op-amp is loaded heavily to ground with R3 and R4.

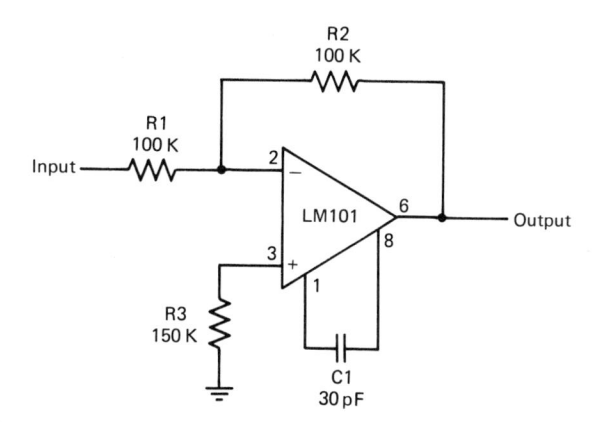

Single-supply, ac-coupled amplifier. Input impedance is approximately 500k and output swing is more than 8V peak-to-peak with a 12V supply.

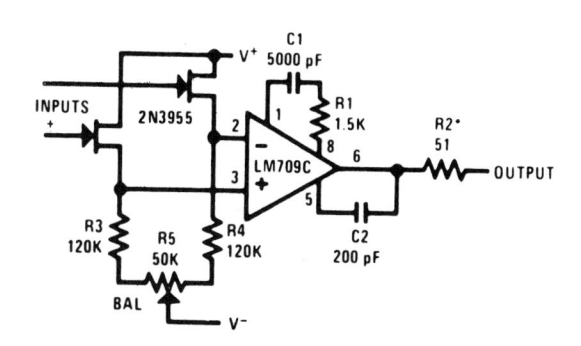

A summing amplifier with bias-current compensation for fixed source resistances. The offset produced by the bias current on the inverting input is canceled by the offset voltage produced across the variable resistor, R3. Bias currents of the two input transistors tend to track well over temperature so that low drift is also achieved. The disadvantage of the method is that a given compensation setting works only with fixed feedback resistors, and the compensation must be readjusted if the equivalent parallel resistance of R1 and R2 is changed.

FET operational amplifier. R2 should be used with any capacitive loading on output.

SECTION **2** *Followers, Buffers, and Special-Purpose Amplifiers* **89**

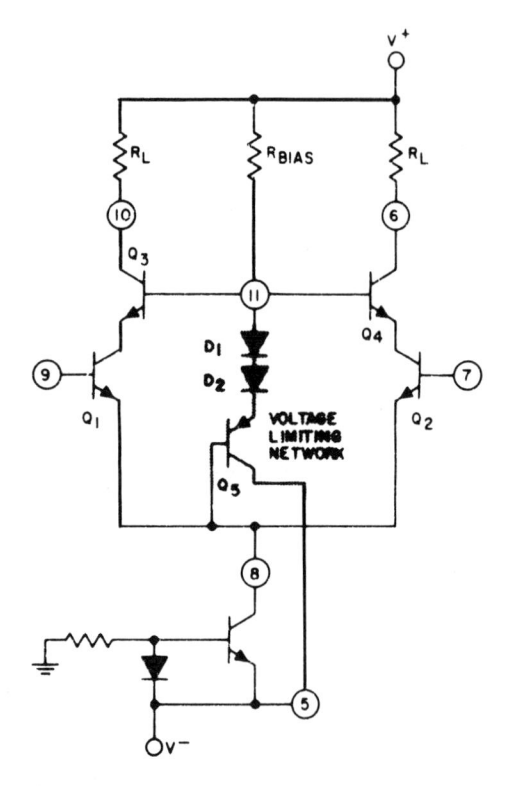

Superbeta differential cascode amplifier. An internal voltage-limiting network (diodes D1, D2, and transistor Q5) incorporated in the differential cascode amplifier assures that the applied collector-to-emitter voltage of each superbeta unit is maintained below 2V. The inset shows a typical bias arrangement of the superbeta differential cascode amplifier. Bias current for this network must be supplied by an exernal source. This bias current can be obtained by simply connecting a resistor from pin 11 to the positive supply of the differential amplifier. The return path for most of the bias current is through the substrate (pin 5) rather than through the common emitter (pin 8). This arrangement provides superior common-mode and power-supply rejection. As a general rule, the current supplied to pin 11 should be 0.04 to 0.1 times the value of the quiescent current of pin 8.

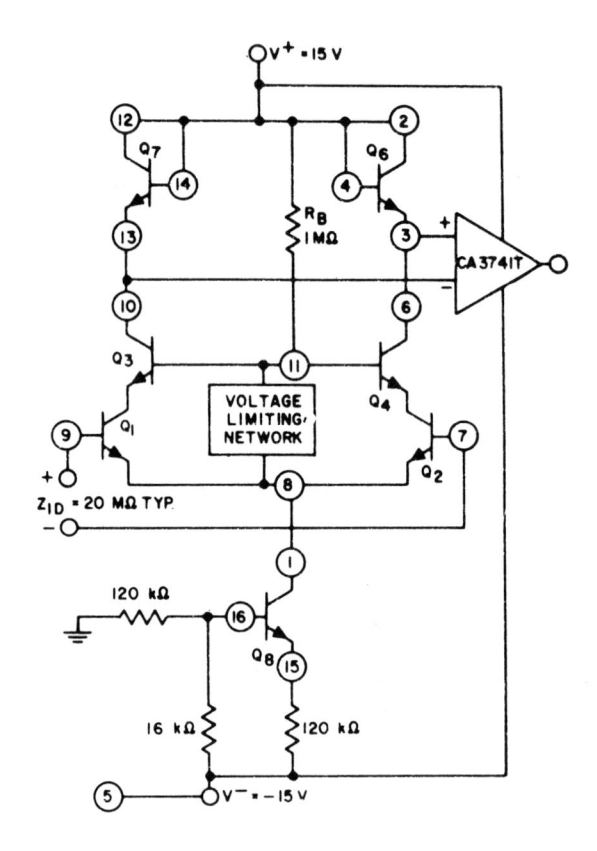

Bias arrangement for operation of the superbeta differential cascode amplifier.

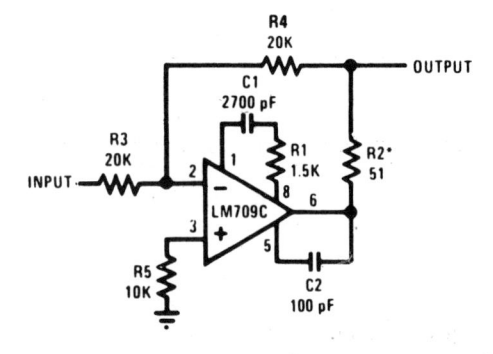

Unity-gain inverting amplifier. R2 should be used with any capacitive loading on output.

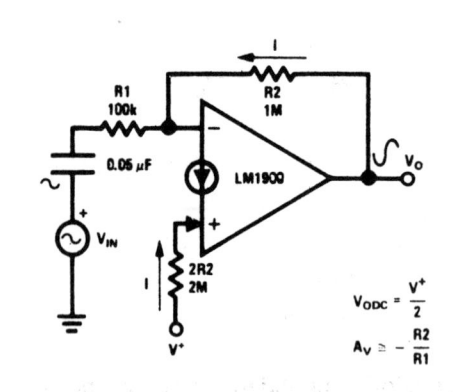

Inverting amplifier. $V^+ = 15V$ dc.

$$V_{ODC} = \frac{V^+}{2}$$

$$A_V \cong -\frac{R2}{R1}$$

90 IC SCHEMATIC SOURCEMASTER

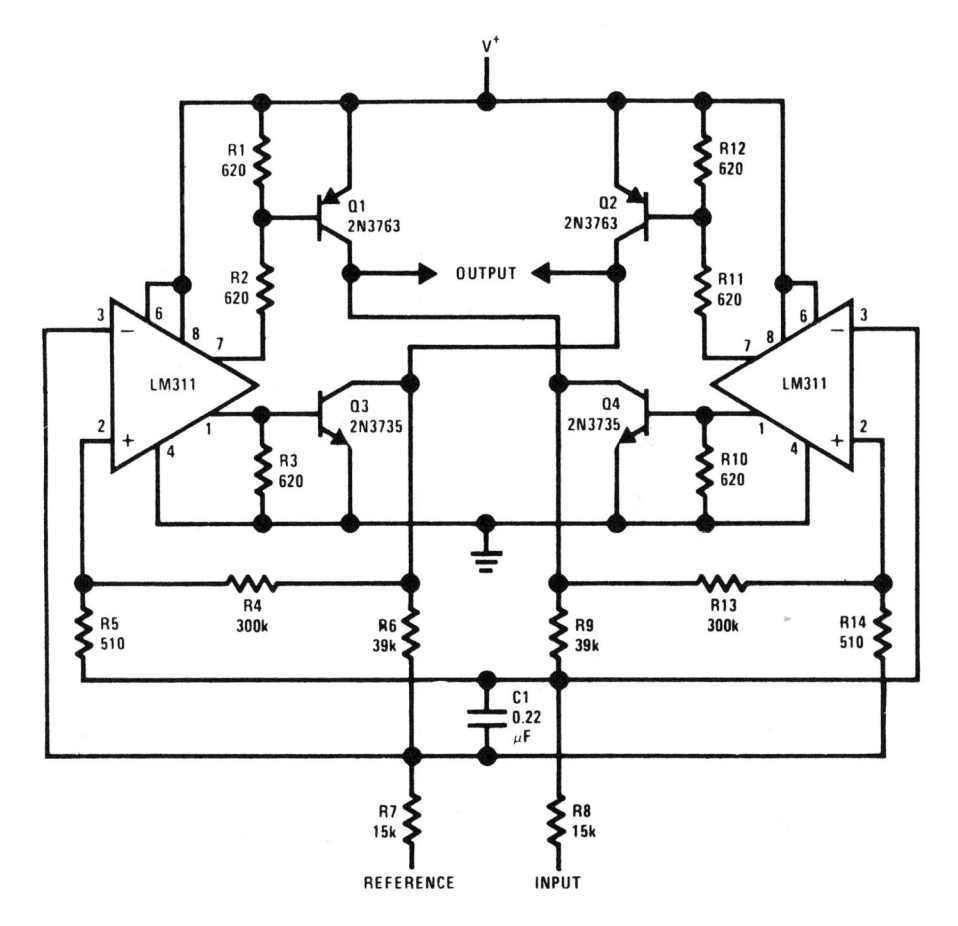

Superbeta op-amp with diode drive network.

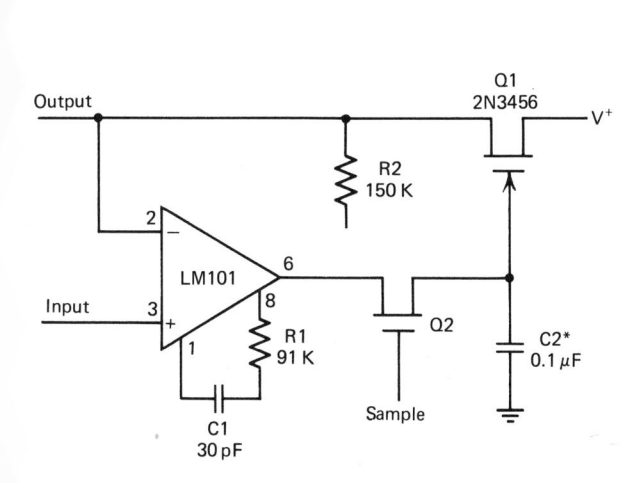

A unity-gain amplifier employing an operational amplifier and FET source follower comprise a capable **sample-and-hold circuit**. In operation, when sample switch Q2 is turned on, it closes the feedback loop to make the output equal to the input, differing only by the offset voltage of the LM101. When the switch is opened, the charge stored on C2 holds the output at a level equal to the last value of the input voltage. C2 should have a polycarbonate dielectric.

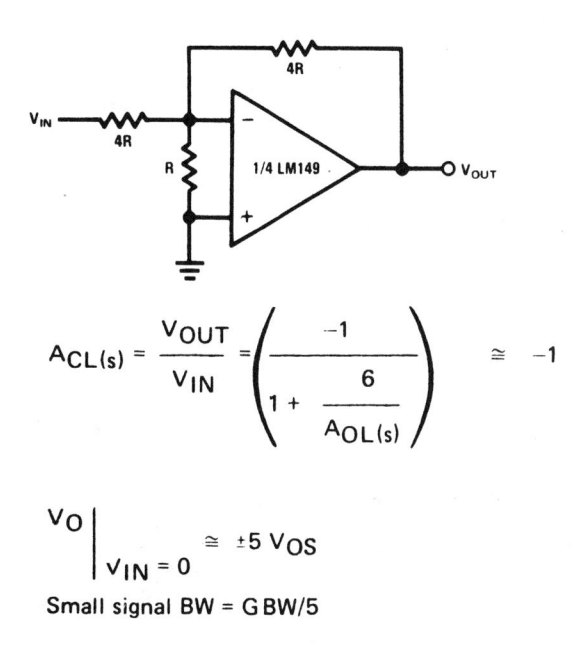

$$A_{CL}(s) = \frac{V_{OUT}}{V_{IN}} = \left(\frac{-1}{1 + \frac{6}{A_{OL}(s)}} \right) \cong -1$$

$$\left. V_O \right|_{V_{IN} = 0} \cong \pm 5\, V_{OS}$$

Small signal BW = GBW/5

Unity-gain inverter using LM149.

SECTION **2** *Followers, Buffers, and Special-Purpose Amplifiers* 91

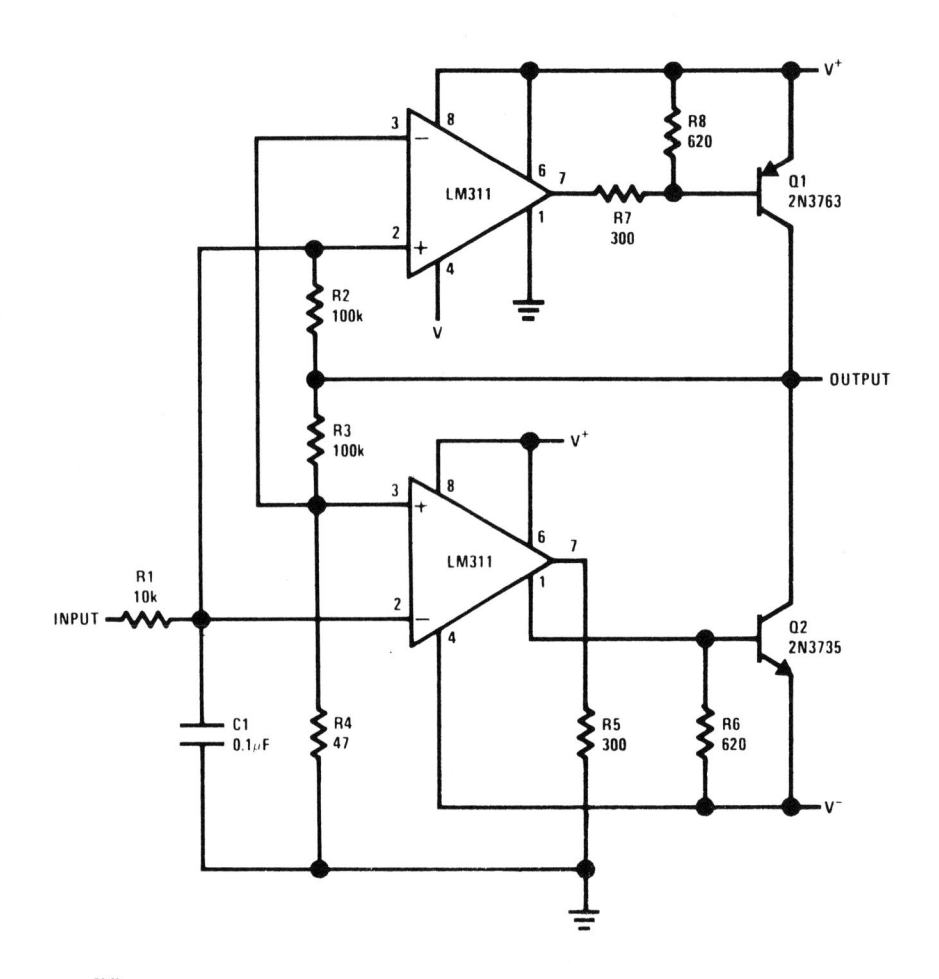

Switching power amplifiers.

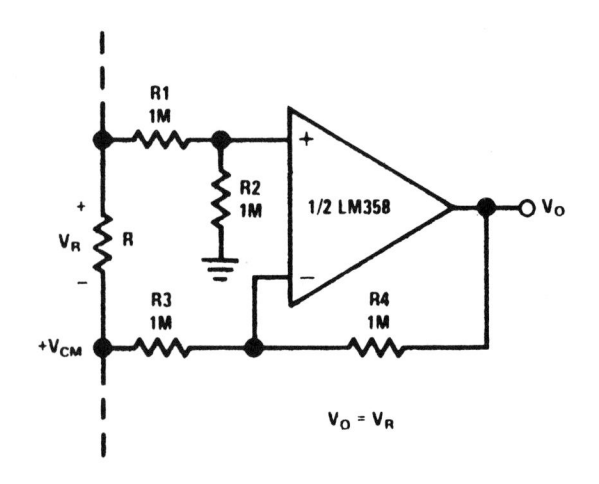

Ground referencing a differential input signal.

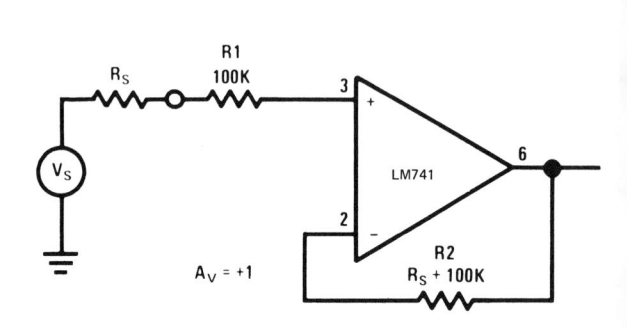

$V_O = V_R$

Unity-gain voltage follower. Resistor R1 limits the current to a safe value of 200 μA. R2 is included to cancel the error voltage due to bias current and should in general be equal to the source resistance plus R1.

92 IC SCHEMATIC SOURCEMASTER

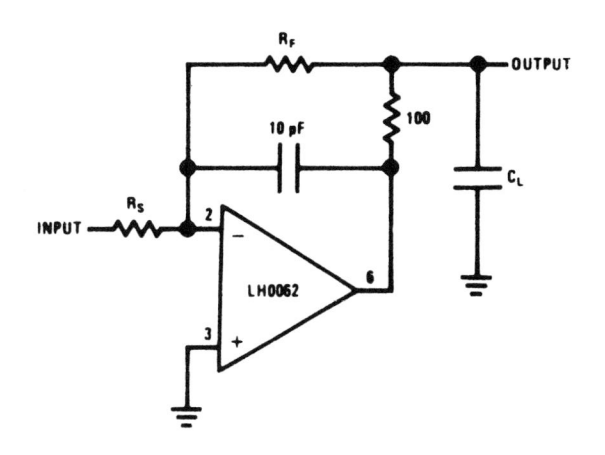

Capacitive load isolator.

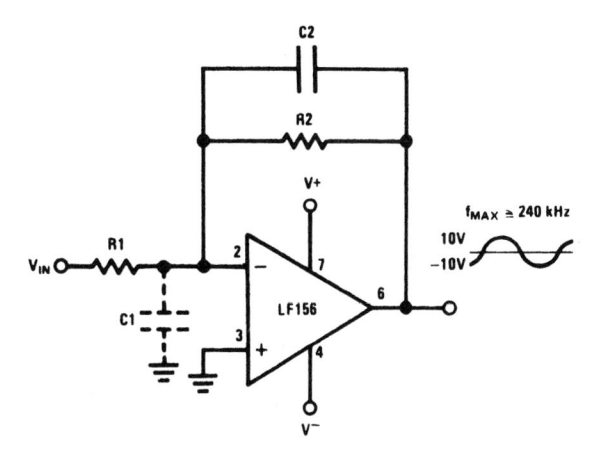

Wide-bandwidth, low-noise, low-drift amplifier. Power bandwidth is $f_{MAX} = S_r/2\pi V_P = 240$ kHz. Parasitic input capacitance C1 $\cong$ (3 pF for LF155, LF156, and LF157 plus any additional layout capacitance) interacts with feedback elements and creates an undesirable high-frequency pole. To compensate, add C2 such that $R2C2 \cong R1C1$.

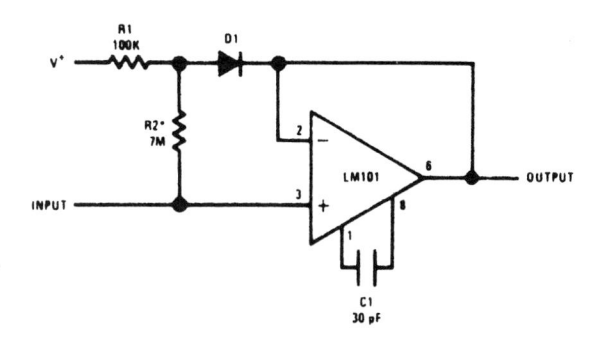

Voltage follower with bias-current compensation. The compensating current is obtained through a resistor connected across a diode bootstrapped to the output. The diode acts as a regulator so that the compensating current does not change appreciably with the signal level, giving input impedances about 1000M. The negative temperature coefficient of the diode voltage also provides some temperature compensation. Select R2 for zero input current.

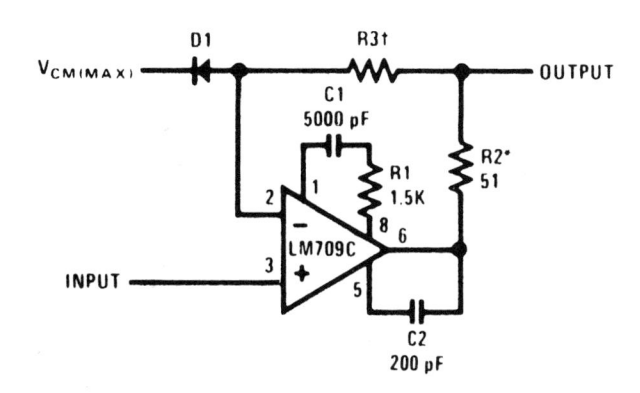

Voltage follower. R2 should be used with any capacitive loading on output. R3 should be equal to dc source resistance on input.

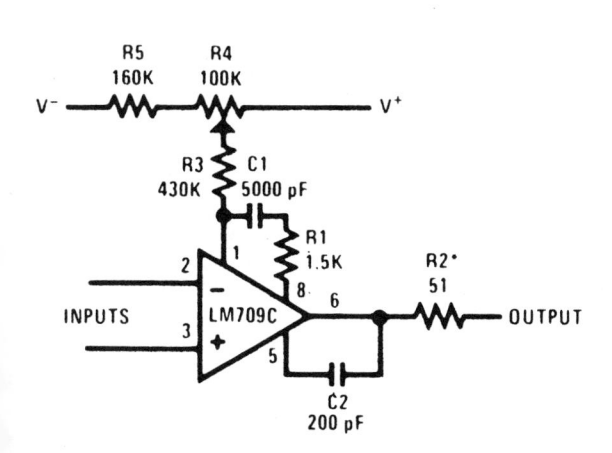

Offset balancing circuit. R2 should be used with any capacitive loading on output.

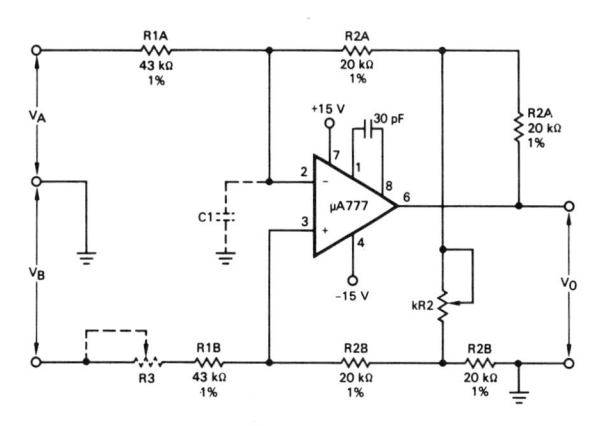

This variable-gain differential amplifier uses the differential form of the resistive T feedback network to achieve a wide variation in gain using a single potentiometer.

SECTION **2** *Followers, Buffers, and Special-Purpose Amplifiers*

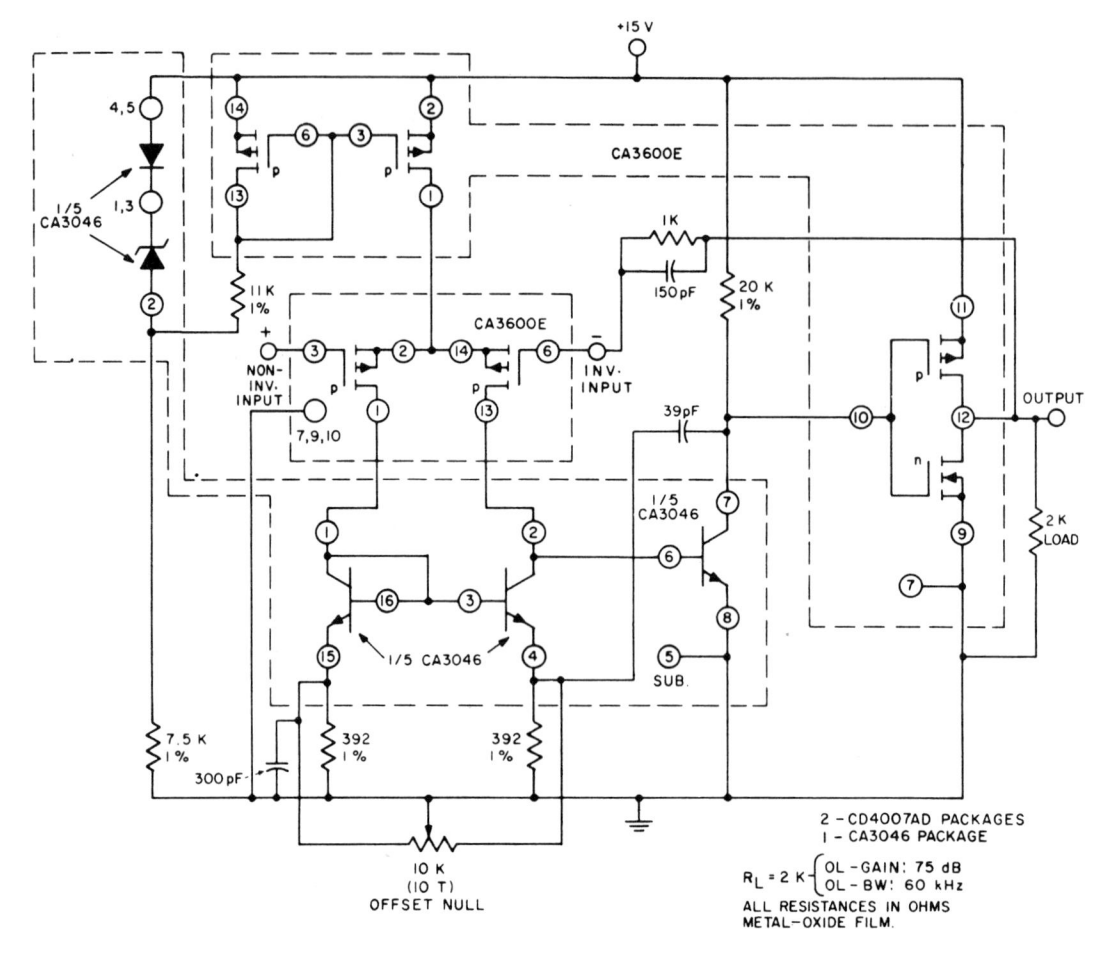

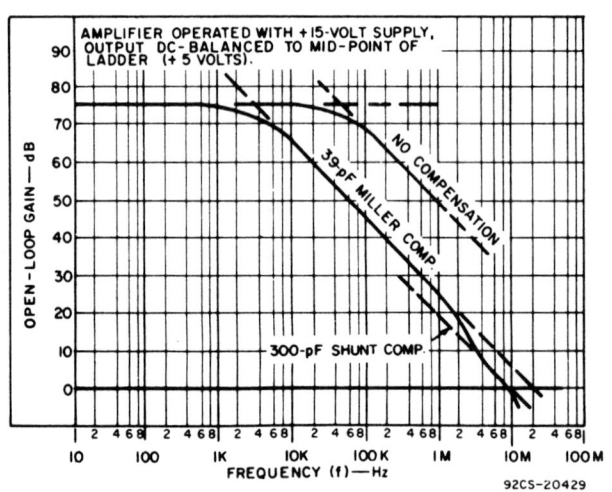

Voltage follower for single-supply operation having a CMOS P-channel input, an NPN second gain stage, and a CMOS inverter output. The IC building blocks are two CA3600Es and a CA3046 transistor array. A zener-regulated leg provides bias for a 400 µA P-channel current source feeding the input stage, which is terminated in an NPN current mirror. Amplifier voltage offset is nulled with the 10k balance potentiometer. The second-stage current level, established by the 20k load, is selected to approximate the first-stage current level to assure similar positive and negative slew rates. The inverter portion forms the final output stage and is terminated in a 2k load.

Voltage gain is affected by the choice of load resistance value. The output stage of this amplifier is easily driven to within 1 mV of the negative supply voltage.

Compensation for the unity-gain noninverting mode is provided by Miller feedback of 39 pF and a 300 pF bypass capacitor shunting half the driving current (1–2 MHz). Unity-gain bandwidth is just under 10 MHz, and the open-loop gain is 75 dB. A potential latch situation at the bipolar mirror is avoided by use of resistor–capacitor network (R = 1k, C = 150 pF), which limits the dc feedback through the P-channel gate-protective diode.

94 IC SCHEMATIC SOURCEMASTER

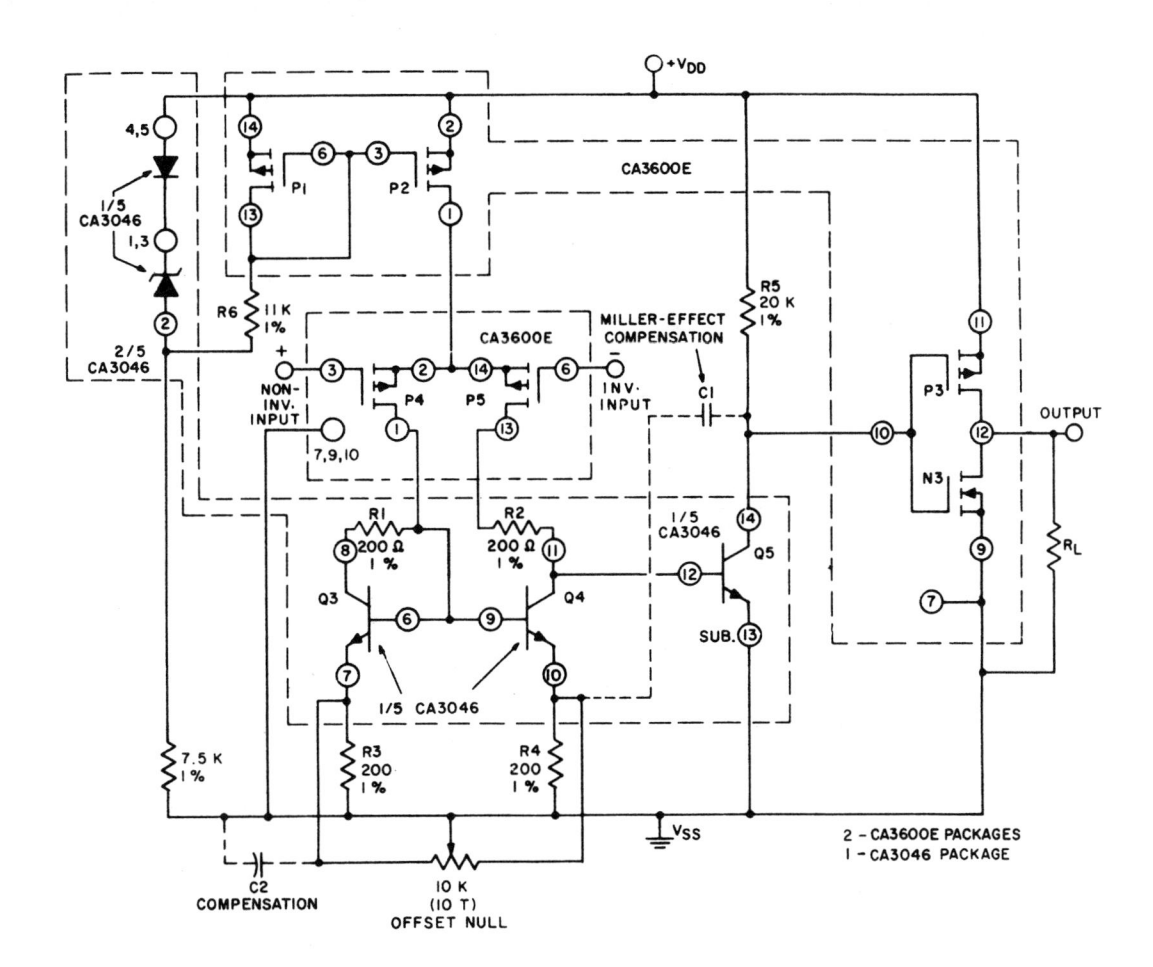

A voltage-follower amplifier for single-supply operation. Resistors should be metal-oxide film.

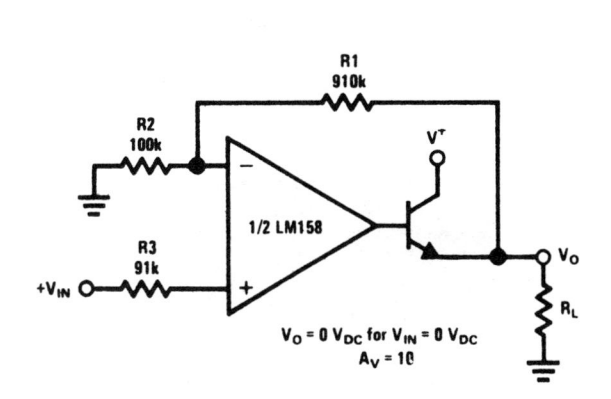

Power amplifier.

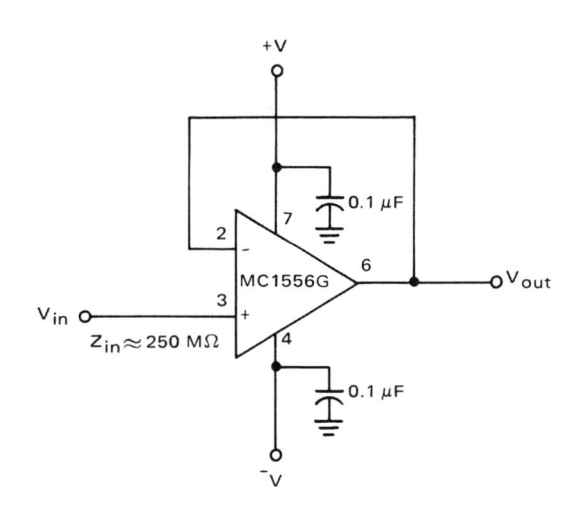

$V_O = 0\ V_{DC}$ for $V_{IN} = 0\ V_{DC}$
$A_V = 10$

This voltage follower has unity gain but makes a great impedance transformer. The use of the MC1556, by virtue of its high input impedance and the value of its input bias current, results in improved performance for virtually all circuit configurations. However, the major performance parameters of some circuits are directly dependent on these factors. The input impedance is theoretically $10^{12}\Omega$ but is, in fact, limited by the common-mode input impedance of 250M. This value represents the low-frequency dynamic input impedance of the follower. At higher frequencies, stray or pin capacitance at the input of the operational amplifier will come into play and probably will be the limiting factor.

SECTION **2** *Followers, Buffers, and Special-Purpose Amplifiers*

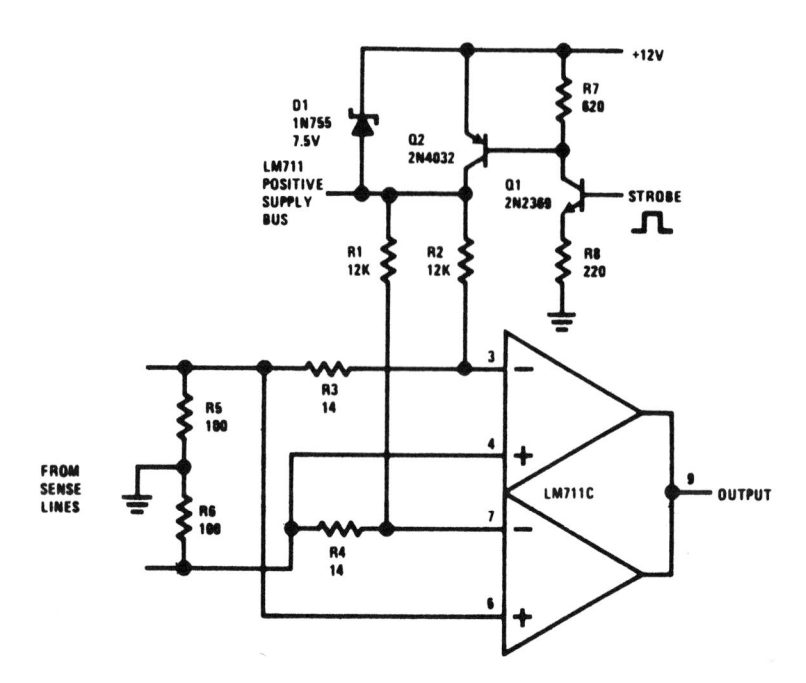

A sense amplifier with supply strobing for reduced power consumption; standby dissipation is about 40 mW.

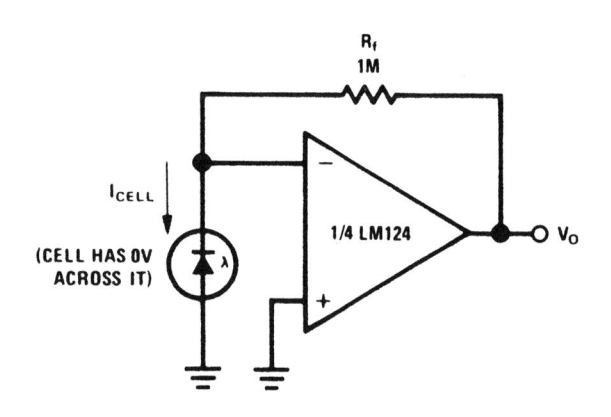

Photovoltaic (solar cell) amplifier.

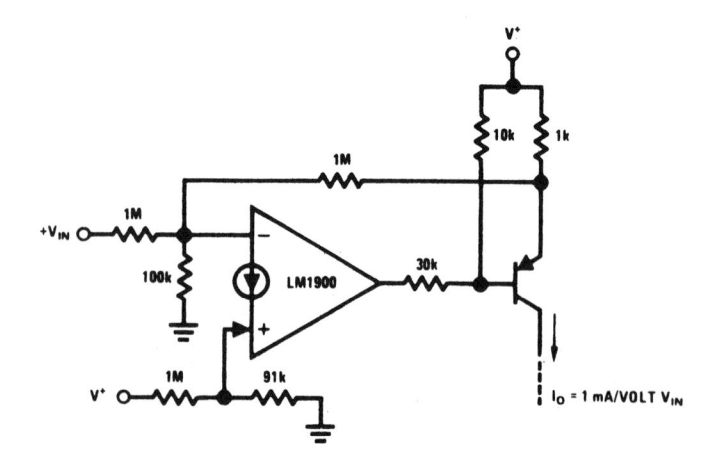

Voltage-controlled current source (transconductance amplifier).

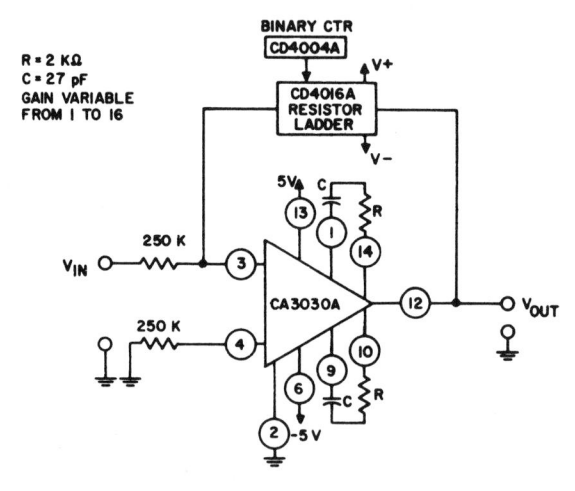

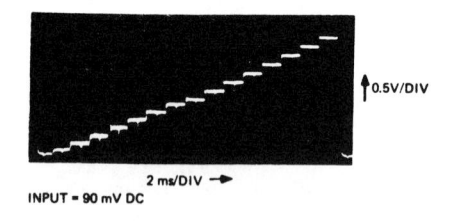

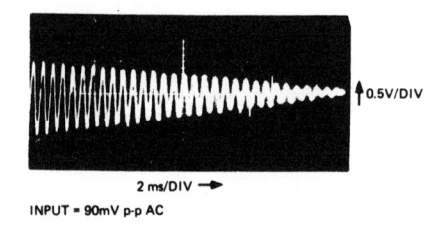

Variable-gain control circuit. Insets show waveform of output voltage for dc input voltage (top) and waveform of output voltage for ac input voltage (bottom).

96 IC SCHEMATIC SOURCEMASTER

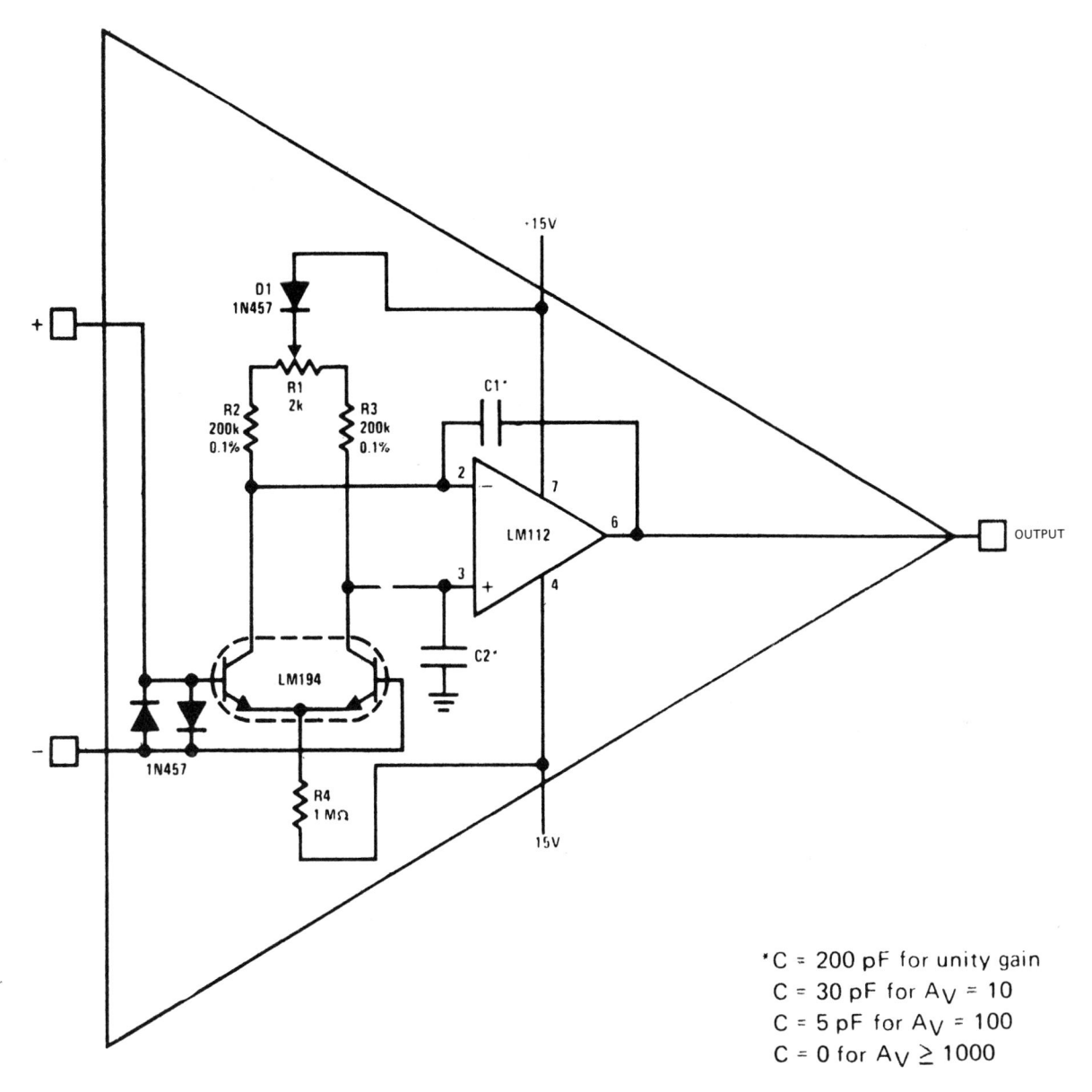

*C = 200 pF for unity gain
C = 30 pF for A_V = 10
C = 5 pF for A_V = 100
C = 0 for $A_V \geq 1000$

Voltage-controlled variable-gain amplifier. Common-mode range is 10V; I_{BIAS}, less than 25 nA; I_{OS}, less than 0.5 nA; V_{OS} (untrimmed), less than 125 µV; ($\Delta V_{OS}/\Delta T$), less than 0.2 µV/°C; CMRR, greater than 120 dB; A_{VOL}, greater than 2.5×10^5.

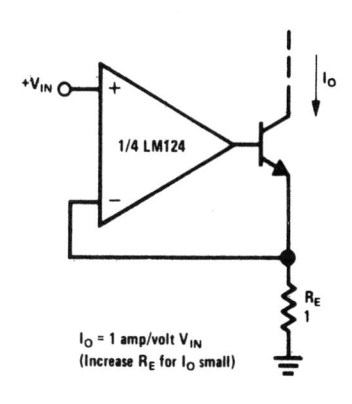

I_O = 1 amp/volt V_{IN}
(Increase R_E for I_O small)

High-compliance current sink.

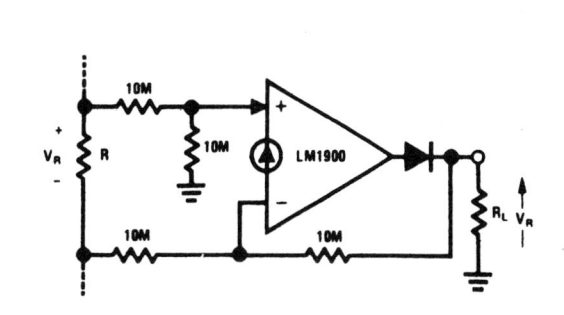

Ground-referencing a differential input signal.

SECTION **2** *Followers, Buffers, and Special-Purpose Amplifiers* **97**

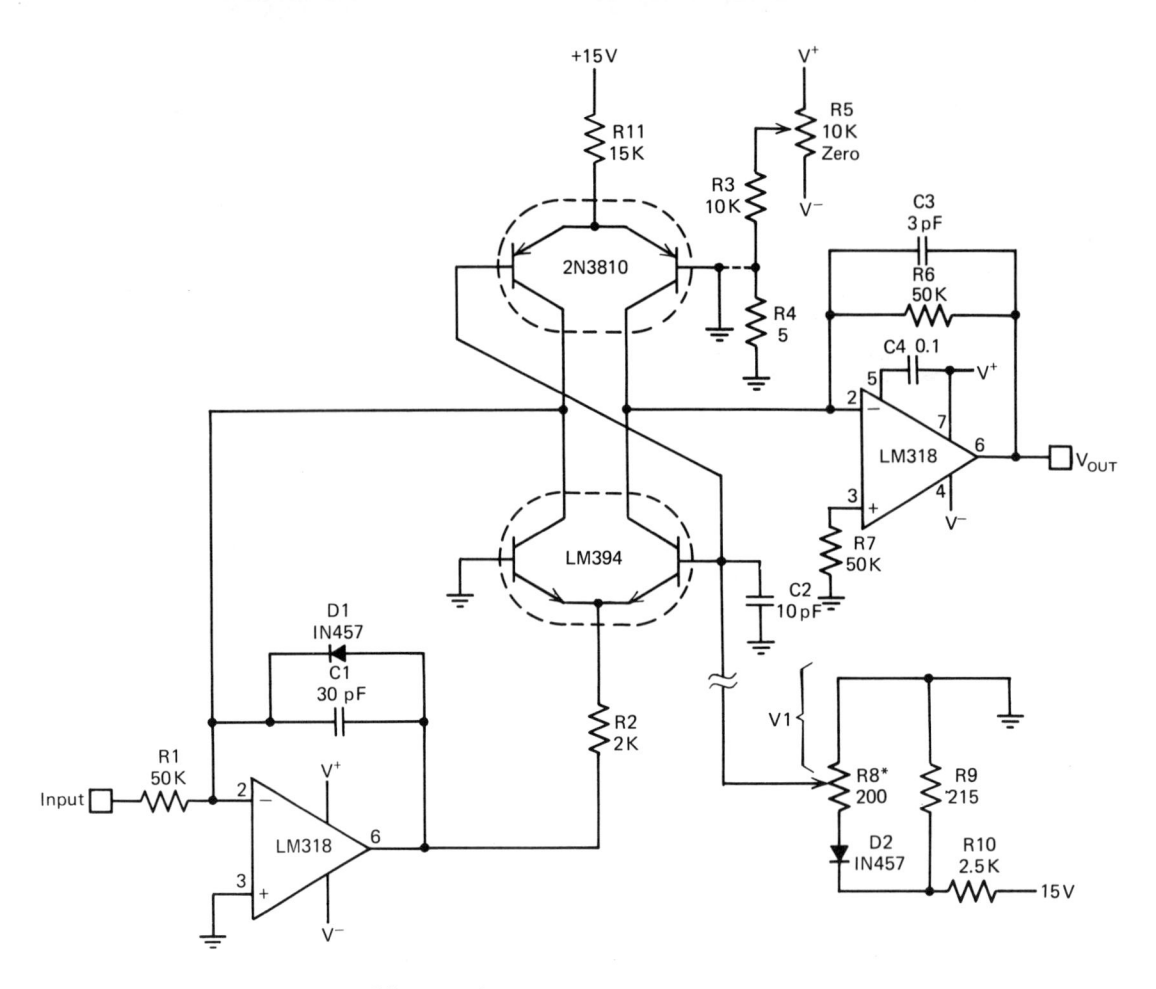

Precision low-drift operational amplifier. Resistors R8–R10 and diode D2 provide a temperature-independent gain control: G = −336 V1 (dB). Distortion is less than 0.1% over the full 1 MHz bandwidth. Gain range is 100 dB.

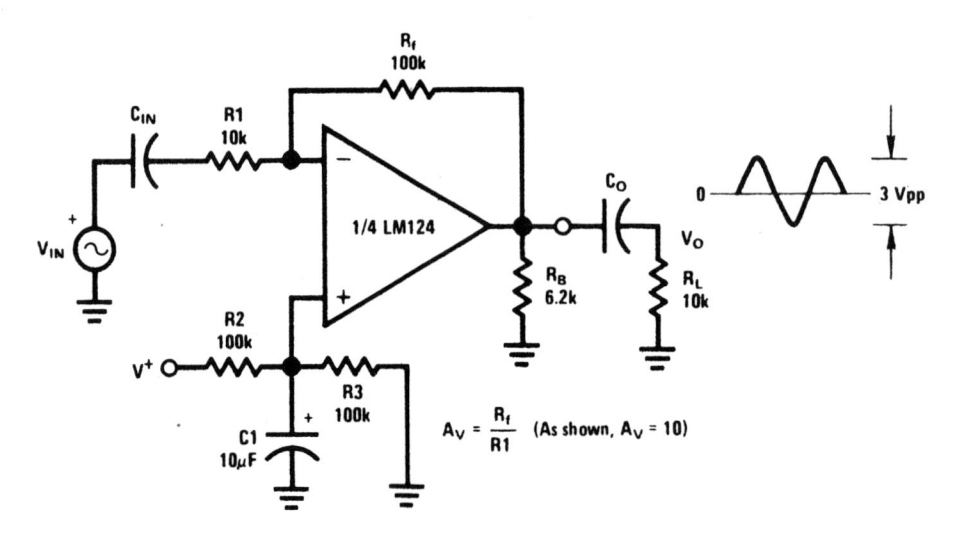

$$A_V = \frac{R_f}{R_1} \quad \text{(As shown, } A_V = 10\text{)}$$

Ac-coupled inverting amplifier.

98 IC SCHEMATIC SOURCEMASTER

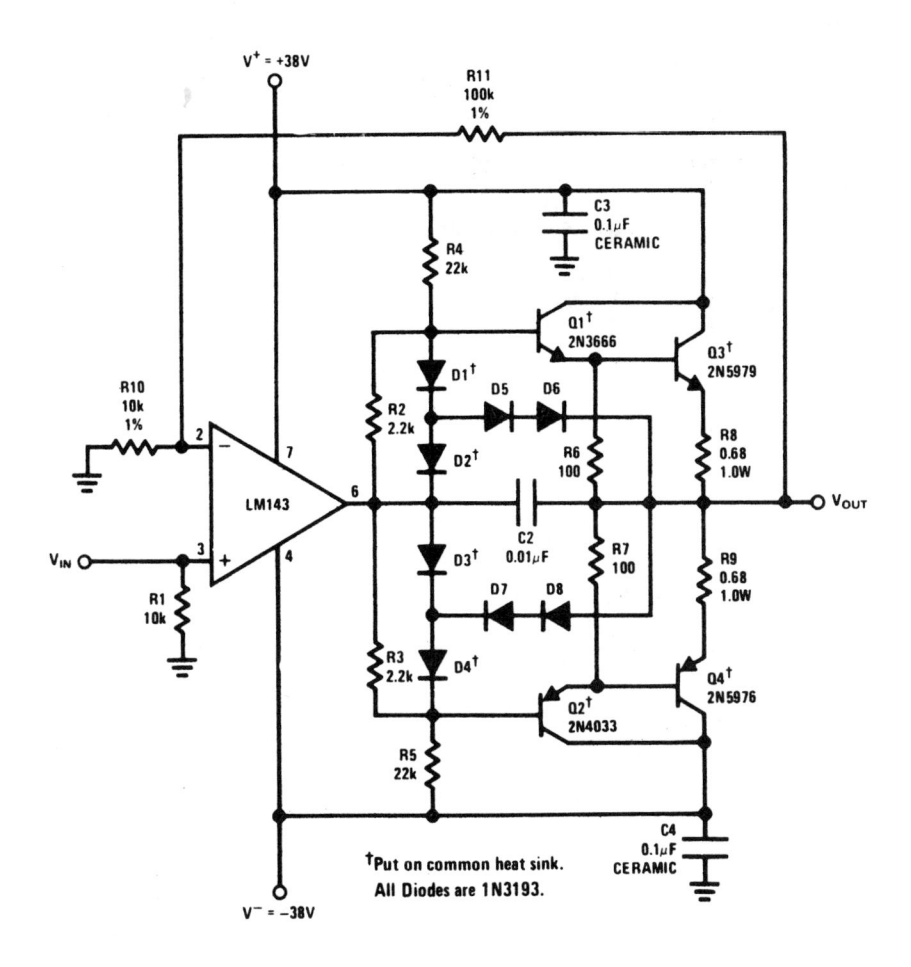

1A power amplifier with short-circuit protection. The 38V supplies allow for a 5% voltage tolerance; all resistors are 0.5W, except as noted.

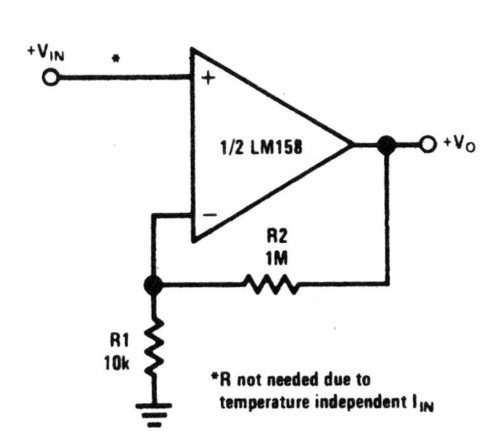

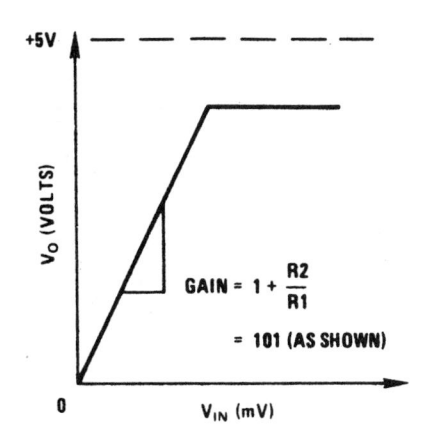

†Put on common heat sink. All Diodes are 1N3193.

Noninverting dc gain control (0V input = 0V output).

*R not needed due to temperature independent I_{IN}

$$\text{GAIN} = 1 + \frac{R2}{R1} = 101 \text{ (AS SHOWN)}$$

SECTION **2** *Followers, Buffers, and Special-Purpose Amplifiers* **99**

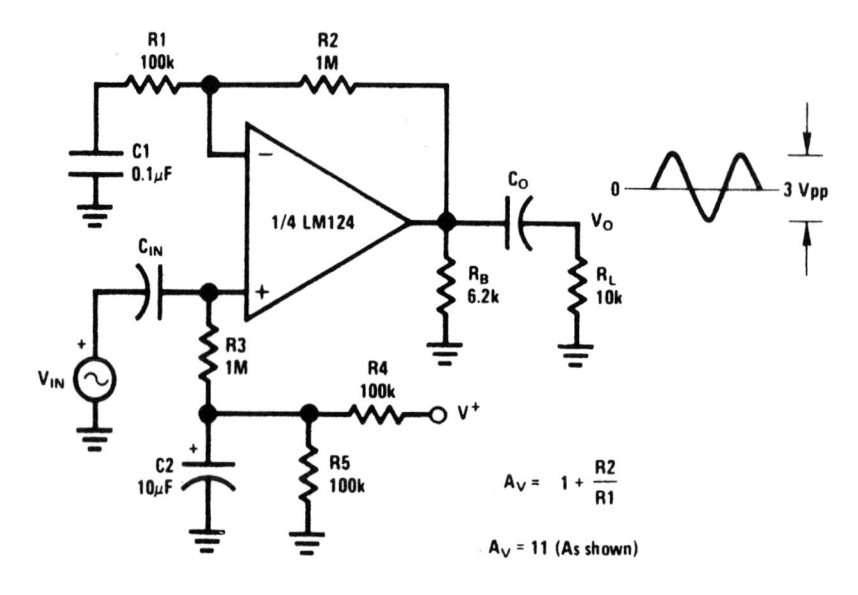

Ac-coupled noninverting amplifier.

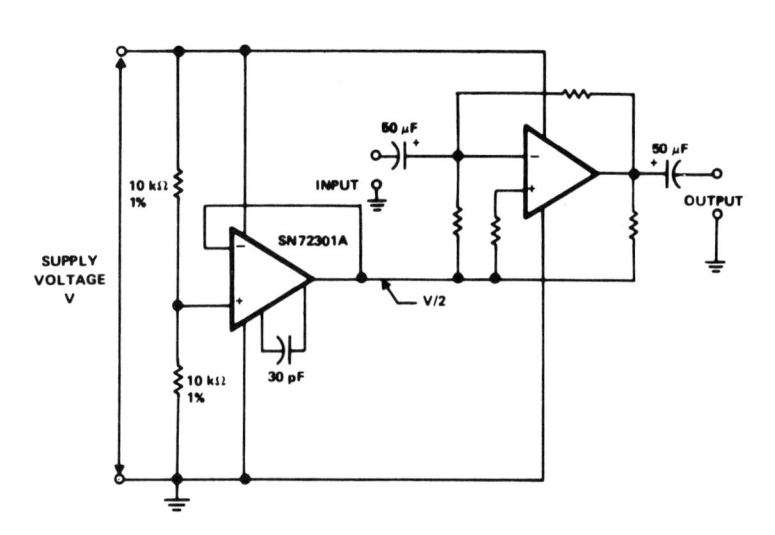

$$A_V = 1 + \frac{R2}{R1}$$

$A_V = 11$ (As shown)

Operating an ac-coupled operational amplifier from a single supply voltage.

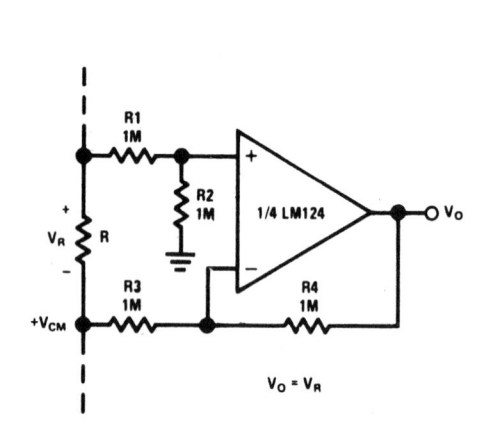

Ground-referencing a differential input signal.

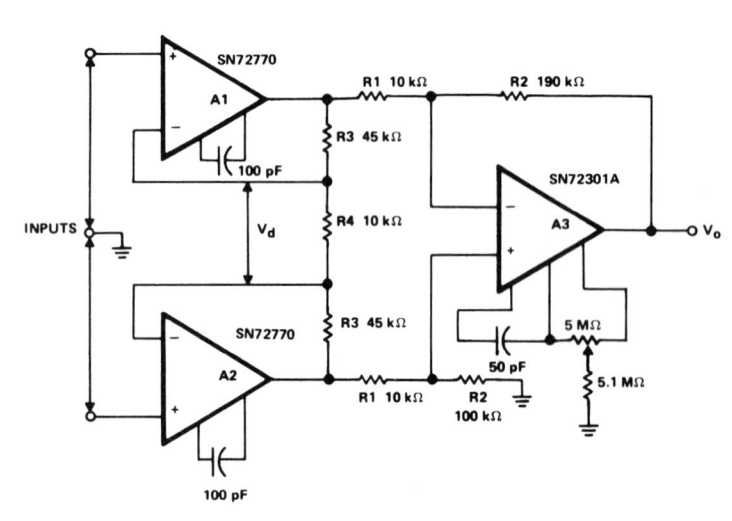

$V_O = V_R$

High-input-impedance differential amplifier with improved common-mode rejection.

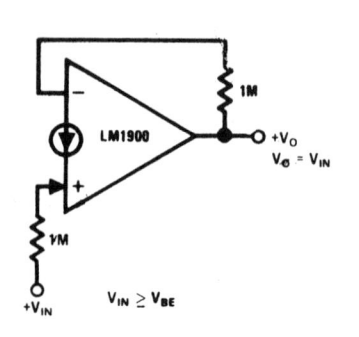

Buffer amplifier.

$V_O = V_{IN}$

$V_{IN} \geq V_{BE}$

100 IC SCHEMATIC SOURCEMASTER

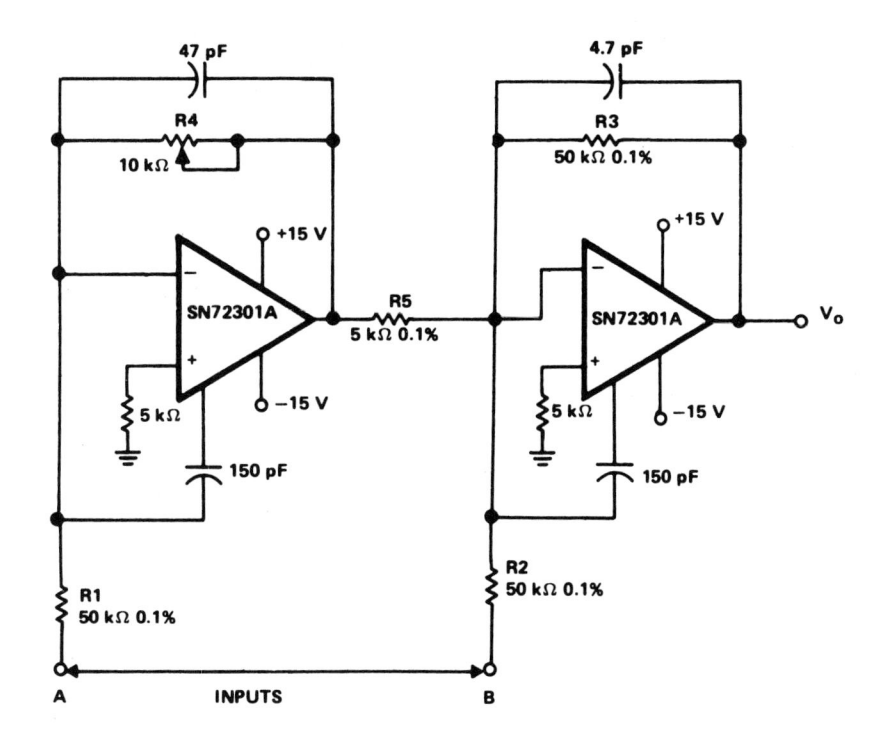

A differential amplifier with ±100V common-mode (input) range. Adjust R4 to the exact value (~5k) required for maximum common-mode rejection.

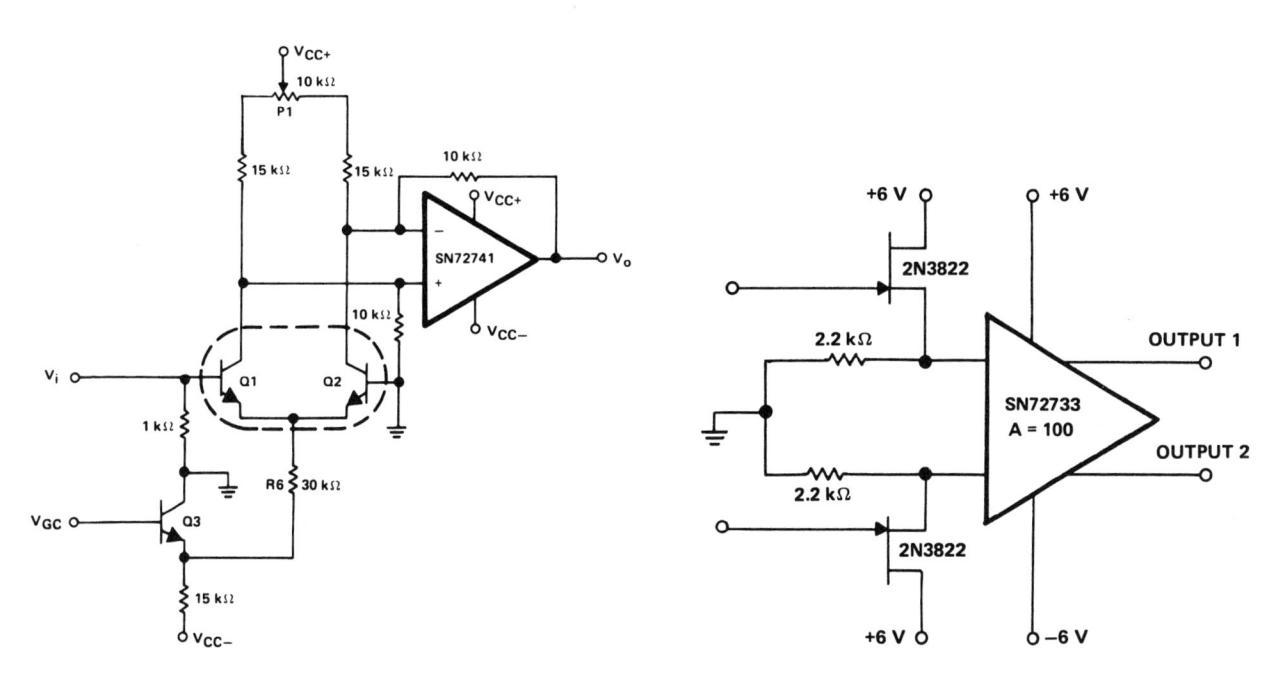

An amplifier controlled by a negative control voltage applied to the base of Q3. Emitter currents for Q1 and Q2 are controlled by V_{GC}, the total current being equal to V_{GC}/R. Controlling this current controls the circuit gain. The potentiometer in the collector circuit of Q1 and Q2 is used to correct for offset voltages. The gain-control voltage applied to the base of Q3 may be used to switch the amplifier on and off or to modulate the input signal.

High-input-impedance wideband amplifier.

SECTION 2 *Followers, Buffers, and Special-Purpose Amplifiers*

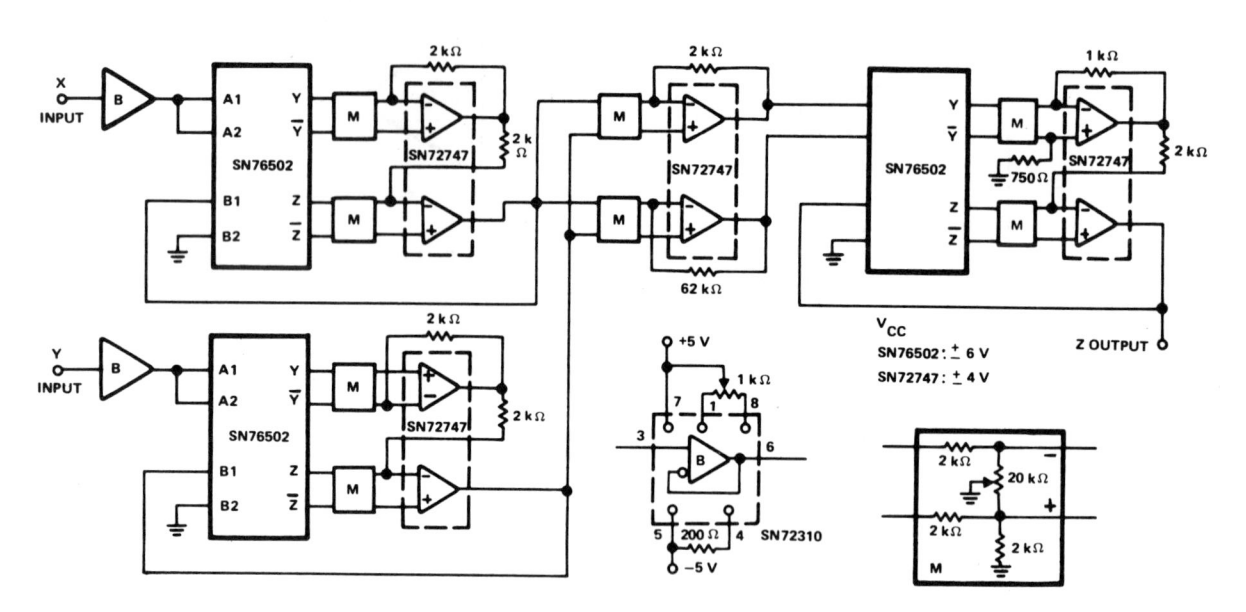

This logarithmic amplifier solves for the hypotenuse of a right triangle when given the two sides. With SN76502 logarithmic amplifier sections it is possible to solve for Z (the hypotenuse), applying the values of X and Y into the input as shown. The output of the circuit is automatically the value for Z. An analysis of the various log functions required is as follows:

$$Z^2 = X^2 + Y^2$$
$$= \text{antilog } 2 \log X + \text{antilog } 2 \log Y$$
$$Z = \text{antilog } \tfrac{1}{2} \log Z^2$$
$$= \text{antilog } \tfrac{1}{2} \log (\text{antilog } 2 \log X$$
$$+ \text{antilog } 2 \log Y)$$

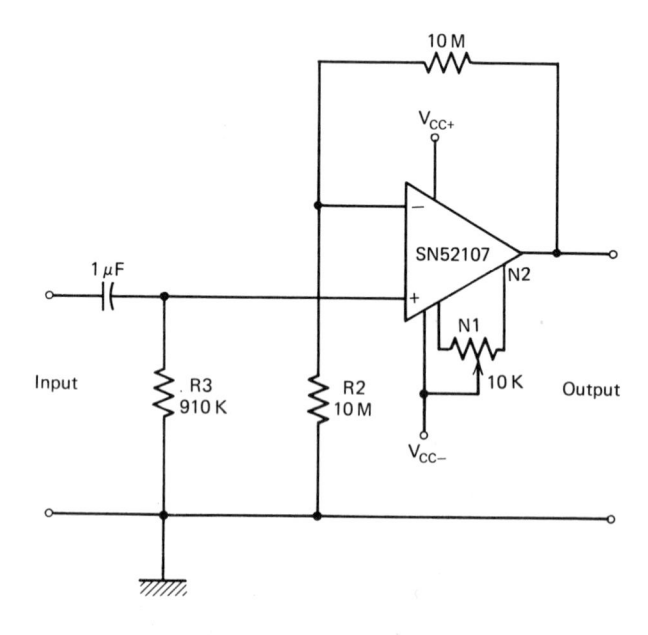

High-input-resistance noninverting amplifier. In this application, a high input resistance (1M) is required in matching a high-resistance source to a low-impedance load. Input offset current should be as low as possible to prevent a large dc output offset. With 1M source resistance, its 10 nA maximum input offset current will generate only 10 mV of input offset.

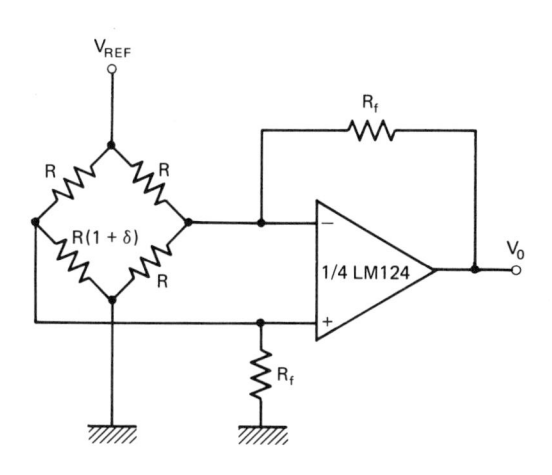

Bridge current amplifier. For $\delta \ll 1$ and $R_f \gg R$,

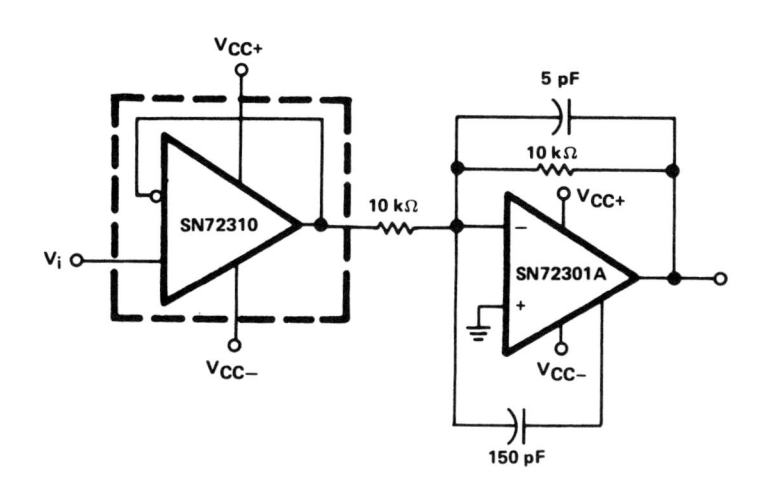

$$V_o \cong V_{REF} \left(\frac{\delta}{2}\right) \frac{R_f}{R}$$

Very high-input-impedance inverter.

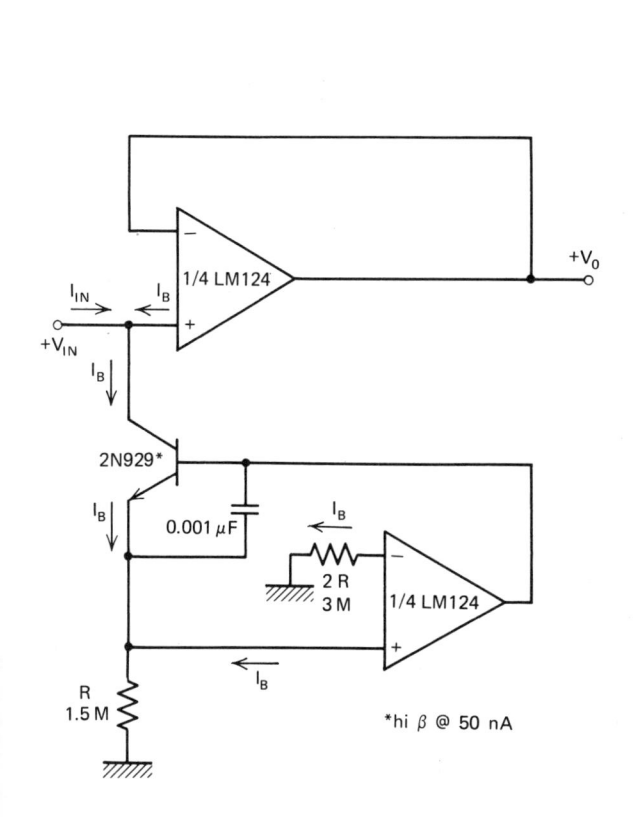

Using symmetrical amplifiers to reduce input current (general concept).

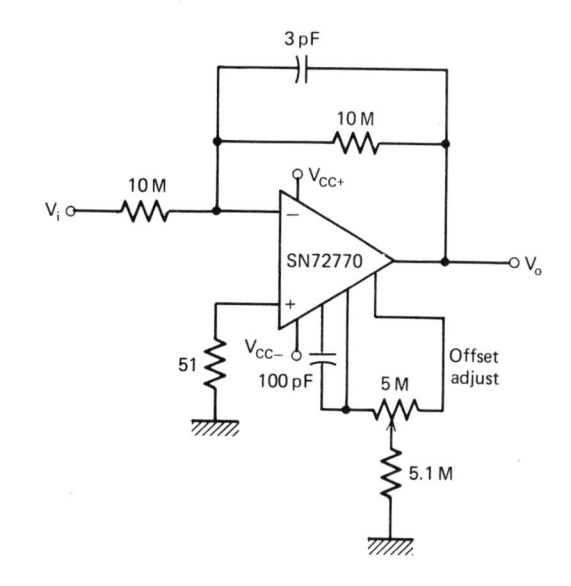

High-input-impedance inverting amplifier. If the source and input termination impedances of an op-amp are high (greater than 1M), the offset voltages generated by input offset bias currents could be prohibitive. However, a superbeta amplifier such as the SN72770 has several advantages. The very low input offset current of 10 nA maximum at 25°C will generate only 50 mV of offset voltage. The device's high input resistance (100M) allows operation from a 10M source, as shown, with only slight loss in signal. In addition, external compensation allows adjustment of the compensation capacitor for optimum frequency response.

SECTION **2** Followers, Buffers, and Special-Purpose Amplifiers

SECTION 3
INSTRUMENTATION AND TEST EQUIPMENT CIRCUITS

Some of the circuits on these pages represent new applications for long-established ICs; others use ICs recently developed for the comparatively new but wildly growing technology of medical electronics. In choosing the schematics for this section, I have tried to choose those that can be considered "precision," leaving the indoor/outdoor thermometers and moisture detectors for a subsequent section. There were two reasons for doing this: (1) it avoids the possibility of confusing first-time IC constructionists with a choice of **apparently** similar circuits that may be significantly disparate in terms of cost and complexity; and (2) it precludes the design engineer who wants a sophisticated instrument of measure from making an attempt to adapt a loose-tolerance circuit that would never be capable of delivering the performance required of it.

Of course, there is no clear line that distinguishes a precision circuit from the garden-variety sensor or its associated amplification system; the application itself will in many cases be the factor that determines the classification of a particular circuit. Since you are the individual who is singularly familiar with your specific circuit application, you should compare the schematics in this section with others in related sections of this book. Section 8, on control and indicating sensors, detectors, and alarms, for example, is the most likely place to "shop" when you've exhausted the possibilities here.

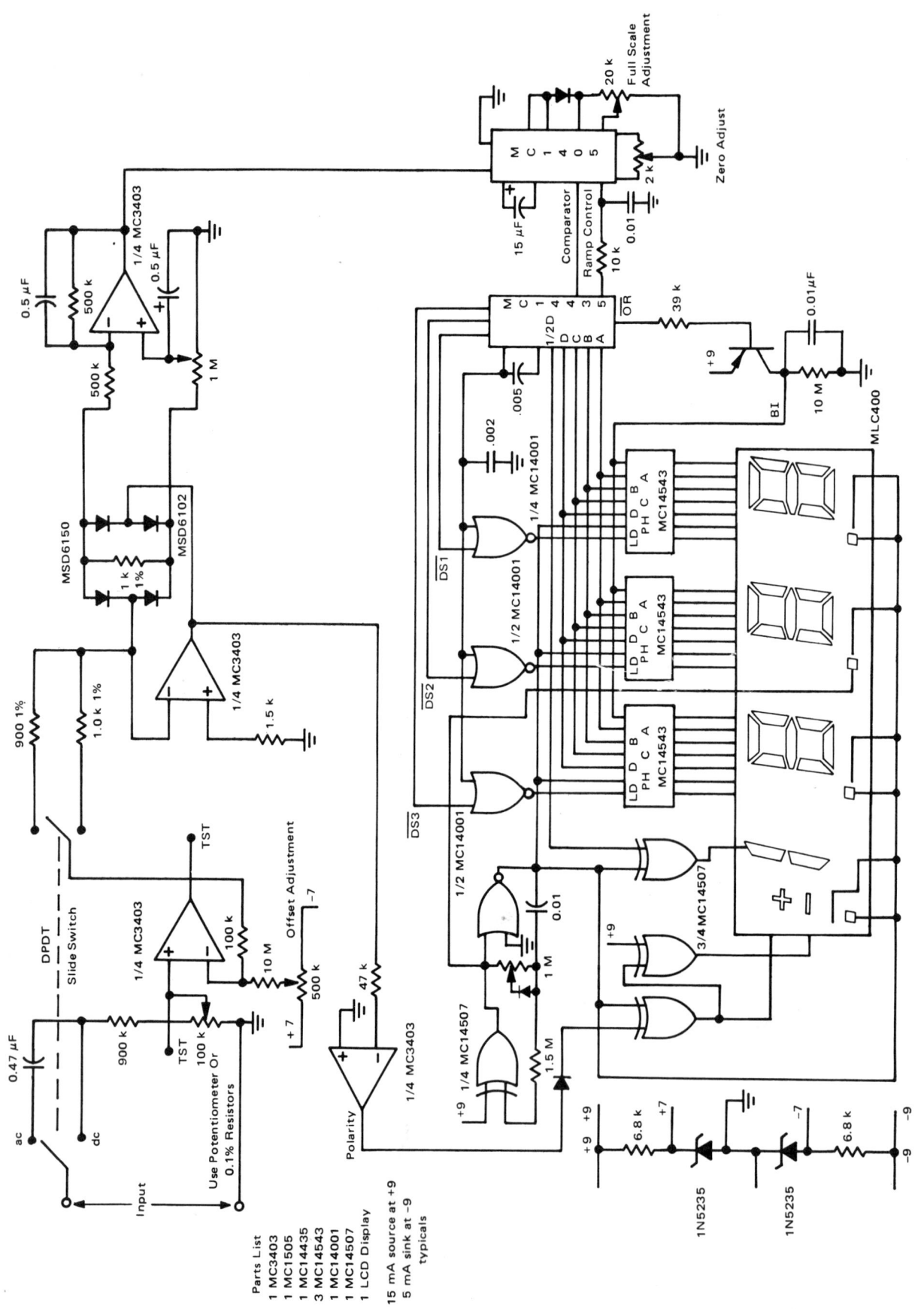

0–20V ac/dc DVM with liquid-crystal display.

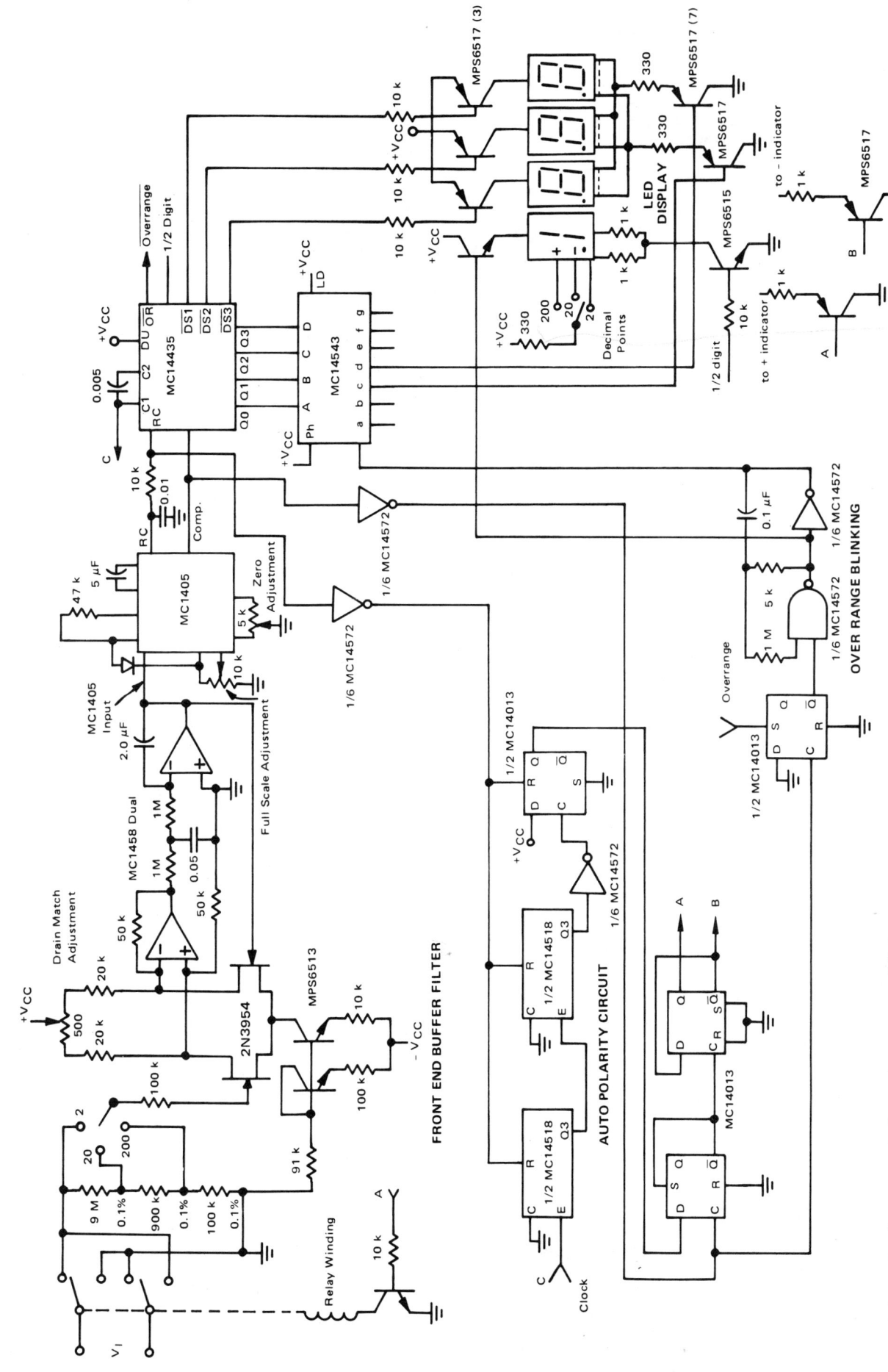

3.5-digit DVM with autopolarity and overrange blinking (Motorola).

SECTION 3 *Instrumentation and Test Circuits*

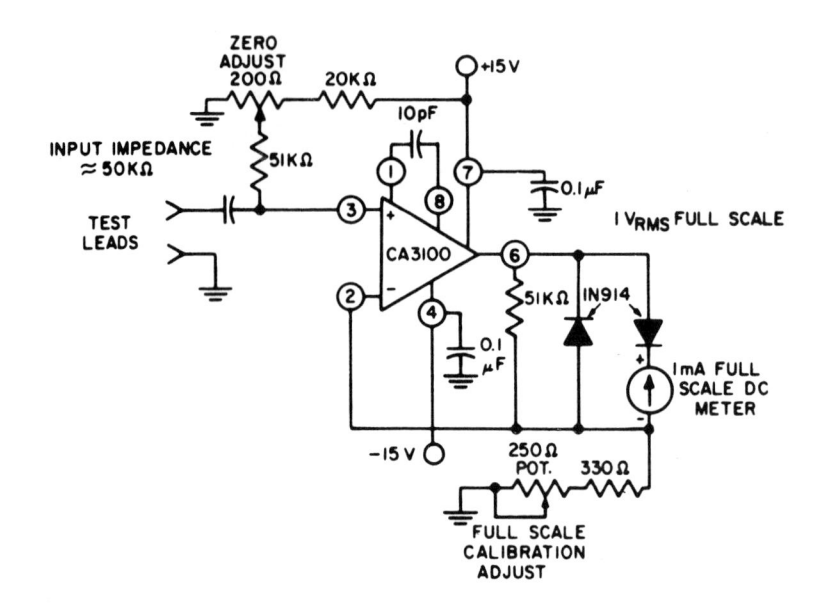

1 MHz meter-driver amplifier.

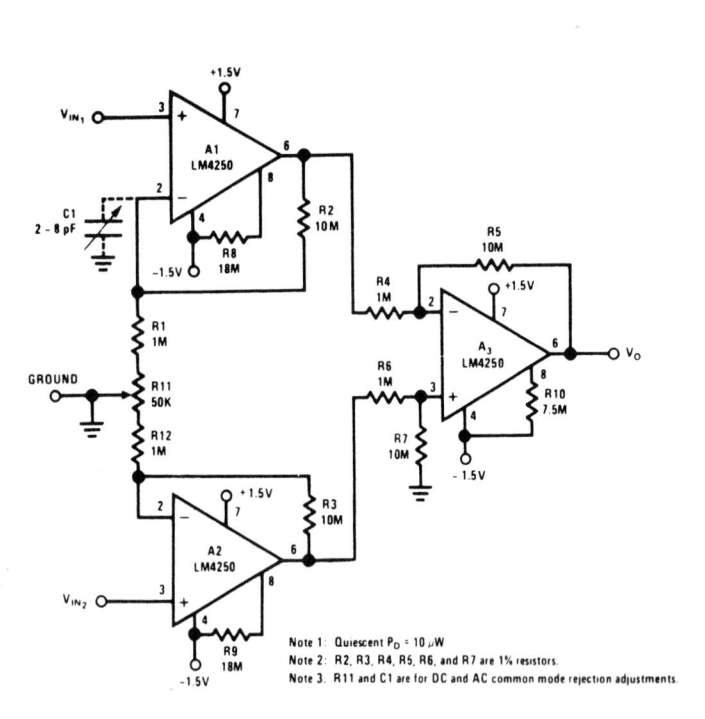

This X100 instrumentation amplifier has a full differential input center-tapped to ground. With the bias current set at approximately 0.1 μA, the impedance looking into either V_{IN1} or V_{IN2} is 100M with respect to ground, and the input bias current at either terminal is 0.2 nA. The entire circuit can run from two 1.5V cells connected directly to the V⁺ and V⁻ terminals. With a total current drain of 2.8 μA, the quiescent power dissipation of the circuit is 8.4 μW. This is low enough to have no significant effect on the shelf life of most batteries.

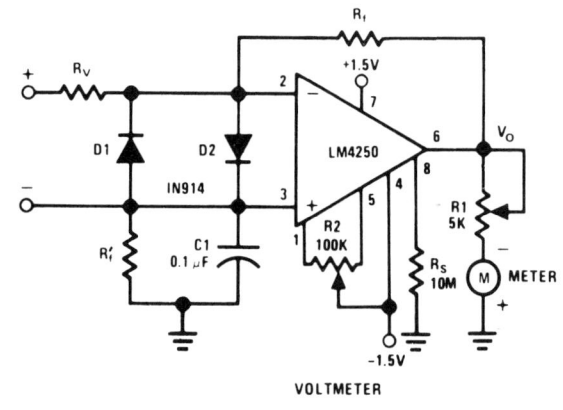

A 10 mV to 100V full-scale voltmeter. This inverting amplifier has a gain varying from −30 for the 10 mV full-scale range to −0.003 for the 100V full-scale range. The chart lists the proper values of R_v, R_f, and R'_f for each range. Diodes D1 and D2 provide complete amplifier protection for input overvoltages as high as 500V on the 10 mV range, but if overvoltages of this magnitude are expected under continuous operation, the power rating of R_v should be adjusted accordingly.

Resistance Values

V FULL SCALE	R_V [Ω]	R_f [Ω]	R'_f [Ω]
10 mV	100k	1.5M	1.5M
100 mV	1M	1.5M	1.5M
1V	10M	1.5M	1.5M
10V	10M	300k	0
100V	10M	30k	0

108 IC SCHEMATIC SOURCEMASTER

4.5-digit DVM digital subsystem (Motorola).

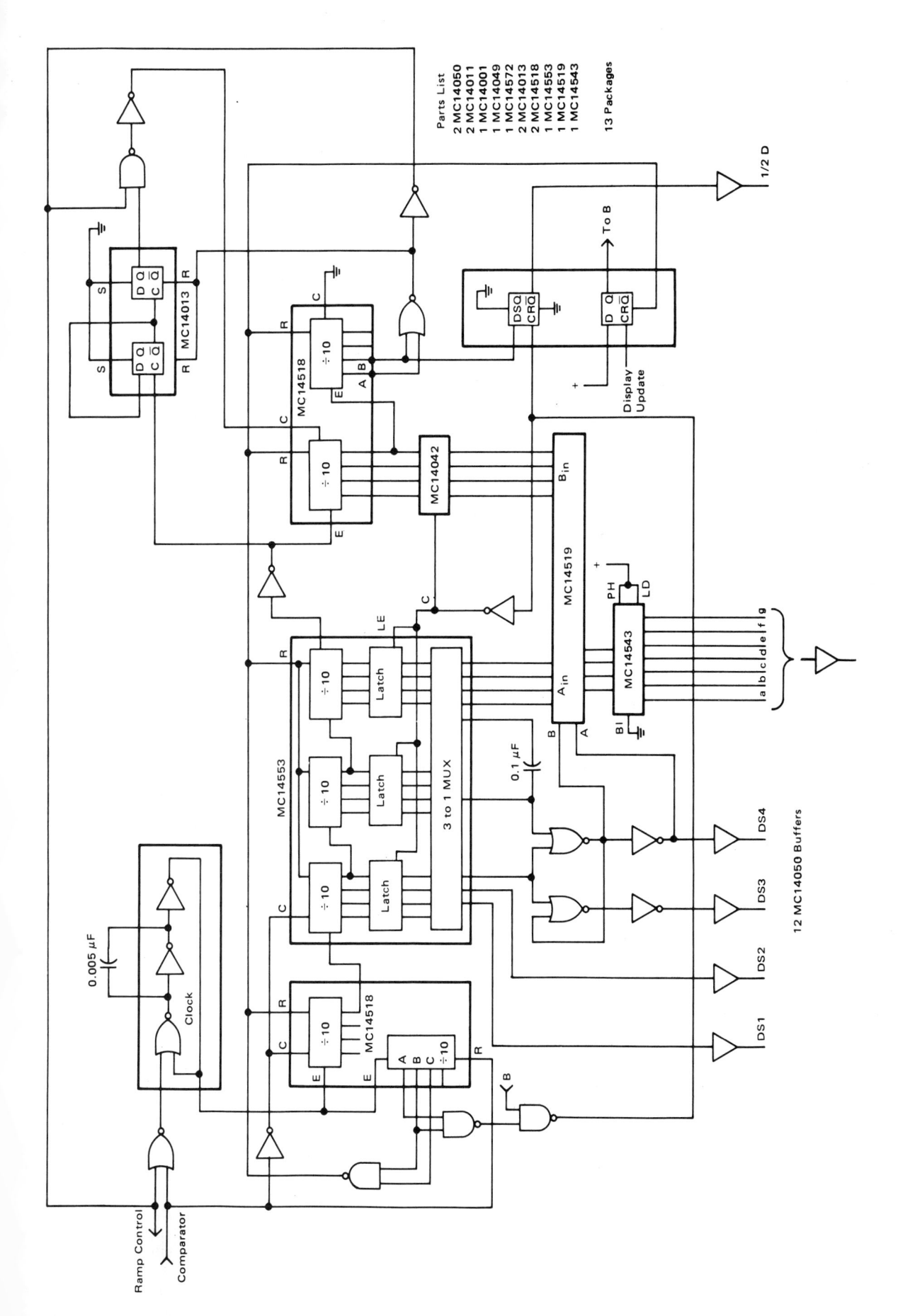

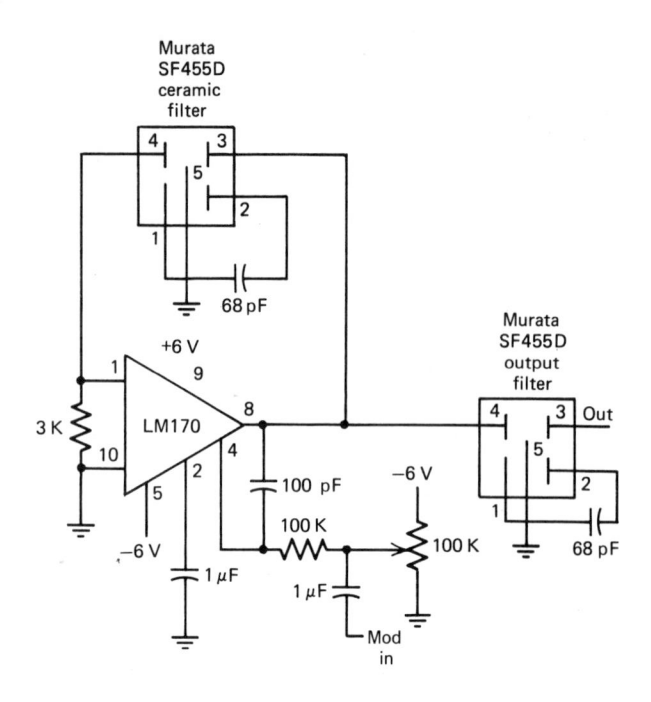

A modulated 455 kHz signal generator has constant output and is ideal for use as an AM IF alignment generator. If the AGC threshold voltage, which determines stabilized output, is varied at a low (audio) rate, the output amplitude will be forced to track the audio modulation.

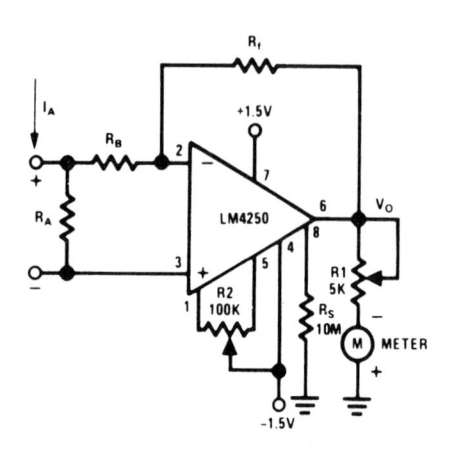

An ammeter for current readings above 100 μA. Resistor R_A develops a voltage drop in response to input current I_A. This voltage is amplified by a factor equal to the ratio of $R_f:R_B$. R_B must be sufficiently larger than R_A so as not to load the input signal. The chart shows the proper values of R_A, R_B, and R_f for full-scale meter deflections of from 1 mA to 10A.

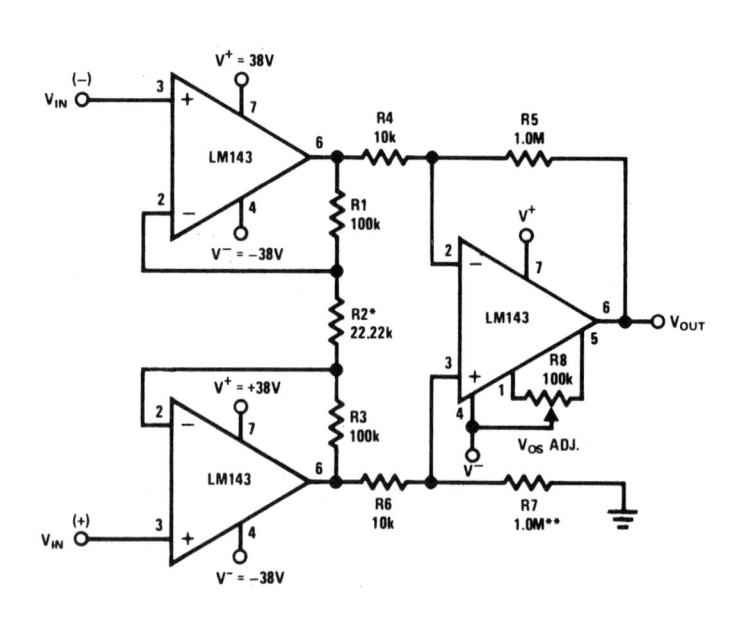

A 34V common-mode instrumentation amplifier. R2 may be adjustable to trim gain. R7 may be adjusted to compensate for the resistance tolerance of R4–R7 for best CMRR.

$$A_V = \left(1 + \frac{2R1}{R2}\right)\frac{R5}{R4}$$

where R4 = R6 and R5 = R7.

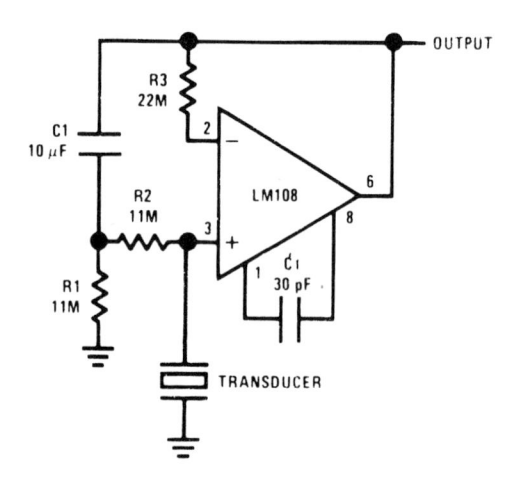

An amplifier for piezoelectric transducers such as accelerometers. These sensors normally require a high-input-resistance amplifier. The LM108 can provide input resistances in the range of 10 to 100M, using conventional circuitry. However, conventional designs are sometimes ruled out either because large resistors cannot be used or because prohibitively large input resistances are needed.

Using this circuit, input resistances that are orders of magnitude greater than the values of the dc return resistors can be obtained. This is accomplished by bootstrapping the resistors to the output. With this arrangement, the lower cutoff frequency of a capacitive transducer is determined more by the RC product of R1 and C1 than by resistor values and the equivalent capacitance of the transducer.

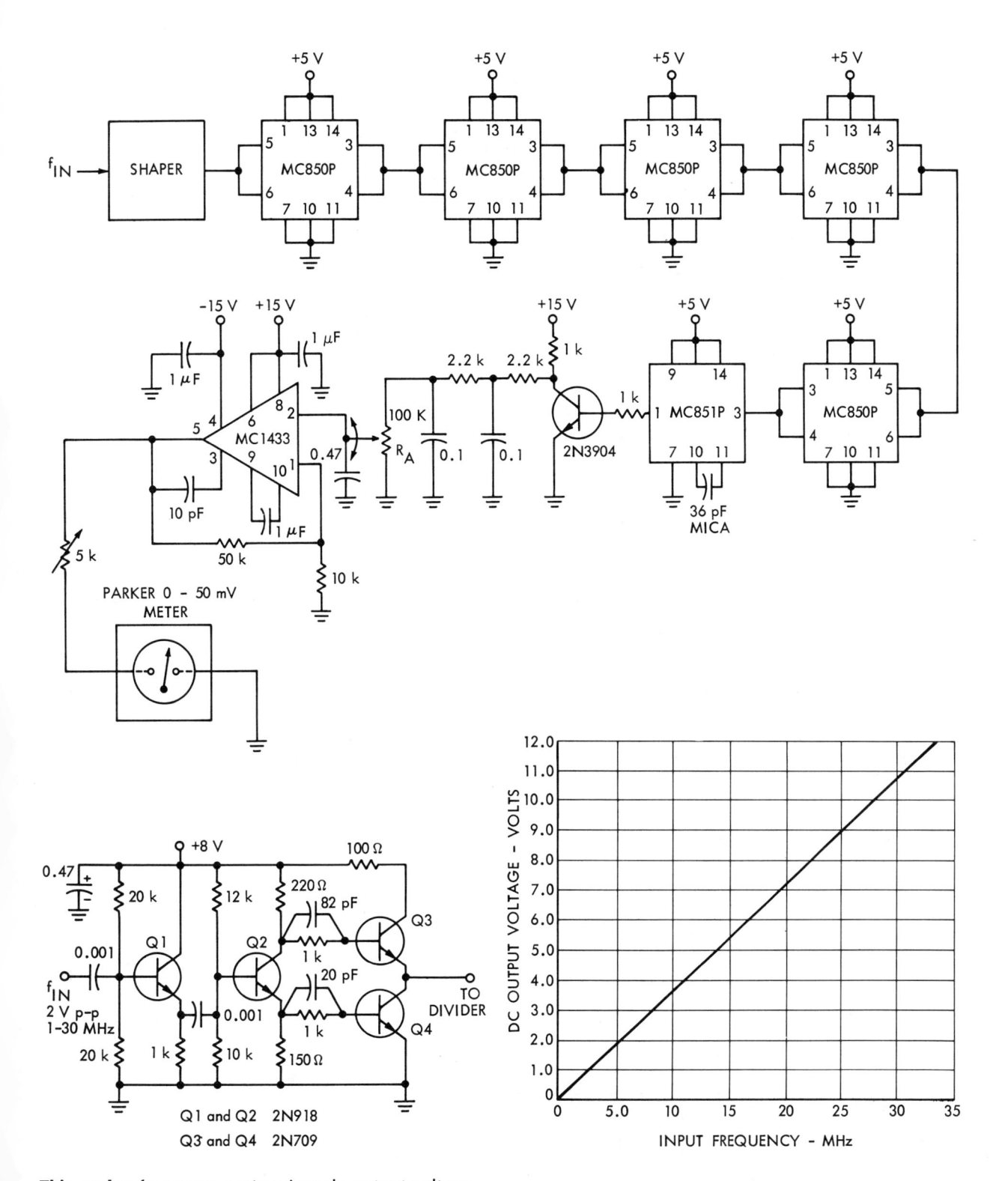

This analog-frequency meter gives dc output voltage proportionate to input frequency from 50 kHz to 40 MHz, with worst-case inaccuracy of 3.2% between 25 and 30 MHz.

SECTION **3** *Instrumentation and Test Circuits* 111

NC = Normally Closed
NO = Normally Open

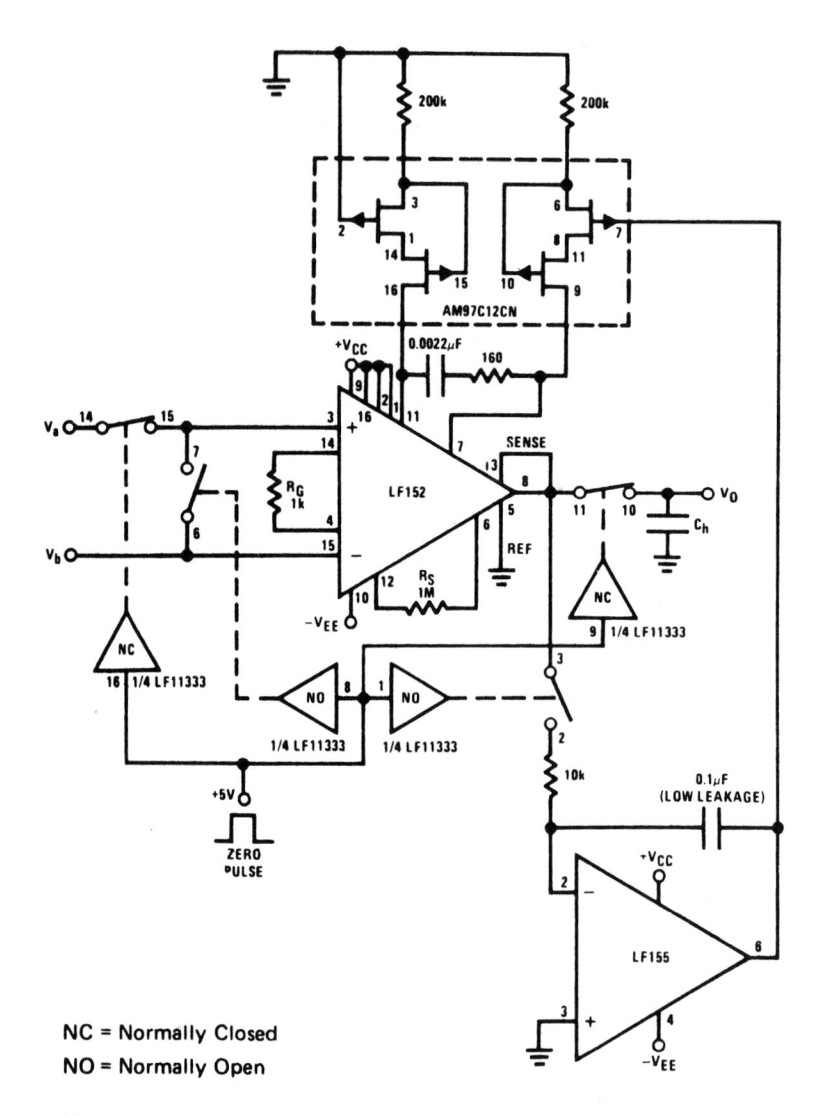

Minimum pulse width to drive V_O to zero is 400µs.

Automatic input-offset voltage adjust ($G \geq 100$).

Compensation Component Values

A_{VCL}	R1 (Ω)	C1 (µF)	R2 (Ω)	C2 (µF)
10,000	10K	50 pF	—	—
1,000	470	.001	—	—
100	47	.01	—	—
10	27	.05	270	.0015
1	10	.05	39	.02

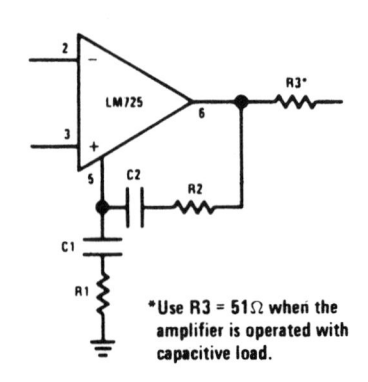

*Use R3 = 51Ω when the amplifier is operated with capacitive load.

Frequency-compensation circuit.

112 IC SCHEMATIC SOURCEMASTER

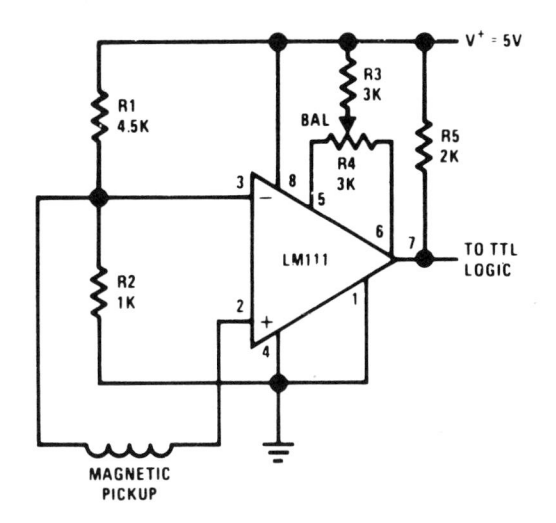

Circuit will **detect zero crossing** in the output of a magnetic transducer within a fraction of a millivolt. The magnetic pickup is connected between the two inputs of the comparator. The resistive divider, R1 and R2, biases the inputs 0.5V above ground, within the common-mode range of the IC. The output will directly drive DTL or TTL. The exact value of pullup resistor R5 is determined by the speed required from the circuit, since it must drive any capacitive loading for positive-going output signals.

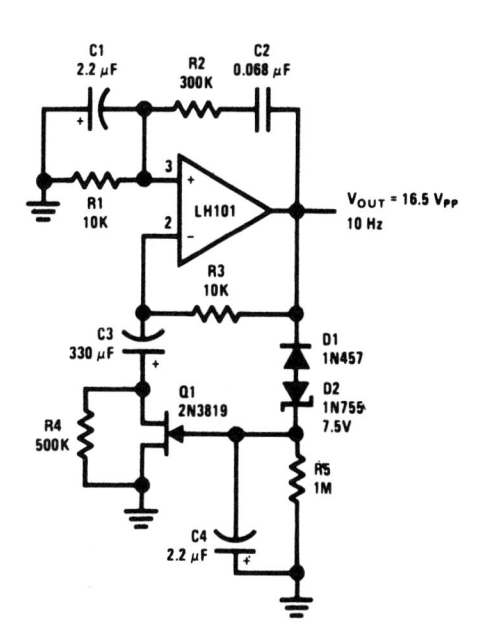

This amplitude-stabilized Wien bridge sine-wave oscillator provides high-purity sine-wave output down to low frequencies with minimum circuit complexity. R4 is chosen to adjust the negative feedback loop, so that the FET is operated at a small negative gate bias. The circuit shown provides optimum values for a general-purpose oscillator.

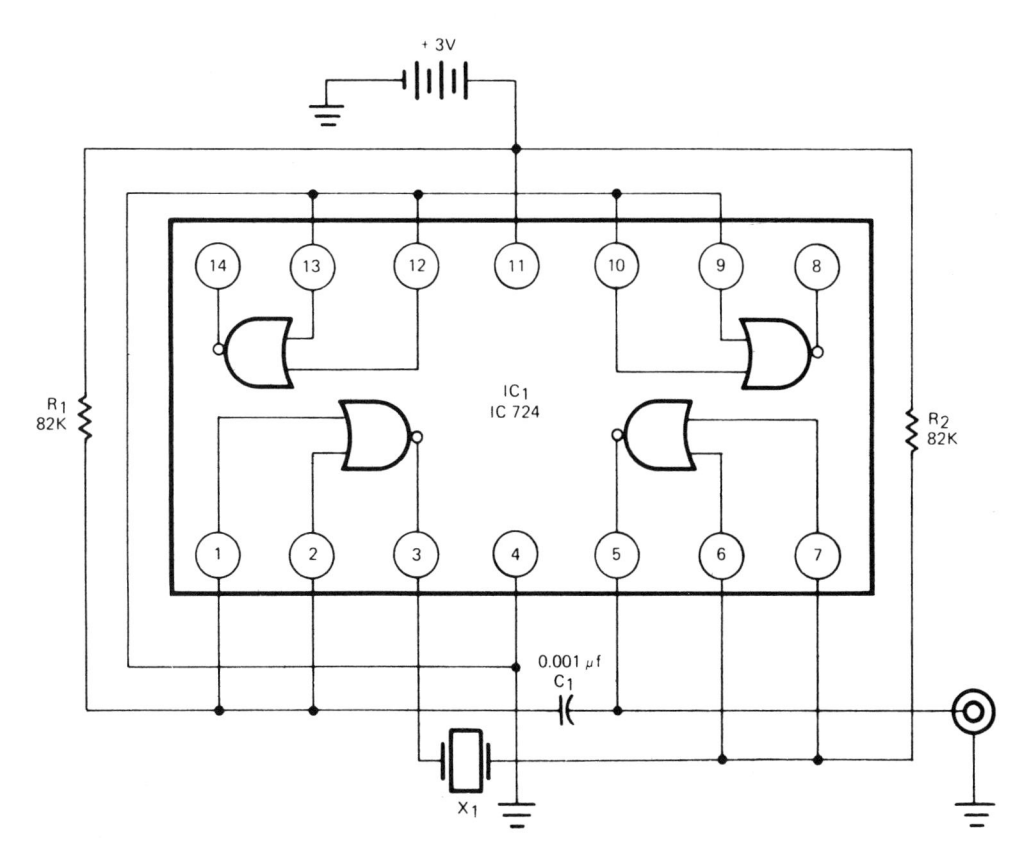

This crystal calibrator circuit offers a secondary frequency standard required in many projects and in radio and TV servicing (IRC/Workman integrated circuit).

SECTION **3** *Instrumentation and Test Circuits* **113**

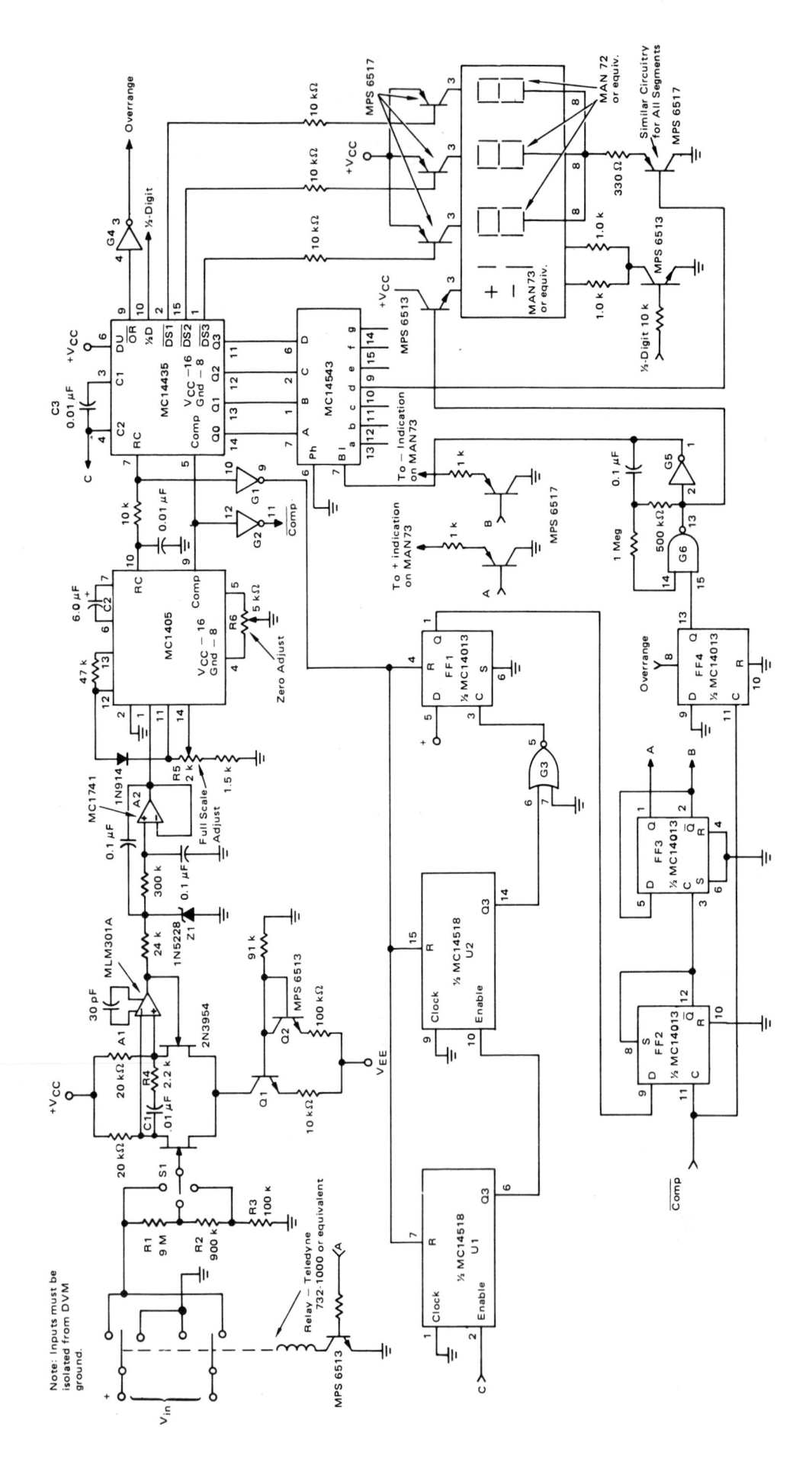

Complete digital voltmeter (3.5 digits) includes the digital readout, autopolarity circuit with indicator, high-impedance input, and overrange indicator. There are three input voltage ranges that allow the DVM to measure a dc voltage of 0–1.999V, 0–19.99V, and 0–199.9V.

114　IC SCHEMATIC SOURCEMASTER

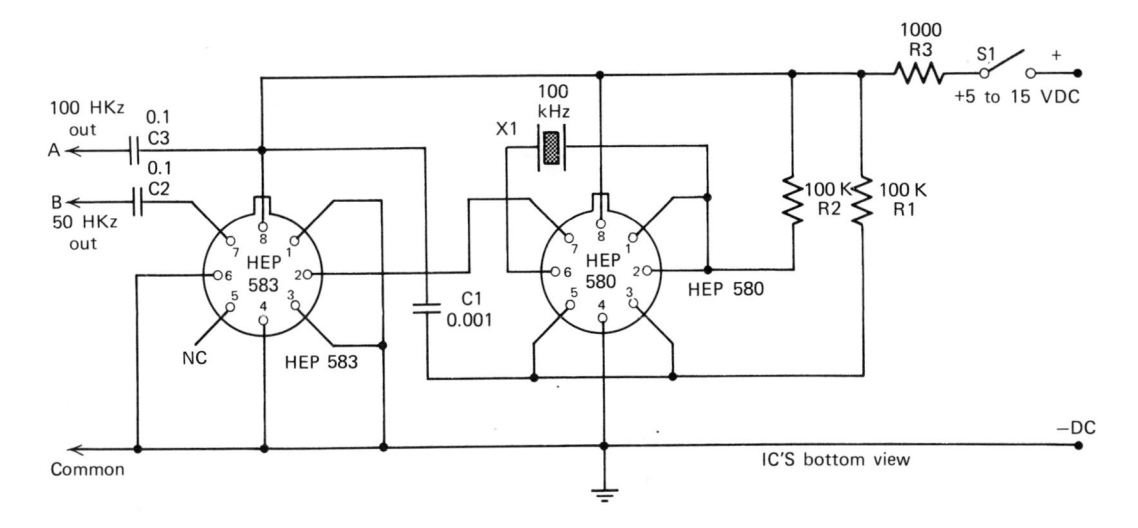

Crystal-controlled square-wave generator. This is an accurate frequency source of 100 or 50 kHz square-wave signals for test purposes or a marker for communications receivers. It operates from 5 to 15V. Drain is 5.5 mA max. The voltage output at "A" or "B" is 1.0V peak-to-peak at +9V power source.

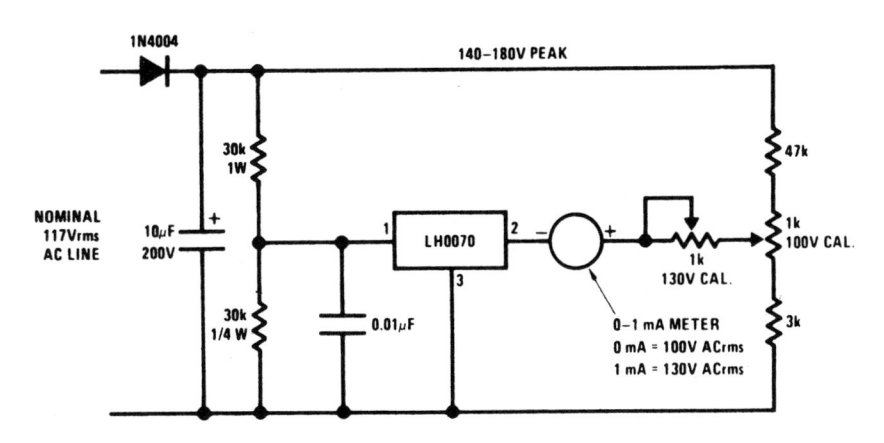

Expanded-scale ac voltmeter.

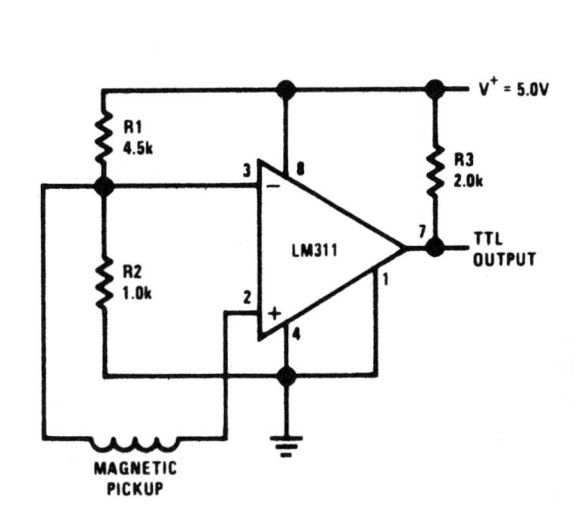

Detector for magnetic transducer.

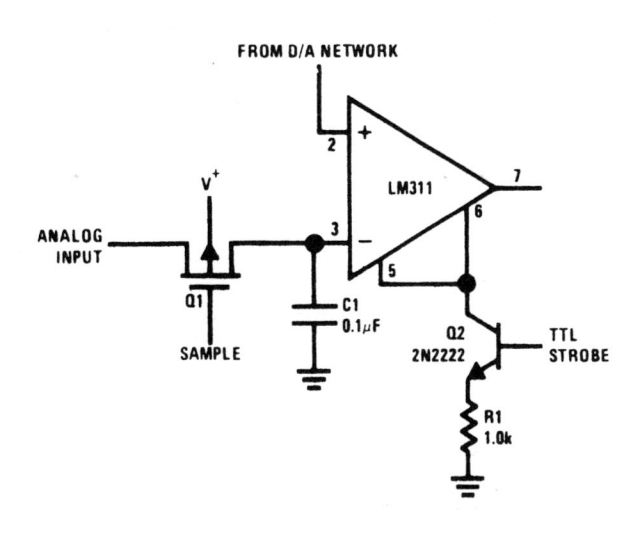

Strobing off both input and output stages. Typical input current is 50 pA with inputs strobed off.

SECTION **3** *Instrumentation and Test Circuits* 115

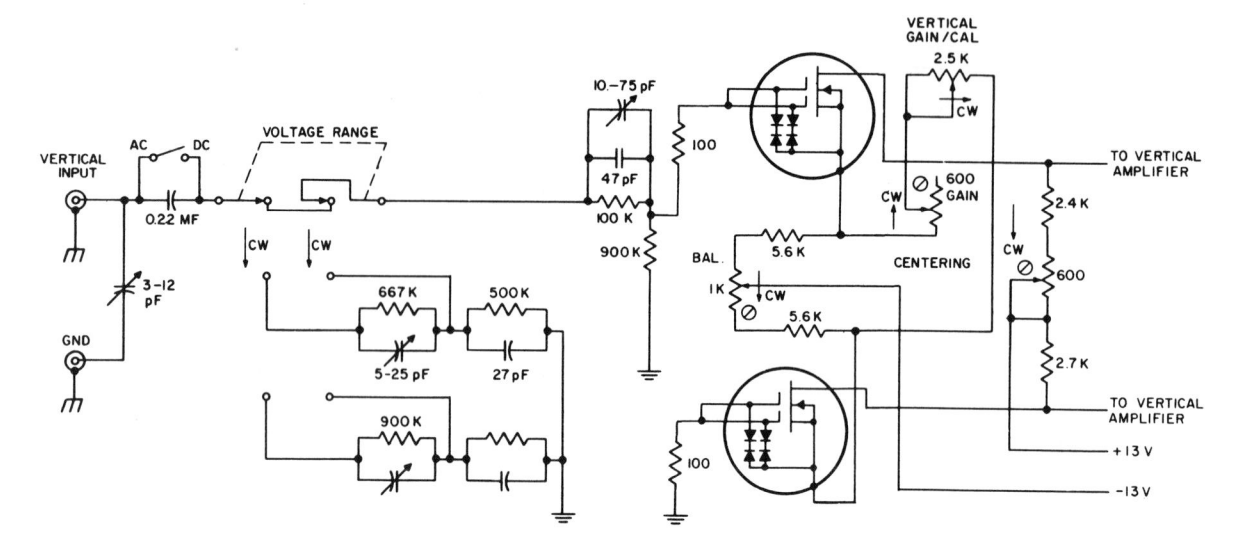

A differential amplifier using the 40841 IC in the vertical input stage of a solid-state oscilloscope.

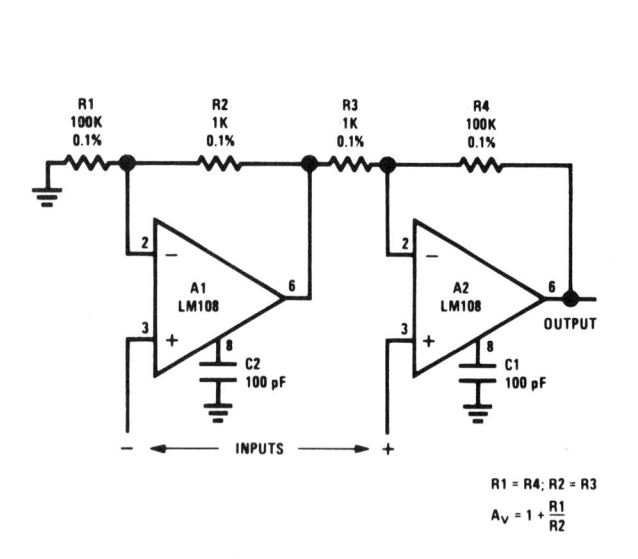

$R1 = R4; R2 = R3$
$A_V = 1 + \dfrac{R1}{R2}$

This differential input instrumentation amplifier has an input resistance greater than $10^{10}\Omega$, yet it does not need large resistors in the feedback circuitry. With the component values shown, A1 is connected as a noninverting amplifier with a gain of 1.01; it feeds into A2, which has an inverting gain of 100. Hence, the total gain from the input of A1 to the output of A2 is 101, which is equal to the noninverting gain of A2. If all the resistors are matched, the circuit responds only to the differential input signal—not to the common-mode voltage.

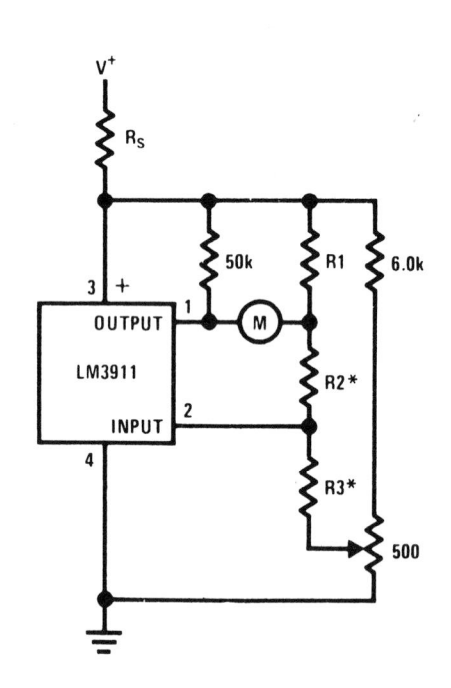

Meter thermometer with trimmed output. T_0 should be 5k more than desired and $I_0 = 100 \, \mu A$. Variable resistor calibrates T_0.

116 IC SCHEMATIC SOURCEMASTER

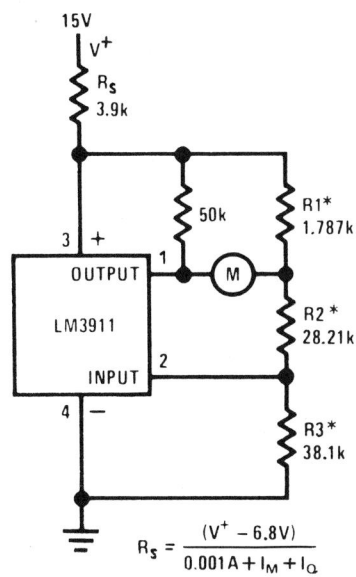

$$R1 = \frac{(V_Z)(10\,mV)(\Delta T)}{I_M(V_Z - 0.01\,T_O)}$$

$$R2 = \frac{0.01\,T_O - I_Q R1}{I_Q}$$

$$R3 = \frac{V_Z}{I_Q} - R1 - R2$$

$$\left(I_Q \leq \frac{2V}{R1}\right)$$

V_Z = Shunt regulator voltage (use 6.85)
ΔT = Meter temperature span (°K)
I_M = Meter full scale current (A)
T_O = Meter zero temperature (°K)
I_Q = Current through R1, R2, R3 at zero meter current (10µA to 1.0 mA) (A)

* Values shown for:
T_O = 300 K, ΔT = 100 K,
I_M = 1.0 mA, I_Q = 100µA

$$R_s = \frac{(V^+ - 6.8V)}{0.001A + I_M + I_Q}$$

Electronic thermometer. As with any temperature sensor, internal power dissipation will raise the sensor's temperature above ambient. The nominal suggested operating current for the shunt regulator is 1.0 mA and causes 7.0 mW of power dissipation. In free air, this raises the package temperature by about 1.2°K. Although the regulator will operate at higher reverse currents and the output will drive loads up 5.0 mA, these higher currents will raise the sensor temperature at about 19°K above ambient-degrading accuracy. Therefore, the sensor should be operated at the lowest possible power level.

Thermometer with meter output. Values shown for T_0 = 300k, I_M = 1.0 mA, I_0 = 100 µA.

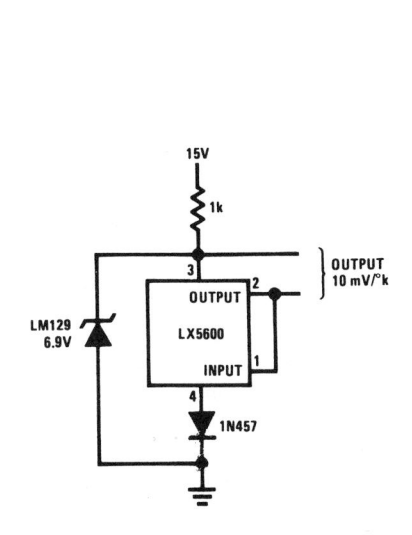

$$V_{OUT} = (10\,mV/°C)\left(\frac{R_1 + R_2}{R_1}\right)(T_2 - T_1)$$

External reference for temperature transducer.

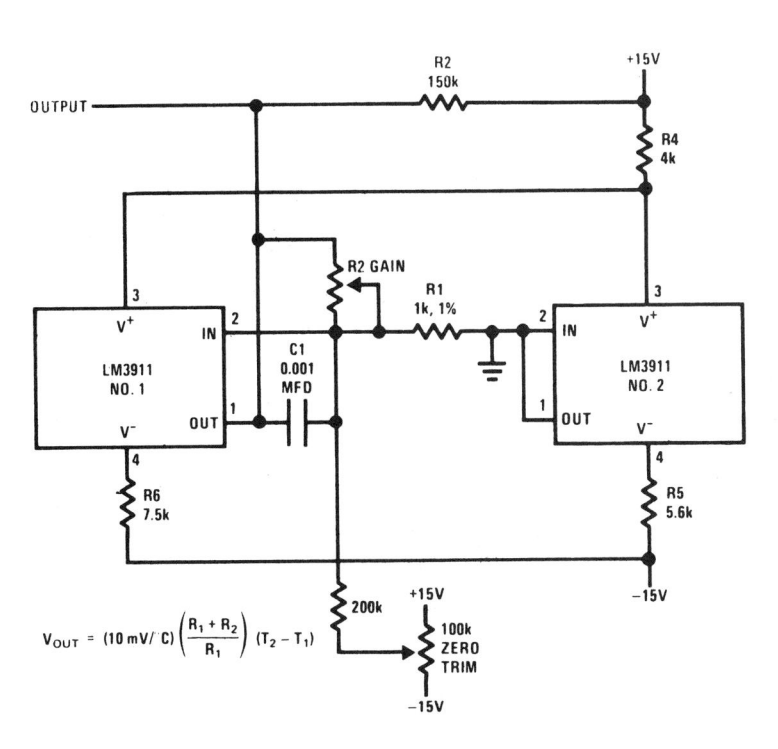

Differential thermometer. Output can swing +3V at +50 µA with low output impedance.

SECTION **3** *Instrumentation and Test Circuits* **117**

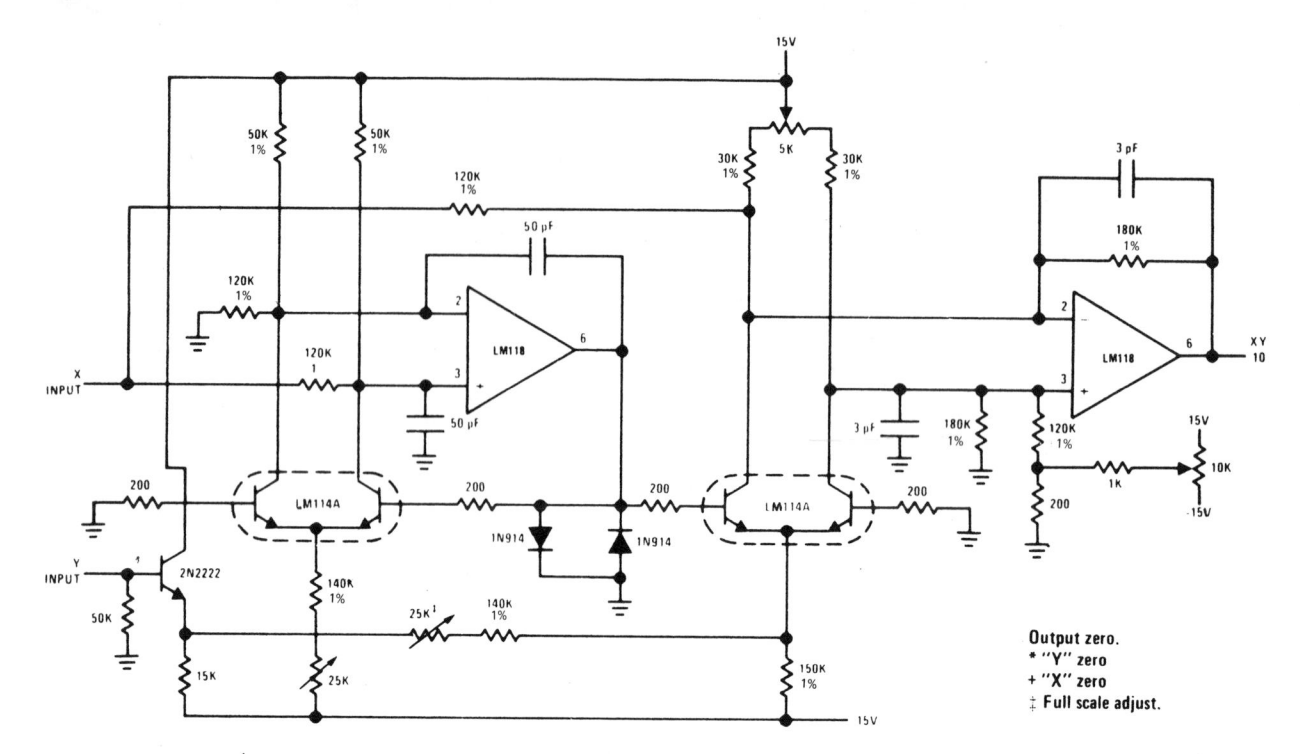

Four-quadrant multiplier.

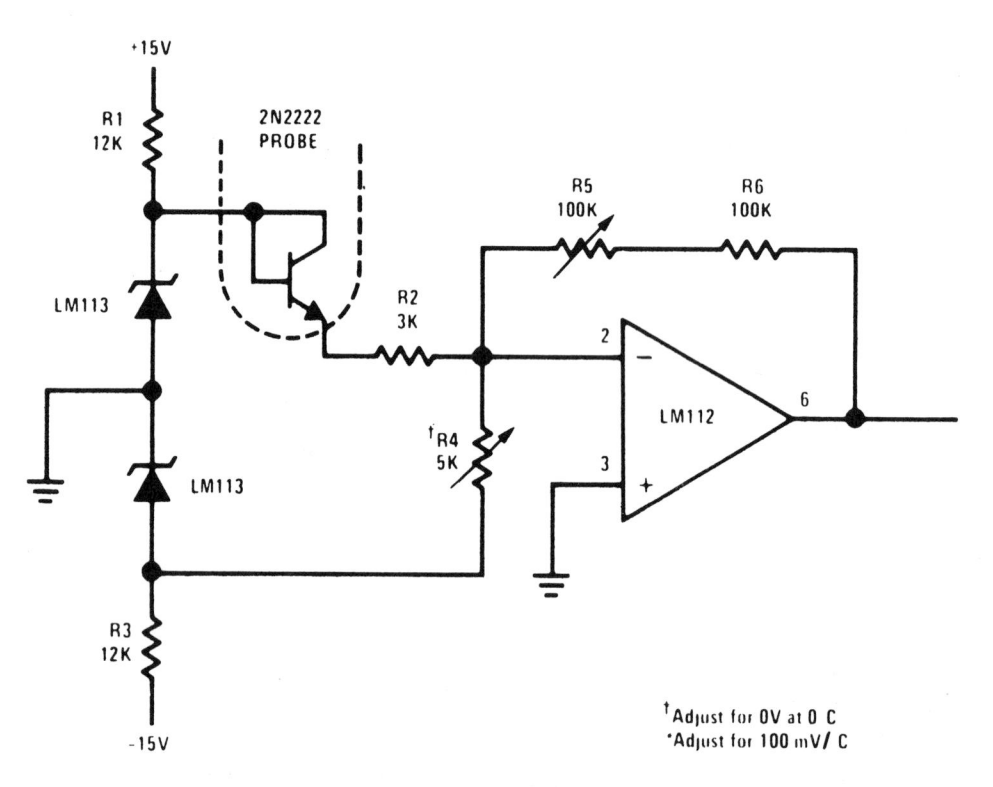

An electronic thermometer using an inexpensive silicon transistor as the temperature sensor. It can provide better than 1°C accuracy over a 100°C range. The emitter–base turn-on voltage of silicon transistors is linear with temperature; if the operating current of the sensing transistor is made proportional to absolute temperature, the nonlinearity of emitter–base voltage can be minimized. Over a −55 to 125°C temperature range, the nonlinearity is less than 2 mV or the equivalent of 1°C temperature change. Resistor R4 biases the output of the amplifier for zero output at 0°C. Feedback resistor R5 is then used to calibrate the output scale factor to 100 mV/°C. Once the output is zeroed, adjusting the scale factor does not change the zero.

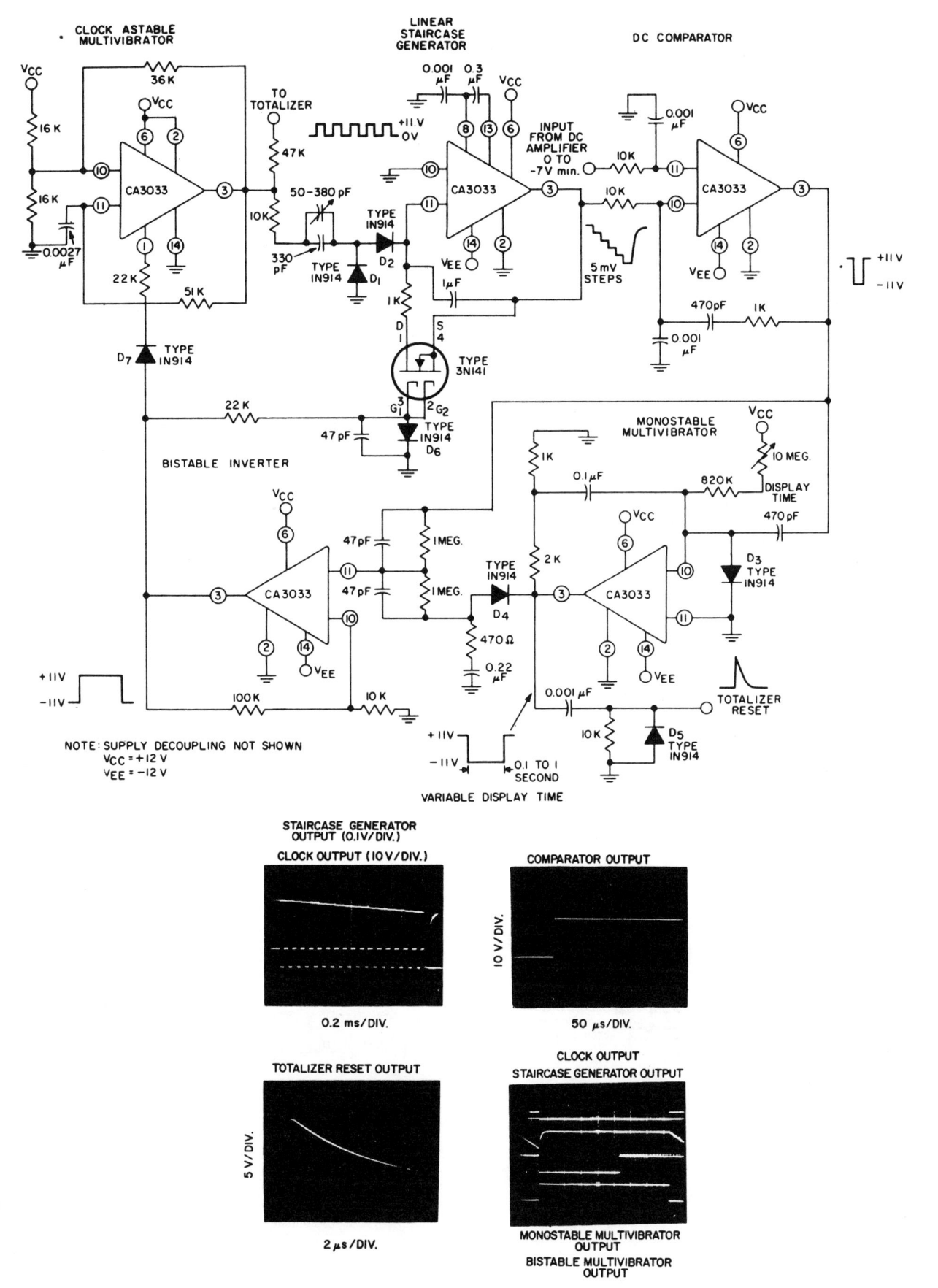

A digital voltmeter using CA3033 ICs. Regeneration is added around the comparator circuit in this system to accelerate the transition time when the two input voltages are equal. Waveforms are shown for critical points. The 470 pF capacitor and 100Ω resistor between terminals 3 and 10 of the CA3033 in the comparator circuit provide the regeneration. Two 0.001 μF capacitors on each input filter externally generated noise.

SECTION 3 *Instrumentation and Test Circuits* 119

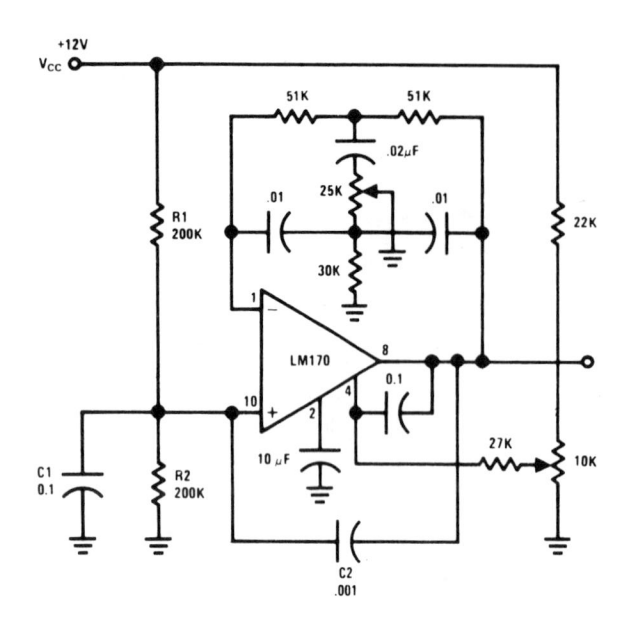

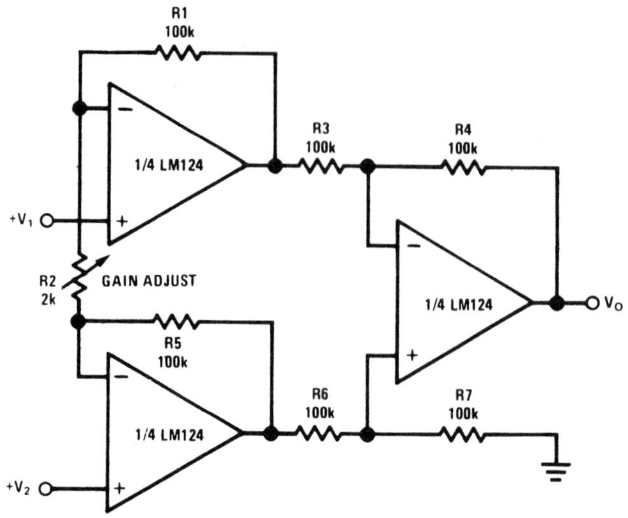

A decade sine-wave oscillator tunable over one decade in frequency. Positive feedback is applied to the noninverting input by the capacitive divider (C1, C2). Capacitor C1 also decouples the input from supply noise. Negative feedback occurs through the twin tee at all frequencies except the null frequency of the tee network, allowing the circuit to oscillate there. The nominal low frequency of oscillation for the circuit is approximately 320 Hz. At this frequency and 1V out, the total harmonic distortion is under 0.25%. At the upper frequency limit (3300 Hz), the output drops less than 1.5 dB and the distortion is 0.45%.

This high-input-impedance, adjustable-gain dc instrumentation amplifier uses single +5V supply. If R1 = R5 + R3 = R4 = R6 = R7 (CMRR depends on match),

$$V_o = .1 + 2R1/R2(V2V1)$$

As shown, $V_o = 101(V2V1)$.

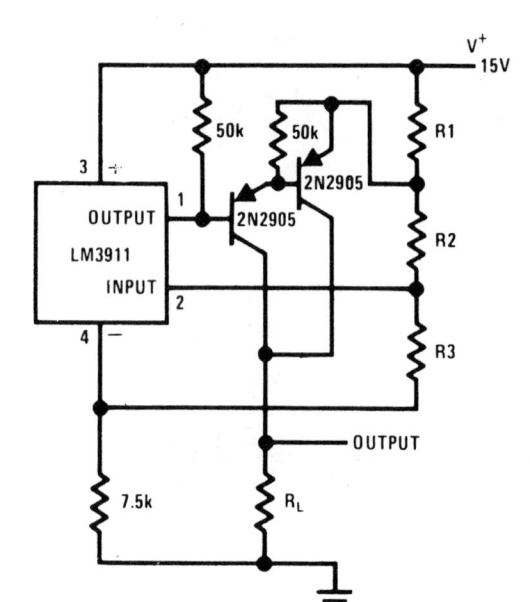

$$R1 = \frac{(V_z)(10\ mV)(\Delta T)}{\frac{V_O}{R_L}(V_z - 0.01\ T_O)}$$

$$R2 = \frac{0.01\ T_O - I_Q R1}{I_Q}$$

$$R3 = \frac{V_z}{I_Q} - R1 - R2$$

V_z = Shunt regulator voltage
ΔT = Temperature span (°K)
T_O = Temperature for zero output (°K)
V_O = Full scale output voltage $\leq$ 10V
I_Q = Current through R1, R2, R3 at zero output voltage (typically 100μA to 1.0 mA)

Ground-referred thermometer.

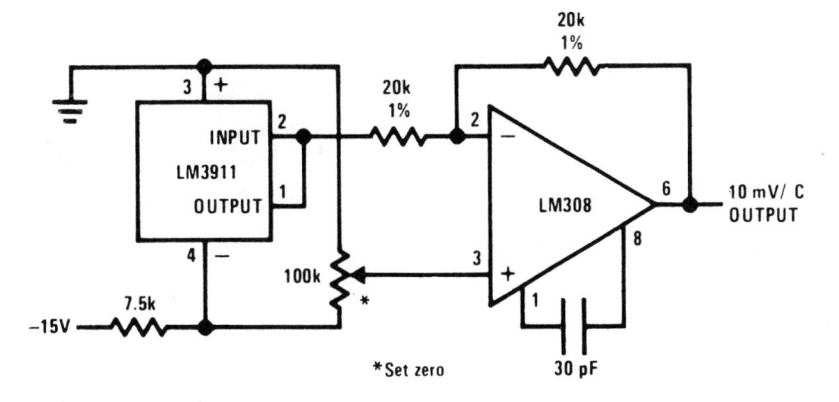

Ground-referred Celsius thermometer.

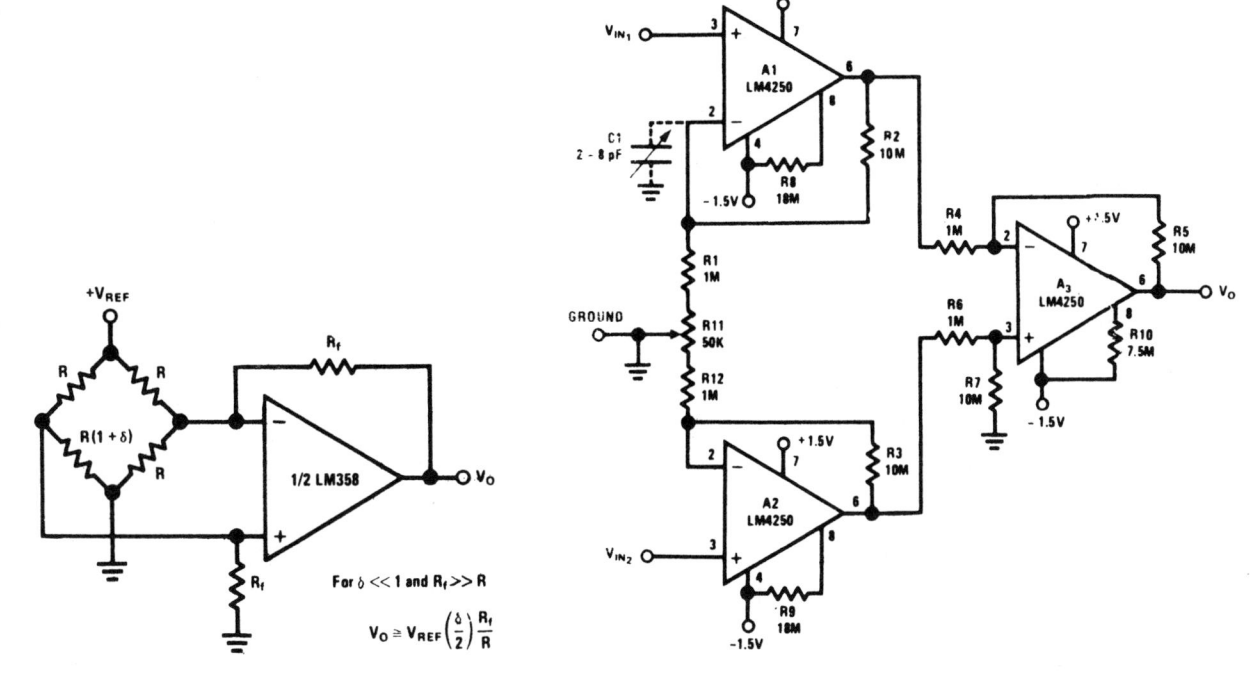

Bridge current amplifier. For $\delta \ll 1$ and $R_f \gg R$, $V_0 \cong V_{REF}(\delta/2)(R_f/R)$.

This X100 instrumentation amplifier has quiescent power dissipation of 10 μW. Use 1% resistors for R2–R7. Resistor R11 and capacitor C1 adjust CMRR.

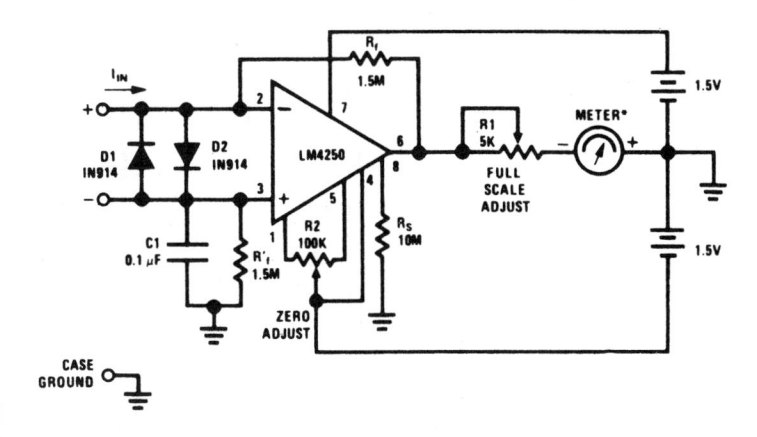

This floating-input meter amplifier gives sensitivity of 100 nA full scale. Meter movement (0–100 μA, 2k) marked for 0–100 nA full scale. Quiescent power dissipation is about 1.8 μW.

SECTION 3 *Instrumentation and Test Circuits*

LOW COST THERMOMETER
−40°C to +100°C
−40°F to +199°F

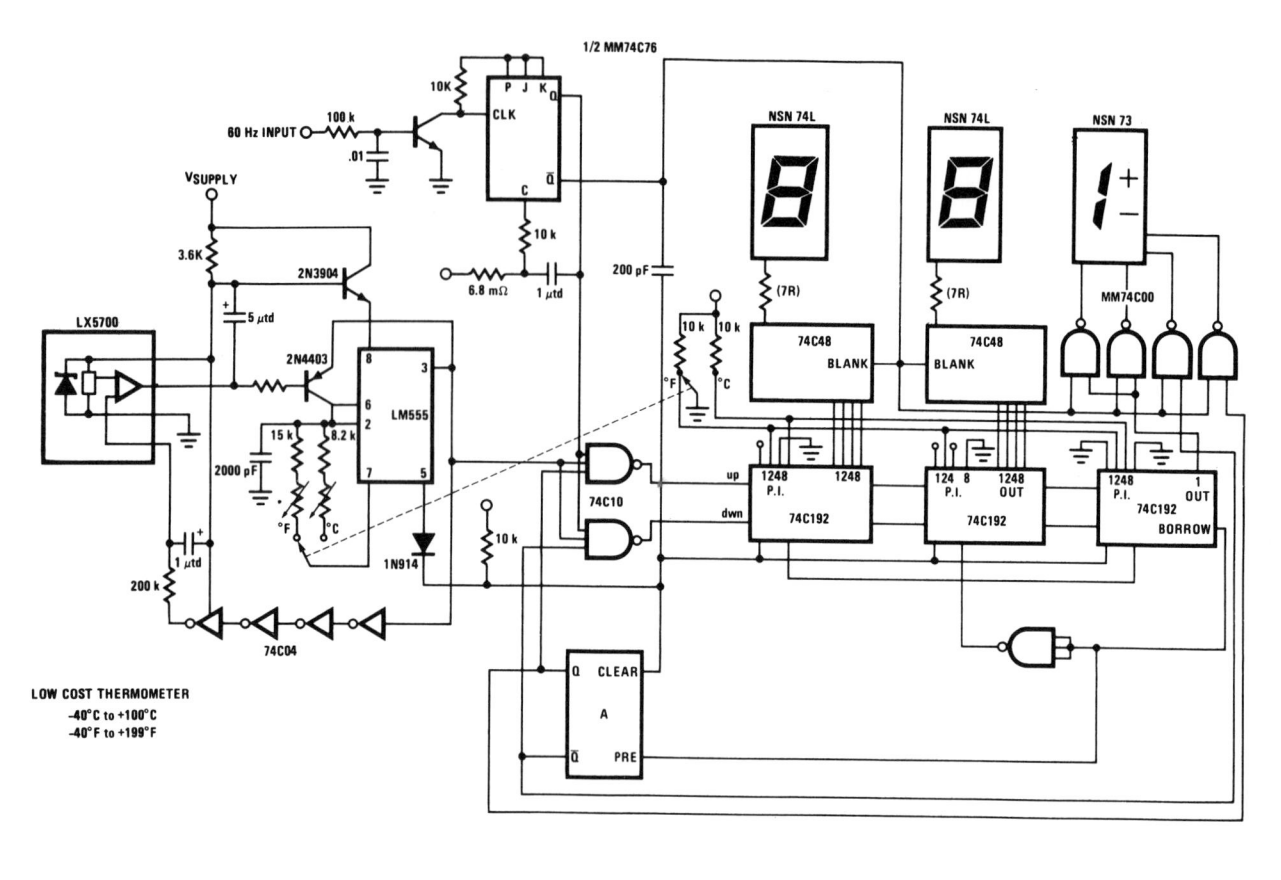

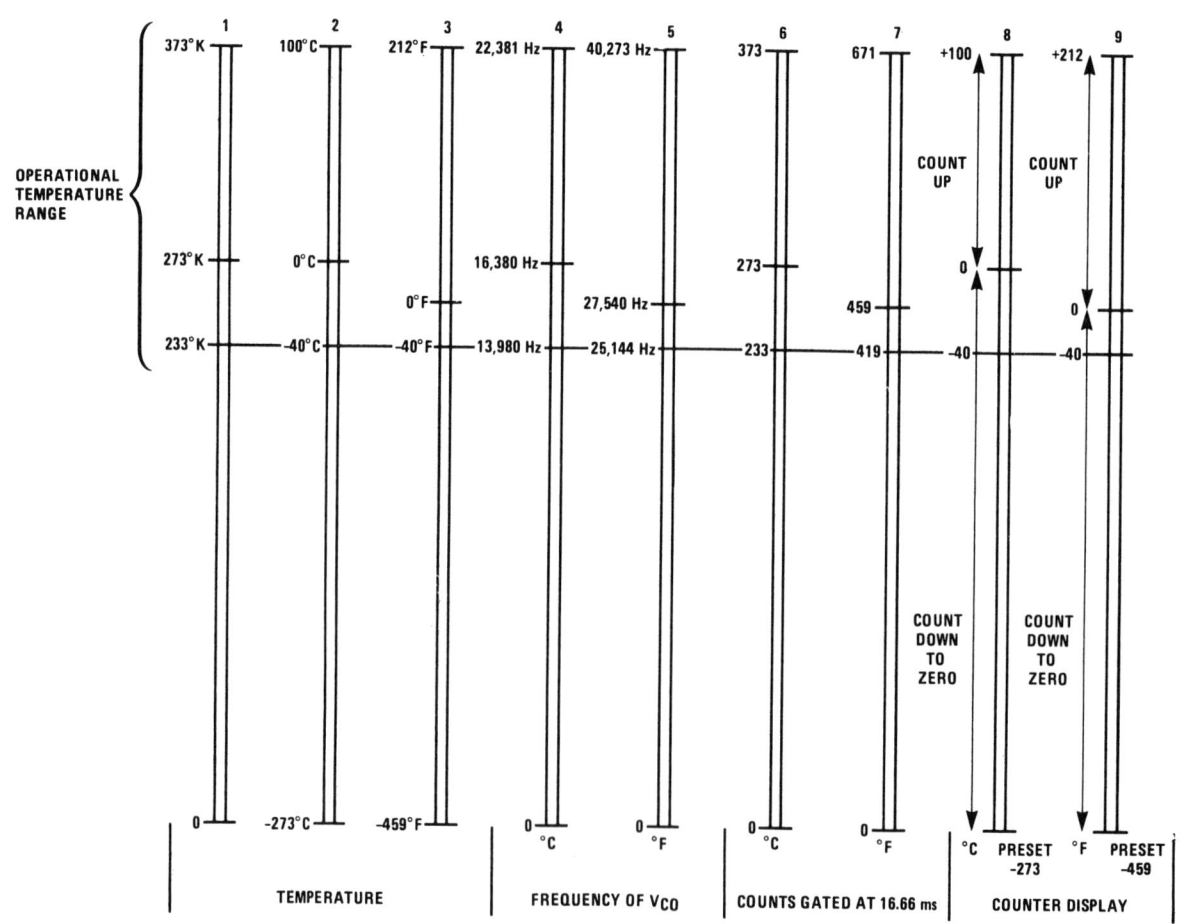

122 IC SCHEMATIC SOURCEMASTER

Fahrenheit/Celsius low-cost LED thermometer. Use of a "±1" digit and unusual code conversion allows a display of −40° to +100°C and −40° to +199°F. One supply is required; critical voltages in the analog subsection are referenced to the zener contained in the LM5700 sensor, making additional regulation unnecessary. To display in degrees Celsius or Fahrenheit, a digital code converter is used. The operation of this converter is represented by the bar graphs. The first three bars show the conversion between degrees Kelvin, Celsius, and Fahrenheit. Graphs 4 and 5 show the actual frequencies the LM555 is calibrated to produce at various temperatures. By gating these frequencies with a 16.66 msec pulse developed from the 60 Hz line, the counts that are fed to the counters are developed. These counts are depicted in graphs 6 and 7.

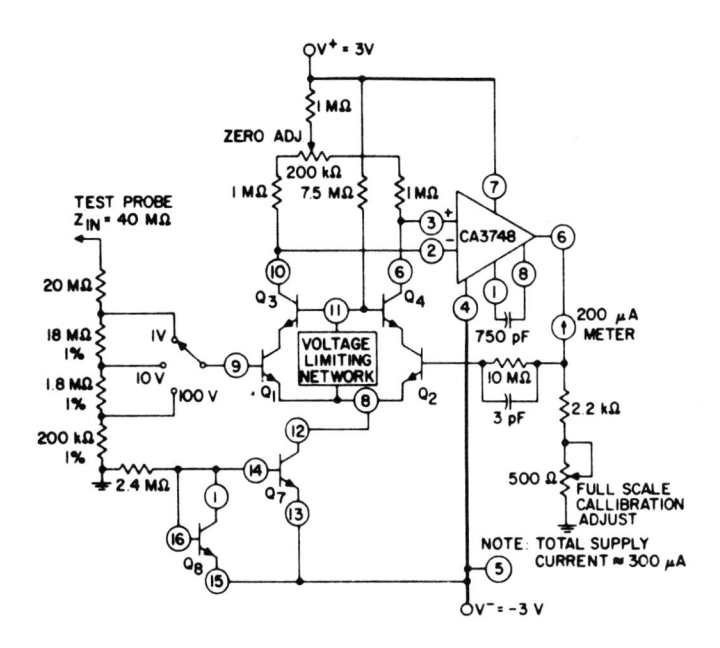

High-impedance voltmeter. Don't be scared off by the block marked "voltage-limiting network." It is just an internal diode-transistor-resistor network that assures that the applied collector-to-emitter voltage of the superbeta IC is maintained below 2V. All you have to do is supply the bias current for this internal network by simply connecting a resistor from pin 11 to the positive supply of the differential amplifier. The current supplied at pin 11 should be about half the value of quiescent current at pin 8.

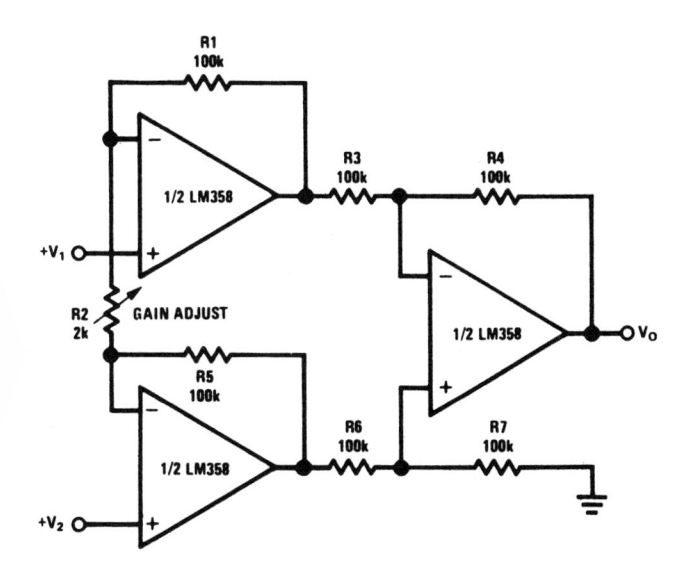

High-input-impedance adjustable-gain dc instrumentation amplifier. If R1 = R5 and R3 = R4 = R6 = R7 (CMRR depends on match),

$$V_o = 1 + \frac{2R1}{R2}(V_2 - V_1)$$

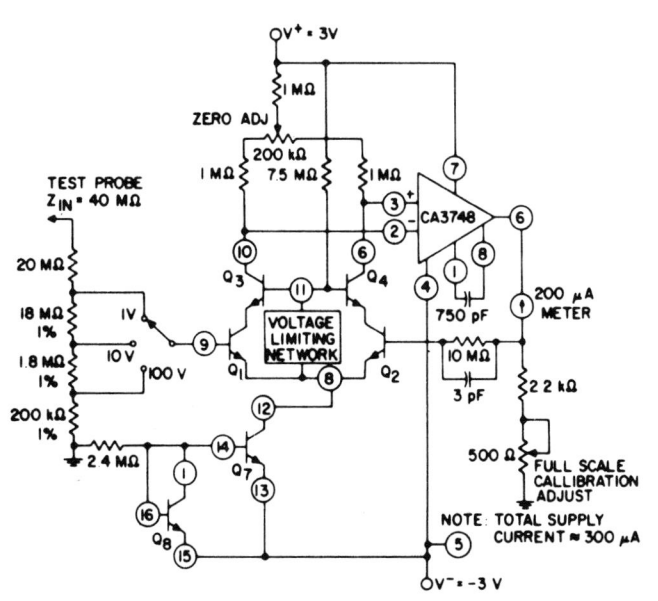

As shown, $V_o = 101(V_2 - V_1)$.

High-input-impedance dc voltmeter circuit. Voltage-limiting network is same as in superbeta differential cascode amplifier and requires an external bias-current supply.

SECTION 3 *Instrumentation and Test Circuits*

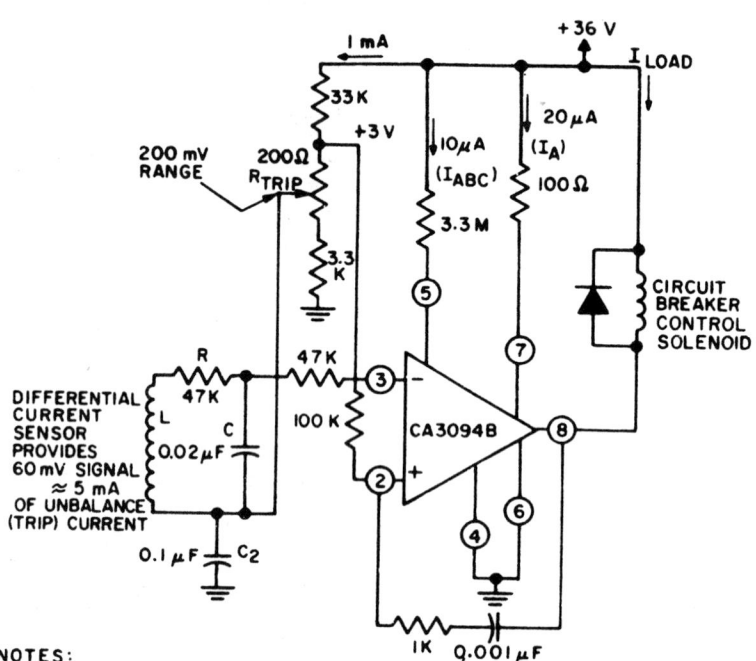

NOTES:
1. ALL RESISTORS IN OHMS, 1/2 WATT, ±10%
2. RC SELECTED FOR 3db POINT AT 200 Hz
3. C_2 = AC BY-PASS
4. OFFSET ADJ. INCLUDED IN R_{TRIP}
5. INPUT IMPEDANCE FROM 2 TO 3 EQUALS 800 K.
6. WITH NO INPUT SIGNAL TERMINAL 8 (OUTPUT) AT +36 VOLTS

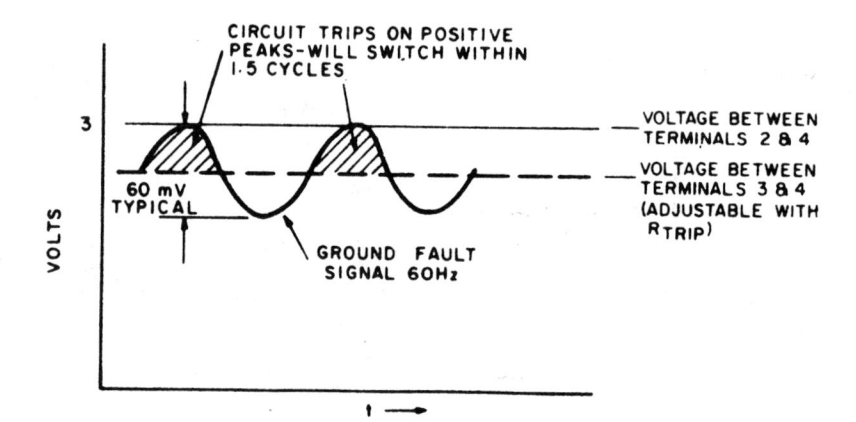

A ground fault interrupter (GFI) and waveform pertinent to ground fault detector. Circuit can be operated from a single supply, has adequate sensitivity, and can drive a relay or thyristor directly to effect power interruption. Vernier adjustment of the trip point is made by the R_{TRIP} potentiometer. When the differential current sensor supplies a signal that exceeds the selected trip-point voltage level (for example, 60 mV), the CA3094 is toggled on, and terminal 8 goes low to energize the circuit breaker trip coil. Under quiescent conditions, the entire circuit consumes approximately 1 mA. Resistor R, connected to one leg of the current sensor, provides current limiting to protect the CA3094 against voltage spikes as large as 100V.

124   IC SCHEMATIC SOURCEMASTER

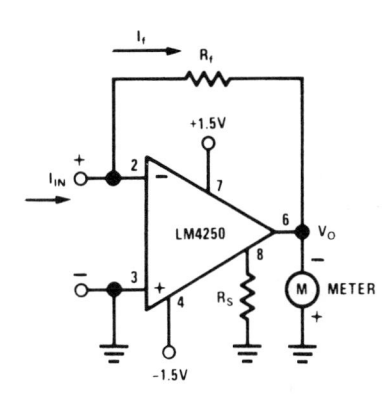

This IC meter amplifier runs on flashlight batteries. Meter amplifiers normally require one or two 9V transistor batteries. Because of the heavy current drain on these supplies, the meters must be switched off when not in use. This meter circuit operates on two 1.5V flashlight batteries and has a quiescent power drain so low that no on-off switch is needed. A pair of D cells will serve for 1 year without replacement. Circuit will provide current ranges as low as 100 nA full-scale. The circuit shown is a current-to-voltage converter. Negative feedback around the amplifier ensures that currents I_{IN} and I_f are always equal, and the high gain of the op-amp ensures that the input voltage between pins 2 and 3 is in the microvolt region. Output voltage V_o is therefore equal to $-I_f R_f$. Considering the ±1.5V sources (±1.2V end-of-life), a practical value of V_o for full-scale meter deflection is 300 mV. With the master bias-current setting resistor (R_S) set at 10M, the total quiescent current drain of the circuit is 0.6 µA for a total power supply drain of 1.8 µW. The input bias current required is in the range of 600 pA.

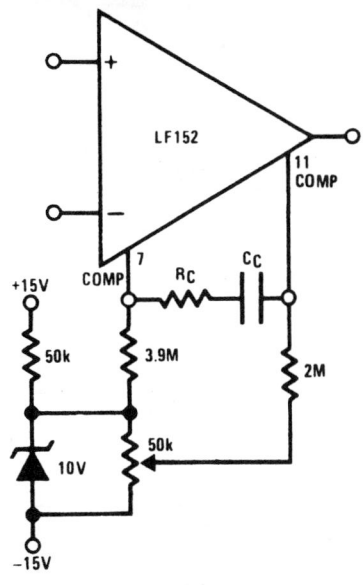

Simple input offset adjust. The output offset is affected by adjustment of the input offset. For every millivolt of input offset adjust, the output offset will change by approximately 32 mV. Adjustment of the output offset has no effect on the input offset, so it should always be done last. Offset adjustment changes the temperature coefficient of the V_{OS} drift. The typical input offset drift of the unadjusted device is -10 µV/°C. If the input offset is adjusted, the V_{IOS} drift increases by approximately V_{IOS} drift ≈ -10 µV/°C + 2 µV/°C/mV of adjustment. The V_{OOS} drift will be improved by output offset adjust because the magnitudes of the current sources adjusted become less sensitive to V_{BE} variations. If V_{OOS} adjust is not used, pins 1, 2, and 16 must be shorted to the positive supply for circuit operation.

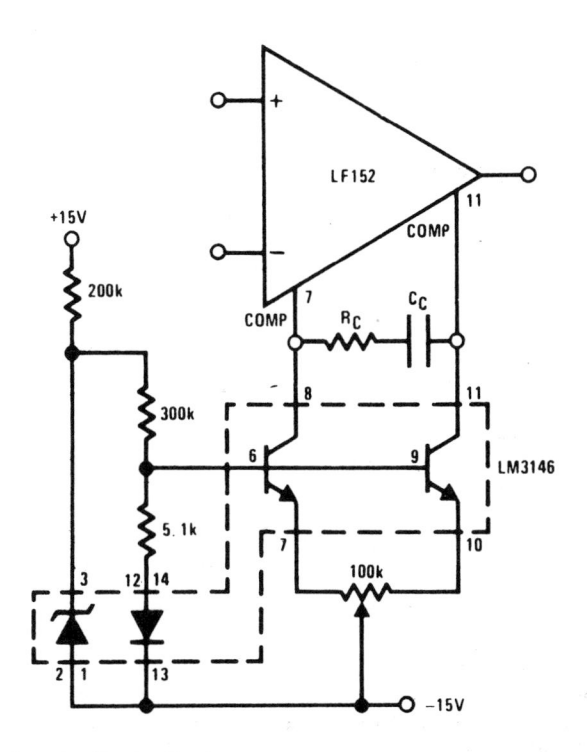

Input offset adjust.

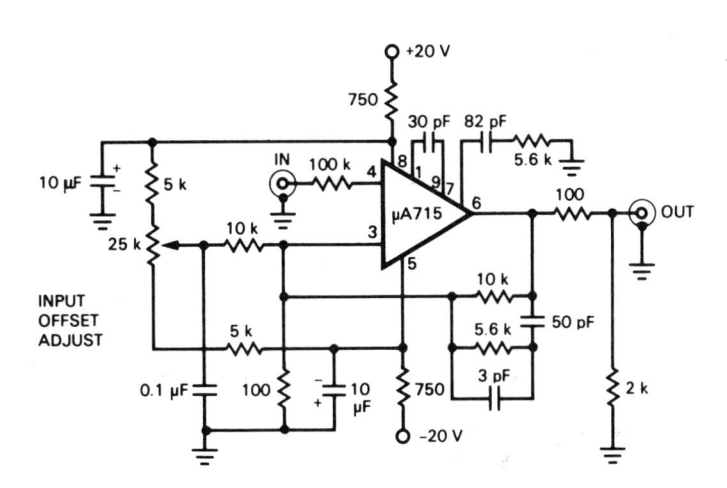

40 dB noninverting high-input-impedance biomedical amplifier. This configuration has a very high input impedance at dc, but it falls with increase in frequency as a result of the input capacitance of the µA715. This restricts the available bandwidth to about 500 kHz with a 100k source resistance. An input offset control is required because of the presence of input offset current flowing through the input resistance, causing an input offset voltage. Thus the higher the input resistance, the higher the offset.

SECTION **3** *Instrumentation and Test Circuits* **125**

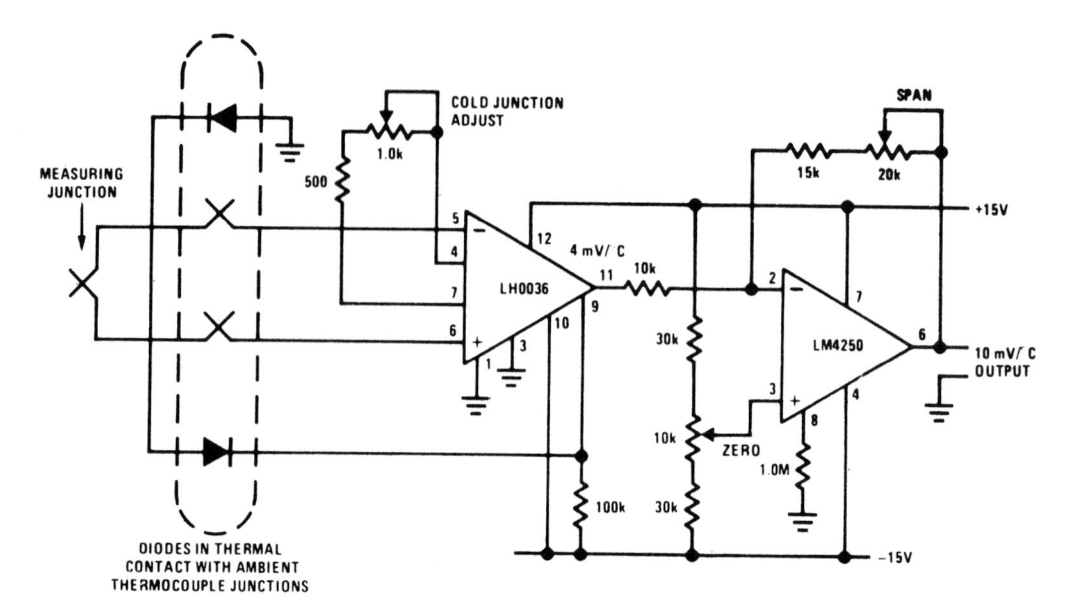

Thermocouple amplifier with cold junction compensation.

Instrumentation circuits. The National integrated circuit used in the schematics on this page is a true micropower instrumentation amplifier designed for precision differential signal processing. The input impedance is 300M and the CMRR is 100 dB. Gain is programmable with a single external resistor from ×1 to ×1000. Power consumption is on the order of 90 µW. Supply voltage can vary over a wide range.

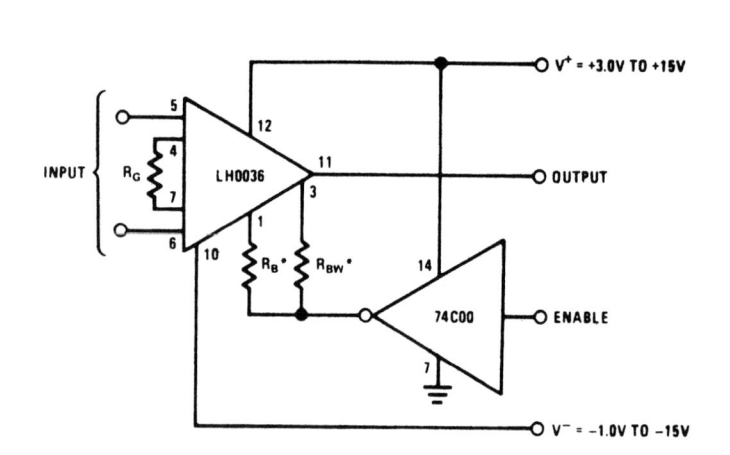

Premultiplex signal conditioning. (All resistors are 0.1%.)

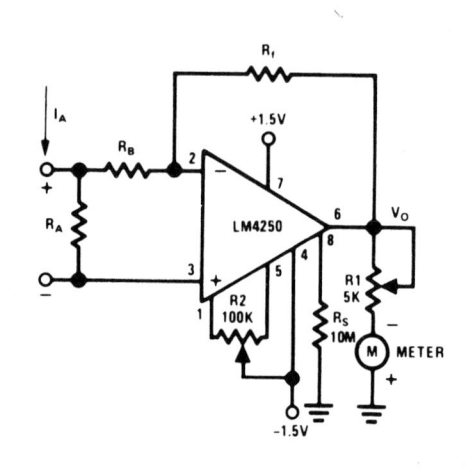

Resistance Values

I FULL SCALE	R_A [Ω]	R_B [Ω]	R_f [Ω]
1 mA	3.0	3k	300k
10 mA	.3	3k	300k
100 mA	.3	30k	300k
1A	.03	30k	300k
10A	.03	30k	30k

A microammeter for dc readings higher than 100 µA. Resistor R_A develops a voltage drop in response to input current I_A. This voltage is amplified by a factor equal to the ratio of $R_f:R_B$. R_B must be sufficiently larger than R_A so as not to load the input signal.

126 IC SCHEMATIC SOURCEMASTER

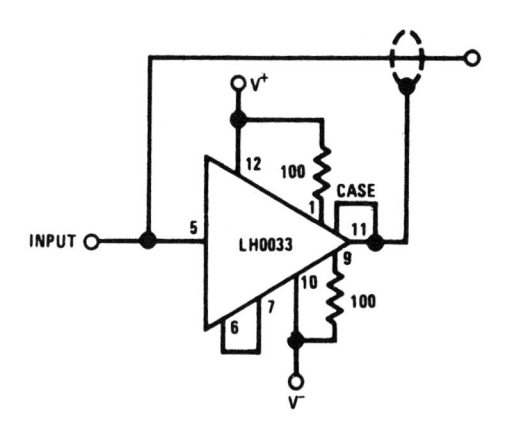

Instrumentation shield/line driver.

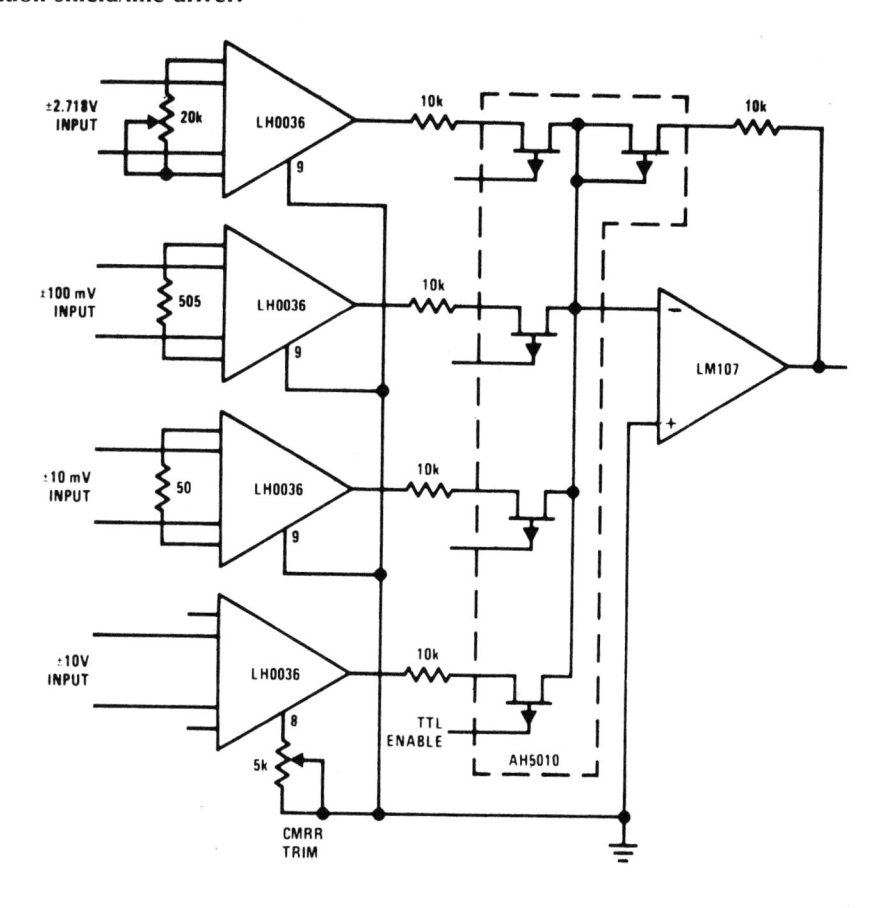

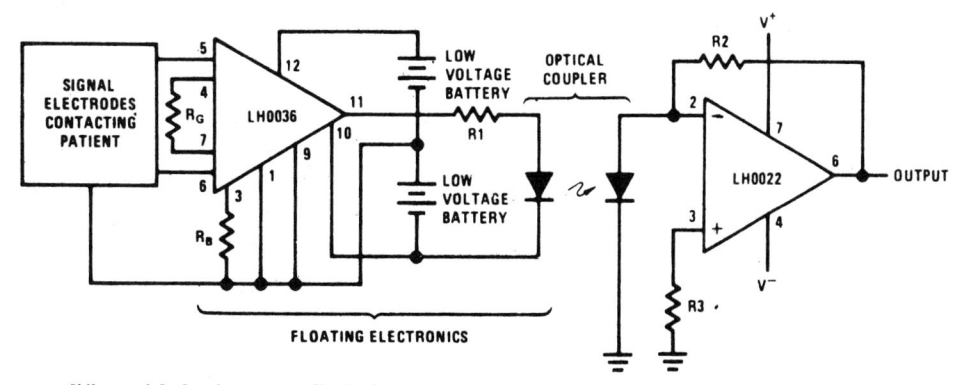

Instrumentation amplifier with logic-controlled shutdown. R_{BW} and R_B are optional bandwidth and input bias current controlling resistors.

SECTION **3** *Instrumentation and Test Circuits* **127**

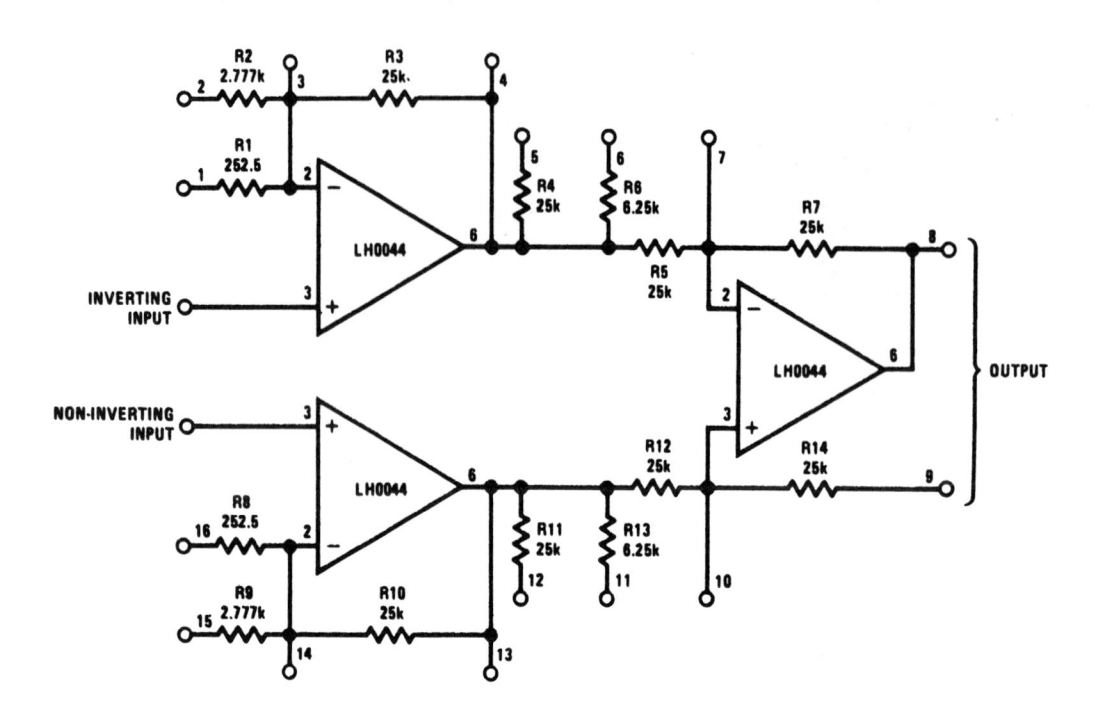

OVERALL GAIN	INPUT STAGE GAIN	OUTPUT STAGE GAIN	JUMPER PINS ON RA201
X1	X1	X1	
X2	X1	X2	5 to 7, 12 to 10
X5	X1	X5	6 to 7, 11 to 10
X10	X10	X1	2 to 15
X20	X10	X2	2 to 15, 5 to 7, 12 to 10
X50	X10	X5	2 to 15, 6 to 7, 11 to 10
X100	X100	X1	1 to 16
X200	X100	X2	1 to 16, 5 to 7, 12 to 10
X500	X100	X5	1 to 16, 6 to 7, 11 to 10
X995	X199	X5	1 to 14, 6 to 7, 11 to 10

Precision instrumentation amplifier. All resistors are part of National's RA201 resistor array.

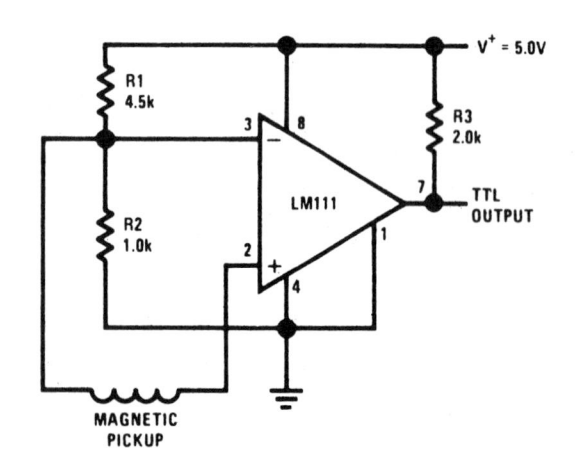

Detector for magnetic transducer.

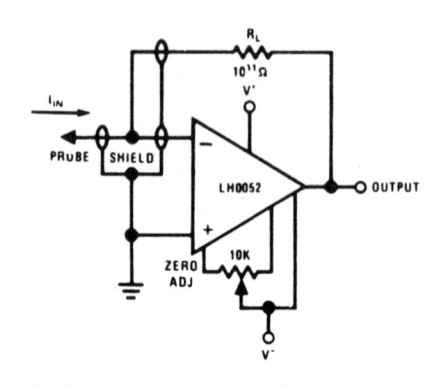

A picoamp amplifier for pH meters and radiation detectors; $e_{out} = I_{IN}(0.1\ V_{OL}{}^{-1}/pA)$.

128

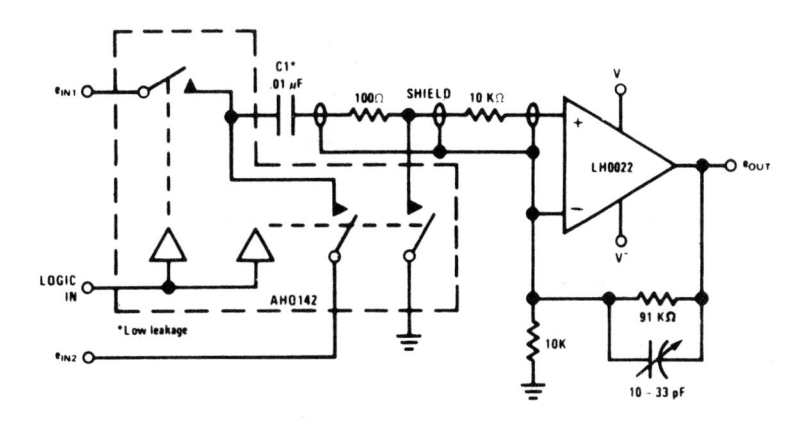

A precision subtractor for automatic test gear. C1 should be a low-leakage capacitor; $e_{OUT} = 10(e_{IN1} - e_{IN2})$.

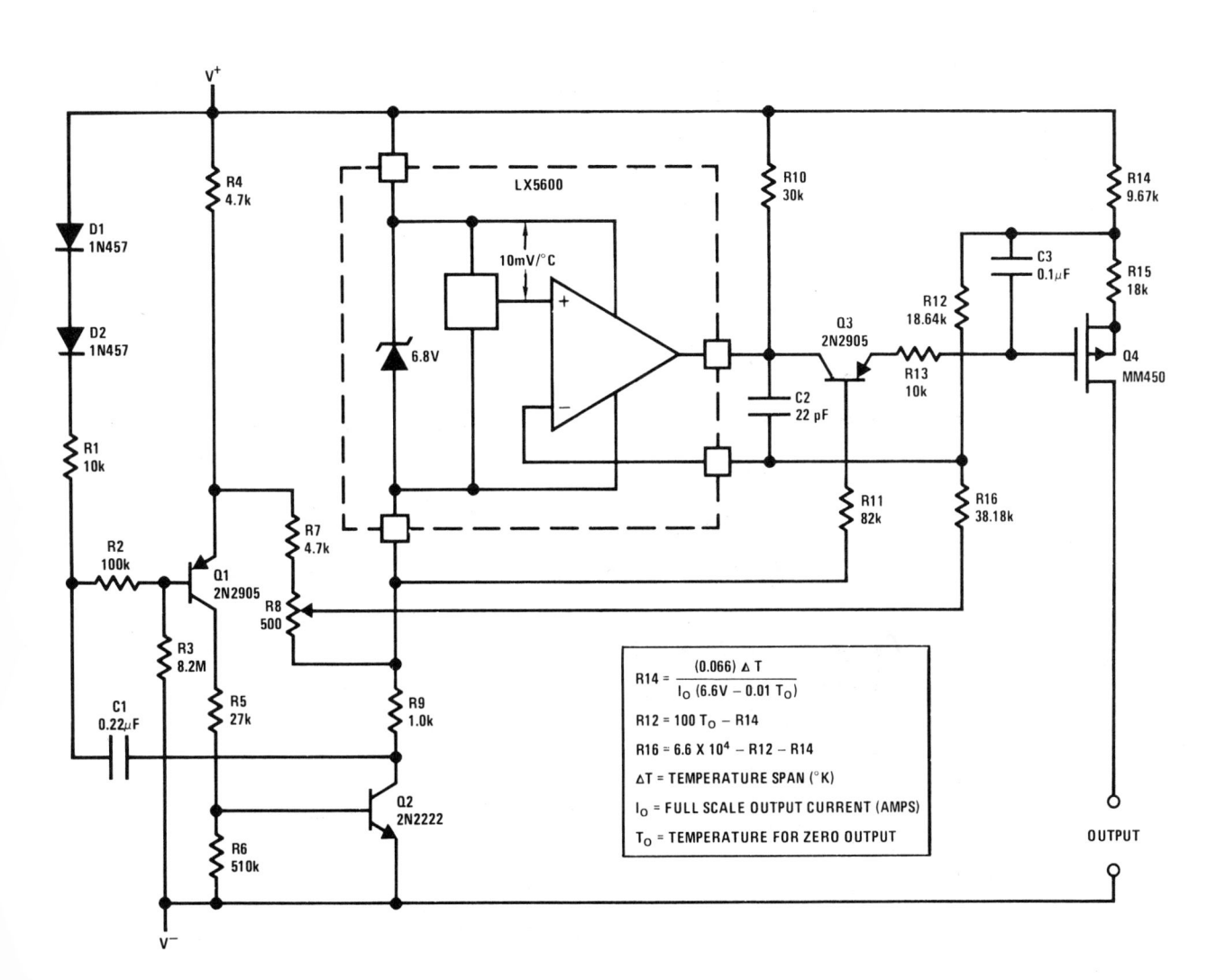

Micropower thermometer. Output is a current proportional to temperature, which can be used to drive a meter for a direct readout. Alternatively, a resistor or op-amp can be used to obtain voltage output. With the components shown, the duty cycle is about 0.2% with a 1-sec sample rate, which gives an average current drain of about 25 µA plus the output current. Designed to operate over a supply voltage of 8.0 to 12V with good results. A small 8.4V mercury battery can give an operational life in excess of 1 year.

$$R14 = \frac{(0.066) \Delta T}{I_O (6.6V - 0.01 T_O)}$$

$$R12 = 100 T_O - R14$$

$$R16 = 6.6 \times 10^4 - R12 - R14$$

ΔT = TEMPERATURE SPAN (°K)
I_O = FULL SCALE OUTPUT CURRENT (AMPS)
T_O = TEMPERATURE FOR ZERO OUTPUT

SECTION 3 *Instrumentation and Test Circuits*

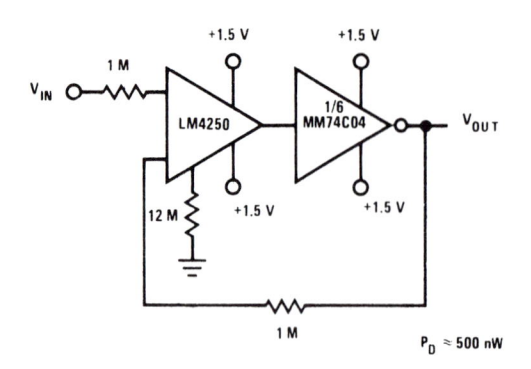

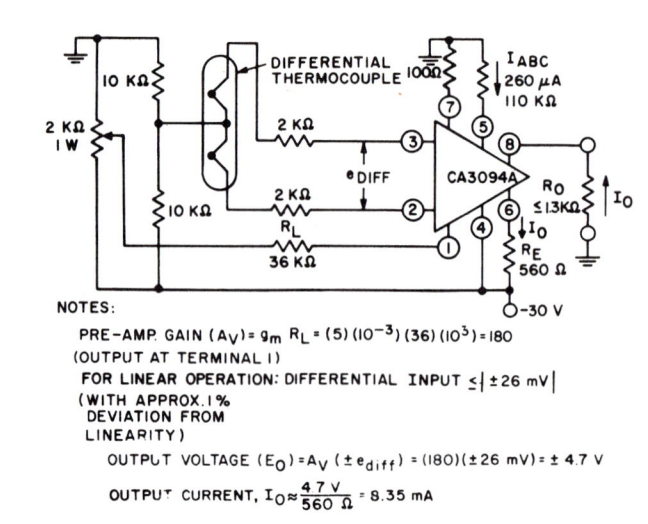

Postamplifier for op-amps. A standard operational amplifier used with a CMOS inverter for a postamplifier has several advantages: (1) the operational amplifier essentially sees no-load condition, since the input impedance to the inverter is very high; (2) the CMOS inverters will swing to within millivolts of either supply. This gives the designer the advantage of operating the operational amplifier under no-load conditions, yet having the full supply swing capability on the output. Shown is the LM4250 micropower op-amp used with a 74C04 inverter for increased output capability while maintaining the low-power advantage of both devices.

An open-loop instrumentation amplifier with differential input and single-ended output. In this circuit an I_{ABC} of 260 µA results in a g_m of 5 mmho. The operating point of the output stage is controlled by the 2k potentiometer. With no differential input signal it is adjusted to obtain a quiescent output current (I_o) of 12 mA. This output current is established by the 560 Ω emitter resistor, R_E, as follows:

$$I_0 \approx \frac{(g_m R_L)(e_{diff})}{R_E}$$

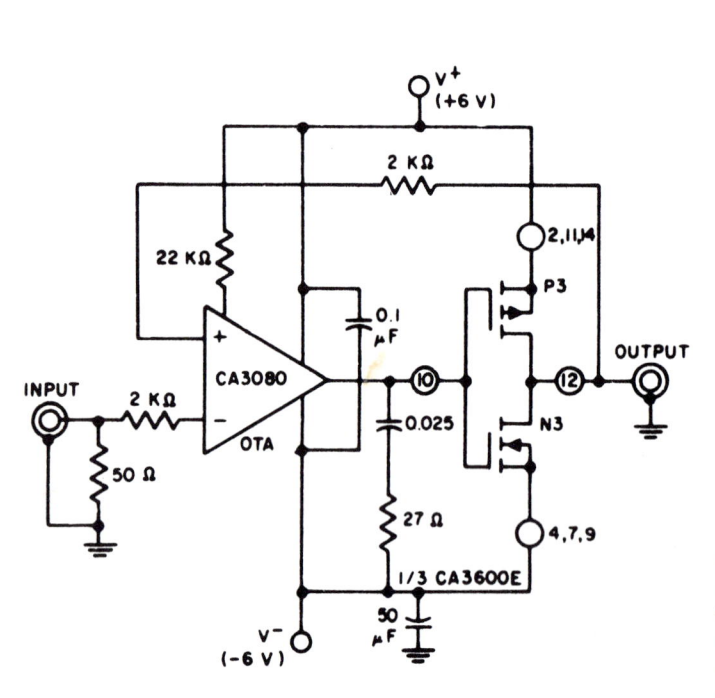

CMOS transistor pair used as postamplifier to op-amp in unity-gain circuit. The open-loop slew rate is 65V/µsec; when compensated for the unity-gain voltage-follower mode shown here, the slew rate is about 1V/µsec. For greater current output, the two remaining transistor pairs of the CA3600E may be connected in parallel with the single stage.

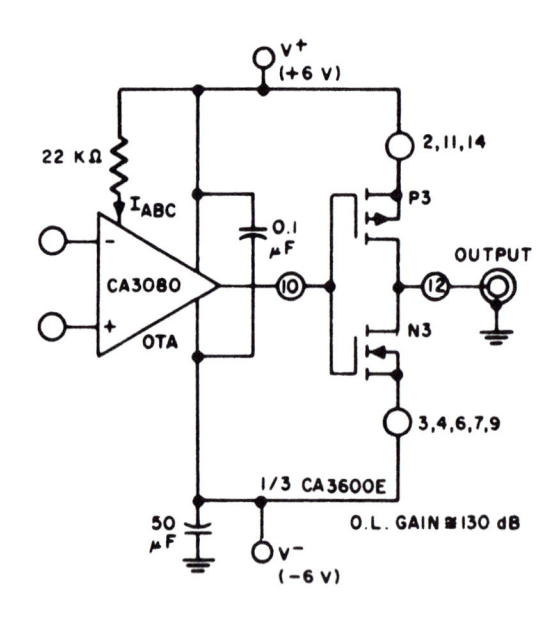

Postamplifiers for op-amps. Because the input impedance of the CMOS pair is comparatively high, the op-amp operates under essentially unloaded conditions. Each CMOS pair can sink and source output current up to about 10 mA. Additionally, the op-amp output can be directly coupled to bias the CMOS pair.

The first schematic shows a CMOS transistor pair serving as a postamplifier to an RCA CA3080 operational transconductance amplifier. The 30 dB gain in a single CMOS transistor pair is an added increment to the 100 dB gain in the CA 3080, yielding a total forward gain of about 130 dB.

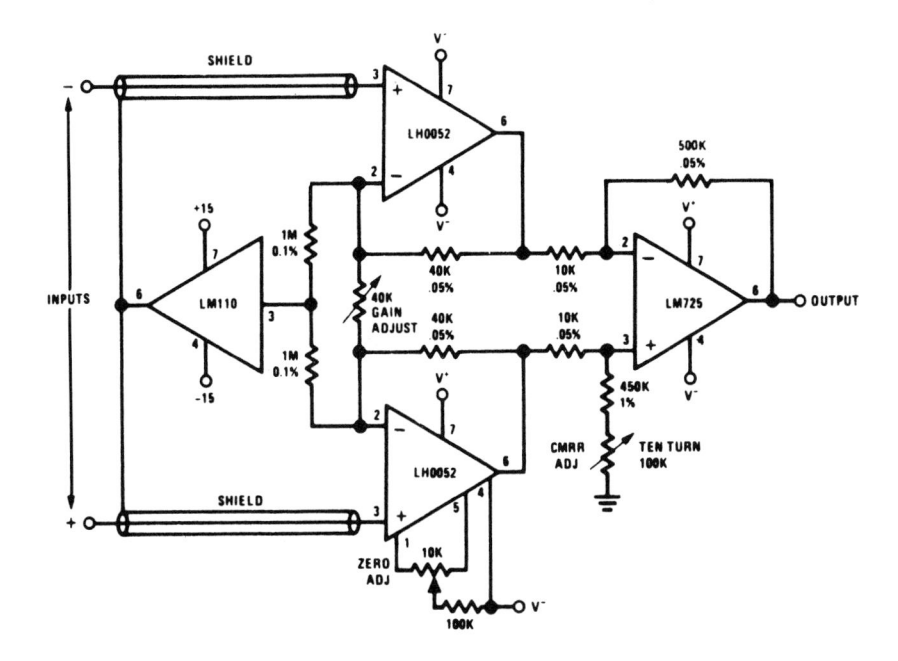

True instrumentation amplifier.

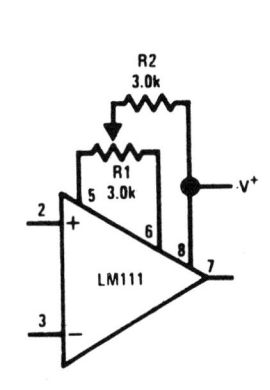

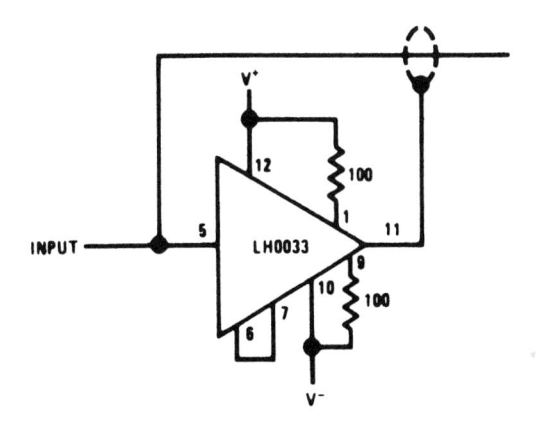

Offset balancing.

Shield or line driver for high-speed automatic test equipment. The LH0033 is mounted close to the test device and drives the cable shield to allow higher speed operation since the test device does not have to charge the cable.

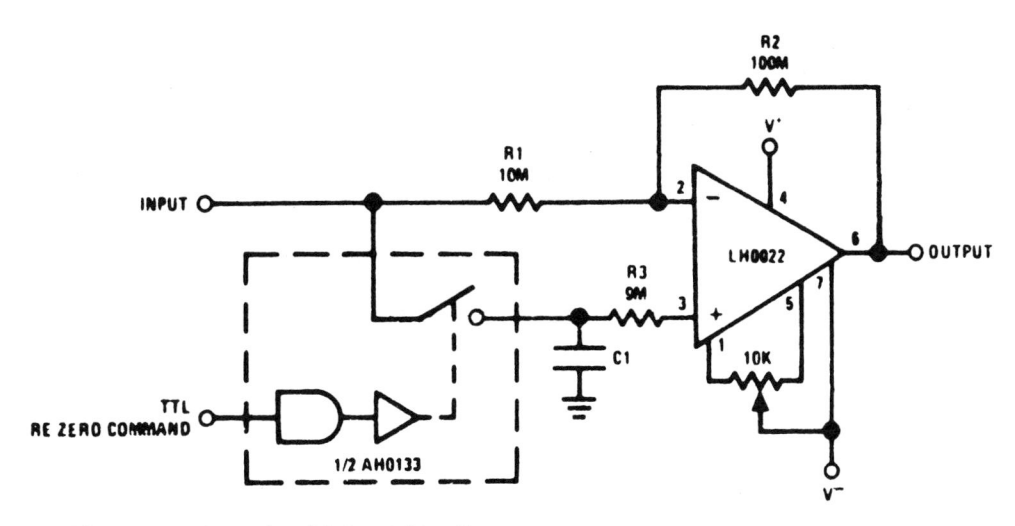

Rezeroing amplifier. Capacitor should be 0.01 µF polystyrene.

SECTION **3** Instrumentation and Test Circuits

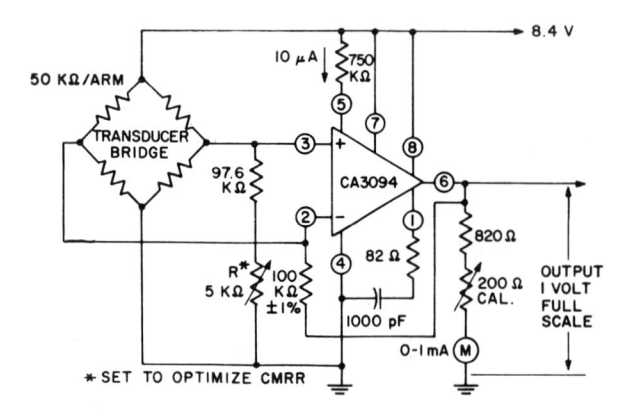

Single-supply differential-bridge amplifier.

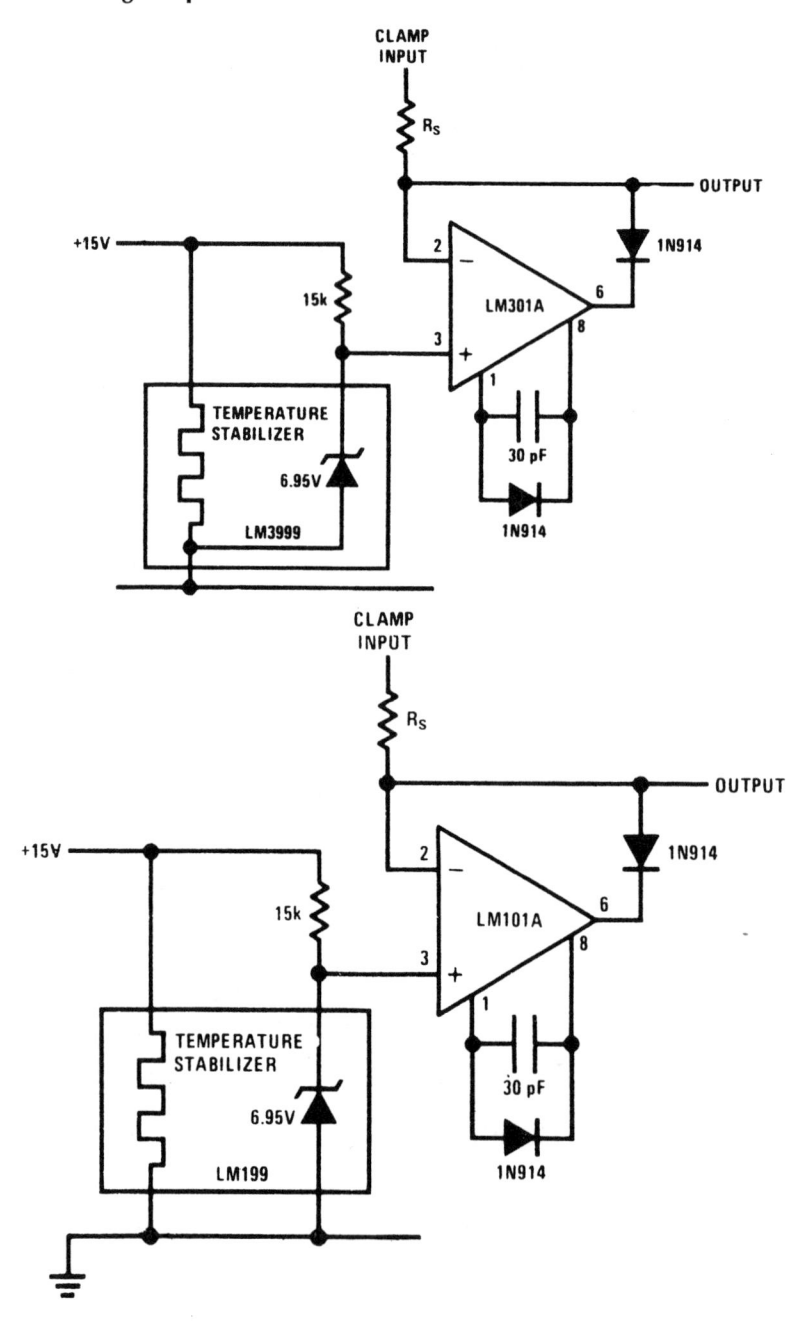

Precision clamps will sink 5 mA when input goes more positive than reference.

132 IC SCHEMATIC SOURCEMASTER

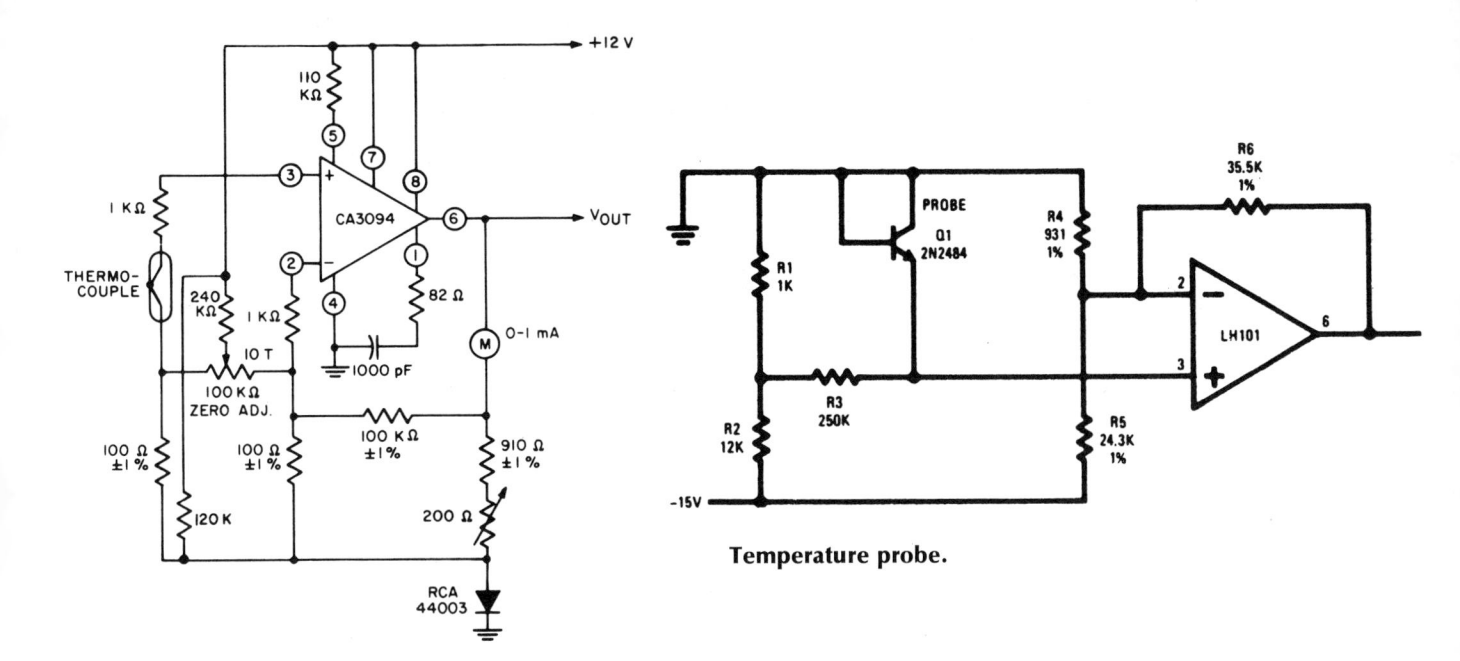

Single-supply amplifier for thermocouple signals: 1 mV from thermocouple produces full-scale output current.

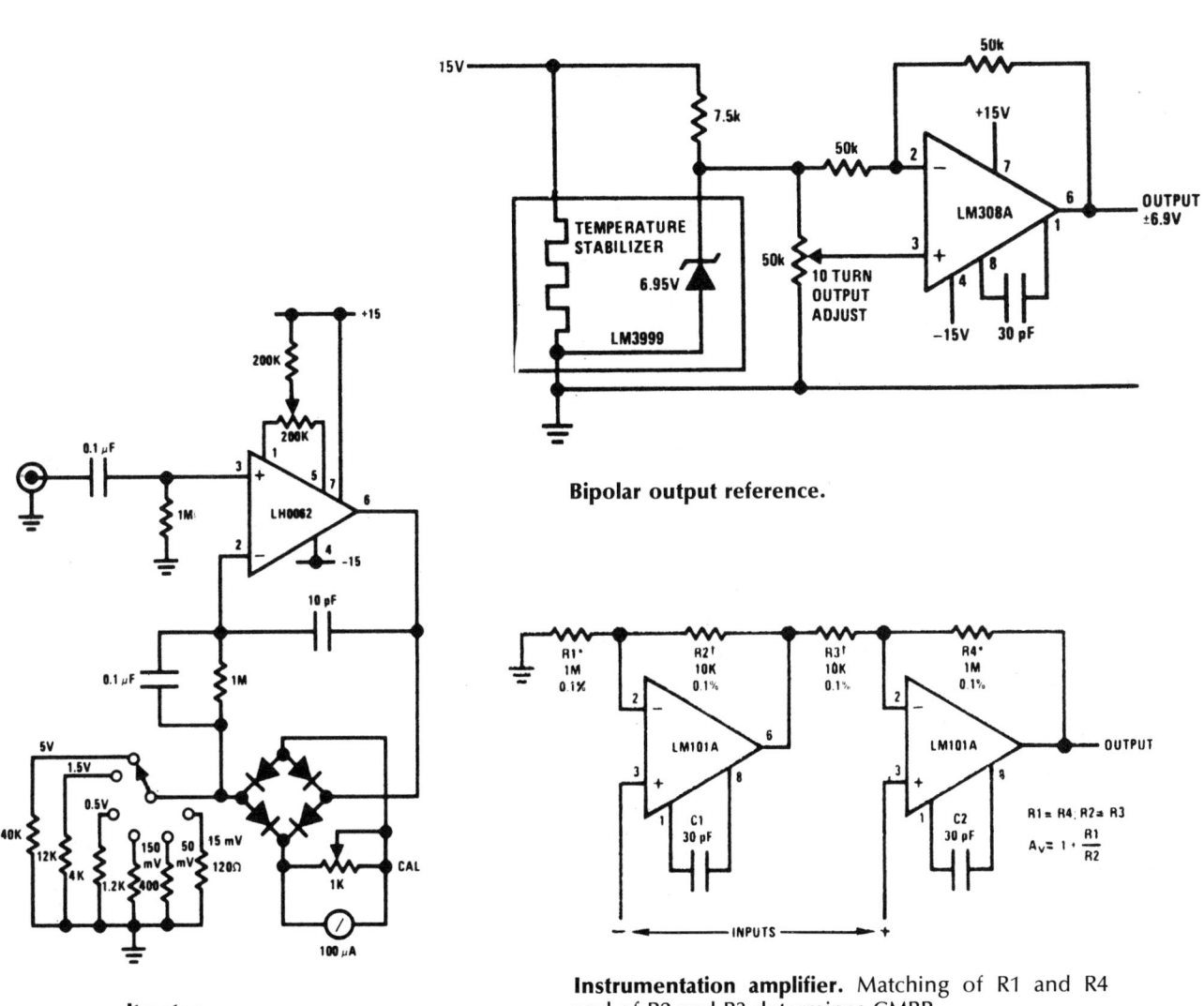

Temperature probe.

Bipolar output reference.

Wide-range ac voltmeter.

Instrumentation amplifier. Matching of R1 and R4 and of R2 and R3 determines CMRR.

$$R1 = R4, R2 = R3$$
$$A_V = 1 + \frac{R1}{R2}$$

SECTION **3** *Instrumentation and Test Circuits* **133**

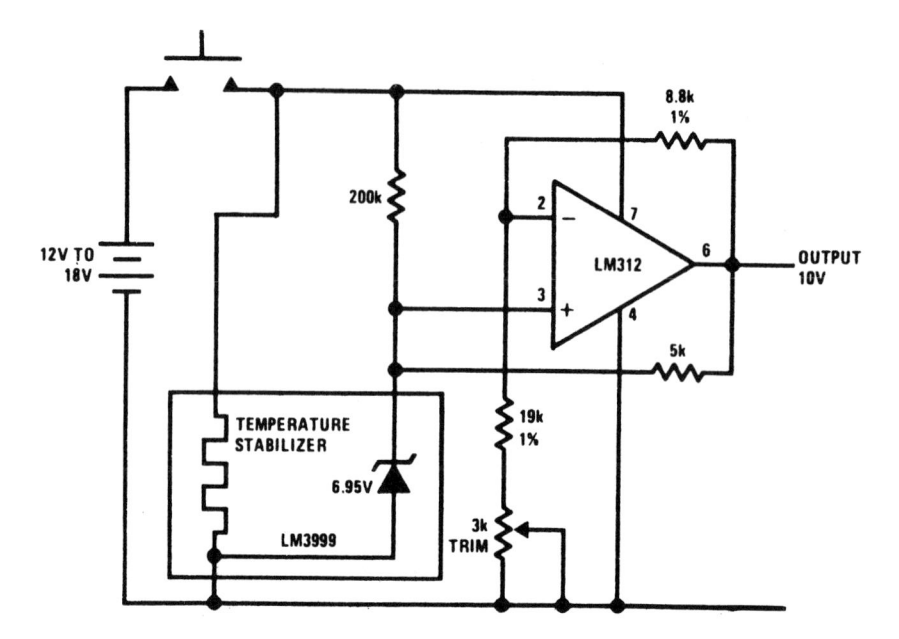

This portable calibrator delivers precisely 10V. Warm-up time is 10 seconds; intermittent operation does not degrade long-term stability.

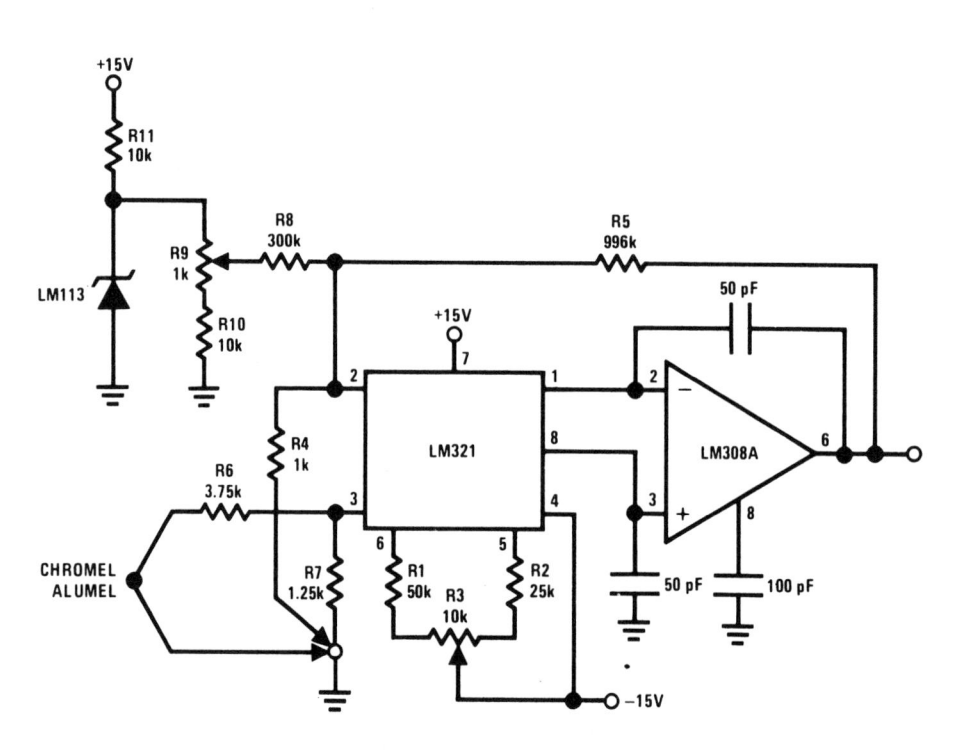

A thermocouple amplifier with cold-junction compensation. Since the output voltage of thermocouples is proportional to the temperature difference, the ambient temperature or measurement end of the thermocouple must be known. Alternatively, compensation can be applied for temperature changes. This is done either by terminating the thermocouple in a temperature-controlled environment or with electrical compensation circuitry. The amplifier shown here provides a direct reading output of 10 mV/°C and automatically compensates for reference-junction temperature changes. Calibration is relatively simple. Resistors R1, R2, and R3 set the operating current of the preamp, and R3 is used to adjust the offset. The offset and drift are amplified by the ratio of the feedback resistors, R4 and R5, and appears at the output. R6 and R7 attenuate the thermocouple's output to 10 μV/°C to match the amplifier drift and set the scale factor at 10 mV/°C. The LM113 provides a temperature-stable reference for offsetting the output to read directly in degrees Celsius.

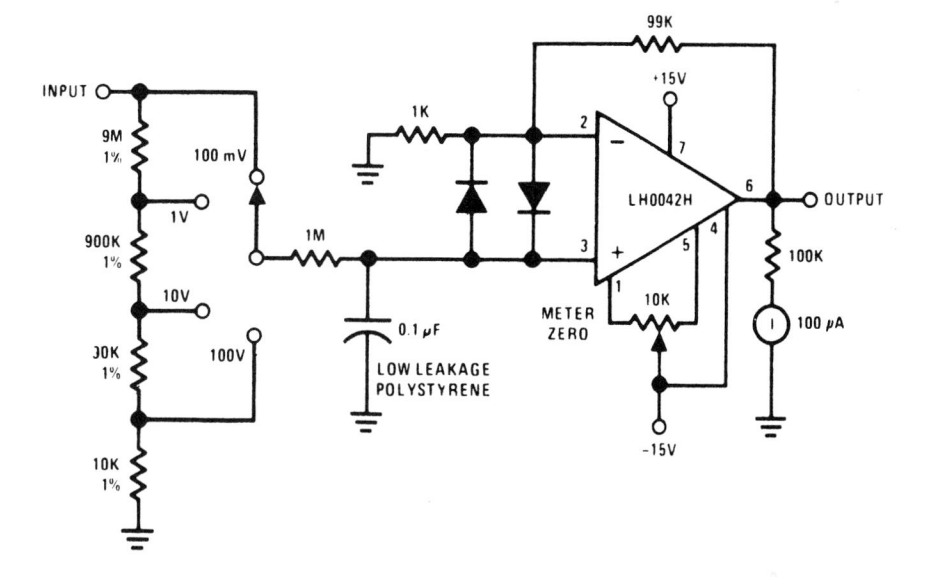

This sensitive low-cost high-impedance electronic voltmeter may be run off of 8 flashlight batteries, drawing only 20 mW of power. Add some more switching to allow operation of the FET op-amp in transconductance mode, thus combining both voltage and current-measuring capability into the same circuit.

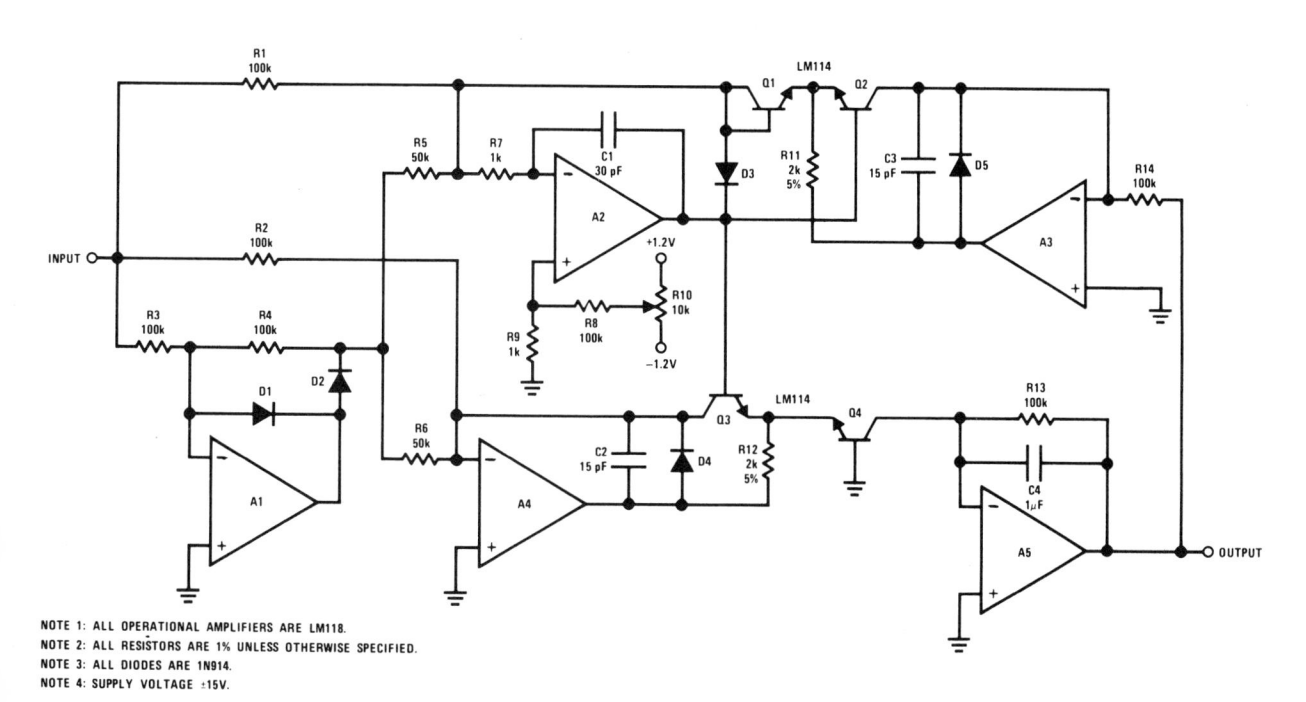

NOTE 1: ALL OPERATIONAL AMPLIFIERS ARE LM118.
NOTE 2: ALL RESISTORS ARE 1% UNLESS OTHERWISE SPECIFIED.
NOTE 3: ALL DIODES ARE 1N914.
NOTE 4: SUPPLY VOLTAGE ±15V.

True-RMS detector provides a dc output equal to the RMS value of the input. Accuracy is typically 2% for a 20V_{P-P} input signal from 50 Hz to 100 kHz, although it's usable to about 500 kHz. The lower frequency is limited by the size of the filter capacitor. Further, since the input is dc coupled, it can provide the true RMS equivalent of a dc and ac signal.

Basically, the circuit is a precision absolute-value circuit connected to a one-quadrant multiplier-divider. Amplifier A1, the absolute-value amplifier, provides a positive input current to amplifiers A2 and A4, independent of signal polarity. If the input signal is positive, A1's output is clamped at −0.6V, D2 is reverse-biased, and no signal flows through R5 and R6. Positive signal current flows through R1 and R2 into the summing junctions of A2 and A4. When the input is negative, an inverted signal appears at the output of A1 (output taken from D2). This is summed through R5 and R6 with the input signal from R1 and R2. Twice the current flows through R5 and R6, and the net input to A2 and A4 is positive.

For best results, transistors Q1 through Q4 should be matched, have high beta, and be at the same temperature. Since dual transistors are common, good results can be obtained if Q1, Q2 and Q3, Q4 are paired. They should be mounted in close proximity or on a common heatsink, if possible. Bypass all op-amps with 0.1 μF (op-amps are LM118, diodes are 1N914).

SECTION **3** *Instrumentation and Test Circuits* **135**

$$\text{GAIN} = \frac{200k}{R8}$$

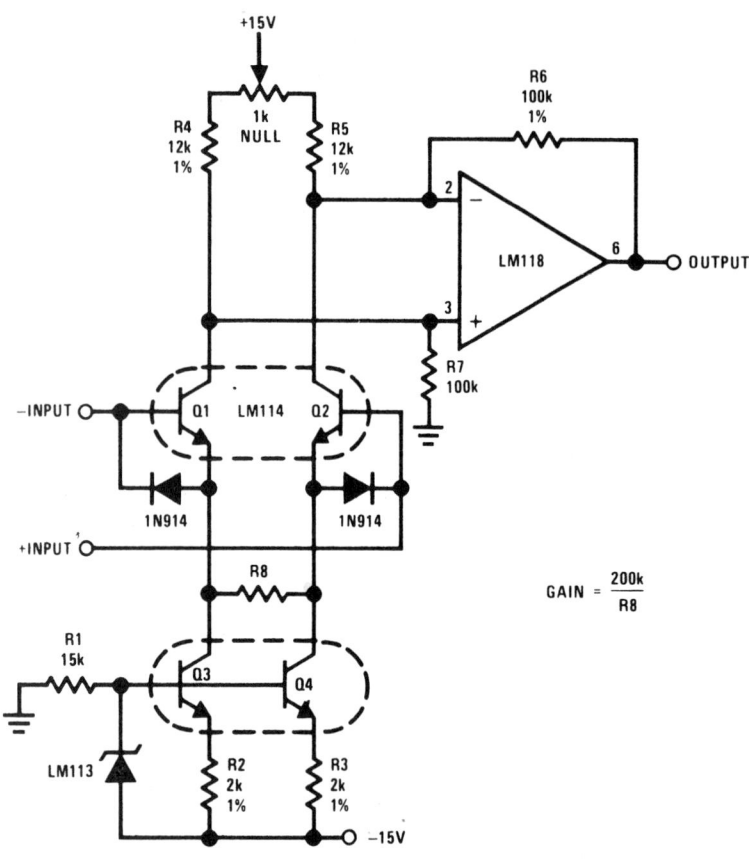

This instrumentation amplifier can faithfully amplify low-level signals in the presence of high common-mode noise. This aspect of its performance makes it especially useful as the input amplifier of a signal-processing system. Other features of the instrumentation amplifier are high input impedance, low input current, and good linearity. It can provide gains from under 1 to over 1000 with a single resistor adjustment. Gain linearity is worst for unity gain at 0.4%, and gain stability is better than 1.5% from $-55°$ to $+125°C$. Typically, over a 0 to $+70°C$ range, gain stability is 0.2%. The common-mode rejection ratio is about 100 dB, independent of gain.

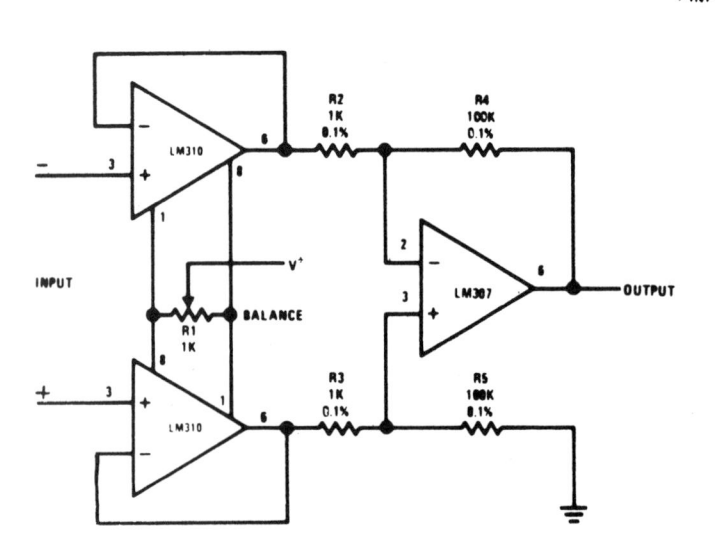

Instrumentation amplifier. Gain $\cong 200k/R_g$ for $1.5k \leq R_f \leq 200k$.

Differential-input instrumentation amplifier.

136 IC SCHEMATIC SOURCEMASTER

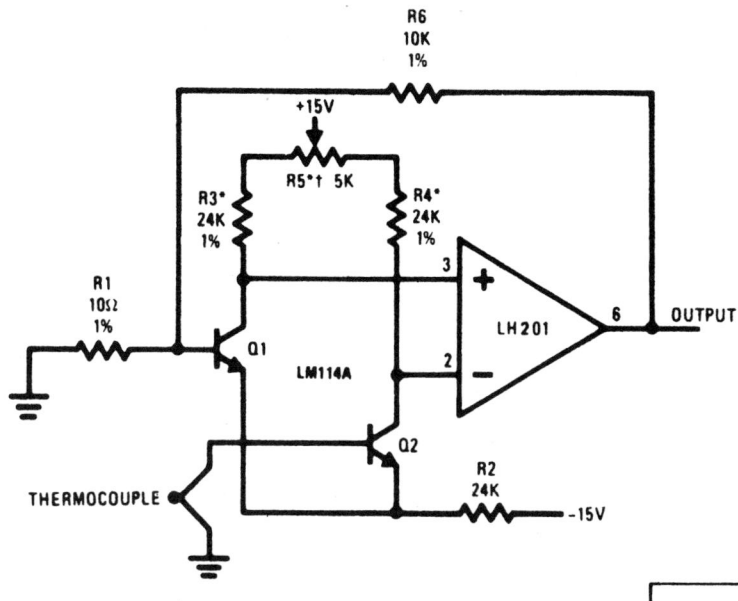

This thermocouple amplifier drifts less than 0.5 µV/°C. Adjust R5 for zero input offset voltage. Resistors R3–R5 must have matched temperature coefficients.

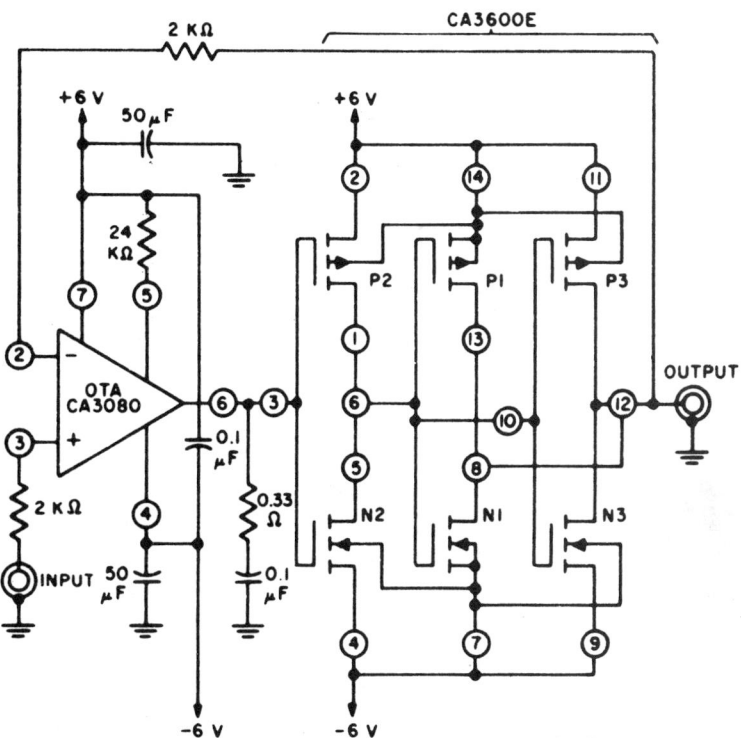

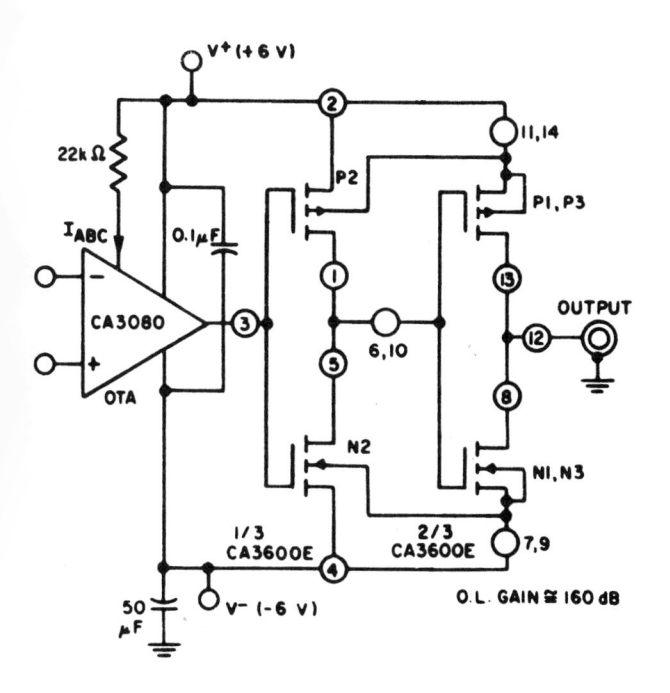

This two-stage CMOS postamplifier increases the total open-loop gain of the system to about 160 dB (×100,000,000). Open-loop slew rate remains at about 65V/µsec. A slew rate of about 1V/µsec is maintained with this circuit connected in the unity-gain voltage-follower mode, as shown.

This unity-gain amplifier uses CMOS transistor pairs as two-stage postamplifier to op-amp.

SECTION 3 *Instrumentation and Test Circuits* 137

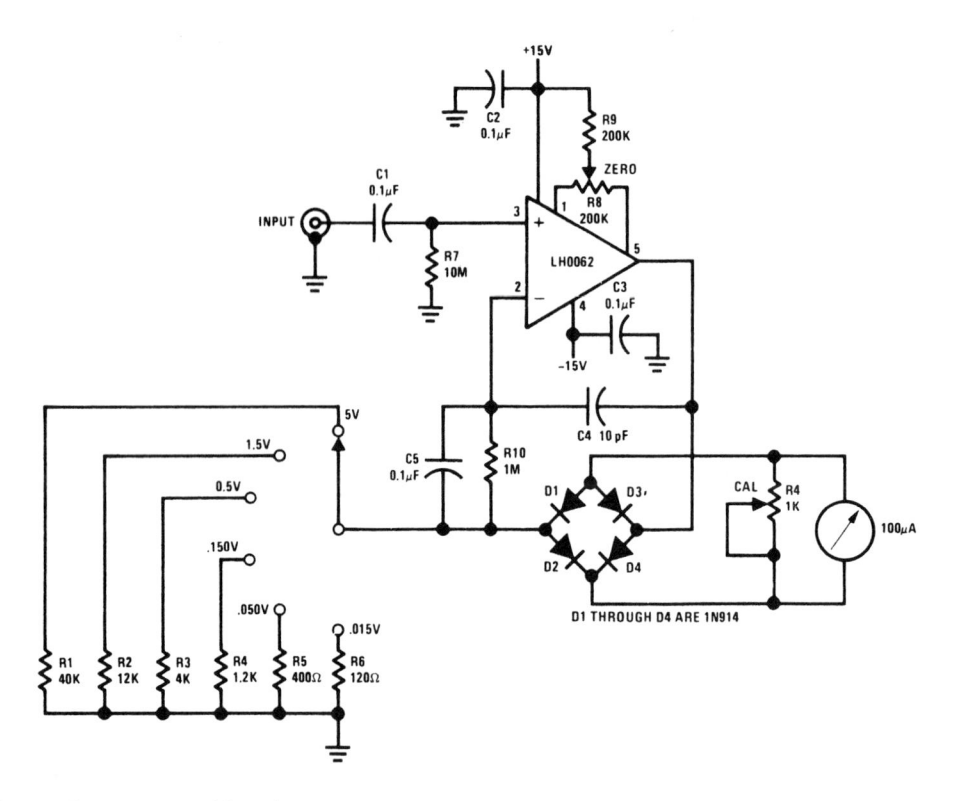

A wideband ac voltmeter capable of measuring ac signals as low as 15 mV at frequencies from 100 Hz to 500 kHz. Full-scale sensitivity may be changed by altering the values R1 through R6 (R ≅ V_{IN}/100 µA).

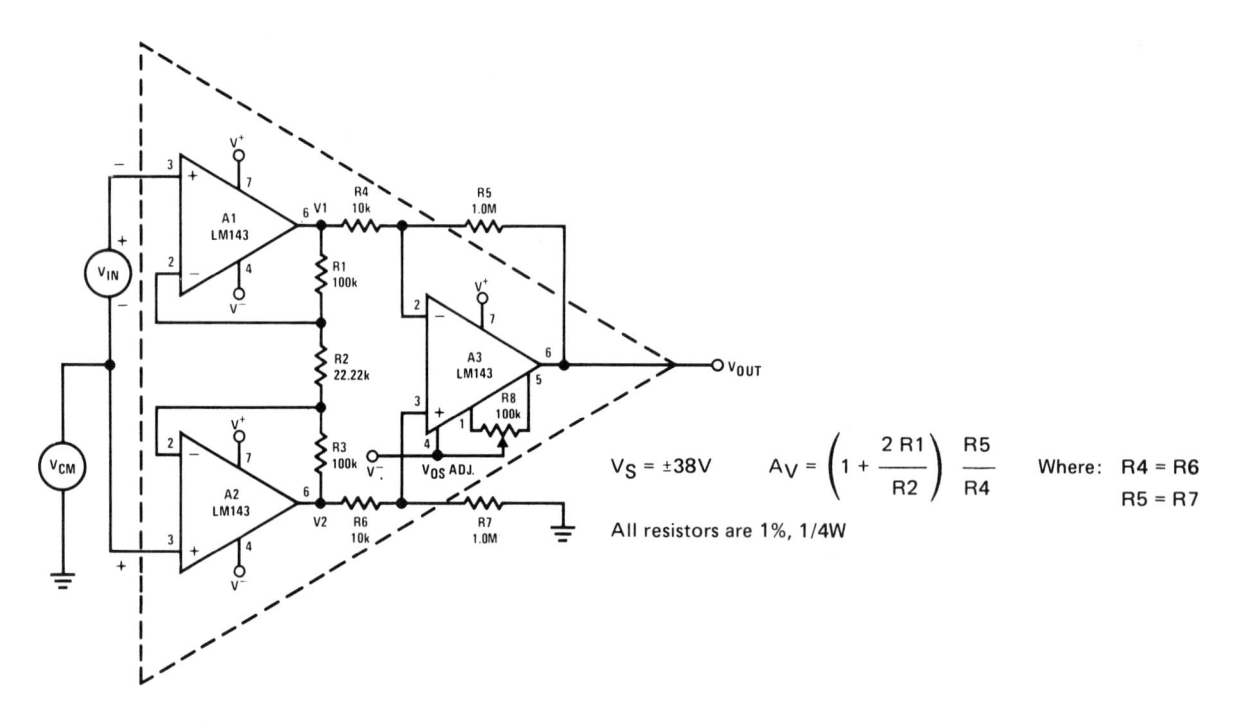

Wide-common-mode-range instrumentation amplifier. For a differential input signal, V_{IN}, A1 and A2 act as noninverting amplifiers of gain $A_V^1 = 1 + (2R1/R2)$, where R1 = R3. However, the gain is unity for common-mode signals, since voltages V1 and V2 are in phase and no current flow is developed through R1, R2, and R3. The second stage is simply an op-amp connected as a simple differential amplifier of gain $A_V^2 = R5/R4$, where R5 = R7 and R4 = R6. The total gain of the instrumentation amplifier is 1000. R7 may be adjusted to take up the resistance tolerances of R4, R5, and R6 for best common-mode rejection (CMR). Also, R2 may be made adjustable to vary the gain of the instrumentation amplifier without degrading the CMRR.

138 IC SCHEMATIC SOURCEMASTER

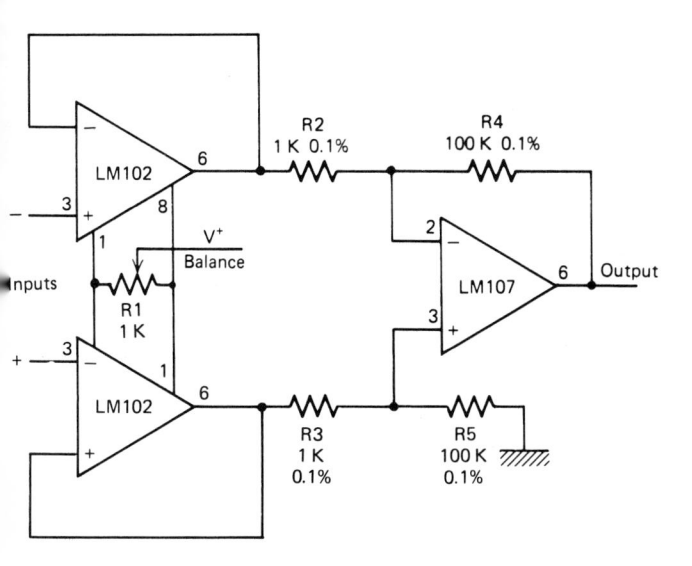

Differential-input instrumentation amplifier. R4/R2 = R5/R3. A_V = R4/R2.

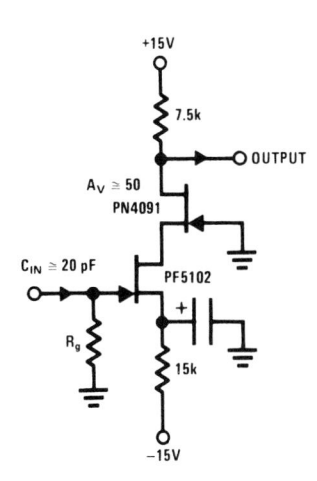

A wideband, low-input-capacitance, very low noise preamplifier with usable bandwidth of 1 MHz.

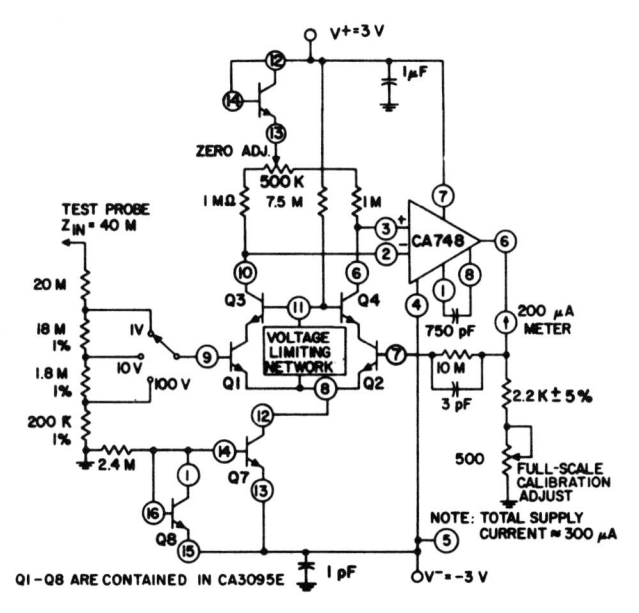

Typical high-input-impedance dc voltmeter circuit. Resistor divider network is provided to develop a dc input signal at terminal 9 of the CA3095E with transistors Q1–Q3 and Q2–Q4 connected in the differential-cascode arrangement. Biasing and dc feedback are applied at terminal 7 of the CA3095E through a 10M resistor. The CA748 op-amp drives a 200 µA meter calibrated in terms of the voltages to be measured. A full-scale reading occurs when the voltage applied to pin 9 is 500 mV dc. The entire circuit is nulled with the 500k zero-adjustment potentiometer. Power requirement is 6V, with a supply current of only 300 µA. The input impedance of this simple circuit is approximately 40M on all scales. The voltage-limiting network is built into superbeta IC. To use it, just connect a resistor from the positive supply of the differential amplifier to pin 11; the value should be such that the current through the resistor is about half that measured at pin 8 under quiescent conditions.

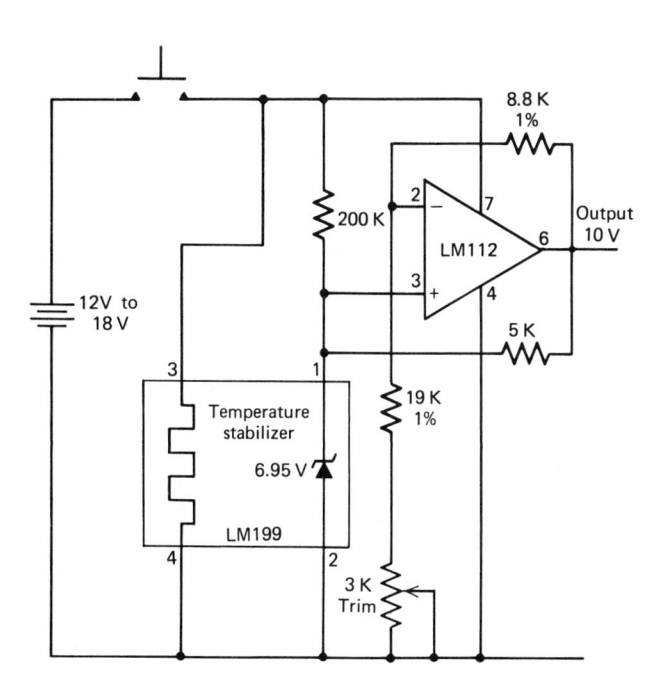

This portable calibrator warms up in 10 seconds. Intermittent operation does not degrade long-term stability.

SECTION 3 *Instrumentation and Test Circuits* **139**

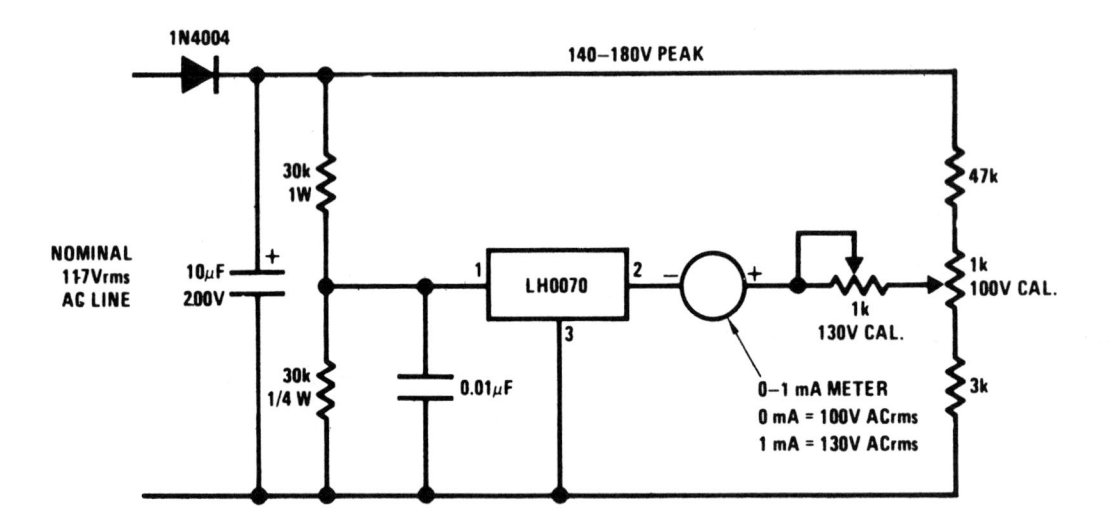

Expanded-scale ac voltmeter.

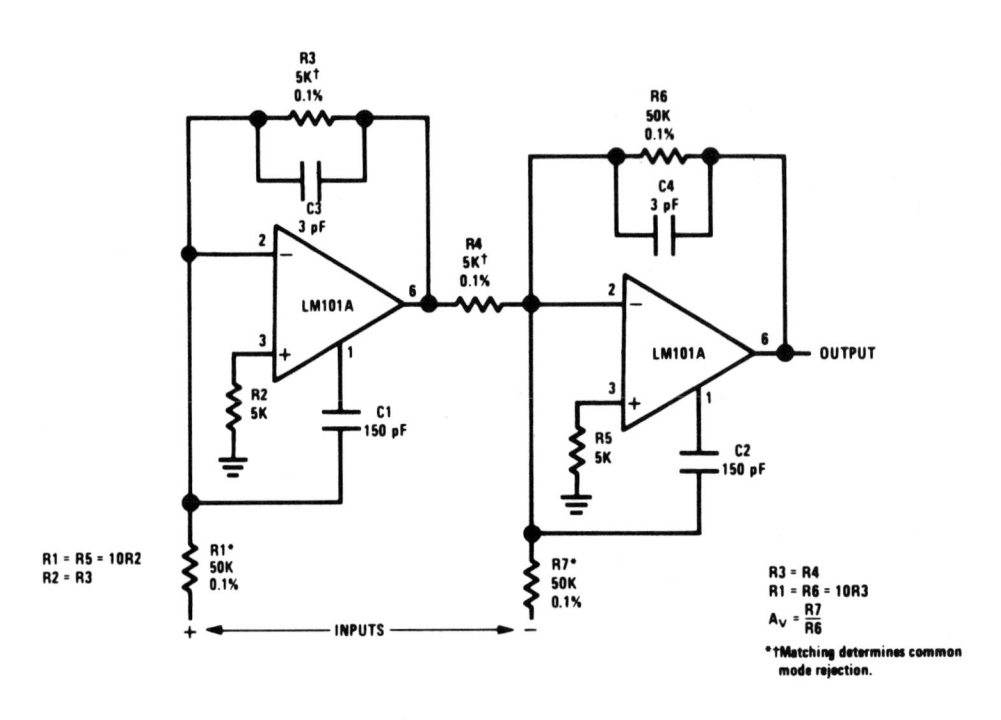

Instrumentation amplifier with ±100V common-mode range.

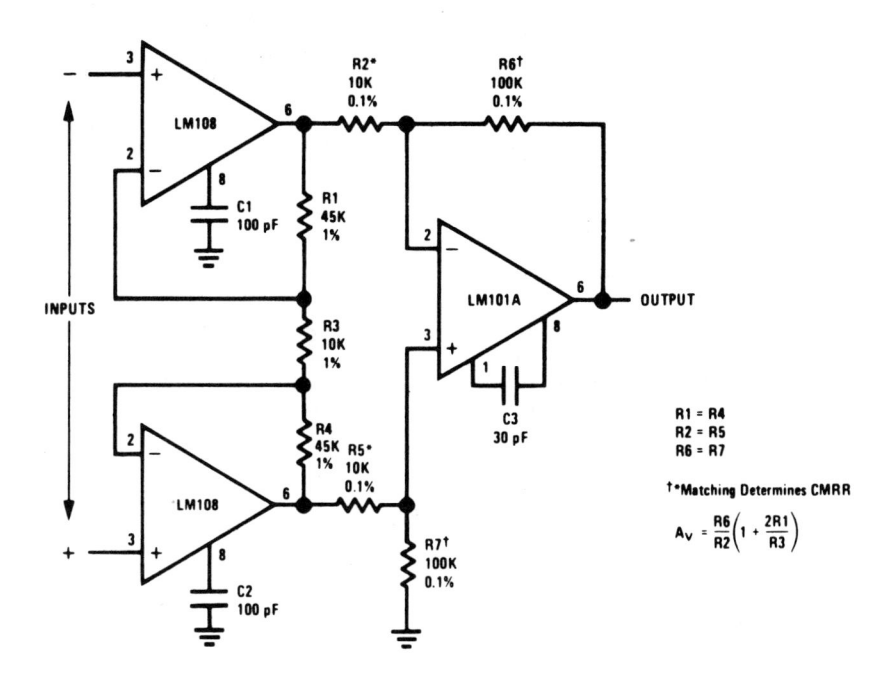

Differential-input instrumentation amplifier with high common-mode rejection.

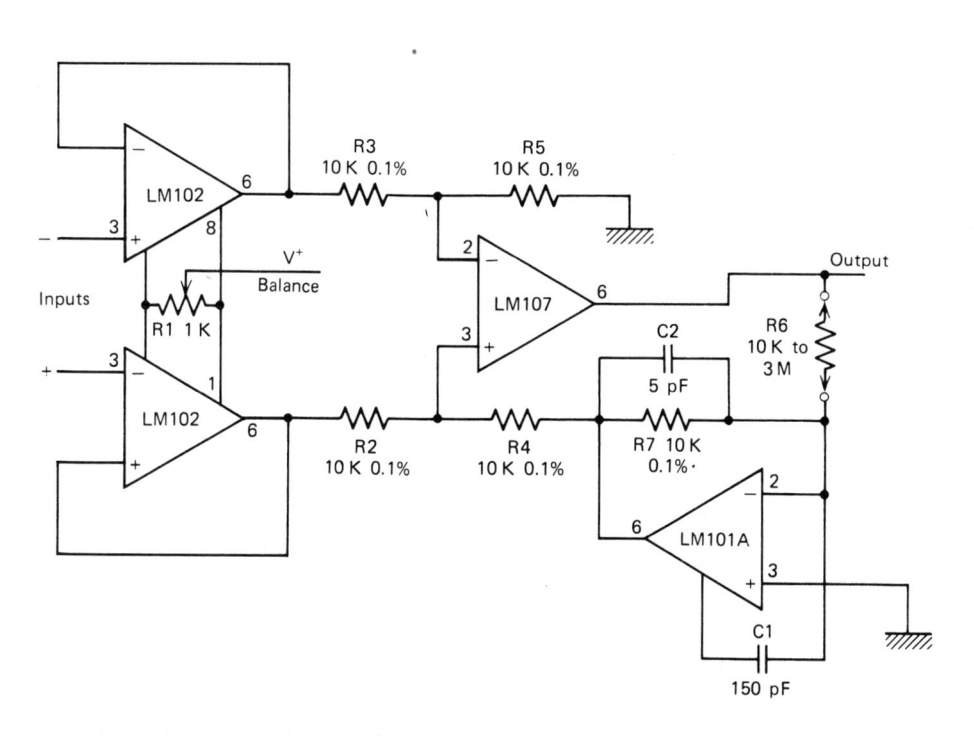

$$A_V = \frac{R6}{R2}\left(1 + \frac{2R1}{R3}\right)$$

R1 = R4
R2 = R5
R6 = R7

†*Matching Determines CMRR

Variable-gain differential-input instrumentation amplifier. R6 adjust gains. $A_V = (10^{-4})R6$.

SECTION 3 *Instrumentation and Test Circuits* 141

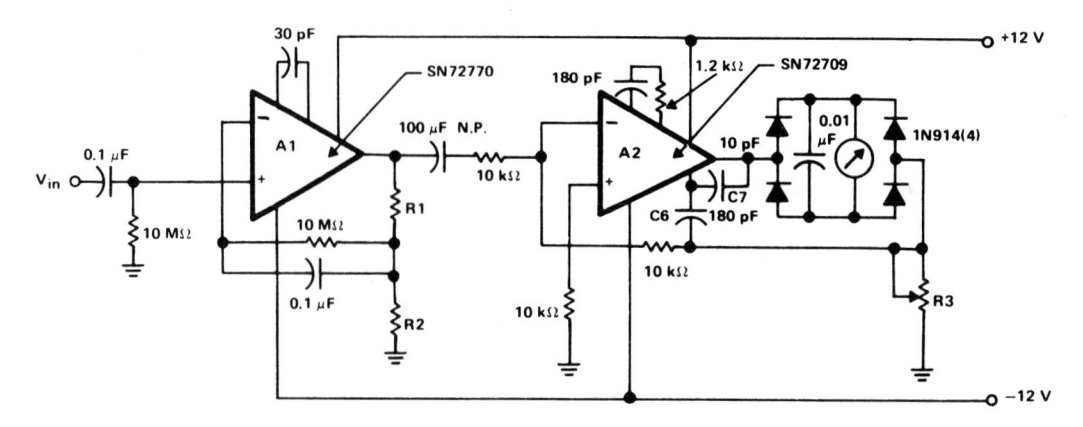

This millivoltmeter combines a high-input-impedance preamplifier (A1) and a voltage-to-current converter (A2). The preamplifier gain may be adjusted in steps, using R1 and R2, for various input sensitivity levels. A2 furnishes the drive for the diode-bridge detector circuit and meter movement. The output current is adjusted by R3 to the desired full-scale meter level.

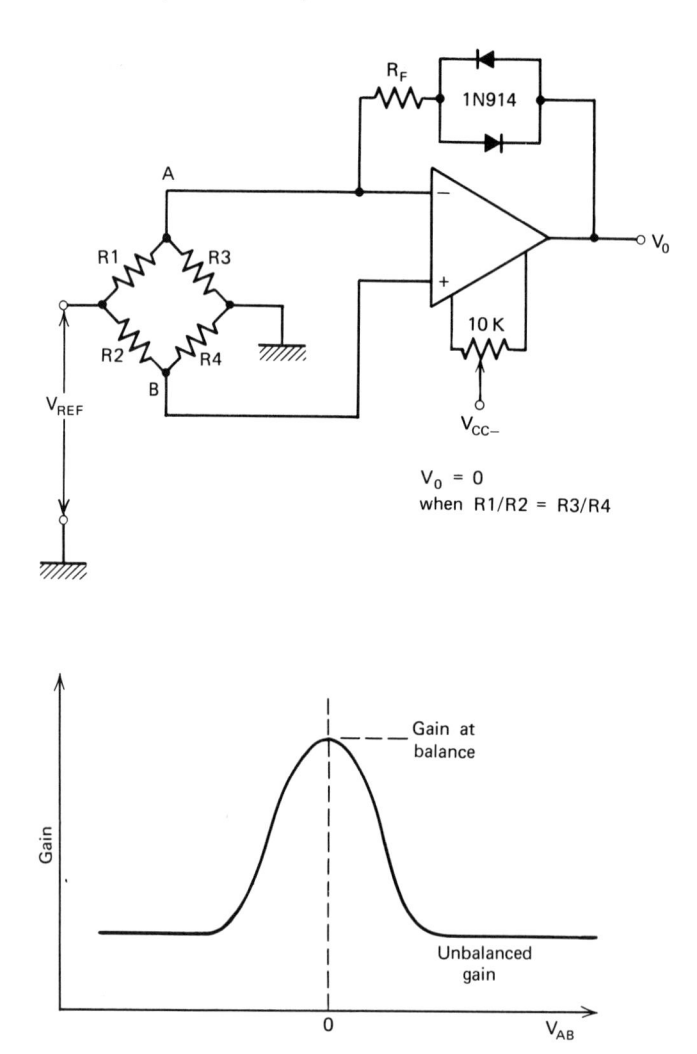

$V_0 = 0$ when R1/R2 = R3/R4

Bridge balance indicator. In this application the 1N914 diodes in the feedback loop result in high sensitivity and accuracy near the point of balance (R1/R2 = R3/R4). When the bridge is unbalanced (R1/R2 ≠ R3/R4), the amplifier's closed-loop gain is approximately R_F/r, where r is the parallel equivalent of R1 and R3. The resulting gain equation is $G = R_F (1/R1 + 1/R3)$. During an unbalanced condition the voltage at point A is different from that of point B. This difference voltage V_{AB}, amplified by the gain factor G, appears as a voltage V_0 at the output. As the bridge approaches a balanced condition (R1/R2 = R3/R4), V_{AB} approaches zero. Under this condition the diodes in the feedback loop lose their forward bias and their resistance increases. This results in an increase in the total feedback resistance, thus increasing the circuit gain and accuracy in detecting a balanced condition.

142 IC SCHEMATIC SOURCEMASTER

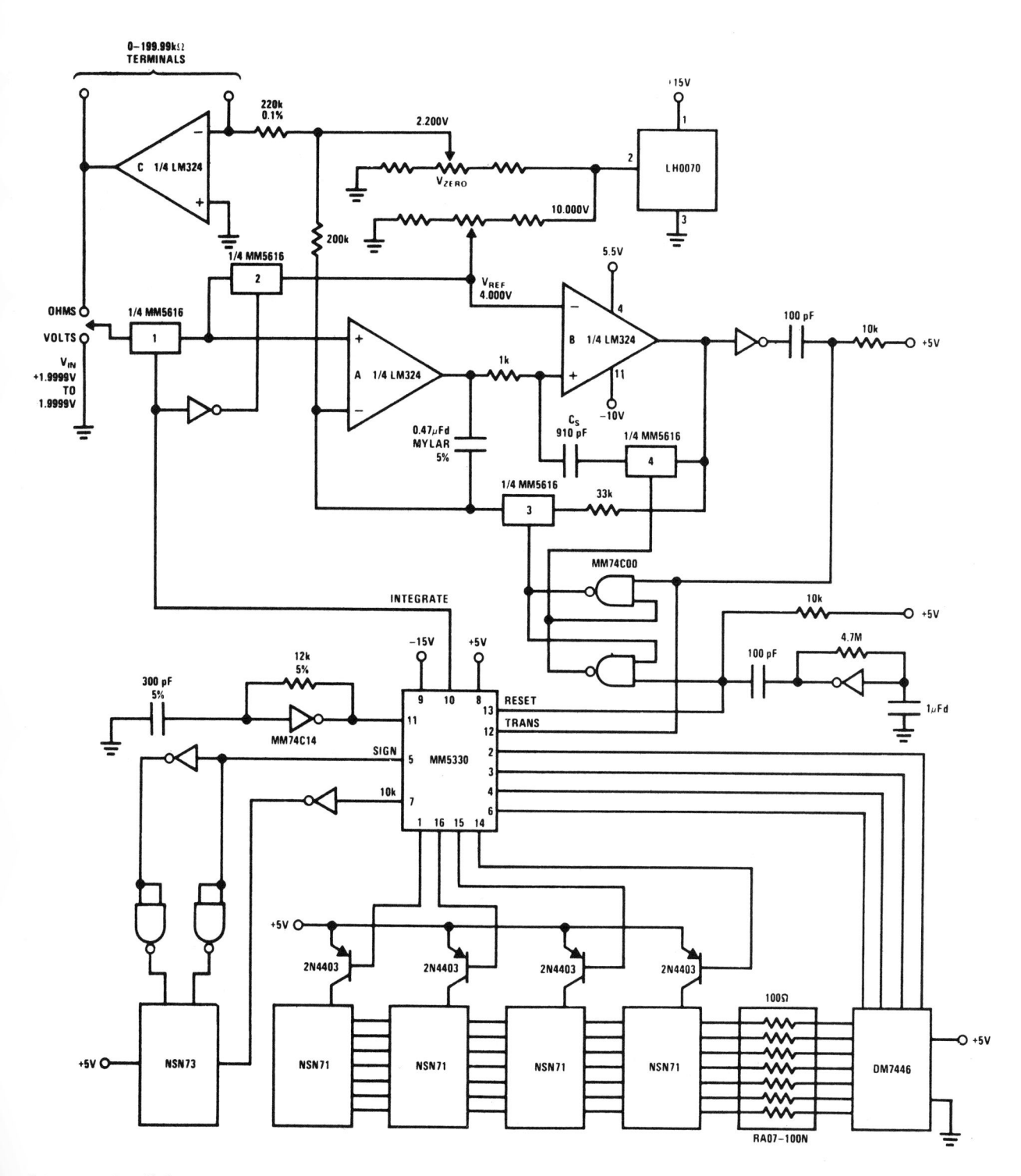

Low-cost 4.5-digit DVM (National). The display interface used is a TTL, 7-segment decoder driver and four P-type transistors. The ±1 digit is driven directly by CMOS. The clock synchronous reset and transfer functions prevent any cyclic digit variations and present a blink-free display. CMOS analog switches are used as reference, zero, and input switches and are incorporated in the comparator slew rate circuit.

SECTION **3** Instrumentation and Test Circuits 143

SECTION 4
WAVEFORM GENERATORS AND SHAPERS

This section presents waveform generators and shapers of every description. No attempt has been made to separate toys from precision devices, but the textual information associated with the schematics should give you the kind of data required to help you make an intelligent decision as to which circuits are most appropriate for your needs.

Oscillators cover an incredibly broad spectrum, from single-device tone makers to full-fledged function generators and from simple shapers (pulse generators and integrators) to power inverters. Bear in mind that the signal-generating element, like the common amplifier, is a very basic part of a great many types of circuits. When you're looking for just the right circuit for a specific job, don't stop with this section; think about the total function you want your circuit to perform, then spend a few extra minutes browsing through other sections of this book that might contain the appropriate diagrams. For example, the section on sensors and alarms (Section 8) contains a large number of circuits that incorporate oscillator elements integrally, and of course, Section 7, containing communications circuits, is also rich in diagrams that show devices employed in signal generation operations.

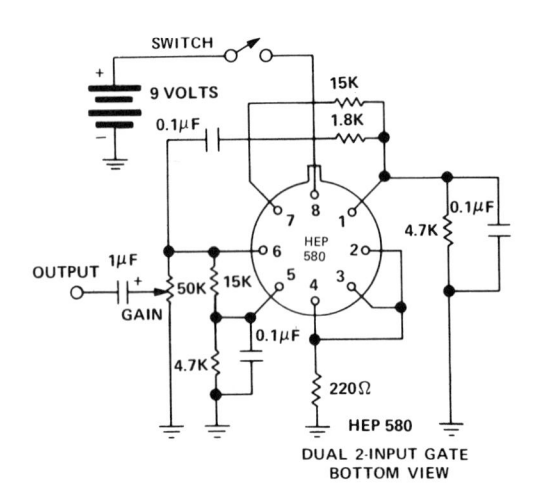

A cheap audio signal generator for general audio testing features adjustable output from 0 to approximately $2V_{p-p}$ with a sine-wave output of fixed frequency of approximately 1 kHz.

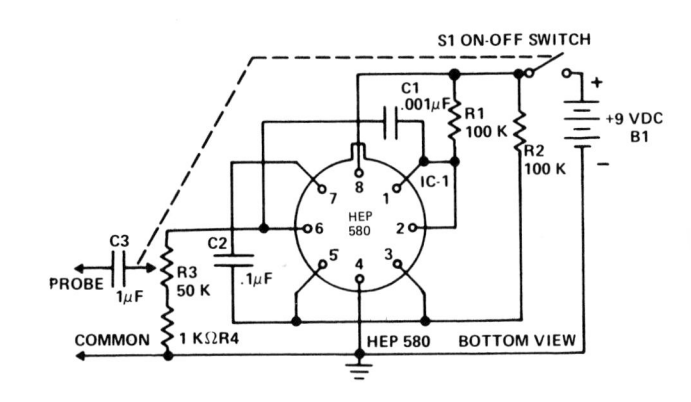

This signal injector is an ideal audio signal probe for signal injection in testing AM receivers or audio amplifiers. Output voltage is variable from 1 to $7V_{p-p}$, 1300 Hz tone. Battery drain at 9V is about 5.5 mA. Frequency of tone may be raised or lowered by lowering or raising the value of C1.

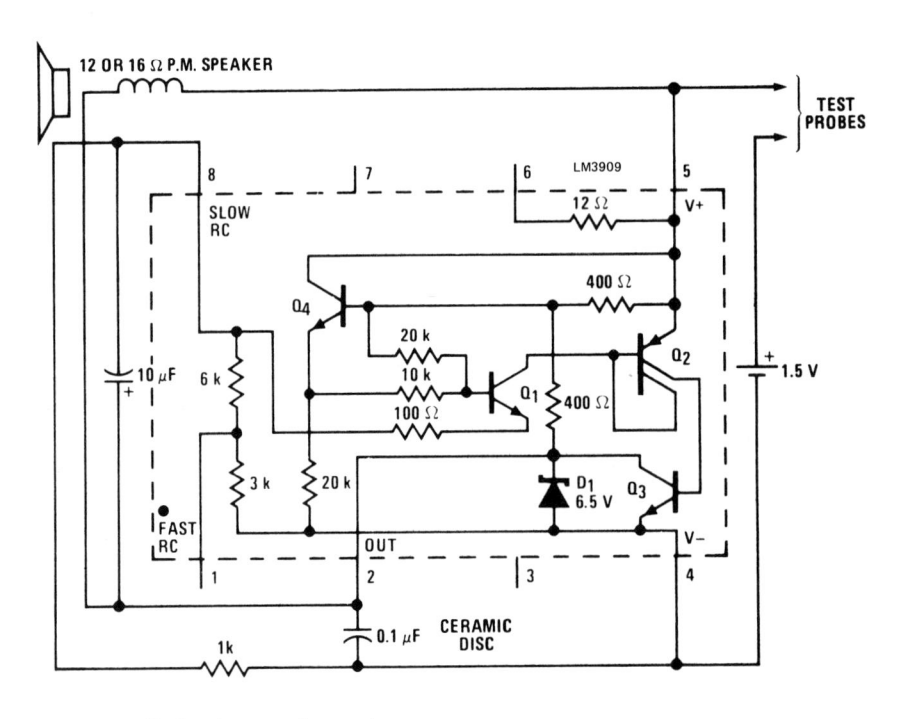

Buzz box continuity and coil checker. A short of up to about 100Ω across the test probes provides enough power for audible oscillation. By probing 2 values in quick succession, small differences, such as between a short and 5Ω, can be detected by differences in tone.

146 IC SCHEMATIC SOURCEMASTER

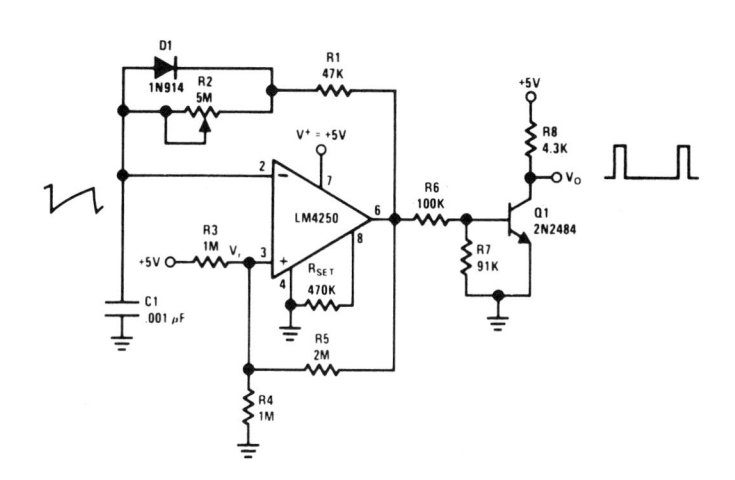

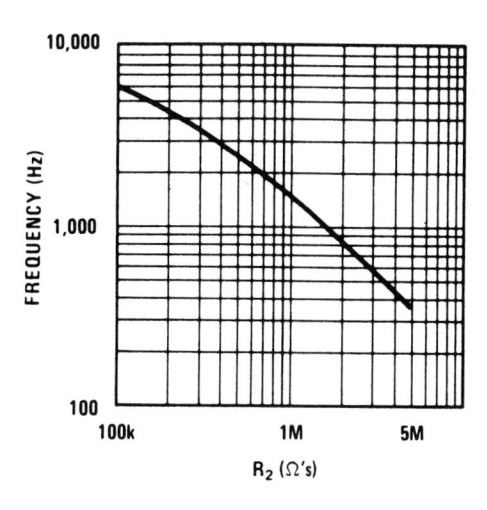

Pulse Frequency vs R₂

This variable-frequency pulse generator provides an example of the LM4250 operated from a single supply. The circuit is a buffered-output free-running multivibrator with a constant-width output pulse occurring with a frequency determined by potentiometer R2.

The change in output frequency as a function of supply voltage is less than ±4% for a V⁺ change of from 4 to 10V. This stability of frequency versus supply voltage results from the fact that the reference voltage V$_r$ and the drive voltage for the capacitor are both direct functions of V⁺.

The power dissipation of the free-running multivibrator is 300 μW, and the power dissipation of the buffer circuit is approximately 5.8 mW.

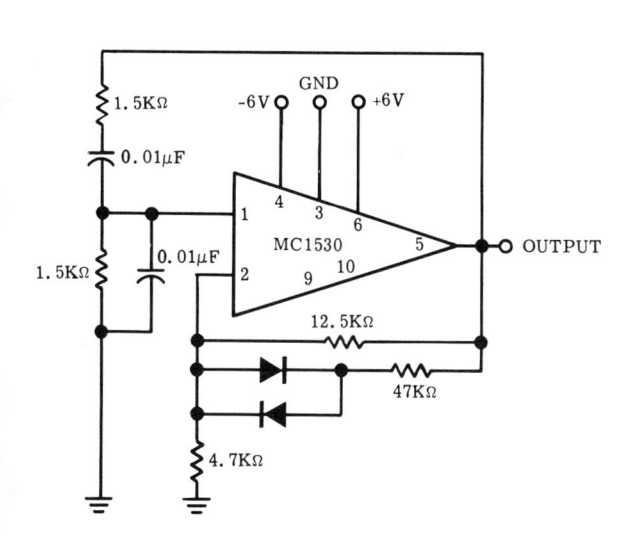

Wien bridge oscillator (Motorola).

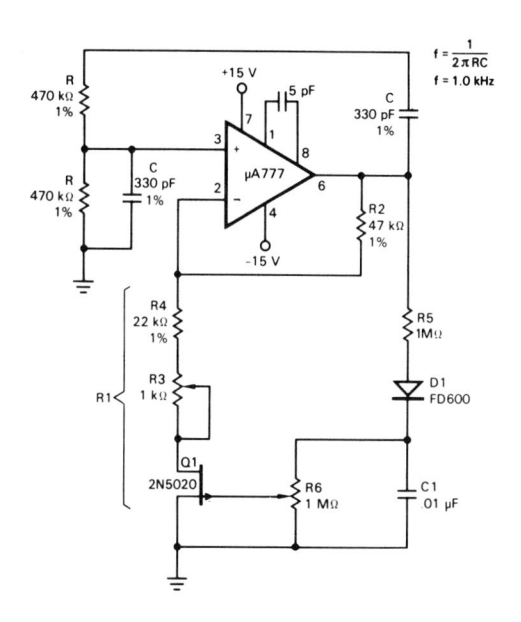

A Wien bridge oscillator (Fairchild), in which field effect transistor Q1 operates in the linear resistive region to provide automatic gain control. Because the attenuation of the RC network is one-third at the zero-phase-shift oscillation frequency, the amplifier gain determined by resistor R2 and equivalent resistor R1 must be just equal to 3 to make up the unity-gain positive-feedback requirement needed for stable oscillation. Resistors R3 and R4 are 1k less than the required R1 resistance. The FET dynamically provides the trimming resistance needed to make R1 half the resistance of R2.

SECTION **4** Waveform Generators and Shapers

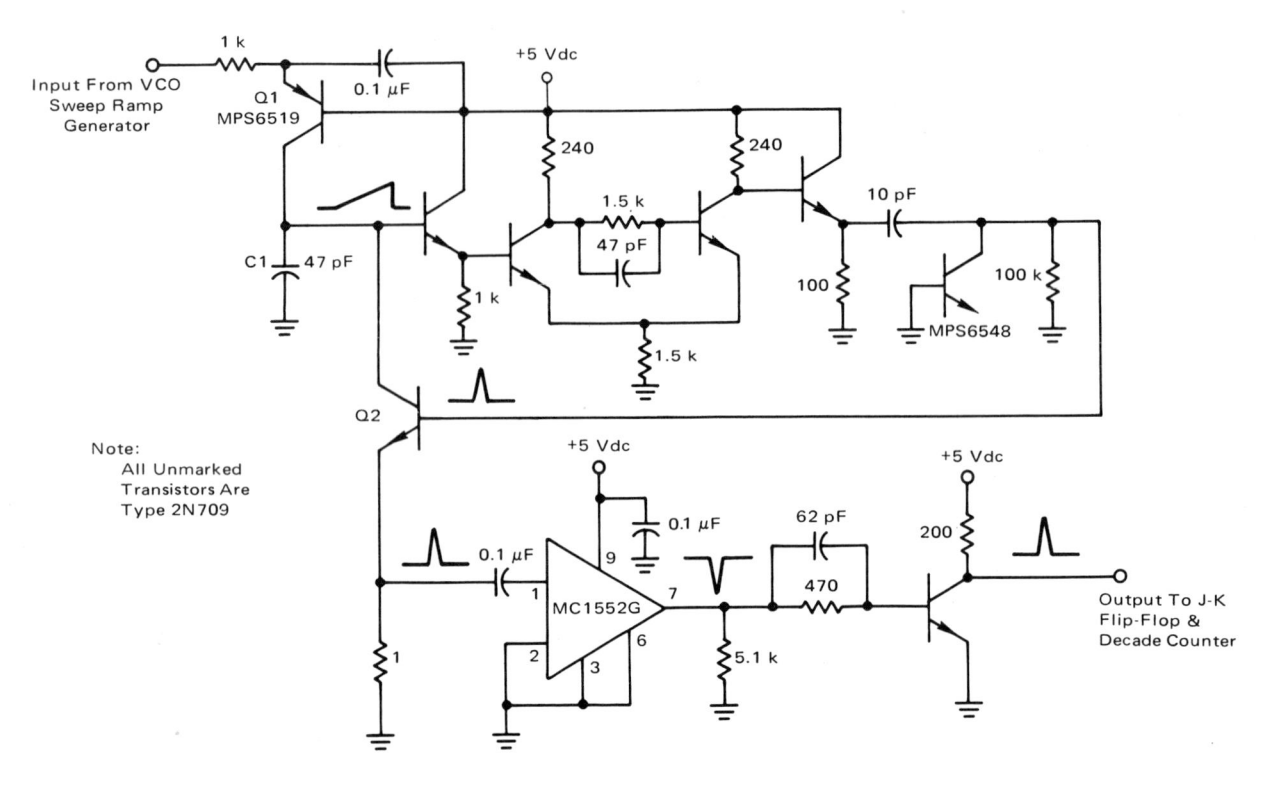

Voltage-controlled oscillator. A voltage-controlled current generator sources current into a timing capacitor. The charge accumulated in the capacitor is monitored by a voltage-sensitive, regenerative threshold detector—a Schmitt trigger. At a predetermined threshold, the detector provides a positive drive signal to a transistor that discharges the timing capacitor to another predetermined level, after which the cycle repeats itself.

Current generator Q1 is voltage controlled by a ramp function that serves to sweep the repetition rate. When the Schmitt trigger threshold is reached, the regenerative slope provides a signal that saturates Q2, discharging the timing capacitor. The discharge is sampled by the emitter resistor of Q2. This pulse is then amplified by an integrated circuit and transistor to provide a power pulse suitably shaped for driving the subsequent logic functions.

The dynamic frequency range is primarily determined by the dynamic range of the voltage-controlled current generator. The absolute values of frequency are determined by the absolute values of current supplied to the timing capacitor, the absolute value of the timing capacitor, and the upper and lower voltage levels between which the timing capacitor is regulated.

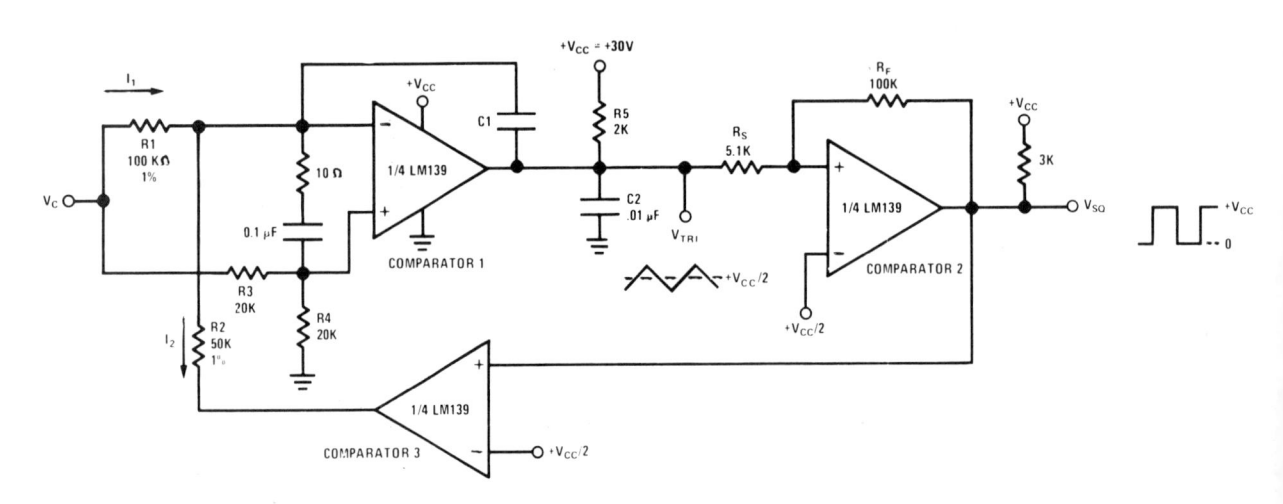

A wide-range stable voltage-controlled oscillator using a minimum of external components. Comparator 1 is used as an integrator, comparator 2 is used as a triangle-to-square-wave converter, and comparator 3 is the switch driving the integrator.

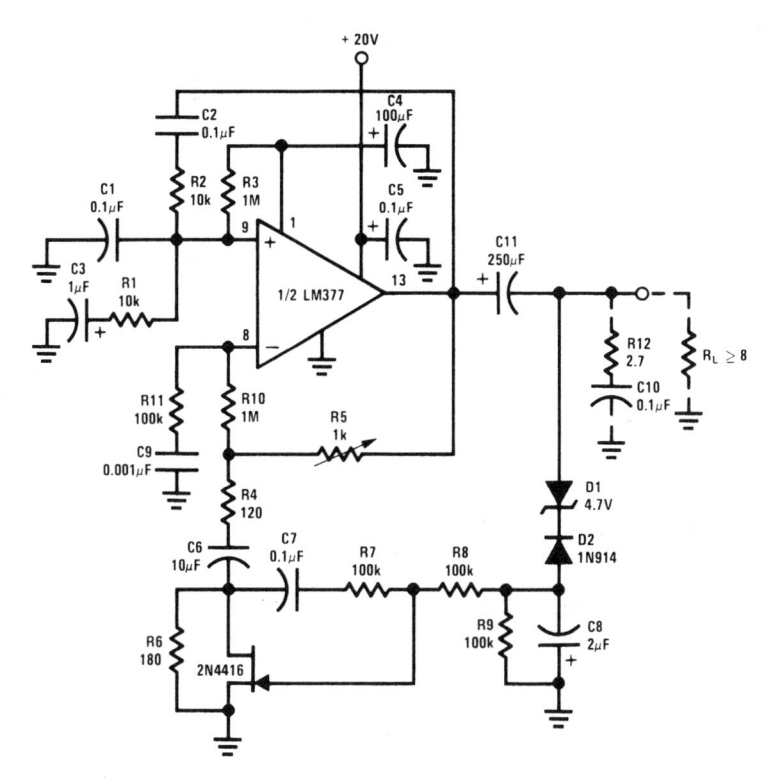

Wien bridge oscillator (National) with FET amplitude stabilization in the negative feedback path. The circuit employs internal biasing and operates from a single supply. C3 and C9 allow unity-gain dc feedback and isolate the bias from ground. Total harmonic distortion is under 1% to 10 kHz and could be improved with careful adjustment of R5. The FET acts as the variable element in the feedback attenuator R4 to R6. Minimum negative feedback gain is set by resistors R4 to R6, while the FET shunts R6 to increase gain in the absence of adequate output signal. The peak detector (D2, C8) senses output level to apply control bias to the FET. Zener D1 sets the output level, although adjustment could be made if R9 were a potentiometer with R8 connected to the slider. Maximum output level with the values shown is 5.3V RMS at 60 Hz. C7 and the attenuator of R7 and R8 couple half the signal of the FET drain to the gate for improved FET linearity and low distortion. The amplitude control loop could be replaced by an incandescent lamp in noncritical circuits, although dc offset will suffer by a factor of about 3 (dc gain of the oscillator). R10 matches R3 for improved dc stability, and the network R11, C9 increases high-frequency gain for improved stability. Without this RC network, oscillation may occur on the negative half-cycle of the output waveform. A low-inductance capacitor, C5, located directly at the supply leads on the package, is important to maintain stability and present high-frequency oscillation on negative half-cycles of the output waveform. C5 may be 0.1 μF ceramic or 0.47 μF Mylar. Layout is important; especially take care to avoid ground loops. If high-frequency instability still occurs, add the R12, C10 network to the output.

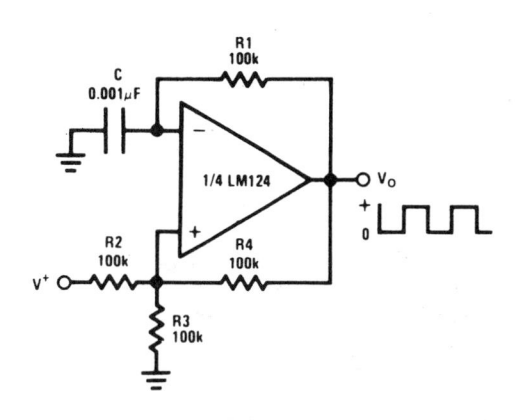

Square-wave oscillator.

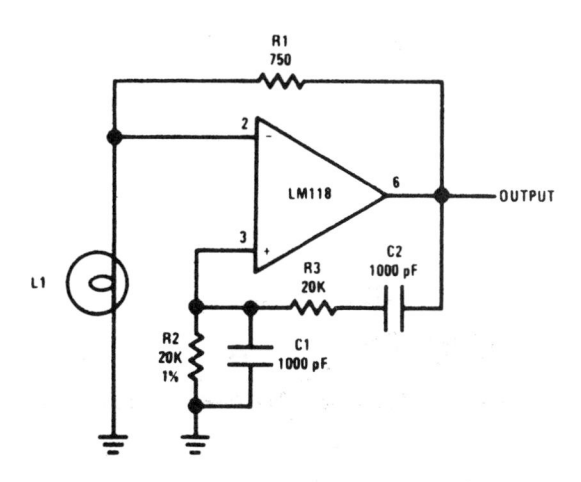

Wien bridge sine-wave oscillator. L1 is 10V 14 mA bulb (Eldema 1869). R1 = R2, C1 = C2; $f = 1/(2\pi R2C1)$.

SECTION **4** Waveform Generators and Shapers

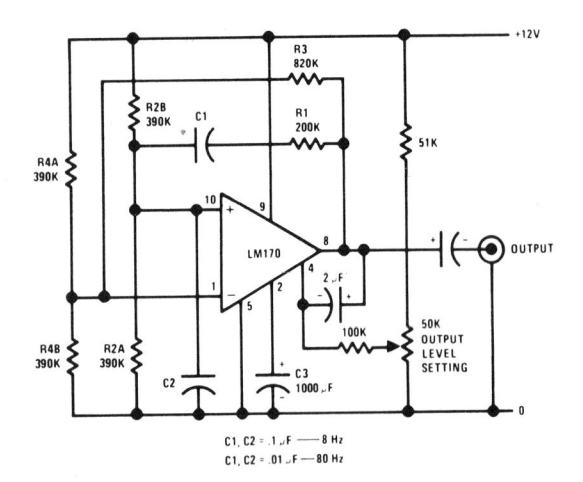

Wien bridge oscillator. The positive feedback loop from pin 8, the output, to pin 10 uses R = 200k and C = 0.1 for 8 Hz. The R of the lag arm is formed from two resistors that provide bias for pin 10. The 820k resistor provies a maximum loop gain of about 4; the system needs a gain of 3 for oscillation, since the attenuation of the positive feedback loop at resonance is 3.

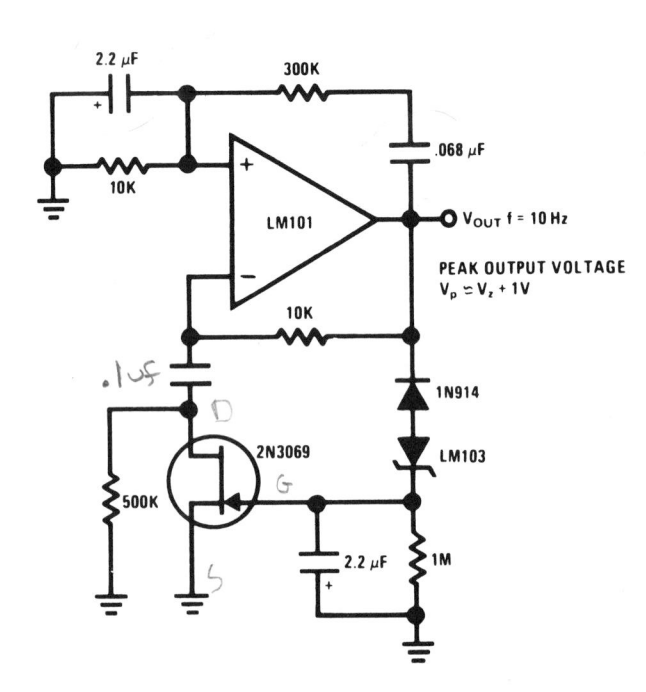

Wien bridge sine-wave oscillator. The major problem in producing a low-distortion, constant-amplitude sine wave is getting the amplifier loop gain just right. However, this can be easily achieved by using the 2N3069 JFET as a voltage-variable resistor in the amplifier feedback loop. The zener provides the voltage reference for the peak sine-wave amplitude; this is rectified and fed to the gate of the 2N3069, thus varying its channel resistance and loop gain.

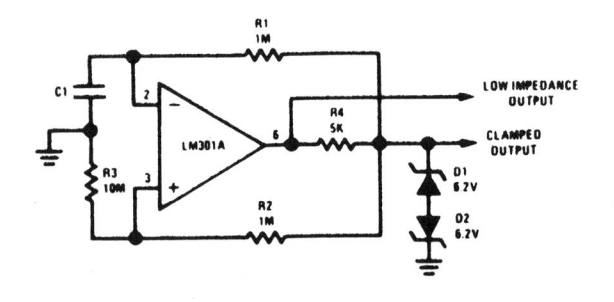

Low-frequency square-wave generator.

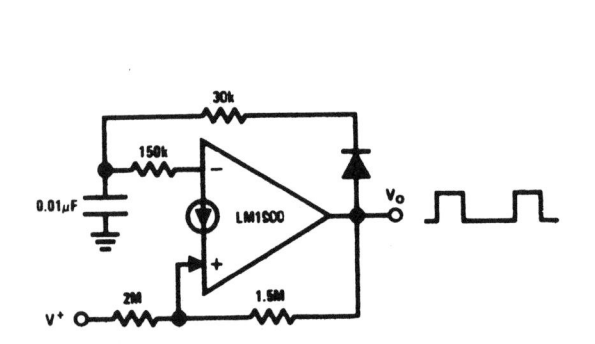

Pulse generator.

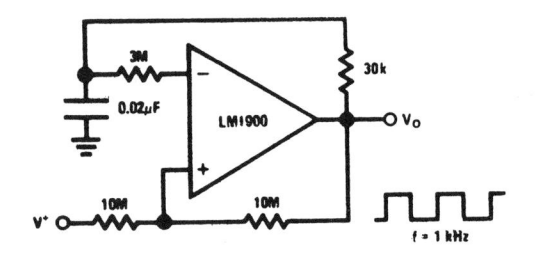

Square-wave oscillator.

150 IC SCHEMATIC SOURCEMASTER

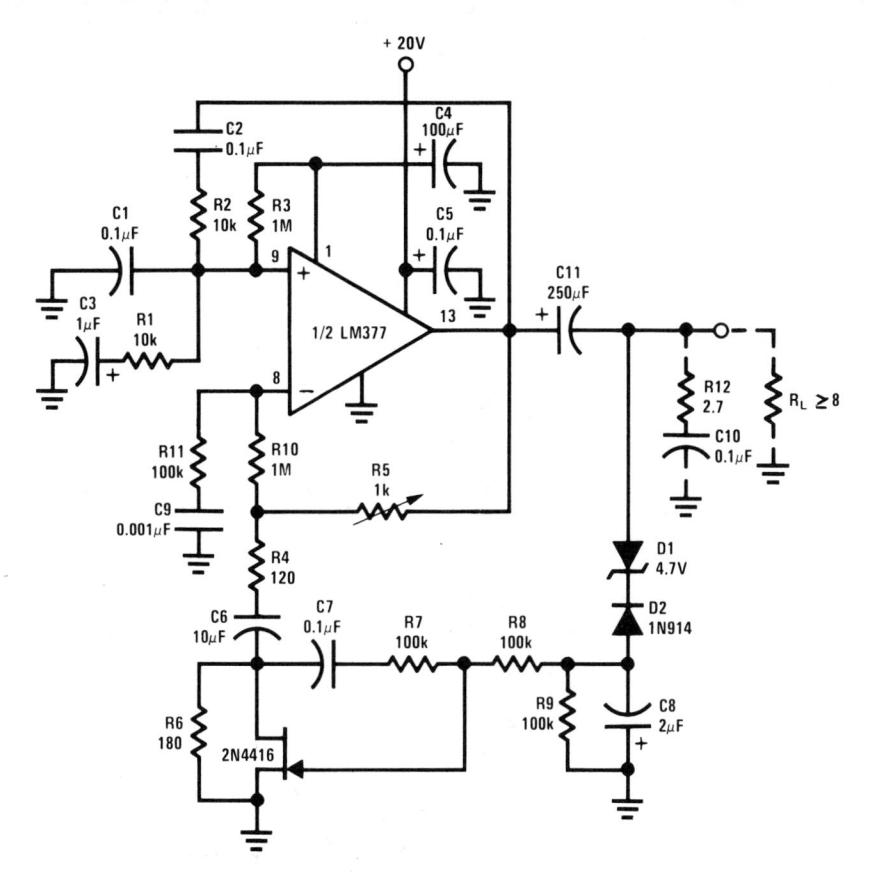

A Wien bridge power oscillator with FET amplitude stabilization in the negative feedback path. The circuit employs internal biasing and operates from a single supply. C3 and C6 allow unity-gain dc feedback and isolate the bias from ground. Total harmonic distortion is under 1% to 10 kHz and could be improved with careful adjustment of R5. The FET acts as the variable element in feedback attenuator of R4 to R6. Minimum negative feedback gain is set by resistors R4 to R6, while the FET shunts R6 to increase gain in the absence of adequate output signal. The peak detector (D2 and C8) senses output level to apply control bias to the FET. Zener D1 sets the output level. Maximum output level with the values shown is 5.3V at 60 Hz.

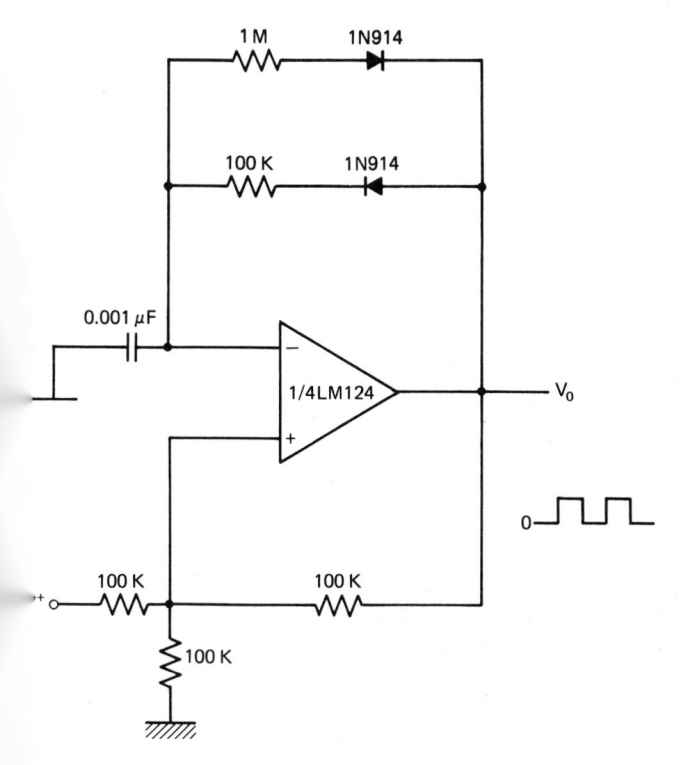

Pulse generator.

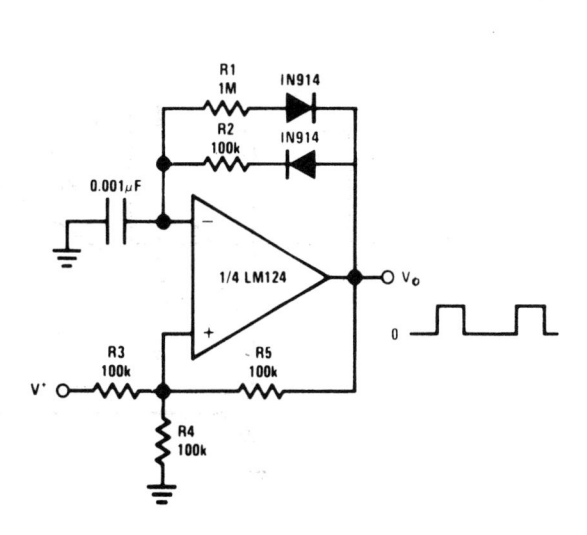

Pulse generator.

SECTION **4** *Waveform Generators and Shapers*

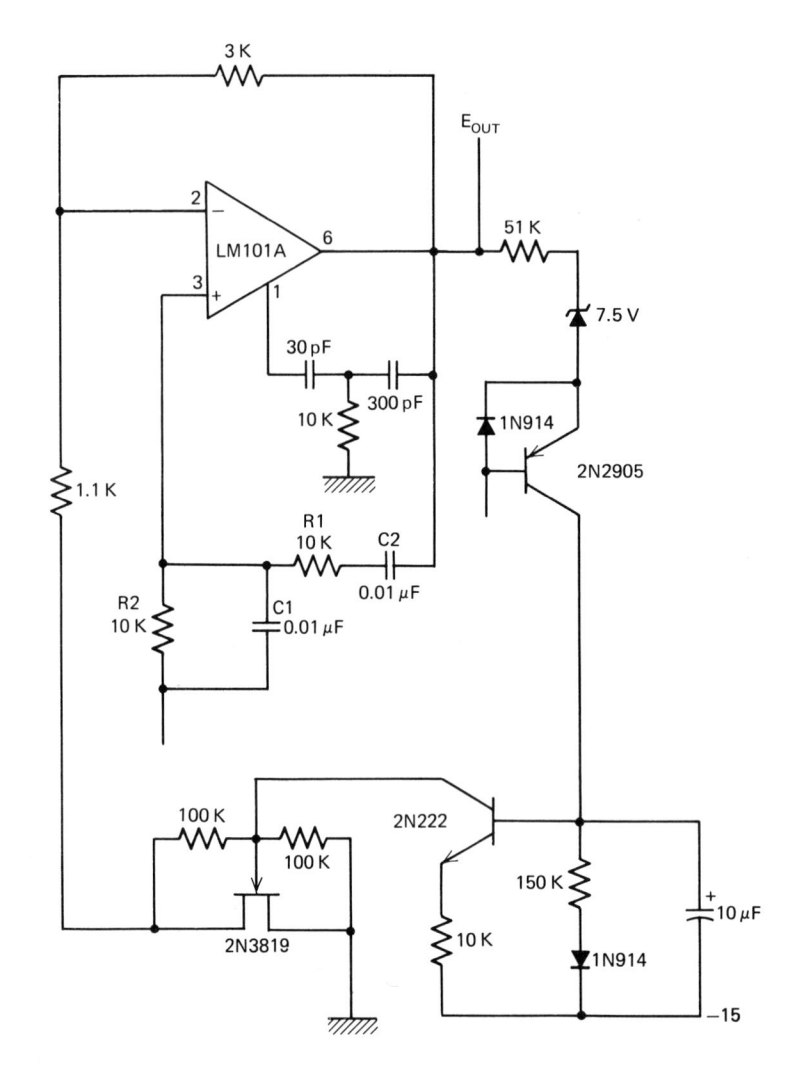

Wien bridge oscillator with FET amplitude stabilization. R1 = R2, C1 = C2, $f = 1/(2\pi R1C1)$.

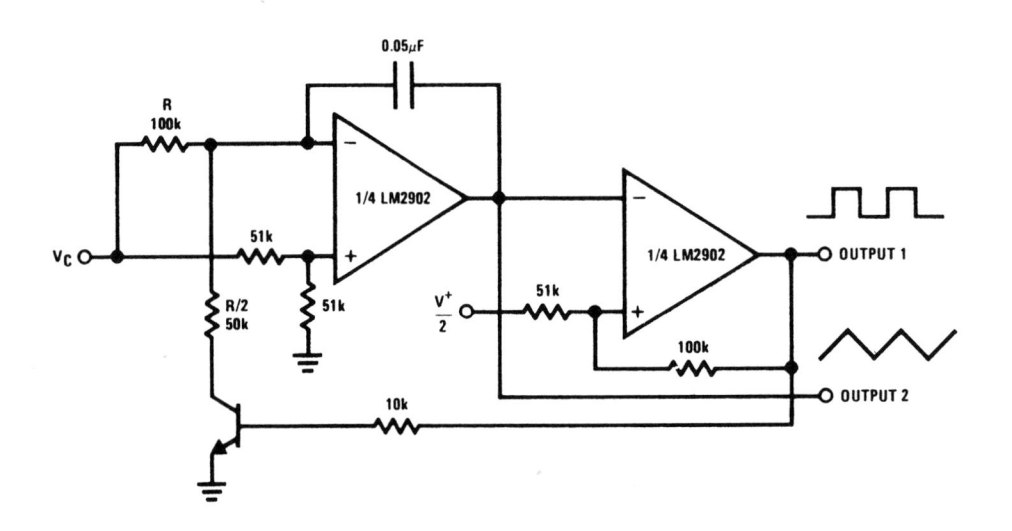

Voltage-controlled oscillator circuit.

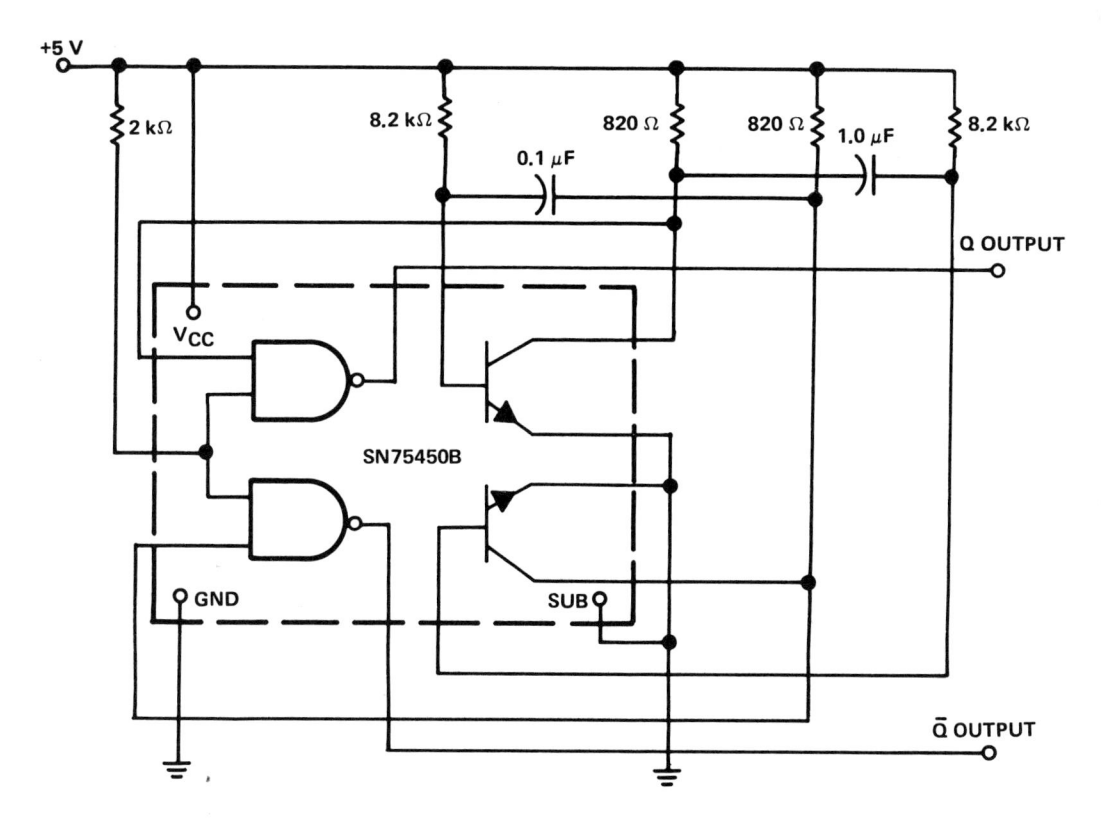

This square-wave generator can drive two-phase TTL clock inputs directly.

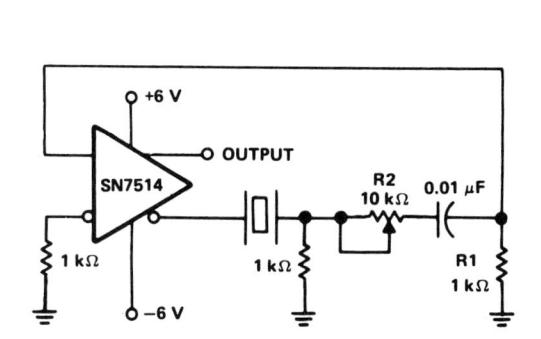

A crystal oscillator with variable feedback provided by a simple voltage divider. The crystal precisely defines the operating frequency, since only at this frequency will significant feedback signal occur. The divider network can be used to adjust the feedback characteristic to provide an undistorted sinusoidal output. Since the SN7514 provides greater than 50 dB gain at frequencies up to 80 MHz, this network could be used to above 200 MHz in an oscillator circuit of this type.

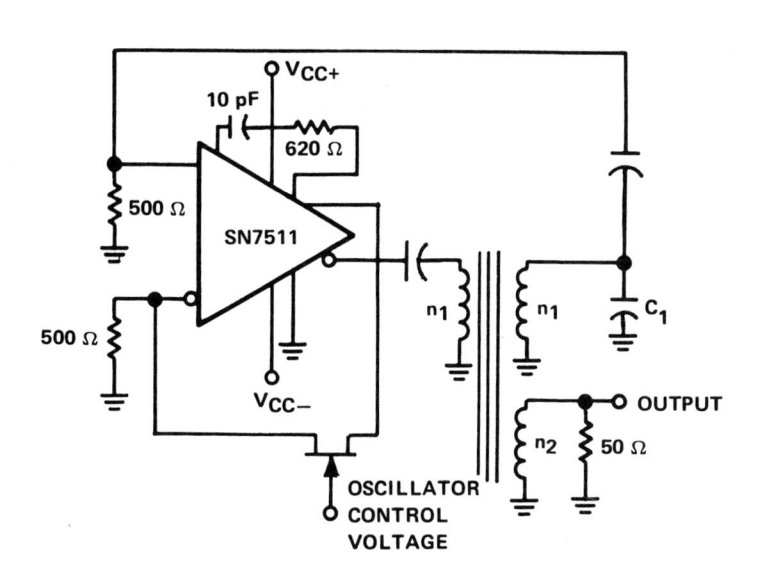

Transformer feedback oscillator. With unity turns ratio and C and L at resonance, the feedback factor is unity, and the gain of the network is the voltage gain of the amplifier, which is the resistance of the FET divided by 500Ω. If the gain and feedback-factor product is greater than 1 (in this case, if G is greater than 1), oscillations will occur at a frequency determined by the LC network. With proper choice of component values, the circuit can be used at frequencies as high as 50 MHz. The feedback loop provides a voltage-controlled method of reducing the amplifier gain to below that required to sustain oscillation. The gate input of the FET can therefore be used as a strobe control to switch the oscillator off and on.

SECTION **4** *Waveform Generators and Shapers* **153**

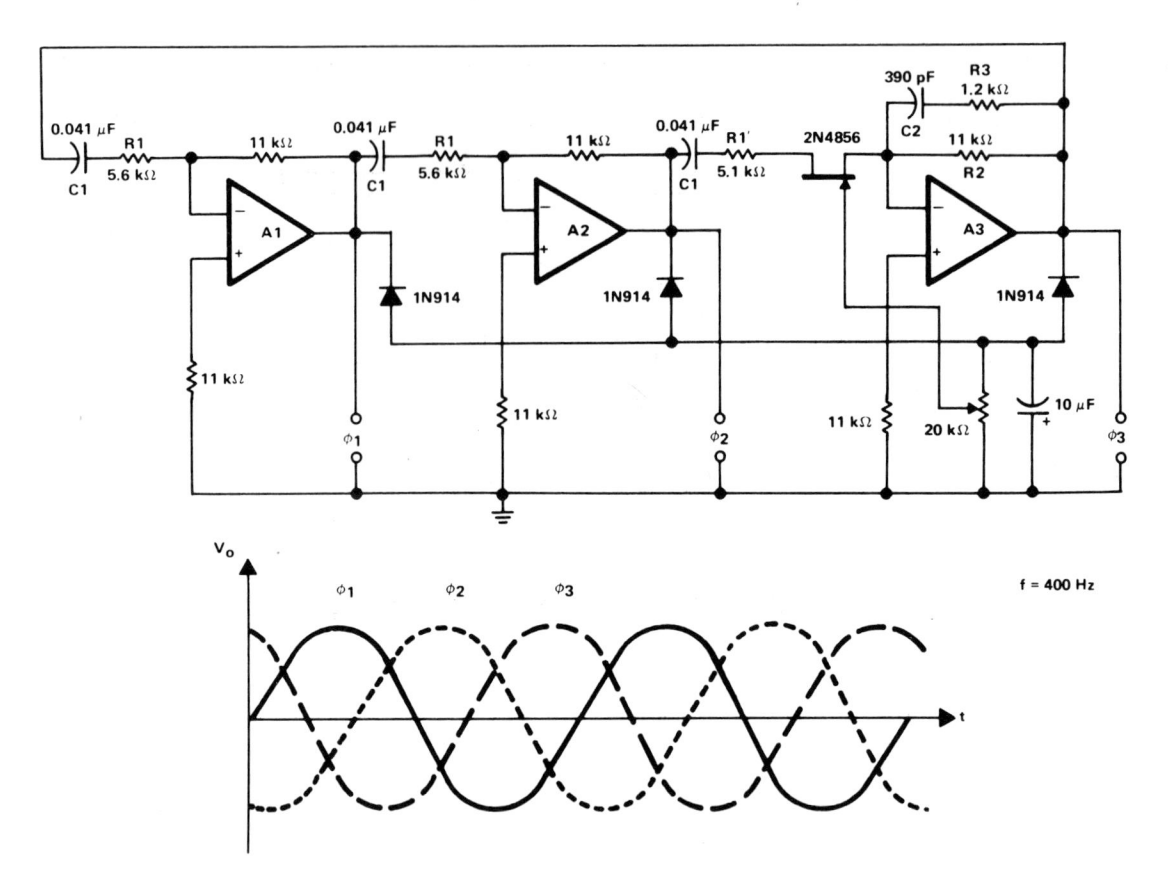

Three-phase signal generator employs three stages with individual phase shifts of 120° for three-phase generation. The first two stages, A1 and A2, are basic active phase-shift networks that must have unity gain at a phase shift of 120°. Therefore, the tangent of 120°, which is $\sqrt{3}$, must equal $X_{C1}/R1$ at the frequency of operation.

Output amplitude is adjusted by a control inserted in the third stage (amplifier A3). A 20k potentiometer is used to set the FET operating level in the inverting input of this amplifier. The FET *on* resistance will be very low initially, giving amplifier A3 a gain in excess of unity to ensure that the oscillator starts. As soon as oscillations reach a desired level, AGC feedback from each amplifier to the FET gate increases its *on* resistance until R1' + R_{ON} = R1 and operational balance is achieved. Some high-frequency rolloff, preventing undesired oscillations and noise, is obtained by the use of C2 and R3.

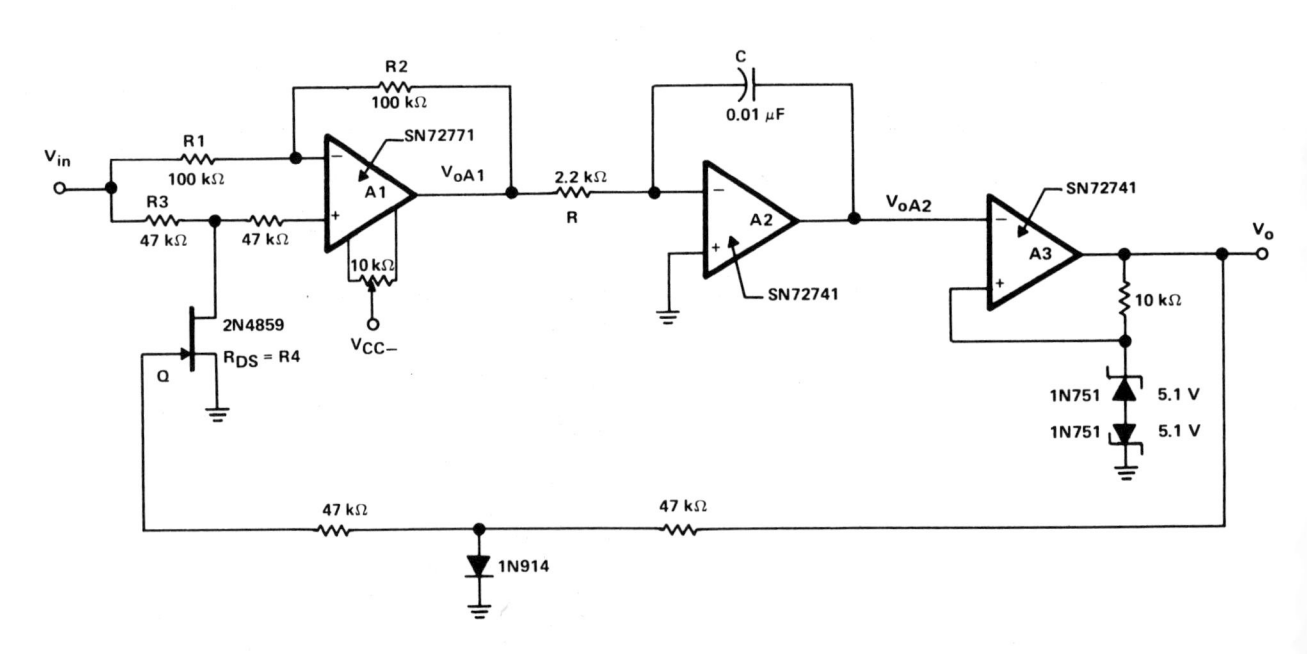

Multivibrator with voltage-controlled frequency.

154 IC SCHEMATIC SOURCEMASTER

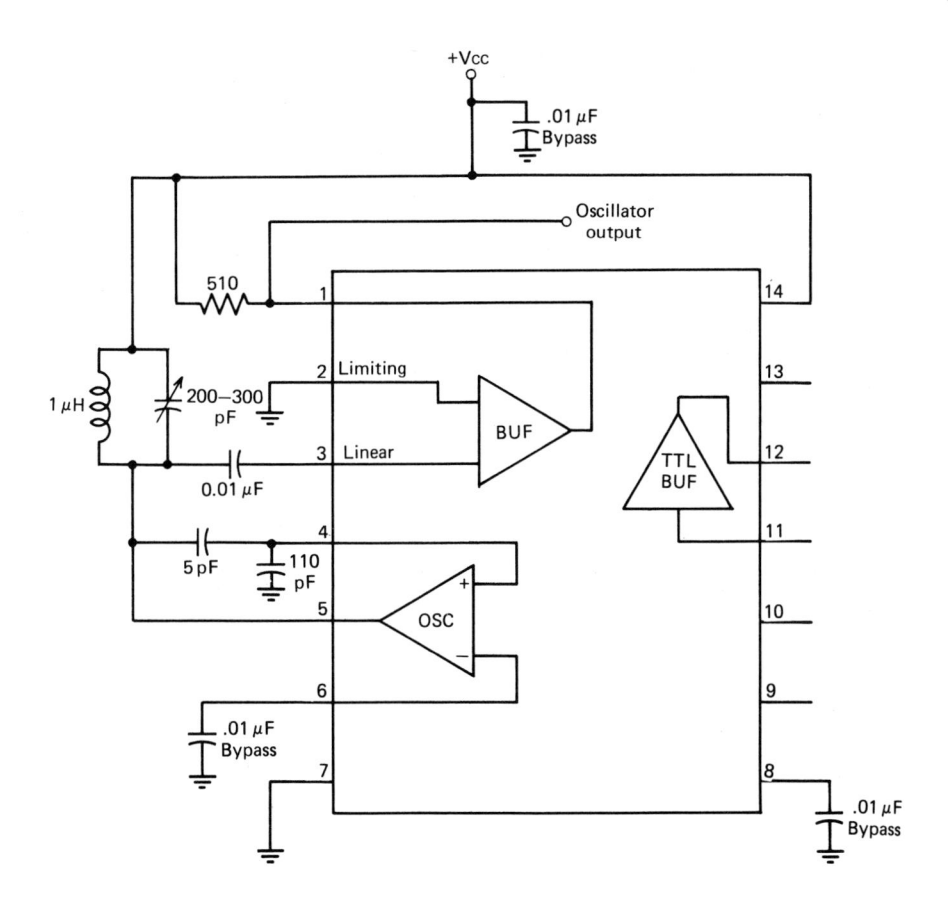

10 MHz LC sine-wave oscillator.

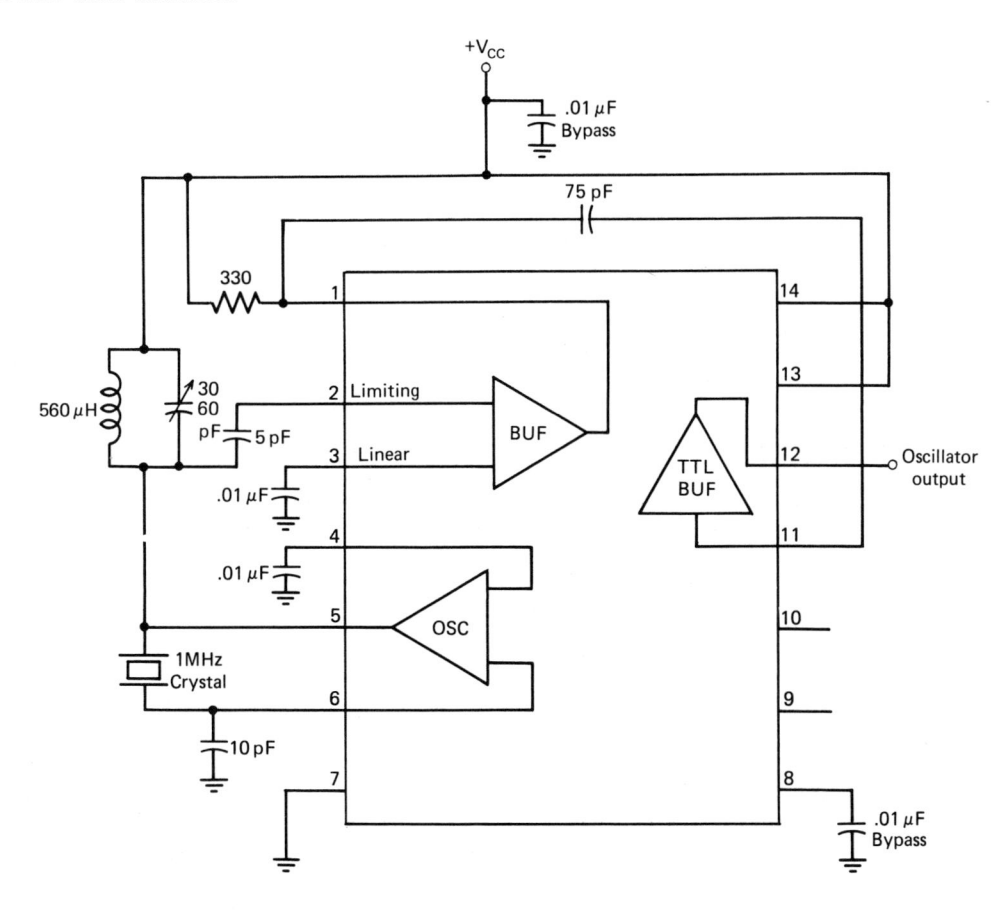

1 MHz crystal oscillator with TTL output.

SECTION **4** *Waveform Generators and Shapers* 155

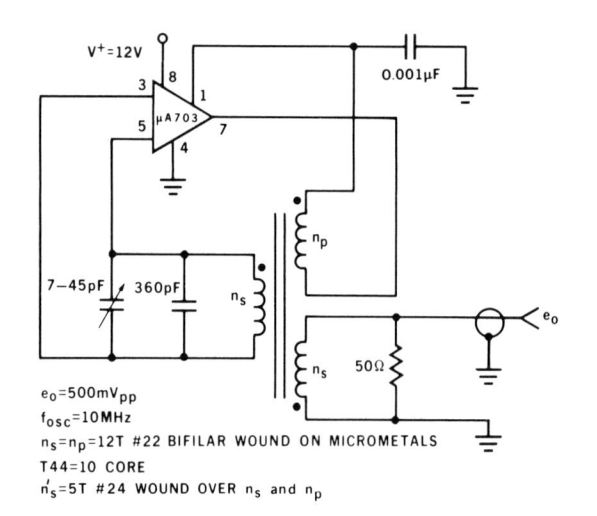

A 10 MHz oscillator circuit with collector voltage swing of $2V_{p-p}$; an effective load resistance of 400Ω is required using the maximum output current of 5 mA.

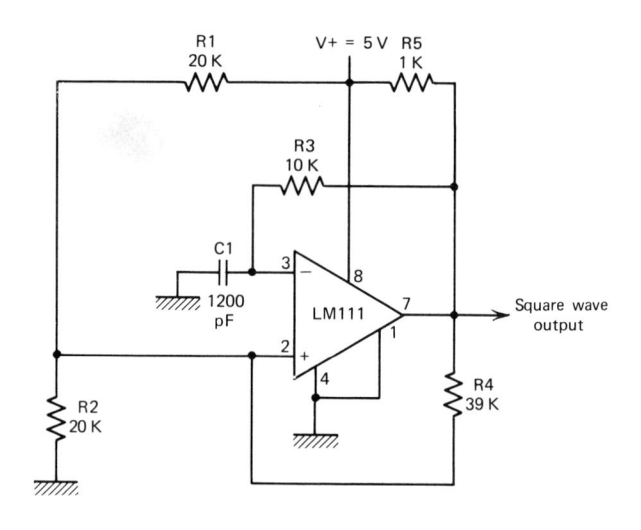

100 kHz free-running multivibrator. Output is TTL or DTL fanout of two.

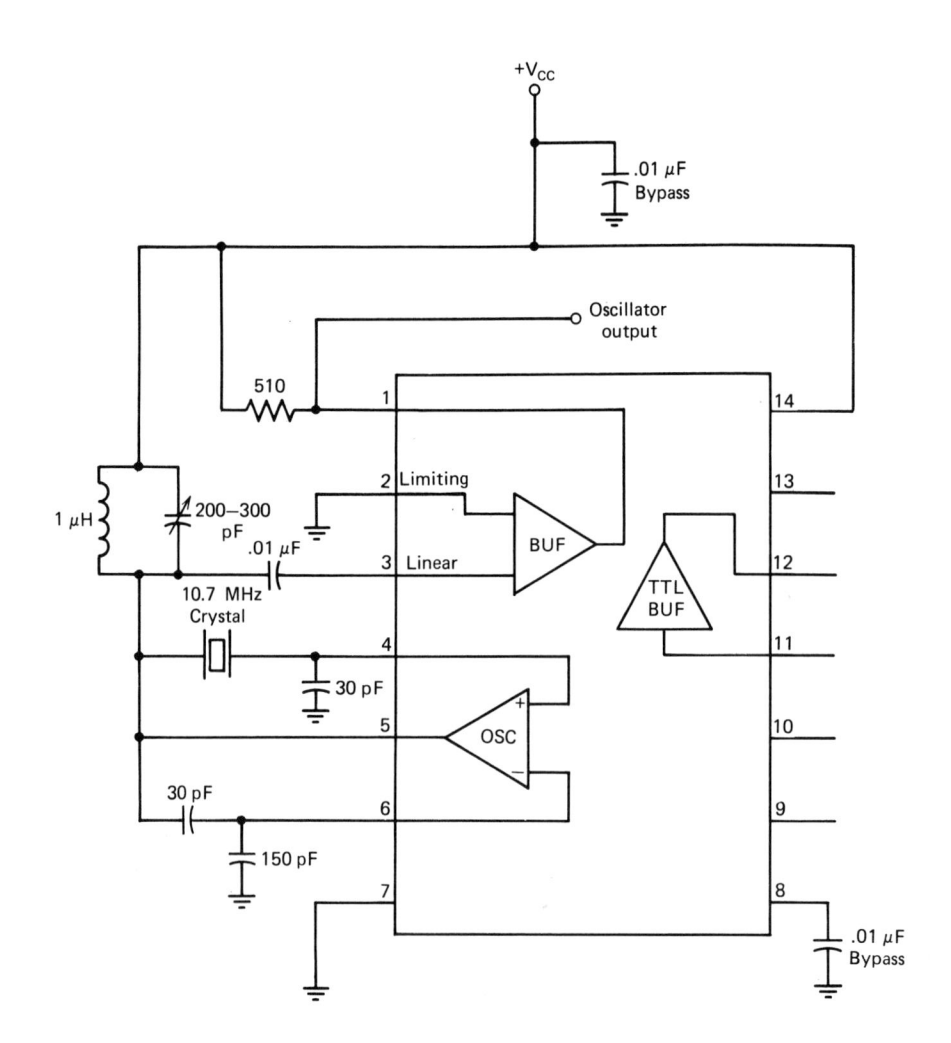

10.7 MHz series-resonant crystal oscillator.

156 IC SCHEMATIC SOURCEMASTER

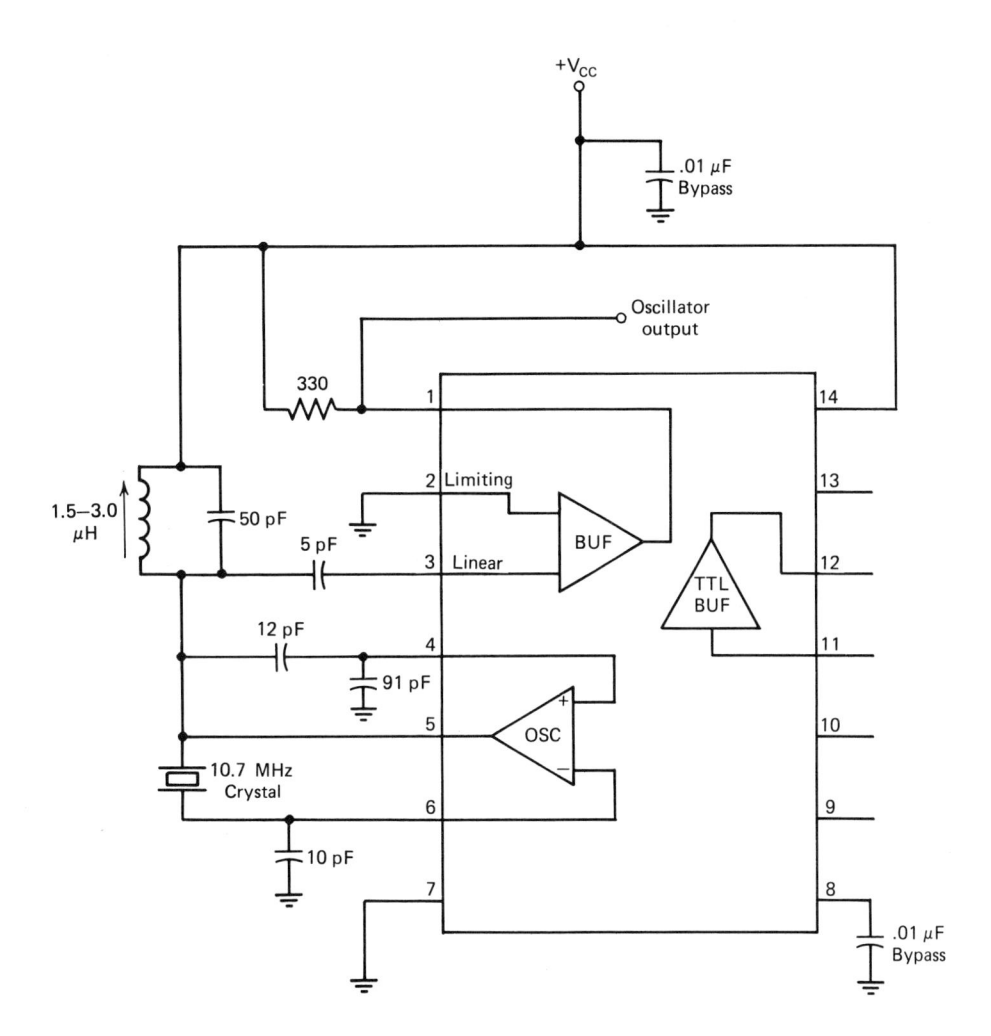

10.7 MHz parallel-resonant crystal oscillator.

FREQUENCY RANGE: 50 MHz TO 100 MHz, DEPENDENT ON CRYSTAL FREQUENCY AND TANK TUNING

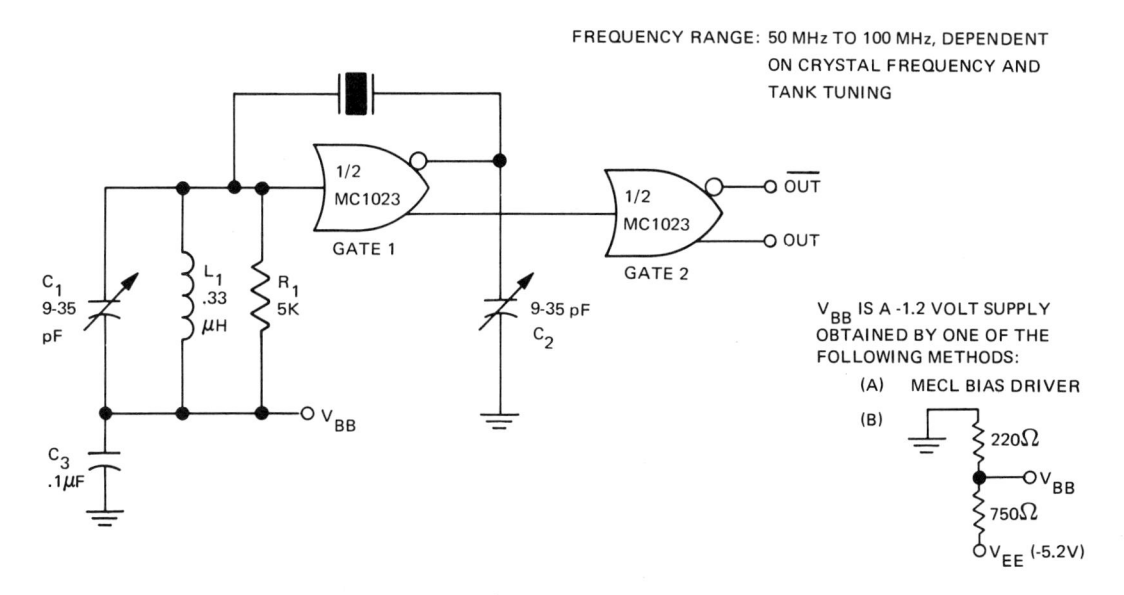

V_{BB} IS A -1.2 VOLT SUPPLY OBTAINED BY ONE OF THE FOLLOWING METHODS:

(A) MECL BIAS DRIVER
(B) [220Ω / 750Ω divider to V_{EE} (-5.2V)]

This crystal-controlled oscillator employs an adjustable resonant tank circuit that ensures operation at the desired crystal overtone. C1 and L1 form the resonant tank circuit which, with the values specified, has a resonant frequency adjustable from 50 to 100 MHz. Overtone operation is accomplished by adjusting the tank circuit frequency at or near the desired frequency. The tank circuit exhibits a low-impedance shunt to off-frequency oscillations and a high impedance to the desired frequency, allowing feedback from the output. Operation in this manner guarantees that the oscillator will always start at the correct overtone.

SECTION **4** Waveform Generators and Shapers

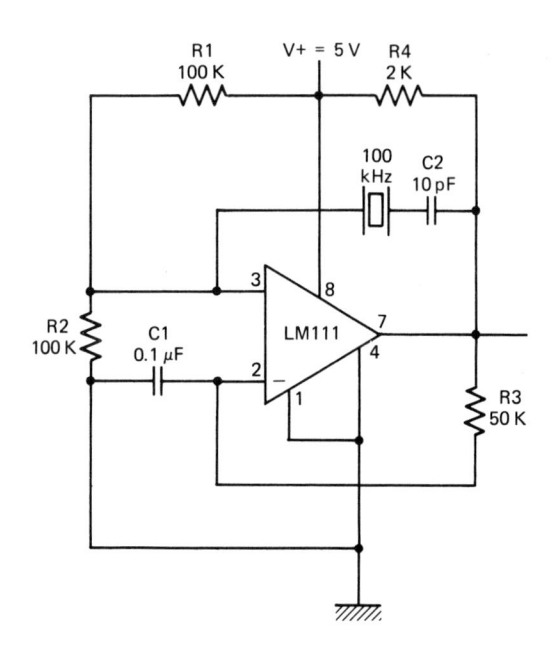

Crystal oscillator.

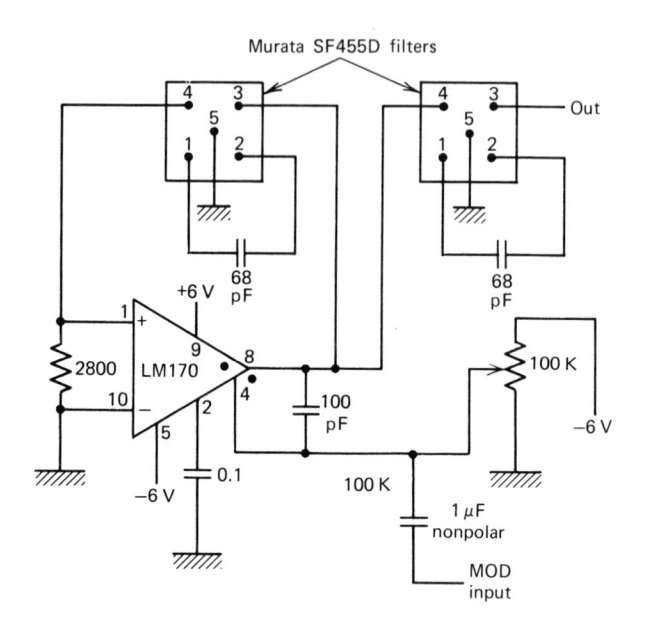

455 kHz modulated signal generator.

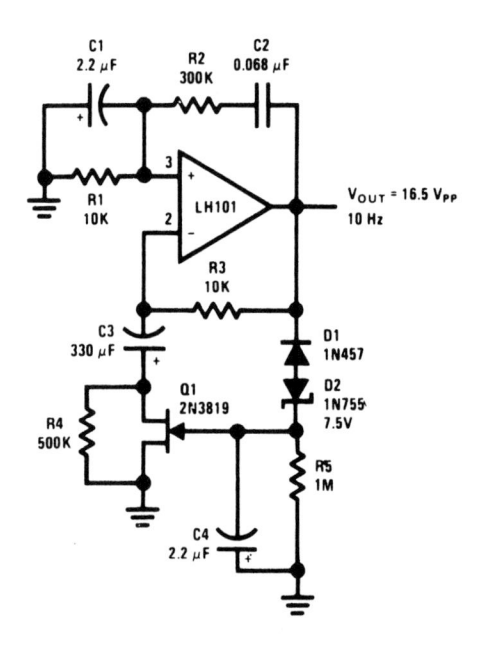

This amplitude-stabilized Wien bridge sine-wave oscillator provides high-purity sine-wave output down to low frequencies with minimum circuit complexity. The traditional tungsten filament lamp amplitude regulator is eliminated, along with its time constant and linearity problems.

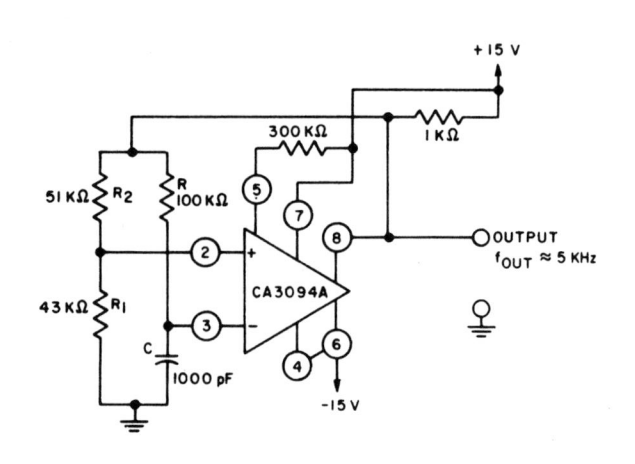

An astable multivibrator using dual supply;

$$f_{OUT} = \frac{1}{2RC \, \ln\left(\frac{2R1}{R2} + 1\right)}$$

If R2 = 3.08R1, f_{OUT} = 1/RC.

158 IC SCHEMATIC SOURCEMASTER

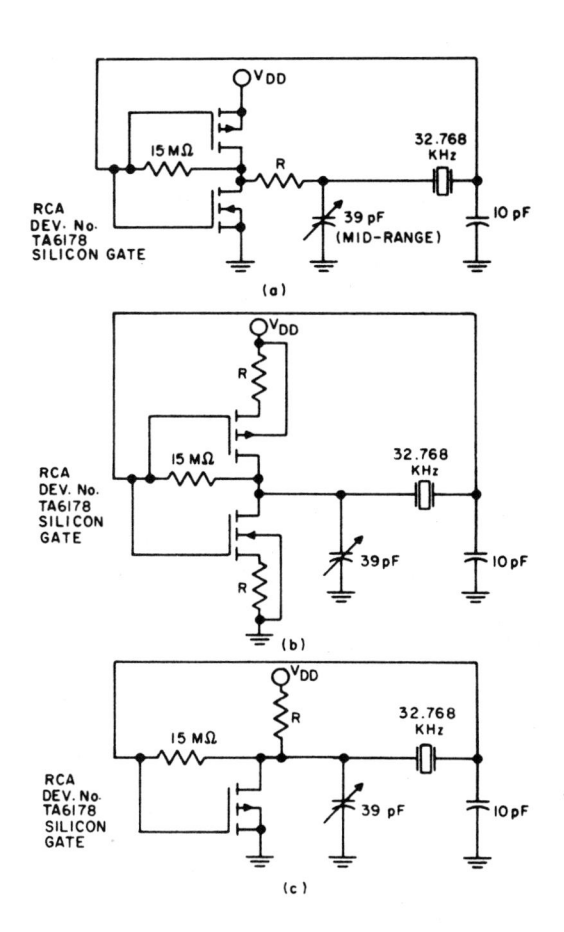

Crystal oscillator circuits.

Typical Oscillator Data

Circuit	Value of R (Ω)	V_{DD} (V)	Current (μA)	Frequency Stability $V_{DD} = 1.45 - 1.6V$
(a)	0	1.60	4.0	2.8
	0	1.45	3.1	
	100k	1.60	3.1	2.6
	100k	1.45	2.4	
	200k	1.60	2.9	2.6
	200k	1.45	2.1	
(b)	100k	1.60	2.3	0.3
	100k	1.45	2.0	
	100k	1.1	1.5	
	150k	1.60	1.8	0.2
	150k	1.45	1.6	
	150k	1.1	0.95	
(c)	200k	1.60	5.0	0.6
	200k	1.45	4.4	
	300k	1.60	3.5	0.5
	300k	1.45	3.0	

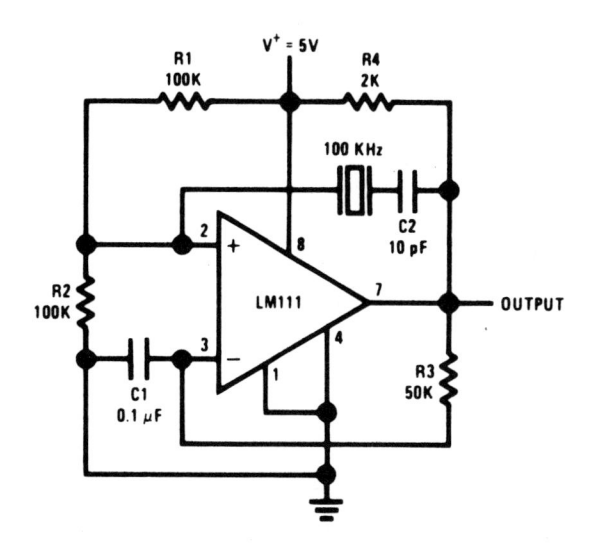

Crystal-controlled oscillator. The high input impedance of the comparator and the isolating capacitor, C2, minimize loading of the crystal and contribute to frequency stability. As shown, the oscillator delivers a 100 kHz square-wave output.

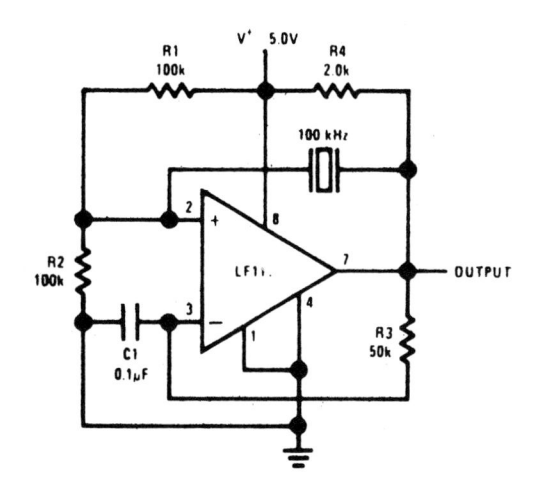

Crystal oscillator.

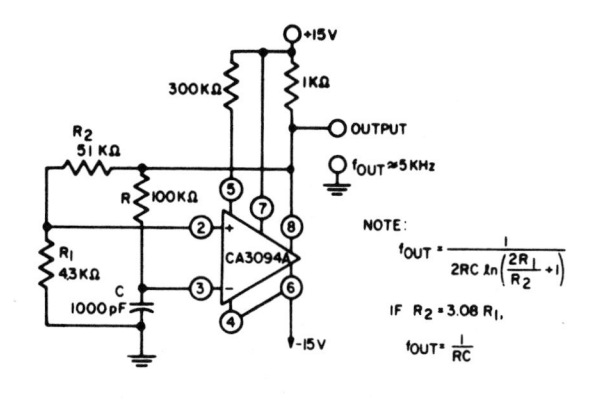

Op-amp astable multivibrator (dual-supply).

NOTE:
$$f_{OUT} = \frac{1}{2RC \ln\left(\frac{2R_1}{R_2}+1\right)}$$

IF $R_2 = 3.08 R_1$,
$$f_{OUT} = \frac{1}{RC}$$

SECTION 4 *Waveform Generators and Shapers* 159

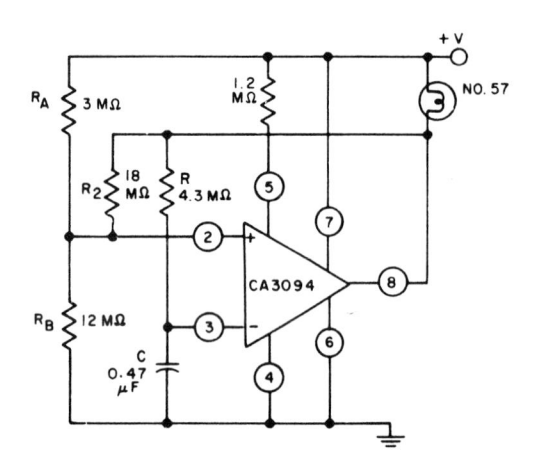

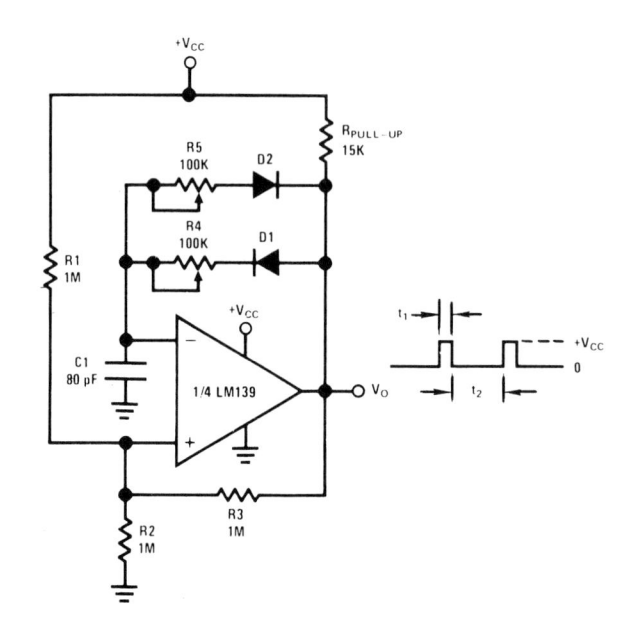

An astable multivibrator flasher using single supply features one flash per second, 25% duty cycle, frequency independent of V⁺ from 6 to 15V dc.

$$f_{osc} = \frac{1}{2RC \ln\left(\frac{2R1}{R2} + 1\right)}$$

where $R1 = R_A R_B / R_A + R_B$.

Basic square-wave generator can be modified to obtain an **adjustable-duty-cycle pulse generator** by providing a separate charge and discharge path for capacitor C1. Path through R4 and D1 will charge the capacitor and set the pulse width (t_1). The other path, R5 and D2, will discharge the capacitor and set the time between pulses (t_2). By varying resistor R5, the time between pulses of the generator can be changed without changing the pulse width. By varying R4, the pulse width will be altered without affecting the time between pulses. However, both controls will change the frequency of the generator.

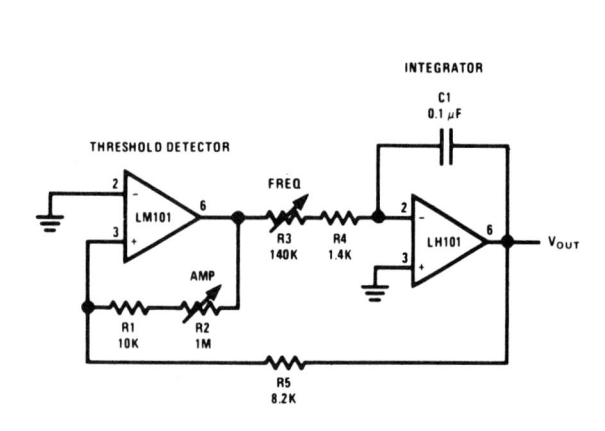

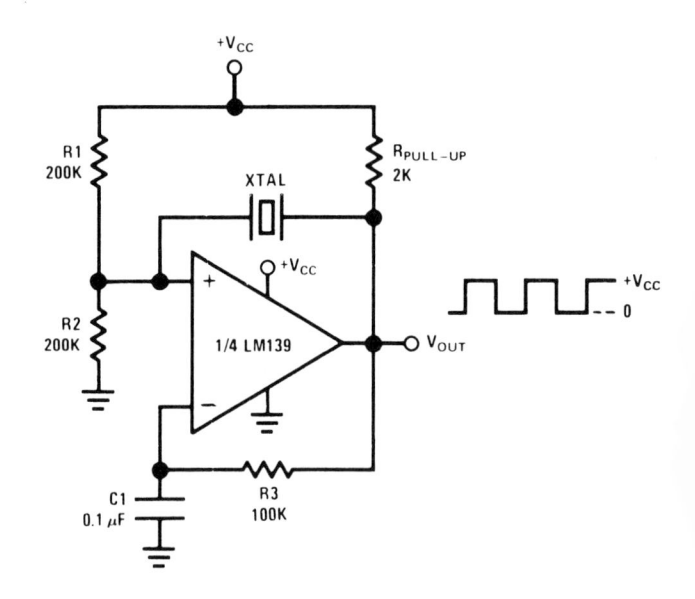

This **constant-amplitude triangular-wave generator** embodies an integrator as a ramp generator and a threshold detector with hysteresis as a reset circuit. Frequency is determined by R3, R4, and C1 and the positive and negative saturation voltages of the amplifier. Amplitude is determined by the ratio of R5 to the combination of R1 and R2 and the threshold detector saturation voltages. Positive and negative ramp rates are equal, and positive and negative peaks are equal if the detector has equal positive and negative saturation voltages. The output waveform may be offset with respect to ground if the inverting input of the threshold detector is offset with respect to ground.

Crystal-controlled oscillator. R1 and R2 are equal so that the comparator will switch symmetrically about $+V_{CC}/2$. The RC time constant of R3 and C1 is set to be several times greater than the period of the oscillating frequency, ensuring a 50% duty cycle by maintaining a dc voltage at the inverting input equal to the absolute average of the output waveform.

When specifying the crystal, be sure to order *series resonant* along with the desired temperature coefficient and load capacitance to be used.

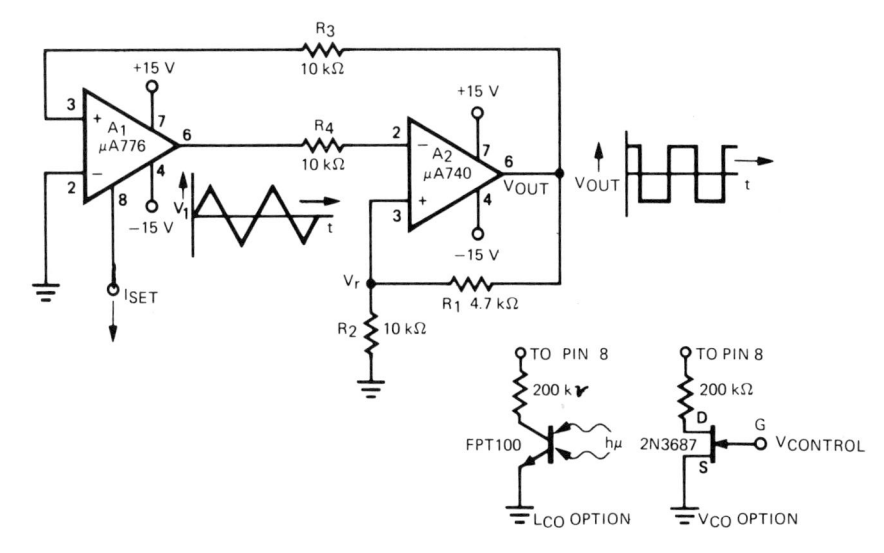

Current-controlled oscillator. The current-controlled oscillator exhibits five decades of frequency range in response to a similar range of set currents fed to pin 8 of the amplifier. This linear change in frequency as a function of set current is possible because the slew rate of the μA776 is directly proportional to the current out of the master bias-setting pin of the device.

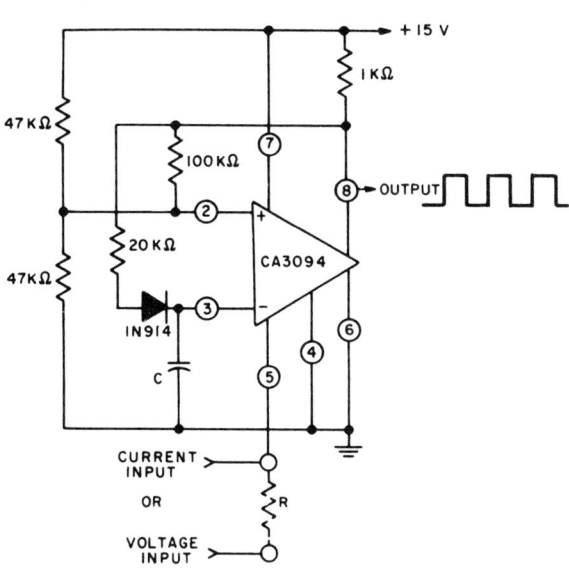

Current- or voltage-controlled oscillator.

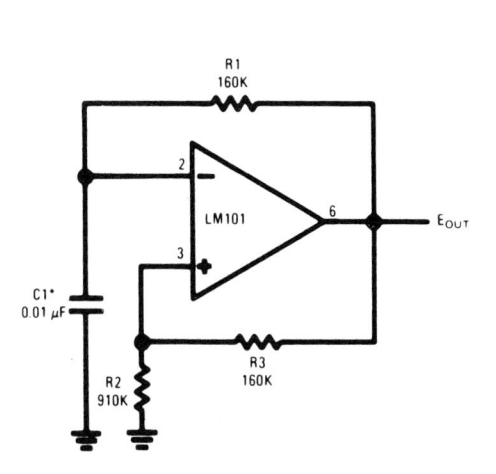

This free-running multivibrator is an excellent example of an application where one does not normally consider using an operational amplifier. However, this circuit operates at low frequencies with relatively small capacitors because it can use a longer portion of the capacitor time constant, since the threshold point of the operational amplifier is well determined. In addition, it has a completely symmetrical output waveform along with a buffered output, although the symmetry can be varied by returning R2 to some voltage other than ground. C1 is chosen for oscillation at 100 Hz.

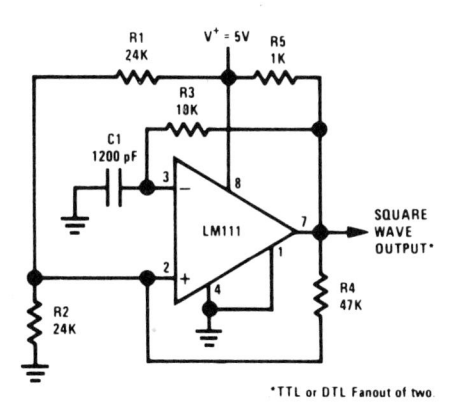

This free-running multivibrator produces a 100 kHz square wave. The frequency of oscillation depends almost entirely on the resistor and capacitor values because of the precision of the comparator. Further, the frequency changes by only 1% for a 10% change in supply voltage. Waveform symmetry is also good, but the symmetry can be varied by changing the ratio of R1 to R2.

SECTION **4** Waveform Generators and Shapers **161**

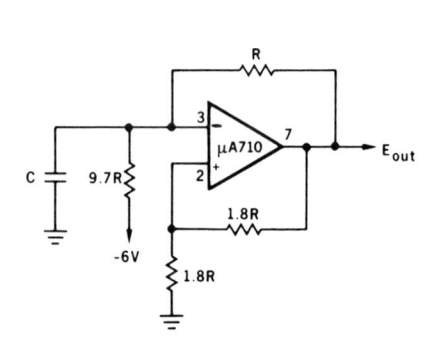

A free-running multivibrator using Fairchild μA710. A large amount of dc negative feedback, along with a resistor connected to the negative supply, is used to set the output of the μA710 in the center of its active region to ensure starting and reasonable symmetry in the output waveform. The capacitor reduces the negative feedback at high frequencies, giving net positive feedback and oscillation.

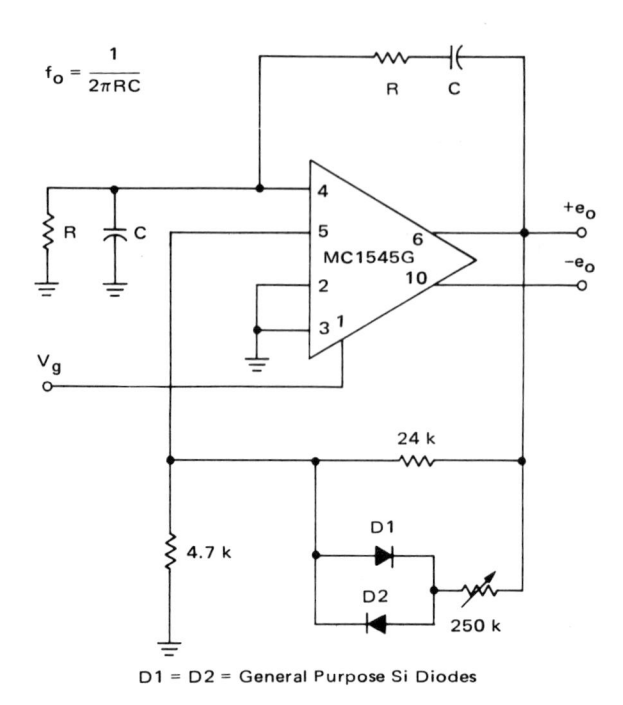

A gated oscillator with an adjustable level control. Rise times of this oscillator are dependent on the RC network, varying from 1 to several cycles of the oscillation frequency. At 160 kHz, t_{rise} is about 30 μsec and at 8 MHz, t_{rise} is greater than 100 nsec. Fall time is the channel select time, equal to 20 nsec.

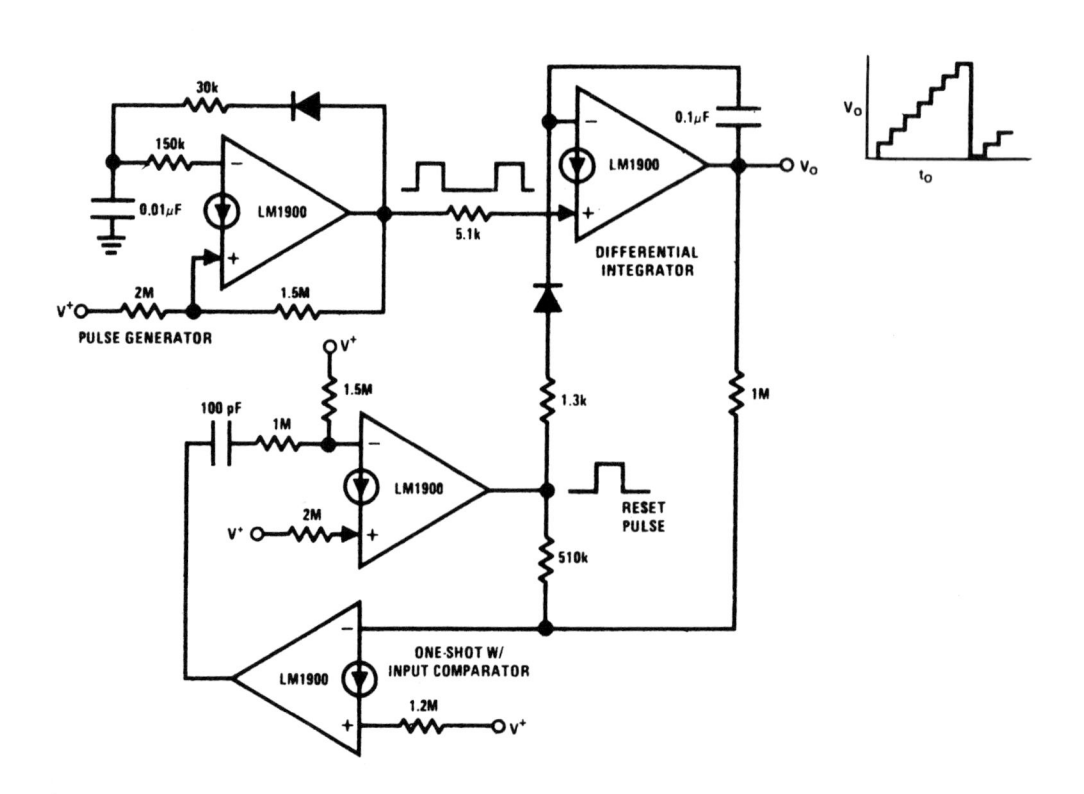

$$f_o = \frac{1}{2\pi RC}$$

Free-running staircase generator/pulse counter.

162   IC SCHEMATIC SOURCEMASTER

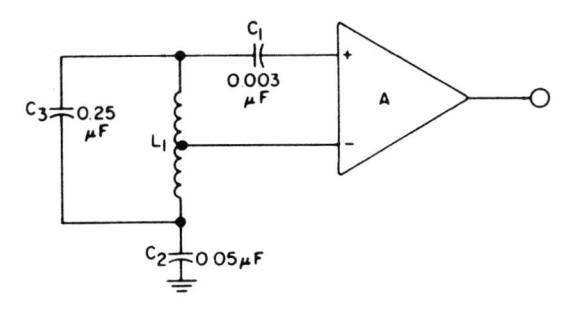

A IS ANY AMPLIFIER OF THE CA3048

This Hartley oscillator is easily designed and constructed using the CA3048 amplifier. No feedback capacitor is required, and it is possible to extract square, sawtooth, or sinusoidal wave shapes. In the circuit, the tap on the coil is located at one-fourth the total turns, capacitors C1 and C2 provide dc blocking, and capacitor C3 tunes with inductor L1 ($\omega_o = 1/\sqrt{L1C3}$). When the circuit is operated from a 12V supply, the output voltage is a clipped sine wave that has a peak-to-peak value of about 7V. The voltage at the inverting input is a sawtooth that has a peak-to-peak value of about 0.300V. If an unclipped sine wave is desired, it is available across coil L1. A sine wave can be obtained in the single-ended connection if the value of C2 is made large with respect to C3 so that it effectively bypasses the sawtooth to ground; the voltage across L1 is then sinusoidal with respect to ground.

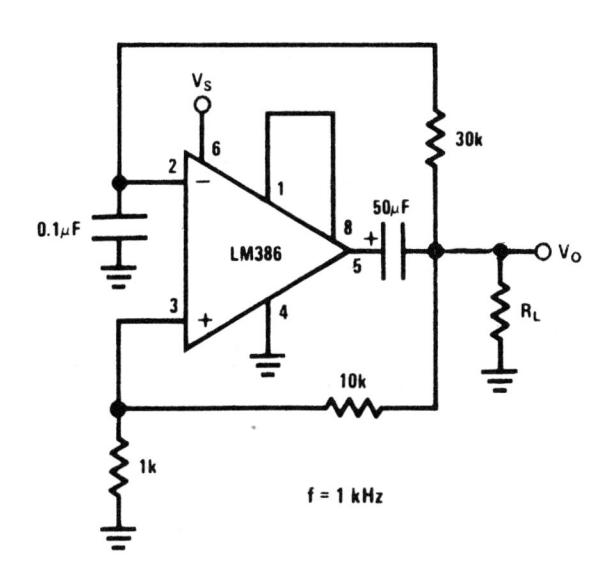

Square-wave oscillator.

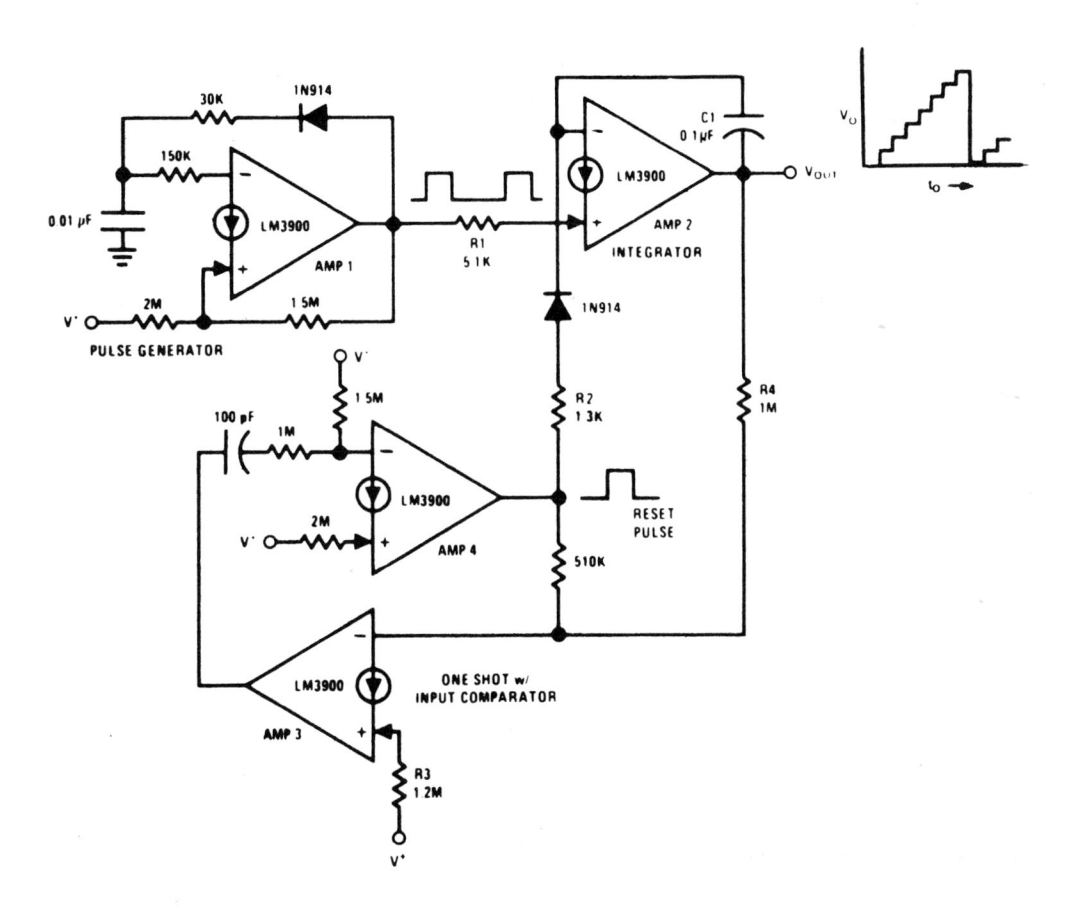

This free-runnig staircase generator uses all four of the amplifiers available in one LM3900 package.

SECTION **4** *Waveform Generators and Shapers* **163**

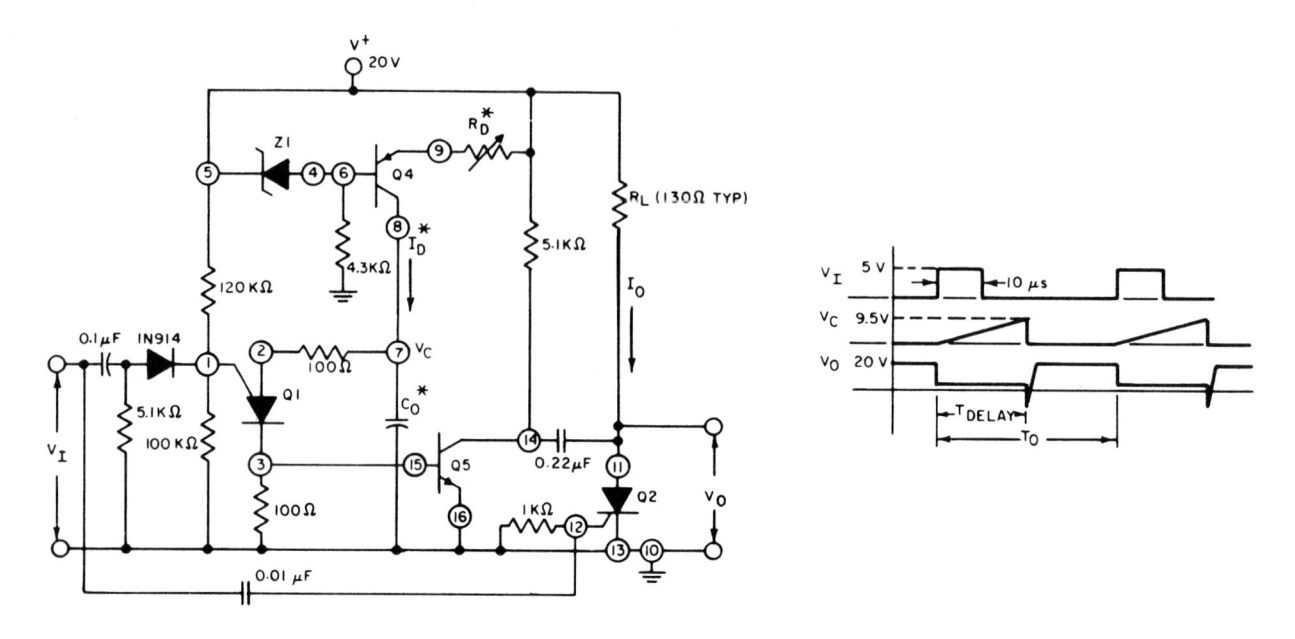

A monostable multivibrator with variable delay. Monostable delay time is set by adjusting I_D (vary R_D) or by C_D. I_D must be greater than I_V of Q1 (PUT) for monostable operation. SCR Q2 switching times: gate-controlled turn-on time (t_{gt}), 50 nsec typical; circuit-commutated turn-off time (t_q), 10 μsec typical.

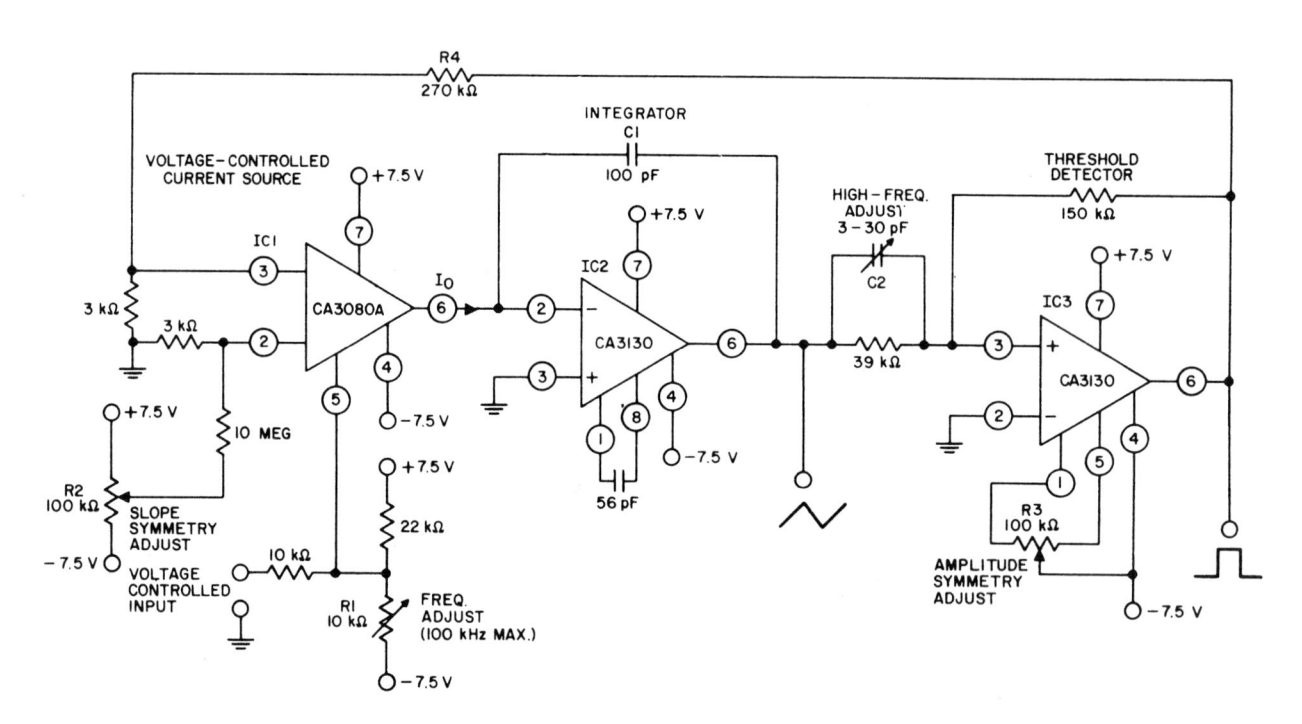

Function generator (RCA). This circuit generates a triangular- or square-wave output that can be swept over a 1,000,000:1 range (0.1 Hz to 100 kHz) by means of a single control, R1. A voltage-control input is also available for remote sweep control. The heart of the frequency-determining system is an operational transconductance amplifier operated as a voltage-controlled current source. The output current, I_o, is applied directly to the integrating capacitor, C1, in the feedback loop of the integrator IC2, using a CA3130, to provide the triangular-wave output. Potentiometer R2 is used to adjust the circuitry for slope symmetry of positive-going and negative-going signal excursions. IC3 is used as a controlled switch to set the excursion limits of the triangular output from the integrator circuit. Capacitor C2 is a peaking adjustment to optimize the square wave. Potentiometer R3 adjusts the amplitude symmetry of the square-wave output. Output from the threshold detector is fed back via resistor R4 to the input of IC1 to toggle the current source from plus to minus in generating the linear triangular wave.

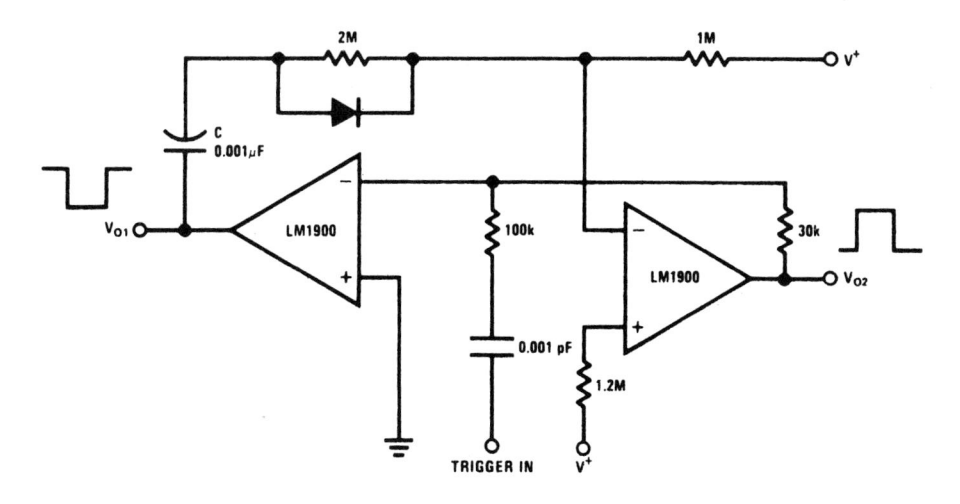

One-shot multivibrator. The pulse width is about $2 \times 10^6 C$. Diode speeds recovery.

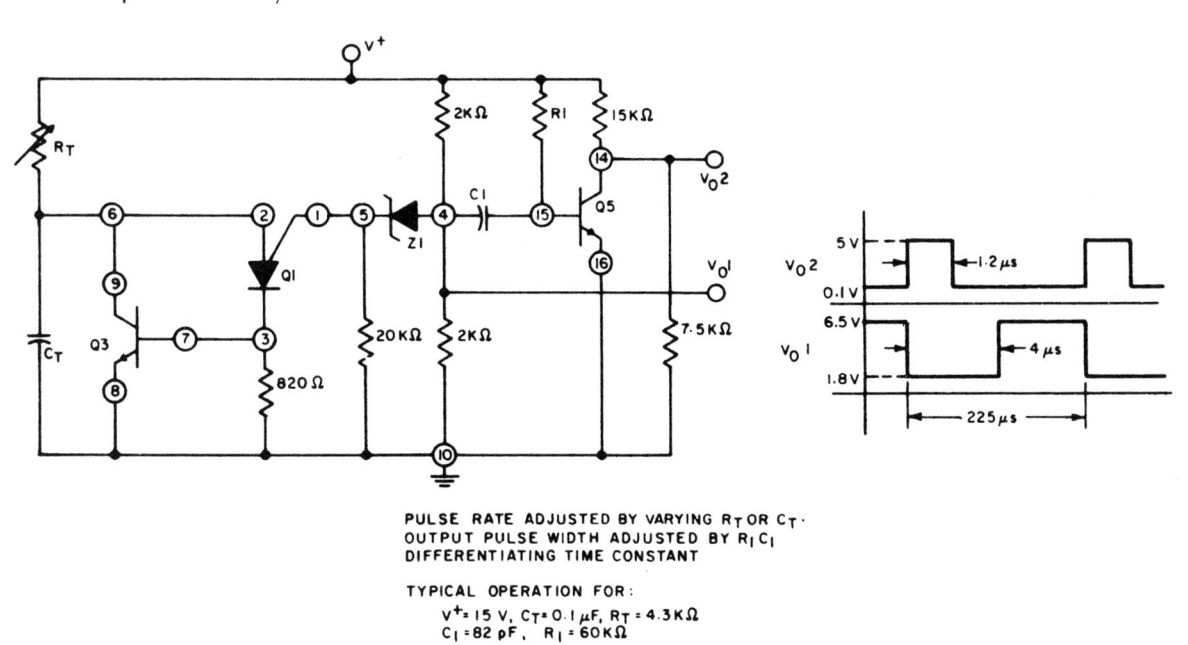

PULSE RATE ADJUSTED BY VARYING R_T OR C_T.
OUTPUT PULSE WIDTH ADJUSTED BY $R_1 C_1$
DIFFERENTIATING TIME CONSTANT

TYPICAL OPERATION FOR:
$V^+ = 15$ V, $C_T = 0.1 \mu F$, $R_T = 4.3 K\Omega$
$C_1 = 82$ pF, $R_1 = 60 K\Omega$

Pulse generator.

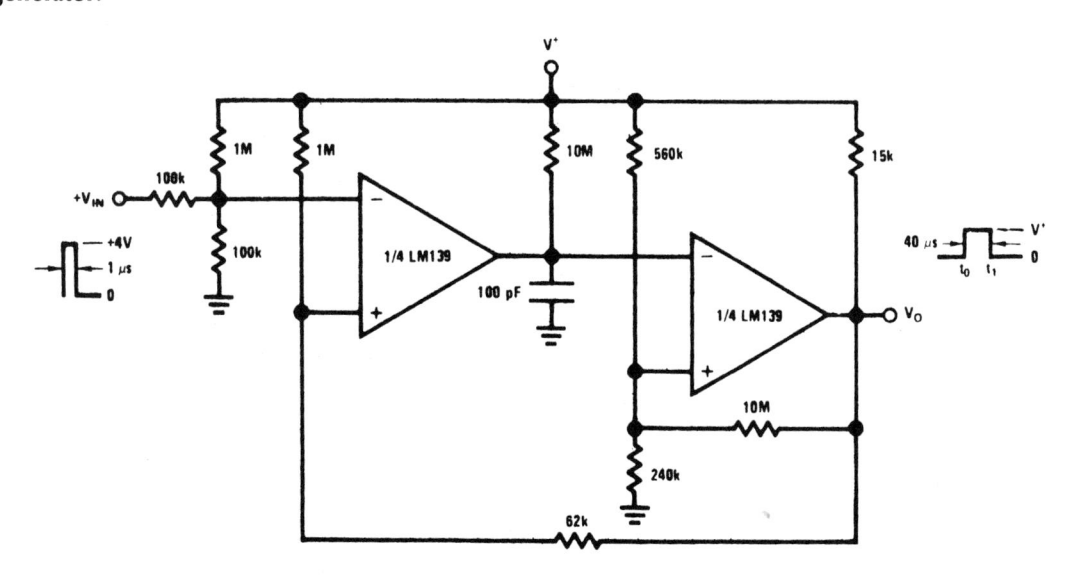

One-shot multivibrator with input lockout. $V^+ = 15V$ dc.

SECTION **4** *Waveform Generators and Shapers* **165**

MC1520 Alternate Integrator

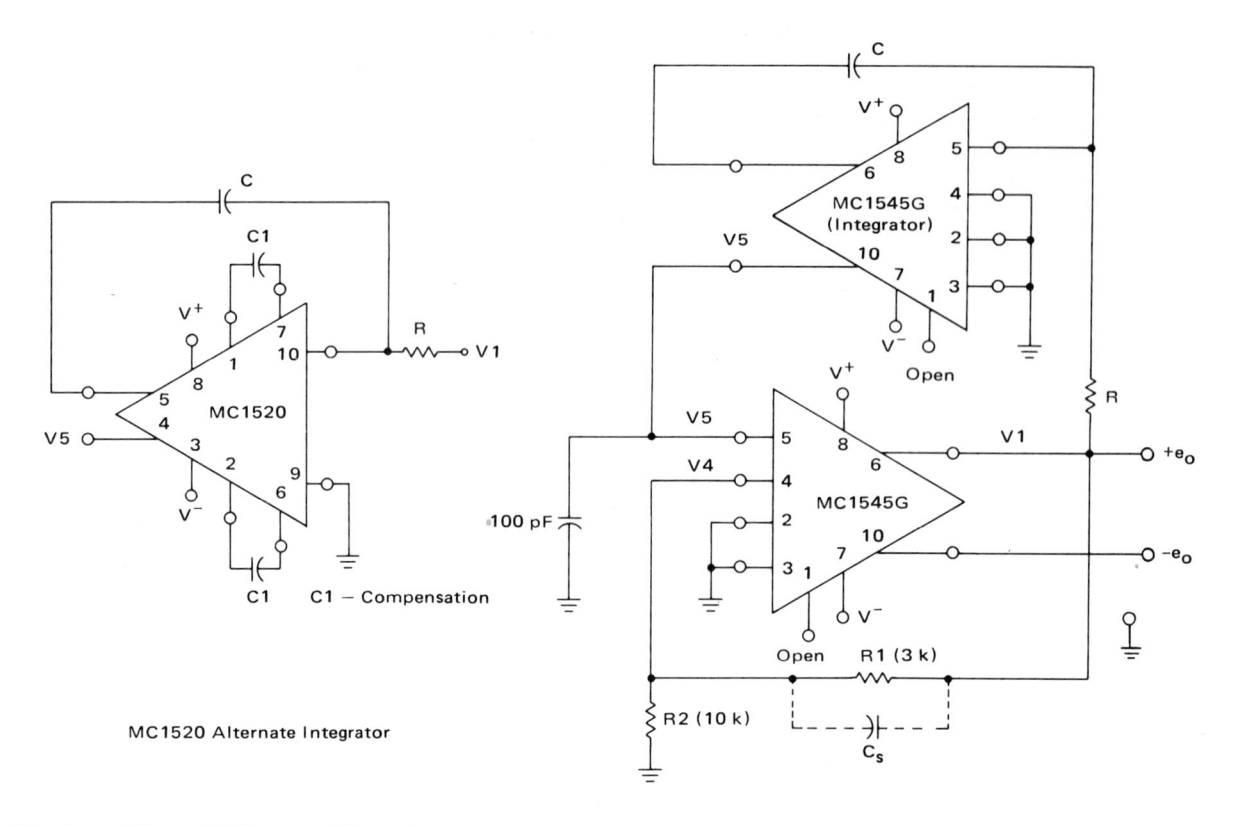

Gated astable multivibrator. The voltage on pin 5 (V5) of the lower unit is slightly higher than that on pin 4 (V4). The output (pin 6) will then be V1 = −A$_V$ (V5 − V4). A portion of this is applied to the noninverting input, pin 4. If R2/R1 + R2 is greater than 1/A$_V$, the gain of the circuit is greater than 1 and the output will quickly be driven to the extreme. The upper device functions as an integrator.

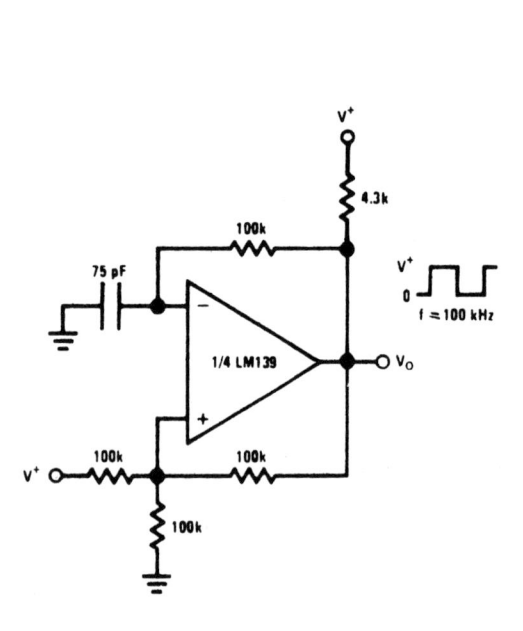

Square-wave oscillator. V$^+$ = 15V dc.

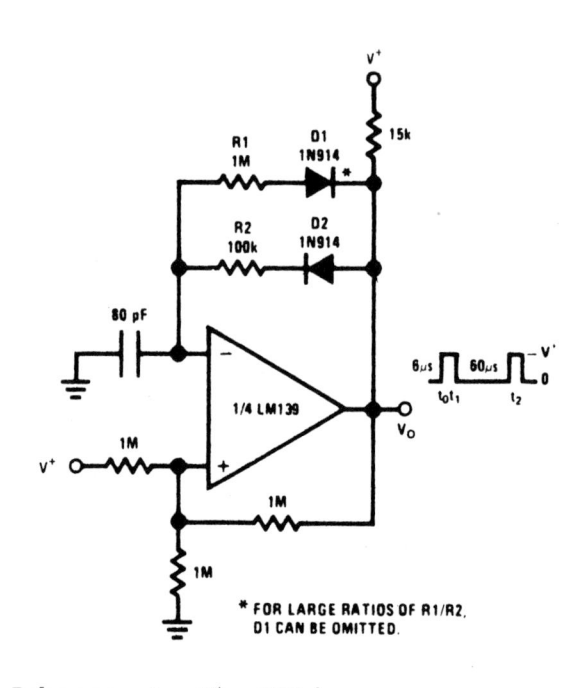

Pulse generator. V$^+$ = 15V dc.

166 IC SCHEMATIC SOURCEMASTER

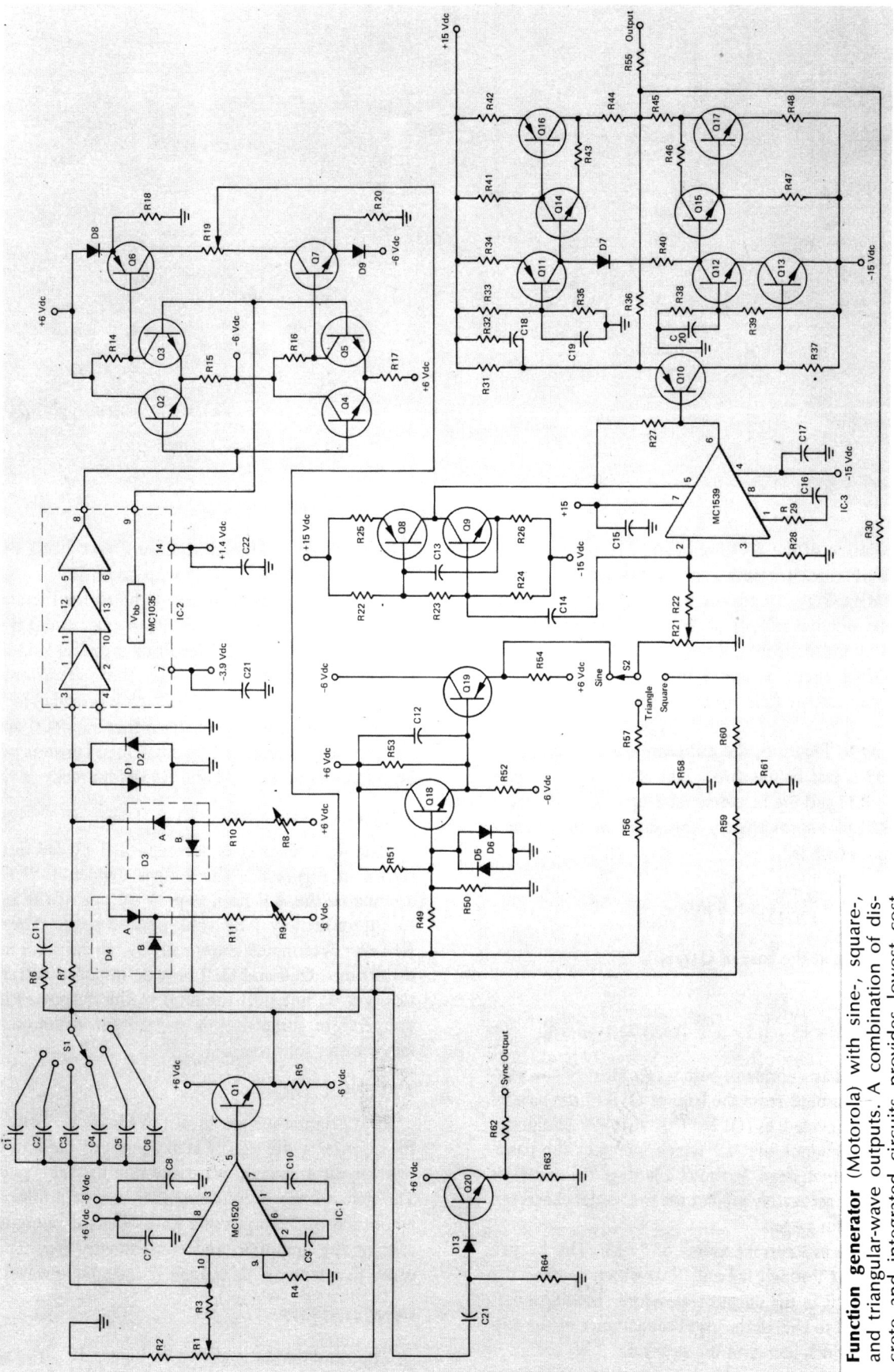

Function generator (Motorola) with sine-, square-, and triangular-wave outputs. A combination of discrete and integrated circuits provides lowest cost without sacrificing good performance. Maximum output amplitude of all waveforms is 20V peak-to-peak from a 50Ω output impedance. Frequency range is 1 Hz to 1 MHz.

SECTION 4 *Waveform Generators and Shapers*

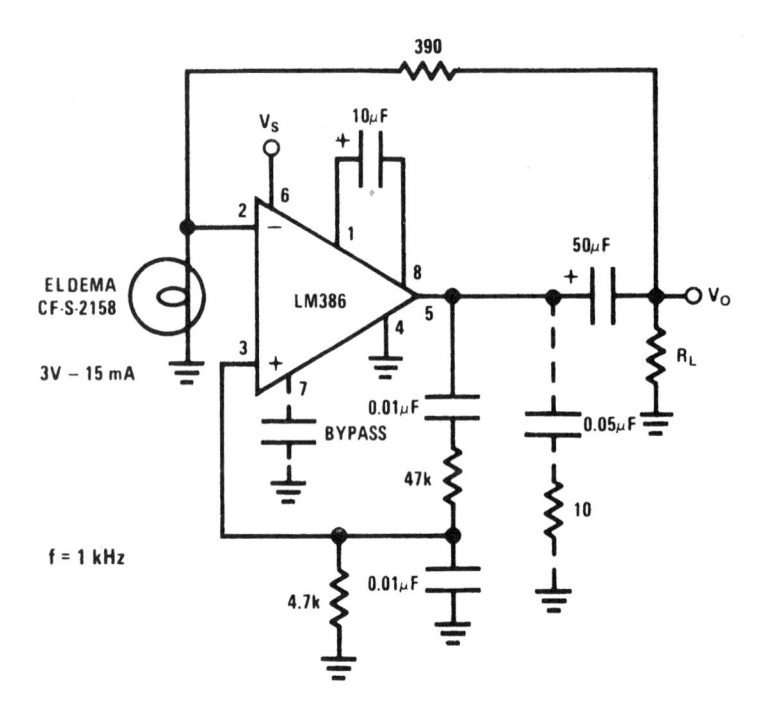

Low-distortion power Wien bridge oscillator.

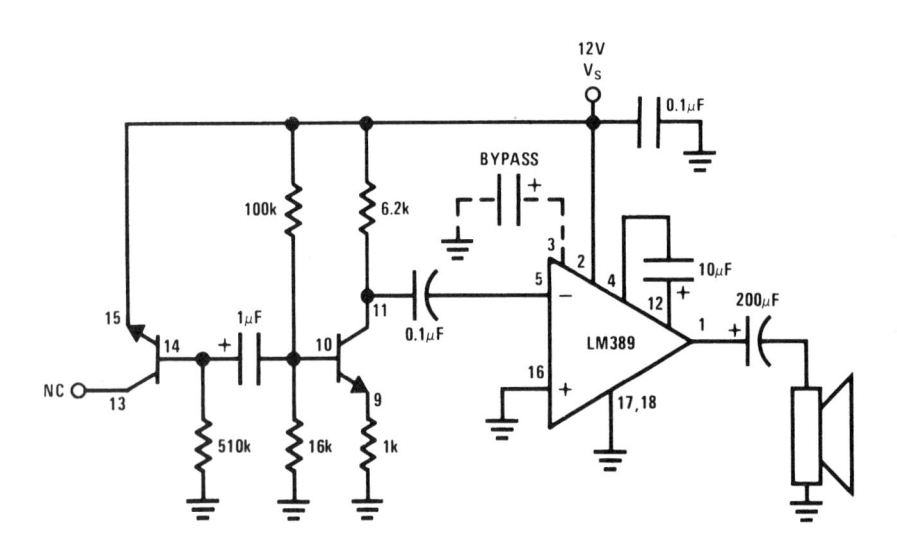

Noise generator. By applying reverse voltage to the emitter of a grounded-base transistor, the emitter-base junction will break down in an avalanche mode to form a handy zener. The reverse voltage characteristic is typically 7.1V and may be used as a voltage reference or a noise source. The noise voltage is amplified by the second transistor and delivered to the power amplifier stage where further amplification takes place before being used to drive the speaker.

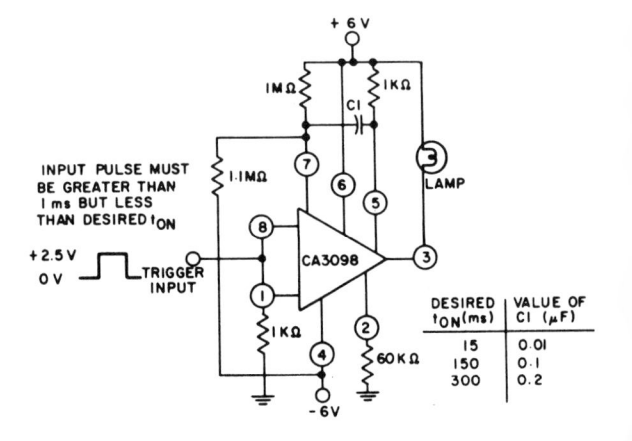

DESIRED t_{ON}(ms)	VALUE OF CI (μF)
15	0.01
150	0.1
300	0.2

One-shot multivibrator.

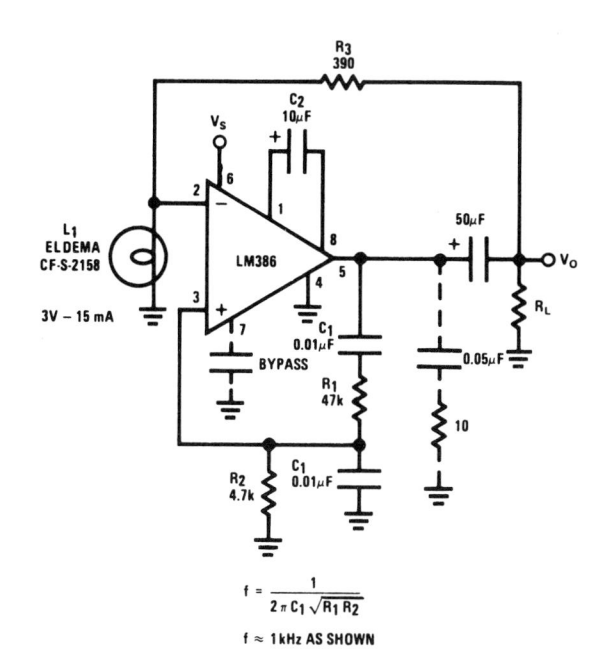

Power Wien bridge oscillator. Capacitor C2 raises the open-loop gain to 200V/V. Closed-loop gain is fixed at approximately 10 by the ratio of R1 to R2. A gain of 10 is necessary to guard against spurious oscillations that may occur at lower gains, since the LM386 is not stable below 9V/V. The frequency of oscillation is given by the equation in the figure and may be changed easily by altering capacitors C1. Resistor R3 provides amplitude-stabilizing negative feedback in conjunction with lamp L1. Almost any 3V, 15 mA lamp will work.

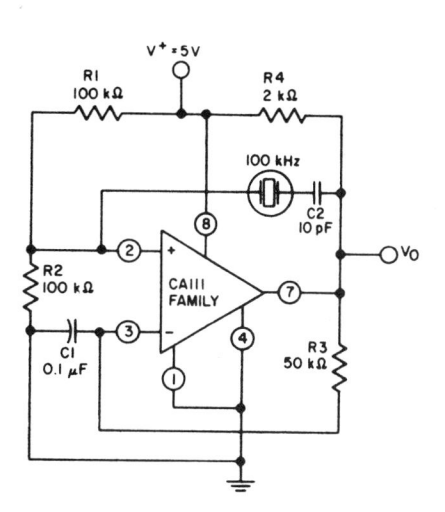

$$f = \frac{1}{2\pi C_1 \sqrt{R_1 R_2}}$$

$f \approx 1\text{kHz}$ AS SHOWN

Crystal oscillator.

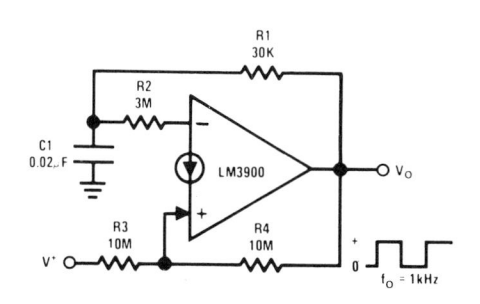

An op-amp square-wave generator, in which capacitor C1 alternately charges and discharges (via R1) between the voltage limits established by R2, R3, and R4. When the output is low, R2 causes the Schmitt trigger to fire when the current through this resistor equals the current at the (+) input. This gives a firing voltage of approximately $R2/(R3)V^+$ (or $V^+/3$). The other trip point, when the output voltage is high, is approximately $[2(R2/R3)]V^+$, as R3 = R4, or $2/2(V^+)$. The voltage across C1 will be the first half of an exponential waveform between these voltage trip limits and will have good symmetry and be essentially independent of the magnitude of the power supply voltage.

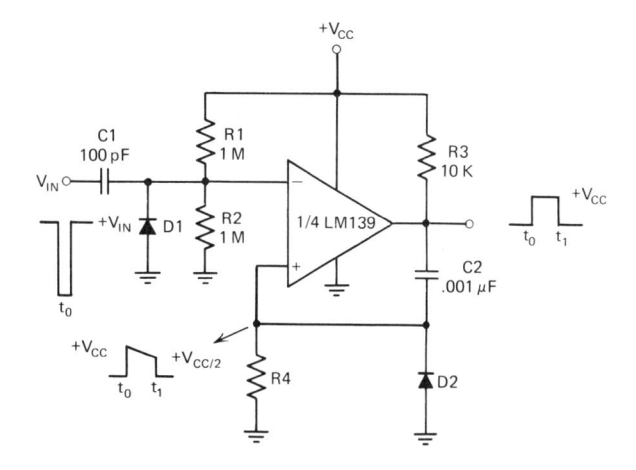

One-shot multivibrator. The output pulse width is set by the values of C2 and R4 (with R4 greater than 10R3 to avoid loading the output). The magnitude of the input trigger pulse required is determined by the resistive divider R1 and R2. Temperature stability can be achieved by balancing the temperature coefficients of R4 and C2 or by using components with very low TC. In addition, the TC of resistors R1 and R2 should be matched so as to maintain a fixed reference voltage of $+V_{CC}/2$. Diode D2 provides a rapid discharge path for capacitor C2 to reset the one-shot at the end of its pulse. It also prevents the noninverting input from being driven below ground. The output pulse width is relatively independent of the magnitude of the supply voltage and will change less than 2% for a 5V change in $+V_{CC}$.

SECTION 4 *Waveform Generators and Shapers*

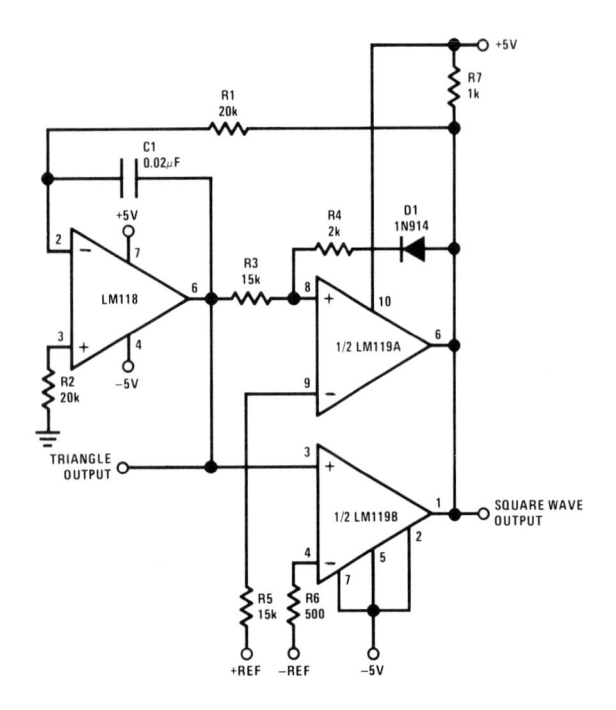

This precise tri-wave function generator consists of an integrator and two comparators. One comparator sets the positive peak and the other sets the negative peak of the tri-wave. The frequency of operation is dependent upon R1, C1, and the reference voltages. Frequency is given by

$$f = \frac{5.0V}{2R1C1(V^+_{REF} = V^-_{REF})}$$

The maximum frequency of operation is limited by circuit delay to about 200 kHz. The maximum difference in reference voltages is 5.0V.

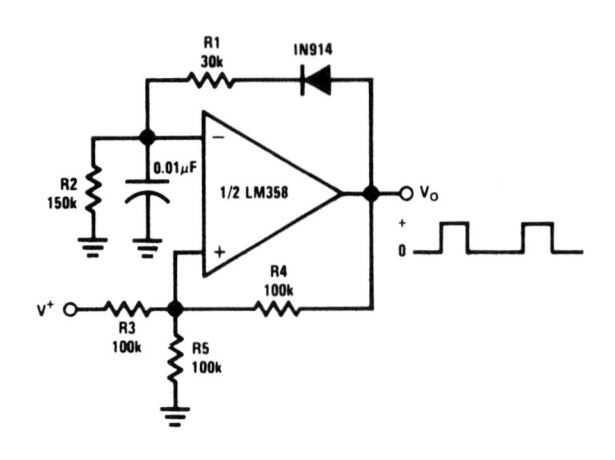

Pulse generator.

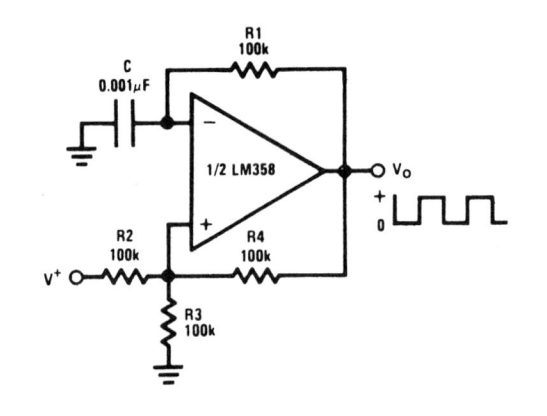

Square-wave oscillator.

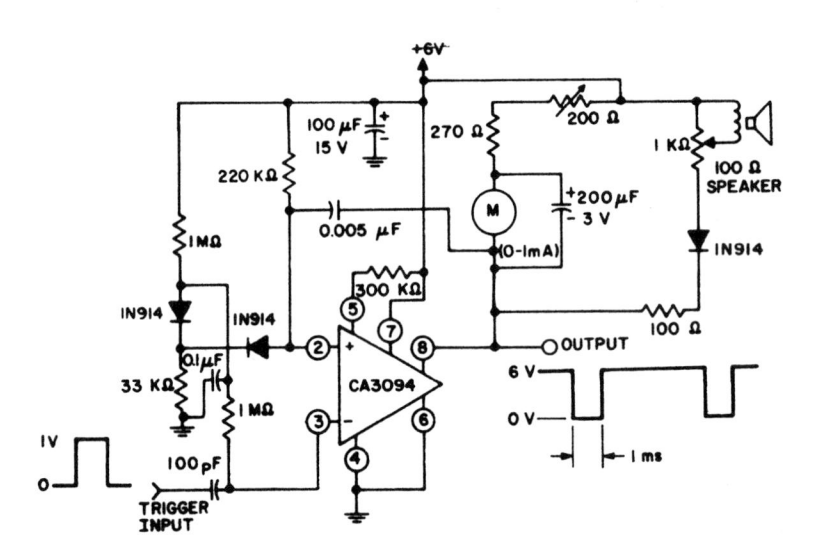

Power monostable multivibrator. Circuit is a pulse counter in which the duration of the output pulses is independent of the trigger-pulse duration. The meter reading is a function of the pulse repetition rate, which can be monitored with the speaker. Full-scale meter deflection ≈83 pulses/sec.

170 IC SCHEMATIC SOURCEMASTER

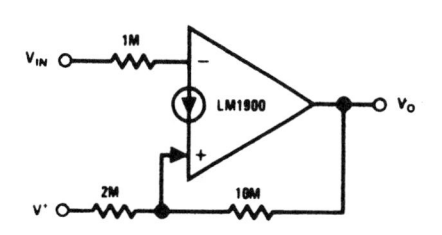

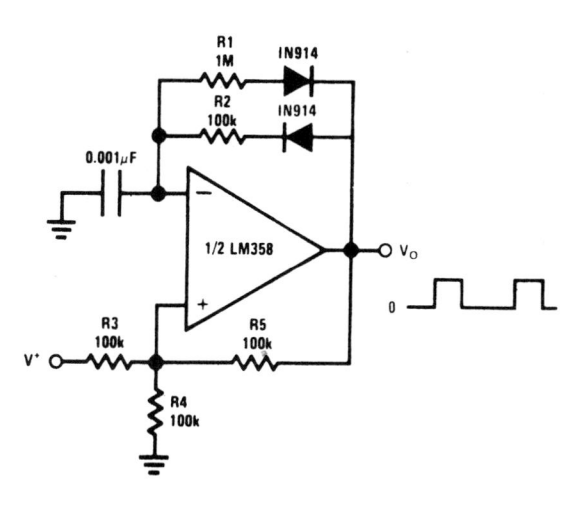

Pulse generator.

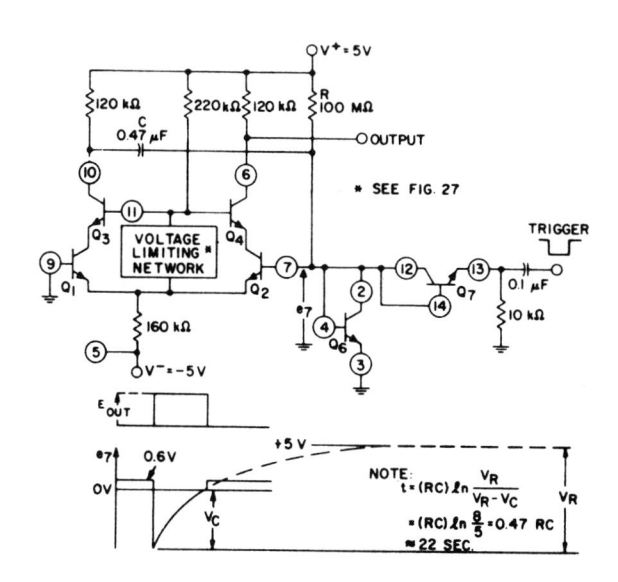

Long-delay monostable multivibrator circuit. Voltage-limiting network is built into superbeta IC. To use it, just connect a resistor from the positive supply of the differential amplifier to pin 11; the value should be such that current through the resistor is about half that measured at pin 8 under quiescent conditions.

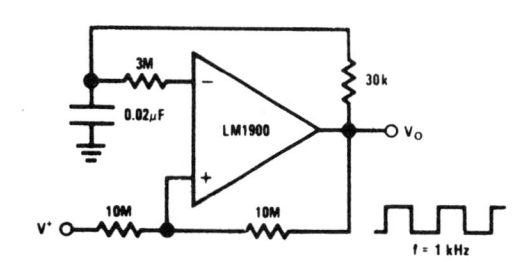

Schmitt trigger

LM 1900 variations include Schmitt trigger and pulse and square-wave generator. Operation can best be understood by noticing that input currents are differenced at the inverting input terminal, and this difference current then flows through the external feedback resistor to produce the output voltage. Common-mode current biasing is generally useful to allow operation with signal levels near ground or even negative, as this maintains the inputs biased at $+V_{BE}$. Internal clamp transistors catch negative input voltages at approximately $-0.3V$ dc, but the magnitude of current flow has to be limited by the external input network. For operation at high temperature, this limit should be approximately 100 μA.

Square-wave oscillator

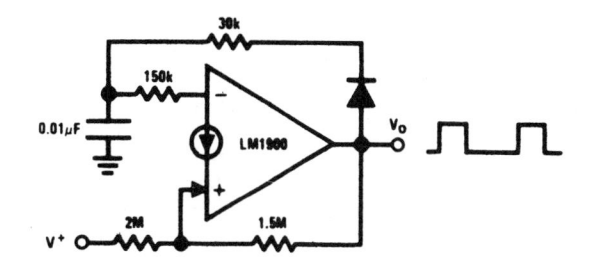

Pulse generator

SECTION 4 *Waveform Generators and Shapers* **171**

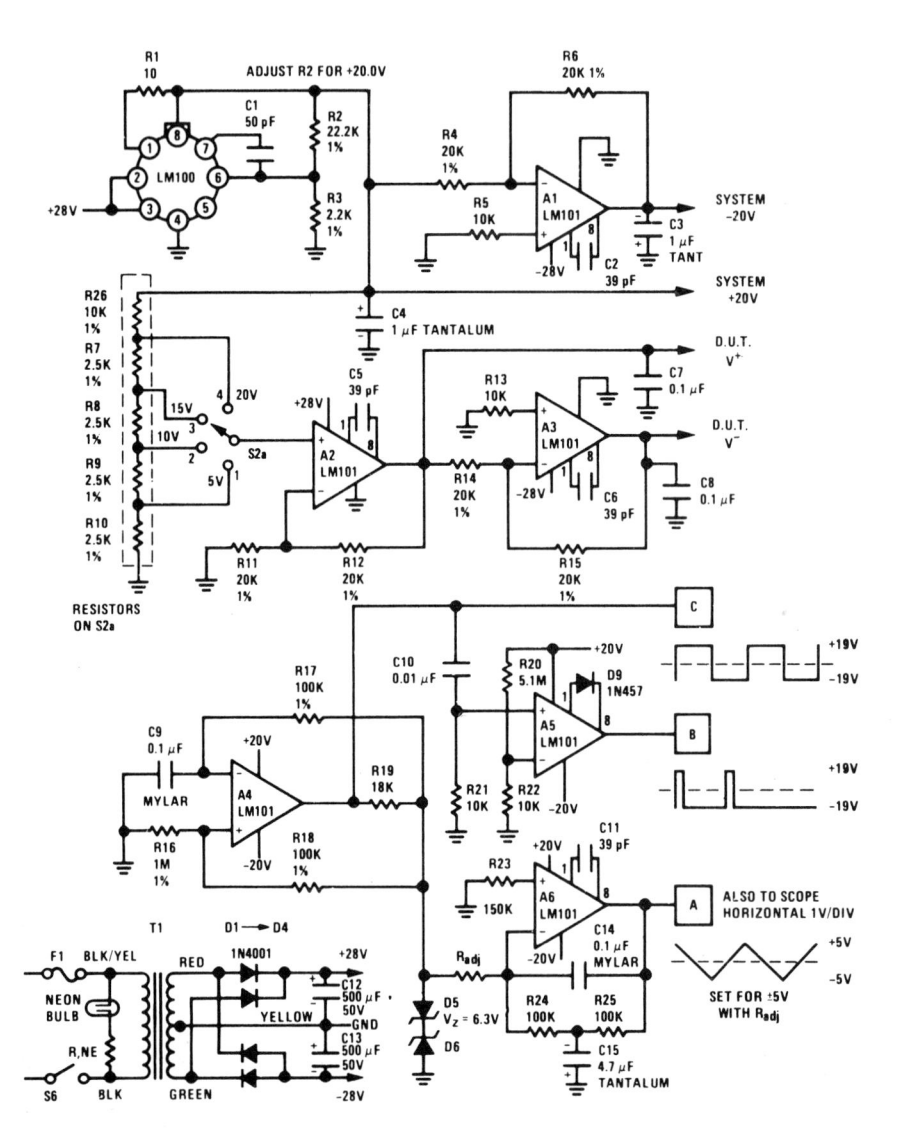

Power supply and function generator (National). The function generator provides three outputs, a ±19V square wave, a −19V to +19V pulse having a 1% duty cycle, and a ±5V triangular wave. The square wave is the basic function from which the pulse and triangular waves are derived, the pulse is referenced to the leading edge of the square wave, and the triangular wave is the inverted and integrated square wave.

Amplifier A4 is an astable multivibrator generating a square wave from positive to negative saturation. The amplitude of this square wave is approximately ±19V. The frequency is determined by the ratio of R18 to R16 and by the time constant of R17 and C9. The operating frequency is stabilized against temperature and power regulation effects by regulating the feedback signal with the dividers R19, D5, and D6.

Amplifier A5 is a monostable multivibrator triggered by the positive-going output of A4. The pulse width of A5 is determined by the ratio of R20 to R22 and by the time constant of R21 and C10. The output pulse of A5 is 1%-duty-cycle pulse from approximately −19V to +19V.

Amplifier A6 is a dc-stabilized integrator driven from the amplitude-regulated output of A4. Its output is a ±5V triangular wave. The amplitude of the output of A6 is determined by the square-wave voltage developed across D5 and D6 and the time constant of R_{adj} and C14. Dc stabilization is accomplished by the feedback network R24, R25, and C15. The ac attenuation of this feedback network is high enough so that the integrator action at the square-wave frequency is not degraded.

Operating frequency of the function generator may be varied by adjusting the time constants associated with A4, A5, and A6 in the same ratio.

172 IC SCHEMATIC SOURCEMASTER

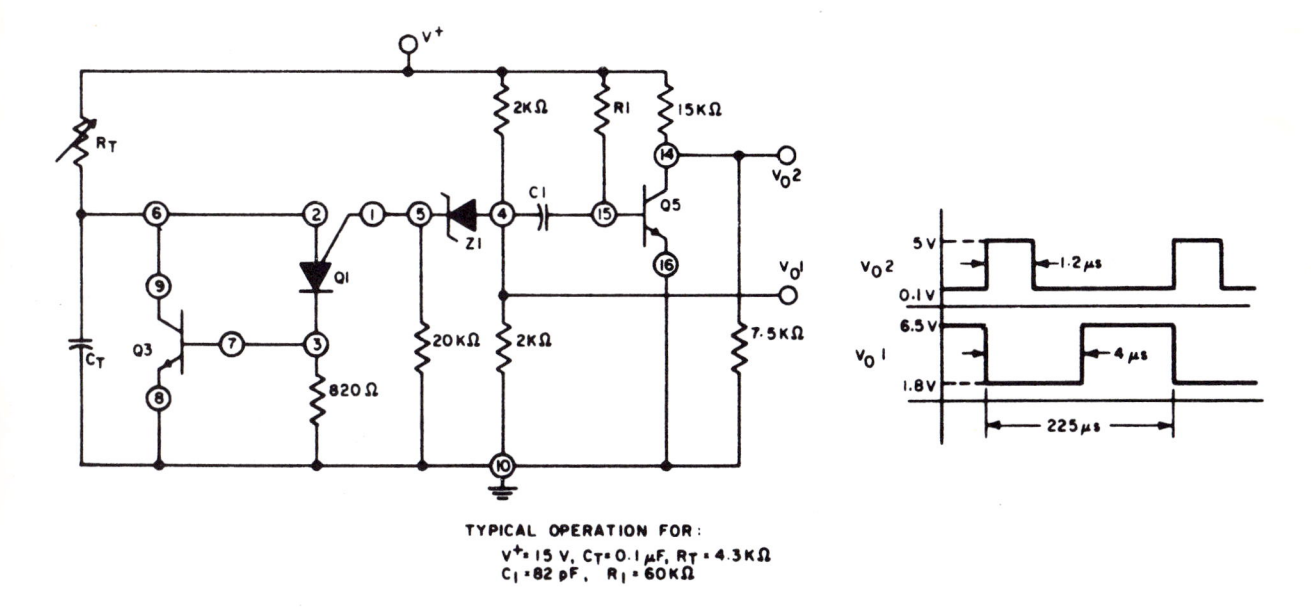

TYPICAL OPERATION FOR:
V+ = 15 V, C_T = 0.1 μF, R_T = 4.3 KΩ
C_1 = 82 pF, R_1 = 60 KΩ

A pulse generator using CA3097E. The pulse rate is adjusted by varying R_T or C_T; the output pulse width is adjusted by R1C1 differentiating time constant.

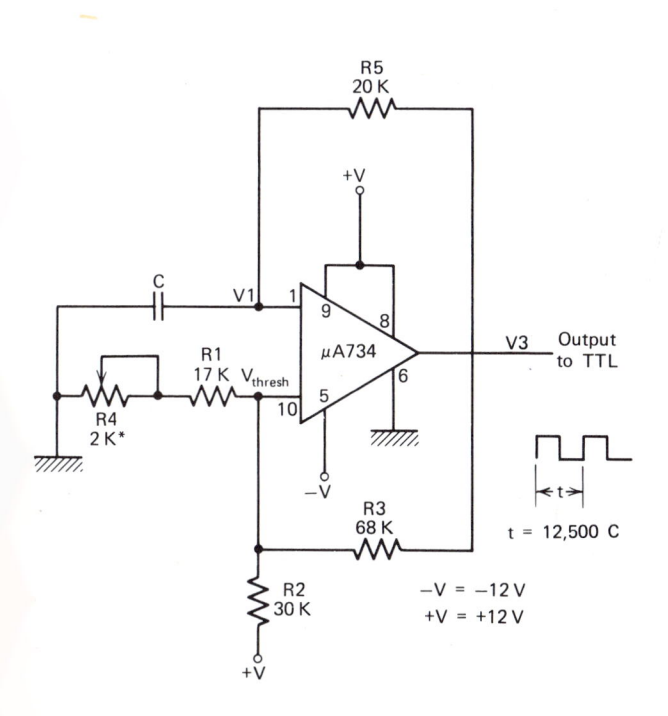

Self-starting pulse oscillator. When power is turned on, the output of the Fairchild μA734 goes high, and capacitor C starts charging. When voltage V1 crosses the threshold voltage set at V_{thresh}, the output of the μA734 goes low, setting a new threshold voltage. Capacitor C discharges until V1 crosses the new threshold voltage and the output goes high again, repeating the cycle.

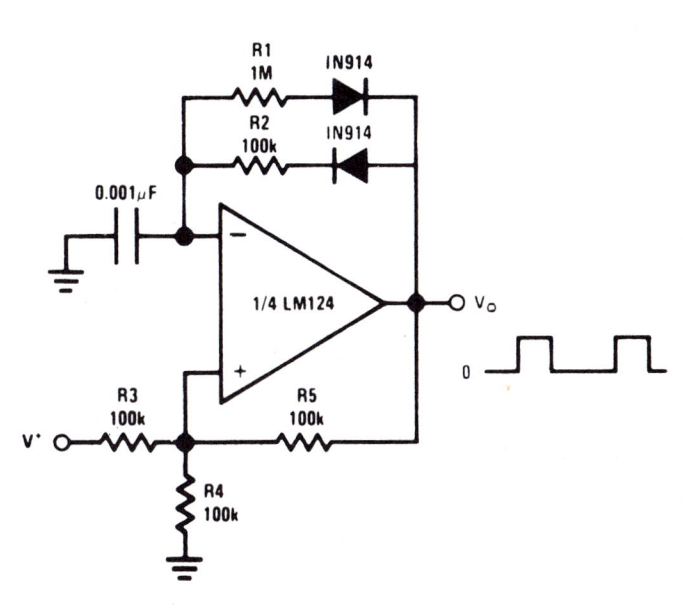

Pulse generator.

SECTION **4** *Waveform Generators and Shapers* **173**

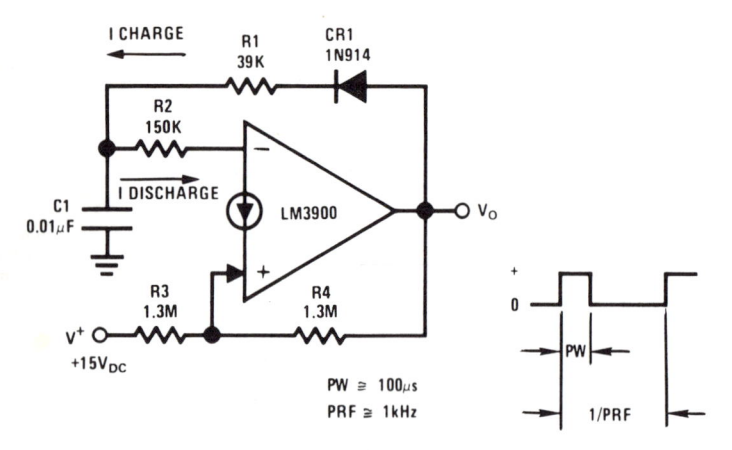

Pulse generator.

Triangle/square waveform generator. V+ = 15V.

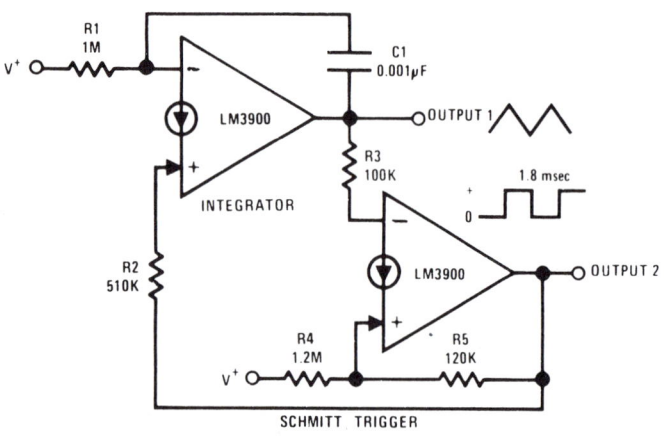

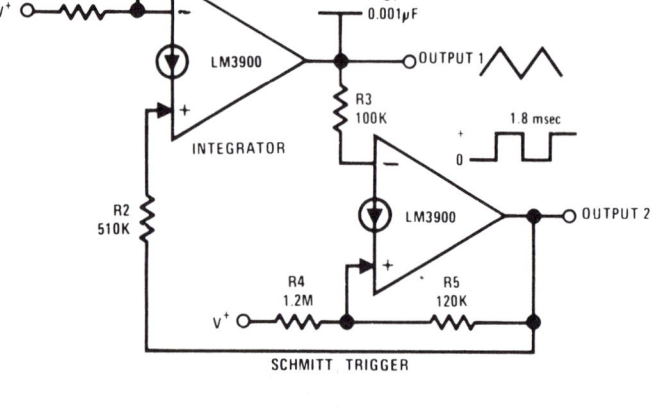

Triangle waveform generator. One amplifier is doing the integrating by operating first with the current through R1 to produce the negative output voltage slope; when the output of the second amplifier (the Schmitt trigger) is high, the current through R2 causes the output voltage to increase. If R1 = 2R2, the output waveform will have good symmetry.

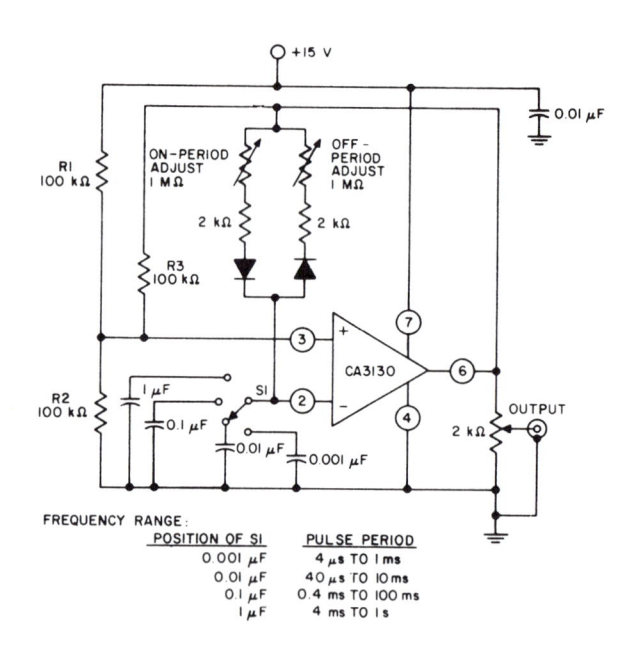

A pulse generator (astable multivibrator) with provisions for independent control of *on* and *off* periods.

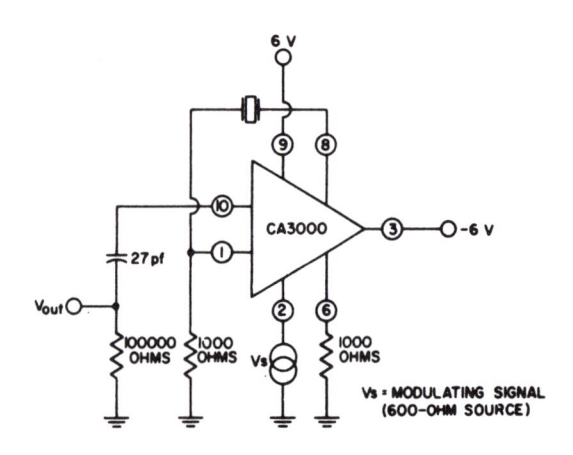

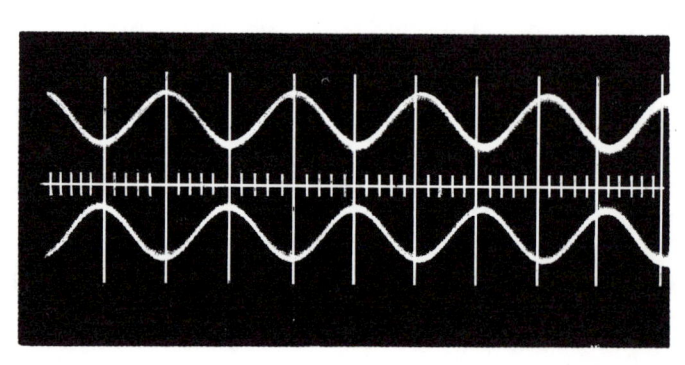

Schematic diagram and output waveform of CA3000 modulated 455 kHz oscillator.

174 IC SCHEMATIC SOURCEMASTER

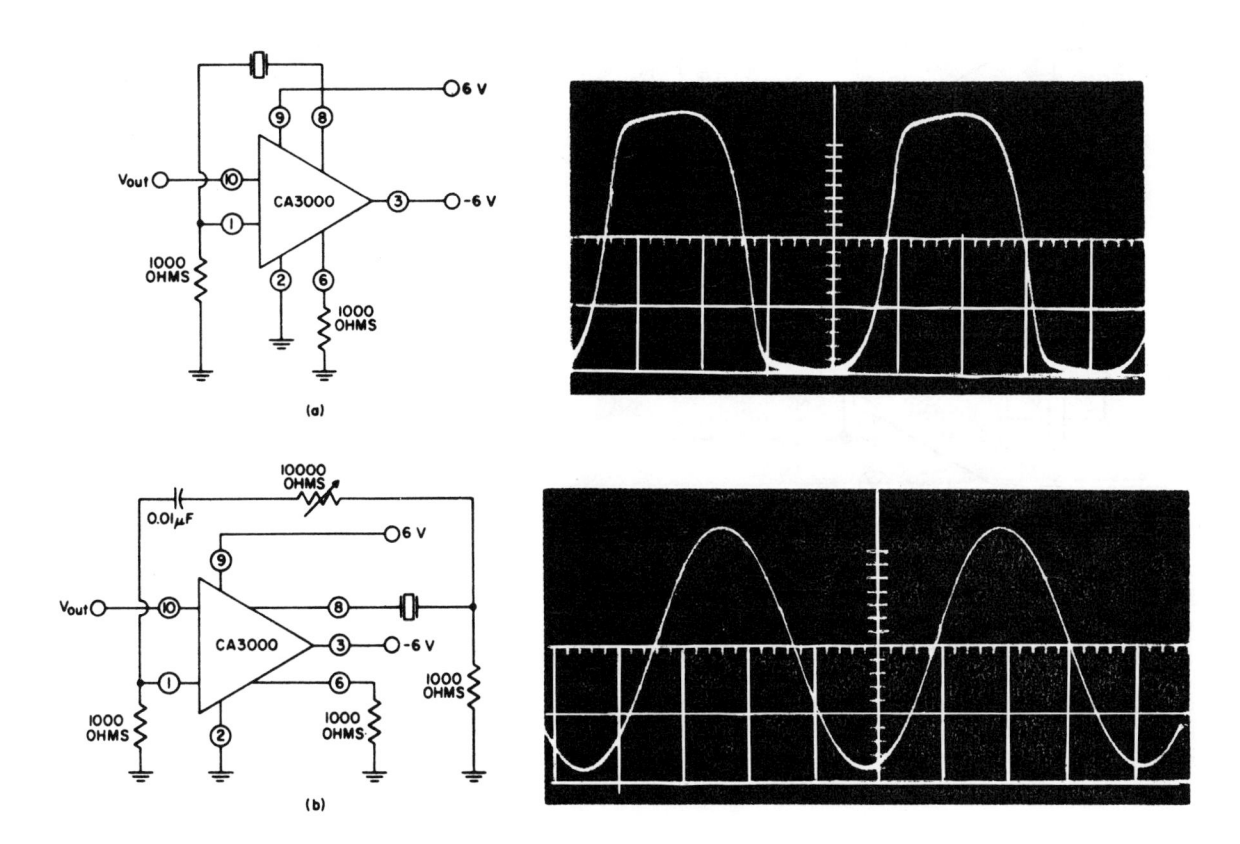

Schematic diagrams and output waveforms of 455 kHz oscillator (top) and 455 kHz crystal oscillator with variable feedback (bottom)

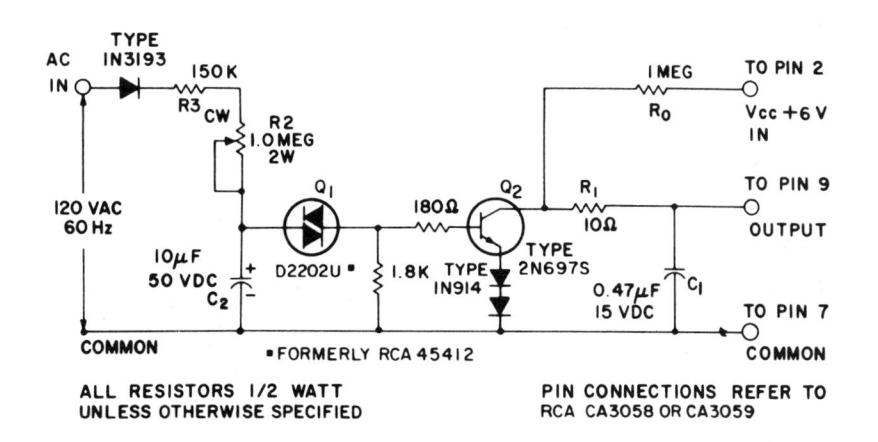

Ramp generator. The ramp voltage is generated when capacitor C1 charges through resistors R0 and R1. The timebase of the ramp is determined by resistors R2 and R3, capacitor C2, and the breakover voltage of the D3202U diac. When the voltage across C2 reaches 32V, the diac switches and turns on the 2N697S and 1N914 diodes. Capacitor C1 then discharges through the collector-to-emitter junction of the transistor. This discharge time is the retrace or flyback time of the ramp. The circuit can generate ramp times ranging from 0.3 to 2.0 seconds through adjustment of R2. For precise temperature regulation, the timebase of the ramp should be shorter than the thermal time constant of the system but long with respect to the period of the 60 Hz line voltage.

SECTION **4** *Waveform Generators and Shapers* **175**

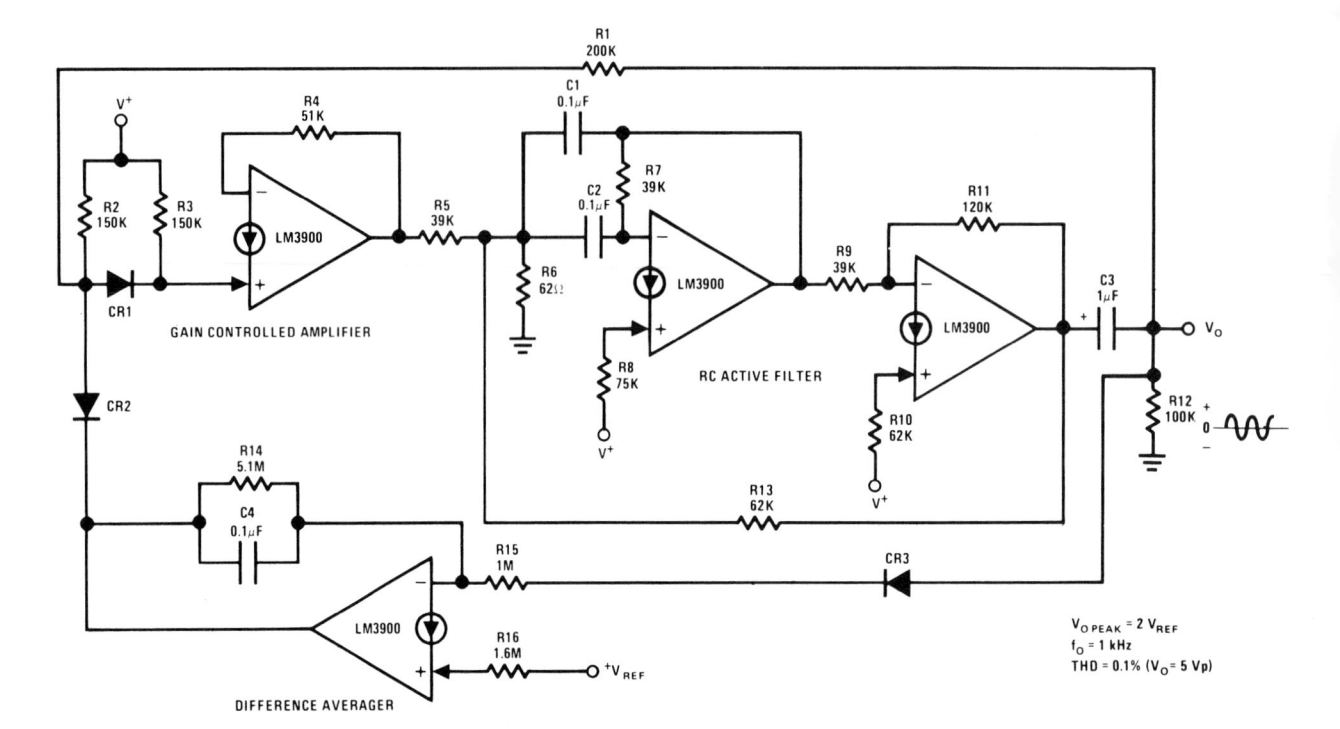

This sine-wave oscillator has both amplitude stability (and predictability) and output waveform purity. An RC bandpass filter is used as a high-Q resonator for the oscillator circuit. The two-amplifier RC active filter is used, as it requires only two capacitors and provides an overall noninverting phase characteristic. The output voltage is sensed and regulated, as the average value is compared to a dc reference voltage, V_{REF}, by use of a differential averaging circuit. The output sinusoid is essentially independent of both temperature and the magnitude of the power supply voltage.

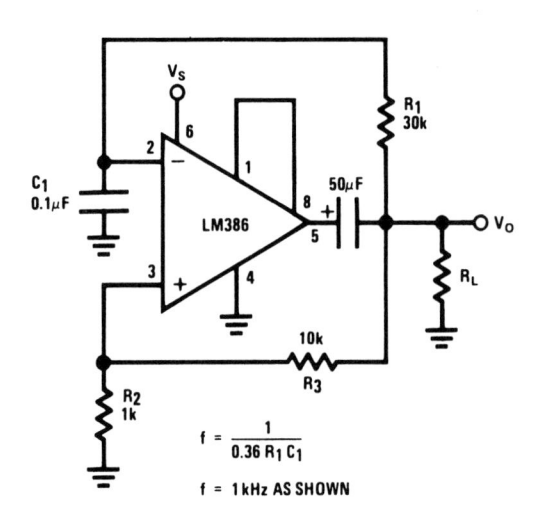

A square-wave oscillator capable of driving an 8Ω speaker with 0.5W from a 9V supply. Altering either R1 or C1 will change the frequency of oscillation by the equation given in the figure. A reference voltage determined by the ratio of R3 to R2 is applied to the positive input from the LM386 output. Capacitor C1 alternately charges and discharges about this reference value, causing the output to switch states. A triangle output may be taken from pin 2 if desired. Since dc offset voltages are not relevant to the circuit operation, the gain is increased to 200V/V by a short circuit between pins 1 and 8, thus saving one capacitor.

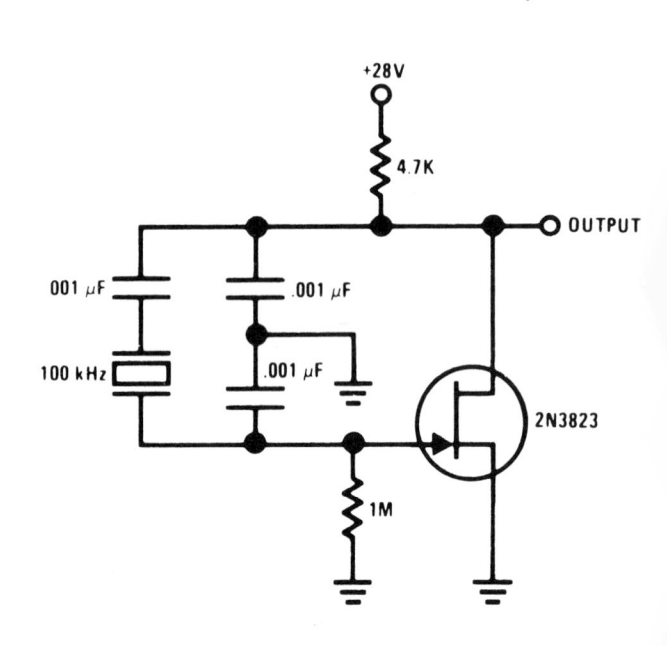

Stable Colpitts crystal oscillator is ideal for low-frequency circuits. Excellent stability is assured because the 2N3823 JFET circuit loading does not vary with temperature.

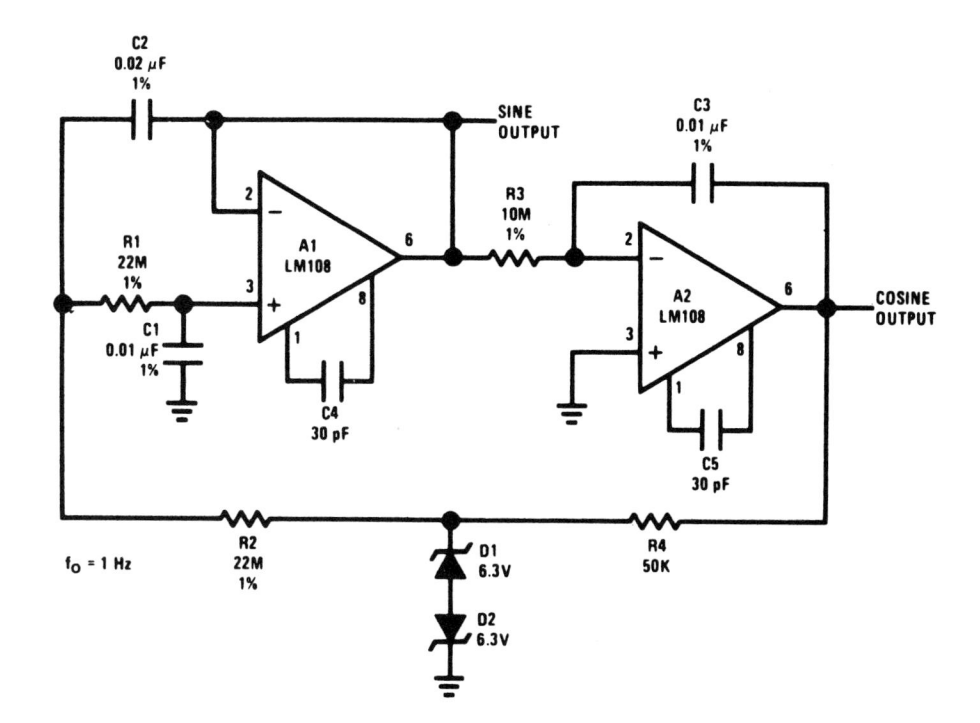

This sine-wave oscillator delivers a high-purity sinusoid with a stable frequency and amplitude. A1 is connected as a two-pole low-pass active filter, and A2 is connected as an integrator. Since the ultimate phase lag introduced by the amplifiers is 270°, the circuit can be made to oscillate if the loop gain is high enough at the frequency where the lag is 180°. The gain is actually made somewhat higher than is required for oscillation to ensure starting. Therefore, the amplitude builds up until it is limited by some nonlinearity in the system.

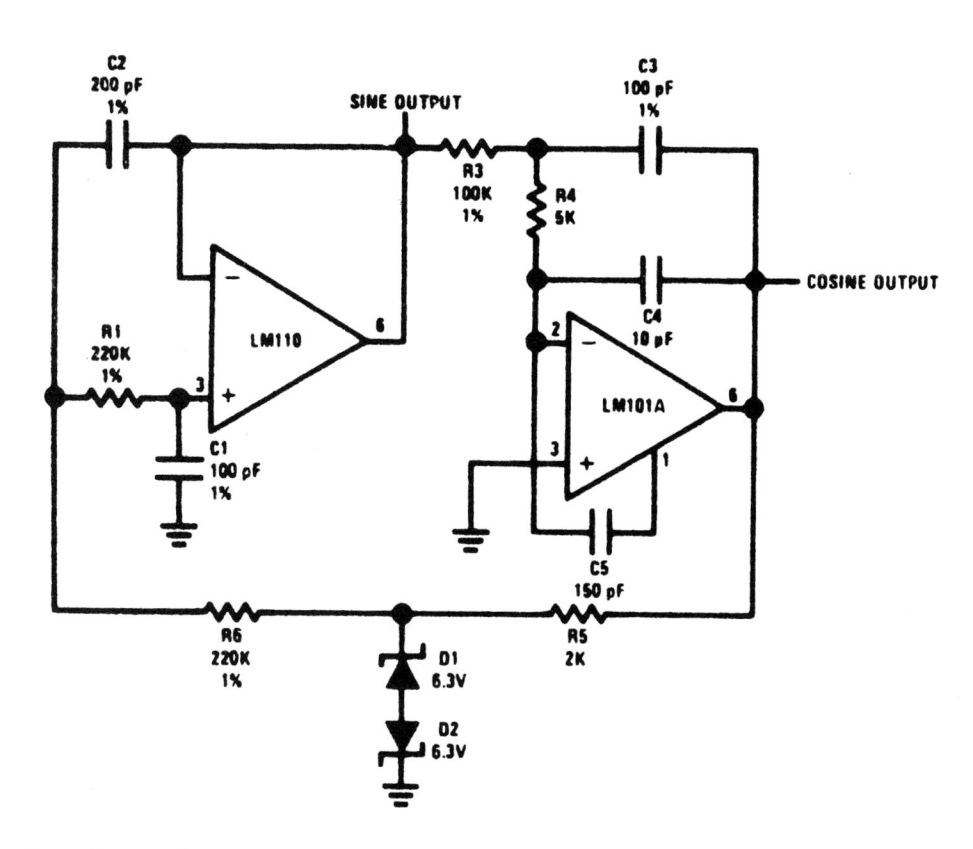

Sine-wave oscillator. $f_o = 10$ kHz.

SECTION 4 *Waveform Generators and Shapers* **177**

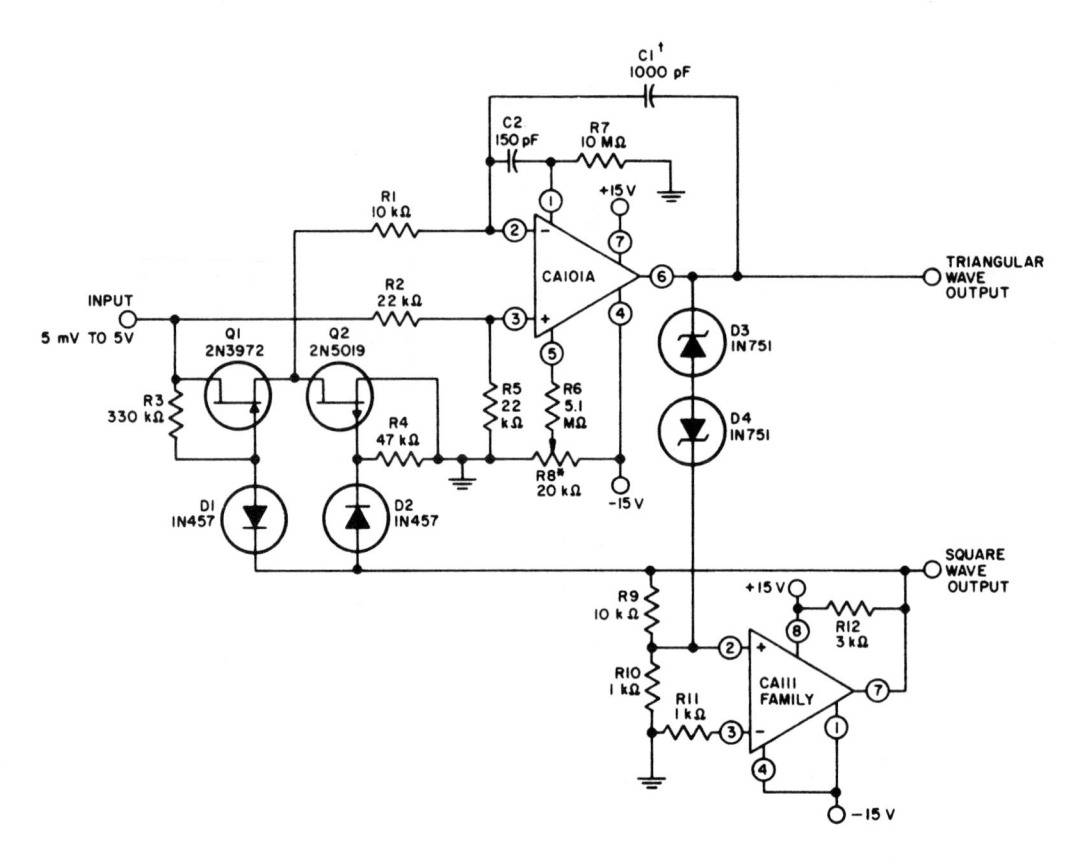

10 Hz to 10 kHz voltage-controlled oscillator. Adjust R8 for symmetrical square-wave time when $V_I = 5$ mV. Minimum capacitace of C1, 20 pF; maximum frequency, 50 kHz.

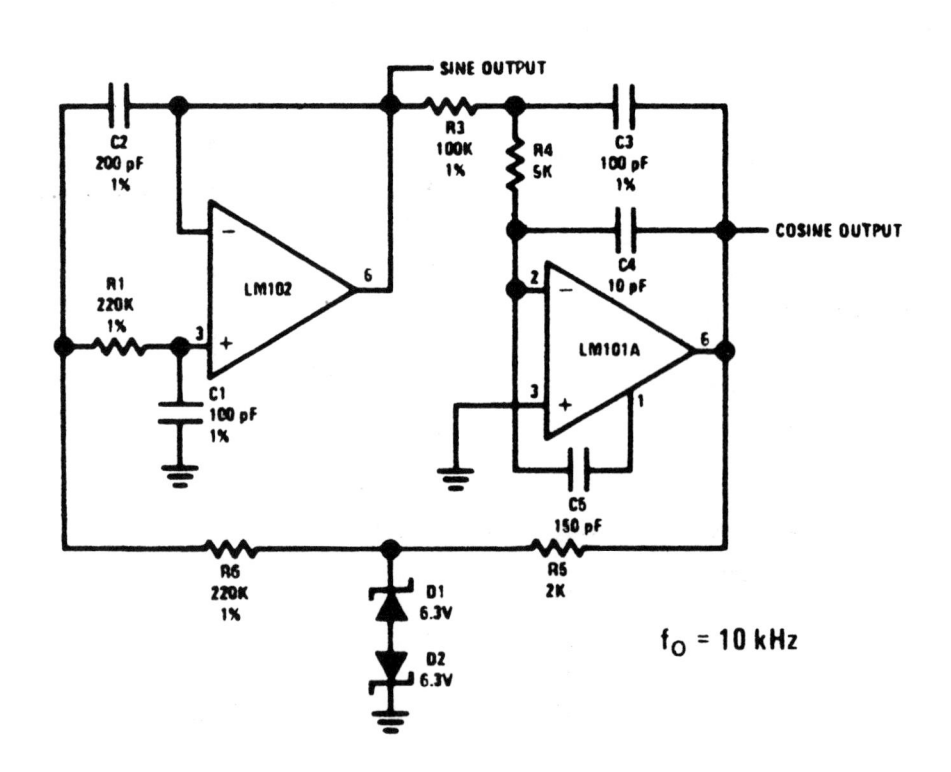

Sine-wave oscillator.

178 IC SCHEMATIC SOURCEMASTER

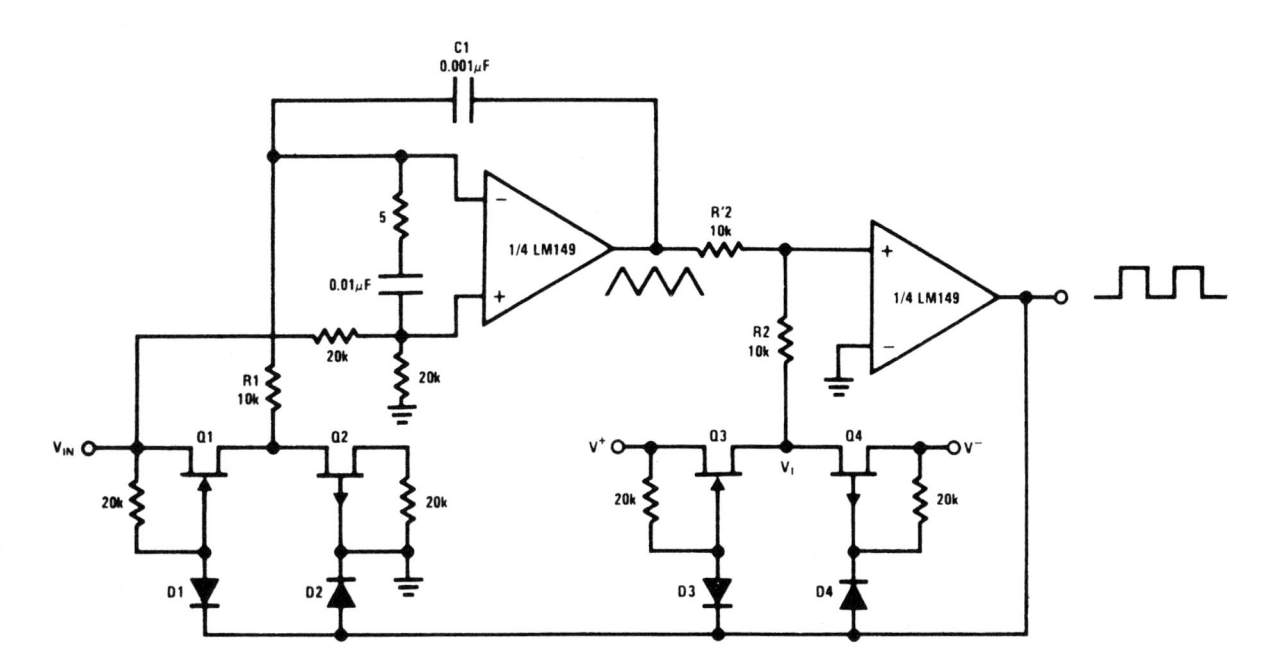

This triangular square-wave generator can be used as a low-frequency V/f for process control. Q1, Q3: KE4393; Q2, Q4: P1087E; diodes are 1N914. $f = K \times V_{IN}/8V^+C1R1$; $K = R2/R'2$; $2V_1/K \leq 25V$; $V^+ = V^-$; $V_S = \pm15V$. (Use LM125 for $\pm15V$ supply.)

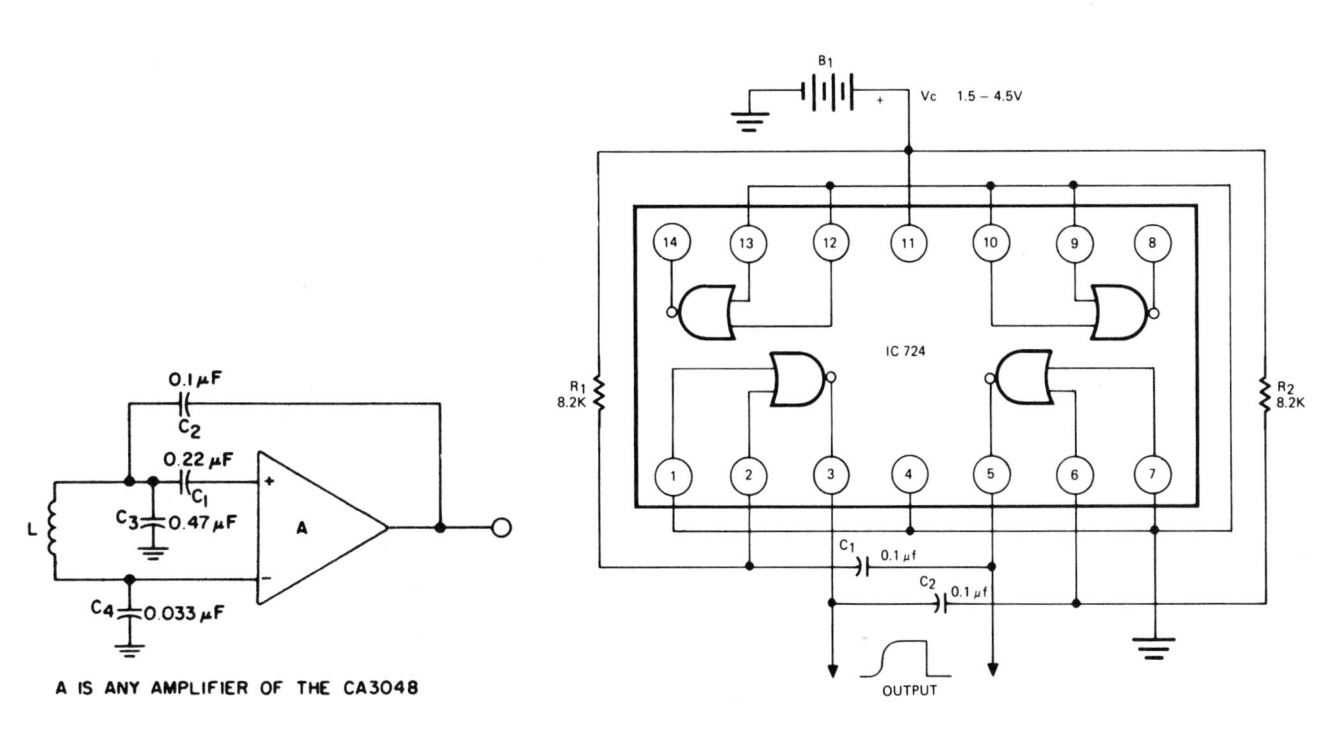

This tunable Colpitts oscillator is readily designed using one of the amplifiers of the CA3048 array. Capacitors C1 and C2 are dc-blocking capacitors; the series combination of capacitors C2 and C4 resonates with coil L. The ratio of C3 to C4 determines the relative amounts of signal fed back to the two inputs and may be chosen on the basis of stability or strength of oscillation. For the component values shown, the frequency of oscillation is 33.536 kHz with a 12V supply; it increases to 33.546 kHz when the supply voltage is reduced 25% to 9V.

Square-wave multivibrator. Operating frequency can be increased by reducing the values of C1 and C2 or reduced by increasing them. (Semiconductors are IRC/Workman.)

SECTION **4** *Waveform Generators and Shapers* **179**

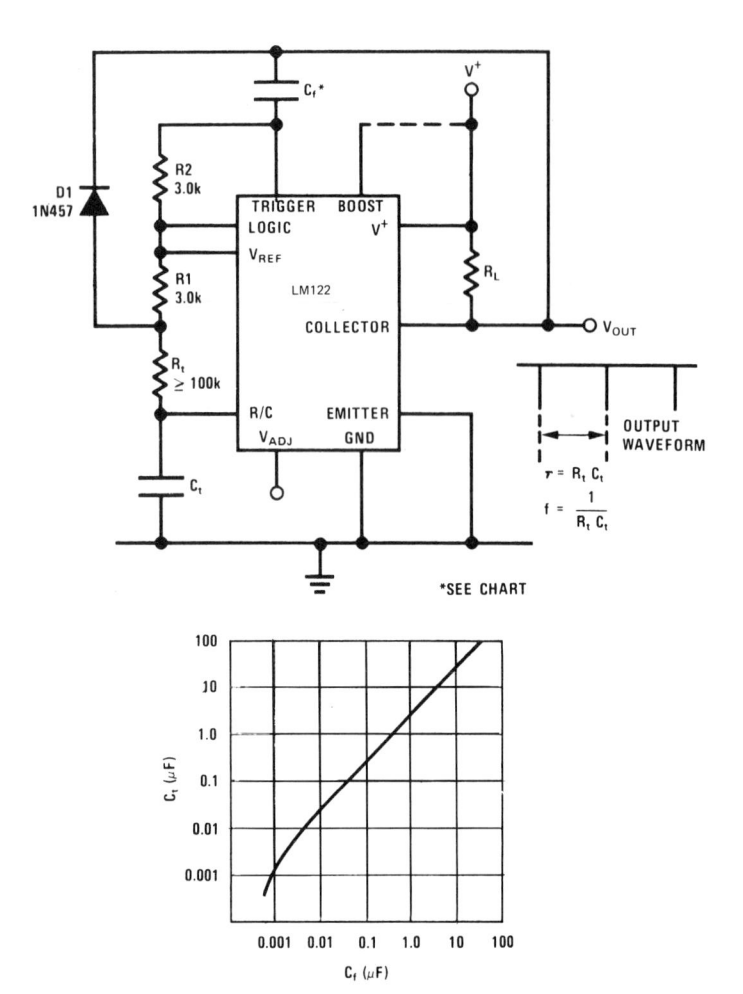

Self-starting oscillator. Operating frequency is $1/R_tC_t$. The output is a narrow negative pulse whose width is approximately $2R2C_f$. For optimum frequency stability, C_f should be as small as possible. The minimum value is determined by the time required to discharge C_t through the internal discharge transistor. A conservative value for C_f can be chosen from the graph. For frequencies below 1 kHz, the frequency error introduced by C_f is 0.3% or less when R_t is greater than 500k.

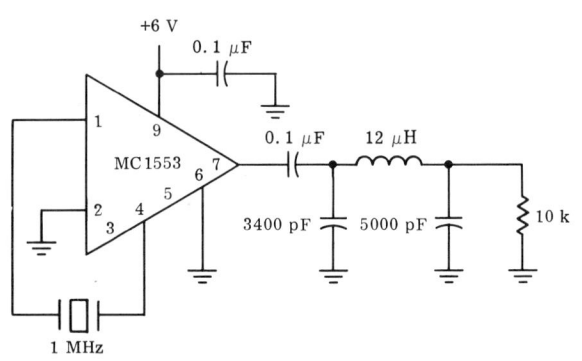

A sine-wave oscillator, in which positive feedback is injected through the 1 MHz series-mode crystal to input pin 1. The bias-decoupling capacitor normally on pin 5 should be omitted in this oscillator design to ensure that the crystal operates into a relatively low impedance. The output of the oscillator is taken from pin 7, which is buffered from the oscillator circuit proper by a stage of gain and an emitter follower.

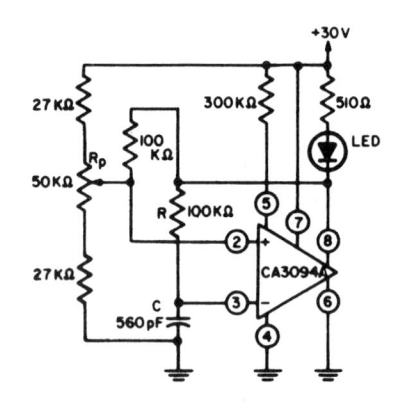

Single-supply astable multivibrator.

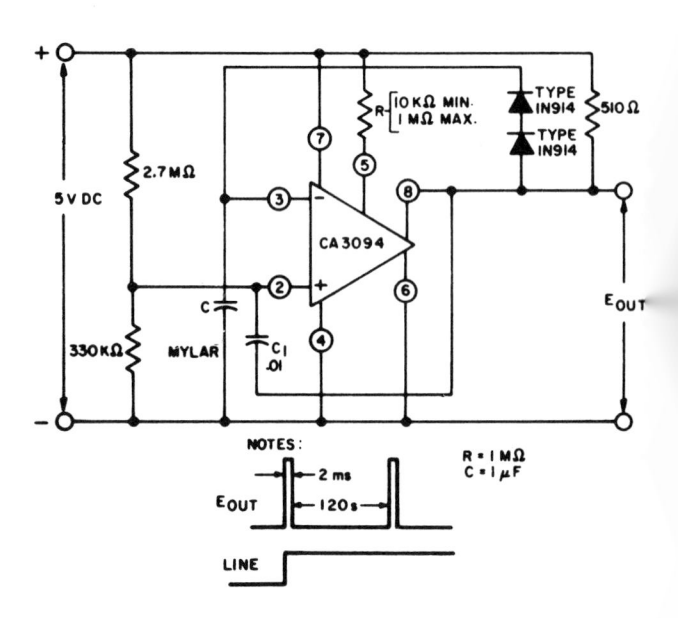

Free-running pulse generator. Time constant $t \approx 120RC$; pulse width $\omega \approx K(C1/C)$.

180 IC SCHEMATIC SOURCEMASTER

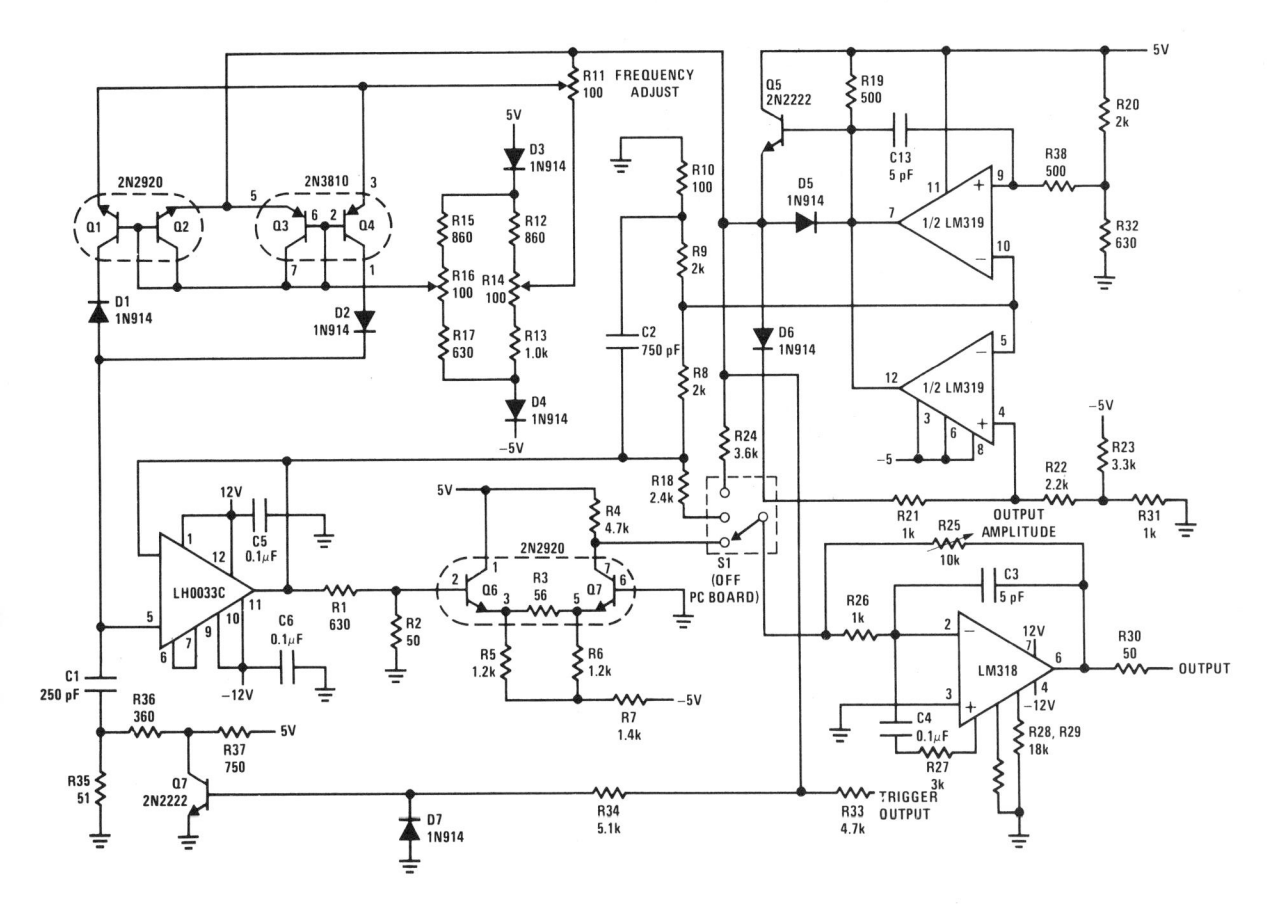

This sine, square, and triangle function generator provides all three waveforms and operates from below 10 Hz to 1 MHz, with usable output to about 2 MHz. An integrator–comparator generates the square and triangle waveforms, with a shaping circuit forming the triangle wave into a sine wave. The triangle wave is generated by switching current-source transistors to alternately charge and discharge the timing capacitor. This generates a linear tri-wave without the use of an op-amp integrator. The FET voltage follower buffers the tri-wave and drives the comparator, output amplifier, and sine converter.

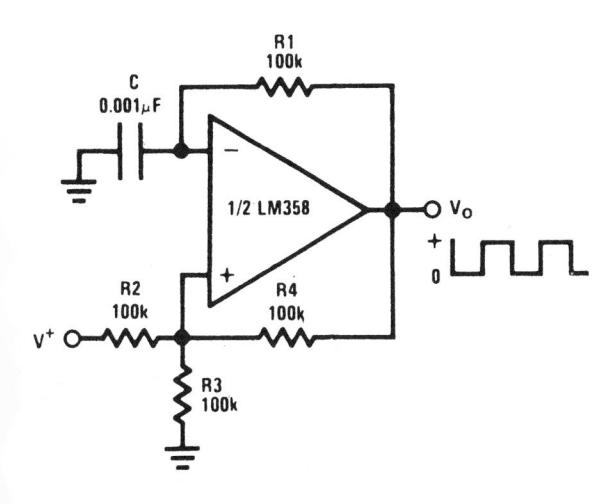

Square-wave oscillator.

SECTION 4 *Waveform Generators and Shapers* **181**

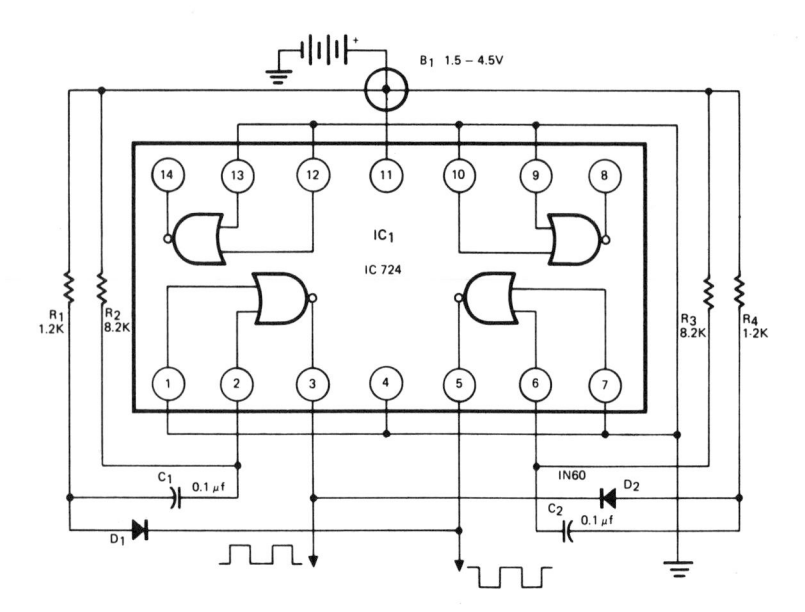

Square-wave generator. Frequency can be modified by changing the values of C1 and C2 (but they must always be of the same value). (Semiconductors are IRC/Workman.)

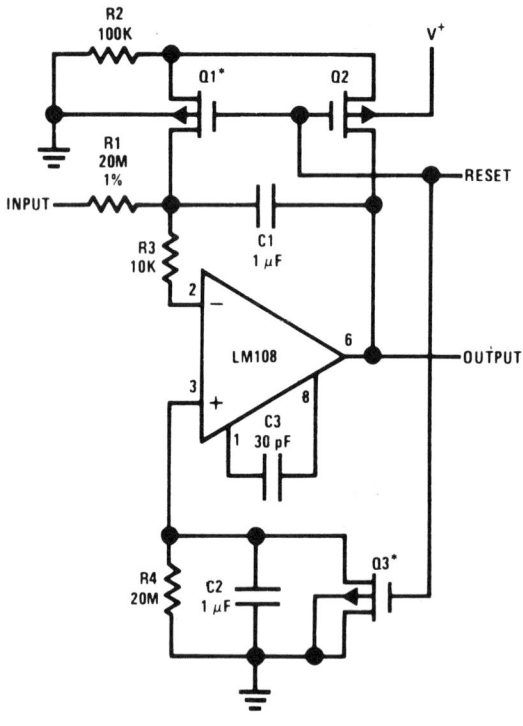

Sine-wave oscillator; $f_o = 10$ kHz.

*Q1 and Q3 should not have internal gate-protection diodes.

This low-drift integrator with reset gets rid of switch leakages. A negative-going reset pulse turns on Q1 and Q2, shorting the integrating capacitor. When the switches turn off, the leakage current of Q2 is absorbed by R2 while Q1 isolates the output of Q2 from the summing node. Q1 has practically no voltage across its junctions because the substrate is grounded; hence, leakage currents are negligible.

During the integration interval, the bias current of the noninverting input accumulates an error across R4 and C2 just as the bias current on the inverting input does across R1 and C1. Therefore, if R4 is matched with R1 and if C2 is matched with C1, the output will drift at a rate proportional to the difference in these currents. At the end of the integration interval, Q3 removes the compensating error accumulated on C2 as the circuit is reset.

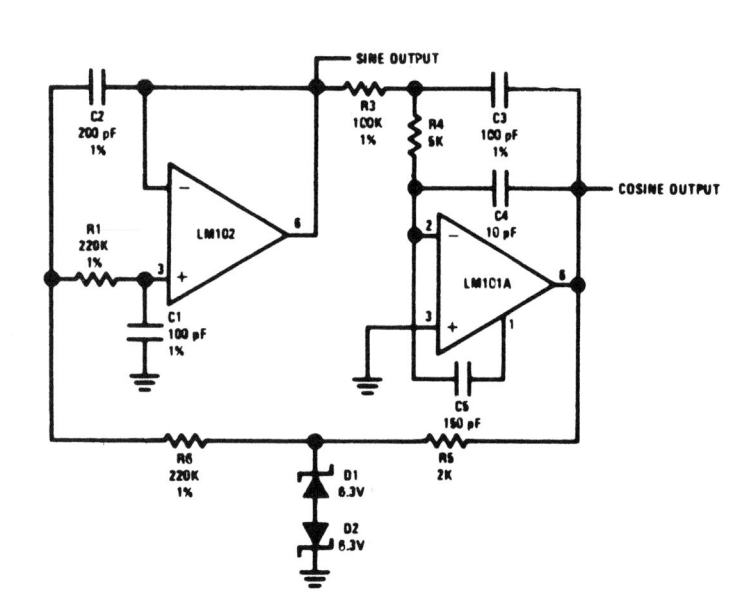

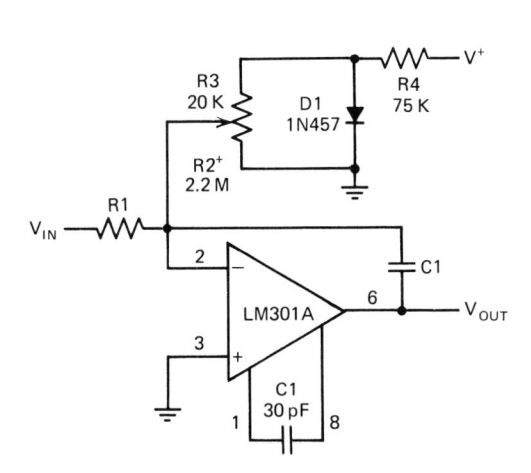

An integrator with bias-current compensation. Adjust R2 for zero integrator drift. Typical current drift is 0.1 nA/°C over a 0 to 70°C range.

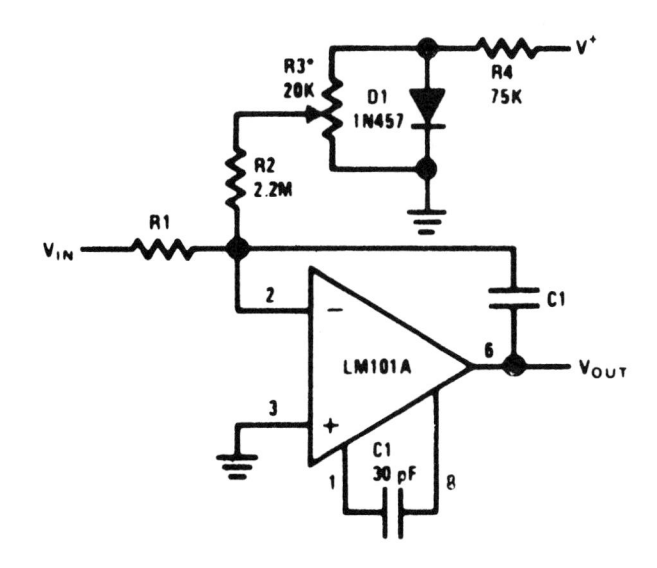

Integrator with bias-current compensation. Current drift is typically 0.1 nA/°C over a −55 to +125°C range. Adjust R3 for zero integrator drift.

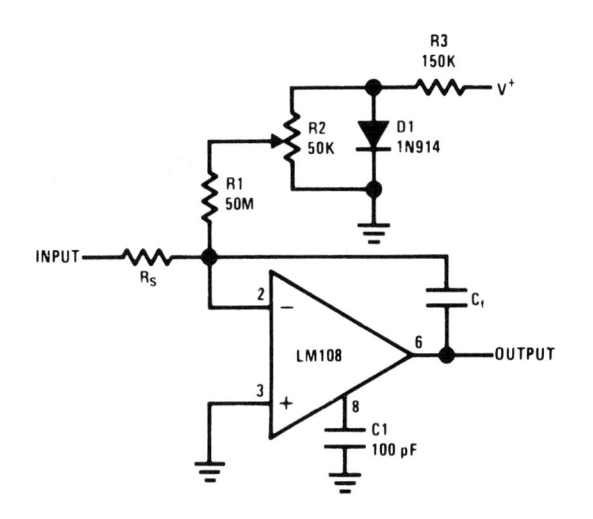

Integrator with bias-current compensation. A current is fed into the summing node through R1 to supply the bias current. The potentiometer, R2, is adjusted so that this current exactly equals the bias current, reducing the drift rate to zero.

The diode is used for two reasons. First, it acts as a regulator, making the compensation relatively insensitive to variations in supply voltage. Second, the temperature drift of diode voltage is approximately the same as the temperature drift of bias current. Therefore, the compensation is more effective if the temperature changes. Over a 0 to 70°C temperature range, the compensation will give a factor-of-10 reduction in input current. Even better results are achieved if the temperature change is less.

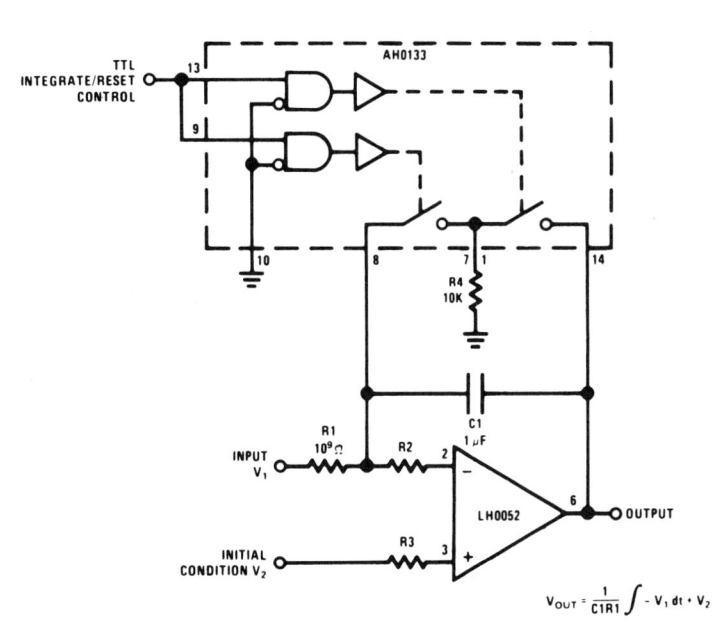

$$V_{OUT} = \frac{1}{C1R1} \int -V_1 \, dt + V_2$$

Precision integrator. Resistor R1 should be selected so that the total leakage current at the summing node is smaller than the signal current (V1/R1) by a margin sufficient to ensure the required accuracy. C1 should be chosen for low leakage, stability, accuracy, and low voltage coefficient. Polystyrene or polycarbonate dielectric is the best choice for capacitances up to about 1 μF; Teflon is good for the lower values. R2 protects the input circuit during the reset transient, although many low-speed applications will not require it.

SECTION **4** *Waveform Generators and Shapers* **183**

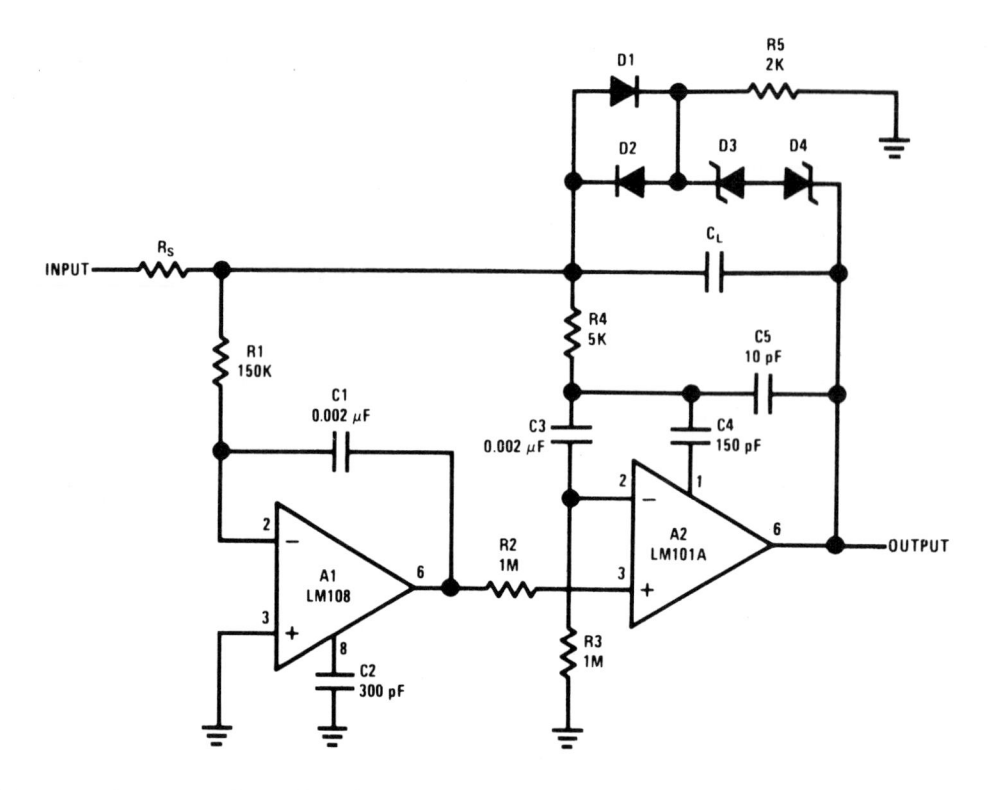

This fast integrator is arranged so that the high-frequency gain characteristics are determined by A2, while A1 determines the dc and low-frequency characteristics. The noninverting input of A1 is connected to the summing node through R1. A1 is operated as an integrator, going through unity gain at 500 Hz. Its output drives the noninverting input of A2. The inverting input of A2 is also connected to the summing node through C3. C3 and R3 are chosen to roll off below 750 Hz. Hence, at frequencies above 750 Hz, the feedback path is directly around A2, with A1 contributing little. Below 500 Hz, however, the direct feedback path to A2 rolls off, and the gain of A1 is added to that of A2.

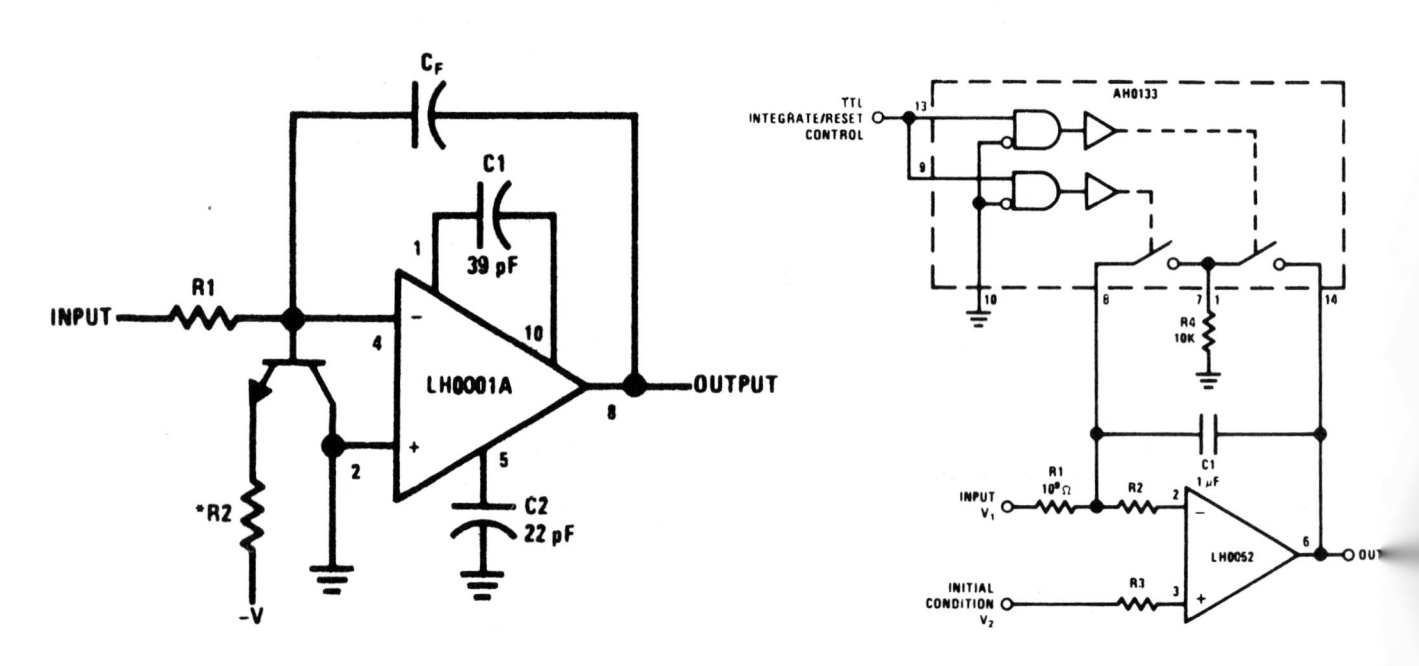

An integrator with bias compensation. Adjust R2 for zero integrator drift.

Precision integrator. $V_{OUT} = 1/C1R1 - V1dt + V2$.

184 IC SCHEMATIC SOURCEMASTER

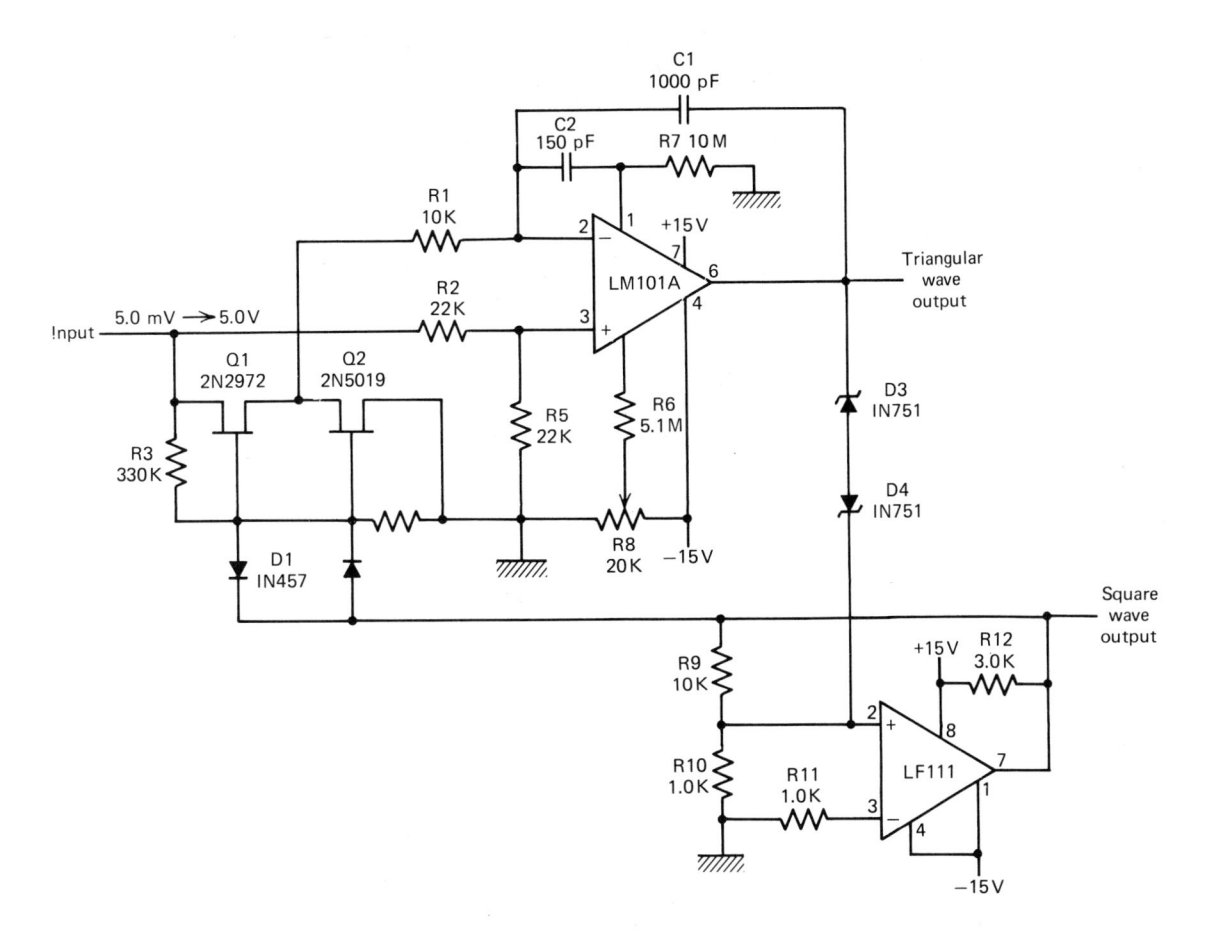

10 to 10,000 Hz voltage-controlled oscillator. Adjust R8 for symmetrical square-wave time when $V_{IN} = 5.0$ mV. Minimum capacitance for C1 is 20 pF; maximum frequency, 50 kHz.

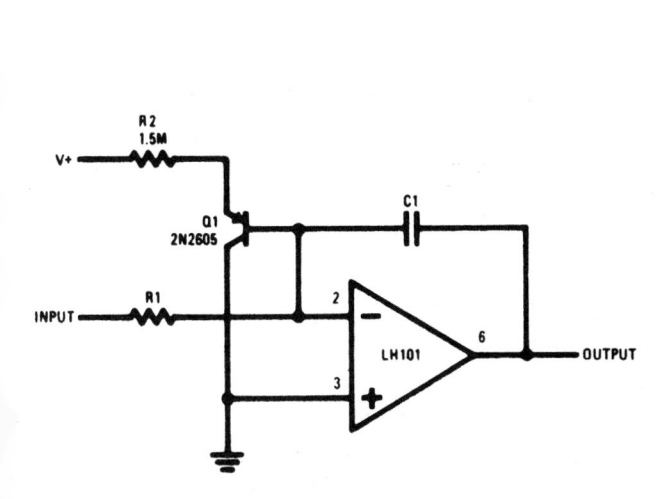

Integrator with bias-current compensation. IC is metal can. Select R2 for zero integrator drift.

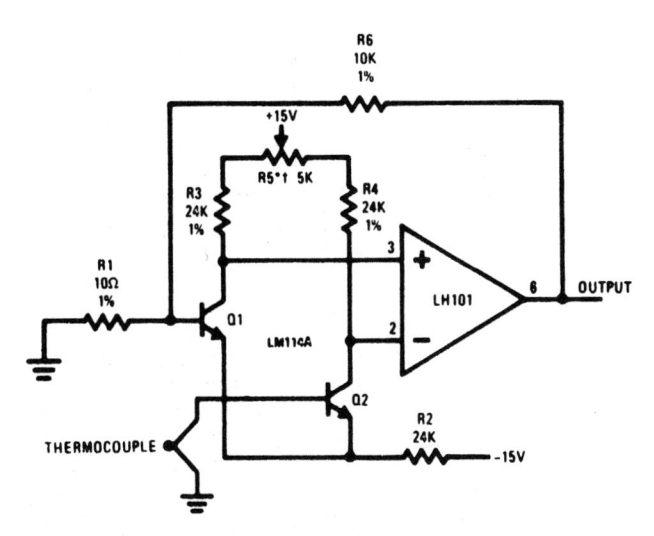

Low-drift thermocouple amplifier. R3 through R5 must match temperature coefficients. Adjust R5 for zero input offset voltage. Device drifts less than 0.5 $\mu V/°C$.

SECTION **4** *Waveform Generators and Shapers* **185**

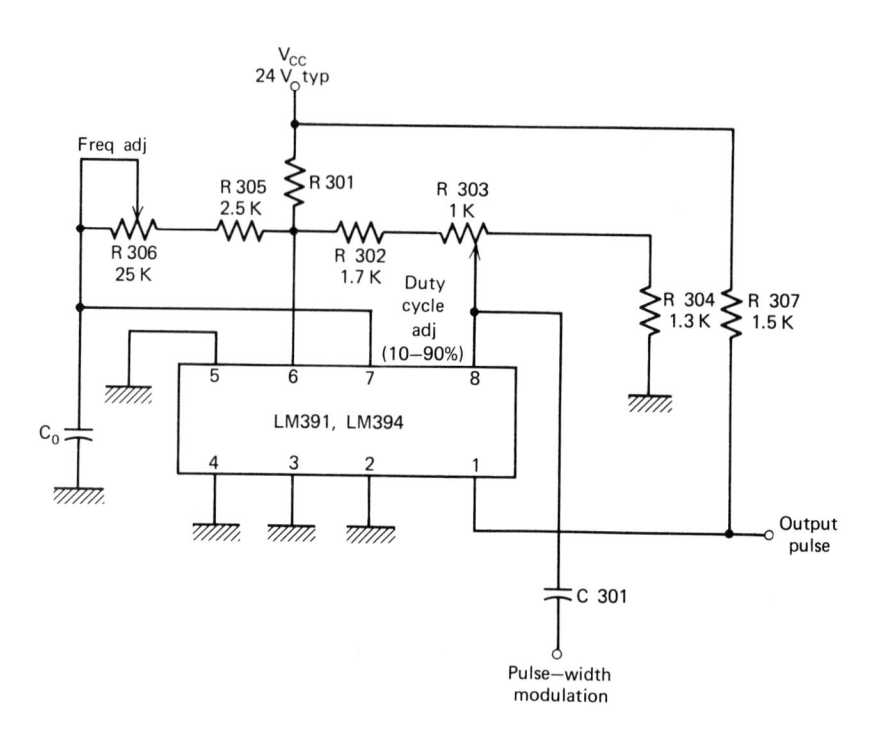

Variable-duty-cycle oscillator.

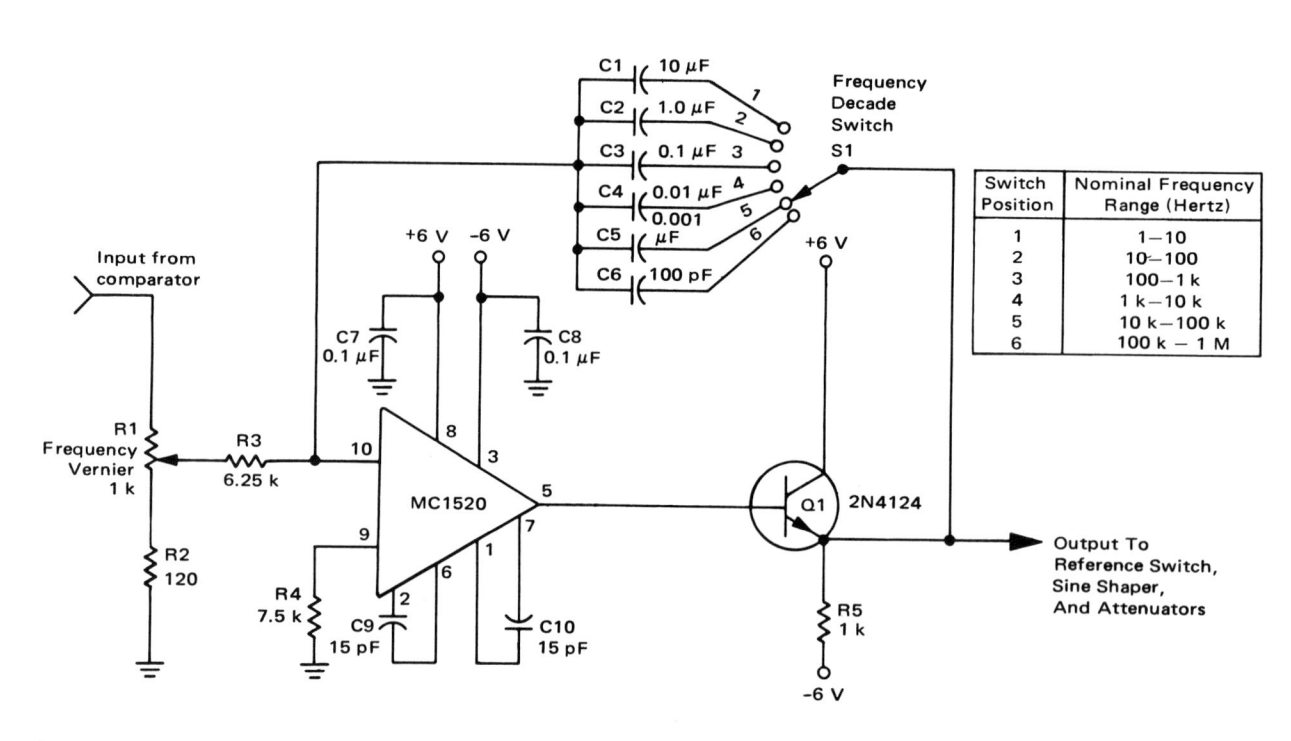

Switch Position	Nominal Frequency Range (Hertz)
1	1–10
2	10–100
3	100–1 k
4	1 k–10 k
5	10 k–100 k
6	100 k–1 M

Integrator is a standard configuration using an MC1520 operational amplifier. The input voltage (whose amplitude is controlled by R1) is a square wave from the comparator. When the input is positive, the output is a negative-going ramp with a velocity determined by the current through R3 and feedback capacitors C1 through C6. Conversely, for a negative input the output is a positive-going ramp. With the peak-to-peak output at 4V at 1 MHz, dv/dt is 8V/μsec. The 100 pF feedback capacitor is used for this frequency. About 0.8 mA flows through the integrating resistor, R3, assuming the offset voltage and bias currents are negligible. Since the output of the comparator is 5V peak,

$$R3 = \frac{5V}{0.8 \text{ mA}} = 6.25k$$

To operate over one decade, the feedback current must change by a factor of 10. With R1 at the low-resistance end, the input voltage is 0.5V, and the input current is 0.08 mA; therefore, the lowest frequency obtainable with a 100 pF feedback capacitor is 100 kHz.

186 IC SCHEMATIC SOURCEMASTER

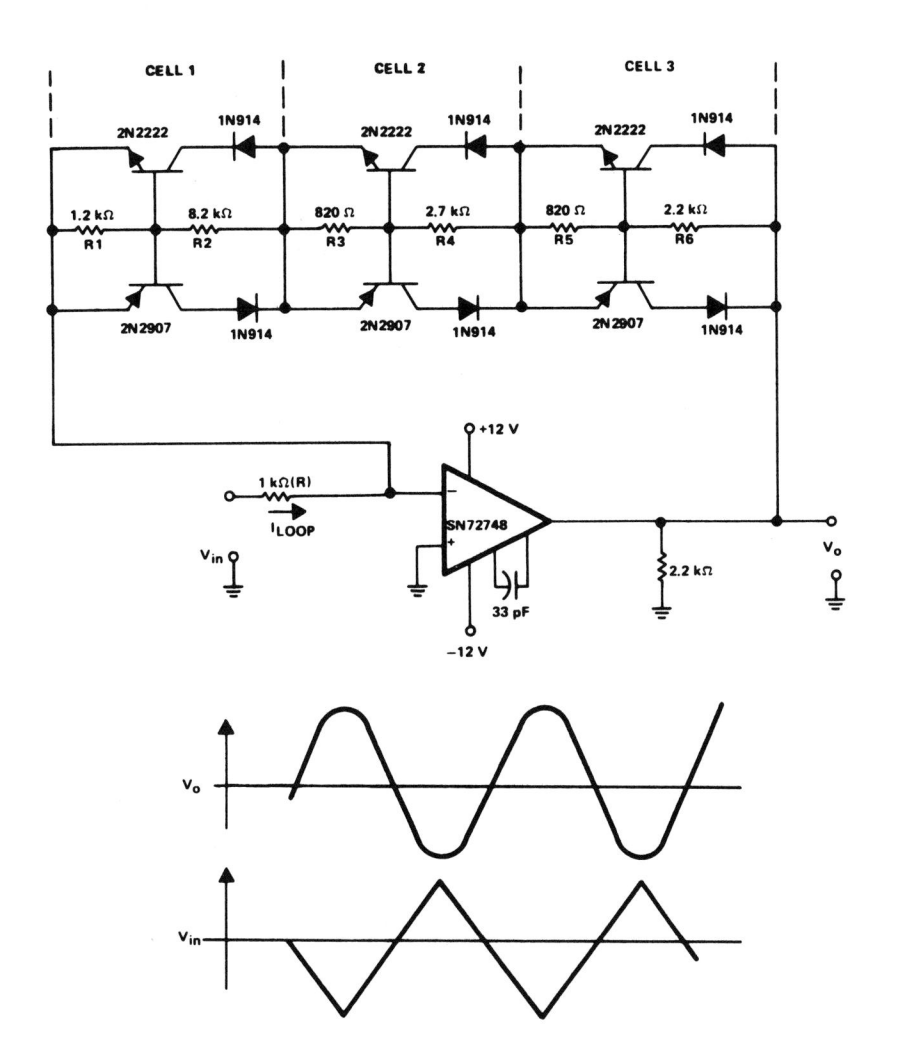

This triangle-to-sine converter uses a nonlinear transfer characteristic to obtain a sine-wave output from a triangular input. To obtain the required transfer function, the amplifier gain must decrease as the input voltage increases. A feedback loop consisting of three cells provides the required gain variations and will handle both positive and negative input signals.

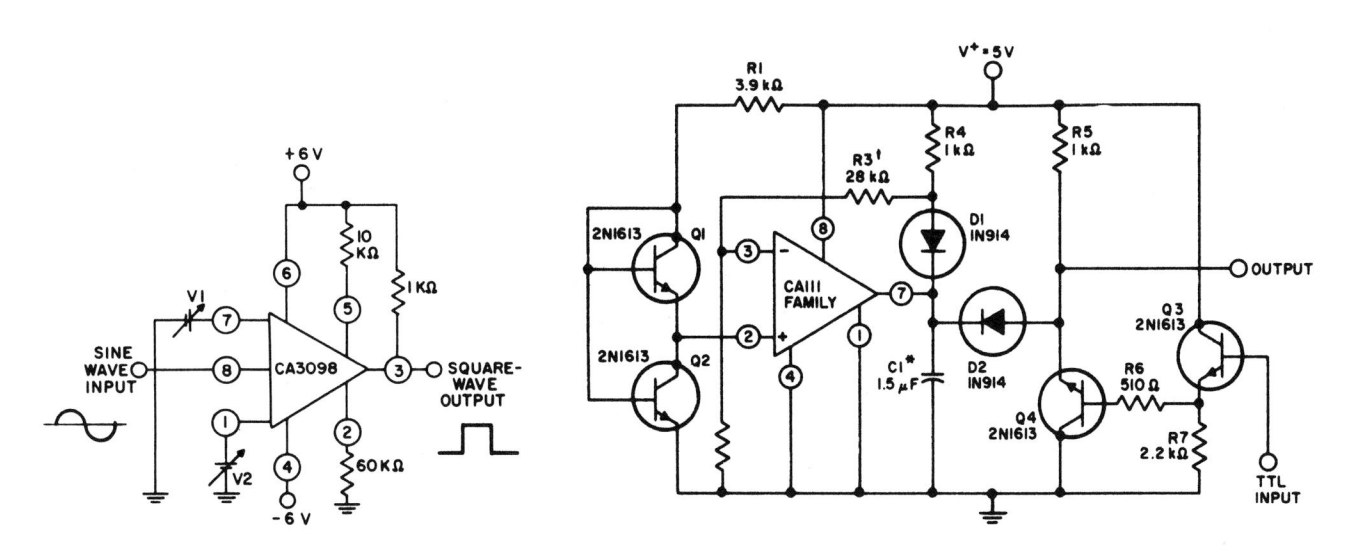

A sine- to square-wave converter with duty-cycle adjustment (V_1 and V_2).

Precision squarer. C1 should be solid tantalum. Adjust R3 to set clamp level.

SECTION 4 *Waveform Generators and Shapers*

SECTION 5
TELEVISION AND VIDEO CIRCUITS

According to Milt Wilcox of National Semiconductors, "Every area of the television (set) which does not have too high a frequency or voltage requirement has been integrated at least once, and many are on second- and third-generation ICs." The use of integrated circuits in television systems was not new in 1975, when Wilcox made that statement—and video applications have increased steadily since then. In this section, we present a gamut of circuits, including one or two of those early hybrid types that successfully employed integrated circuits in concert with vacuum tubes.

The circuits in this section do not tell the whole story. For related circuit-function diagrams, look in the logic section (Section 15) for multipliers, log amplifiers and attenuators, and other "video" circuits that have multiple functions; see Section 14 for appropriate data transmission circuits. And do not overlook the communications section (Section 7), the audio circuits section (Section 6), and the amplifier section (Section 2).

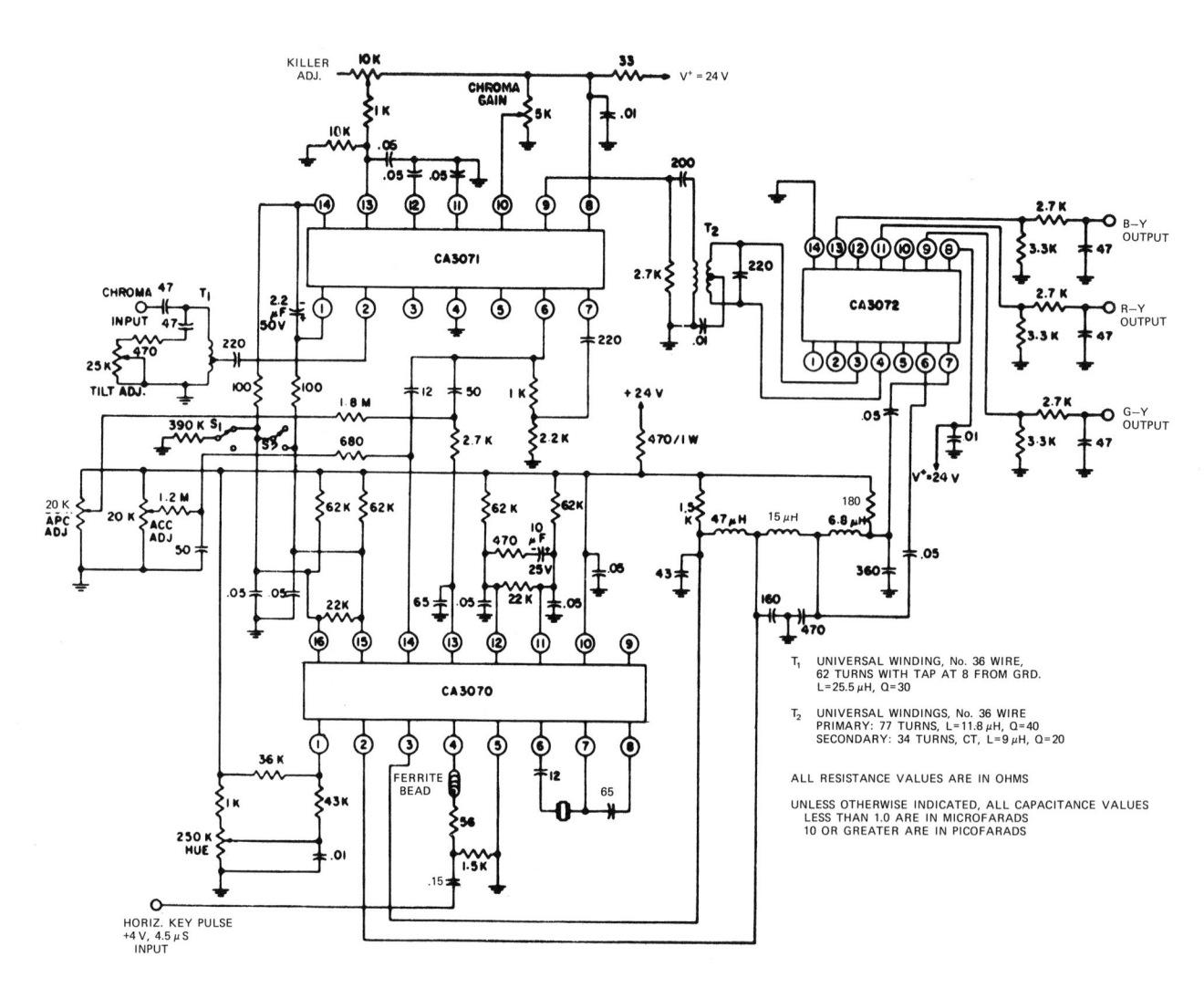

Complete signal-processing system for color TV. RCA CA3070, CA3071, and CA3072 monolithic integrated circuits are used as the subcarrier regenerator, chroma amplifier, and chroma demodulator, respectively. The chroma signal input, taken from the first or second video stage, is coupled into the CA3071 chroma amplifier through a bandpass filter. The outputs from the system are the color difference signals that are intended to drive high-level amplifiers. Luminance mixing may be external to the picture tube, or the difference signals may be amplified and applied to the picture tube grid or cathode where they are internally mixed with the luminance signal.

Other input requirements to the system are the power supply voltage of +24V and the horizontal keying pulse. The power supply voltage should be maintained within ±3V of the recommended value of +24V. The total current for the system is approximately 70 mA. The horizontal keying pulse input to the subcarrier regenerator is approximately +4V peak and centered on the burst as seen at pins 13 and 14 of the CA3070. The pulse width should be maintained as close as possible to the recommended value of 4.5 μsec.

190 IC SCHEMATIC SOURCEMASTER

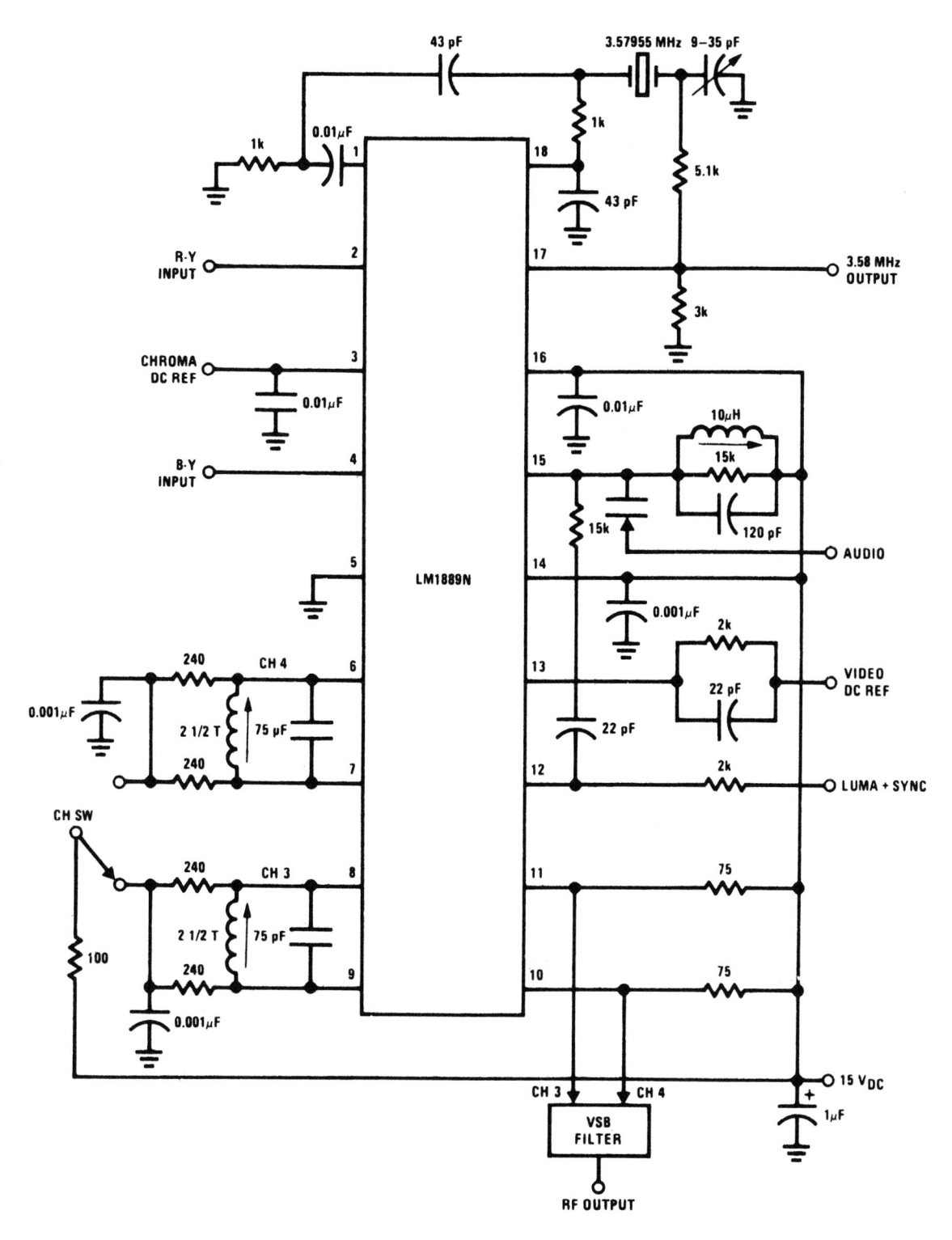

This TV video modulator is designed to interface audio, color difference, and luminance signals to the antenna terminals of a TV receiver. It consists of a sound subcarrier oscillator, chroma subcarrier oscillator, quadrature chroma modulators, and RF oscillators and modulators for two low VHF channels. The LM1889 allows video information from VTRs, games, test equipment, or similar sources to be displayed on black-and-white or color TV receivers. When used with MM57100 and MM53104, a complete TV game is formed.

SECTION **5** *Television and Video Circuits* **191**

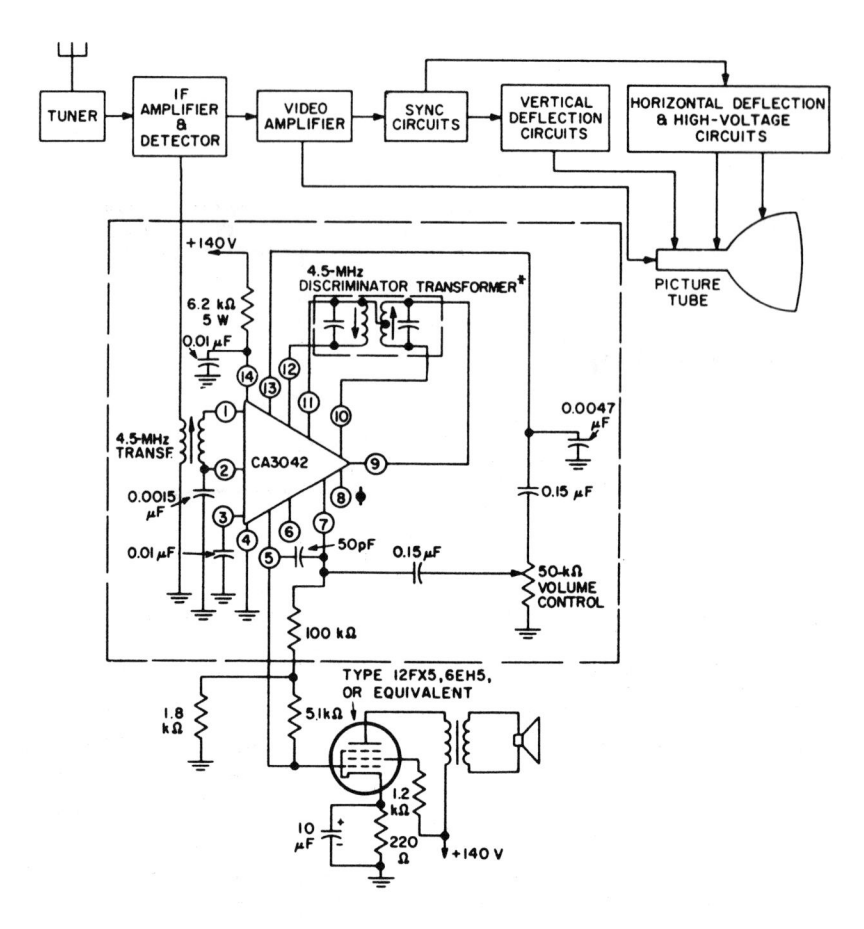

A TV receiver using CA3042 and a 12FX5, 6EH5, or equivalent. Muting can be accomplished by grounding terminal 8.

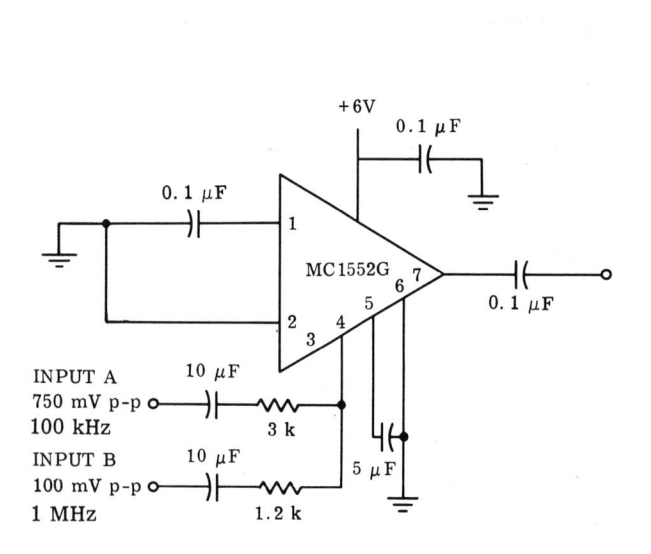

MC 1552G video amplifier connected as a **summing/scaling amplifier.** In this noninverting configuration, the summation of input signal currents is accomplished at the summing point, pin 4, through the input resistors. Scale factor considerations are accounted for by adjustment of the design values of the input summing resistors.

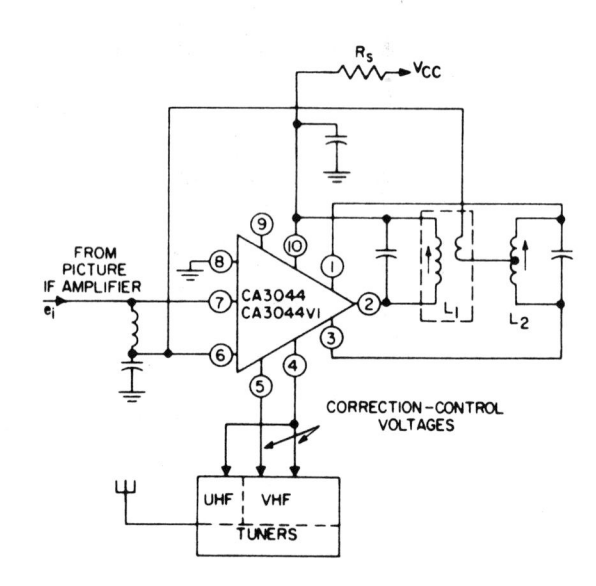

Automatic fine tuning (AFT) application using CA3044 or CA3044V1 in color TV receiver shown in block diagram. L1 and L2 serve as a phase detector transformer.

192   IC SCHEMATIC SOURCEMASTER

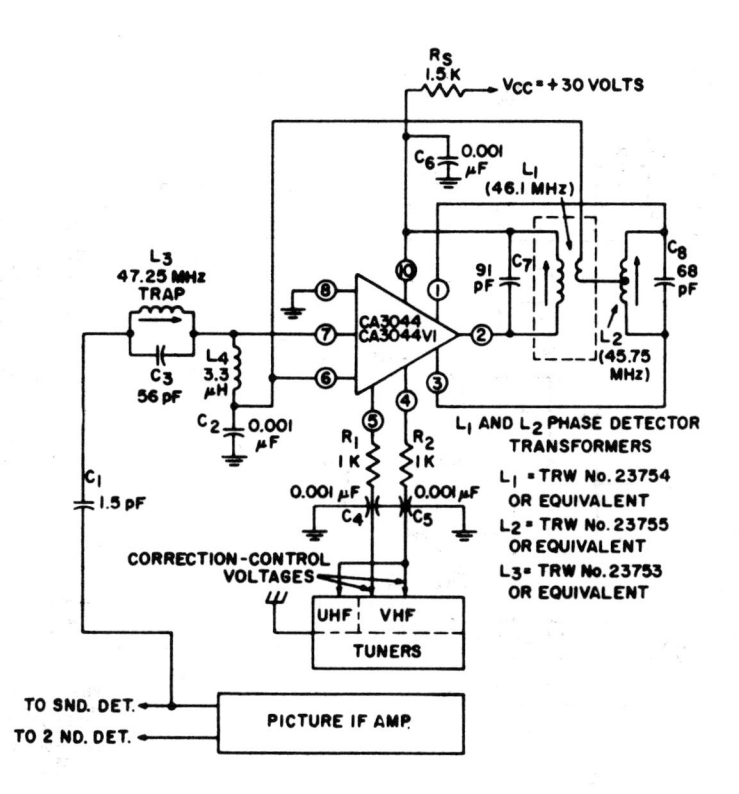

A typical automatic fine-tuning (AFT) application showing the CA3044 or CA3044VI in use in a color TV receiver. Load impedance at the center frequency is about 1800 Ω. RF bypassing is required for both terminals 6 and 10, the latter of which is connected through the primary winding of the detector transformer to terminal 2.

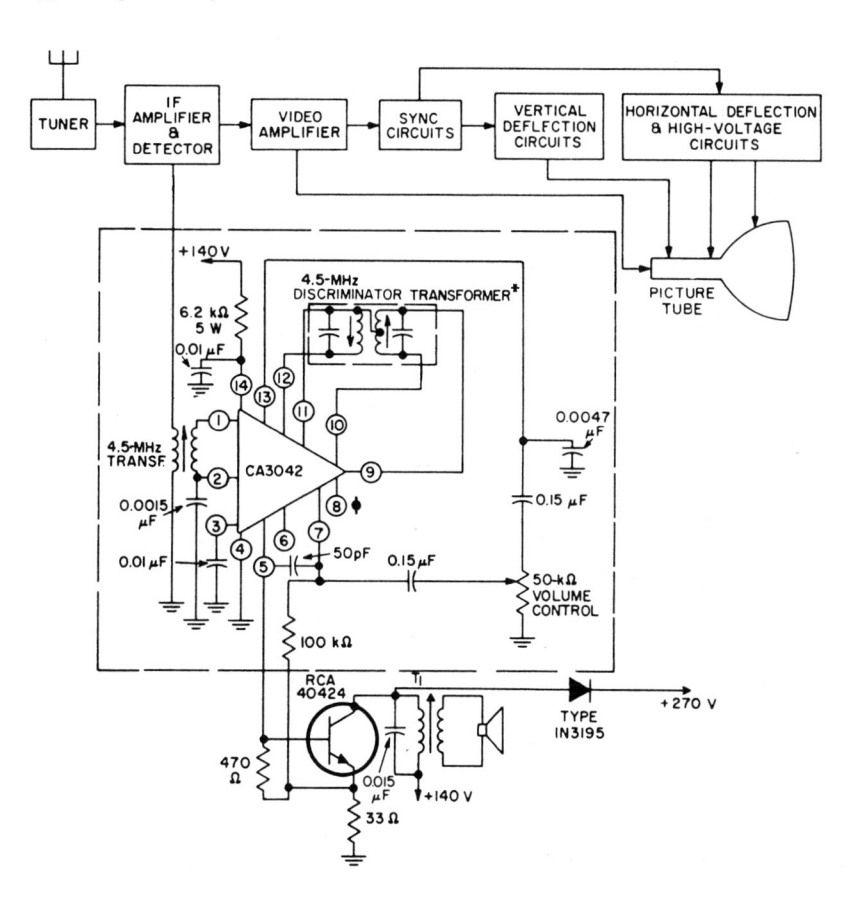

A typical TV receiver using RCA 40425. Muting can be accomplished by grounding terminal 8. Discriminator transformer is TRW E023874 or equivalent. T1 is 2.5W, 3500 Ω to speaker; primary dc is 40 mA.

SECTION 5 *Television and Video Circuits* 193

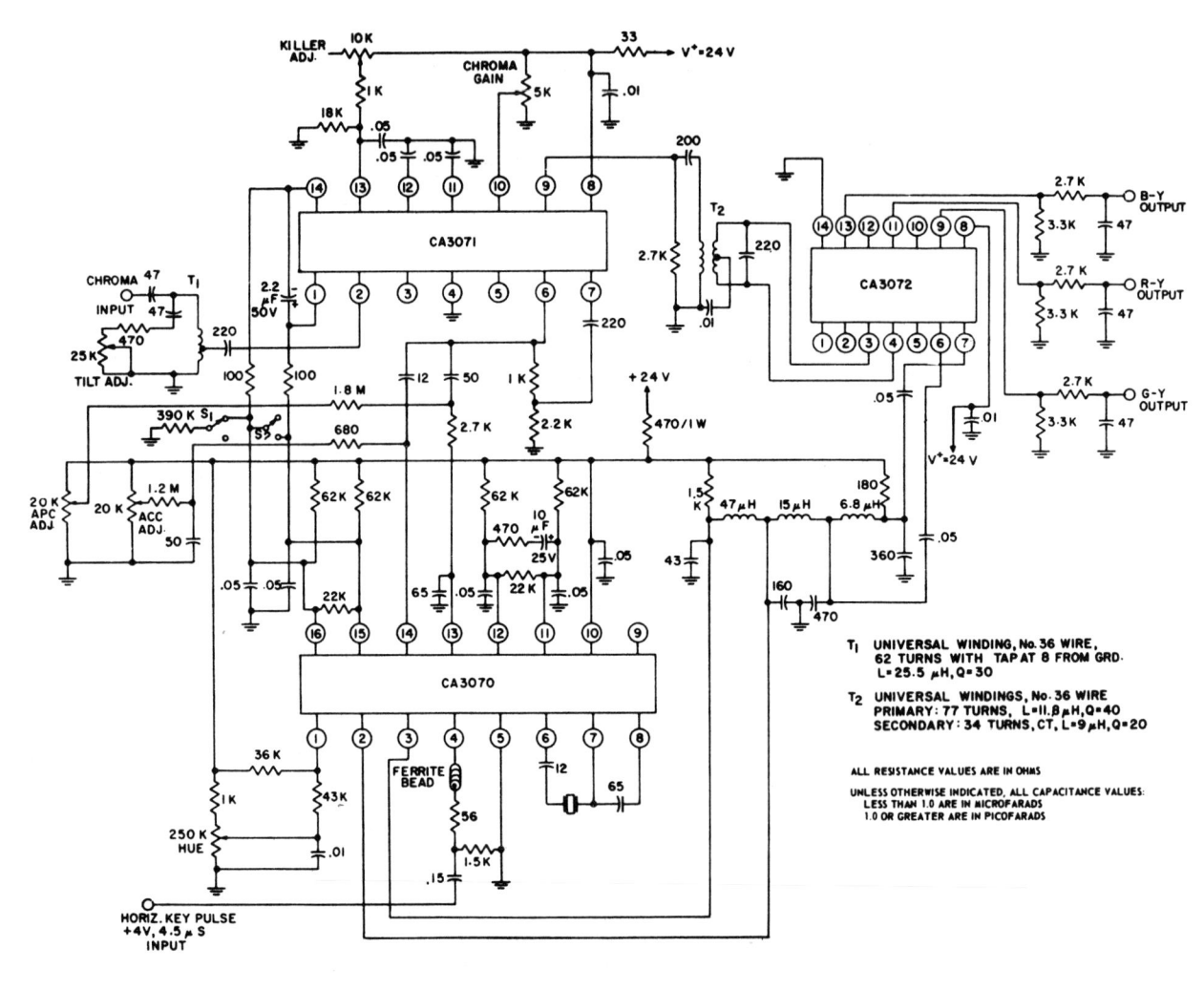

A chroma system for color TV receivers using RCA CA3070, CA3071, and CA3072.

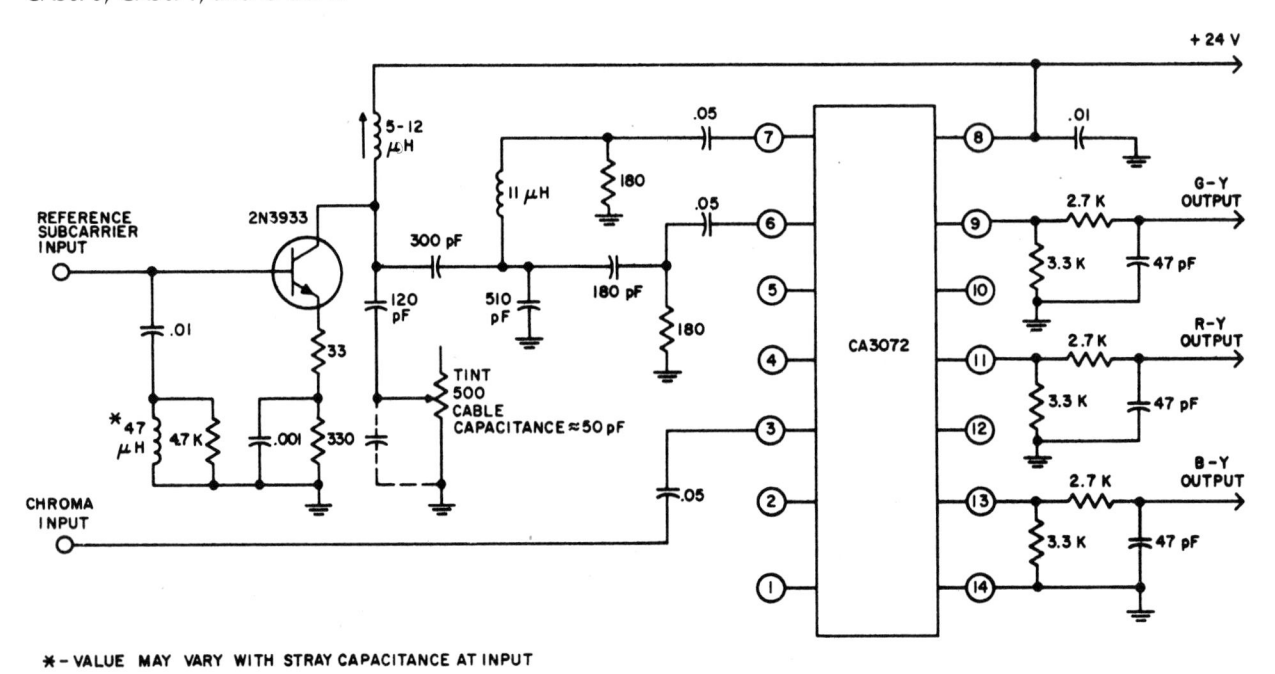

A chroma demodulator with 2N3933 tint control amplifier circuit using CA3072.

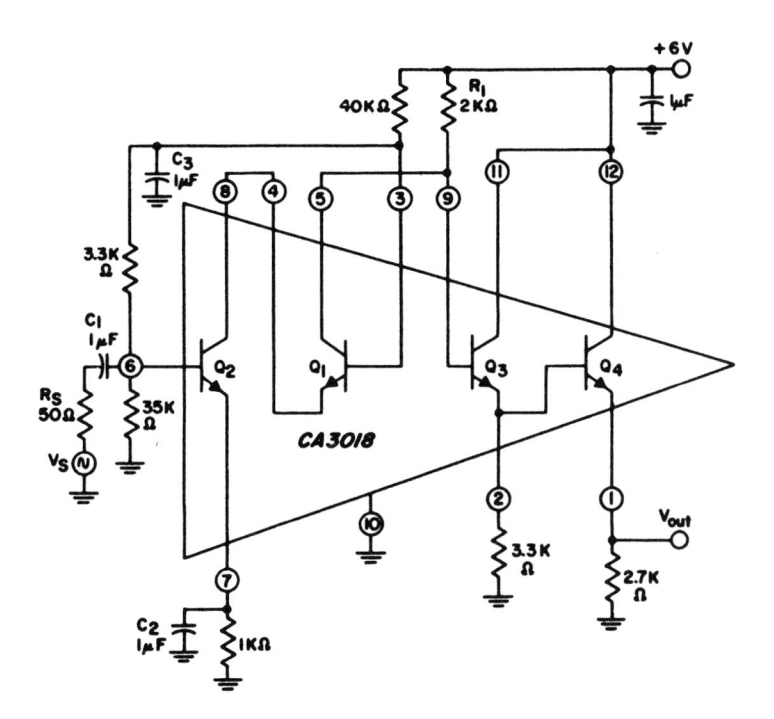

A cascode video amplifier using CA3018. Idling-current bias is provided to Q1 and Q2 by use of transistor Q3 as a diode (with collector and base shorted) and connection of a series resistor to the supply. The idling current for each transistor in the class B output is equal to the current established in the resistance-diode loop. Because resistor R1 is the predominant factor in controlling the current in the bias loop, the bias current is relatively independent of temperature. Because the devices have nearly equal characteristics and are at the same temperature, the idling current is nearly independent through the full military temperature range. Ac feedback as well as dc feedback, can be obtained by substituting, respectively, two resistors R2 and R3 in place of R1, the cascode. The common-base unit is followed by cascaded emitter followers (Q3 and Q4) that provide a low output impedance to maintain bandwidth for iterative operation.

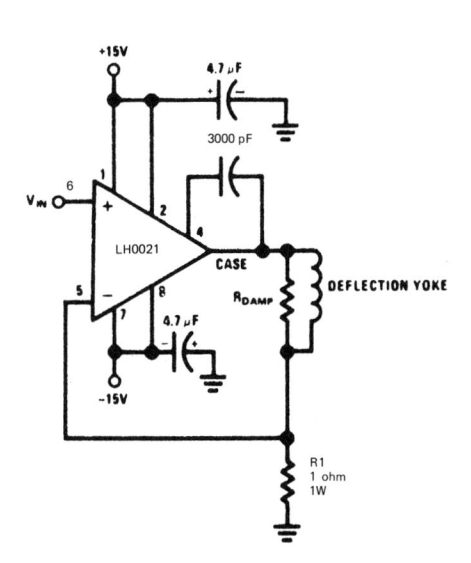

CRT deflection yoke driver. $I_{OUT} = V_{IN}/R1$.

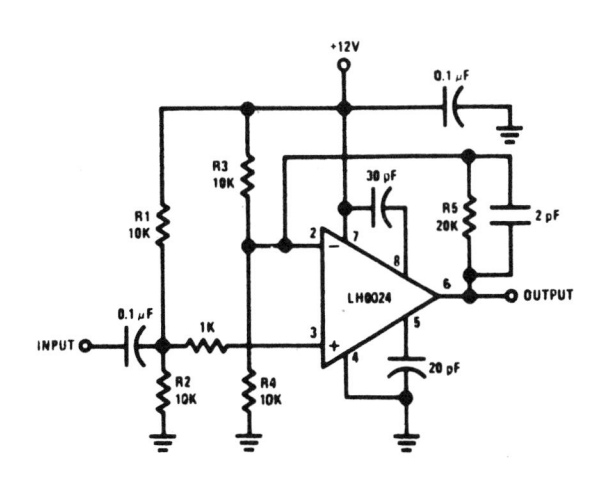

Video amplifier. R1 = R2 = R3 = R4; $A_V = R5 + (R3 \| R4)/(R3) \| (R4) = 5$.

SECTION 5 Television and Video Circuits **195**

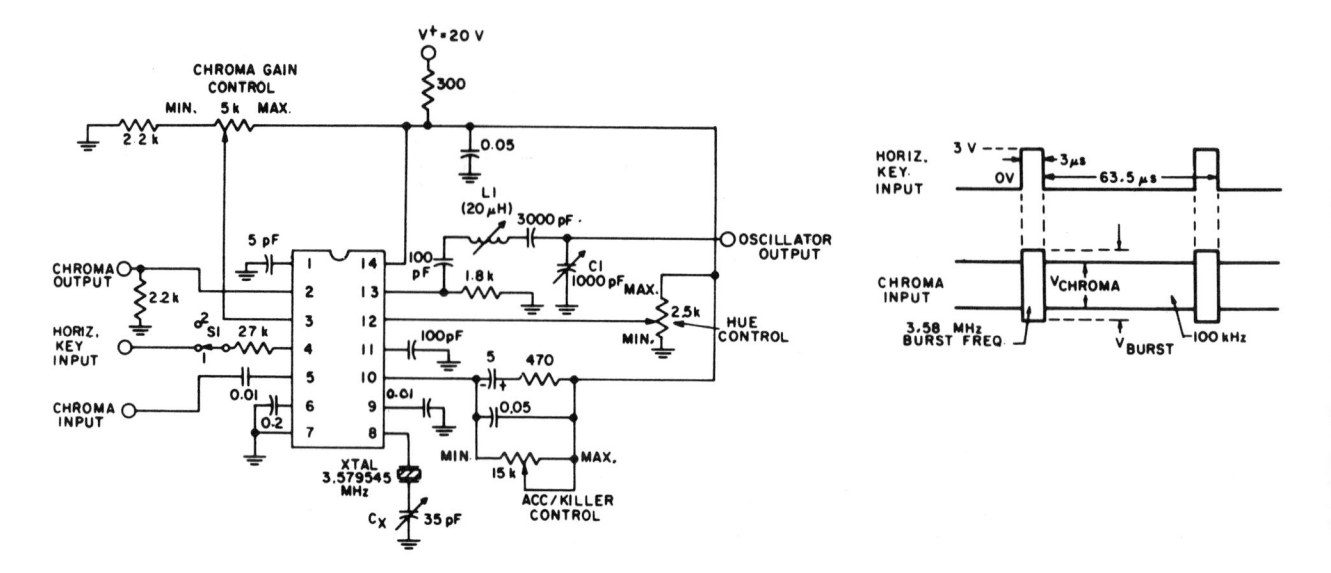

This television chroma processor uses RCA CA1398E, a monolithic silicon chroma processor containing chroma amplifier and gain control, color killer, color subcarrier oscillator, hue control, and ACC circuitry. It has been designed for interchangeability with other "1398"-type chroma processor devices. It functions compatibly with the RCA CA3125E chroma demodulator, as well as with other commercially available chroma demodulators in color TV receivers.

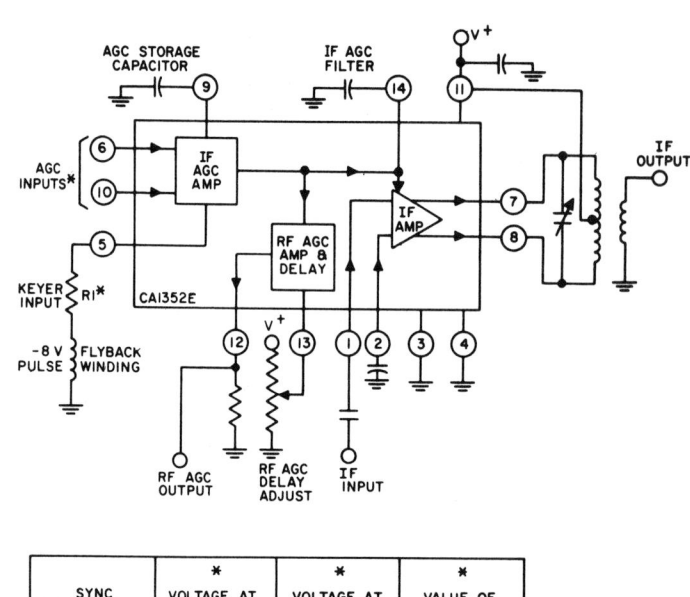

SYNC POLARITY	VOLTAGE AT TERMINAL 6	VOLTAGE AT TERMINAL 10	VALUE OF R1-Ω
NEGATIVE	5.5 V / -2 V / 0 V	1 TO 4 V NOM = 2 V	0
POSITIVE	1 TO 8 V NOM = 4.5	4.5 V / 0 V	3.9k

This TV video IF amplifier with AGC and keyer circuit features a high-gain gated AGC system with a 68 dB range. A delayed forward AGC output is adjustable by means of a potentiometer. Either positive- or negative-going sync may be used for this system, which typically features a high 45 MHz gain of 53 dB.

196 IC SCHEMATIC SOURCEMASTER

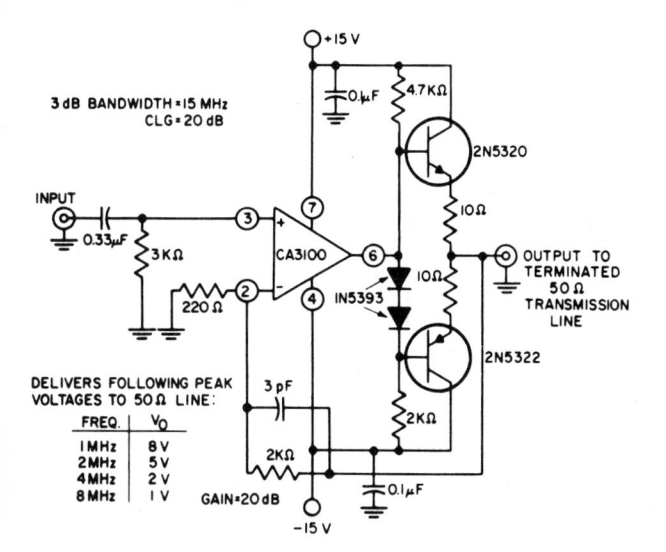

20 dB video line driver.

3 dB BANDWIDTH = 15 MHz
CLG = 20 dB

DELIVERS FOLLOWING PEAK VOLTAGES TO 50 Ω LINE:

FREQ.	V_O
1 MHz	8 V
2 MHz	5 V
4 MHz	2 V
8 MHz	1 V

GAIN = 20 dB

Interstage phase-shift network.

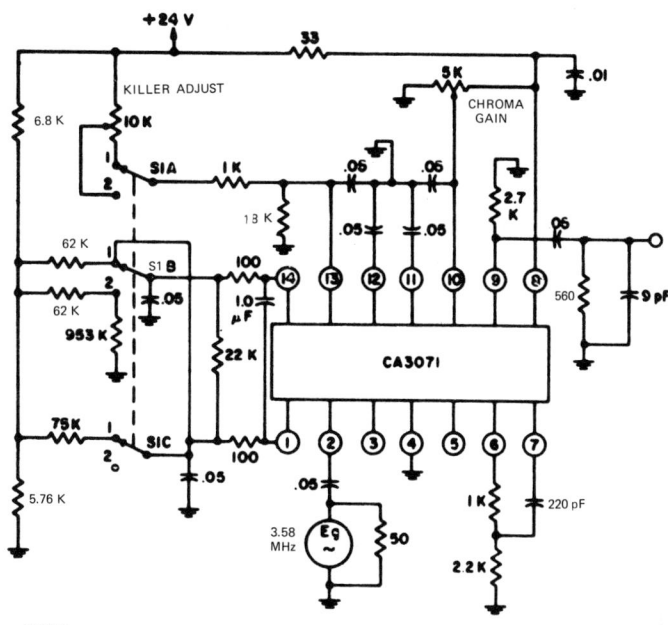

NOTES:
1. SWITCH S1 IN POSITION 1 UNLESS OTHERWISE NOTED IN TABLE OF DYNAMIC CHARACTERISTICS
2. CHROMA GAIN CONTROL SET TO GROUND UNLESS OTHERWISE NOTED IN TABLE OF DYNAMIC CHARACTERISTICS
3. ALL RESISTANCES IN OHMS
4. ALL CAPACITANCES ARE IN MICROFARADS UNLESS OTHERWISE SPECIFIED

Two-stage chroma amplifier and functional control circuit. The input signal is received from the video amplifier and is applied to pin 2 of the input amplifier stage. The first amplifier stage is part of the ACC system and is controlled by differential adjustment from ACC input pins 1 and 14. The output of the first amplifier is directed to pin 6 from where the signal may be applied to the ACC detection system of the CA3070 or an equivalent circuit. The output at pin 6 is also applied to 7, which is the input to the second amplifier stage. Another output of the first amplifier (pin 13) is directed to the killer adjustment circuit. The dc voltage level at pin 13 rises as the ACC differential voltage decreases with a reduction in the burst amplitude. At a preset condition determined by the killer adjustment resistor, the killer circuit is activated and causes the second chroma amplifier stage to be cut off. The second chroma amplifier stage is also gain controlled by the adjustment of dc voltage at pin 10. The output of the second chroma amplifier stage is available at pin 9. The typical output termination circuit provides a differential chroma drive signal to the demodulator circuit. Both amplifier outputs utilize emitter followers with short-circuit protection.

SECTION 5 *Television and Video Circuits* **197**

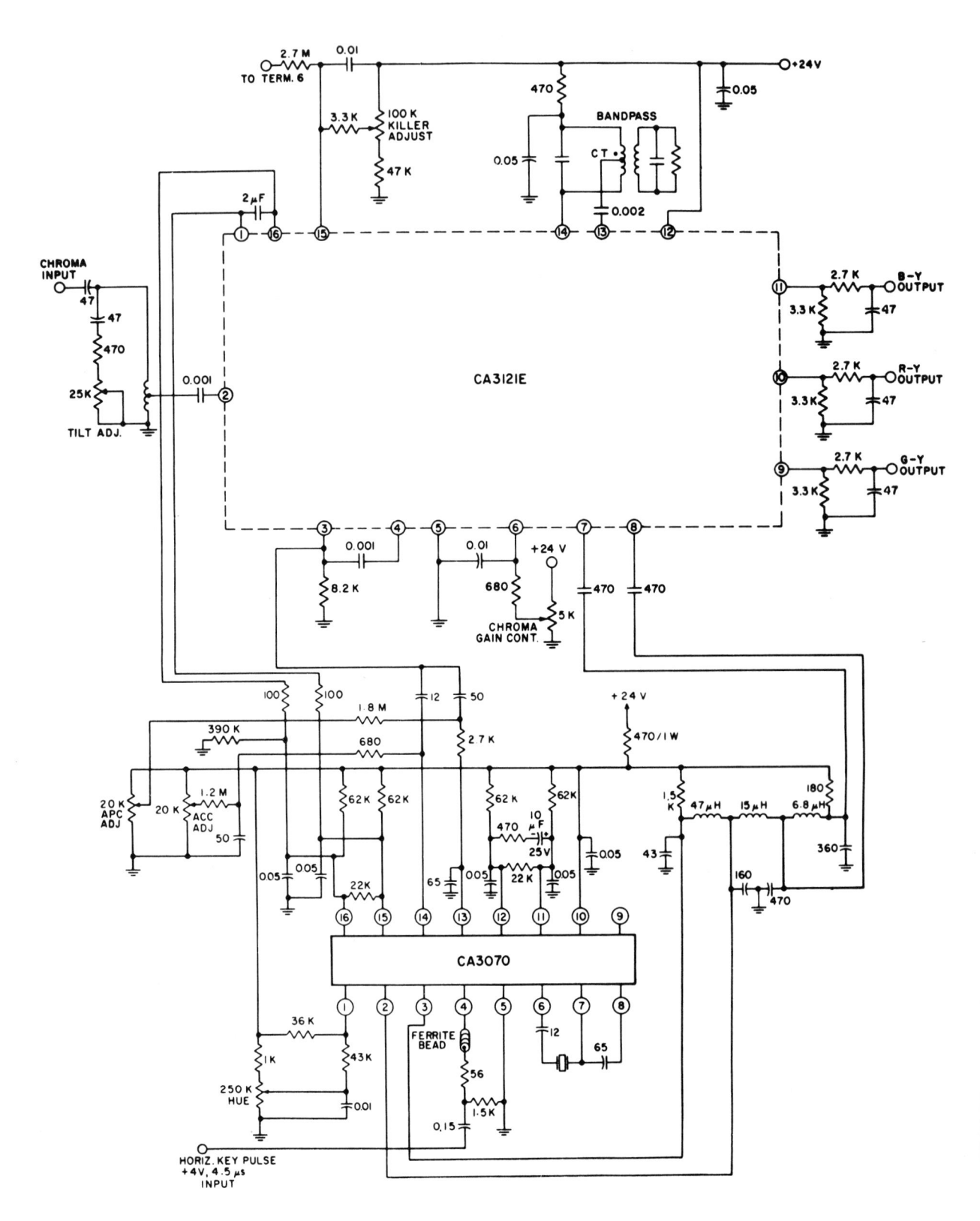

Outboard circuitry of a typical **two-package chroma system** for color-TV receivers using the CA3121E and CA3070.

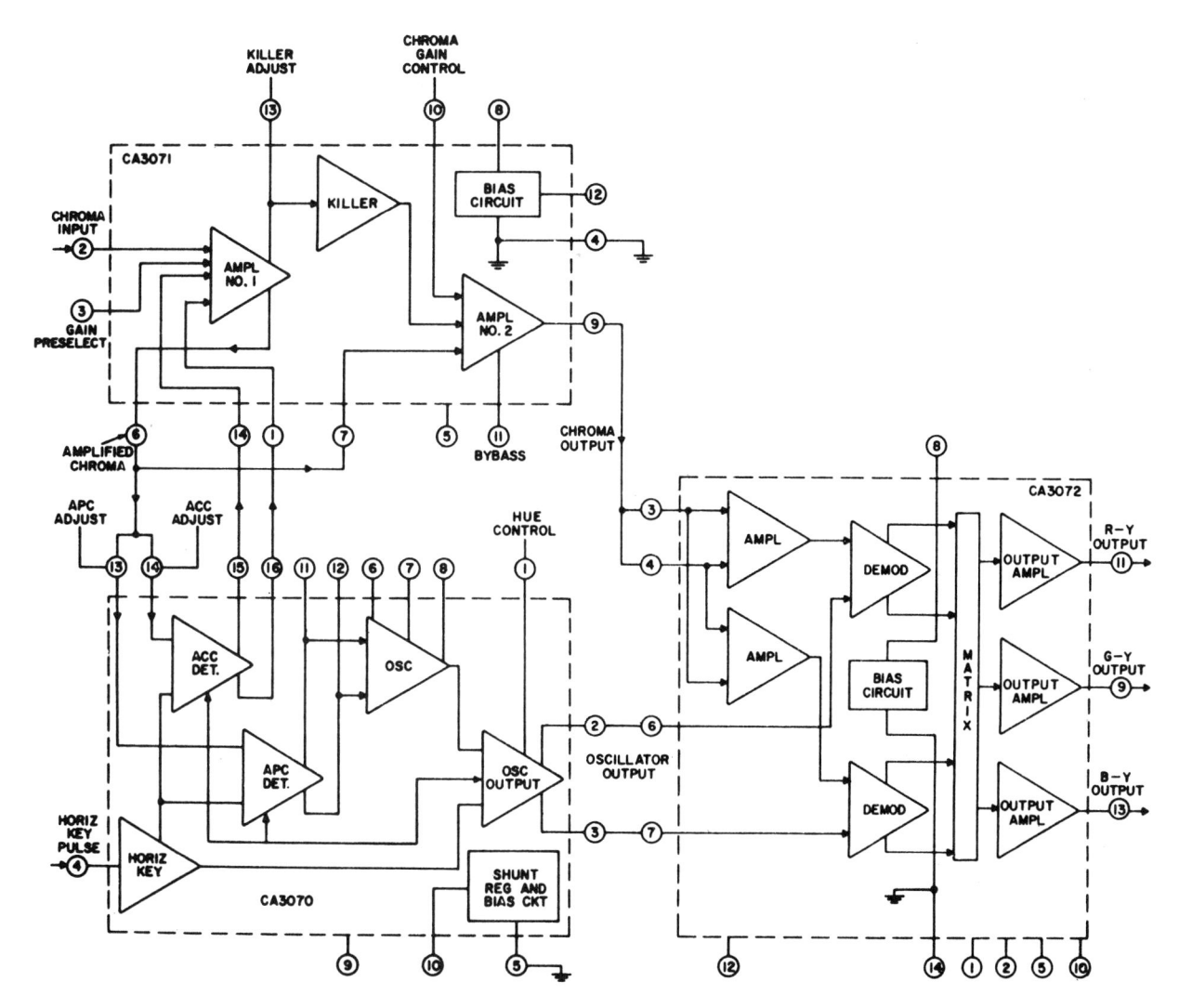

Chroma signal processor uses RCA's CA3070, a complete subcarrier regeneration system with automatic phase control applied to the oscillator. An amplified chroma signal from the CA3071 is applied to terminals 13 and 14, which are the automatic phase control (APC) and the automatic chroma control (ACC) inputs. APC and ACC detection is keyed by the horizontal pulse that also inhibits the oscillator output amplifier during the burst interval.

The ACC system uses a synchronous detector to develop a correction voltage at the differential output, pins 15 and 16. This control signal is applied to input pins 1 and 14 of the CA3071. The APC system also uses a synchronous detector. The APC error voltage is internally coupled to the 3.58 MHz oscillator at balance; the phase of the signal at pin 13 is in quadrature with the oscillator.

To accomplish phasing requirements, an RC phase-shift network is used between the chroma input and terminals 13 and 14. The feedback loop of the oscillator is from terminals 7 and 8 back to 6. The same oscillator signal is available at pins 7 and 8, but the dc output of the APC detector controls the relative signal levels at 7 or 8. Because the output at pin 8 is shifted in phase compared to the output at 7, which is applied directly to the crystal circuit, control of the relative amplitudes at terminals 7 and 8 alters the phase in the feedback loop, thereby changing the frequency of the crystal oscillator. Balance adjustments of dc offsets are provided to establish an initial no-signal offset control in the ACC output. The oscillator output stage is differentially controlled at pins 2 and 3 by the hue control input to pin 1. The hue phase shift is accomplished by the external R, L, and C components that couple the oscillator output to the demodulator input terminals. The CA3070 includes a shunt regulator to establish a 12V dc supply.

SECTION **5** *Television and Video Circuits*

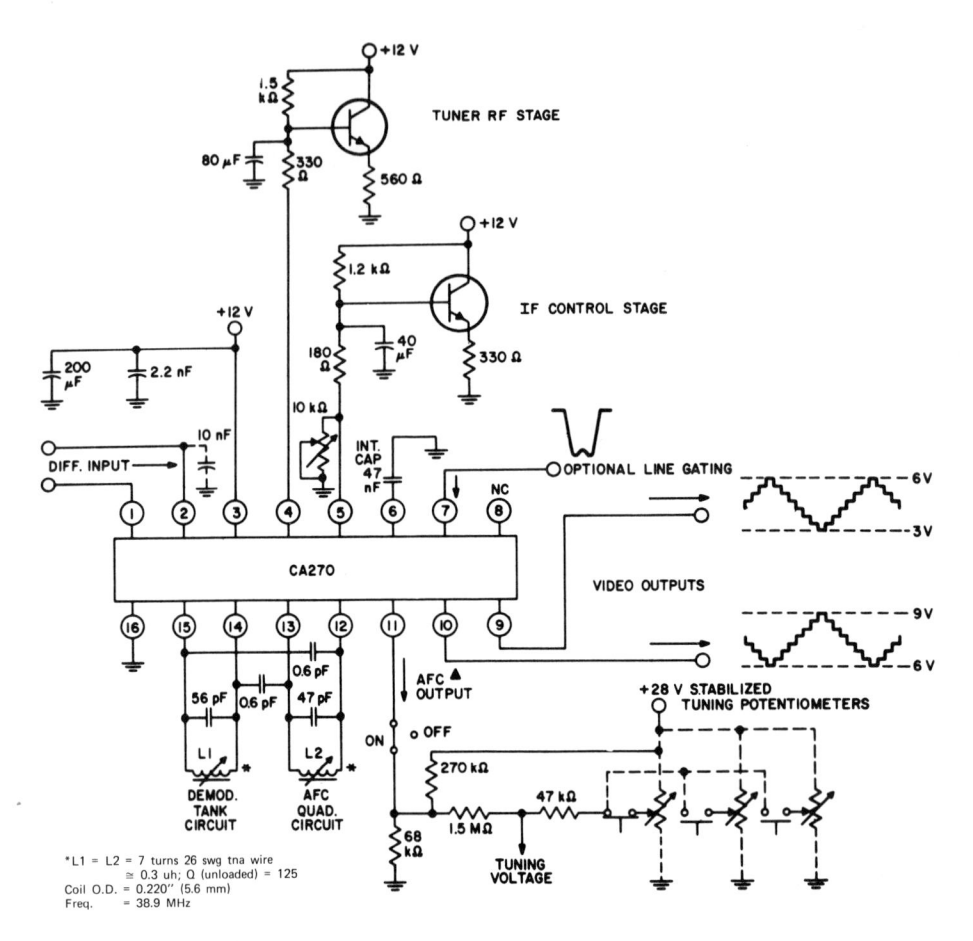

This TV synchronous demodulator for color and black-and-white TV systems performs the functions of synchronous detection of the TV IF video amplification and buffering and noise inversion on dual-polarity waveforms. Both positive and negative polarities of video output are available, a feature that provides great flexibility by permitting the use of either output for deriving the video and sound channels.

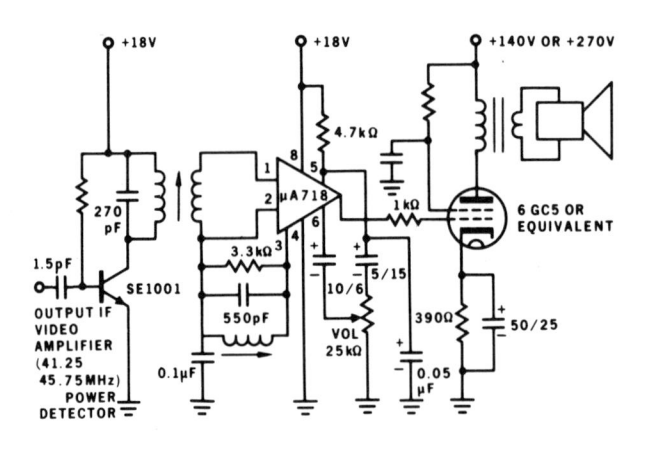

Color TV receiver sound IF system for hybrid models.

200 IC SCHEMATIC SOURCEMASTER

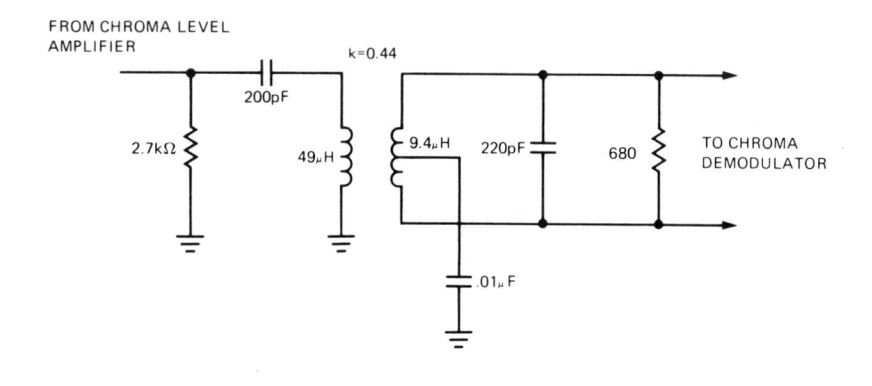

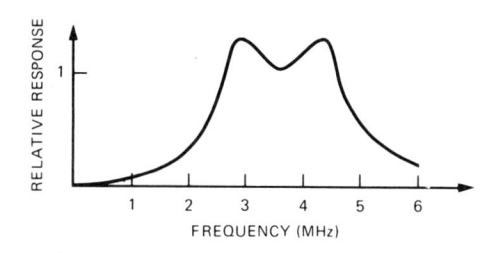

Chroma output network and frequency-response curve.

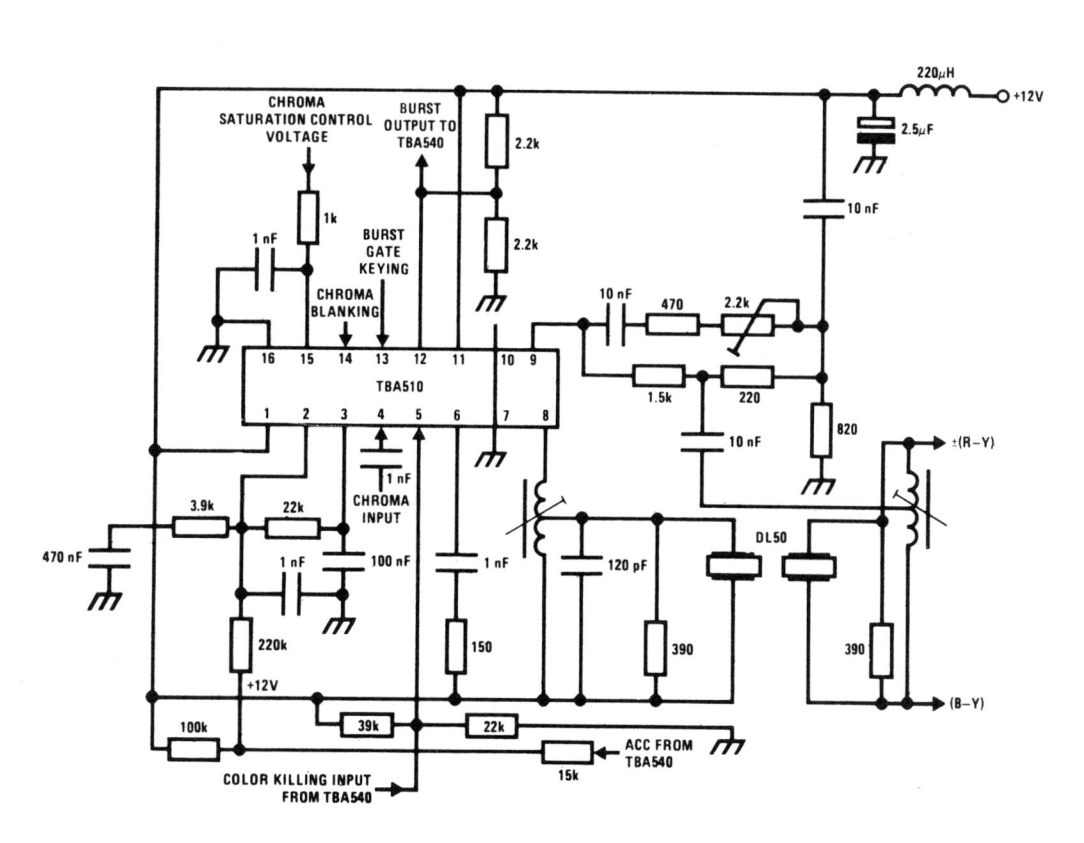

Note: The A.C.C. loop gain can be defined by inserting a suitable resistor between pins 2 & 3. (Example 22 kΩ).

A chrominance amplifier circuit for color TV receivers incorporating a variable gain ACC circuit, a dc control for chroma saturation that can be ganged to the receiver contrast control, chroma blanking and burst gating functions, a burst output stage, a color killer, and a PAL delay line driver (National Semiconductors).

SECTION 5 *Television and Video Circuits* 201

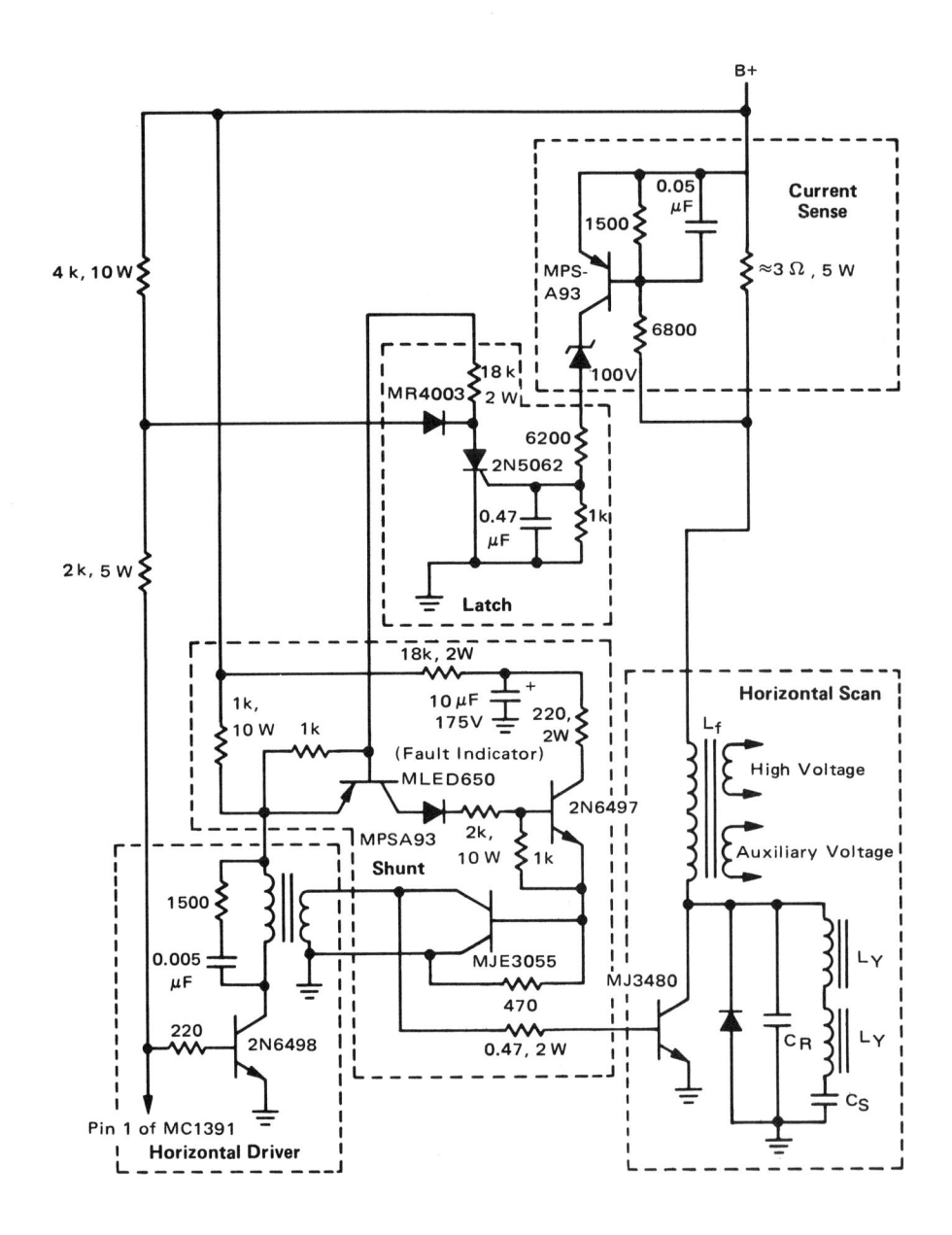

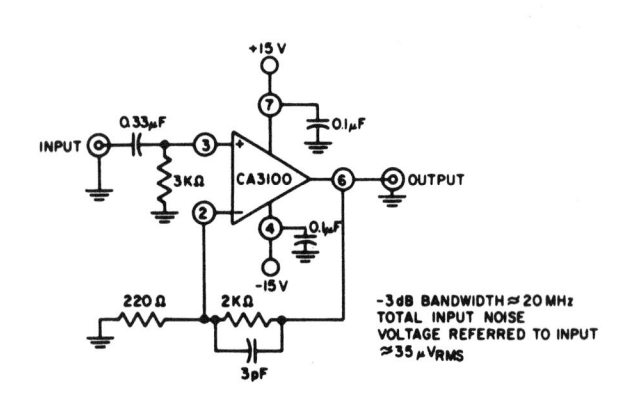

20 dB video amplifier.

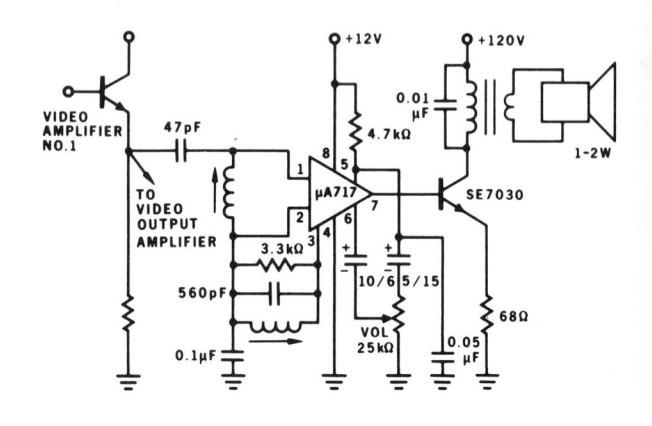

Black-and-white solid-state TV sound IF system.

202 IC SCHEMATIC SOURCEMASTER

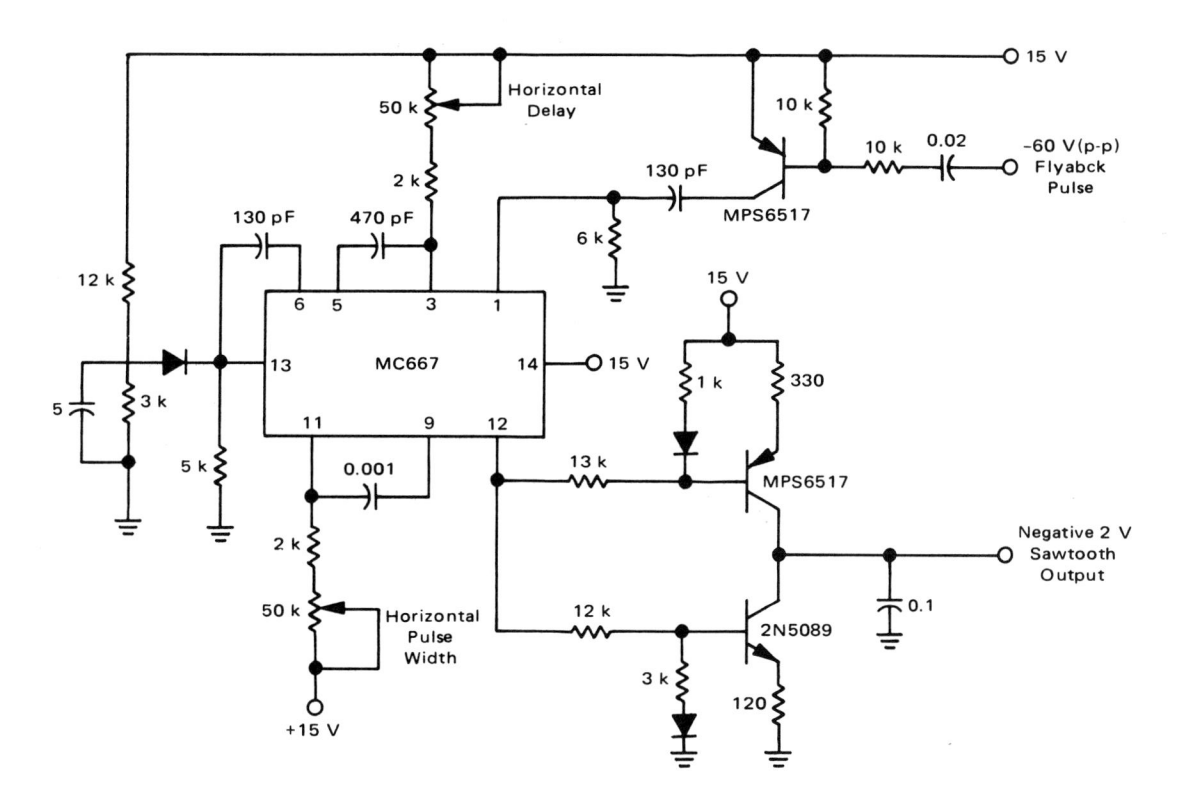

Delayed horizontal sawtooth circuit.

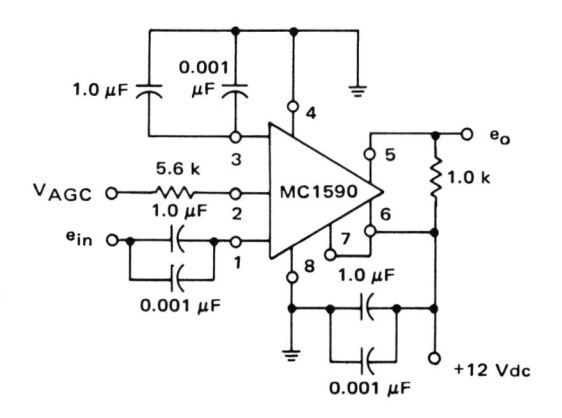

Video amplifier. The MC1590G can be used as a wideband video amplifier. In addition to its high gain, the AGC capability makes the MC1590 quite attractive for video amplifier applications. Several MC1590G circuits can be cascaded to increase the gain. To cascade two MC1590G circuits, the two input pins need only be capacitively coupled to the output pins of the previous stage.

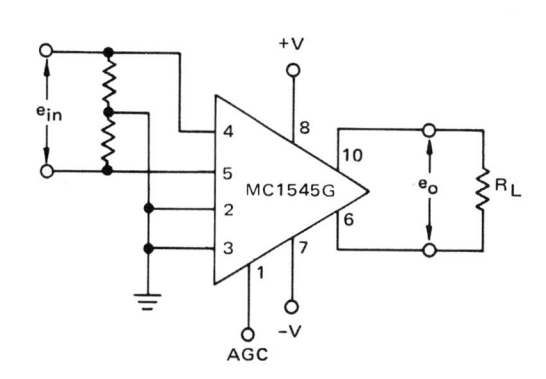

Wideband differential amplifier with AGC. With a dc voltage applied to the gate pin, as much as 100 dB of AGC can be obtained. Since there is essentially no dc level shift with AGC, the output waveform will collapse symmetrically about zero with little or no distortion.

SECTION **5** *Television and Video Circuits*

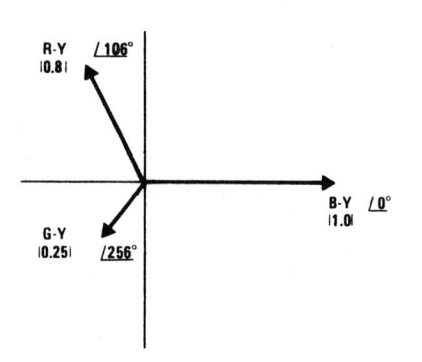

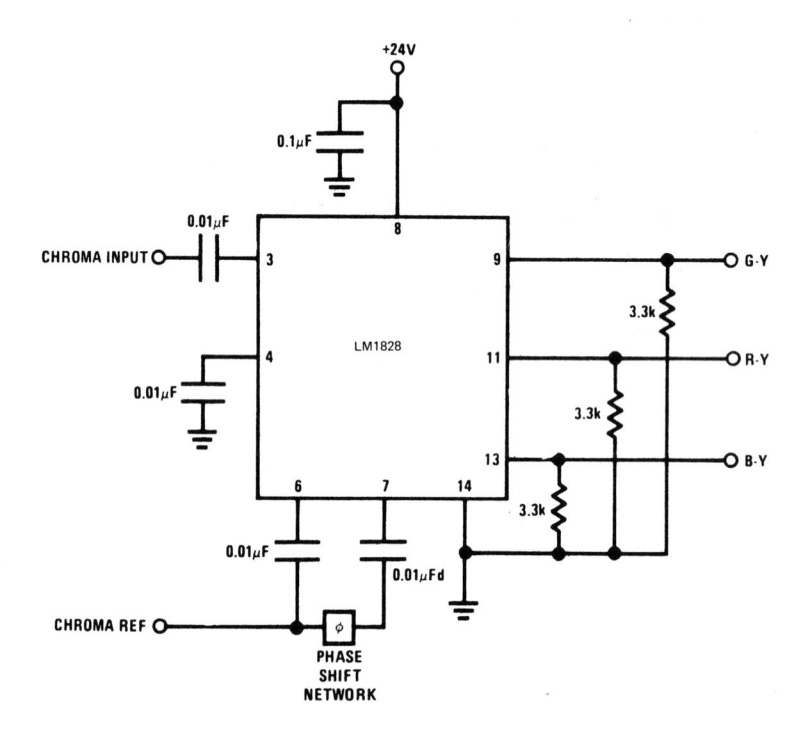

LM1828 standard matrix.

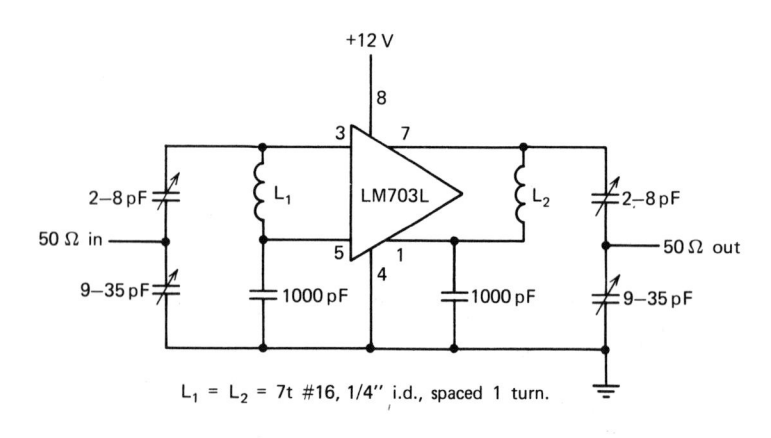

100 MHz narrow-band amplifier. L1 = L2 = 7 turns No. 16, ¼-inch (ID) wire spaced 1 turn.

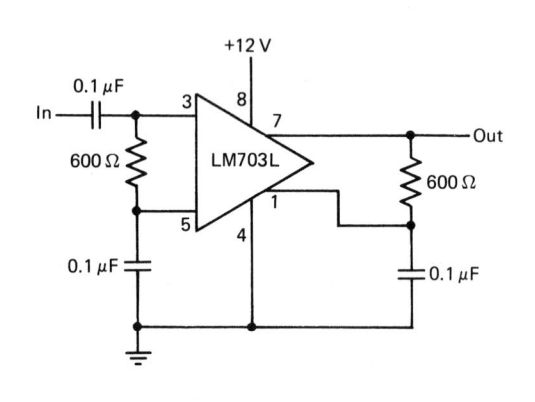

RC-coupled video amplifier.

204 IC SCHEMATIC SOURCEMASTER

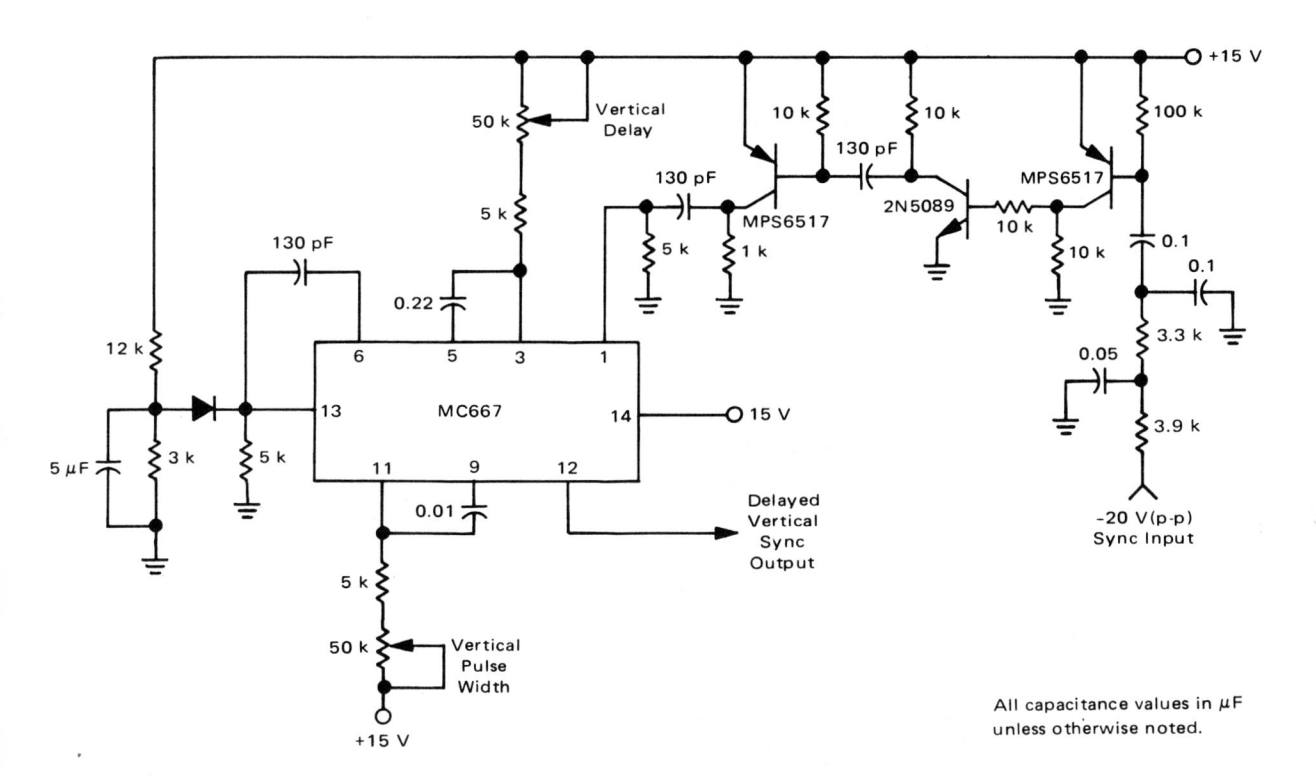

Vertical sync delay circuit.

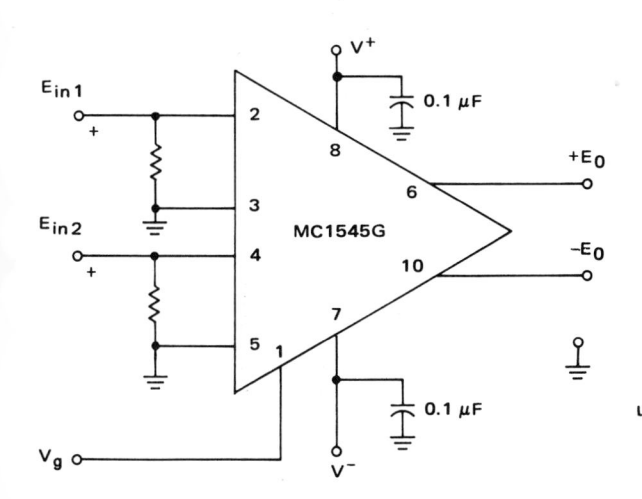

Two-channel wideband amplifier, video switch. The single-ended gain in this configuration is 18 dB (typical).

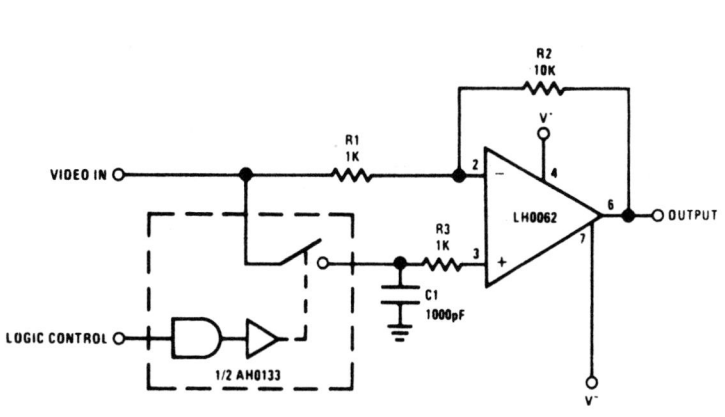

Video dc-restoring amplifier.

SECTION 5 *Television and Video Circuits* 205

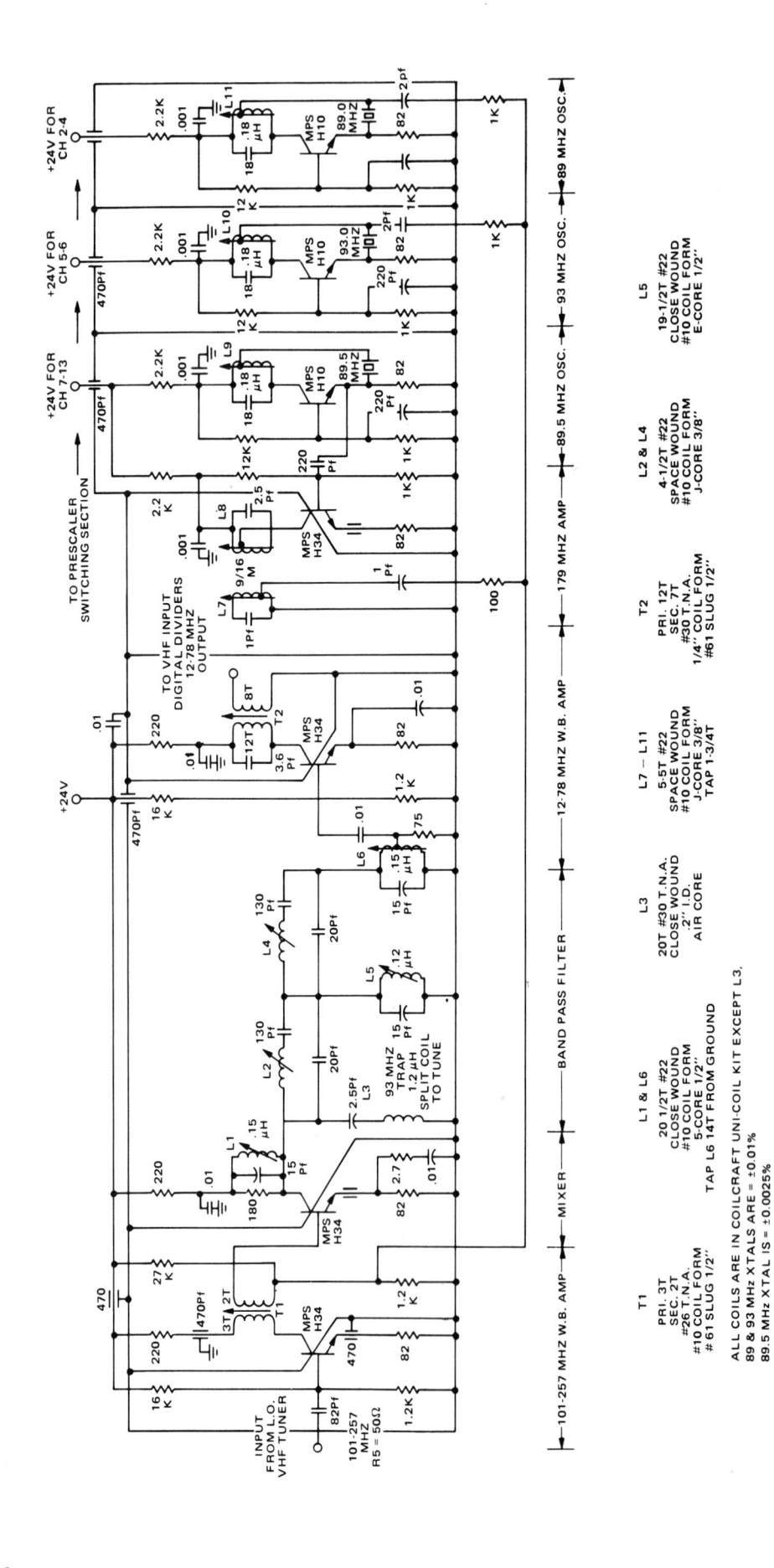

VHF offset oscillators, wideband mixer, and amplifier.

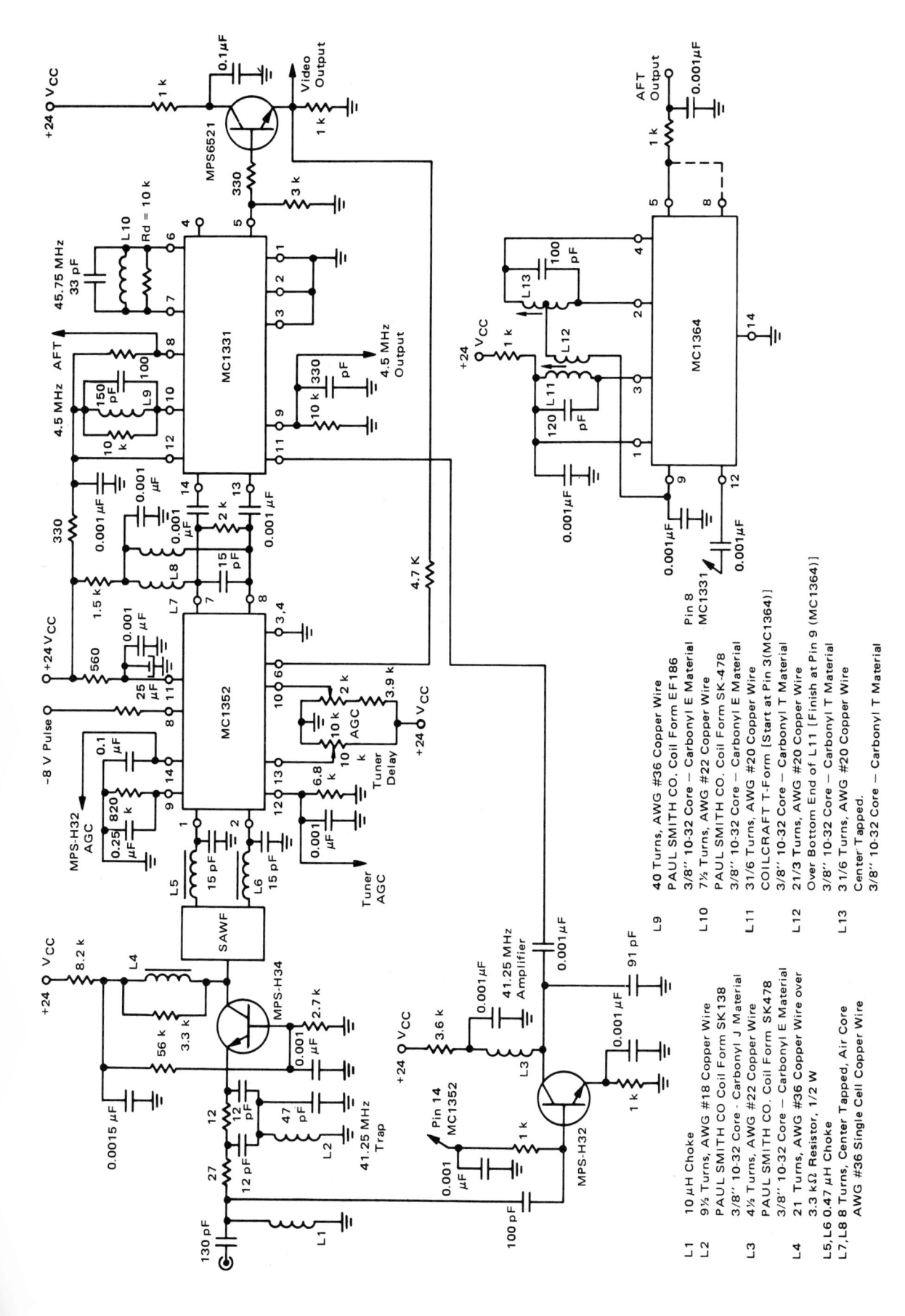

Complete video IF schematic.

SECTION 5 Television and Video Circuits

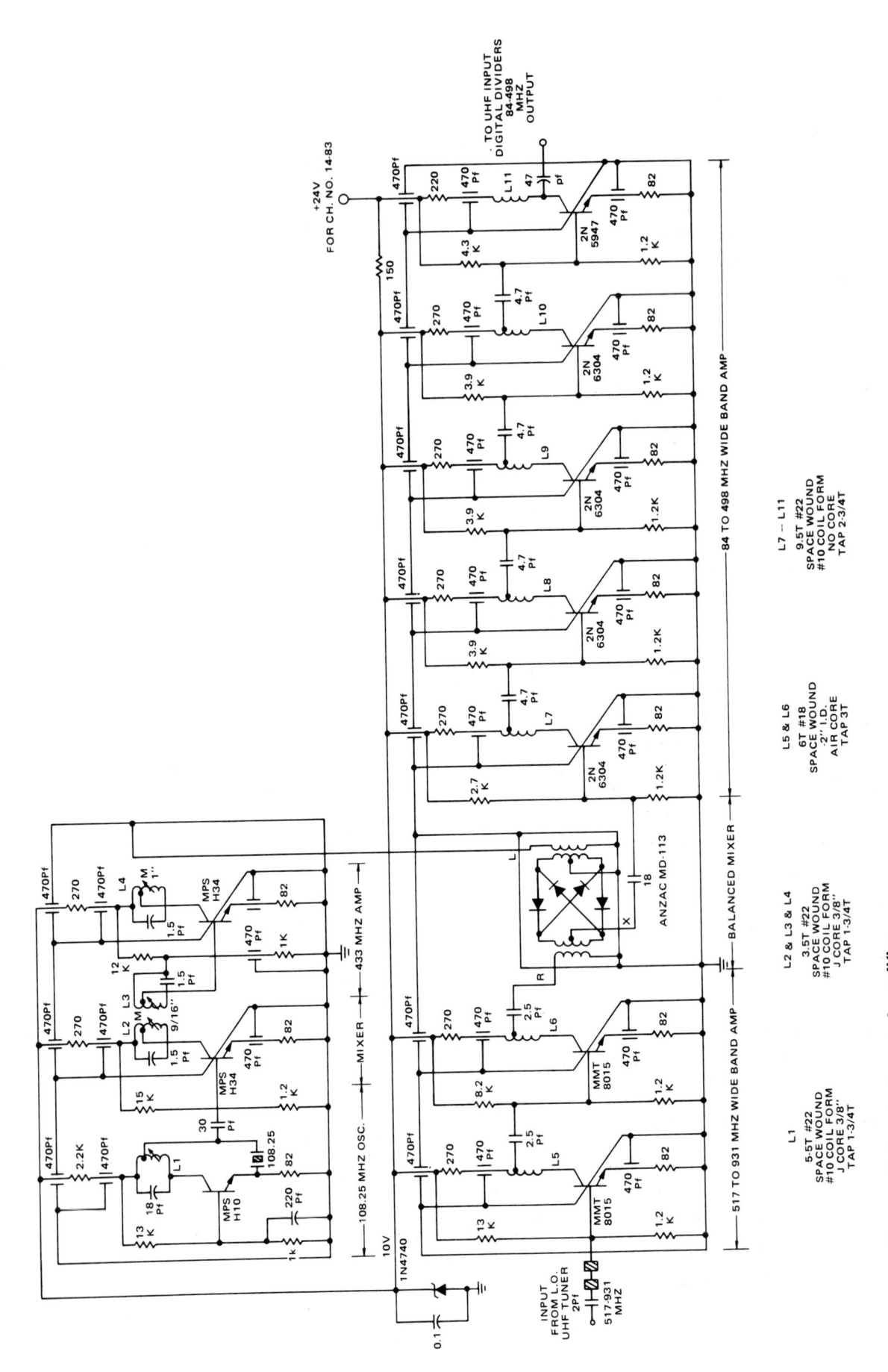

UHF offset oscillator, wideband mixer, and amplifier.

TABLE I

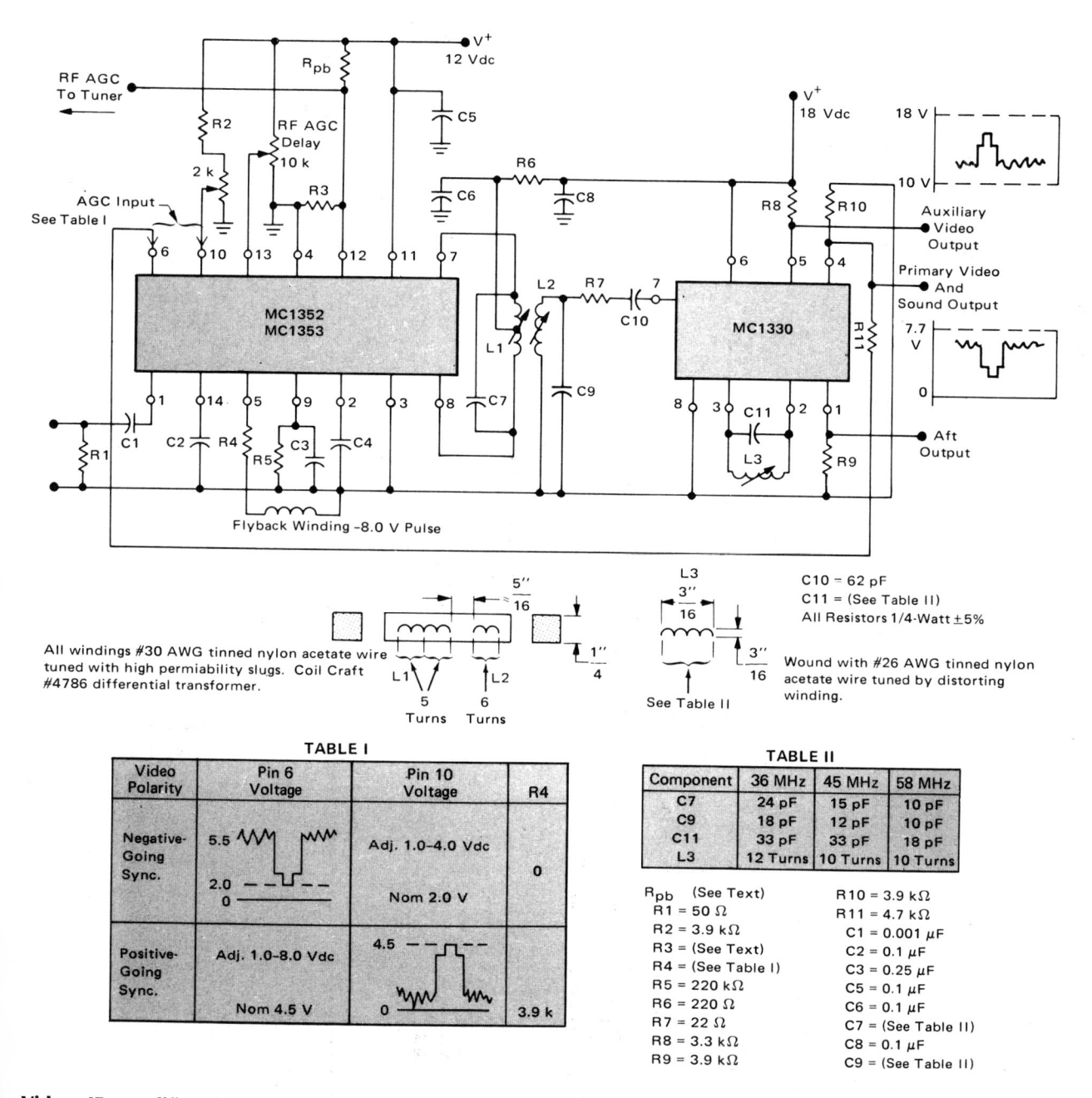

Video Polarity	Pin 6 Voltage	Pin 10 Voltage	R4
Negative-Going Sync.	5.5 / 2.0 / 0 (waveform)	Adj. 1.0–4.0 Vdc / Nom 2.0 V	0
Positive-Going Sync.	Adj. 1.0–8.0 Vdc / Nom 4.5 V	4.5 / 0 (waveform)	3.9 k

TABLE II

Component	36 MHz	45 MHz	58 MHz
C7	24 pF	15 pF	10 pF
C9	18 pF	12 pF	10 pF
C11	33 pF	33 pF	18 pF
L3	12 Turns	10 Turns	10 Turns

C10 = 62 pF
C11 = (See Table II)
All Resistors 1/4-Watt ±5%

R_{pb} (See Text)
R1 = 50 Ω
R2 = 3.9 kΩ
R3 = (See Text)
R4 = (See Table I)
R5 = 220 kΩ
R6 = 220 Ω
R7 = 22 Ω
R8 = 3.3 kΩ
R9 = 3.9 kΩ
R10 = 3.9 kΩ
R11 = 4.7 kΩ
C1 = 0.001 µF
C2 = 0.1 µF
C3 = 0.25 µF
C5 = 0.1 µF
C6 = 0.1 µF
C7 = (See Table II)
C8 = 0.1 µF
C9 = (See Table II)

All windings #30 AWG tinned nylon acetate wire tuned with high permiability slugs. Coil Craft #4786 differential transformer.

Wound with #26 AWG tinned nylon acetate wire tuned by distorting winding.

Video IF amplifier for use at 45 MHz using an MC1330 low-level detector and either an MC1352 or MC1353. If the RF amplifier transistor in the tuner is an NPN device, an MC1352 is used. To set the maximum gain prebias on a forward AGC NPN transistor, a fixed resistor R_{pb} must be selected. This forms a voltage divider with 6.8k resistor (R3). In the maximum-gain condition, the voltage on pin 12 would be zero without the voltage divider. To prebias a forward AGC PNP transistor, R_{pb} is set at 6.8k, and R3 is selected for proper prebias. To obtain maximum gain without AGC for alignment purposes, connect a 22k resistor between pins 9 and 11 of the MC1352 or MC1353. Connecting a 200k variable resistor between pin 14 and ground and the 22k resistor between pins 9 and 11 provides a method of obtaining any particular gain desired.

SECTION 5 *Television and Video Circuits*

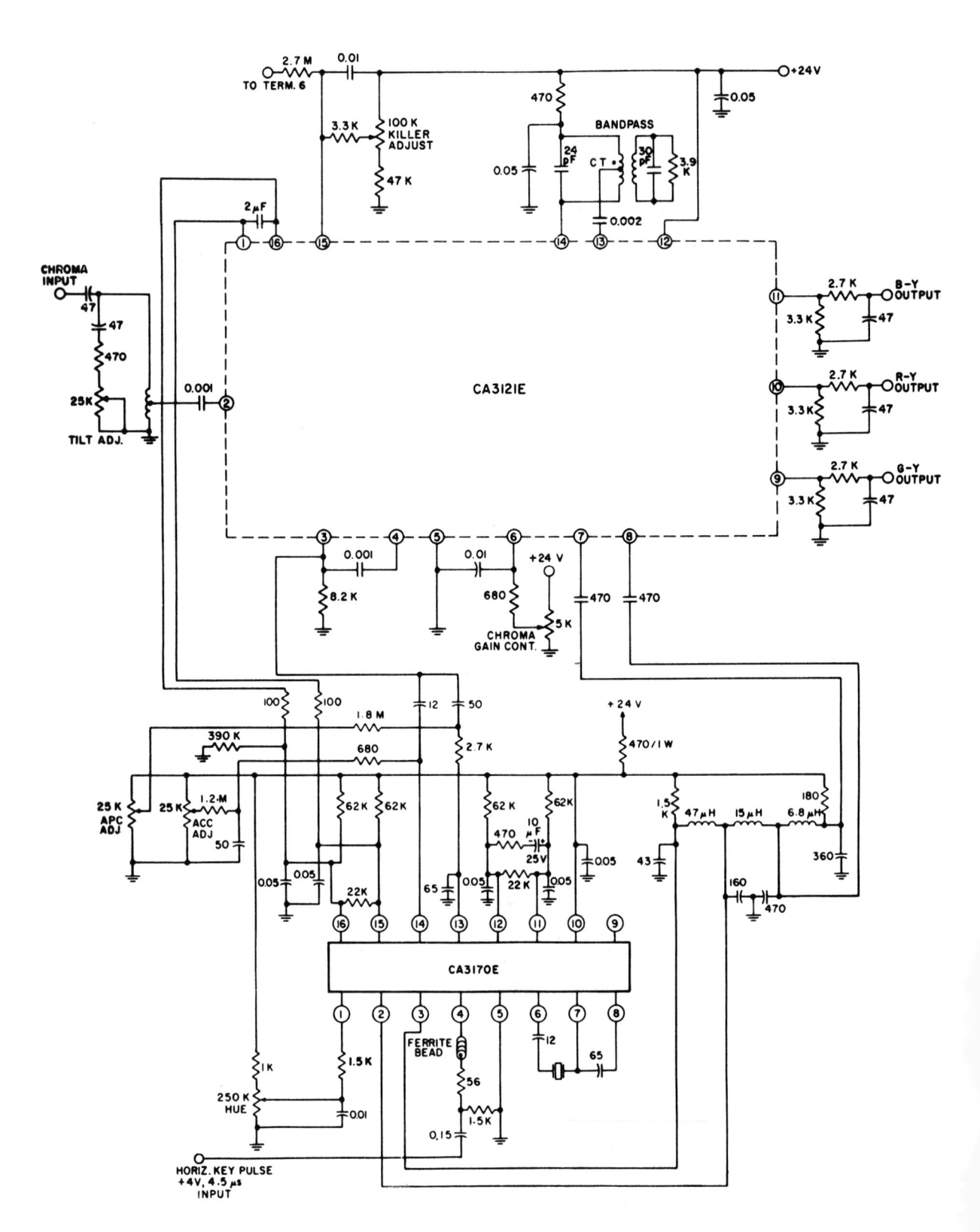

Outboard circuitry of a typical **two-package chroma system** for color-TV receivers using the CA3121E and CA3170E.

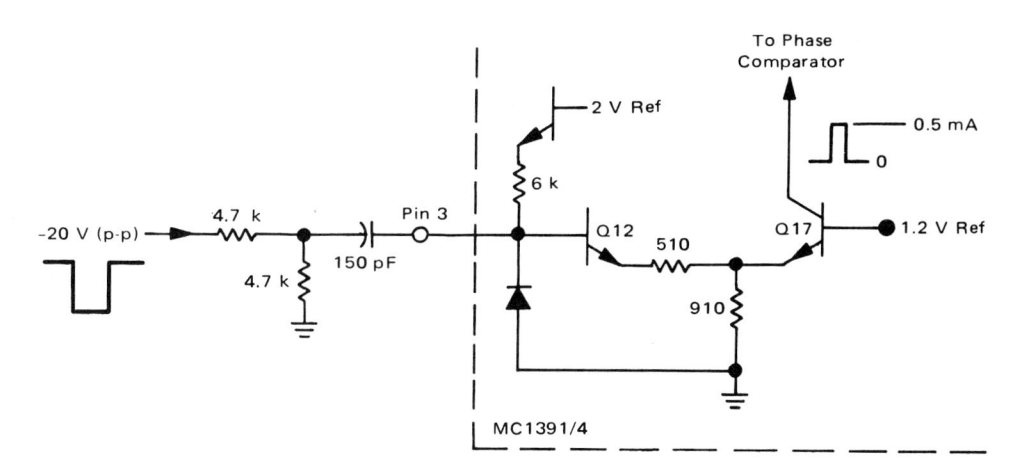

TV sync pulse differentiation.

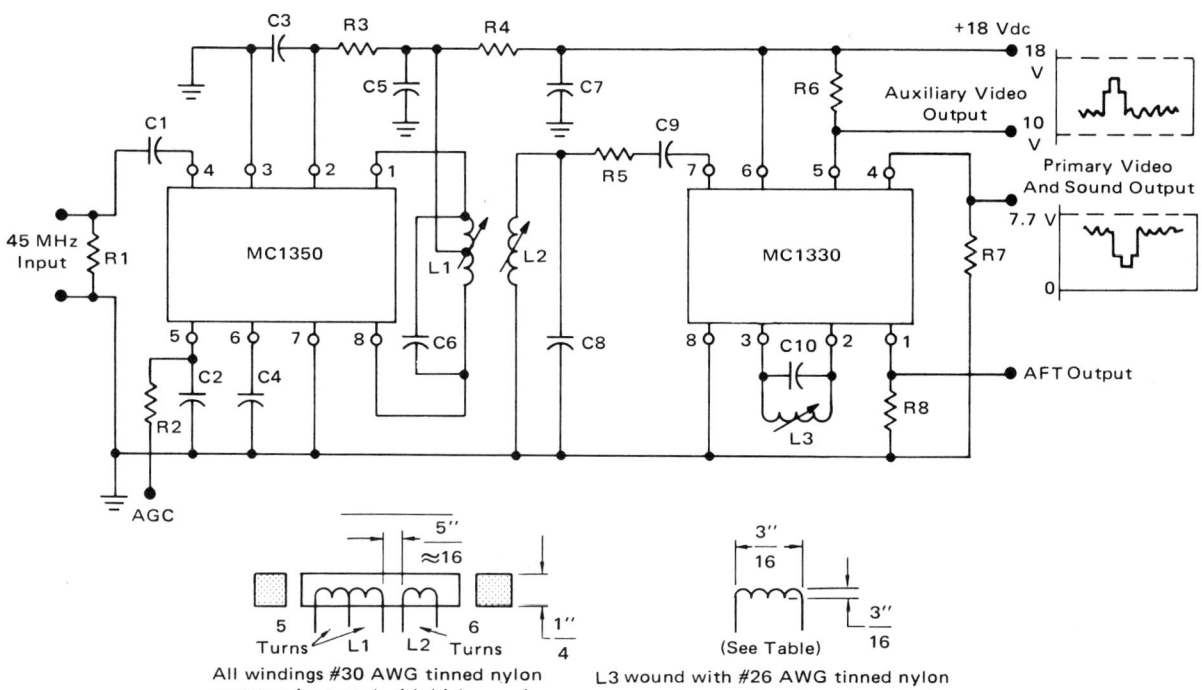

C1 = 0.001 µF	C6 = See Table	R1 = 50 Ω	R6 = 3.3 kΩ
C2 = 0.002 µF	C7 = 0.1 µF	R2 = 5 k	R7 = 3.9 kΩ
C3 = 0.002 µF	C8 = See Table	R3 = 470 Ω	R8 = 3.9 kΩ
C4 = 0.002 µF	C9 = 68 pF	R4 = 220 Ω	All Resistors
C5 = 0.002 µF	C10 = See Table	R5 = 22 Ω	1/4-W ±10%

All Caps Marked µF Ceramic HiK
All Caps Marked pF Silver Mica 5%

Table of Component Values

Component	36 MHz	45 MHz	58 MHz
C6	24 pF	15 pF	10 pF
C8	18 pF	12 pF	10 pF
C10	33 pF	33 pF	18 pF
L3	12 Turns	10 Turns	10 Turns

Practical video IF amplifier and detector. This circuit has a typical voltage gain of 84 dB and a typical AGC range of 80 dB. It gives very small changes in bandpass shape, usually less than 1 dB tilt for 60 dB compression. There are no shielded sections. The detector uses a single tuned circuit (L3 and C10).

Coupling between the two integrated circuits is achieved by a double-tuned transformer (L1 and L2). No block filters or traps have been designed for the front end of this amplifier. The sound intercarrier information may be taken from the detected video output.

SECTION 5 *Television and Video Circuits* 211

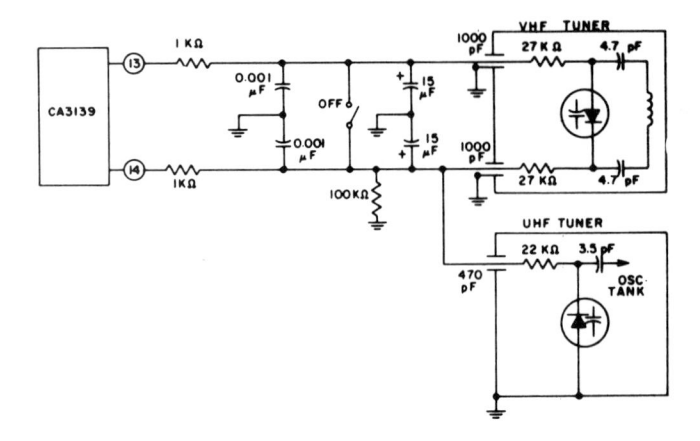

Typical **TV tuner connection** using CA3139 AFT circuit.

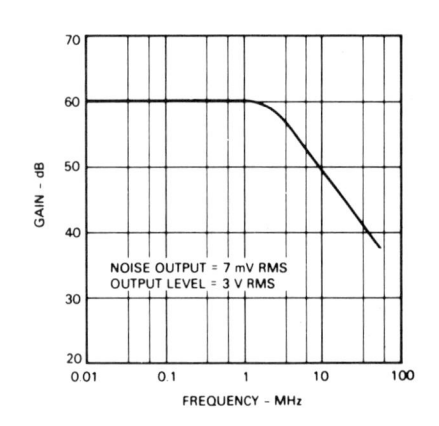

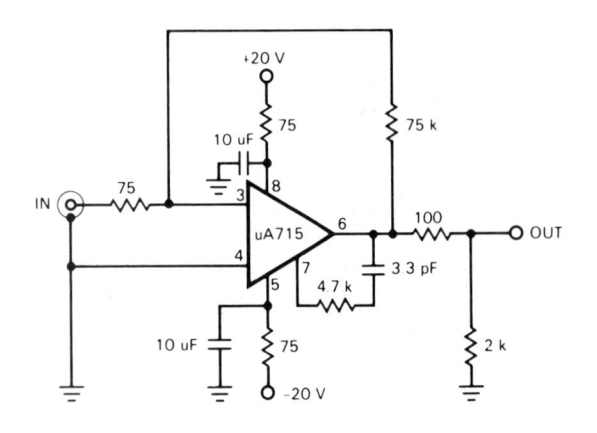

A 75Ω wideband amplifier with 60 dB of gain. The high gain results in a slightly lower bandwidth because the high-frequency rolloff characteristic is governed by the open-loop rolloff.

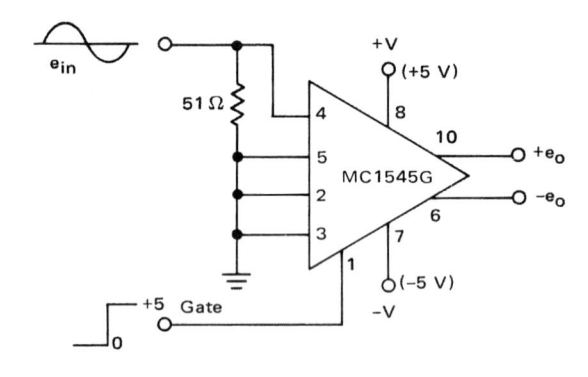

Video switch. The signal (analog or digital) is applied to the amplifier at pin 4. With the logic signal at pin 1 at a logic 1 state (positive voltage), the input signal is amplified and passed through the amplifier. However, if the logic signal at pin 1 is at a logic 0 state, the amplifier is turned off, and no signal will pass through the device. If the opposite logic levels had to pass or block the signal, the input signal can just as easily be applied to pin 2 or 3 with pins 4 and 5 grounded. In this case, a high logic level would block transmission and a low logic level would pass the signal, making the use of inverters unnecessary. Taking "channel select time" as the time delay from the 50% point of the gate pulse to the 50% point of the full output swing, it is observed to be approximately 20 nsec. During the time that the gating logic is in the low state, the circuit that gates the MC1545 must sink a maximum of 2.5 mA, which most forms of saturated logic can do easily. When the gating logic is in the high state, the circuit that gates the MC1545 must source only the leakage current of a reverse-biased diode, which is 2 µA maximum. These requirements are quite similar to the input requirements of a standard DTL or TTL logic gate.

212 IC SCHEMATIC SOURCEMASTER

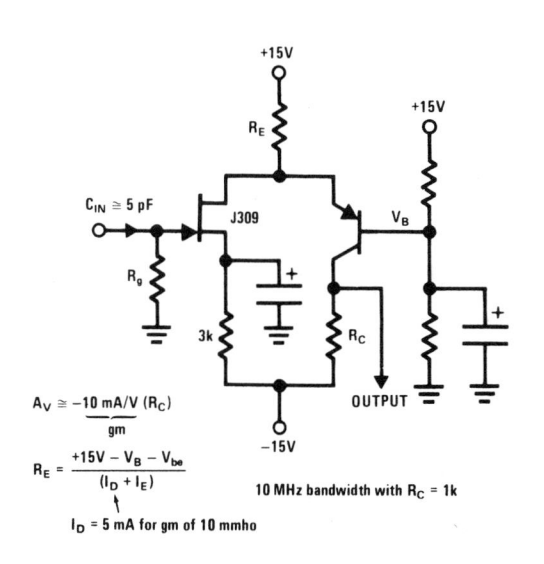

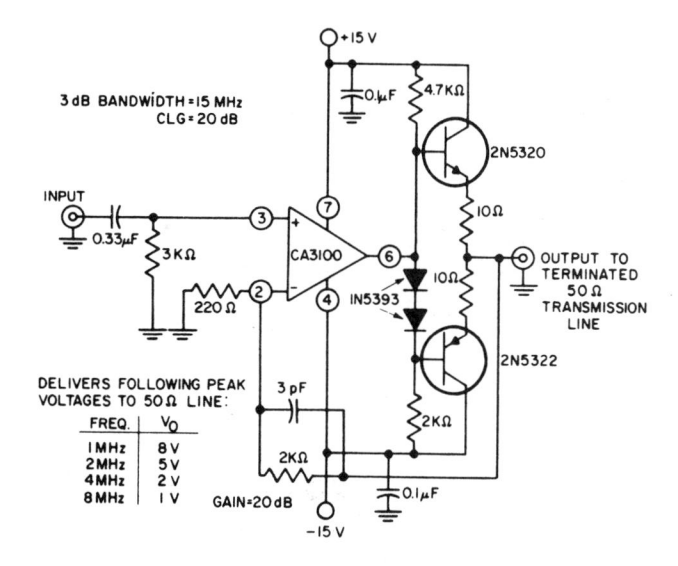

Low noise, very low input-capacitance video amplifier.

$$A_V \cong \underbrace{-10\ mA/V}_{g_m}\ (R_C)$$

$$R_E = \frac{+15V - V_B - V_{be}}{(I_D + I_E)}$$

$I_D = 5\ mA$ for g_m of 10 mmho

10 MHz bandwidth with $R_C = 1k$

20 dB video line driver.

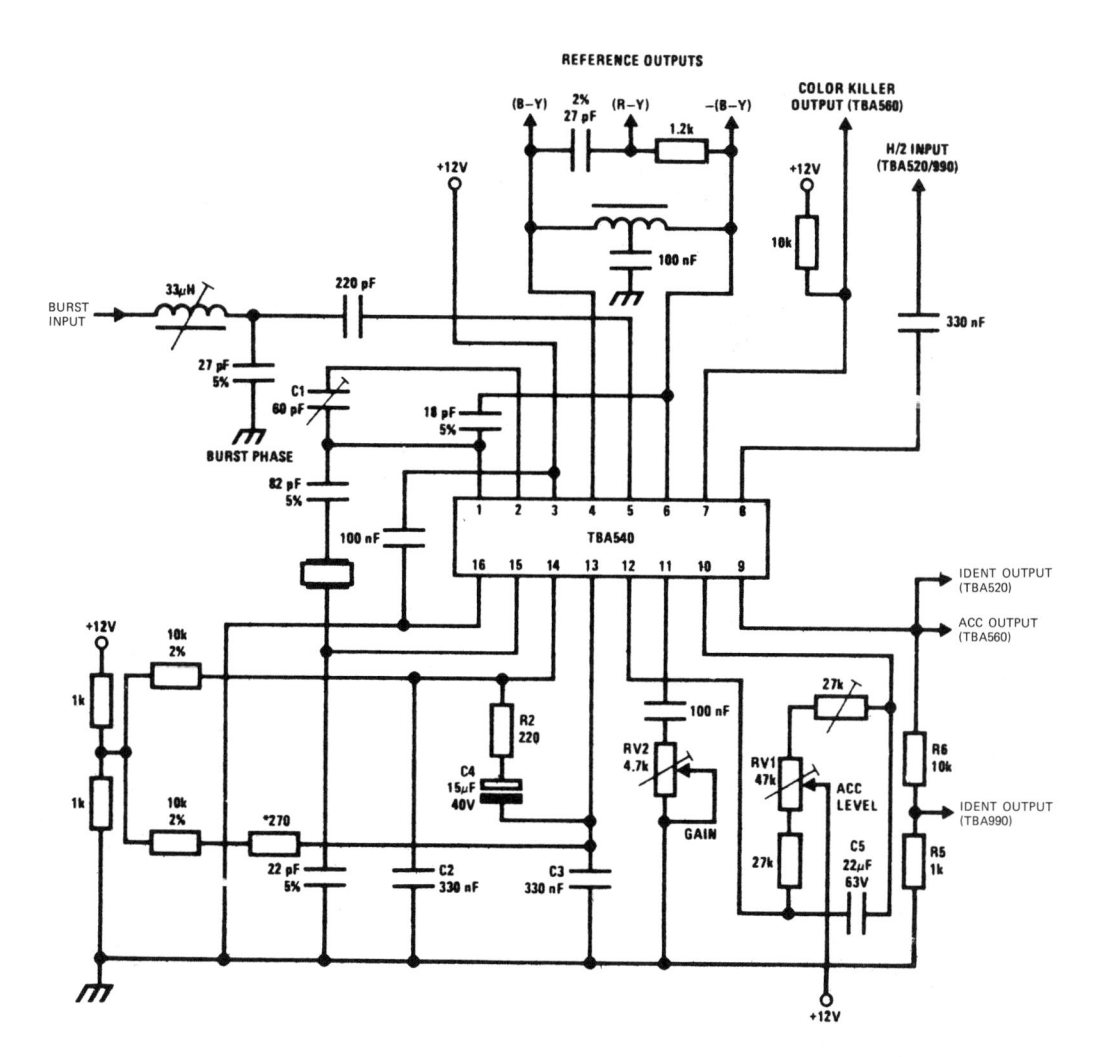

3 dB BANDWIDTH = 15 MHz
CLG = 20 dB

DELIVERS FOLLOWING PEAK VOLTAGES TO 50 Ω LINE:

FREQ.	V_O
1 MHz	8 V
2 MHz	5 V
4 MHz	2 V
8 MHz	1 V

GAIN = 20 dB

Color reference oscillator circuit for PAL TV receivers. The oscillator employs a quartz crystal and incorporates automatic phase and amplitude control. A synchronous demodulator is used to compare the phase and amplitude of the swinging burst ripple with the PAL flip-flop waveform and generates appropriate ACC color killer and identification signals. A high standard of noise immunity has been obtained by using synchronous demodulation.

SECTION 5 *Television and Video Circuits* 213

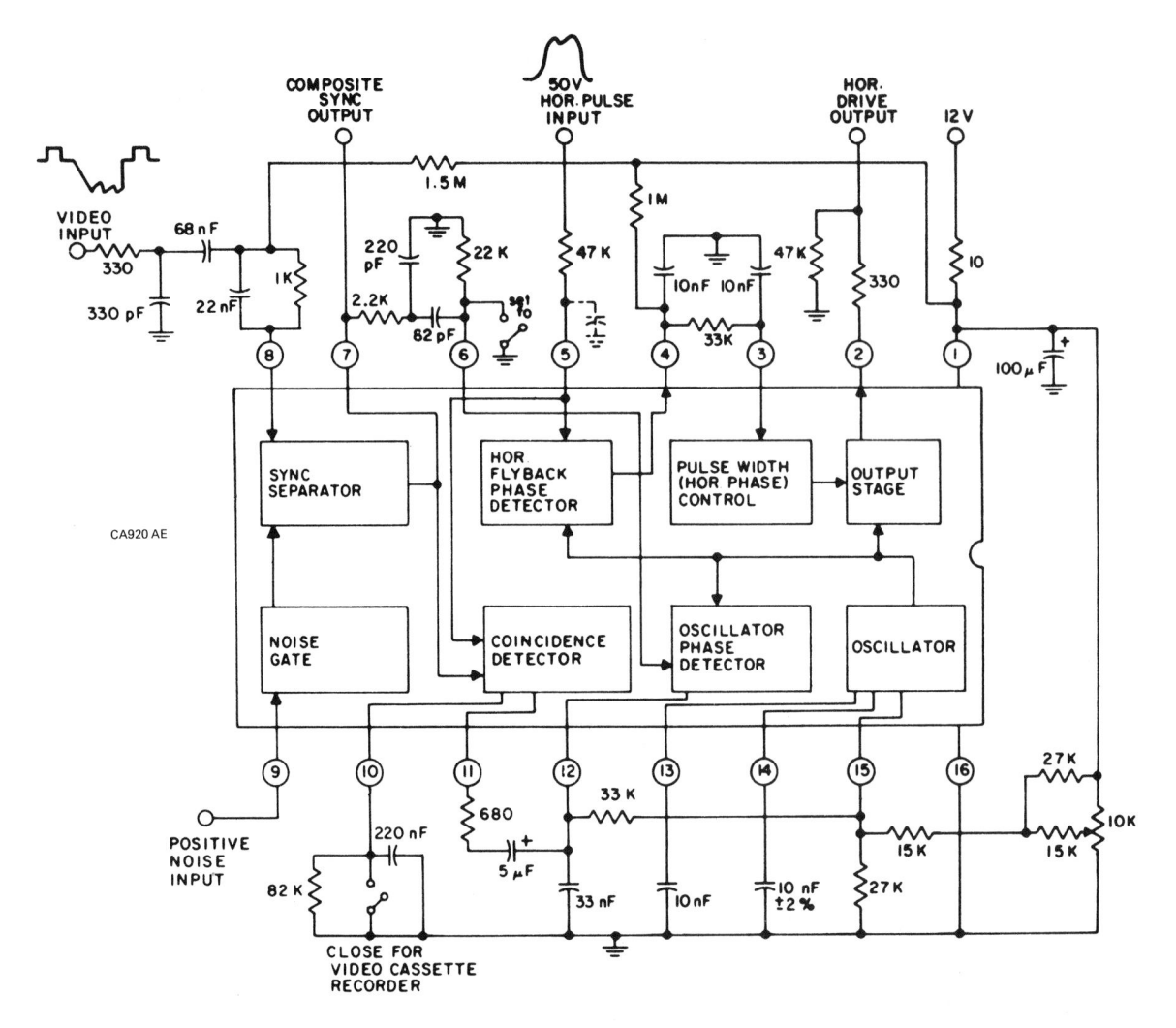

This TV horizontal oscillator for color and monochrome receivers performs the functions of a sync separator, noise gate, and horizontal oscillator with dual time-constant switching in the flywheel loop. It also generates automatic phase control between horizontal flyback pulses and the horizontal oscillator frequency and provides fast edge switching drive for transistor or thyristor horizontal output stages.

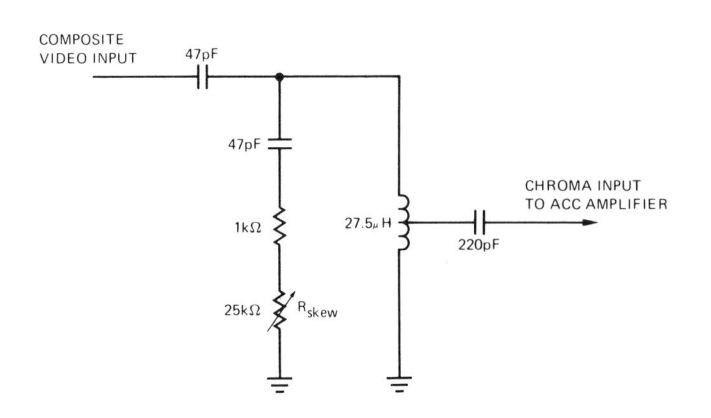

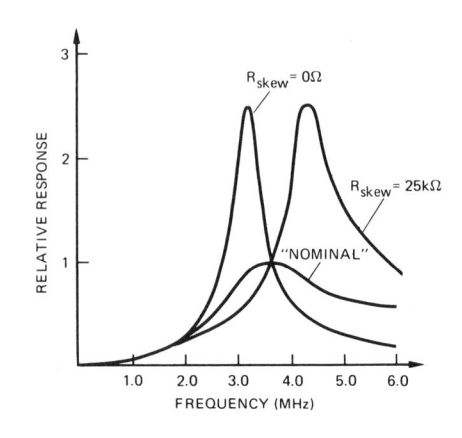

This color TV receiver chroma bandpass filter provides a sloping response that is low at 3.1 MHz to compensate for rolloff of video IF amplifier. Tilt is adjustable to accommodate statistical variations in IF response, allowing the system to be trimmed for best quadrature crosstalk performance.

214 IC SCHEMATIC SOURCEMASTER

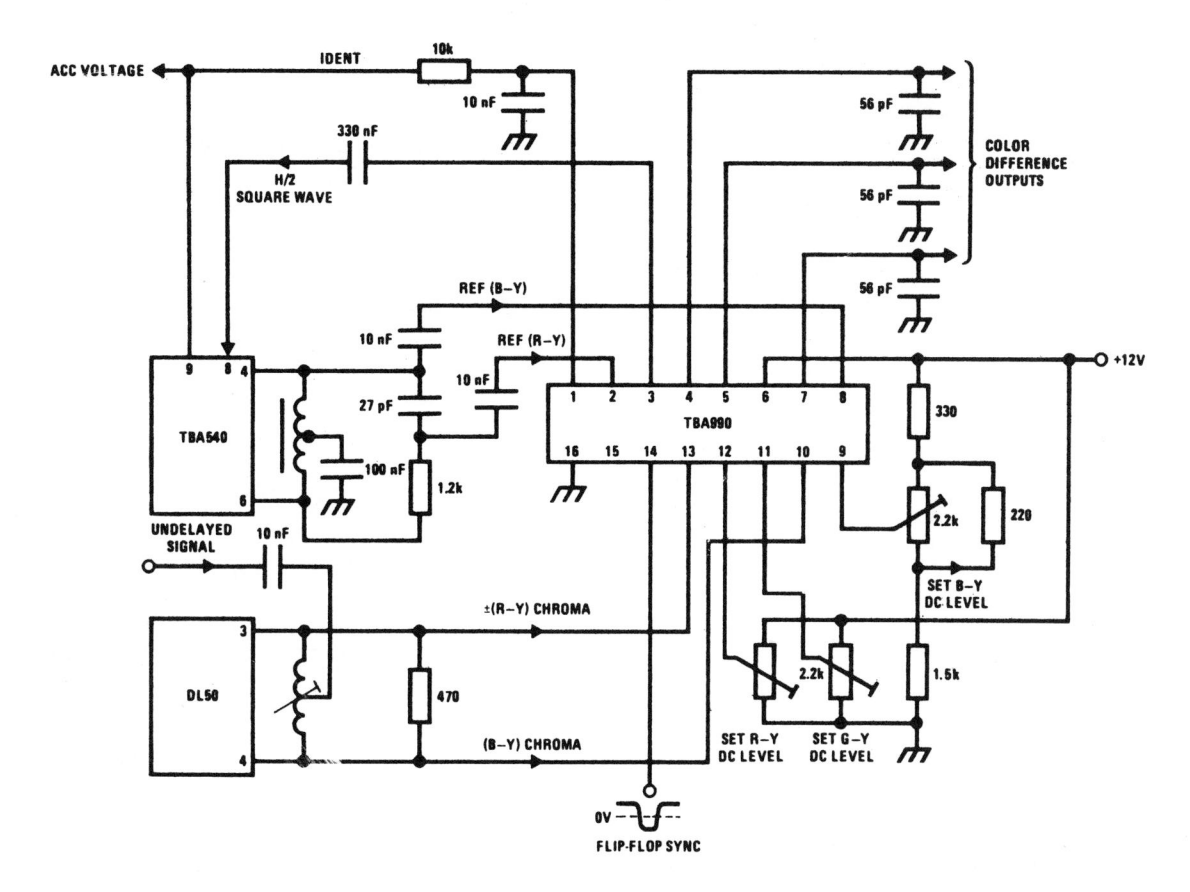

A color demodulator circuit for color television receivers incorporating two active synchronous demodulators for the R-Y and B-G chrominance signals, a matrix (producing the G-Y color difference signal), PAL phase switch, and flip-flop. It is suitable for dc-coupled drive to the picture tube when associated with the matrix integrated circuit (TBA530) and R-G-B output stages.

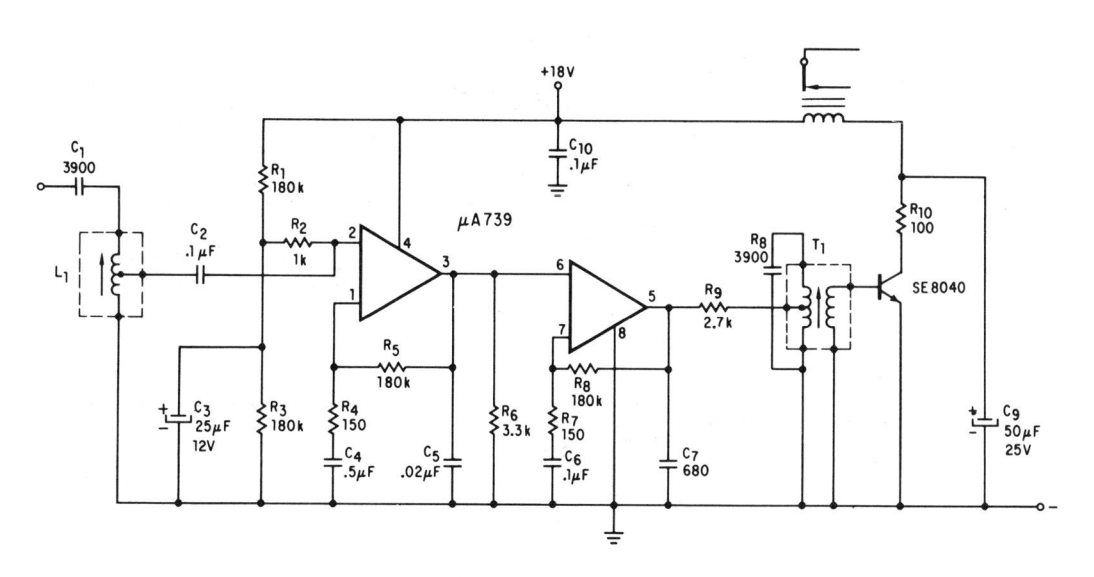

A TV remote control receiver, which consists of a high-gain amplifier, a transformer and coil for high selectivity (high Q), and a relay driver. The transmitted signal is detected by a condenser microphone and fed to the cascade amplifier. The amplified signal is applied to the base of the relay driver through the frequency-selective transformer. Operation is unaffected by fluctuations in line voltage, temperature, or signal strength.

SECTION 5 Television and Video Circuits   **215**

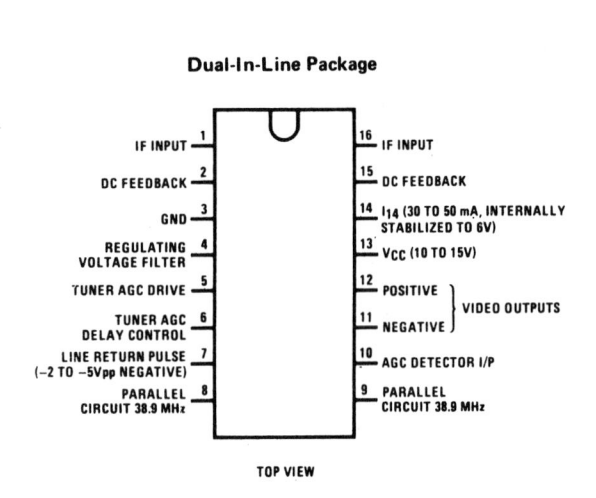

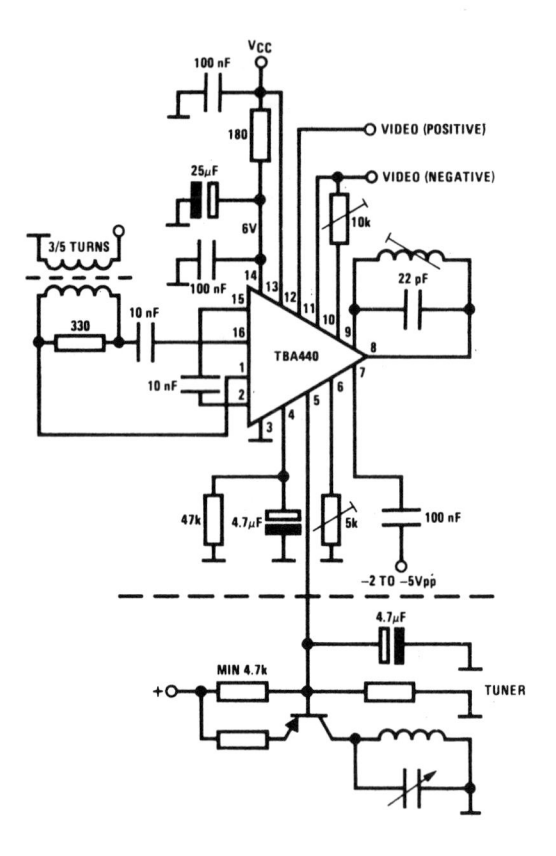

Video IF amplifier for color and monochrome television receivers. The circuit includes three IF amplifier stages, a balanced video IF detector, and a gated AGC section for the IF amplifier and PNP tuner.

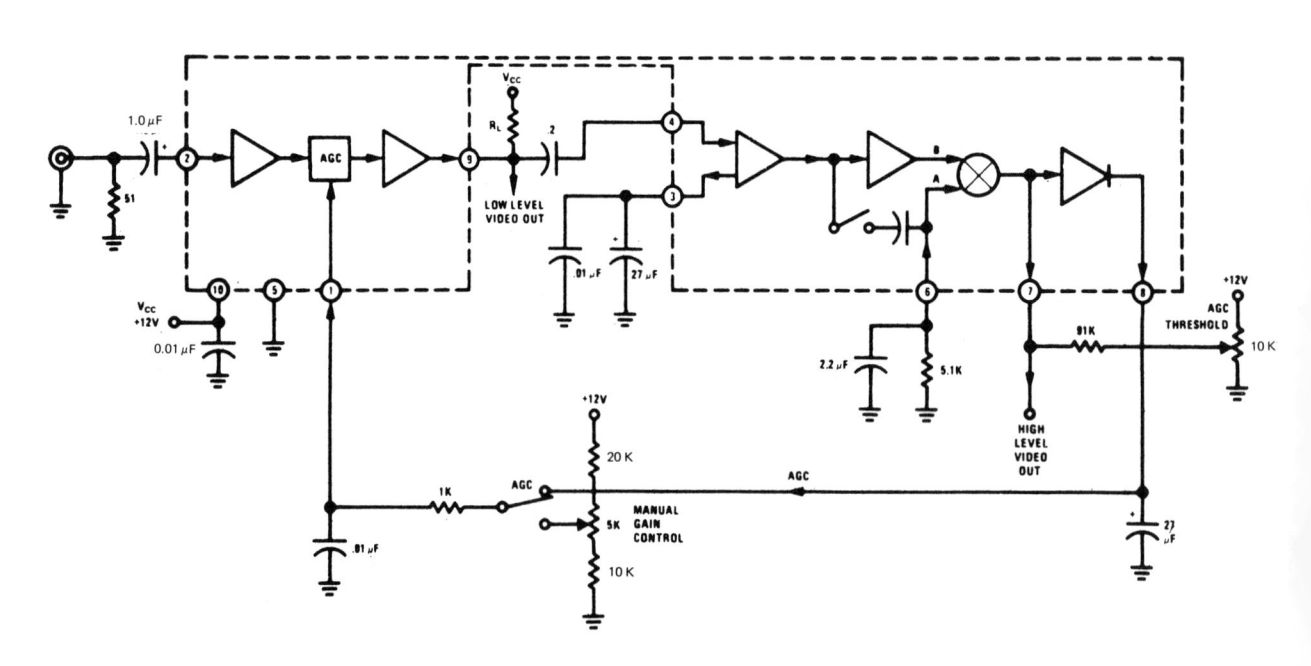

An LM274/LM374 video amplifier configuration. Proper layout and minimum lead length should be observed. The first gain block, pins 2 to 9, shows a typical gain of 32 dB; the second gain block, pins 4 to 7, shows a typical gain of 37 dB. But the device does not require any shielding between stages. Construction on a copperclad printed board is adequate. A power supply bypassing directly at pin 10 and a dc-feedback bypassing at pin 3 are necessary.

216 IC SCHEMATIC SOURCEMASTER

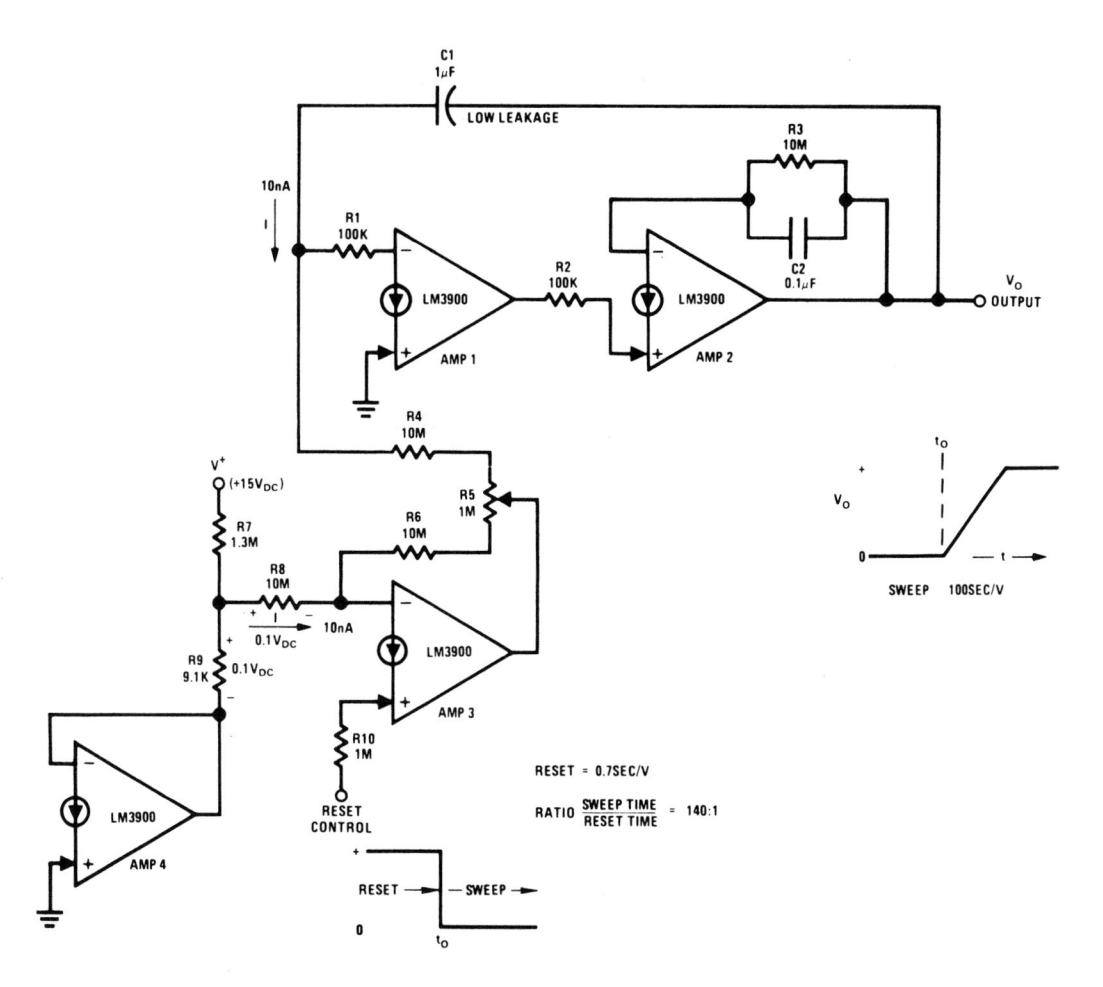

Very slow sawtooth waveform generator. Amps 1 and 2 are cascaded to increase the gain of the integrator; the output is the very slow sawtooth waveform. Amp 3 is used to exactly supply the bias current to amp 1.

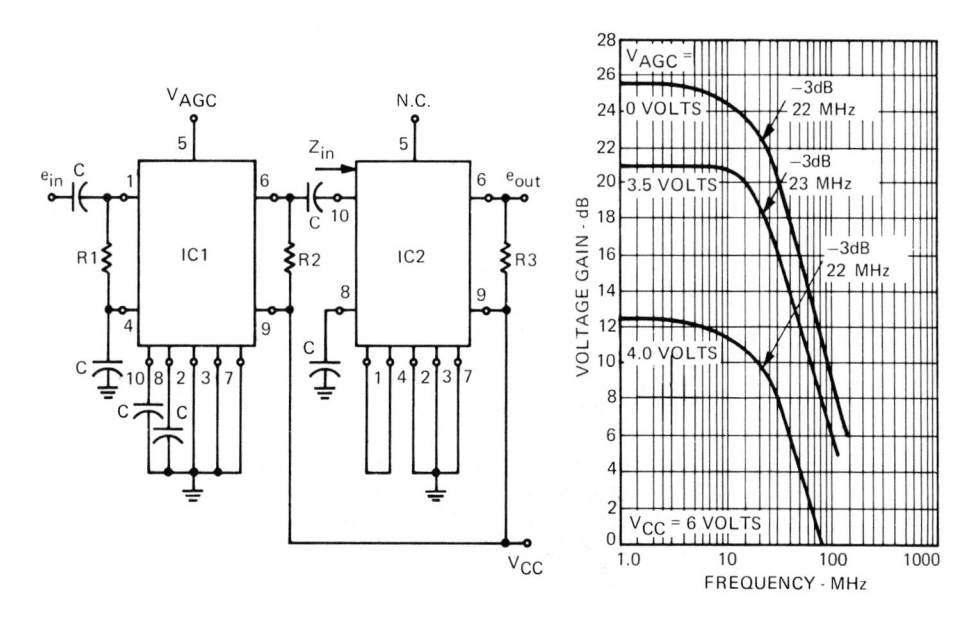

This two-stage video amplifier requires no tuned circuits and features automatic gain control of IC amplification. Video amplifier response with AGC is shown on inset graph. IC1 and IC2 are Motorola HEP 590s, R1 is 47Ω, R2 is 1800Ω, and R3 is 3300Ω. The coupling and bypass capacitors are all 0.1 μF—preferably Mylar.

SECTION 5 *Television and Video Circuits* **217**

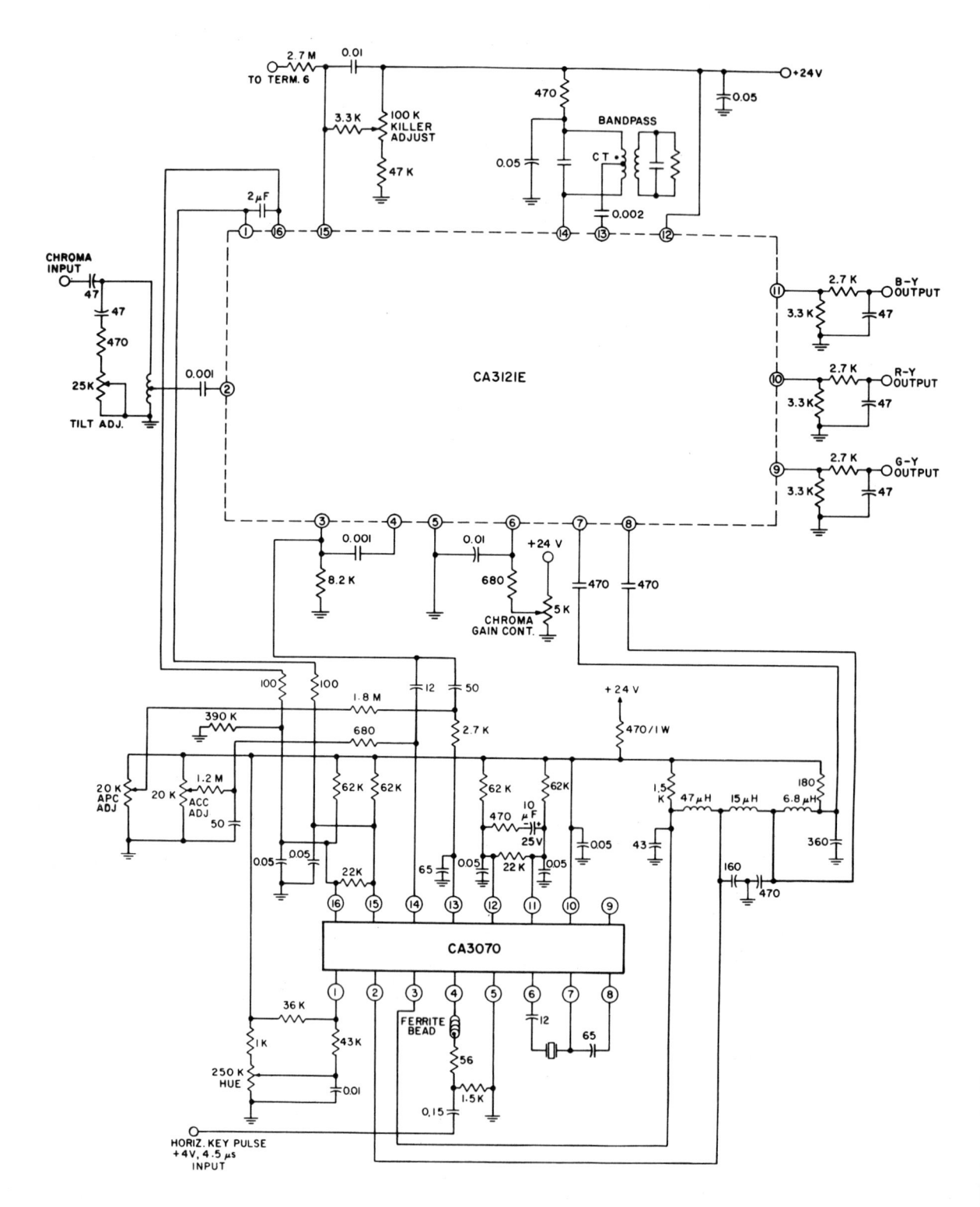

Outboard circuitry of a typical two-package chroma system for color TV receivers using a CA3121E and CA3070.

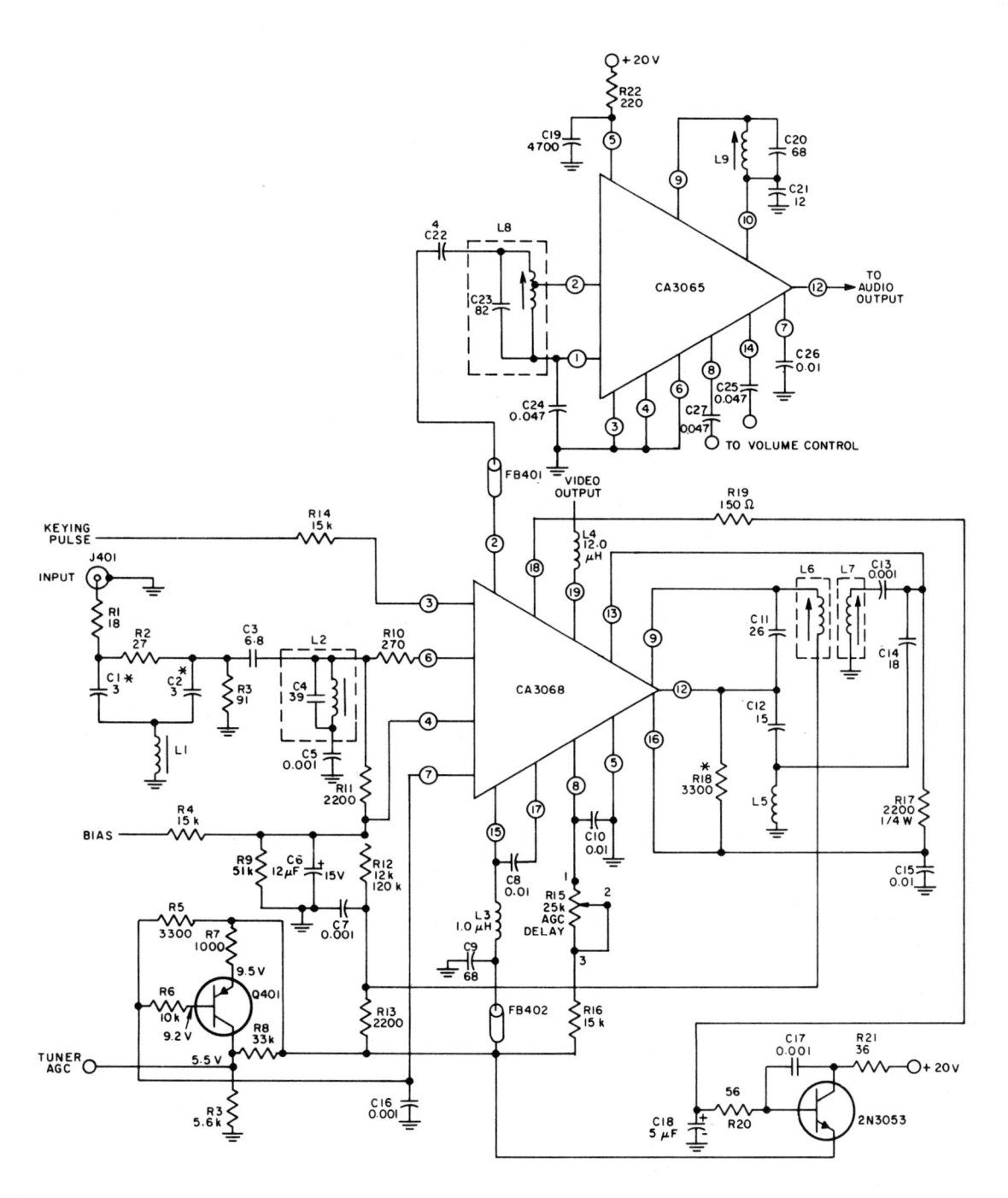

A picture IF circuit for a monochrome TV system using a CA3068. Component tolerances: C11, C12, C14, R3, R9, R11, R17, R18, 5%; R2, 2%.

SECTION 5 *Television and Video Circuits*

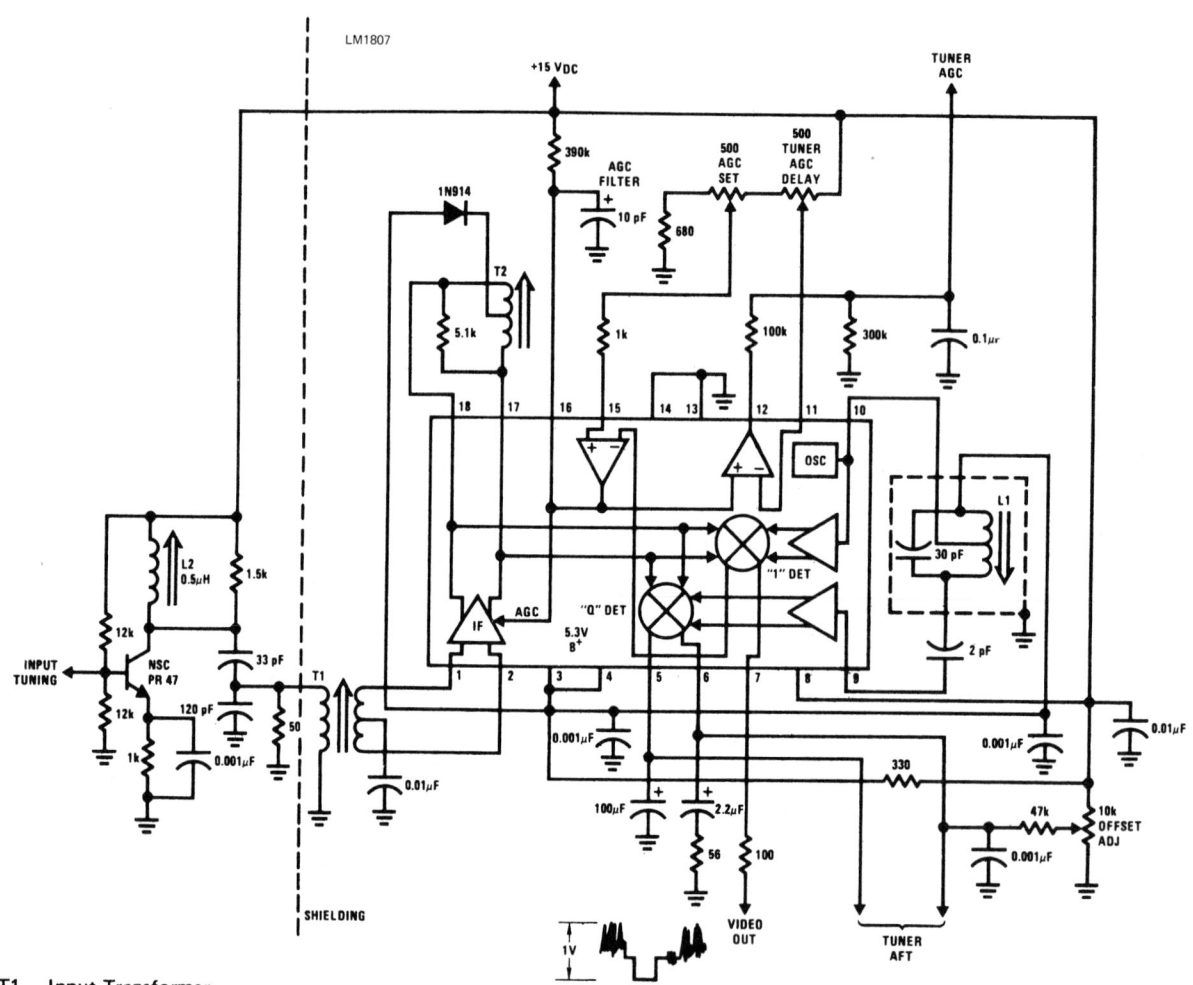

T1 — Input Transformer
2 turns of #32 enameled wire. Windings are tightly coupled and wound on a ferrite bead or toroid.

T2 — Interstage Transformer
7 turns of #32 enameled wire. Windings are tightly coupled and wound on a 7/32" O.D. coil form with high freq. slug.

L1 — Oscillator Coil
7 turns of #20 bare wire wound 0.025" apart on an 11/32" O.D. coil form with high freq. slug. Tap at 2 1/4 turns from the ac ground end.

This TV phase-locked video IF and detector performs all the video IF functions required in a television receiver. It includes a gain-controlled IF amplifier and a true synchronous detector with the reference supplied by a local oscillator on the ship. A phase-locked loop locks the tuner to the IF frequency, providing an AFT function. The AGC section gain-reduces the IF amplifier and supplies a delayed control voltage to the tuner.

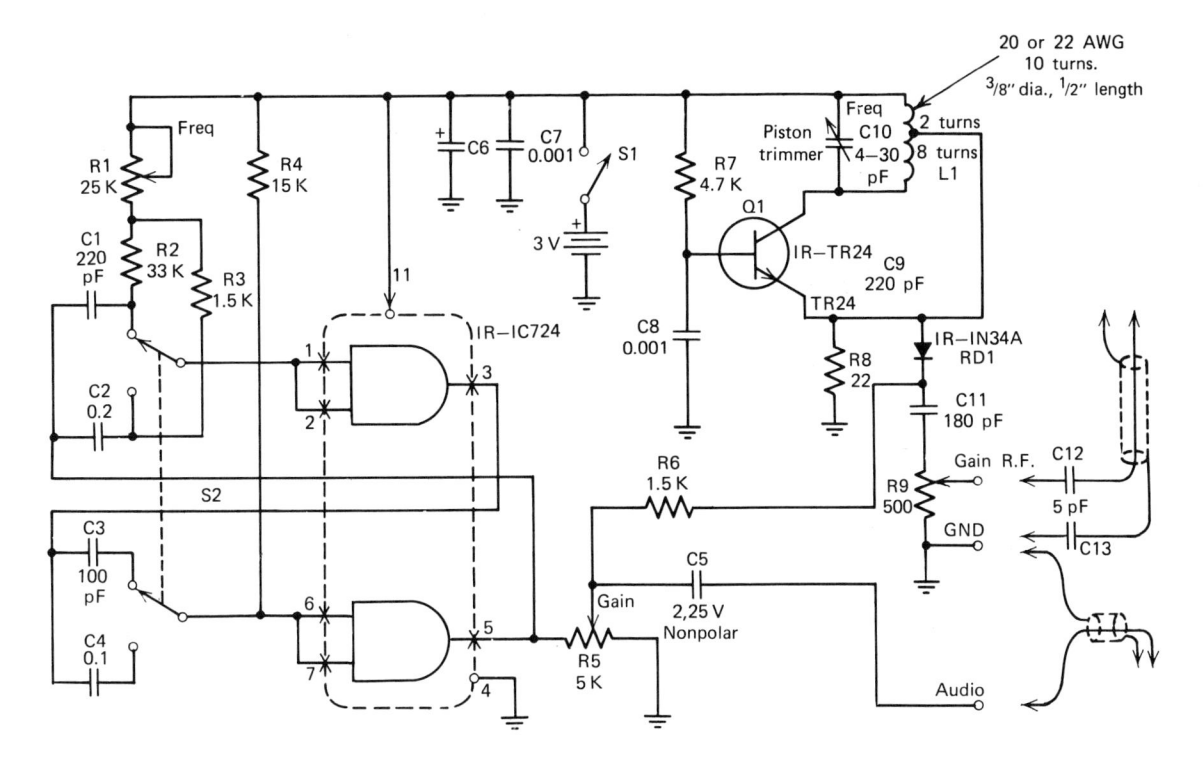

This IC TV pattern generator for TV maintenance will provide horizontal or vertical bars for color TV convergence and linearity. The entire unit can be mounted in a 4 × 3 × 2-inch box. Variable capacitor C10 can be manufactured by mounting a bolt to the capacitor if no suitable piston trimmer is available. Also, coil L1 can be made. The diagram illustrates how the tap connection is formed. (Semiconductors are IRC/Workman.)

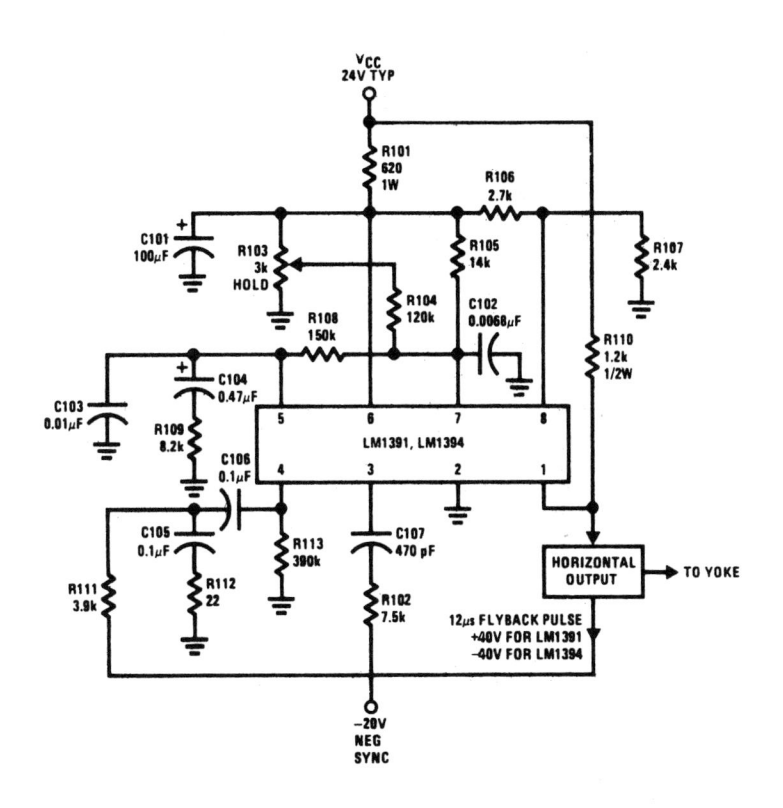

TV horizontal processor.

SECTION **5** *Television and Video Circuits* **221**

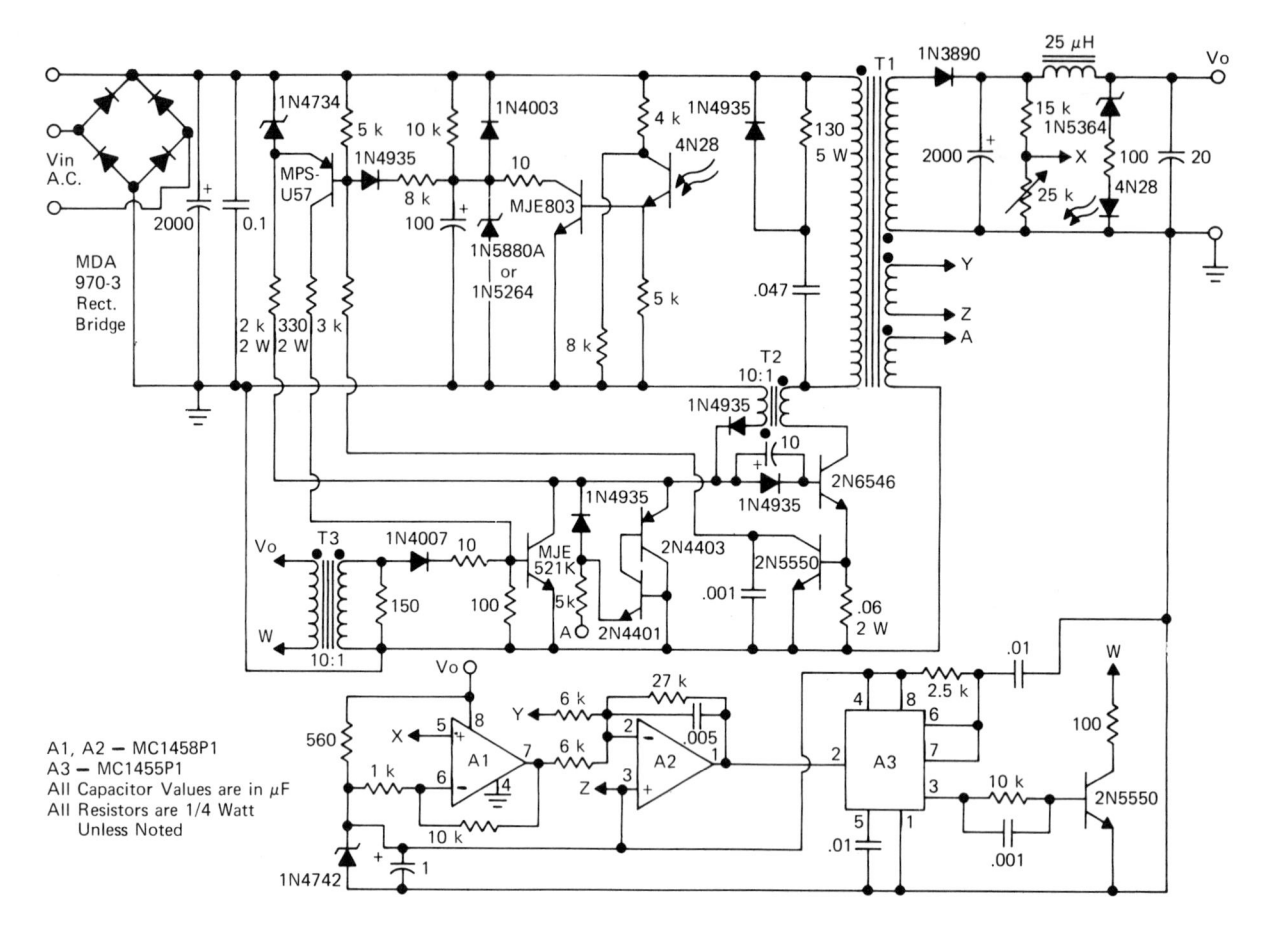

80W CATV switching supply. When the output voltage of op-amp A2 goes below the MC1455's threshold voltage, the MC1455 fires, turning on a control transistor, which pulls the base of the 2N6546 to ground and turns it off. A pulse transformer is used between this control circuitry and the power transistor to maintain input/output isolation. The *off* time of the 2N6546 is controlled by the MC1455 to allow complete transfer of energy under worst-case conditions (heavy load and low input voltage). When the MC1455 times out, the 2N6546 is again allowed to turn on, and the cycle repeats.

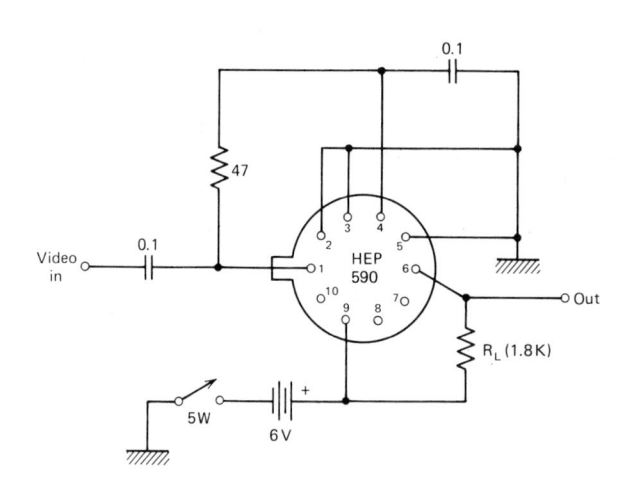

Simple video amplifier using Motorola hobby IC has a 20 dB gain to 20 MHz and as much as a 10 dB gain at 160 MHz. Unique circuitry requires no tuning or other circuit optimization. Just build the circuit as shown and connect the input and output leads.

222 IC SCHEMATIC SOURCEMASTER

SECTION 6
AUDIO CIRCUITS

Audio circuits come in many forms, which is the basis for the exceptionally large number of diagrams contained in this section. Presented here are stereo and four-channel schematics; circuits for preamplifiers of every description, including microphone, phonograph and tape, and general control units for hi-fi systems; amplifiers from a few milliwatts to auditorium powerhouses; and various mixers and audio signal conditioners.

Like other sections in this book, this one is not the exclusive source for audio circuits. If you plan to build a preamplifier or a graphic equalizer, for example, you should also consult Section 11 (filters); Section 2, on voltage followers and amplifiers, contains many circuits that could have been included here. See also Section 10 (automotive) and Section 7 (communications) for more applicable circuits.

GE integrated circuits are no longer being manufactured, but they are plentiful on the surplus market.

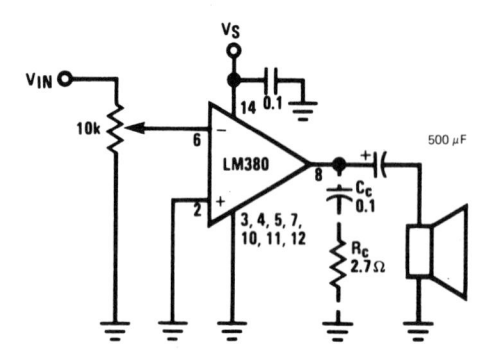

Simple IC amplifier. With the internal biasing and compensation of the LM380, the simplest and most basic circuit configuration requires only an output coupling capacitor. The input pot is a volume control. (Of course, the dc line is bypassed according to good design practice.)

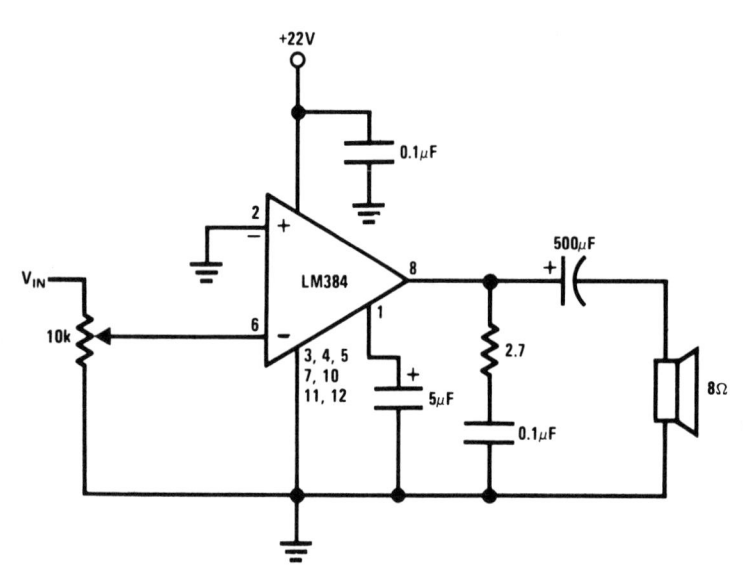

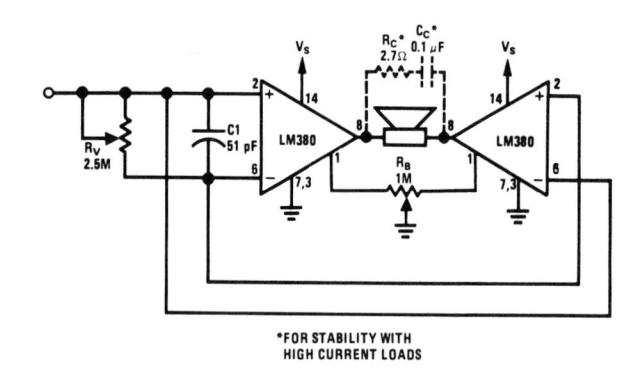

A bridge-type audio amplifier with quiescent balance control. Adjust for zero dc on speaker with amplifier working but no signal applied to input.

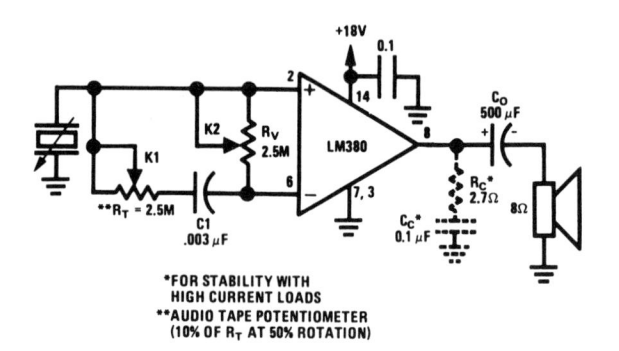

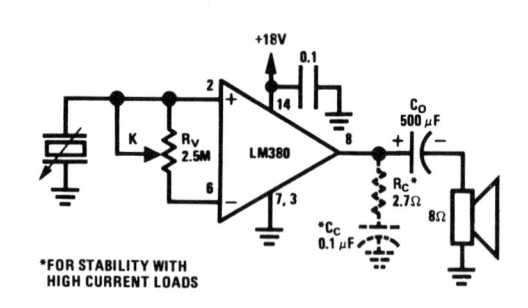

Common-mode volume and tone control.

Common-mode volume control.

5W audio power amplifier. Higher allowed operating voltage means higher output power, and this is what distinguishes the LM384 from the LM380 audio amplifier. Typical power levels of 7.5W (10% THD) into 8Ω are possible when operating from a supply voltage of 26V. Output power with 22V input is 5W. For applications where output ripple and high-frequency oscillations are not a problem, all capacitors except the 500 μF output capacitor may be eliminated, along with the 2.7Ω resistor. This creates a complete amplifier with only one external capacitor and no resistors!

Audio amplifier with common-mode volume and tone controls. When maximum input impedance is required or the signal attenuation of a voltage-divider volume control is undesirable, a common-mode volume control may be used. With this volume control the source loading impedance is only the input impedance of the amplifier when in the full-volume position. This reduces to half the amplifier input impedance at the zero volume position. The common-mode volume control can be combined with a common-mode tone control as shown. This circuit has a distinct advantage when transducers of high source impedance are used, in that the full input impedance of the amplifier is realized. It also has an advantage with transducers of low source impedance, since the signal attenuation of the input voltage divider is eliminated.

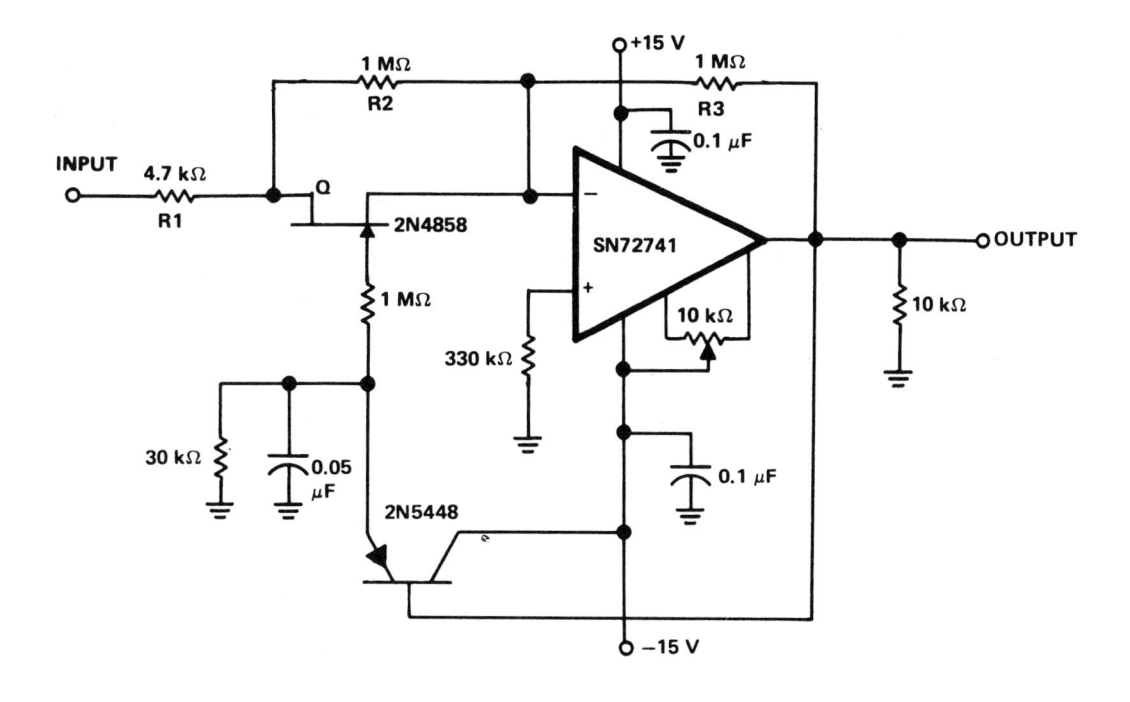

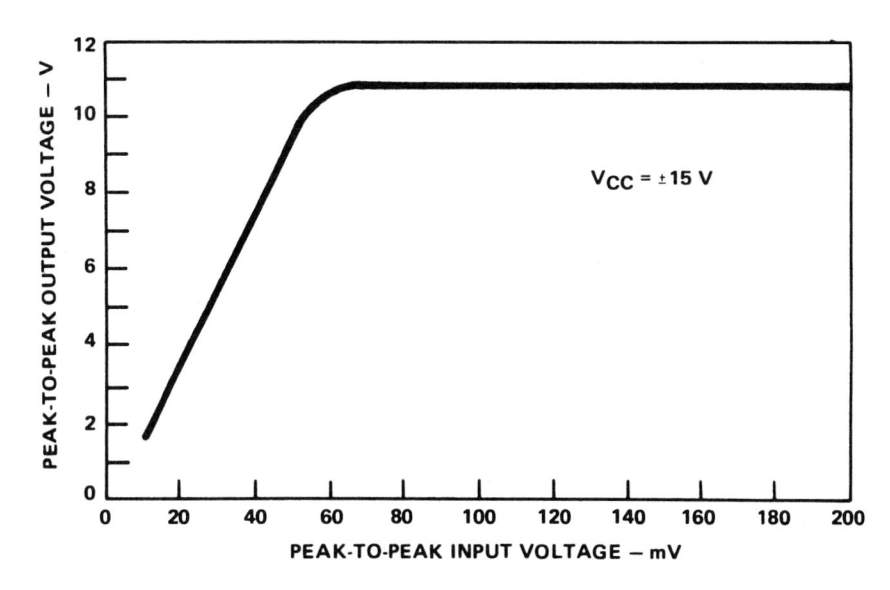

Audio amplifier with automatic gain control, accomplished by detection of the output signal with a 2N5448 silicon PNP transistor and by feeding the filtered result to the gate of a 2N4858 silicon FET controlling the termination resistance of the inverting input. Any increase in output signal results in an increase in the negative dc voltage fed to the gate of the FET. This negative voltage results in less conduction through the FET, increasing the termination resistance and reducing the circuit gain.

With the FET turned off under large-input-signal conditions, the circuit gain is R3/(R1 + R2), or 1M/1.005M ≈ 1. As the signal input decreases, the FET will be turned on, increasing the gain to maintain the output level. At very low input levels the FET is fully on, with very low forward resistance.

Although there is high gain at low signal levels (46 dB in this case), it is possible to handle large input signals without going into distortion.

SECTION **6** *Audio Circuits*

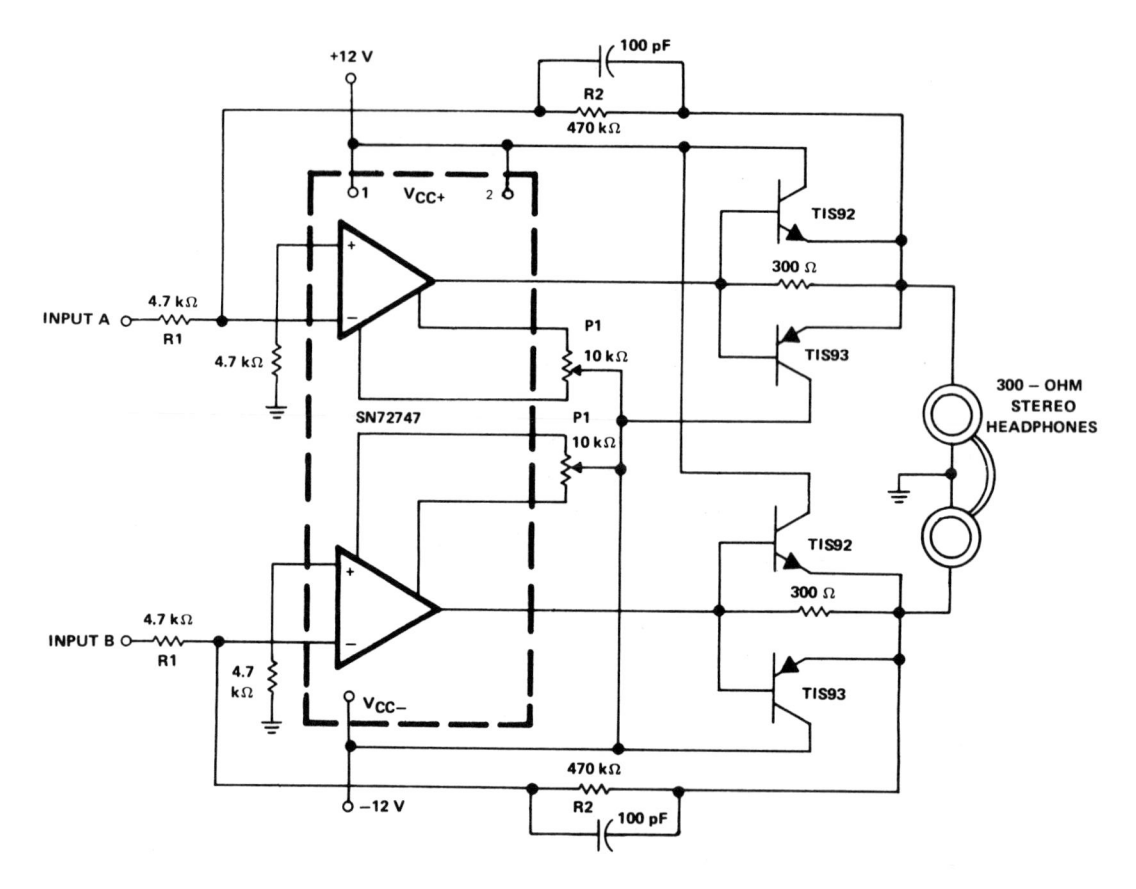

This stereo headphone amplifier combines an economical dual-channel op-amp with equally economical Silect transistors. This combination produces a stereo amplifier capable of driving low-impedance (typically 300Ω) stereo headphones with sufficient power for good fidelity. TIS92 and TIS93 complementary transistors provide the desired impedance matching to the outputs of the SN72747.

This circuit was designed to provide a 40 dB gain using source impedances of 4.7k (R1). Different source impedances will result in different circuit gains unless the value of R2 is adjusted to maintain an R2/R1 ratio of 100. High-frequency rolloff is provided by 100 pF capacitors around the 470k feedback resistors. Offset voltage is adjusted by a separate control (P1) for each channel.

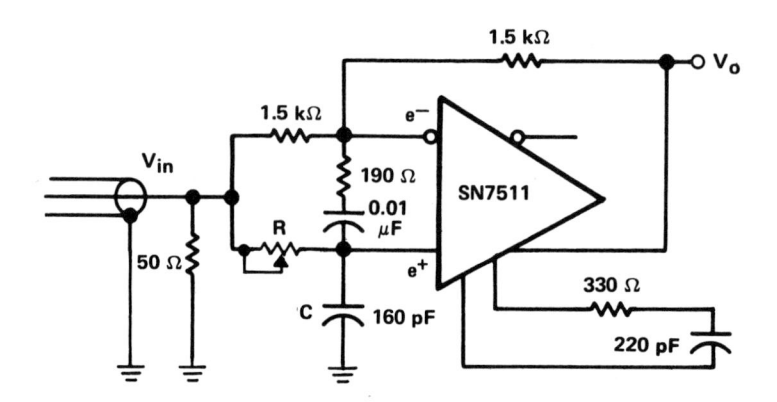

This variable phase shifter is controllable between 0° and 180°, depending on the values of R and C and the frequency of operation. Normally, this circuit is used to phase-shift a fixed frequency.

226 IC SCHEMATIC SOURCEMASTER

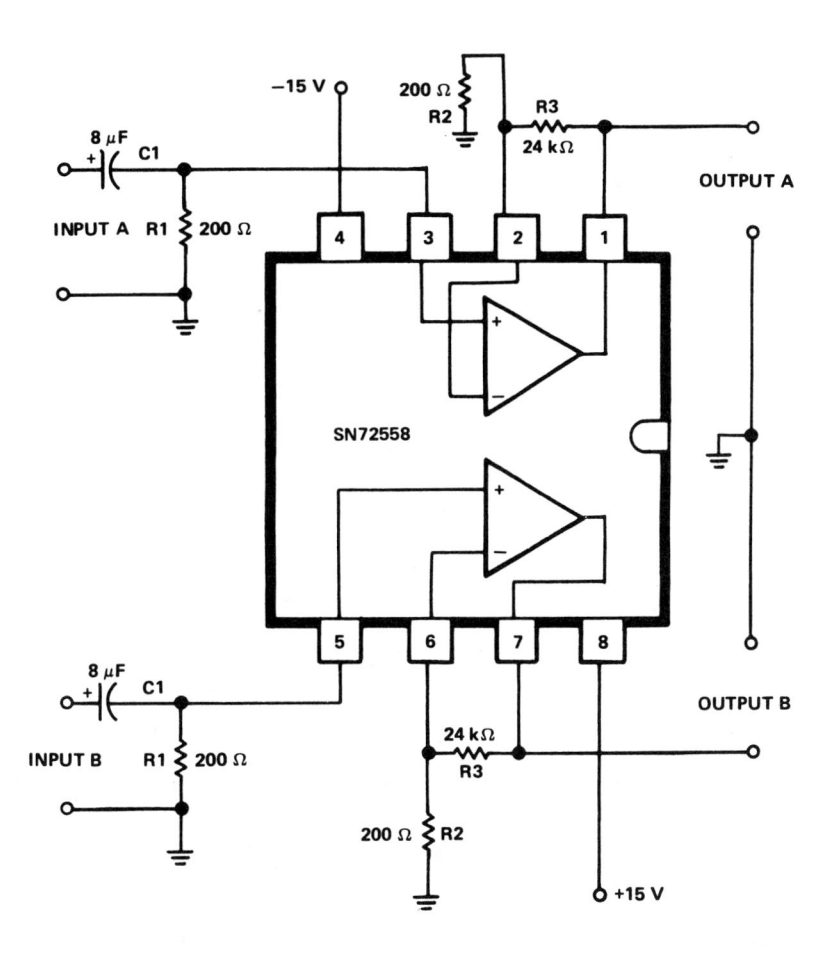

Stereo preamplifier. Performance characteristics are:
 Equivalent input noise (over 10 kHz BW)—20 µV
 Input impedance (set by R1)—200 Ω
 −3 dB bandwidth—100 Hz to 20 kHz
 Gain (determined by R2 and R3)—40 dB
 Interchannel crosstalk rejection—greater than 60 dB

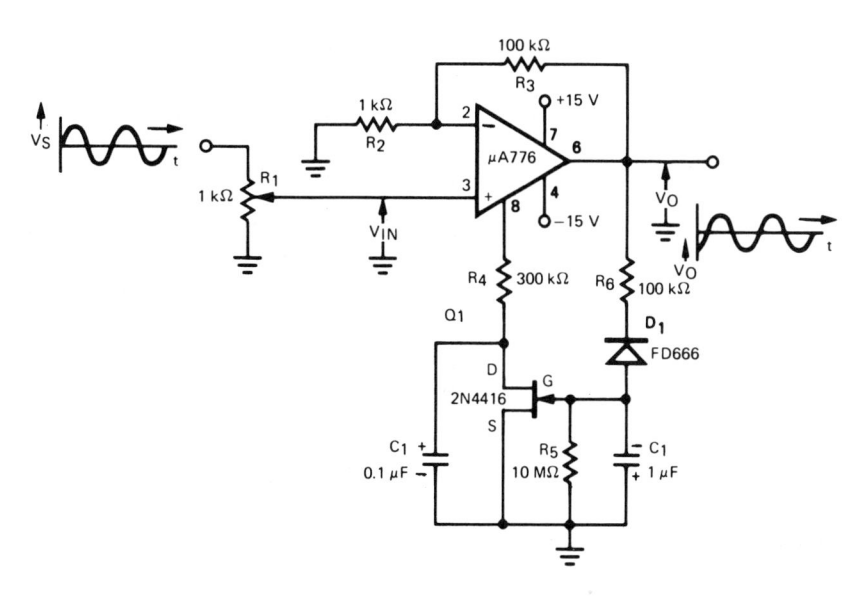

−45° phase shifter, 1–10 kHz. The circuit gives a constant phase shift and constant amplitude at the output for a range of input frequencies from 1 to 10 kHz.

SECTION **6** Audio Circuits 227

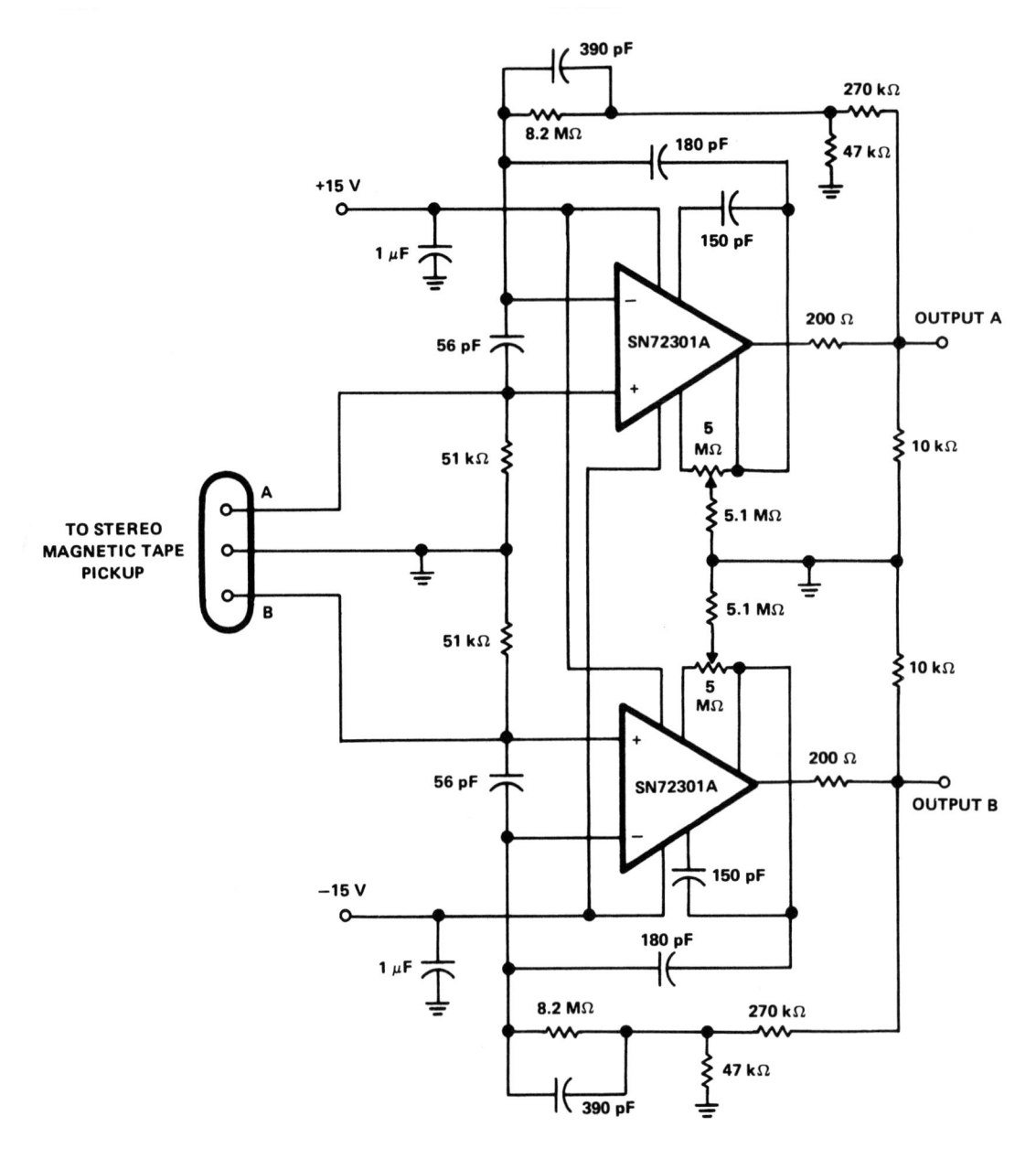

Stereo preamplifier for a magnetic tape recorder. Feedback networks demonstrate a technique for providing uniform gain throughout the audio range. Symmetrically balanced output signals are provided by separate offset controls for each channel. Amplifier gain is 15 dB; with output nose down it is 70 dB, and crosstalk rejection is greater than 60 dB.

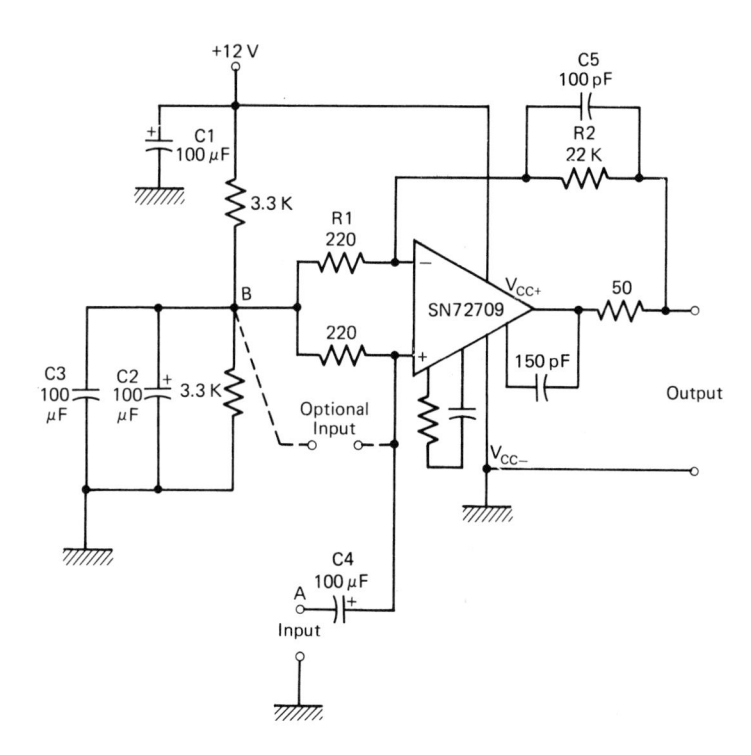

Low-noise audio preamplifier wired to operate from a single +12V supply, with +V_{CC} tied to the supply and −V_{CC} connected to ground. The amplifier's quiescent point is determined by the resistive divider network R3 and R4. It will be about half the +V_{CC} level and is bypassed by C2 and C3 to provide good low- and high-frequency filtering. The input signal may be applied in one of two ways:

1. Between point A and ground, with dc isolation provided by C4. This capacitor will determine the low-frequency rolloff point.
2. Between points B and C, in which case C4 would not be necessary. In this case the input signal must be floating (not terminated to ground), as neither point B nor C can be connected to ground. The input signal could be transformer coupled into points B and C to maintain a floating condition.

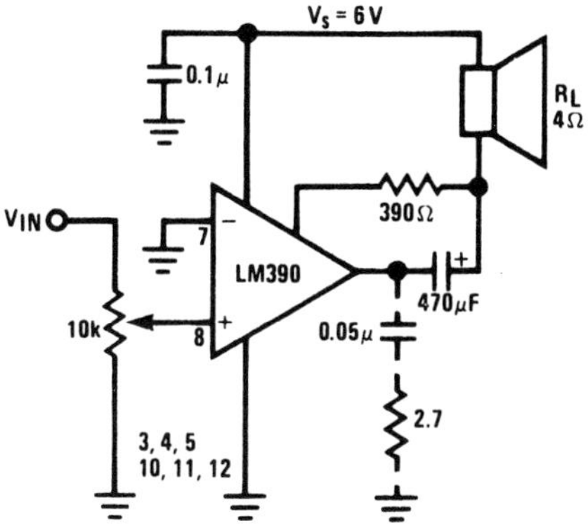

1W battery-operated audio power amplifier. Battery-operated consumer products often employ 4Ω speaker loads for increased power output. The LM390 meets the stringent output voltage swings and higher currents demanded by low-impedance loads. Bootstrapping of the upper output stage maximizes positive swing, while a unique biasing scheme used on the lower half allows negative swings down to within one saturation drop above ground. Special processing techniques are employed to reduce saturation voltages to a minimum. The result is a monolithic solution to the difficulties of obtaining higher power levels from low-voltage supplies.

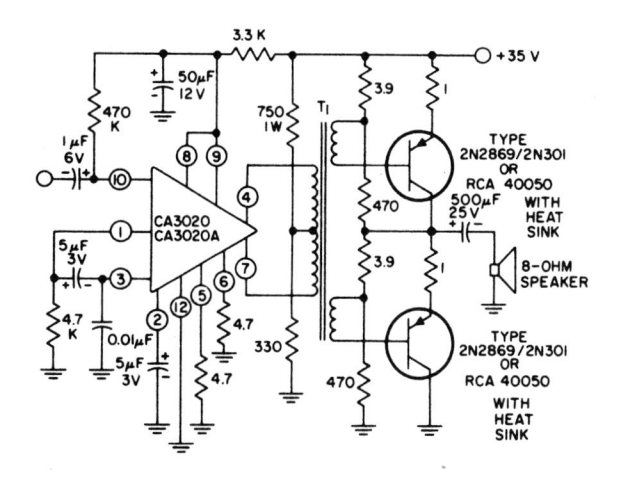

A 10W single-ended class B audio amplifier using the CA3020 or CA3020A as a driver amplifier. T1 details: primary impedance, 4000Ω center-tapped; secondary impedance, 600Ω center-tapped, split; Thordarson TR-454 or equivalent.

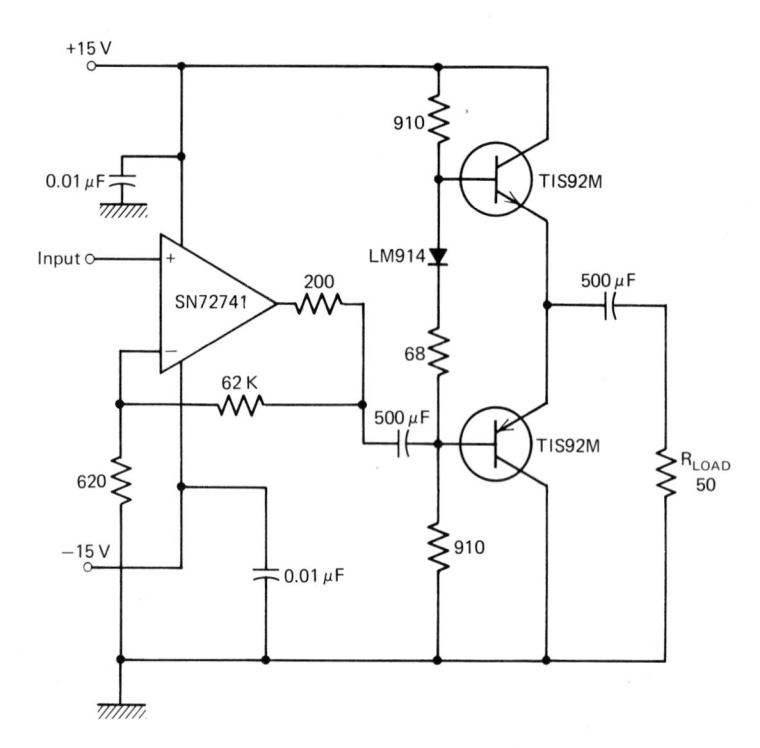

A 0.5W audio amplifier circuit combines the internally compensated SN72741 with the TIS92M and TIS93M matched complementary silicon transistors.

A 200Ω resistor is used in series with the op-amp output and its capacitive load for stability. For coupling between the output and load, 500 μF capacitors are used to provide dc isolation, so that slight offset voltages do not appear at the output. A 1N914 diode and 68Ω resistor provide proper compensation for transistor thresholds, preventing crossover distortion in the output stages.

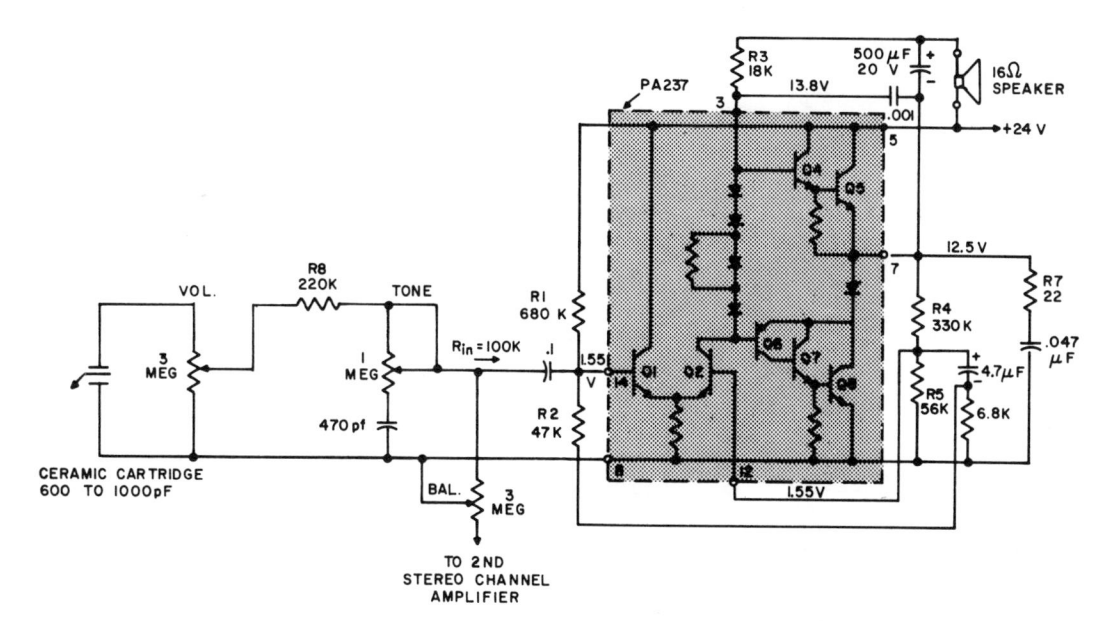

A 2W integrated phonograph amplifier using GE's PA237 IC. The phono cartridge output level will determine the optimum resistance for R8. The circuit shown is designed for a 700 mV ceramic cartridge with a capacitance of 600 to 1000 pF. A 500 mV cartridge will give 2W power output. The total harmonic distortion at 1 kHz is typically between 1.5 and 2%. In normal operation, the volume control setting will decrease the cartridge loading (compared to maximum output) and increase the bass-frequency response. The tone control at maximum treble cut attenuates a 10 kHz signal by 10 dB or more (depending on the volume setting) with respect to 1 kHz. A 1 kHz signal is changed by 1 dB or less at all tone control settings.

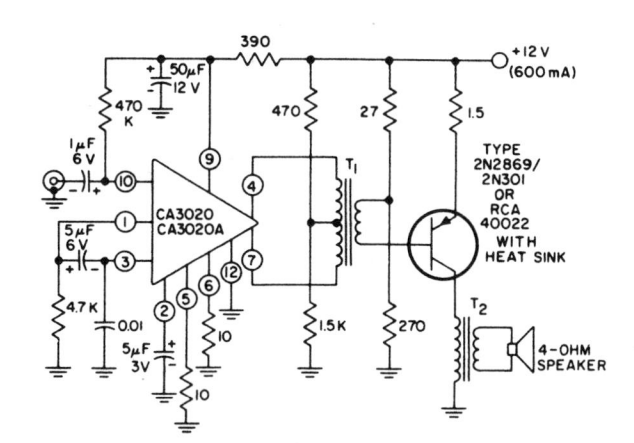

T_1: primary impedance, 10,000 ohms; center-tapped at 160 ohms; primary direct current, 2 milliamperes; Thordarson TR-207 (entire secondary), or equiv.

T_2: primary impedance, 20 ohms; primary direct current, 0.6 ampere; secondary, 4 ohms; Thordarson TR-304, Stancor TP62, or equiv.

A 2.5W class A audio amplifier using the CA3020 or CA3020A as a driver amplifier. Sensitivity for an output of 2.5W is 3 mV; this figure can be improved at a slight increase in distortion by reduction of the 4.7 Ω resistors between terminals 5 and 6 and ground.

SECTION **6** Audio Circuits

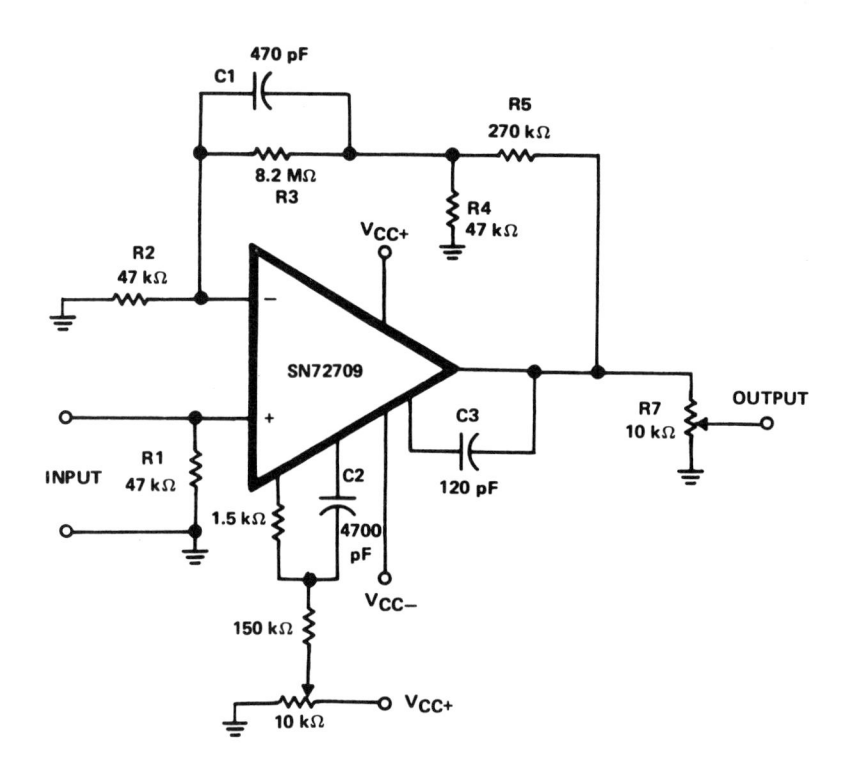

Magnetic tape preamplifier. Closed-loop gain is established by two different networks in the feedback loop. The first network in the feedback path, consisting of R4 and R5, determines the loop gain affecting all frequencies of operation. The second, consisting of R3, C1, and R2, provides a gain segment that is frequency dependent. Capacitor C1 (470 pF), shunting the 8.2M feedback resistor, reduces the gain at high frequencies. This compensation results in the bass boost desired in tape recorder amplifiers.

C1 is effective at or above the frequency where the impedance (R3 in parallel with C1) is 0.707 × R3. At this point the loop gain has been reduced by about 3 dB. Circuit calculations will show that the half-power point is about 40 Hz.

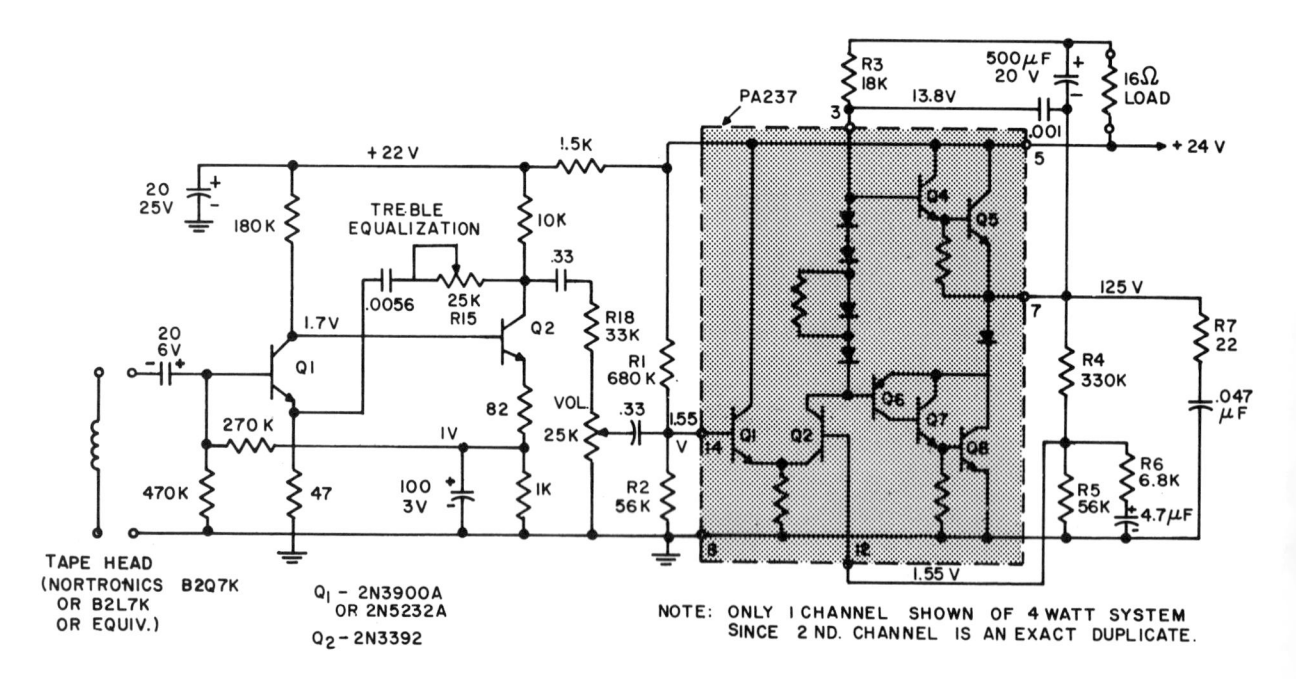

4W stereo 8-track tape amplifier system. Two-transistor preamp drives the GE power amplifier IC (PA-237). Preamp is equalized for 1⅞ or 3¾ in./sec tape speed, with R15 set at approximately 15k. The treble equalization control (R15) can be adjusted to compensate for variations in program material, tape head, or speaker; also, it can function as the normal treble-cut control. The sensitivity of the system can be adjusted by the resistor value for R18. The system as shown will give 2W output with 0.7 mV input signal from the tape head. This sensitivity is adequate for an 8-track (cartridge) stereo tape playback system.

232 IC SCHEMATIC SOURCEMASTER

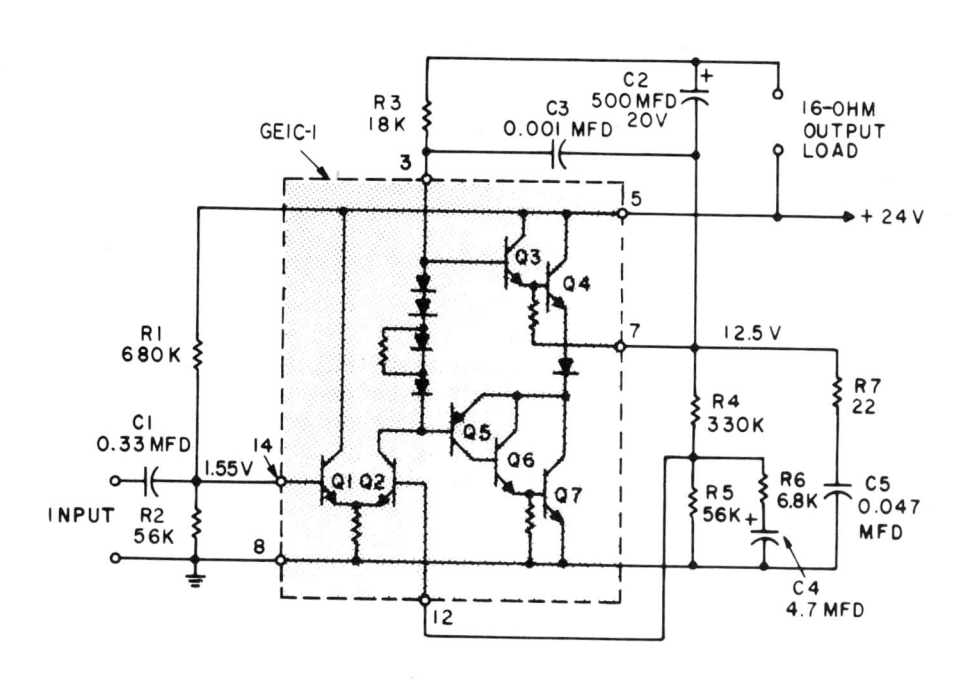

Parts List

IC audio amplifier — General Electric GEIC-1
C1 — 0.33-mfd, 50-volt capacitor
C2 — 500-mfd, 20-volt electrolytic capacitor
C3 — 0.001-mfd, 50-volt capacitor
C4 — 4.7-mfd, 25-volt electrolytic capacitor
C5 — 0.047-mfd, 50-volt capacitor

R1 — 680K-ohm, 1/2-watt resistor
R2 — 56K-ohm, 1/2-watt resistor
R3 — 18K-ohm, 1/2-watt resistor
R4 — 330K-ohm, 1/2-watt resistor
R5 — 56K-ohm, 1/2-watt resistor
R6 — 6.8K-ohm, 1/2-watt resistor
R7 — 22-ohm, 1/2-watt resistor

Sensitivity and Resistance for Input Impedance Matching

Resistor added in series with Capacitor C1	Input Resistance (ohms)	Two-watt Sensitivity
None	40K	120 Millivolts
68K	108K	300 Millivolts
120K	160K	450 Millivolts
330K	370K	1.0 Volts
470K	510K	1.4 Volts
680K	720K	2.0 Volts

2W audio amplifier. Broken lines enclose GE's IC.

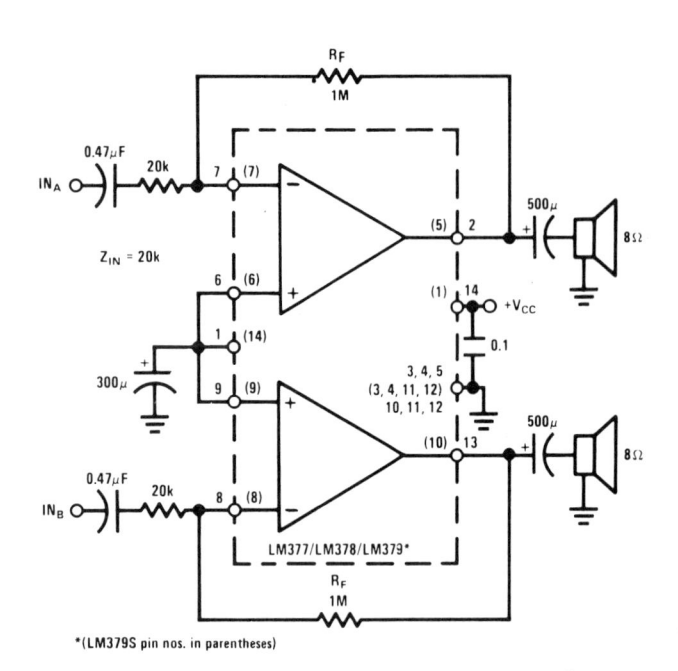

	LM377	LM377/LM378	LM379
P_O =	2W/CH	3W/CH	4W/CH
e_i =	80mV MAX	98mV MAX	113mV MAX
A_v =	50	50	50
V_{CC} =	18V	24V	28V

A 2–4W stereo amplifier may be operated in either the noninverting or the inverting mode. The inverting circuit has the lowest parts count, so it is most economical when driven by relatively low-impedance circuitry. The feedback resistor value of 1M is about the largest practical value because of an input bias current maximum of approximately ½ μA (100 nA typical). This will cause a −1.0 to 0.5V shift in dc output level, thus limiting peak negative signal swing. This output voltage shift can be corrected by the addition of series resistors (equal to the R_F in value) in the + input lines. However, when this is done, a potential exists for high-frequency instability owing to capacitive coupling of the output signal to the + input. Bypass capacitors could be added at + inputs to prevent such instability, but this increases the parts count.

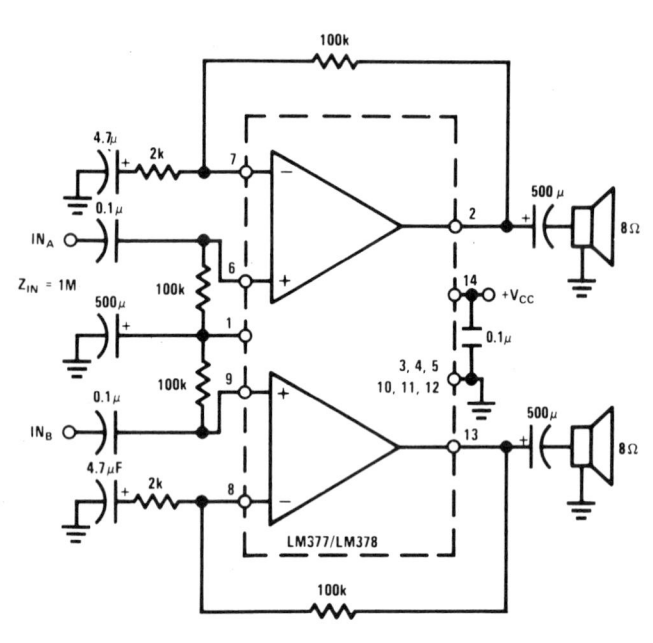

Noninverting stereo amplifier.

Inverting stereo amplifier.

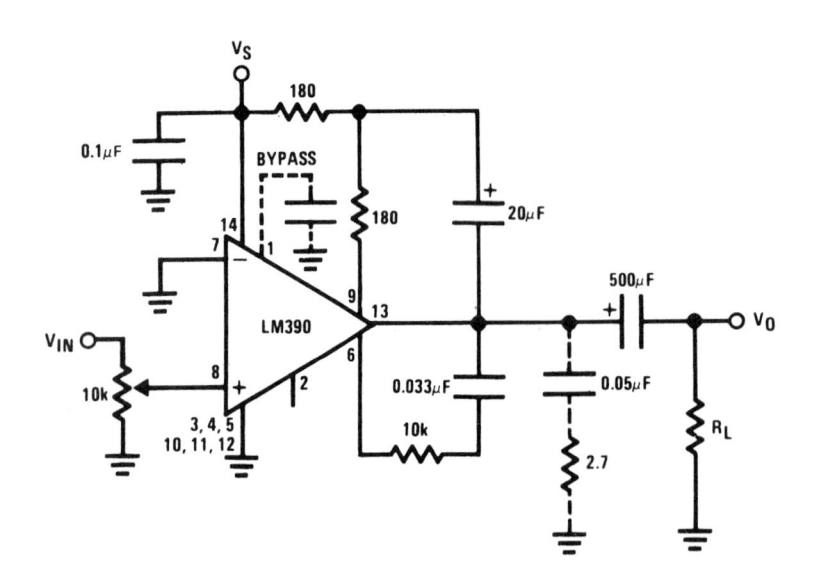

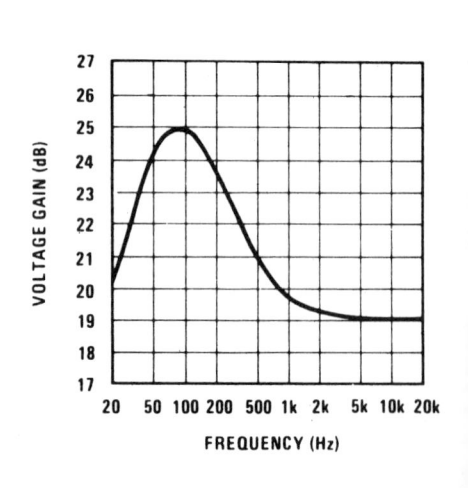

Amplifier with bass boost.

234   IC SCHEMATIC SOURCEMASTER

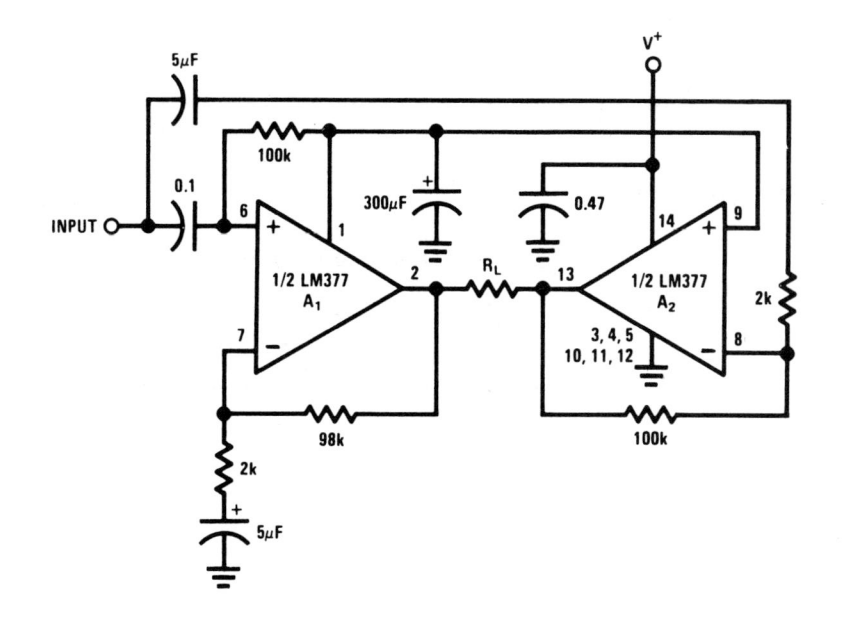

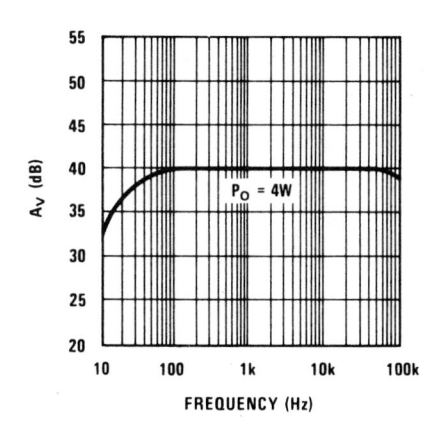

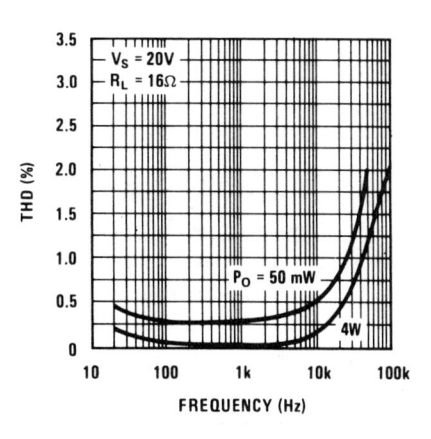

4W bridge amplifier.

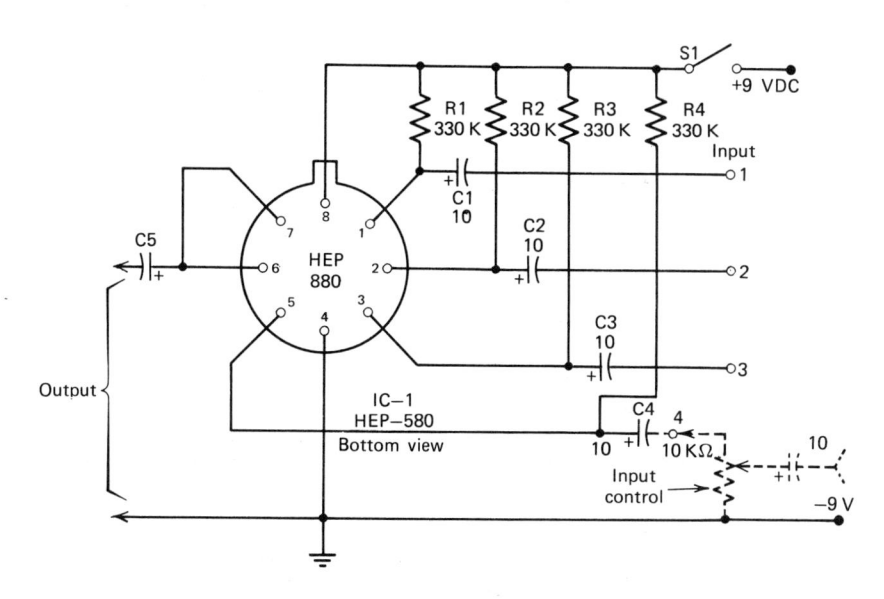

Four-channel preamp/mixer lets you mix four low-level audio signals for one output. At 0.25V input, output is constant 4.5V, 20 Hz to 1 MHz ± 1 dB (input voltage must not exceed 0.25V over the frequency range specified). Current drain at 9V is 3.0 mA.

SECTION **6** Audio Circuits **235**

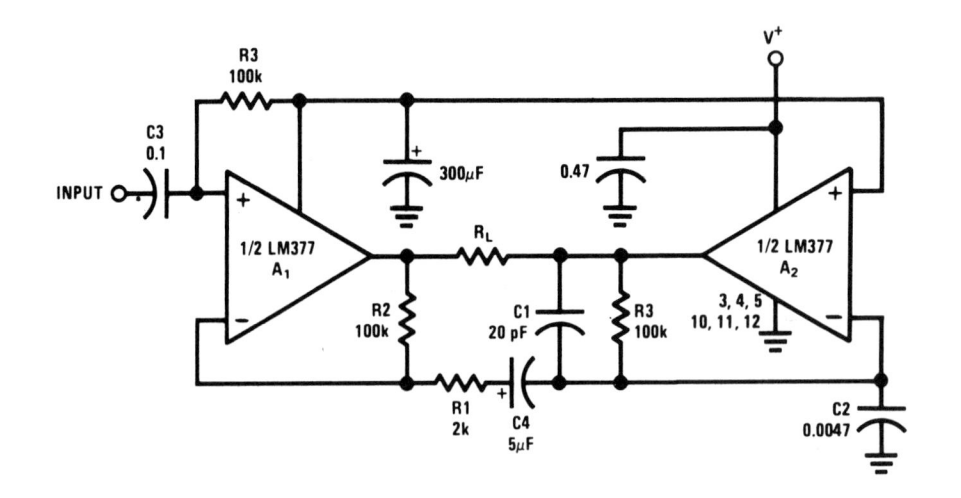

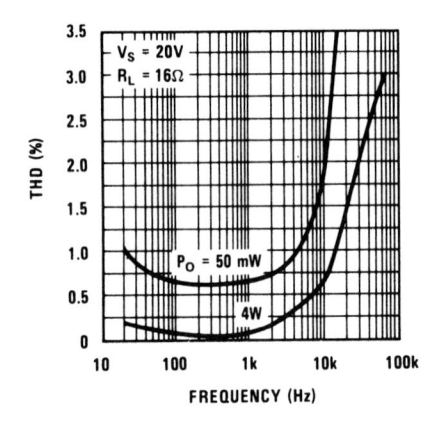

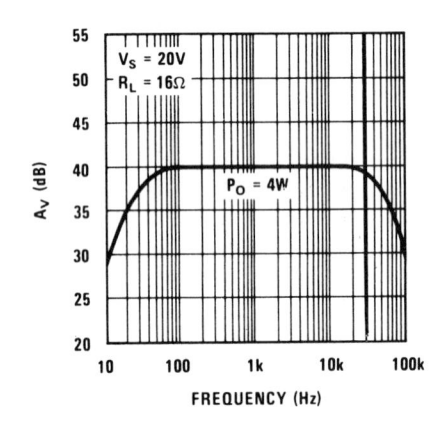

4W bridge amplifier with high input impedance.

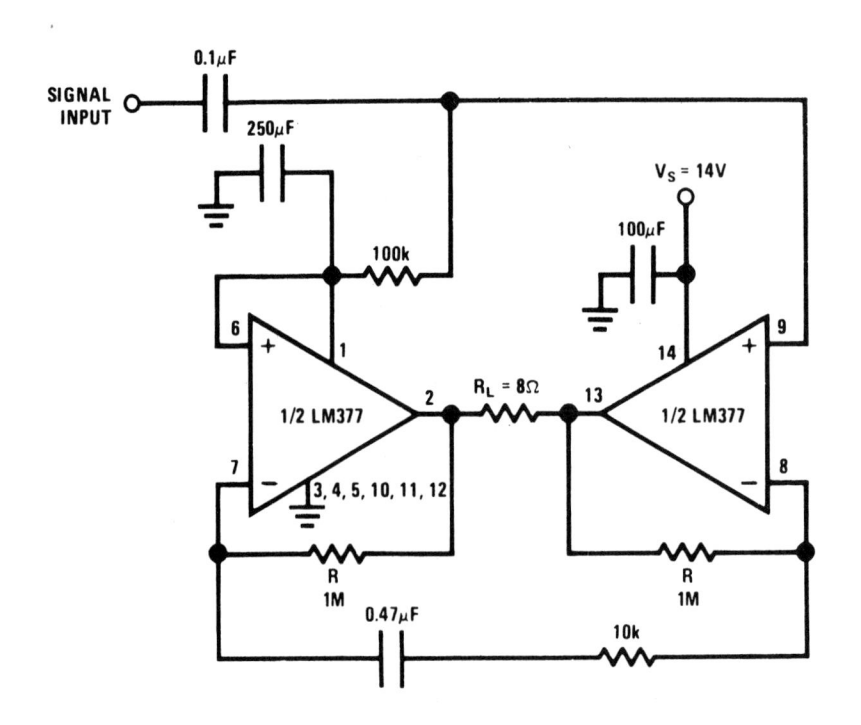

A 4W bridge amplifier keeps speaker off ground by driving voice coil from either end.

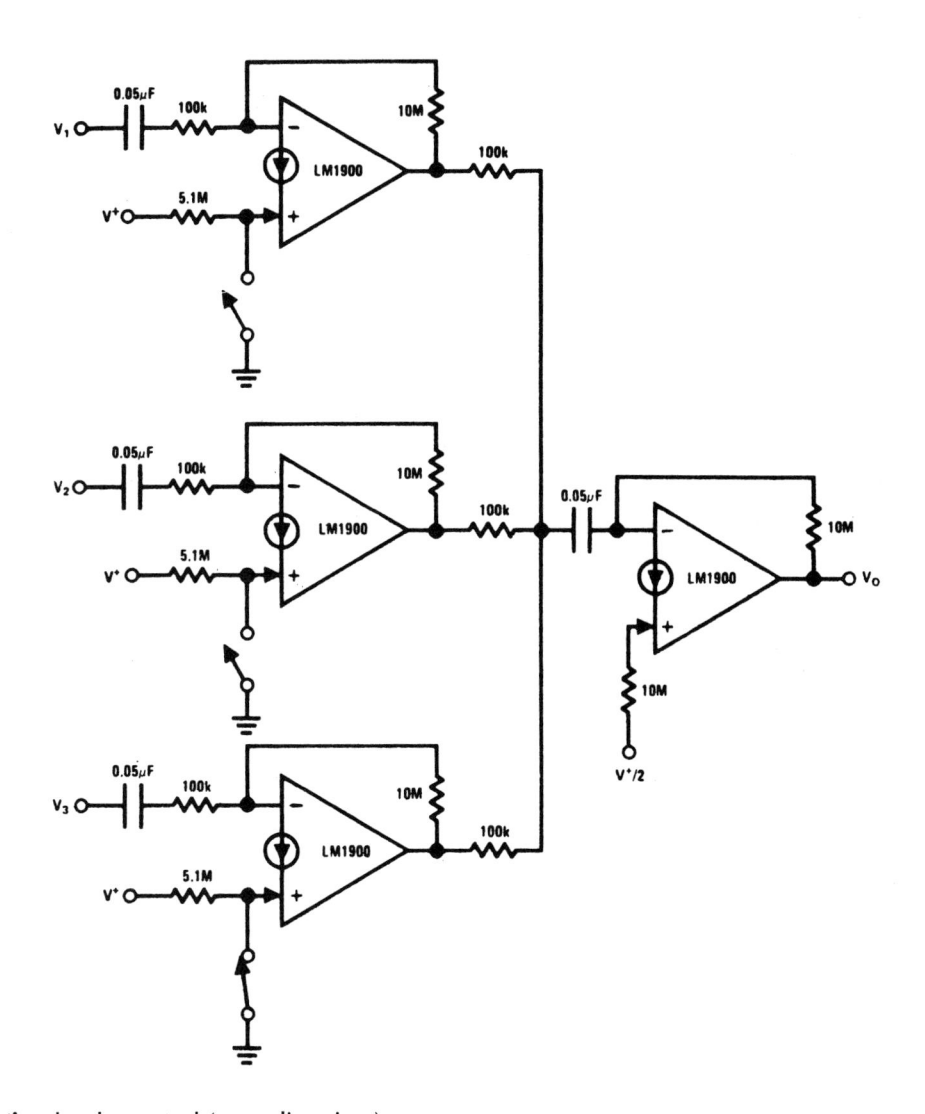

Channel selection by dc control (or audio mixer).

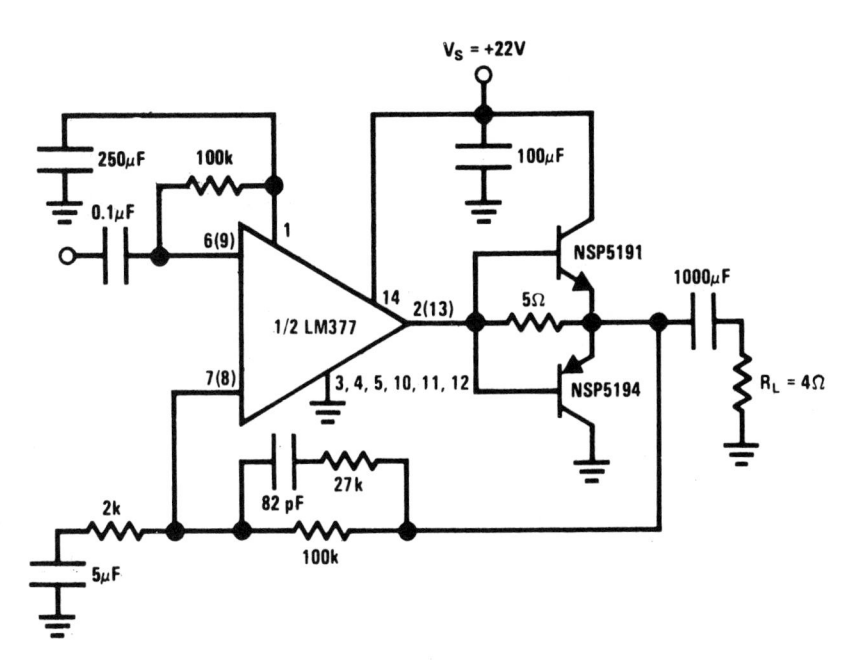

10W per channel stereo amplifier. Only one of stereo pair is shown. It requires a 22V supply.

SECTION **6** *Audio Circuits*

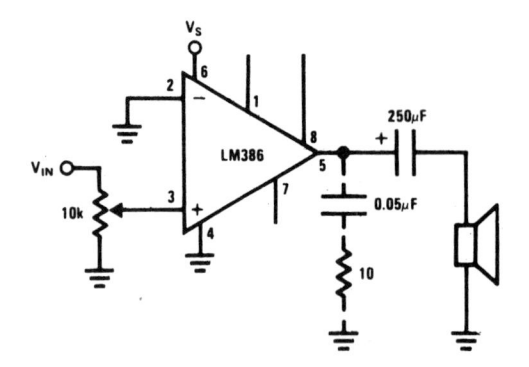

Amplifier with a gain of 20 using a minimum of parts

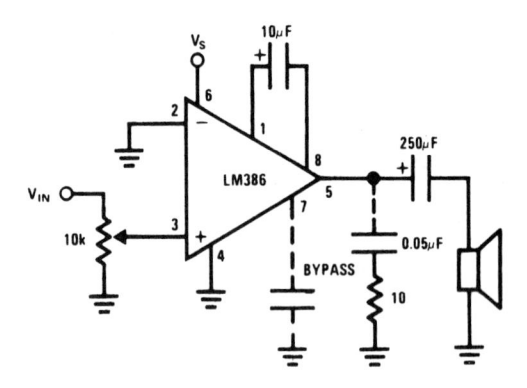

Amplifier with a gain of 200

Audio amplifiers based on design simplicity without undue performance compromise. Pins 1 and 8 are provided for gain control. With pins 1 and 8 open, the 1.35k resistor sets the gain at 20 (26 dB). If a capacitor is placed from pin 1 to pin 8, bypassing the 1.35k resistor, the gain will go up to 200 (46 dB). If a resistor is placed in series with the capacitor, the gain can be set to any value from 20 to 200. Gain control can also be accomplished by capacitively coupling a resistor (or FET) from pin 1 to ground.

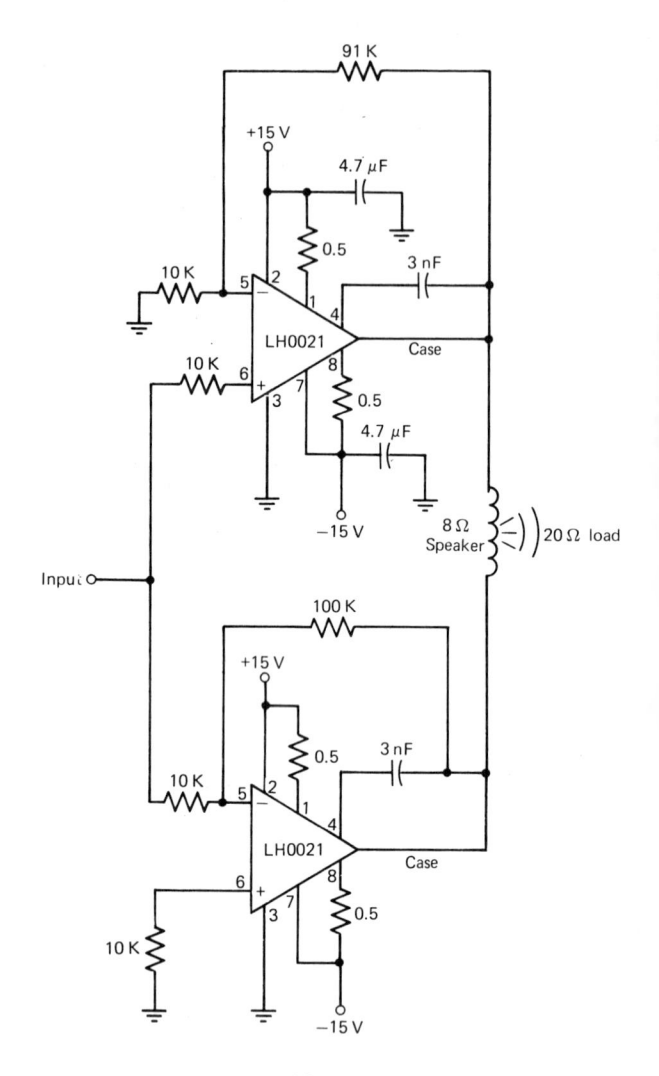

10W (RMS) audio amplifier.

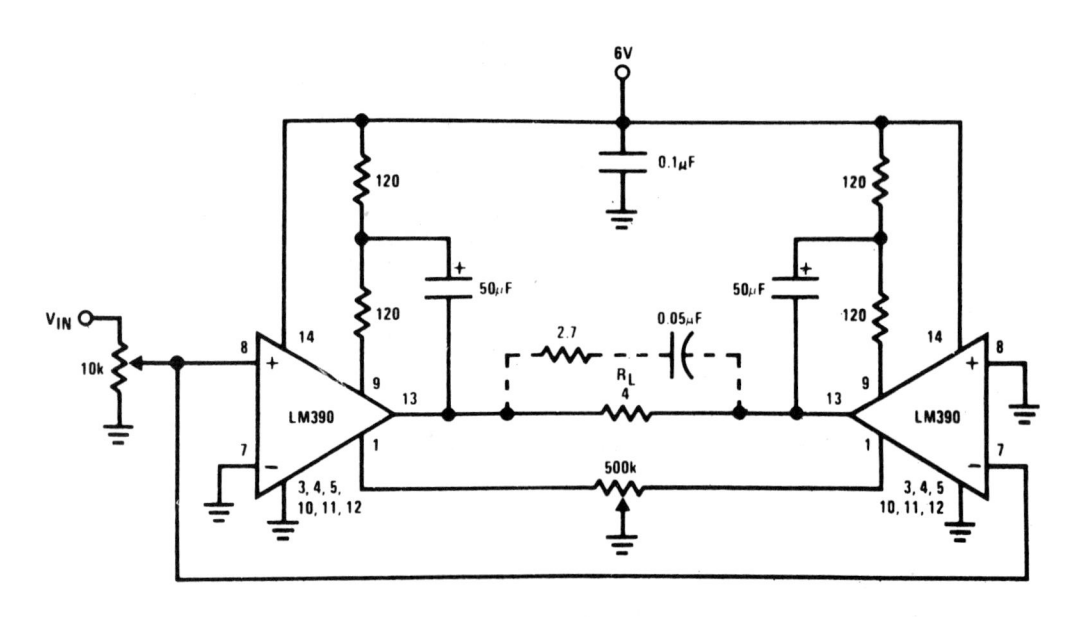

2.5W bridge amplifier.

238 IC SCHEMATIC SOURCEMASTER

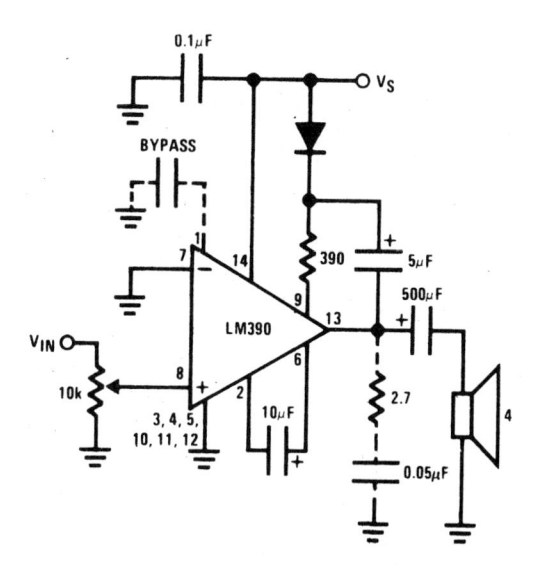

Amplifier with gain of 200 and minimum C_B.

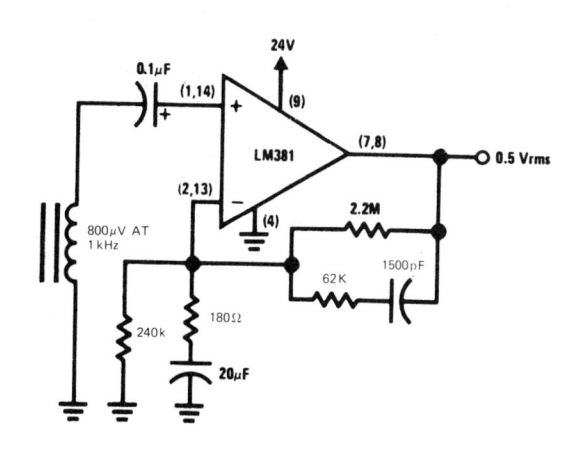

Typical tape playback amplifier, fully NAB-compensated.

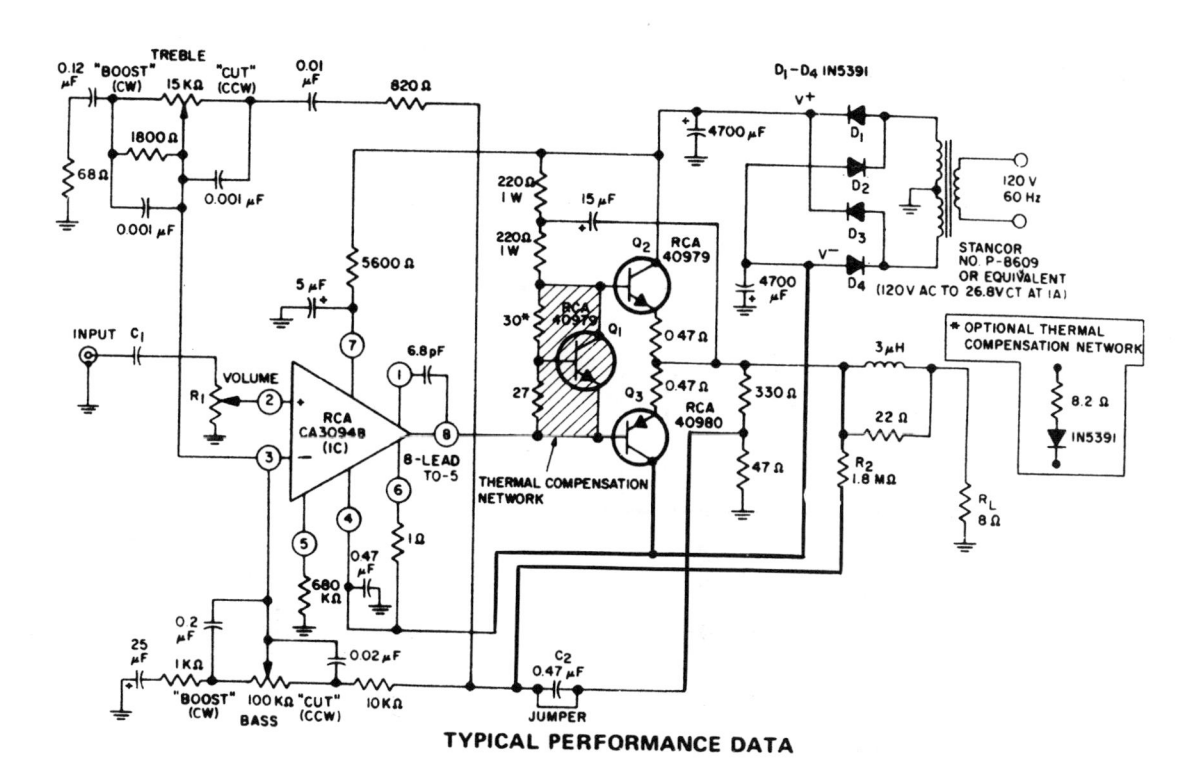

TYPICAL PERFORMANCE DATA

Power Output (8Ω load, Tone Control set at "Flat")		
Music (at 5% THD, regulated supply)	15	W
Continuous (at 0.2% IMD, 60 Hz & 2 kHz mixed in a 4:1 ratio, unregulated supply) See Fig. 8	12	W
Total Harmonic Distortion		
At 1 W, unregulated supply	0.05	%
At 12 W, unregulated supply	0.57	%
Voltage Gain	40	dB
Hum and Noise (Below continuous Power Output)	83	dB
Input Resistance	250	kΩ

12W amplifier circuit featuring true complementary-symmetry output stage with CA3094 in driver stage. For standard input: short C2; R1 = 250k, C1 = 0.047 μF; remove R2. For ceramic cartridge input: C1 = 0.0047 μF, R1 = 2.5M; remove jumper from C2, leave R2.

SECTION **6** *Audio Circuits* **239**

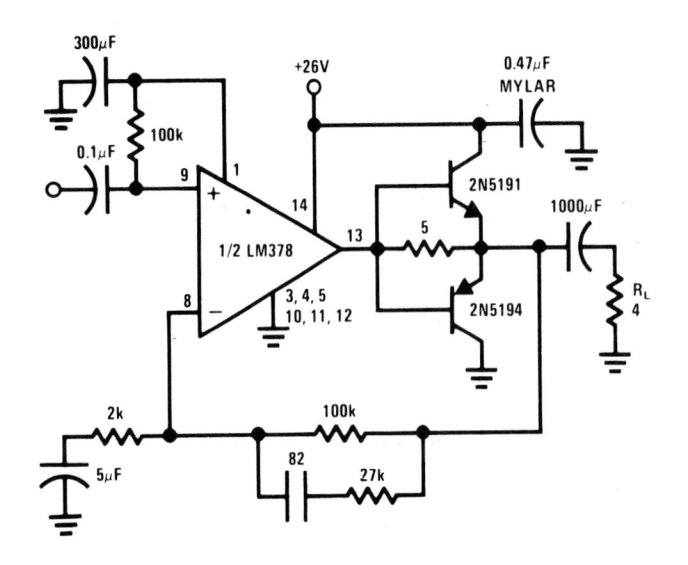

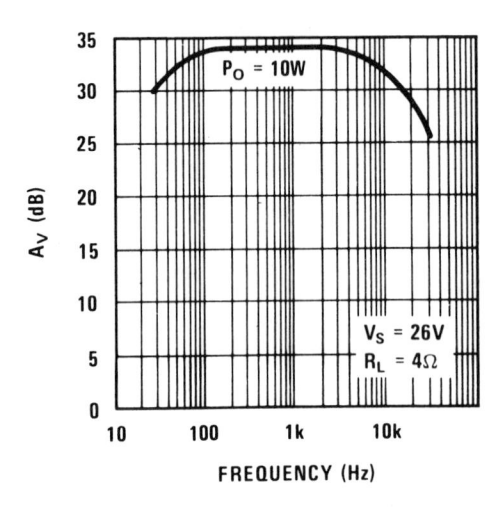

This audio booster amplifier allows power output of 10W per channel when driven from a low-power automotive tape player. The circuit is exceptionally simple, and the output exhibits very low levels of crossover distortion owing to the inclusion of the booster transistors within the feedback loop. At signal levels below 20 mW, the LM378 supplies the load directly through the 5Ω resistor to about 100 mA peak current. Above this level, the booster transistors are biased on by the load current through the same 5Ω resistor.

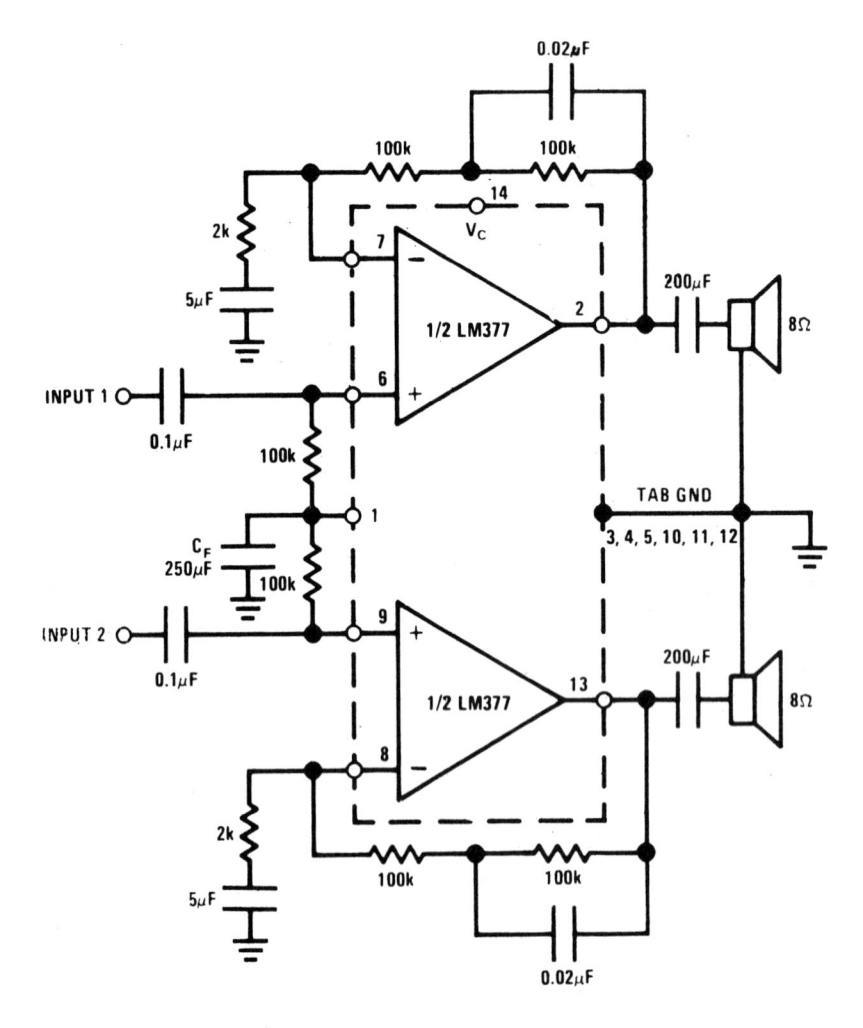

A simple 10W per channel stereo amplifier with bass boost.

240 IC SCHEMATIC SOURCEMASTER

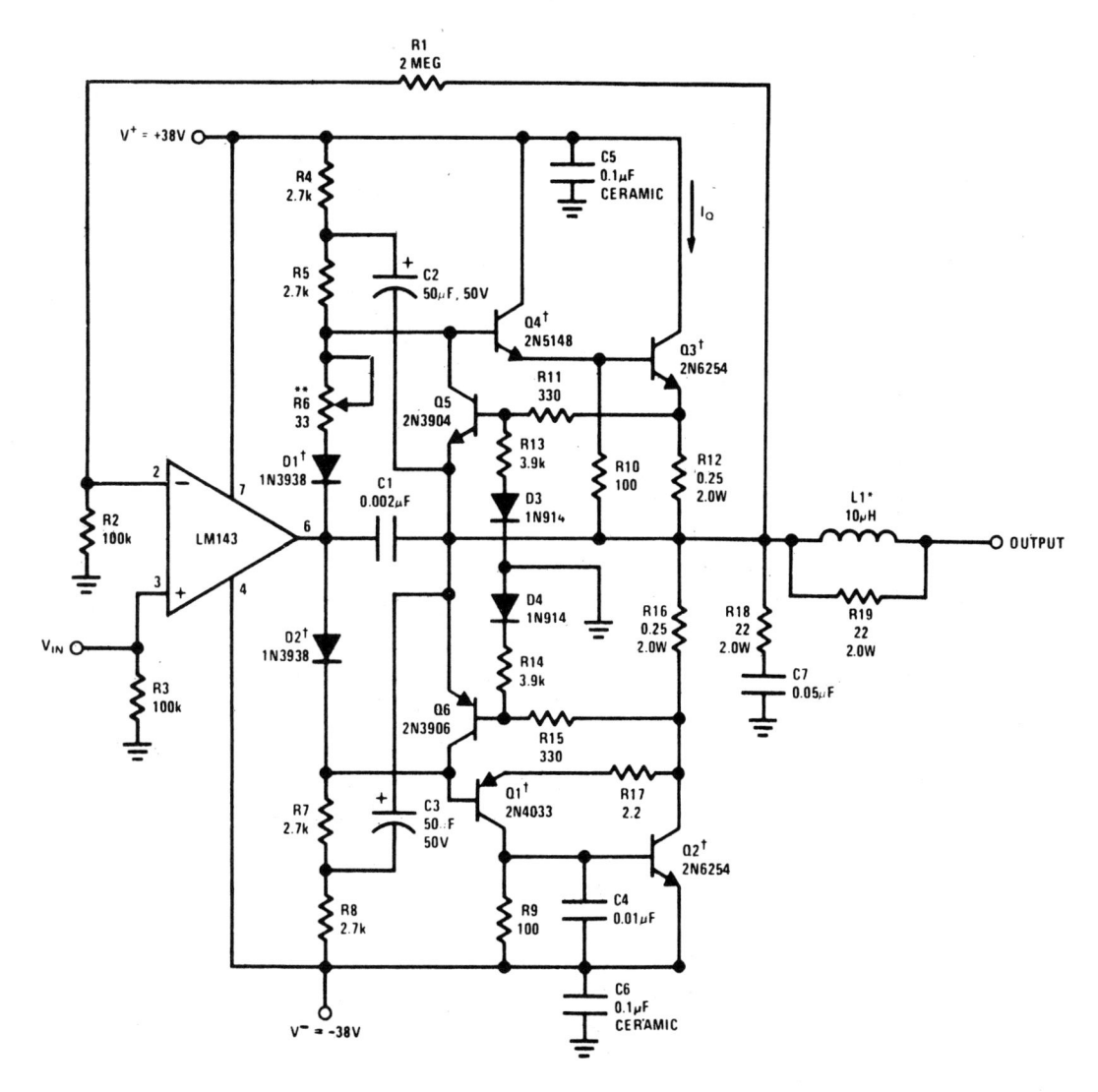

A 90W audio power amplifier with safe-area protection. Precautions should be taken to ensure that the power supplies never become reversed in polarity, even under transient conditions. With reverse voltage, the IC will conduct excessive current, fusing the internal aluminum interconnects. Voltage reversal between the power supplies will almost always result in a destroyed LM143.

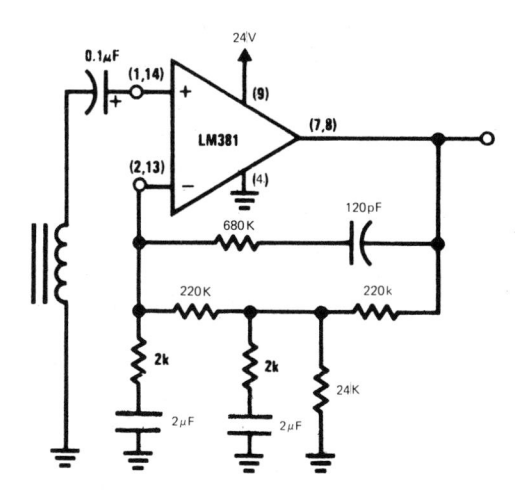

Two-pole fast turn-on NAB tape preamp.

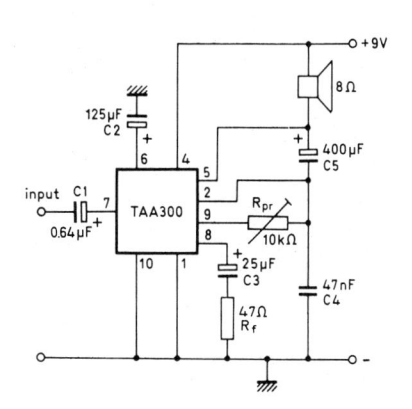

Basic circuit arrangement of **1W amplifier.**

SECTION **6** Audio Circuits 241

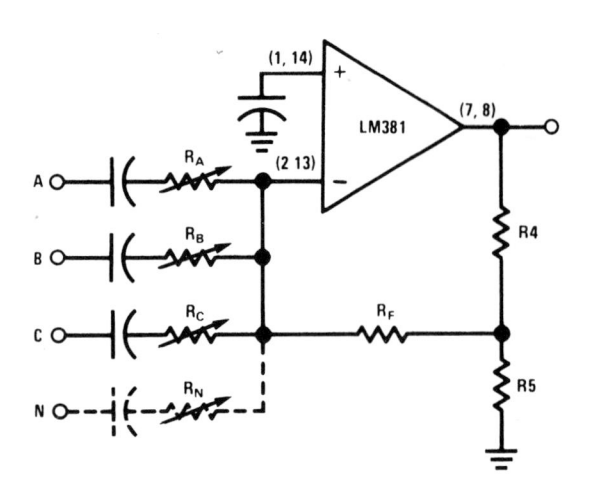

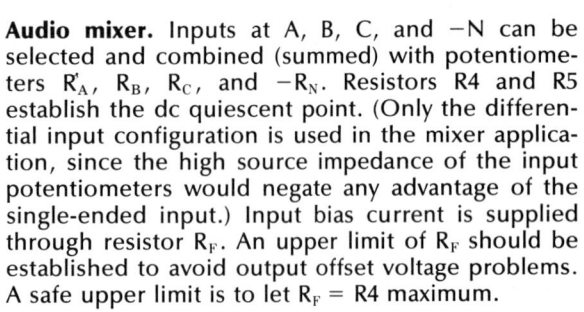

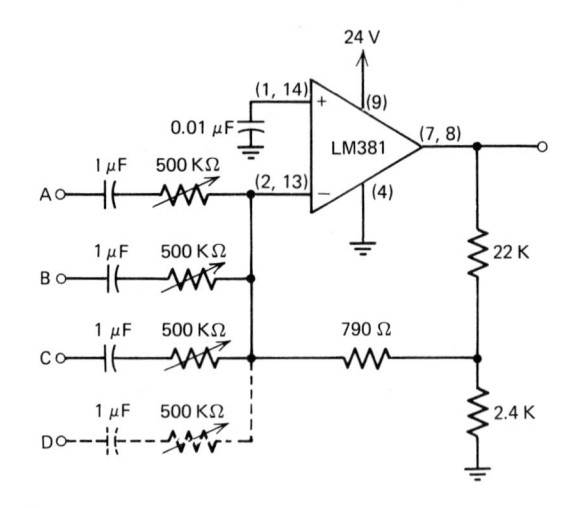

Audio mixer. Inputs at A, B, C, and −N can be selected and combined (summed) with potentiometers R'_A, R_B, R_C, and −R_N. Resistors R4 and R5 establish the dc quiescent point. (Only the differential input configuration is used in the mixer application, since the high source impedance of the input potentiometers would negate any advantage of the single-ended input.) Input bias current is supplied through resistor R_F. An upper limit of R_F should be established to avoid output offset voltage problems. A safe upper limit is to let R_F = R4 maximum.

Audio mixer. Each of the two amplifiers is completely independent, with an individual internal power supply decoupler–regulator providing a 120 dB supply rejection and a 60 dB channel separation. Other outstanding features include high gain (112 dB), large output voltage swing (V_{cc} −2V peak-to-peak), and wide power bandwidth (75 kHz, 20V peak-to-peak). The LM381/LM382A operates from a single supply of 9 to 40V.

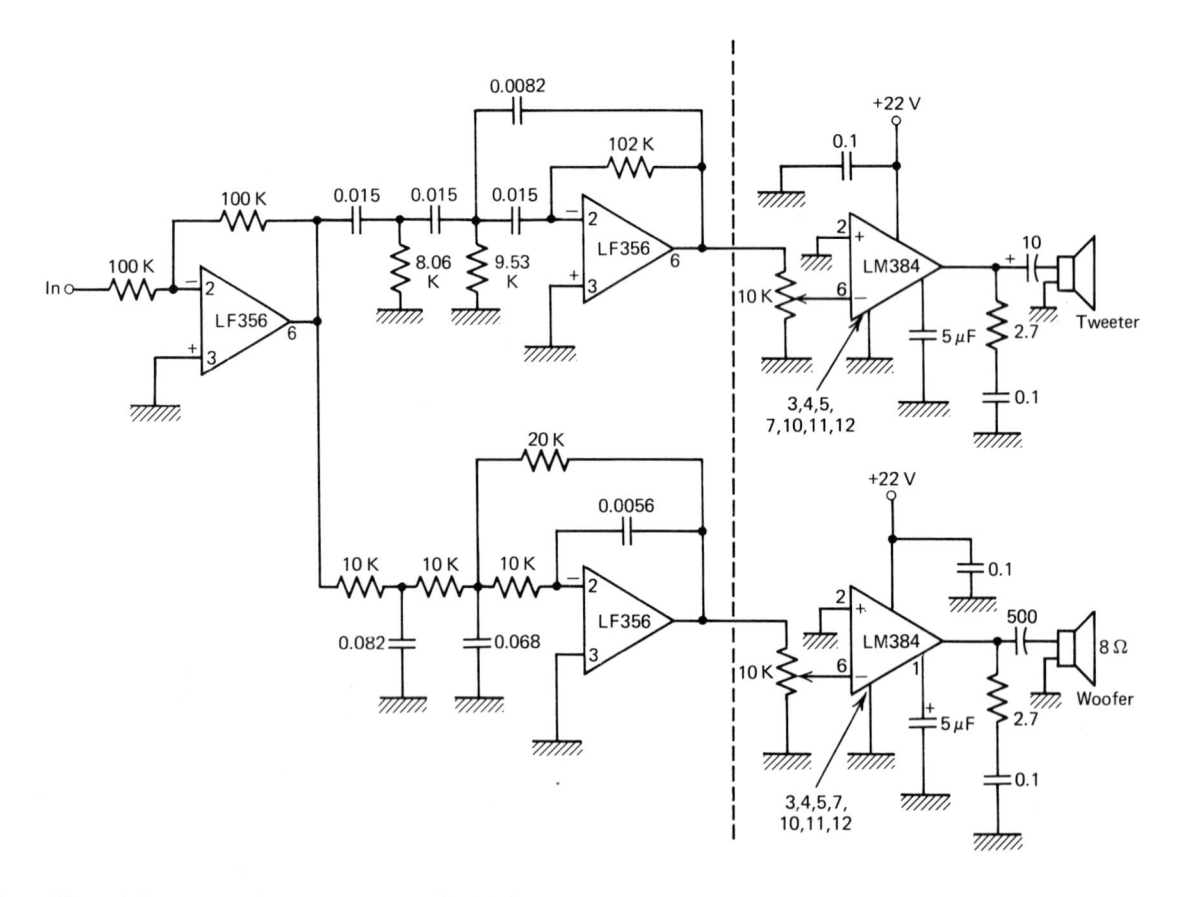

Biamplifier delivers 5W into tweeter and 5W into woofer. Distortion is noticeably lower than obtainable with conventional passive crossover network. Circuit shown is one channel only; for stereo, double up on all components. Broken line shows separation between filtering system and amplifiers.

242   IC SCHEMATIC SOURCEMASTER

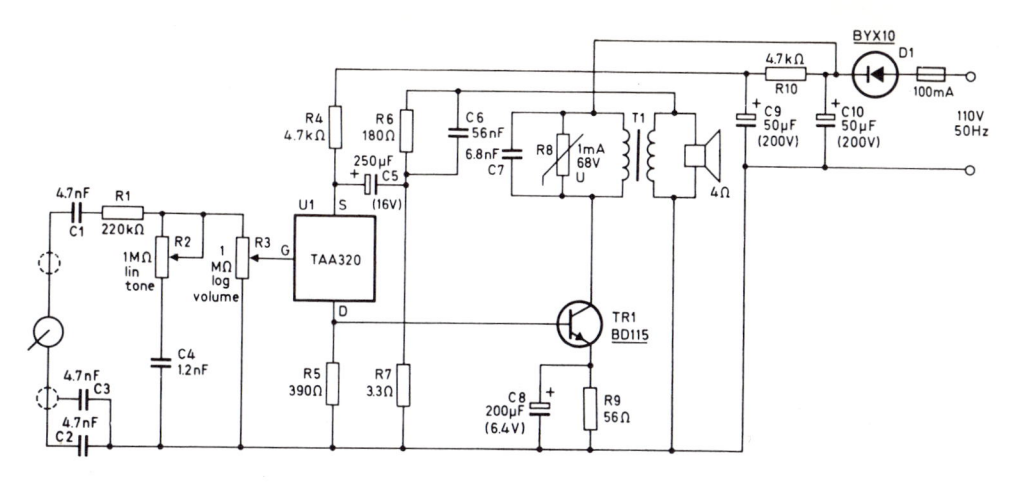

Circuit diagram of pickup amplifier.

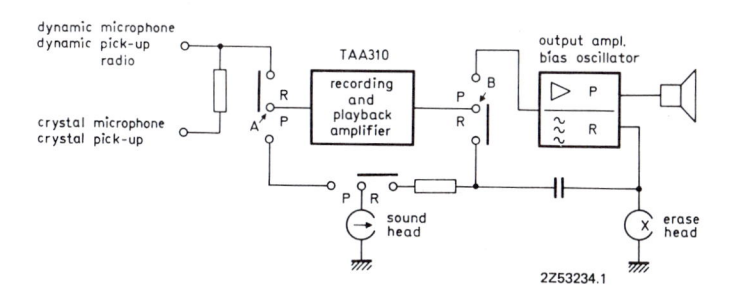

Block diagram of tape recorder using the TAA310. R = record, P = playback. Points A and B refer to input and output (respectively) of the circuit schematic.

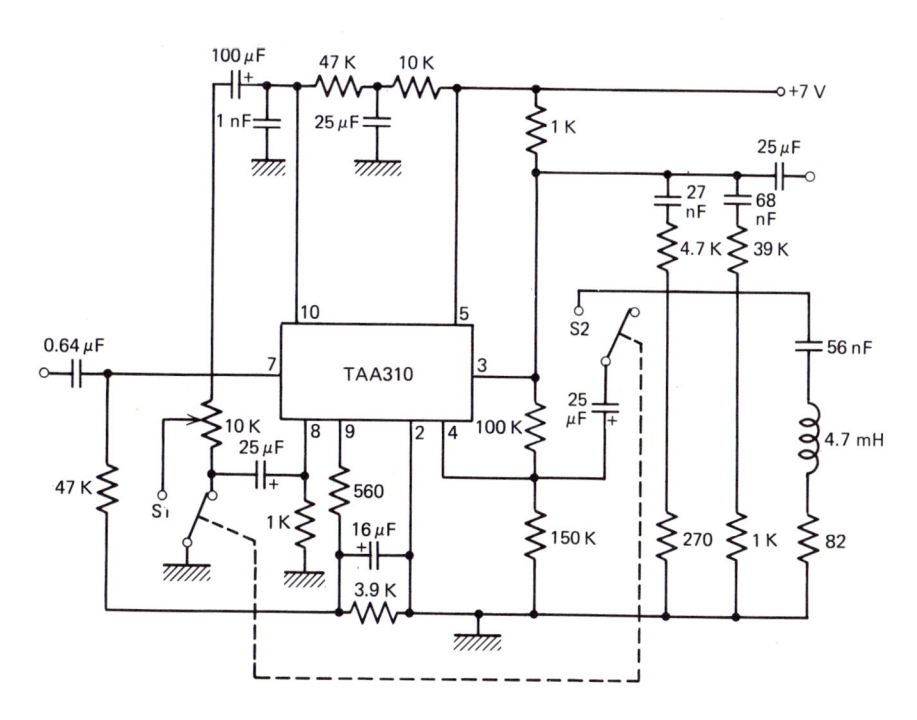

Recording and playback amplifier (cassette tape). Switches S2 and S1 are in position 1 for recording and position 2 for playback.

SECTION 6 Audio Circuits 243

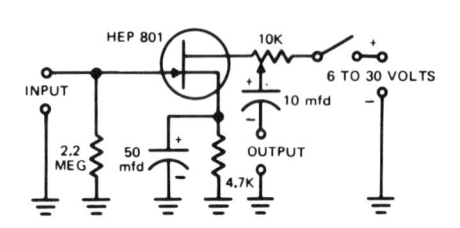

Audio preamplifier for ceramic or crystal phono cartridges or microphone. This circuit exhibits excellent frequency response, operates on a wide range of supply voltages, and can be built in minutes. If other than the Motorola FET is used, you may have to make an adjustment in the 4.7k source resistor, which can be as high as 10k in some cases.

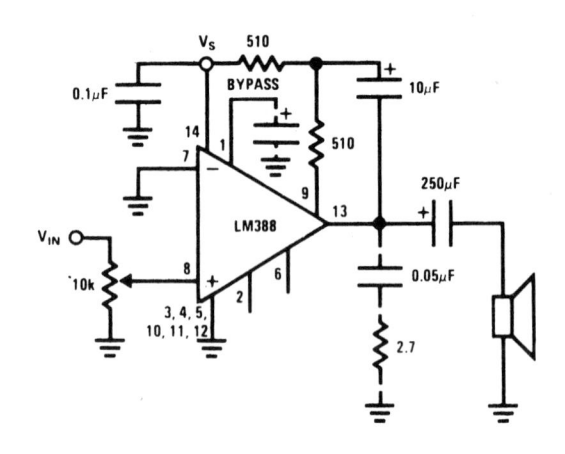

Gain-of-20 amplifier with load returned to ground.

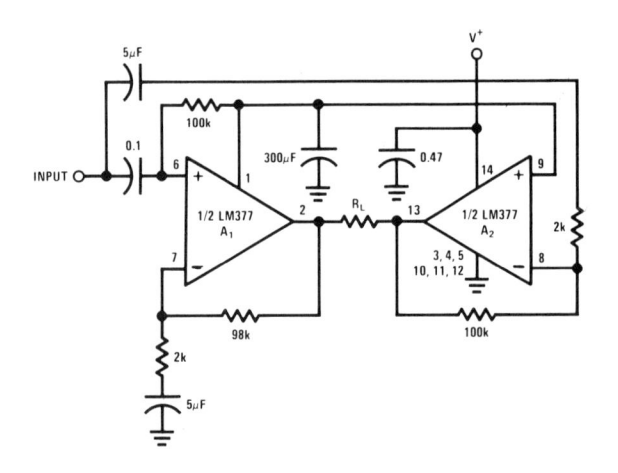

This bridge amplifier drives floating loads, which may be loadspeakers, servo motors, etc. Twice the power output of a single IC can be obtained in this connection, and output coupling capacitors are not required. Load impedance may be either 8 or 16Ω. Response is 20 Hz to 160 kHz, and distortion is 0.1% midband at 4W, rising to 0.5% at 10 kHz.

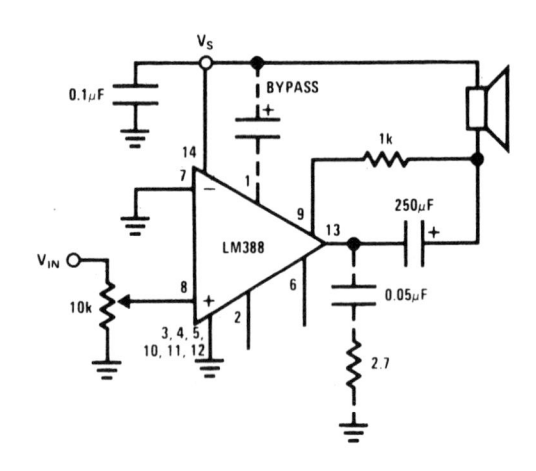

Gain-of-20 amplifier with load returned to V_S.

244 IC SCHEMATIC SOURCEMASTER

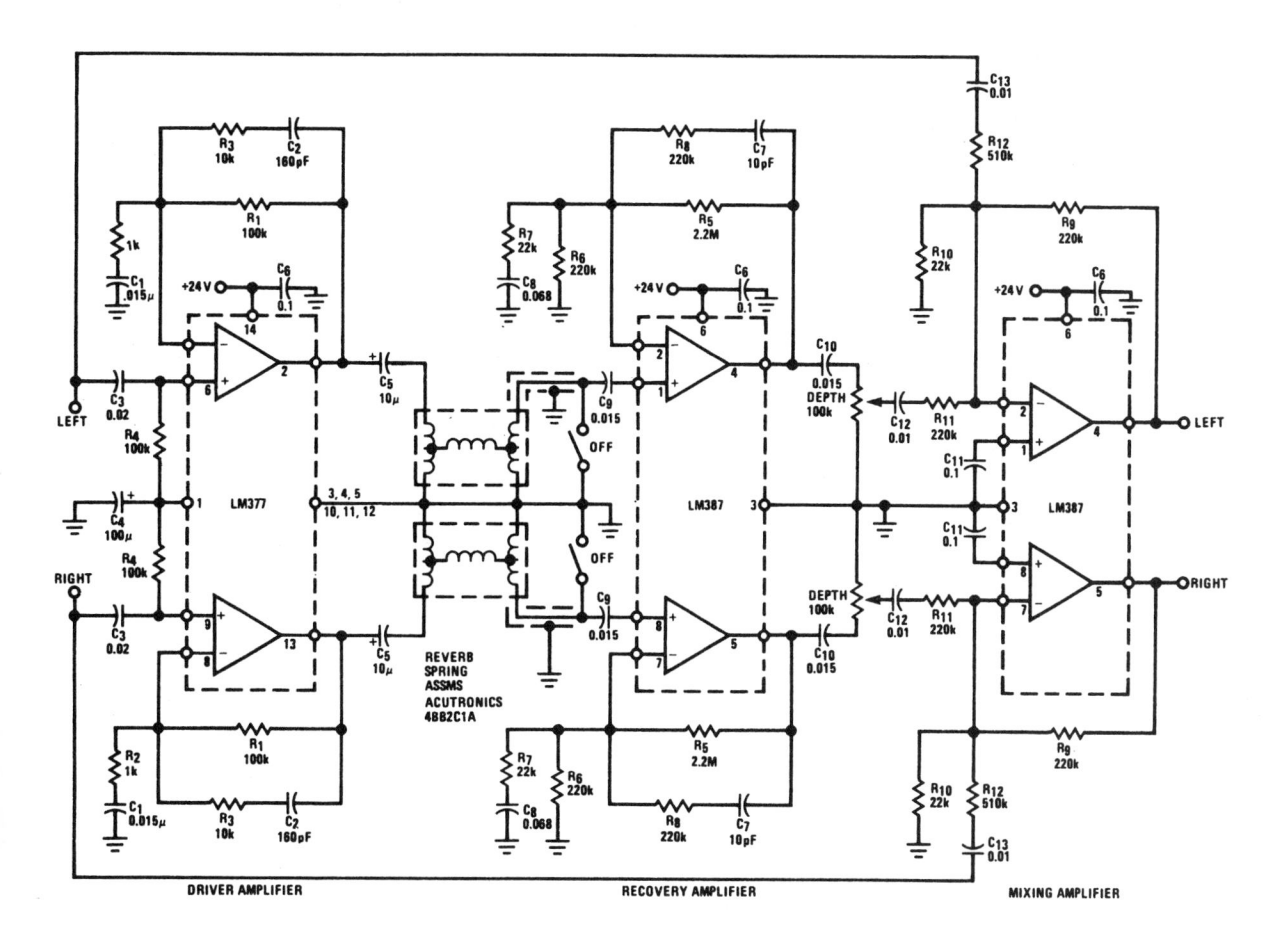

Complete stereo reverb system. The LM377 dual power amplifier is used as the spring driver because of its ability to deliver large currents into inductive loads. Some reverb assemblies have input transducer impedance as low as 8Ω and require drive currents of 30 mA. (There is a preference among certain users of reverbs to drive the inputs with as much as several hundred milliamps.) The recovery amplifier is easily made by using the LM387 low-noise dual preamplifier, which gives better than a 75 dB signal-to-noise performance at 1 kHz (10 mV recovered signal). Mixing of the delayed signal with the original is done with another LM387 used in an inverting summing configuration.

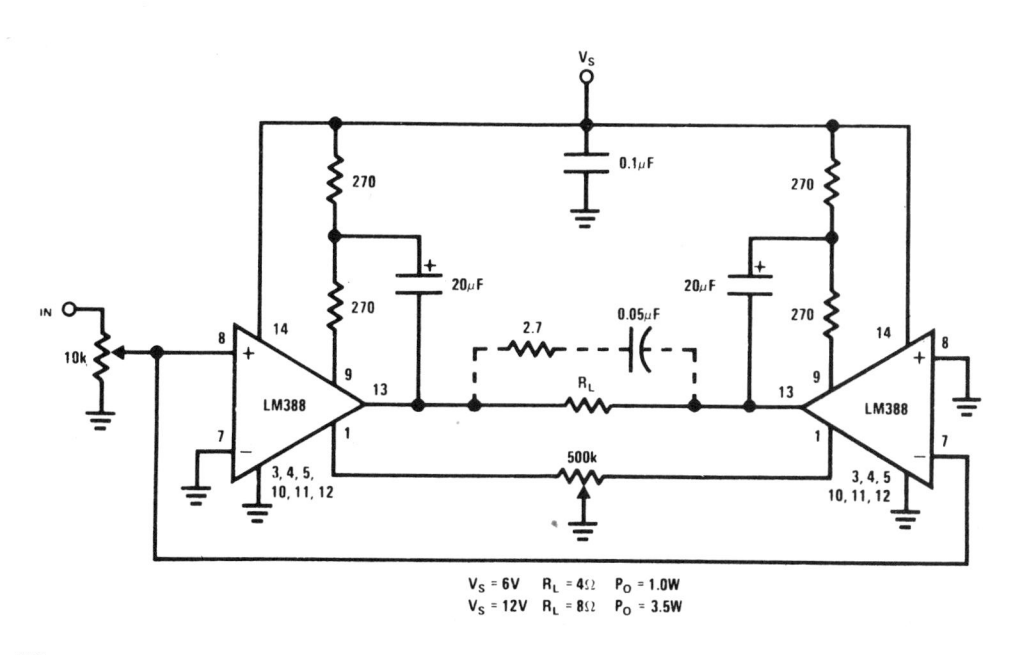

Bridge amplifier.

SECTION **6** Audio Circuits   **245**

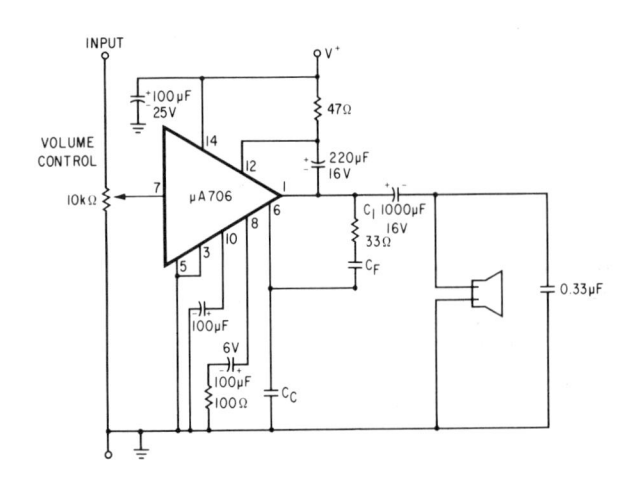

G_V	34 dB		46 dB	
BW	10 kHz	20 kHz	10 kHz	20 kHz
R_B	100Ω		0Ω	
C_C	10 nF	6.8 nF	2.7 nF	1.5 nF
C_F	1 nF	470 pF	330 pF	150 pF

This audio output amplifier has less than 1% total harmonic distortion at 2W output.

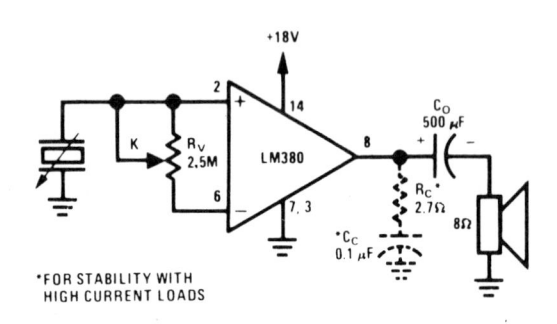

This phono amplifier for crystal/ceramic cartridges uses common-mode volume control whereby the source loading impedance is only the input impedance of the amplifier when in the full-volume position. This reduces to half the amplifier input impedance at the zero volume position.

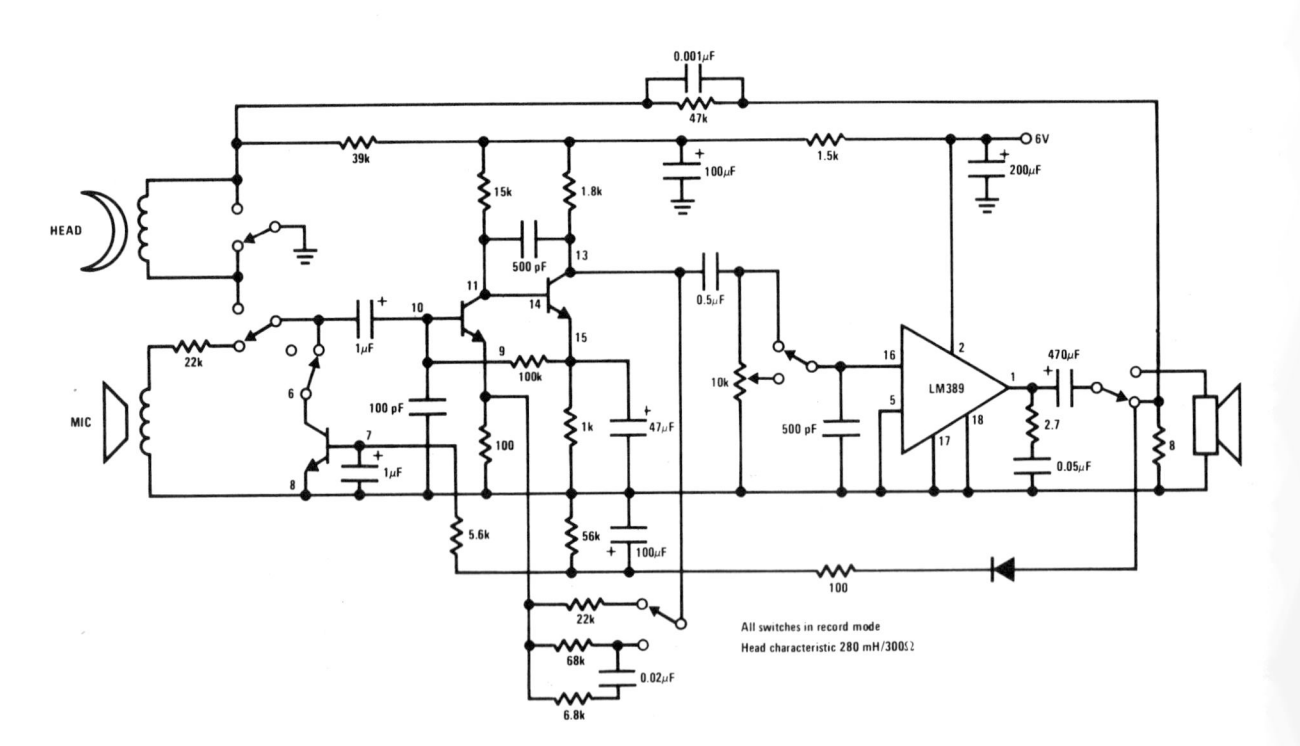

Complete record/playback cassette tape machine amplifier. Two of the transistors act as signal amplifier, with the third used for automatic level control during the "record" mode. The complete circuit consists of only the LM389 plus one diode and the passive components. Transistors are integral to IC.

246 IC SCHEMATIC SOURCEMASTER

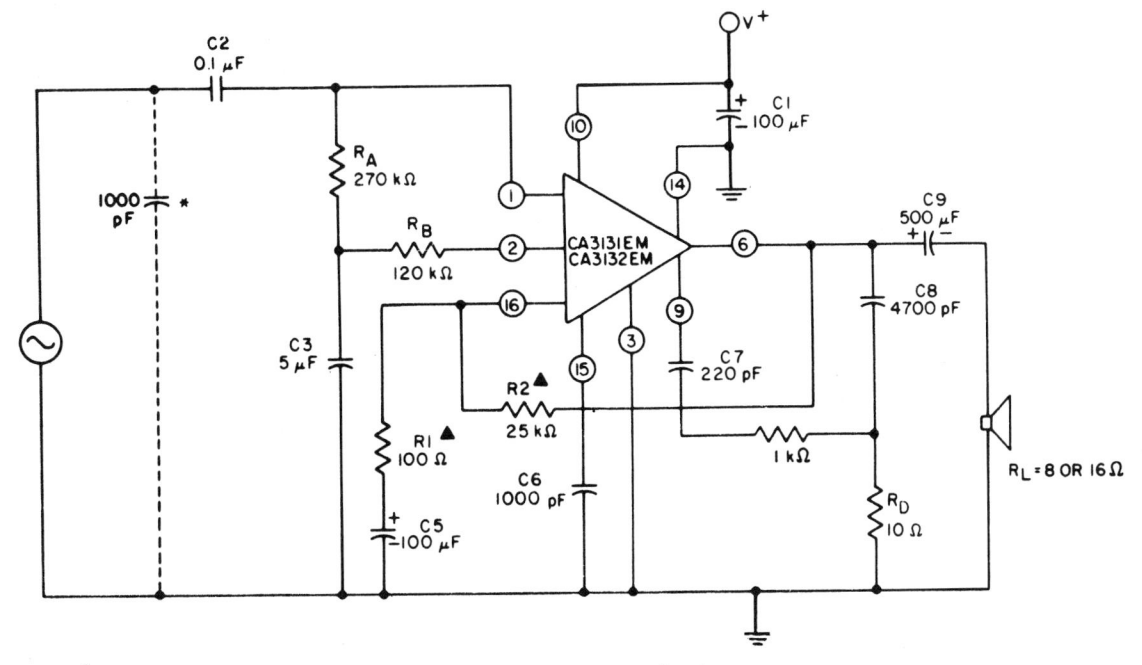

* A 1000-pF capacitor is required if input has an open circuit.

▲ External resistors R1 and R2 are used only with the CA3132EM. When testing the CA3131EM, omit R1 and R2 and connect the (+) termination of C5 to Terminal 16.

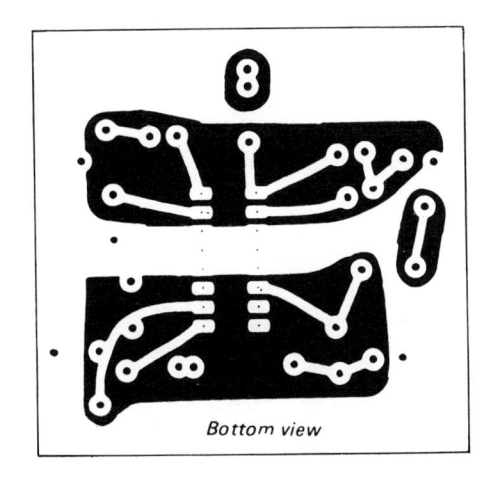

Circuit and full-size PC board layout for **5W audio amplifier** with integral heatsink. The dc quiescent output voltage is set by the voltage at pin 1. This voltage, in turn, is set by the internal voltage at pin 2 less I_1 (input current, fixed by $R_A + R_B$, for Q4). The voltage at pin 2 is set slightly above half the supply voltage to allow for the voltage drop across $R_A + R_B$. Filter $R_B C_3$ attenuates any ac ripple injected from the supply line and prevents positive feedback to pin 1. The rejection of supply voltage is a direct function of the filter attenuation.

The input impedance of the audio amplifiers is a function of the closed-loop gain and the magnitude of the Q8 current. In practice the input impedance is well above 1M. The input signal, applied through C2, sees an impedance equivalent to the resistance of R_A connected in parallel with the amplifier input impedance. Hence, the value of R_A in most cases is dominant in establishing the input signal impedance.

The value of C1 depends on the regulation of the power supply. It is possible for the amplifier to work with a value of C1 as low as 0.1 µF to attenuate high-frequency signals in the supply line. Ideally, C1 should be placed as near pin 10 as possible. An electrolytic capacitor should be used for C1 if the power supply is poorly regulated to avoid ripple at the output.

If a 1000 pF capacitor is used for C6, the first breakpoint for a 46 dB closed-loop gain occurs at 200 kHz. Higher capacitance values will cause the constant current to charge C6 on the positive voltage swing and thus limit the slew rate at high signal levels.

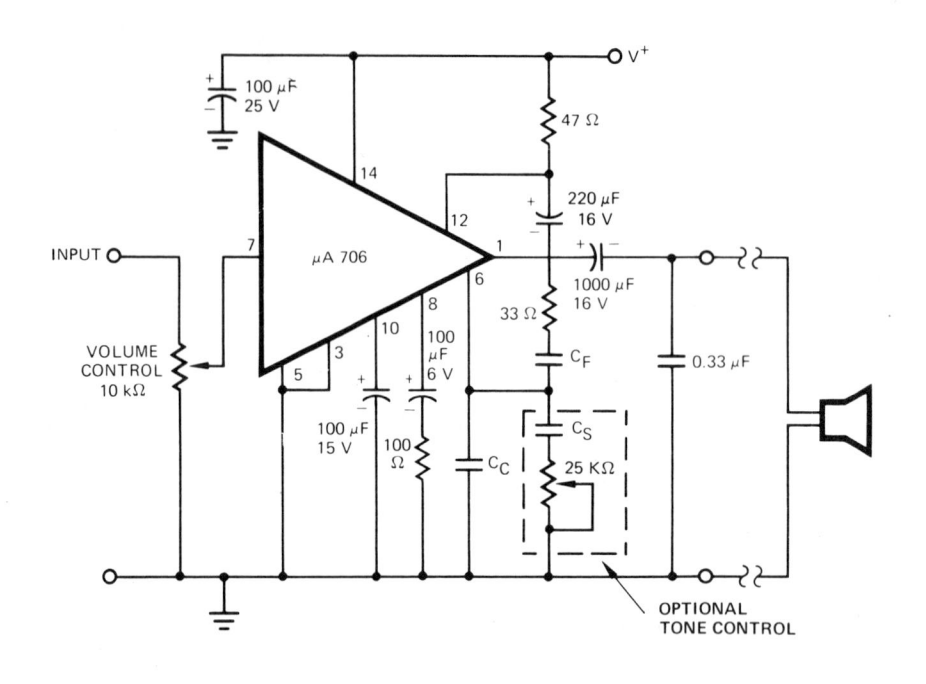

G_V	34 dB		46 dB	
BW	10 kHz	20 kHz	10 kHz	20 kHz
R_B	100 Ω		0 Ω	
C_C	10 nF	6.8 nF	2.7 nF	1.5 nF
C_C	1 nF	470 pF	330 pF	150 pF
C_S	27 nF		5.6 nF	

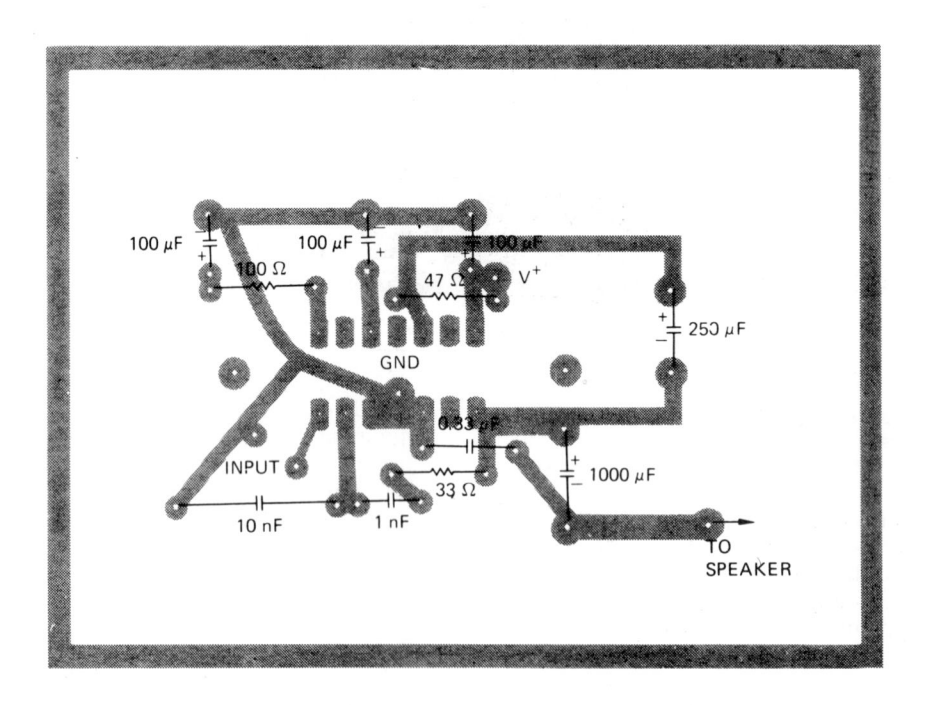

Good-fidelity 5W audio amplifier. Table is a guide to component selection for different gain levels and bandwidths. The value of C_s is selected for a 3 dB gain falloff at 4 kHz. The insert is a printed-board layout for this amplifier.

248 IC SCHEMATIC SOURCEMASTER

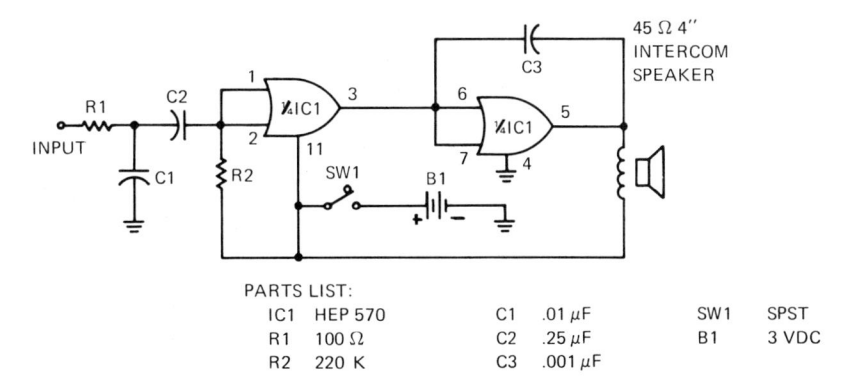

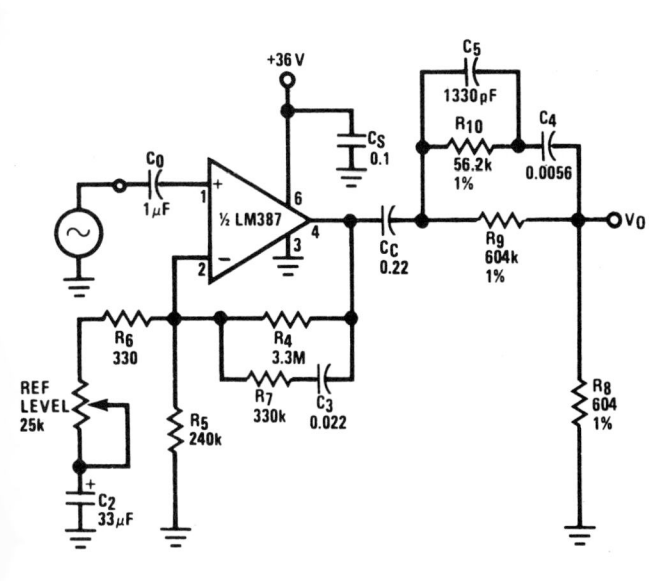

PARTS LIST:
IC1	HEP 570	C1	.01 µF	SW1	SPST
R1	100 Ω	C2	.25 µF	B1	3 VDC
R2	220 K	C3	.001 µF		

This high-gain audio amplifier can be used for microphone preamp or any other class A audio amplification function. Integrated circuit is a digital device being used in a linear application. By replacing the 45Ω speaker with an interstage transformer (50:500Ω), this circuit can be used as power-amplifier driver.

Inverse RIAA response generator. A useful test box to have handy while designing and building phono preamps is one that will yield the opposite of the playback characteristic, that is, an inverse RIAA (or record) characteristic. The circuit is achieved by adding a passive filter to the output of an LM387, which is used as a flat-response adjustable gain block. Gain is adjustable over a range of 24 to 60 dB and is set in accordance with the 0 dB reference gain (1 kHz) of the phono preamp under test. For example, assume the preamp being tested has +34 dB gain at 1 kHz. Connect a 1 kHz generator to the input. The passive filter has a loss of −40 dB at 1 kHz, which is corrected by the LM387 gain; so if a 1 kHz test output level of 1V is desired from a generator input level of 10 mV, then the gain of the LM387 is set at +46 dB (+46 dB − 40 dB + 34 dB = ×100; 10 mV × 100 = 1V). Break frequencies of the filter are determined by

$$f_1 = 50 \text{ Hz} = \frac{1}{2\pi R9 C4}$$

$$f_2 = 500 \text{ Hz} = \frac{1}{2\pi R10 C4}$$

$$f_3 = 2120 \text{ Hz} = \frac{1}{2\pi R10 C5}$$

The R7-C3 network is necessary to reduce the amount of feedback for ac and is effective for all frequencies beyond 20 Hz.

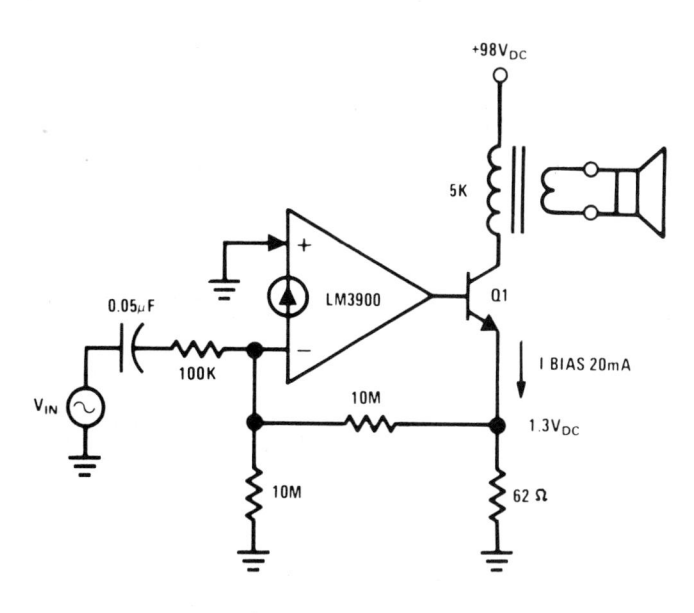

A line-operated audio amplifier can be powered from rectified line voltage. The external high-voltage transistor is biased and controlled by the LM3900. The magnitude of the dc bias voltage, which appears across the emitter resistor of Q1, is controlled by the resistor from the (−) input to ground.

SECTION 6 *Audio Circuits* 249

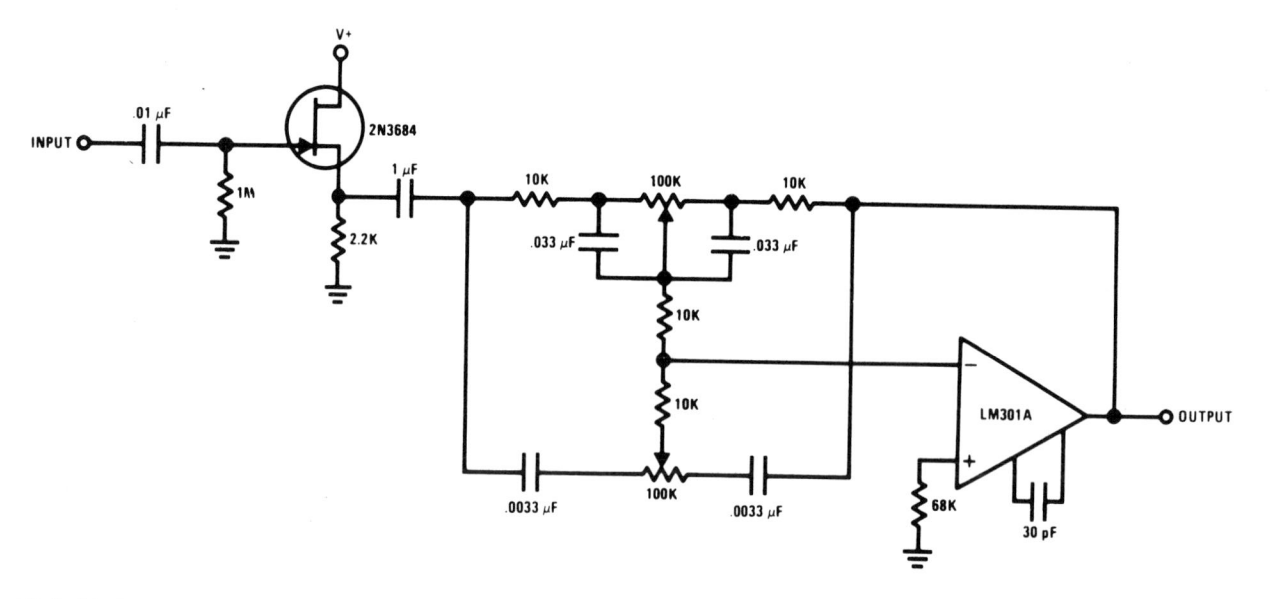

High-fidelity tone control circuit. The 2N3684 JFET provides a high input impedance and low-noise characteristics to buffer an op-amp operated feedback tone control circuit. The upper 100k potentiometer is the bass control; the lower one is the treble.

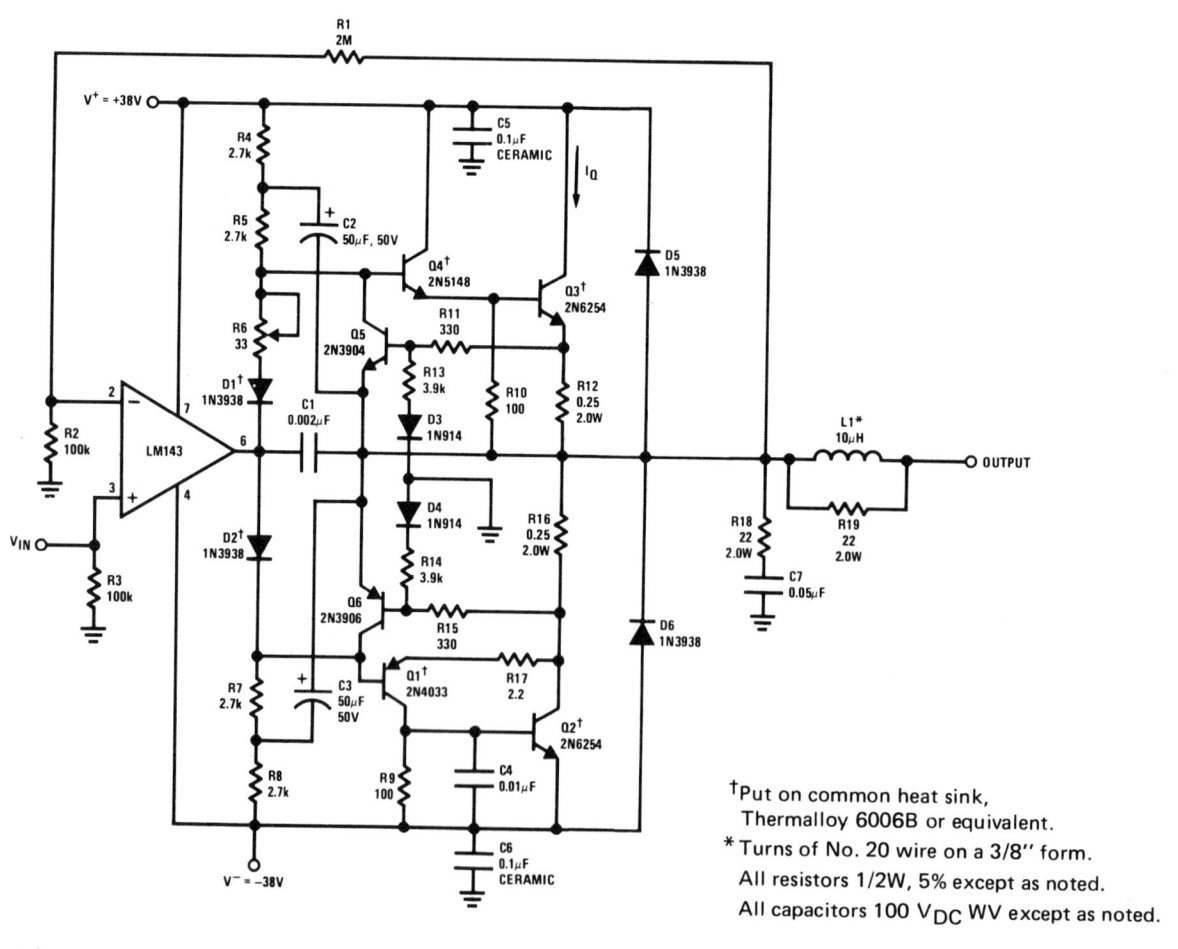

This high-power audio amplifier delivers 90W into 4Ω load or 70W into 8Ω. Circuit features safe-area, short-circuit, and overload protection; harmonic distortion less than 0.1% at 1.0 kHz; and an all-NPN output stage.

The output of the LM143 drives a quasi-complementary output stage made up of Q1, Q2, Q3, and Q4.

†Put on common heat sink, Thermalloy 6006B or equivalent.
*Turns of No. 20 wire on a 3/8" form.
All resistors 1/2W, 5% except as noted.
All capacitors 100 V_{DC} WV except as noted.

250 IC SCHEMATIC SOURCEMASTER

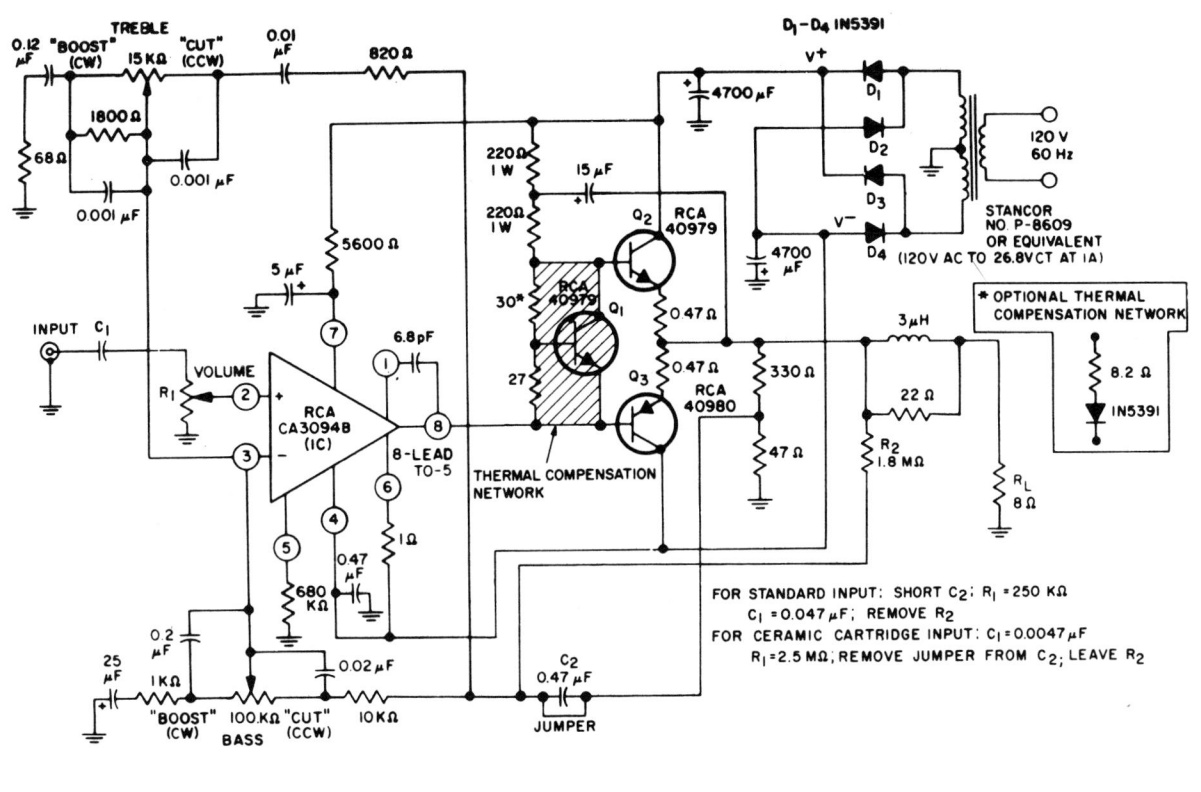

Power Output (8Ω load, Tone Control set at "Flat")		
Music (at 5% THD, regulated supply)	15	W
Continuous (at 0.2% IMD, 60 Hz & 2 kHz mixed in a 4:1 ratio, unregulated supply) See Fig. 8 In ICAN-6048	12	W
Total Harmonic Distoration		
At 1 W, unregulated supply	0.05	%
At 12 W, unregulated supply	0.57	%
Voltage Gain	40	dB
Hum and Noise (Below continuous Power Output)	83	dB
Input Resistance	250	kΩ

This high-fidelity 15W audio amplifier features a true complementary output stage with CA3094 in the driver stage. High-impedance input. Noise is 83 dB down (from rated output). Low-power harmonic distortion is 0.05%, and it's only 0.6% at full-rated output power.

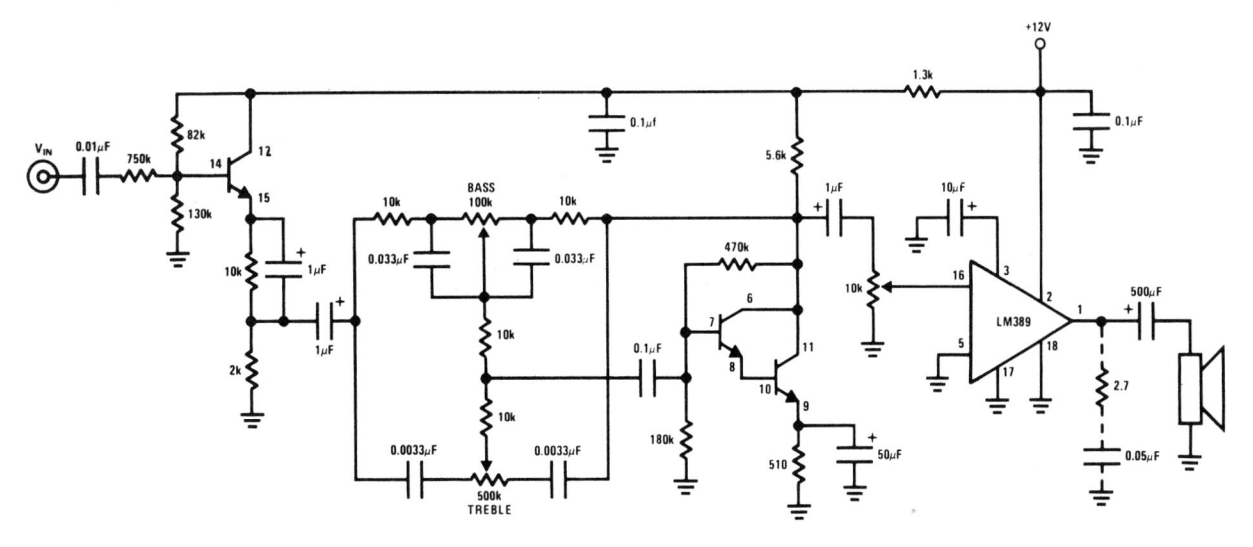

A ceramic phono amplifier with tone controls. For proper frequency responses (particularly at the low end), ceramic cartridges require a high termination impedance. In this low-cost single-IC phono amplifier, one of the LM389 transistors is used as a high-input-impedance emitter follower to provide the required cartridge load. The remaining transistors form a high-gain Darlington pair, which is used as the active element in a low-distortion Baxandall tone control circuit.

SECTION **6** Audio Circuits **251**

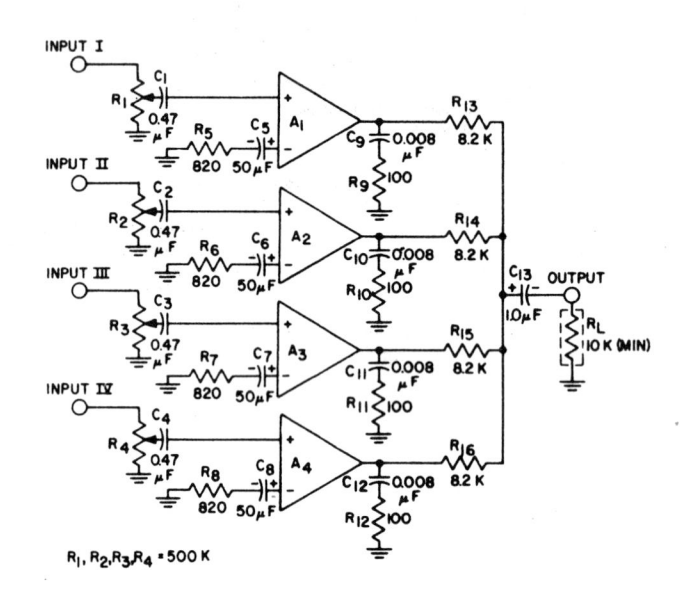

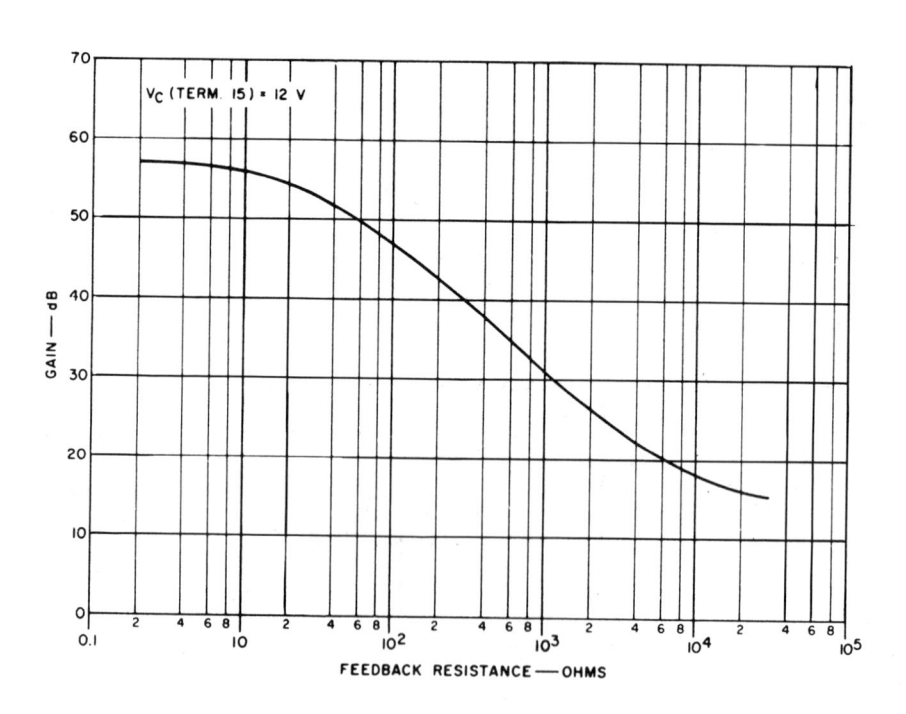

Linear mixer. The gain of any input to the corresponding output is 20 dB for the circuit values shown and a load impedance of 10k or greater. Resistors R5 through R8 program the gain of the system and may be varied to provide more or less gain, depending on the requirements of the application. The curve illustrates the effect of variation in the resistance in the feedback circuit of the CA3048.

Gain as a function of feedback resistance

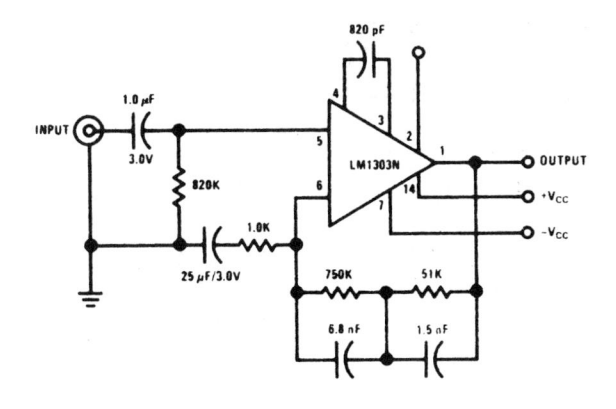

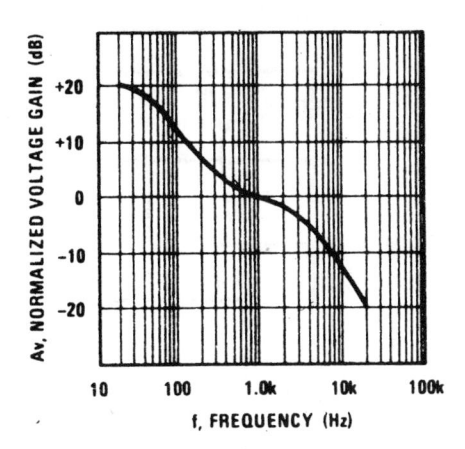

Voltage gain . . . . 34 dB at 1 KHz
Input overload point . . . . 100 mVrms at 1 KHz
Output voltage swing . . . . 5.0 Vrms at 1 KHz and 0.1% THD
Output noise level . . . . Better than 70 dB below 10 mV phono input (input shorted)

This magnetic phono playback preamplifier, RIAA equalized, has large output voltage swing (4V RMS min), high open-loop voltage gain (6000 min), and channel separation of 60 dB min at 10 kHz.

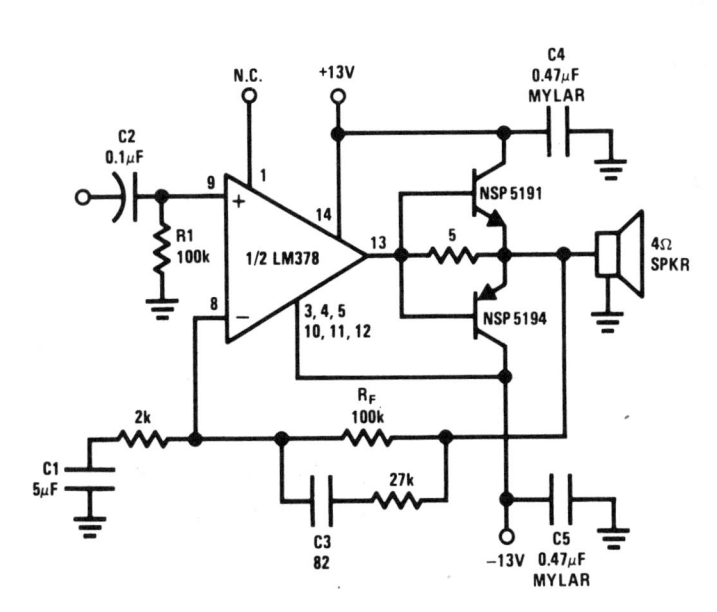

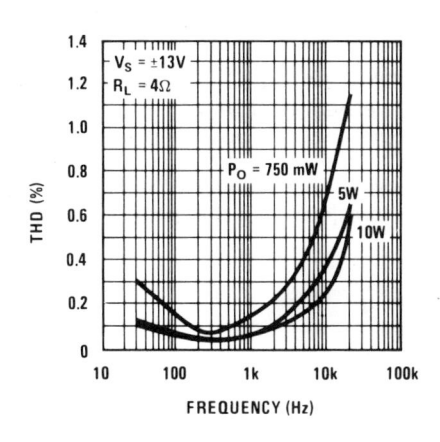

Low-distortion power amplifier achieves about 12W per channel output prior to clipping. Power output is increased because there is no power loss owing to effective series resistance and capacitive reactance of the output coupling capacitor required in the single supply circuit. At power up to 10W per channel, the output is extremely clean, containing less than 0.2% THD midband at 10W. The bandwidth is also improved owing to the absence of the output-coupling capacitor. The frequency response and distortion are plotted for low and high power levels. Note that the input coupling capacitor is still required, even though the input may be ground referenced, in order to isolate and balance the dc input offset resulting from input bias current. Feedback-coupling capacitor C1 maintains dc loop gain at unity to ensure zero dc output voltage and zero dc load current. Capacitors C1 and C2 both contribute to decreasing gain at low frequencies. Either or both may be increased for better low-frequency bandwidth. C3 and the 27k resistor provide increased high-frequency feedback for improved high-frequency distortion characteristics. C4 and C5 are low-inductance Mylar capacitors connected within 2 inches of the IC terminals to ensure high frequency stability. R1 and R_f are made equal to maintain V_{OUT} dc = 0. The output should be within 10 to 20 mV of zero volts dc. The internal bias is unused; pin 1 should be open circuit. When experimenting with this circuit, use the amplifier connected to terminals 8, 9, and 13. If using only the amplifier on terminals 6, 7, and 2, connect terminals 8 and 9 to ground (split supply) to cause the internal bias circuits to disconnect.

SECTION **6** Audio Circuits 253

*L = 7–14 µHy
Qu = 60
f_o = 4.5 MHz

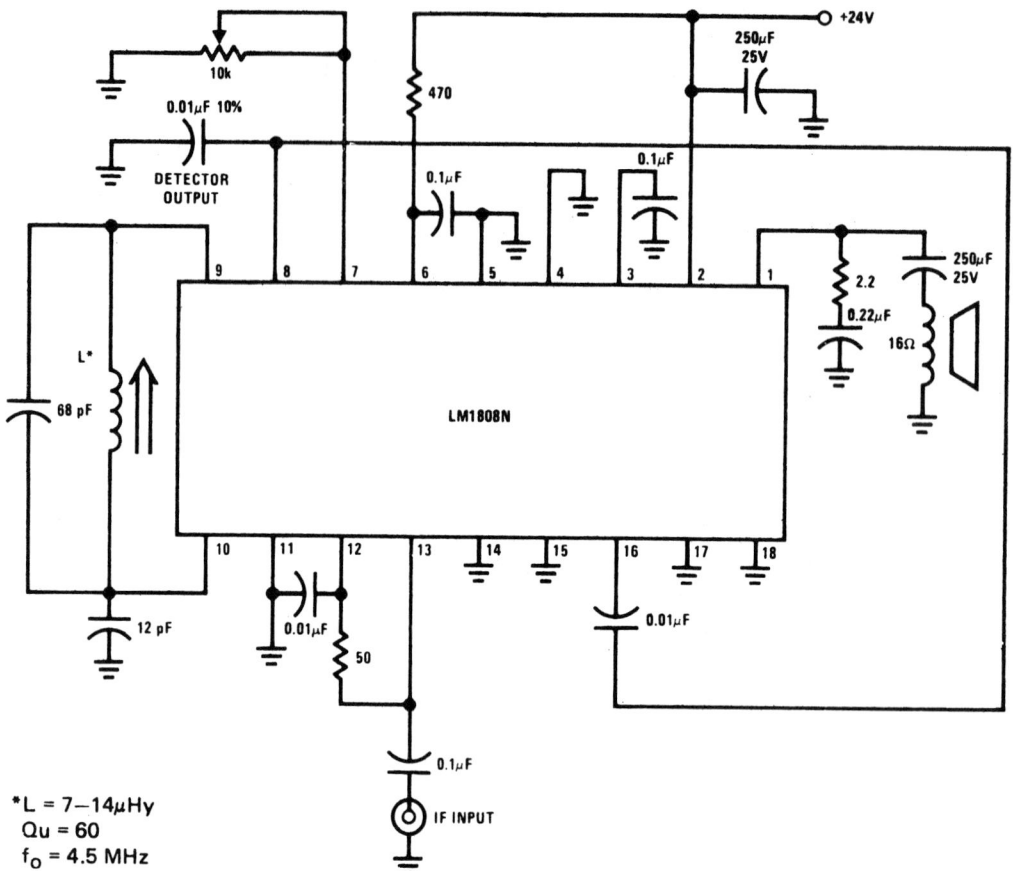

A monolithic 2W sound system designed for television and related applications. The circuit consists of two independent functions: a sound IF and an audio power amplifier. An improved volume control circuit is included so that recovered audio is a linear function of the resistance of the control potentiometer.

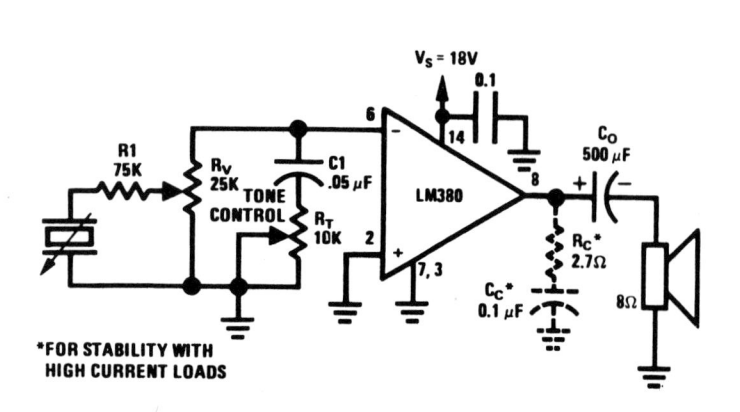

*FOR STABILITY WITH HIGH CURRENT LOADS

This ceramic/crystal cartridge phono amplifier uses LM380 with a voltage-divider volume control and high-frequency rolloff tone control.

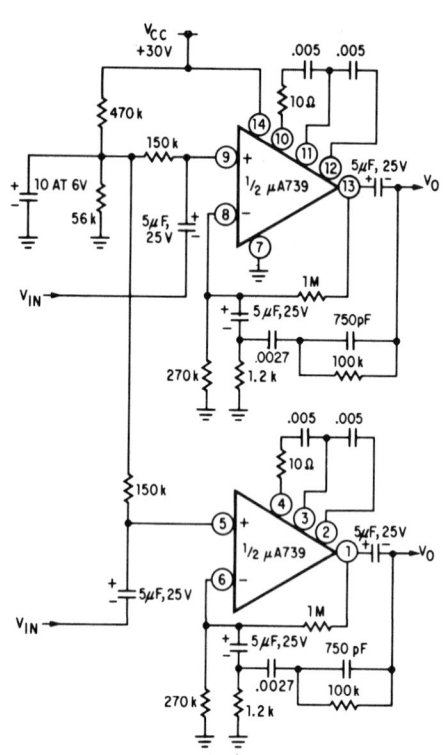

This magnetic-cartridge stereo phono preamp uses a single 30V supply. Total harmonic distortion is less than 0.2% at 50 Hz and less than 0.1% at 20,000 Hz. Input noise is lower than most designs incorporating discrete components—typical performance at 1.2V input noise gives 78 dB signal-to-noise ratio at 1 kHz gain of 40 dB.

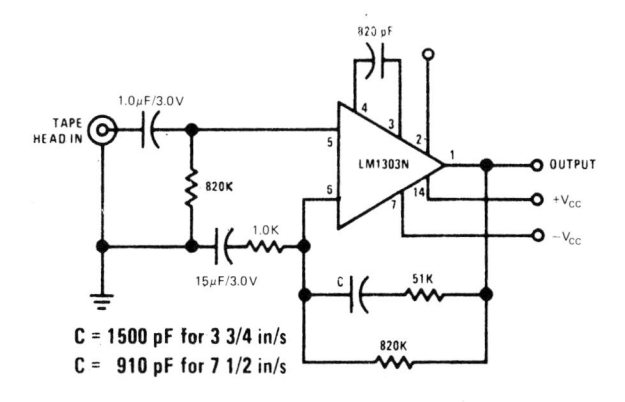

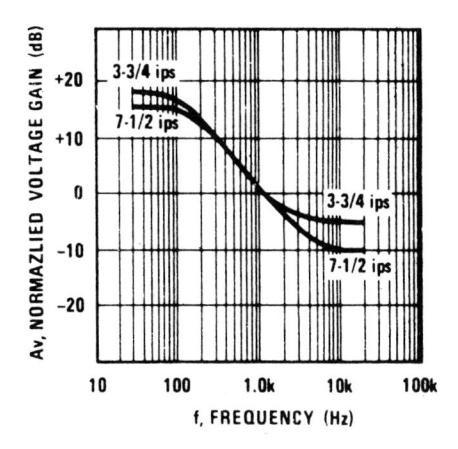

A tape head playback preamplifier with NAB equalization. Voltage gain is 35 dB at 1 kHz. Output voltage swings up to 5V. C = 1500 pF for 3¾ in./sec, 910 pF for 7½ in./sec.

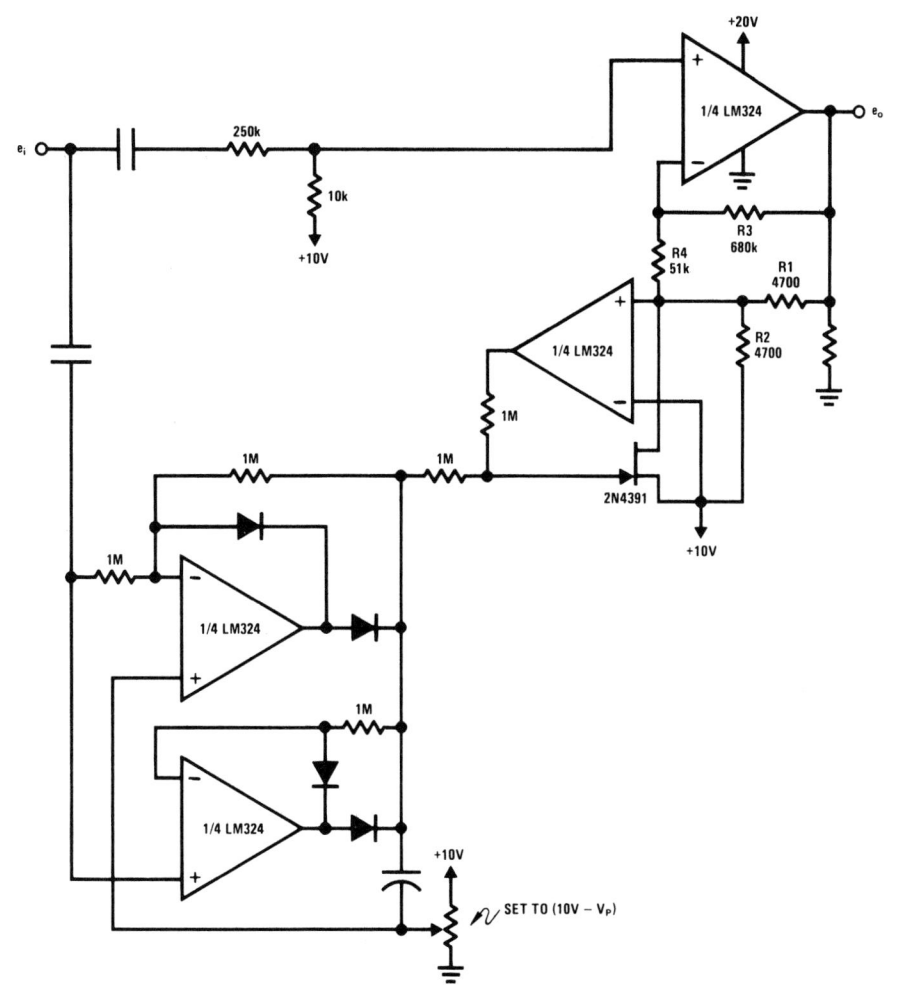

This dynamic range expander for audio signals uses gain-controlled amplifier. The control is linear, the required control range is about 1:4, and the input signal is small for the low-gain condition (when distortion would otherwise be most apparent). The gain-controlled amplifier exhibits a 12 dB variation in gain, being lowest for small signals. The slope of (log) gain in decibels versus (log) signal in decibels is linear. The peak detector is linear down to very small signals, exhibits a fast attack of a millisecond or less and a discharge time constant of about 2 seconds, and operates on the first half-cycle. Resistors R3 and R4 modify the linear control curve to the desired log curve. The input signal is attenuated before amplification to reduce distortion and maintain an overall gain of approximately 0 dB at midrange of expansion. The noise with the LM124 over a 20 kHz bandwidth is a function of signal; but the maximum signal-to-noise ratio is 80 dB.

SECTION **6** Audio Circuits

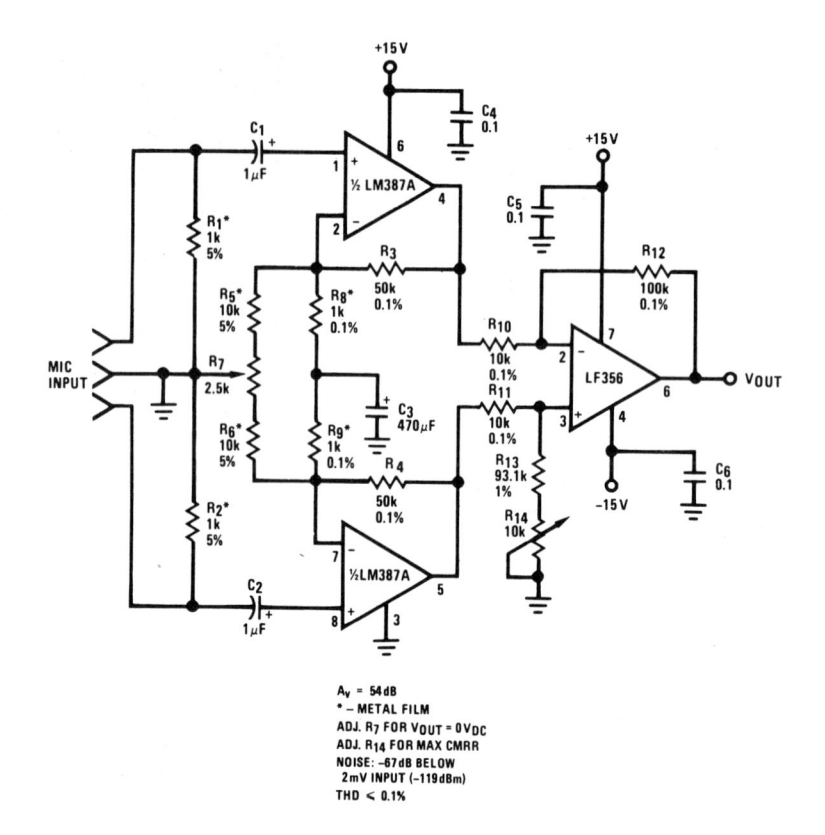

$A_V = 54$ dB
* – METAL FILM
ADJ. R_7 FOR $V_{OUT} = 0 V_{DC}$
ADJ. R_{14} FOR MAX CMRR
NOISE: –67 dB BELOW
  2 mV INPUT (–119 dBm)
THD ≤ 0.1%

Low-noise transformerless balanced-line microphone preamp. An improvement in noise performance is possible by using a LM387A in front of the LF356 (or LF357) as shown. This configuration is known as an instrumentation amplifier, after its main usage in balanced bridge instrumentation applications. In this design, each half of the LM387A is wired as a noninverting amplifier. Resistors R1 and R2 set the input impedance at 2k (balanced). Potentiometer R7 is used to set the output level at zero volts by matching the dc levels of pins 4 and 5 of the LM387A, which allows direct coupling between the stages, thus eliminating the coupling capacitors and the associated matching problem for optimum CMRR. Ac gain resistors R8 and R9 are grounded by the common capacitor, C3, eliminating another capacitor and assuring ac gain match. Close resistor tolerance is necessary around the LM387A in order to preserve common-mode signals appearing at the input. The function of the LM387A is to amplify the low-level signal, adding as little noise as possible.

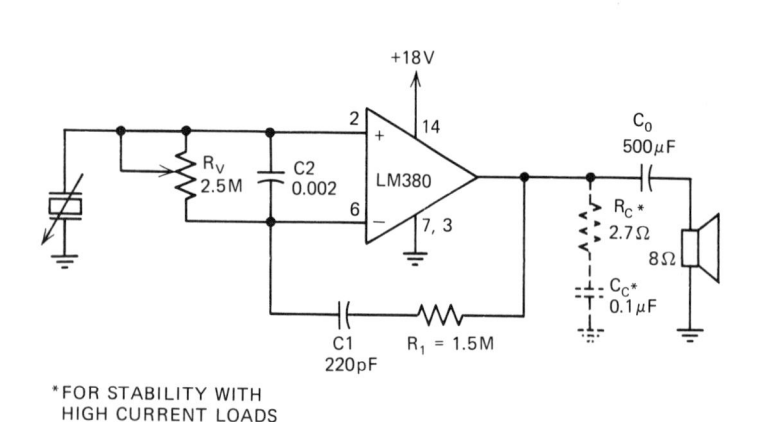

*FOR STABILITY WITH HIGH CURRENT LOADS

RIAA phono amplifier.

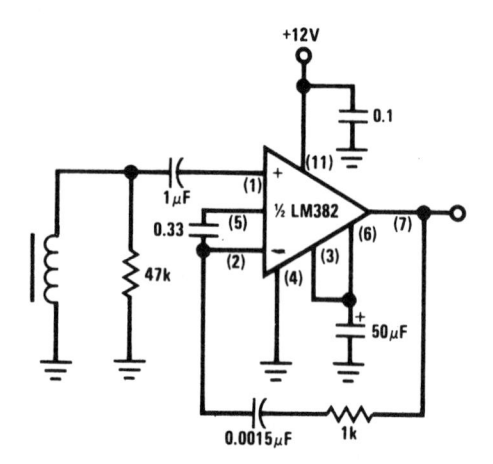

A minimum-parts-count low-noise phono preamp using the LM382. The circuit has been optimized for 12 to 14V. The midband 0 dB reference gain equals 46 dB (200V/V) and cannot easily be altered. For designs requiring either gain or supply voltage changes, the required extra parts make selection of an LM381 or LM387 more appropriate.

256 IC SCHEMATIC SOURCEMASTER

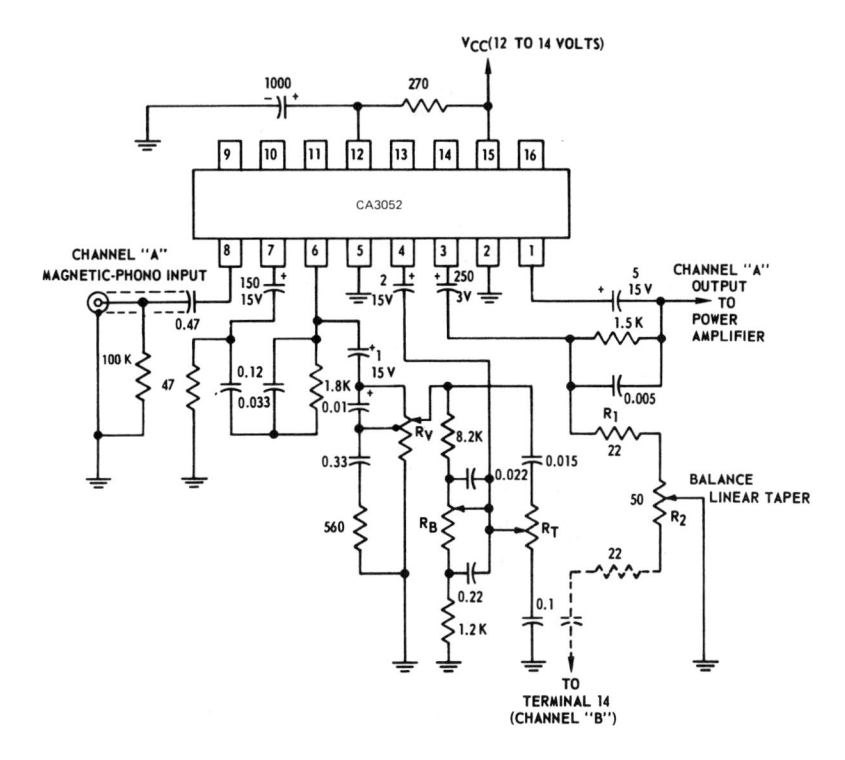

Magnetic phono preamplifier using CA3052. R1 and R2 resistor values are selected for a sensitivity of 3 mV input at 1 kHz. R_V, volume control potentiometer is 15,000 Ω tapped at 6000 Ω with audio taper. R_T and R_B are 25k tone controls.

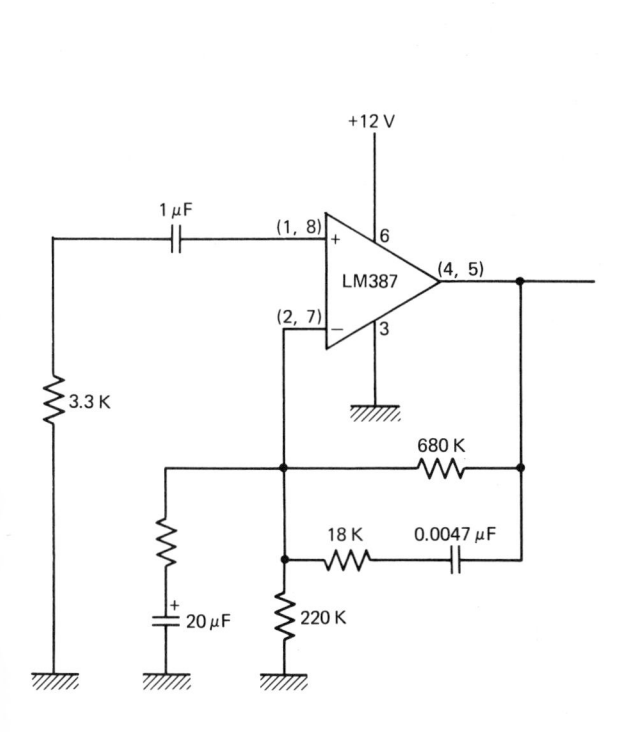

NAB tape circuit.

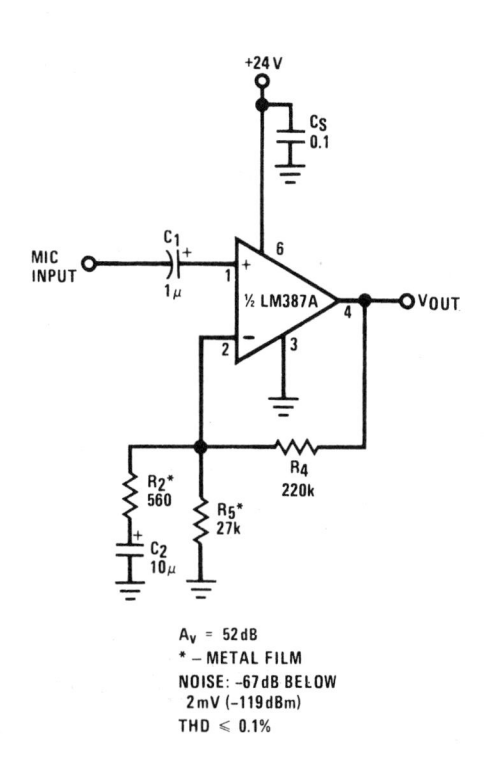

Low-impedance mike preamp. The noise degradation referenced to the LM381A is only +2 dB, making it a desirable alternative for designs where space and cost are dominant factors. Biasing and gain resistors are similar to LM381A.

SECTION **6** Audio Circuits **257**

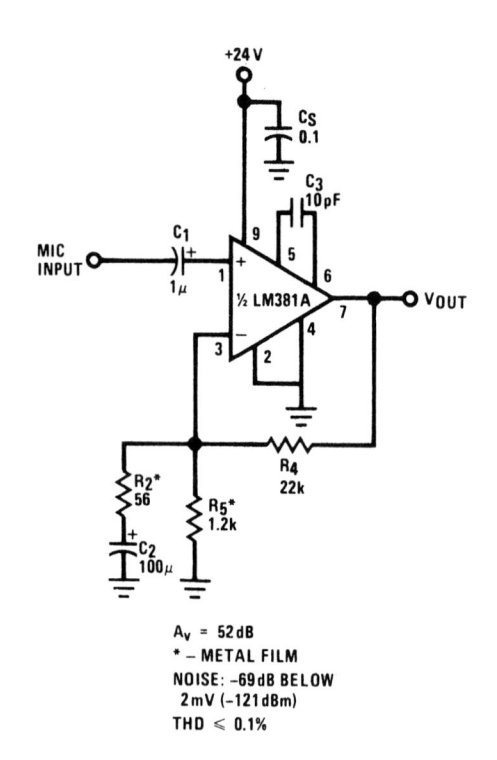

Microphone preamps. Microphones fall into two groups: high impedance (20K), high output (~200 mV); and low impedance (~200 Ω), low output (~2 mV). The first category places no special requirements upon the preamp; amplification is done simply and effectively with the standard noninverting or inverting amplifier configurations. The frequency response is reasonably flat, and no equalization is necessary. Hum and noise requirements of the amplifier are minimal because of the large input levels. So, if everything is so easy, where is the hook? It surfaces with regard to hum and noise pickup of the microphone itself. Being a high-impedance source, these mikes are very susceptible to stray magnetic field pickup (for example, 60 Hz), and their use must be restricted to short distances (typically less than 10 feet of cable length). Because of this problem, high-impedance mikes are rarely used.

Low-impedance microphones also have a flat frequency response, requiring no special equalization in the preamp section. Their low output levels do, however, impose rather stringent noise requirements upon the preamp. For a signal-to-noise ratio of 65 dB with a 2 mV input signal, the total equivalent input noise (EIN) of the preamp must be 1.12 μV (10–10,000 Hz). National's line of low-noise dual preamps with their guaranteed EIN of 0.7 μV (LM381A) and 0.9 μV (LM387A) make excellent mike preamps, giving at least 67 dB S/N (LM387A) performance.

There are two types of low-impedance mikes: unbalanced two-wire output, one of which is ground, and balanced three-wire output, two signal and one ground. Balanced mikes are used most often, since the three-wire system facilitates minimizing hum and noise pickup by using differential input schemes. This takes the form of a transformer with a center-tapped primary (grounded) or use of a differential op-amp.

Low-impedance, transformerless unbalanced-mike preamps, with noise performance −69 dB below a 2 mV input reference point. Resistors R3 and R5 provide negative input bias current and establish the dc output level at one-half supply. Gain is set by the ratio of R4 to R2, while C2 establishes the low-frequency −3 dB corner. High-frequency rolloff is accomplished with C3. Capacitor C1 is made large to reduce the effects of 1/f noise currents at low frequencies.

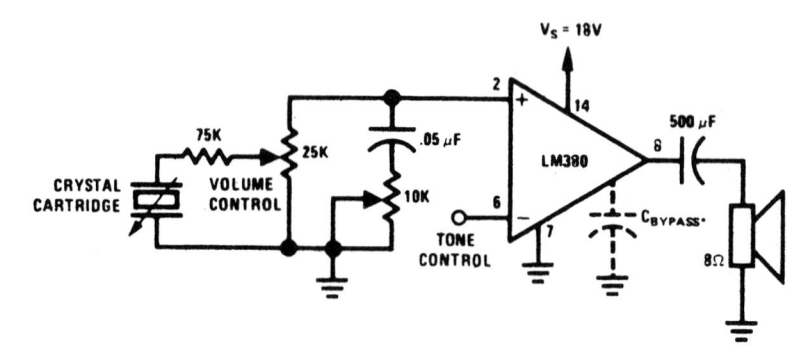

A phono amplifier with 34 dB gain. Input impedance is about 35k.

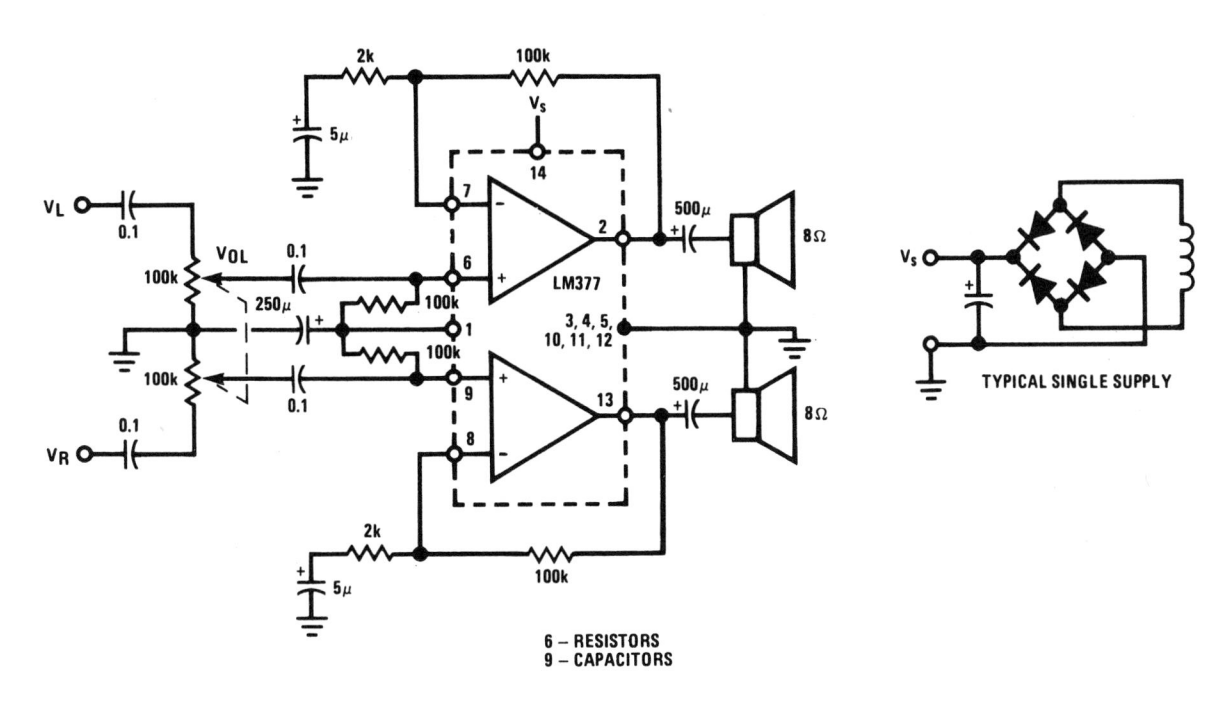

A noninverting amplifier using single supply.

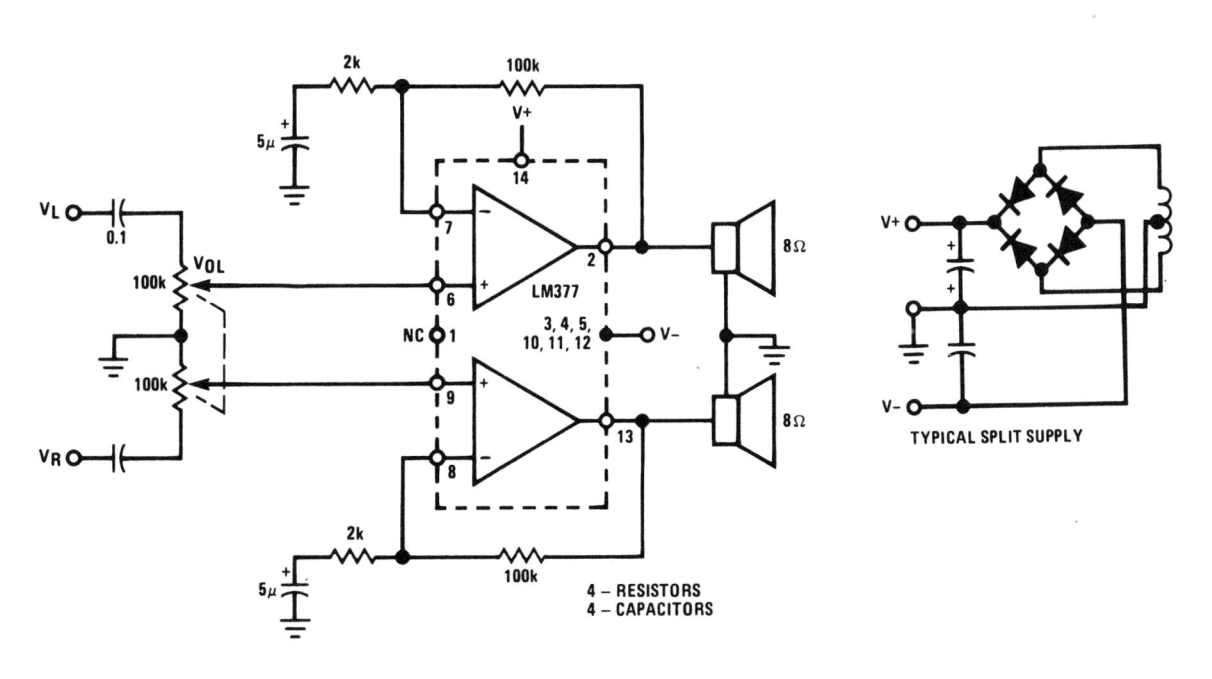

A noninverting amplifier using split supply.

SECTION **6** *Audio Circuits*

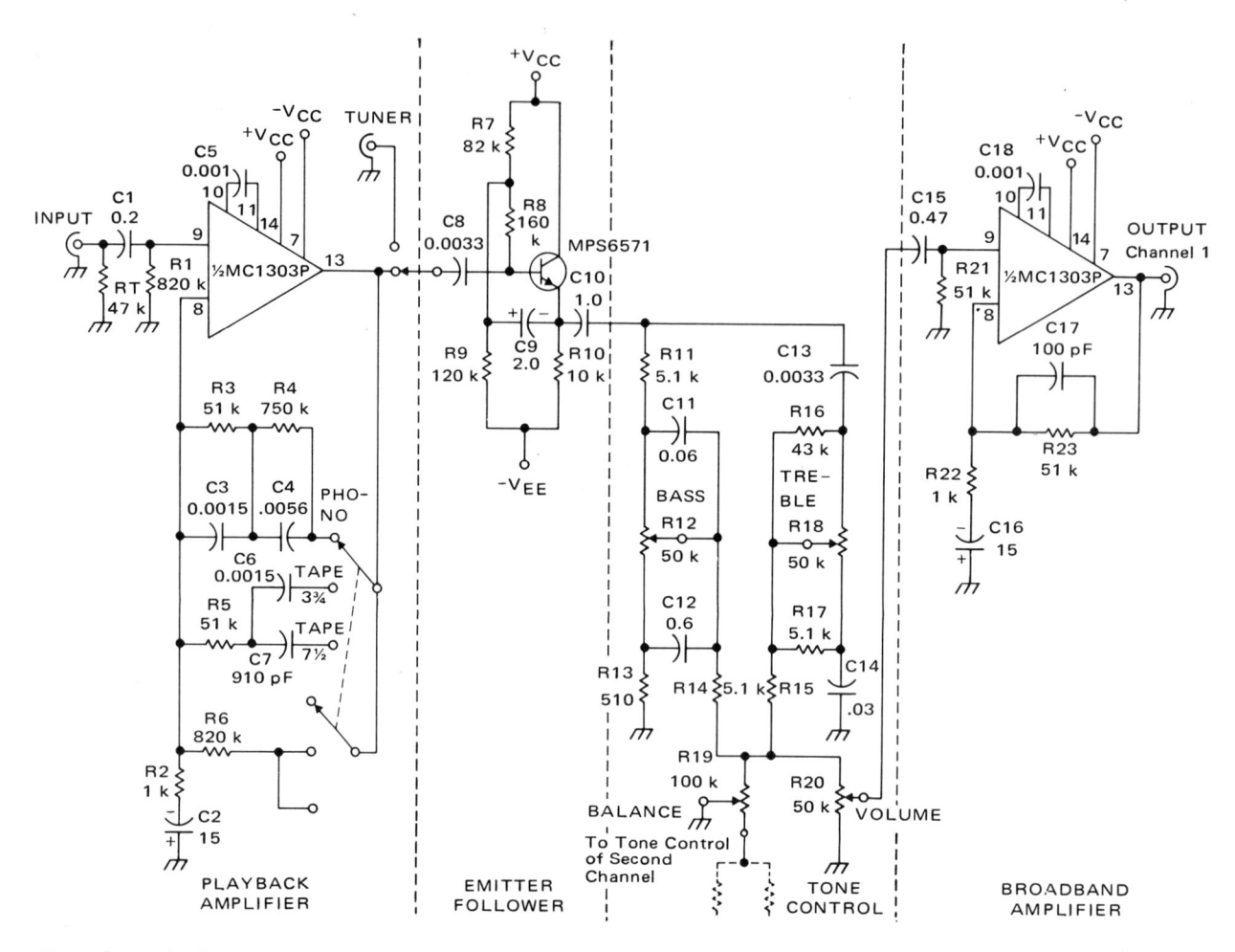

One channel of a stereo preamplifier designed with Motorola MC1303P. A single MC1303P is used for both playback (compensated) stages, and another single unit is used for the broadband (flat gain) stages in the stereo preamplifier. Each channel of the amplifier has a differential input amplifier followed by a second differential stage with single-ended output and two emitter-follower stages.

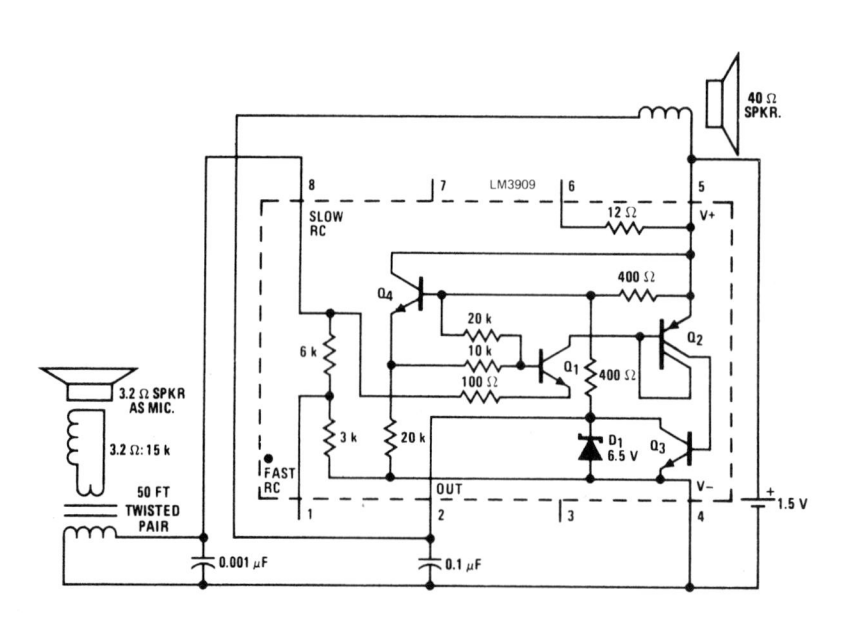

This mini PA system uses single IC with enough sensitivity to bring in speaker input signal from 50 feet away.

260 IC SCHEMATIC SOURCEMASTER

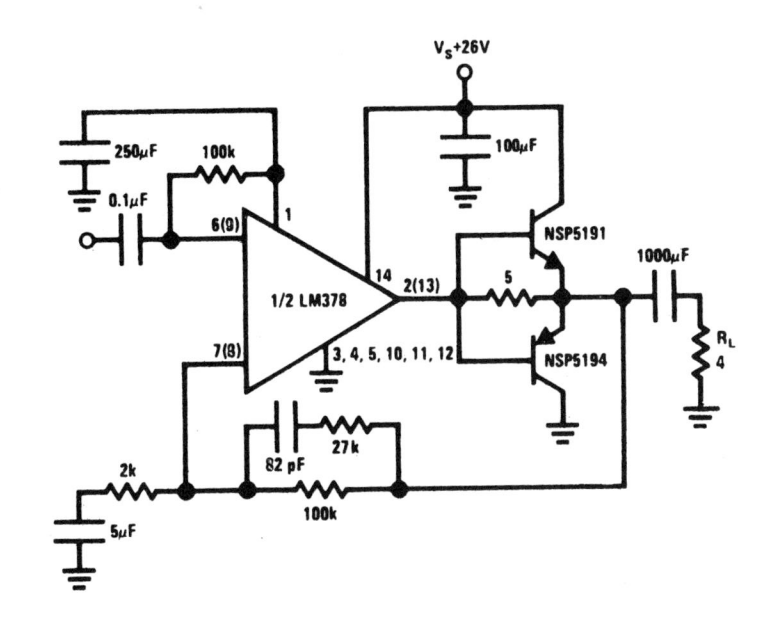

One channel of a 30W stereo amplifier.

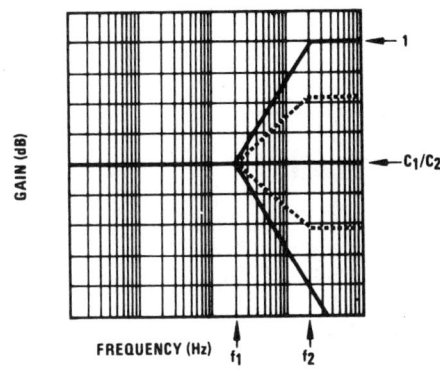

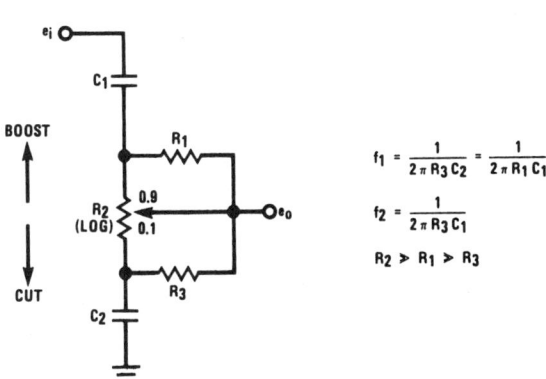

Passive treble control, in which the amount of boost or cut is set by the following ratios: R3/R1 = C1/C2 = treble boost or cut amount, assuming R2 ≫ R1 ≫ R3. Treble turnover frequency f_1 occurs when the reactance of C1 equals R1 and the reactance of C2 equals R3:

$$C1 = \frac{1}{2\pi f_1 R1}$$

$$C2 = \frac{1}{2\pi f_1 R3}$$

The amount of available boost is reached at frequency f_2 and is determined when the reactance of C1 equals R3.

$$f_2 = \frac{1}{2\pi R3 C1}$$

$$f_1 = \frac{1}{2\pi R_3 C_2} = \frac{1}{2\pi R_1 C_1}$$

$$f_2 = \frac{1}{2\pi R_3 C_1}$$

$$R_2 \gg R_1 \gg R_3$$

$$f_1 = \frac{1}{2\pi R_2 C_2}$$

$$f_2 = \frac{1}{2\pi R_2 C_1}$$

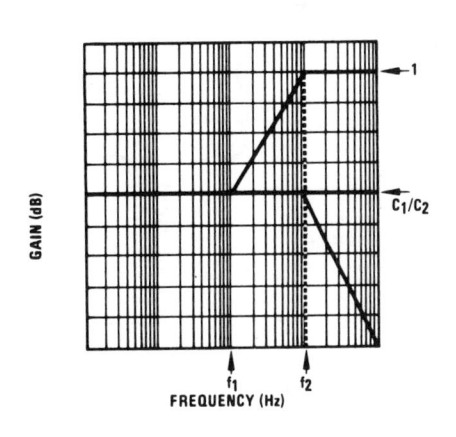

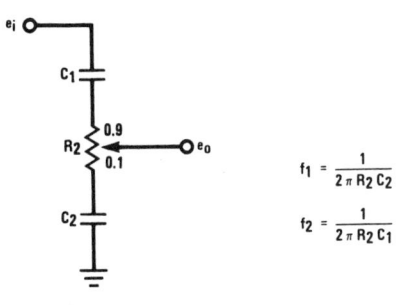

Minimum-parts treble tone control.

SECTION **6** Audio Circuits **261**

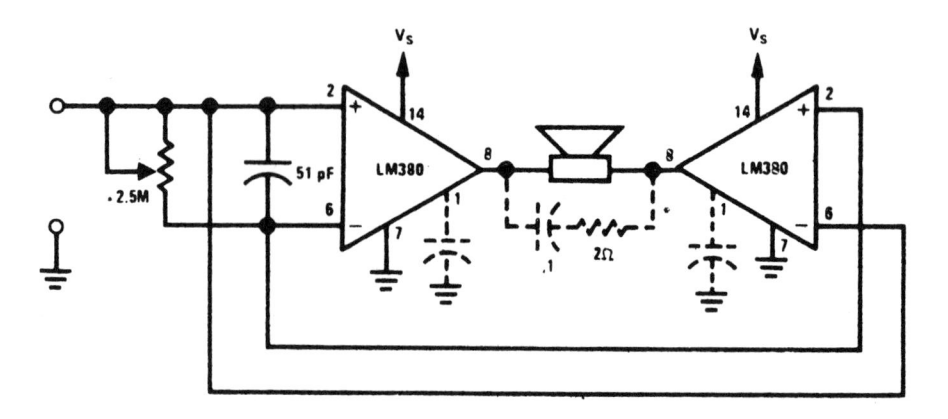

A bridge amplifier with short-proof output. Gain is fixed at 34 dB per stage.

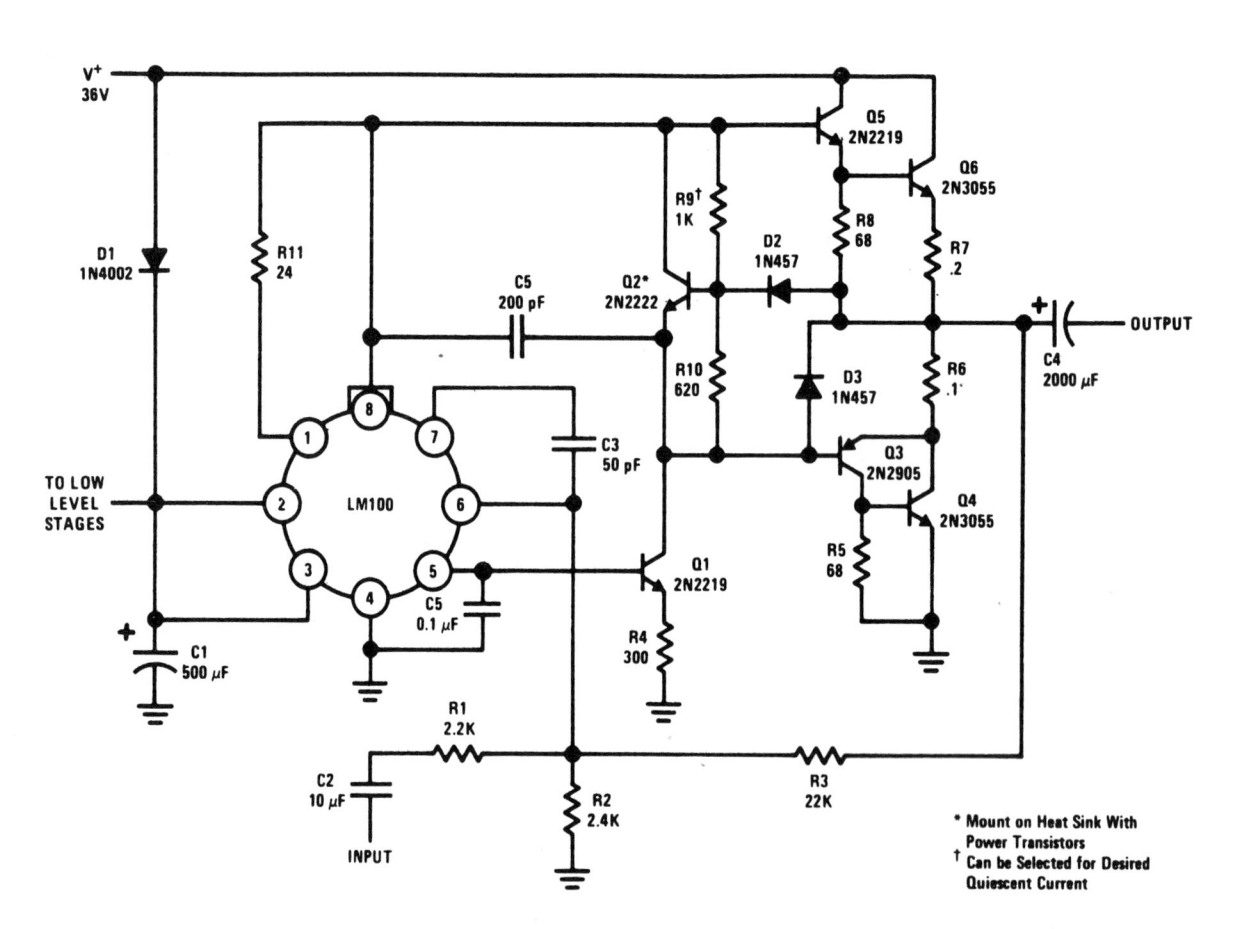

Power amplifier with current limiting. The LM100 is used as a high-gain amplifier and connected to a quasi-complementary power output stage. Feedback around the entire circuit stabilizes the gain and reduces distortion. In addition, the regulation characteristics of the LM100 are used to stabilize the quiescent output voltage and minimize ripple feedthrough from the power supply.

The LM100 drives the output transistors, Q5 and Q6, for positive-going output signals, while Q1, operating as a current source from the 1.8V on the reference terminal of the LM100, supplies base drive to Q3 and Q4 for negative-going signals. Q2 eliminates the dead zone of the class B output stage, and it is bypassed by C5 to present a lower driving impedance to Q3 at high frequencies. The voltage drop across Q2 will be a multiple of its emitter–base voltage, determined by R9 and R10. These resistors can therefore be selected to give the desired quiescent current in Q4 and Q6. It is important that Q2 be mounted on the heatsink with the output and driver transistors to prevent thermal runaway.

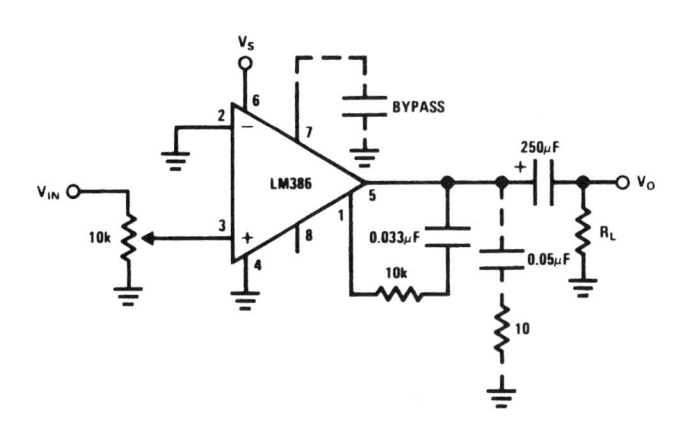

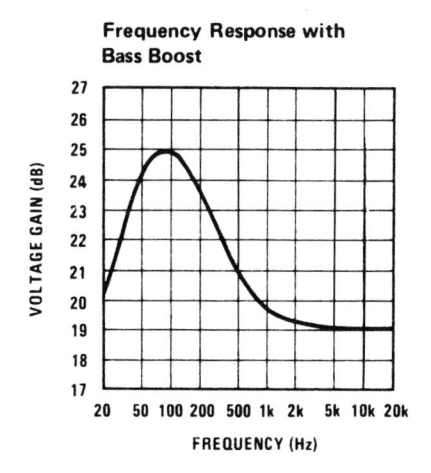

An amplifier with bass boost.

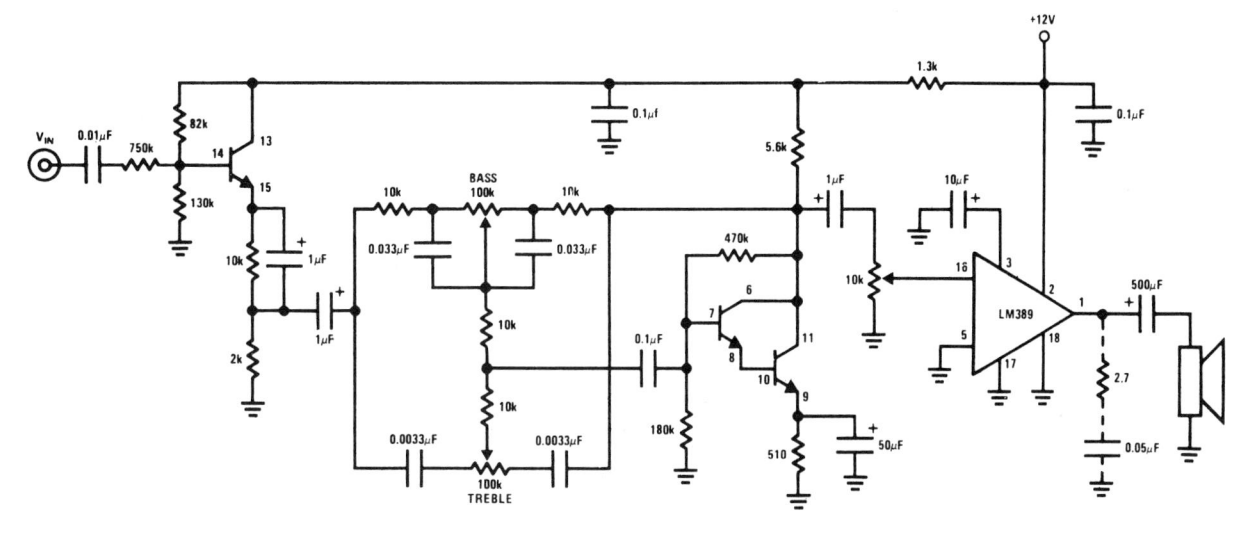

A ceramic phono amplifier with tone controls. The transistors are built into the IC package.

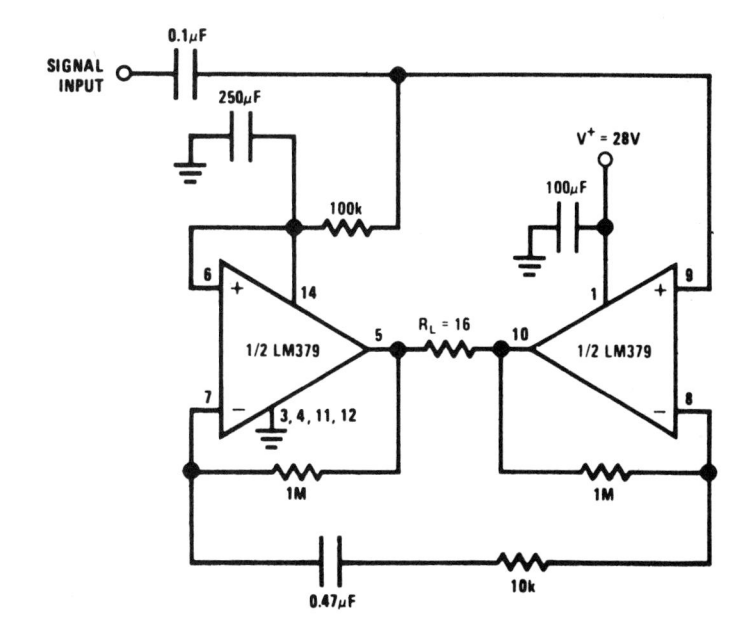

12W bridge amplifier.

SECTION **6** *Audio Circuits* 263

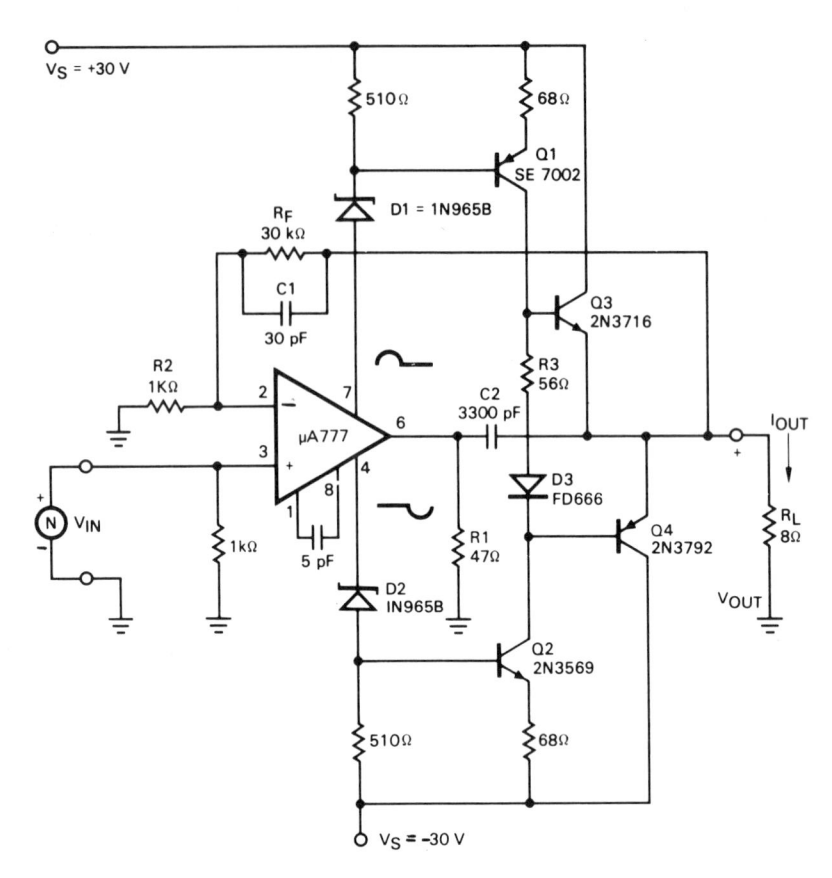

This power amplifier drives ampere-level currents at supply voltages of ±30V or more. Amplifier response is flat from dc to about 100 kHz, with distortion values ranging from 0.05% to 0.1% at 1 kHz, depending on the output current required. Applications include servo drivers, audio amplifiers, and many control circuits where flat response and medium power characteristics are an advantage.

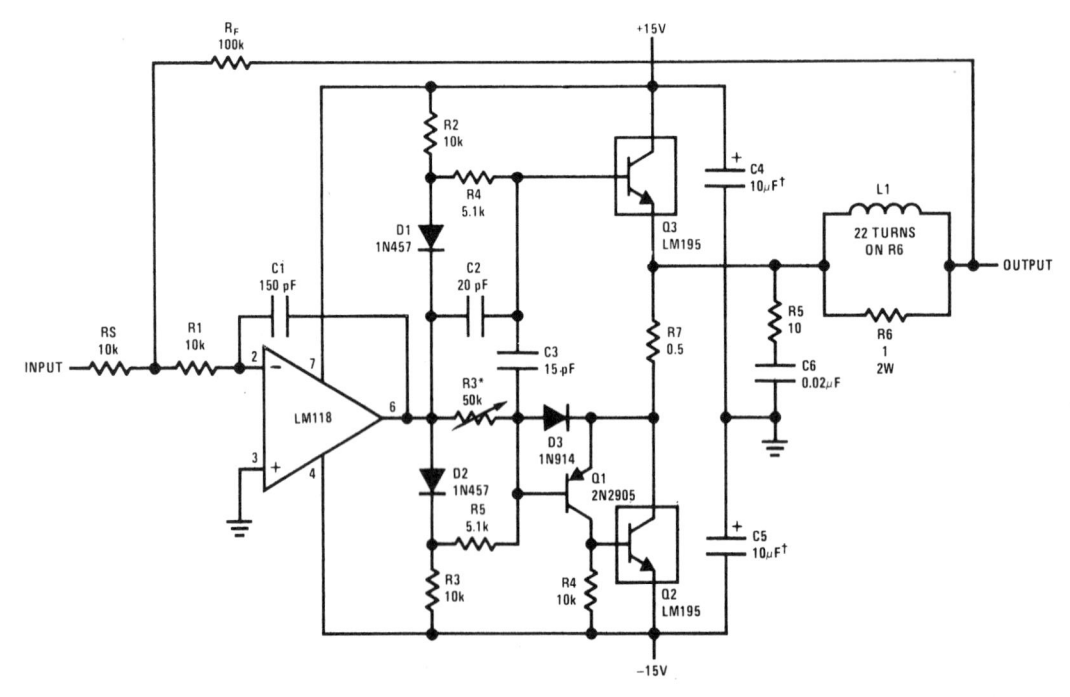

Power op-amp with a quasi-complementary output stage. Q1 and Q2 form the equivalent of a power PNP. The circuit is simply an op-amp with a power output stage. As shown, the circuit is stable for almost any load. Better bandwidth can be obtained by decreasing C1 to 15 pF (to obtain 150 kHz full output response), but capacitive loads can cause oscillation. If the quasi-complementary loop oscillates, collector-to-base capacitance on Q1 will stabilize it. Adjust R3 for 50 mA quiescent current. C5 should be solid tantalum.

264   IC SCHEMATIC SOURCEMASTER

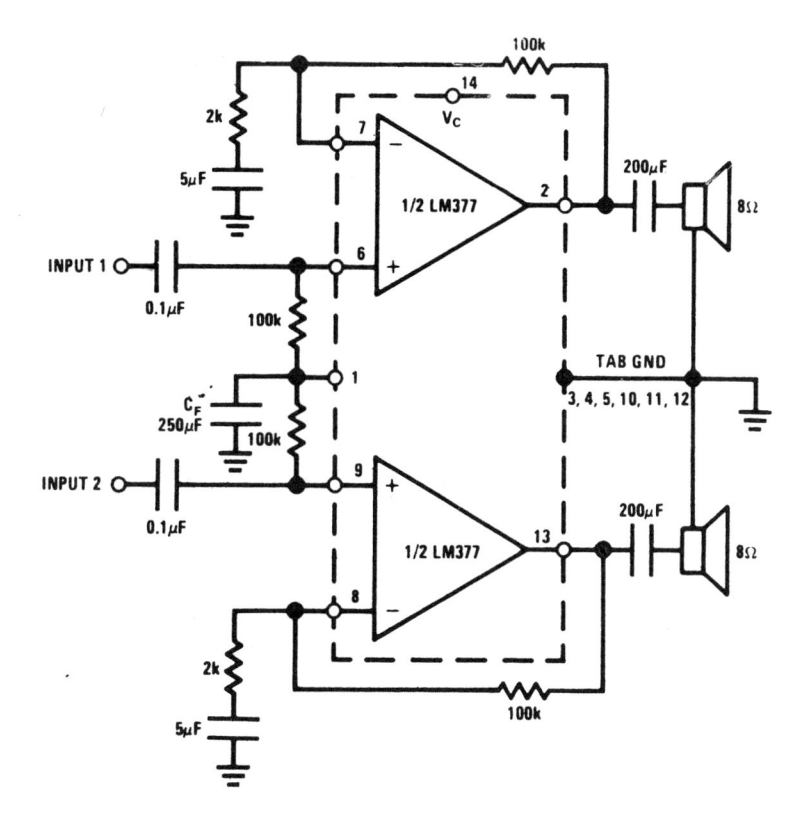

A simple stereo amplifier delivers 2W per channel into 8 or 16Ω loads. The amplifier is designed to operate with a minimum of external components and contains an internal bias regulator to bias each amplifier. Device overload protection consists of both internal current limit and thermal shutdown.

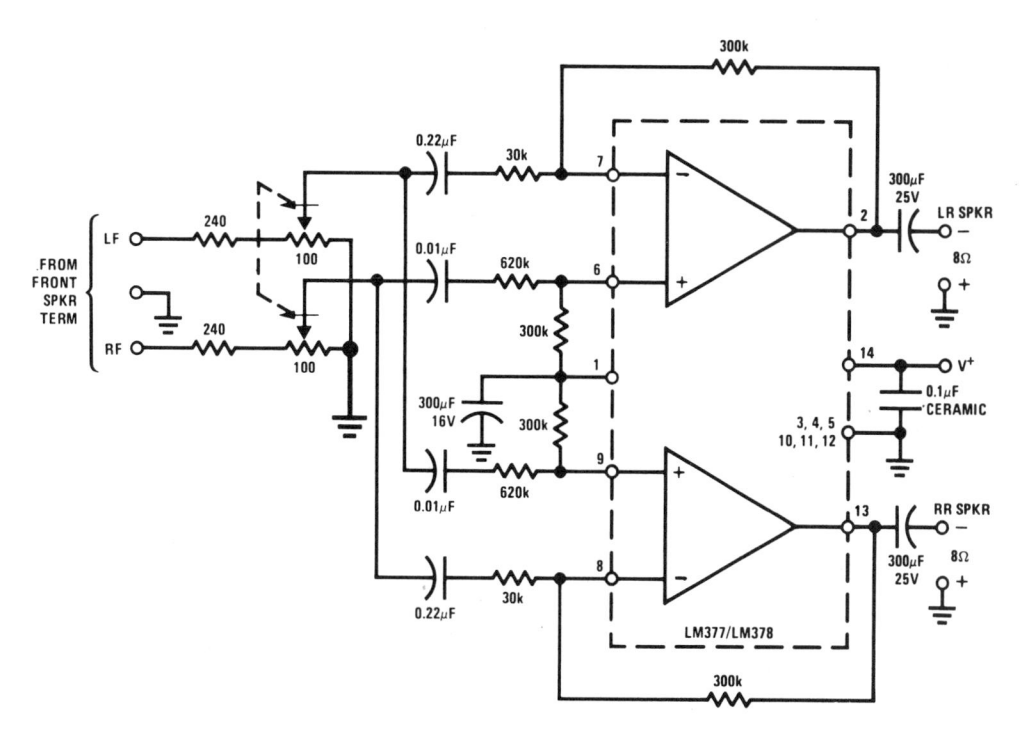

This rear-channel ambience amplifier can be added to an existing stereo system to extract a difference signal (R − L or L − R) which, when combined with some direct signal (R or L), adds some fullness or "concert hall" realism to reproduction of recorded music. Very little power is required at the rear channels; hence, an LM377 will suffice for most "ambience" applications. The inputs are merely connected to the existing speaker output terminals of a stereo set, and two more speakers are connected to the ambience circuit outputs. Note that the rear speakers should be connected in opposite phase to those of the front speakers, as indicated by the +/− signs.

SECTION **6** Audio Circuits 265

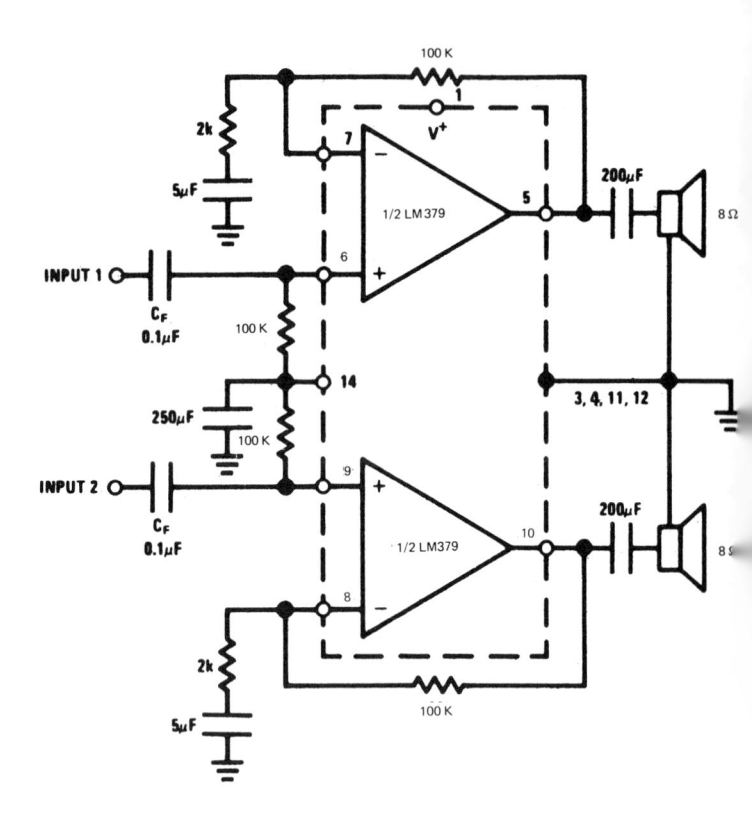

A simple stereo amplifier delivers 6W per channel into 8Ω loads.

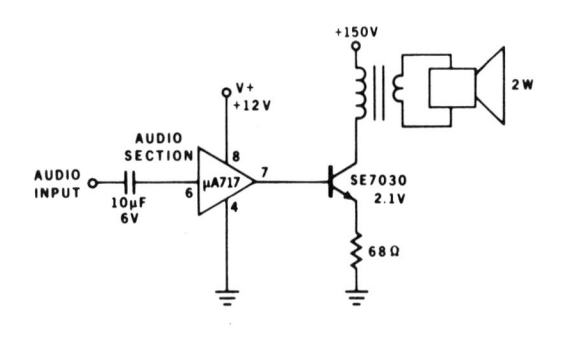

High-voltage audio output stage.

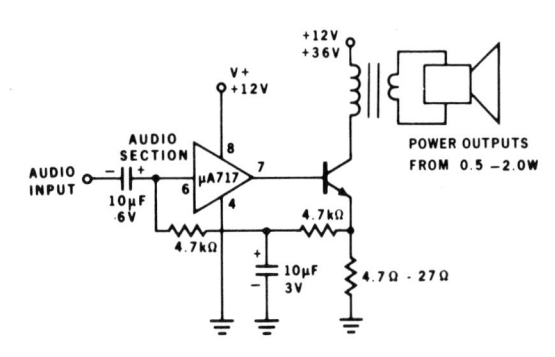

Low-voltage audio output stage.

*FOR STABILITY WITH HIGH CURRENT LOADS
**AUDIO TAPE POTENTIOMETER (10% OF R_T AT 50% ROTATION)

Simple audio output stages. The high-voltage scheme permits the most economical operation and can provide 2W of power to the loudspeaker. The low-voltage scheme employs a dc feedback circuit to reestablish the output dc voltage from the microcircuit, allowing more of the supply voltage to be used to provide audio power with less voltage across the emitter resistor of the output transistor.

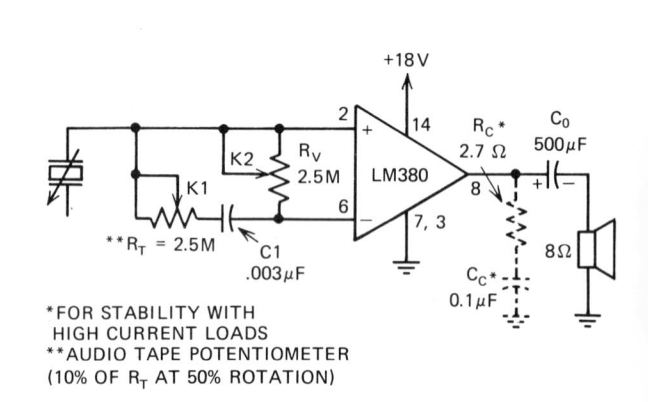

This phono amplifier incorporates common-mode volume and tone control. This circuit has a distinct advantage when transducers of high source impedance are used, in that the full input impedance of the amplifier is realized. It also has an advantage with transducers of low source impedance, since the signal attenuation of the input voltage divider is eliminated.

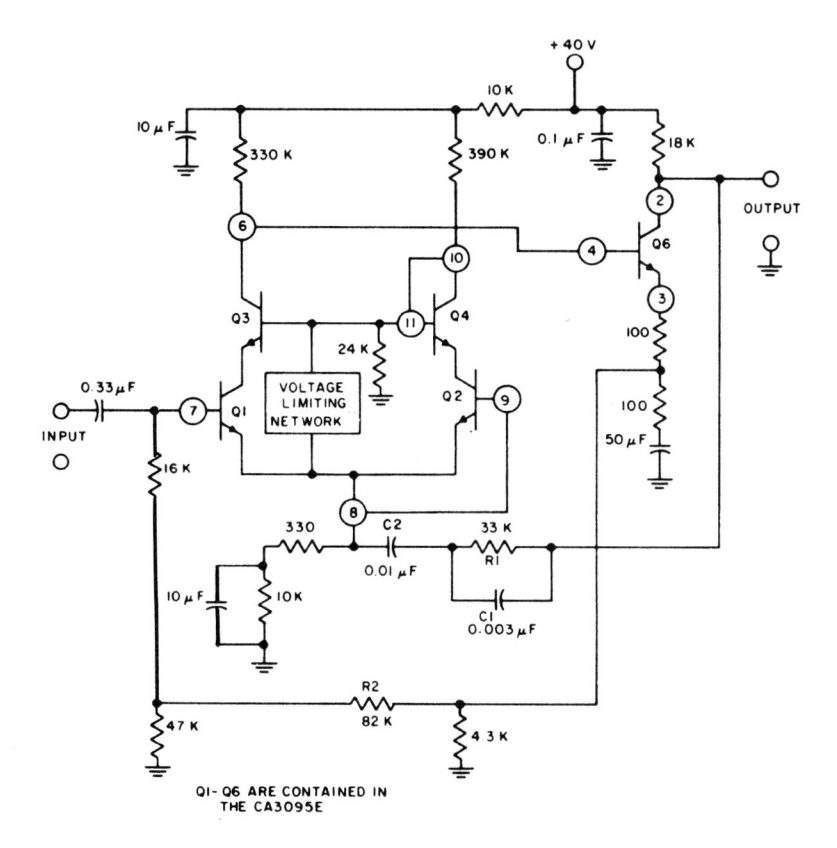

A preamplifier equalized for RIAA standards applicable to magnetic phono cartridges. Transistors Q1 and Q3 are cascode connected as the input stage, and transistor Q6 is connected as a common-emitter postamplifier. Transistors Q2 and Q4 are nonconductive because the emitter–base junction in Q2 and the base-collector junction in Q4 are shunted by external wiring. Equalization for the RIAA phono frequency-response characteristics is provided by the R1-C1 network connected in the ac feedback path.

Dc feedback stabilization is provided by the path through resistor R2. The amplifier has an overall gain of about 40 dB at 1 Hz and can deliver output voltages in the order of 25V peak-to-peak. A voltage-limiting network is built into the superbeta IC. To use it, just connect a resistor from the positive supply of the differential amplifier to pin 11; the value should be such that current through the resistor is about half that measured at pin 8 under quiescent conditions.

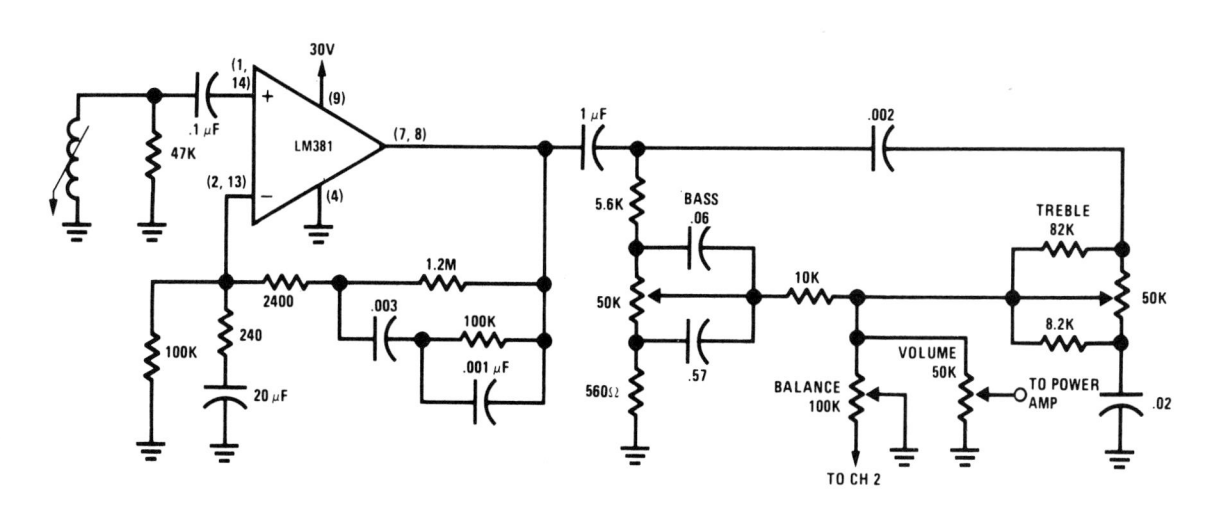

Single channel of complete **phono preamp** (RIAA) contains bass, treble, and balance controls plus volume (log taper).

SECTION **6** Audio Circuits 267

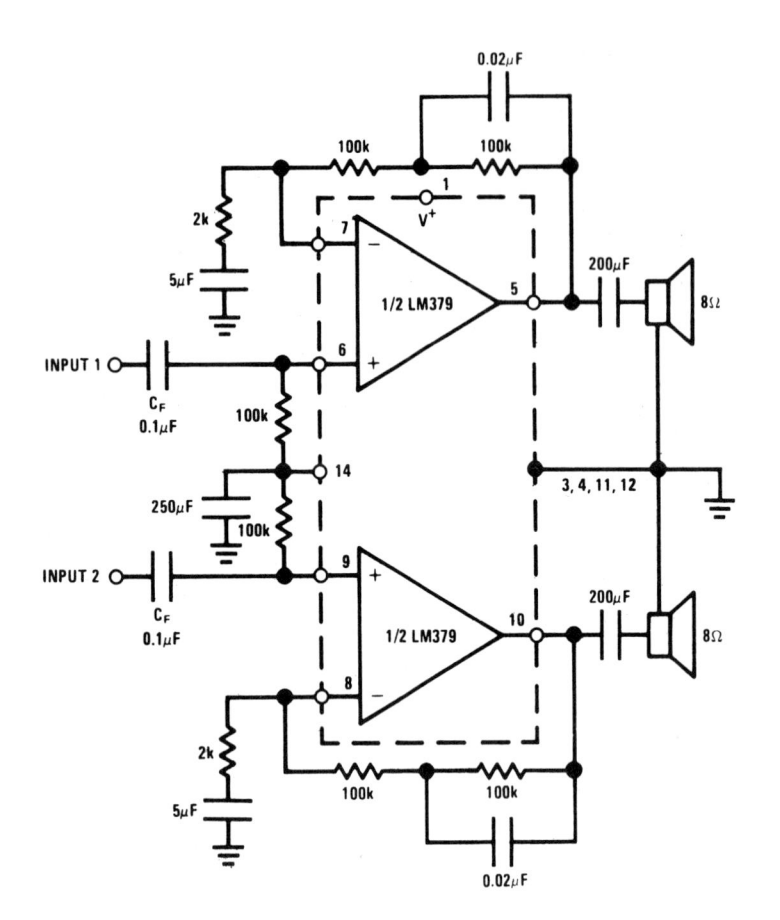

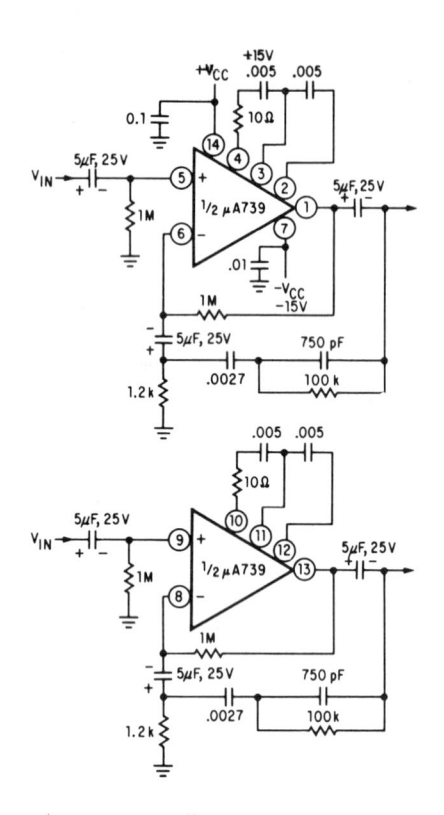

Simple stereo amplifier with bass boost.

Stereo preamp for magnetic phono cartridge. Gain and playback equalization are fixed by the feedback components. This circuit is designed for use with a split supply (+15V and −15V).

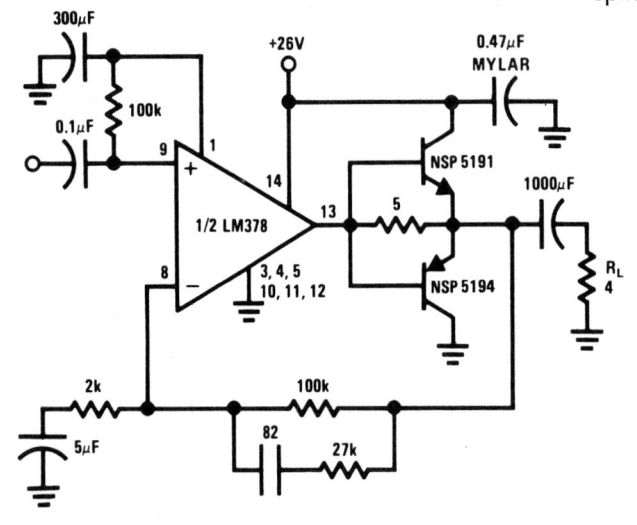

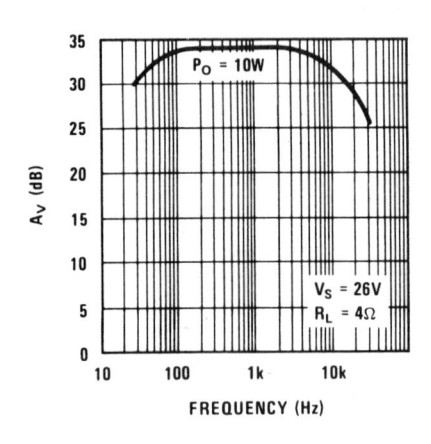

A simple audio booster allows power output of 10W per channel when driven from the LM378. The circuit is exceptionally simple, and the output exhibits lower levels of crossover distortion than does the LM378 alone due to the inclusion of the booster transistors within the feedback loop. At signal levels below 20 mW, the LM378 supplies the load directly through the 5Ω resistor to about 100 mA peak current. Above this level, the booster transistors are biased on by the load current through the same 5Ω resistor.

The response of the 10W boosted amplifier is indicated for power levels below clipping. Distortion is below 2% from about 50 Hz to 30 kHz. 15W RMS power is available at 10% distortion; however, this represents extreme clipping. Although the LM378 delivers little power, its heatsink must be adequate for about 3W package dissipation. The output transistors must also have an adequate heatsink.

268 IC SCHEMATIC SOURCEMASTER

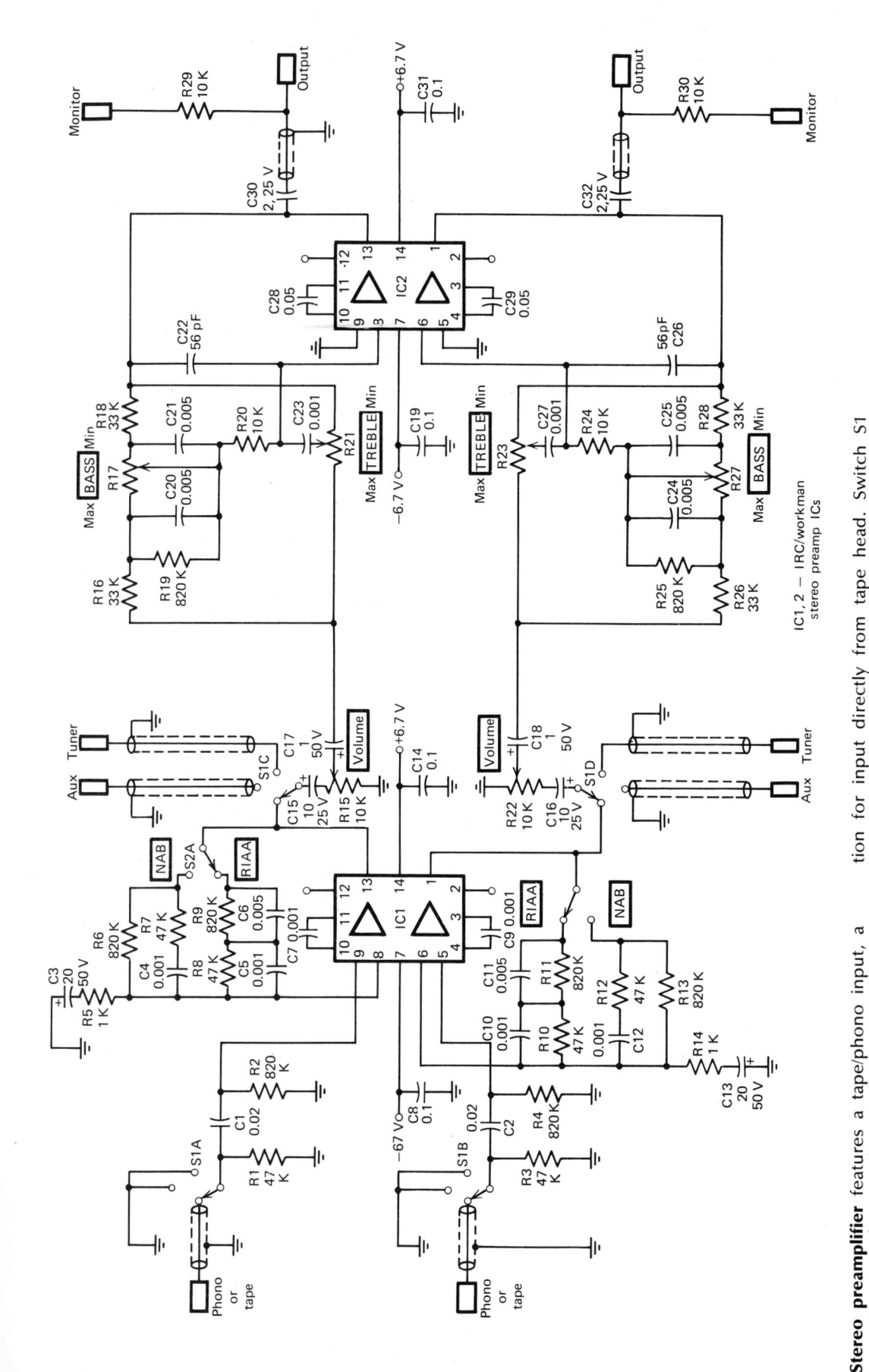

Stereo preamplifier features a tape/phono input, a tuner input, and an auxiliary input, and includes a monitor output for headphones or VU meter. During operation, IC1 is an amplifier for phono or tape input. Switch S2 provides selection of RIAA equalization for magnetic phono cartridges or NAB equalization for input directly from tape head. Switch S1 selects output from IC1, from tuner, or from the auxiliary input. IC2 provides active tone control. Unit uses very low voltages, but is extremely critical. (Semiconductors are RC/Workman.)

SECTION **6** *Audio Circuits* **269**

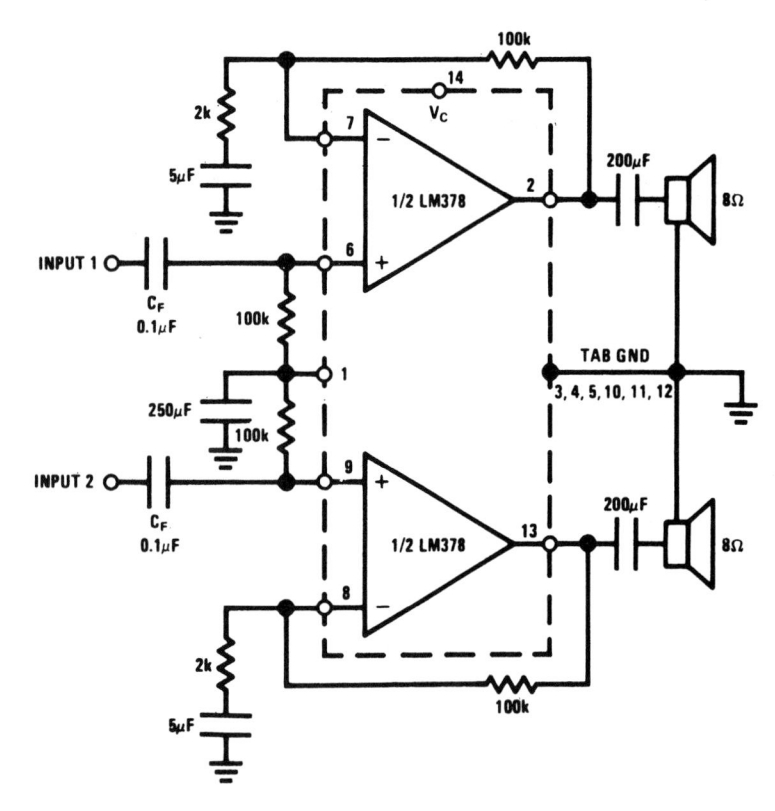

This simple stereo amplifier will deliver 4W per channel into 8 or 16Ω loads. The amplifier is designed to operate with a minimum of external components and contains an internal bias regulator to bias each amplifier. Device overload protection consists of both internal current limit and thermal shutdown.

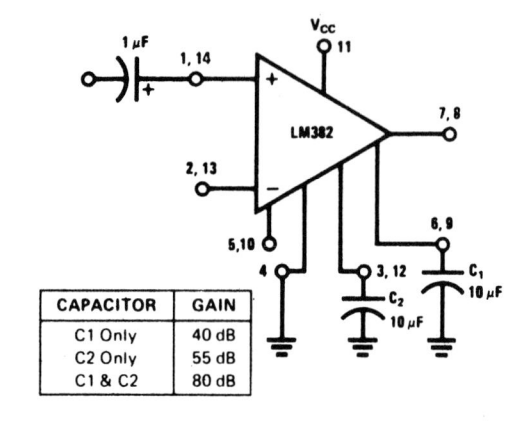

Flat response, fixed-gain-preamplifier.

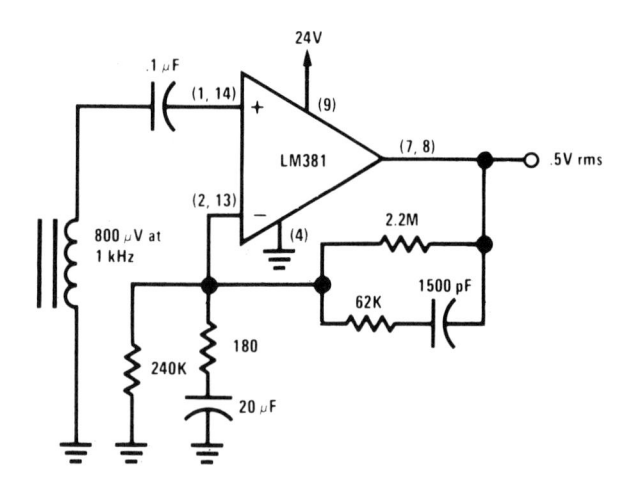

CAPACITOR	GAIN
C1 Only	40 dB
C2 Only	55 dB
C1 & C2	80 dB

Tape playback amplifier.

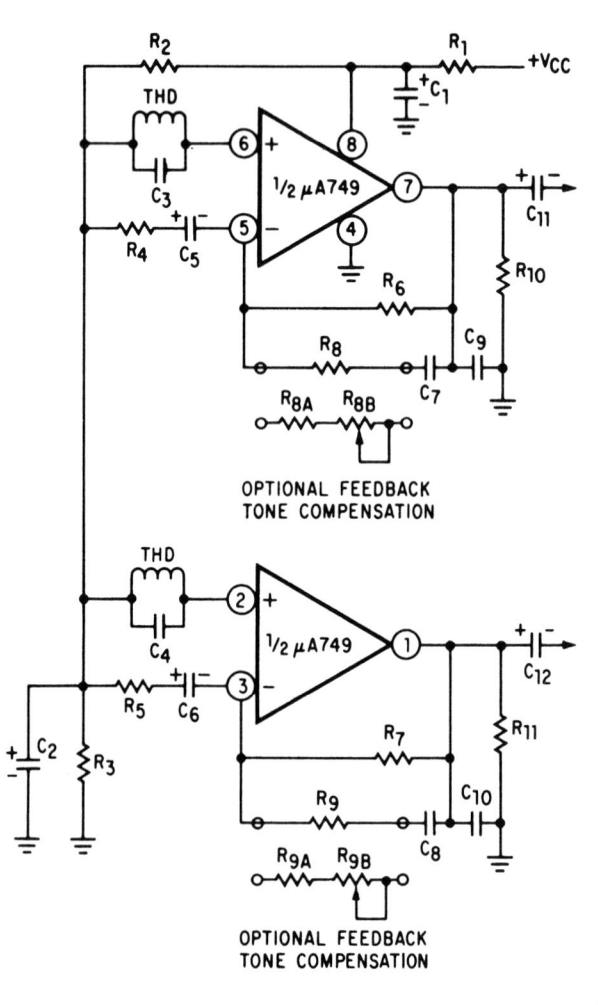

Tape preamp. R1 and C1 are power supply decoupling for the circuit. Since the power supply rejection of the μA749 is typically 80 dB, no further filtering is needed for the supply. R2, R3, and C2 bias the plus input of the μA749. The bias current for the input (less than 2.5 μA) flows through the playback heads. R6 and R7 bias the minus input with feedback from the output of each amplifier. R10 and R11 are the collector loads to ground for the output stage.

270   IC SCHEMATIC SOURCEMASTER

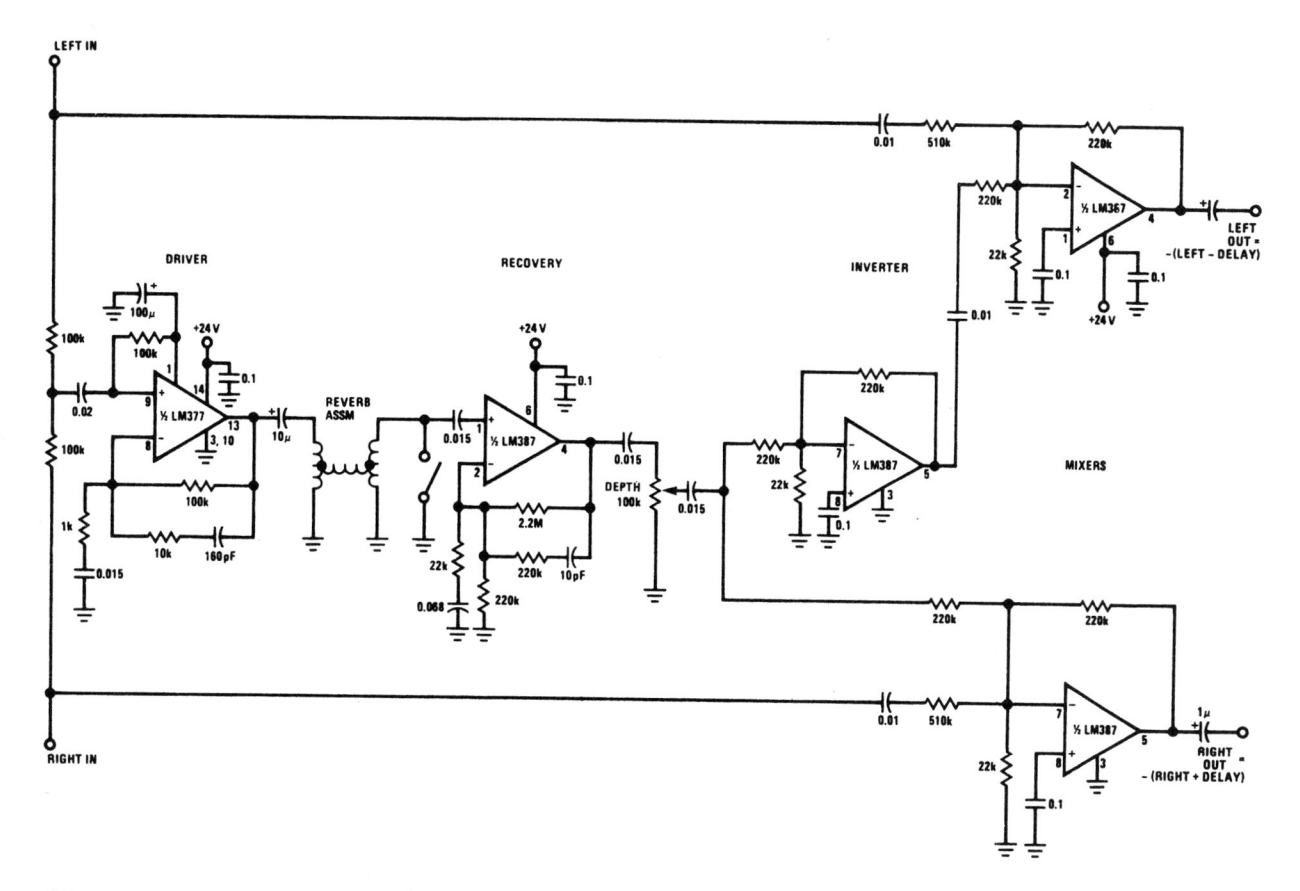

This stereo reverb enhancement system can also be used to synthesize a stereo effect from a monaural source such as AM radio or FM mono broadcast, or it can be added to an existing stereo (or quad) system where it produces an exciting spatial effect that is truly impressive.

The second half of the LM387 recovery amplifier is used as an inverter, and a new LM387 is added to mix both channels together. The outputs are inverted scaled sums of the original and delayed signals such that the left output is composed of LEFT minus DELAY and the right output is composed of RIGHT plus DELAY.

When applied to mono source material, both inputs are tied together, and the two outputs become INPUT minus DELAY and INPUT plus DELAY, respectively. If the outputs are to be used to drive speakers directly (as in an automotive application or small home systems), then the LM387 may be replaced by one of the LM377/378/379 dual 2W/4W/6W amplifier family wired as an inverting power summer.

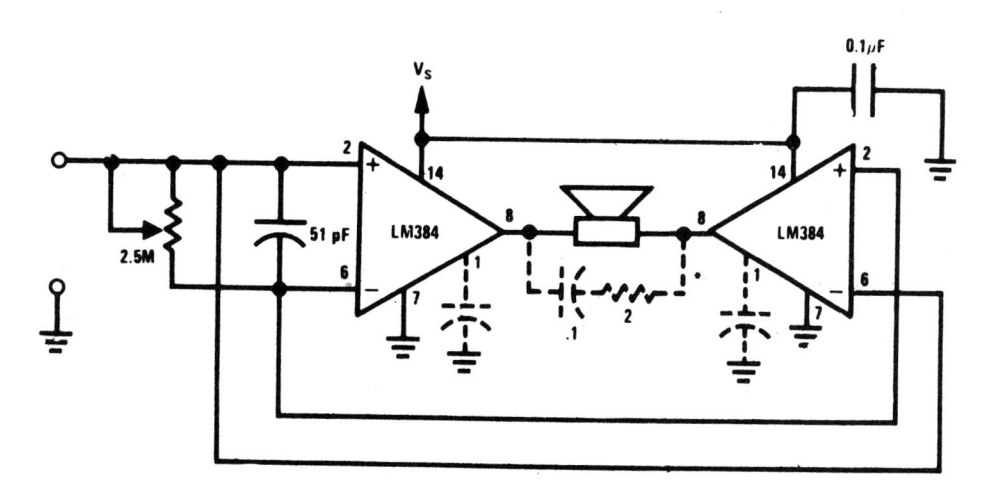

Bridge amplifier.

SECTION 6 Audio Circuits 271

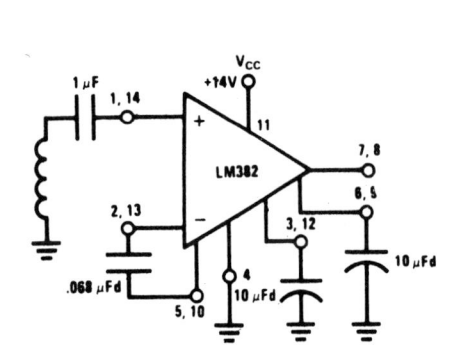

Tape preamp (NAB equalization).

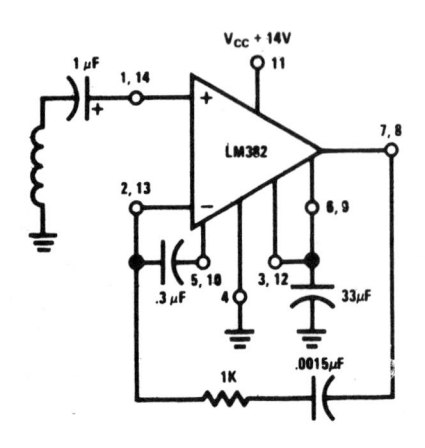

Phono preamp (RIAA equalization).

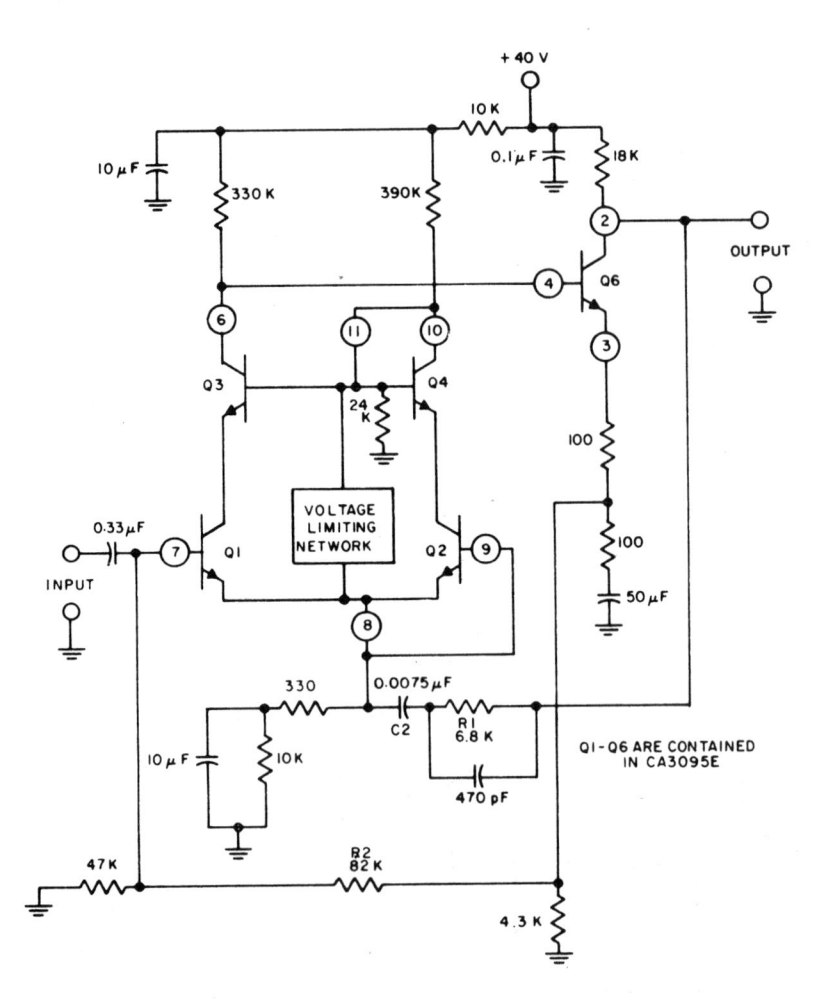

A tape playback preamplifier equalized for NAB standards (7.5 in./sec). Dynamic range is typically about 95 dB with gain indicated. A voltage-limiting network is built into the superbeta IC. To use it, just connect a resistor from the positive supply of the differential amplifier to pin 11; the value should be such that current through the resistor is about half that measured at pin 8 under quiescent conditions.

272 IC SCHEMATIC SOURCEMASTER

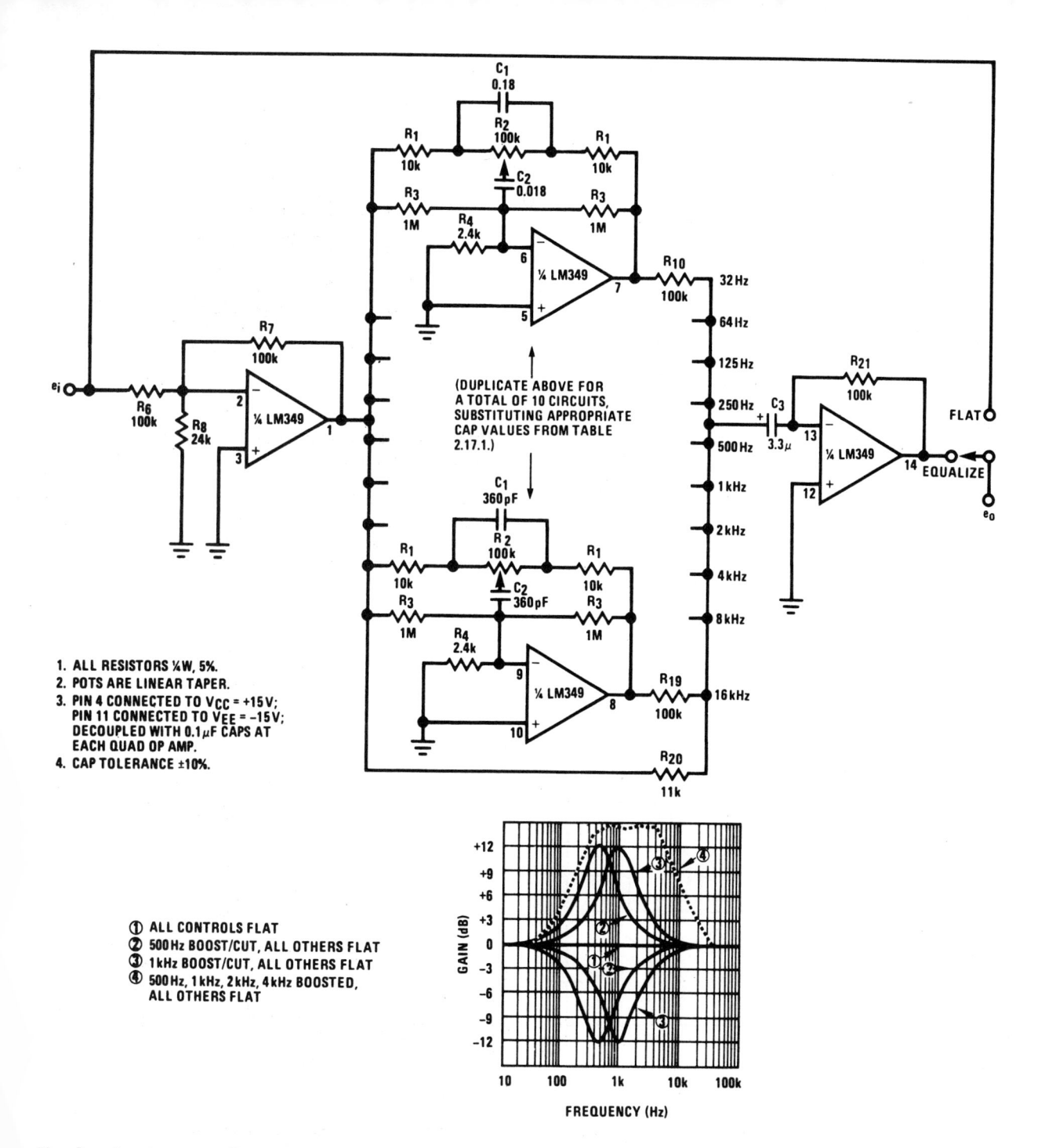

Ten-band octave equalizer. While it appears complicated, it is really just repetitious. By using quad amplifier ICs, the whole thing consists of only three integrated circuits. This one channel would be duplicated for a stereo system. The input buffer amplifier guarantees a low source impedance to drive the equalizer and presents a large input impedance for the preamplifier. Resistor R8 is necessary to stabilize the LM349 while retaining its fast slew rate (2V/μsec). The output amplifier is a unity-gain inverting summer used to add each equalized octave of frequencies back together again. One odd aspect of the summing circuit is that the original signal is subtracted from the sum via R20. (It is subtracted rather than added because each equalizer section inverts the signal relative to the output of the buffer, and R20 delivers the original signal without inverting.) The reason this subtraction is necessary is to maintain a unity-gain system. Without it, the output would equal 10 times the input; for example, an input of 1V, with all pots flat, would produce 1V at each equalizer output—the sum of which is 10V. By scaling R20 such that the input signal is multiplied by 9 before the subtraction, the output now becomes 10V − 9V = 1V output, that is, unity gain. The addition of R4 to each section is for stability. Capacitor C3 minimizes possibly large dc offset voltages from appearing at the output. If the driving source has a dc level, then an input capacitor is necessary (0.1 μF), and similarly, if the load has a dc level, then an output capacitor is required.

Typical frequency response of equalizer.

SECTION **6** *Audio Circuits* 273

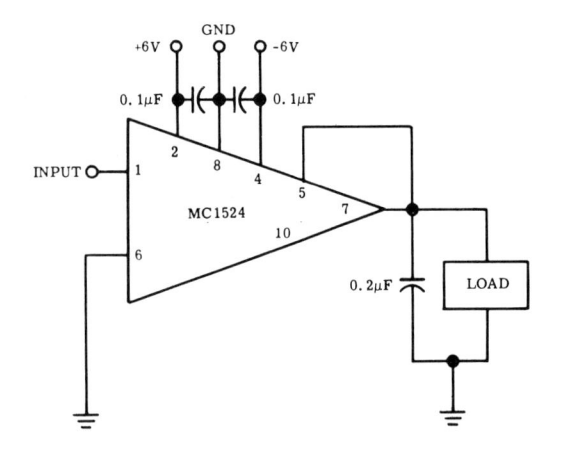

Typical **audio amplifier** using Motorola MC1524. A frequency rolloff capacitor of 0.2 µF is used at the load, and 0.1 µF capacitors are used for power supply bypassing.

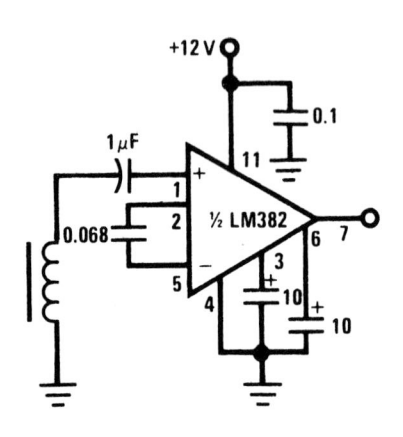

This tape playback preamp is NAB-equalized. The circuit is optimized for automotive use; that is, V_S = 10 − 15V. The wideband 0 dB reference gain is equal to 46 dB (200V/V) and is not easily altered. For designs requiring either gain or supply voltage changes, the required extra parts make selection of a LM387 a more appropriate choice.

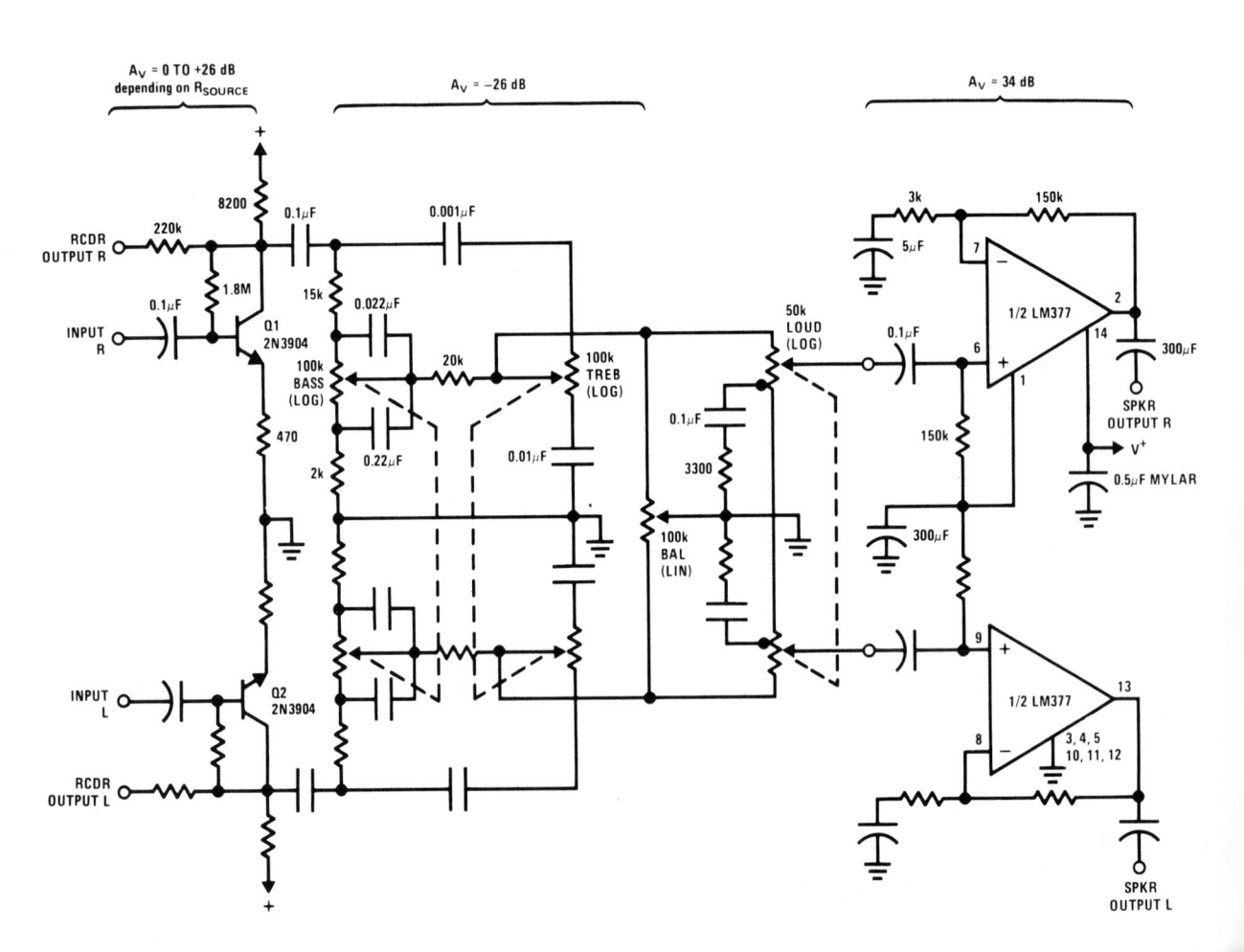

A two-channel sound system with inputs for AM radio, stereo FM radio, phono, and tape playback. It combines power amplifier pair with loudness, balance, and tone controls. The tone controls allow boost or cut of bass and treble. Transistors Q1 and Q2 act as input line amplifiers with the triple function of (1) presenting a high input impedance to the inputs, especially ceramic phono; (2) providing an amplified output signal to a tape recorder; and (3) providing gain to make up for the loss in the tone controls.

274   IC SCHEMATIC SOURCEMASTER

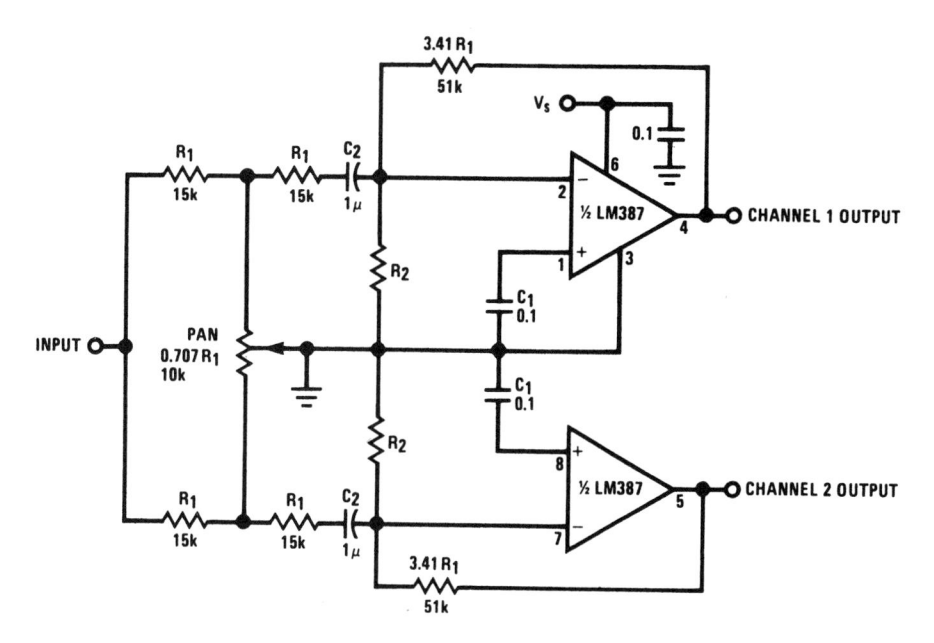

A two-channel panning circuit with the ability to move the apparent position of one microphone's input between two output channels. Panning is how recording engineers manage to pick up your favorite pianist and "float" the sound over to the other side of the stage and back again. The output of a pan circuit is required to have unity gain at each extreme of pot travel (that is, all input signal delivered to one output channel with the other output channel zero) and −3 dB output from each channel with the pan-pot centered. Normally, panning requires two oppositely wound controls ganged together; however, this circuit provides smooth and accurate panning with only one linear pot. With the pot at either extreme, the effective input resistance equals 3.41R1, and the gain is unity. Centering the pot yields an effective input resistance on each side equal to 4.83R1, and both gains are −3 dB. Using standard 5% resistor values as shown, gain accuracies within 0.4 dB are possible; replacing R1 with 1% values (for example, input resistors equal 14.3k and feedback resistors equal 48.7k) allows gain accuracies of better than 0.1 dB. Biasing resistor R2 is selected as a function of supply voltage. Capacitor C1 is used to decouple the positive input, while C2 is included to prevent shifts in output dc level due to the changing source impedance.

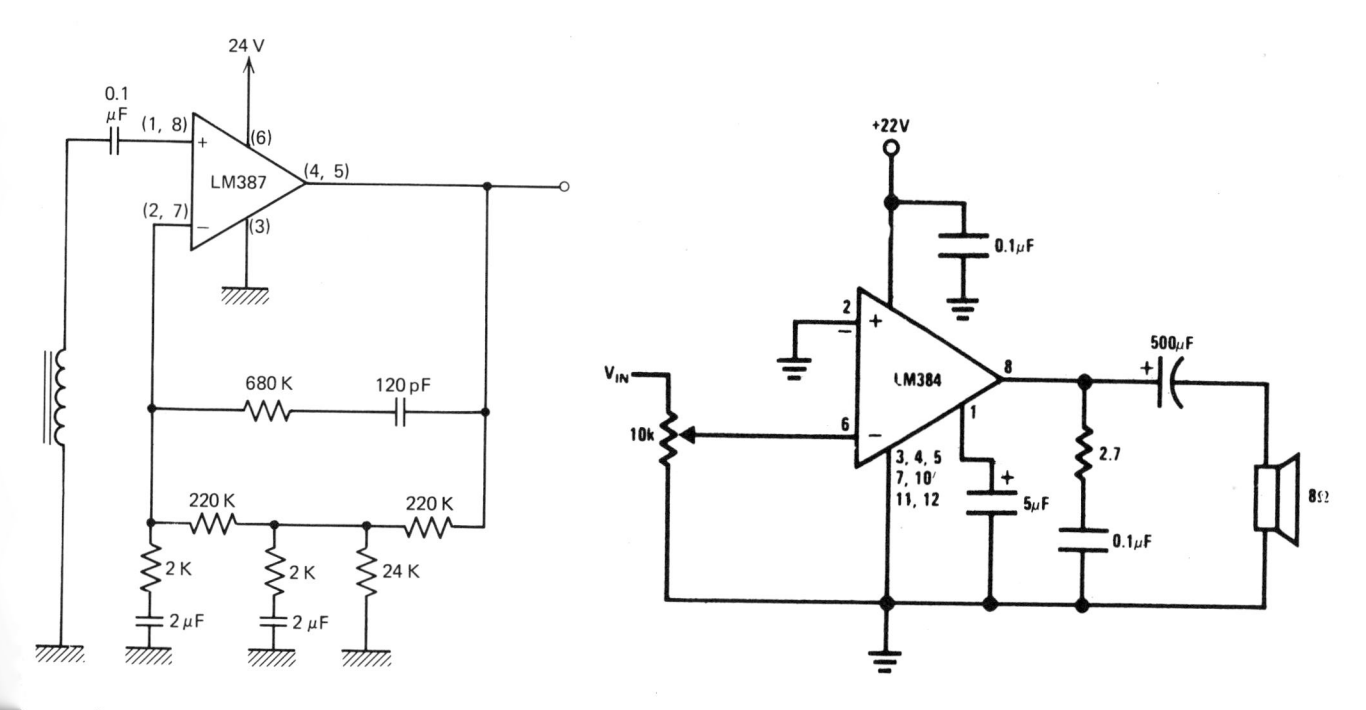

Two-pole, fast-turn-on NAB tape preamplifier.

Typical 5W amplifier.

SECTION **6** Audio Circuits

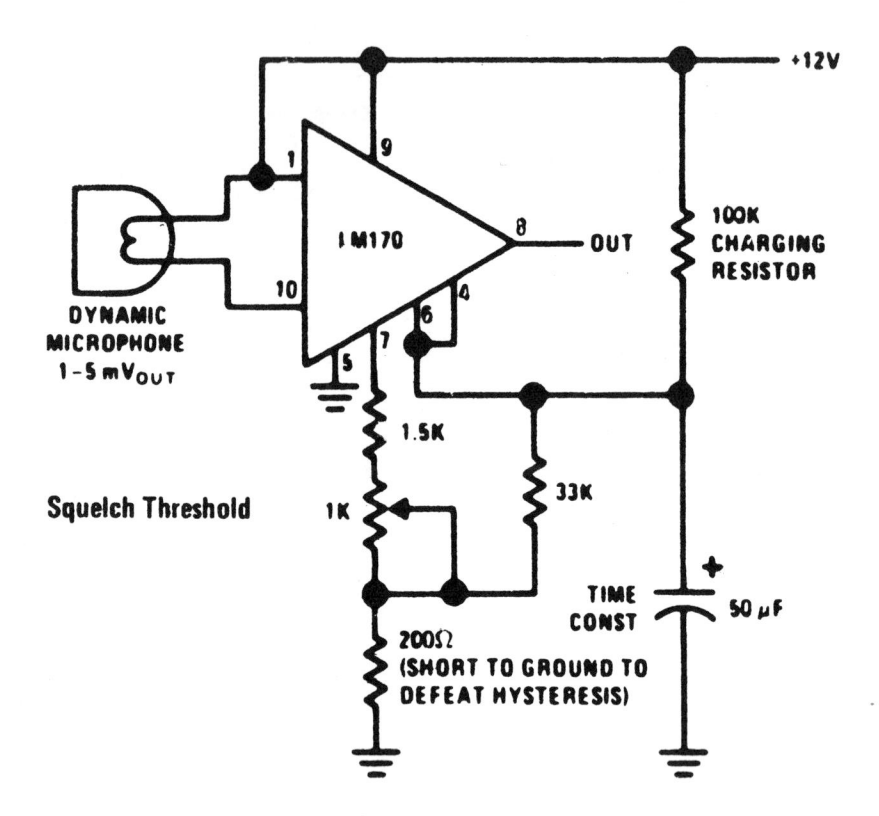

A squelched mike preamplifier with hysteresis. Voltage gain is continuously variable from a maximum value, dependent upon supply voltage, to minimum value, by application of a dc control voltage at pin 3 or 4. Dc output voltage is substantially independent of gain changes, provided that differential dc input voltage is minimized, so that direct-coupled or fast gain-control operation is possible with minimum disturbance of succeeding amplifiers.

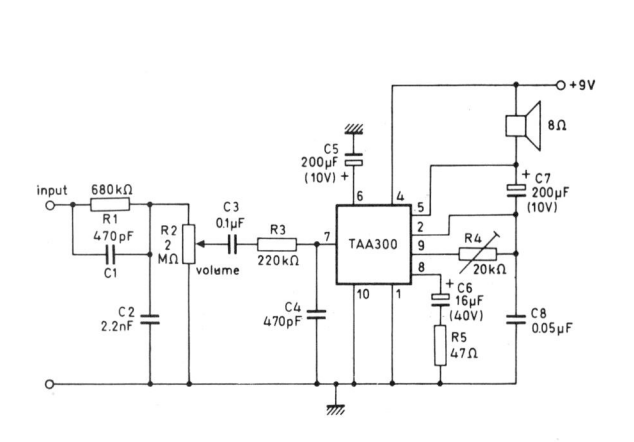

A 1W Amperex IC amplifier for use with a ceramic pickup.

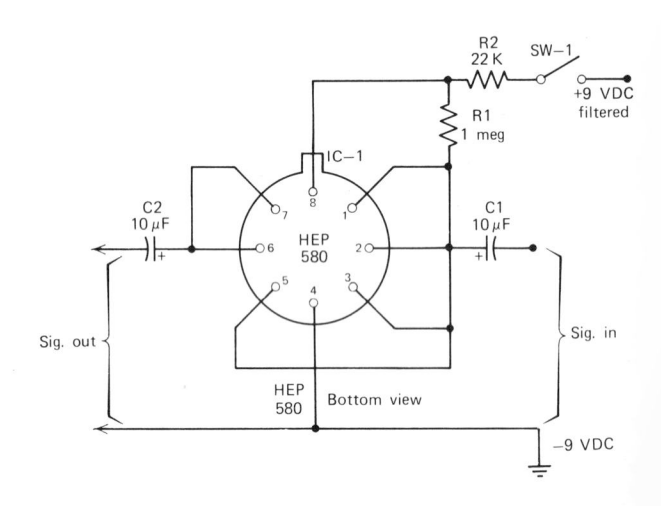

A wideband preamplifier using dual 2-input NOR gate. With 9.0V supply, output is approximately 1V from 10 Hz to 1 MHz. Current drain is about 0.2 mA. The gain factor over specified bandwidth is 100. It may be used as a stereo (2 channel) preamp or a 2- or 4-channel mixer. Refer to the IC chip schematic for input and output connections.

276 IC SCHEMATIC SOURCEMASTER

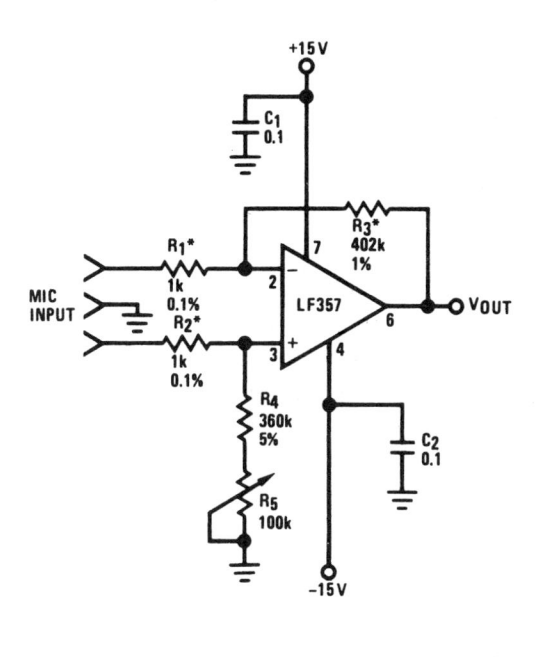

Transformerless FET-input balanced-line mike preamp. Transformer input designs offer the advantage of nearly noise-free gain and do indeed yield the best noise performance for microphone applications; however, when the total performance of the preamplifier is examined, many deficiencies arise. Even the best transformers will introduce a certain amount of harmonic distortion; they are very susceptible to hum pickup; common-mode rejection is not optimum; and high-quality input transformers are quite expensive. By utilizing the inherent ability of an operational amplifier to amplify differential signals while rejecting common-mode siginals, it becomes possible to eliminate the input transformer.

$A_V = 52\,dB$
* – METAL FILM
NOISE: –64 dB BELOW 2 mV (–115 dBm)
THD $\leq 0.1\%$

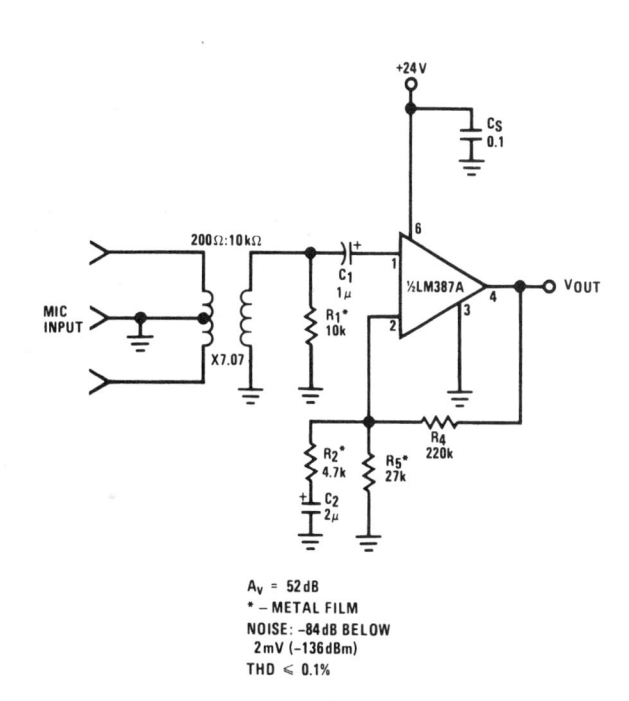

Transformer-input balanced-line mike preamp. Balanced microphones are used where hum and noise must be kept at a minimum. This is achieved by using a three-wire system—two for signal and a separate wire for ground. The two signal wires are twisted tightly together, with an overall shield wrapped around the pair acting as the ground. Proper grounding of microphones and their interconnecting cables is crucial since all noise and hum frequencies picked up along the way to the preamplifier will be amplified as signal. The rationale behind the twisted-pair concept is that all interference will be induced equally into each signal wire and will thus be applied to the preamp common mode, while the actual transmitted signal appears differential. Balanced-input transformers with center-tapped primaries and single-ended secondaries dominate balanced mike preamp designs. By grounding the center tap, all common-mode signals are shunted to ground, leaving the differential signal to be transformed across to the secondary winding, where it is converted into a single-ended output.

$A_V = 52\,dB$
* – METAL FILM
NOISE: –84 dB BELOW 2 mV (–136 dBm)
THD $\leq 0.1\%$

SECTION 6 *Audio Circuits* 277

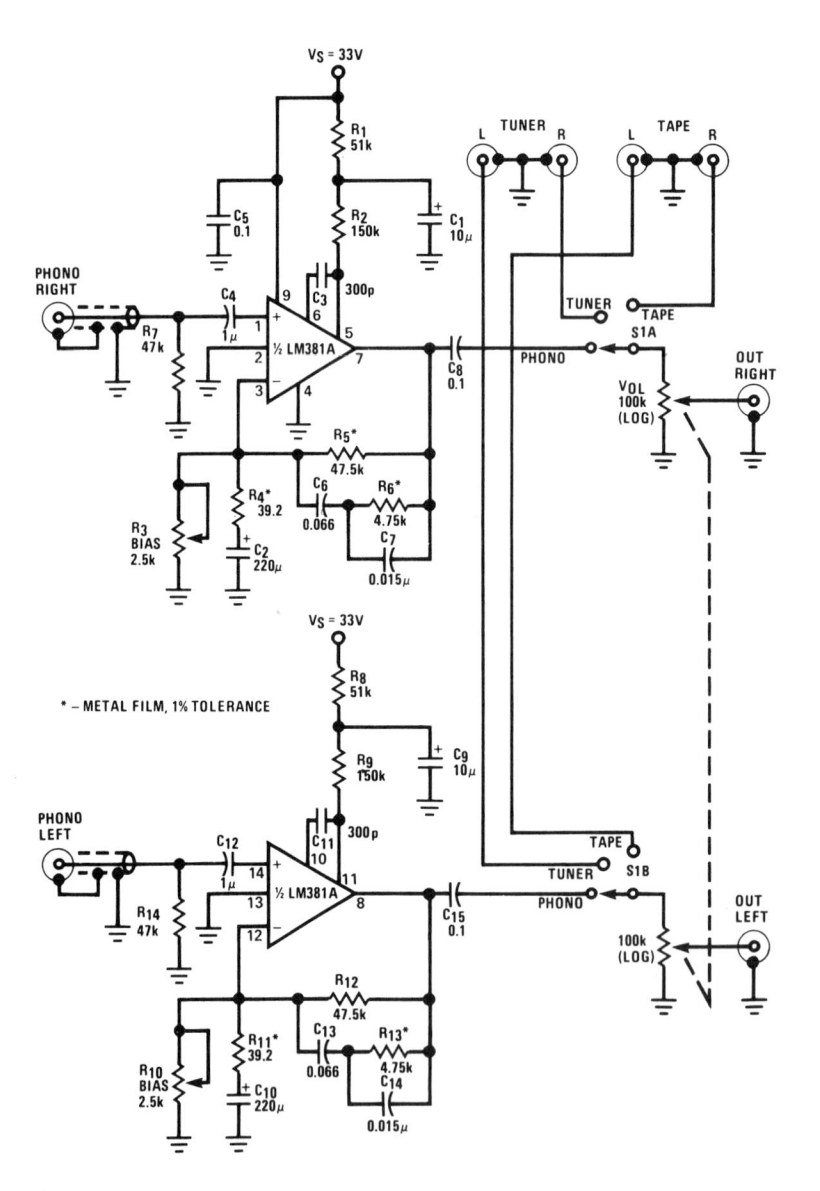

Ultralow-noise mini preamp, with provisions for tuner and tape inputs, selector switch, and ganged volume control. Tone controls are omitted but may be easily added. The RIAA frequency response is within ±0.6 dB of standard values. The 0 dB reference gain at 1 kHz is 41.6 dB (120V/V), producing 1.5V output from a nominal 12.5 mV input. With the given dc supply voltage of 33V, this gives better than +25 dB headroom (dynamic range) for a typical 5 mV input at 1 kHz. Input overload limit equals 91 mV at midband frequencies. Signal-to-noise ratio is better than −85 dB referenced to a 10 mV input level, with unweighted total output noise less than 100 μV (input shorted). Metal film resistors and close-tolerance capacitors should be used to minimize excess noise and maintain RIAA frequency accuracy.

278 IC SCHEMATIC SOURCEMASTER

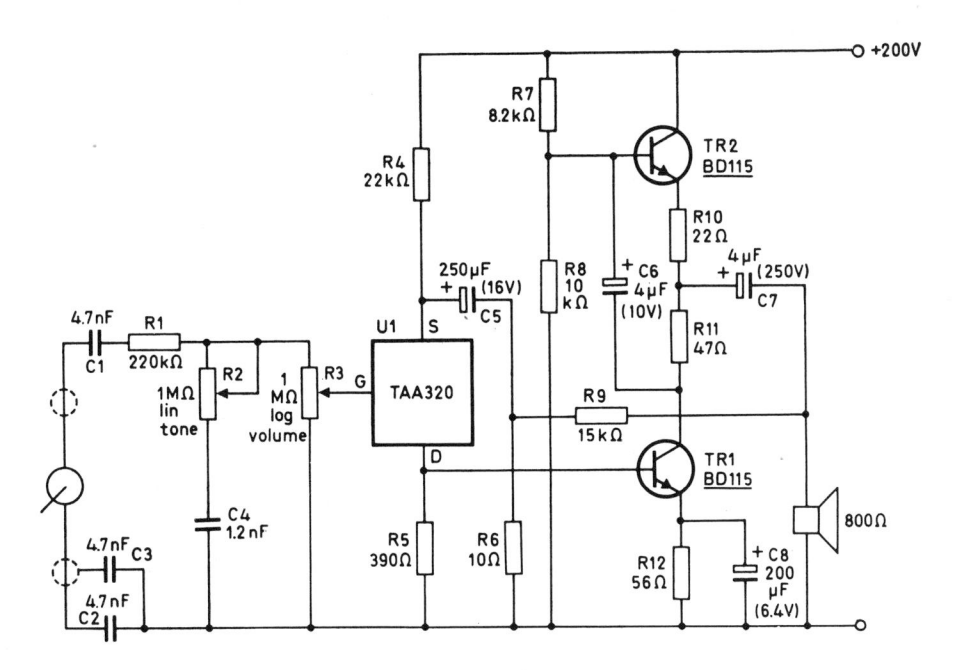

4W pickup amplifier.

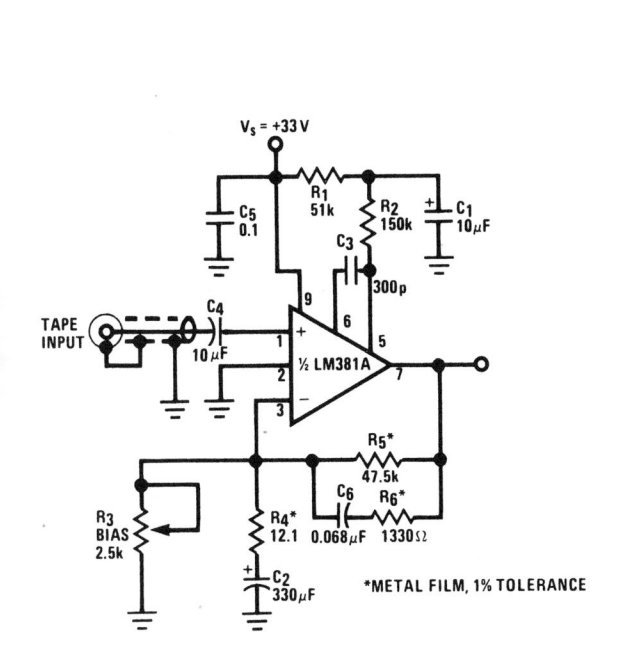

An ultralow-noise tape playback preamp for popular tape speeds of 1⅞ and 3¾ in./sec. Metal film resistors should be used where indicated to reduce excess noise. The 0 dB reference gain is 41 dB and produces an output level equal to 200 mV from a head output of 1 mV at 1 kHz. Turn-on is inherently fast owing to the low voltage required at pin 3 (~0.5V compared to ~1.2V for differential scheme).

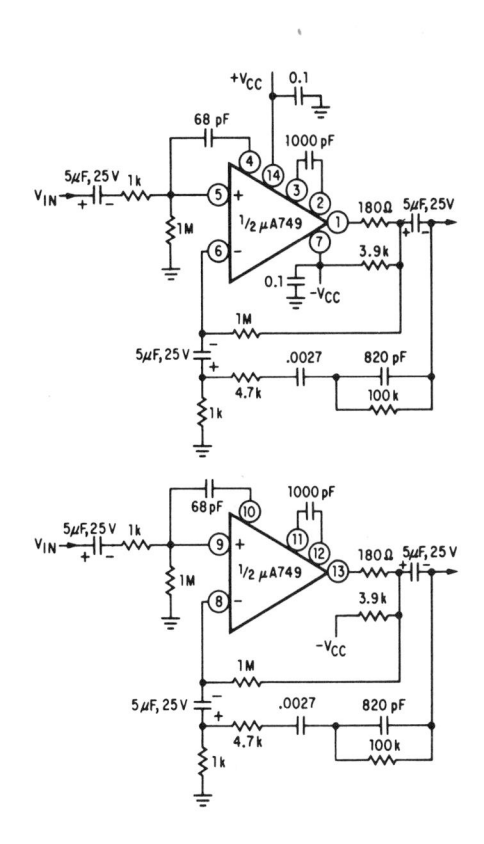

Wideband magnetic phono preamp.

SECTION **6** Audio Circuits **279**

SECTION 7
COMMUNICATIONS CIRCUITS

To date, no clever circuit design engineer has managed to pack a complete transceiver into a tiny IC package; indeed, very few transmitter functions are performed with integrated circuits, and most receivers contain a considerable number of peripheral discrete devices. But there are still many applications of ICs in communications circuits. This section presents a very large number of circuits for a wide variety of seemingly unrelated functions. You'll find simple intercoms, wireless microphones, and other "fun"-type circuits alongside the more serious intermediate-frequency amplifiers, frequency multipliers, modulators, and squelch circuitry.

Let this section be the first stop in your quest for good circuits to adapt for any communications application, but also check other appropriate sections. Oscillators, for example, are integral to almost every communications system, and you will find many more of them in Section 4. Audio circuits in general are presented in Section 6. If your communications system is to be logic-oriented or part of a machinated transmission system, be sure to check the schematics in Sections 14 and 15.

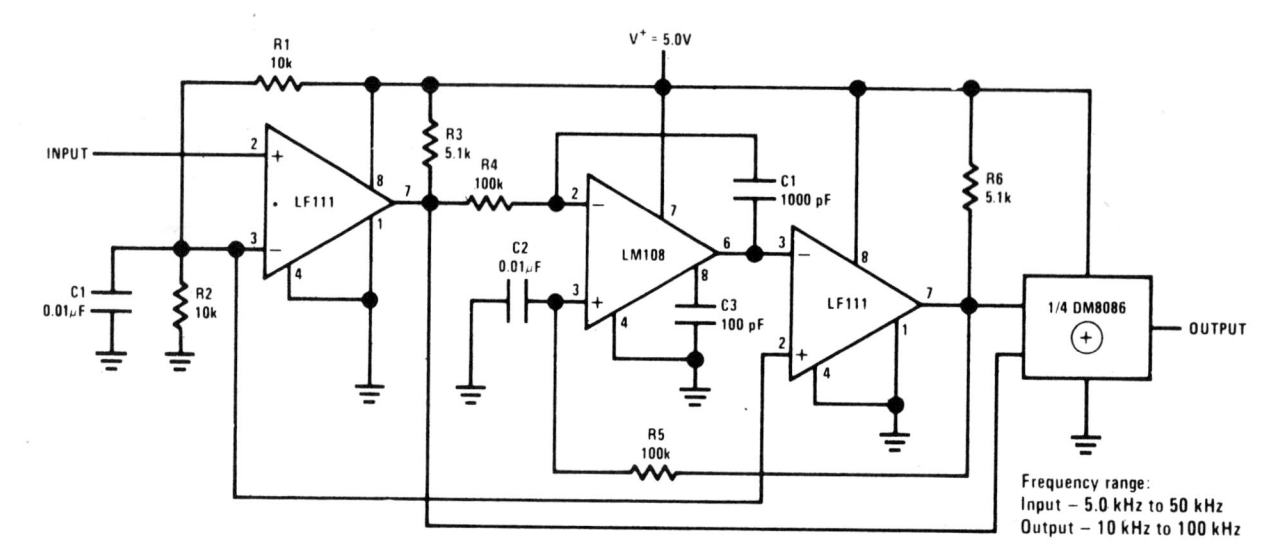

Frequency doubler.

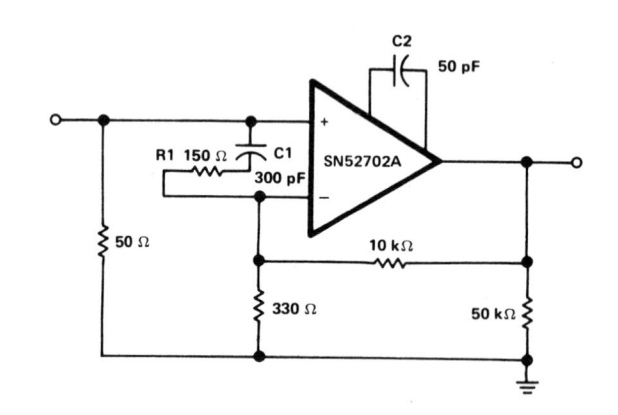

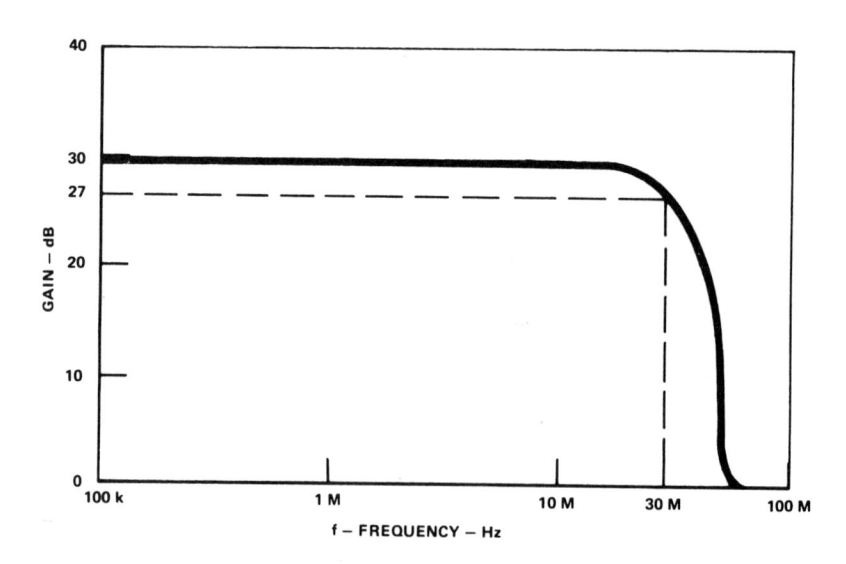

Although the Texas Instruments SN52702A is an operational amplifier and is generally considered an audio amplifier, its broad bandwidth allows it to be used in the RF ranges as well. The schematic shows a broadband application in which the unity-gain bandwidth is about 50 MHz.

A 10k feedback resistor and 330Ω input termination set the low-frequency gain at 30 dB. R1 and C1 at the input provide phase-delay compensation, and C2 supplies some phase-advance compensation.

Suppressed-carrier modulator.

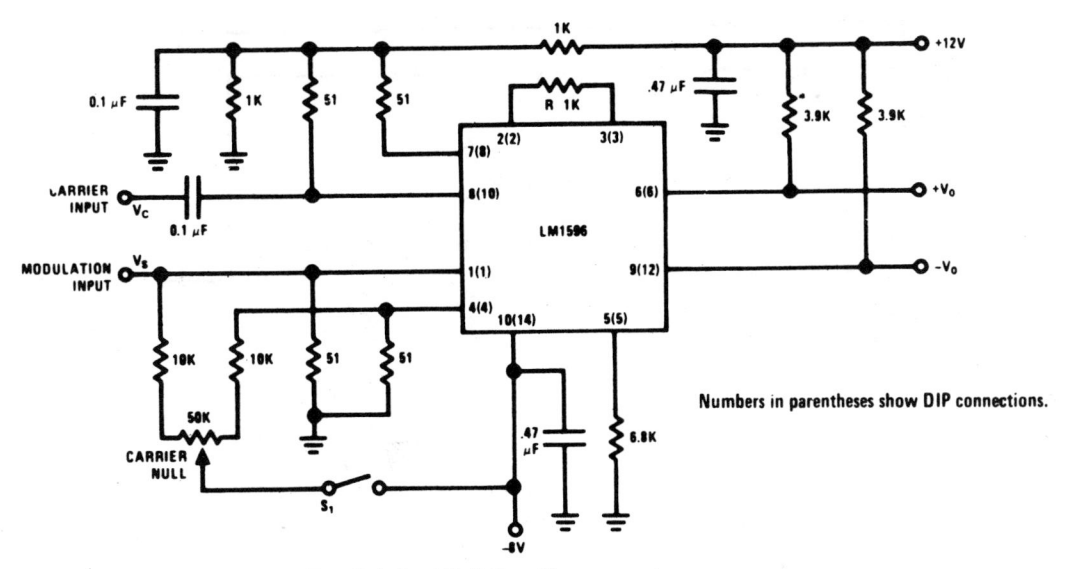

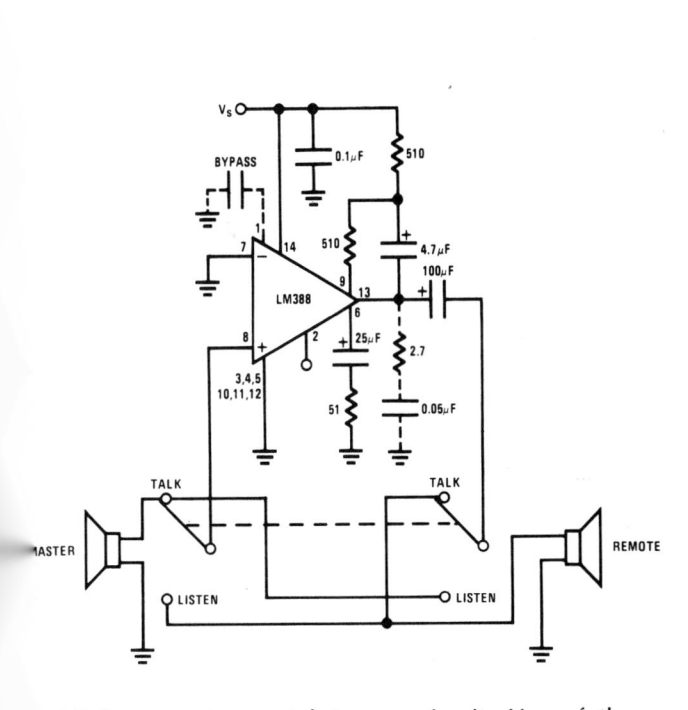

Numbers in parentheses show DIP connections.

Note: S₁ is closed for "adjusted" measurements.

Minimum-parts-count **intercom** circuit. Use of the gain control pin to set the ac gain to approximately 300V/V ($A_V \approx 15k/51\Omega$) allows elimination of the step-up transformer normally used in intercom designs. The optional 2.7Ω–0.05 µF RC network suppresses spurious oscillations.

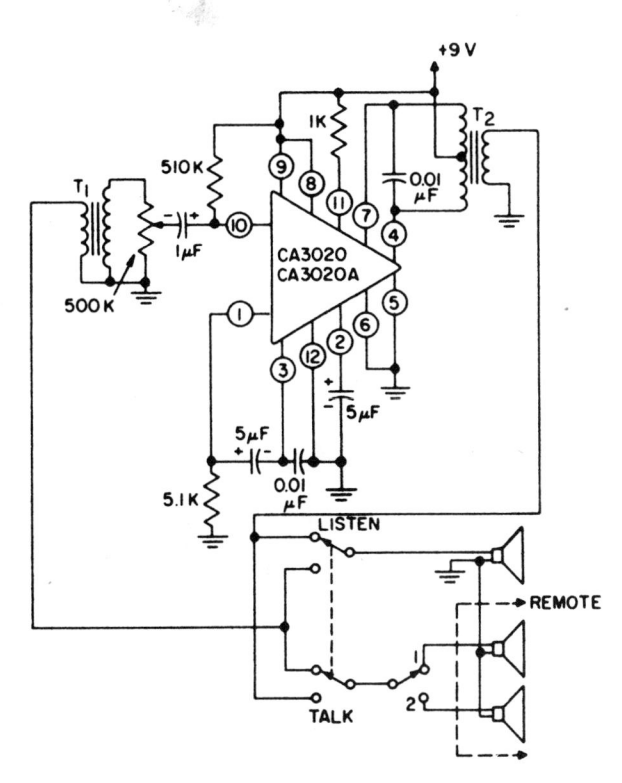

T_1: Primary 4 ohms, Secondary 25,000 ohms; Stancor A4744 or equiv.

T_2: Better Coil and Transformer DF1084, Thordarson TR-192, or equiv.

Speakers: 4 ohms

Intercom using CA3020 or CA3020A. Listen–talk position switch controls two or more remote positions. Only the speakers, the switch, and the input transformer are added to the basic audio amplifier circuit. A suitable power supply for the intercom could be a 9V battery used intermittently rather than continuously.

SECTION 7 *Communications Circuits* **283**

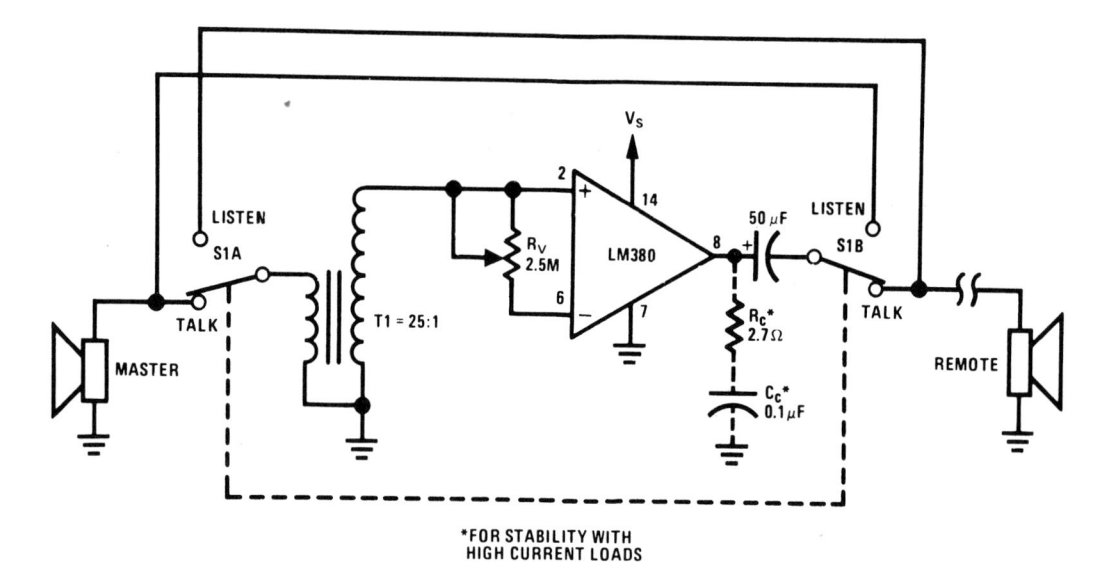

Minimum-component intercom. With switch S1 in the **talk** position, the speaker of the master station acts as the microphone with the aid of stepup transformer T1. A turns ratio of 25 and a device gain of 50 allows a maximum loop gain of 1250. R_V provides a "common-mode" volume control. Switching S1 to the **listen** position reverses the role of the master and remote speakers.

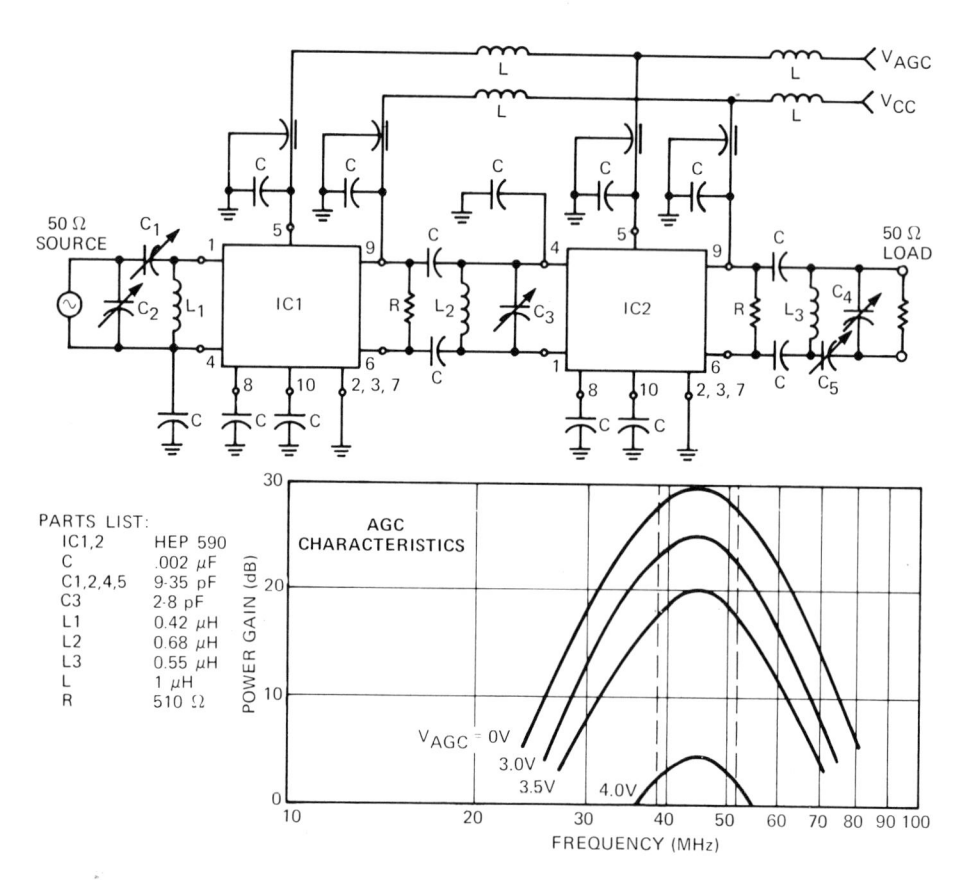

PARTS LIST:
IC1,2   HEP 590
C   .002 µF
C1,2,4,5 9-35 pF
C3   2·8 pF
L1   0.42 µH
L2   0.68 µH
L3   0.55 µH
L   1 µH
R   510 Ω

This wideband 45 MHz amplifier features AGC and extremely wide bandwidth at very high signal gain. AGC characteristics are shown on inset graph. Both ICs are Motorola HEP 590s.

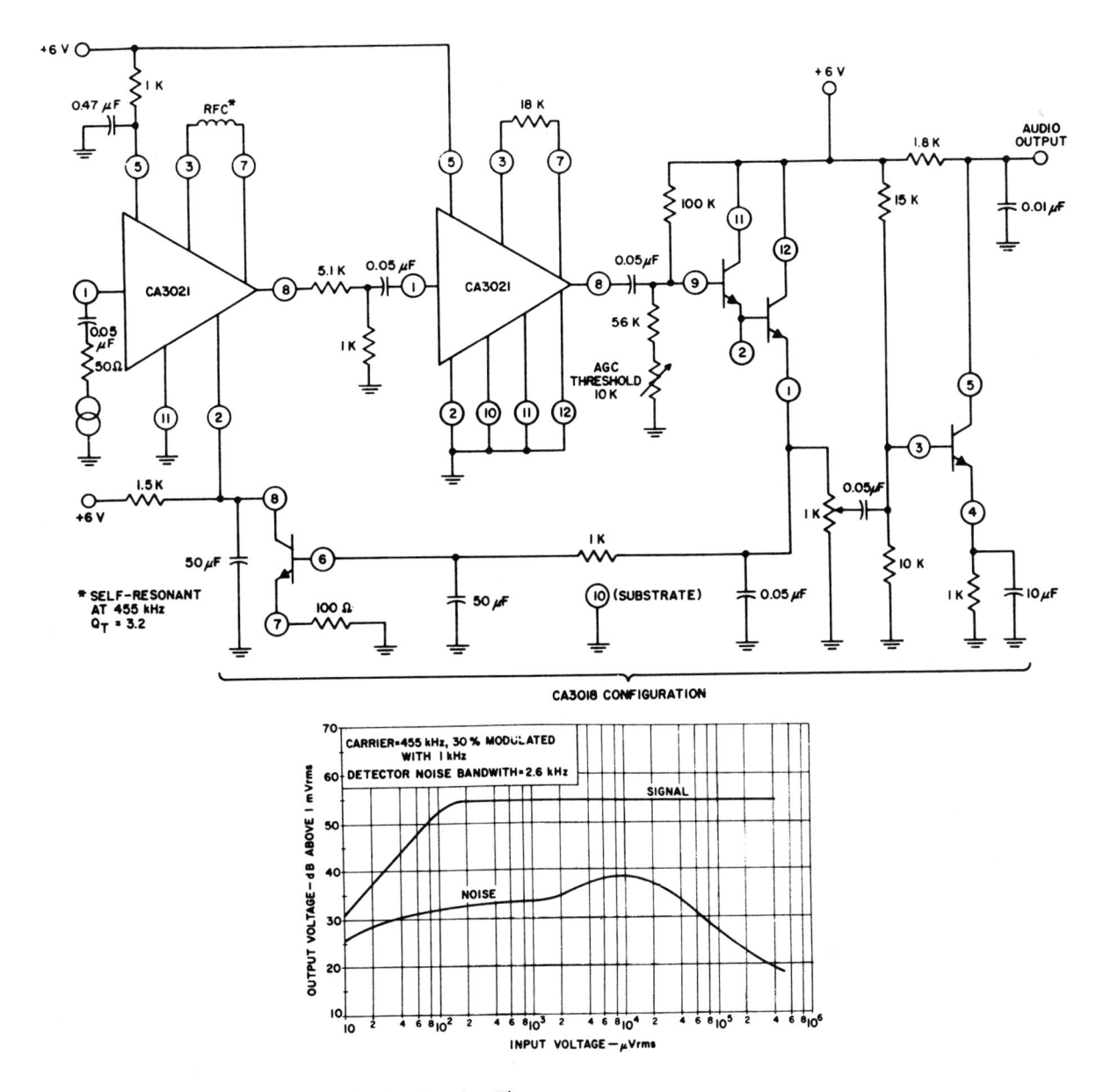

A 455 kHz IF amplifier using two CA3021 circuits. The RF feedback choke is self-resonant at 455 kHz and has a Q of 3.2 in the circuit. The second stage is a video amplifier. Input filtering would normally be provided to obtain the desired IF response. For the particular choice of stage gain and AGC loop gain, an interstage pad network is used to maintain stability and achieve an acceptable signal-to-noise ratio with gain control.

SECTION 7 Communications Circuits

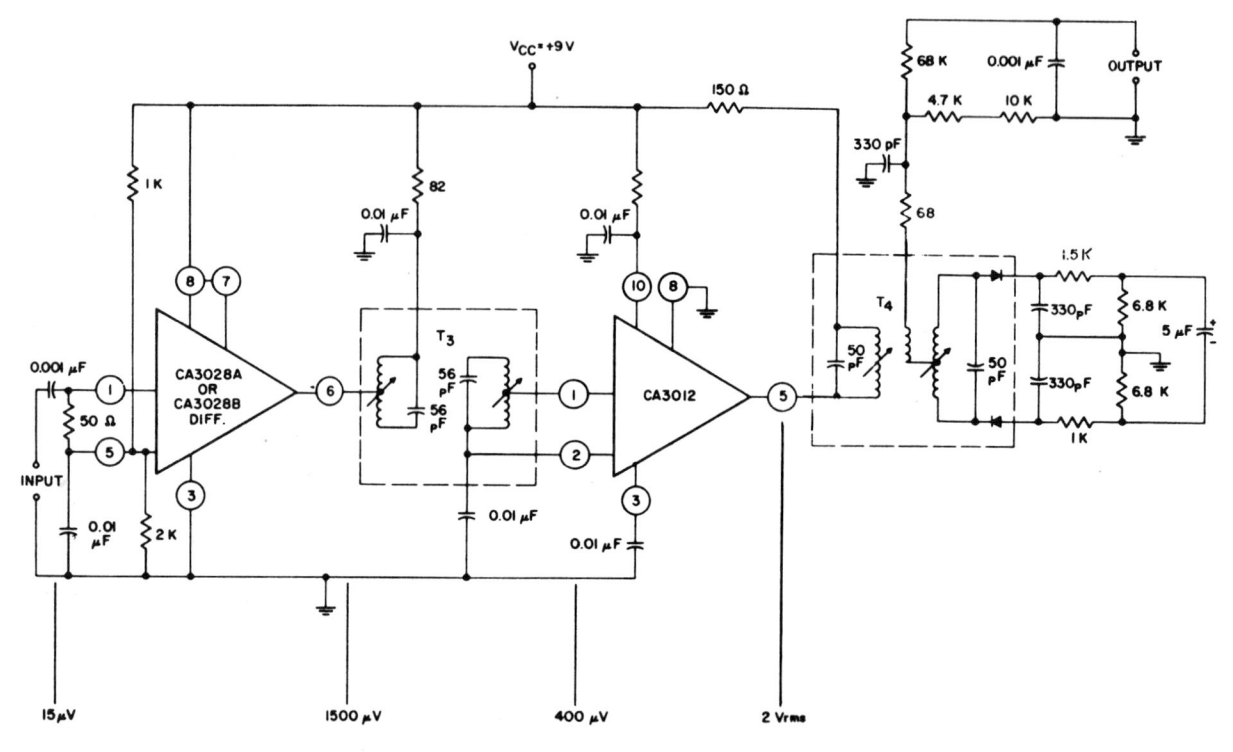

T₃: Interstage transformer TRW #22486 or equiv.
T₄: Ratio detector TRW #22516 or equiv.
Audio output: 155 mV rms for 15 μV ± 75 kHz input 3 dB below knee of transfer characteristic.

A 10.7 MHz IF strip using a CA3028A or CA3028B in the differential mode. An input of approximately 1500 μV is required to the interstage filter. The differential-mode voltage gain of the CA3028A or CA3028B into a 3000Ω load is 39.3 dB. This voltage gain requires that an input of approximately 15 μV be available at the base of the CA3028A or CA3028B differential amplifier. Even if a triple-tuned filter having a voltage insertion loss of 28 dB is used in a low-gain front end, a receiver having an IHFM sensitivity of 5 μV results. If 26 dB second-channel attenuation is permissible, a 3 μV-sensitivity IHFM receiver can be realized.

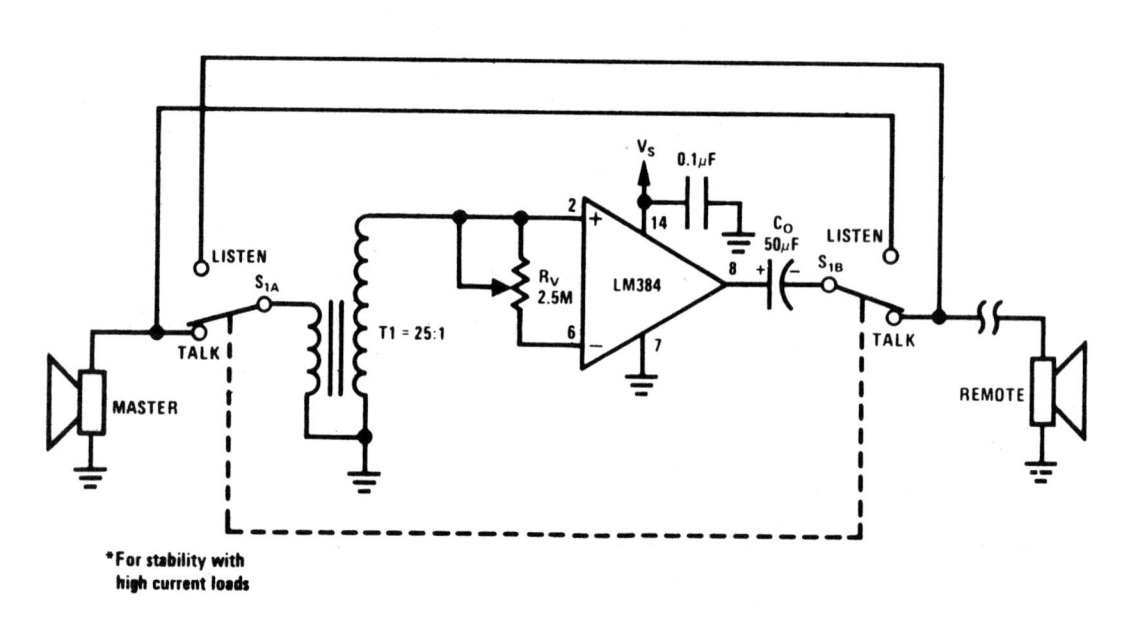

Intercom.

286 IC SCHEMATIC SOURCEMASTER

*Capacitors noted by asterisk are 0.1 at 455 kHz. L is Miller No. 43A105CBI for 455 kHz; 8 turns No. 26 AWG on Micrometals T25-2 Carbonyle Core (0.255 OD x 0.180 ID x 0.096W) for 10.7 MHz; 3 1/2 turns No. 20 AWG 5/16" dia x 1/4" long for 27 MHz.

f	C1	C2	C3	C4	L
455 kHz	.01	1000	.012	.001	10.5 µH
10.7 MHz	1000	250	1000	500	.22 µH
27 MHz	1000	180	300	500	.12 µH

An LM273/LM373 AM IF connection.

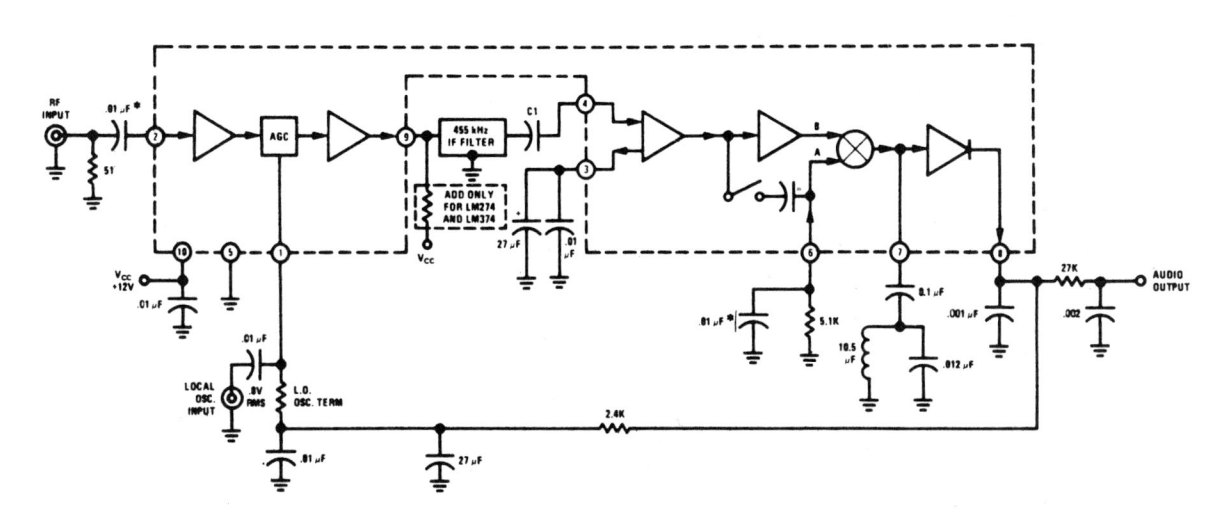

*Capacitors noted by asterisk are 0.1 at 455 kHz. L is Miller No. 43A105CBI for 455 kHz; 6 turns No. 26 AWG on Micrometrls T25-2 Carbonyle Core (0.255 OD x 0.180 ID x 0.096W) for 10.7 MHz; 3-1/2 turns No. 20 AWG 5/16" dia x 1/4" long for 27 MHz.

An LM274/LM374, LM273/LM373 first-stage converter operation for AM signal detection at 455 kHz.

Ac-couple an LC tank from the input of the active peak detector to ground. A lossy filter from pin 9 to pin 4 should be avoided, as this will greatly reduce the audio output and AGC figure of merit. The tank on pin 7 should have a high enough Q to limit noise yet be low enough to pass the full IF signal. It should also have a high enough impedance (greater than 5k) to avoid affecting the gain of that stage. Proper audio output is attained by a small capacitor at pin 8 to peak detect the RF envelope, followed by a series RC rolloff to shape the audio response. Excessive loading will reduce available output. There is a tradeoff between audio level out and AGC range, so the feedback resistor from pin 8 to pin 1 should be adjusted to give the desired results. Pin 1 must be filtered well with a 15 µF capacitor or larger to prevent any ac variation from causing erratic AGC action.

SECTION 7 Communications Circuits

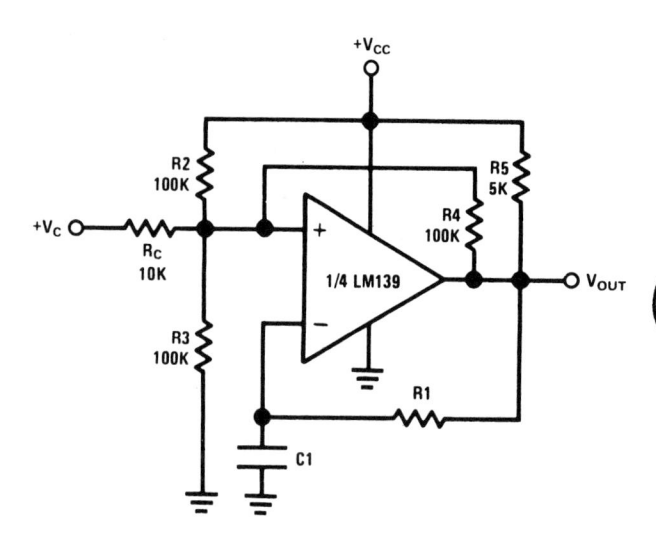

Pulse-width modulator. The input control voltage is equal to $+V_{CC}/2$. If the input control voltage is moved above or below $+V_{CC}/2$, the duty cycle of the output square wave will be altered because the addition of the control voltage at the input has now altered the trip points.

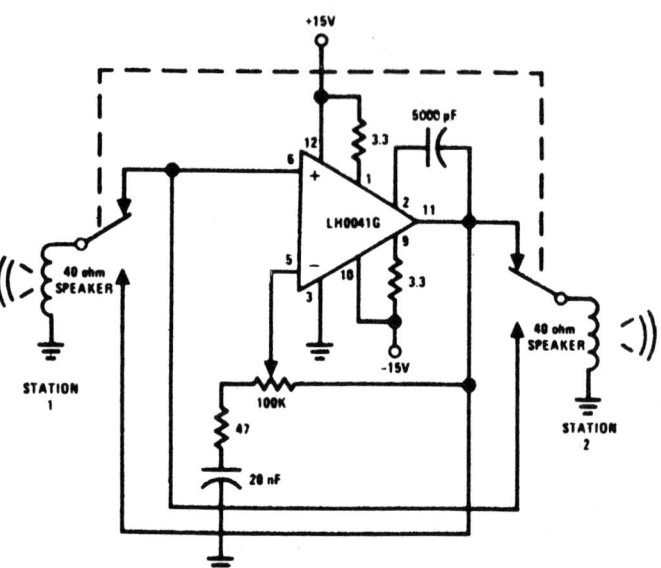

Two-way intercom.

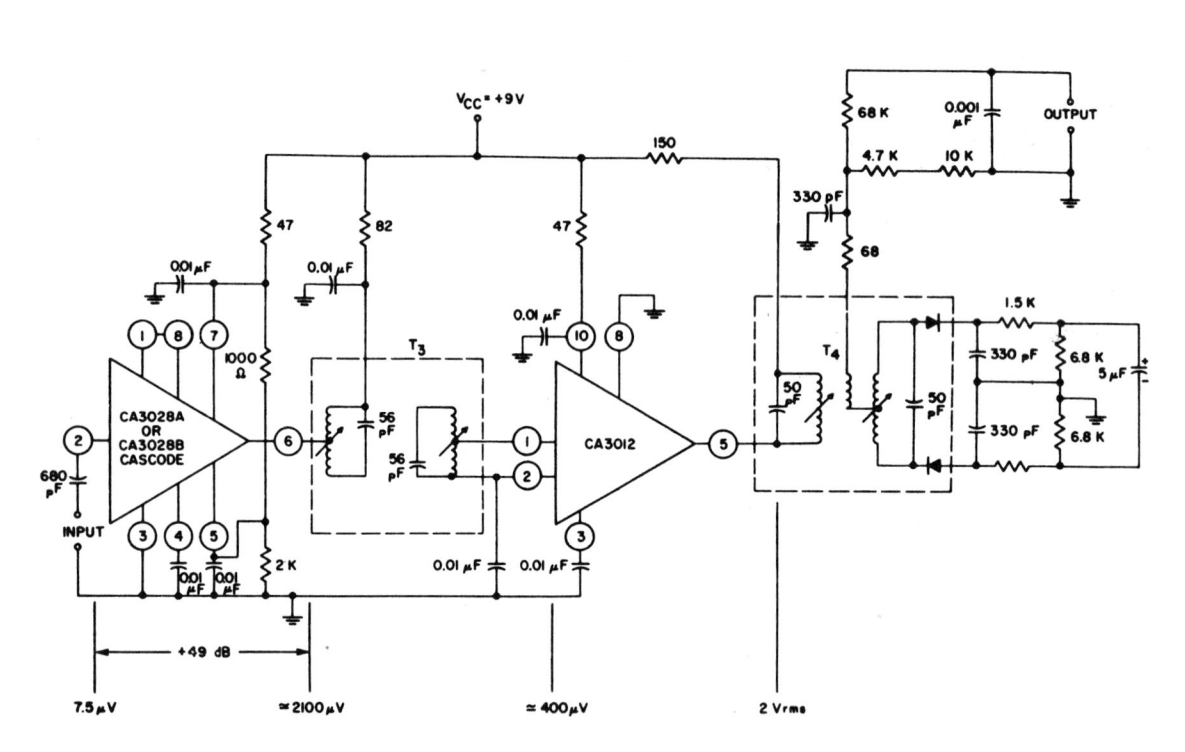

T_3: Interstage transformer TRW #22486 or equiv.
T_4: Ratio detector TRW #22516 or equiv.
Audio Output: 155 mV rms for 7.5 μV ± 75 kHz input 3 dB below knee of transfer characteristic.

A 10.7 MHz IF amplifier using a CA3028A or CA3028B in cascode. An input of approximately 400 μV is required at the base of the CA3012 for -3 dB below full limiting. An impedance-transfer device and filter must be connected between the CA3012 base (terminal 1) and the output of the CA3024A or CA3028B (terminal 6). The insertion loss of this filter should be kept near 6 dB (1:2 ratio of loaded to unloaded Q), so that all possible gain can be realized up to the CA3012 base. In addition to this insertion loss, a voltage stepdown loss of 5.8 dB in the interstage filter is unavoidable. Therefore, the total voltage loss is 9 to 14 dB, and an output of 1500 to 2000 μV must be available from the CA3028A or CA3028B to provide the required 400 μV input to the CA3012.

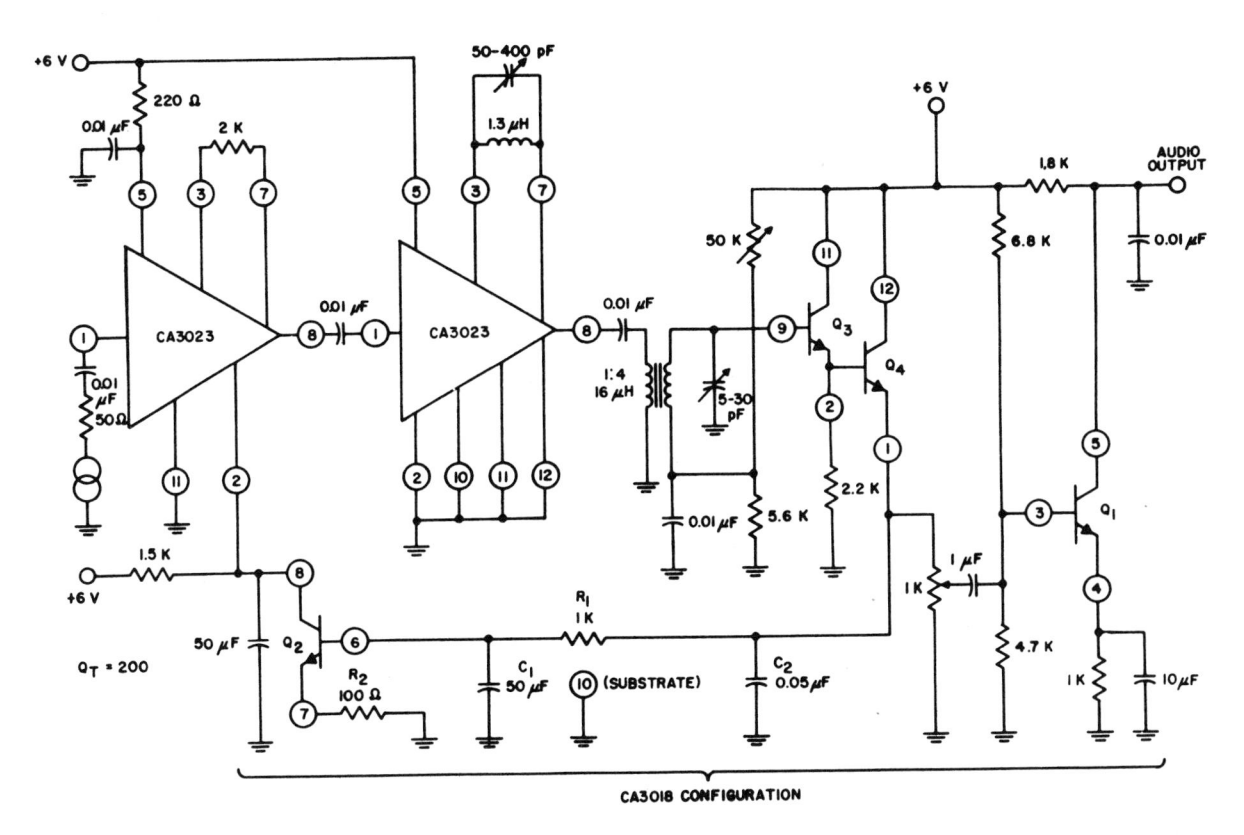

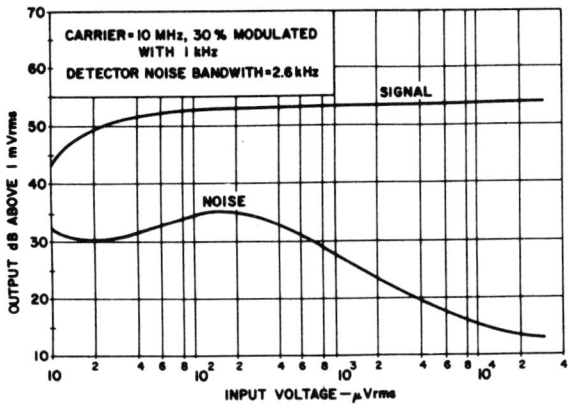

A 10 MHz IF amplifier using two CA3023 circuits. The first stage is operated in a broadband mode with a 2000Ω feedback resistor between terminals 3 and 7. The second stage is a tuned IF amplifier. Because the sinusoidal output capability of the CA3023 at 10 MHz is in the 200 mV range, it is necessary to step up the voltage to drive the envelope detector; therefore, a tuned transformer that has a 1:4 turns ratio is used at the second-stage output. The total effective circuit Q for this IF configuration is 200, and the full voltage gain is 86 dB from the input of the first stage to the output of the stepup transformer.

SECTION **7** Communications Circuits

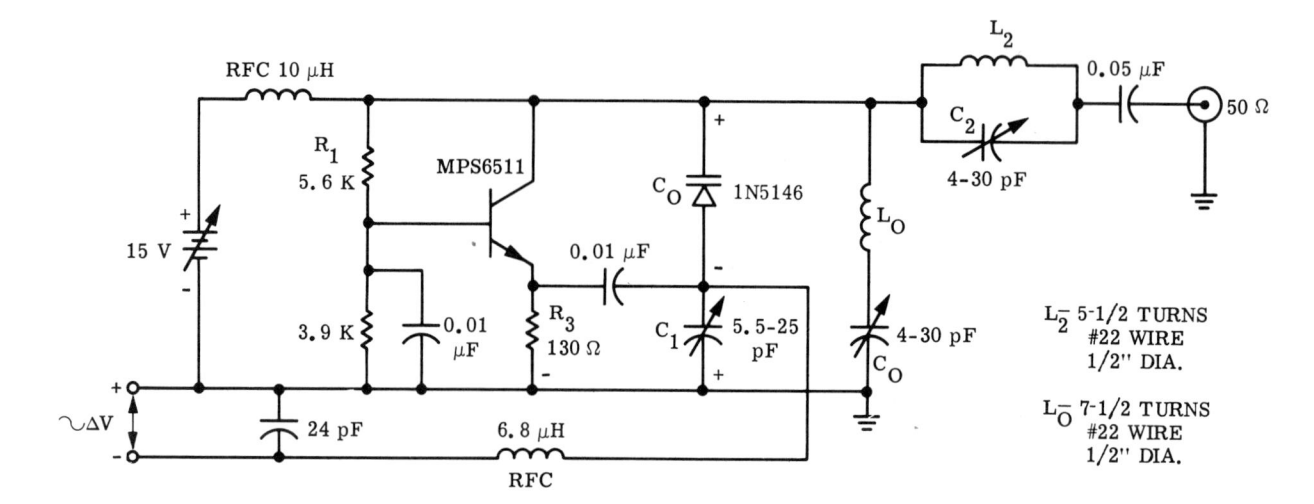

This 52 MHz frequency-modulated oscillator uses a varactor and a Motorola MPS6511 transistor designed especially for oscillator applications. The varactor is a Motorola 1N5146, a silicon epitaxial passivated diode rated at 33 pF for a reverse-bias voltage of −4V. L2 and C2 are used to tune out unwanted harmonics; their combined effect approaches a short circuit at the frequency of oscillation. The 6.8 μH RFC and the 24 pF capacitor in the input circuit are used to isolate the external input circuit from the RF circuit of the oscillator. Circuit operation of prototypes was limited to voltage inputs of ±200 mV or less, with the frequency deviations on the order of ±60 kHz or less.

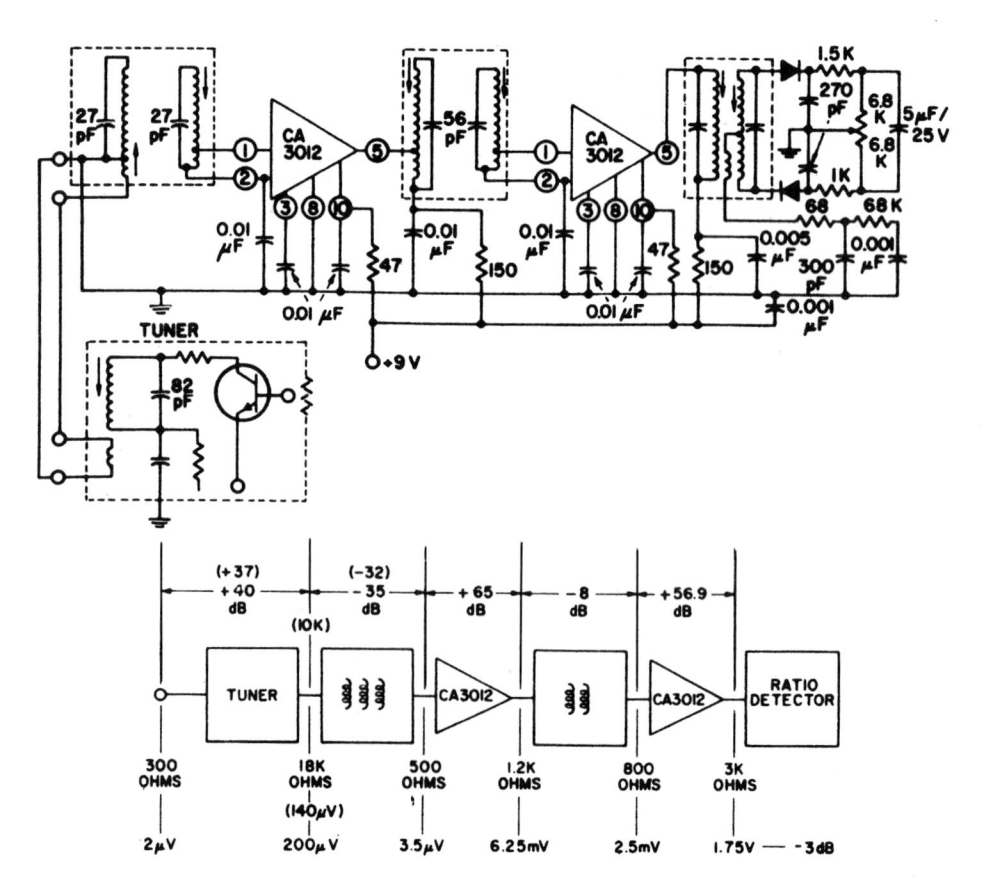

A complete 10.7 MHz IF amplifier strip using two CA3012 ICs. Inset shows voltage gain and impedance values. Values are given for two levels of mixer output impedance. All other impedance levels shown have exhibited good stability. The capture ratio varies from 5 dB at 2 μV to 1.2 dB above 500 μV. With careful adjustment, values as low as 0.8 dB can be obtained.

290 IC SCHEMATIC SOURCEMASTER

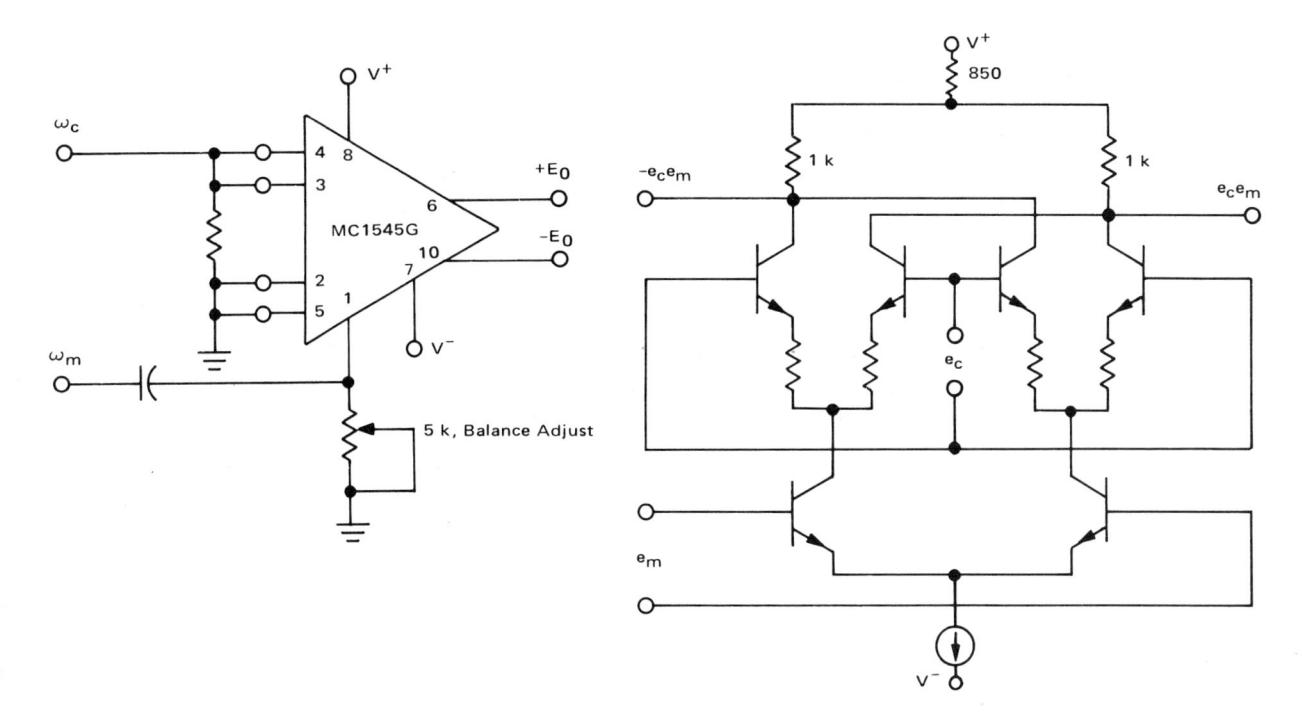

When the MC1545 is connected as shown, the function of **balanced modulation** can be obtained. Here, the internal differential amplifiers have been connected in a manner that cross-couples the collectors. Considering a carrier level adequate to switch the cross-coupled pair of differential amplifiers, the modulation signal applied to the gate will be switched between collector loads at the carrier rate, resulting in multiplying the modulation by a symmetrical switching function.

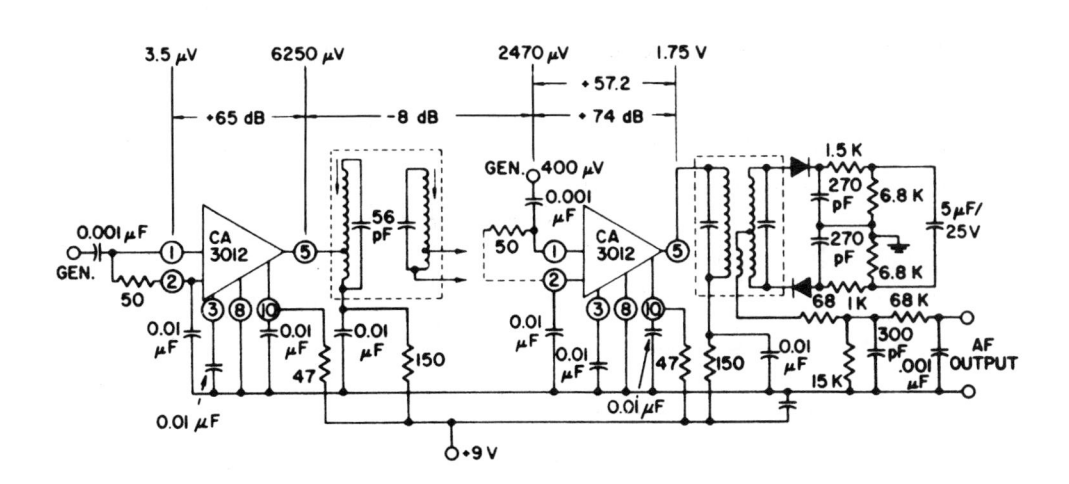

A 10.7 MHz IF amplifier strip using CA3012 integrated circuit. A double-tuned filter that has a voltage insertion loss of 8 dB is located between the two CA3012 units to provide a filter input of approximately 1000 μV (at terminal 5 of the first CA3012). For an IF strip sensitivity of 4 μV, a gain of 48 dB is required. However, if the CA3012 used has a load impedance of 1200Ω, the available gain is 65 dB, or approximately 17 dB more than required. The extra gain is not wasted but drives the second CA3012 harder, causing it to limit so that its gain is reduced by approximately 17 dB.

SECTION 7 Communications Circuits

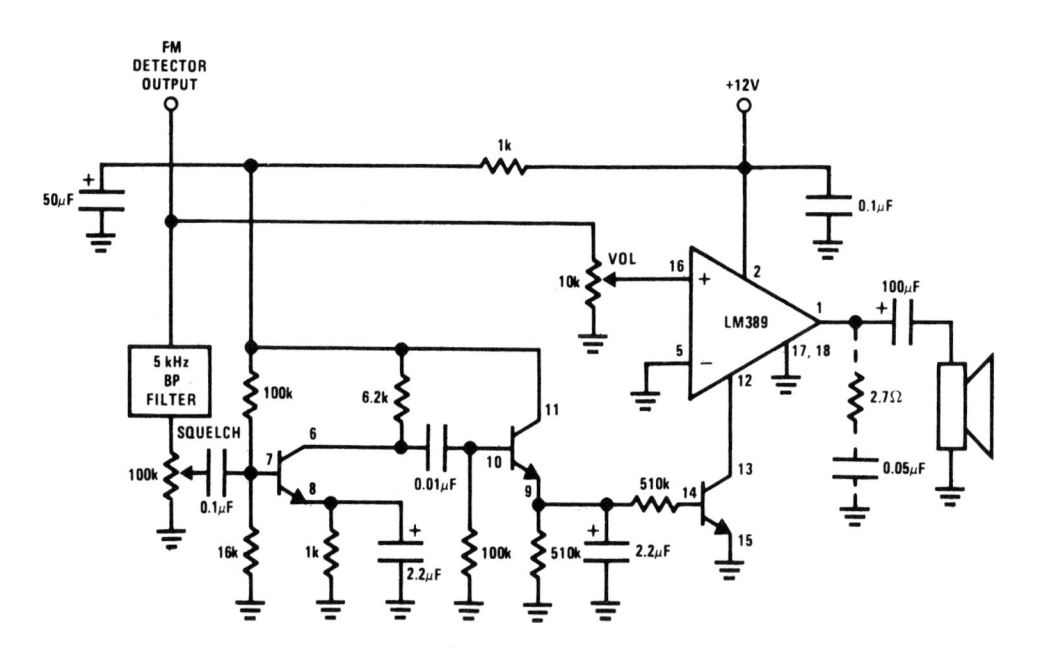

FM scanner noise squelch circuit. Transistors are in IC package.

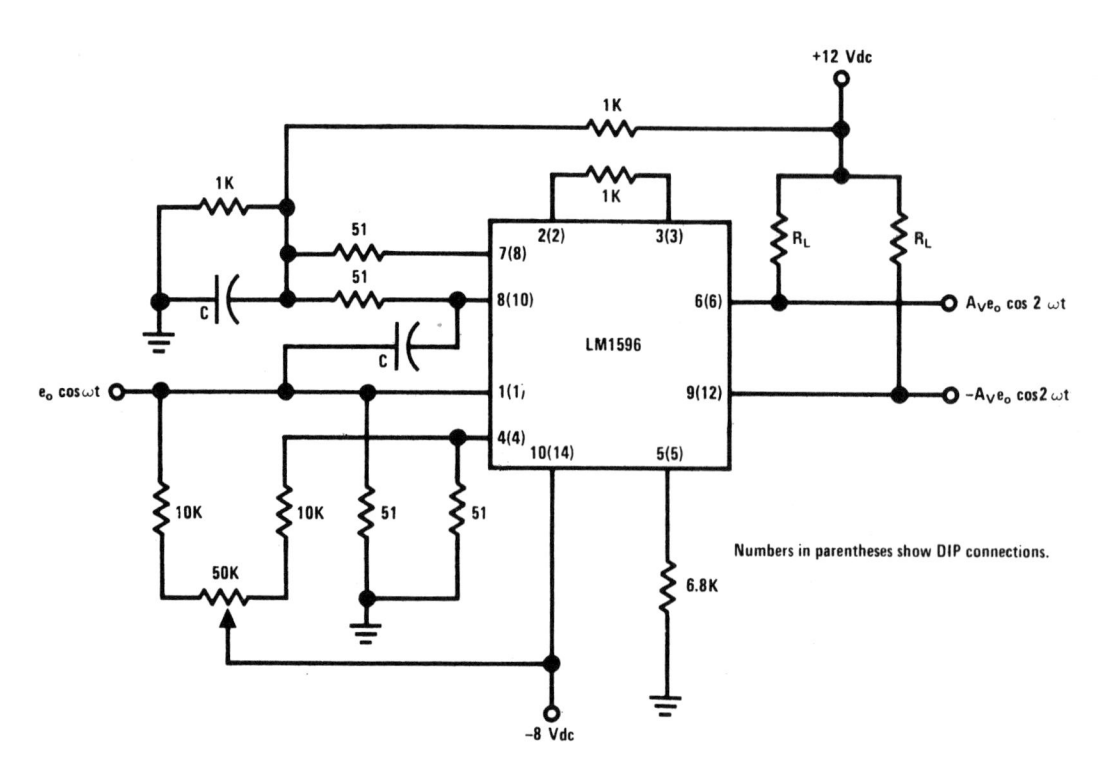

Frequency-doubler circuit doubles low-level signals with low distortion. The value of C should be chosen for low reactance at the operating frequency. Signal level at the carrier input must be less than 25 mV peak to maintain operation in the linear region of the switching differential amplifier. Levels to 50 mV peak may be used with some distortion of the output waveform. If a larger input signal is available, a resistive divider may be used at the carrier input, with full signal applied to the signal input.

292 IC SCHEMATIC SOURCEMASTER

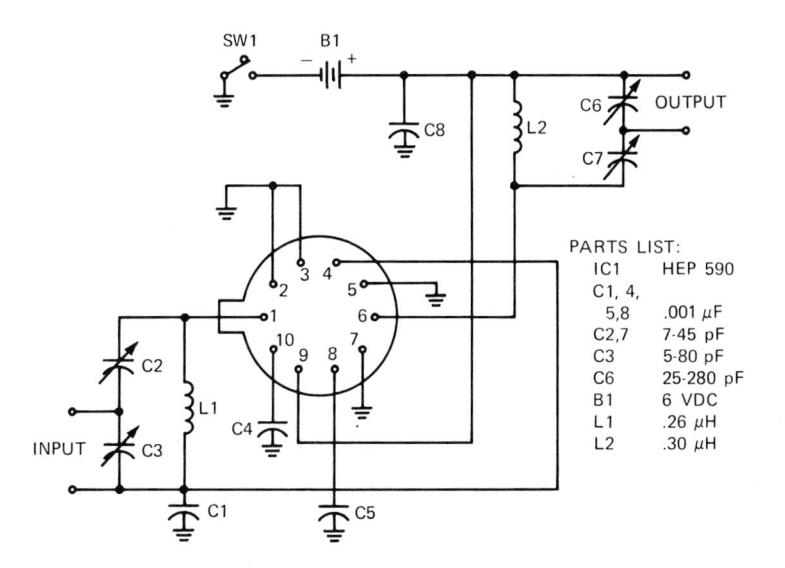

A 6-meter RF preamplifier offers full 30 dB of signal gain with a 600 kHz bandwidth. Tubable capacitors at input and output are used for matching impedance of preamp to requirements of antenna and receiver. AGC can be effectively added to pin 5 of the HEP 590. Coils should be "grid-dipped" for center frequency of about 20.25 MHz.

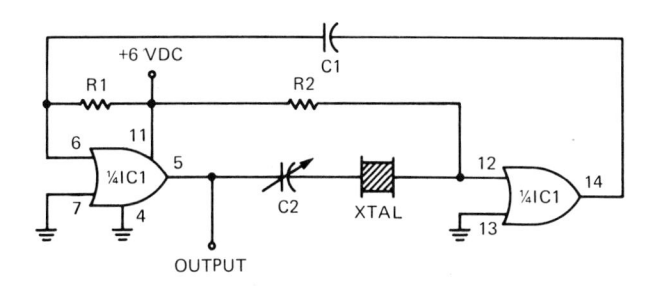

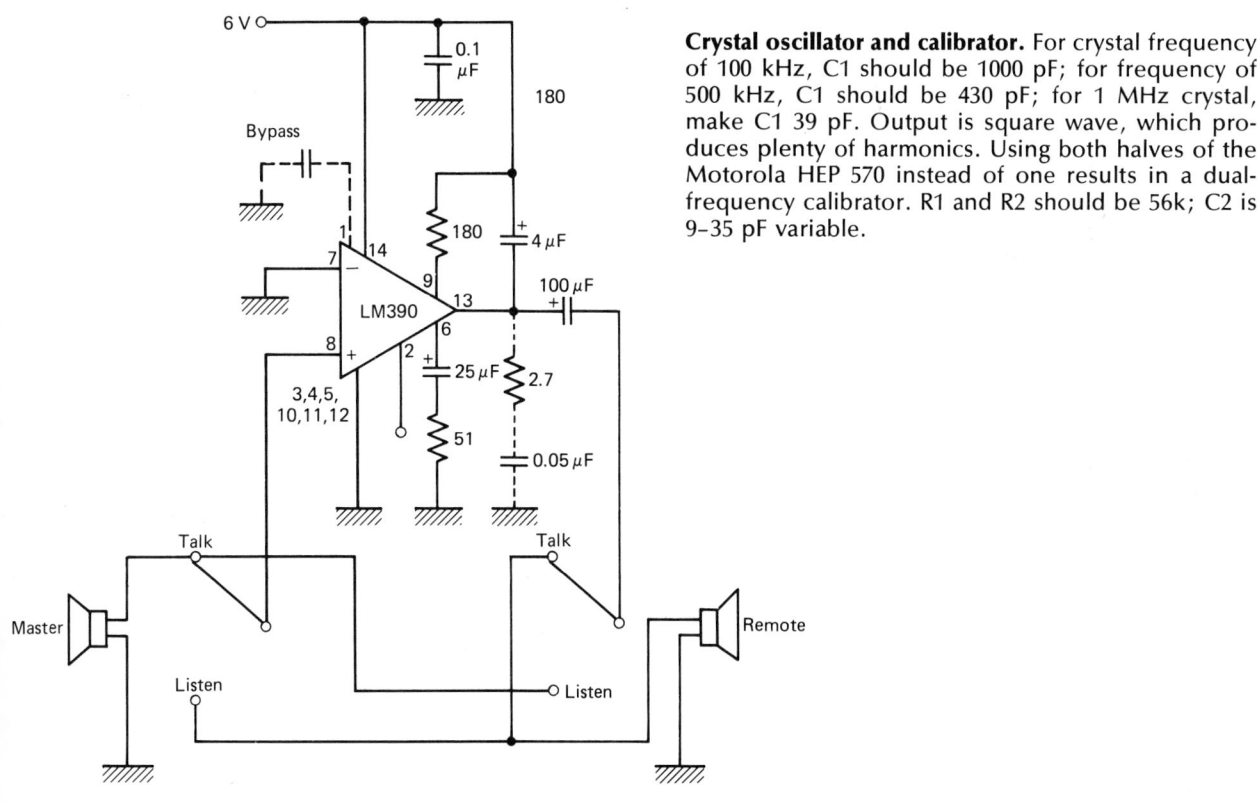

PARTS LIST:
IC1   HEP 590
C1, 4, 5, 8   .001 μF
C2, 7   7-45 pF
C3   5-80 pF
C6   25-280 pF
B1   6 VDC
L1   .26 μH
L2   .30 μH

Crystal oscillator and calibrator. For crystal frequency of 100 kHz, C1 should be 1000 pF; for frequency of 500 kHz, C1 should be 430 pF; for 1 MHz crystal, make C1 39 pF. Output is square wave, which produces plenty of harmonics. Using both halves of the Motorola HEP 570 instead of one results in a dual-frequency calibrator. R1 and R2 should be 56k; C2 is 9–35 pF variable.

Intercom.

SECTION **7** Communications Circuits   **293**

Circuit

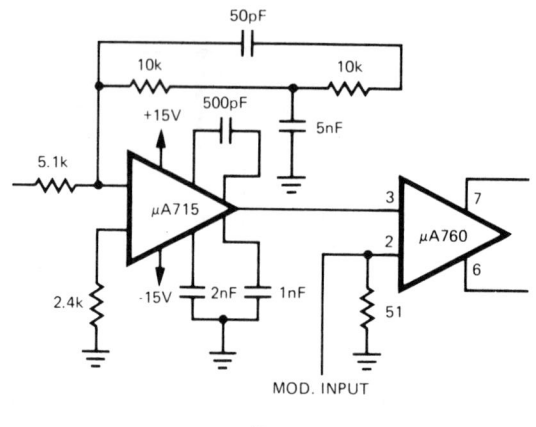

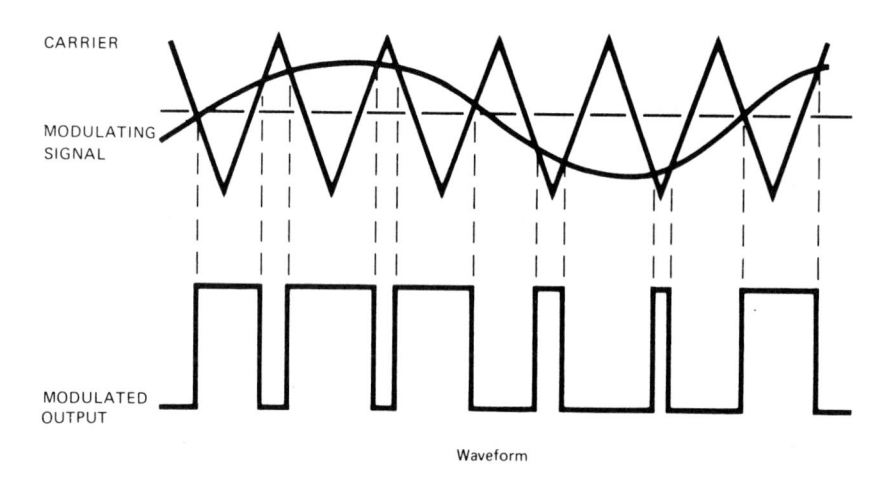

Waveform

This pulse-width modulator uses a Fairchild μA760. The μA715 is connected as a ramp generator, and the ramp output is fed into one input of the μA760 while the other input of the μA760 is fed from an audio signal. The μA760 outputs will change state each time the audio input equals the ramp voltage. The μA760 outputs vary in pulse width as a function of the instantaneous value of the audio input.

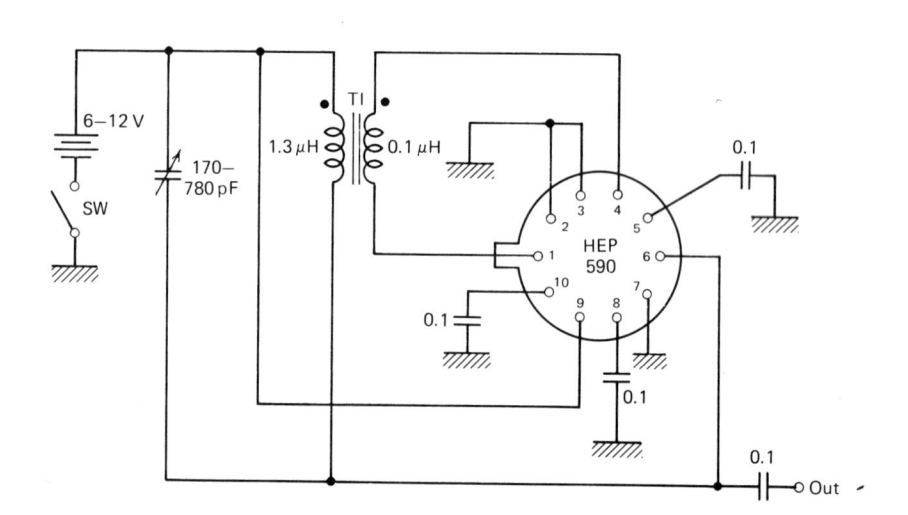

This variable-frequency oscillator tunes from 5 to 10 MHz and can be used as a complete 40-meter transmitter without modification. Output impedance is 50Ω. Output over full range is a near-ideal sine wave.

T1 can be wound with 36 AWG wire on Micrometals T-12-2 core; for primary, use 21 turns, and for secondary, use 7 turns—space all turns about the diameter of wire.

294 IC SCHEMATIC SOURCEMASTER

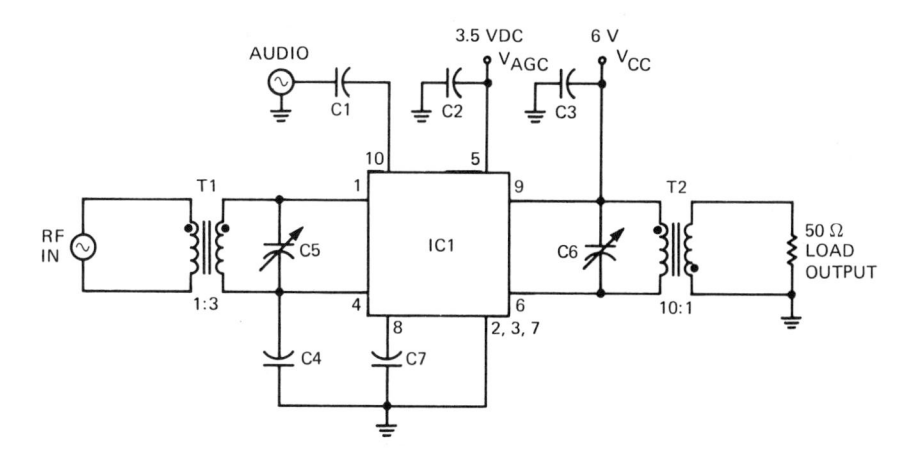

This 6-meter (50–54 MHz) amplitude modulator gives 90% modulation with very little distortion for real RF punch! T1 and T2 can be wound on Micrometal T-12-2 cores: T1 ratio is 6 turns: 18 turns of 32-gage solid wire spaced at wire diameter. T2 ratio is 28:3 of 36-gage wire at same spacing. For tuneup, set C5 and C6 to 52.25 MHz. Components are as follows: IC1 = HEP590; C1 = 1 μF; C2 and 3 = 0.47 μF; C4 and 7 = 0.1 μF; C5 and 6 = 1.5 to 10 pF variables.

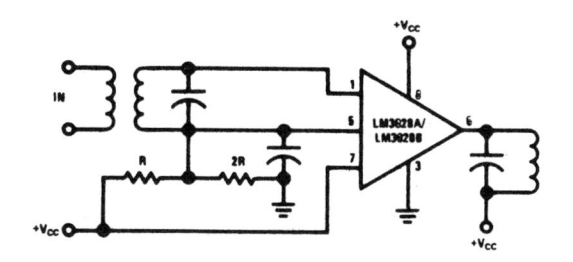

A balanced differential amplifier with a controlled constant-current-source drive and AGC capability.

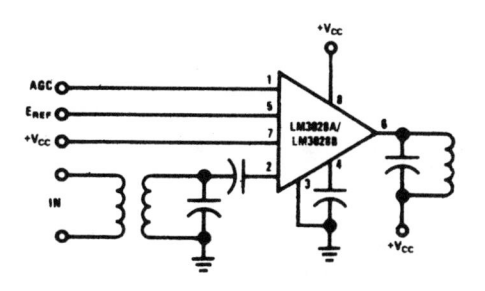

A cascode amplifier with a constant-impedance AGC capability

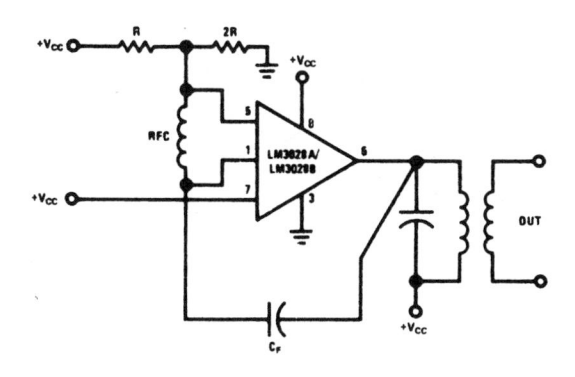

Oscillator

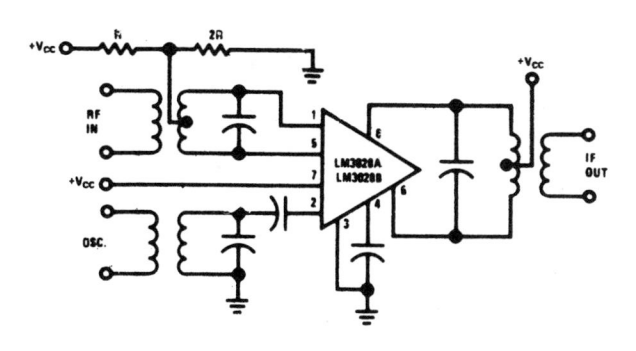

Mixer

A monolithic RF/IF amplifier intended for emitter-coupled (differential) or cascode amplifier operation from dc to 120 MHz in industrial and communications equipment. The LM3028A/LM3028B and LM3053 are plug-in replacements for the CA3028A/CA3028B and CA3053, respectively. The LM3028B is similar to the LM3028A but has premium performance with tighter limits in offset voltage and current, bias current, and voltage gain. The LM3053 is similar to the LM3028A/LM3028B but is recommended for IF amplifier operation with less critical dc parameters.

SECTION **7** Communications Circuits

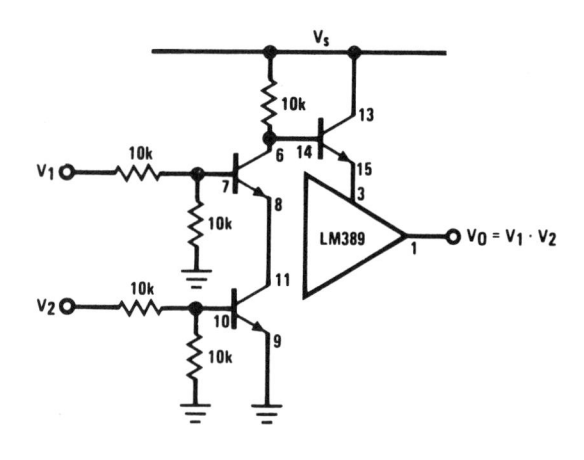

AND Muting

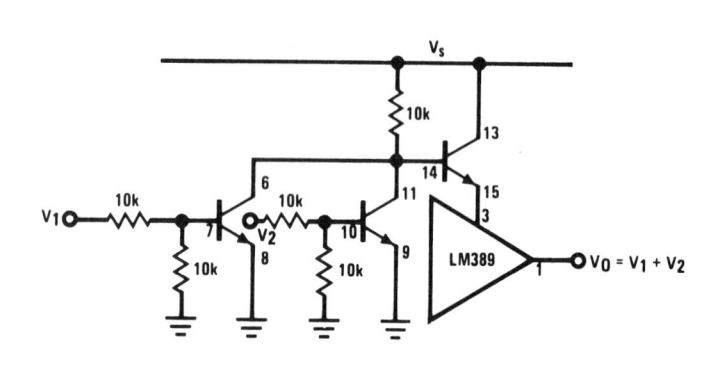

OR Muting

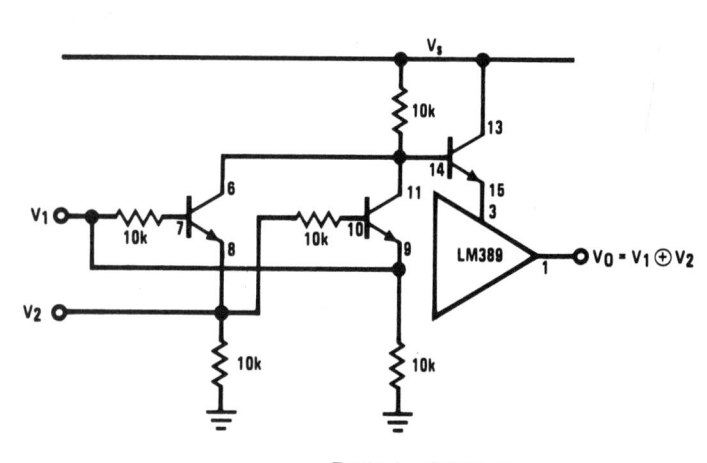

Exclusive-OR Muting

Logic-controlled mute. Various logic functions are possible with the three NPN transistors of the LM389, making logic control of the mute function possible. Here, standard AND, OR, and exclusive-OR circuits are shown for controlling the muting transistor. Using the optional mute scheme of shorting pin 12 to ground gives NAND, NOR, and exclusive-NOR operation.

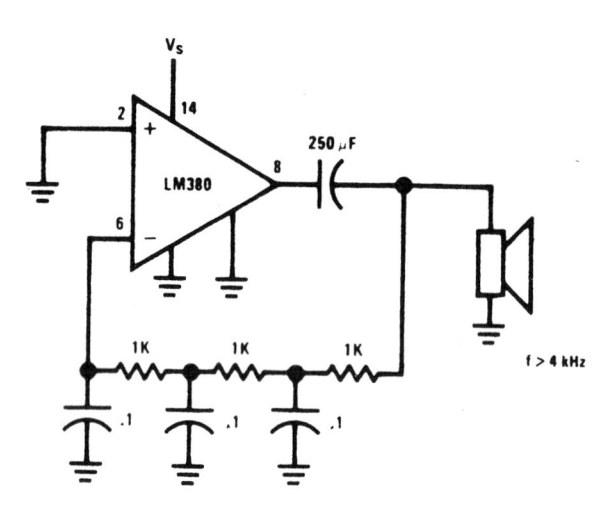

This phase-shift oscillator produces a nice sine wave for code practice or tone encoder applications.

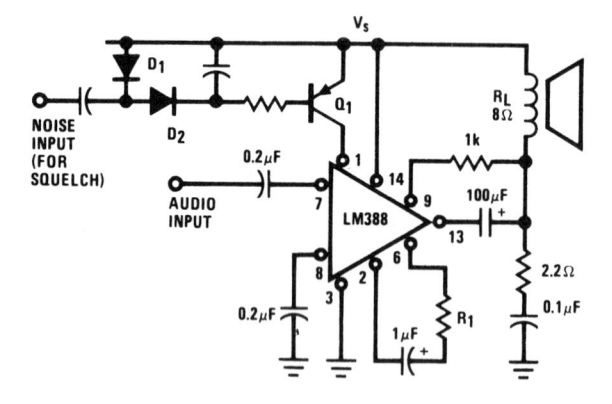

Hand-held portable squelched amplifier for FM two-way units. Diodes D1 and D2 rectify noise from the limiter or the discriminator of the receiver, producing dc to turn on Q1, which clamps the LM388 off. As shown, the following performance is obtained:

Voltage gain equals 20 to 200 (selectable with R1).
Noise (output squelched) equals 20 μV.
P_o = 0.53W (V_S = 7.5V, R_L = 8Ω, THD = 5%).
P_o = 0.19W (V_S = 4.5V, R_L = 8Ω, THD = 5%).
Current consumption (V_S = 7.5V): squelched, 0.8 mA; P_o = 0.5W, 110 mA.

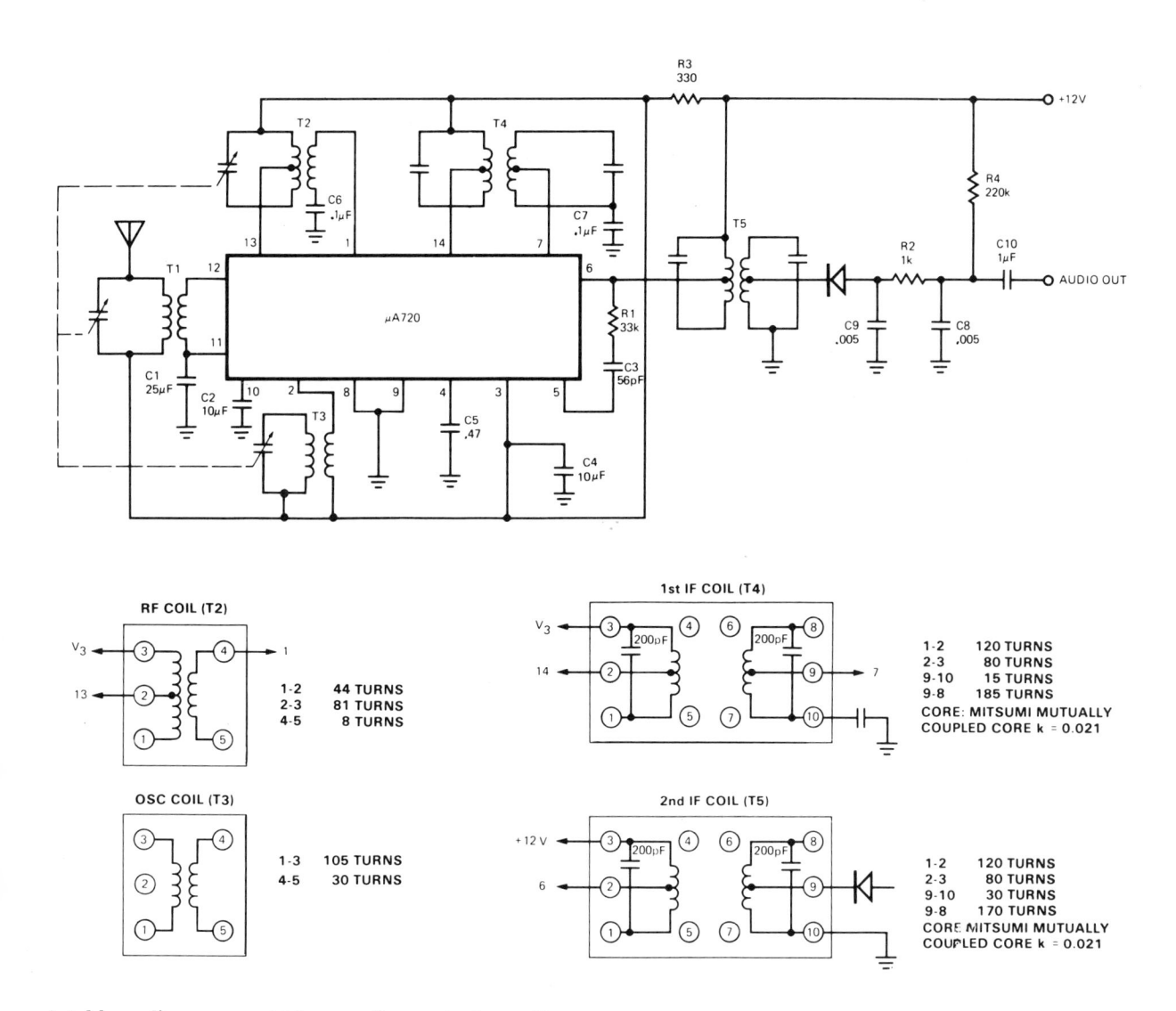

A table radio uses variable-capacitance tuning with an IF of 455 kHz.

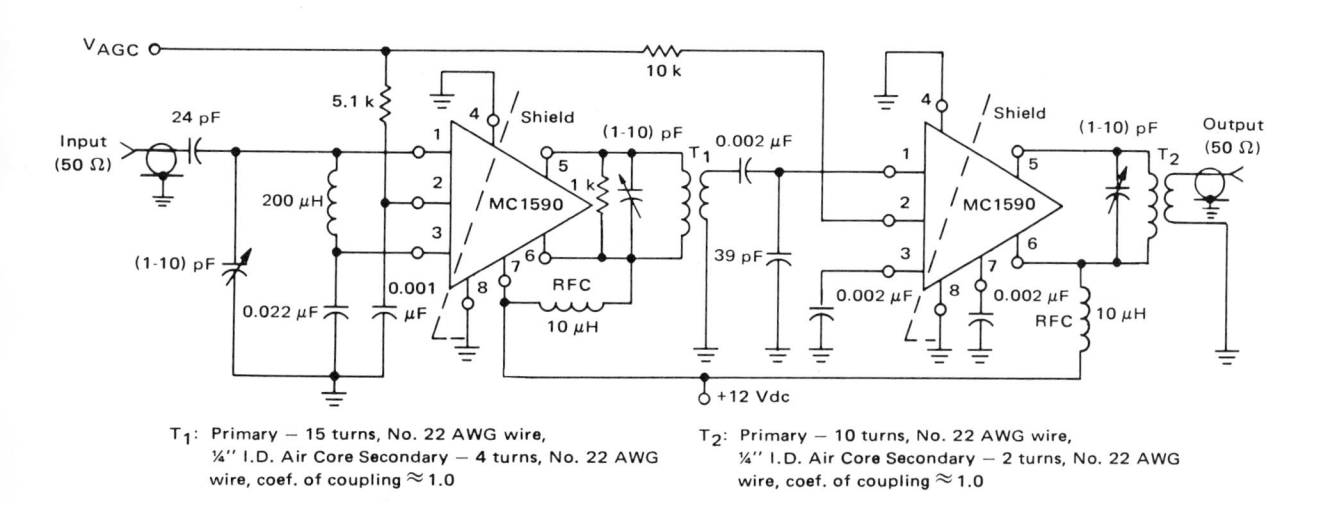

A two-stage 60 MHz IF amplifier with power gain of 80 dB and bandwidth of about 1.5 MHz.

SECTION **7** *Communications Circuits* **297**

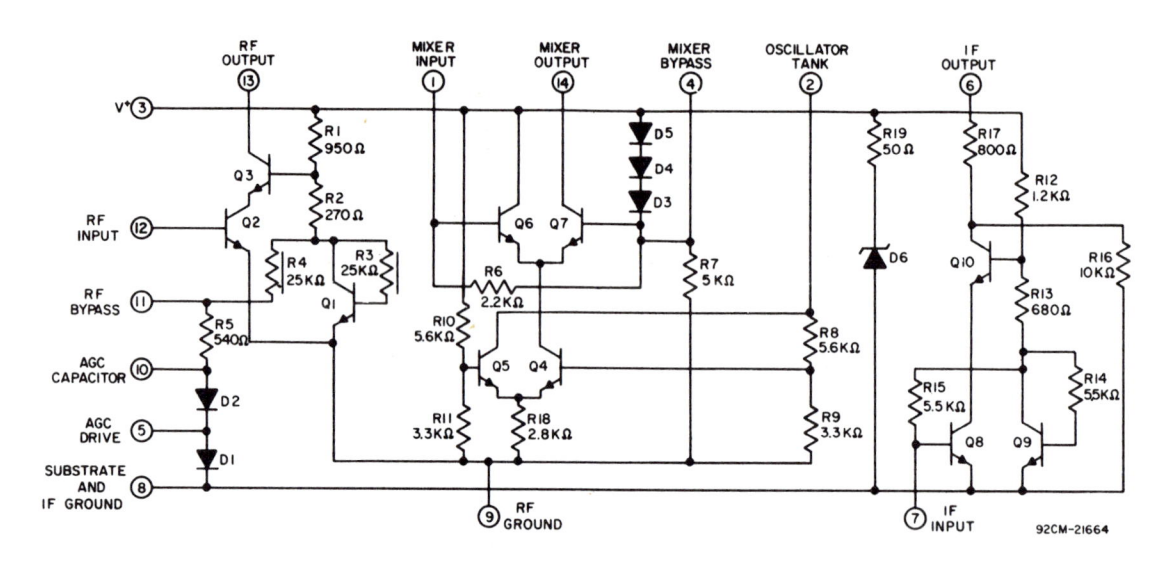

Schematic diagram of CA3123E.

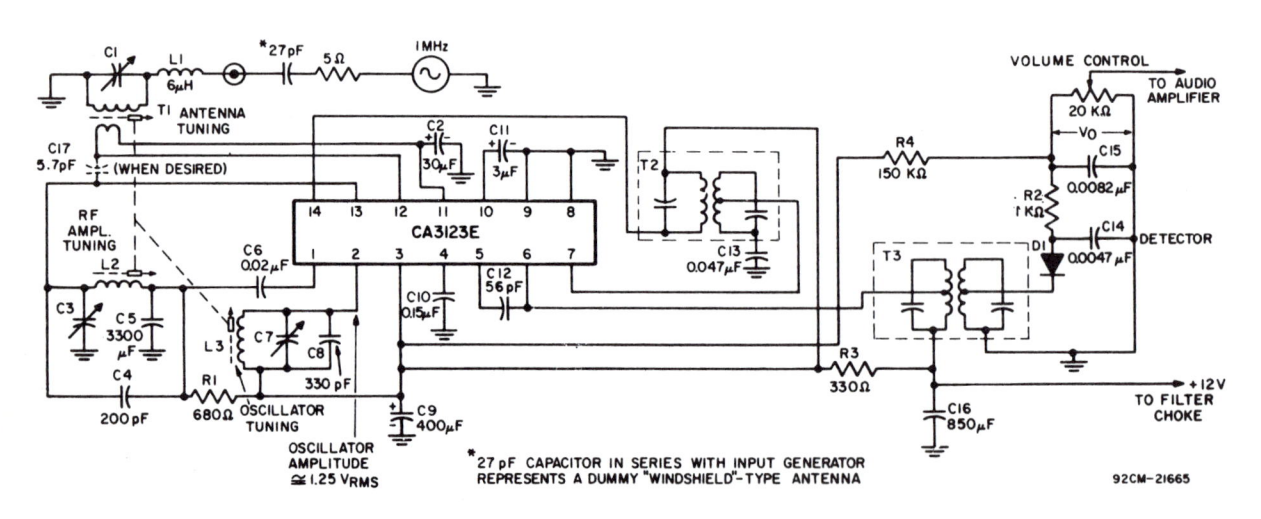

27 pF CAPACITOR IN SERIES WITH INPUT GENERATOR REPRESENTS A DUMMY "WINDSHIELD"-TYPE ANTENNA

Transformer	Symbol	Frequency	Inductance µh ($\approx$)	Capacitance pF ($\approx$)	Q ($\approx$)	Total Turns To Tap Turns Ratio	Coupling
First IF: Primary	T_2	262 kHz	2840	130	60	none	critical
Secondary			2840	130	60	or 30:1 31:1	$\approx 0.017 \approx 1/Q$
Second IF: Primary	T_3	262 kHz	2840	130	60	8.5:1	—
Secondary			2840	130	60	8.5:1	critical $\approx 0.017 \approx 1/Q$
Antenna: Primary	T_1	1 MHz	195	(C_1)–130	65		
Secondary		Adjusted to an impedance of 75 Ω with primary resonant at 1 MHz. Coupling should be as tight as practical. Wire should be wound around end of coil away from tuning core.					
Coils	L_1	7.9 MHz	6		50		
	L_2	1 MHz	55		50		
	L_3	1.262 MHz	41		40		

Schematic diagram of **AM radio receiver** using RCA CA3123E.

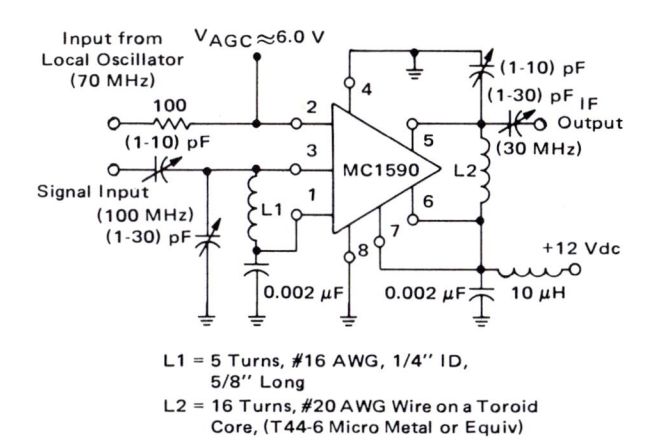

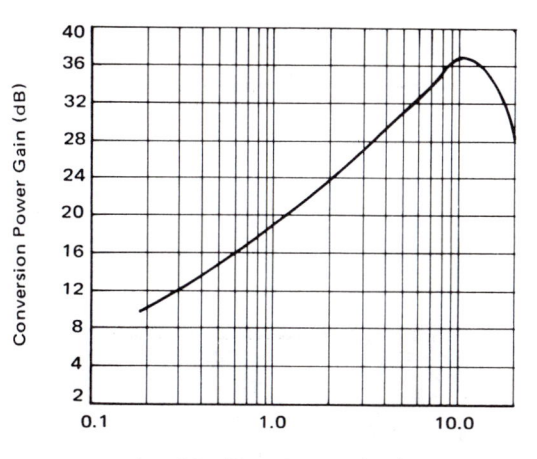

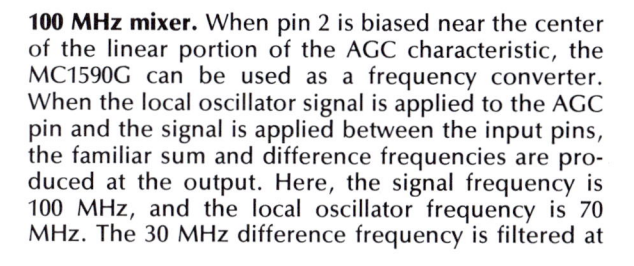

100 MHz mixer. When pin 2 is biased near the center of the linear portion of the AGC characteristic, the MC1590G can be used as a frequency converter. When the local oscillator signal is applied to the AGC pin and the signal is applied between the input pins, the familiar sum and difference frequencies are produced at the output. Here, the signal frequency is 100 MHz, and the local oscillator frequency is 70 MHz. The 30 MHz difference frequency is filtered at the output. Also shown is the mixer conversion gain versus local oscillator voltage for the test circuit. In addition to the exceptionally high conversion gain, the circuit also provides excellent isolation of the signal source from the local oscillator. With this frequency-conversion capacity, it is possible to construct a two-stage mixer–IF amplifier combination with a total power gain of 70 dB.

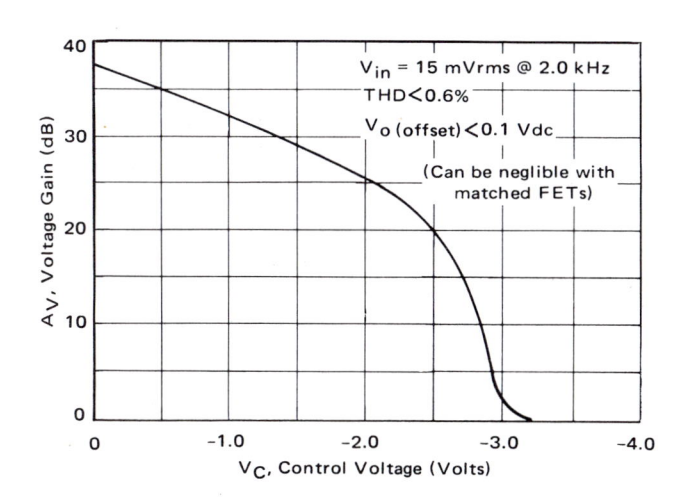

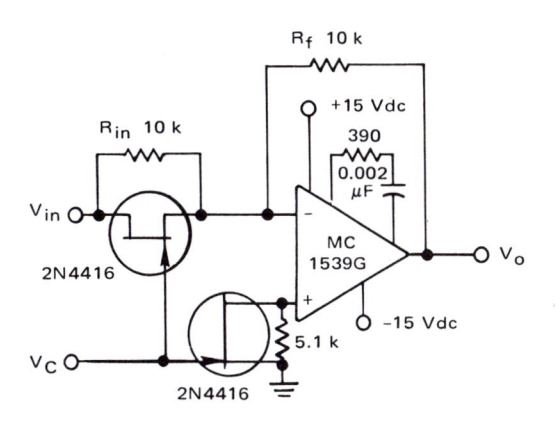

AGC amplifier. Here, the FET is used in conjunction with an operational amplifier; the gain is made variable by changing the ratio of R_f/R_{in}. This is a simple AGC circuit but with improved distortion characteristics. The FET in the inverting input line is sufficient for AGC, but owing to the changing impedance and the input bias current required by the op-amp, the dc level of the output also varies. Offset voltage in the output owing to input bias currents is normally minimized by placing a resistance equivalent to the parallel combination of R_f and R_{in} in the noninverting leg to ground. An FET is added, so that this resistance varies accordingly with the variation in R_{in}; this minimizes the change in the output offset voltage due to the changing R_{in}.

SECTION **7** *Communications Circuits* **299**

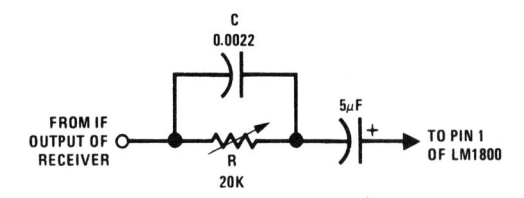

Compensation for receiver IF rolloff

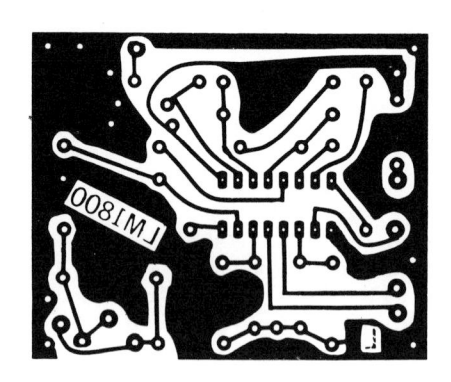

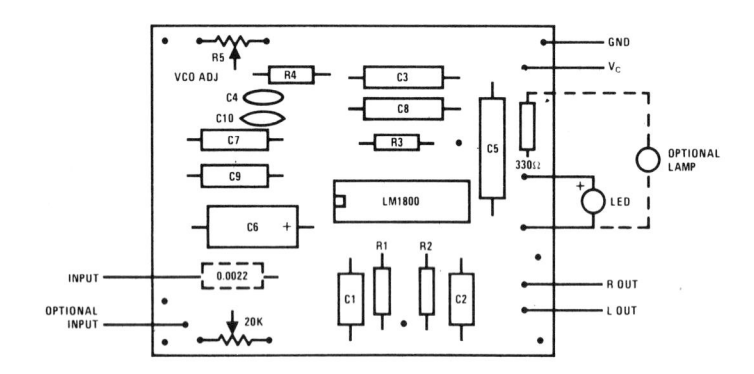

PCB layout for FM demodulator.

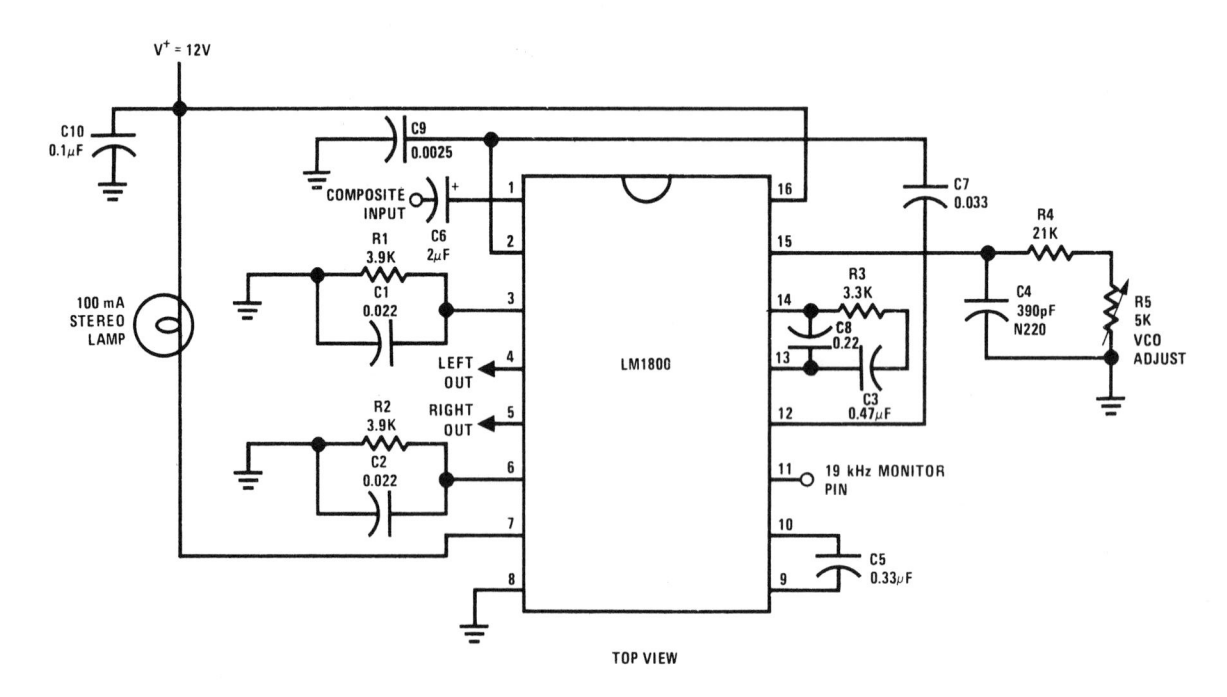

Phase-locked-loop FM stereo demodulator. In addition to separating left (L) and right (R) signal information from the detected IF output, the LM1800 features automatic stereo/monaural switching, 45 dB power supply rejection, and a 100 mA stereo indicator lamp driver. Particularly attractive is the low external part count and total elimination of coils. A single inexpensive potentiometer performs all tuning. The resulting FM stereo system delivers high-fidelity sound while still meeting the cost requirements of inexpensive stereo receivers.

300 IC SCHEMATIC SOURCEMASTER

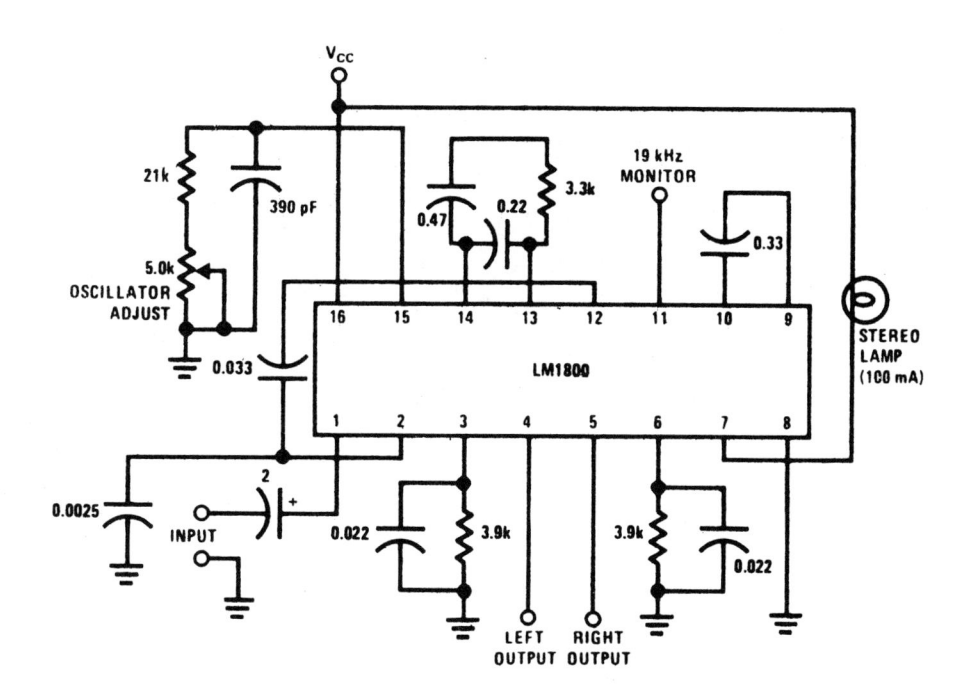

Phase-locked-loop FM stereo demodulator. The numerous features integrated on the die make possible a system delivering high-fidelity sound while still meeting the cost requirements of inexpensive stereo receivers. Features include automatic stereo/monaural switching; 45 dB power supply rejection; no coils, since all tuning is performed with single potentiometer; wide operating supply voltage range; excellent channel separation; and emitter-follower output buffers.

SECTION **7** *Communications Circuits*

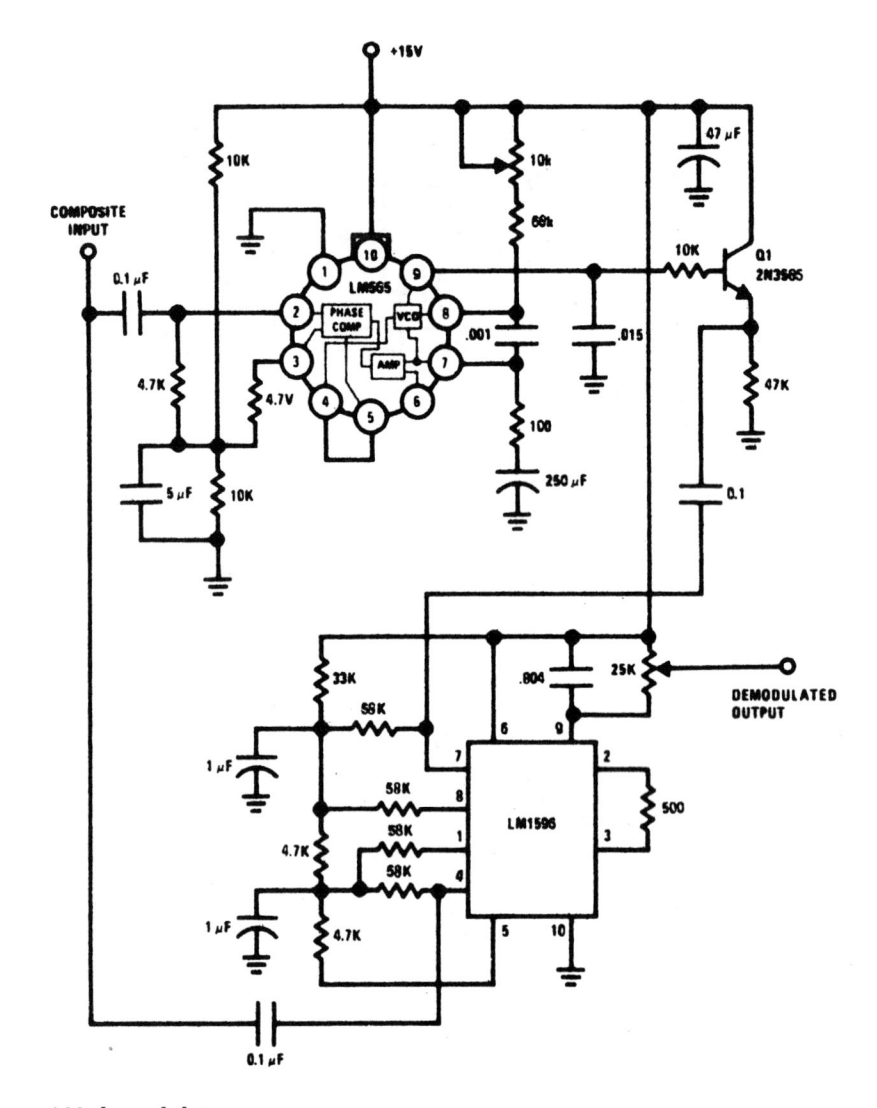

2400 Hz synchronous AM demodulator.

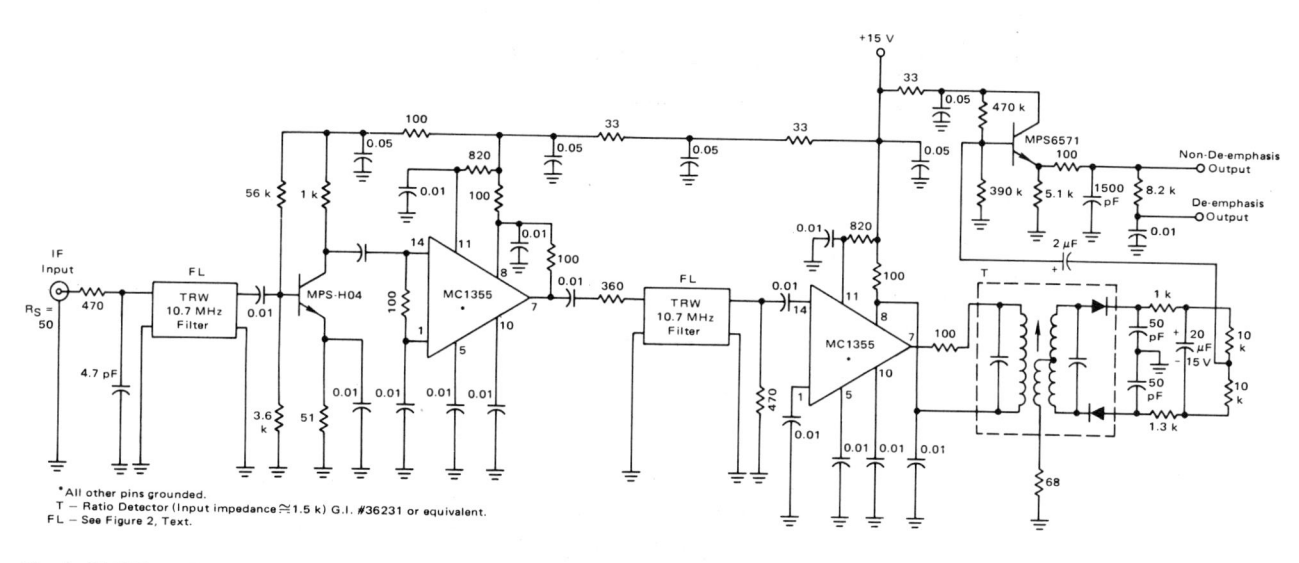

Dual MC1355 FM IF strip with high-performance capability. Specifications are as follows:

Supply:	+15V, 35 mA
Sensitivity (using deemphasis output):	5.6 µV for 3% THD, 5 µV for −3 dB limiting
Selectivity (V_{IN} = 4 µV):	70 dB at ±400 kHz
Total harmonic distortion:	0.18% for 100 µV input <0.5% for 16 µV to >100 mV input (using deemphasis output)
Frequency response:	<5 Hz to 50 kHz for < −1 dB (using nondeemphasized output)
Signal-to-noise ratio:	76 dB for 1000 µV input, Δf = 75 kHz (using deemphasis output)
AM rejection:	48 dB for V_{IN} = 18 µV (V_{IN} = 10 dB above that required for 3% THD)

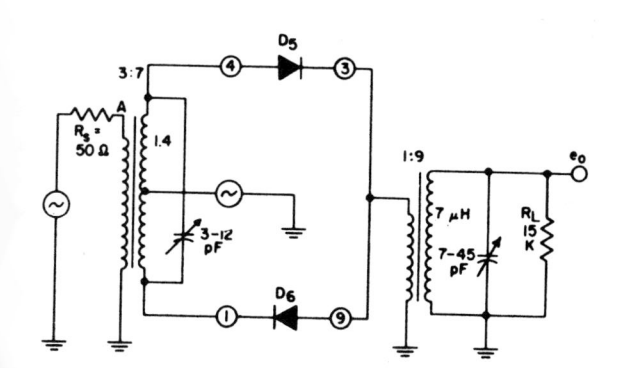

A balanced mixer using the CA3019. 45 MHz in; 55 MHz oscillator.

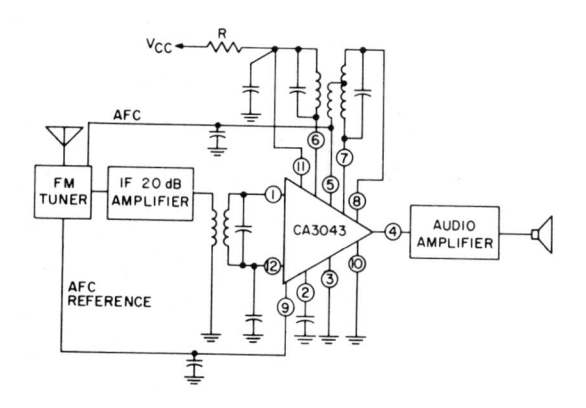

A high-gain limiter and amplifier/detector for FM receiver using CA3043.

SECTION 7 *Communications Circuits* 303

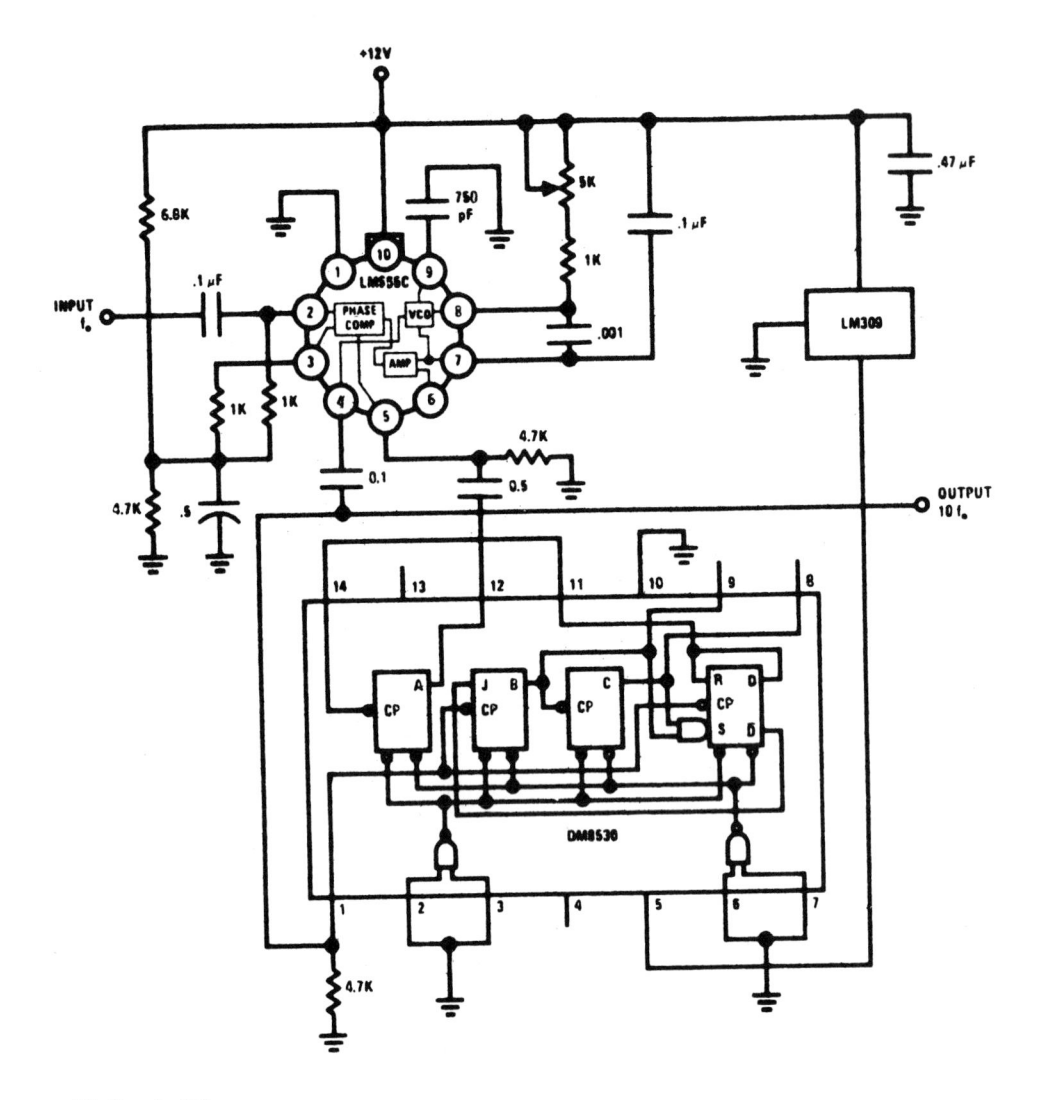

Frequency multiplier (×10).

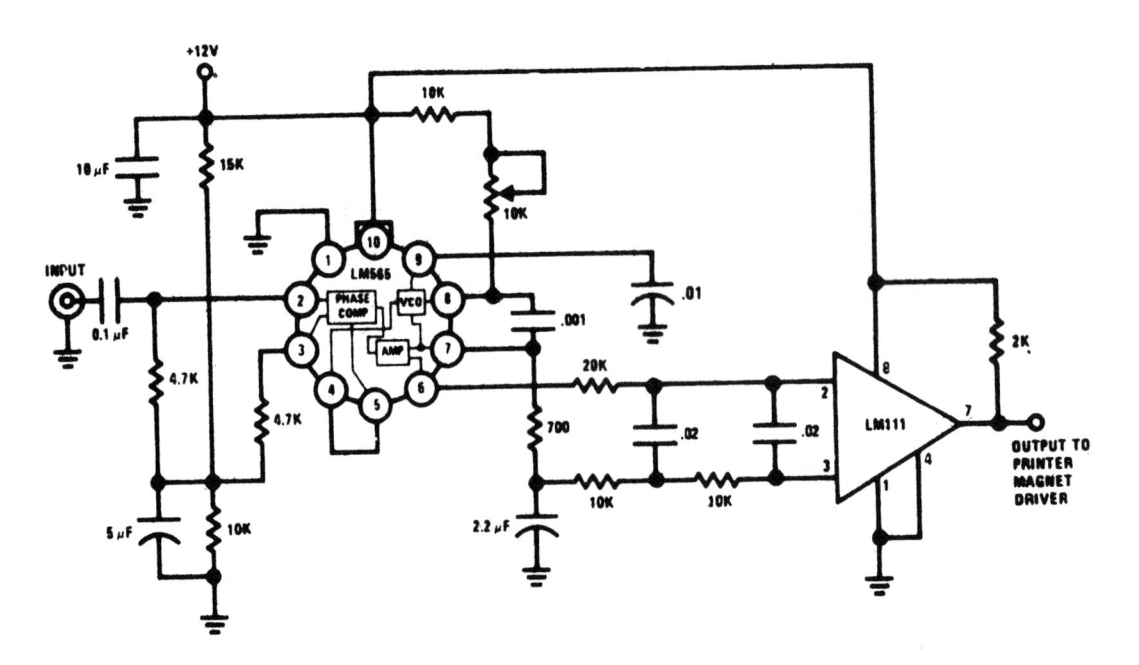

FSK demodulator (2025–2225 Hz).

304 IC SCHEMATIC SOURCEMASTER

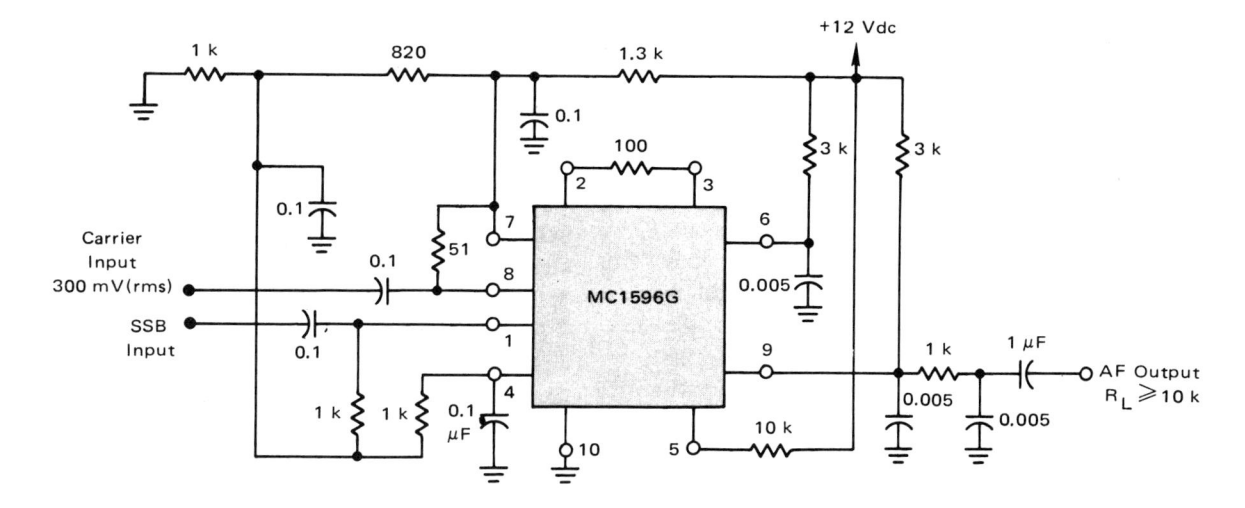

SSB product detector. All frequencies except the desired demodulated audio are in the RF spectrum and can be easily filtered in the output. As a result, the carrier null adjustment need not be included.

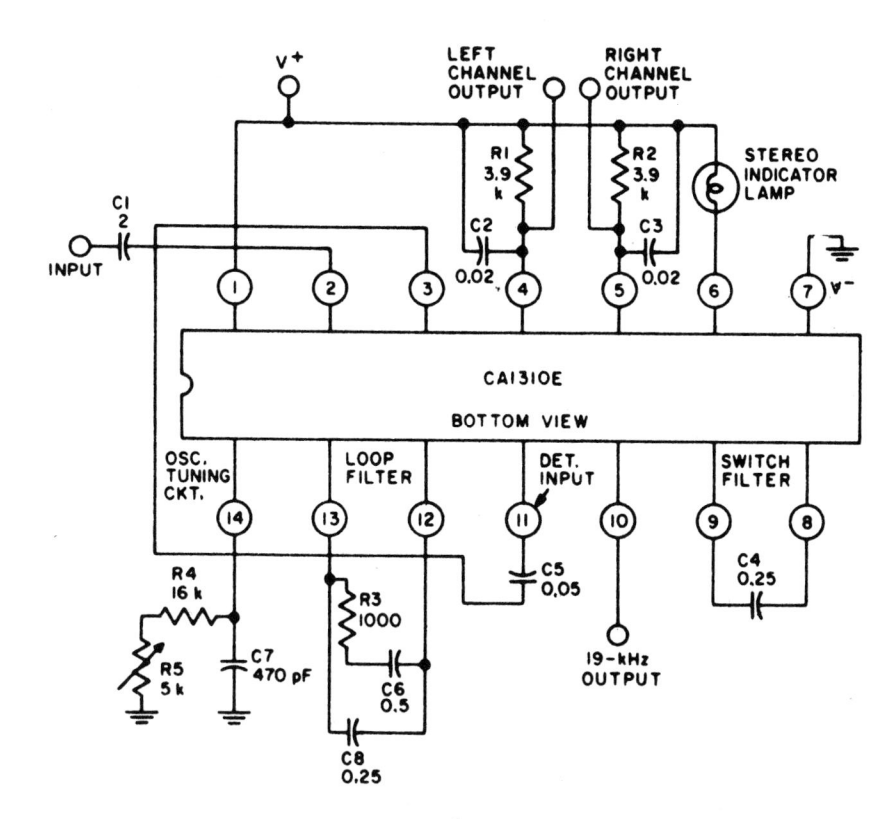

RC phase-locked-loop stereo decoder for FM multiplex systems features low THD (0.3%), excellent SCA rejection (75 dB), and high audio channel separation (40 dB). It requires only one adjustment for complete alignment. Despite these high-performance features, this decoder uses a minimum of external components and requires only one adjustment (oscillator frequency) for complete alignment.

A buffered 3V positive-going square wave is available at IC pin 10. The alignment of the free-running oscillator frequency may be checked at this point with a frequency counter. A lower value input coupling capacitor (C1) may be used in place of the 2 μF value if reduced separation at low frequencies is acceptable.

The time constant for the stereo switch level detector circuit is calculated by C4 × 53,000Ω ±30%, with a maximum dc voltage drop across C4 of 0.25V (pin 8 positive) and a pilot level voltage of 100 mV RMS. Signal voltage across C4 is negligible.

The recommended 0.05 μF capacitor for C5 provides a 1.75° phase lead at 19 kHz. Load resistance values are related to supply voltage as follows:

Minimum Supply Voltage	8V	10V	12V
R1, R2 maximum load resistance (Ω)	2700	4300	6200

If a capture range greater than ±3% is required, reduce the value of C7 and increase the values of R4 and R5 proportionally. However, beat-note distortion is increased at high signal levels because of oscillator phase jitter.

SECTION 7 Communications Circuits **305**

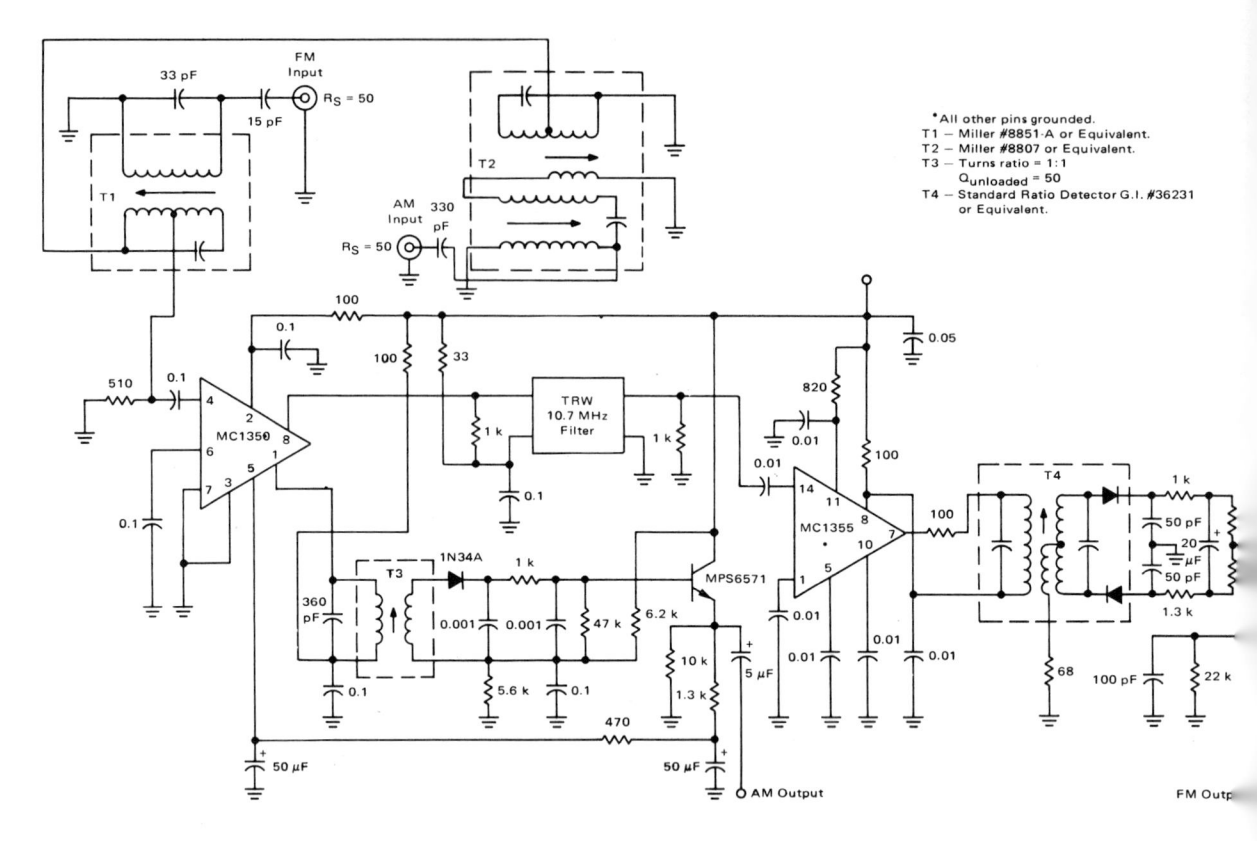

Composite AM/FM IF amplifier/detector. The two input filters are standard double-tuned Miller coils, while the FM block filter is a TRW linear phase filter. The AM detector coil is hand wound, although a standard AM detector coil could be used.

The source impedance of the MC1350 is set by the filters used, and the 10.7 MHz load impedance is set by the block filter used. For a Stern stability factor of 5 at 10.7 MHz, the required source and load impedances are 300 and 13,000Ω respectively. These load conditions are met by using the TRW filter and the input coils.

The two double-tuned tank circuits used for the input to the MC1350 may be wired in series, since the 10.7 MHz coil exhibits a low impedance at 455 kHz, and the 455 kHz coil appears as a low impedance at 10.7 MHz.

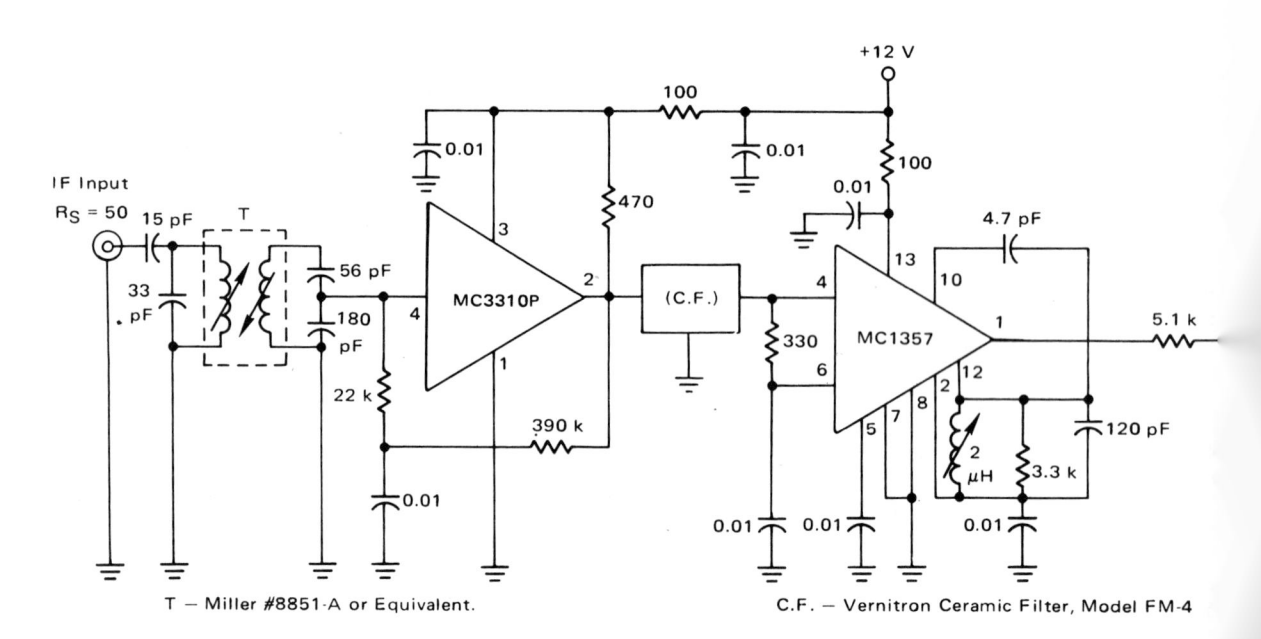

This quadrature-detector FM IF strip is ideal for automotive use because of the wide latitude of operating voltages (8 to 16V) that can be used. The four-pole ceramic filter provides most of the filtering in the IF amplifier.

306 IC SCHEMATIC SOURCEMASTER

Notes:

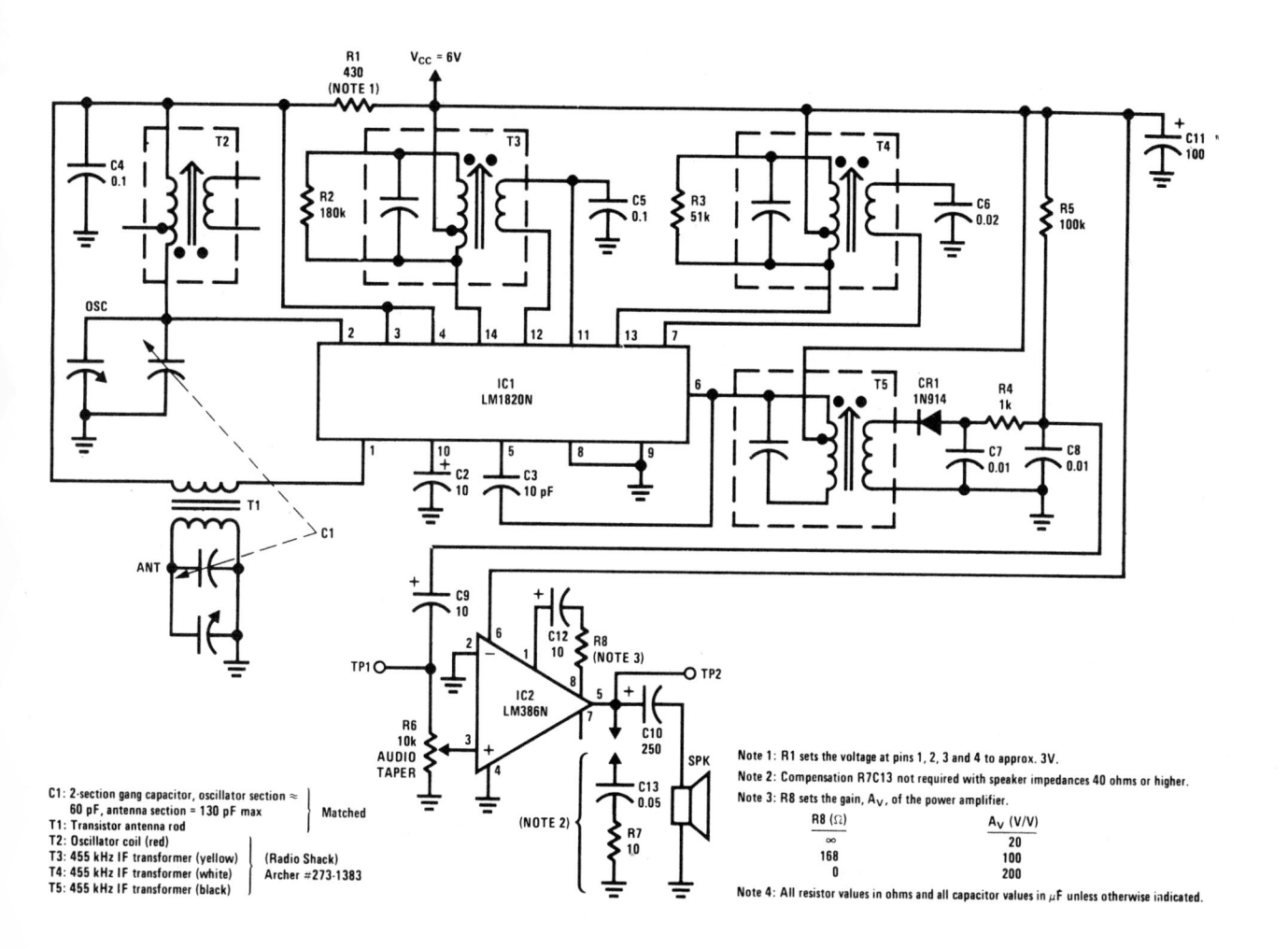

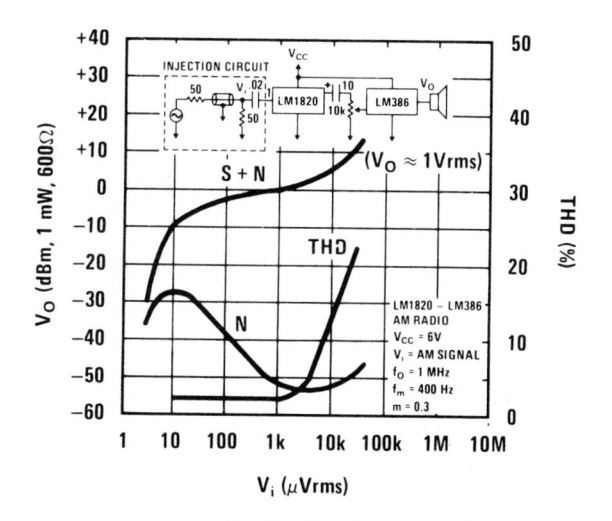

Note 1: R1 sets the voltage at pins 1, 2, 3 and 4 to approx. 3V.
Note 2: Compensation R7C13 not required with speaker impedances 40 ohms or higher.
Note 3: R8 sets the gain, A_V, of the power amplifier.

R8 (Ω)	A_V (V/V)
∞	20
168	100
0	200

Note 4: All resistor values in ohms and all capacitor values in μF unless otherwise indicated.

C1: 2-section gang capacitor, oscillator section ≈ 60 pF, antenna section = 130 pF max — Matched
T1: Transistor antenna rod
T2: Oscillator coil (red)
T3: 455 kHz IF transformer (yellow)
T4: 455 kHz IF transformer (white) (Radio Shack)
T5: 455 kHz IF transformer (black) Archer #273-1383

Radio Performance Plots

A low-cost AM radio system using LM1820 and LM386. Power output of 0.25W into an 8Ω speaker is obtained by the LM386, the gain of which is externally set to 200. Pins 1, 2, 3, and 4 are biased from the same supply through a 430Ω dropping resistor. This reduces the total current consumption to approximately 10 mA, making operation from a 6V battery feasible. The dc return of pins 1 and 4 to pin 3 improves component count and prevents transistor Q4 in the oscillator section from saturating. Large swings are preserved by returning the collectors at pins 14, 13, and 6 to V_{CC} via the primary windings of transformers T3, T4, and T5. For better linearity, detector 1N914 is biased slightly in the forward direction.

SECTION **7** Communications Circuits **307**

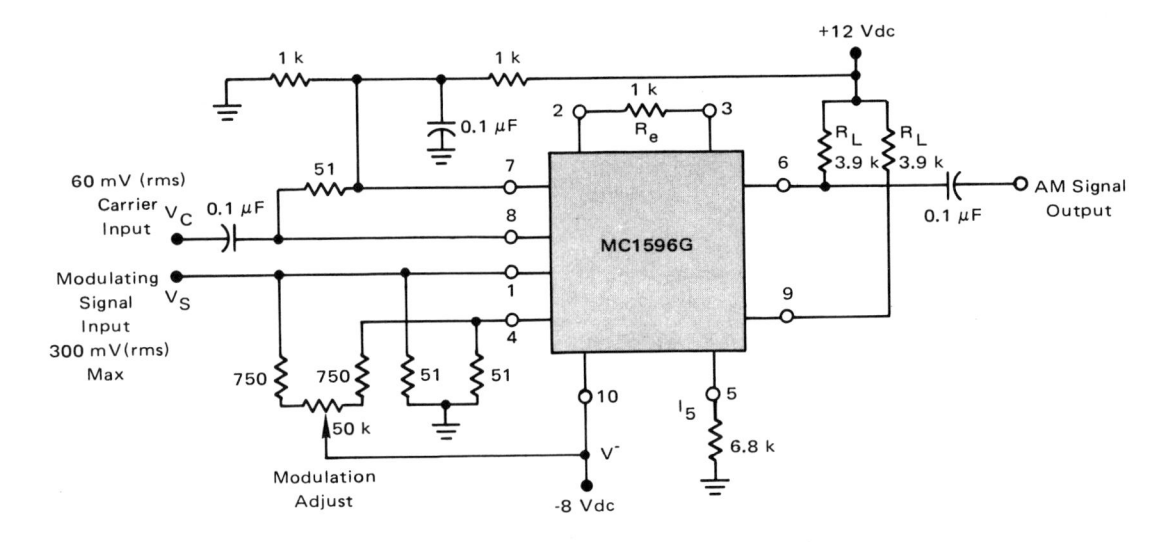

An amplitude modulator will provide excellent modulation at any percentage from zero to greater than 100%.

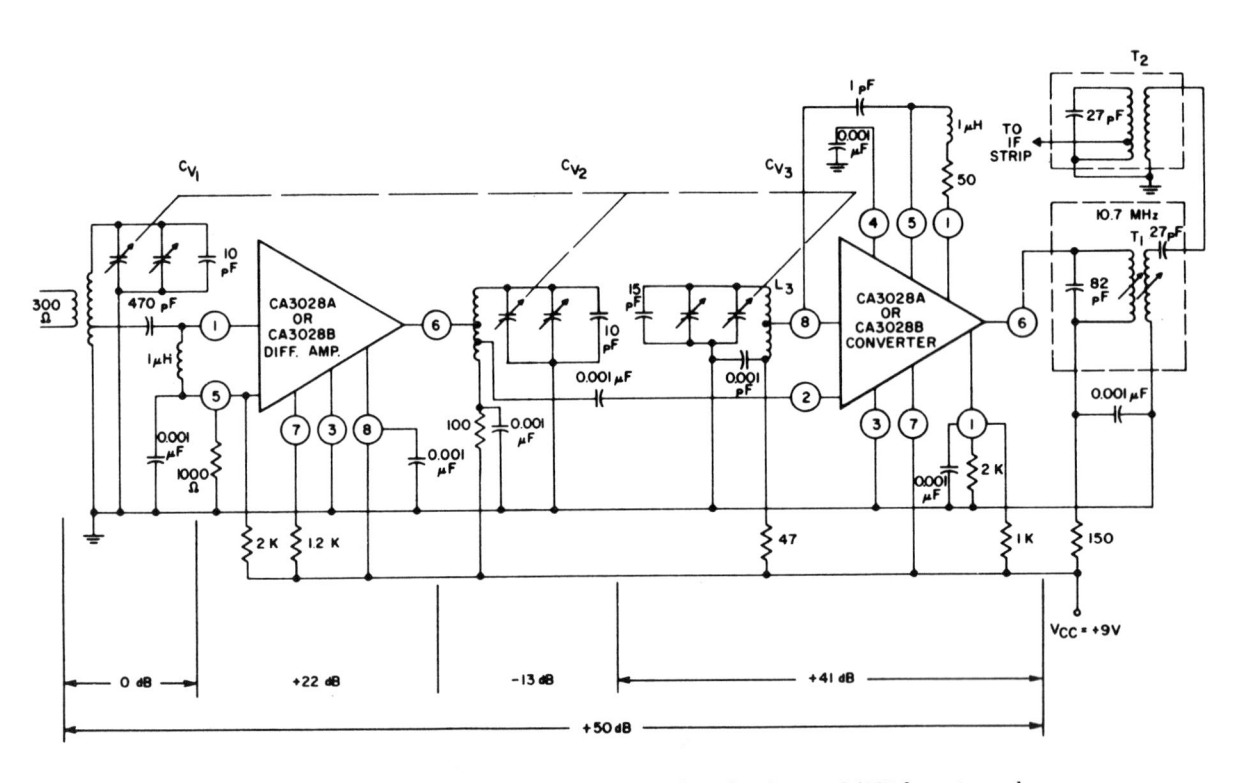

L₁: 3-3/4 T #18 tinned copper wire; winding length 5/16" on 9/32" form; tapped at 1-3/4 T; primary — 2 turns #30 SE.

L₂: 3-3/4 T #18 tinned copper wire; winding length 5/16" on 9/32" form; tapped at 6 2-1/4 T, A 3/4 T.

C_{v1-2}: variable $\Delta C \approx 15$ pF

T₁: Mixer transformer TRW #22484 or equiv.

T₂: Input transformer TRW #22485 or equiv.

L₃: 3-1/2 T #18 tinned copper wire; winding length 5/16" on 9/32" form.

C_{v1-3}: variable, $\Delta C \approx 15$ pF.

88 to 108 MHz FM front end.

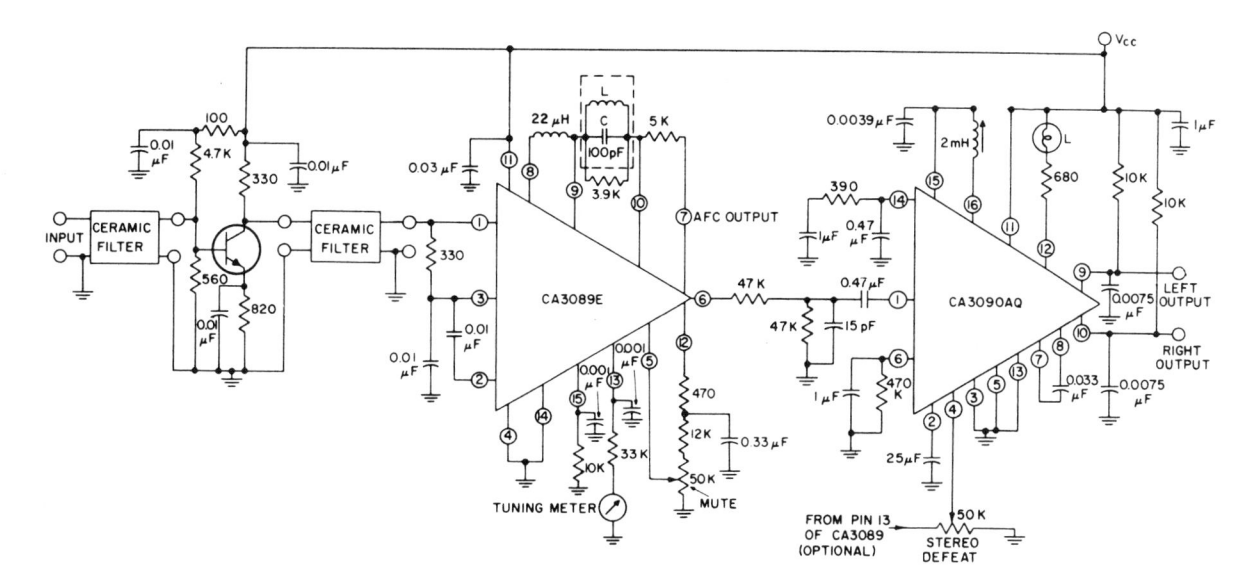

IF amplifier/detector system and stereo decoder.

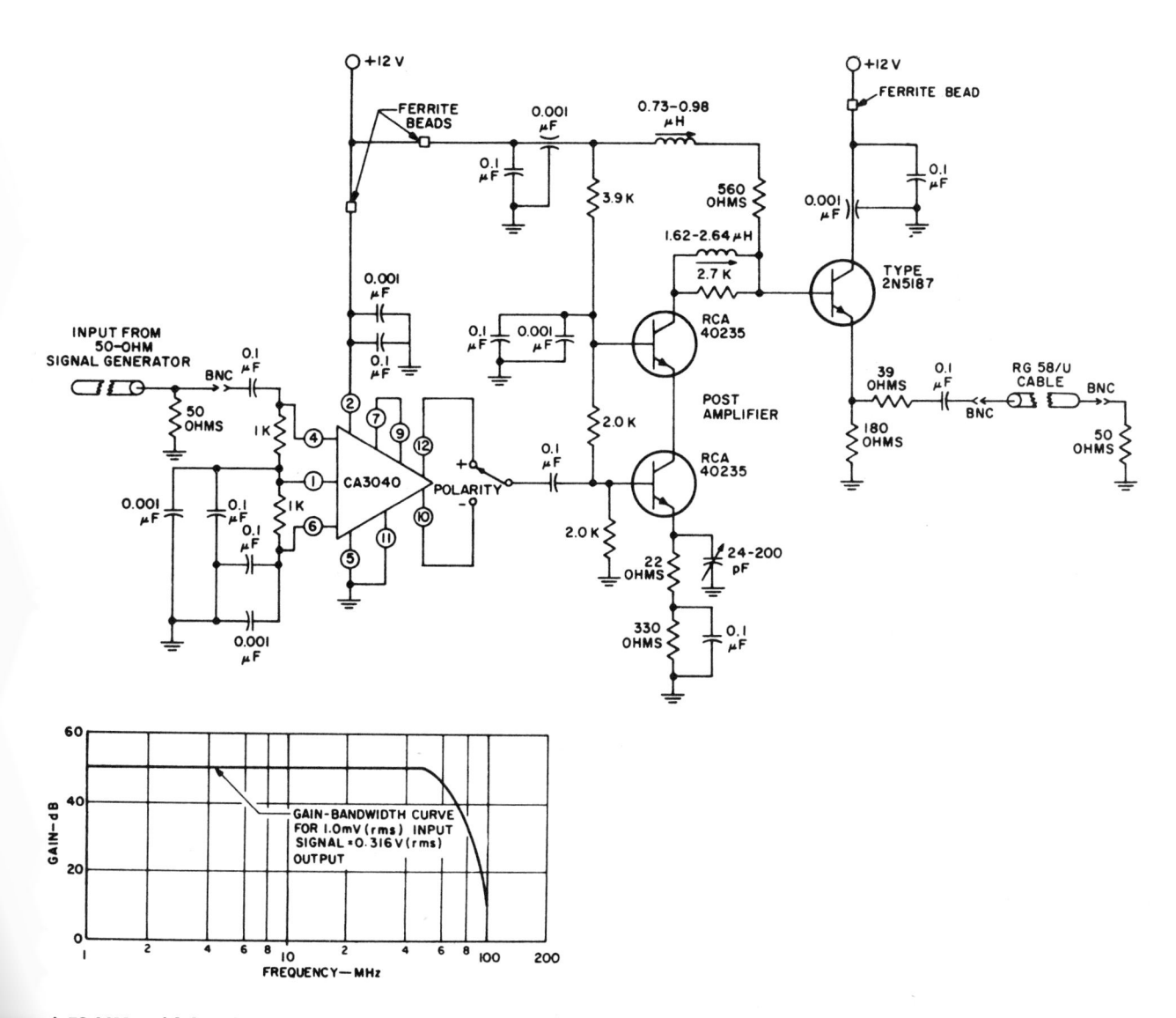

A 58 MHz wideband amplifier including postamplifier and frequency response characteristic.

SECTION 7 *Communications Circuits* 309

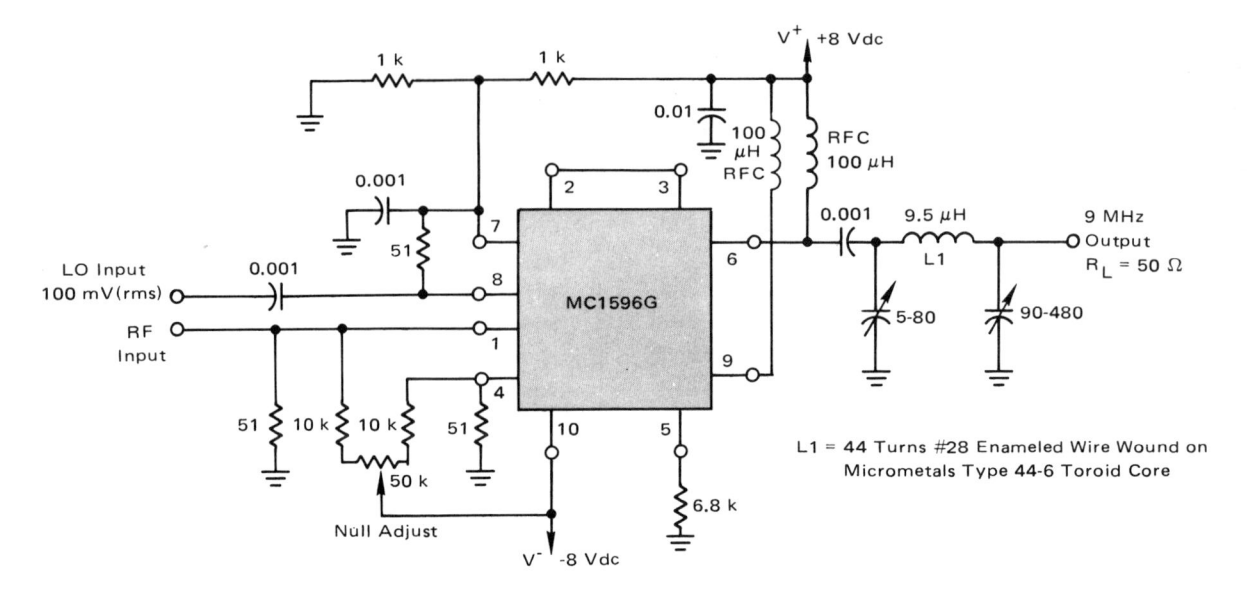

A double-balanced mixer with a broadband input and a tuned output at 9 MHz. The 3 dB bandwidth of the 9 MHz output tank is 450 kHz.

The local oscillator (LO) signal is injected at the upper input port with a level of 100 mV RMS. The modulated signal is injected at the lower input port, with a maximum level of about 15 mV RMS. For maximum conversion gain and sensitivity the external emitter resistance on the lower differential amplifier pair has been reduced to zero.

For a 30 MHz input signal and a 39 MHz local oscillator, the mixer has a conversion gain of 13 dB and an input signal sensitivity of 7.5 μV for a 10 dB (S + N)/N ratio in the 9 MHz output signal.

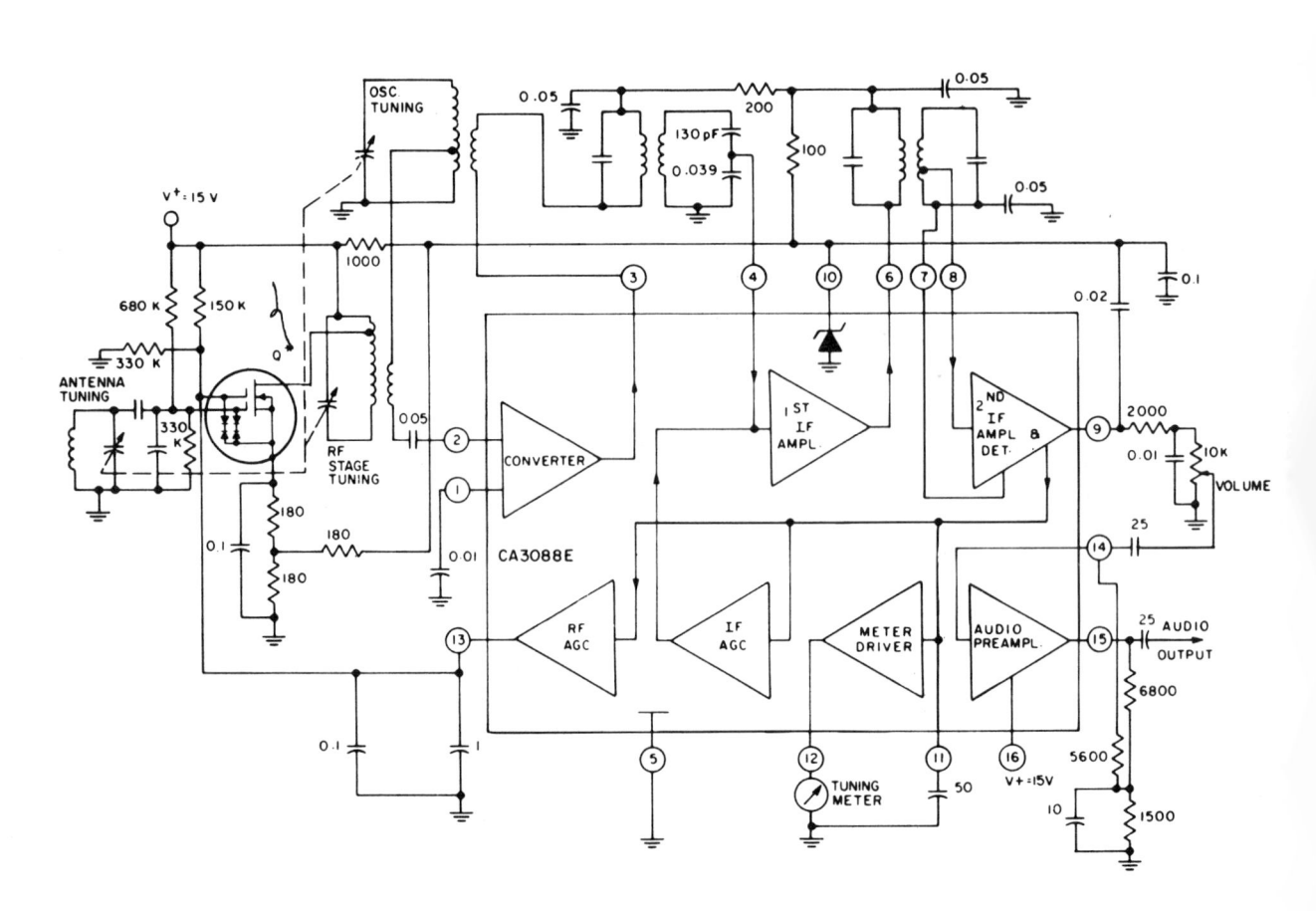

An AM broadcast receiver using CA3088E with optional RF amplifier stage. MOSFET is RCA 40841. Transformers are standard IF cans.

310 IC SCHEMATIC SOURCEMASTER

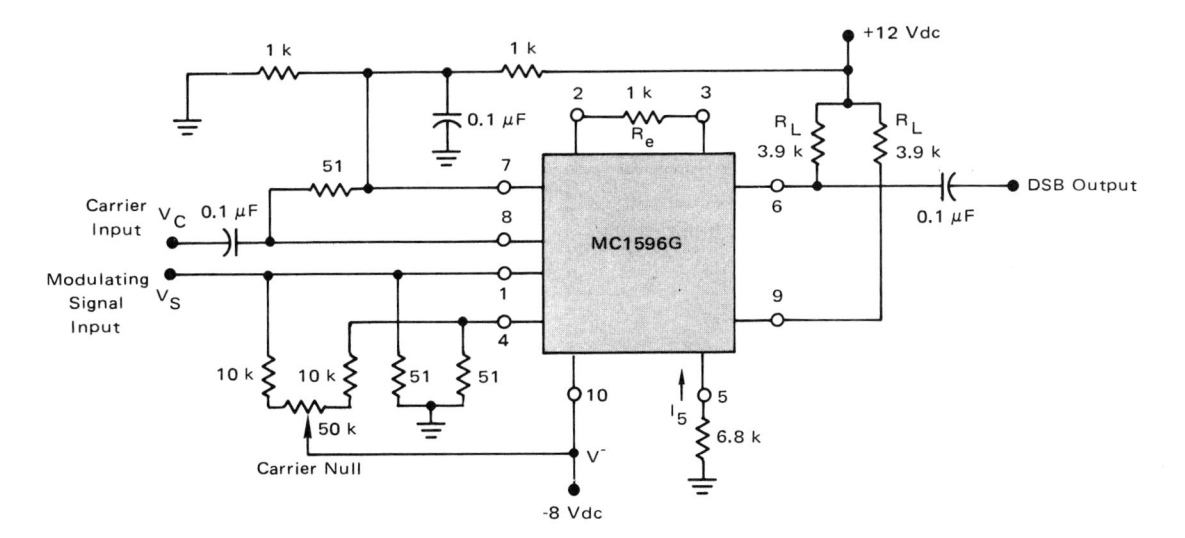

A broadband double-sideband suppressed-carrier balanced modulator without transformers or tuned circuits. Excellent gain and carrier suppression can be obtained with this circuit by operating the upper (carrier) differential amplifiers at a saturated level and the lower differential amplifier in a linear mode. The recommended input signal levels are 60 mV for the carrier and 300 mV for the maximum modulating signal levels.

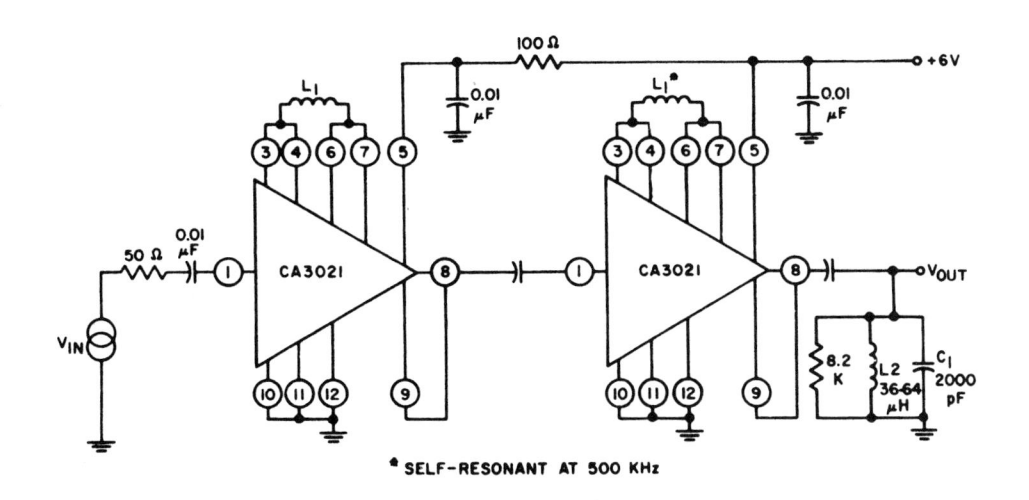

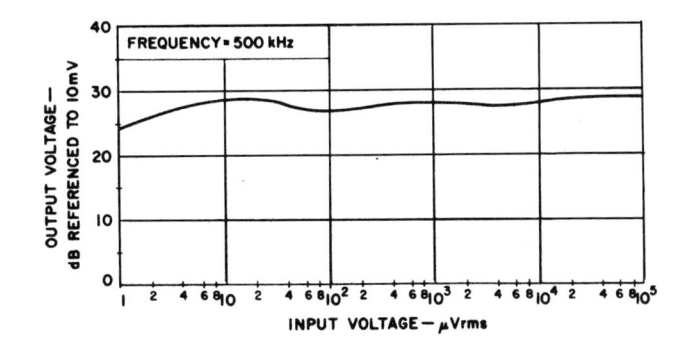

Schematic diagram and limiting performance of two-stage 500 kHz limiter amplifier using the CA3021. Two 500 kHz self-resonant chokes are used in the feedback path. A tuned circuit is included in the output to obtain a sine-wave output. The limiting characteristics of this amplifier are shown in the inset. Although limiting occurs for noise, a limited signal is apparent above the noise at an input signal of 1 μV. Because of the noise and early limiting, voltage gain can only be estimated; however, it is at least 100 dB. Good limiting performance was obtained for an input signal up to 3V RMS. Total power drain for the circuit with a 6V supply is approximately 4 mW.

SECTION **7** Communications Circuits

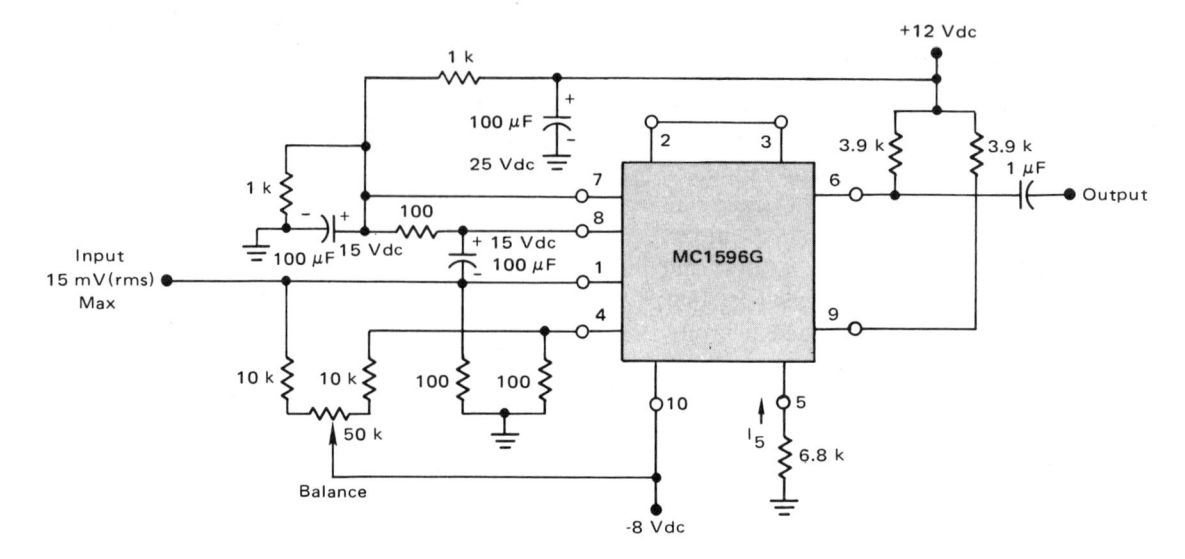

Low-frequency doubler. This circuit will double in the audio- and low-frequency range below 1 MHz with all spurious outputs greater than 30 dB below the desired $2f_{IN}$ output signal. For optimum output-signal spectral purity, both upper and lower differential amplifiers should be operated within their linear ranges. This corresponds to a maximum input signal level of 15 mV.

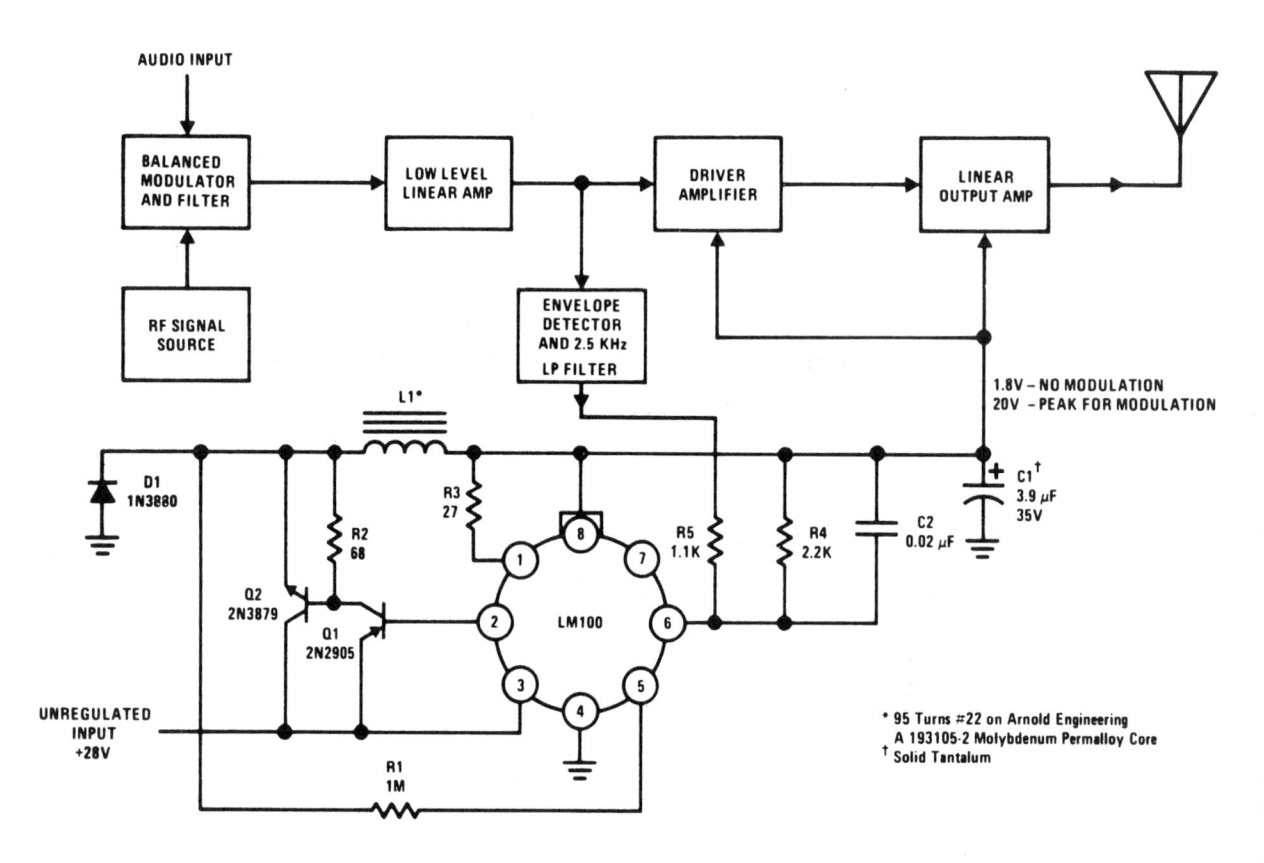

High-efficiency modulation scheme for single-sideband transmitter. A switching regulator operates the linear output amplifiers of a conventional single-sideband transmitter at a voltage just higher than that required to accommodate the envelope of the RF output signal. With no modulating signal, the driver and output amplifiers are operated at 1.8V, which is the reference voltage of the LM100. When modulation is present, the envelope of the RF waveform is detected and used to drive the regulator, so that its output voltage follows the shape of the envelope. Hence, the amplifiers are always supplied just enough voltage to keep them from saturating.

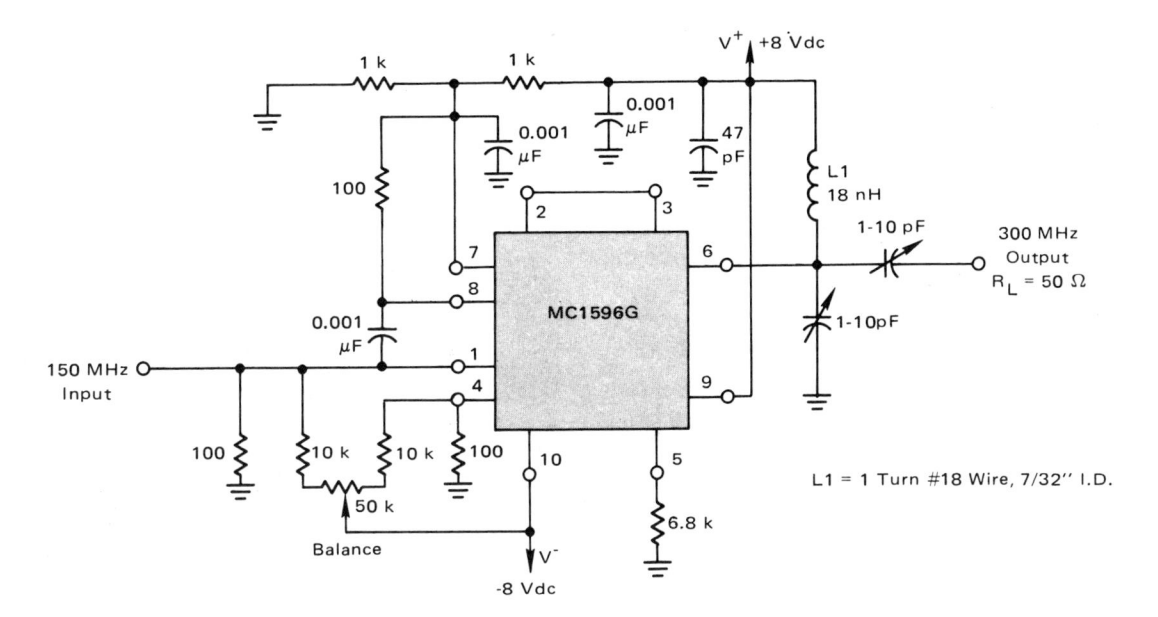

VHF (150 to 300 MHz) doubler. Even at this frequency, the MC1596 is still superior to a conventional transistor doubler before output filtering. All spurious outputs are 20 dB or more below the desired 300 MHz output.

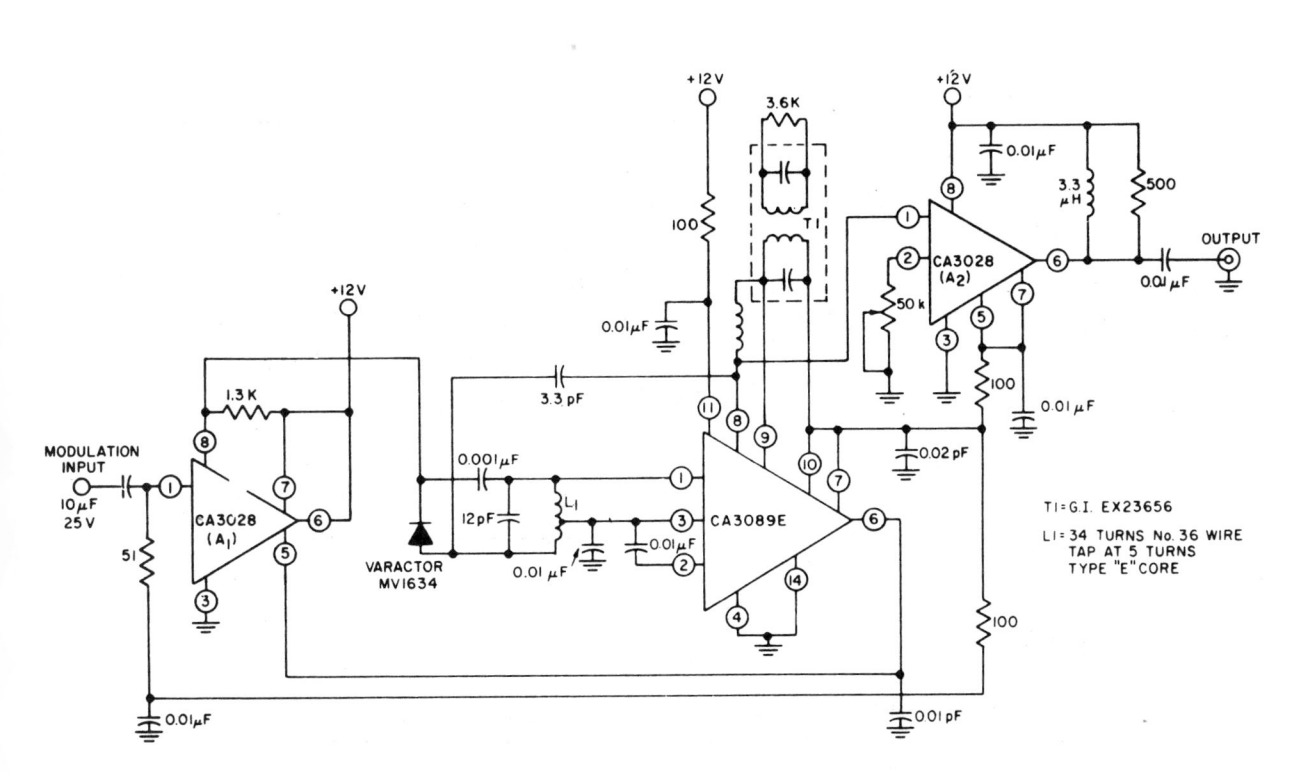

FM generator circuit. Carrier is generated by the CA3089E with the introduction of feedback from the output terminal, pin 8. The carrier is modulated by the varactor connected across the tuned circuit at the input of the CA3089E. The varactor is driven by the output of the differential amplifier, A1, using a CA3028 IC. This differential amplifier stage is driven at one of its input terminals by the audio modulating signal. Negative feedback of the audio signal is provided by driving the other differential amplifier input from the recovered audio output of the CA3089E at pin 6. The detector circuit uses a double-tuned transformer to produce audio with very little distortion at pin 6. This feedback technique results in very low modulation distortion. The RF output of the CA3089E at pin 8 is essentially a square wave and is fed to a tuned amplifier stage to buffer the signal and restore the sine-wave-shaped RF output signal.

SECTION 7 Communications Circuits

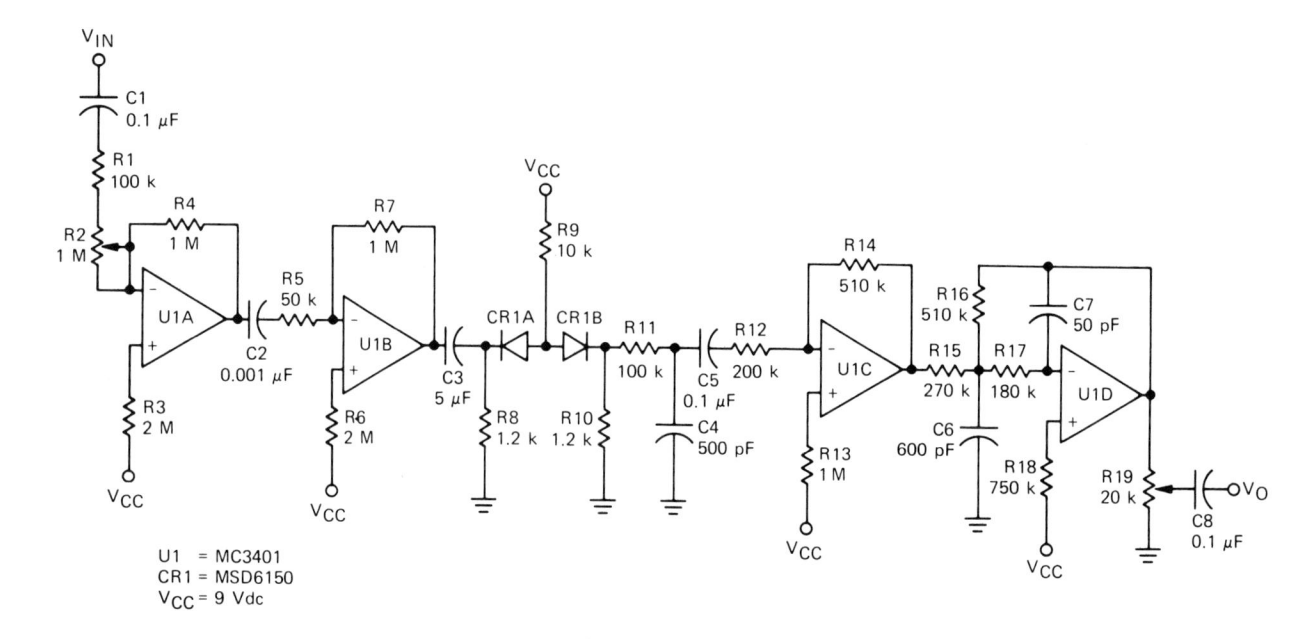

This voice processor offers preemphasis response of 6 dB/octave between 300 Hz and 3 kHz with +1 and −3 dB tolerances. The amplitude limiter prevents transmitter peak deviation from exceeding the allowed maximum. A low-pass filter (12 dB/octave above 3 kHz) is located between the limiter and the modulator. There is less than 10% harmonic distortion at 1000 Hz.

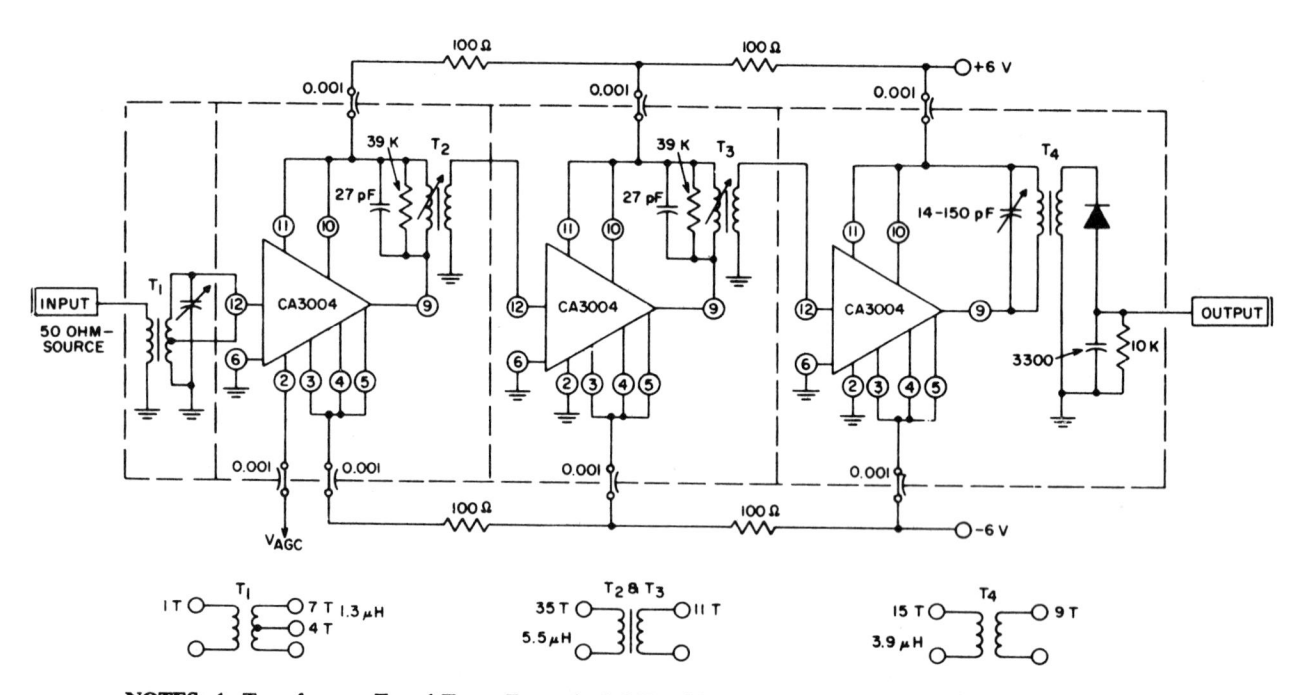

NOTES: 1. Transformers T_1 and T_4 are Ferramic Q-2 Toroid Types (unloaded Q = 200).
2. Transformers T_2 and T_3 are slug-tuned with Carbonyl IT-71 material (unloaded Q = 70).

A three-stage, 12 MHz gain-controlled AM amplifier using CA3004 circuits. The circuit uses three CA3004s and provides a stage gain of 25 dB. The source resistance to the input circuit is 800Ω, a satisfactory compromise for gain, noise figure, and modulation-distortion performance. Input and output transformers T1 and T4 have high unloaded Qs (37) to achieve the required gain. The second detector has a 3 dB bandwidth of 5.0 kHz. The typical overall performance characteristics are:

Power drain:	83 mW
Power gain (to second detector input):	76 dB
AGC range (1st stage):	60 dB
Noise figure:	4.5 dB
3 dB bandwidth:	160 kHz

314 IC SCHEMATIC SOURCEMASTER

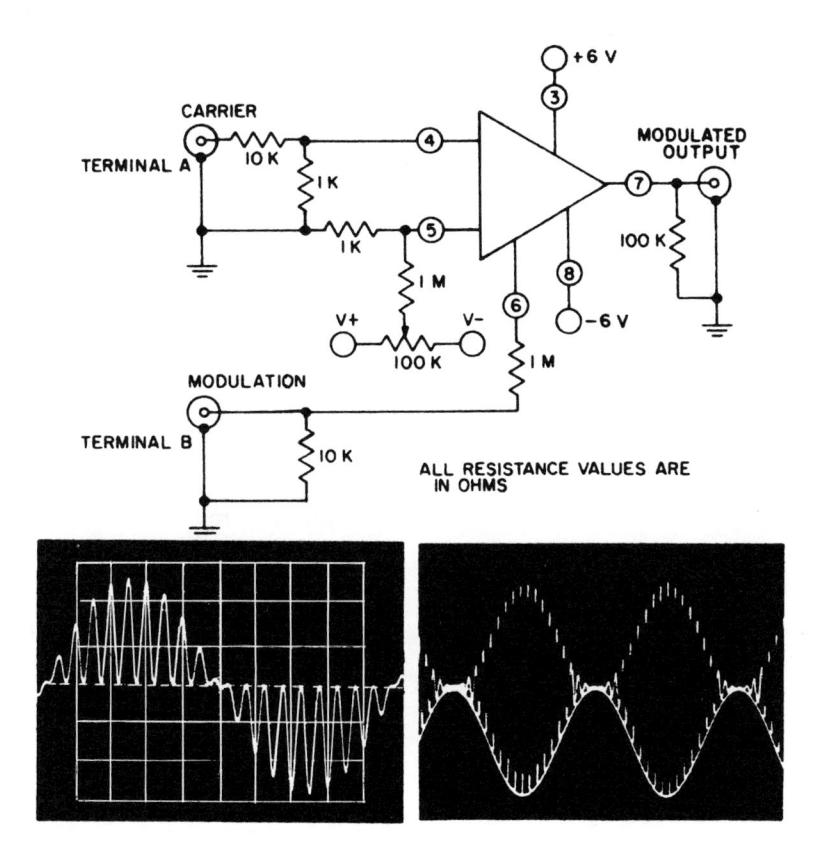

AM modulator. When modulation is applied to the amplifier bias input, terminal B, and the carrier frequency is applied to the differential input, terminal A, the waveform shown below is obtained. The linearity of the modulator is indicated by the solid trace of the superimposed modulating frequency. The maximum depth of modulation is determined by the ratio of the peak input modulating voltage to V.

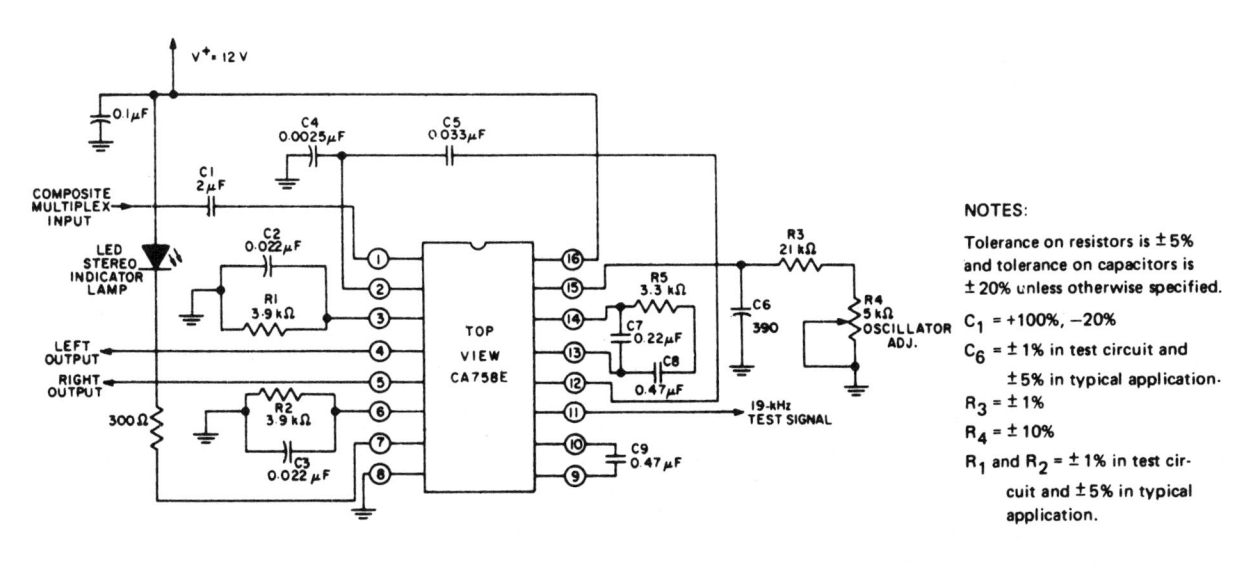

A phase-locked-loop multiplex decoder for FM solid-state stereo systems. The RCA CA758E decodes the multiplexed stereo input signal into left- and right-channel audio output signals. The decoder also suppresses SCA (storecast) transmissions when present in the composite stereo signal. The decoder uses a minimum of external components and requires one adjustment (oscillator frequency) for complete alignment. The CA758E provides automatic mono/stereo mode switching and energizes a stereo indicator lamp.

SECTION **7** Communications Circuits **315**

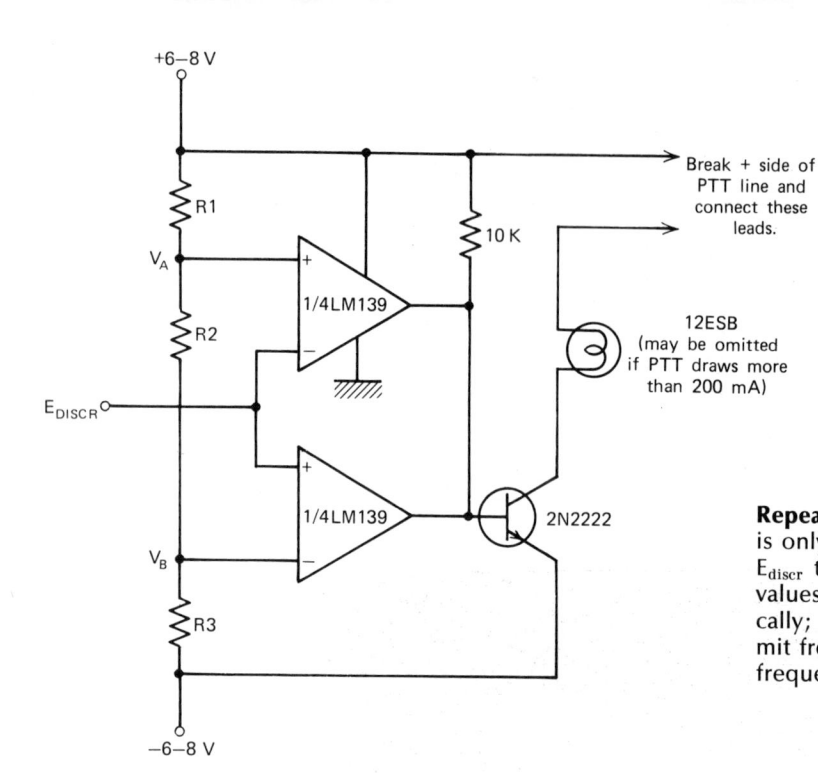

Repeater discriminator window. Push-to-talk circuit is only activated with on-frequency signals. Connect E_{discr} to discriminator meter source. Determine the values of voltage-divider resistors R1, 2, 3 empirically; when discriminator voltage exceeds V_A (transmit frequency too high) or drops below V_B (transmit frequency too low), repeater will be disabled.

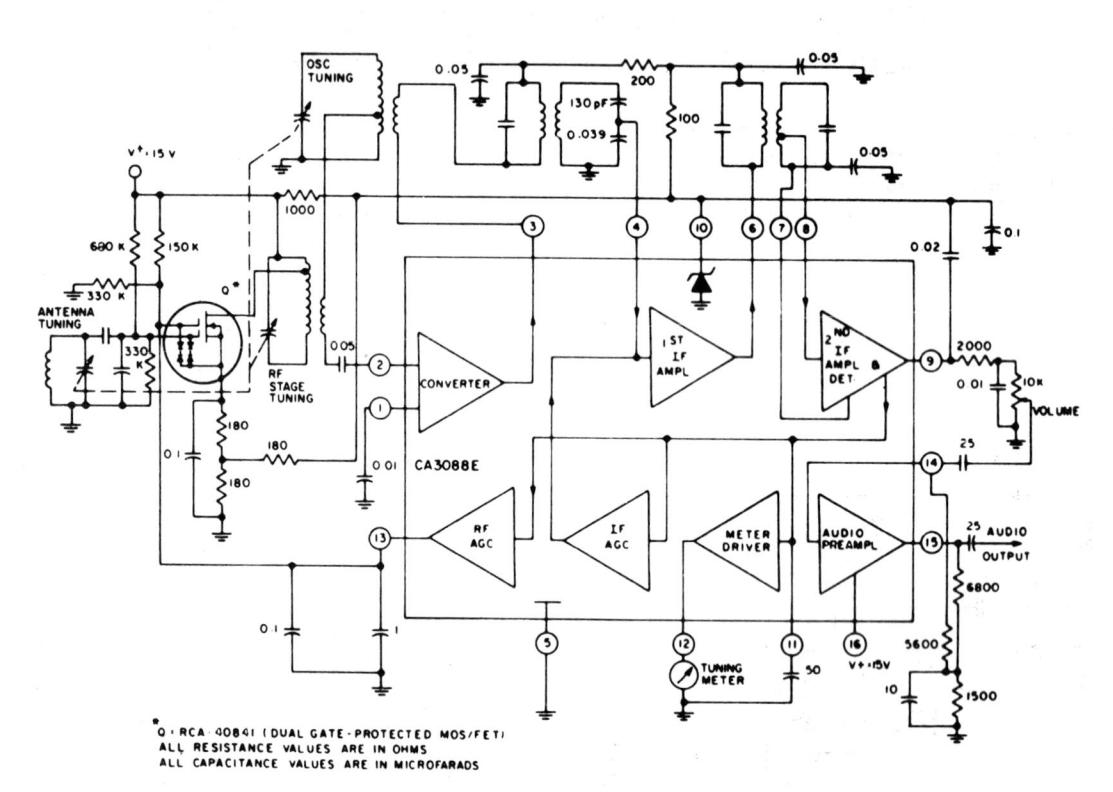

AM receiver subsystem includes AM converter, IF amplifiers, detector, and audio preamplifier for AM broadcast and communications receivers. Features include:

    Excellent overload characteristics
    AGC for IF amplifier
    Buffered output signal for tuning meter
    Internal voltage regulation
    Two IF amplifier stages
    Low-noise converter and first IF amplifier
    Low harmonic distortion

Delayed AGC for RF amplifier
Terminals for optional inclusion of tone control
Operation from wide range of power supplies (V^+ = 6–16V).
Optional ac or dc feedback on wideband amplifier
Array of amplifiers for general-purpose applications
Suitability for use with optional external RF stage—either MOS or bipolar

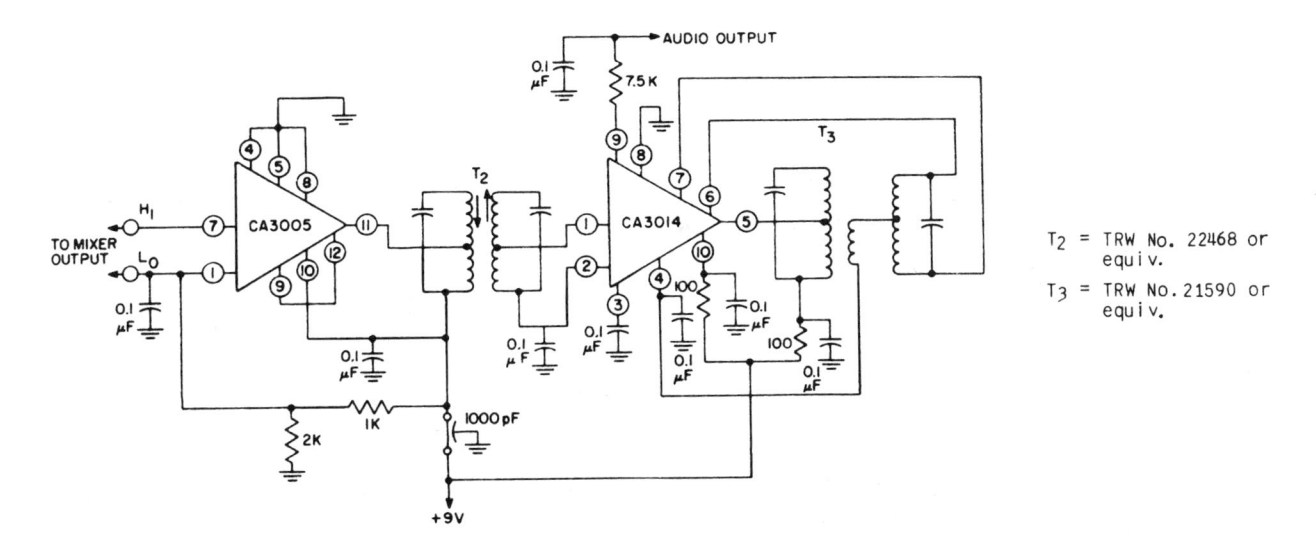

Two-chip 10.7 MHz IF amplifier, limiter, and discriminator using CA3005, CA3014, and interstage transformer. Insets show selectivity and detector "S" curves.

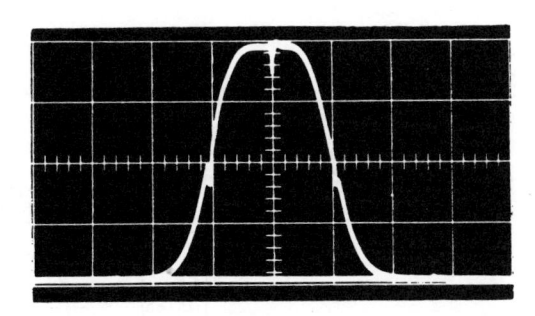

T_2 = TRW No. 22468 or equiv.
T_3 = TRW No. 21590 or equiv.

A 10.7 MHz IF selectivity curve; markers are 100 kHz apart with center at 10.7 MHz.

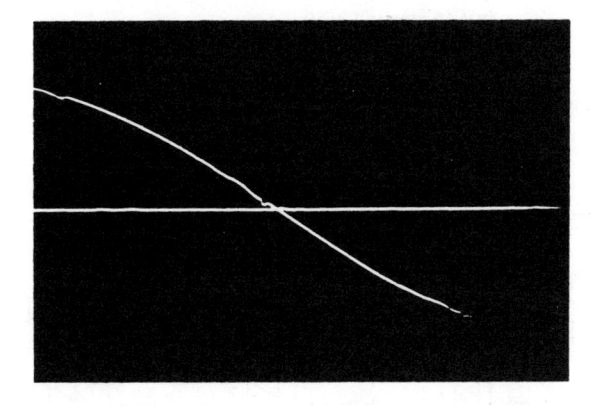

Detector "S" curve; markers are 100 kHz apart with center at 10.7 MHz.

SECTION **7** *Communications Circuits*

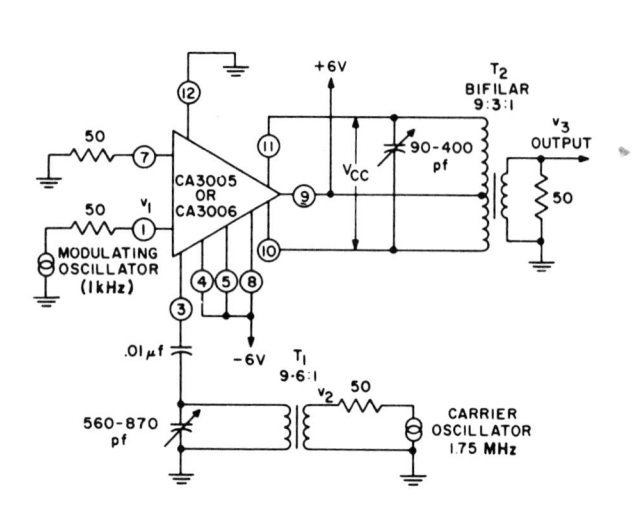

A double-sideband, suppressed-carrier modulator using CA3005 or CA3006.

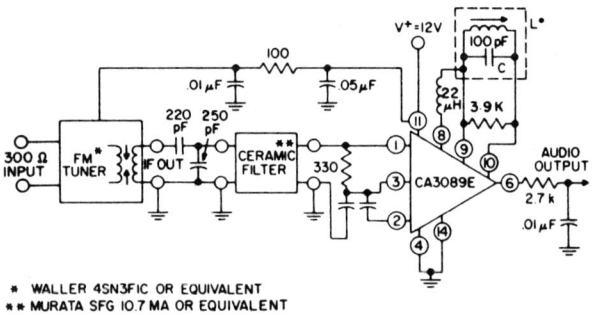

Single-stage tuned amplifier for various VHF ranges.

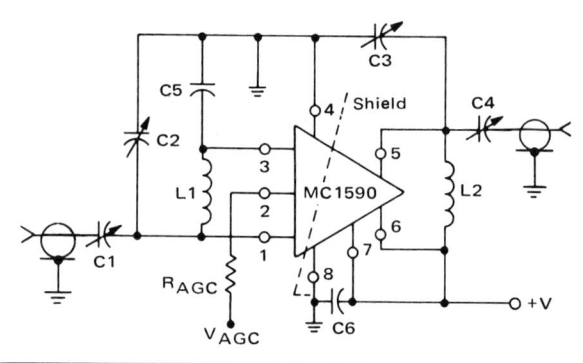

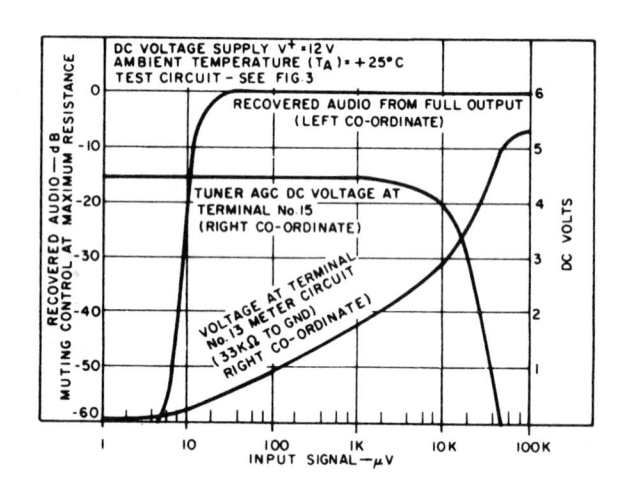

Parameter		30 MHz	60 MHz	100 MHz
Power Gain (dB)		50.8–54	44.2–46.7	31.6–35.7
BW (MHz)		0.7–1.4	1.9–2.4	7.8–9.2
C1	pF	38	1–30	1–30
C2	pF	1–30	1–30	1–10
C3	pF	1–10	1–30	1–15
C4	pF	1–30	1–10	1–10
C5	μF	0.002	0.001	470
C6	μF	0.002	0.001	470
L1	μH	0.6	0.17	0.07
L2	μH	1.35	0.28	0.13
V+	Vdc	12	12	12

Performance data at f_O = 98 MHz, f_{MOD} = 400 Hz, Deviation = ±75 kHz:
- -3dB Limiting Sensitivity . . . . . 2μV (Antenna Level)
- 20dB Quieting Sensitivity . . . . . 1μV (Antenna Level)
- 30dB Quieting Sensitivity . . . . 1.5μV (Antenna Level)

An FM tuner using the CA3089E with a single-tuned detector coil. Chart shows muting action, tuner AGC, and tuning meter output as a function of input signal voltage.

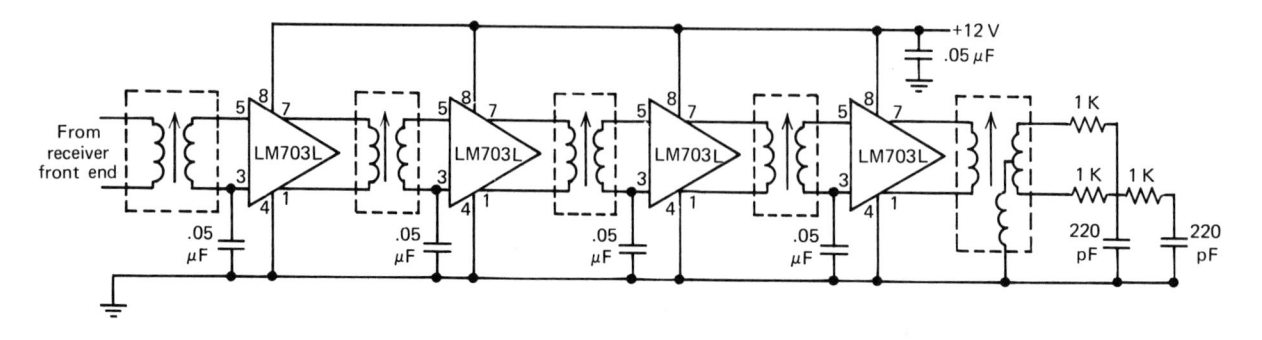

Four-stage 10.7 MHz FM IF amplifier.

318 IC SCHEMATIC SOURCEMASTER

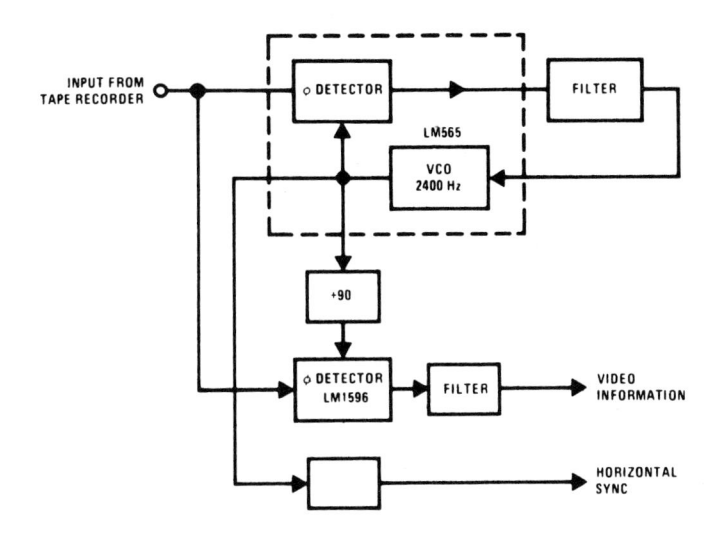

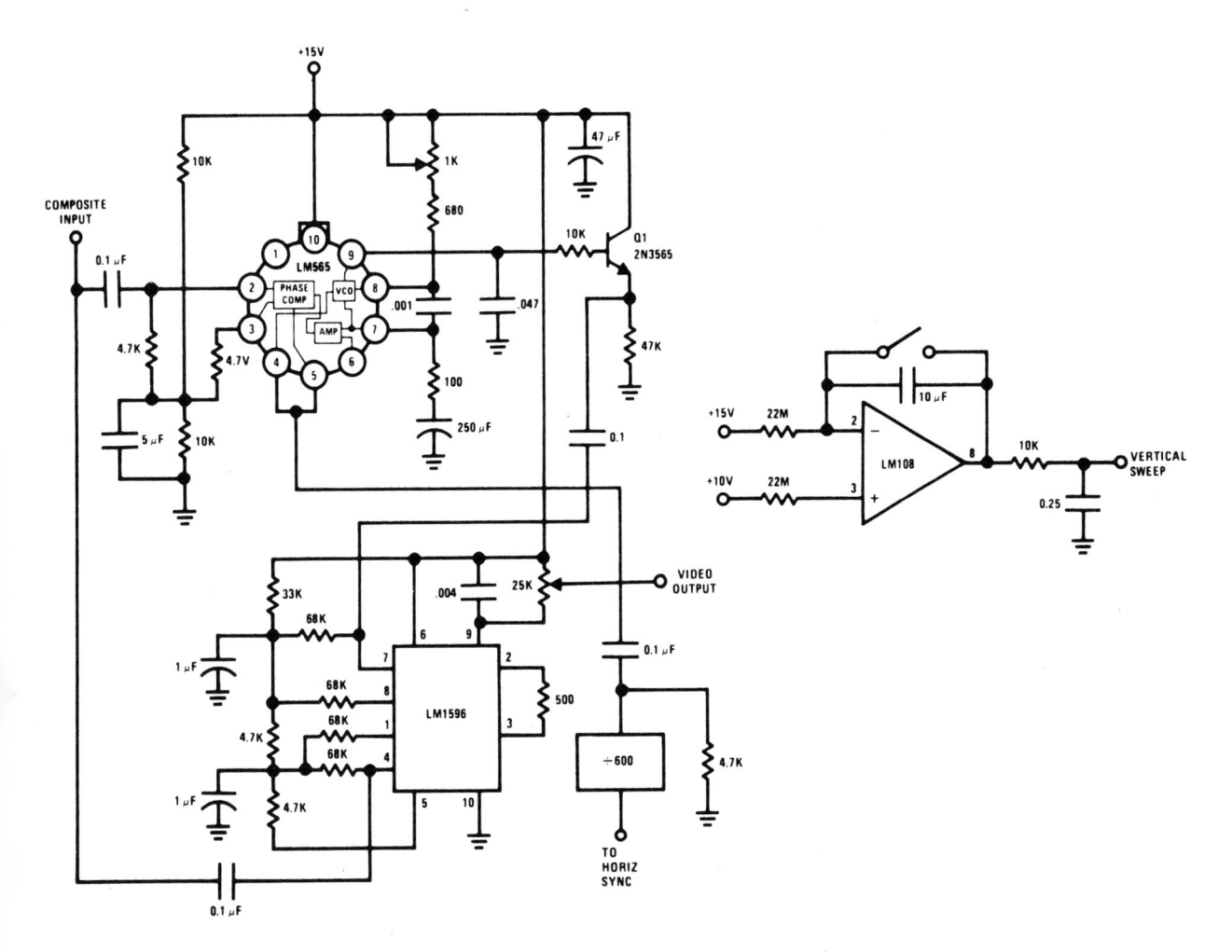

This weather satellite picture demodulator incorporates PLL. Weather satellites of the Nimbus, ESSA, and ITOS series continually photograph the earth from orbits of 100 to 800 miles. The pictures are stored immediately after exposure in an electrostatic storage vidicon and read out during a succeeding 200-second period. The video information is AM on a 2.4 kHz subcarrier, which is on a 137.5 MHz FM carrier.

Upon reception, the output from the receiver FM detector will be the 2.4 kHz tone containing AM video information. It is common practice to record the tone on an audio-quality tape recorder for subsequent demodulation and display. The 2.4 kHz subcarrier frequency may be divided by 600 to obtain the horizontal sync frequency of 4 Hz.

SECTION 7 Communications Circuits

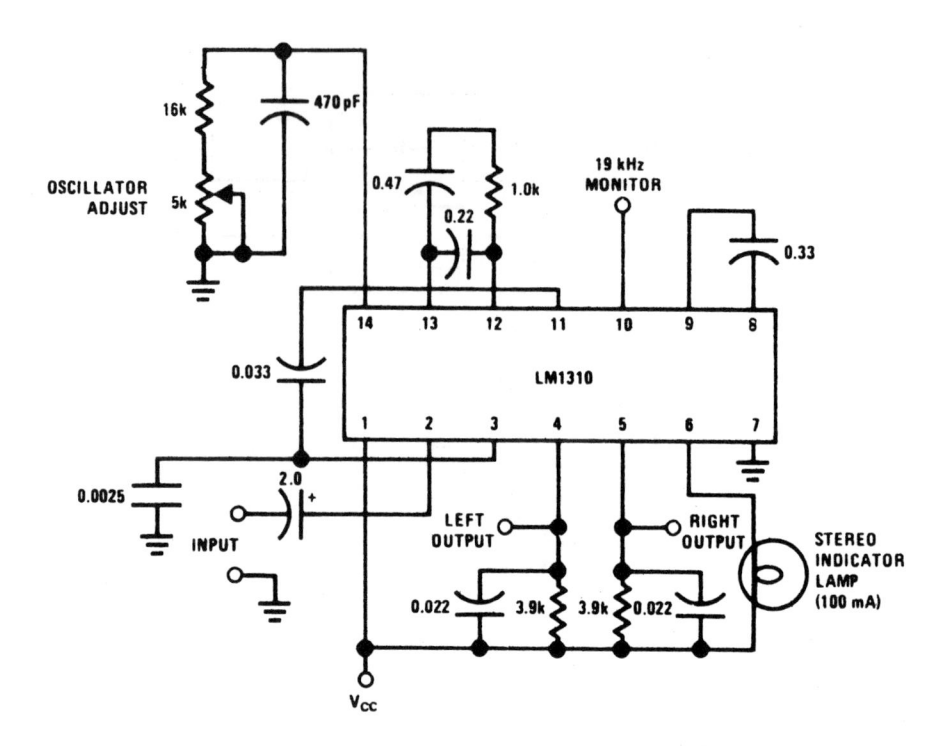

An integrated FM stereo demodulator using phase-locked-loop techniques to regenerate the 38 kHz subcarrier. Delivers high-fidelity sound within the cost restraints of inexpensive stereo receivers.

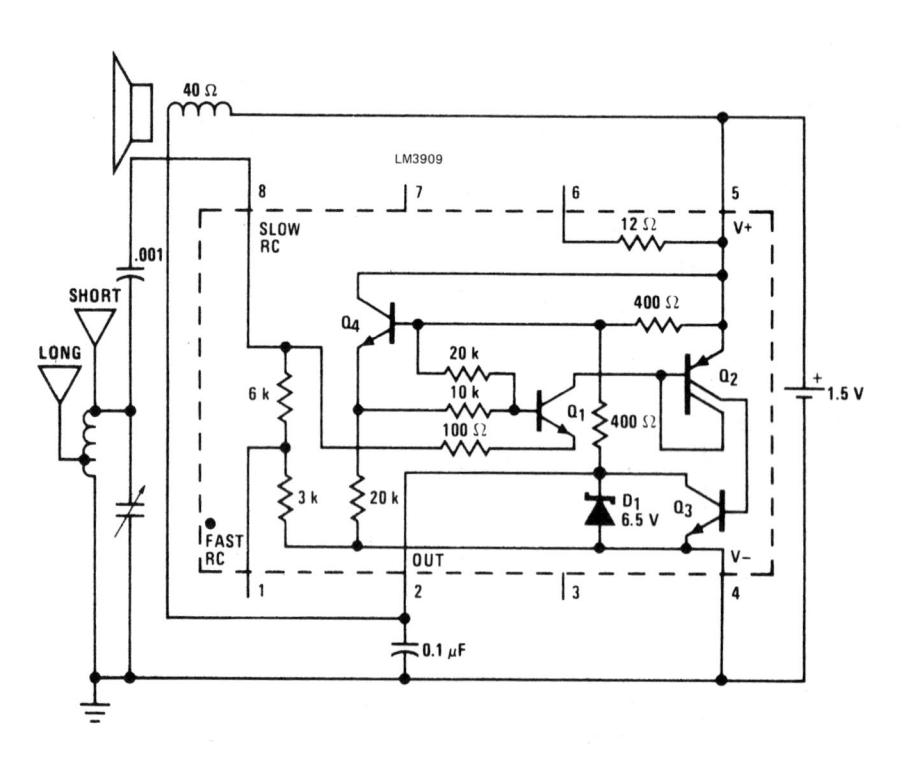

AM broadcast radio. Tuning ability is only as good as a simple crystal set, but a local radio station can provide listenable volume with an efficient 6-inch speaker. Extremely low power drain allows a month of continuous radio operation from a single D cell.

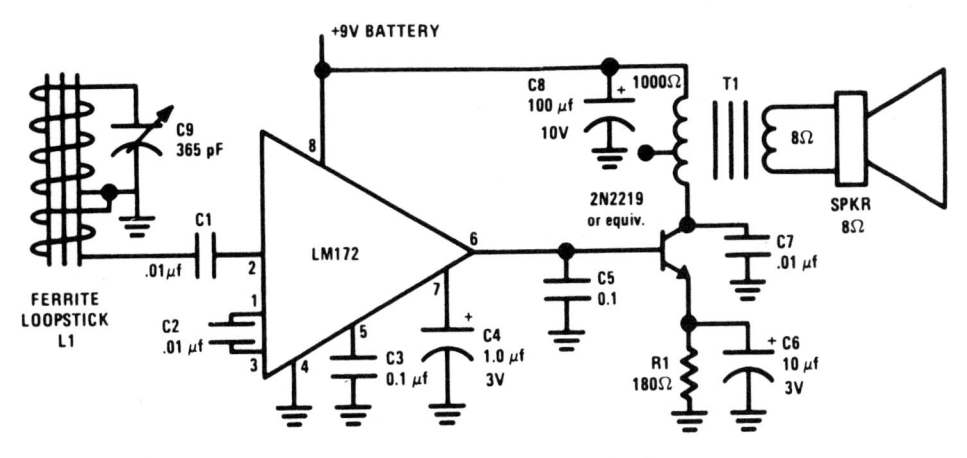

L1 — Ferrite Loopstick — Philmore FF15 (packaged as set of 3 sizes)
C9 — Sub-miniature variable capacitor — Philmore 1949G — 365 pF max.
T1 — Midget Audio Transformer, 1000Ω:8Ω — Archer 273-1380 (Radio Shack, Inc.)
SPKR — 2" PM Speaker, 8Ω, 0.1 watt — Philmore TS20

Low-frequency TRF receiver. Because the LM172 is a broadband functional module, it may be used to amplify and detect signals below 2 MHz directly, without the more complex frequency conversion of superheterodyne receivers. In the AM broadcast band, the strip has sufficient sensitivity to operate alone in urban reception areas, since AGC action is useful down to about 50 µV at pin 2. With additional gain preceding the module or inserted between pins 1 and 3, it may also be useful in monitoring the numerous directional and informational channels below 550 kHz.

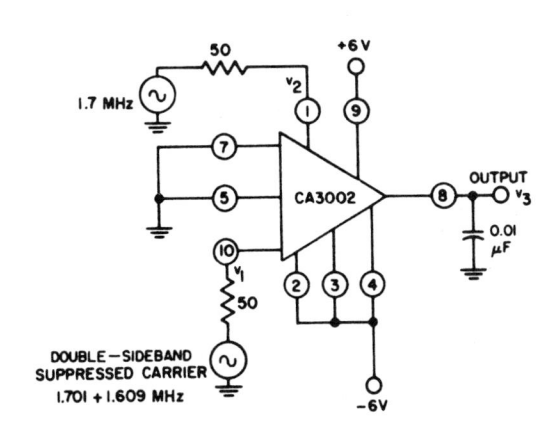

Product detector. A differential pair driven by a constant-current transistor can be used as a product detector if a suppressed-carrier signal is applied to the differential pair and the regenerated carrier is applied to the constant-current transistor. There are two requirements for linearity: (1) the circuit must be operated in a linear region; and (2) the current from the constant-current transistor must be linear with respect to the reinserted carrier voltage. In the circuit shown a double-sideband suppressed-carrier signal is applied at terminal 10, and the 1.7 MHz carrier is applied to terminal 1. Because of the single-ended output, a high-frequency bypass capacitor (0.01 µF) is connected between terminal 8 and ground to provide filtering for the high-frequency components of the oscillator signal at the output.

SECTION 7 *Communications Circuits* 321

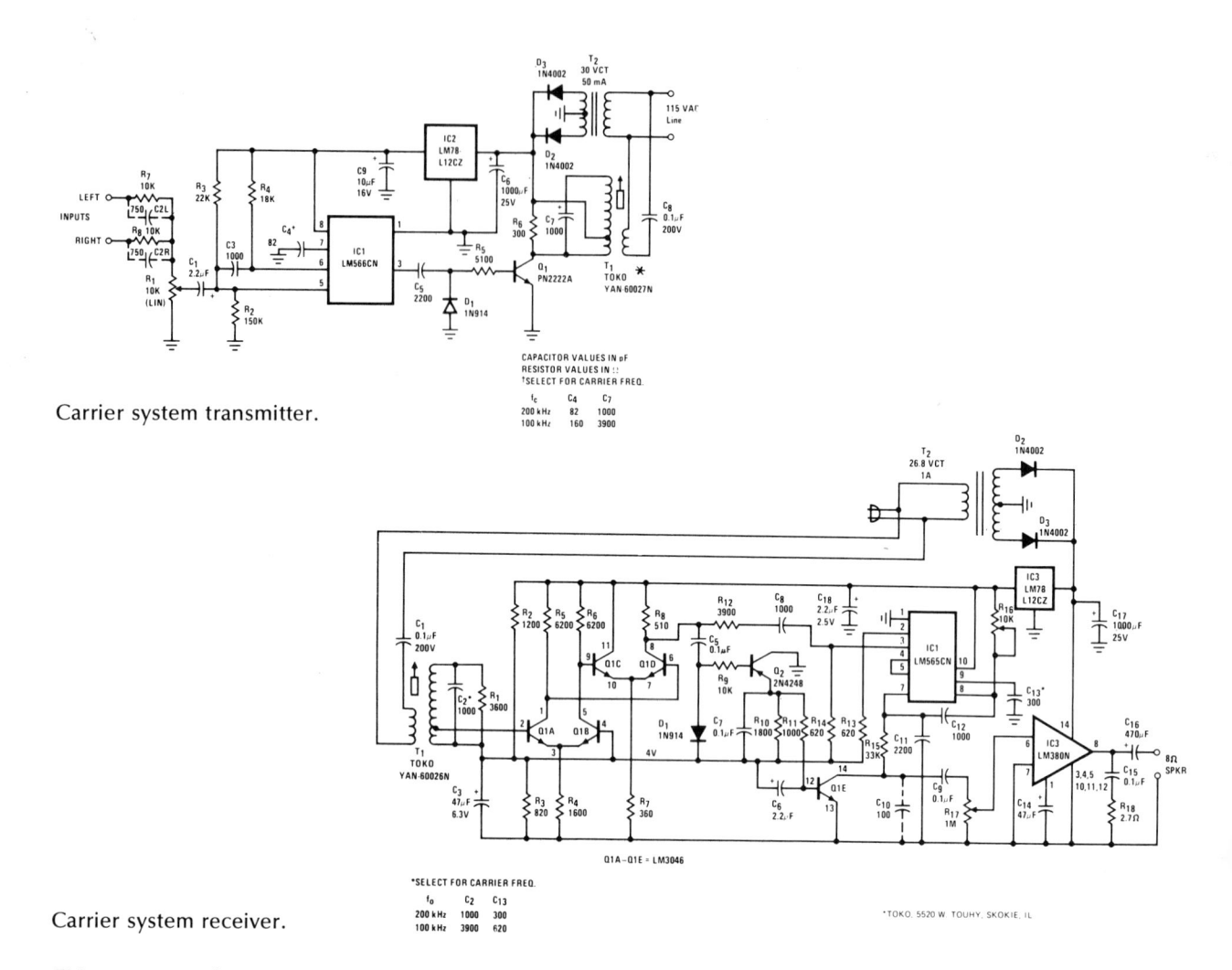

Carrier system transmitter.

f_c	C_4	C_7
200 kHz	82	1000
100 kHz	160	3900

CAPACITOR VALUES IN pF
RESISTOR VALUES IN Ω
†SELECT FOR CARRIER FREQ.

Carrier system receiver.

*SELECT FOR CARRIER FREQ.

f_0	C_2	C_{13}
200 kHz	1000	300
100 kHz	3900	620

Q1A–Q1E = LM3046

*TOKO, 5520 W. TOUHY, SKOKIE, IL

FM remote speaker system using high-quality, noise-free, wireless FM transmitter/receiver. The LM566 VCO is used to convert the program material into FM format, which is then transformer coupled to standard power lines. At the receiver end the material is detected from the power lines and demodulated by the LM565. The complete system is suitable for high-quality transmission of speech or music and will operate from any ac outlet anywhere on a 1-acre homesite. Frequency response is 20–20,000 Hz, and THD is under 0.5% for speech and music program material. The transmission distance along a power line is at least adequate to include all outlets in and around a suburban home and yard. The transmitter is plugged into the ac line at a radio or stereo system source. The signal for the transmitter is ideally taken from the TAPE OUT connectors. The carrier system receiver need only be plugged into the ac line at the remote listening location. The design includes a 2.5W power amplifier to drive a speaker directly.

The transmitter input signal level is adjustable by R1 to prevent overmodulation of the carrier. Adding C2 across each input resistor (R7 and R8) improves the frequency response to 20 kHz. The VCO carrier frequency f_c, determined by R4 and C4, is set at 200 kHz—high enough to be effectively coupled to the ac line. VCO sensitivity under the selected bias conditions with $V_S = 12V$ is about $\pm 0.66 f_C/V$. For minimum distortion, the deviation should be limited to $\pm 10\%$; thus maximum input at pin 5 of the VCO is $\pm 0.15V$ peak. A reduction due to the summing network brings the required input to about 0.2V RMS for $\pm 10\%$ modulation of f_C, based on nominal output levels from stereo receivers. R1 is provided to set the required level. The output at pin 3 of the LM566, being a frequency-modulated square wave of 6V amplitude, is amplified by Q1 and coupled to the ac line via T1. Because T1 is tuned to f_C, it appears as a high-impedance collector load, so Q1 need not have additional current limiting. The collector signal may be as much as $40-50V_{p-p}$. Capacitor C8 isolates the transformer from the line at 60 Hz.

The receiver amplifies, limits, and demodulates the received FM signal in the presence of line transient interference sometimes as high as several hundred volts peak. In addition, it provides audio mute in the absence of carrier and 2.5W output to a speaker. The carrier signal is capacitively coupled from the line to tuned transformer T1. Loaded Q of the secondary tank T1C2 is decreased by shunt resistor R1 to enable acceptance of the $\pm 10\%$ modulated carrier and to prevent excessive tank circuit ringing on noise spikes. The secondary of T1 is tapped to match the base input impedance of Q1A. Recovered carrier at the secondary of T1 may be anywhere from 0.2 to $45V_{p-p}$; the base of Q1A may see signal levels of from 12 mV to 2.6V. Q1A–Q1D operates as a two-stage limiter amplifier whose output is a symmetrical square wave of about 7V, with rise and fall times of 100 nsec. The output of the limiting amplifier is applied directly to the mute peak detector but is reduced to 1V for driving the PLL detector.

322 IC SCHEMATIC SOURCEMASTER

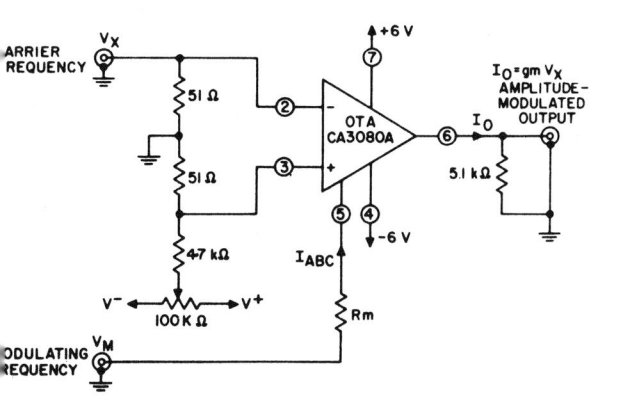

An amplitude modulator using RCA operational transconductance amplifier CA3080A. Output voltage adjustment is provided by 100k pot.

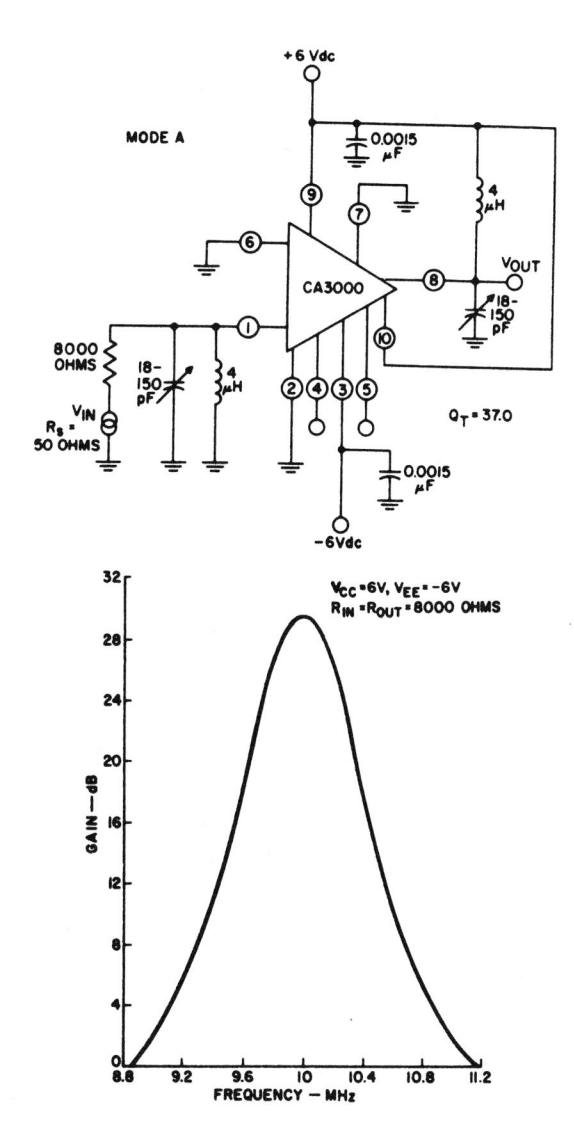

A narrow-band, tuned-input, tuned-output amplifier for operation at 10 MHz with an input Q of 26 and an output Q of 25. The 10 MHz voltage gain is 29.6 dB, and the total effective circuit Q is 37. There is very little feedback skew in the response curve.

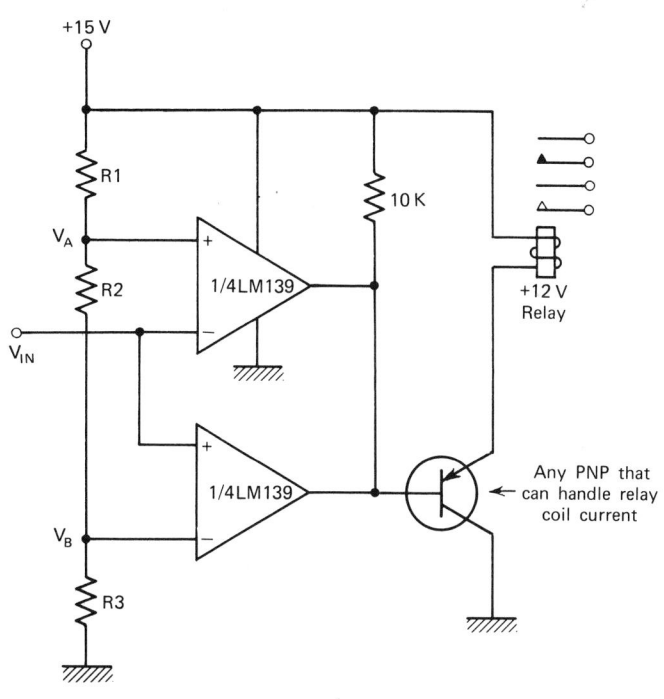

Universal repeater control module. Relay closes when input voltage V_{IN} exceeds V_A or drops below V_B. For carrier-operated relay, apply squelch signal to V_{IN} and establish values for R1, 2, 3 experimentally after determining voltage change of squelch with incoming signal. It can be used for positive- or negative-going squelch signals as long as ground references are independent.

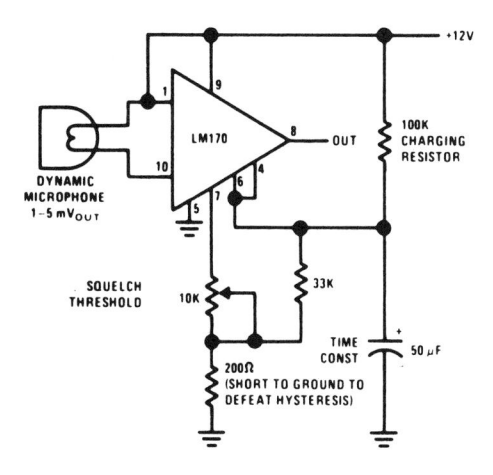

Squelch preamplifier with hysteresis. Audio squelch is useful in noisy acoustic environments to suppress background microphone noises and in receiving systems where the constant clatter of an unused channel must be removed until useful information is received. This circuit includes a number of refinements that make it smooth-acting and easy on the ear of the listener.

The **threshold** pot at pin 7 is manually set to cut in at any desired input level. The large capacitor at pin 6 and its associated charging resistor may be chosen to give squelch release times of as much as several seconds while the large current sinking capability into pin 6 assures fast attack, so that first speech syllables are not lost.

SECTION **7** Communications Circuits **323**

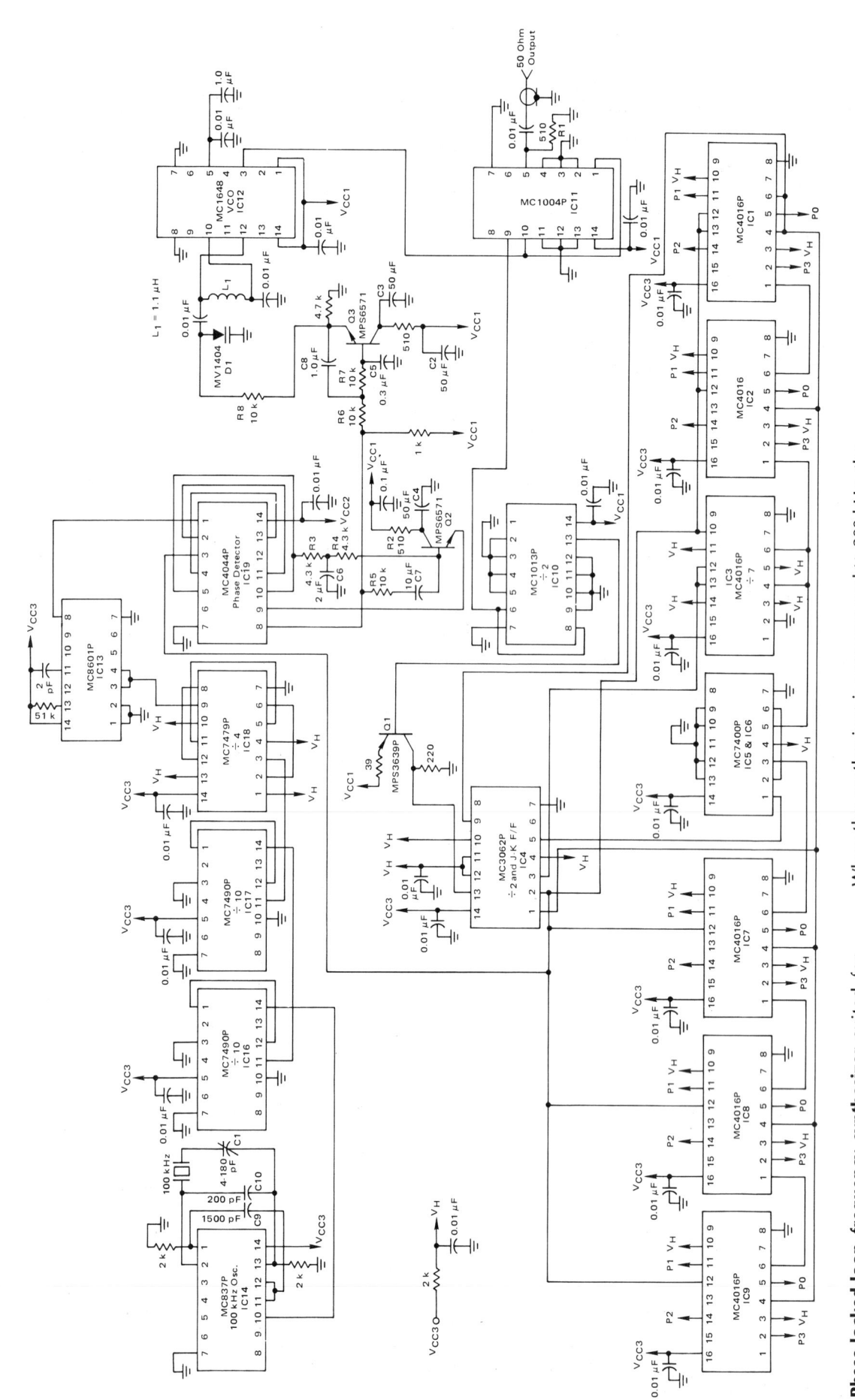

Phase-locked-loop frequency synthesizer suited for the local-oscillator function in aircraft ADF equipment. An ADF oscillator must provide continuous 1 kHz channel spacing over the required frequency range. The required range of ADF equipment is 200–1699 kHz—usually in three bands. The three bands are 200–399 kHz, 400–829 kHz, and 830–1699 kHz.

When the synthesizer is programmed to 200 kHz, the actual output frequency is 10.7 MHz higher, or 10.9 MHz. Since the actual frequency range is relatively small (10.9–12.399 MHz), a simple low-pass filter provides filtering over one band rather than the usual three. (Board art available from Motorola; ask for AN-654.)

324 IC SCHEMATIC SOURCEMASTER

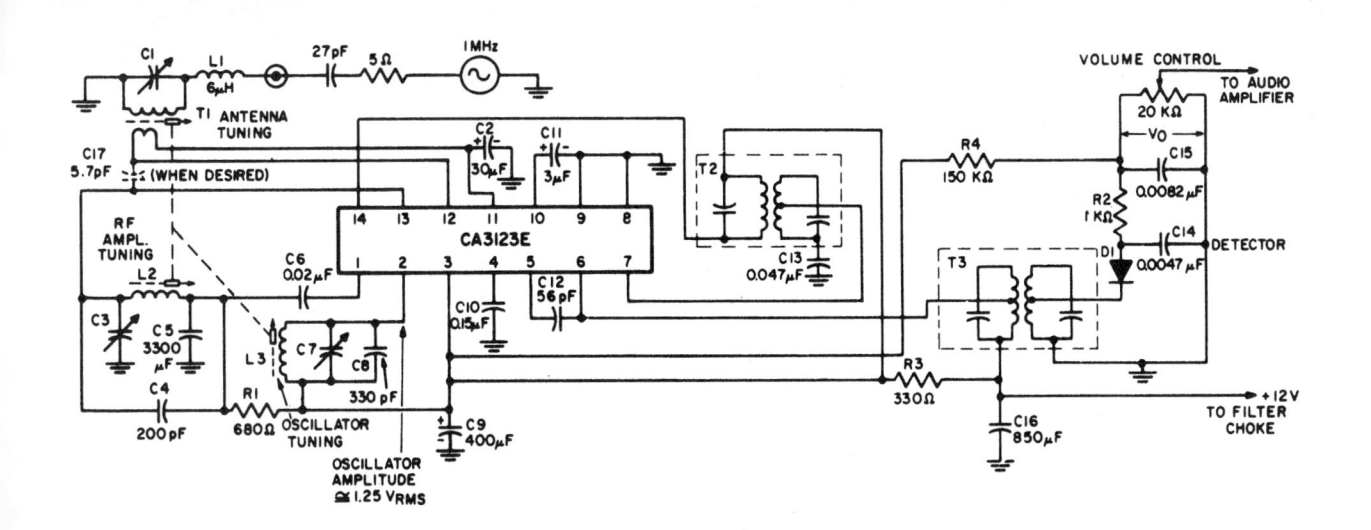

Transformer	Symbol	Frequency	Inductance μh (≈)	Capacitance pF (≈)	Q (≈)	Total Turns To Tap Turns Ratio	Coupling
First IF: Primary	T_2	262 kHz	2840	130	60	none	critical
Secondary			2840	130	60	or 30:1 31:1	≈ 0.017 ≈ 1/Q
Second IF: Primary	T_3	262 kHz	2840	130	60	8.5:1	—
Secondary			2840	130	60	8.5:1	critical ≈ 0.017 ≈ 1/Q
Antenna: Primary	T_1	1 MHz	195	(C_1)–130	65		
Secondary		Adjusted to an impedance of 75 Ω with primary resonant at 1 MHz. Coupling should be as tight as practical. Wire should be wound around end of coil away from tuning core.					
Coils	L_1	7.9 MHz	6		50		
	L_2	1 MHz	55		50		
	L_3	1.262 MHz	41		40		

An AM radio receiver using the CA3123E. The 27 pF capacitor in series with the input generator represents a dummy windshield-type antenna.

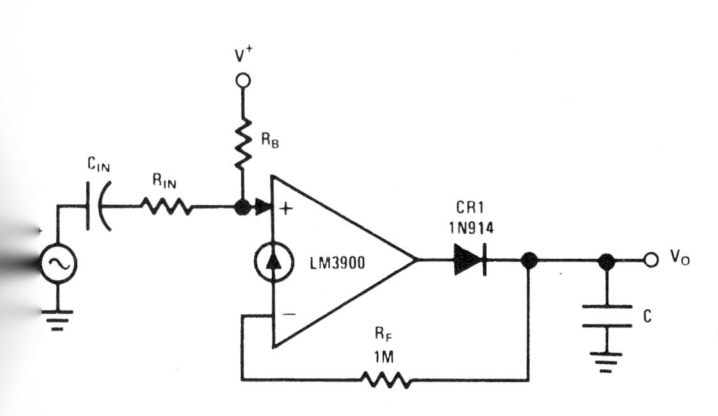

* Contributed by B. Chandler Shaw of Bendix Electrodynamics, North Hollywood, California.

A peak detector rapidly charges a capacitor to the peak value of an input waveform. The voltage drop across the rectifying diode is placed within the feedback loop of an op-amp to prevent voltage losses and temperature drifts in the output voltage. Feedback resistor R_f is kept small (1M) so that the 30 nA base current will cause only a +30 mV error in V_o.

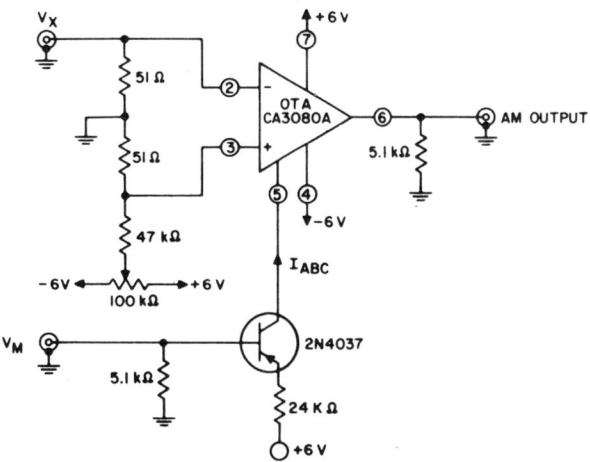

An amplitude modulator using RCA operational transconductance amplifier controlled by PNP transistor. If an NPN is added to drive the PNP, variations attributed to base–emitter characteristics are reduced by the complementary nature of the devices. Adjust 100k pot to set output voltage to vary symmetrically about zero.

SECTION **7** Communications Circuits

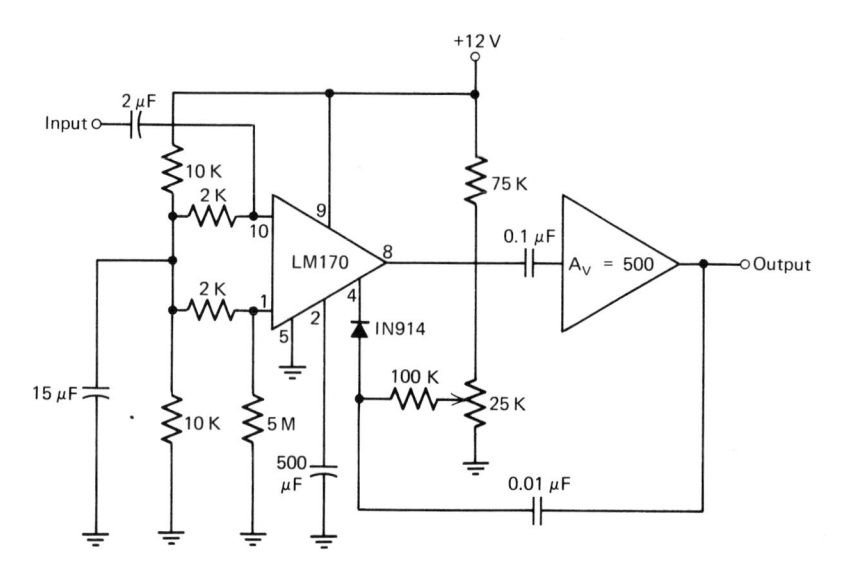

AGC using built-in detector. In systems having widely varying load impedances, AGC derived from the system output can automatically compensate for additional output loading. The emitter follower at pin 4 is used as a high-impedance detector, with detector smoothing performed by a capacitor at pin 2. The dc threshold for the detector is set at any desired level by a potentiometer, determining the positive peak output voltage that initiates gain regulation.

When operating the LM170 with an external gain stage to provide very high AGC loop gains (on the order of several hundred), proper layout is essential. As with any high-gain circuitry, good power supply regulation is a necessity. Multiple bypass capacitors give effective wideband filtering. This should prevent any undesirable ripples or transient spikes from being transferred from one device to another through V_{cc}.

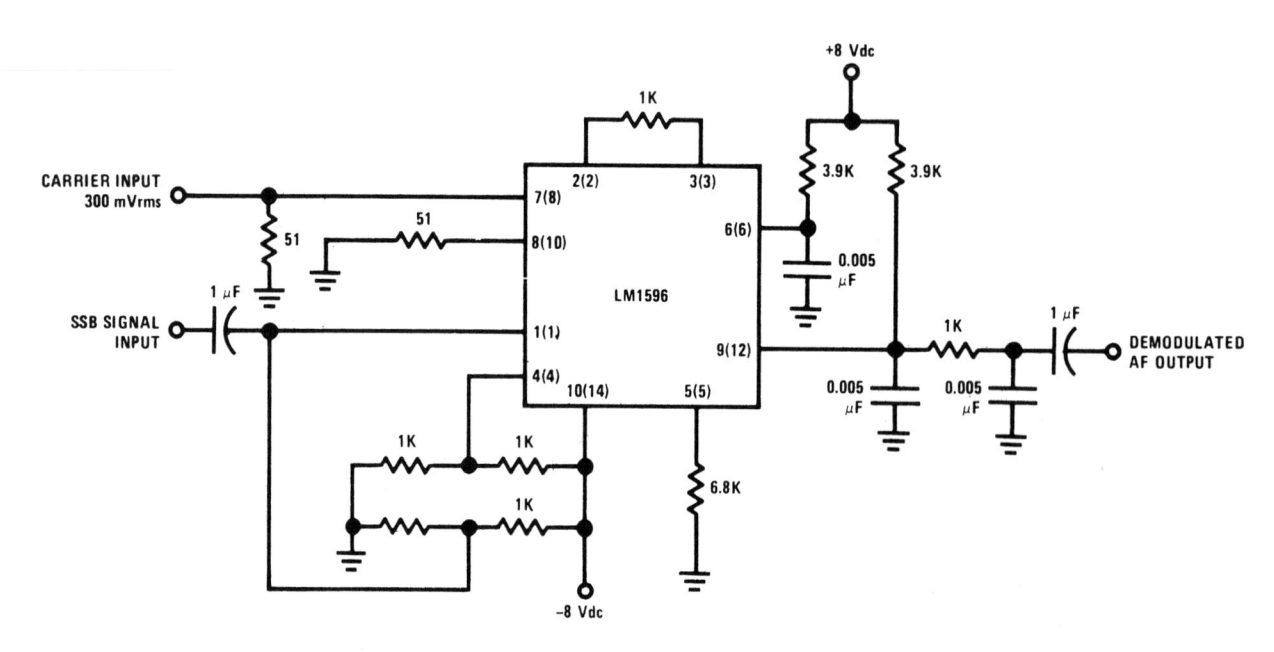

Single-sideband (SSB) suppressed-carrier demodulator (product detector). The carrier signal is applied to the carrier input port with sufficient amplitude for a switching operation. A carrier input level of 300 mV RMS is optimum. The composite SSB signal is applied to the signal input port with an amplitude of 5.0 to 500 mV RMS. All output signal components except the desired demodulated audio are filtered out, so that an offset adjustment is not required. This circuit may also be used as an AM detector by applying composite and carrier signals in the same manner as described for product detector operation. (Numbers in parentheses show DIP connections.)

326 IC SCHEMATIC SOURCEMASTER

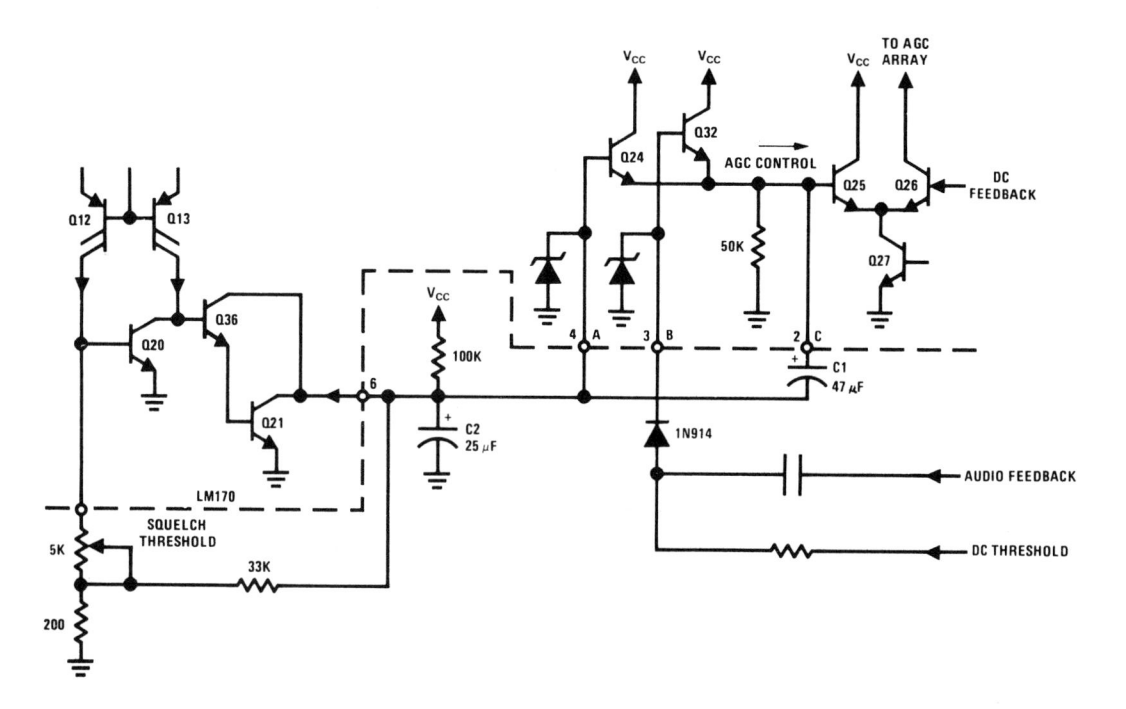

Simultaneous squelch and AGC. An interesting application of the LM170 involves simultaneously obtaining AGC and a fast-attack, slow-release squelch.* In normal AGC operation, a filter capacitor is required on pin 2 (LM170) to store the peak AGC signal. But if a filter capacitor were connected from pin 2 to ground, it would not be possible to lower the voltage at pin 2 fast enough to obtain fast-attack squelch. However, with C1 connected to pin 6 as shown, pin 2 is drawn down rapidly when unsquelching, and the low-impedance path through Q21 provides the ground for the filtering action required for AGC with signal and threshold level applied at pin 3. With no signal, Q21 turns off, and the voltage at pin 4 rises normally, slowly squelching the amplifier. Note that C1 becomes reverse-biased with no signal. Since the voltage between pins 4 and 2 is only one diode drop, this is insufficient to forward-bias the capacitor, so no deforming occurs. Hysteresis is provided by the positive feedback to the bottom end of the threshold control through the 33k resistor.

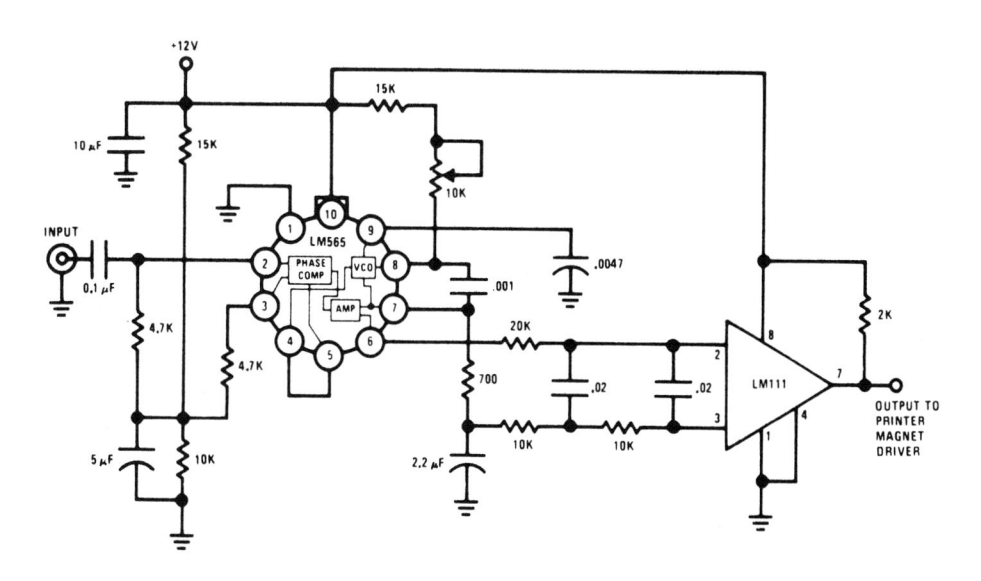

FSK demodulator (2025 to 2225 Hz). The values of the loop filter components (C1 = 2.2 μF and R1 = 700 Ω) were changed to accommodate a keying rate of 300 bauds (150 Hz), since calculated values above cause too much rolloff of 150 Hz. The two 10k resistors and 0.02 μF capacitors at the input to the LM111 comparator provide filtering of the carrier.

SECTION 7 *Communications Circuits* 327

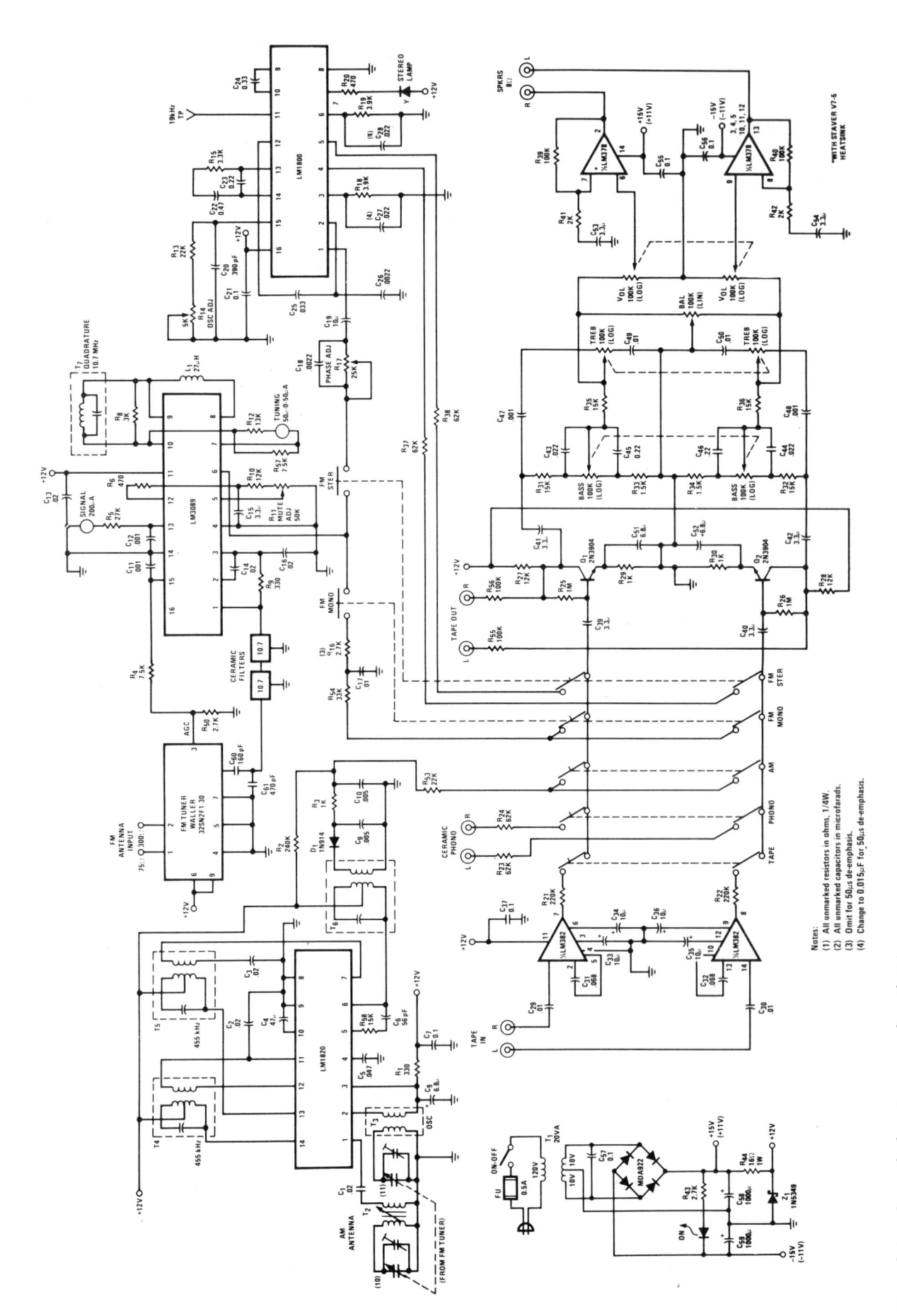

Complete schematic for stereo receiver that incorporates PLL demodulator.

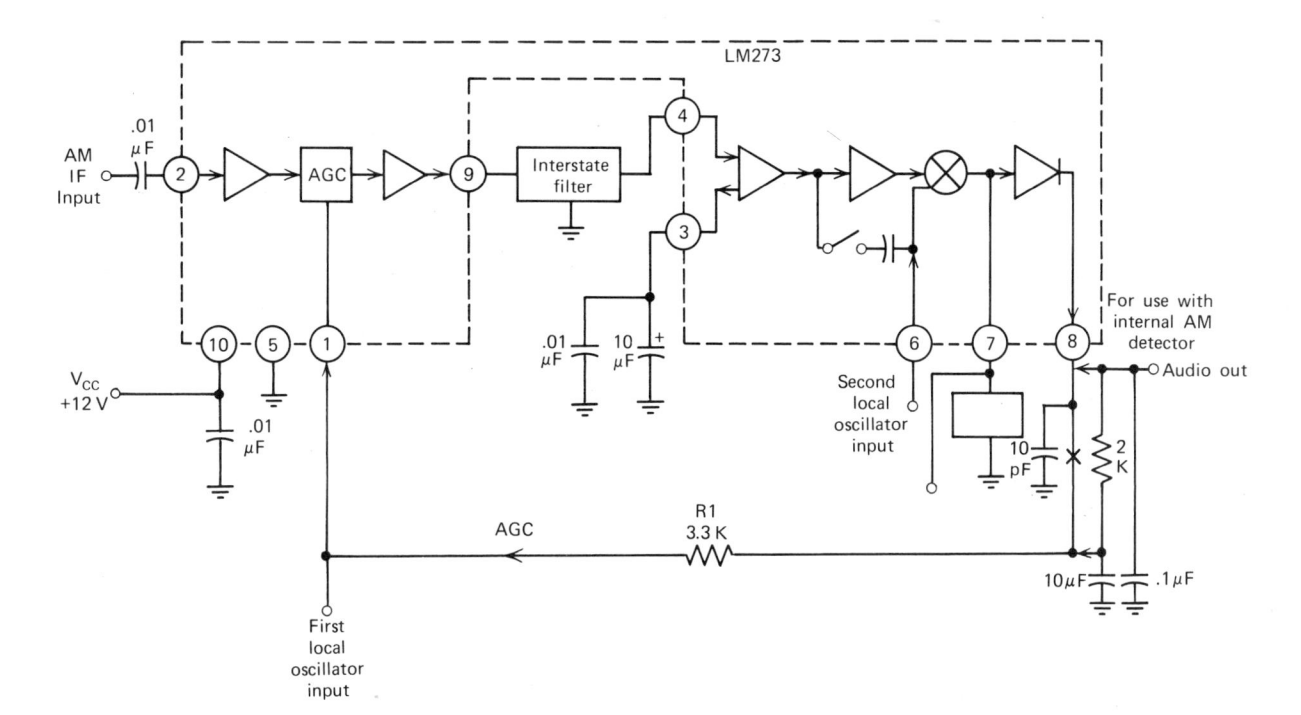

A monolithic double-conversion IF strip is possible with the LM273 and LM274 by using what is normally the AGC block as a balanced mixer. The incoming signal is fed into pin 2, and after passing through one gain stage, it is converted down in frequency the first time by inserting the first local-oscillator signal into pin 1. A bandpass network from pin 9 to pin 4 selects the desired frequency and passes it to successive gain stages. The second frequency conversion occurs in the balanced mixer, with the second local oscillator output injected into pin 6. Additional filtering is connected at pin 7 to shunt all but the desired signal from the succeeding stage. Simultaneous AGC is available by feeding back dc voltage from pin 8 to pin 1. Frequency conversion with the AGC block requires a local oscillator signal level on the order of 800 mV, while the signal level into pin 6 can be 60 mV.

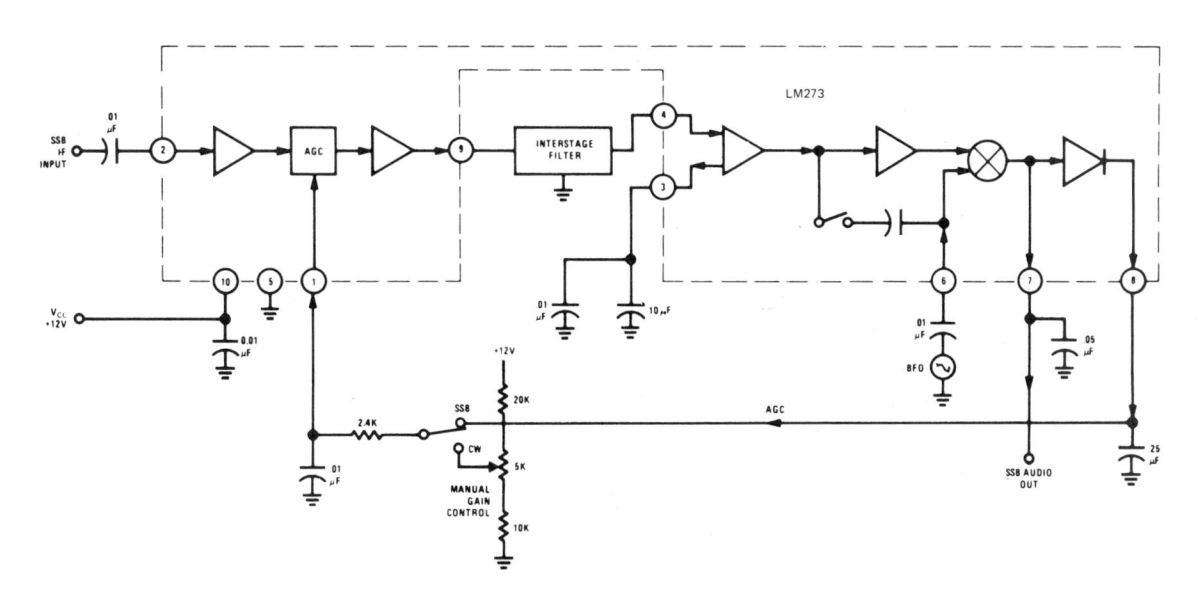

SSB IF strip. The balanced mixer is used as a product detector instead of as a simple gain stage. The local oscillator is fed into pin 6, with the optimum level of switching signal to the upper port being approximately 60 mV. The AGC operates in the same manner as for AM, with the feedback resistor from pin 8 to pin 1 determining output level and AGC figure of merit. The audio is taken from pin 7, so it should have all RF shunted to ground through a capacitor. The impedance looking into pin 7 is well defined at approximately 1k, so this facilitates making a simple RC rolloff with a well-defined cutoff frequency. Additional AGC filtering may be added by a capacitor from pin 8 to ground.

SECTION **7** *Communications Circuits* **329**

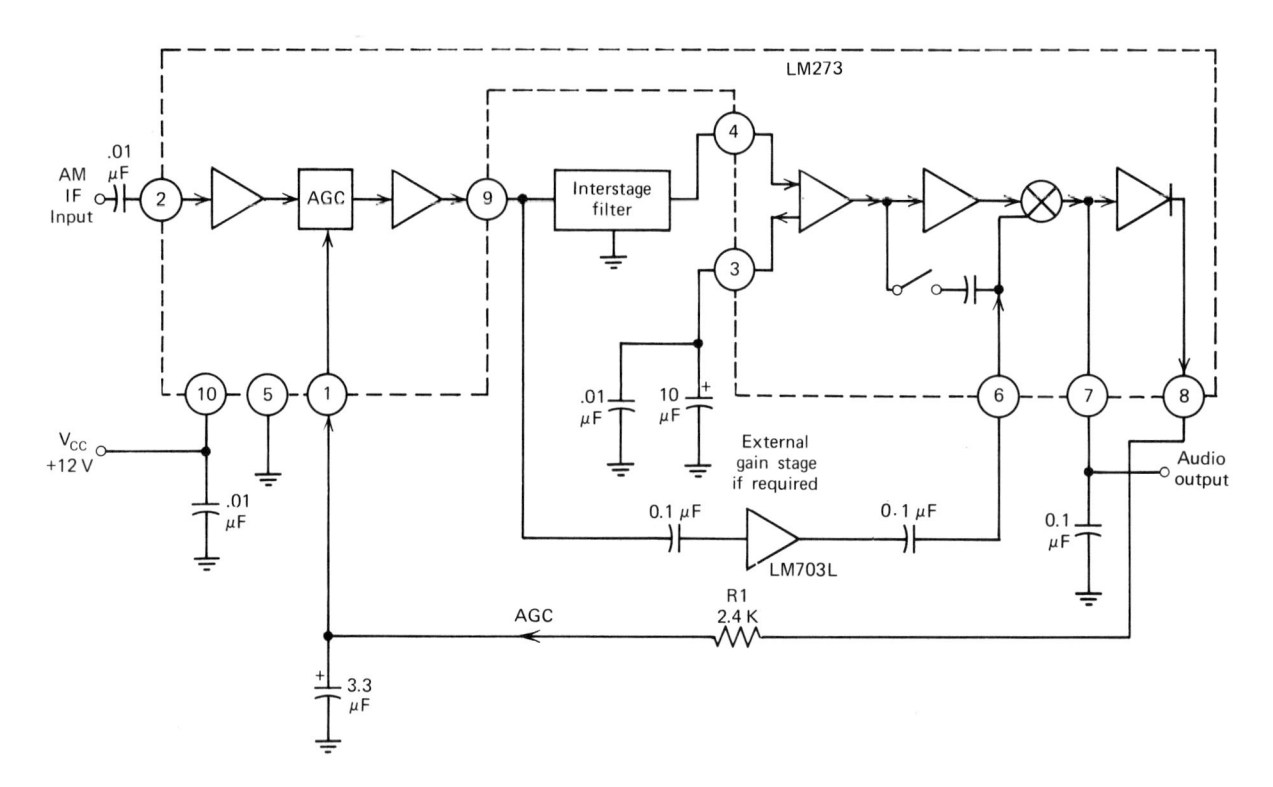

Synchronous AM detector. Detection takes place in the product detector rather than in the peak detector. For current synchronous detection, the signal level at pin 6 should be sufficient to drive the upper port into a full switching mode. A level near 60 mV is optimum. If enough signal is not available at the input of the LM273 to develop this large a signal swing at pin 6, then an additional gain stage may be inserted from pin 9 to pin 6. The audio output is taken at pin 7; additional AGC filtering may be added at pin 8. The interstage filter is optional and need only be used if receiver bandwidth requirements necessitate its use.

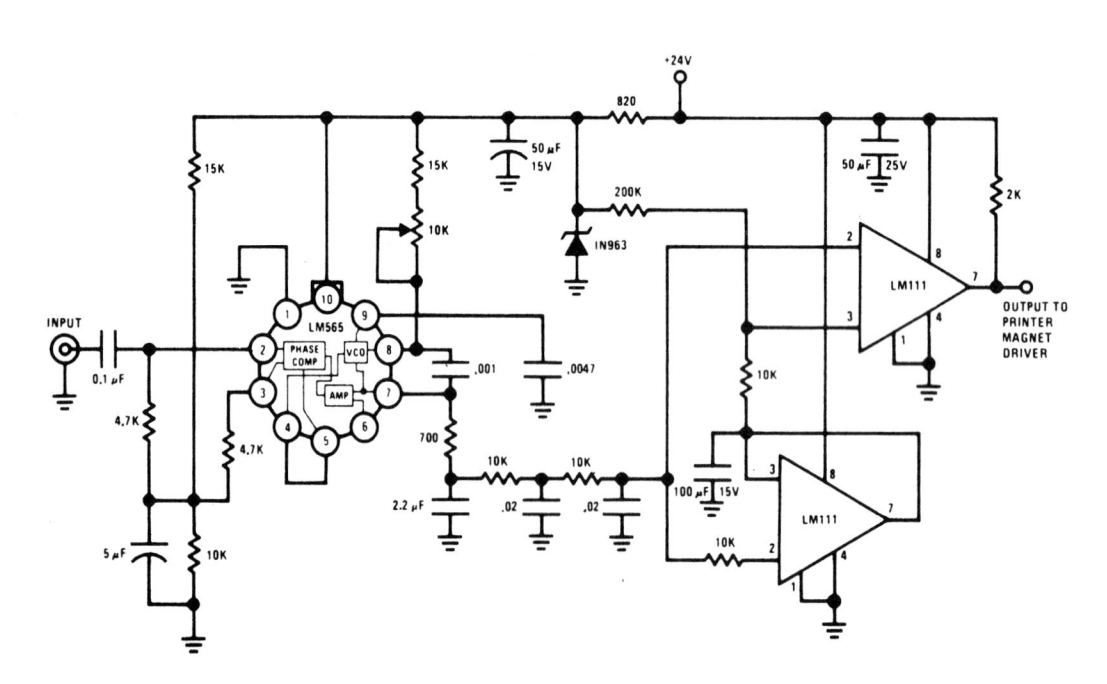

FSK demodulator with dc restoration. An LM111 is used as an accurate peak detector to provide a dc bias for one input to the comparator. When a **space** frequency is transmitted, the output at pin 7 of the LM565 goes negative and switching occurs; the detected and filtered voltage of pin 3 to the comparator will not follow the change.

Horiz. 10 ns/cm
Vert. 1 V/cm

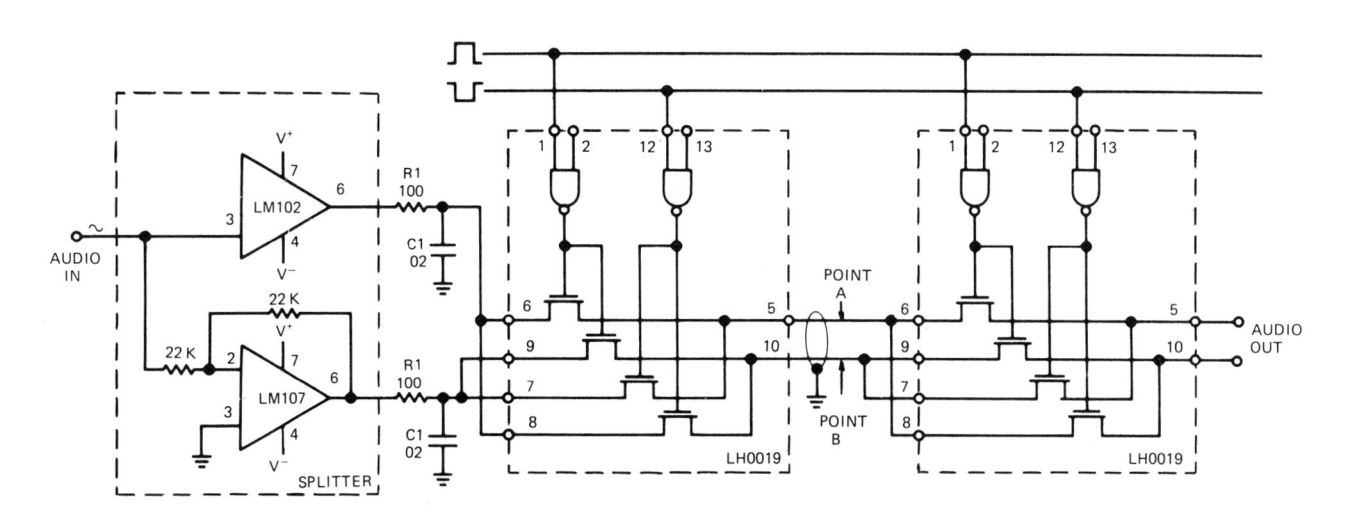

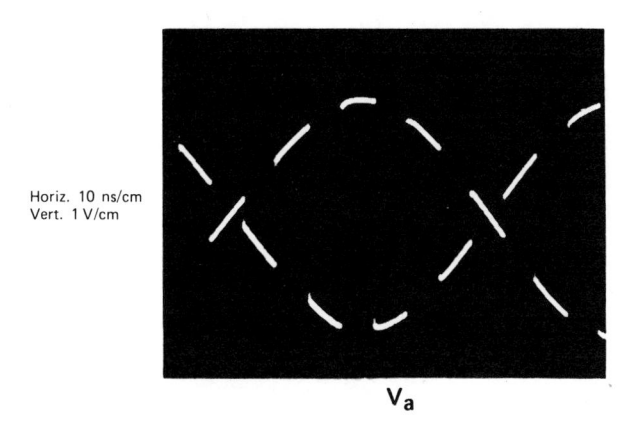

V_a

Double-sideband modulator. The LH0019 functions as a double-pole switch that alternately reverses the polarity of the modulating signal at the chopper frequency. Both points A and B are DSB-modulated outputs. The point-A waveform is illustrated for a carrier frequency of 100 kHz and an audio frequency of 12.5 kHz. Point B is equal and out of phase.

SECTION **7** Communications Circuits

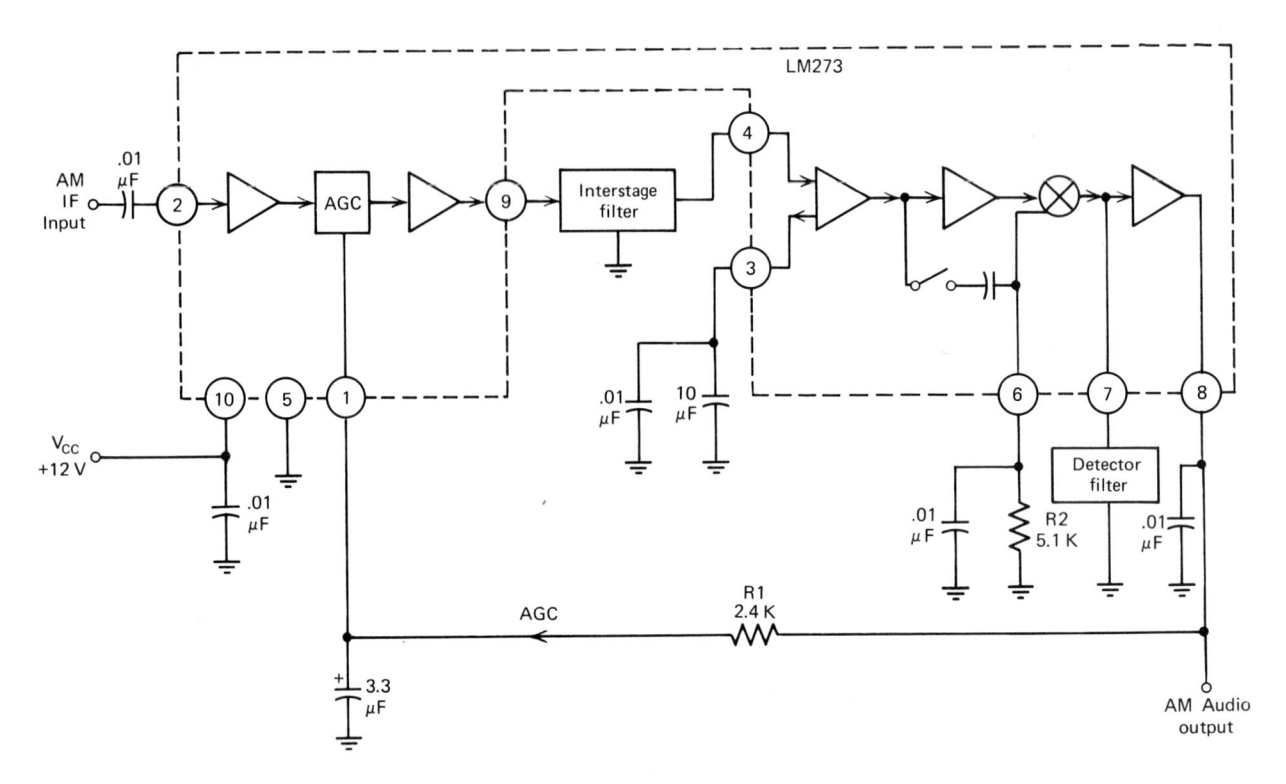

455 kHz AM IF strip. AM operation is achieved by connecting R2 from pin 6 to ground to offset the product detector and by connecting R1 from the AM detector output at pin 8 to the AGC input, pin 1. The value of R1 may be modified to obtain different tradeoffs between output voltage and AGC range. The filter from pin 7 to ground is used to shape the bandwidth of the passband to the active peak detector. The ac-coupled tank passes the desired signal while shunting all wideband noise to ground, preventing undesirable AGCing on noise spikes. The interstage filter shown from pin 9 to pin 4 may also be used to provide additional bandpass shaping. It may be an LC, ceramic, crystal, or any other type of filter.

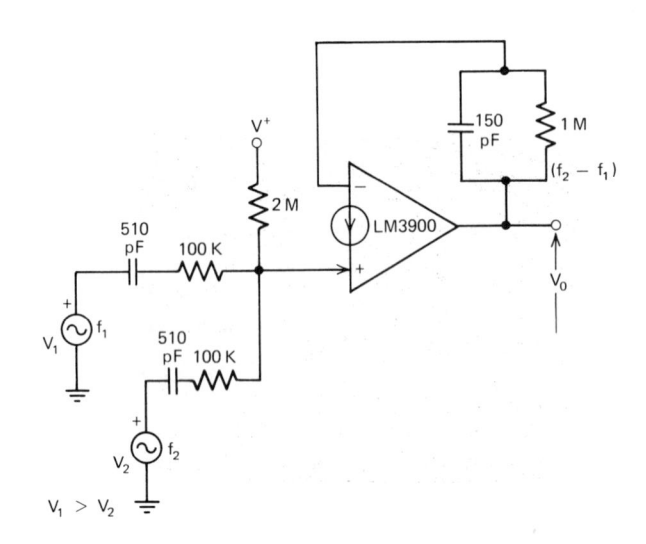

This low-frequency RF mixer has a gain of 10 and contains a low-pass single-pole filter (1M and 150 pF feedback elements) with a corner frequency of 1 kHz. With one signal larger in amplitude to serve as the local oscillator input (V_1), the transconductance of the input diode is gated at this rate (f_1). A smaller signal (V_2) can now be added at the second input, and the difference frequency is filtered from the composite resulting waveform and made available at the output. Relatively high frequencies can be applied at the inputs as long as the desired difference frequency is within the bandwidth capabilities of the amplifier and the RC low-pass filter.

332 IC SCHEMATIC SOURCEMASTER

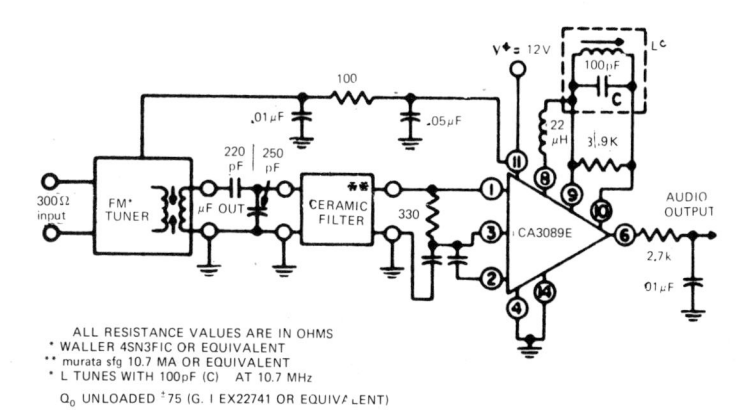

ALL RESISTANCE VALUES ARE IN OHMS
* WALLER 4SN3FIC OR EQUIVALENT
** murata sfg 10.7 MA OR EQUIVALENT
* L TUNES WITH 100pF (C) AT 10.7 MHz
  Q_0 UNLOADED $\pm$75 (G. I EX22741 OR EQUIVALENT)

Performance data at f_0 = 98 MHz, f_{MOD} = 400 Hz,
Deviation = $\pm$75kHz:

.3dB Limiting Sensitivity . . . . . 2µV (Antenna Level)
20dB Quieting Sensitivity . . . . 1µV (Antenna level)
30dB Quieting Sensitivity . . . . 1.5µV (Antenna Level)

This FM IF system incorporates RCA CA3089E, which includes IF amplifier, quadrature detector, AF preamplifier, and specific circuits for AGC, AFC, muting (squelch), and tuning meter. It is particularly suited for FM IF amplifier applications in high-fidelity, automotive, and communications receivers. Distortion in a CA3089E FM IF system is primarily a function of the phase linearity characteristic of the outboard detector coil.

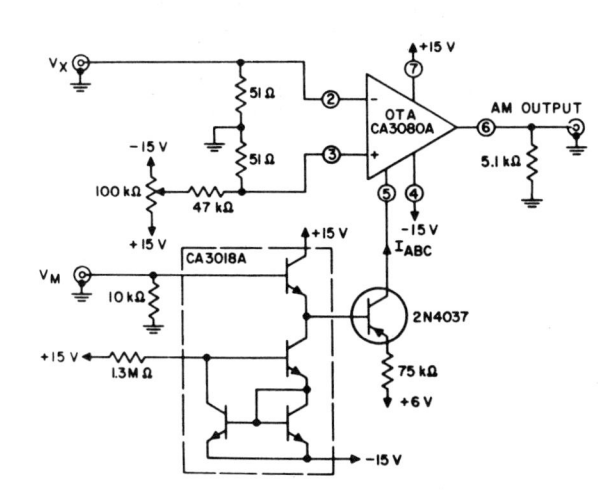

An amplitude modulator using RCA operational transconductance amplifier controlled by PNP and NPN transistors. One transistor in the CA3018 array is used as an emitter follower, and the three other transistors of this array serve as a current source. The 100k pot nulls the effects of amplifier input offset voltage. Adjust the pot to set the output voltage symmetrically about zero.

SECTION **7** *Communications Circuits* **333**

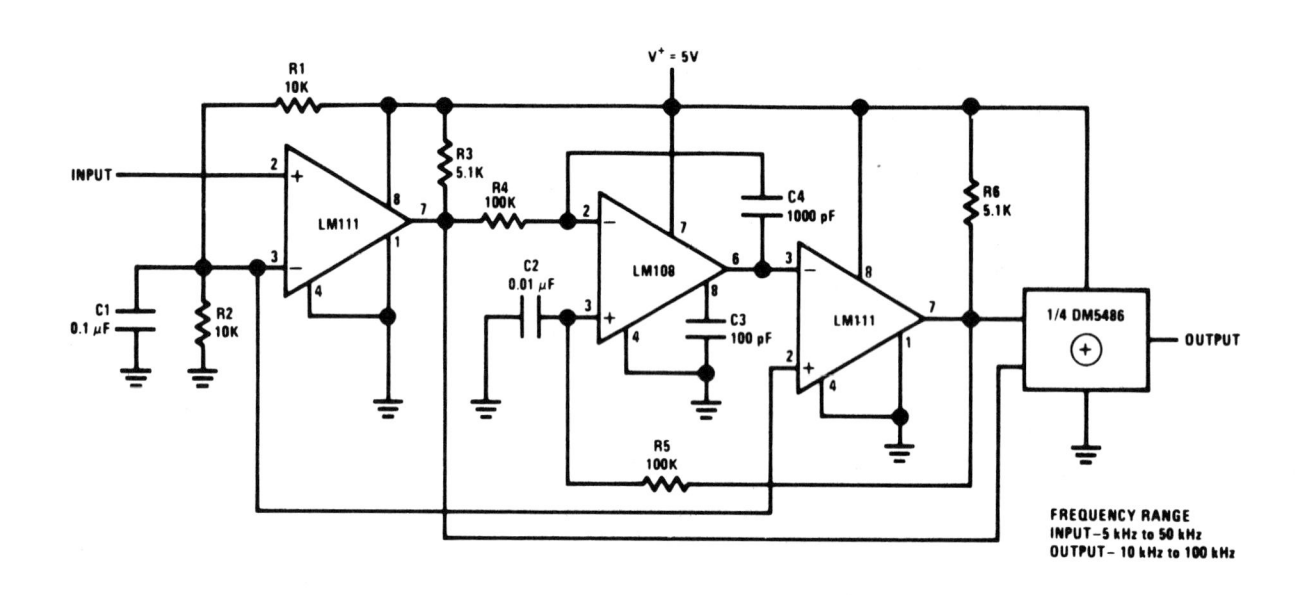

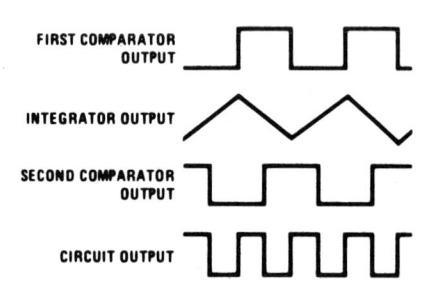

The frequency doubler uses a comparator to shape the input signal and feed it to an integrator. The shaping is required because the input to the integrator must swing between the supply voltage and ground to preserve symmetry in the output waveform. An op-amp that works from the 5V logic supply serves as the integrator. This feeds a triangular waveform to a second comparator that detects when the waveform goes through a voltage equal to its average value. The output of the second comparator is delayed by half the duration of the input pulse. The two comparator outputs can then be combined through an *exclusive*-OR gate to produce the double-frequency output.

With the component values shown, the circuit operates at frequencies from 5 to 50 kHz. Lower frequency operation can be secured by increasing both C1 and C2.

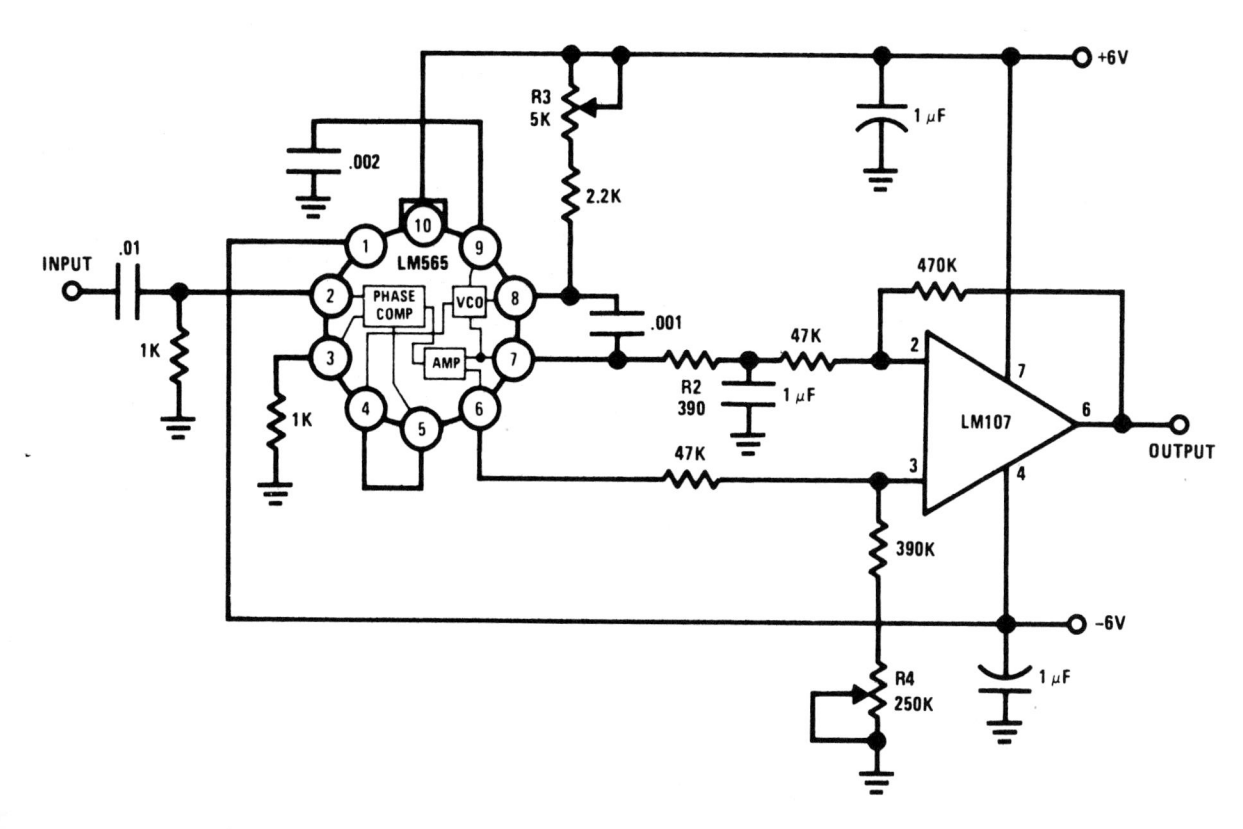

IRIG channel 13 demodulator. In the field of missile telemetry it is necessary to send many channels of narrowband FM data via a radio link on a set of subcarriers with center frequencies in the range of 400 Hz to 200 kHz. Standardization of these frequencies has resulted in several sets of subcarrier channels, some based on deviations that are a fixed percentage of center frequency and others that have a constant deviation regardless of center frequency. For channel 13:

Center frequency:	14.5 kHz
Max. deviation:	±7.5%
Frequency response:	220 Hz
Deviation ratio:	5

With a deviation of 7.5%, the LM565 will produce approximately 225 mV output. It is desirable to amplify and level-shift this signal to ground so that plus and minus output voltages can be obtained for frequency shifts above and below center frequency.
R4 is used to set the output at zero with no input. The frequency of the VCO can be adjusted with R3 to provide zero output voltage when an input signal is present.

SECTION **7** *Communications Circuits* **335**

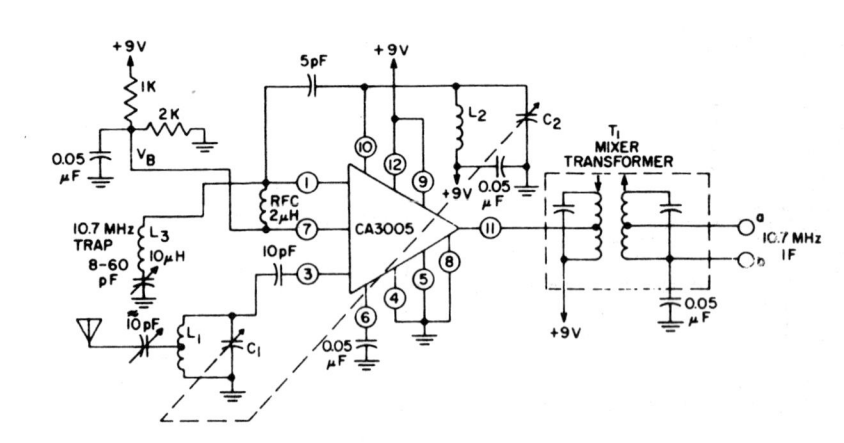

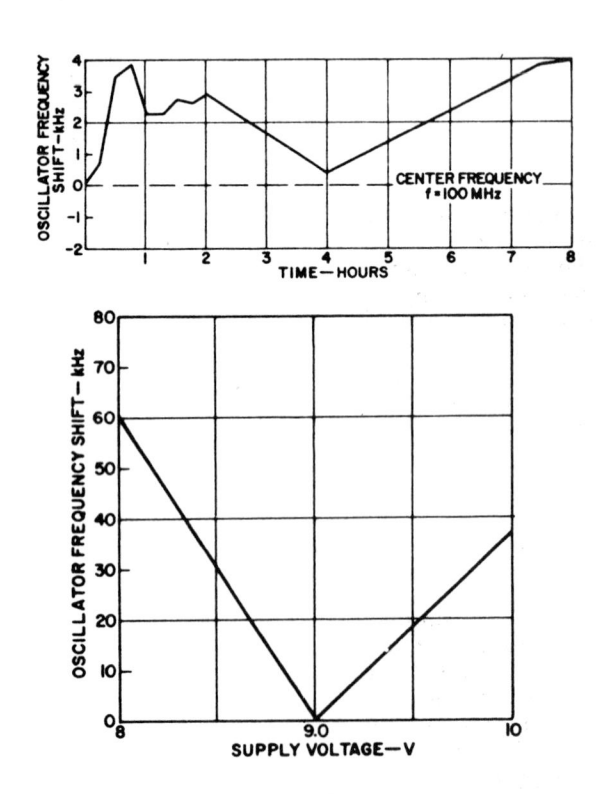

A single-chip front end using the CA3005. Curves show oscillator stability with time and supply voltage.

L_1 — 4 turns of No. 22 wire, center-tapped; 1/4" O.D. coil form, "E" mat'l slug

L_2 — 6 turns No. 32 wire on toroid core; Radio Industries Inc., 1/4" O.D., No. 8 mat'l

T_1 — TRW No. 21629, or equiv.

336   IC SCHEMATIC SOURCEMASTER

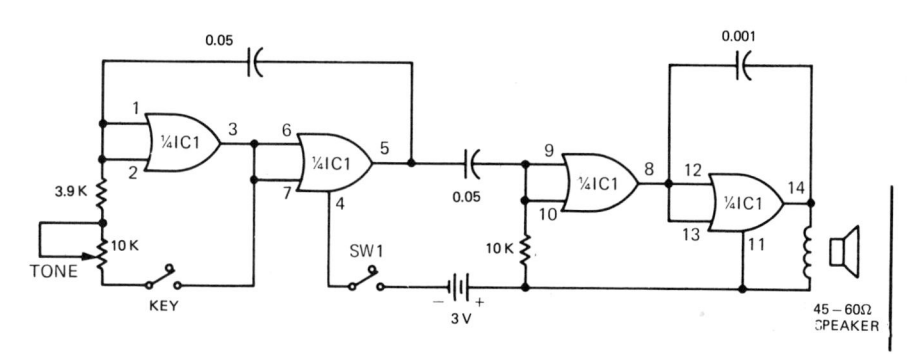

A code practice oscillator is simple to construct and offers loud audio signal and relatively stable tone. IC is Motorola HEP 570, a quad 2-input positive-logic NOR gate.

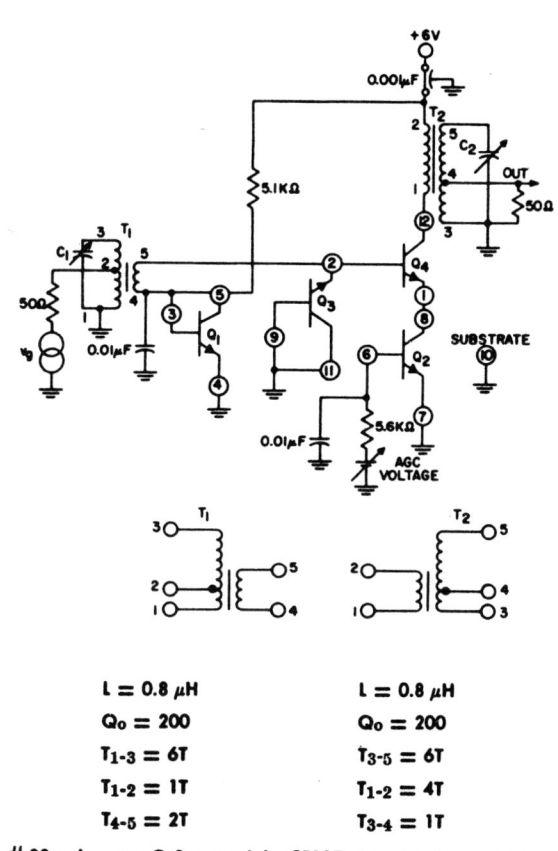

$L = 0.8\ \mu H$
$Q_0 = 200$
$T_{1\text{-}3} = 6T$
$T_{1\text{-}2} = 1T$
$T_{4\text{-}5} = 2T$

$L = 0.8\ \mu H$
$Q_0 = 200$
$T_{3\text{-}5} = 6T$
$T_{1\text{-}2} = 4T$
$T_{3\text{-}4} = 1T$

#22 wire on Q-2 material, CF107 Torroid from Indiana General.

C_1, C_2 = Arco 425 or equiv.

A 15 MHz RF amplifier for receivers in 2 to 30 MHz range using CA3018. This circuit was designed for a midband frequency of 15 MHz to demonstrate its capability. Gain is obtained in a common-emitter stage (Q4). Transistor Q2 is used as a variable resistor the emitter of Q4 to provide improved signal-handling capability with AGC. Transistor Q1 is used as a bias diode to stabilize Q4 with temperature, and the reverse breakdown of Q3 as a diode is used to protect the common-emitter stage from signal overdrive of adjacent transmitters.

SECTION 7 Communications Circuits 337

SECTION 8
SENSING, MONITORING, CONTROLLERS, AND POWER DRIVERS

This section contains many circuits that might logically be grouped into other categories; the classification is necessarily arbitrary, since so many different types of electronic devices perform sensing and monitoring functions. The circuits for thermometers and temperature-detection switches, for example, might well have been included in Section 3 along with voltage- and current-monitoring instruments. The voice-operated relay might have been included in the section with communications circuits.

When you're looking for circuits fitting the description of the section title, do think about other possible sections in this book where other candidates might be found; many times the only difference between a circuit of one class and that of another is in the form of transducer employed. Look for zero-crossing detectors, peak detectors, and analog switches in Section 14 on analog and digital transmission; look for sound detectors, continuity sensors, and touch switches in Section 16, which contains hobby, fun, and experimenter circuits.

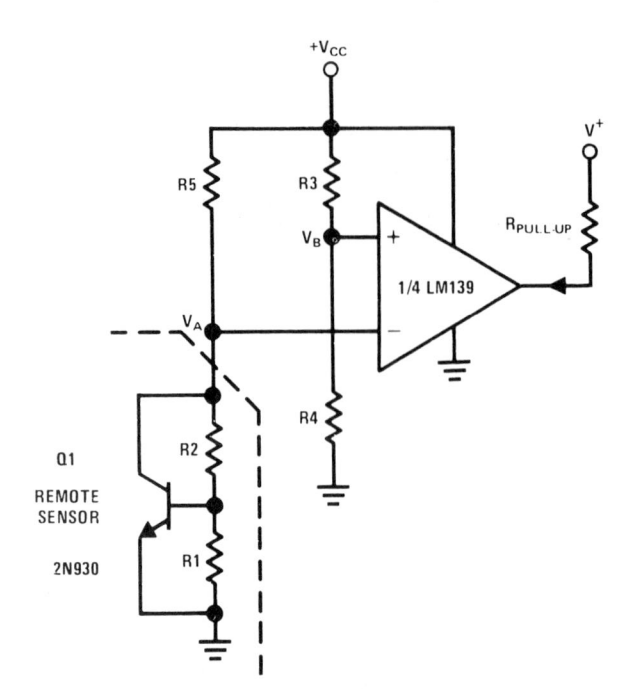

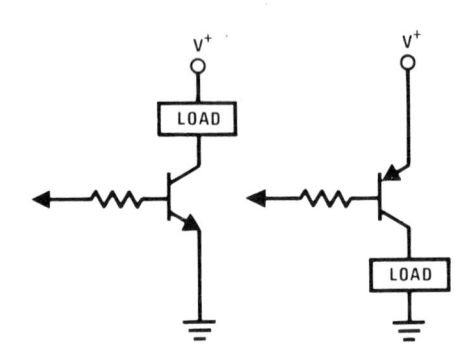

Remote temperature sensor alarm. The 2N930 is a National process 07 silicon NPN transistor connected to produce a voltage reference equal to a multiple of its base–emitter voltage along with a temperature coefficient equal to a multiple of 2.2 mV/°C. That multiple is determined by the ratio of R1 to R2. With transistor Q1 biased up, its base-to-emitter voltage will appear across resistor R1. Assuming a reasonably high beta ($\beta \geq 100$), the base current can be neglected so that the current that flows through resistor R1 must also be flowing through R2. This base-to-emitter voltage is strongly temperature dependent (−2.2 mV/°C for a silicon transistor). This temperature coefficient is also multiplied by the resistor ratio R1/R2. This provides a highly linear, variable temperature coefficient reference ideal for use as a temperature sensor over a temperature range from approximately −65 to +150°C. When this temperature sensor is connected as shown, it can be used to indicate an alarm condition of either too high or too low a temperature excursion. Resistors R3 and R4 set the trip point reference voltage, V_B, with switching occurring when $V_A = V_B$. Resistor R5 is used to bias Q1 at some low value of current simply to keep quiescent power dissipation to a minimum (10 μA is acceptable).

Using one LM139, four separate sense points are available. The outputs of the four comparators can be used to indicate four separate alarm conditions or the outputs can be ORed to indicate an alarm condition at any one of the sensors. For the circuit shown, the output will go high when the temperature of the sensor goes above the present level. This could easily be inverted by simply reversing the input leads. For operation over a narrow temperature range, the resistor ratio R2/R1 should be large to make the alarm more sensitive to temperature variations.

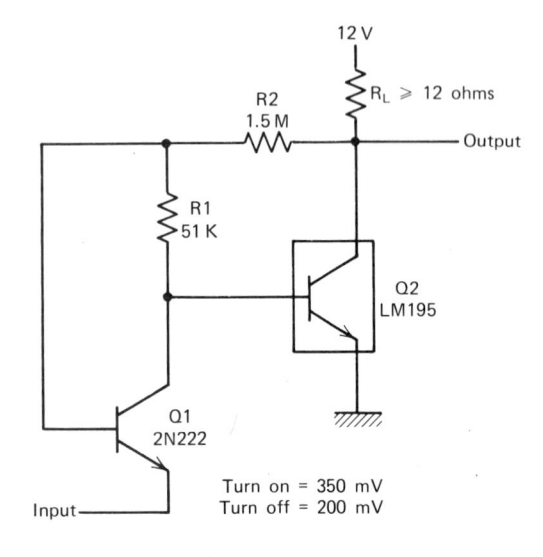

Low-level power switch.

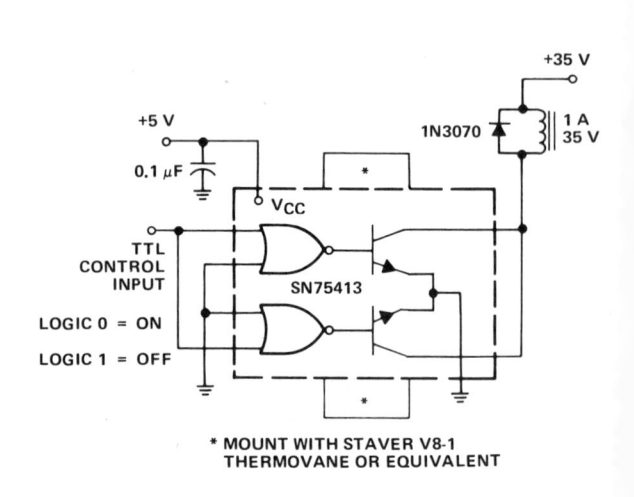

Power solenoid driver.

340 IC SCHEMATIC SOURCEMASTER

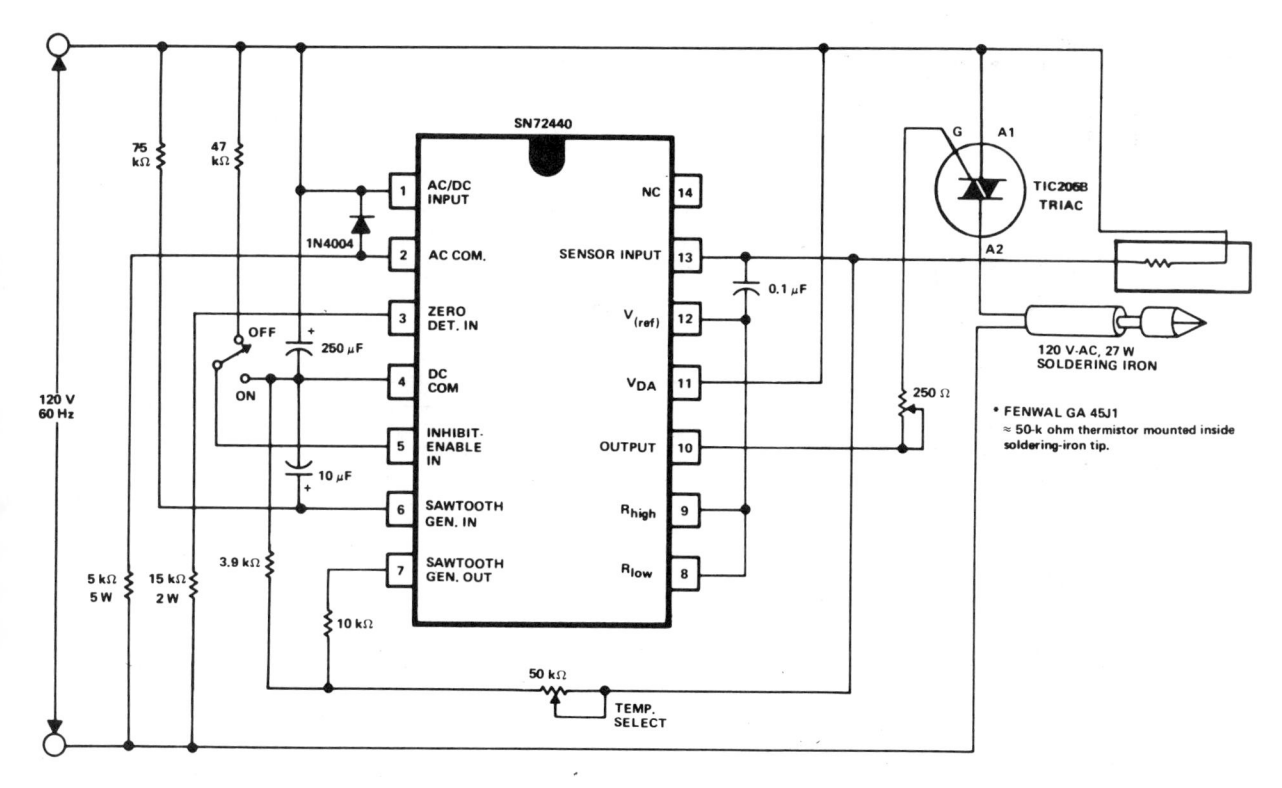

A soldering iron thermal probe uses the SN72440 in a proportionally controlled power application, where the temperature of the probe, a 27W soldering iron, is maintained at a preset level. The probe temperature is sensed by a thermistor built into its tip. By adjusting the temperature to a desired level and applying the tip of the iron to a transistor, integrated circuit, resistor, or other component, individual reaction of the component to that temperature may be determined. With the temperature-select control calibrated, it is possible to slowly raise the temperature of a component and easily determine the temperature at which it fails or changes characteristics.

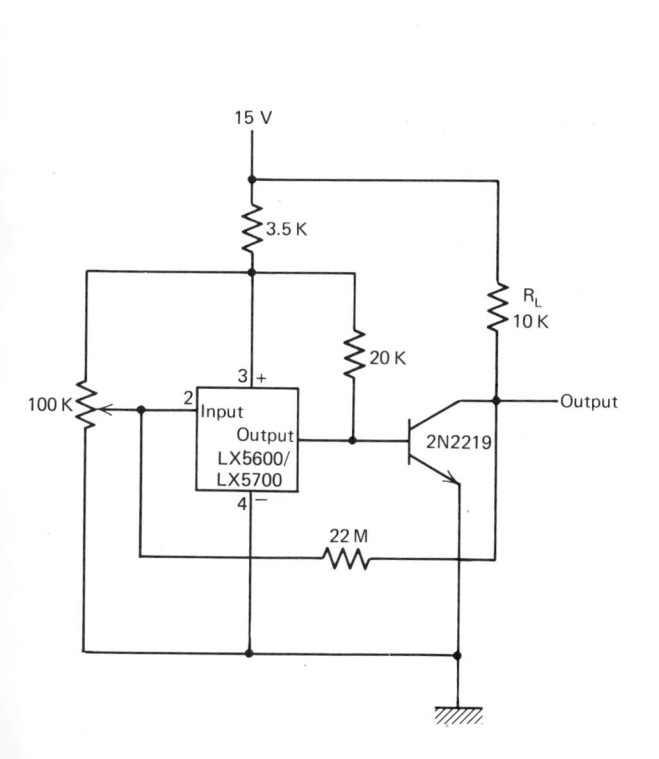

Temperature controller with hysteresis. Output goes positive on temperature increase.

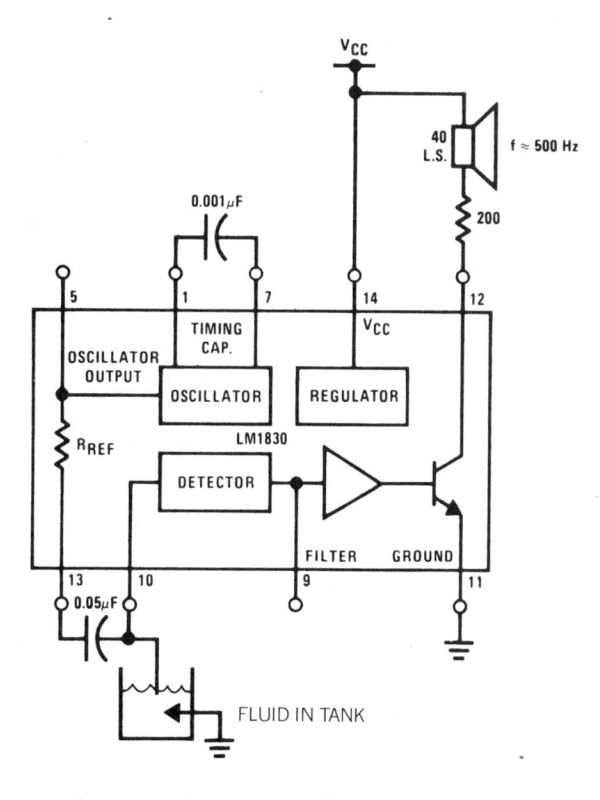

Low-level warning with audio output.

SECTION **8** *Sensing, Monitoring, Controllers, and Power Drivers* **341**

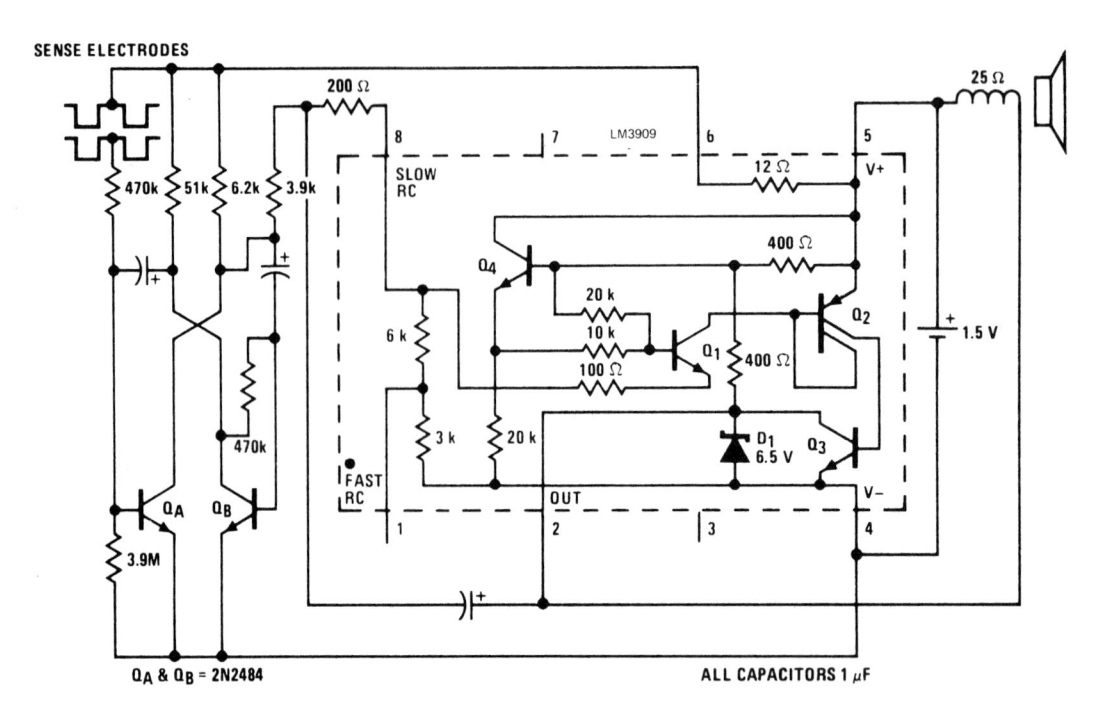

A **water seepage alarm** is safe on potentially damp floors since there is no connection to the power line. Standby battery drain of 100 µA yields a battery life close to shelf life. Without moisture, Q_a is completely off, and its collector load (6.2k) provides enough current to hold pin 8 of the LM3909 above 0.75V (where it can't oscillate). When the sense electrodes pass about 0.25 µA (due to moisture), Q_a starts turning on; since Q_b is already partially biased on, positive feedback occurs. Q_a and Q_b are now an astable multivibrator that oscillates faster as more leakage passes across the sense electrodes. The sensor, part of the base of the box the alarm circuitry is packaged in, consists of two 6- or 8-inch electrodes about ⅛ inch apart. Two strips of stainless steel on insulators or the appropriate zigzag path cut in the copper cladding of a circuit board will work well. The bare circuit board between the copper sensing areas should be coated with warm wax so that moisture on the floor, **not** that absorbed by the board, will be detected. The circuit and sensor can be tested by just touching a damp finger to the electrode gap.

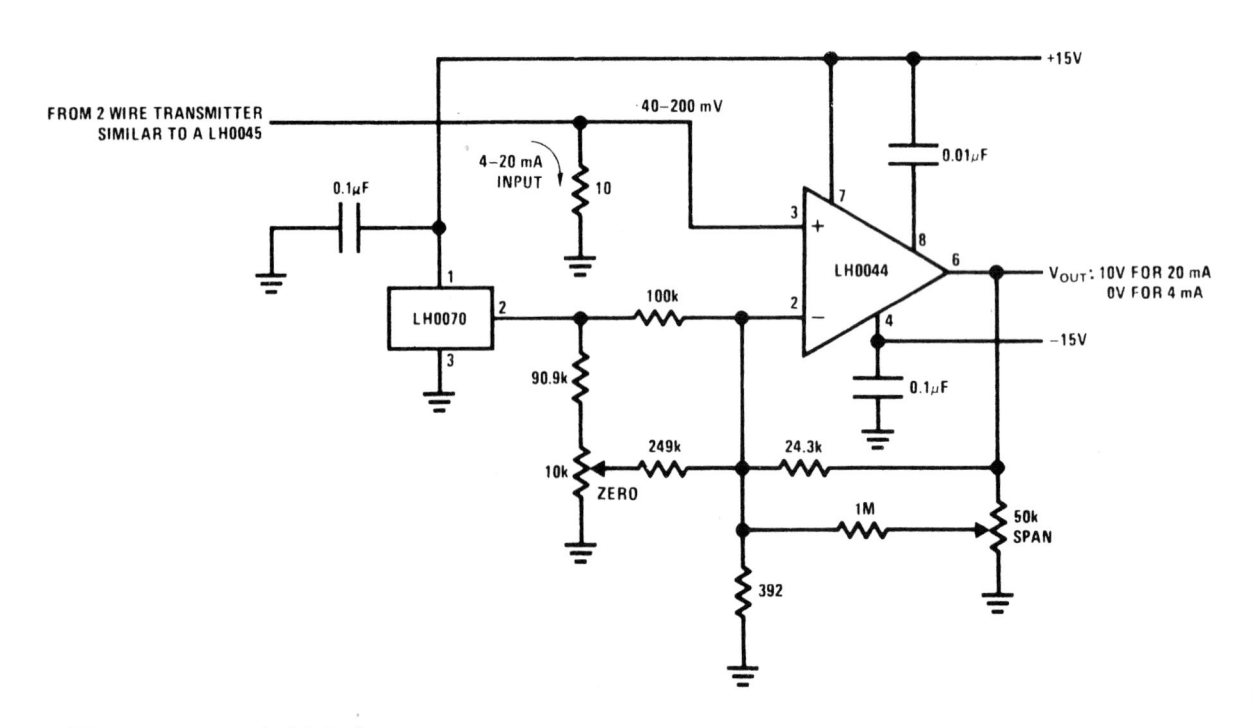

Precision process control interface.

342 IC SCHEMATIC SOURCEMASTER

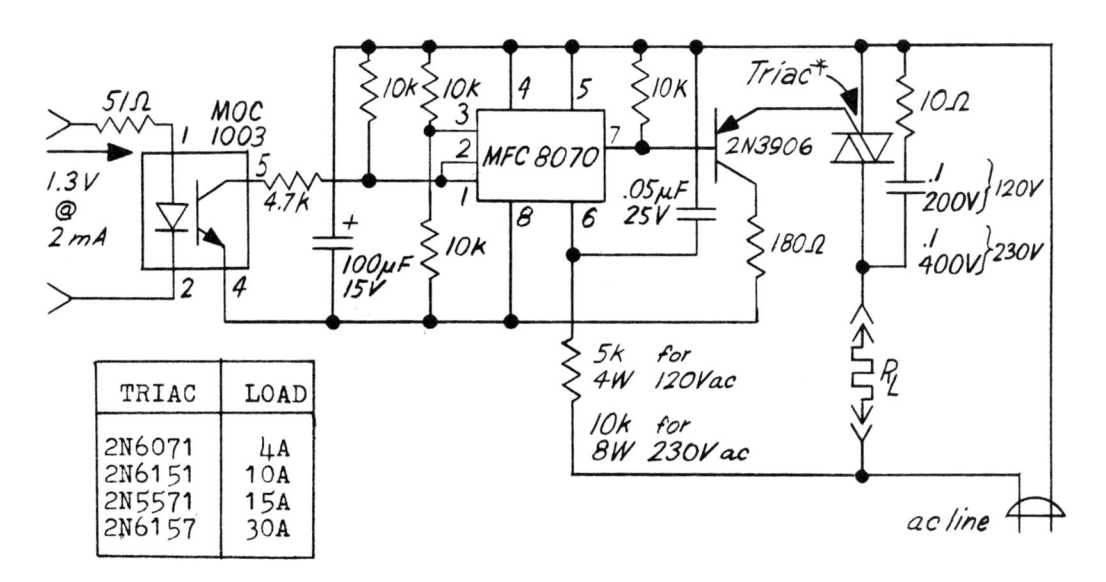

TRIAC	LOAD
2N6071	4A
2N6151	10A
2N5571	15A
2N6157	30A

Motorola's **solid-state relay** used to drive triacs in control circuits. It features saturated logic controllable input requiring 1.5V at 10 mA, optical coupling for isolation to 500V, zero-voltage-crossing switching for interference elimination, 120 or 130V line operation. Circuit employs MOC1003 optical coupler to provide the logic-compatible drive and isolation. The MFC8070 senses a differential input signal adjacent to the line's zero-voltage crossing and provides an output pulse (when the coupler is energized) of about 50 mA. The 2N3906 provides a power boost to ensure operation at lower temperatures. The load can be slightly inductive with the inclusion of the RC snubber as shown.

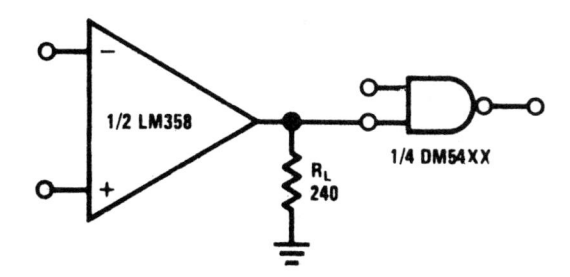

Driving TTL.

SECTION **8** Sensing, Monitoring, Controllers, and Power Drivers

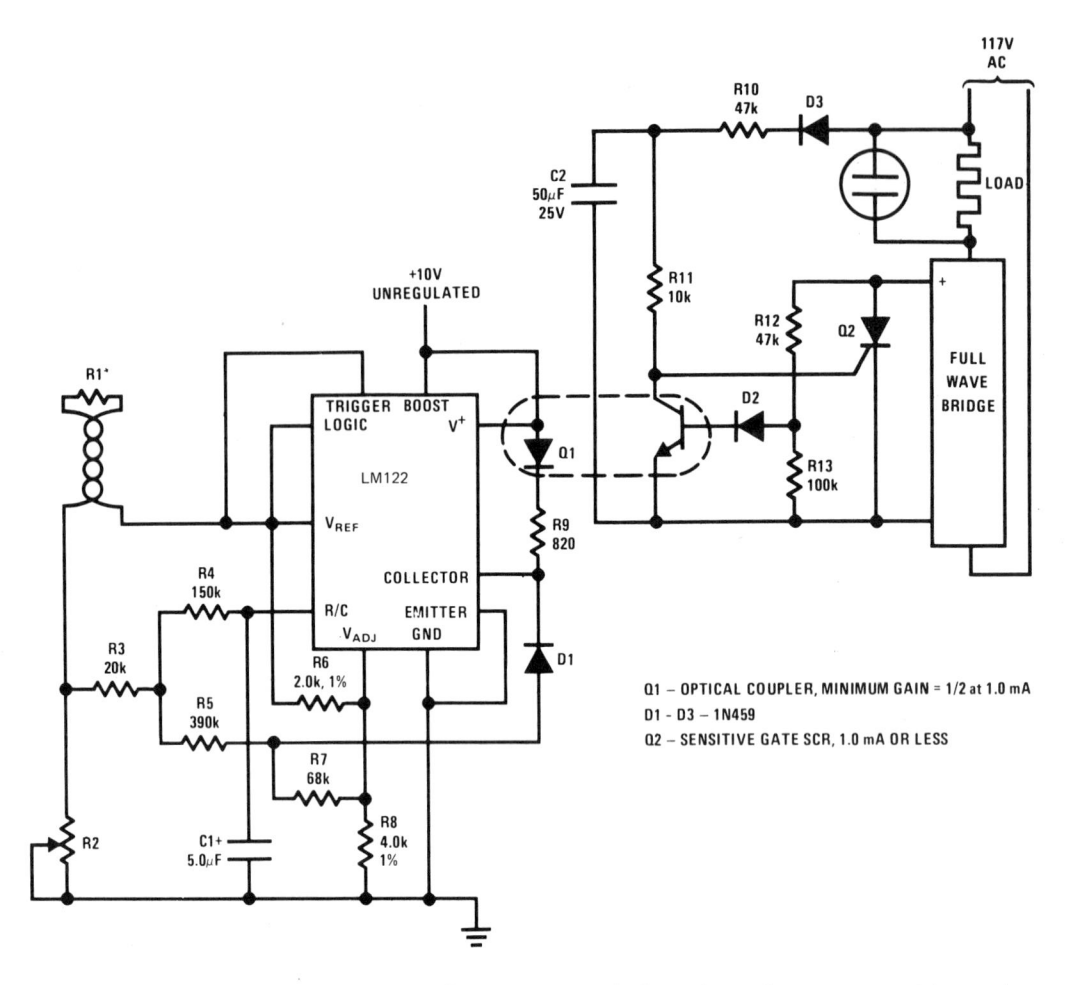

Q1 – OPTICAL COUPLER, MINIMUM GAIN = 1/2 at 1.0 mA
D1 – D3 – 1N459
Q2 – SENSITIVE GATE SCR, 1.0 mA OR LESS

Proportioning temperature controller with synchronized zero crossing. The IC's timing function is not used; instead, the trigger terminal is held high, and the LM122 is used as a high-gain comparator with a built-in reference. R1 is a thermistor (−4%/°C temperature coefficient). R2 is used to set the temperature to be controlled by R1. R3 through R8 set up the proportioning action. R3 raises the impedance of the R1/R2 divider so that R5 sees a relatively constant impedance independent of the set-point temperature. R6 and R8 reduce the V_{ADJ} impedance so that internal variations in divider impedance do not affect proportioning action. R5 and R7 set the actual width of the proportioning band and can be scaled as necessary to alter the width of the band. Larger resistors make the band narrower. The values shown give approximately a 1°C band. R4 and C1 determine the proportioning frequency, which is about 1 Hz with the values shown. C1 or R4 can change to alter frequency, but R4 should be between 50 and 500k, and C1 must be a low-leakage type to prevent temperature shifts. D1 prevents supply voltage fluctuations from affecting set point or proportioning band. Any unregulated supply between 6 and 15V is satisfactory.

Q1 is an optical isolator with a minimum gain of 0.5. With the values shown for R9, R10, and R11, Q1 is overdriven by at least 3 to 1 to ensure deep saturation for reliable turnoff of the SCR. Q2 must be a sensitive-gate device with a worst-case gate firing current of 0.5 mA. R12, R13, and D2 implement the synchronized zero-crossing feature by preventing Q1 from turning off after the voltage across Q2 has climbed above 2.5V. D3, R10, and C2 provide a source of semifiltered dc for SCR gate drive. D3 and Q2 must have a minimum breakdown of 20V.

344 IC SCHEMATIC SOURCEMASTER

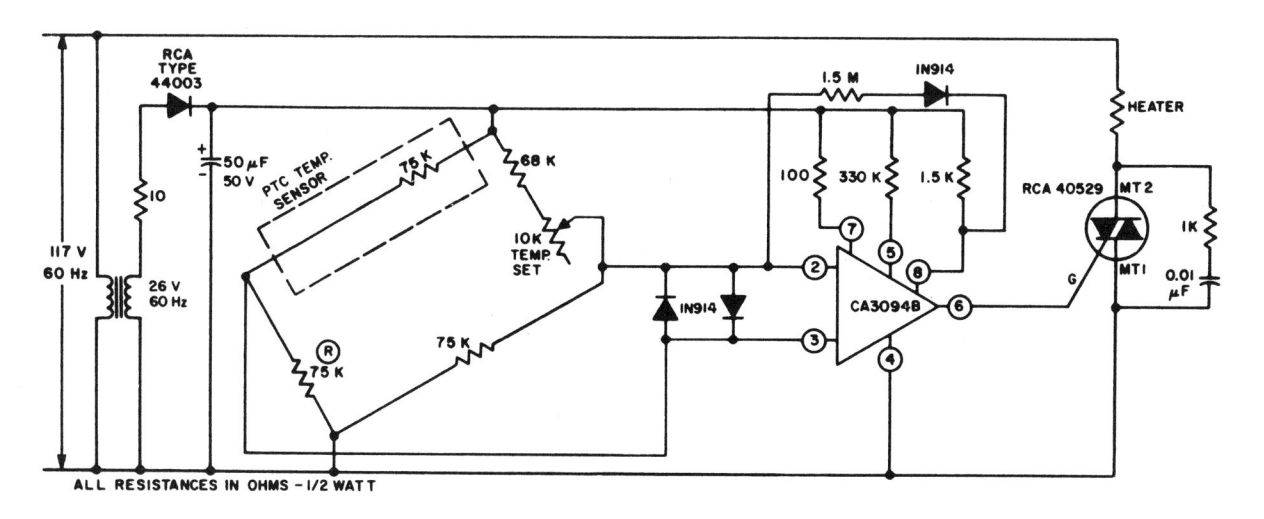

Temperature controller.

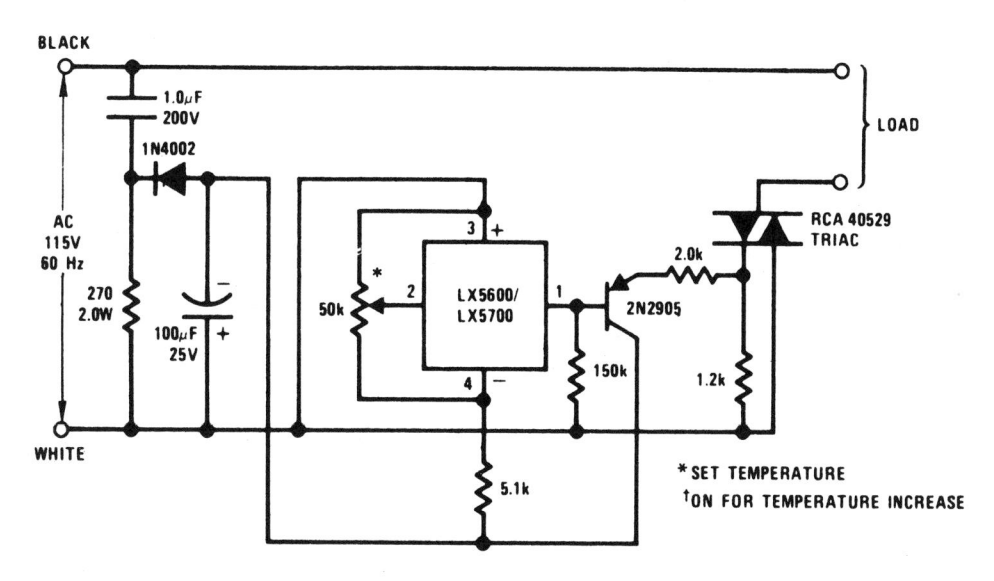

Temperature controller driving triac.

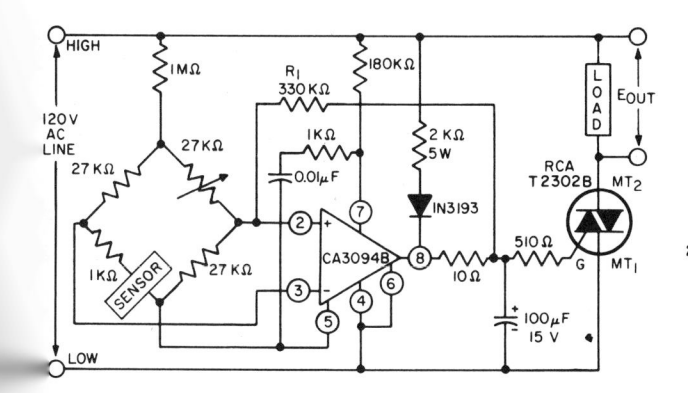

Line-operated thyristor-firing circuit controlled by ac-bridge sensor.

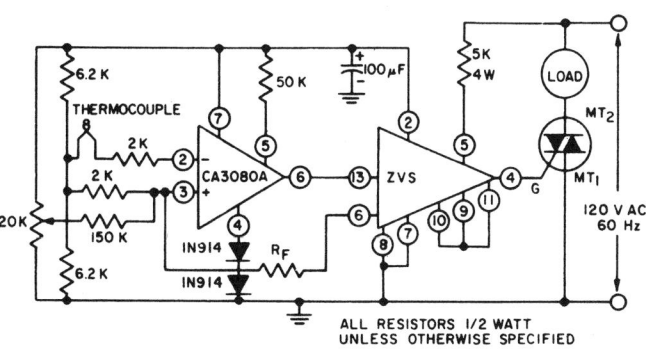

Thermocouple temperature control with zero-voltage switching for use with thermocouple sensor.

SECTION **8** Sensing, Monitoring, Controllers, and Power Drivers

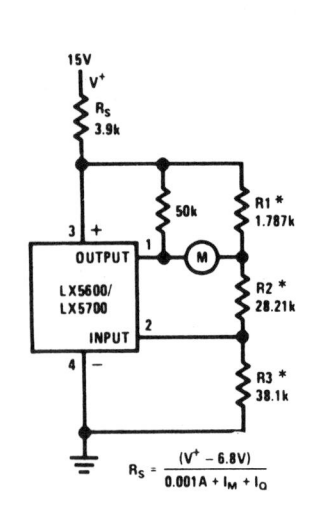

* VALUES SHOWN FOR:
$T_O = 300°K$, $\Delta T = 100°K$,
$I_M = 1.0$ mA, $I_Q = 100\mu A$

$$R1^* = \frac{(V_Z)(10\text{ mV})(\Delta T)}{I_M(V_Z - 0.01\,T_O)}$$

$$R2 = \frac{0.01\,T_O - I_Q R1}{I_Q}$$

$$R3 = \frac{V_Z}{I_Q} - R1 - R2$$

$$\left(I_Q \leq \frac{2V}{R1}\right)$$

V_Z = SHUNT REGULATOR VOLTAGE (USE 6.85)
ΔT = METER TEMPERATURE SPAN (°K)
I_M = METER FULL SCALE CURRENT (A)
T_O = METER ZERO TEMPERATURE (°K)
I_Q = CURRENT THROUGH R1 R2 R3 AT ZERO METER CURRENT (10µA TO 1.0 mA) (A)

$$R_S = \frac{(V^+ - 6.8V)}{0.001A + I_M + I_Q}$$

Thermometer with meter output.

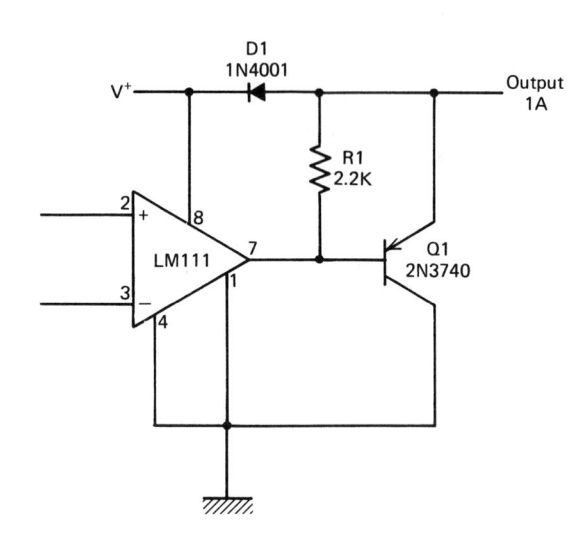

Comparator and solenoid driver.

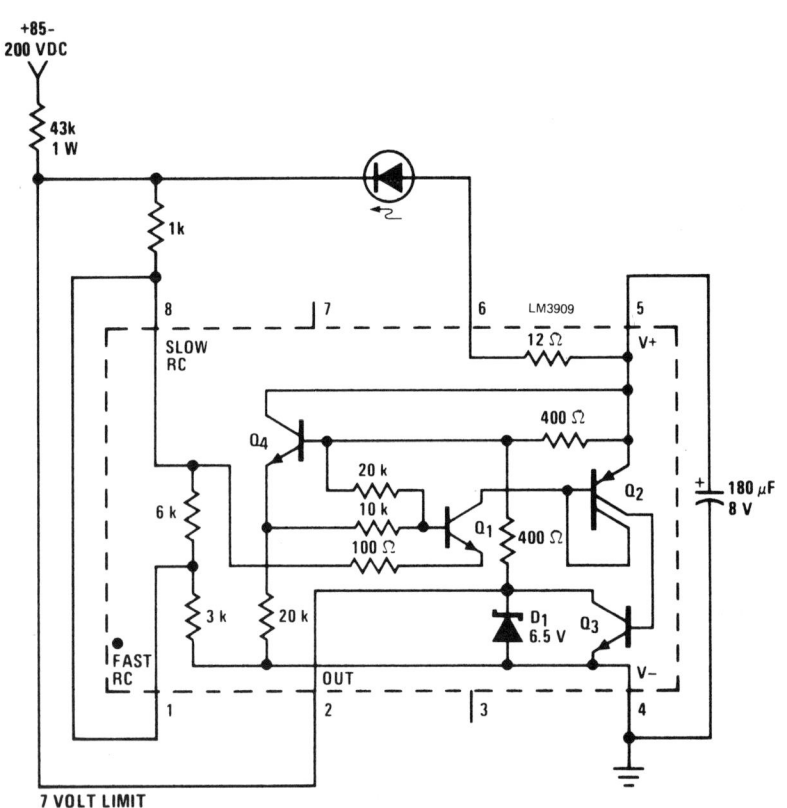

Safe high-voltage flasher. Indication or monitoring of a high-voltage power supply at a remote location can be done much more safely than with neon lamps. If the dropping resistor (43k) is located at the source end, all other voltages on the line, the IC, and the LED will be limited to less than 7V about ground. The timing capacitor is charged through the dropping resistor and the two 400Ω collector loads between pins 2 and 5 of the IC. When capacitor reaches 5V, there is enough voltage across the 1k resistor (pin 8) to turn on the whole IC to discharge the capacitor through the LED.

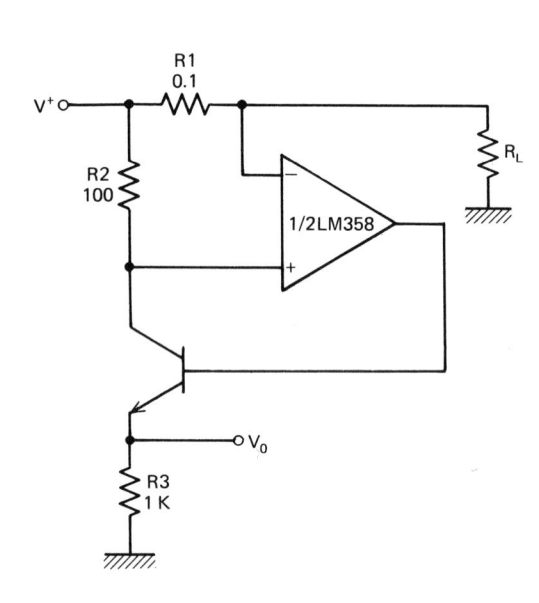

Current monitor. $V_o = 1V(I_L)/1A$; increase R1 for small I_L values.

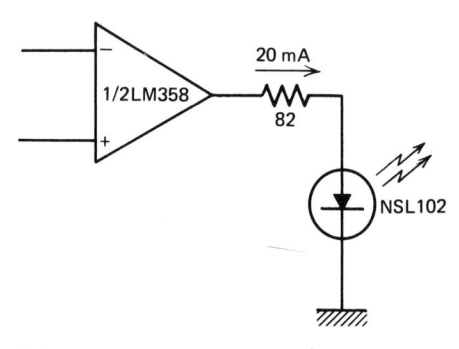

LED driver.

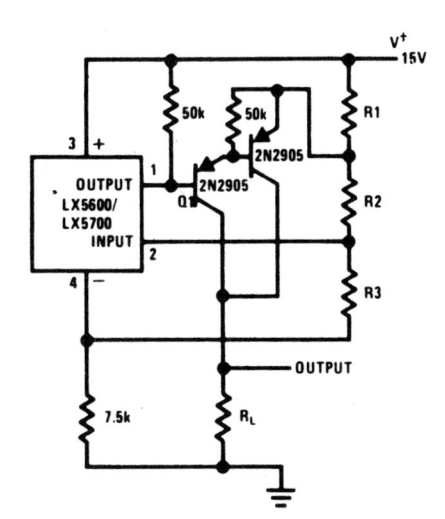

$$R1 = \frac{(V_Z)(10\,mV)(\Delta T)}{\frac{V_O}{R_L}(V_Z - 0.01\,T_O)}$$

$$R2 = \frac{0.01\,T_O - I_Q R1}{I_Q}$$

$$R3 = \frac{V_Z}{I_Q} - R1 - R2$$

V_Z = SHUNT REGULATOR VOLTAGE
ΔT = TEMPERATURE SPAN (K)
T_O = TEMPERATURE FOR ZERO OUTPUT (K)
V_O = FULL SCALE OUTPUT VOLTAGE $\leq 10V$
I_Q = CURRENT THROUGH R1, R2, R3, AT ZERO OUTPUT VOLTAGE (TYPICALLY 100μA TO 1.0 mA)

Ground-referred thermometer.

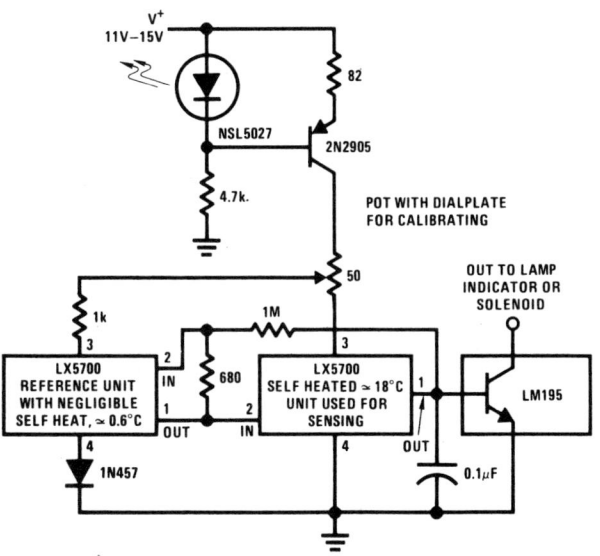

Output "OFF" if sensing unit becomes hot, i.e., out of liquid or airstream.
Reference unit is 1 inch from the sensing unit in airstreams, and below the sensor in liquid sensing systems.

Relay driver with strobe. D1 absorbs inductive kickback of relay and protects IC from severe voltage transients on V^{++} line.

Liquid or moving-air detection circuit. Basic two-sensor configuration for performing a number of differential temperature functions. One sensor is heated considerably by internal dissipation and the other is biased slightly lower at pin 3, so that if the "hot" sensor loses half its excess heat, its output (operating essentially open-loop) will switch to a *high* state. This turns on the power transistor. The feedback network, consisting of a 1M and a 680Ω resistor, provides slightly less than ½° hysteresis or snap action, which prevents partial outputs or oscillation near the set point.

Adjustments are not too critical owing to the large temperature differences that are created. Cooling of the sensing unit by an electrically nonconducting liquid (such as oil) is about 16°C. Air at about 400 cpm (a little under 5 mph) provides about 10°C cooling.

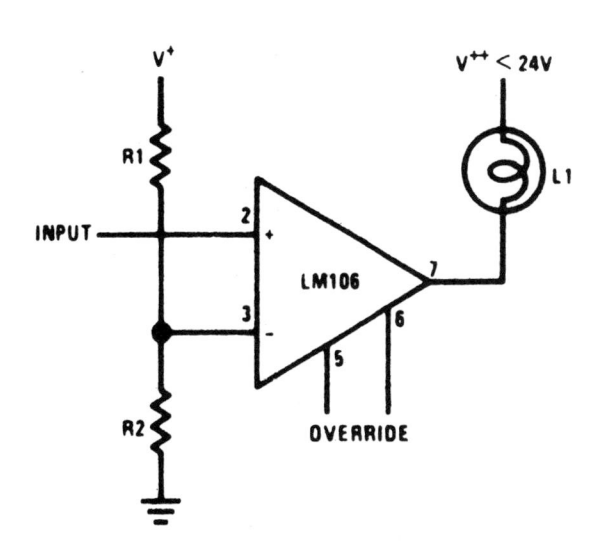

Level detector and lamp driver.

SECTION **8** *Sensing, Monitoring, Controllers, and Power Drivers* **347**

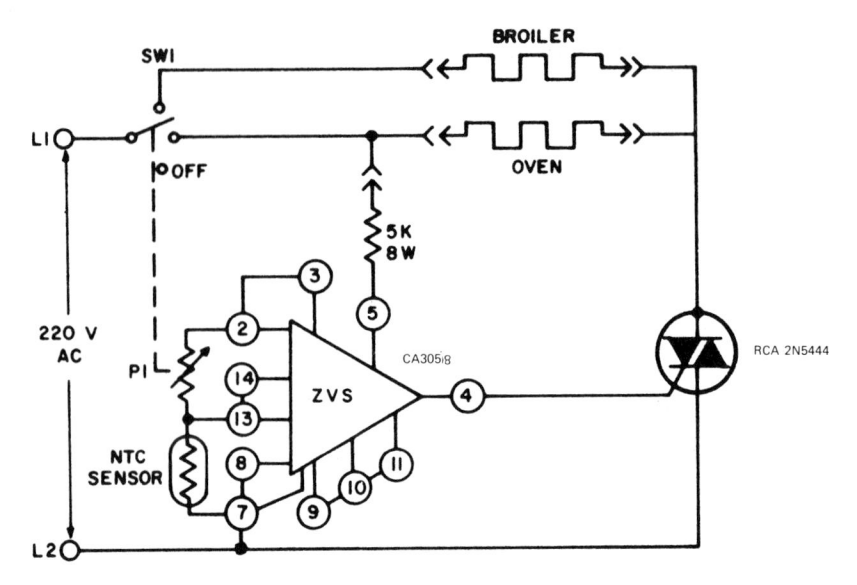

Oven control. A sensor element is included in the oven to provide a closed-loop system for accurate control of temperature: the element is cycled on and off to maintain the desired temperature to nearly ±2°C of the value set by adjusting potentiometer R1.

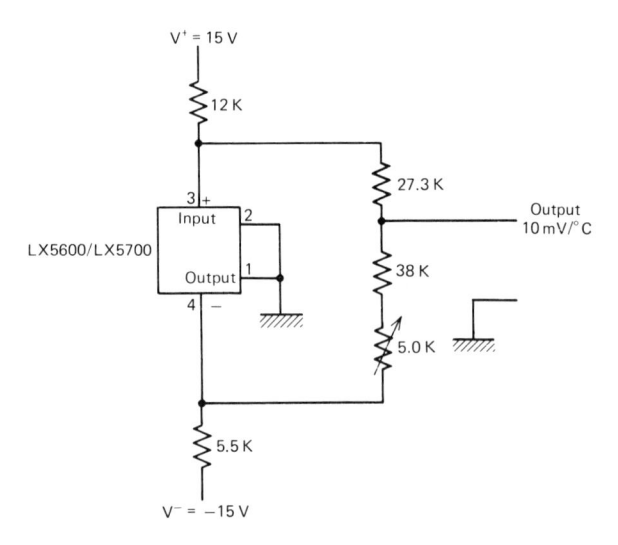

Ground-referred Celsius (centigrade) thermometer. R1 = 1000(V^+ − 3.0V); R2 = 500($V^−$ − 4.0V). Pot trims out initial zener tolerance; set output to read C.

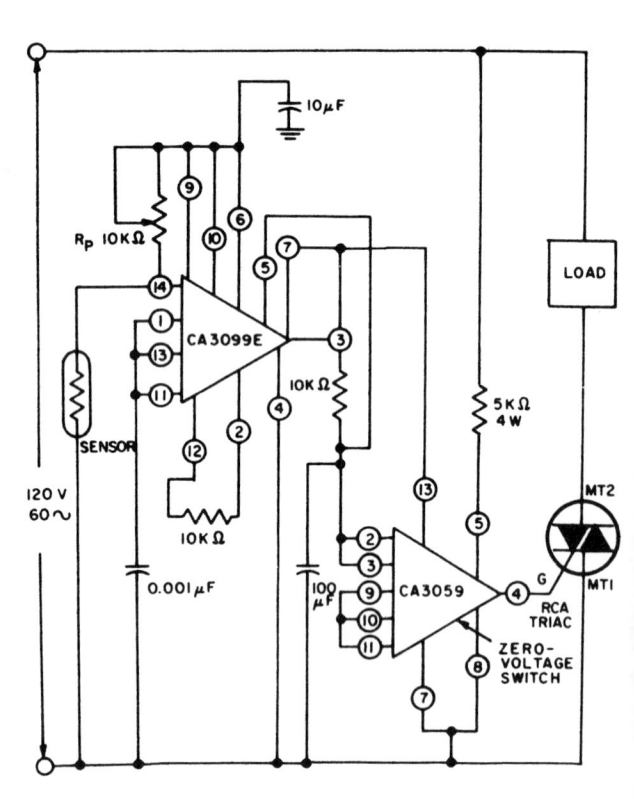

Sensitive temperature control. This circuit, which changes state when there is a change of approximately 1Ω in a 5k sensor, does not exhibit spurious half-cycle conduction effects.

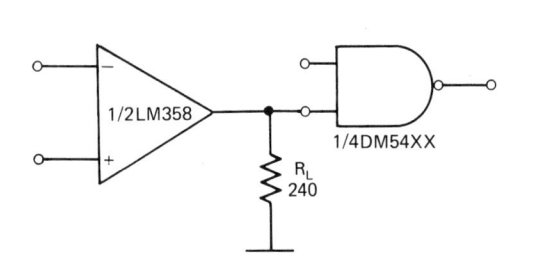

Driving TTL.

348 IC SCHEMATIC SOURCEMASTER

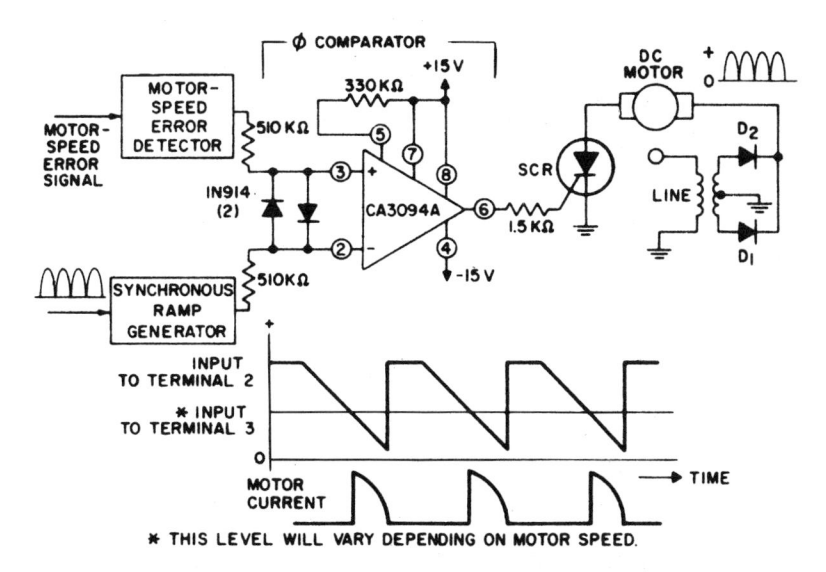

Motor speed controller system. Circuitry associated with rectifiers D1 and D2 comprises a full-wave rectifier that develops a train of half-sinusoid voltage pulses to power the dc motor. The motor speed depends on the peak value of the half-sinusoids and the period of time (during each half-cycle) the SCR is conductive. The SCR conduction, in turn, is controlled by the time duration of the positive signal supplied to the SCR by the phase comparator. The magnitude of the positive dc voltage supplied to terminal 3 of the phase comparator depends on motor-speed error.

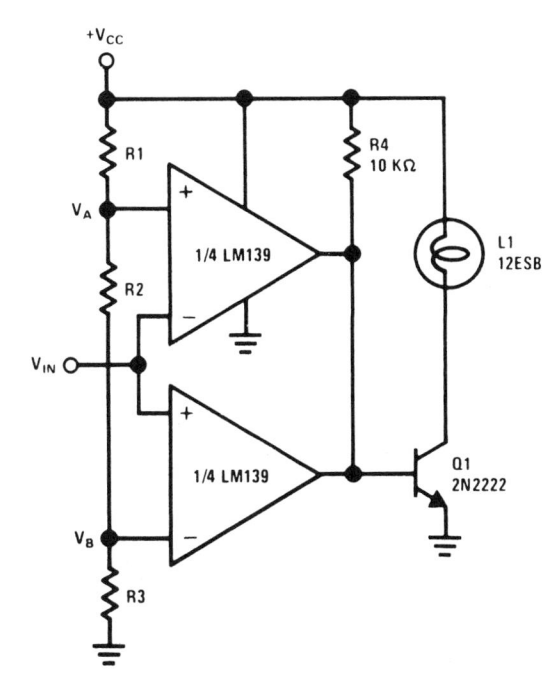

A limit comparator with lamp driver provides a range of input voltages between which the output devices of both LM139 comparators will be off. This will allow base current for Q1 to flow through pullup resistor R4, turning on Q1, which lights the lamp. If input voltage changes to a value greater than V_A or less than V_B, one of the comparators will switch on, shorting the base of Q1 to ground and causing the lamp to go off. If a PNP transistor is substituted for Q1 (with emitter tied to $+V_{CC}$), the lamp will light when the input is above V_A or below V_B. V_A and V_B are arbitrarily set by varying resistors R1, R2, and R3.

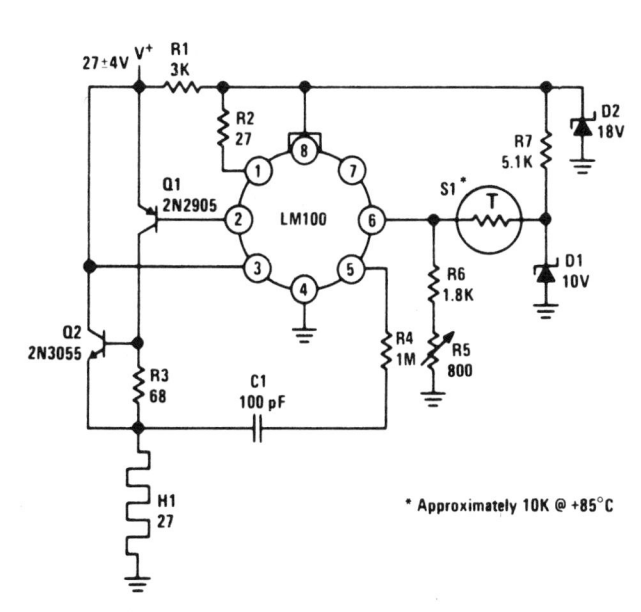

Temperature controller. Temperature changes in the oven are sensed by a thermistor. This signal is fed to the LM100, which controls power to the heater by switching the series-pass transistor, Q2, on and off. Since the pass transistor will be nearly saturated in the *on* condition, its power dissipation is minimized.

If the oven temperature should increase, the thermistor resistance will drop, increasing the voltage on the feedback terminal of the regulator. This action shuts off power to the heater. The opposite would be true if the temperature dropped.

SECTION **8** *Sensing, Monitoring, Controllers, and Power Drivers*

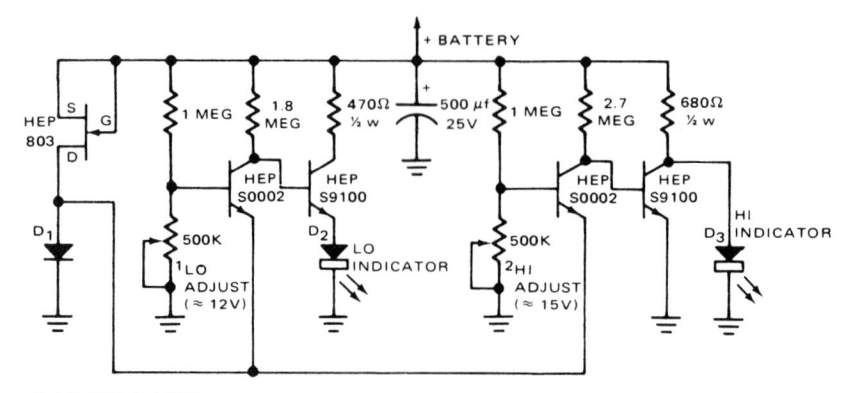

PARTS LIST:

- 1 — P-Channel FET HEP 803
- 2 — Silicon Transistor HEP S0002
- 2 — Darlington Transistor HEP S9100
- D₁ — 1 — Silicon Diode HEP R0050
- D₂ & D₃ — 2 — Light Emitting Diode HEP P2004
- 2 — Resistor 1 meg ¼ or ½ watt
- 1 — Resistor 1.8 meg ¼ or ½ watt
- 1 — Resistor 2.7 meg ¼ or ½ watt
- 1 — Resistor 470Ω ½ watt
- 1 — Resistor 680Ω ½ watt
- 2 — Variable Resistor 500K (Calectro B1-648)
- 1 — Capacitor 500 μf 25V

Battery voltage LO/HI monitor. The LO indicator LED comes on when the battery being monitored drops in output (as from an excessive current drain); a HI indicating LED comes on when battery voltage is excessive, as when the voltage regulator fails. Both LEDs can be adjusted with the 500k pots so that the actual point of turn-on can be varied according to individual requirements. For automotive applications, set LO so that D2 lights when the battery voltage drops below 12V and set HI so that D3 lights when the battery voltage exceeds 15V.

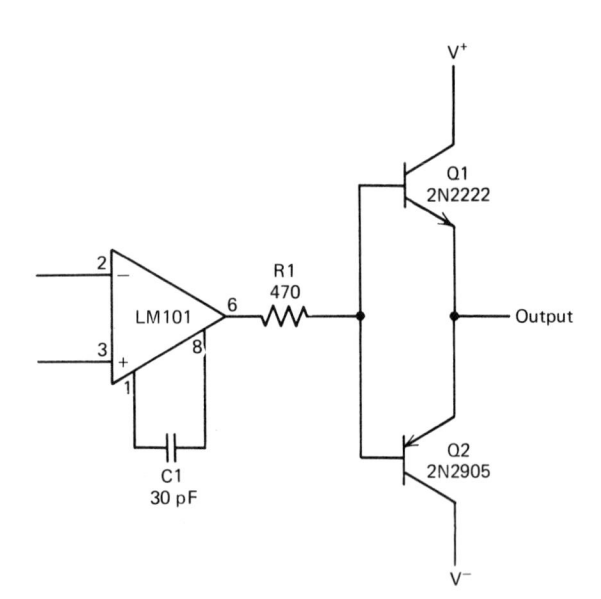

High-current output buffer. A pair of complementary transistors are used on the output of the LM101 to get the increased current swing. Although this circuit does have a dead zone, it can be neglected at frequencies below 100 Hz because of the high gain of the amplifier. R1 is included to eliminate parasitic oscillations from the output transistors. In addition, adequate bypassing should be used on the collectors of the output transistors to ensure that the output signal is not coupled back into the amplifier. This circuit does not have current limiting, but it can be added by putting 50Ω resistors in series with the collectors of Q1 and Q2.

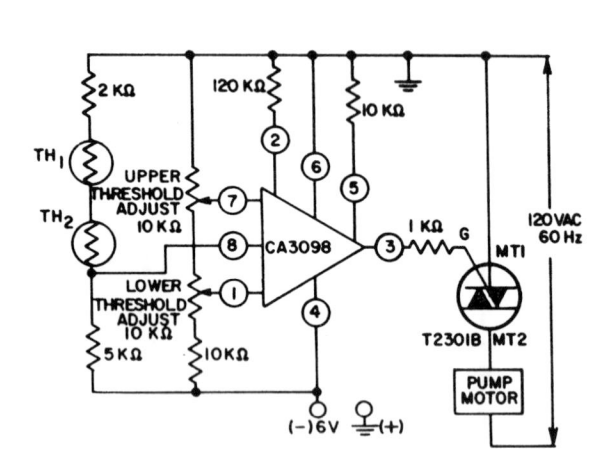

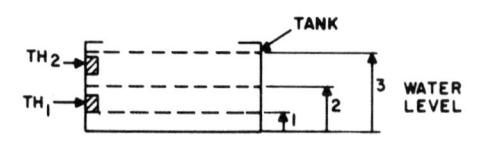

Water-level control circuit. Motor pump is on when the water level rises above thermistor TH_2, and it remains on until the water level falls below thermistor. Thermistors operate in self-heating mode. Triac is RCA T2301B.

350 IC SCHEMATIC SOURCEMASTER

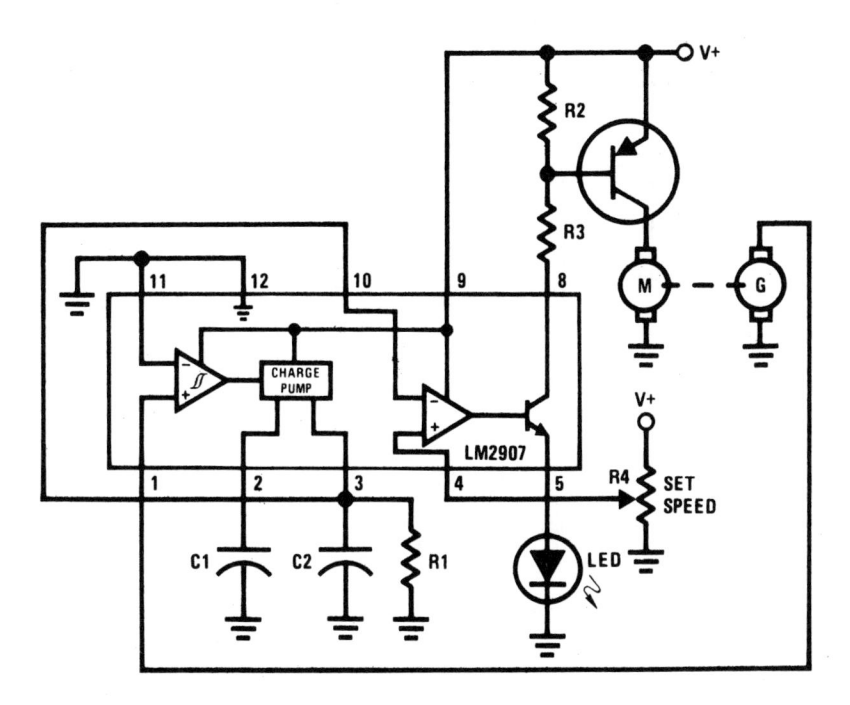

Motor speed control. Dc motors with or without brushes can be purchased with ac tachometer outputs already provided by the manufacturer (TRW Globe Motors, 2275 Stanley Ave., Dayton OH 45404). With these motors in combination with the LM2905, a very low-cost speed control can be constructed. Here, the tachometer drives the noninverting input of the comparator up toward the preset reference level, when the output is turned off and the power is removed from the motor. As the motor slows, the voltage from the charge pump output falls and power is restored. Thus, speed is maintained by operating the motor in a switching mode.

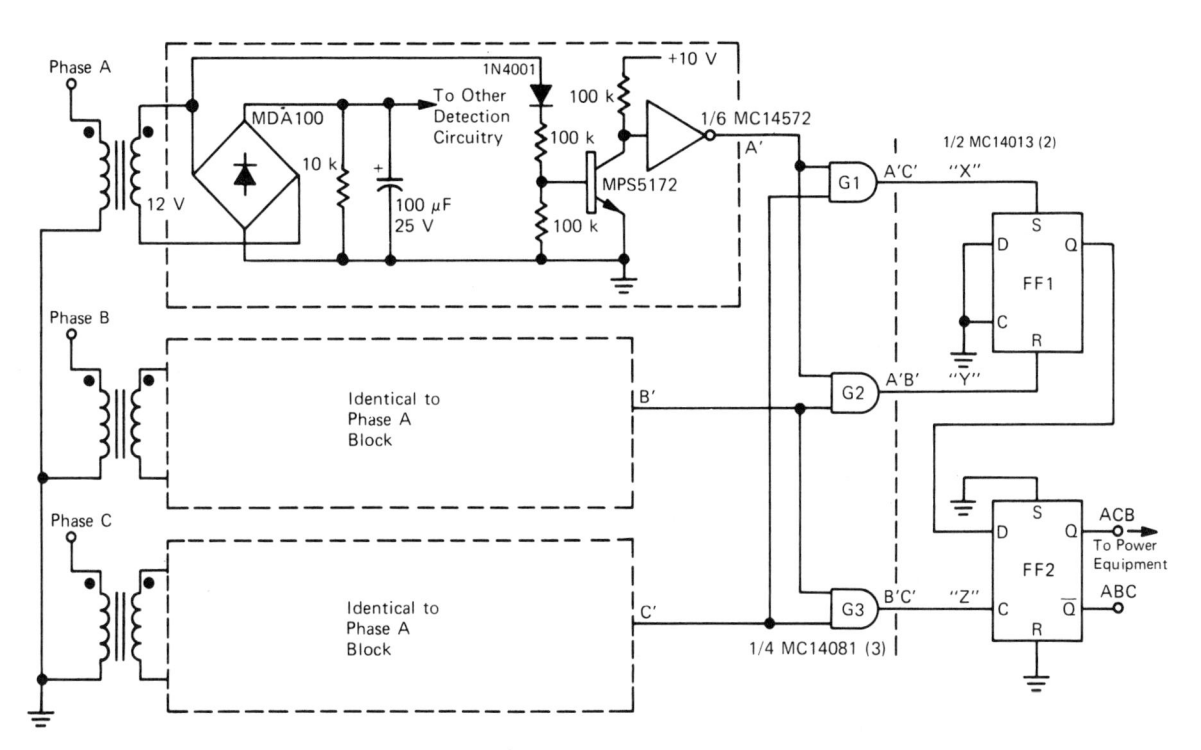

Line-phase-reversal detection circuit. Many three-phase, line-powered applications are sensitive to the direction of rotation of the three phases. If two of the connections to a three-phase motor are inadvertently reversed for example, it will reverse direction, causing disaster in applications such as pump and air-conditioner compressor drives. The circuit shown can be easily added to a line-undervoltage or line-unbalanced detection scheme. Circuit uses and interfaces readily with CMOS logic.

SECTION **8** Sensing, Monitoring, Controllers, and Power Drivers

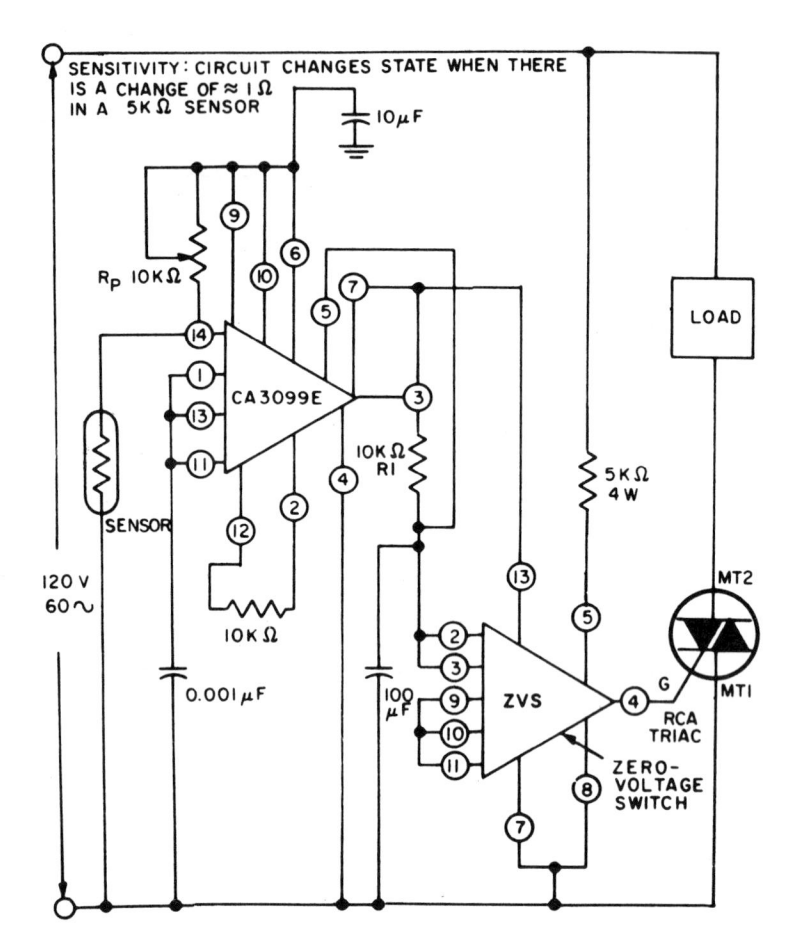

Sensitive temperature control circuit does not exhibit spurious half-cycle conduction effects.

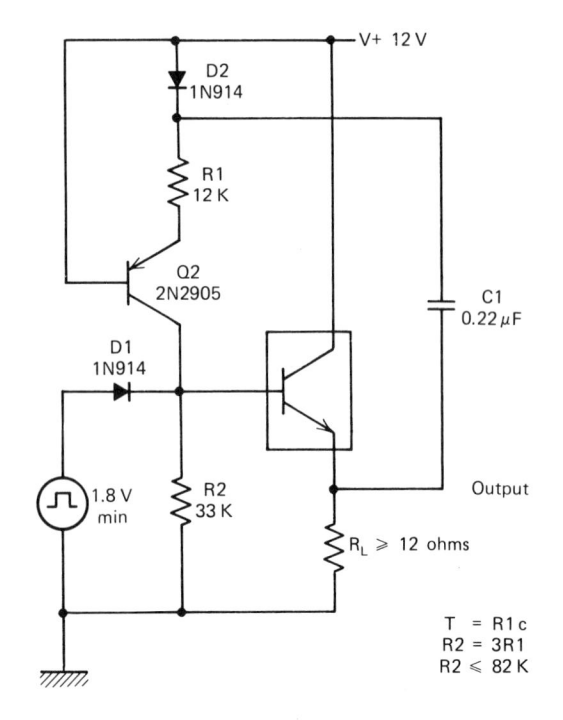

Power one-shot.

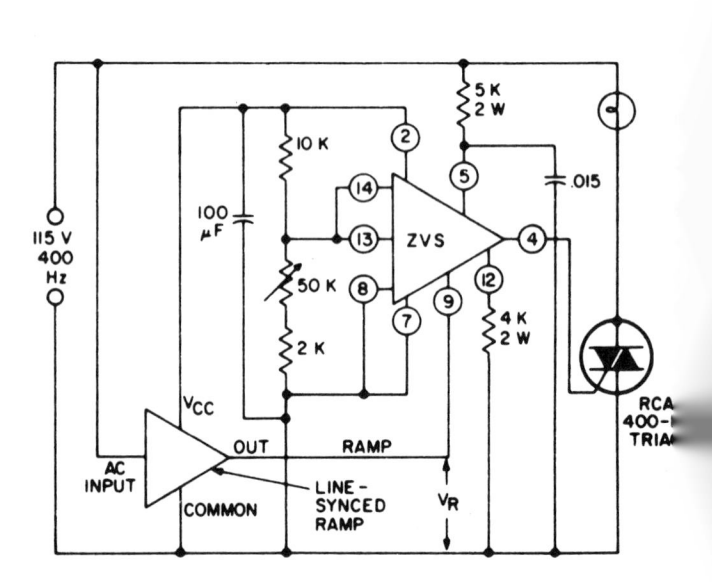

T = R1c
R2 = 3R1
R2 ≤ 82K

A zero-voltage-switched lamp dimmer provides 14 different levels of lamp intensity without noticeable flicker, using 400 Hz power. Light levels are changed by setting the 50k potentiometer. IC is RCA's CA3058. (This device requires extensive RFI filtering.)

352 IC SCHEMATIC SOURCEMASTER

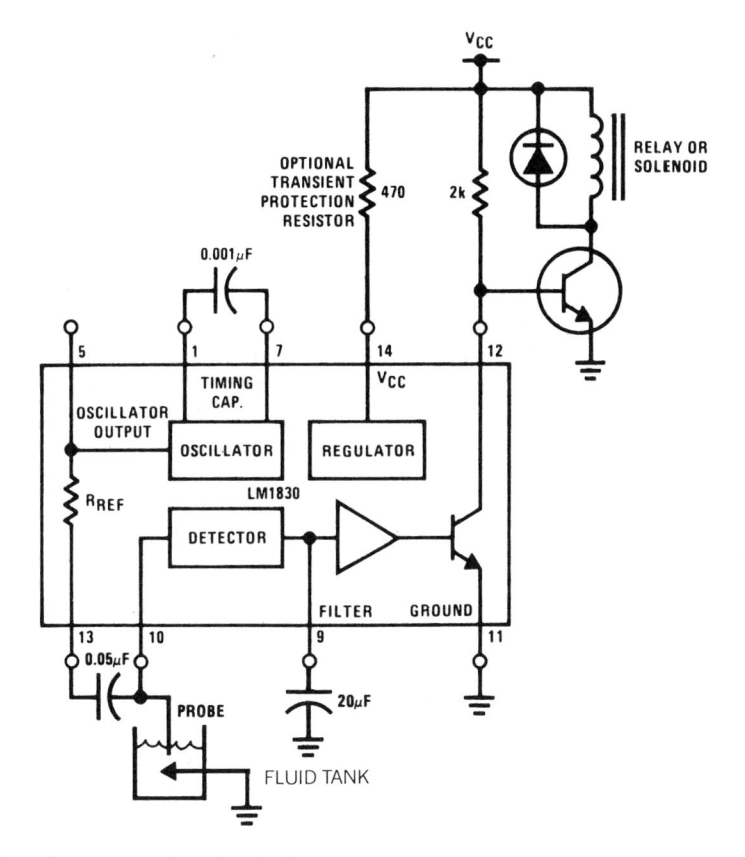

High-level warning device. The output is suitable for driving a sump pump or opening a drain valve and so forth.

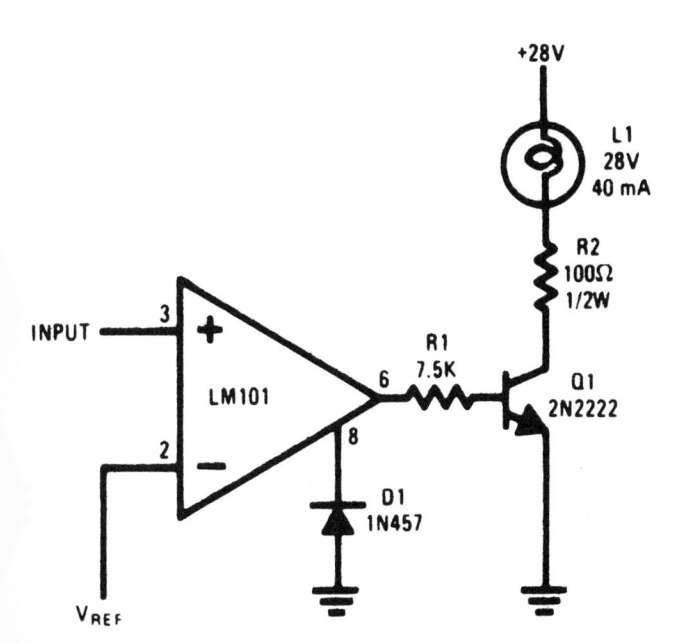

Comparator and lamp driver. Q1 switches the lamp with R2 limiting the current surge resulting from turning on a cold lamp. R1 determines the base drive to Q1, while D1 keeps the amplifier from putting excessive reverse-bias on the emitter–base junction of Q1 when it turns off.

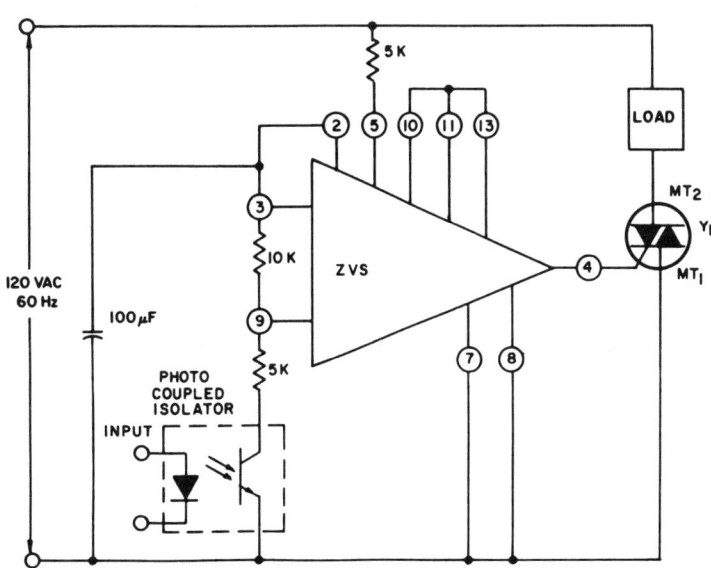

A zero-voltage-switch on-off controller with photocoupler. When a logic 1 is applied at the input of the photocoupler, the triac controlling the load will be turned on whenever the line voltage passes through zero; when a logic 0 is applied to the photocoupler, the triac will turn off and remain off until a logic 1 appears at the input of the photocoupler.

SECTION **8** Sensing, Monitoring, Controllers, and Power Drivers 353

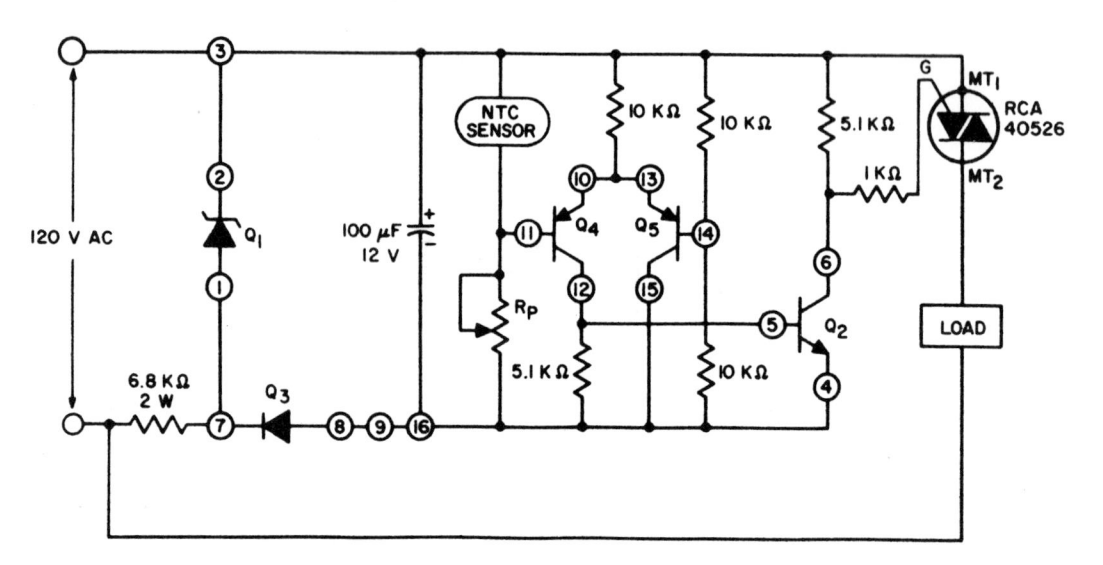

A line-operated level switch using CA3096AE or CA3096E. "NTC sensor" is any thermistor device with a negative temperature coefficient.

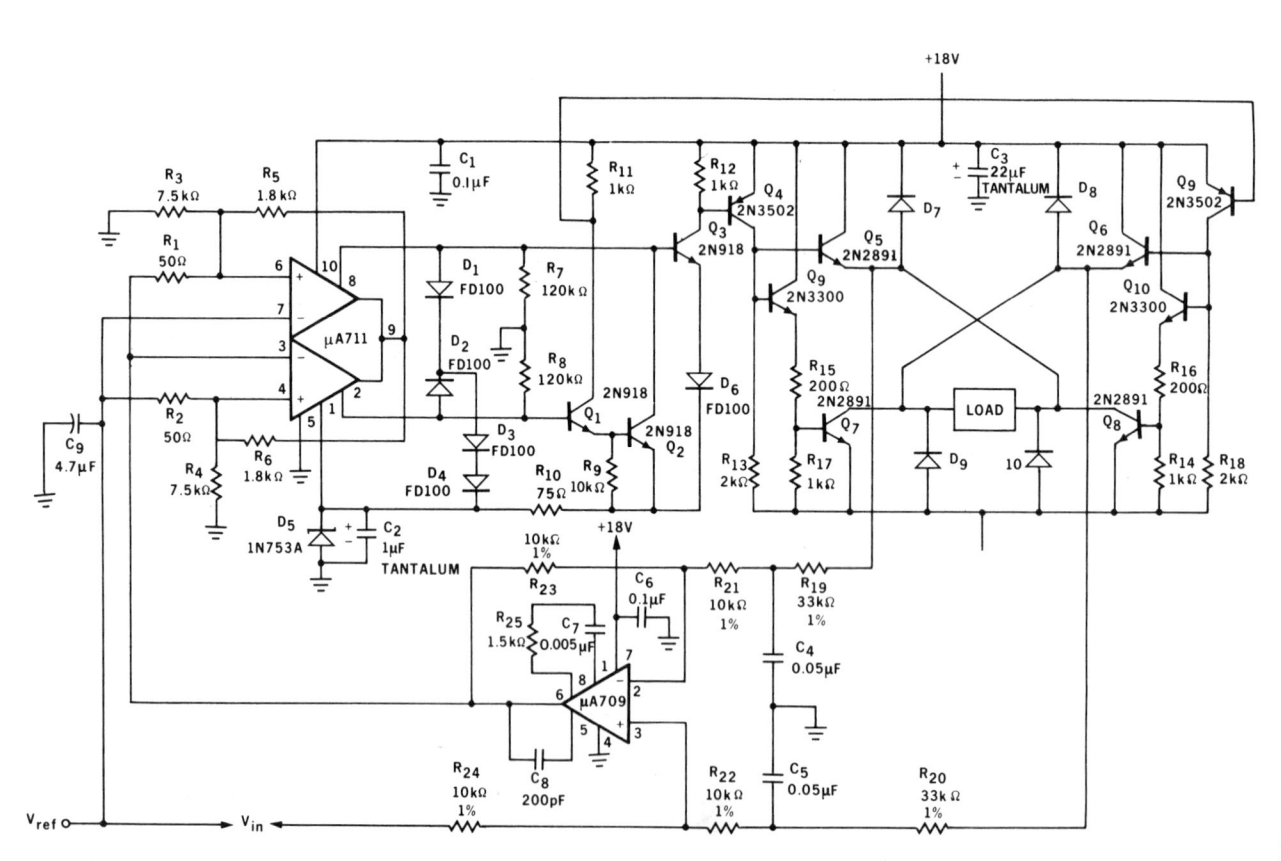

A switching power amplifier circuit operates from a single supply voltage. Emitter followers give an output current capability of several amperes. The bases of the bottom transistors in the bridge are driven from separate emitter followers (Q_9 and Q_{20}). This base drive could be obtained from the emitters of Q_5 and Q_6 except that excessive saturation voltage of one of the lower power transistors could cause the other lower transistor to become turned on. Diodes D_7, D_8, D_9, and D_{20} clamp the current spikes obtained with inductive loads.

354 IC SCHEMATIC SOURCEMASTER

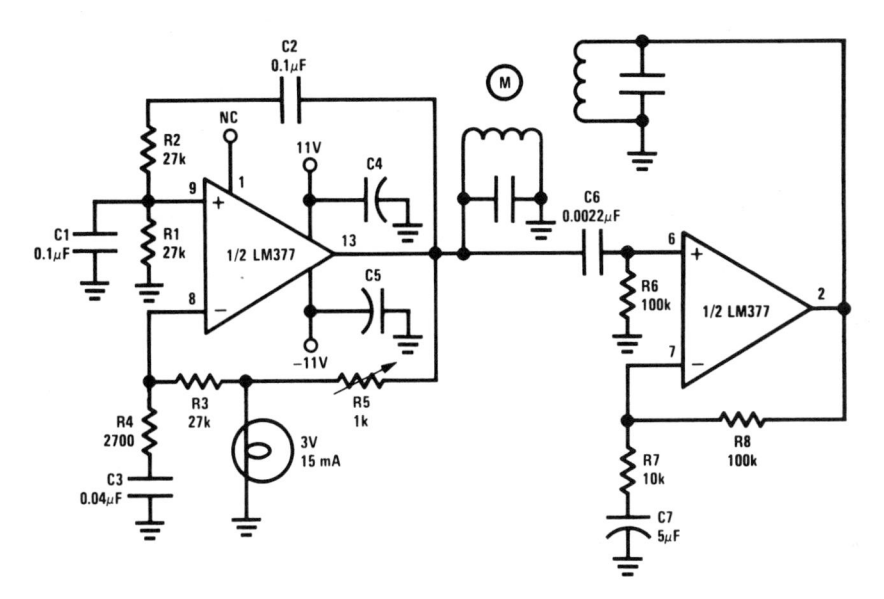

A two-phase motor drive for a small 60 Hz servo motor up to 3W per phase. Applications such as a constant- (or selectable-) speed phonograph turntable drive are adequately met by this circuit. A split supply is used to simplify the circuit, reduce parts count, and eliminate several large bypass capacitors. An incandescent lamp is used in a simple amplitude stabilization loop. Input dc is minimized by balancing dc resistance at (+) and (−) amplifier inputs (R1 = R3 and R6 = R8). High-frequency stability is assured by increasing closed-loop gain from approximately 3 at 60 Hz to about 30 above 40 kHz, with the network consisting of R3, R4, and C3. The interstage coupling C6R6 network shifts phase by 85° at 60 Hz to provide the necessary motor drive signal. The gain of the phase shift network is purposely low so that the buffer amplifier will operate at a gain of 10 for adequate high-frequency stability. The motor windings are tuned to 60 Hz with shunt capacitors. This circuit will drive 8Ω loads to 3W each.

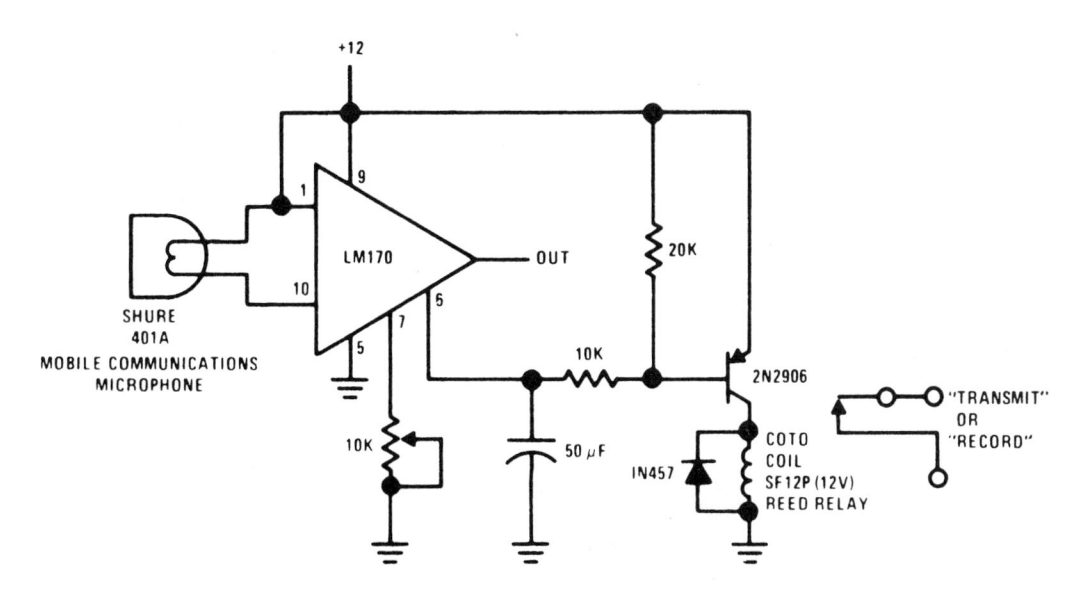

A transmitter or tape recorder voice-operated relay control, or VOX, to switch high powered electronic or electromechanical devices. Automatic transmit/receive operation is possible in two-way communication systems, or tape recorder motors may be switched on at the first syllable of infrequent speech, such as in dictation, thereby conserving tape. To handle large amounts of power, all that is needed is a small PNP power transistor driving a relay, which can have multiple poles. An amplifier may simultaneously be used as a continuous-running preamplifier, may be squelched along with the relay, or may even operate with an independent AGC signal into pin 3 or 4.

SECTION **8** *Sensing, Monitoring, Controllers, and Power Drivers*

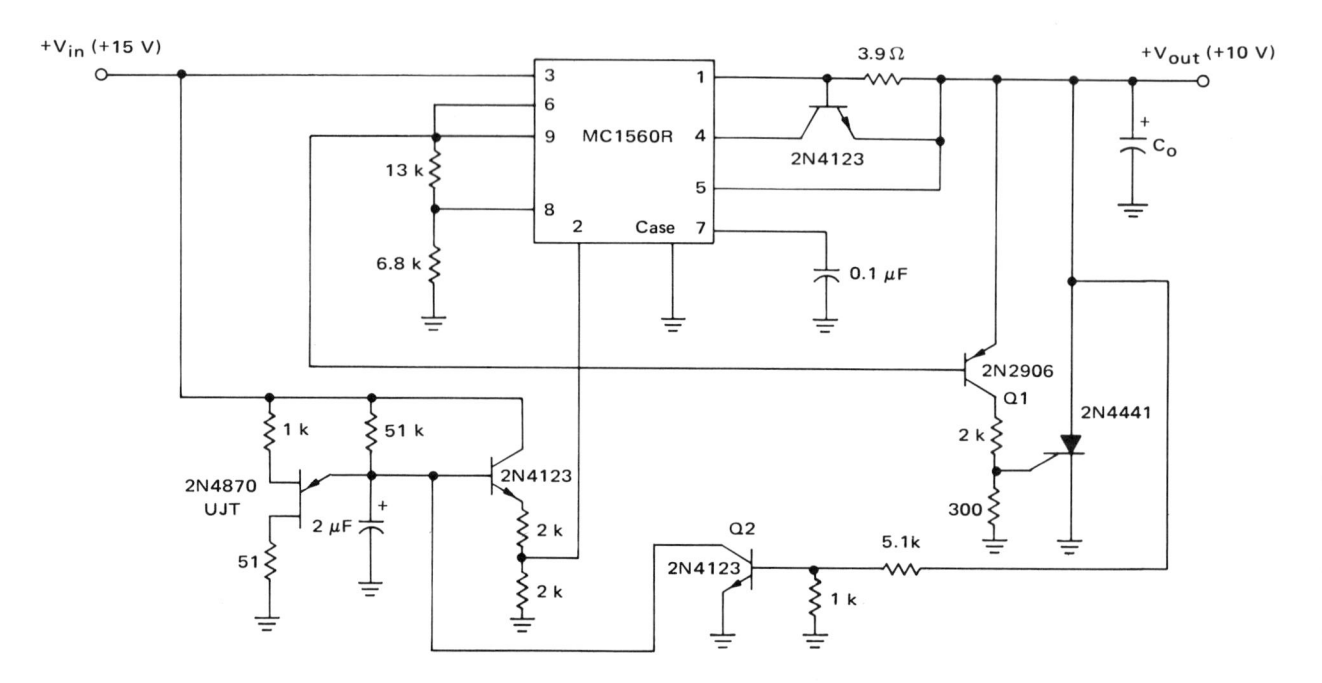

An SCR "crowbar" overvoltage protection with automatic shutdown and recycling.

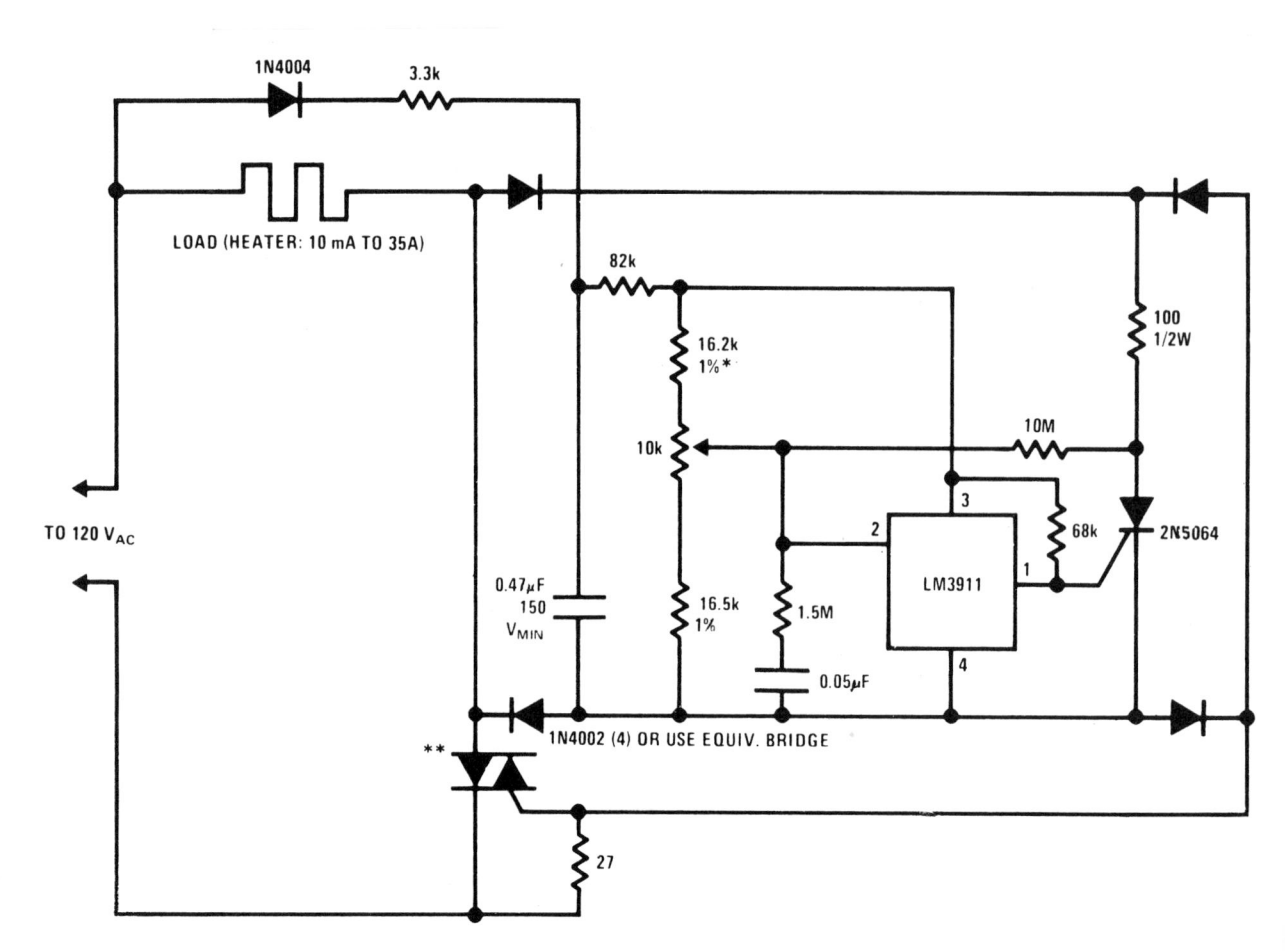

Three-wire electronic thermostat. Divider is set for a nominal 0° to 125 °C range. Wire-wound resistors will provide maximum temperature stability. Almost any triac rated at 1 to 35A is usable with an appropriate load.

356 IC SCHEMATIC SOURCEMASTER

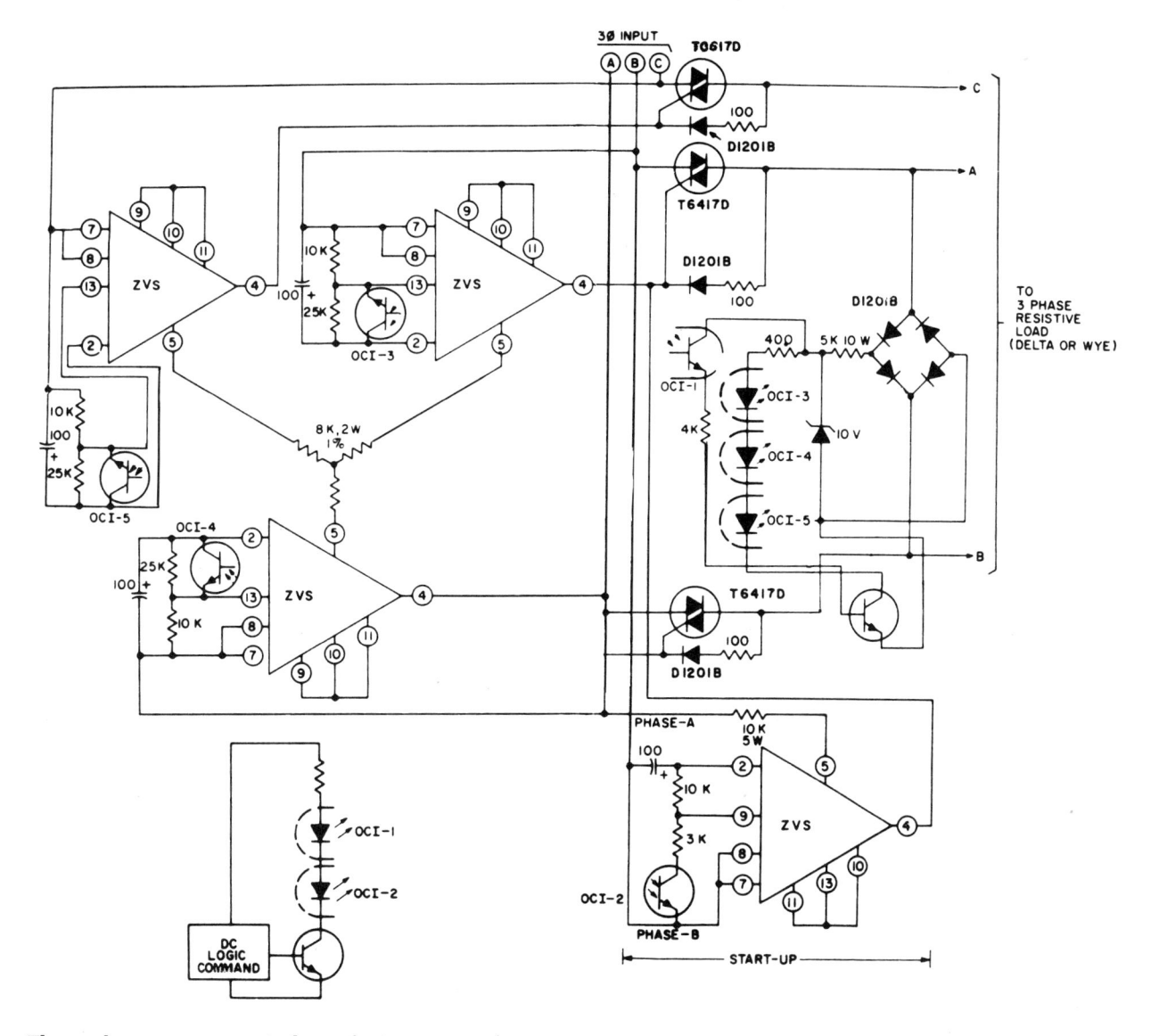

Three-phase power control employing zero-voltage synchronous switching both for steady-state operation and for starting.

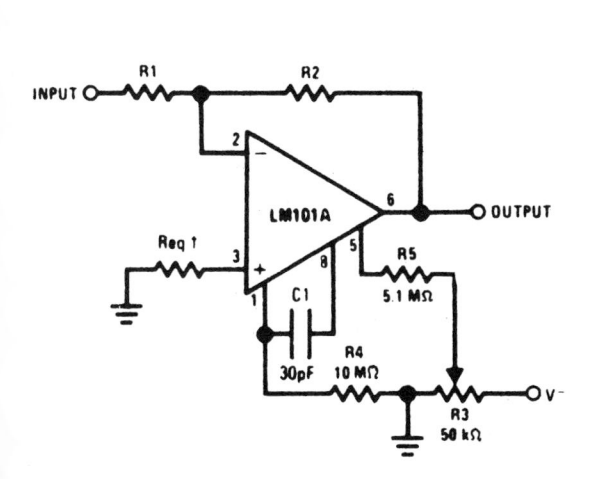

Inverting amplifier with balancing circuit. Resistor grounding pin 3 may be zero or equal to parallel combination of R1 and R2 for minimum offset.

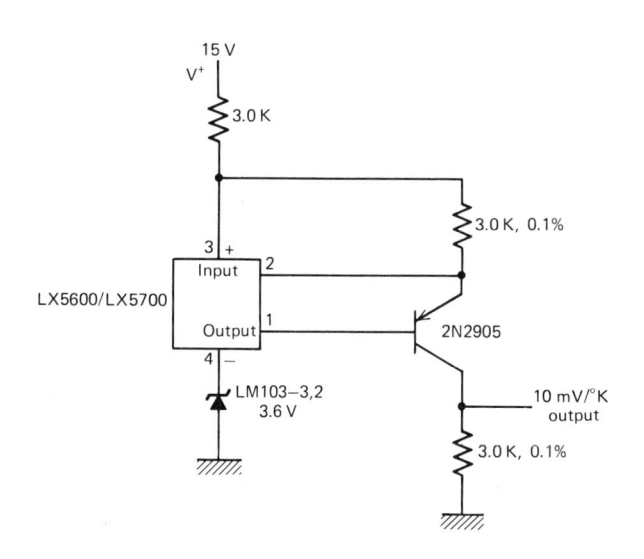

Kelvin thermometer with ground-referred output. $R_S = V_S^+ - 6.8V \times 10^3 \Omega/2$.

SECTION **8** Sensing, Monitoring, Controllers, and Power Drivers

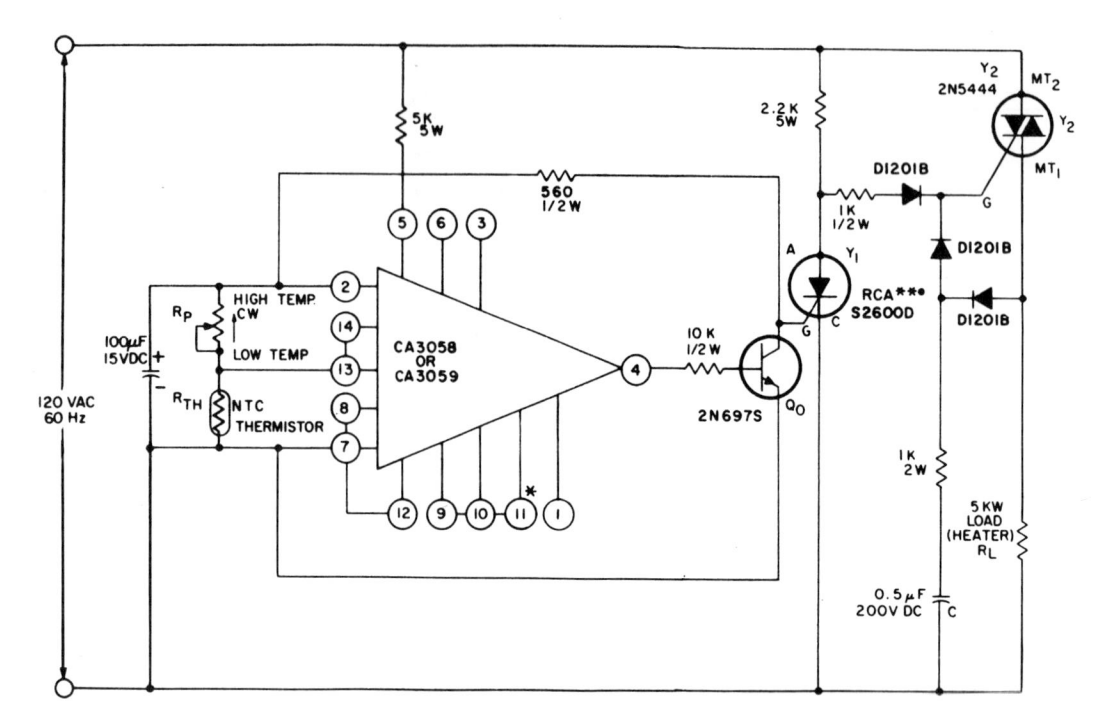

An integral-cycle temperature controller using CA3058 or CA3059 features a protection circuit and no half-cycling effect. The NTC sensor is connected between terminals 7 and 13, and transistor Q_0 inverts the signal output at terminal 4 to nullify the phase reversal introduced by the SCR (Y1). The internal power supply of the zero-voltage switch supplies bias current to transistor Q_0. For proportional operation, open terminals 9, 10, and 11 and connect positive ramp voltage to terminal 9. RCA S2600D selected for I_{GT} = 6 mA maximum.

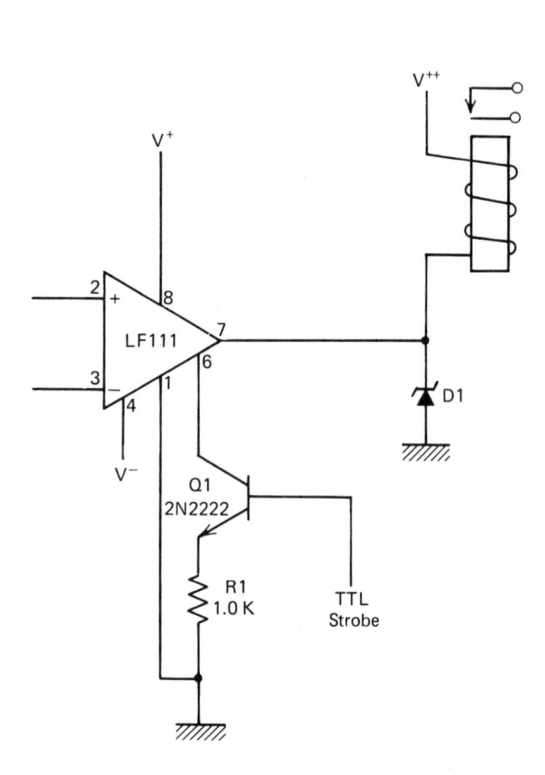

Relay driver with strobe. D1 absorbs inductive kickback of relay and protects IC from severe voltage transients on V^{++} line.

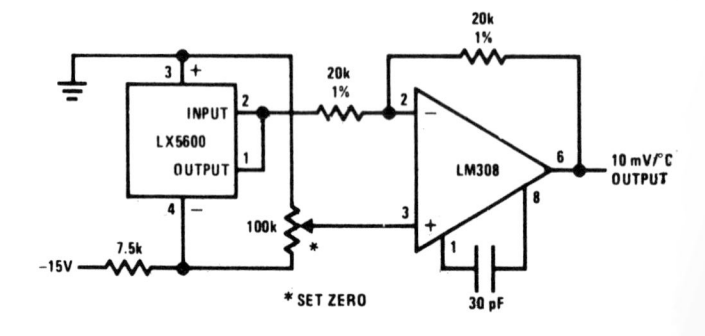

Ground-referred Celsius (centigrade) thermometer.

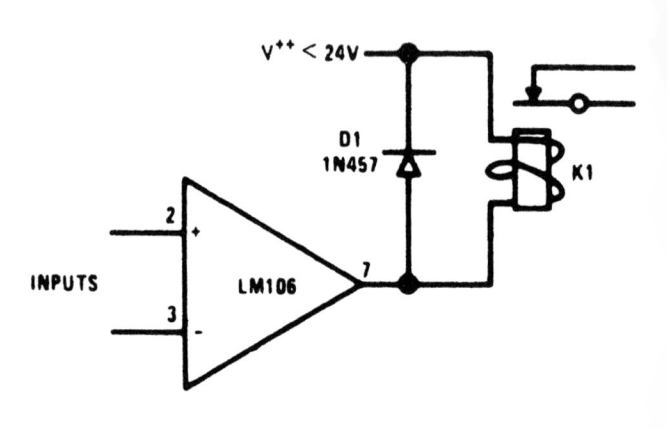

Relay driver.

358 IC SCHEMATIC SOURCEMASTER

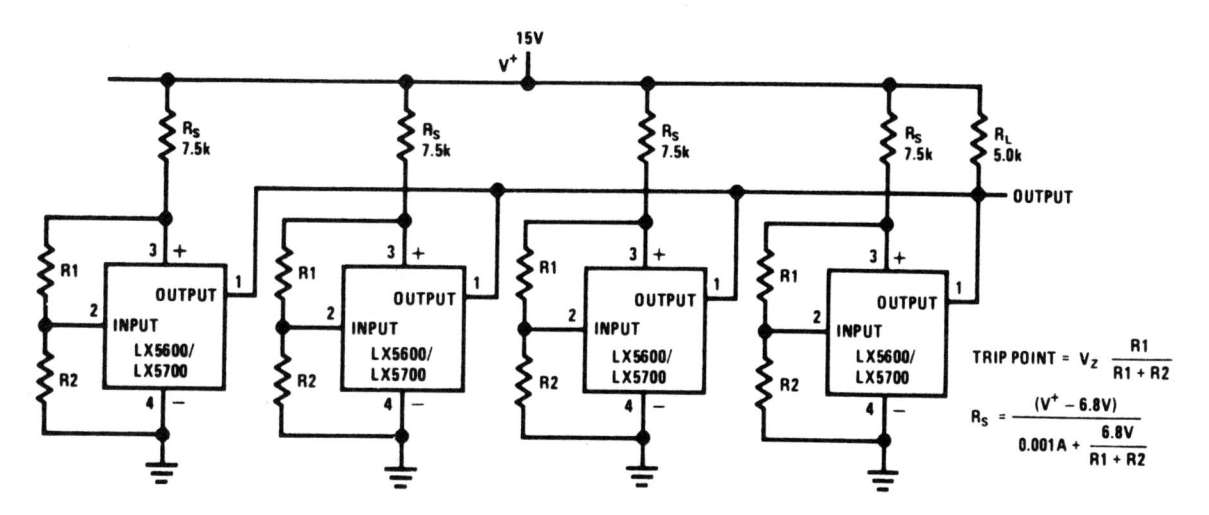

Overtemperature detectors with common output.

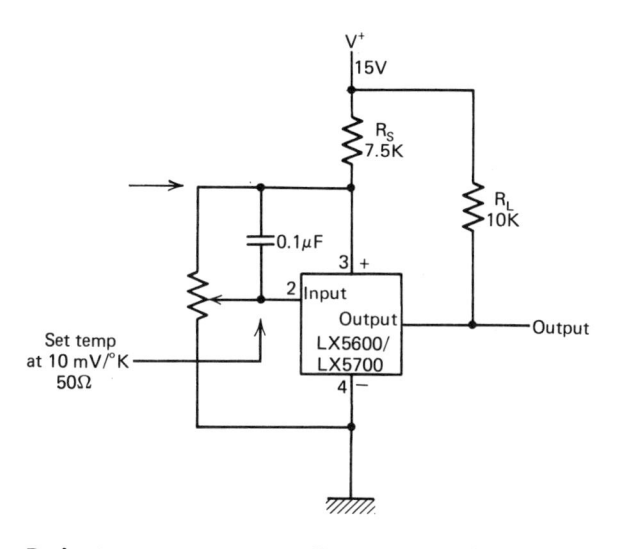

$$\text{TRIP POINT} = V_z \frac{R1}{R1 + R2}$$

$$R_S = \frac{(V^+ - 6.8V)}{0.001A + \frac{6.8V}{R1 + R2}}$$

Basic temperature controller. $R_S = (V^+ - 6.8V)$k. Output goes negative on temperature increase.

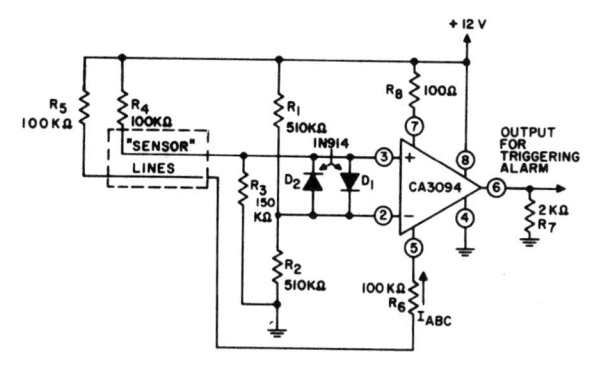

Alarm system uses two sensor lines. In the no-alarm state, the potential at terminal 2 is lower than the potential at terminal 3, and terminal 5 (I_{ABC}) is driven with sufficient current through resistor R5 to keep the output voltage high. If either sensor line is opened, shorted to ground, or shorted to the other sensor line, the output goes low and activates some type of alarm system.

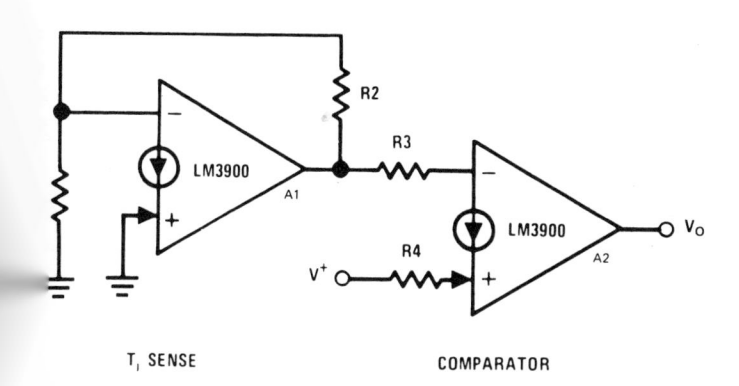

Temperature sensing. The LM3900 can be used to monitor the junction temperature of the monolithic chip. Amplifier A will generate an output voltage that can be designed to undergo a large negative temperature change by design of R1 and R2. The second amplifier compares this temperature-dependent voltage with the power supply voltage and goes high at a designed maximum junction temperature.

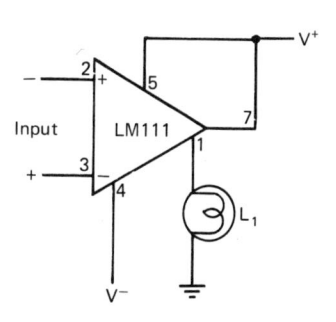

Driving ground-referred load. The input polarity is reversed because the ground terminal is used as the output. An incandescent lamp, which is the load here, has a cold resistance eight times lower than it is during normal operation. This produces a large inrush current when it is switched on, which can damage the switch. However, the current limiting of the LM111 holds this current to a safe value. Input polarity is reversed when using pin 1 as output.

SECTION **8** Sensing, Monitoring, Controllers, and Power Drivers

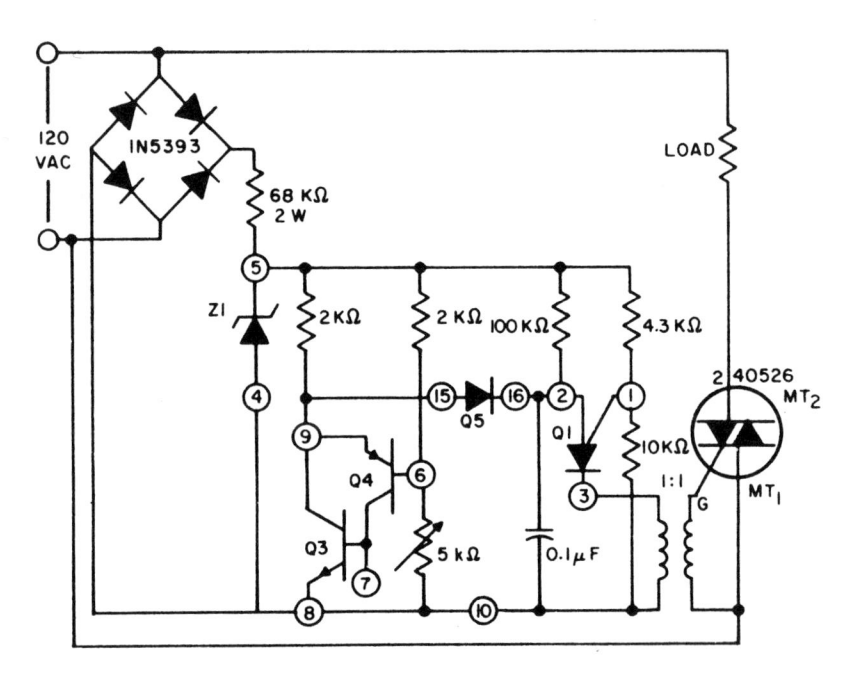

Phase control circuit. To use Q5 as a diode, short terminal 15 to terminal 14.

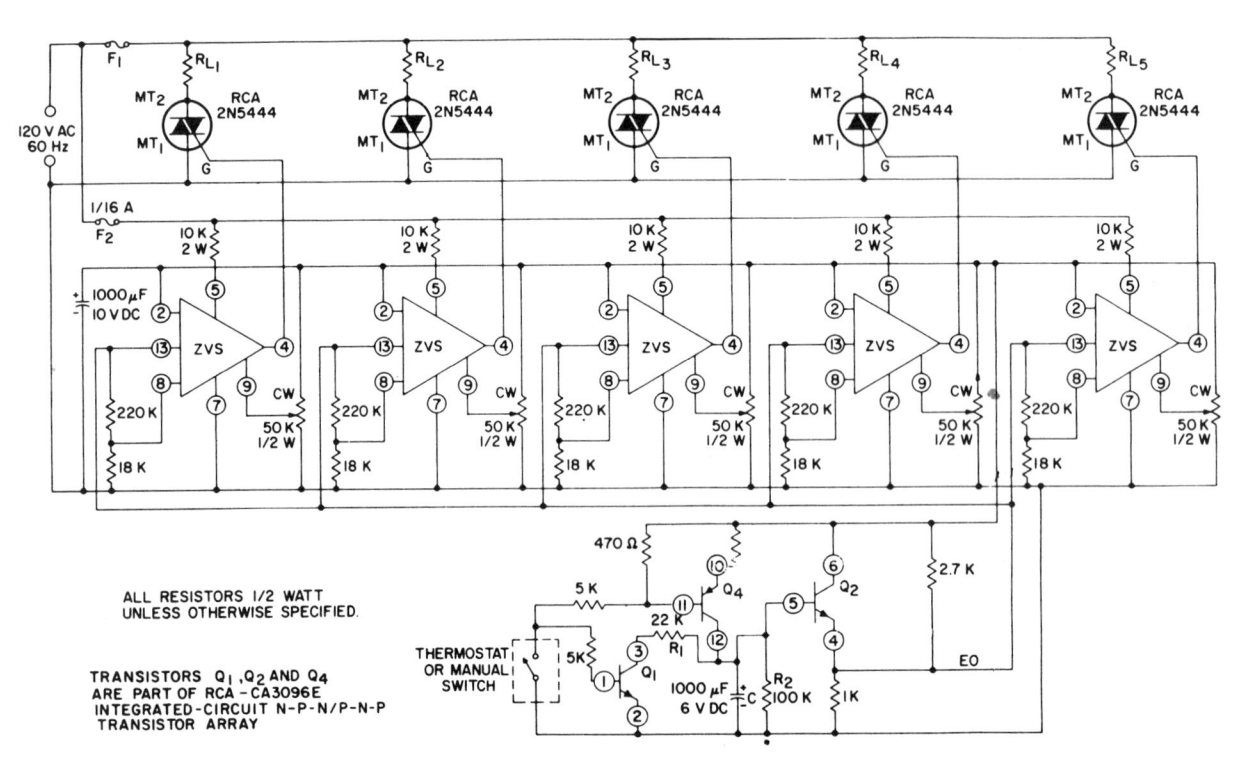

A synchronous-switching heat-staging controller using a series of CA3058 zero-voltage switches. Loads as heavy as 5 kW are switched sequentially at zero voltage to eliminate RFI and to prevent a dip in line voltage that would be caused by switching the full 25 kW simultaneously. At approximately 3-second intervals, each heating element is switched into the power system by its triac; without a demand for heat, the thermostat opens and capacitor C discharges through R1 and R2, causing the triacs to turn off in a reverse sequence.

360 IC SCHEMATIC SOURCEMASTER

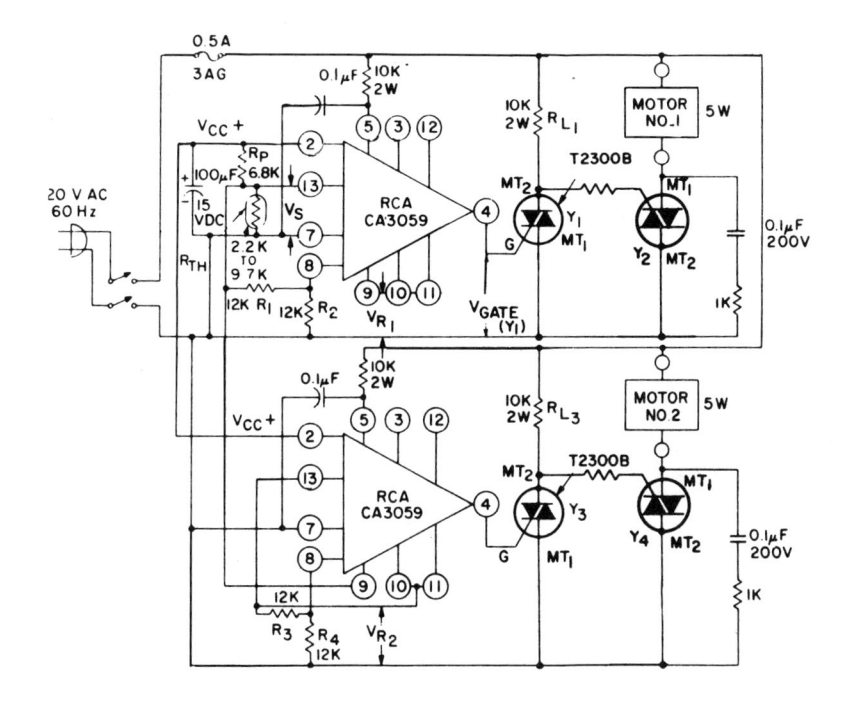

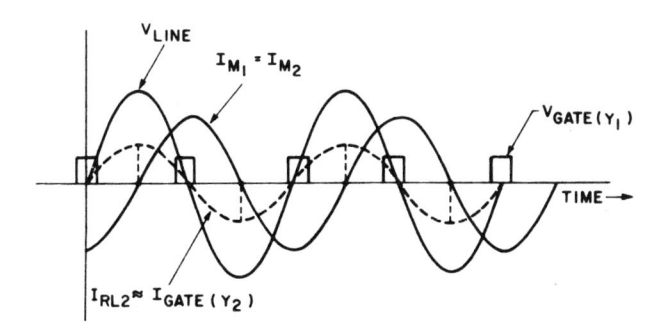

A dual-output temperature controller drives two triacs. When voltage V_S, developed across the temperature-sensing network, exceeds the reference voltage V_{R1}, motor 1 turns on. When the voltage across the network drops below the reference voltage V_{R2}, motor 2 turns on. Because the motors are inductive, the currents I_{M1} lag the incoming line voltage. The motors are switched by the triacs at zero current.

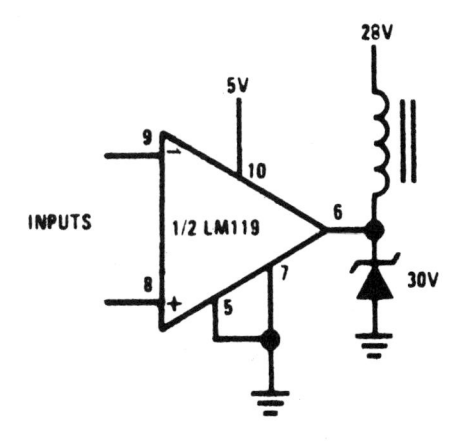

Relay driver.

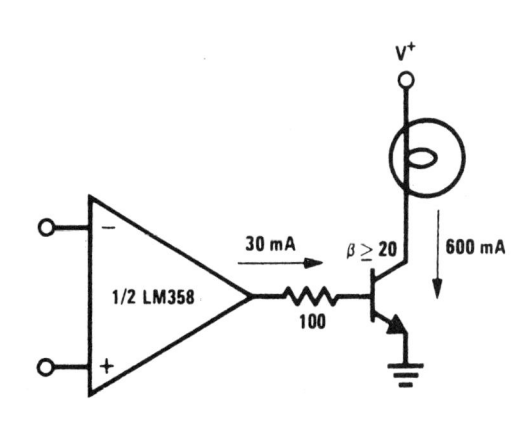

Lamp driver.

SECTION **8** *Sensing, Monitoring, Controllers, and Power Drivers* 361

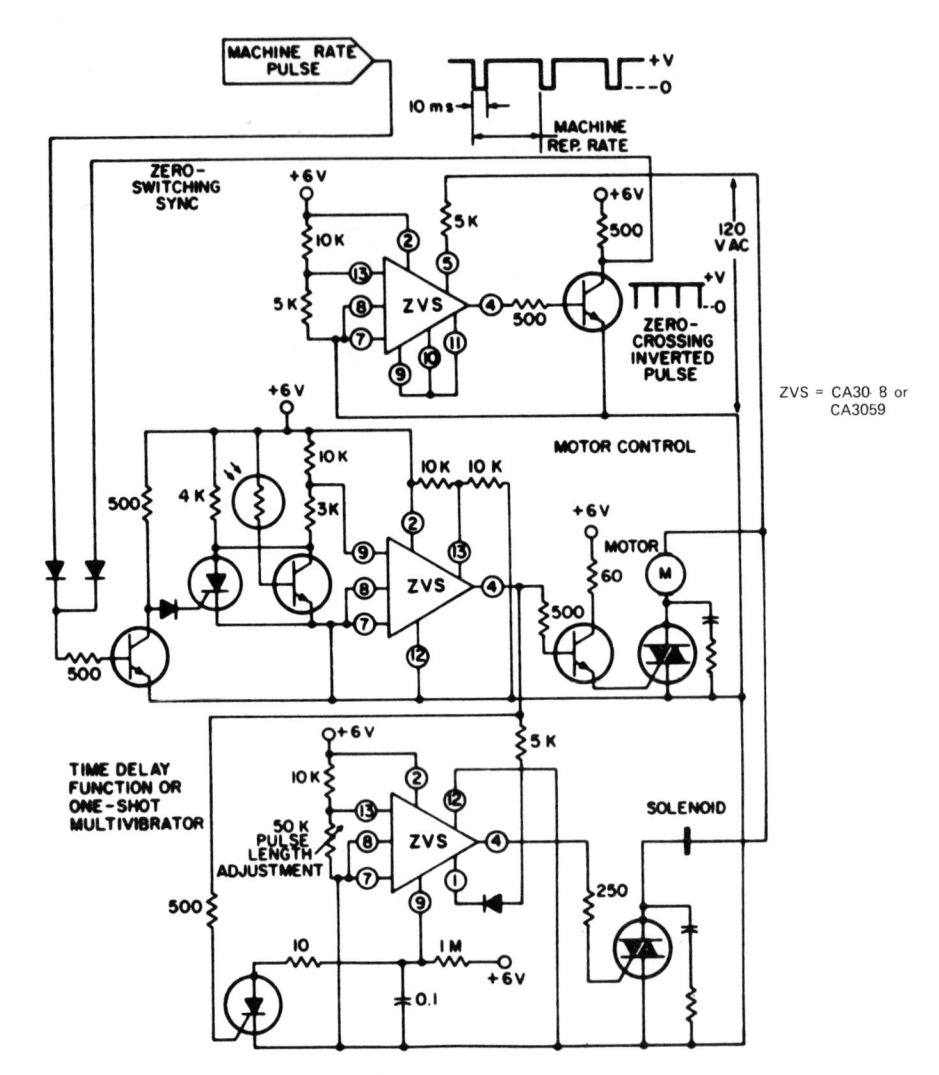

Solid-state machine control. The gate drive to the motor triac is continuous dc starting at zero-voltage crossing to reduce the chance of generating RFI. The inhibit input at terminal 1 ensures that the solenoid will not operate while the motor is running; the motor is started when the machine rate pulse and zero-voltage sync are at low voltage. The time delay is adjustable by the 50k potentiometer at terminal 13. Under dc operation, limit the output to 20 mA.

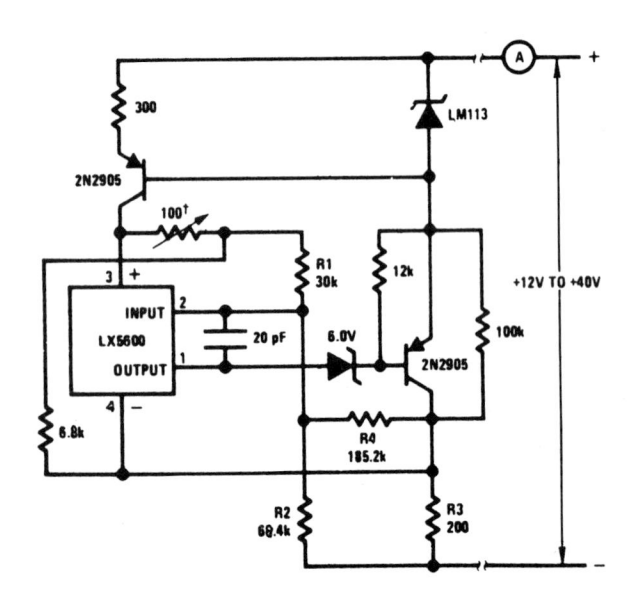

$$R2\ (\Omega) = \frac{\left(V_Z - 0.01\ T_L\right)\left(I_H - \dfrac{0.01\ T_H}{R1}\right) + \left(V_Z - 0.01\ T_H\right)\left(\dfrac{0.01\ T_L}{R1} - I_L\right)}{\dfrac{0.01}{R1\ R3}\left[T_H\left(V_Z - 0.01\ T_L\right) - T_L\left(V_Z - 0.01\ T_H\right)\right]}$$

$$R3\ (\Omega) \geqq \frac{V_Z\left(\dfrac{T_H}{T_L} - 1\right)}{I_H - \dfrac{I_L\ T_H}{T_L}}$$

$$\frac{1}{R4} = \frac{1}{(V_Z - 0.01\ T_L)(R2)}\left[\frac{(R2)(0.01\ T_L)}{R1} + \frac{\left(\dfrac{V_Z - 0.01\ T_L}{R2} - I_L\right)}{\dfrac{1}{R2} + \dfrac{1}{R3}}\right]\frac{1}{R2}$$

T_L = TEMPERATURE FOR I_L (°K)
T_H = TEMPERATURE FOR I_H (°K)
V_Z = ZENER VOLTAGE (VOLTS)
I_L = LOW TEMPERATURE OUTPUT CURRENT (mA)
I_H = HIGH TEMPERATURE OUTPUT CURRENT (mA)

*VALUES SHOWN FOR I_{OUT} = 1 mA TO 10 mA FOR 10°F TO 100°F
†SET TEMPERATURE

Two-terminal temperature-to-current transducer.

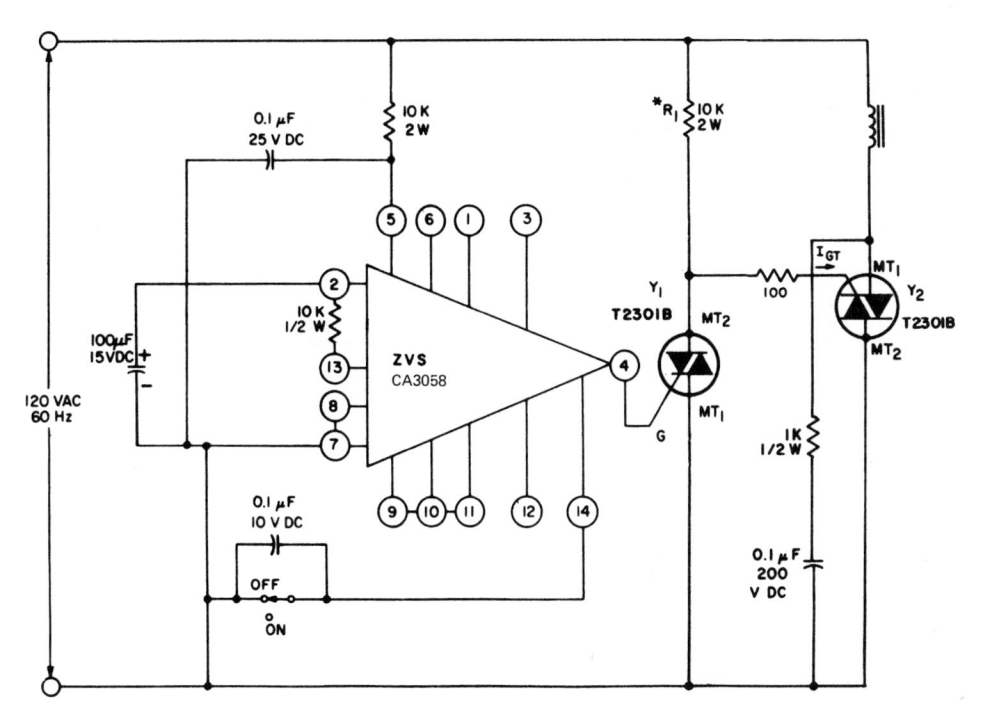

*IF Y₂, FOR EXAMPLE, IS A 40-AMPERE TRIAC, THEN R₁ MUST BE DECREASED TO SUPPLY SUFFICIENT I_{GT} FOR Y₂.

Zero-voltage-switching transient-free switch controller. Power is supplied to the load when the switch is open. After control terminal 14 is opened, the electronic logic waits until the line voltage reaches a zero crossing before power is applied to load Z_L. When the control terminals are shorted, the load current continues until it reaches a zero crossing. The circuit can switch a resistive or inductive load at zero current.

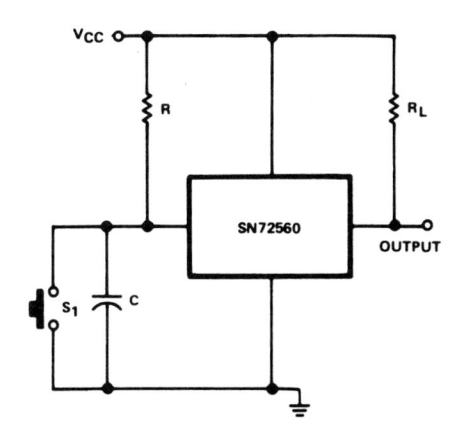

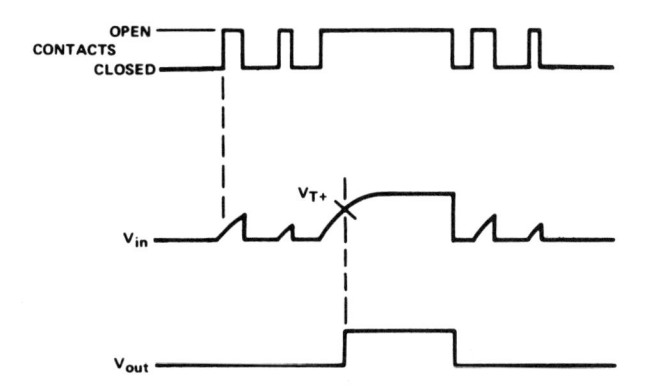

Bounceless switch. An application that demonstrates the basic operating characteristic of the SN72560 is this bounceless switch. As soon as switch S1 is opened, capacitor C will be charged through resistor R, and when V_{T+} is reached, the output will go high. If, during the charge time, switch S1 bounces and closes, capacitor C will discharge, and nothing will be seen on the output until bouncing ceases. When the switch is closed, the output changes directly to zero. When a bounce occurs, nothing is seen on the output if C is not sufficiently charged. This is illustrated by the waveforms.

SECTION **8** Sensing, Monitoring, Controllers, and Power Drivers

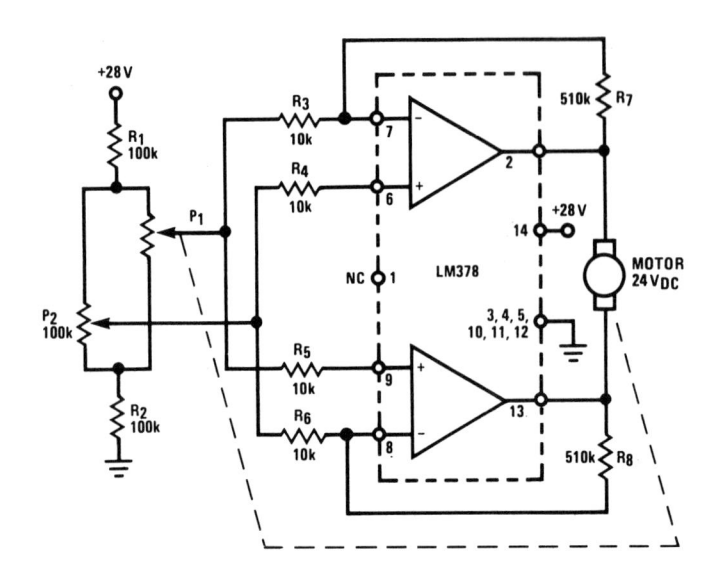

A low-cost proportional speed controller for use with 12 to 24V dc motors at continuous currents up to several hundred milliamps. This circuit allows remote adjustment of angular displacements in drive shaft. Typical applications include rooftop rotary antennas and motor-controlled valves.

Proportional control results from an error signal developed across the Wheatstone bridge comprised of resistors R1, R2, and potentiometers P1 and P2. Control P1 is mechanically coupled to the motor shaft as depicted by the dotted line and acts as a continuously variable feedback sensor. Setting position control P2 creates an error voltage between the two inputs that is amplified by the LM378 (wired as a difference bridge amplifier); the magnitude and polarity of the output signal of the LM378 determines the speed and direction of the motor. As the motor turns, potentiometer P1 tracks the movement, and the error signal—that is, difference in positions between P1 and P2—becomes smaller and smaller until ultimately the system stops when the error voltage reaches zero volts.

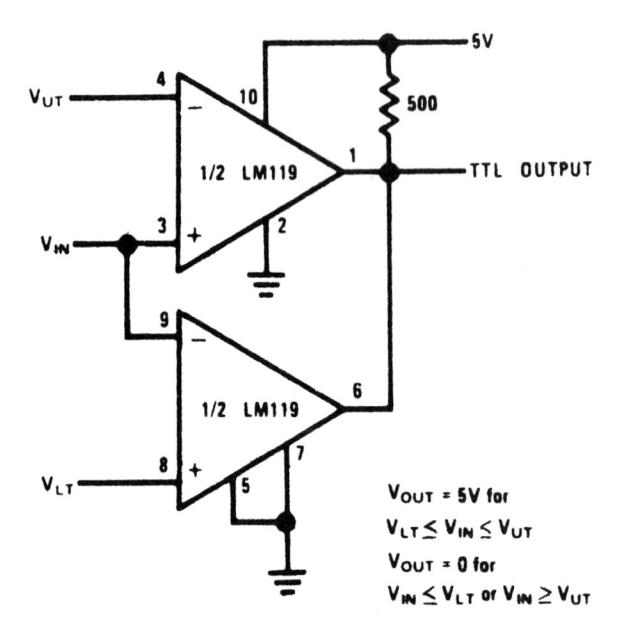

Window detector.

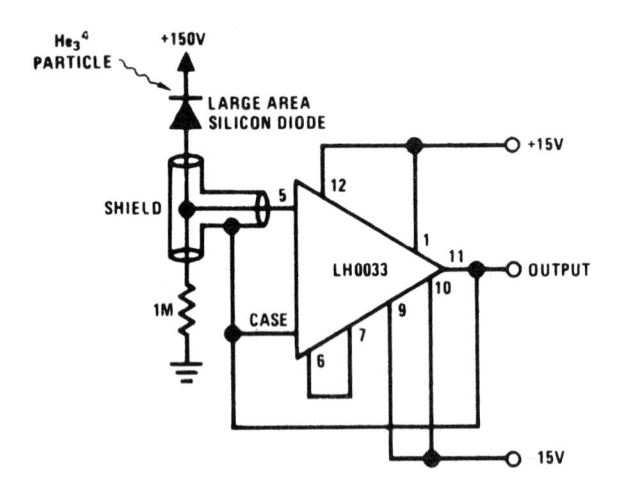

$V_{OUT} = 5V$ for $V_{LT} \leq V_{IN} \leq V_{UT}$
$V_{OUT} = 0$ for $V_{IN} \leq V_{LT}$ or $V_{IN} \geq V_{UT}$

Nuclear particle detector.

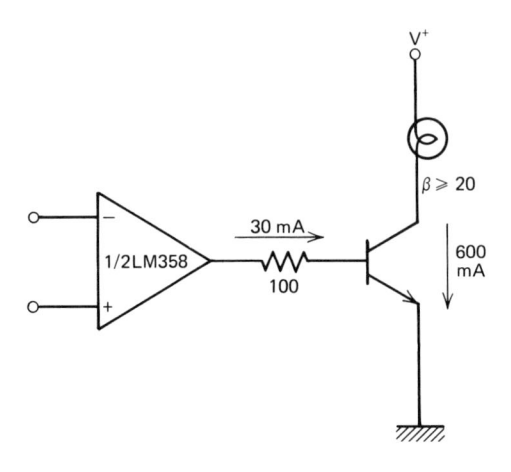

Circuit diagram for the power one-shot control.

Lamp driver.

364 IC SCHEMATIC SOURCEMASTER

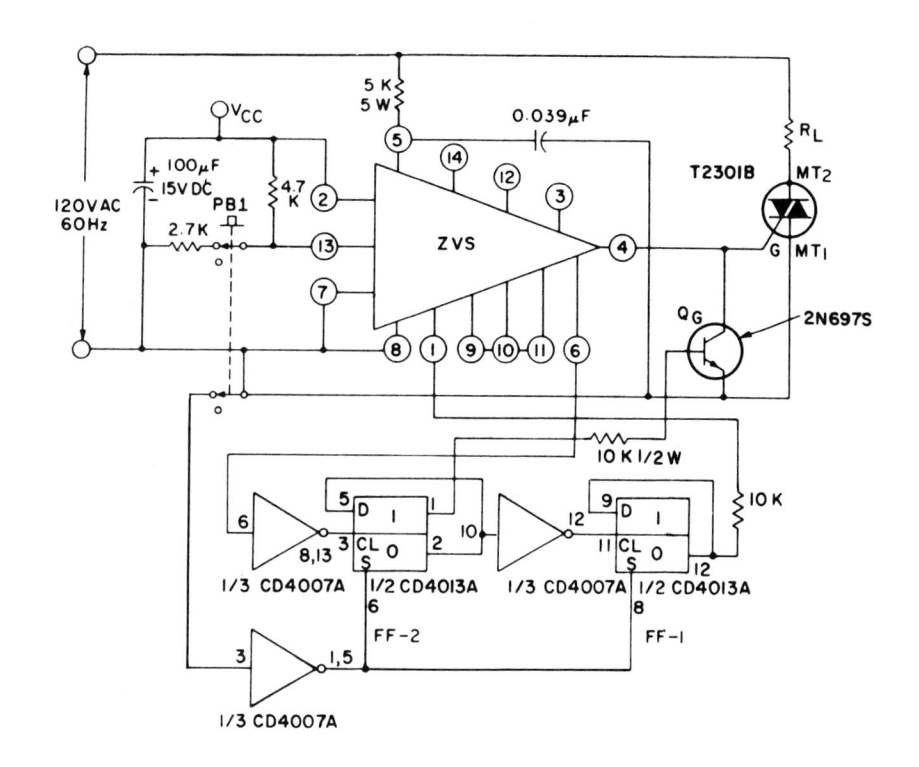

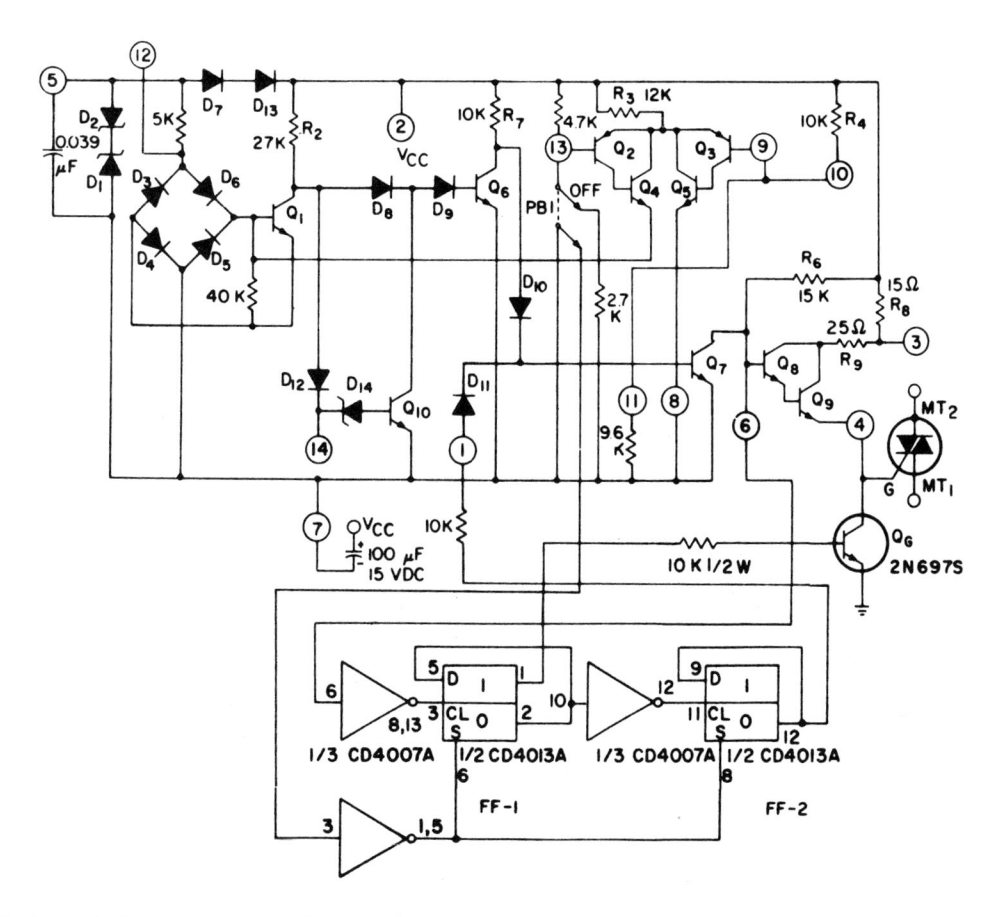

Power one-shot control using a zero-voltage switch. The differential comparator (part of the zero-voltage switch) is initially biased to inhibit output pulses. When the pushbutton is pressed, pulses are generated, but the state of Q_G determines the requirement for their supply to the triac gate. The first pulse generated serves as a "framing pulse" and does not trigger the triac but toggles FF-1. The second pulse triggers the triac and FF-1, which toggles flip-flop FF-2. The output of FF-2 turns on transistor Q7. When the pushbutton is released, the circuit resets itself until the process is repeated. (ZVS is CA3058.)

SECTION **8** *Sensing, Monitoring, Controllers, and Power Drivers*

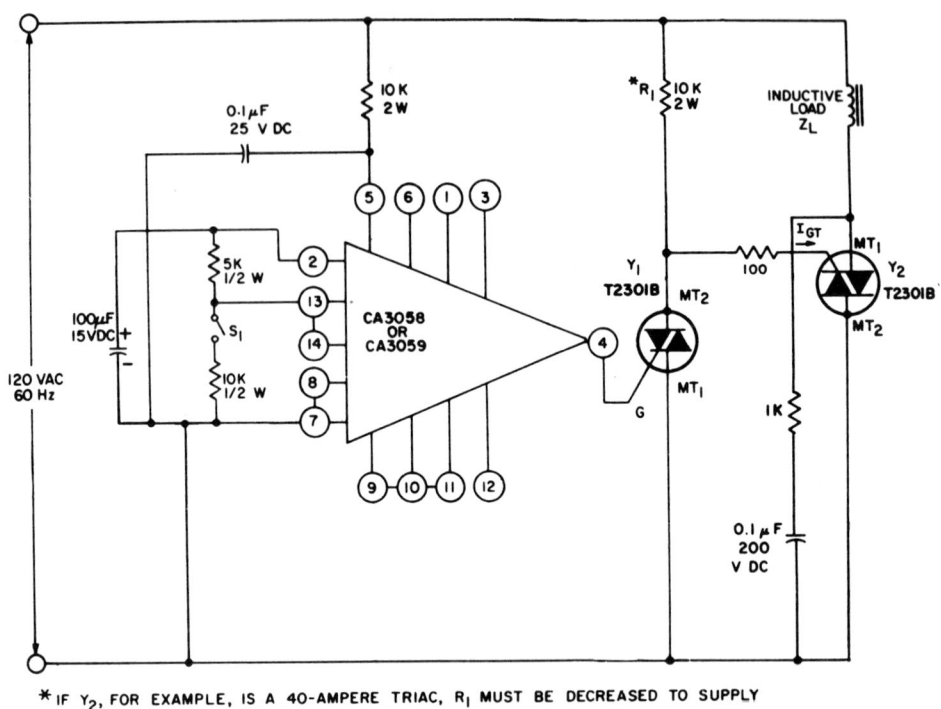

*IF Y_2, FOR EXAMPLE, IS A 40-AMPERE TRIAC, R_1 MUST BE DECREASED TO SUPPLY SUFFICIENT I_{GT} FOR Y_2.

A zero-voltage switch transient-free switch controller using control logic to supply power to the load when the switch is closed.

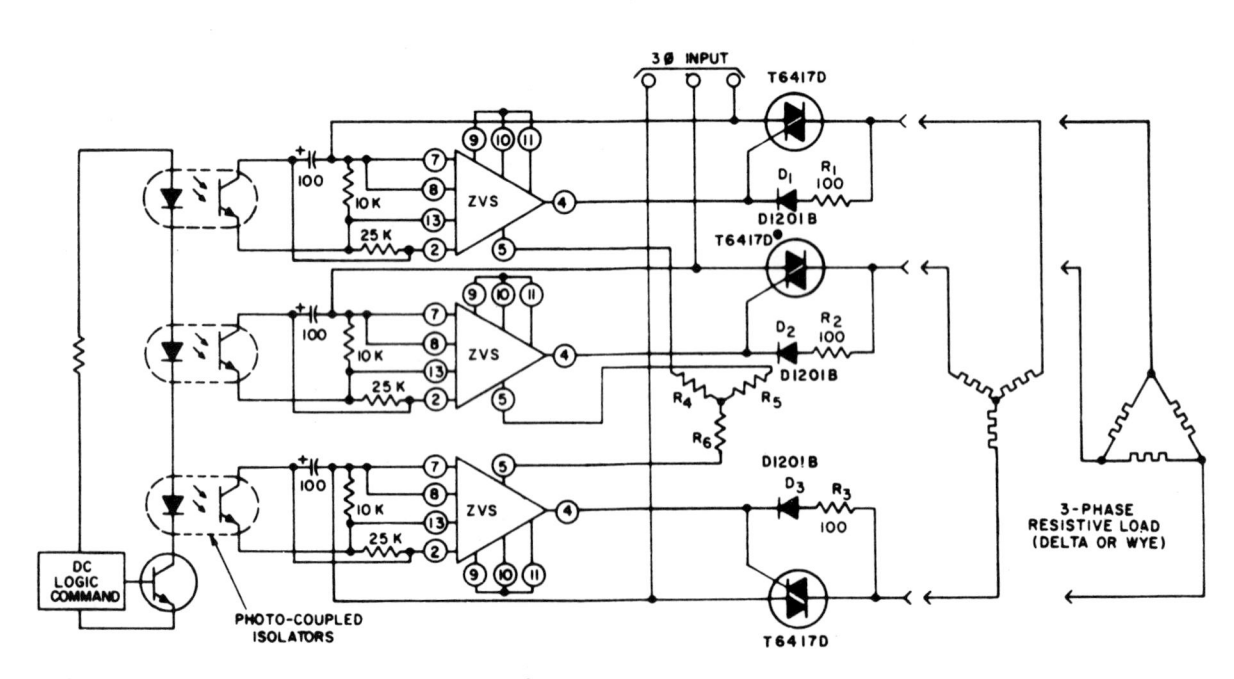

Three-phase heater control employing zero-voltage synchronous switching in the steady-state operating conditions.

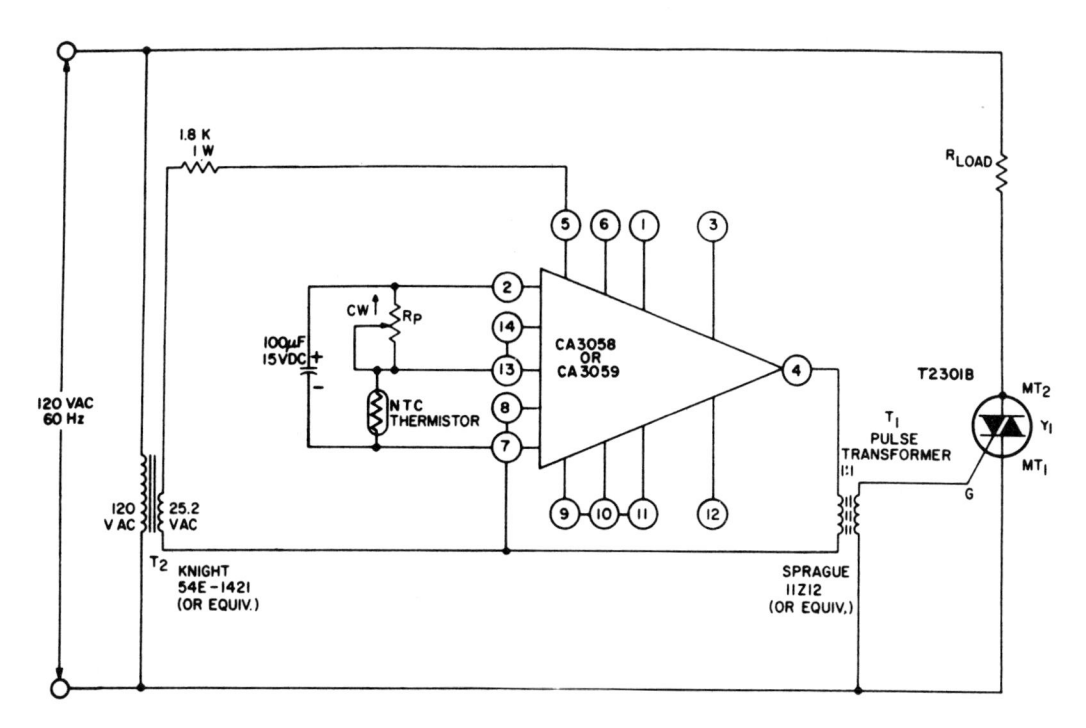

A zero-voltage-switch on-off controller with an isolated sensor. A pulse transformer is used to provide electrical isolation of the sensor from incoming ac power lines.

Relay driver with strobe.

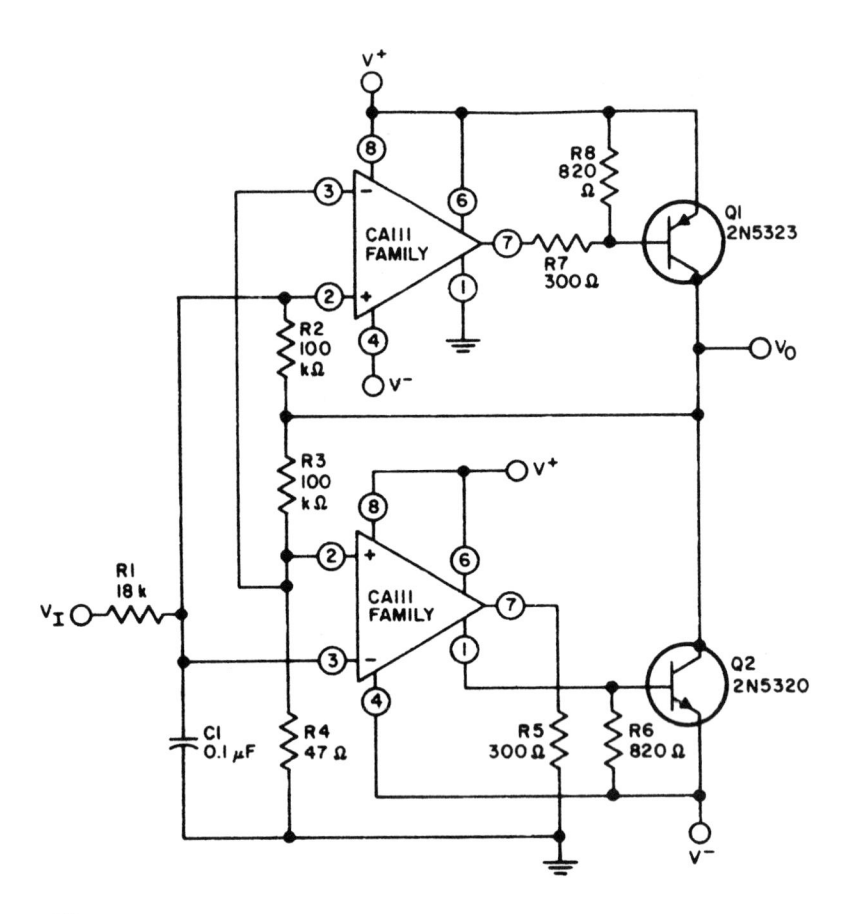

Switching power amplifier.

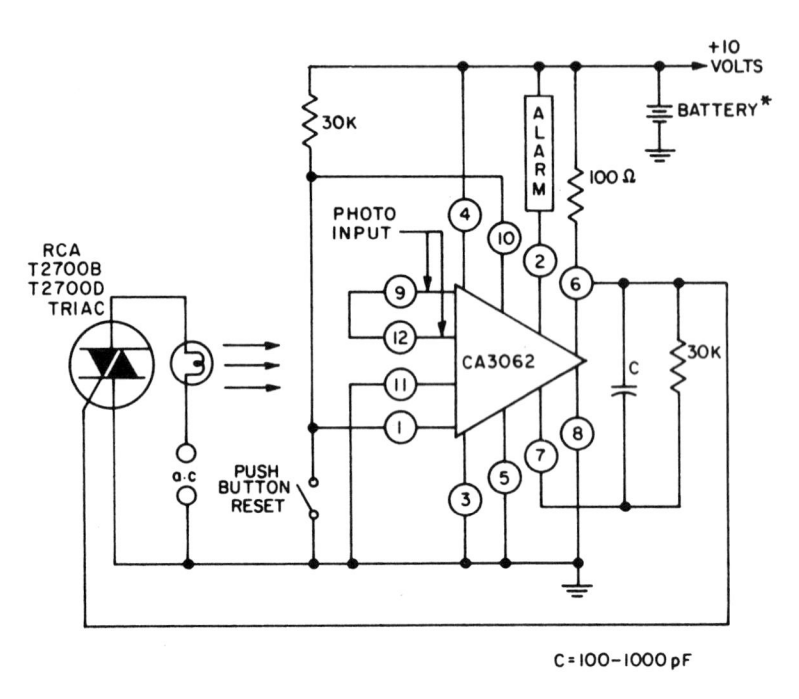

Triac-controlled intrusion alarm. If the light path is broken or the supply ac is interrupted, the alarm is activated, provided the battery is adequately charged. Use any 10 to 12V alarm between V_{CC} and pin 2 of RCA CA3062. The 10V supply voltage serves to charge the battery while the circuit is operating from the ac line.

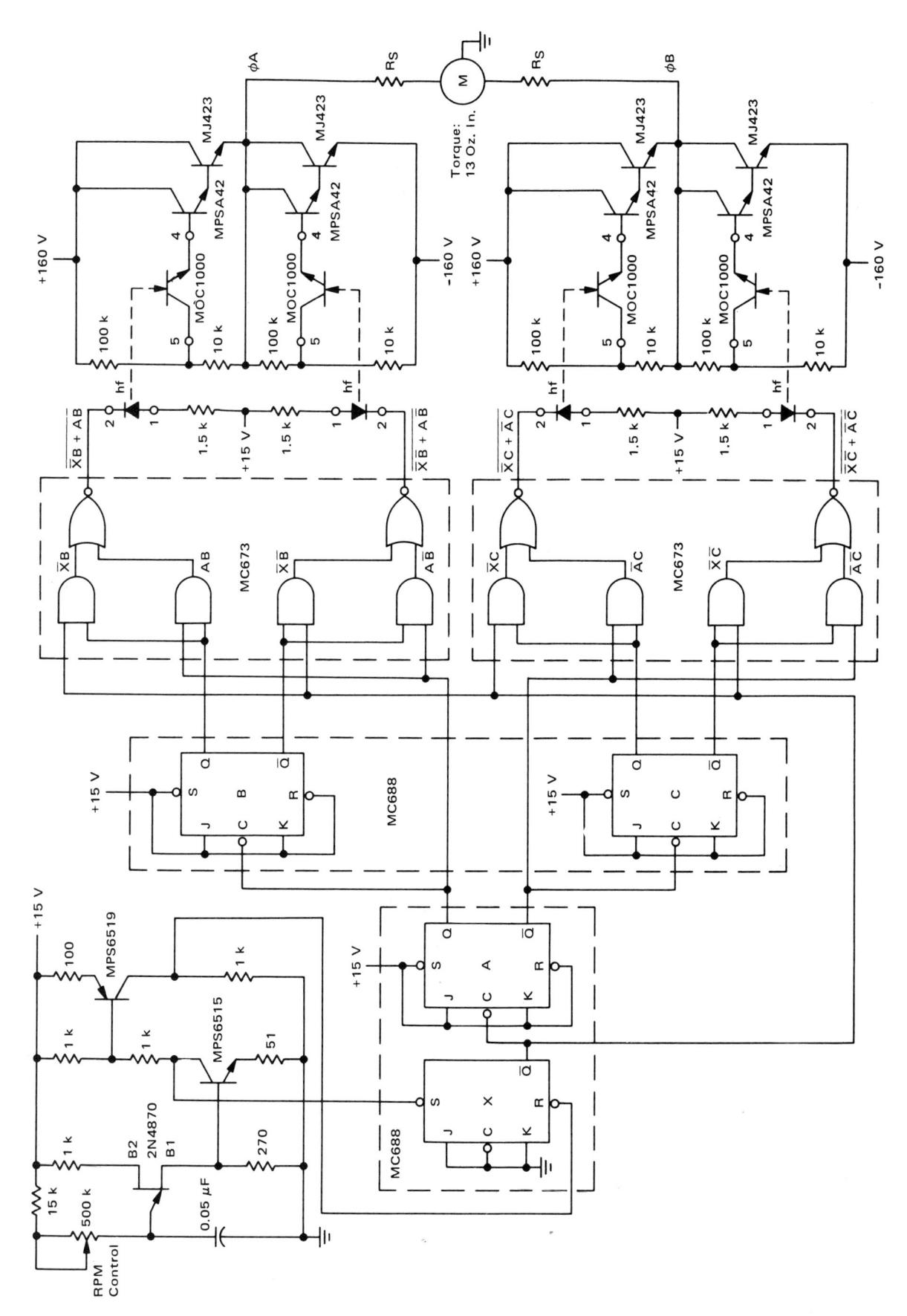

Variable speed control circuit for induction motors.

SECTION **8** *Sensing, Monitoring, Controllers, and Power Drivers* **369**

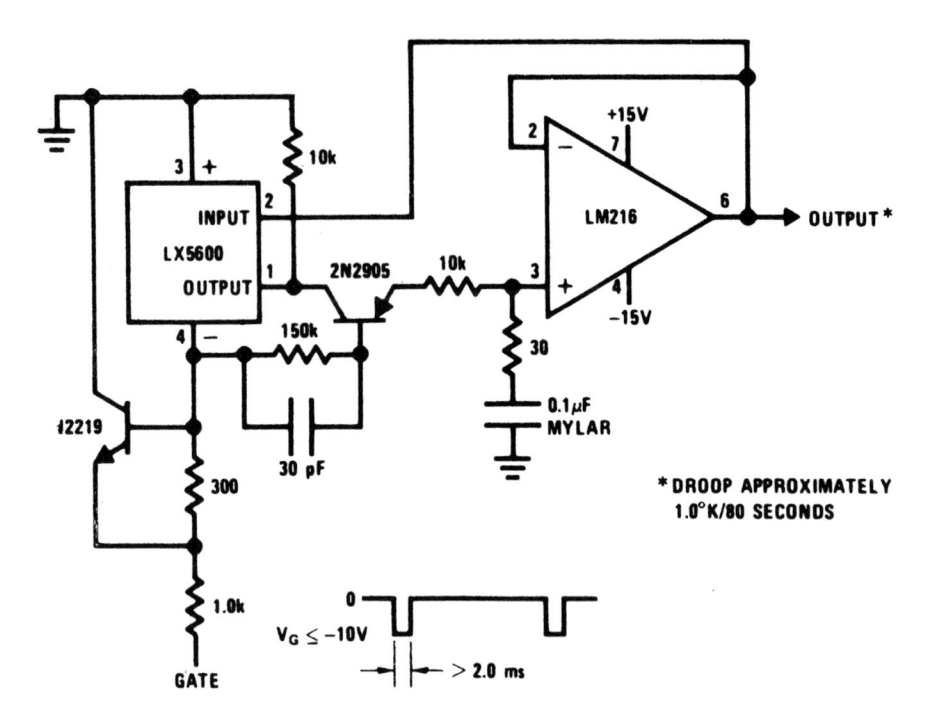

Low-duty-cycle thermometer.

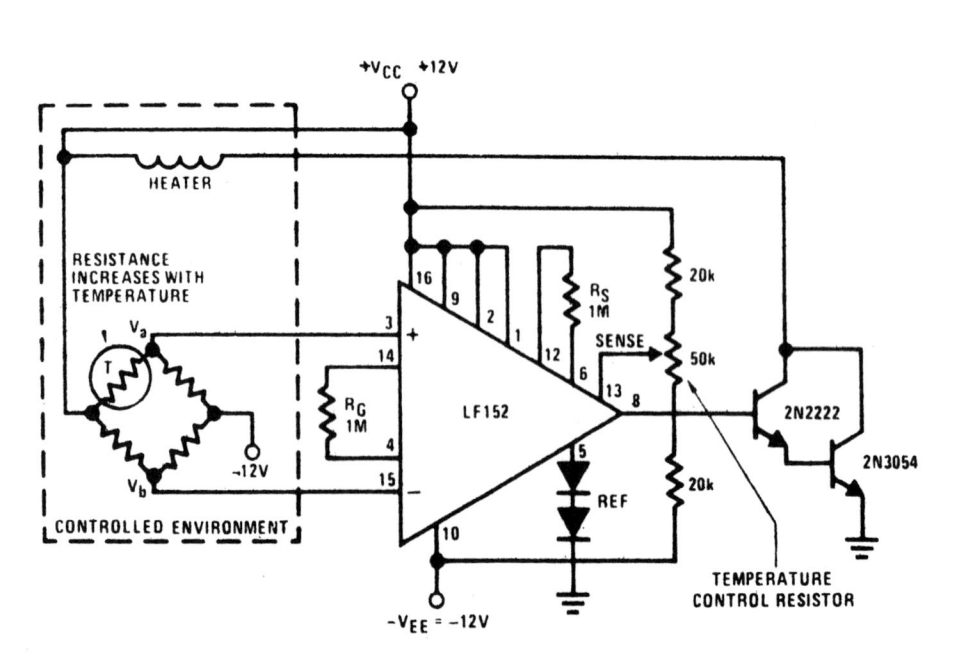

Temperature controller. Under balanced conditions, $V_{sense} - V_{ref}$ appears across R_s and $V_a - V_b$ appears across R_g, $I_{rg} = I_{rs}$. V_{sense} is fixed by the temperature control resistor and R_g/R_s is constant. The *National LF152* is used as a comparator with a feedback loop that is closed through the heater and the temperature-dependent resistor. If $V_a - V_b$ is greater than $V_{sense} R_g/R_s$, the output goes high, which turns *on* the heater. If it is lower, the output goes low, which turns *off* the heater. The closed-loop gain is R_s/R_g.

370 IC SCHEMATIC SOURCEMASTER

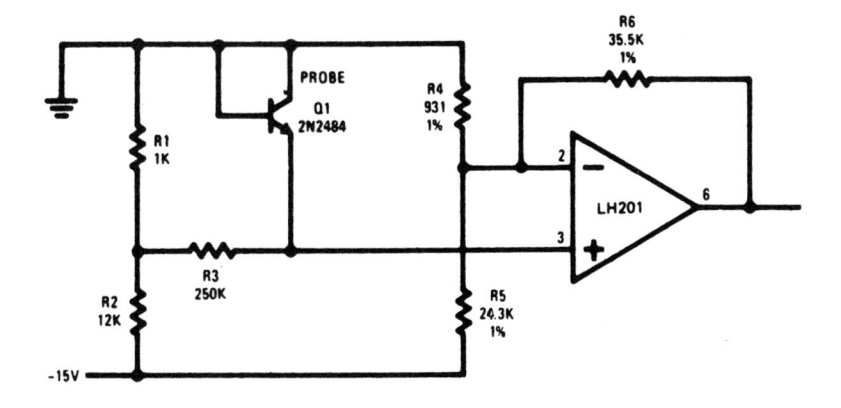

Temperature probe.

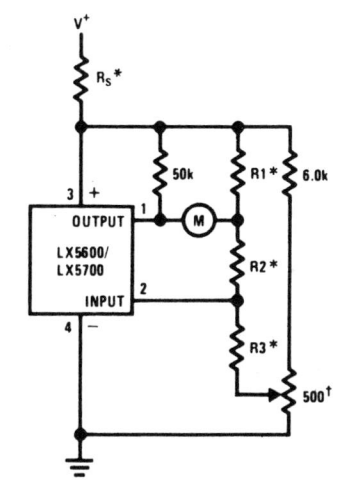

Meter thermometer with trimmed output.

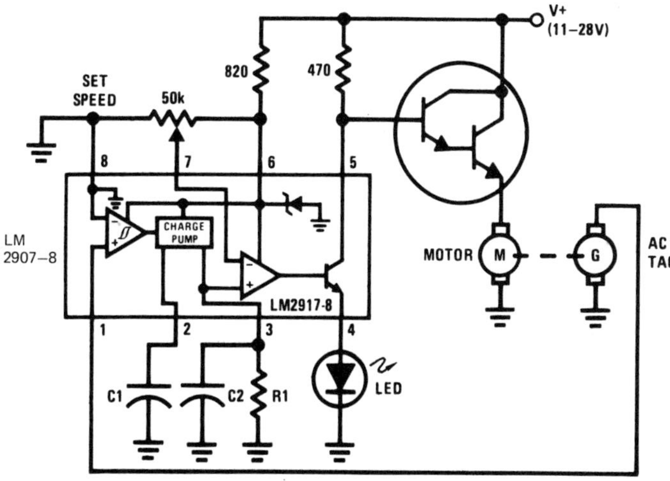

*SELECTED AS FOR METER THERMOMETER EXCEPT T_O SHOULD BE 5°K MORE THAN DESIRED AND $I_Q = 100\mu A$.
†CALIBRATES T_O.

Motor speed control employs an LM2907-8 or LM2917 acting as a shunt mode regulator. It also features an LED to indicate when the device is in regulation.

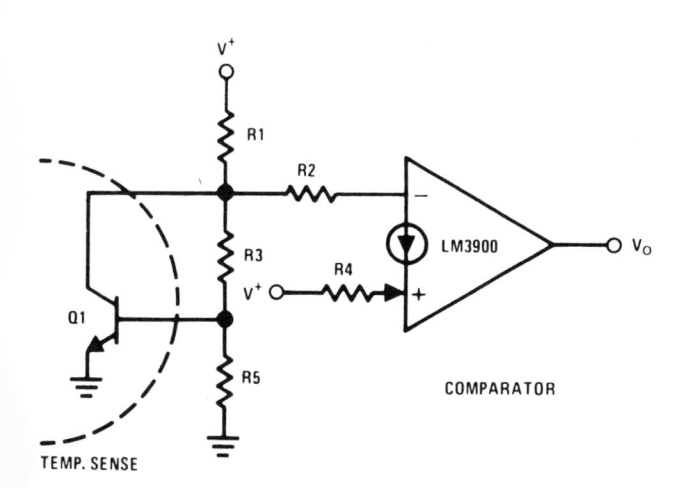

For remote temperature sensing, an NPN transistor, Q1, is connected as an N V_{BE} generator (with R3 and R5) and biased via R1 from the power supply voltage. The LM3900 compares this temperature-dependent voltage with the supply voltage and can be designed to have the output voltage go high at a maximum temperature of remote temperature sensor Q1.

SECTION **8** Sensing, Monitoring, Controllers, and Power Drivers **371**

SECTION 9
TIMERS

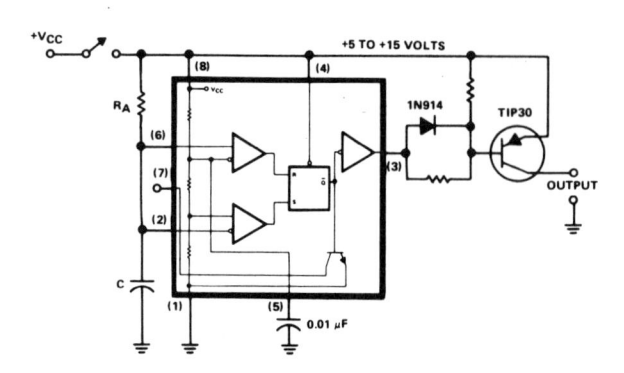

Delayed turn-on power sequence

Time delays. The R_AC time constant may be adjusted to provide the desired power-on delay time. Delaying the turn-off of power to some part of a system requires slightly different circuitry, as shown. As long as the power switch is on, the output is maintained on owing to output-pass transistor drive from R1 and

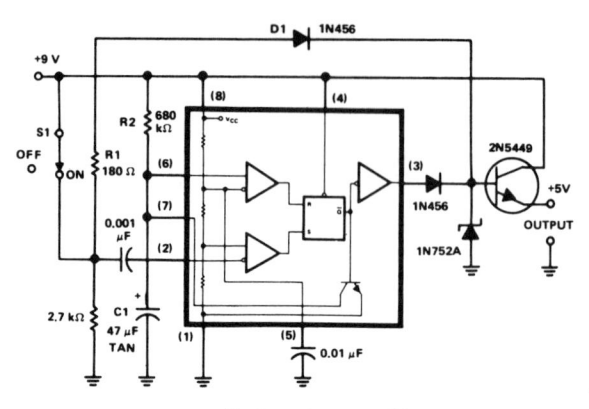

Delayed turn-off power sequence

D1. When switch S1 is turned off, a negative transient coupled into pin 2 starts the timing function. R2 (680k) and C1 (47 μF) will time out in slightly less than 30 seconds, at which point the output at pin 3 goes low, turning off the 2N5449 series-pass transistor.

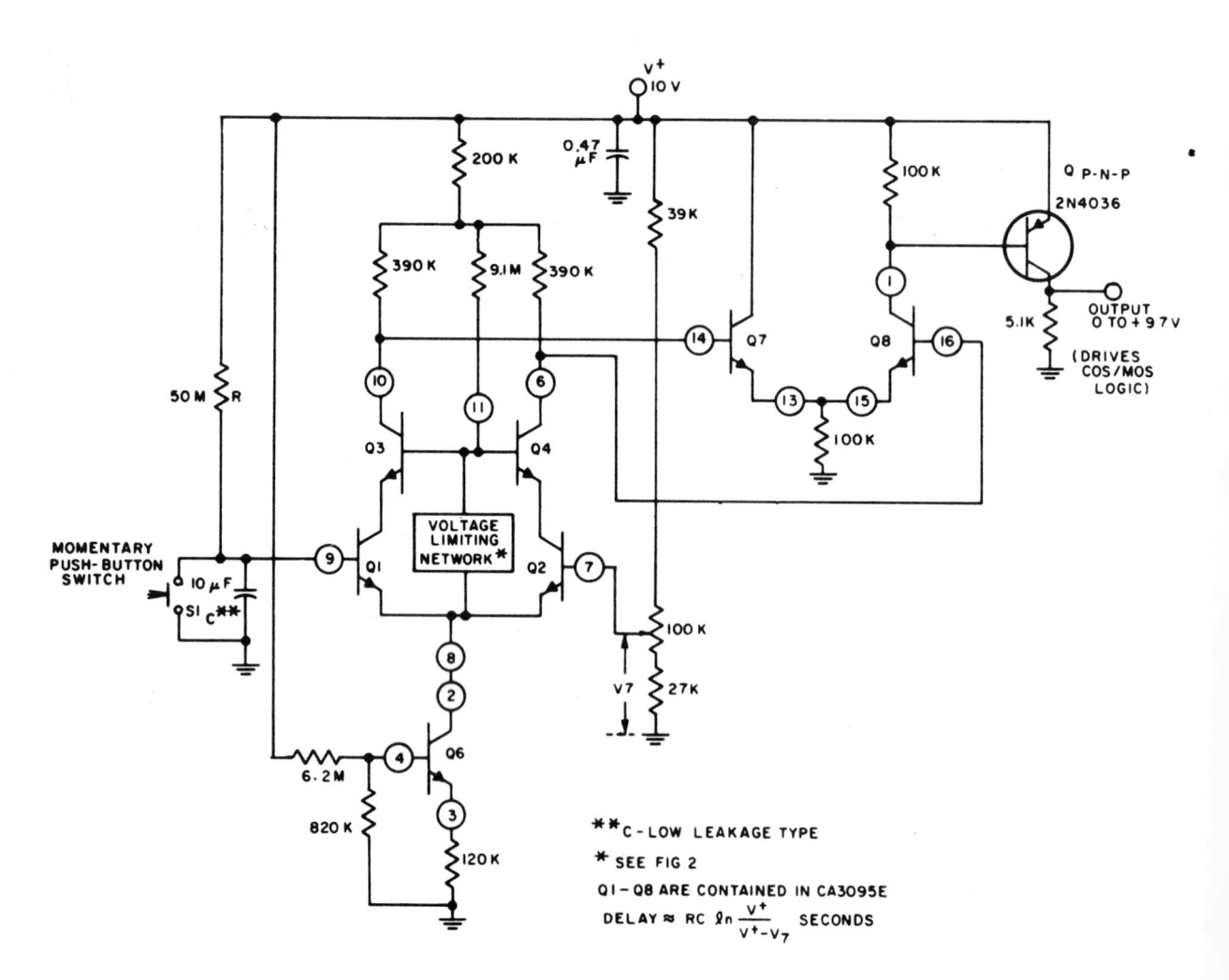

**C-LOW LEAKAGE TYPE
* SEE FIG 2
Q1-Q8 ARE CONTAINED IN CA3095E
DELAY $\approx RC \ln \dfrac{V^+}{V^+ - V_7}$ SECONDS

Analog timer for long delays. Voltage-limiting network is built into superbeta IC. To use it, just connect a resistor from the positive supply of the differential amplifier to pin 11; the value should be such that current through the resistor is about half that measured at pin 8 under quiescent conditions.

374 IC SCHEMATIC SOURCEMASTER

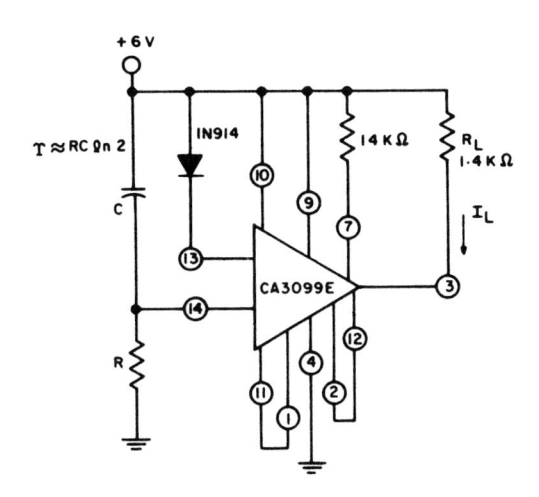

Time-delay circuit. Terminal 3 sinks after t seconds.

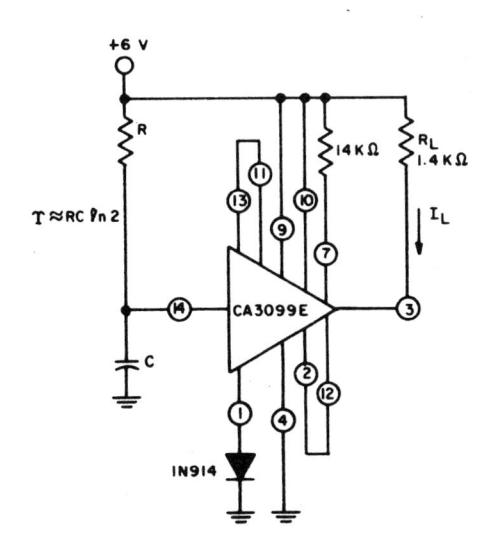

Time-delay circuit. Sink current is interrupted after t seconds.

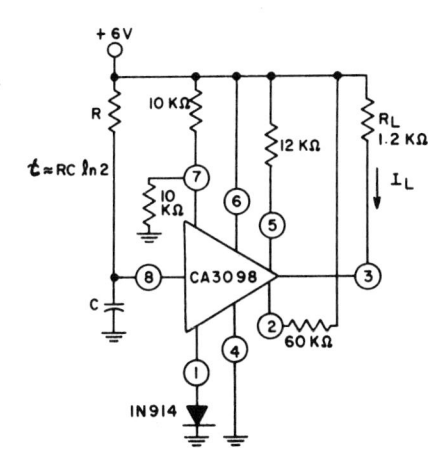

Time-delay circuit. Sink current interrupted after t seconds.

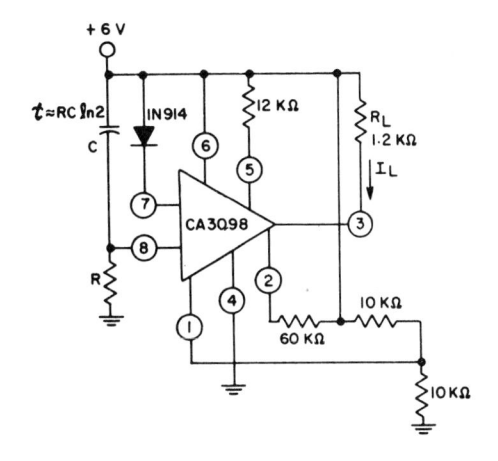

Time-delay circuit. Terminal 3 sinks after t seconds.

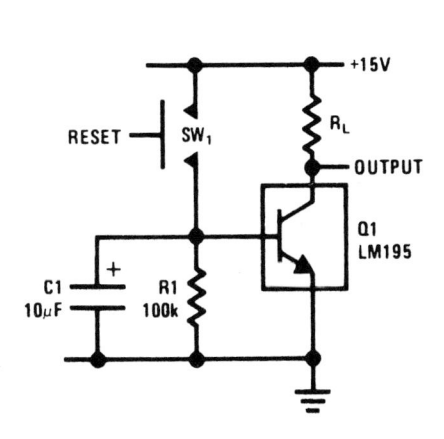

Time delay.

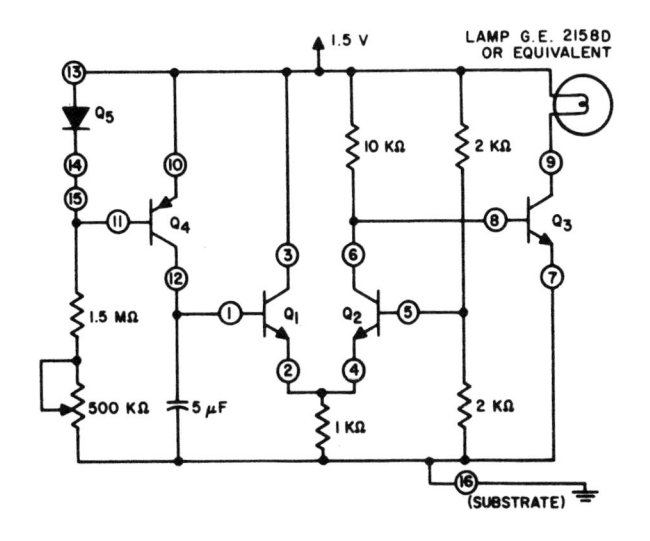

A 10-second timer operated from 1.5V supply using CA3096E.

SECTION 9 *Timers* 375

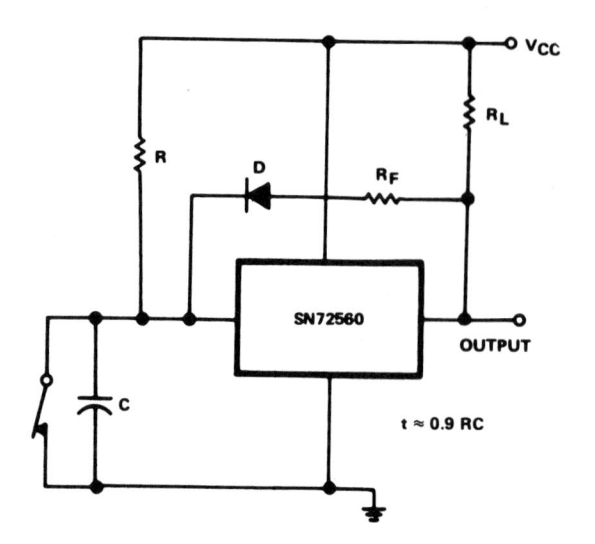

Long-delay timer. Resistor R can be much higher than 1M because, if the output is low, diode D is reverse-biased, and the delay time is determined only by the values of C and R. As soon as the switch is opened and voltage V_{T+} is reached, the output goes high, and diode D will be forward-biased to deliver the extra input current necessary to keep the input high. The upper limitation for resistor R is influenced by the leakage resistance of capacitor C and the reverse resistance of the diode and R_F.

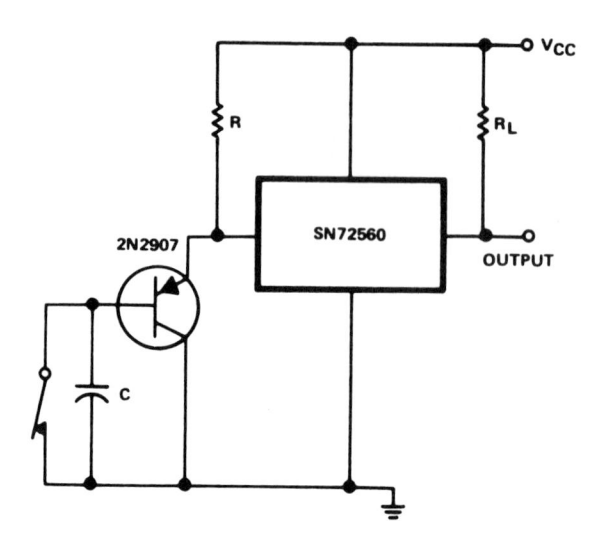

Long-delay timer. When the switch is opened, capacitor C is charged by the base current of the transistor. This makes the charging time equal to beta times CR. The timing period is beta dependent, but the circuit has the advantage of small resistor and capacitor values.

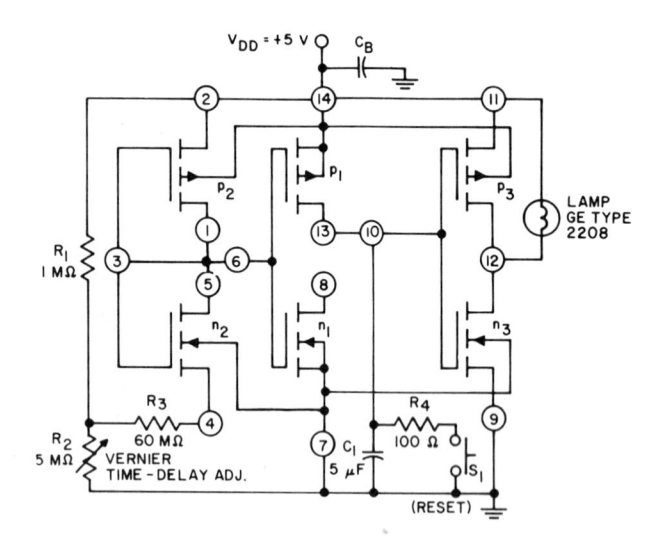

An analog timer using CA3600E. After time t, capacitor C1 is charged sufficiently in the positive direction so that transistor N3 is driven into conduction by its gate, and the lamp is lighted to signify the end of the time-delay period. The circuit is reset by momentarily closing switch S1 to discharge capacitor C1 through R4. Resistor-divider network R1–R2 establishes the supply voltage to a constant-current network comprised of resistor R3 and the series-connected CMOS pair N2–P2 biased for linear operation by resistor R_b. With the circuit values shown, time delay is 1 hour.

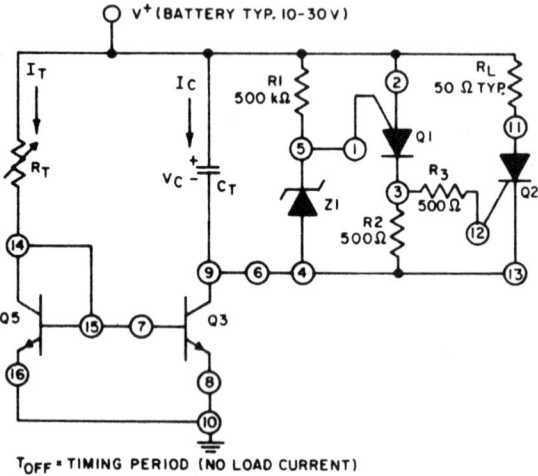

Low-current-drain, battery-operated, long-interval astable timer built around CA3097E.

376 IC SCHEMATIC SOURCEMASTER

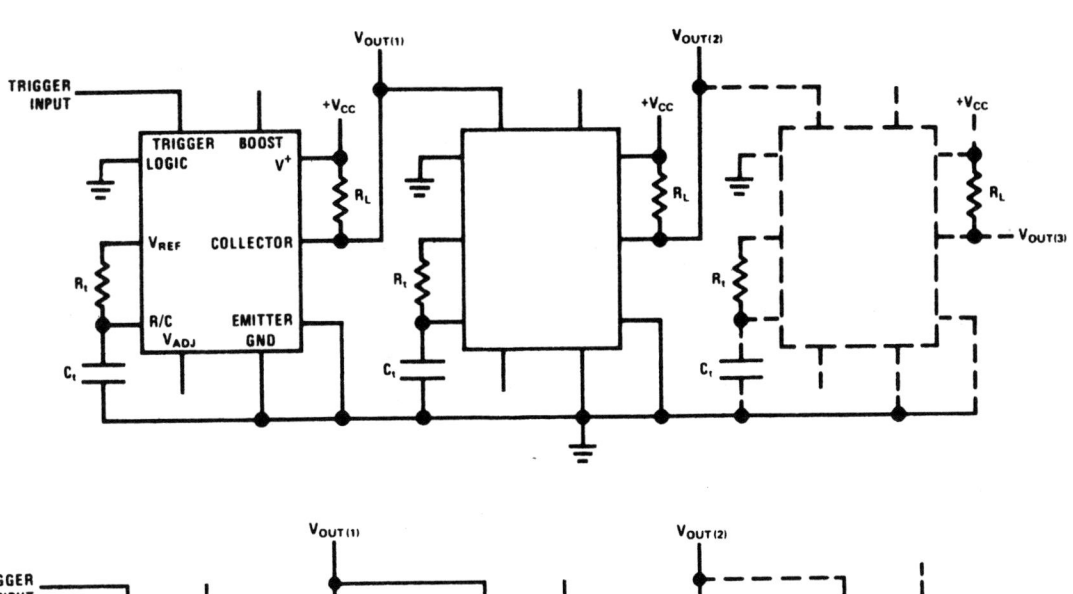

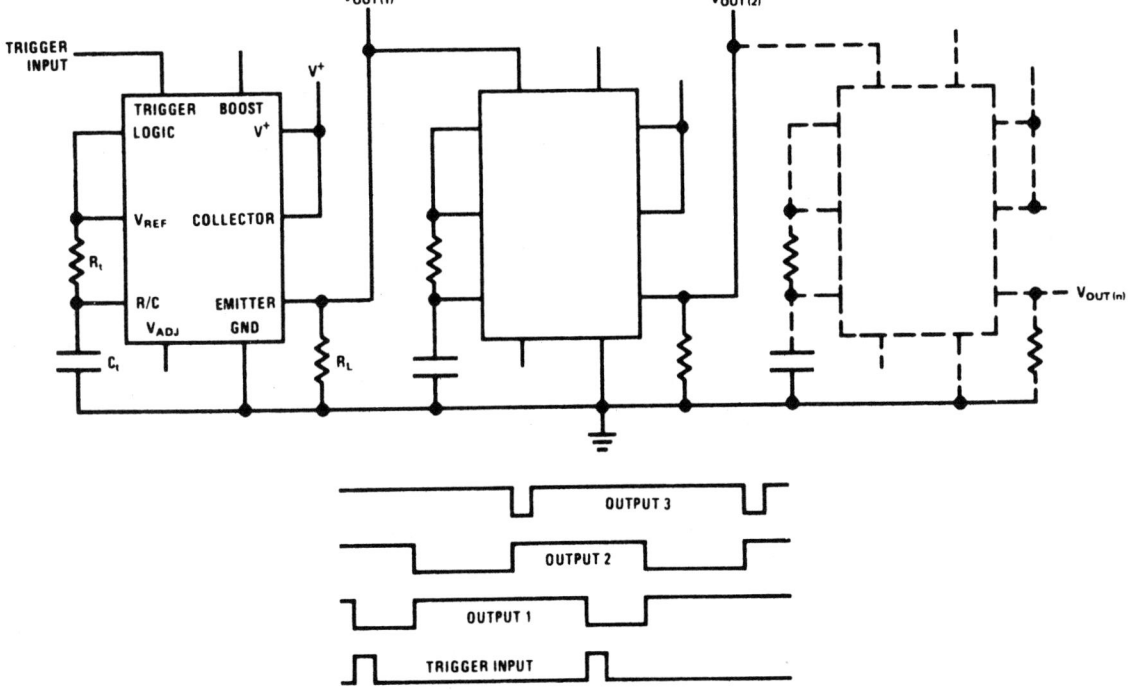

The National LM122 can be connected as a chain of timers quite easily with no interface required. Here, two possible connections are shown. In both cases, the output of the timer is low during the timing period so that the positive-going signal at the end of timing period can trigger the next timer. There is no limitation on the timing period of one timer with respect to any other timer before or after it because the trigger input to any timer can be high or low when that timer ends its timing period.

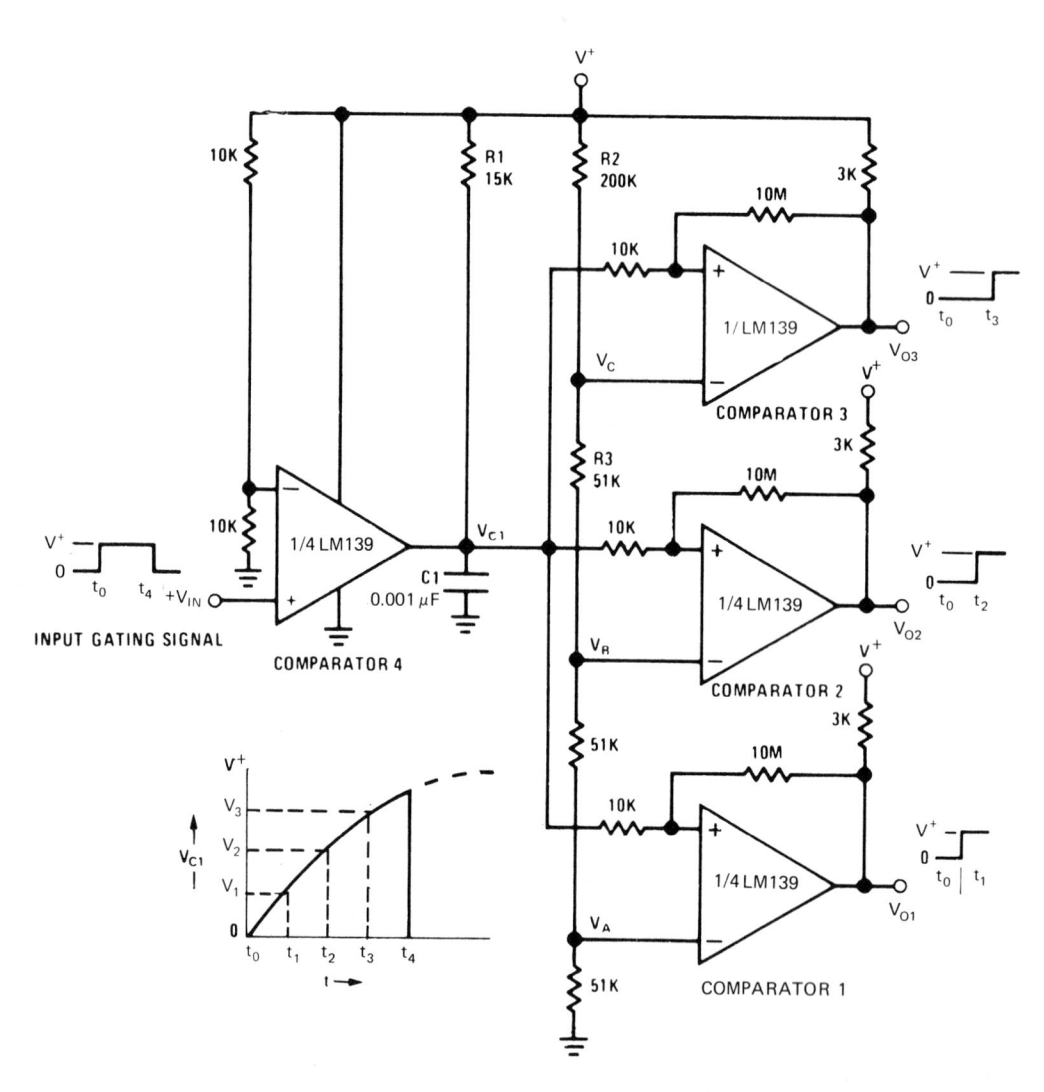

A time-delay generator will provide output signals at prescribed time intervals from a time reference t_0 and will automatically reset when the input signal returns to ground. For circuit evaluation, consider the quiescent state ($V_{IN} = 0$) where the output of comparator 4 is on, which keeps the voltage across C1 at zero. This keeps the outputs of comparators 1, 2, and 3 in their *on* state (V_{OUT} = ground). When an input signal is applied, comparator 4 turns off, allowing C1 to charge at an exponential rate through R1. As this voltage rises past the preset trip points V_A, V_B, and V_C of comparators 1, 2, and 3, the output voltage of each of these comparators will switch to the high state ($V_{OUT} = +V_{CC}$). A small amount of hysteresis has been provided to ensure fast switching for the case where the RC time constant has been chosen large to give long delay times. It is not necessary that all comparator outputs be low in the quiescent state. Several or all may be reversed as desired simply by reversing the inverting and noninverting input connections. Hysteresis again is optional.

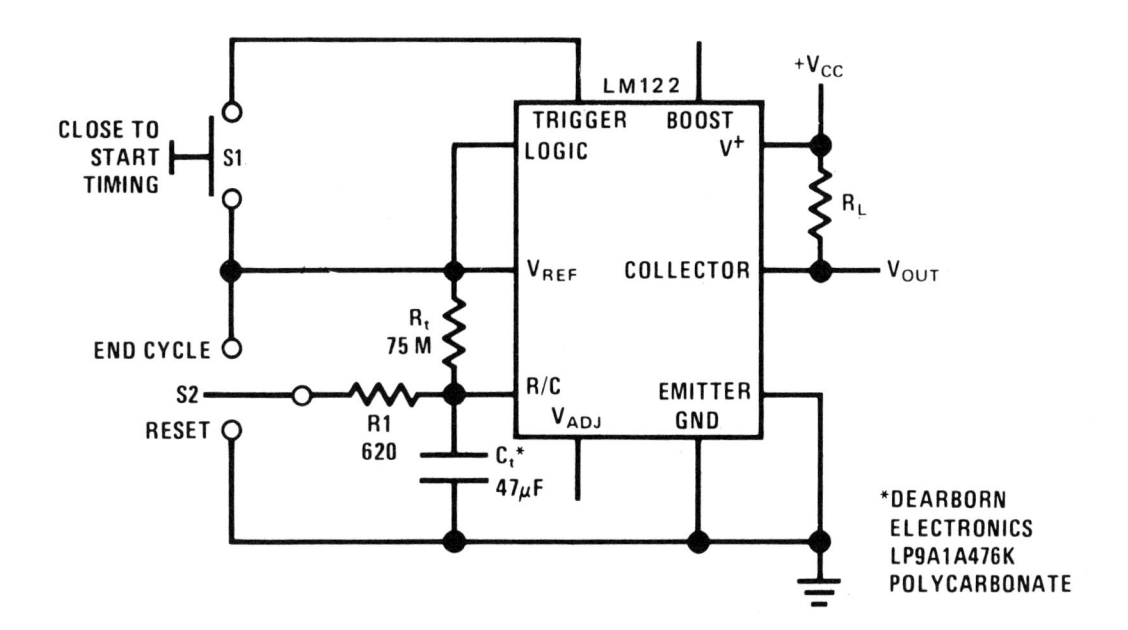

A 1-hour timer with manual controls for start, reset, and cycle end. S1 starts timing but has no effect after timing has started. S2 is a center-off switch that can either end the cycle prematurely (with the appropriate change in output state and discharging of C_t) or cause C_t to be reset to 0V without a change in output. In the latter case, a new timing period starts as soon as S2 is released. The average charging current through R_t is about 30 nA, so some attention must be paid to parts layout to prevent stray leakage paths. The suggested timing capacitor has a typical time constant of 300 hours and a guaranteed minimum of 25 hours at +25°C. Other capacitor types may be used if sufficient data is available on their leakage characteristics.

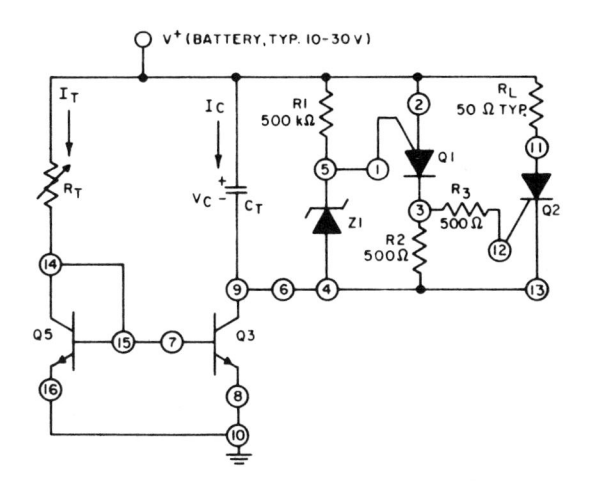

A low-current battery-operated astable timer for long intervals. T_{OFF} = timing period (no load current). PUT fires when $V_C \approx 8V$. $V_C = I_C(T_{OFF})/C_T$. $I_C \approx I_T$ (Q3, Q5 matched). I_T set by adjusting R_T, $I_T \approx V^+ - 0.7/R_T$. T_{ON} = capacitor discharge time through load. Load turns off when SCR anode current falls below holding current I_{HO} (typical I_{HO} = 1.2 mA); for example, for a timing period of 8.3 min, C_T = 1000 μF, I_T = 16 μA, $R_T = V^+ - 0.7/I_T$ (for V^+ = 16V, $R_T \approx$ 1M).

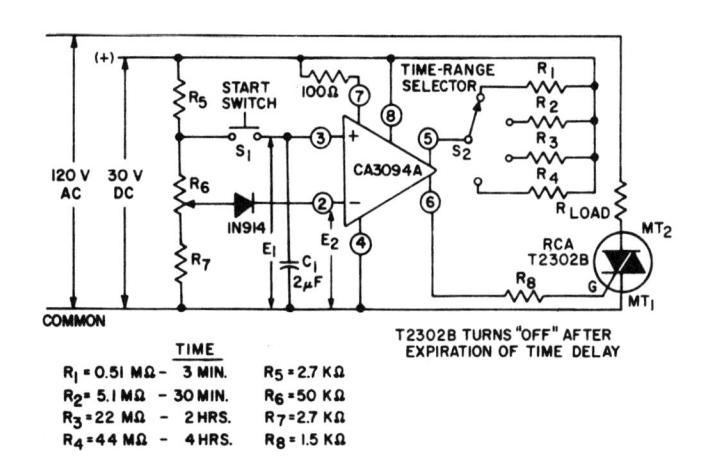

Presettable analog timer. Long timing intervals (up to 4 hours) are achieved by discharging timing capacitor C1 into the signal-input terminal of the CA3094. This discharge current is controlled precisely by the magnitude of the amplifier bias current (I_{ABC}) programmed into terminal 5 through a resistor selected by switch S2. Operation of the circuit is initiated by charging capacitor C1 through the momentary closing of switch S1. Capacitor C1 starts discharging and continues discharging until voltage E1 is less than voltage E2. The differential input transistors in the CA3094 then change state, and terminal 2 draws sufficient current to reverse the polarity of the output voltage (terminal 6).

SECTION 9 Timers **379**

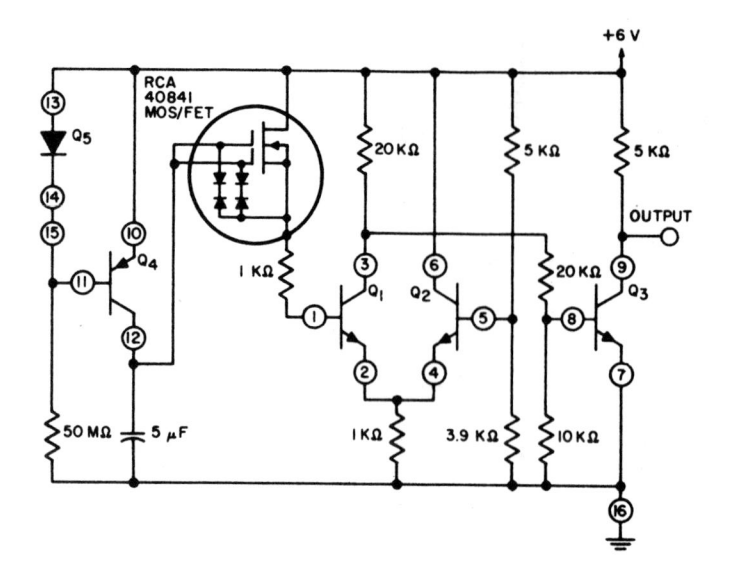

A 1-minute timer using CA3096AE and a MOSFET. Time delay changes ±7% for supply voltage change of ±10%.

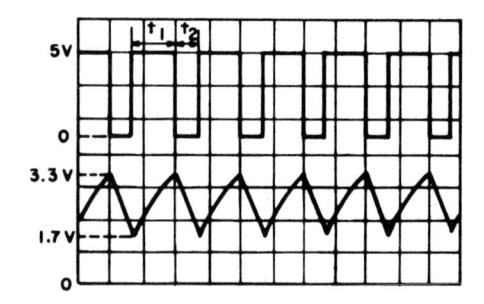

Top Trace: Output voltage (2V/div. and 0.5 ms/div.)
Bottom Trace: Capacitor voltage (1 V/div. and 0.5 ms/div.)

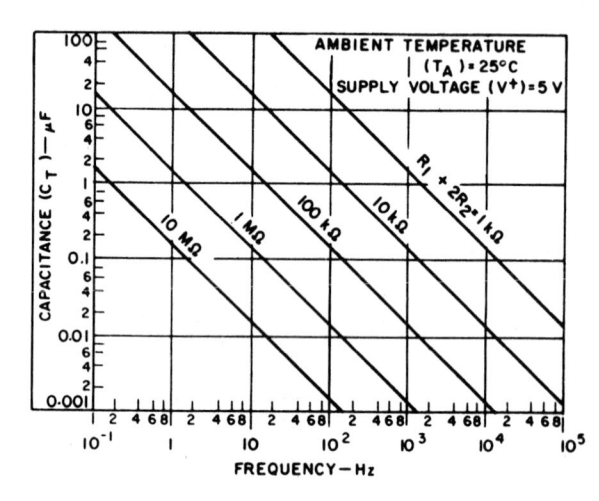

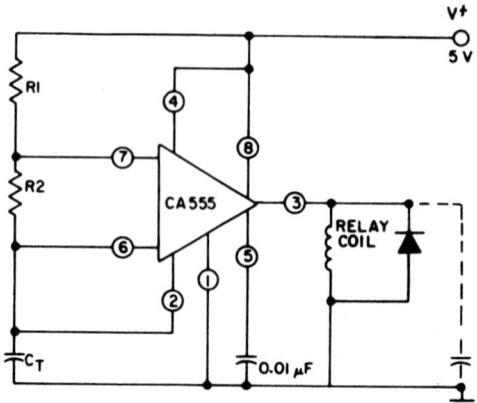

Repeat cycle timer (astable operation). In this mode of operation, the total period is a function of both R1 and R2; $T = 0.693(R1 + 2R2)C_T = t_1 + t_2$, where $t_1 = 0.693(R1 + R2)C_T$ and $t_2 = 0.693(R2)C_T$. The duty cycle is $t_2/t_1 + t_2 = R2/R1 + 2R2$. Typical waveforms generated during this mode of operation are shown along with curves of free-running frequency with variations in the value of (R1 + 2R2) and C_T.

380 IC SCHEMATIC SOURCEMASTER

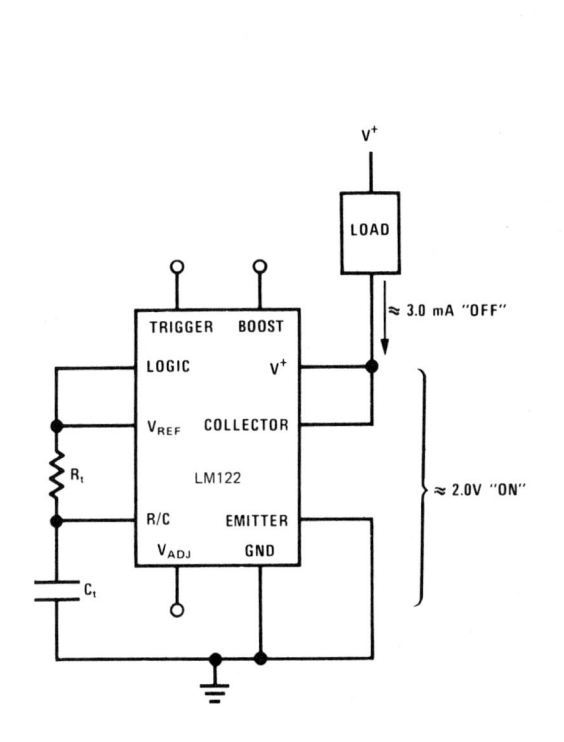

Two-terminal time-delay switch. The timer is used to turn on and drive a relay R_tC_t seconds after application of power. *Off* current of the switch is 4 mA maximum and *on* current can be as high as 50 mA.

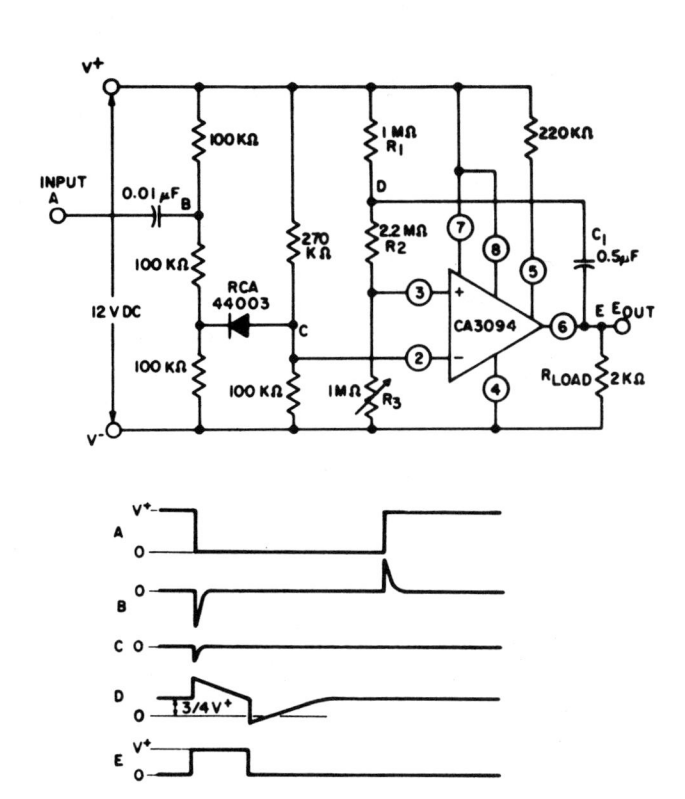

An RC timer triggered by external negative pulse. On a negative-going transient at input (A), a negative pulse at C will turn on the CA3094, and the output (E) will go from a low to a high level.

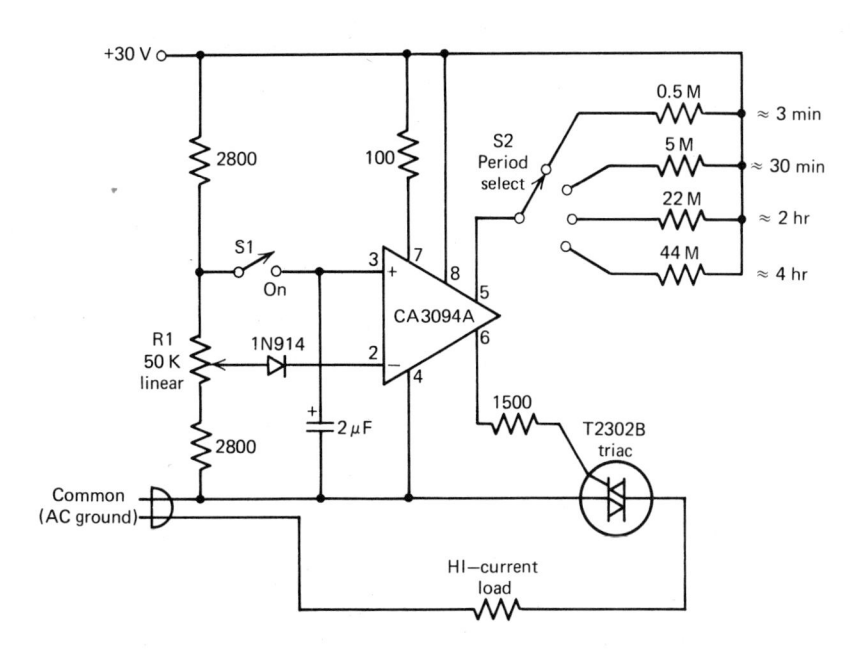

Wide-range timer. Basic timer period is established by S2, and the period can be varied substantially with the 50k adjustment (R1). Closing S1 begins the timing operation; at the end of the period, power is supplied to the load through the triac.

SECTION **9** Timers 381

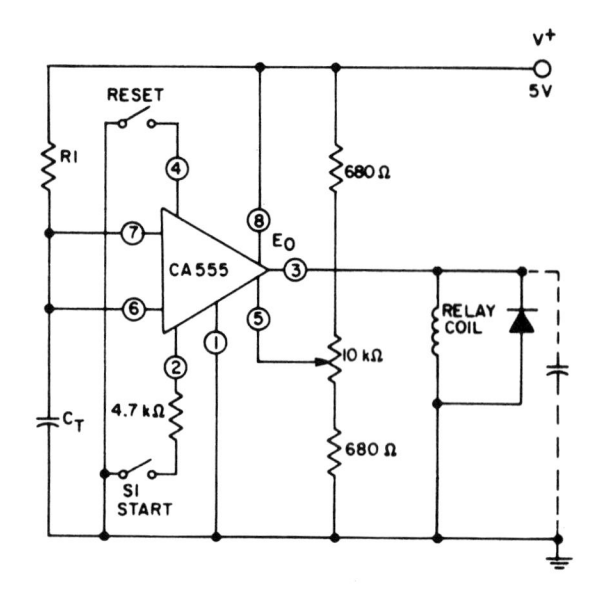

Reset timer (monostable operation). In this mode of operation, capacitor C_T is initially held discharged by a transistor on the integrated circuit. Upon closing the *start* switch or applying a negative trigger pulse to terminal 2, the integral timer flip-flop is set and releases the short circuit across C_T, which drives the output voltage high (relay energized). The action allows the voltage across the capacitor to increase exponentially with the time constant $t = R1C_T$. When the voltage across the capacitor equals $^2/_3V^+$, the comparator resets the flip-flop, which in turn discharges the capacitor rapidly and drives the output to its low state. Since the charge rate and threshold level of the comparator are both directly proportional to V^+, the timing interval is relatively independent of supply voltage variations. Typically, the timing varies only 0.05% for a 1V change in V^+. Applying a negative pulse simultaneously to the reset terminal (4) and the trigger terminal (2) during the timing cycle discharges C_T and causes the timing cycle to restart. Momentarily closing only the reset switch during the timing interval discharges C_T, but the timing cycle does not restart.

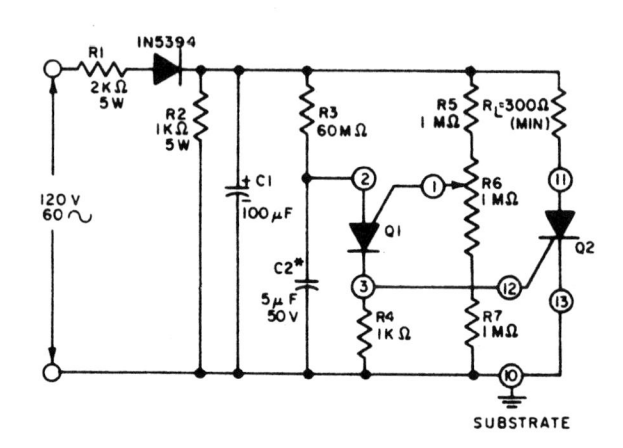

Ac line-operated one-shot timer. Timing period is approximately equal to 200 seconds with 1M pot centered. Timing cycle begins when ac is applied. C2 is Sprague type 4308 (5 μF at 50V), type 6308 (5 μF at 50V), or equivalent.

382 IC SCHEMATIC SOURCEMASTER

SECTION 10
AUTOMOTIVE CIRCUITS

It seems like only yesterday that electronics was foreign to the automobile. The scene is changing drastically, as this section will indicate. Electronics has been steadily making inroads in the automotive market, from simple go/no-go indicators to complex analog and digital parametric-value readouts that sometimes involve quite sophisticated computation. What you see in this section is a small sample of what is currently being done: a clutch control to eliminate slip and prolong transmission and clutch life, electronic high-speed warning devices of various types, wind velocity and temperature indicators, tachometers and other frequency-to-voltage conversion devices, flashers, and even a comparatively simple antiskid device.

The circuits featured here might be considered "thought prodders"—many of them can be adapted readily to functions other than the ones cited; the same applies to those circuits presented in other sections of this book, a great many of which can be applied with little or no modification to the automobile or garage.

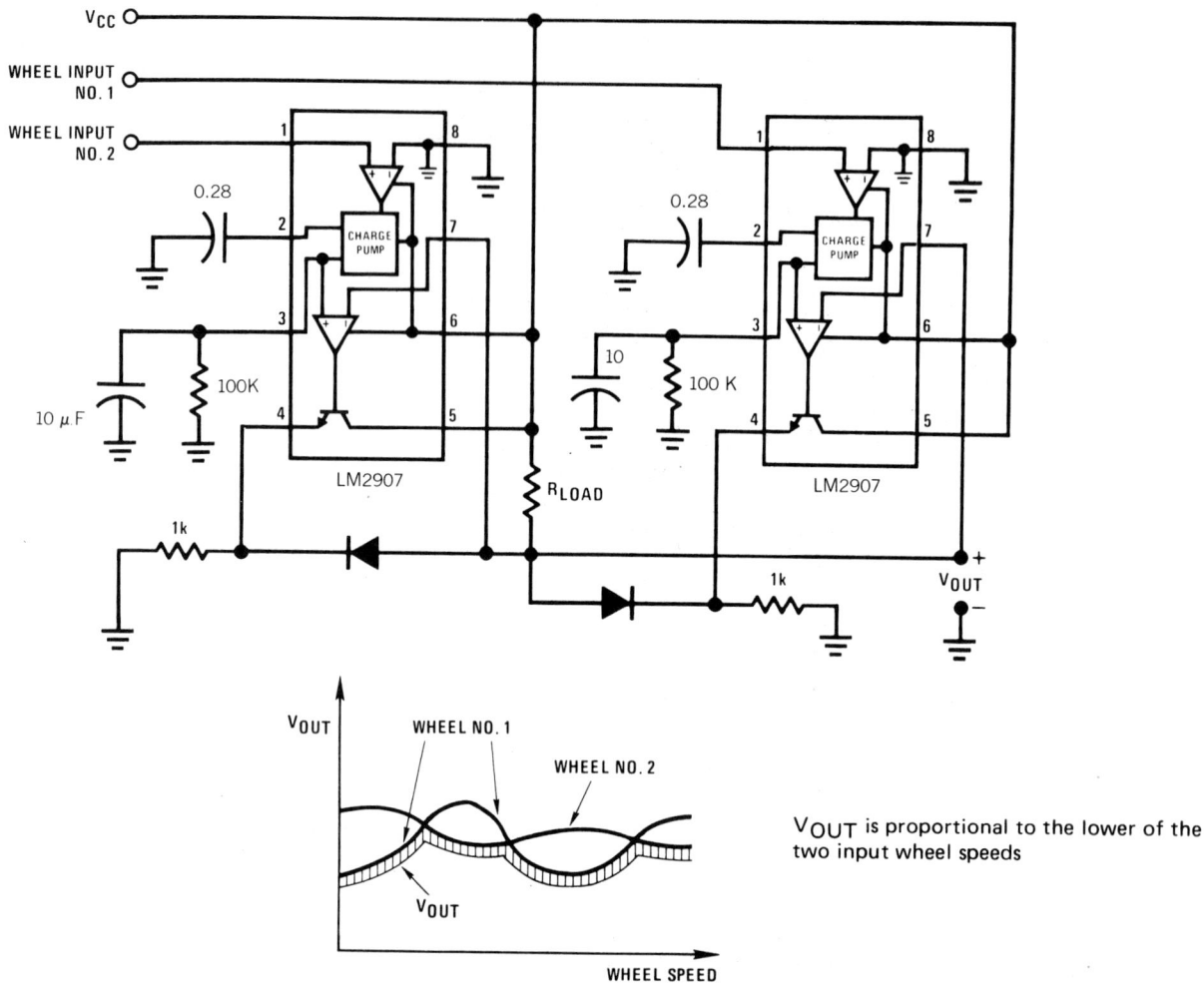

V_{OUT} is proportional to the lower of the two input wheel speeds

Antiskid circuit. Motor Vehicle Standard 121 places certain stopping requirements on heavy vehicles, which necessitate the use of electronic antiskid control devices. These devices generally use variable-reluctance magnetic pickup sensors on the wheels to provide inputs to a control module. Here, the input frequency from each wheel sensor is converted to a voltage in the normal manner. The op-amp/comparator is connected with negative feedback with a diode in the loop so that the amplifier can only pull down on the load and not pull up. In this way, the outputs from the two devices can be joined together, and the output will be the lower of the two input speeds.

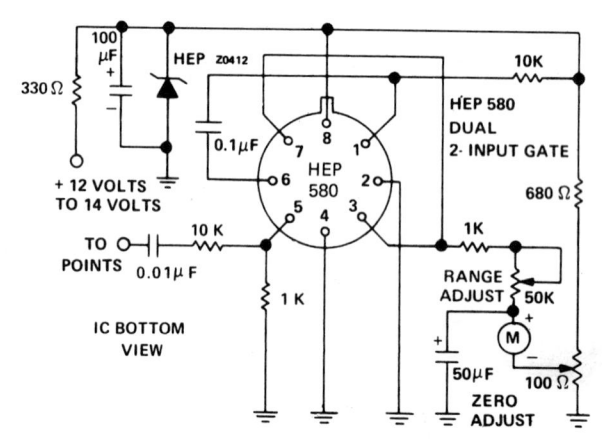

A precision tachometer can be used on automobiles, boats, motorcycles, etc. It features very flexible circuit arrangement, in that *range* and *number of cylinders* can be compensated for by one adjustment. To calibrate, connect lines to ground, 12V, and points with engine turned off but with ignition switch on. Zero meter with zero adjust. To set the range, check auto speed using a tachometer from a service station or garage. (Range maximum can be whatever you desire; calibrate the rest of the scale accordingly.) As an arbitrary adjustment, most automobiles idle at about 600 rpm. With engine idling, adjust the range pot until the needle is in approximately the 600 rpm position.

384 IC SCHEMATIC SOURCEMASTER

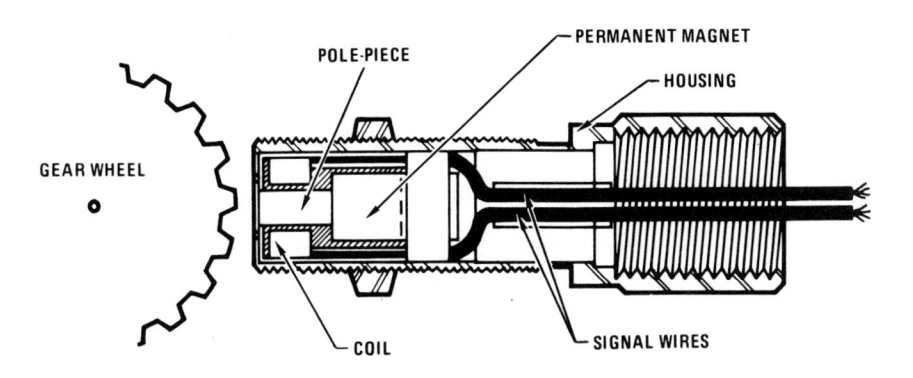

Typical Magnetic Pickup

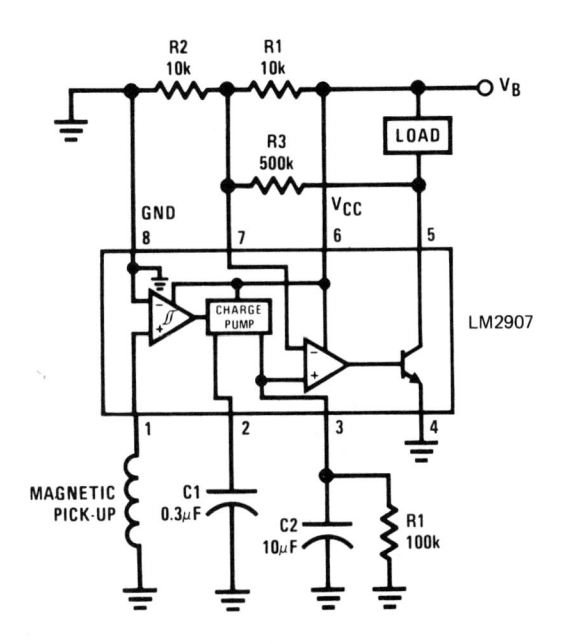

Speed switch. Many antipollution devices recently included on several makes of automobile include a speed switch to disable the vacuum advance function until a certain speed is attained. A circuit that will perform this kind of function is shown here. A typical magnetic pickup for automotive applications provides 1000 pulses per mile, so that at 60 mph the incoming frequency will be 16.6 Hz. If the reference level on the comparator is set by two equal resistors (R1 and R2), then the desired value of C1 and R1 can be determined from the simple relationship

$$\frac{V_{CC}}{2} = (V_{CC})(C1)(R1)(f)$$

or

$$C1 R1 f = 0.5$$

and hence

$$C1 R1 = 0.03$$

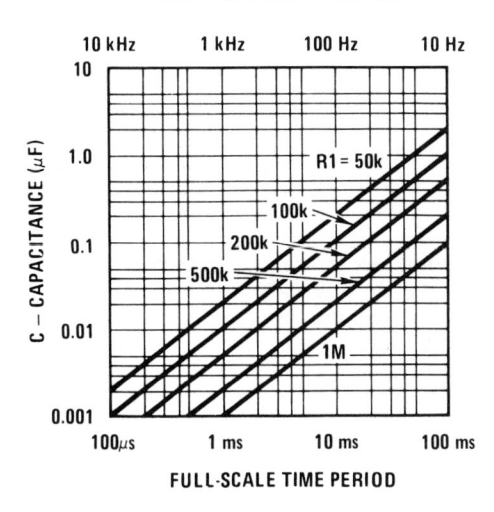

From the RC selection chart, we can choose suitable values for R1 and C1. Examples are 100k and 0.3 µF. The circuit will then switch at approximately 60 mph with the stated input frequency relationshp to speed. To prevent ripple from causing chattering of the load, a certain amount of hysteresis is added by including R3. This will provide typically 1% of supply as a hysteresis or 1.2 mph. Since the reference to the comparator is a function of supply voltage, as is the output from the charge pump, there is no need to regulate the power supply. The frequency at which switching occurs is independent of supply voltage.

SECTION **10** *Automotive Circuits* **385**

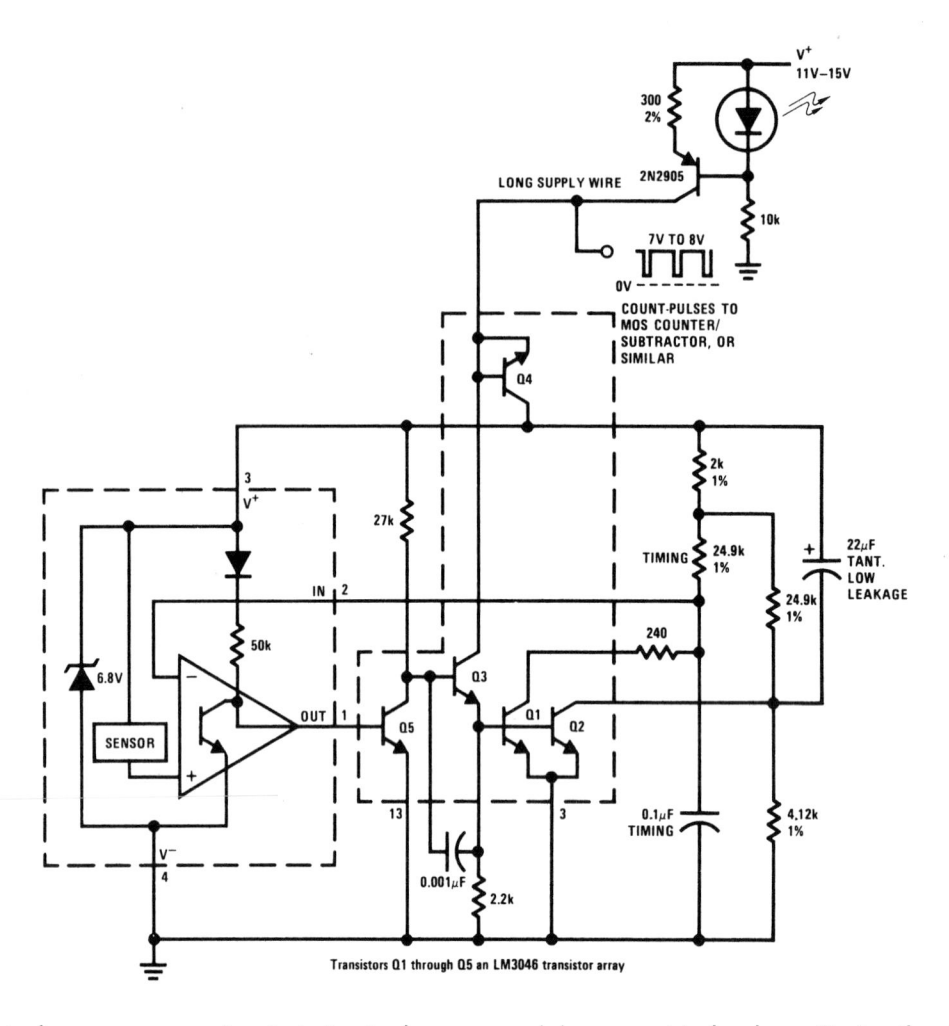

Temperature-to-frequency converter. In trying to determine the lowest-cost package for an automotive external-temperature indicator, a novel system evolved. A simple circuit in the same protective enclosure as the transducer minimizes cost and wiring. The single positive supply wire carries the output signal frequency, so it represents an irreducible minimum cost to hook up. During the short negative output pulses on the supply wire, an internal 22 μF capacitor keeps the converter electronics running. Pulse repetition rate is directly proportional to absolute temperature. An LED/transistor constant-current source provides an ideal supply from a varying voltage source.

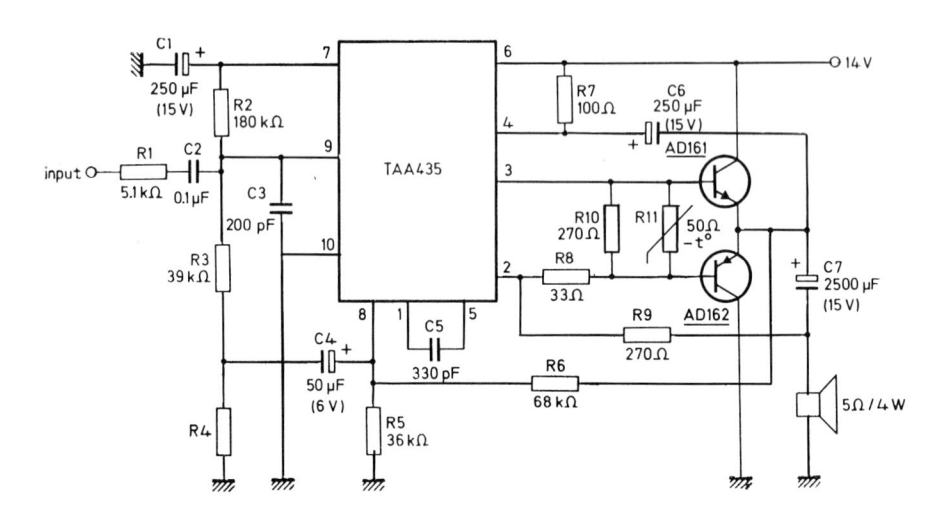

4W car radio amplifier.

386 IC SCHEMATIC SOURCEMASTER

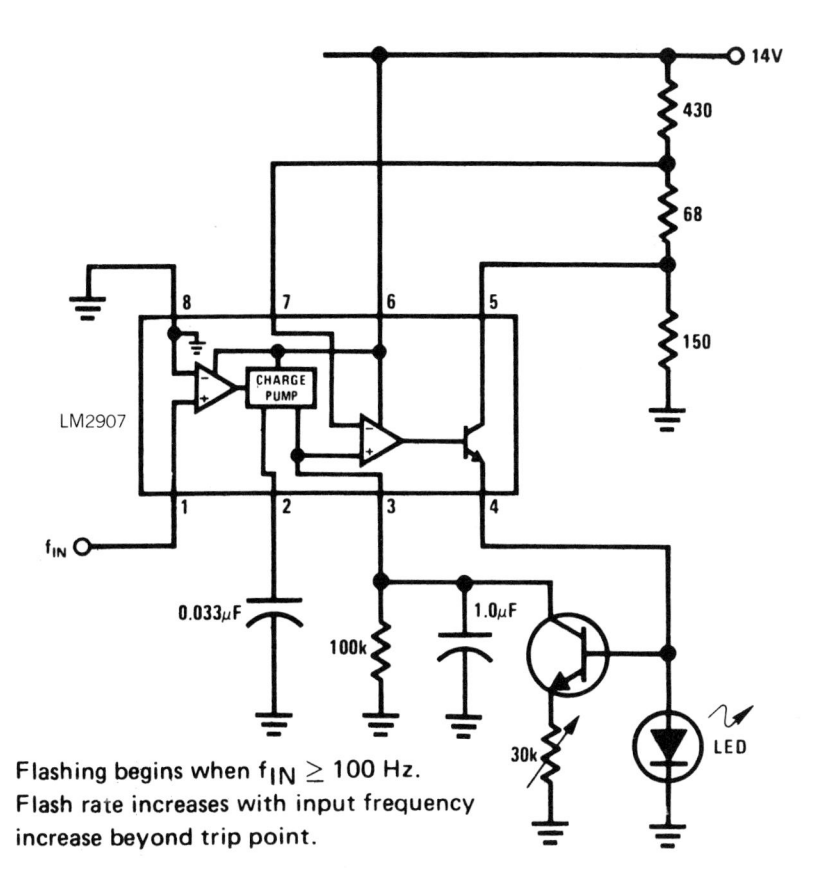

Flashing begins when $f_{IN} \geq 100$ Hz.
Flash rate increases with input frequency
increase beyond trip point.

Overspeed indicator using flashing LED.

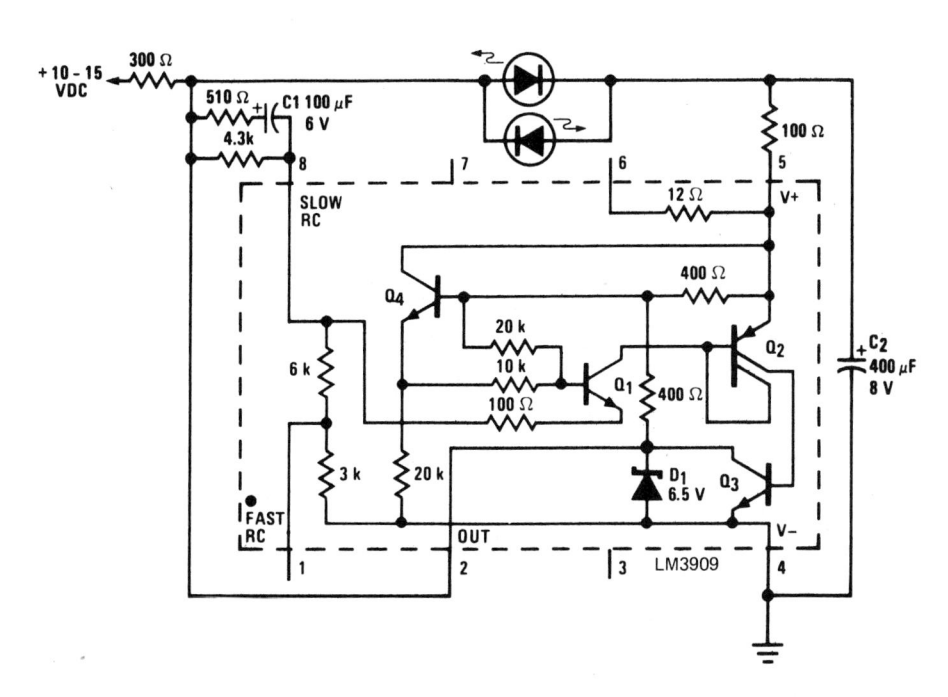

This alternating flasher uses a relaxation-type oscillator that flashes 2 LEDs sequentially. With a 12V supply, repetition rate is 2.5 Hz. C2, the timing and storage capacitor, alternately charges through the upper LED and is discharged through the other by the IC's power transistor, Q3. If a red/green flasher is desired, the green LED should have its anode or plus lead toward pin 5 (like the lower LED); a shorter but higher-voltage pulse is available in this position.

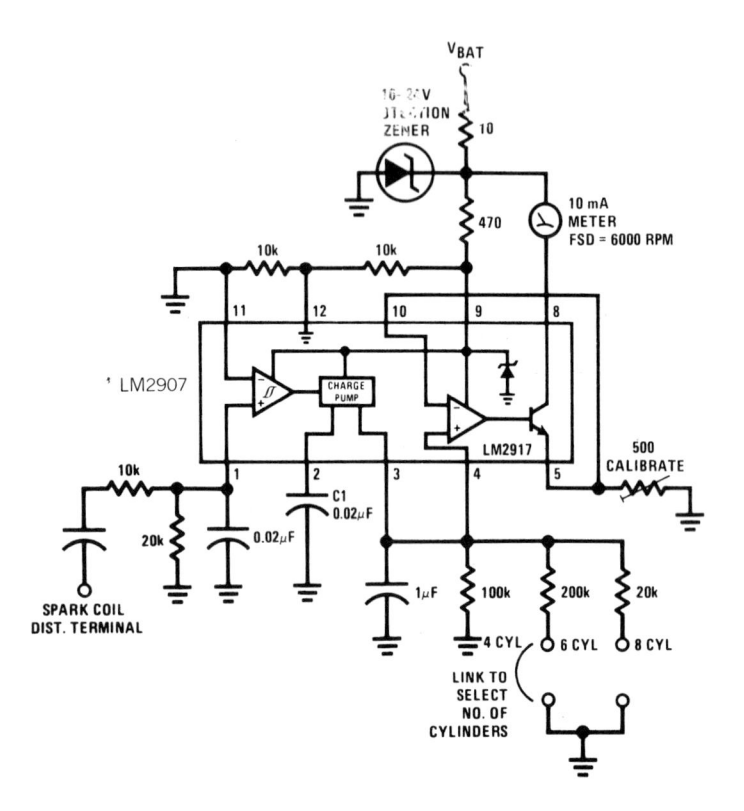

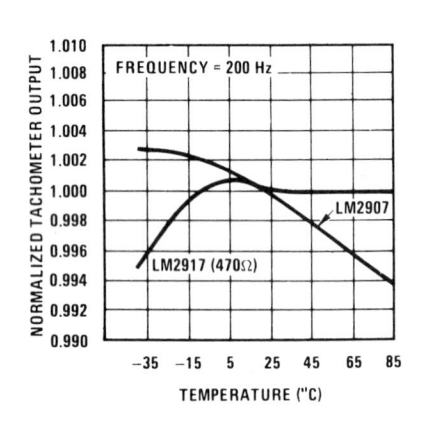

A gasoline engine tachometer can be set up for any number of cylinders by linking the appropriate timing resistor as illustrated. A 500Ω trim resistor can be used to set up final calibration. A protection circuit (10Ω resistor and zener) is precaution against transients found in automobiles.

TUNING MECHANISM: TRW

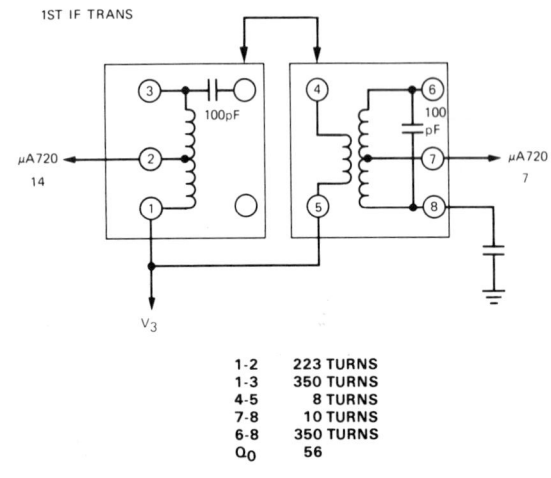

1ST IF TRANS

1-2	223 TURNS
1-3	350 TURNS
4-5	8 TURNS
7-8	10 TURNS
6-8	350 TURNS
Q_0	56

CAN & CORE MITSUMI 10 X 10 MM IF TRANS

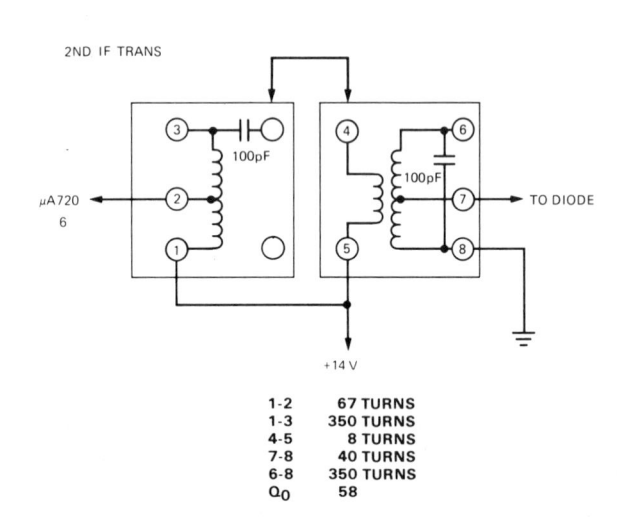

2ND IF TRANS

1-2	67 TURNS
1-3	350 TURNS
4-5	8 TURNS
7-8	40 TURNS
6-8	350 TURNS
Q_0	58

388 IC SCHEMATIC SOURCEMASTER

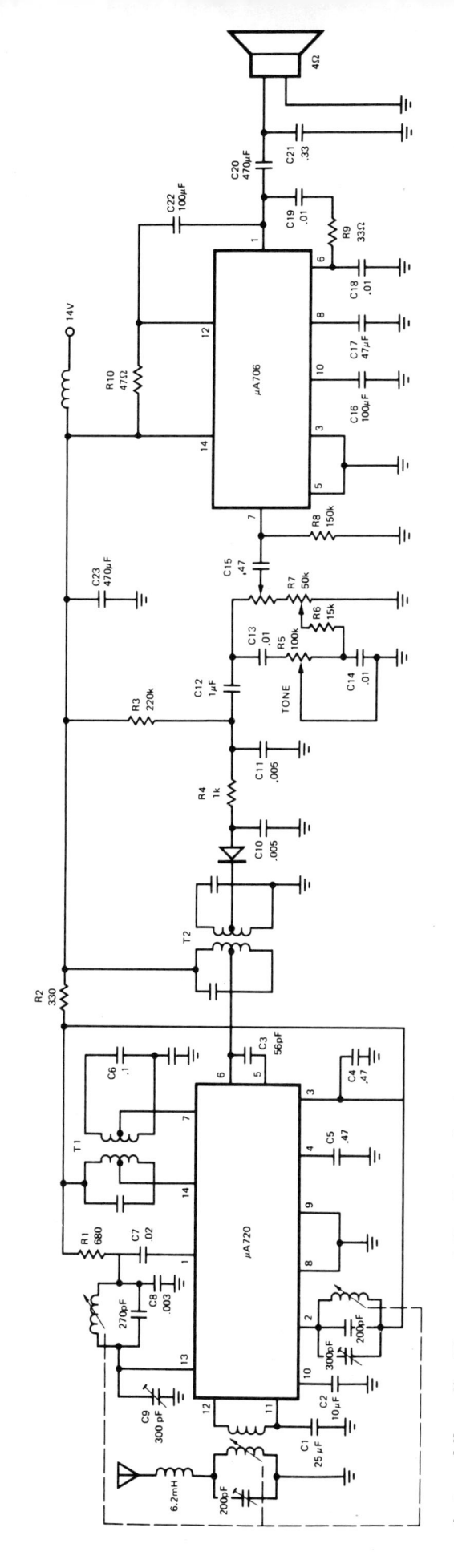

Automobile radio. The second IF coil, T2, should be selected to provide maximum power transfer. Selectivity should be provided by the first IF coil, T1. The audio amplifier and power output stages of this design are provided by the µA706. This device is particularly suited to low-supply-voltage operation and delivers 5W into a 4Ω load from a supply voltage of 14V.

SECTION **10** *Automotive Circuits* 389

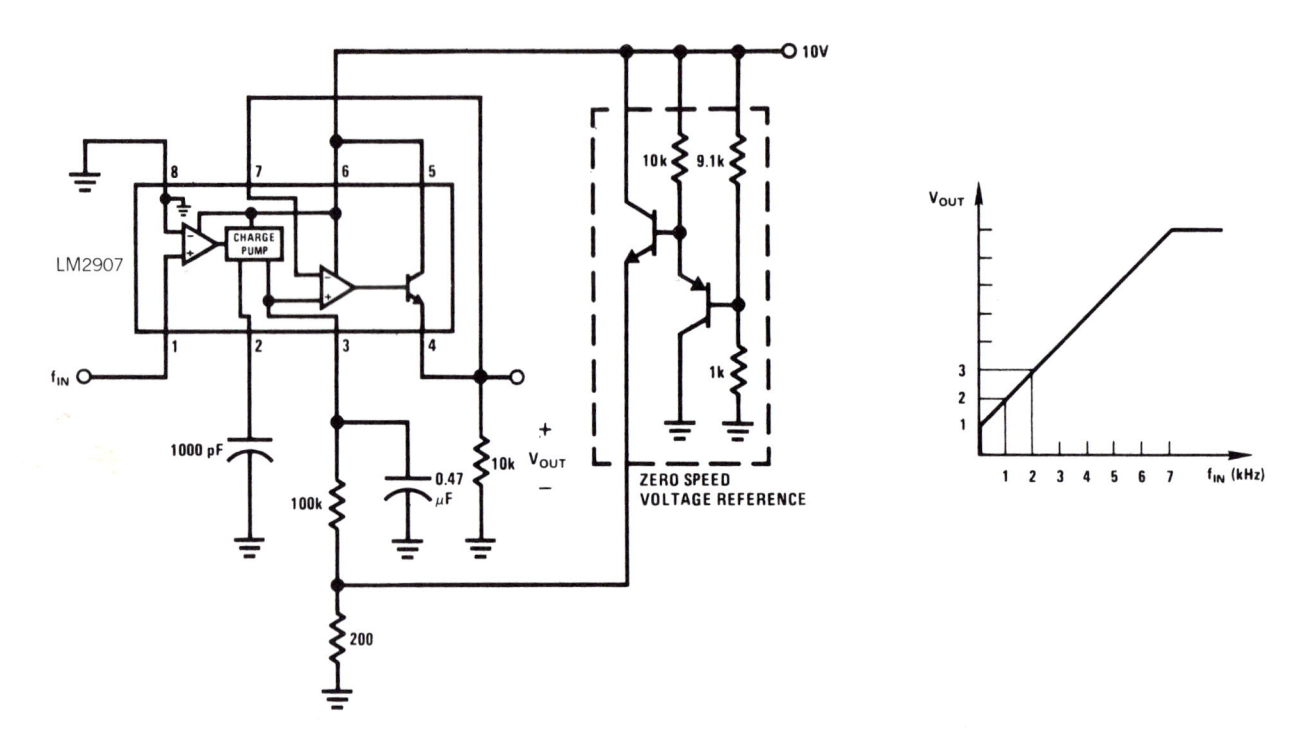

Changing the output voltage for an input frequency of zero.

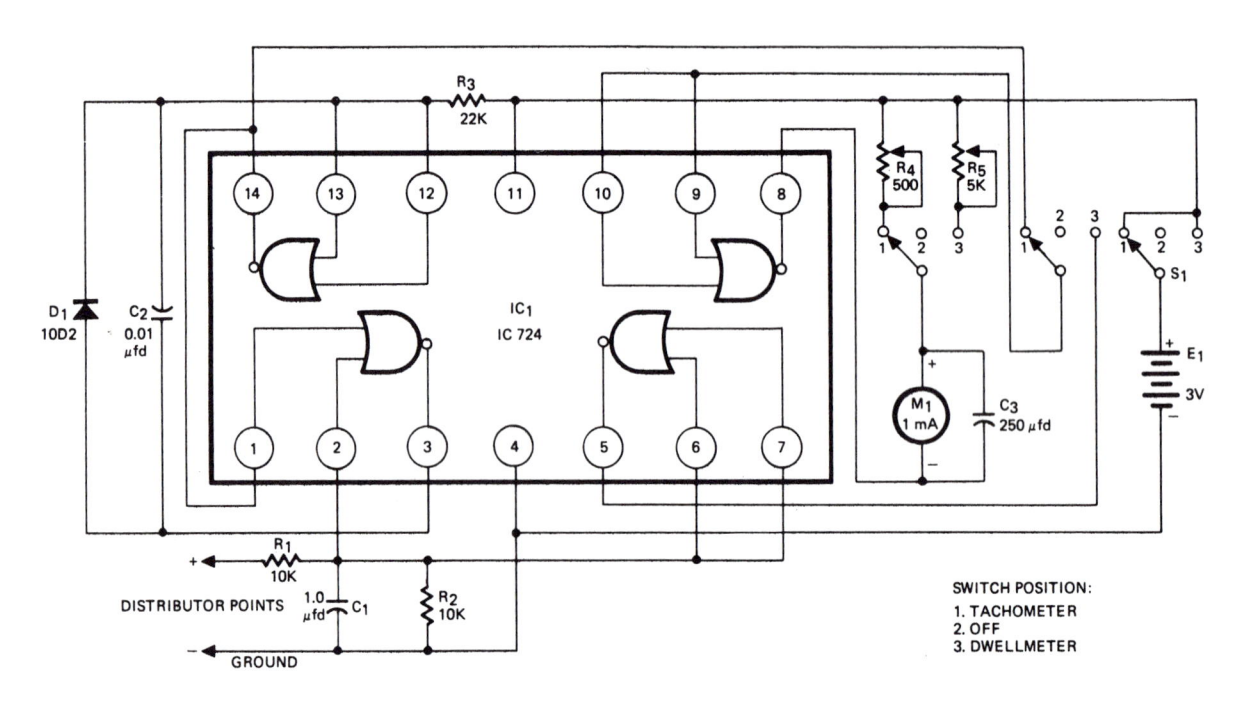

A tachometer and dwellmeter for 4-, 6-, and 8-cylinder automobiles. Circuit converts pulses from distributor points to steady voltage across meter. In *tach* mode, the logic functions at pins 14 and 3 of the International Rectifier IC (Workman Electronics) are serving as a one-shot. For *dwell* mode, pulse amplitude is derived from the dwell angle of the distributor cam; the voltage indicated on the meter is the result of pulse width (duration). (All semiconductors shown are IRC/Workman.)

390 IC SCHEMATIC SOURCEMASTER

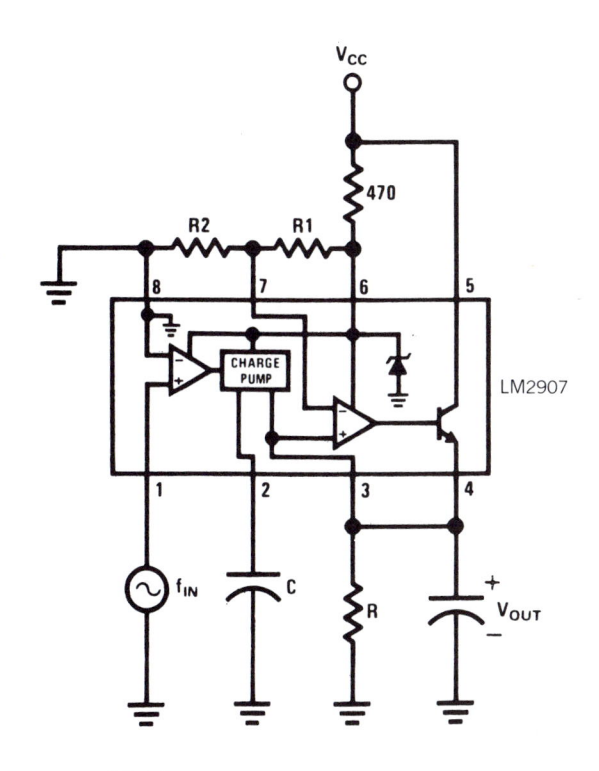

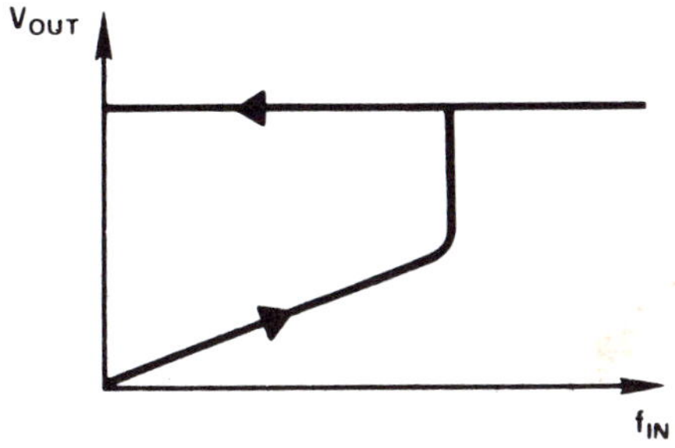

Overspeed latch.

Output latches when
$$f_{IN} = \frac{R2}{R1 + R2} \cdot \frac{1}{RC}$$
Reset by removing V_{CC}.

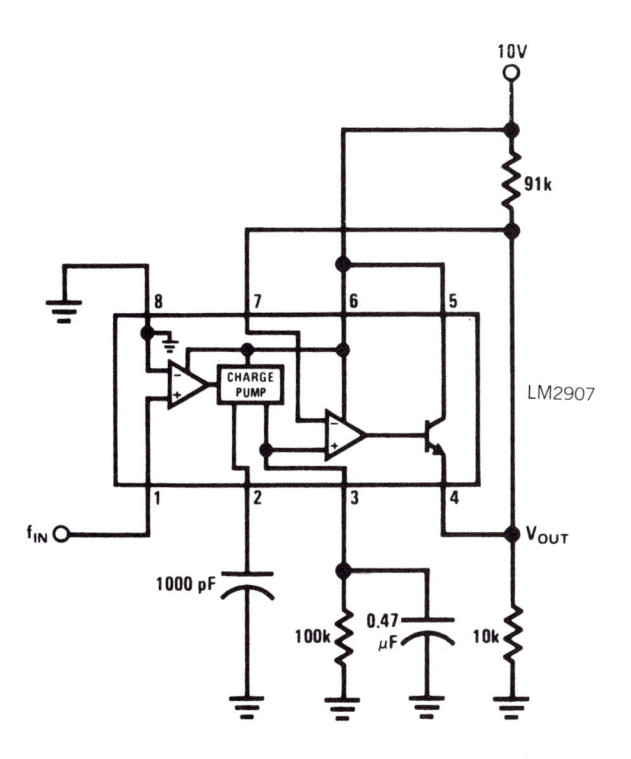

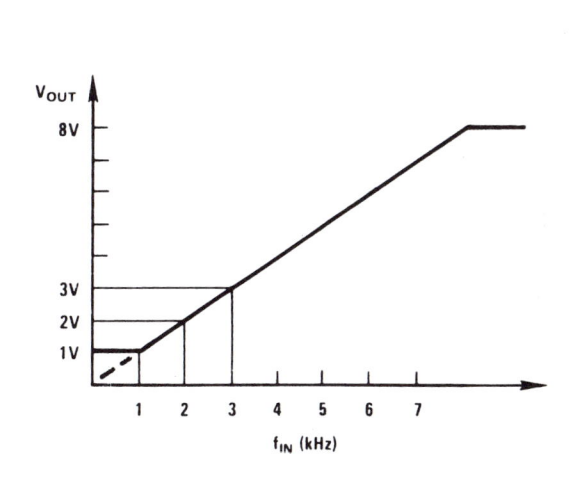

Changing tachometer gain curve or clamping the minimum output voltage.

SECTION **10** *Automotive Circuits* **391**

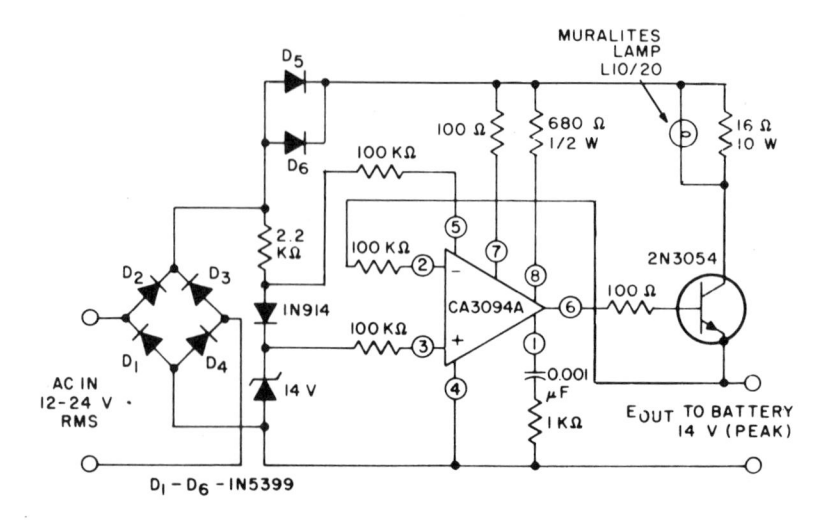

A battery charger regulator circuit accurately limits the peak output voltage to 14V, as established by the zener diode connected across terminals 3 and 4. When the output voltage rises slightly above 14V, signal feedback through a 100k resistor to terminal 2 reduces the current drive supplied to the 2N3054 pass transistor from terminal 6 of the CA3094. An incandescent lamp serves as the indicator of charging current flow. Adequate limiting provisions protect the circuit against damage under load-short conditions. The advantage of this circuit over certain other types of regulator circuits is that the reference voltage supply does not drain the battery when the power supply is disconnected. This feature is important in portable service applications, such as in a trailer where a battery is kept on charge when the trailer is parked, and power is provided from an ac line.

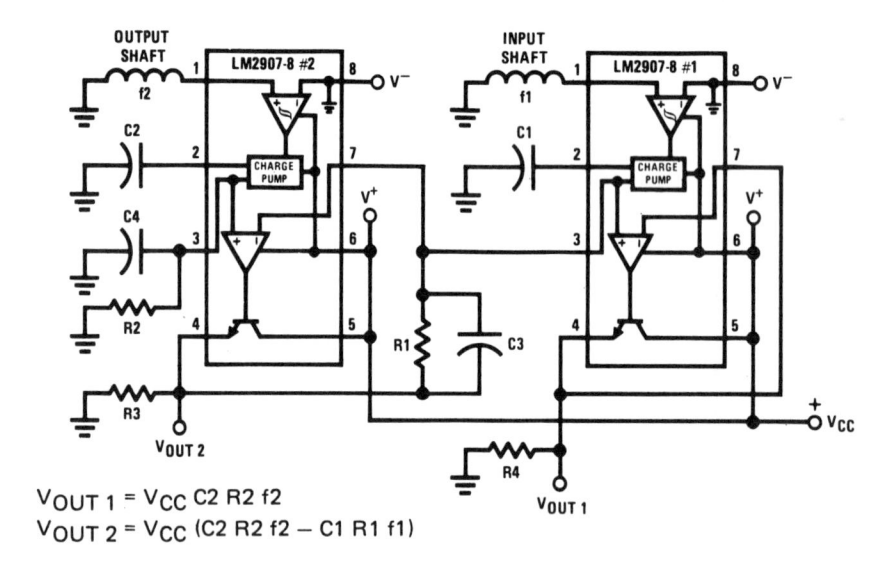

$V_{OUT1} = V_{CC} \, C2 \, R2 \, f2$

$V_{OUT2} = V_{CC} \, (C2 \, R2 \, f2 - C1 \, R1 \, f1)$

Transmission and clutch control functions. Electric clutches can be added to automotive transmissions to eliminate the 6% slip that typically occurs during cruise and that results in a 6% loss in fuel economy. These devices could be operated by a pair of LM2907s as illustrated. Magnetic pickups are connected to input and output shafts of the transmission and provide frequency inputs f_1 and f_2 to the circuit. Frequency f_2, being the output shaft speed, is also a measure of vehicle road speed. Thus the LM2907-8 No. 2 provides a voltage proportional to road speed at pin 3. This is buffered by the op-amp in LM2907-8 No. 1 to provide a speed output V_{OUT1} on pin 4. The input shaft provides charge pulses at the rate of $2f_1$ into the inverting node of op-amp 2. This node has the integrating network R2–C3 going back to the output of the op-amp so that the charge pulses are integrated and provide an inverted output voltage proportional to the input speed. The output V_{OUT2} is proportional to the difference between the two input frequencies. With these two signals—the road speed and the difference between road speed and input shaft speed—it is possible to develop a number of control functions including the electronic clutch and a complete electronic transmission control. In the configuration shown, it is not possible for V_{OUT2} to go below zero, so that there is a limitation to the output swing in this direction. This may be overcome by returning R3 to a negative bias supply instead of to ground.

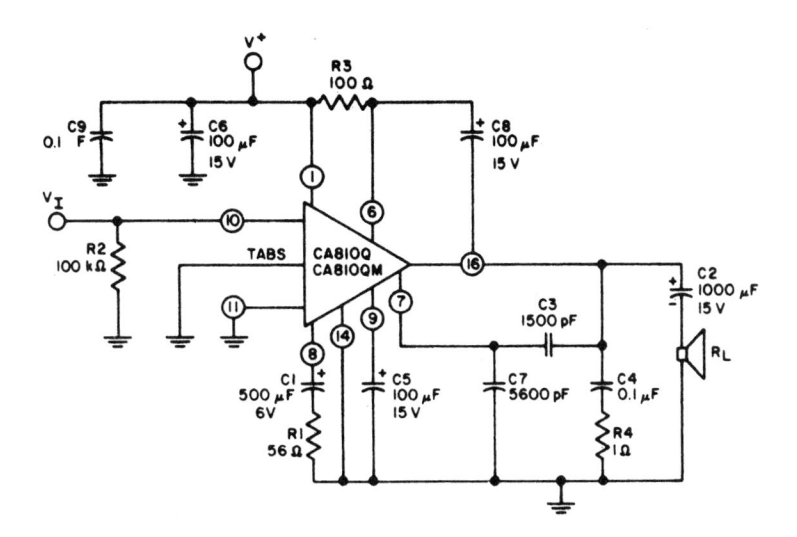

This 7W hi-fi audio power amplifier with thermal shutdown is specifically designed for mobile use. The maximum repetitive peak output current is 2.5A and an integral thermal-limiting circuit shuts the device down in case of output overload or excessive package temperature. The integral IC's supply voltage range is 4 to 10V. The package features very low harmonic and crossover distortion. Output is 4Ω.

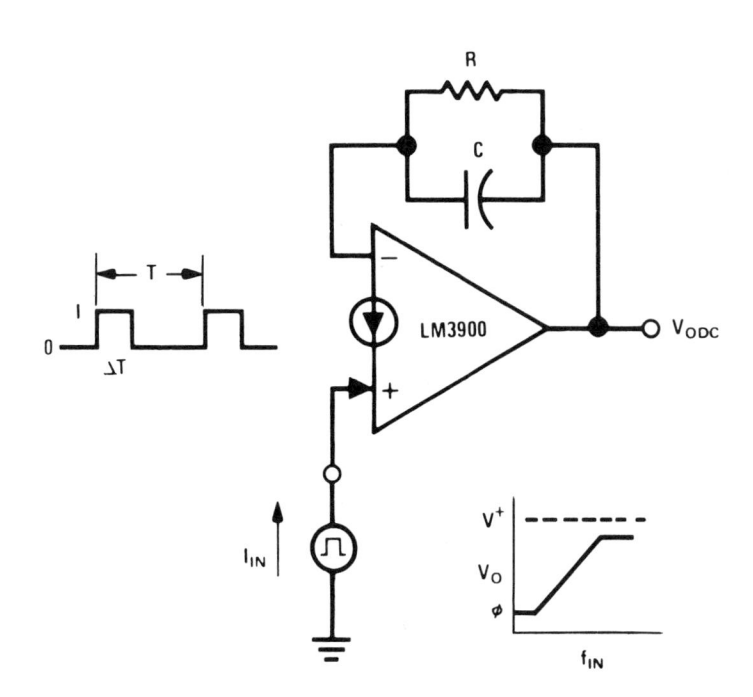

Basic tachometer. Current pulse inputs will provide the desired transfer function shown. Each input current pulse causes a small change in the output voltage. Neglecting the effects of R, we have $\Delta V_{O4} \cong I\Delta t/C$. The inclusion of R gives a discharge path, so the output voltage does not continue to integrate but rather provides the time dependency that is necessary to average the input pulses. If an additional signal source is placed in parallel with the one shown, the output becomes proportional to the sum of these input frequencies. If this additional source were applied to the (−) input, the output voltage would be proportional to the difference between these input frequencies. Voltage pulses can be converted to current pulses by using an input resistor. A series-isolating diode should be used if a signal is applied to the (−) input to prevent loading during the low-voltage state of this input signal.

SECTION **10** *Automotive Circuits* **393**

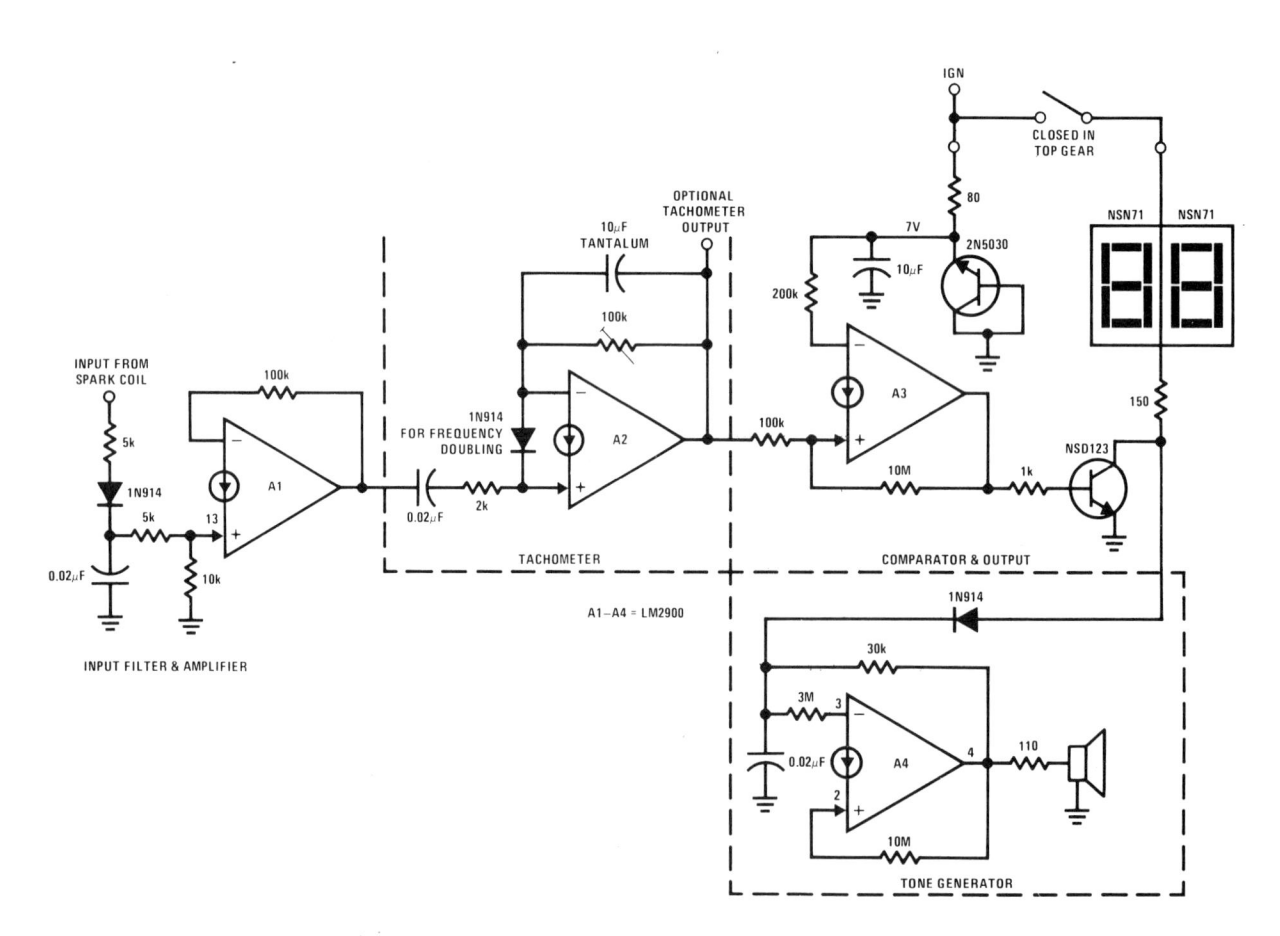

High-speed warning device for automobiles. This circuit uses the engine speed signal available at the primary of the spark coil and a switch in the transmission that is closed only in high gear. Lowest cost display would be a light-emitting diode driven directly from the integrated circuit. Add a Durawatt transistor, and an incandescent lamp can be driven. Two NSN71 1/3-inch high, 7-segment numeric displays can be hardwired to display the speed at which the limit is set. For best effect, the visual warning should be accompanied by an audible alarm. The circuit employs an LM2900 quad Norton op-amp to perform all functions. Specifically, A1 amplifies and regulates the signal from the spark coil. A2 converts frequency to voltage so that its output is a voltage proportional to engine speed. This signal could be used to directly drive a tachometer if desired. A3 compares the tachometer voltage with the reference voltage and turns on the output transistor at the set speed. A small amount (2%) of positive feedback is provided to prevent annoying intermittent operation. Amplifier A4 is used to generate an audible tone whenever the set speed is exceeded. R1 is adjusted according to the gear and axle ratios, number of cylinders, wheel and tire size, and so forth. Note that the 2900 is capable of directly driving the loudspeaker.

In operation, the circuit is powered so that the tachometer drive is always available. When the transmission moves to top gear, switch S1 closes and connects the output light and speaker displays to the power source. When the vehicle speed exceeds the set value (56 mph U.S. or 82 km/hr South American) the light and tone will be energized. To extinguish these warnings, the driver will have to slow the vehicle to below the value set by the hysteresis (say 55 mph or 80 km/hr). The integrating 10 μF capacitor on the frequency-to-voltage converter, A2, could be increased so that the alarm is not sounded during momentary excursions above the set speed.

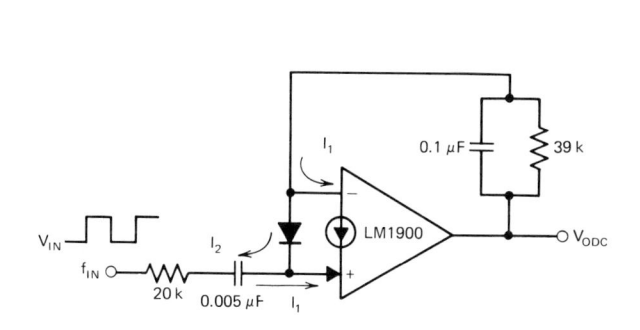

Frequency-doubling tachometer. $V^+ = 15V$.

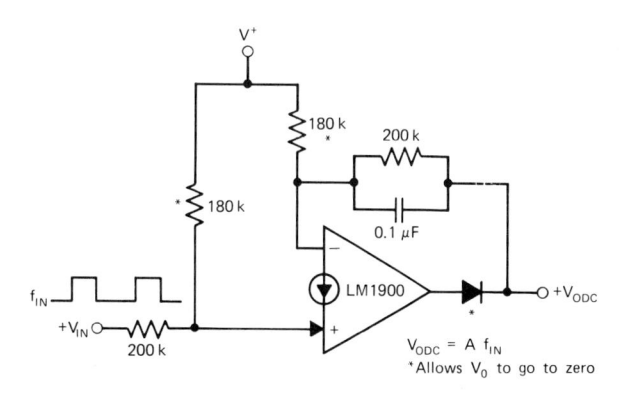

A frequency-doubling tachometer provides a simple circuit that reduces the ripple voltage on a tachometer output dc voltage.

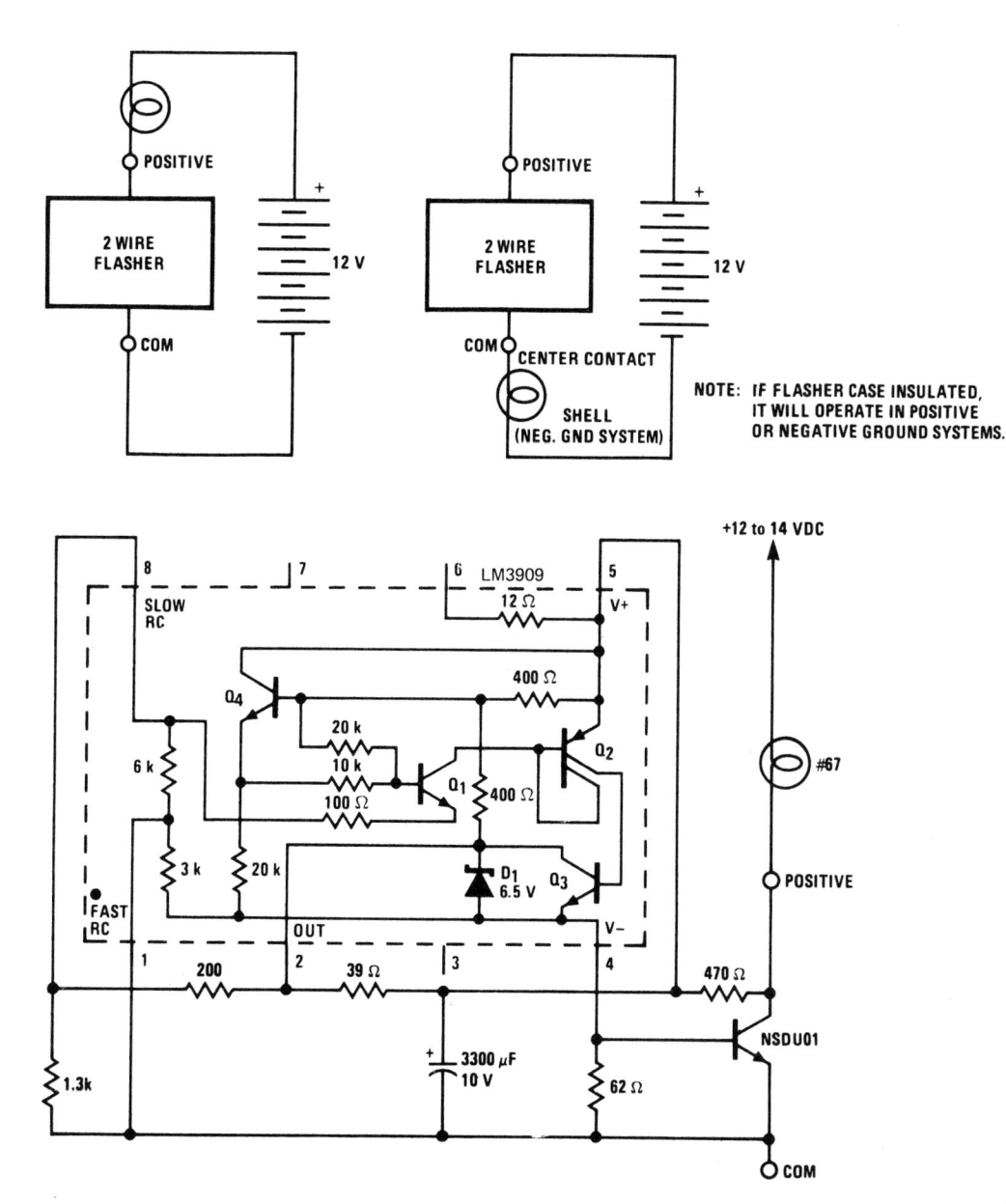

NOTE: IF FLASHER CASE INSULATED, IT WILL OPERATE IN POSITIVE OR NEGATIVE GROUND SYSTEMS.

A 12V flasher uses an automotive storage battery for power. It provides a 1 Hz flash rate and powers a lamp drawing 600 mA. Circuit has only two external wires and thus may be hooked up in either of the two ways shown in the schematic. Circuit failure cannot cause a battery drain greater than that of the bulb itself, continuously lit. The 3300 μF capacitor makes the LM3909 immune to supply spikes and provides the means of limiting the IC's supply voltage.

SECTION **10** *Automotive Circuits* **395**

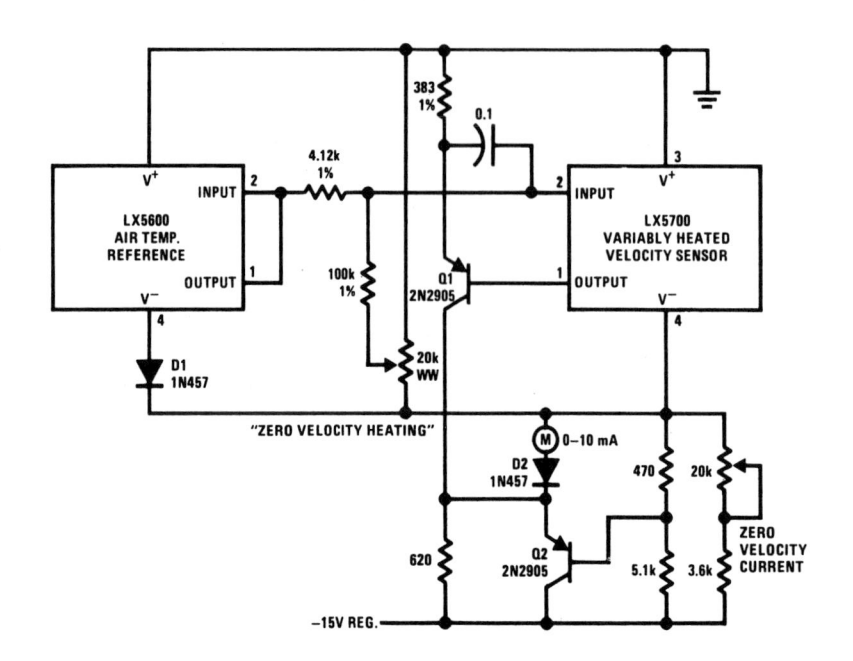

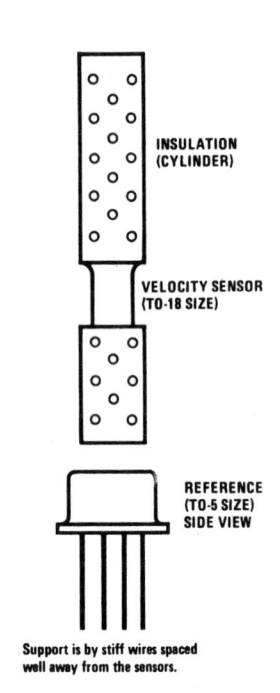

Wind velocity indicator for automobiles. Actual air speed has more effect on gasoline mileage than road speed at above 40 mph. A differential between a heated and unheated transducer is used in this sensing job.

The set point input (pin 2) of the velocity-sensing transducer is biased so that its voltage is always equivalent to a number of degrees above ambient. This voltage is obtained by adding a small dc level to the output of an ambient sensing transducer. The output of the velocity transducing LX5700 is used to control its own heating, holding it the specified number of degrees above ambient. The current to do this is a measure of the amount of heating required. As wind velocity rises, so does this current, as the transducer serves its temperature to a constant amount above ambient. Here, this amount of heating is set by the wiper voltage of the 20k pot into the 100k/4.12k divider. The bottom of this divider is driven by the LX5600 air temperature reference, connected in unity-gain mode. The output drives the base of Q1. If this sensor cools, Q1 base goes positive, reducing its emitter current and collector current. The current drawn by the 620Ω collector load does not change significantly, however. The current now **not** supplied by Q1 is drawn instead through D2, through the meter, and eventually through the reference zener of the velocity sensor. This heats the sensor, closing the feedback loop. The system is made electrically stable by the 0.1 μF compensating capacitor and the fact that Q1 exhibits practically no collector voltage gain. Q2 is not normally conducting, but it and the divider on its base provide a clamp voltage for the Q1 collector. This prevents a latchup mode that occurs if Q1 happens to draw a transient current pulse. The "zero velocity current" adjustment allows this heating current needed at zero velocity to bypass the meter. Zero the meter at about 1% of full scale.

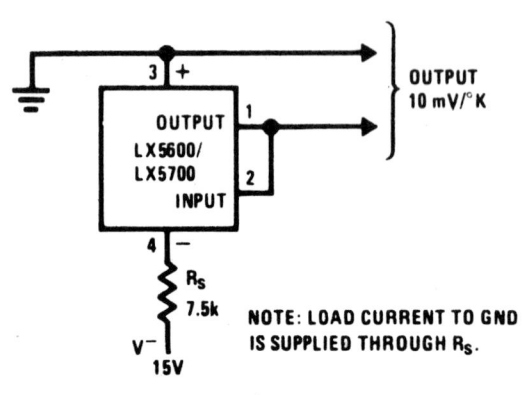

$R_S = (V^- - 6.8V) \times 10^3 \Omega$

$R_S = (V^+ - 6.8V) \times 10^3 \Omega$

Basic thermometer for negative supply.

Basic thermometer for positive supply.

396 IC SCHEMATIC SOURCEMASTER

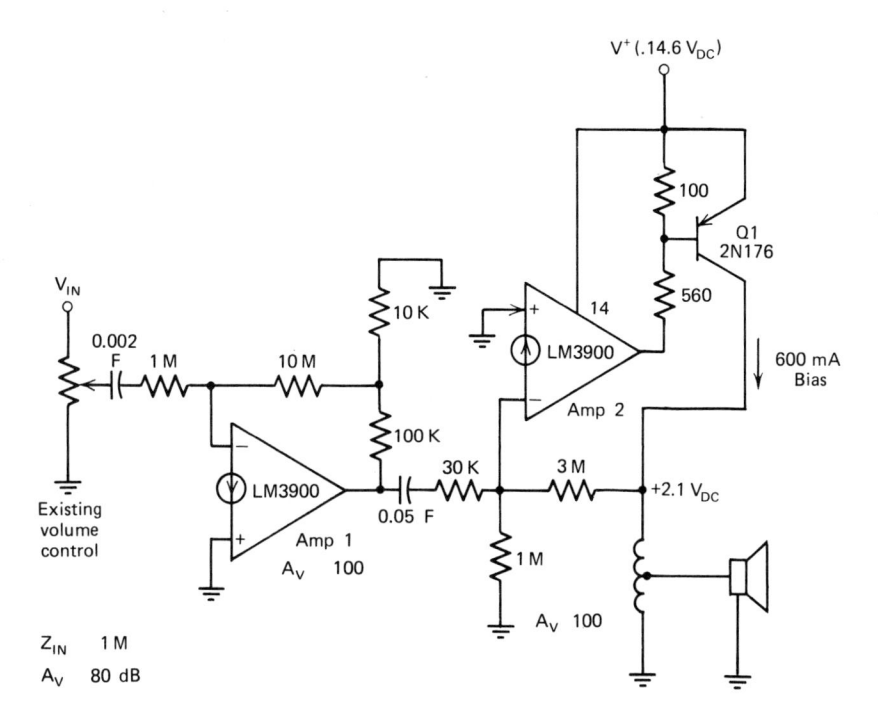

Dual-channel class A driver for auto radios. A germanium power transistor is widely used in automotive class A audio amplifiers. As shown, two amplifiers can be cascaded to bias and to control the 2N176 power transistor. Advantages over the standard discrete circuit:

No electrolytic capacitors are used.
The ripple on the power supply line is rejected.

The input impedance is high (1M).
Large closed-loop gain is achieved (80 dB).
Slow startup delay is eliminated.
Two channels are available in one package.

Pin 14 voltage must be at least as high as the power supply used at the emitter of Q1 to guarantee an *off* control for Q1.

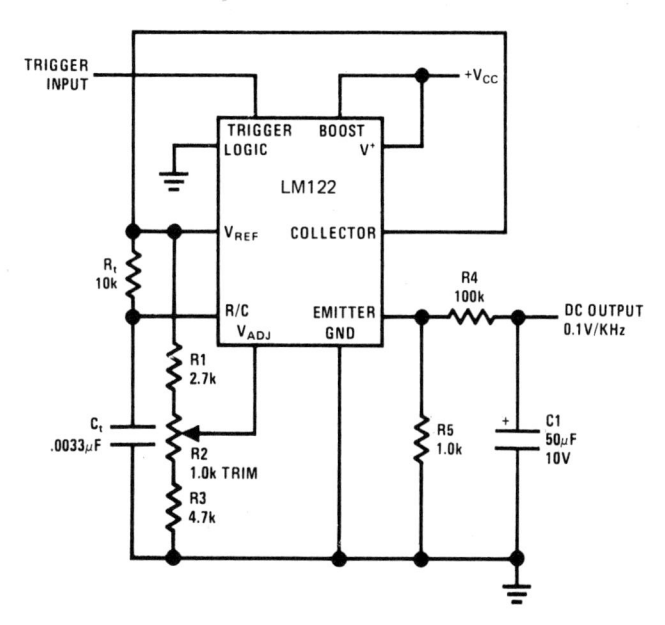

An accurate frequency-to-voltage (tach) converter can be made with the LM122 by averaging output pulses with a simple one-pole filter as shown here. Pulse width is adjusted with R2 to provide initial calibration at 10 kHz. The collector of the output transistor is tied to V_{REF}, giving constant-amplitude pulses equal to V_{REF} at the emitter output. R4 and C1 filter the pulses to give a dc output equal to $(R_t)(C_t)(V_{REF})(f)$. Linearity is about 0.2% for a 0 to 1V output. If better linearity is desired, R5 can be tied to the summing node of an op-amp that has the filter in the feedback path. If a low output impedance is desired, a unity-gain buffer such as the LM110 can be tied to the output. An analog meter can be driven directly by placing it in series with R5 to ground. A series RC network across the meter to provide damping will improve response at very low frequencies.

SECTION **10** Automotive Circuits

SECTION 11
ACTIVE FILTERS

This section presents filter circuits of every kind—bandpass, bandstop, high-pass, and low-pass. As a general rule, the frequency of operation of a filter—whether it is for a radio or an audio application—depends only on the value of the components used in the RC networks; consequently, many of the circuits featured here include formulas to allow construction to your own requirements. The circuit variations depict filters with both steep and gently sloping skirts and notches of varying depth.

The audio constructionist should also consult Section 6, which contains several types of commonly used equalization networks and tone controls, both active and passive in design. Additional radio-frequency filter circuits are incorporated in many of the schematics of Section 7.

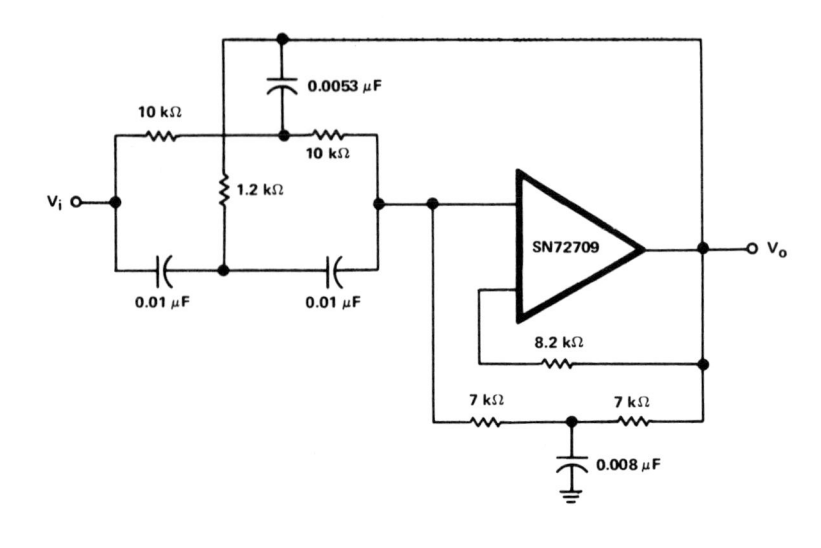

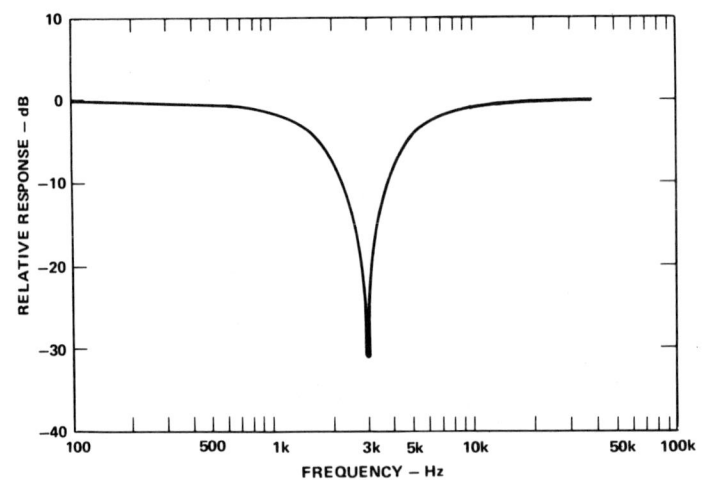

An active bandstop or notch filter uses an SN72709 in a unity-gain configuration. This is a second-order band-reject filter with a notch frequency of 3 kHz. The resulting Q of this filter is about 23, with a notch depth of 31 dB. Although three passive tee networks are used in this application, the operational amplifier has permitted obtaining a sharply tuned low-frequency filter without the use of inductors or large-value capacitors.

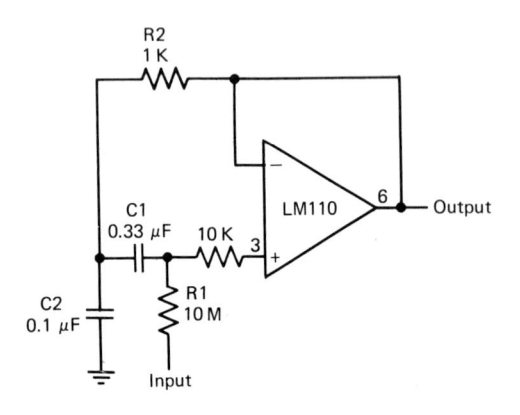

Bandpass filter.

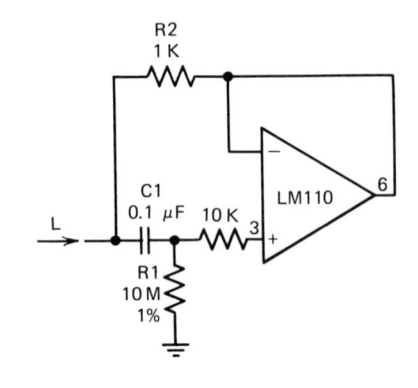

Simulated inductor. $L \cong R1R2C1$; $R_S = R2$; $R_P = R1$.

400 IC SCHEMATIC SOURCEMASTER

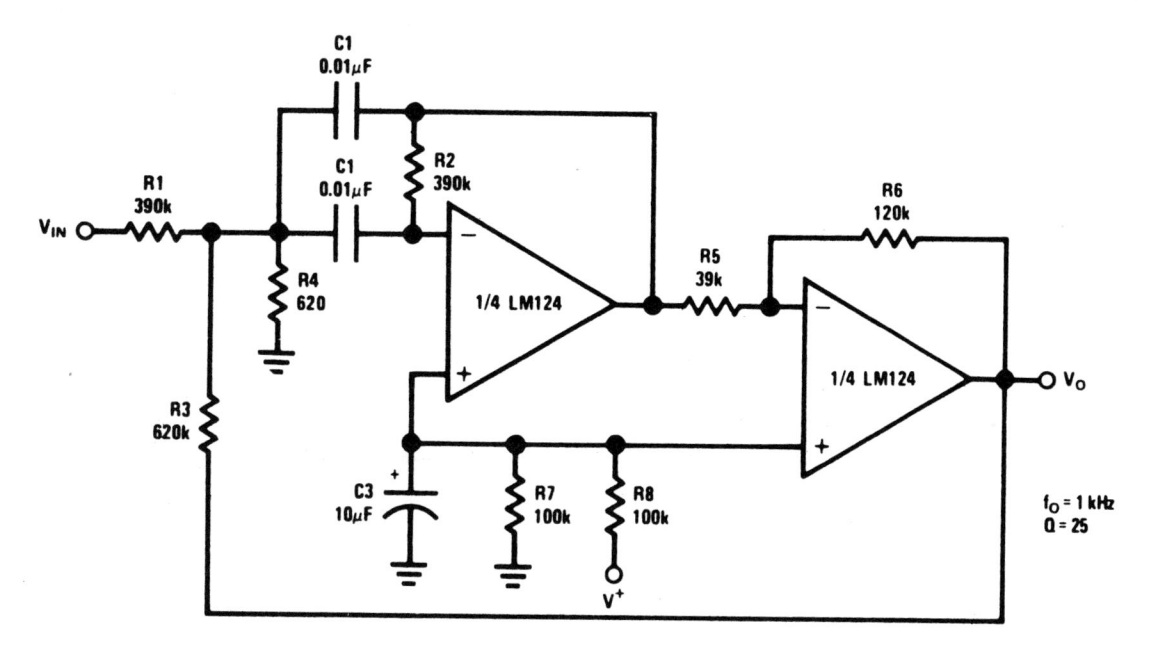

Bandpass active filter. Q = 25; f_0 = 1 kHz.

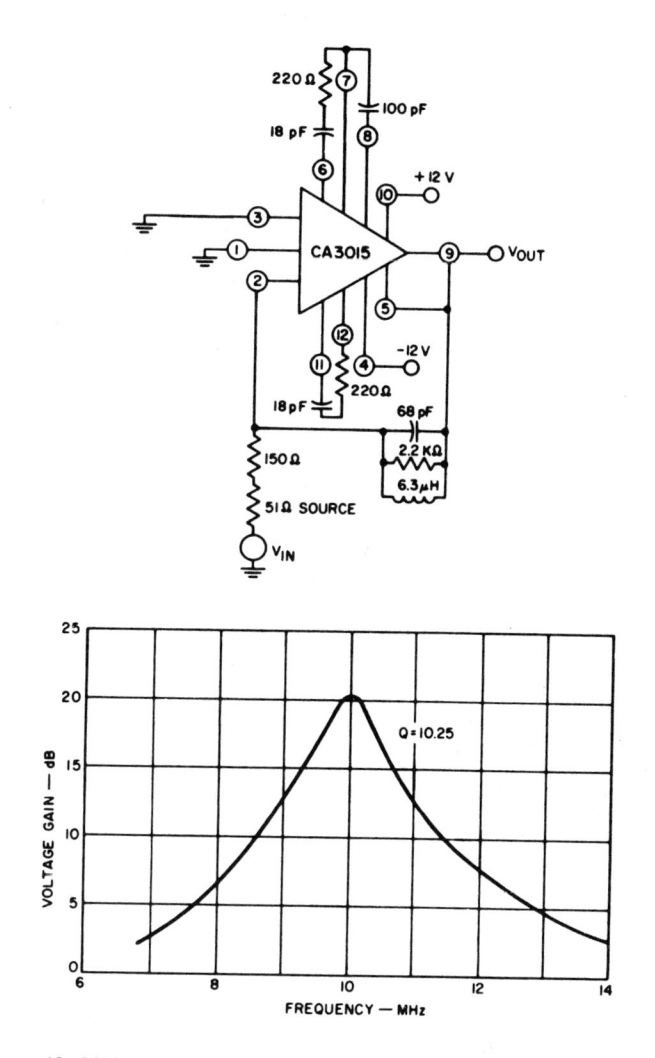

10 MHz RLC bandpass amplifier. Inset shows response curve.

SECTION **11** *Active Filters* **401**

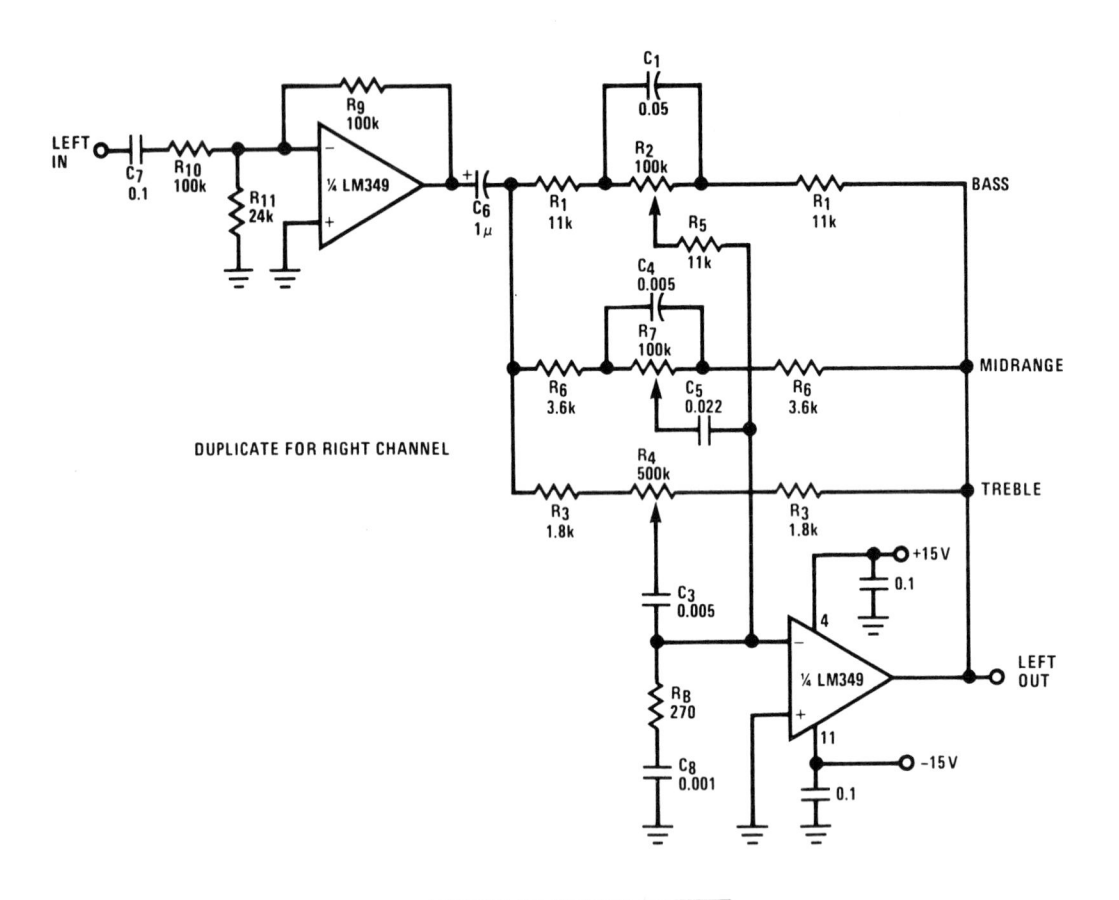

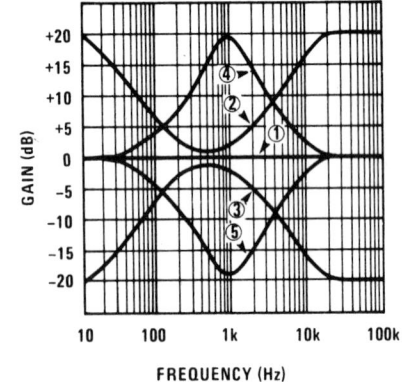

① ALL CONTROLS FLAT
② BASS & TREBLE BOOST, MID FLAT
③ BASS & TREBLE CUT, MID FLAT
④ MID BOOST, BASS & TREBLE FLAT
⑤ MID CUT, BASS & TREBLE FLAT

Three-band active tone control (bass, midrange, and treble). To increase (or decrease) midrange gain, decrease (increase) R6. This will also shift the midrange center frequency higher (lower). (This change has minimal effect upon bass and treble controls.) To move the midrange center frequency (while preserving gain and with negligible change in bass and treble performance), change both C4 and C5. Maintain the relationship that C4 ≈ 5C4. Increasing (decreasing) C5 will decrease (increase) the center frequency. The amount of shift is approximately equal to the inverse ratio of the new capacitor to the old one. For example, if the original capacitor is C5 and the original center frequency is f_0, and the new capacitor is C5' with the new frequency being f_0', then

$$\frac{C5'}{C5} \approx \frac{f_0}{f_0'}$$

402 IC SCHEMATIC SOURCEMASTER

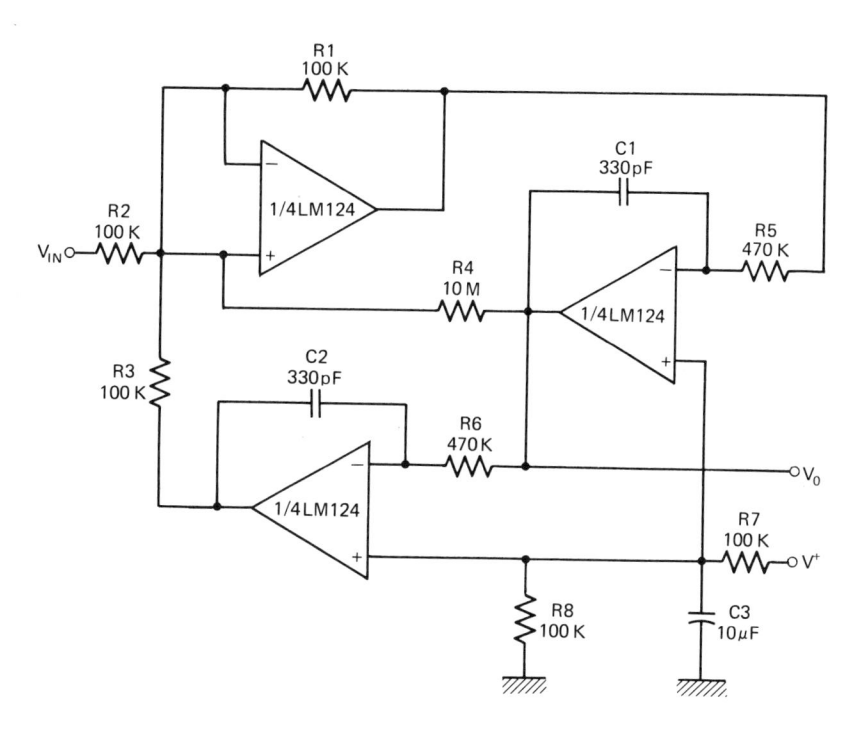

Biquad RC active bandpass filter. Q = 50; f_0 = 1 kHz; A_V = 100 (40 dB).

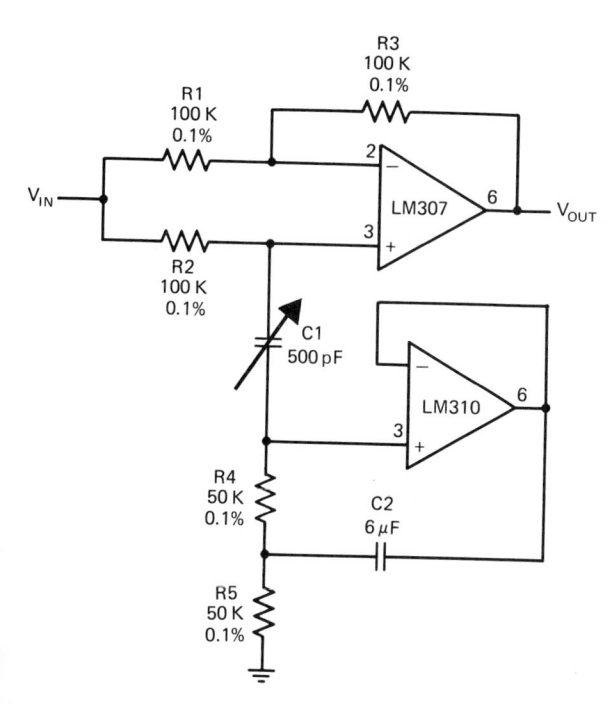

Tunable notch filter. R1 = R2 = R3, R4 = R5 = R1/2;

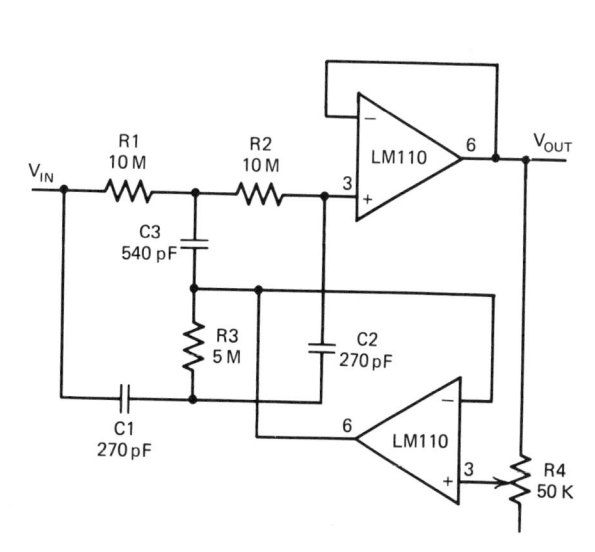

$$f_o = \frac{1}{2\pi\sqrt{C1C2(R2)^2}} = 60 \text{ Hz}.$$

Adjustable-Q notch filter. R1 = R2 = 2R3, C1 = C2 = C3;

$$f_o = \frac{1}{2\pi R1C1} = 60 \text{ Hz}.$$

SECTION 11 Active Filters

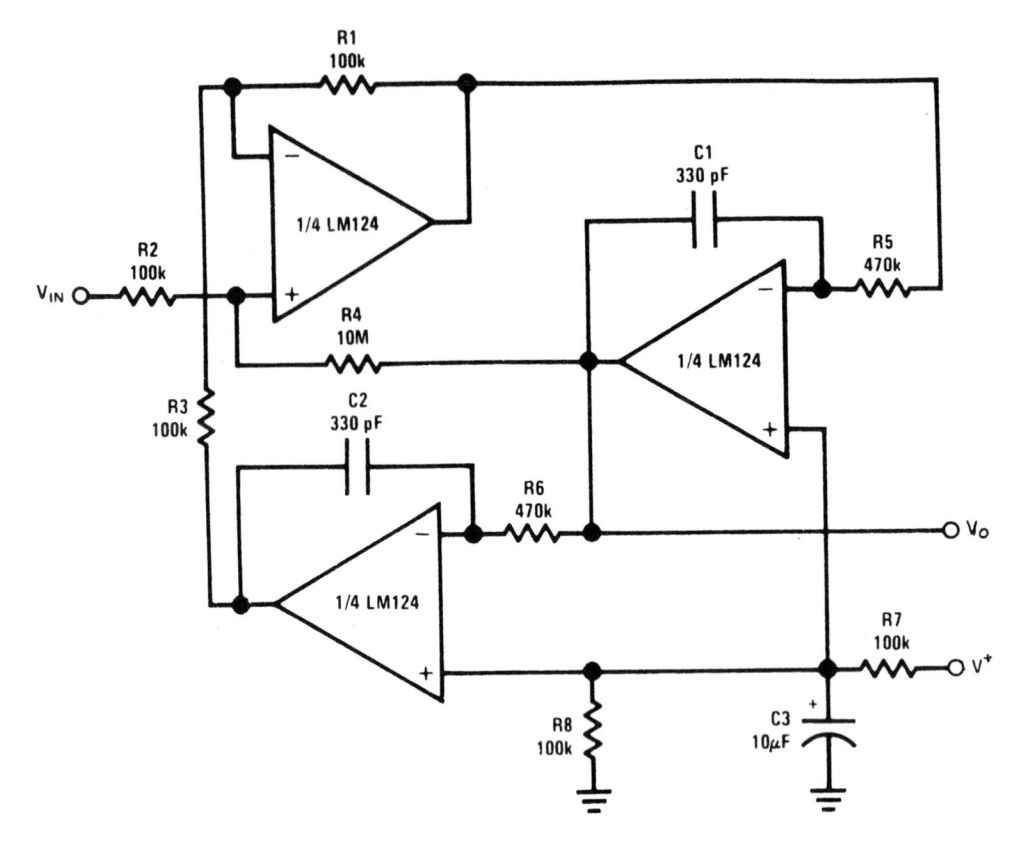

RC active bandpass filter. $f_0 = 1$ kHz, $Q = 50$, $A_V = 100$ (40 dB).

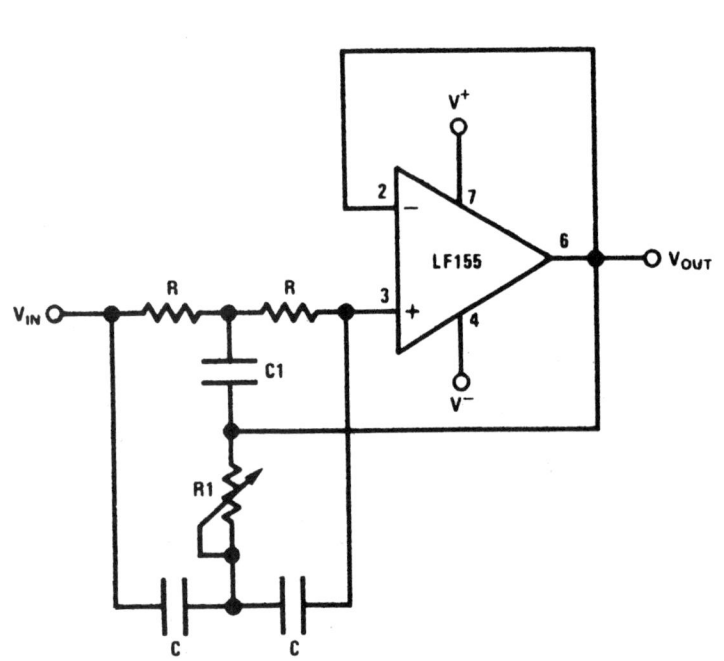

High-Q notch filter. $2R1 = R = 10M$; $2C = C1 = 300$ pF. Capacitors should be matched to obtain high Q. $f_{NOTCH} = 120$ Hz; notch = -55 dB; Q is greater than 100.

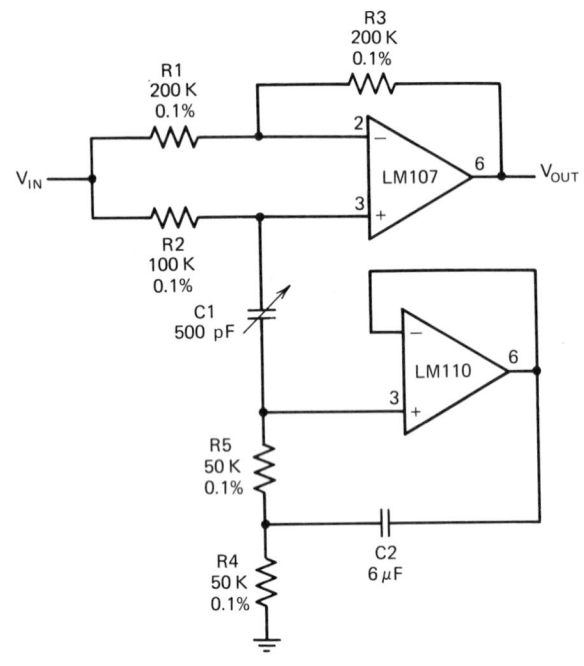

Tunable notch filter. $R1 = R3$, $R4 = R5 = R2/2$, and

$$f_0 = \frac{1}{2\pi R5(C1C2)} = 60 \text{ Hz}.$$

404 IC SCHEMATIC SOURCEMASTER

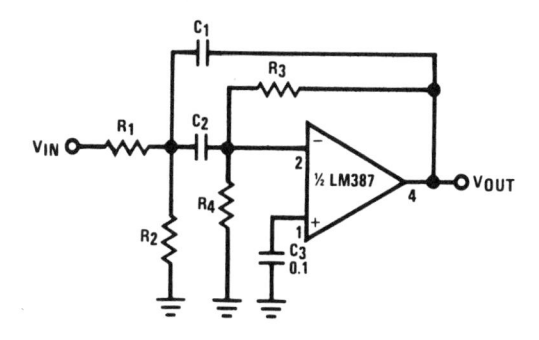

LM387 bandpass active filter.

Bandpass active filters. Narrow-bandwidth active filters do not require cascading of low- and high-pass sections. A single-amplifier bandpass filter using the LM387 is capable of Q ≤ 10 for low-distortion audio applications. The wide-gain bandwidth (20 MHz) and large open-loop gain (104 dB) allow high-frequency, low-distortion performance unobtainable with conventional op-amps.

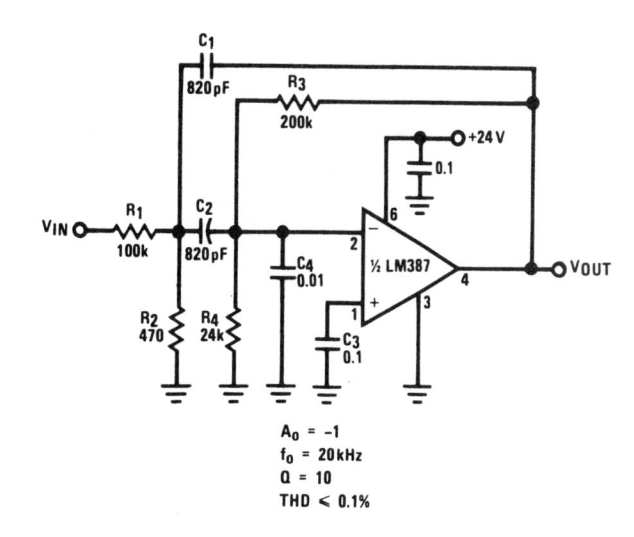

$A_0 = -1$
$f_0 = 20 \text{ kHz}$
$Q = 10$
$THD < 0.1\%$

20 kHz bandpass active filter.

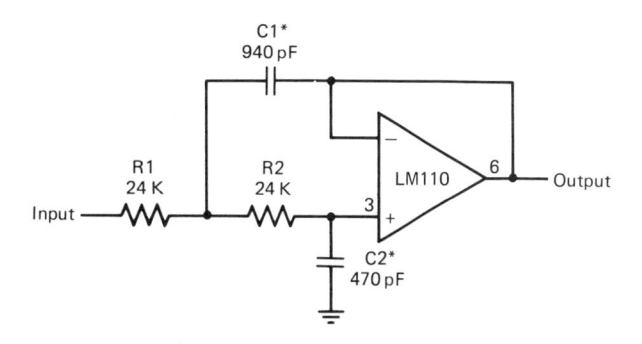

Low-pass active filter. Values are for 10 kHz cutoff. Use silvered mica capacitors for good temperature stability.

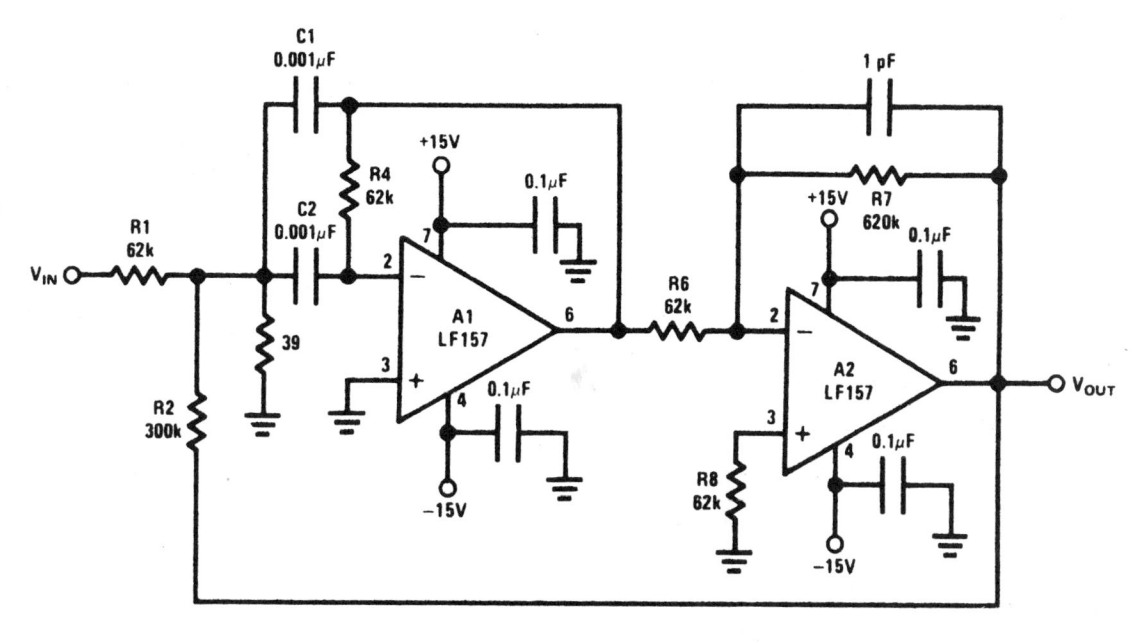

High-Q bandpass filter. By adding positive feedback (R2), Q increases to 40; $f_{BP} = 100$ kHz. Response to a 1V peak-to-peak tone burst is 300 μsec.

SECTION **11** Active Filters 405

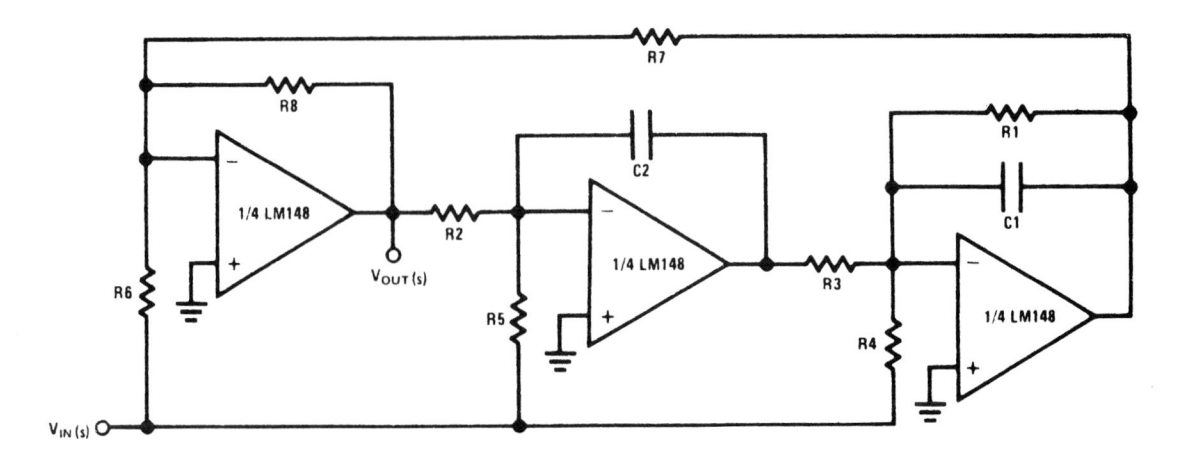

$$Q = \sqrt{\frac{R8}{R7}} \times \frac{R1C1}{\sqrt{R3C2R2C1}} \qquad f_o = \frac{1}{2\pi}\sqrt{\frac{R8}{R7}} \times \frac{1}{\sqrt{R2R3C1C2}} \qquad f_{NOTCH} = \frac{1}{2\pi}\sqrt{\frac{R6}{R3R5R7C1C2}}$$

Three-amplifier biquad notch filter. Necessary condition for notch: $1/R6 = R1/R4R7$; for example, f_{NOTCH} = 3 kHz, Q = 5, R1 = 270k, R2 = R3 = 20k, R4 = 27k, R5 = 20k, R6 = R8 = 10k, R7 = 100k, C1 = C2 = 0.001 μF.

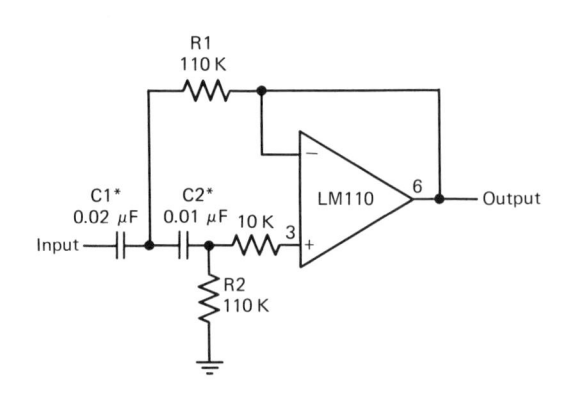

High-pass active filter. Capacitor values are for 100 Hz cutoff. Use metalized polycarbonate capacitors for good temperature stability.

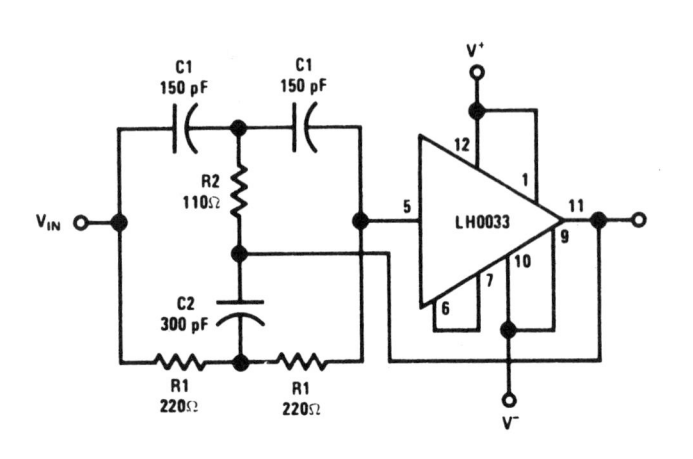

4.5 MHz notch filter. R1 = 2R2, C1 = C2/2; and

$$f_o = \frac{1}{2\pi R1C1}$$

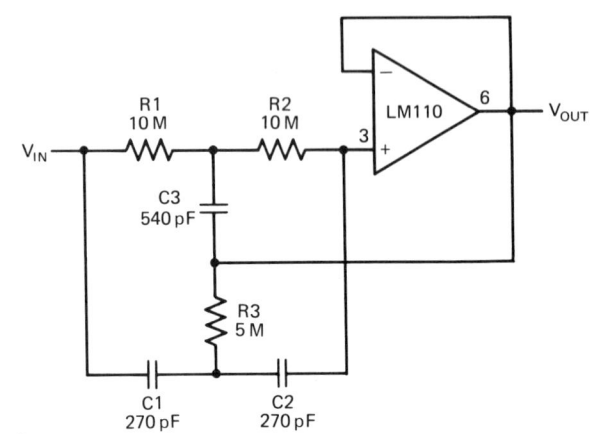

High-Q notch filter. R1 = R2 = 2R3, C1 = C2 = C3/2;

$$f_o = \frac{1}{2\pi R1C1} = 60 \text{ Hz}.$$

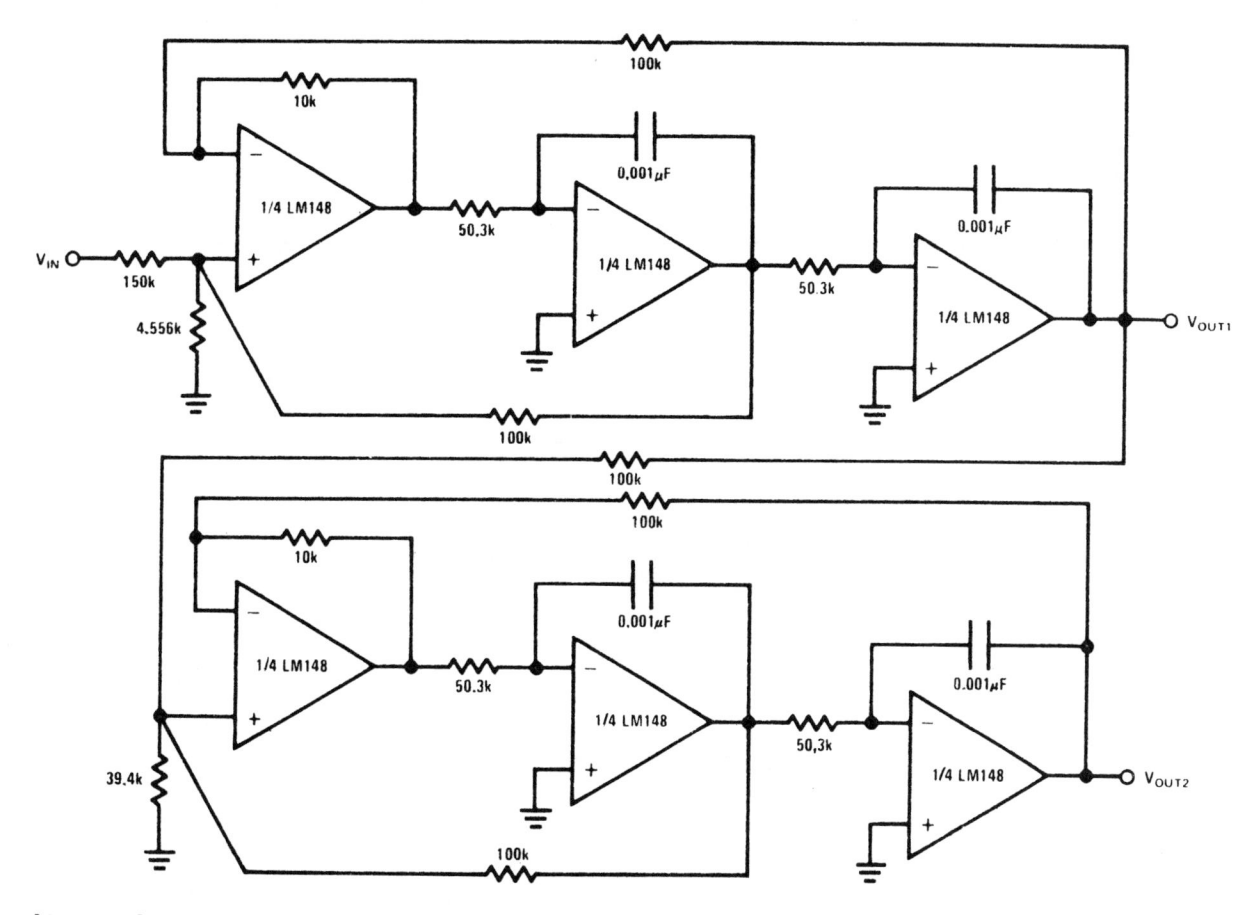

1 kHz 4-pole Butterworth filter. Use general equations and tune each section separately. First section Q = 0.541; second section Q = 2.306. The response should have 0 dB peaking.

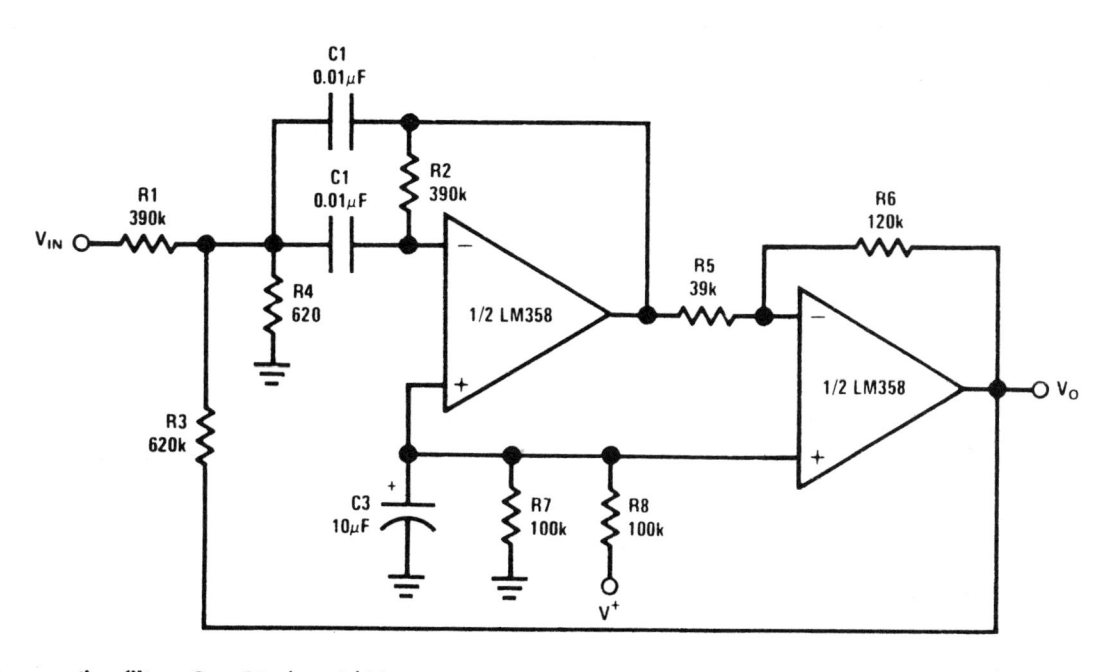

Bandpass active filter. Q = 25; f_0 = 1 kHz.

SECTION **11** *Active Filters*

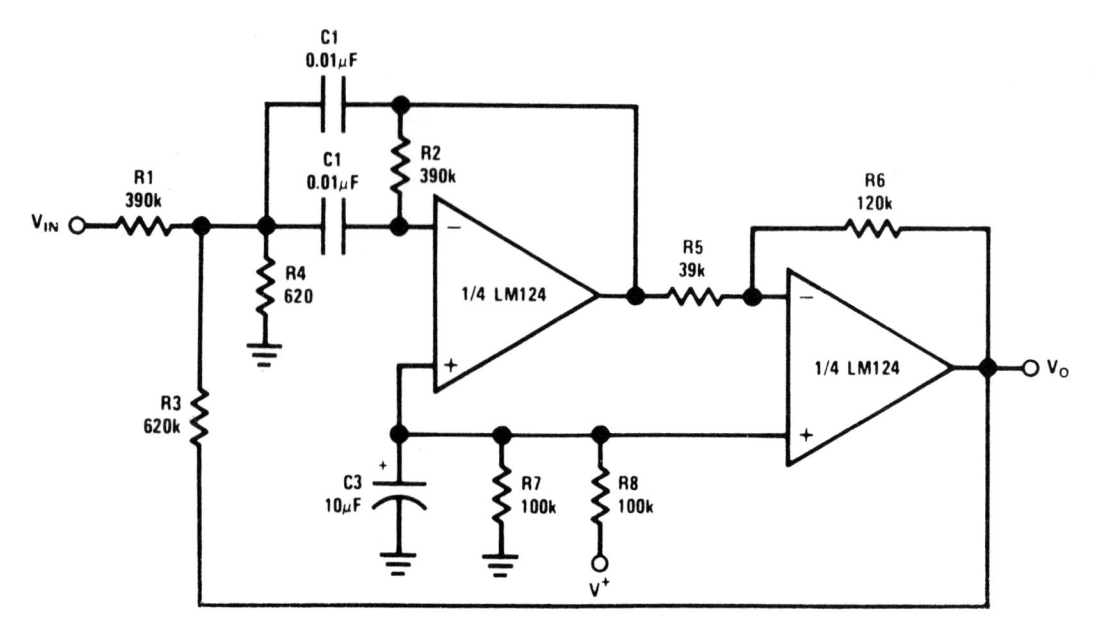

Bandpass active filter. $f_0 = 1$ kHz; Q = 25.

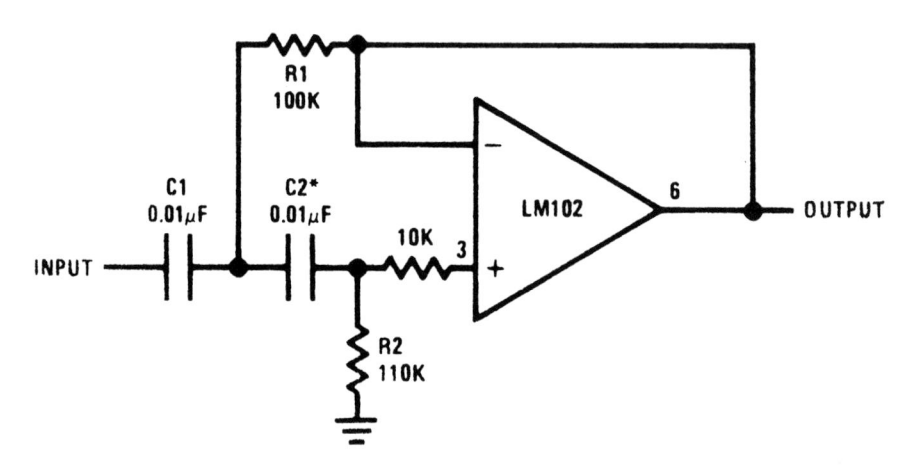

High-pass active filter. C2–R2 values are for 100 Hz cutoff. Use metalized polycarbonate capacitors for good temperature stability. Pin connections shown are for metal can.

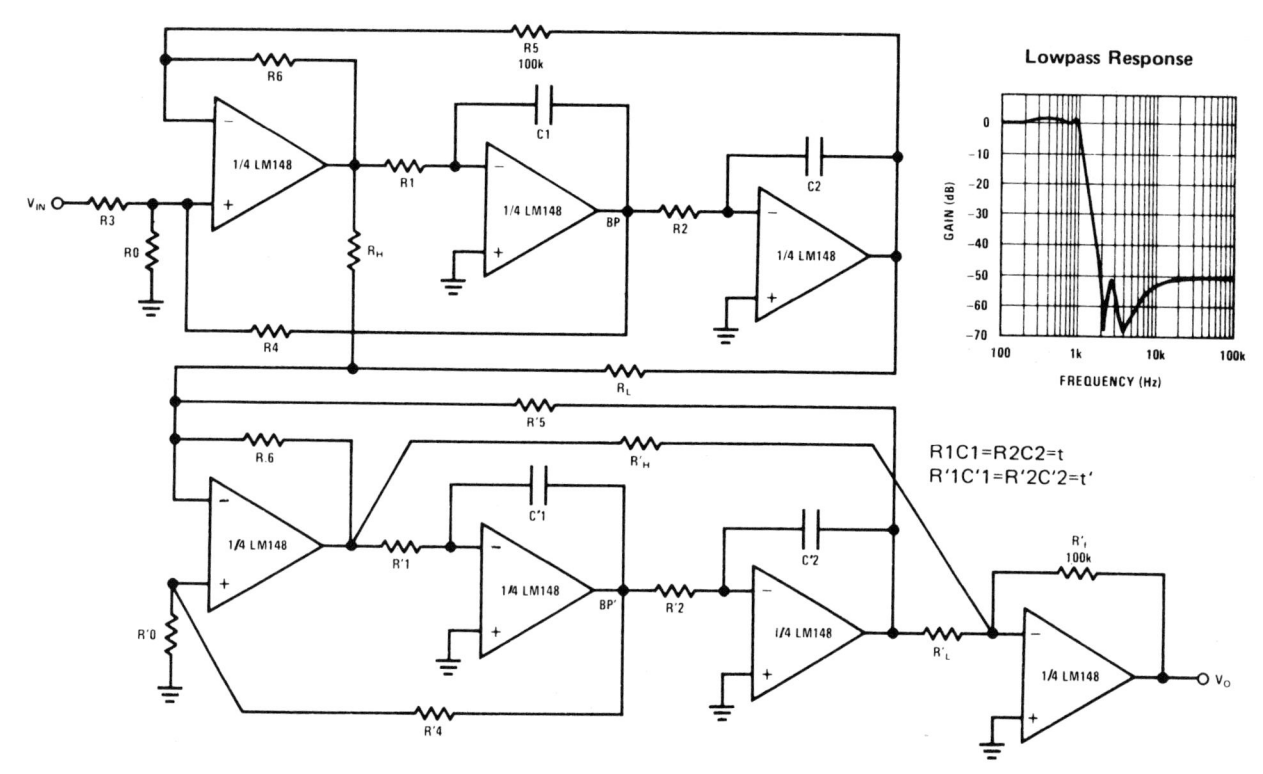

f_C = 1 kHz, f_S = 2 kHz, f_P = 0.543, f_Z = 2.14, Q = 0.841, f'_P = 0.987, f'_Z = 4.92, Q' = 4.403, normalized to ripple BW

$$f_P = \frac{1}{2\pi}\sqrt{\frac{R6}{R5}} \times \frac{1}{t}, \quad f_Z = \frac{1}{2\pi}\sqrt{\frac{R_H}{R_L}} \times \frac{1}{t}, \quad Q = \left(\frac{1 + R4|R3 + R4|R0}{1 + R6|R5}\right) \times \sqrt{\frac{R6}{R5}}, \quad Q' = \sqrt{\frac{R'6}{R'5}} \frac{1 + R'4|R'0}{1 + R'6|R'5 + R'6|R_P}$$

$$R_P = \frac{R_H R_L}{R_H + R_L}$$

Use the BP outputs to tune Q, Q', tune the 2 sections separately
R1 = R2 = 92.6k, R3 = R4 = R5 = 100k, R6 = 10k, R0 = 107.8k, R_L = 100k, R_H = 155.1k,
R'1 = R'2 = 50.9k, R'4 = R'5 = 100k, R'6 = 10k, R'0 = 5.78k, R'_L = 100k, R'_H = 248.12k, R'f = 100k. All capacitors are 0.001μF.

Fourth-order 1 kHz elliptic filter (4 poles, 4 zeros).

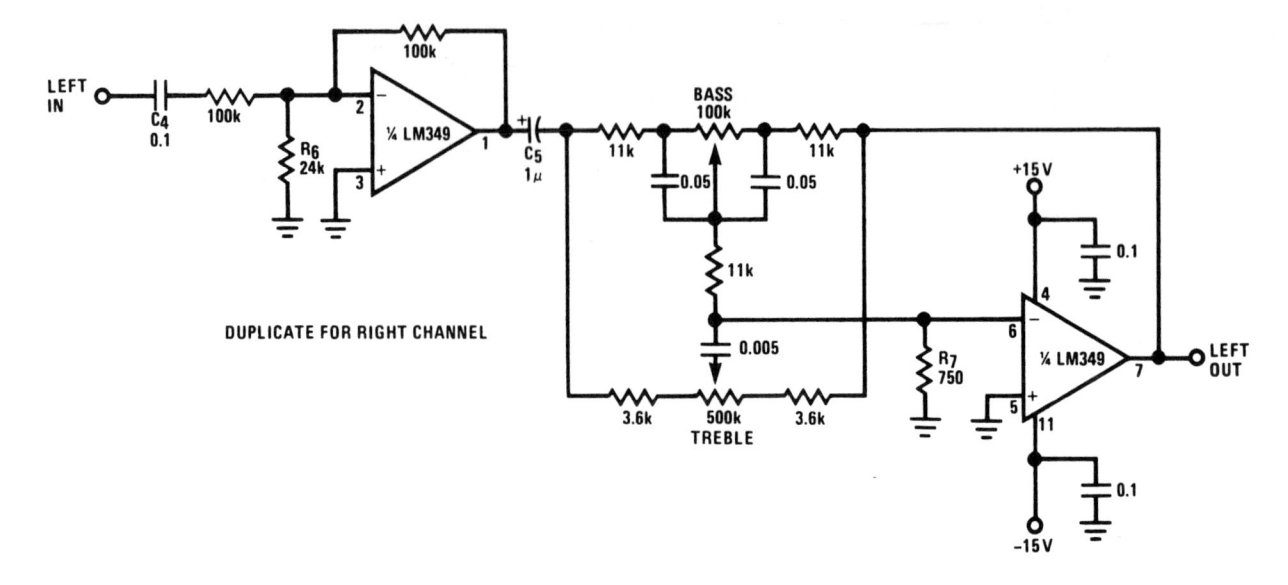

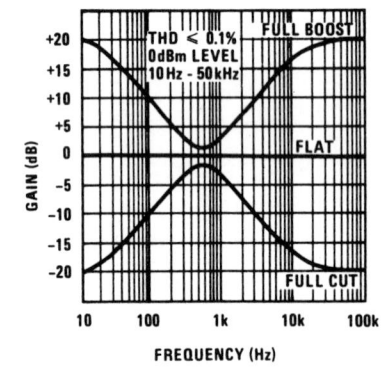

Active tone controls, where the quad op-amp LM349 has been chosen for the active element. The use of a quad makes for a single-IC stereo tone control that is compact and economical. The buffer amplifier is necessary to ensure a low driving impedance for the tone control circuit and creates a high input impedance (100K) for the source. The LM349 was chosen for its fast slew rate (2.5V/μsec), allowing undistorted, full-swing performance out to greater than 25 kHz. Measured THD was typically 0.05% at 0 dBm. (0.77V) across the audio band. Resistors R6 and R7 were added to ensure stability at unity gain, since the LM349 is internally compensated for voltage dividers at high frequencies such that the actual input-to-output gain is never less than five (four if used inverting). Coupling capacitors C4 and C6 serve to block dc and establish low-frequency rolloff of the system; they may be omitted for direct-coupled designs.

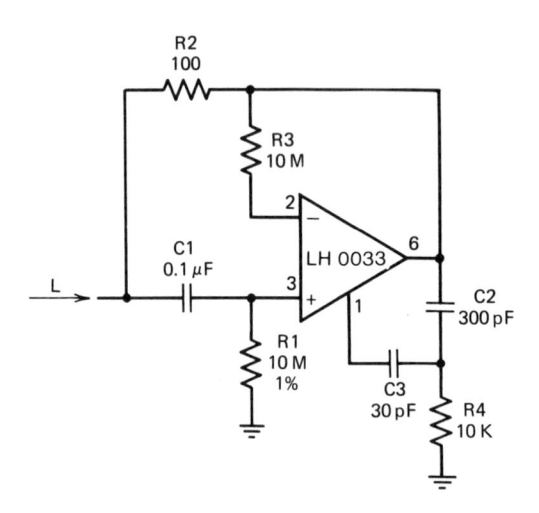

Simulated inductor. L = R1R2C1; R_S = R2; R_P = R1.

410 IC SCHEMATIC SOURCEMASTER

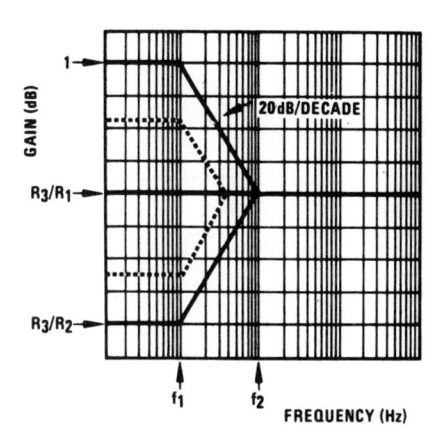

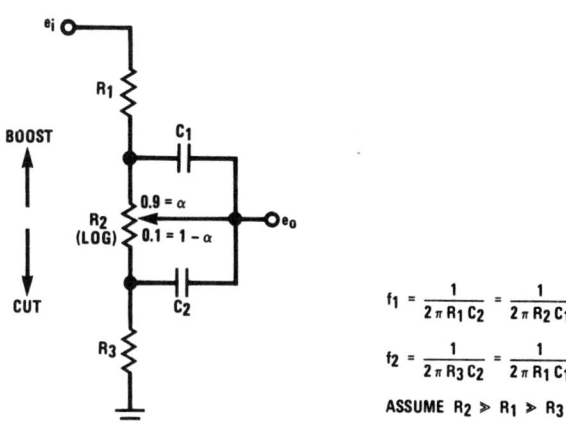

Bass tone control—general circuit.

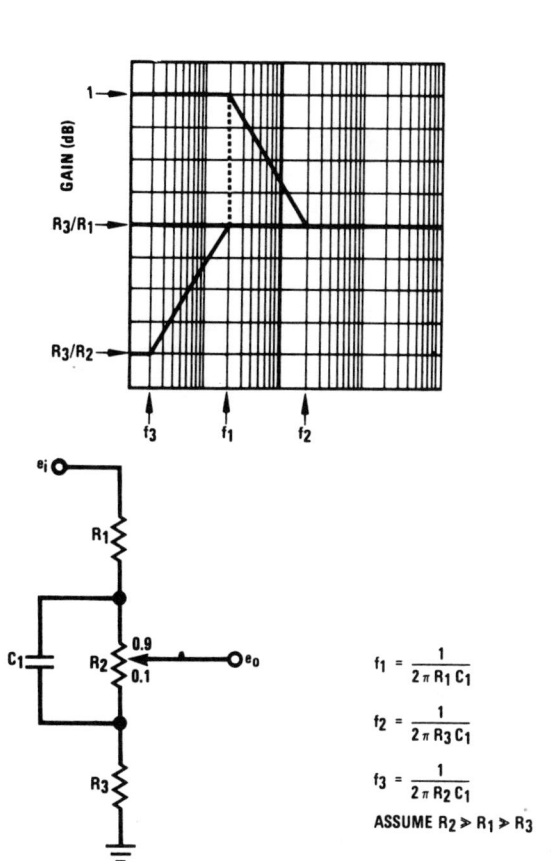

$$f_1 = \frac{1}{2\pi R_1 C_2} = \frac{1}{2\pi R_2 C_1}$$

$$f_2 = \frac{1}{2\pi R_3 C_2} = \frac{1}{2\pi R_1 C_1}$$

ASSUME $R_2 \gg R_1 \gg R_3$

Minimum-parts bass tone control.

$$f_1 = \frac{1}{2\pi R_1 C_1}$$

$$f_2 = \frac{1}{2\pi R_3 C_1}$$

$$f_3 = \frac{1}{2\pi R_2 C_1}$$

ASSUME $R_2 \gg R_1 \gg R_3$

Passive bass controls offer the advantages of lowest cost and minimum parts count while suffering from severe insertion loss, which often creates the need for a tone recovery amplifier. The insertion loss is approximately equal to the amount of available boost, then they will have about −20 dB insertion boost, then they will have about −20-dB insertion loss. This is because passive tone controls work as ac voltage dividers and really only cut the signal.

The most popular bass controls appear here along with associated frequency-response curves. The curves shown can only be approximated. The corner frequencies f_1 and f_2 denote the half-power points and therefore represent the frequencies at which the relative magnitude of the signal has been reduced (or increased) by 3 dB.

Passive tone controls require audio-taper (logarithmic) potentiometers; at the 50% rotation point, the slider splits the resistive element into two portions equal to 90% and 10% of the total value. This is represented in the figures by "0.9" and "0.1" about the wiper arm.

SECTION **11** *Active Filters* **411**

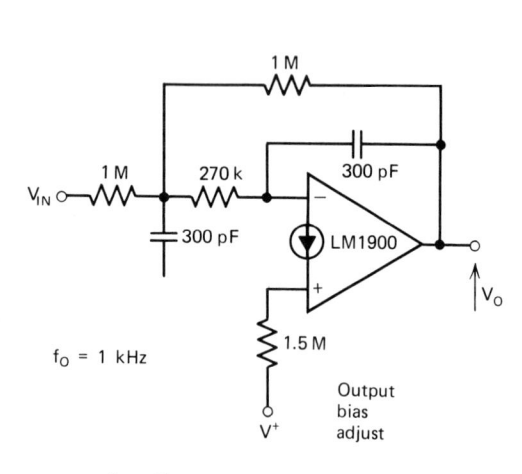

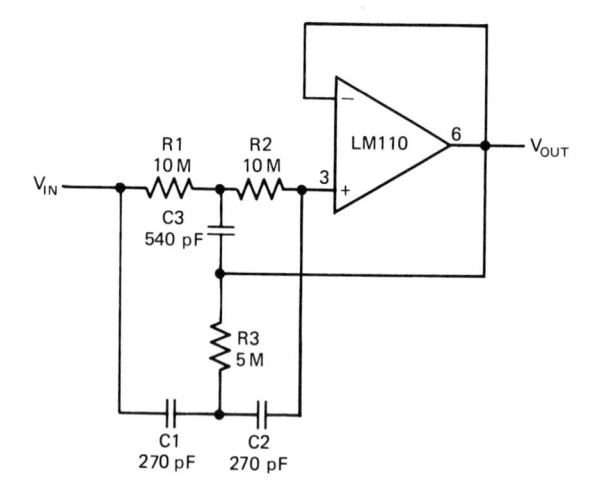

Low-pass active filter. $f_O = 1$ kHz

High-Q notch filter. R1 = R2 = 2R3; C1 = C2 = C3/2;

$$f_o = \frac{1}{2\pi R1C1} = 60 \text{ Hz.}$$

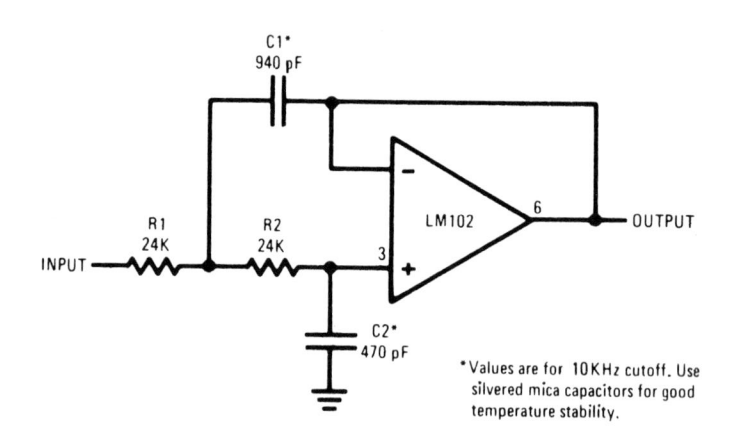

*Values are for 10 KHz cutoff. Use silvered mica capacitors for good temperature stability.

A low-pass filter is one of the simplest forms of active filters. The circuit has the filter characteristics of two isolated RC sections and has a buffered, low-impedance output. The attenuation is roughly 12 dB at twice the cutoff frequency and the ultimate attenuation is 40 dB/decade. A third low-pass RC section can be added on the output of the amplifier for an ultimate attenuation of 60 dB/decade, although this means that the output is no longer buffered.

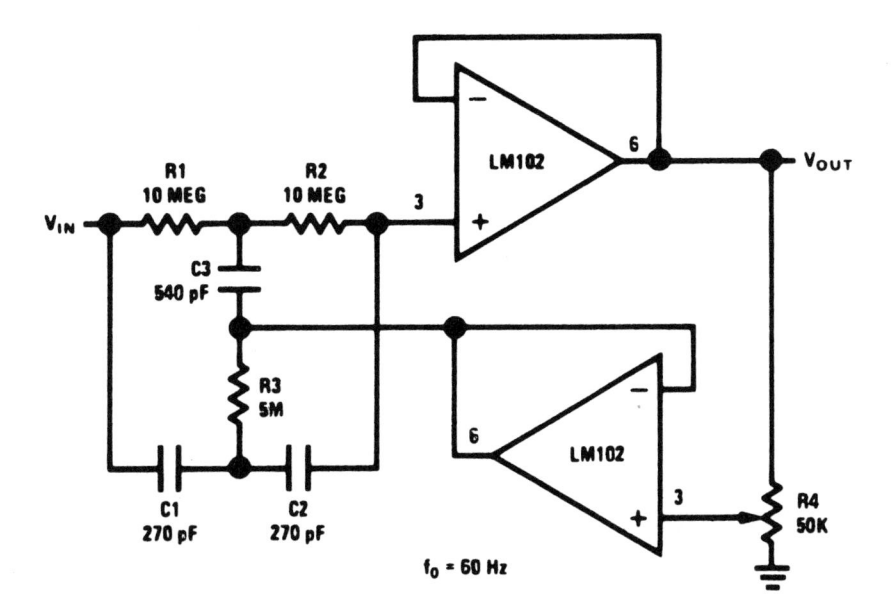

Adjustable-Q notch filter. The circuit is easy to use, and only a few items need be considered for proper operation. To minimize notch frequency shift with temperature, silvered mica or polycarbonate capacitors should be used with precision resistors. Notch depth depends on component match; therefore, 0.1% resistors and 1% capacitors are suggested to minimize the trimming needed for a 60 dB notch. To ensure stability of the LM102, the power supplies should be bypassed near the integrated circuit package with 0.01 µF disc capacitors.

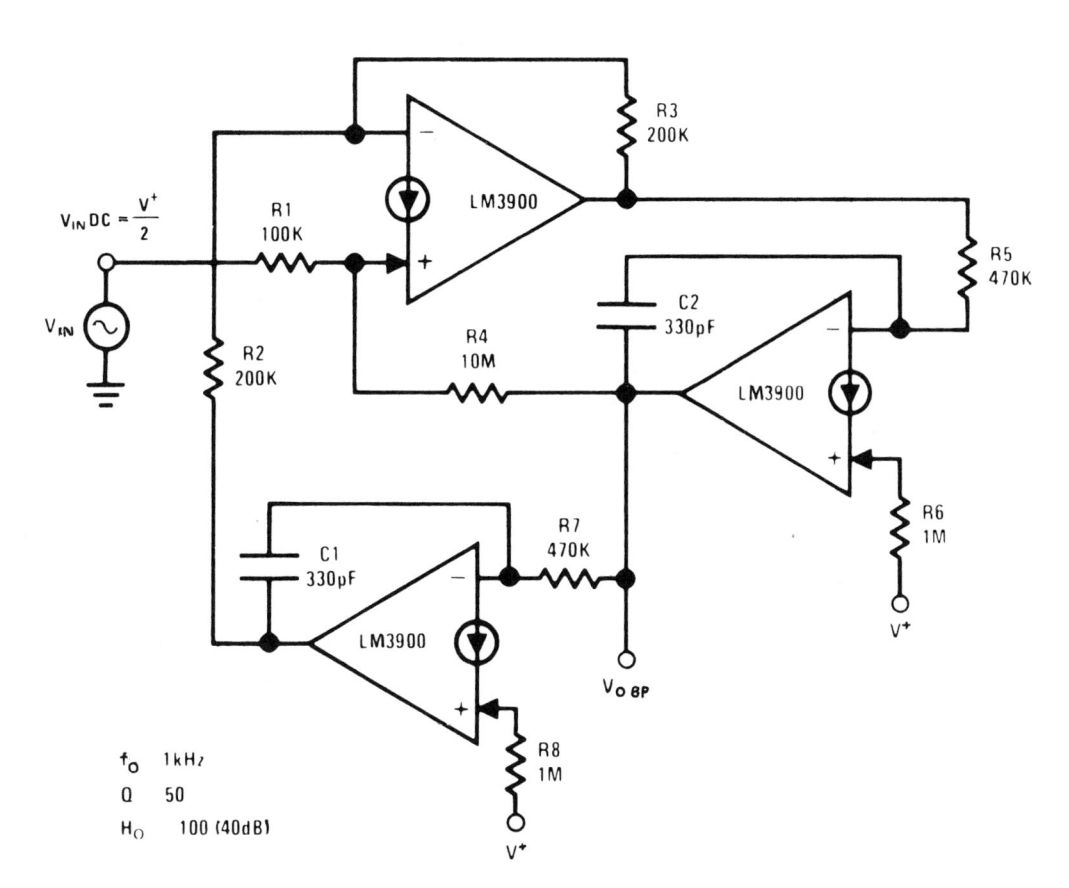

Three-amplifier bandpass active filter. It has been called the "biquad" as it can produce a transfer function that is "quadradic" in both numerator and denominator. Outputs can be taken at any of three points to give low-pass, high-pass, or bandpass response characteristics.

SECTION **11** *Active Filters* **413**

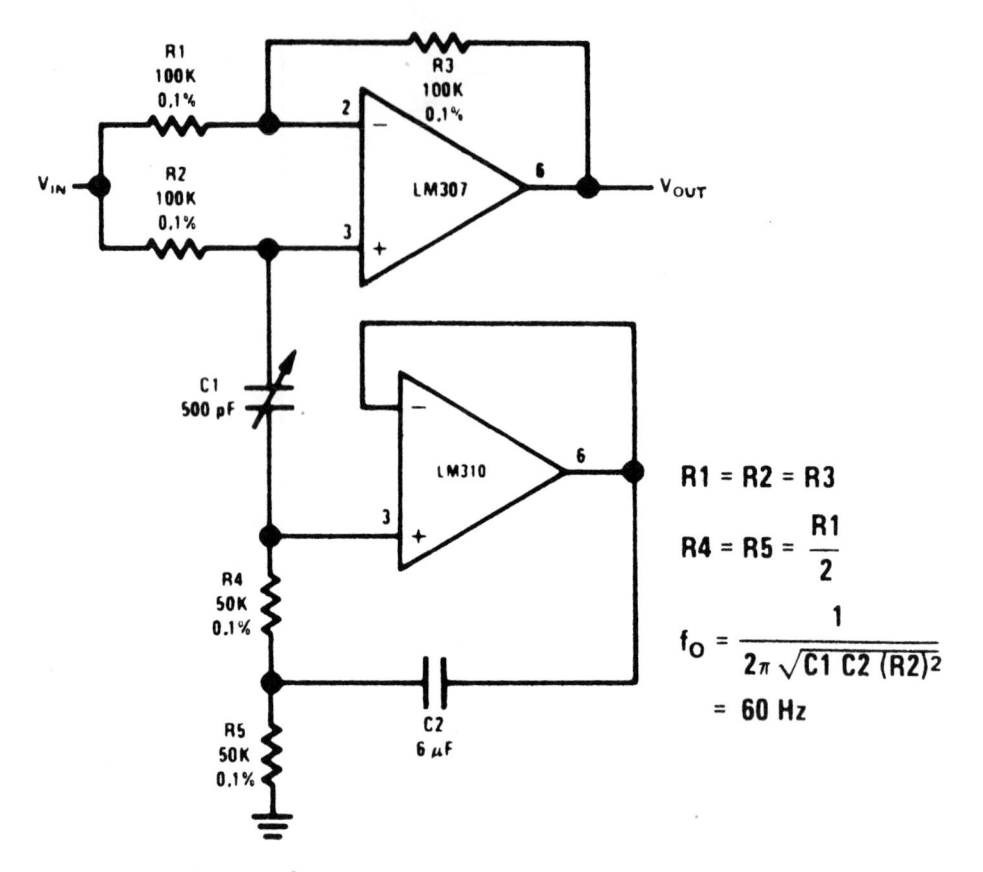

Tunable notch filter.

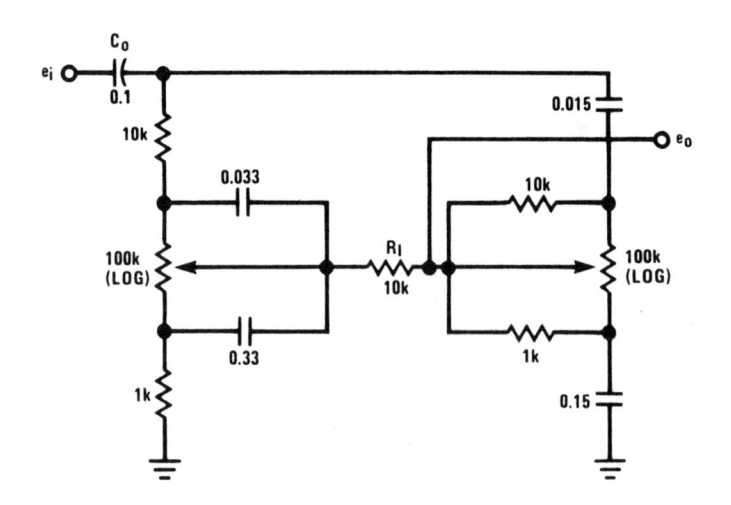

$R1 = R2 = R3$

$R4 = R5 = \dfrac{R1}{2}$

$f_O = \dfrac{1}{2\pi \sqrt{C1\,C2\,(R2)^2}}$

$= 60\text{ Hz}$

Bass & Treble Tone Control Response

Passive tone controls, where R1 has been included to isolate the two control circuits; C_0 is provided to block all dc voltages from the circuit, ensuring the controls are not "scratchy," which results from dc charge currents in the capacitors and on the sliders. C_0 is selected to agree with system low-frequency response:

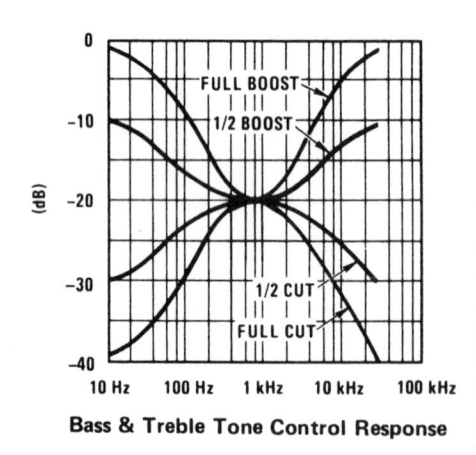

$C_0 = \dfrac{1}{(2\pi)(20\text{ Hz})(10k + 100k + 1k)}$

$= 7.17 \times 10^{-8}$

Use $C_0 = 0.1\ \mu F$.

$$f_O = \frac{1}{2\pi R1 C1}$$
$$R1 = 2 R2$$
$$C1 = \frac{C2}{2}$$

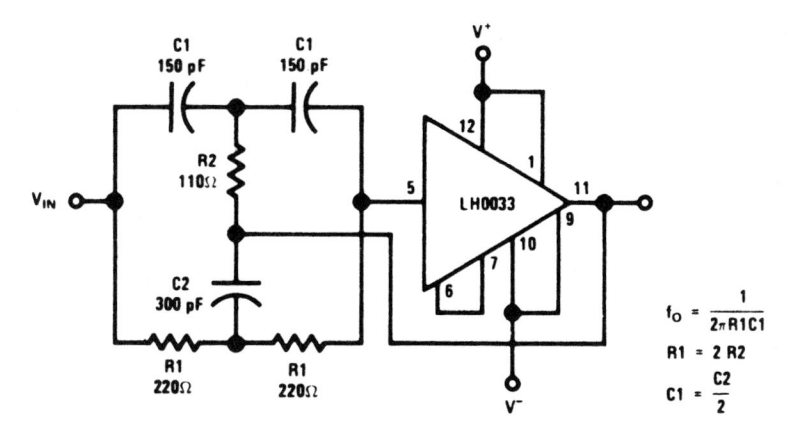

This high-Q notch filter takes advantage of the LH0033's wide bandwidth. For the values shown, the center frequency is 4.5 MHz.

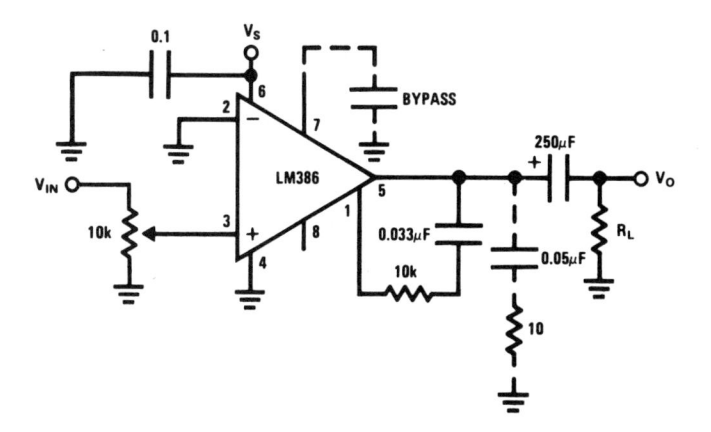

Bass boost circuit. Additional external components can be placed in parallel with internal feedback resistors to tailor gain and frequency response for individual applications. For example, compensate poor speaker bass response by frequency-shaping the feedback path. This is done with a series RC from pin 1 to 5 (paralleling the internal 15k resistor). For 6 dB effective bass boost: R ≈ 15k (the lowest value for good stable operation is R ≈ 10k if pin 8 is open). If pins 1 and 8 are bypassed, then R as low as 2k can be used. This restriction is because the amplifier is only compensated for closed-loop gains greater than 9.

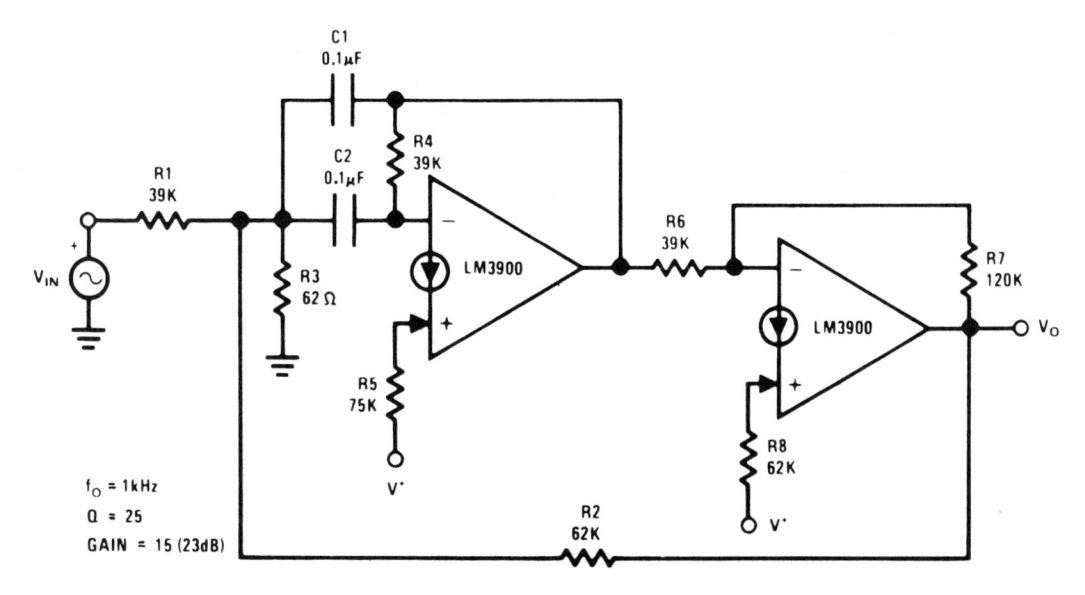

$f_O = 1 kHz$
$Q = 25$
$GAIN = 15 (23dB)$

Two-amplifier bandpass active filter. To allow high Q (between 10 and 50) and higher gain, a two-amplifier filter is required. Resistors R5 and R8 are used to bias the output voltage of the amplifiers at $V^+/2$. R5 is "twice R4," and R8 must be selected after R6 and R7 have been assigned values.

SECTION **11** Active Filters

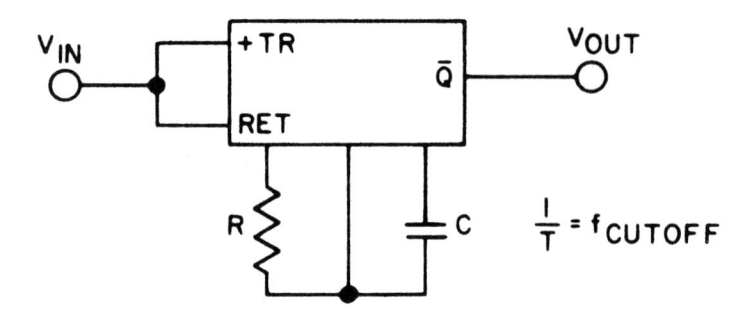

A simple circuit using the CD4047A as a **low-pass filter**. The time constant chosen for the multivibrator will determine the upper cutoff frequency for the filter. The circuit essentially compares the input frequency with its own reference and produces an output that follows the input for frequencies less than f_{CUTOFF}, and a low output for frequencies greater than f_{CUTOFF}.

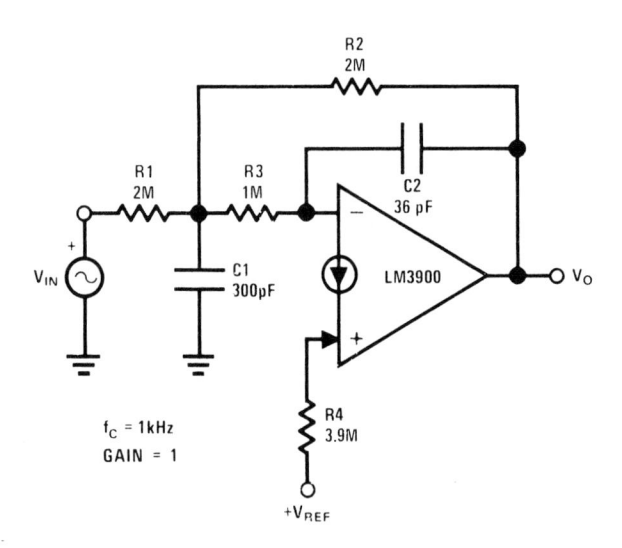

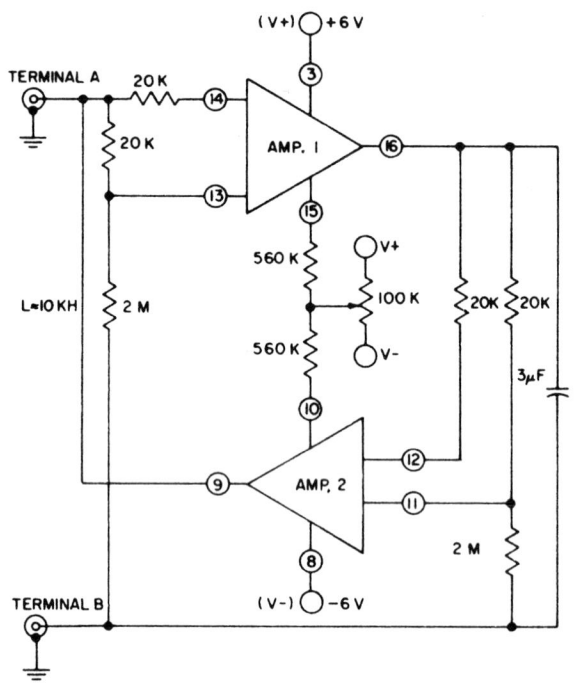

Gyrator in an active filter circuit derived by connecting two operational transconductance amplifiers of the CA3060.

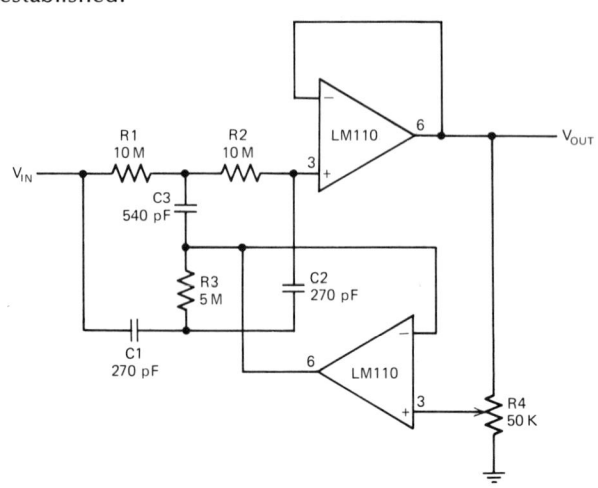

Single-amplifier low-pass active filter. Resistor R4 is used to set the output bias level and is selected after the other resistors have been established.

Adjustable-Q notch filter. R1 = R2 = 2R3; C1 = C2 = 2C3;

$$f_o = \frac{1}{2\pi R1 C1} = 60 \text{ Hz}.$$

416 IC SCHEMATIC SOURCEMASTER

$$f_0 = \frac{1}{2\pi R1 C1}$$
$$= 60 \text{ Hz}$$
$$R1 = R2 = 2 R3$$
$$C1 = C2 = \frac{C3}{2}$$

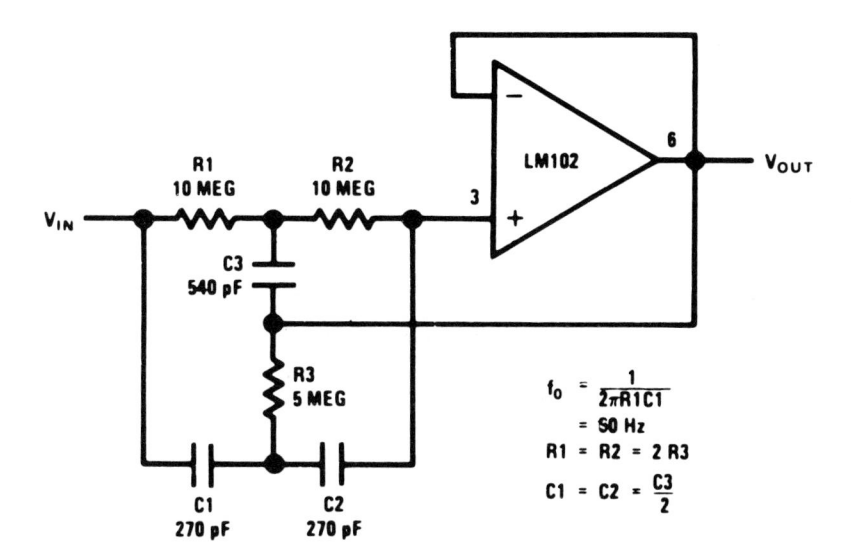

High-Q 60 Hz notch filter. The twin-tee network is one of the few RC filter networks capable of providing an infinitely deep notch. By combining the twin-tee with an LM102 voltage follower, the usual drawbacks of the network are overcome. The Q is raised from the usual 0.3 to something greater than 50. Further, the voltage follower acts as a buffer, providing a low output resistance; the high input resistance of the LM102 makes it possible to use large resistance values so that only small capacitors are required, even at low frequencies. The fast response of the follower allows the notch to be used at high frequencies. Neither the depth nor the frequency of the notch are changed when the follower is added.

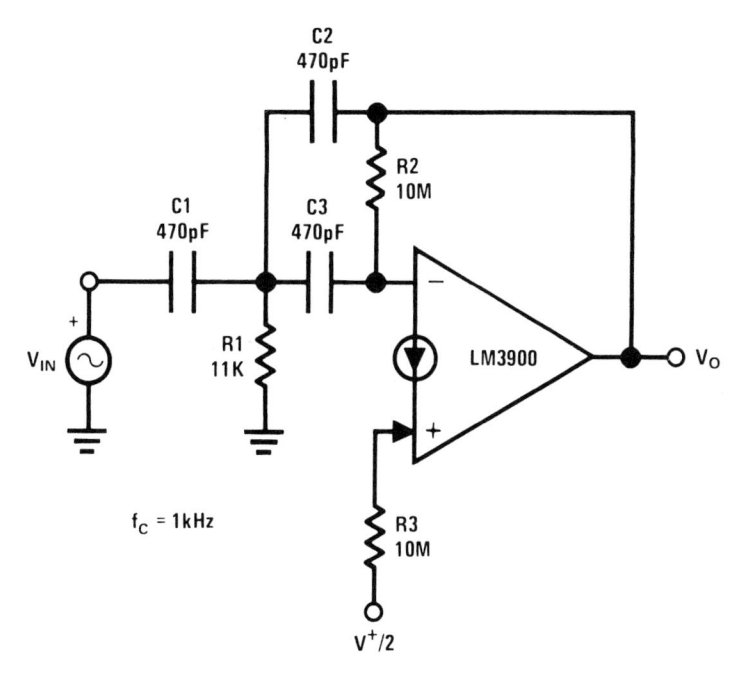

$f_C = 1 \text{kHz}$

A single-amplifier high-pass RC active filter is easily biased using the (+) input of the LM3900. Resistor R3 can be made equal to R2 and a bias reference of $V^+/2$ will establish the output Q point at this value ($V^+/2$). The input is capacitively coupled (C1) and there are therefore no further dc biasing problems.

The design procedure for this filter is to select the pass-band gain, H_0, the Q, and the corner frequency, f_0. A Q value of 1 gives only a slight peaking near the band edge (less than 2 dB) and smaller Q values decrease this peaking. The slope of the skirt of this filter is 12 dB/octave (or 60 dB/decade). If the gain, H_0, is unity, all capacitors have the same value.

SECTION 11 *Active Filters*

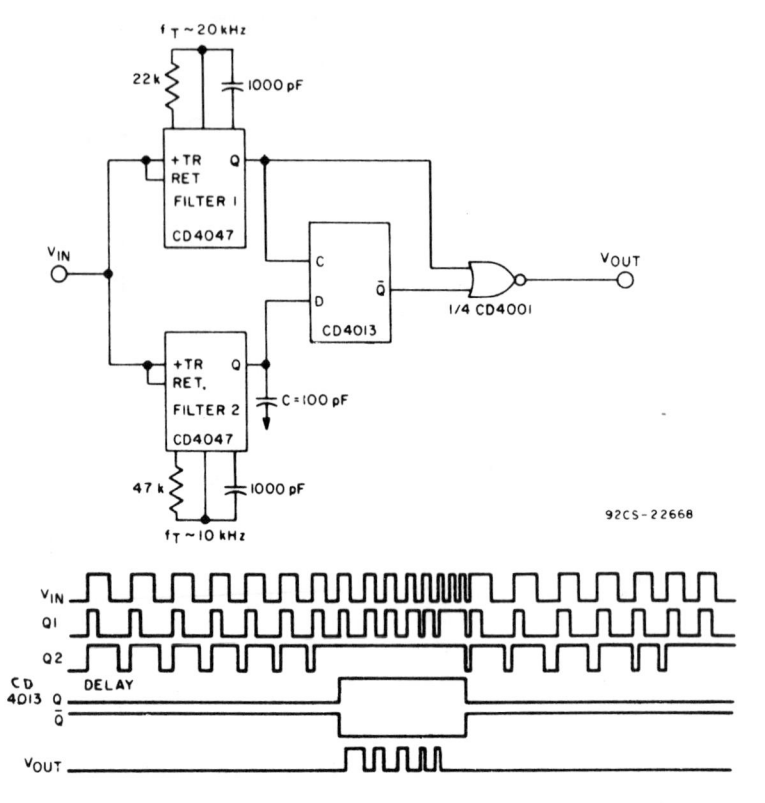

Bandpass filter circuit and waveforms. The pass band is determined by the time constants of the two filters. If the output of filter 2 is delayed by C1, the CD4013A flip-flop will clock high only when the cutoff frequency of filter 2 has been exceeded; this point is illustrated in the timing diagram. The Q output of the CD4013A is gated with the output of filter 1 to produce the desired output.

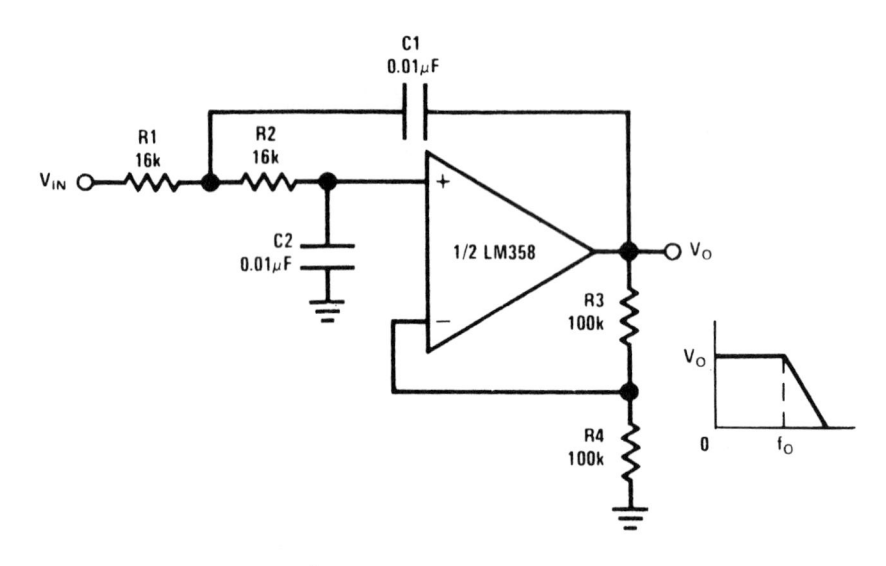

Dc-coupled low-pass RC active filter. $Q = 1$, $f_0 = 1$ kHz, $A_v = 2$.

418 IC SCHEMATIC SOURCEMASTER

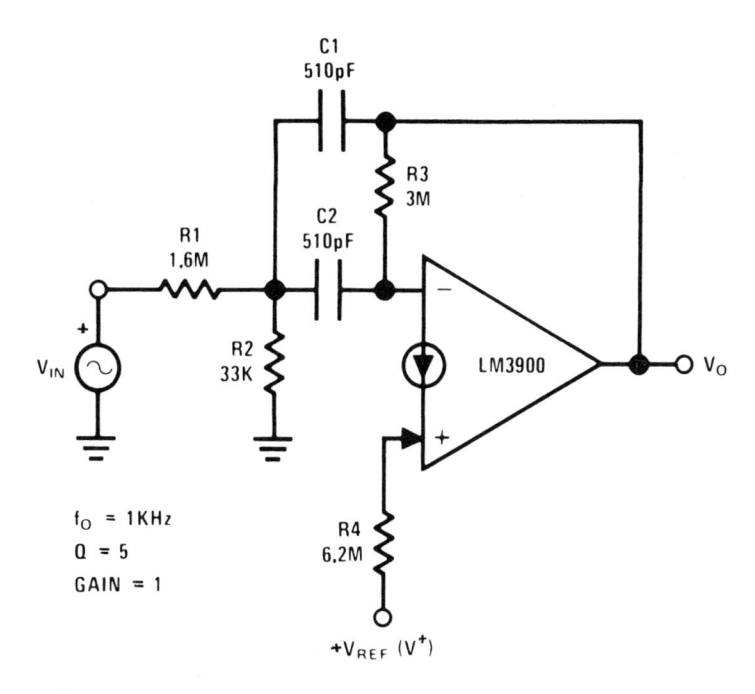

Single op-amp bandpass filter.

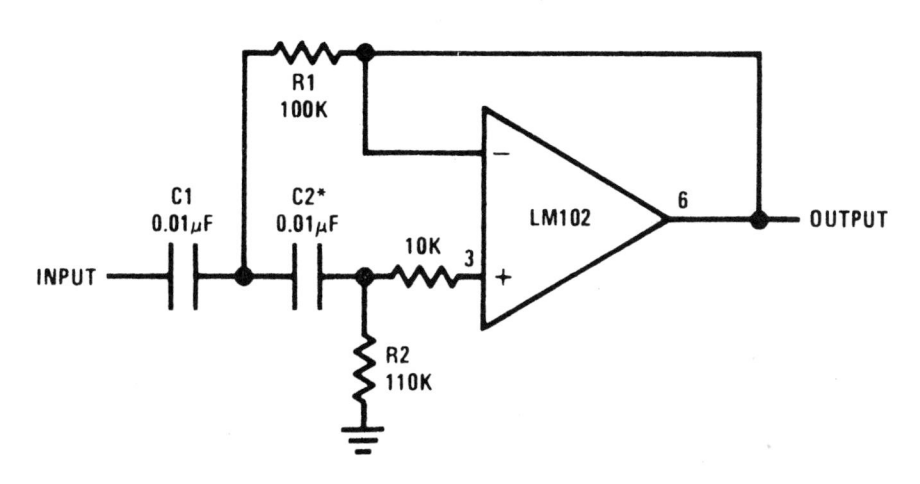

$f_O = 1$ KHz
$Q = 5$
GAIN = 1

High-pass active filter. Capacitive values are for 100 Hz cutoff. Use metalized polycarbonate capacitors for good temperature stability.

SECTION 11 *Active Filters* 419

SECTION 12
OPTOELECTRONIC CIRCUITS

Most optoelectronics circuits involve the use of some form of photo device or devices along with a conventional high-gain integrated-circuit amplifier. However, RCA's CA3062 is an integrated optical sensing amplifier—a unique package shown here performing several different functions.

Combining optics and electronics is a relatively new concept, and it has only recently begun to be exploited to its deserved potential. The applications presented in this section cover almost the entire range of circuit classes in this book, from power control to logic.

While some of these circuits have been designed for specific functional applications—the light-tracking analog multiplier is a typical example—most of them can be considered "basic" in the sense that they can be applied quite simply to any electronic system that requires a digital (on/off) or analog light-proportional voltage as the result of light intensity.

Industrial control applications are the most obvious candidates for opto-electronic implementation, but the text accompanying the circuits shown will doubtless suggest other areas where light can be used to advantage in the sensing, monitoring, or controlling of various electronic circuit functions.

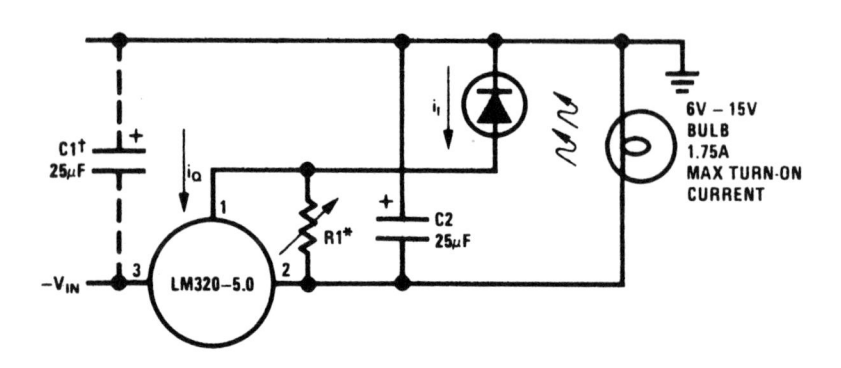

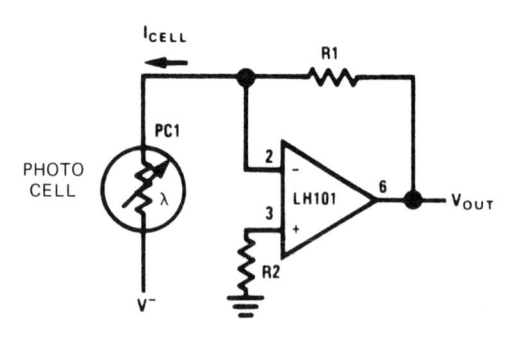

A light controller using silicon photocell. Lamp brightness increases until $i_l = i_o (\approx 1 \text{ mA}) + 5V/R1$. C1 is only necessary if the raw supply filter capacitor is more than 2 inches from LM320.

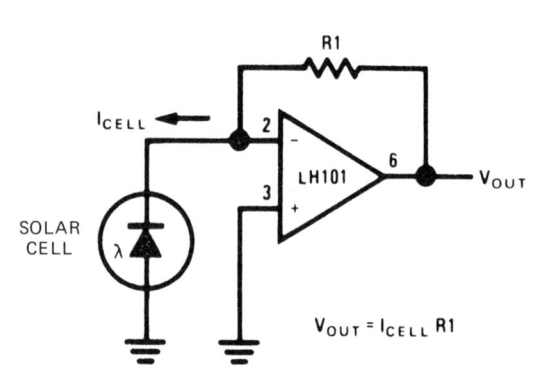

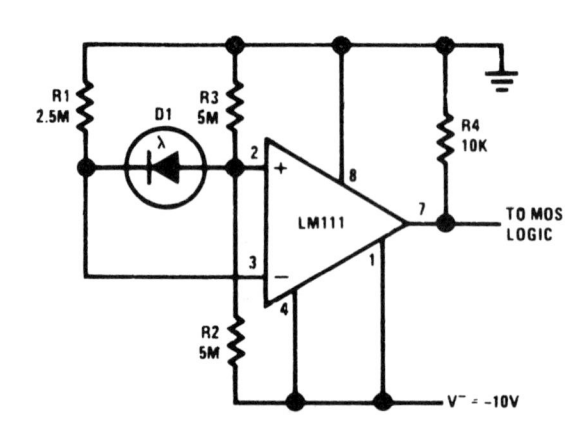

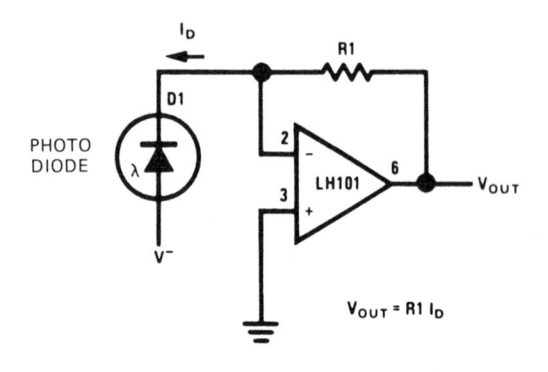

A level detector for a photodiode that operates off a −10V supply. The output changes state when the diode current reaches 1 μA. Even at this low current, the error contributed by the comparator is less than 1%. Higher threshold currents can be obtained by reducing R1, R2, and R3 proportionally. At the switching point, the voltage across the photodiode is nearly zero, so its leakage current does not cause an error. The output switches between ground and −10V, driving the data inputs of MOS logic directly.

In these "photo" amplifiers, R1 (the feedback resistance) is dependent on cell sensitivity and should be chosen for either maximum dynamic range or for a desired scale factor. R2 is elective: if used, it should be chosen to minimize bias current error over the operating range.

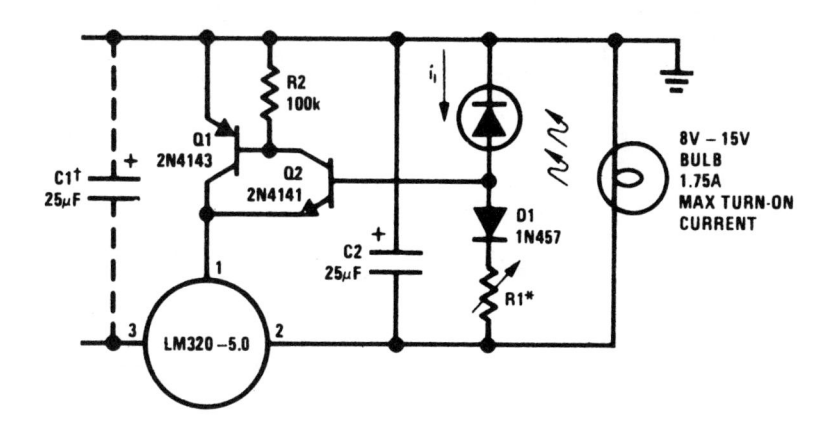

A light controller using silicon photocell. Lamp brightness increases until $i_1 = 5V/R1$; i_1 can be set as low as 1 μA. C1 is only necessary if the raw supply filter capacitor is more than 2 inches from LM320MP.

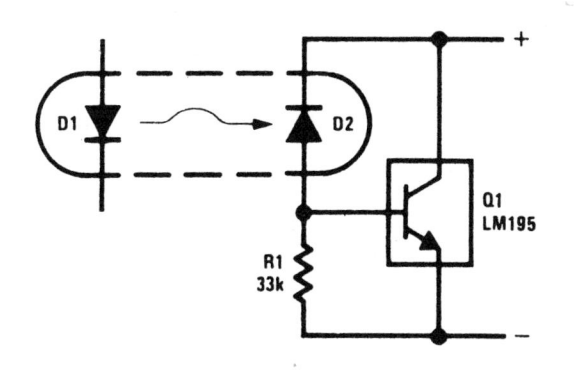

Optically isolated power transistor.

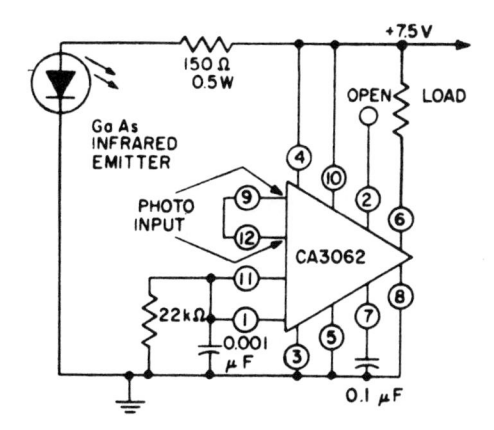

Circuit diagram for on-off photoelectric control applications.

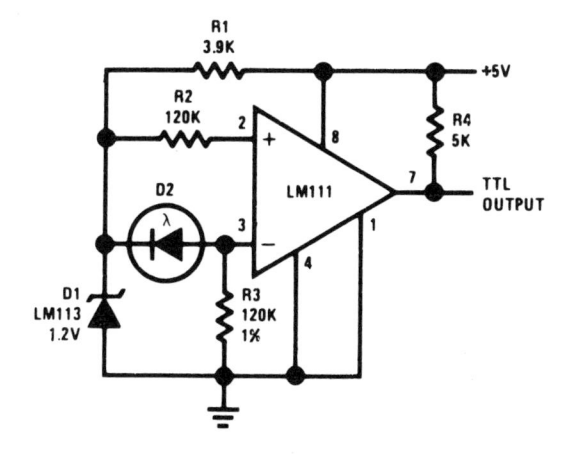

Precision level detector for photodiode. D1 is a temperature-compensated reference diode with a 1.23V breakdown voltage. It acts as a shunt regulator and delivers a stable voltage to the comparator. When the diode current is large enough (about 10 μA) to make the voltage drop across R3 equal to the breakdown voltage of D1, the output will change state. R2 is added to make the threshold error proportional to the offset current of the comparator rather than the bias current; it can be eliminated if the bias-current error is not significant.

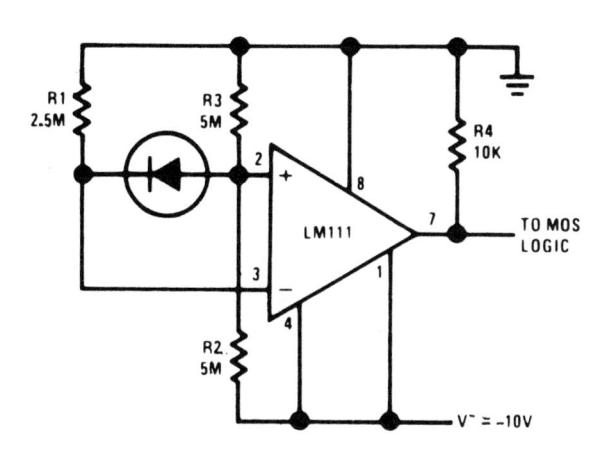

Comparator for a low-level photodiode operating with MOS logic. The output changes state when the diode current reaches 1 μA. At the switching point, the voltage across the photodiode is nearly zero, so its leakage current does not cause an error. The output switches between ground and −10V, driving the data inputs of MOS logic directly.

SECTION **12** Optoelectronic Circuits **423**

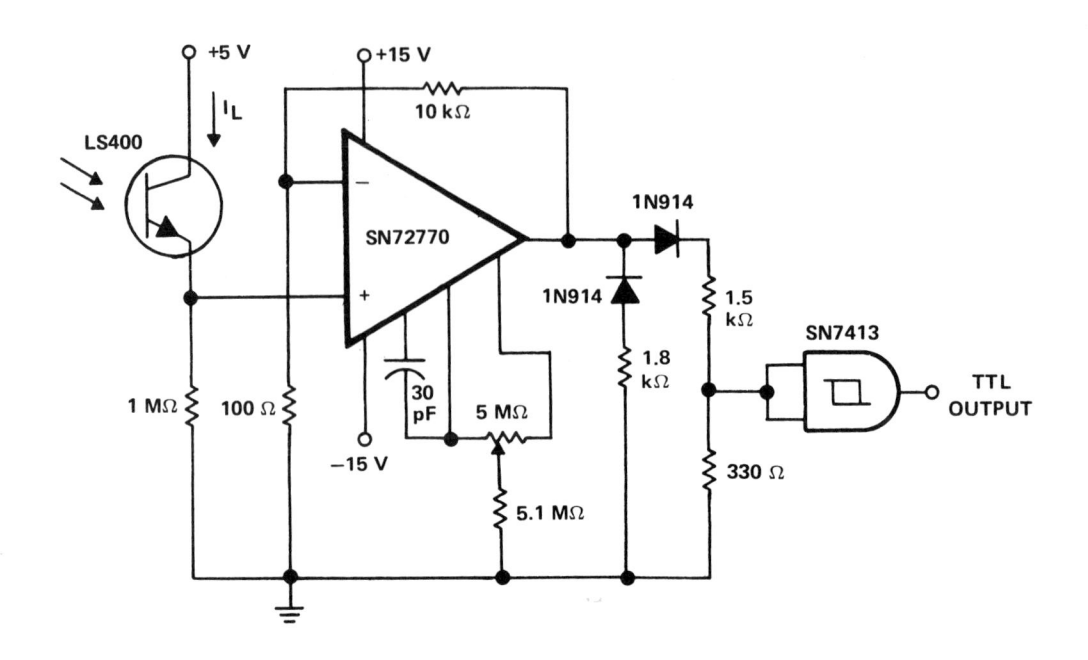

CONDITION	TIL400 CURRENT	OUTPUT LOGIC
LIGHT ON	≥ 250 nA	0
LIGHT OFF	< 25 nA	1

Optical-sensor-to-TTL interface. In low-light-level optoelectronic applications, small signals from the detectors must be amplified. Here the *on* current from the LS400 is assumed to be 250 nA, so it is necessary to use an op-amp with very low input bias currents and high input resistance to successfully detect the *on* condition.

The 250 nA signal results in a +250 mV level at the noninverting input. With a gain of 100, the output is in positive saturation. This output level is fed through a loading network to provide the basic TTL level. Because of the slow speed at which an optical circuit may operate, it might be desirable to follow the network with an SN7413 Schmitt trigger to shape the TTL signal.

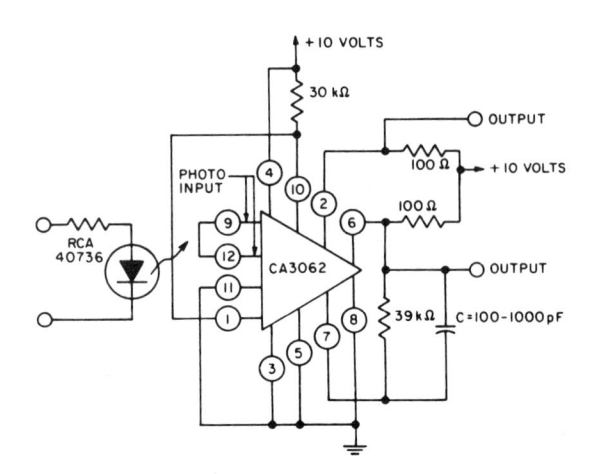

An optically coupled isolator circuit used to transfer signals that are at substantially different voltage levels. Both polarities are available at the output. Current transfer ratios as high as 10:1 can be achieved.

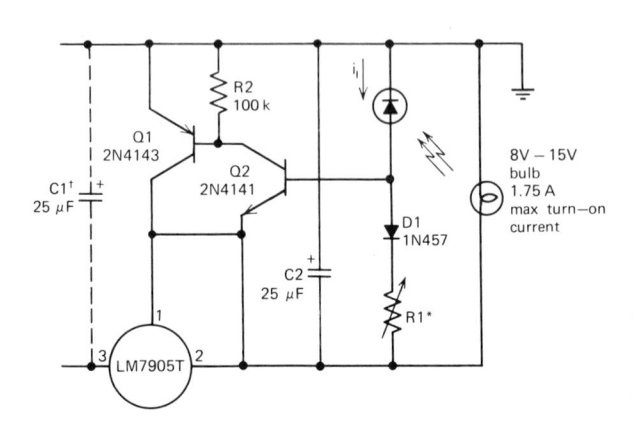

A light controller using silicon photocell. Lamp brightness increases until i_l = 5V/R1 (i_l can be set as low as 1 μA). C1 is only necessary if the raw supply filter capacitor is more than 2 inches from LM7905T.

$$R5 = R1\left(\frac{V^-}{10}\right)$$
$$V_1 > 0$$
$$V_{OUT} = \frac{V_1 V_2}{10}$$

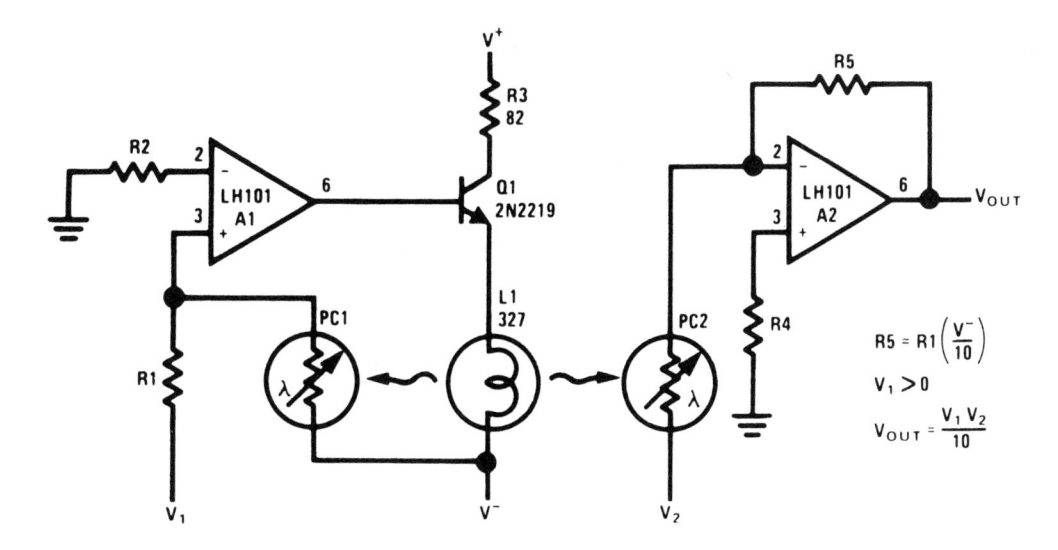

A light-tracking analog multiplier circumvents many of the problems associated with the log/antilog circuit and provides three-quadrant analog multiplication that is relatively temperature insensitive and not subject to the bias-current errors that plague most multipliers.

A2 is a controlled-gain amplifier of V2, whose gain is dependent on the ratio of the resistance of PC2 to R5. A1 is a control amplifier that establishes the resistance of PC2 as a function of V1. In this way, V_{OUT} is a function of both V1 and V2.

A2 acts as an inverting amplifier whose gain is equal to the ratio of the resistance of PC2 to R5. If R5 is chosen equal to the product of R1 and V^-, then V_{OUT} becomes simply the product of V1 and V2. R5 may be scaled in powers of 10 to provide any required output scale factor.

PC1 and PC2 should be matched for best tracking over temperature since the temperature coefficient of resistance is related to resistance match for cells of the same geometry. Small mismatches may be compensated by varying the value of R5 as a scale factor adjustment. The photoconductive cells should receive equal illumination from L1.

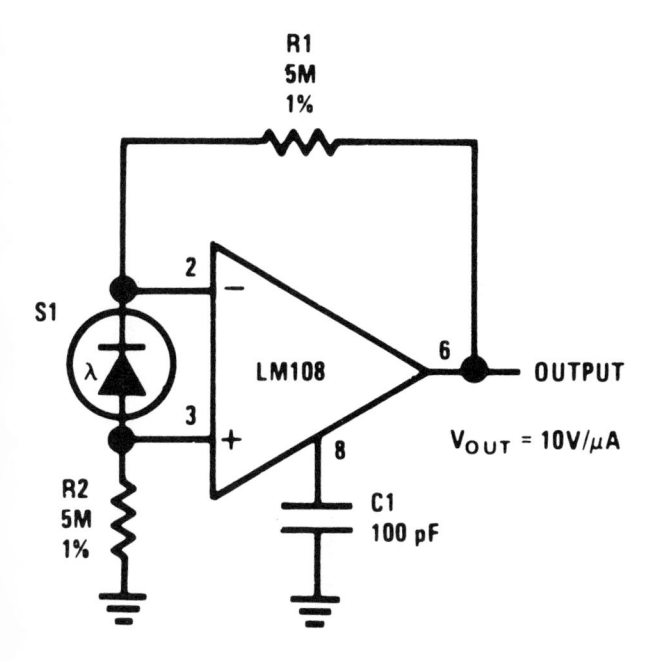

An amplifier for photodiode sensor responds to the short-circuit output current of the photodiode. Since the voltage across the diode is only the offset voltage of the amplifier, inherent leakage is reduced by at least two orders of magnitude. Neglecting the offset current of the amplifier, the output current of the sensor is multiplied by R1 plus R2 in determining the output voltage.

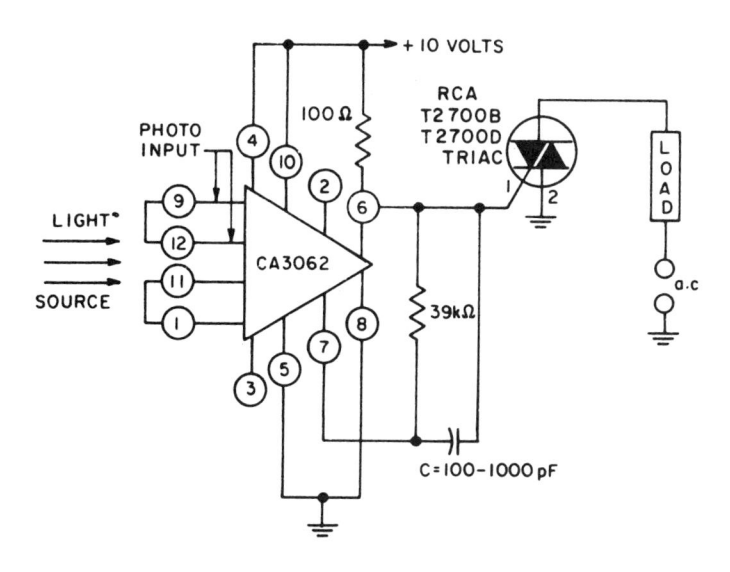

Light-activated triac-controlled load. Supply voltage is furnished to the load when the light source turns on the RCA CA3062, and it is removed when the light path is broken.

SECTION **12** *Optoelectronic Circuits* **425**

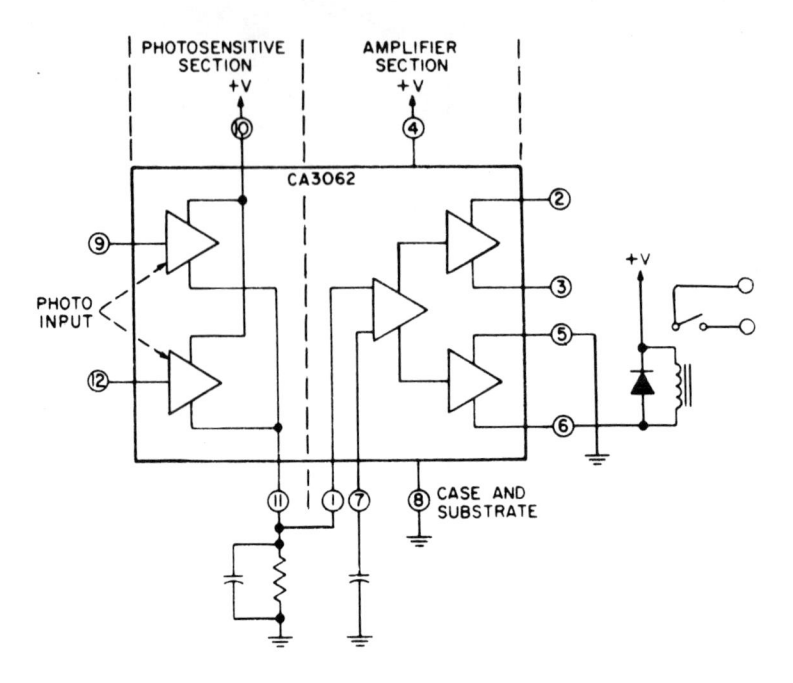

This photodetector and power amplifier for photoelectric control applications features 100 mA output current capability. It can drive a relay or thyristor directly with dc supply voltage of 5 to 15V. The complete system is enclosed in a TO-5 package. The power amplifier has a differential configuration that provides complementing outputs in response to a light input—normally *on* and normally *off*. The separate photodetector, amplifier, and high-current switch provide flexibility of circuit arrangement.

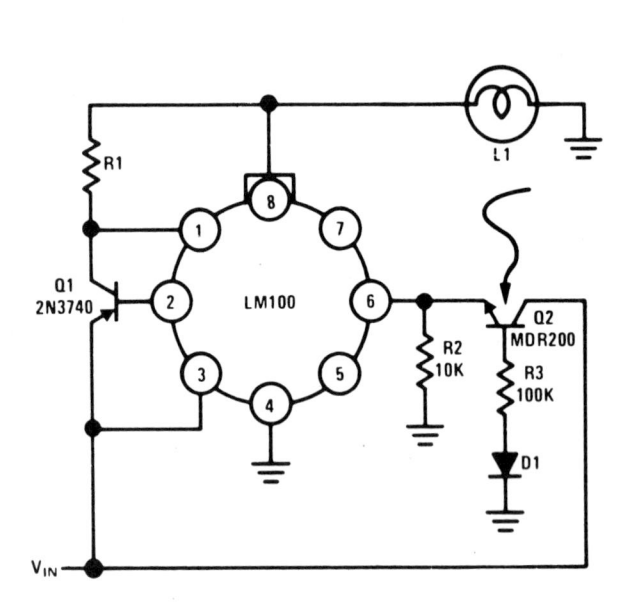

Light-intensity regulator. A phototransistor senses the light level and drives the feedback terminal of the LM100 to control current flow into an incandescent bulb. R1 serves to limit the inrush current to the bulb when the circuit is first turned on.

The current gain of the phototransistor is fixed at 10 to make it less temperature sensitive. The input voltage does not have to be regulated, as the sensitivity of a phototransistor is not greatly affected by the voltage drop across it.

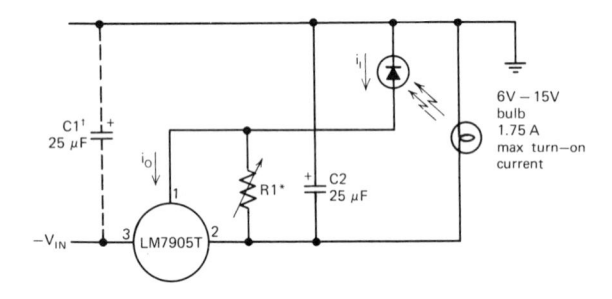

A light controller using silicon photocell. Lamp brightness increases until $i_I = i_Q(\approx 1 \text{ mA}) + 5V/R1$. C1 is only necessary if the raw supply filter capacitor is more than 2 inches from LM5905T.

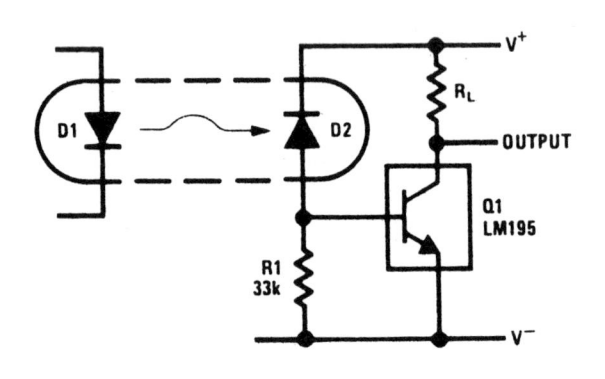

Fast optically isolated switch.

426 IC SCHEMATIC SOURCEMASTER

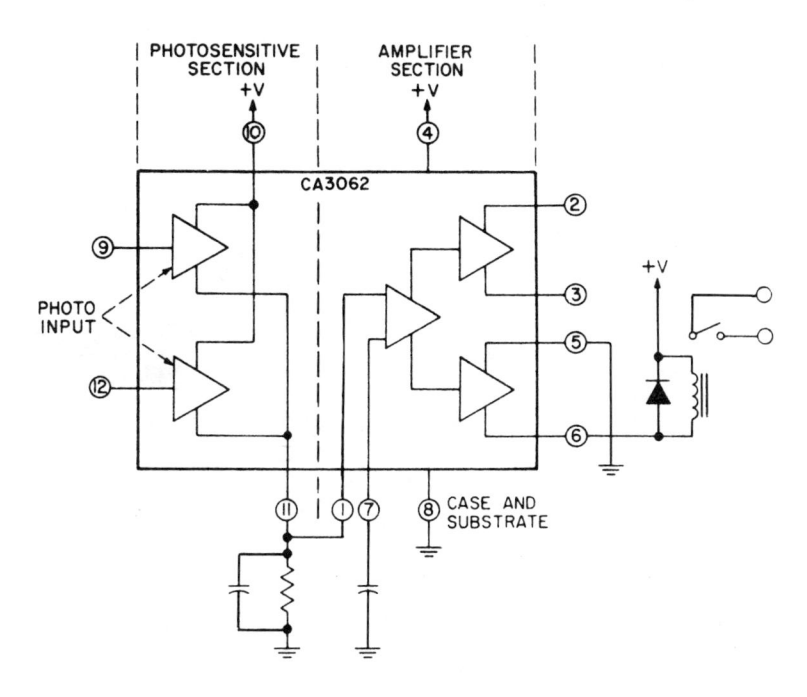

A light-operated relay using CA3062.

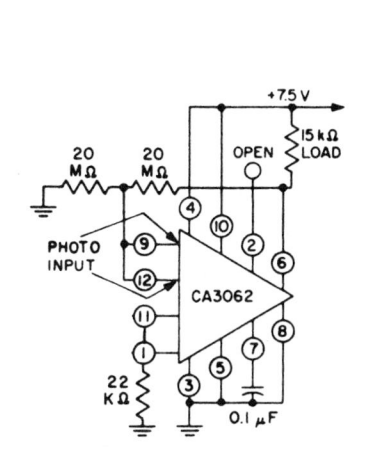

Circuit for linear output photoelectric applications.

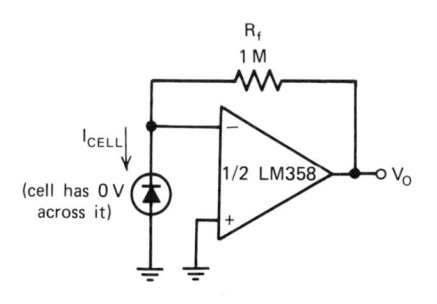

Photovoltaic cell amplifier. Cell has 0V across it.

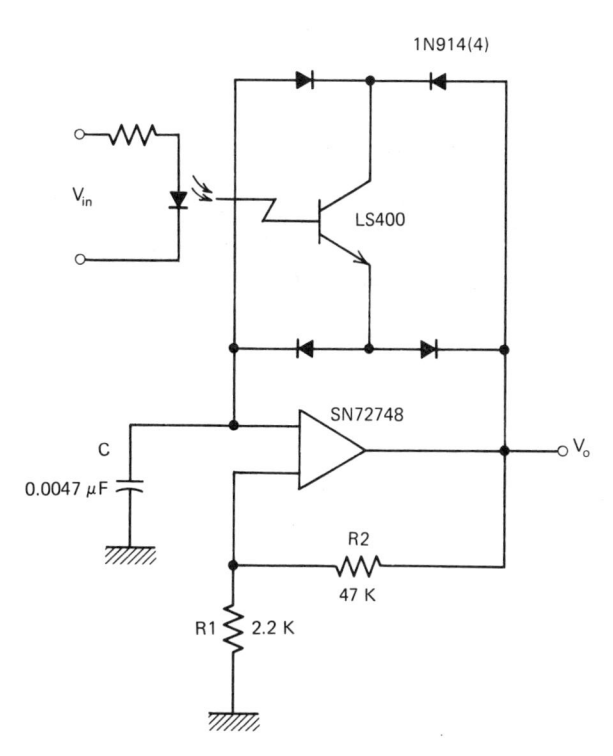

Light-controlled oscillator. A Texas Instruments phototransistor is used in the negative feedback loop of a basic multivibrator. Capacitor C will be charged by a current from the LS400 acting as a current generator. Charging (or discharging) current is proportional to the light level applied. A diode bridge ensures correct flow of charging or discharging currents through the phototransistor. Positive feedback through R2 and R1 provides a Schmitt-trigger type of switching action, and a reference voltage V_{REF} at the noninverting input that is equal to $(V_{0MAX})R1/(R1 + R2)$. The frequency of operation is simply

$$f = \frac{i_{Light}}{4CV_{REF}}$$

SECTION **12** Optoelectronic Circuits

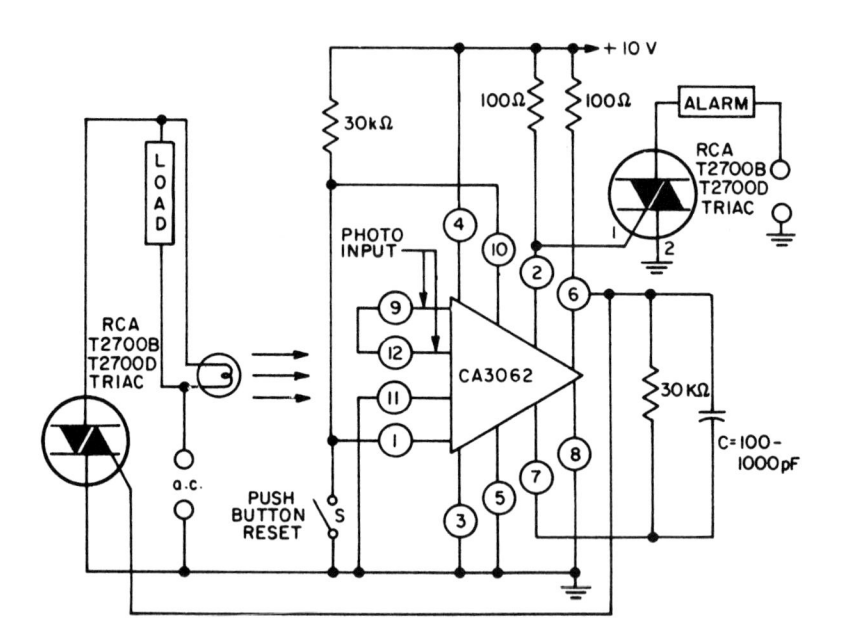

A triac control circuit with safety feature providing **automatic shutoff and alarm**. Alternating voltage is supplied to the load as long as the light source is on. When the light path is broken, the alarm circuit is activated. The left-hand triac controls the load voltage; the alarm is triggered by the other triac.

SECTION 13
ANALOG AND DIGITAL CONVERTERS

Analog-to-digital and digital-to-analog conversion can be achieved in many ways, from cheap-and-dirty (but adequate for a specific purpose) to expensive and dead accurate. The approach you take in a conversion operation will depend on many factors, not the least of which is the speed and accuracy requirements of the system in which such conversion is employed. This section presents but a handful of approaches using specially designed integrated circuits, along with a couple of buffer, comparator, and ladder-network driver circuits.

Converters are assembled from various modular circuits, and these can be found in other sections of this book—the most promising being the logic section (15) the analog and digital transmission section (16), and the section containing oscillators and waveform generators (4).

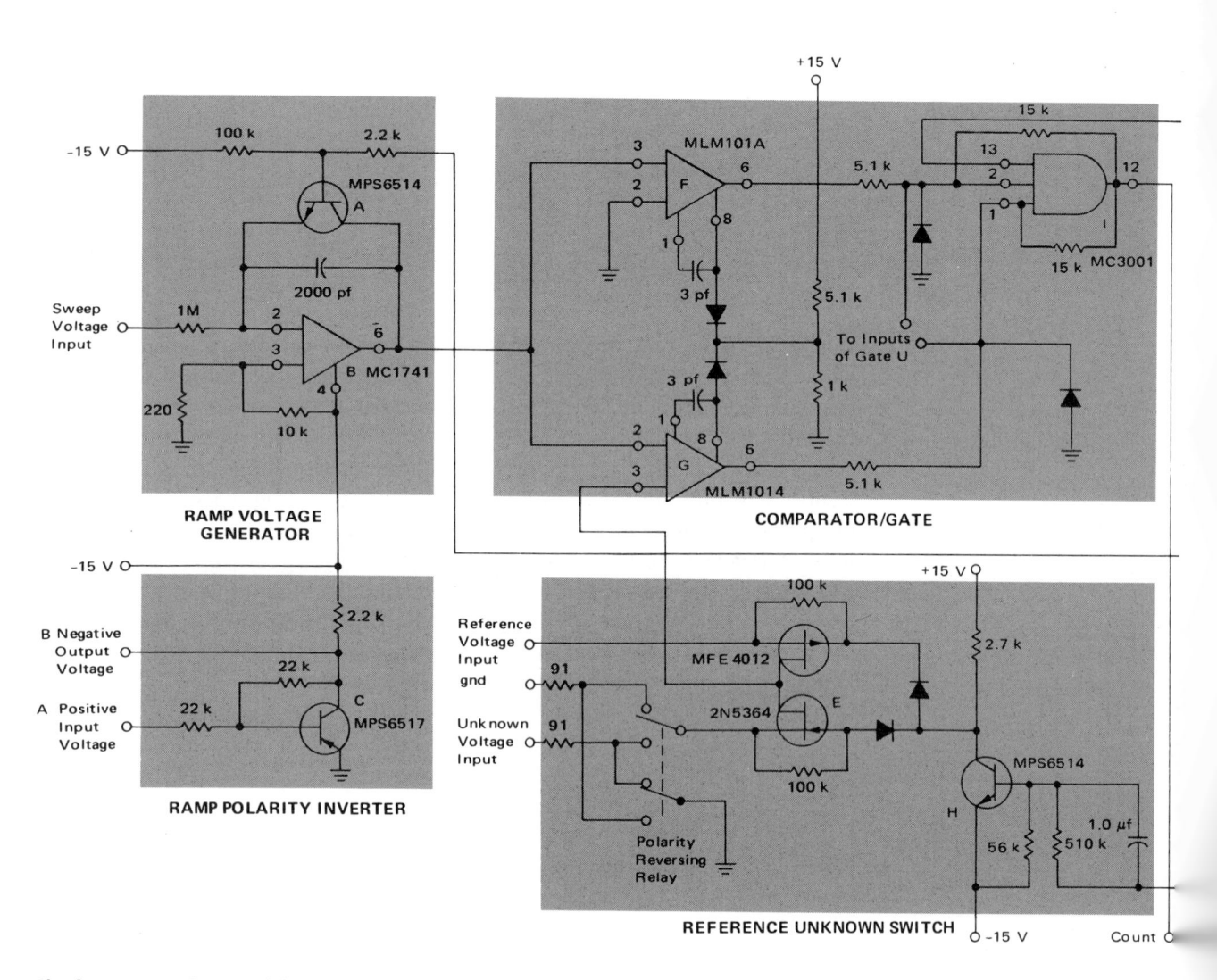

Single-ramp analog-to-digital (A/D) converter.

Clock oscillator frequency:	2 MHz
Gate oscillator frequency:	1.3 kHz
Clock tuning range:	±2.5% or ±50 kHz
Tuning voltage range:	±15V
VCO sensitivity:	3.3 kHz/V
Digital word:	12 bits
Correction increment:	122 Hz
Full-scale reading:	+10V
Calibrate voltage:	+5V
Calibrate time:	1.024 msec
Sweep time (ramp):	3.1 msec
Overrange:	1.052 msec or 50%
Correction rate:	10 Hz/msec
Maximum correction time:	10 seconds
Voltage ramp slope:	5V/msec

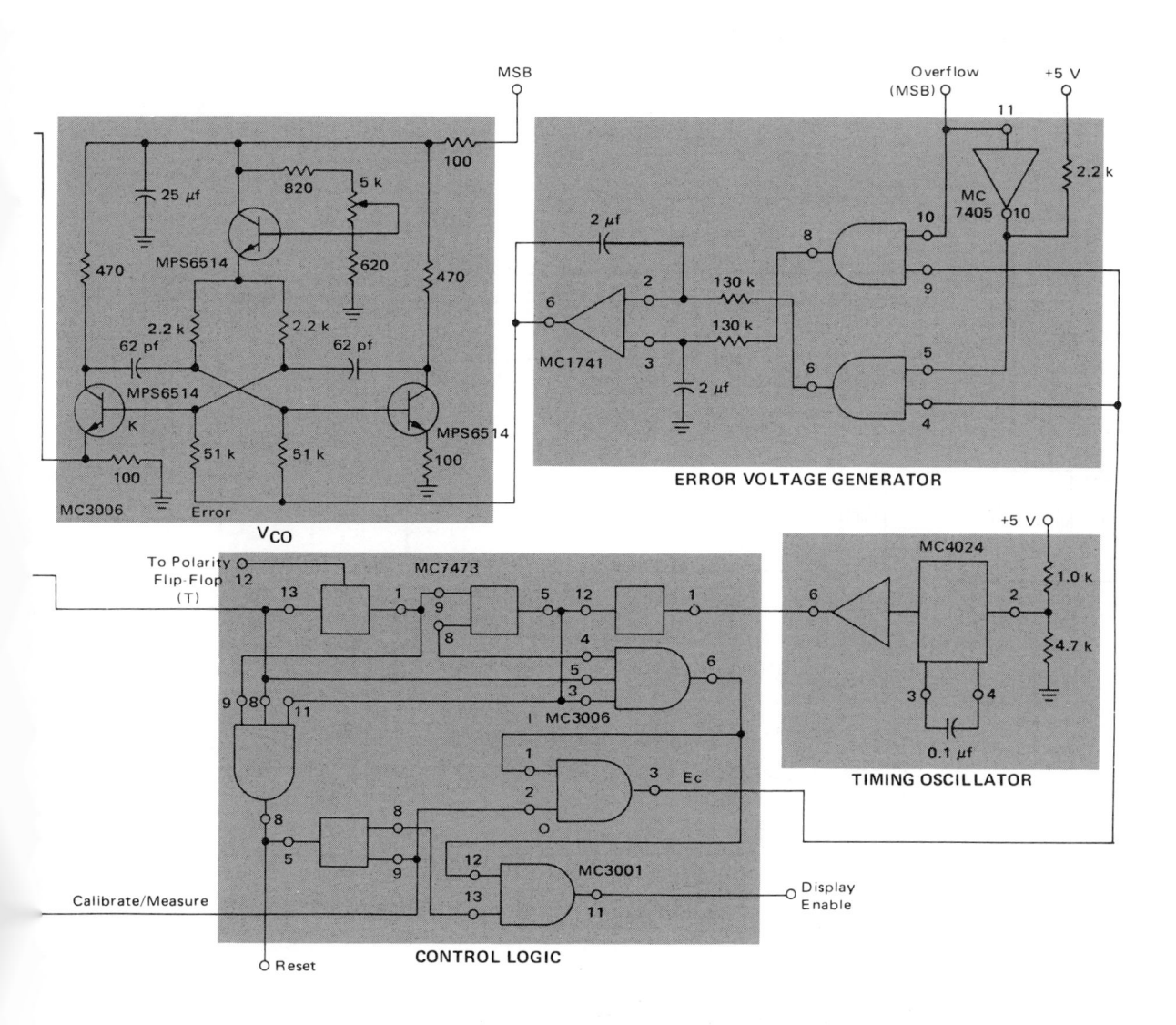

SECTION 13 *Analog and Digital Converters* 431

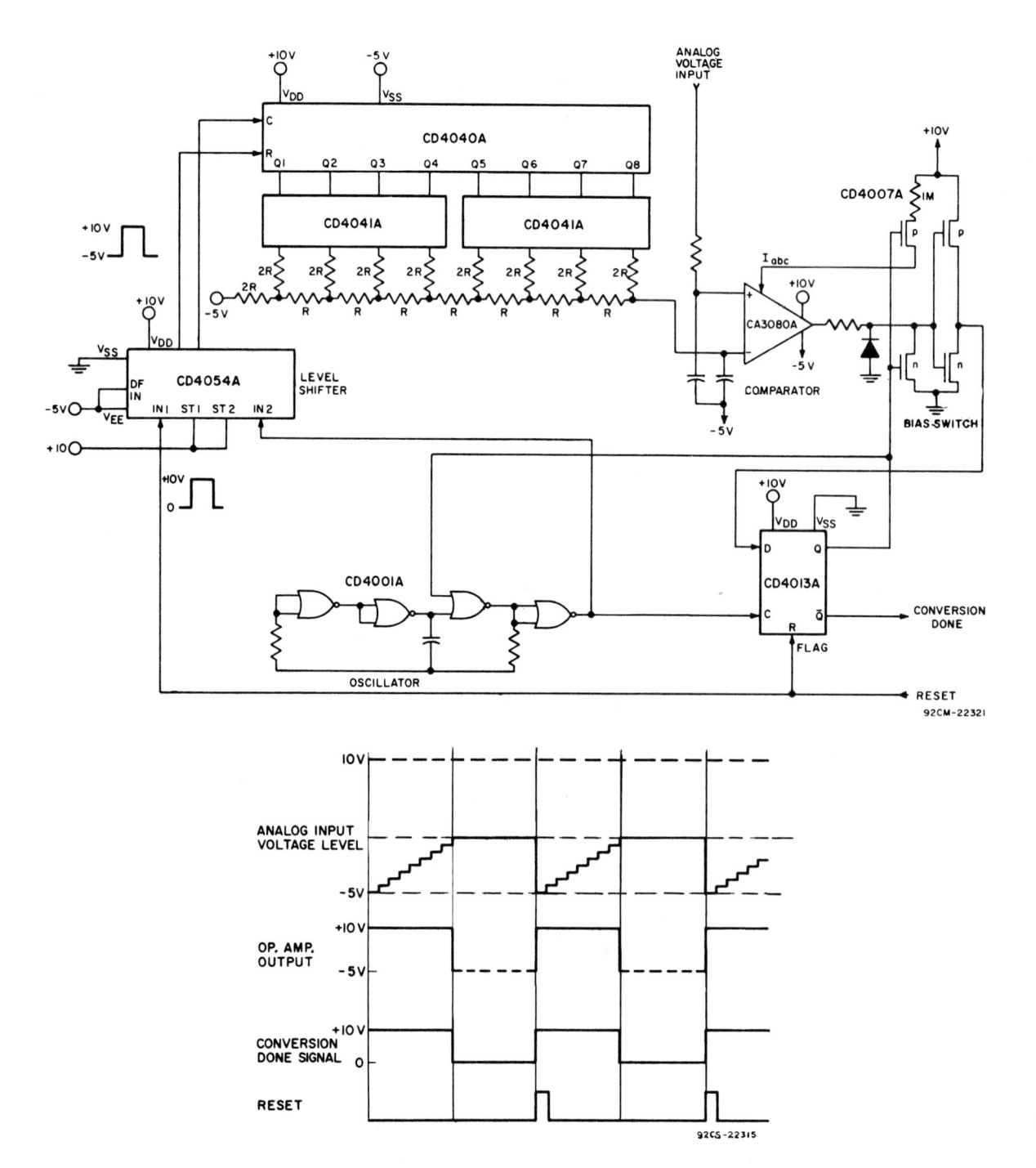

Analog-to-digital converter assembled with standard RCA CMOS parts.

432 IC SCHEMATIC SOURCEMASTER

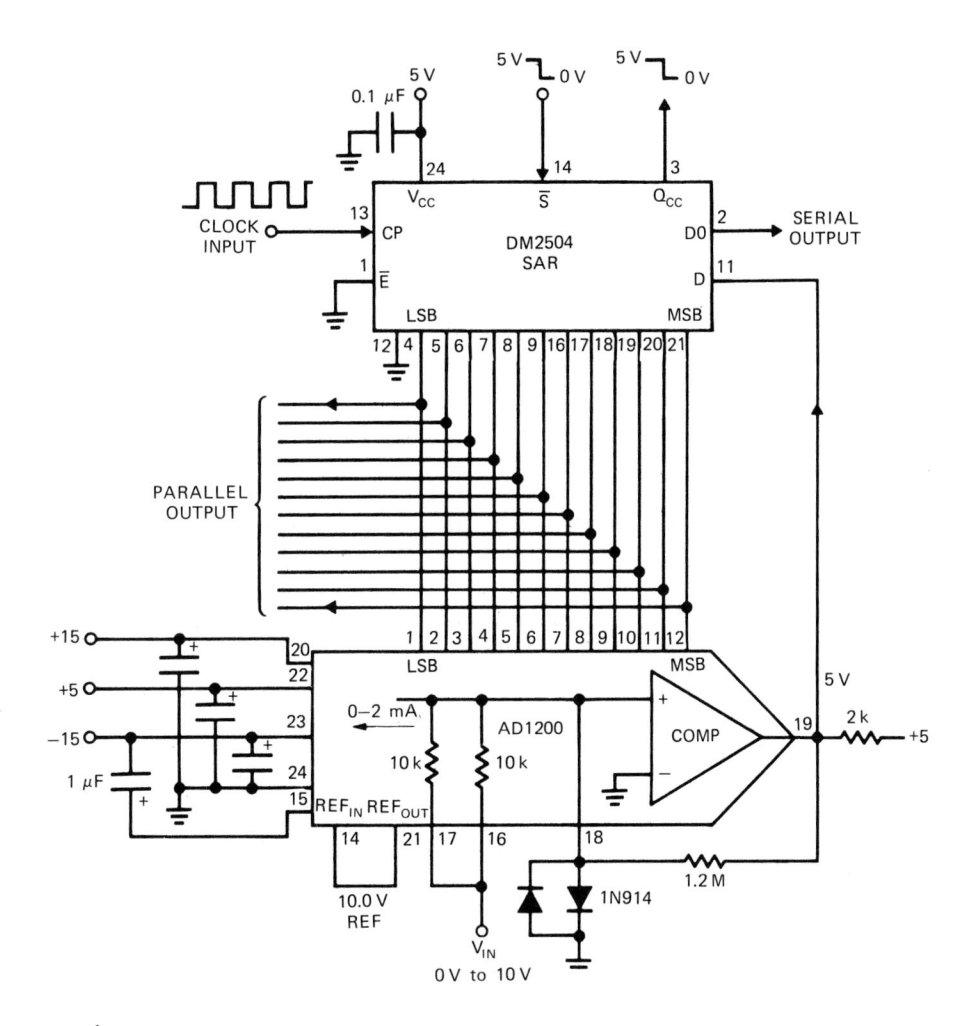

Switching time waveforms

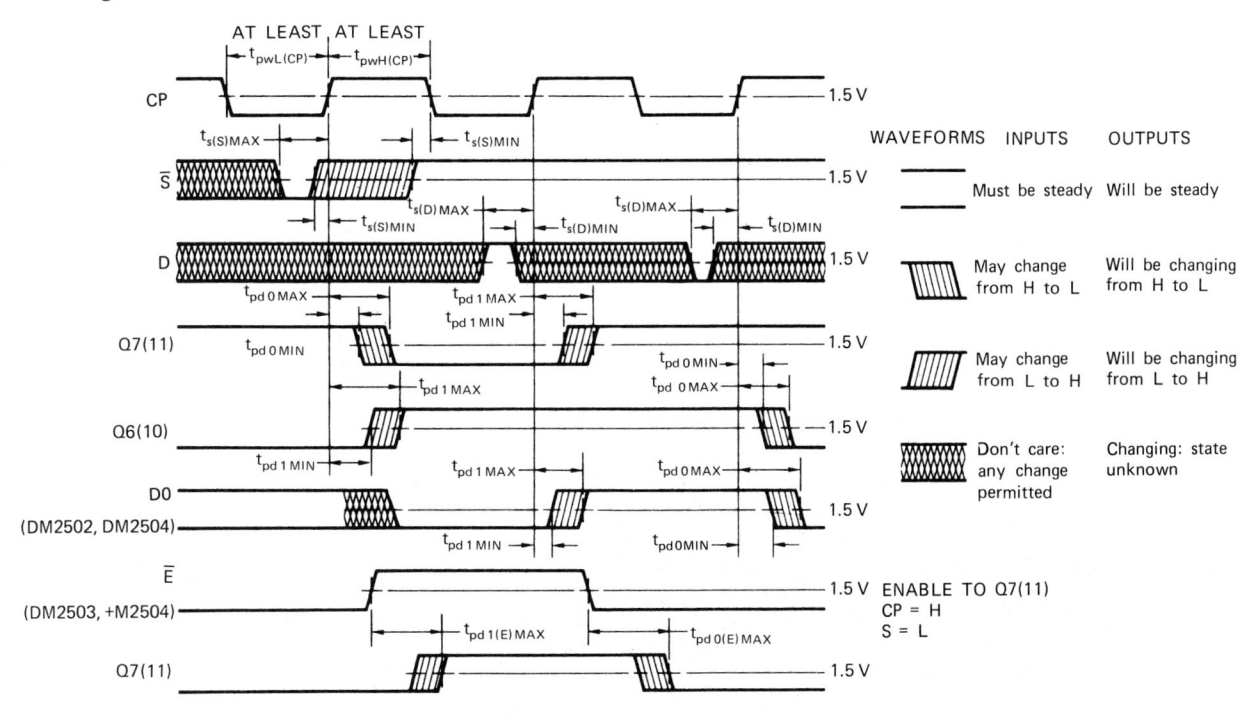

High-speed 12-bit analog-to-digital converter.

SECTION **13** *Analog and Digital Converters*

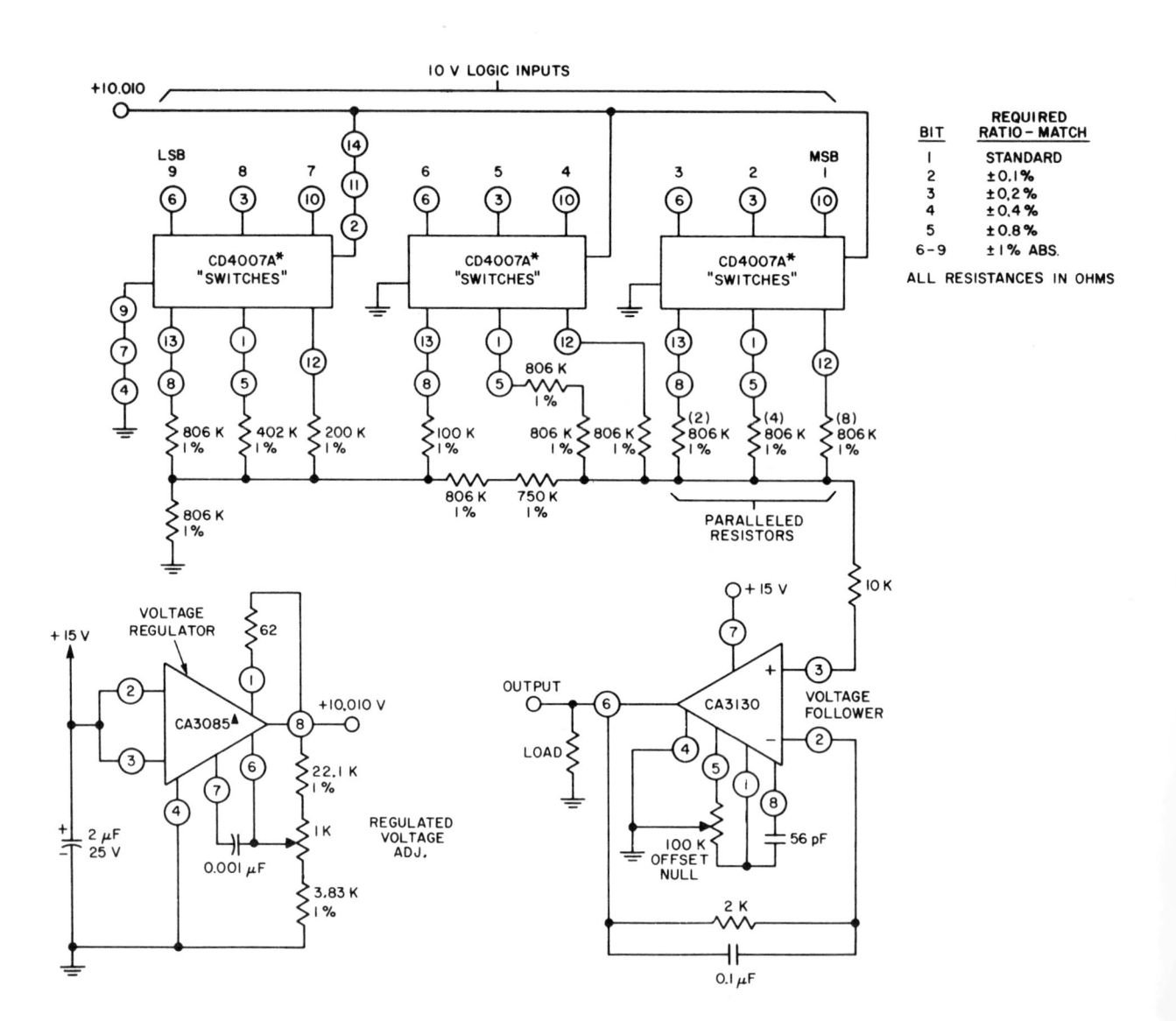

BIT	REQUIRED RATIO – MATCH
1	STANDARD
2	±0.1%
3	±0.2%
4	±0.4%
5	±0.8%
6–9	±1% ABS.

ALL RESISTANCES IN OHMS

A 9-bit digital-to-analog converter (DAC) combines the concepts of multiple-switch CMOS ICs, a low-cost ladder network of discrete metal-oxide film resistors, a CA3130 op-amp connected as a follower, and an inexpensive monolithic regulator in a simple single-supply arrangement. An additional feature of the DAC is that it is readily interfaced with CMOS input logic; for example, 10V logic levels are used in the circuit.

The circuit uses an R/2R voltage-ladder network, with the output potential obtained directly by terminating the ladder arms at either the positive or the negative power supply terminal. Each CD4007A contains three inverters, each functioning a switch to terminate an arm of the R/2R network at either the positive or negative supply terminal. The resistor ladder is an assembly of 1% metal-oxide film resistors. The five arms requiring the highest accuracy are assembled with series and parallel combinations of 806,000 Ω resistors from the same manufacturing lot.

A single 15V supply provides a positive bus for the CA3130 follower amplifier and feeds the CA3085 voltage regulator. A "scale adjust" function is provided by the regulator output control, set to a nominal 10V level in this system. The line-voltage regulation (approximately 0.2%) permits a 9-bit accuracy to be maintained with variations of several volts in the supply. The flexibility afforded by the CMOS building blocks simplifies the design of DAC systems tailored to particular needs.

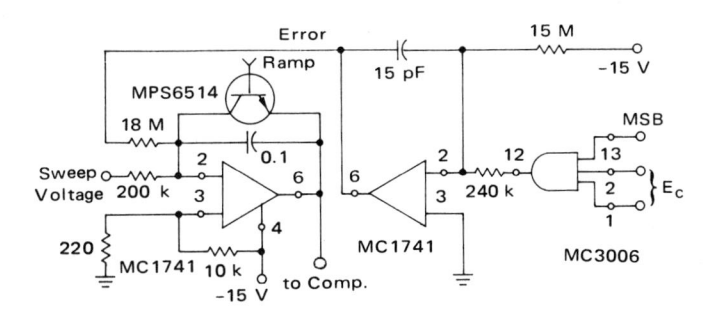

Ramp and error generator. The error input to the integrating amplifier is single-ended; that is, error pulses are supplied only when the gate count is too high. With no error pulses, the integrated error signal drifts slowly positive; with continuous error pulses, it steps more negative at the same average rate.

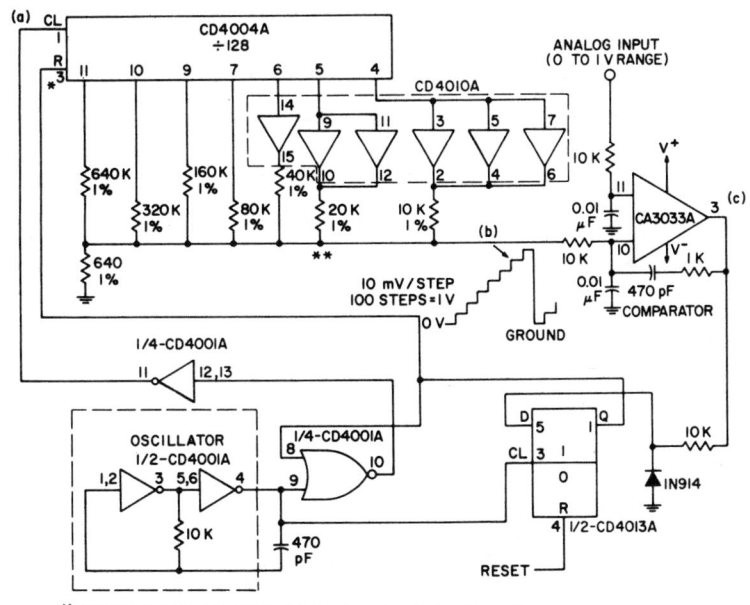

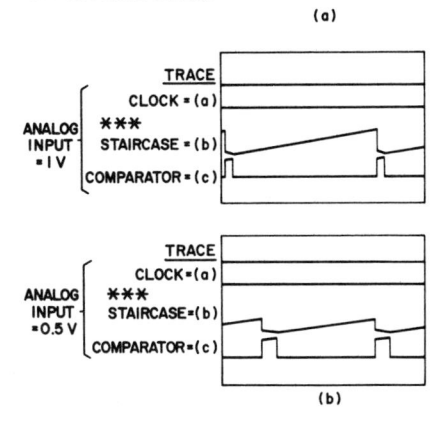

RCA's CD4004A CMOS IC and weighted resistor network in an **analog-to-digital conversion** application. The waveforms show voltage levels at various circuit points.

SECTION **13** *Analog and Digital Converters* 435

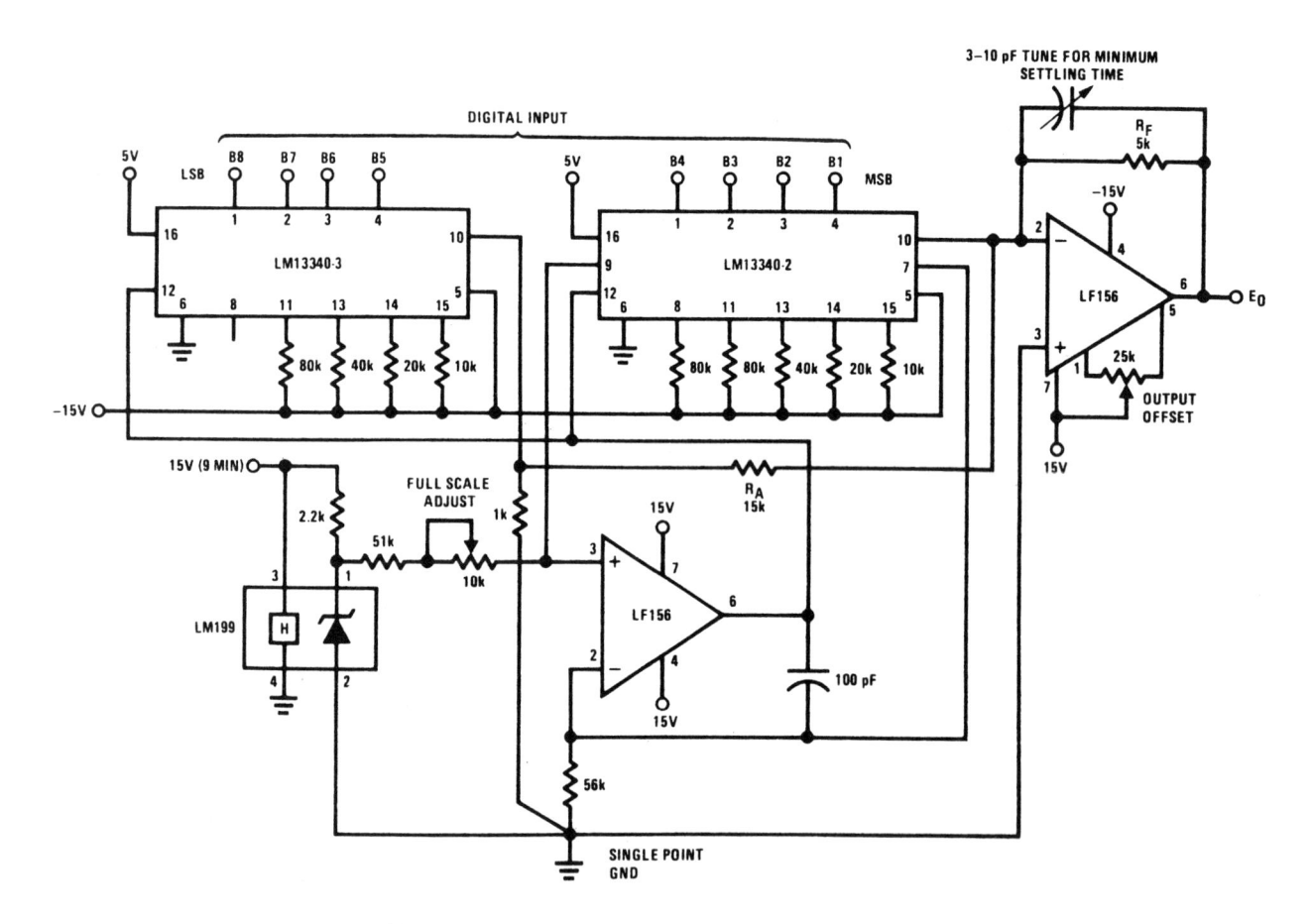

Temperature-compensated 8-bit digital-to-analog converter. Use LF155/6 for fast settling time, low V_{OS} drift, and good stability. For BCD weighting, R_F should be 4k and R_A, 9k. (Resistors R1–R4 form binary ladder.)

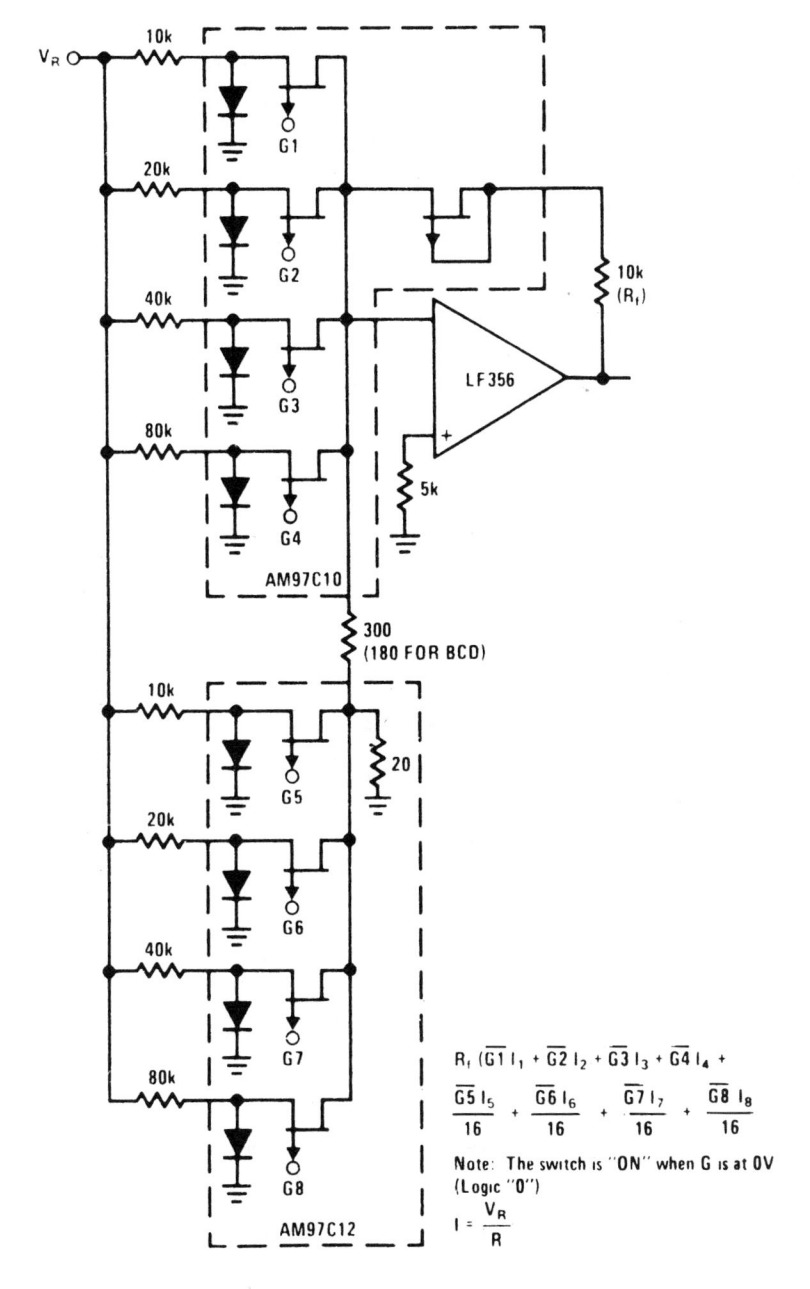

$$R_1 \left(\overline{G1}\,I_1 + \overline{G2}\,I_2 + \overline{G3}\,I_3 + \overline{G4}\,I_4 + \frac{\overline{G5}\,I_5}{16} + \frac{\overline{G6}\,I_6}{16} + \frac{\overline{G7}\,I_7}{16} + \frac{\overline{G8}\,I_8}{16} \right)$$

Note: The switch is "ON" when G is at 0V (Logic "0")

$$I = \frac{V_R}{R}$$

An 8-bit binary (BCD) multiplying digital-to-analog converter.

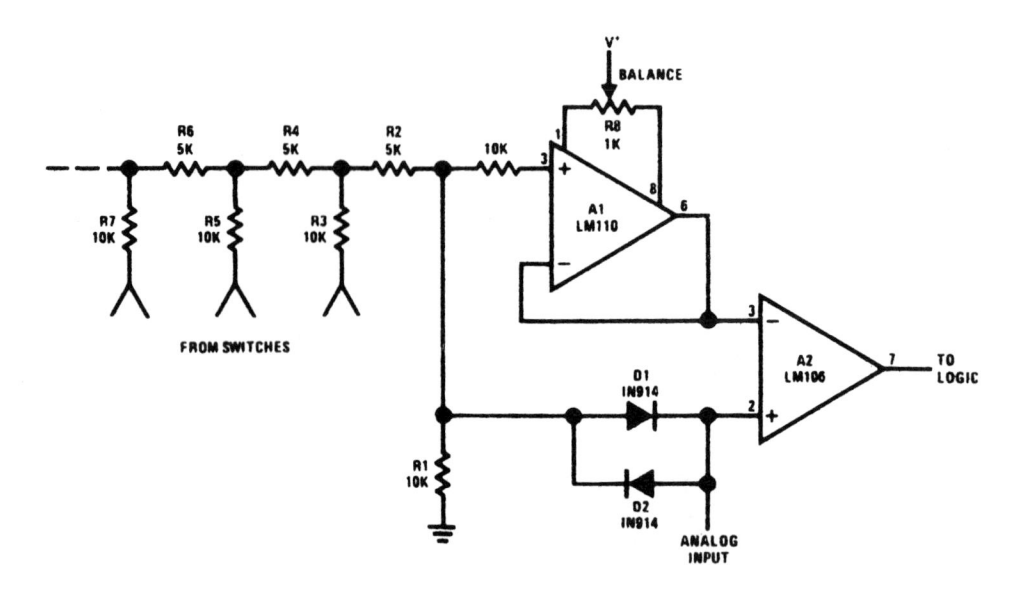

Comparator for A/D converter using a ladder network.

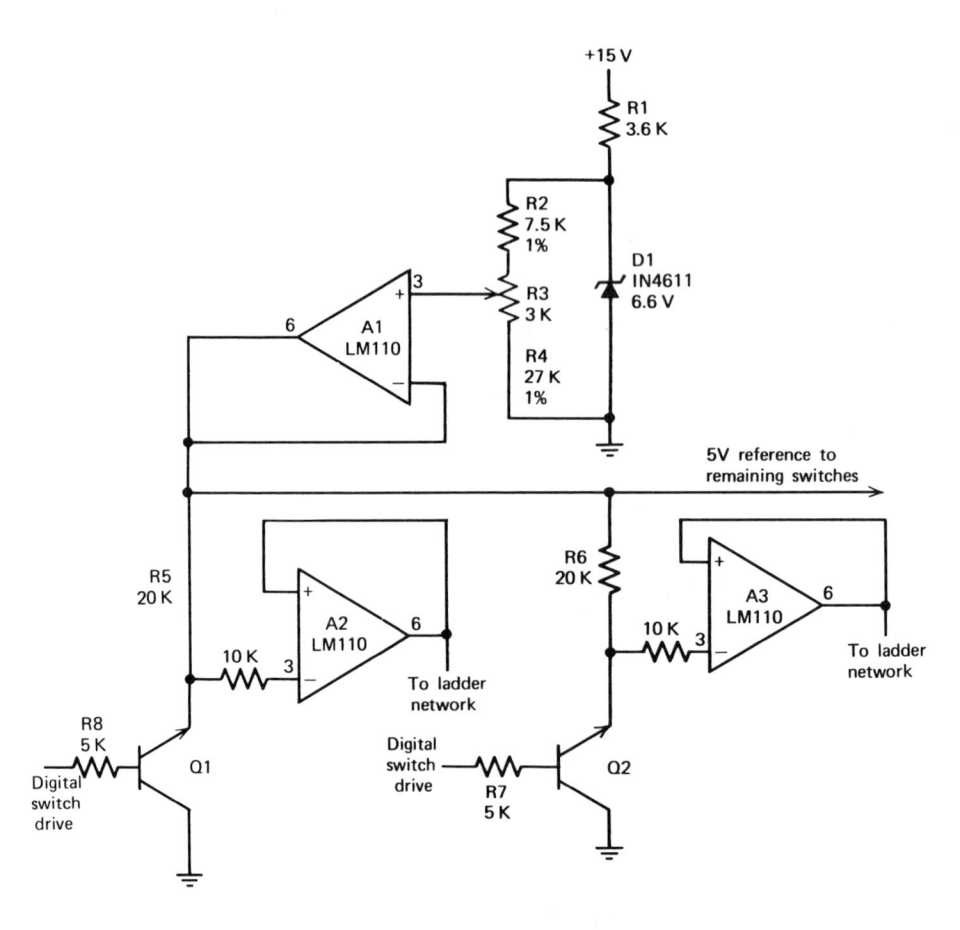

A/D ladder network driver.

438 IC SCHEMATIC SOURCEMASTER

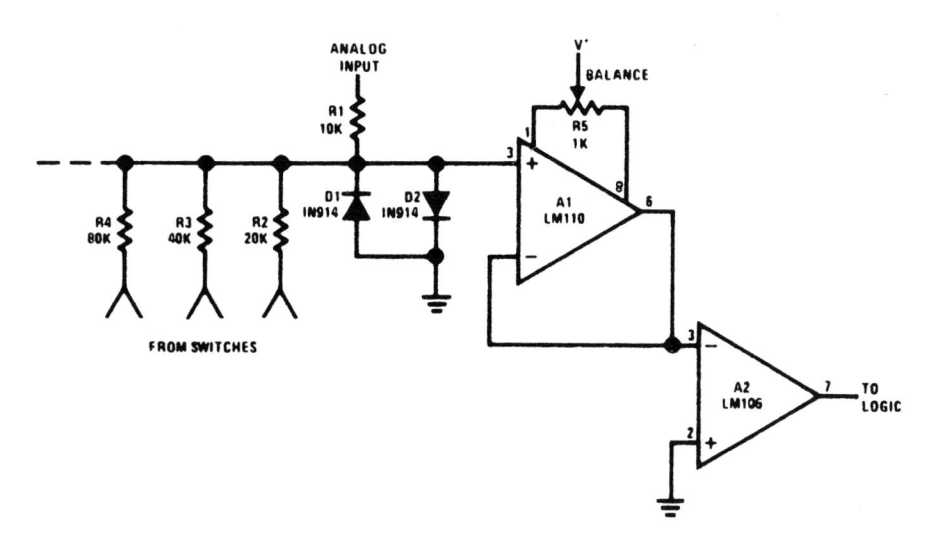

Comparator for analog-to-digital (A/D) converter using a binary-weighted network.

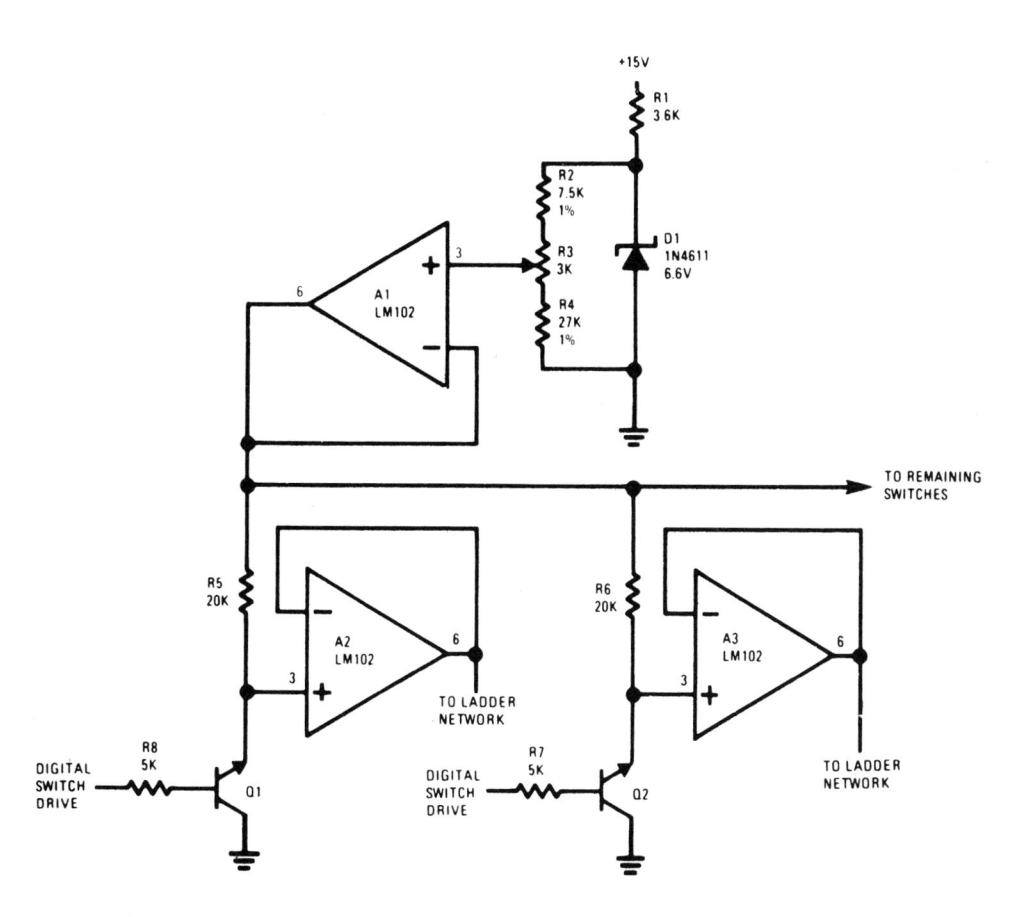

Using the LM102 to drive the ladder network in an A/D converter.

SECTION **13** *Analog and Digital Converters*

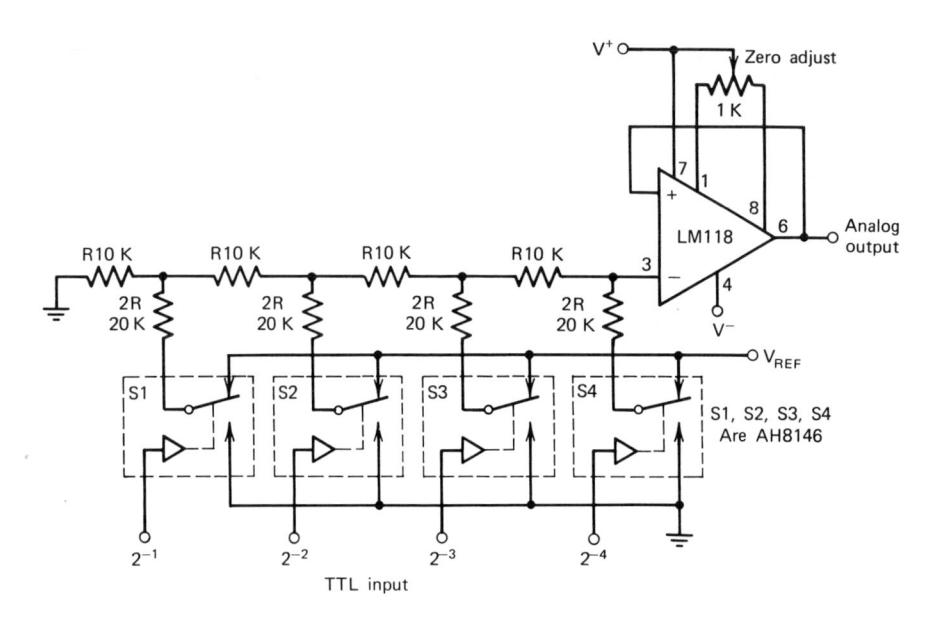

4-to-10-bit digital-to-analog converter (4 bits shown). Settling time, 1 µsec; accuracy, 0.2%. All resistors are 0.1%.

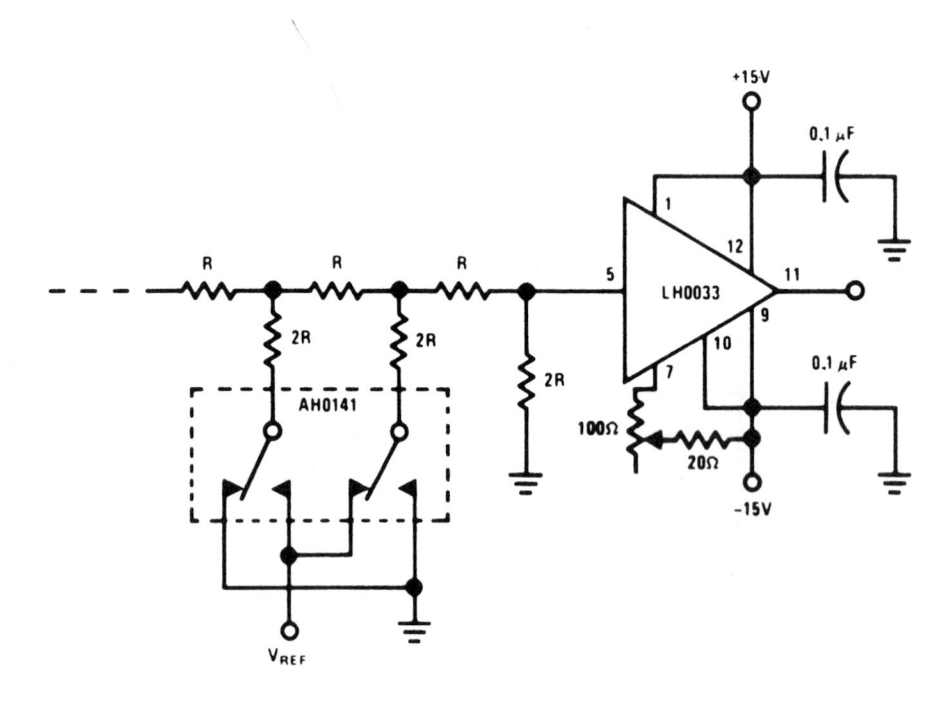

Medium-speed digital-to-analog converter buffer. Offset null is accomplished by connecting a 100Ω pot between pin 7 and V⁻. The 20Ω resistor in series with the pot prevents excessive power dissipation in the LH0033 when the pot is shorted out. In noncritical or ac-coupled applications, pin 6 should be shorted to pin 7. The resulting output offset is typically 5 mV at 25°C.

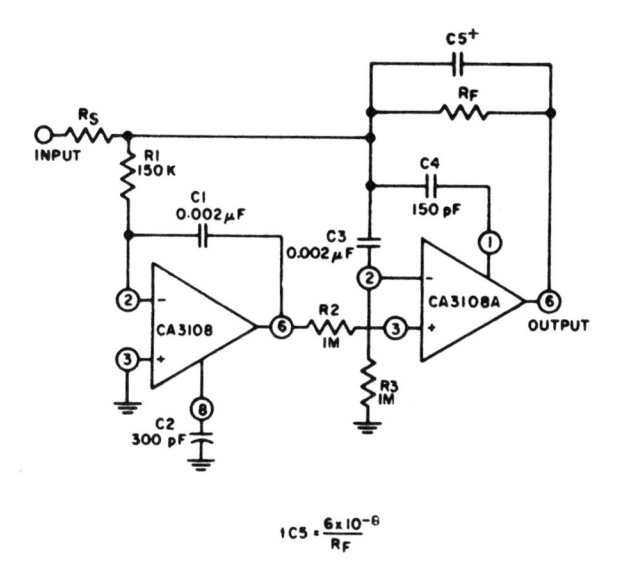

Fast summing amplifier circuit. In this circuit, the power bandwidth is 250 kHz and the small-signal bandwidth is 3.5 MHz. Slew rate is 10V/μsec.

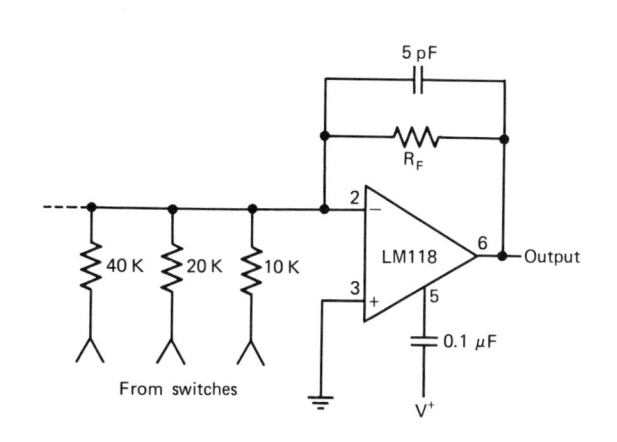

$$1C5 = \frac{6 \times 10^{-6}}{R_F}$$

A digital-to-analog (D/A) converter using binary weighted network. The 0.1 μF capacitor is optional for reduced settling time.

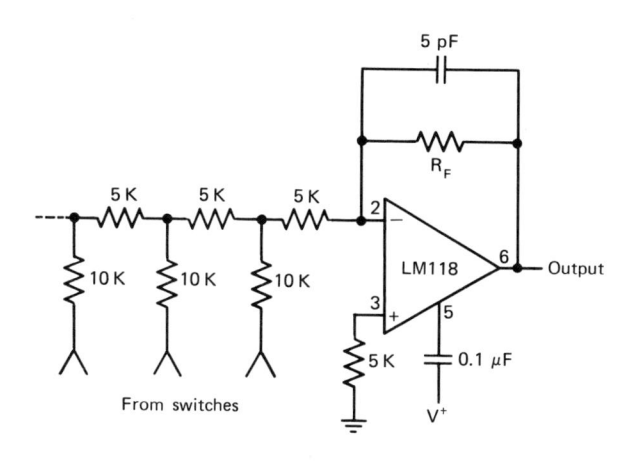

A D/A converter using a ladder network. The 0.1 μF capacitor is optional for reduced settling time.

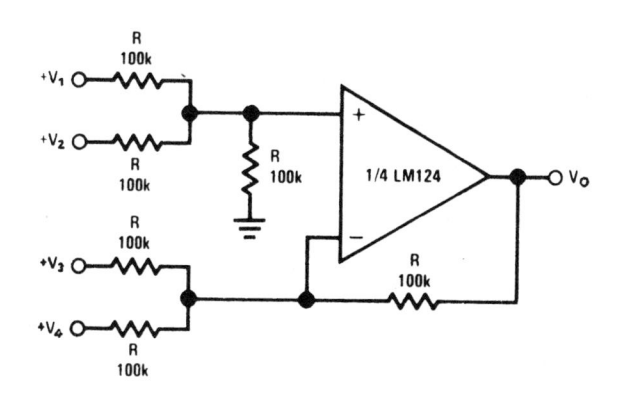

Dc-summing amplifier. $V_{IN'S} \geq$ 0V dc and $V_o \geq$ 0V dc, where V_o = V1 + V2V3V4; (V1 + V2) ≥ (V3 + V4) to keep $V_o >$ 0V dc.

SECTION **13** Analog and Digital Converters

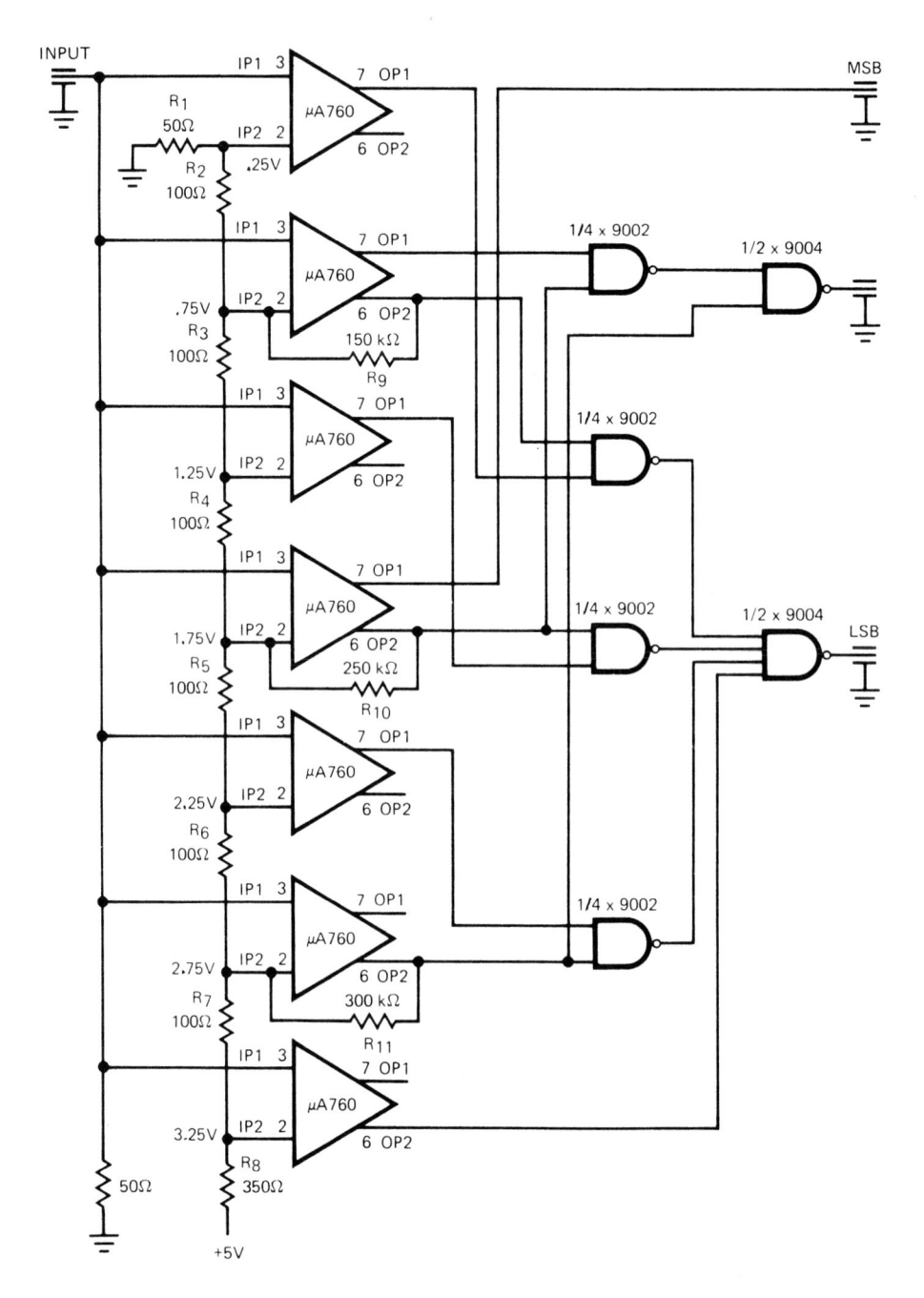

High-speed analog-to-digital converter. Each Fairchild μA760 has one input biased to a voltage level at which the digital equivalent of the input signal should change. The voltages shown along the resistor chain R_1 to R_8 are the theoretical voltages at which each comparator should switch. When comparators 1, 4, and 6 switch, they may affect more than one bit. The switching levels set by the resistor chain are slightly modified by the effect of bias currents: When comparator 3 is about to switch, the input supplies the total bias current of comparators 1 and 2. This is equivalent to the sum of the two bias currents, since there is a large differential voltage between the inputs. The input also will supply half the total bias current of comparator 3, which has zero differential input voltage.

442 IC SCHEMATIC SOURCEMASTER

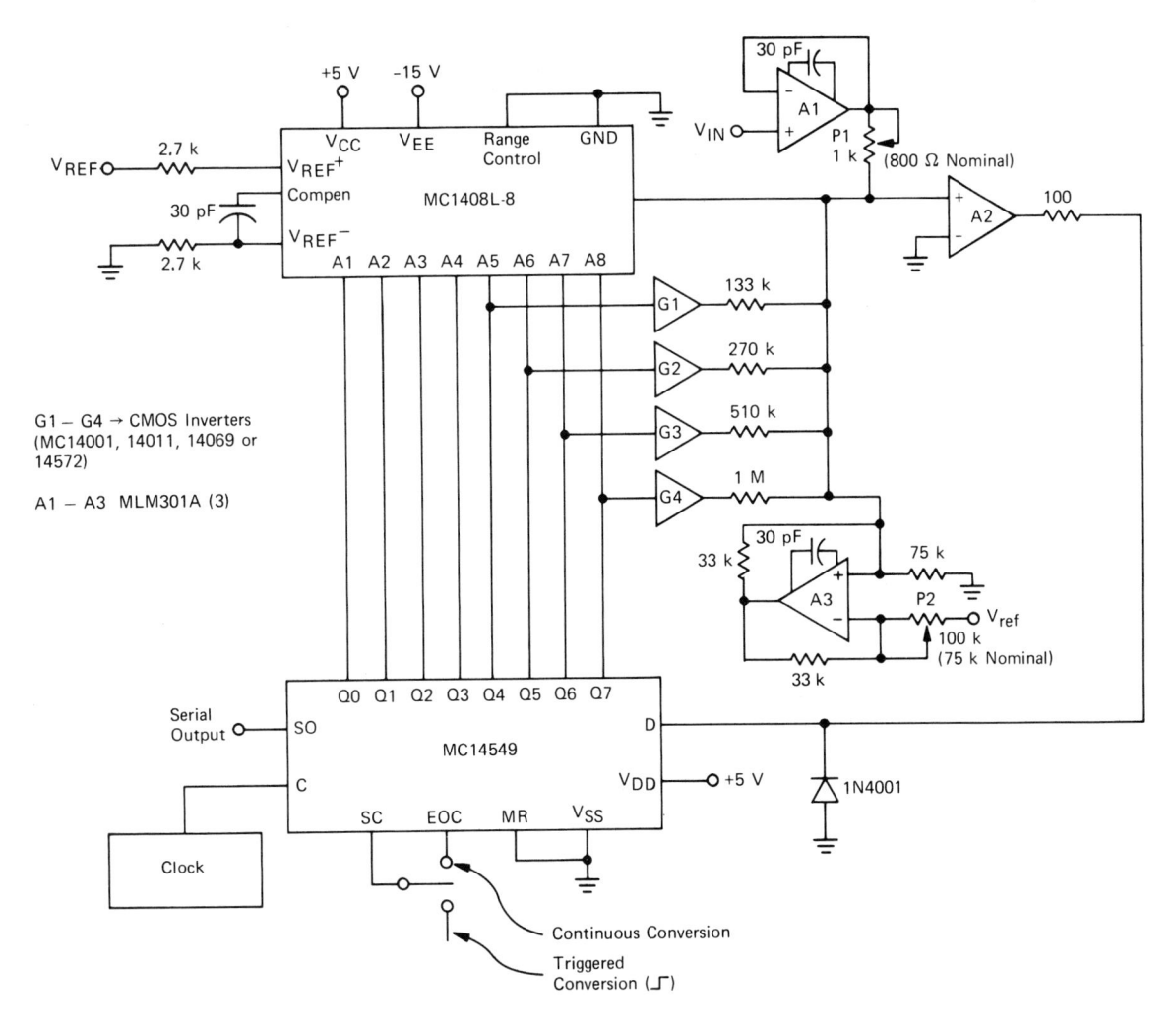

Successive-approximation BCD analog-to-digital converter. A digital-to-analog converter (DAC) is used in a feedback loop to produce a digital output from an analog input. The internal resistor ladder network of the MC1408L-8, an 8-bit binary DAC, has been modified externally to produce a BCD-weighted DAC. A voltage-controlled current source consisting of CMOS inverting gates G1–G4, resistors R1–R4, and op-amp A3, with associated resistors, is used to alter the DAC weighting from binary to BCD. Full scale for the analog-to-digital conversion is 0.99V in 10 mV increments. Its input is buffered by op-amp A1 connected as a voltage follower with potentiometer P1 to supply the output of the MC1408L-8 with a current proportional to the unknown input voltage. This current is compared to that required by the total BCD DAC which includes the MC1408L-8 current and the ladder modifying network current. Potentiometer P1 is used for full-scale calibration, and potentiometer P2, in the op-amp current source, is used for calibration of the least significant digit.

SECTION **13** *Analog and Digital Converters*

SECTION 14
DATA ACQUISITION, MULTIPLEXING, AND TRANSMISSION

This section contains sample-and-hold circuits, a variety of multiplexer system circuitry, signal level and zero-crossing detectors, and various drivers, isolators, and interface circuits. Since many of these circuits are multifunctional they could logically have been grouped under other section heads within this volume. For a complete overview of related circuits, be sure to check the section containing communications circuits and the logic circuits section. Also applicable: Sections 1, 4, and 12.

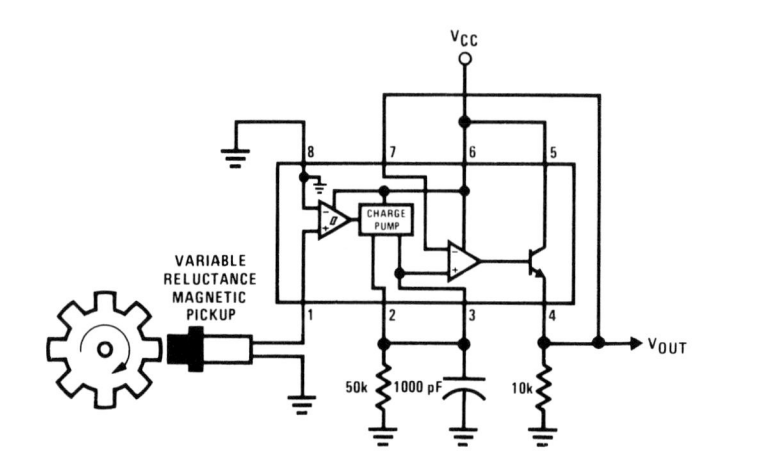

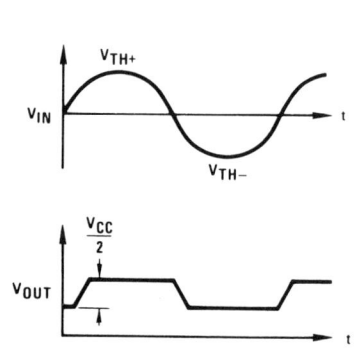

A zero-crossing detector and line drivers for applications where frequency doubling is not required and a square-wave output is preferred. In this case, the output is a swing of half supply voltage starting at 1V_{BE} below three-quarters of supply. This can be increased up to the full output swing capability by reducing or removing the negative feedback around the op-amp.

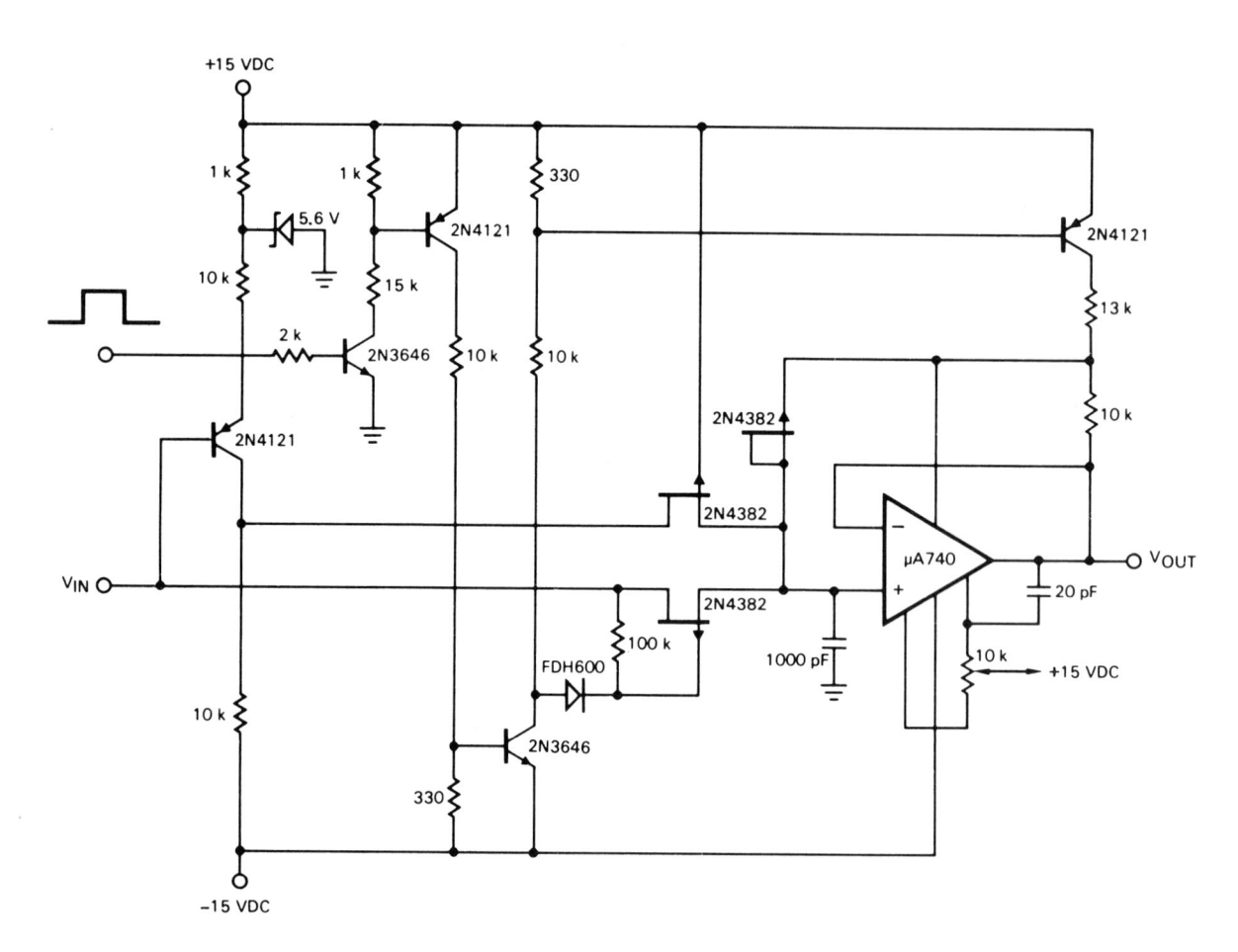

A sample-and-hold circuit with sampling offset compensation and *hold*-mode transient compensation. The 20 pF capacitor between the output of the µA740 and its offset null terminal dampens the overshoot of the amplifier, which occurs following a large (±5V) output swing of the µA740.

446 IC SCHEMATIC SOURCEMASTER

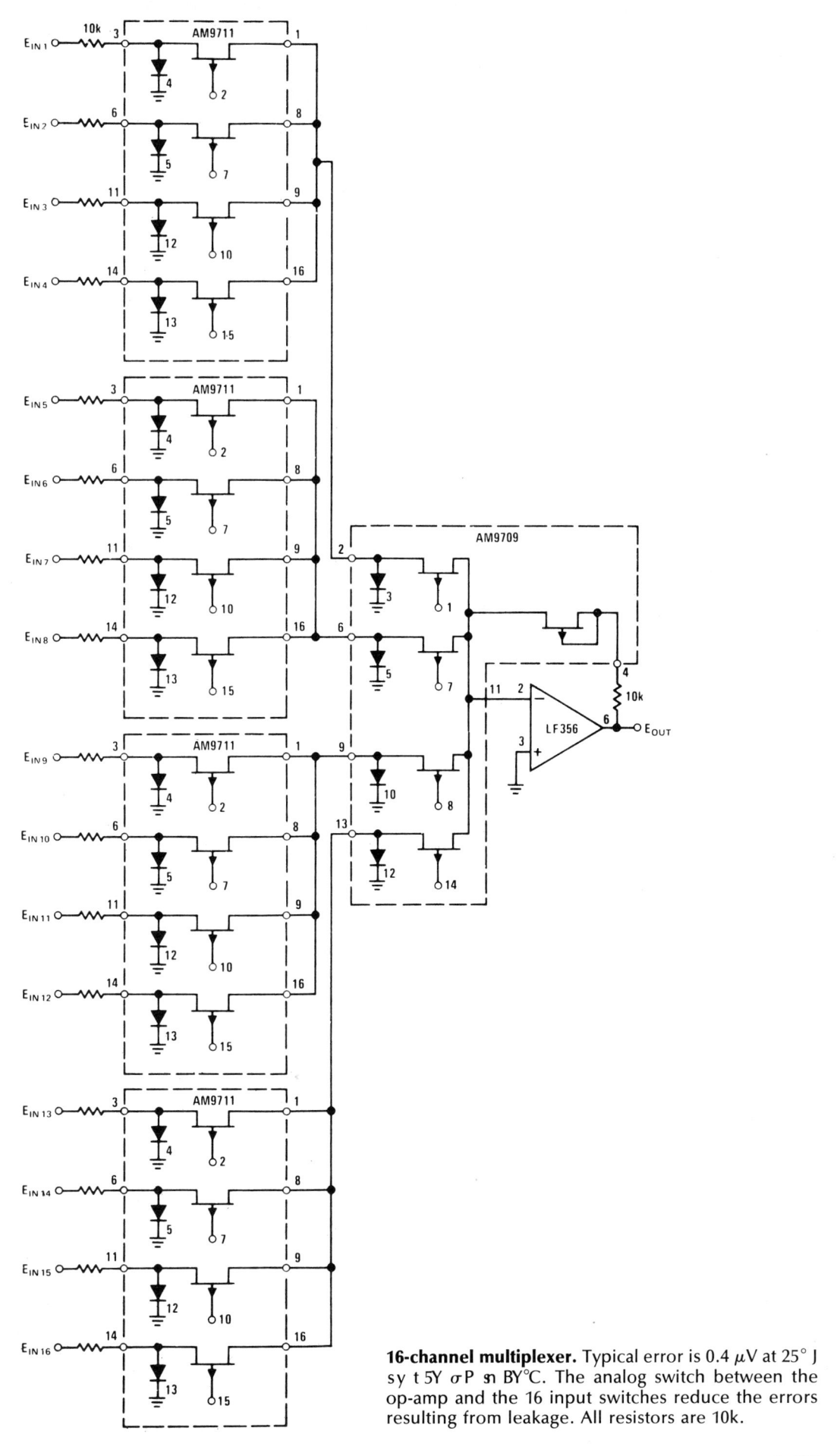

16-channel multiplexer. Typical error is 0.4 μV at 25° J sy t 5Y σP sn BY°C. The analog switch between the op-amp and the 16 input switches reduce the errors resulting from leakage. All resistors are 10k.

SECTION 14 *Data Acquisition, Multiplexing, & Transmission* 447

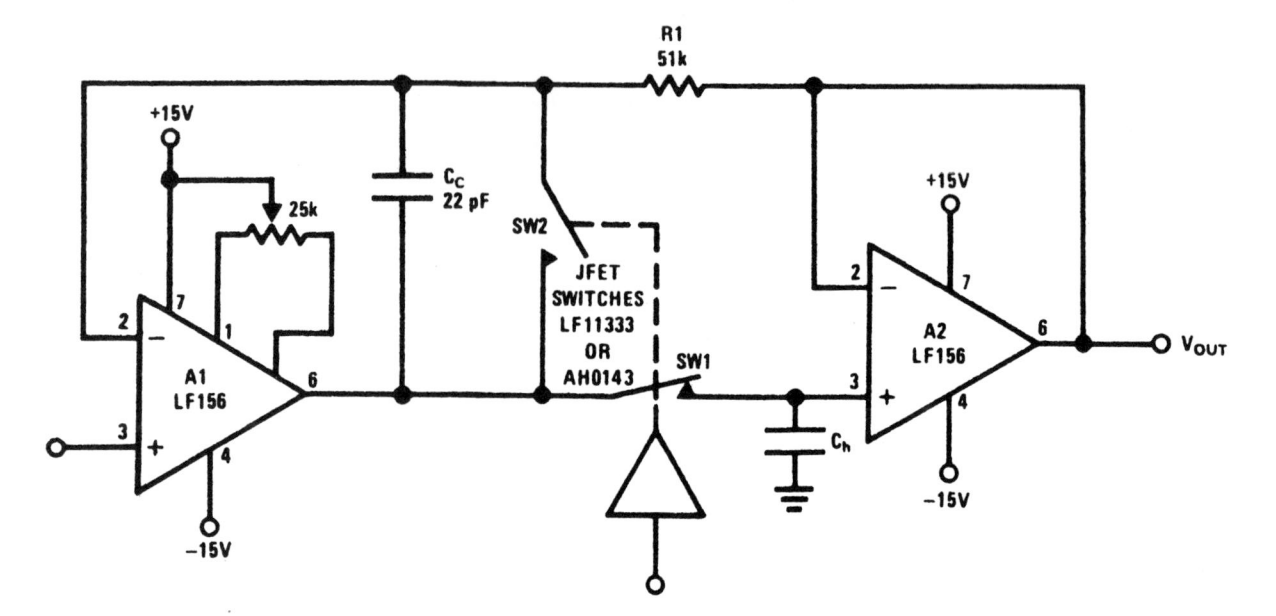

High-accuracy sample-and-hold circuit. By closing the loop through A2, the V_{OUT} accuracy will be determined uniquely by A1. No V_{OS} adjust is required for A2. T_A can be estimated, but because of the added propagation delay in the feedback loop (A2), the overshoot is not negligible. Overall system is slower than the fast sample-and-hold circuit.

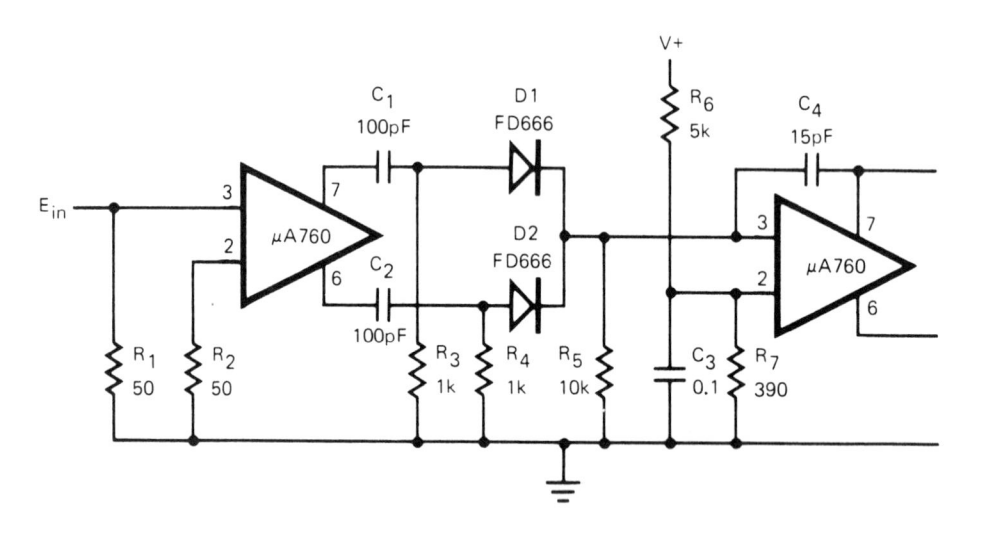

Zero-crossing detector. The outputs of the first µA760 change state each time the input signal passes through zero. Diodes D_1 and D_2, capacitors C_1 and C_2, and resistors R_3, R_4, and R_5 generate a short, positive pulse that is applied to input 1 of the second µA760. A positive pulse results at output 1 of the second µA760. This pulse represents the timing of the input zero crossing. The timing measurement error is equal to the circuit delay, typically 30 nsec. Negative pulses are obtained from the other output. The output pulse widths can be adjusted by varying R_6. The circuit functions at frequencies from 300 Hz to 4 MHz with a signal of about 3 mV RMS at 4 MHz, rising to about 7V at 300 Hz.

448 IC SCHEMATIC SOURCEMASTER

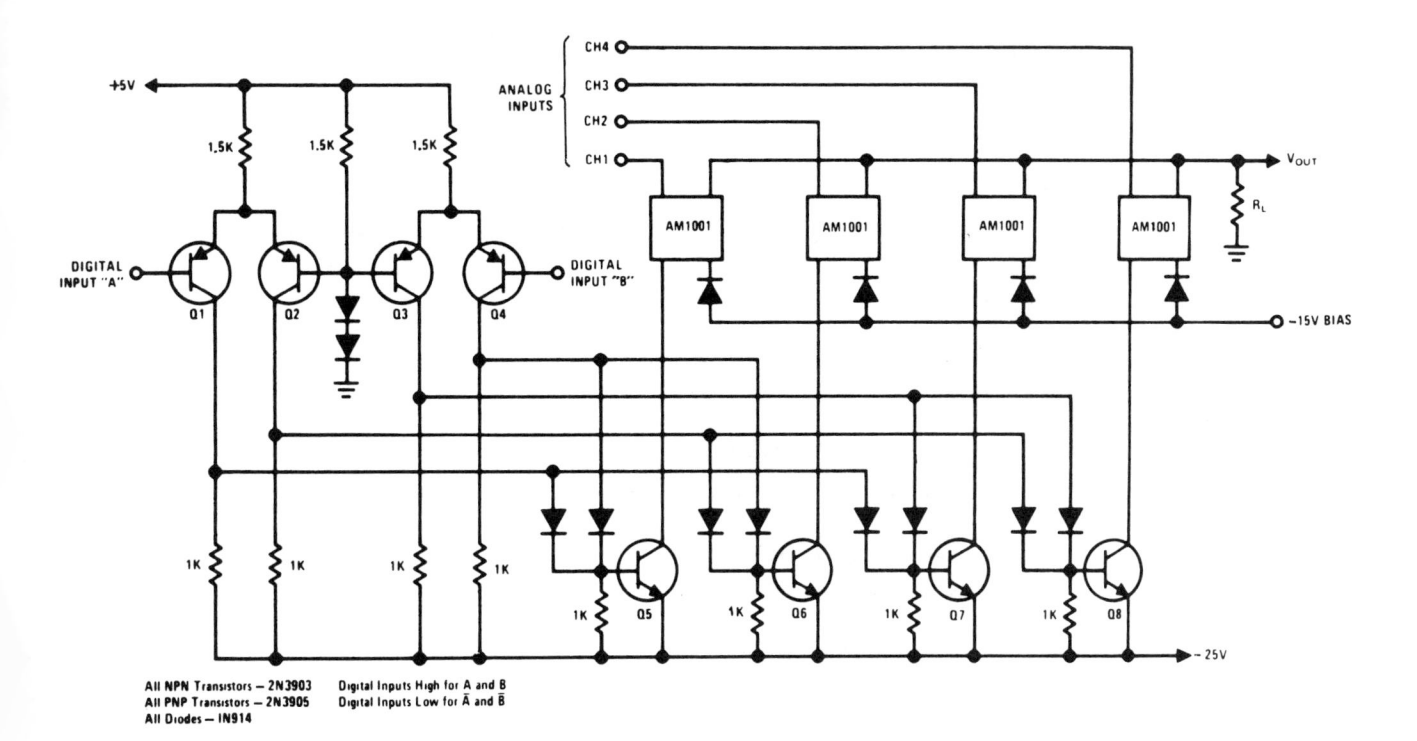

Binary-controlled four-channel multiplexer. The *on* channel is selected by binary coding and is DTL/TTL compatible. If A and B are *high*, drive is removed from Q5, allowing channel 1 AM1001 to pull up and turn on. Q6, Q7, and Q8 pull down on CH2, 3, and 4, thus turning them off. The voltages and devices indicated allow ±15V analog signals to be handled. If large analog voltages must be handled, current-mode multiplexing must be used; the toggle rate is reduced because accurate current-voltage converters are not as fast as noninverting voltage amplifiers.

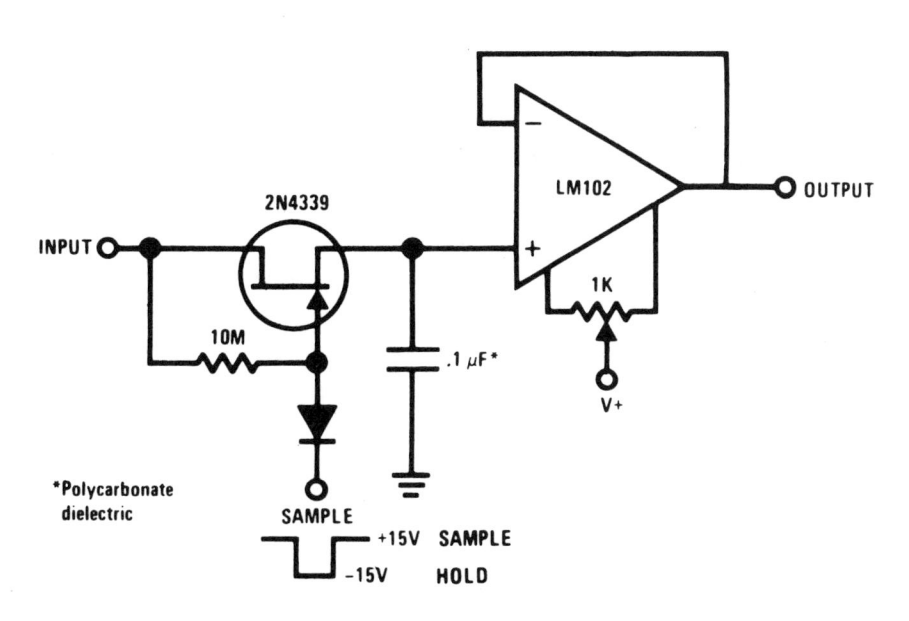

In this sample-and-hold circuit with offset adjustment, the 2N4339 JFET was selected because of its low I_{GSS}, very low $I_{D(OFF)}$ (less than 50 pA), and low pinchoff voltage. Leakages of this level put the burden of circuit performance on clean, solder-resin-free, low-leakage circuit layout.

SECTION **14** *Data Acquisition, Multiplexing, & Transmission* **449**

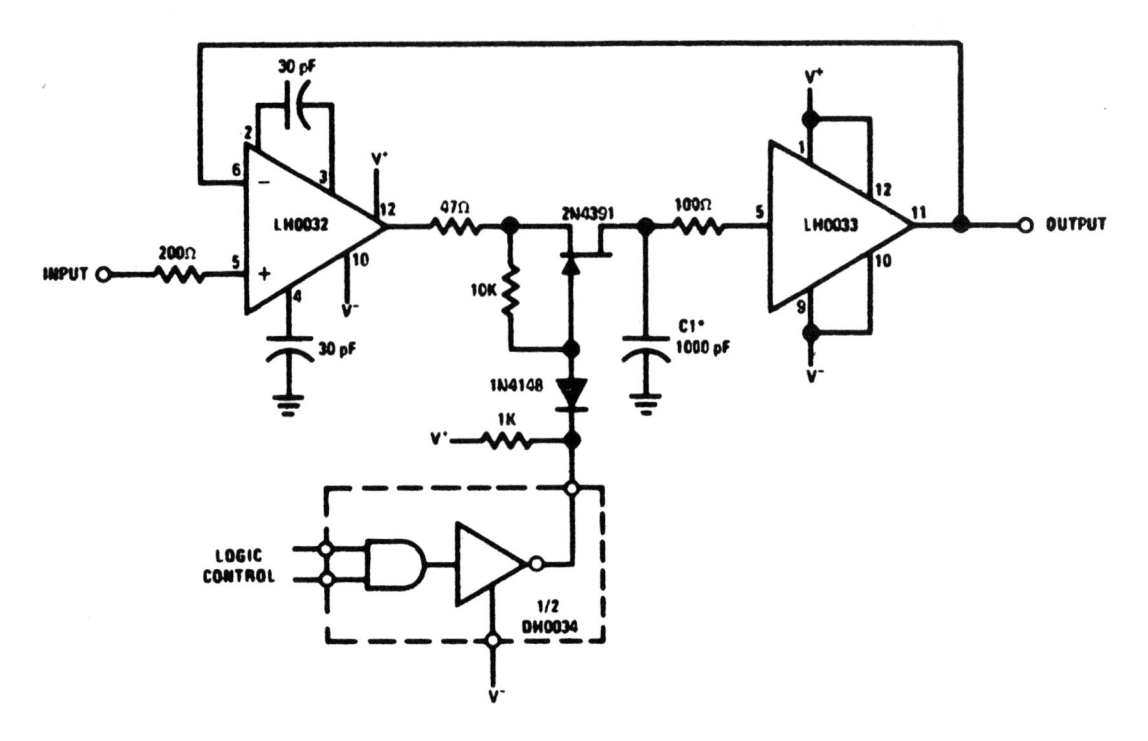

High-speed sample-and-hold circuit. Use polystyrene dielectric for minimum drift.

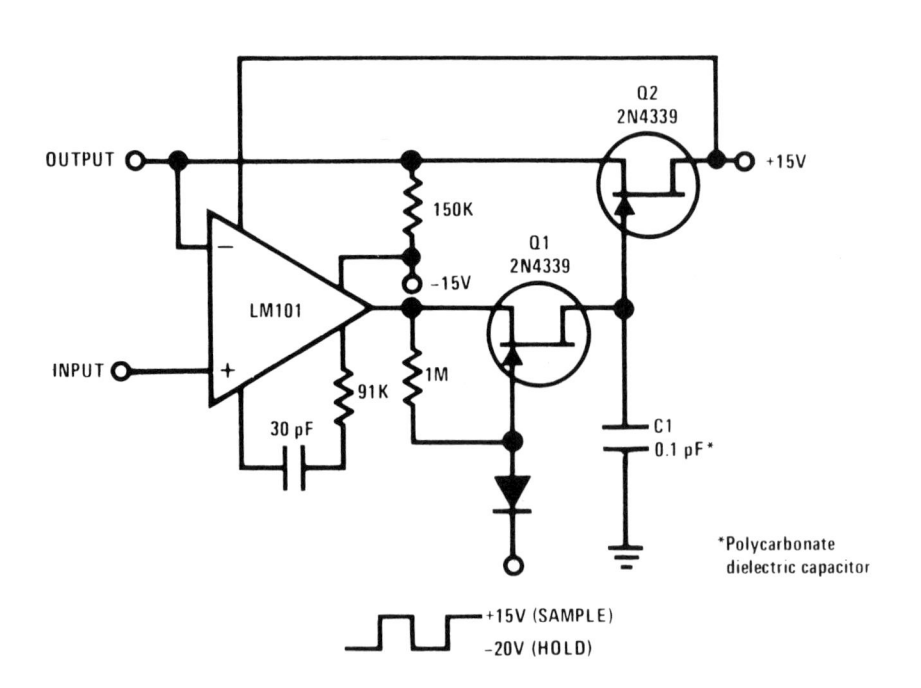

In this low-drift sample-and-hold circuit, JFETs Q1 and Q2 provide complete buffering to C1, the sample-and-hold capacitor. During *sample*, Q1 is turned on and provides a path, $R_{DS(ON)}$, for charging C1. During *hold*, Q1 is turned off, leaving Q1 $I_{D(OFF)}$ (less than 50 pA) and Q2 I_{GSS} (less than 100 pA) as the only discharge paths. Q2 serves a buffering function so feedback to the LM101 and output current are supplied from its source.

450 IC SCHEMATIC SOURCEMASTER

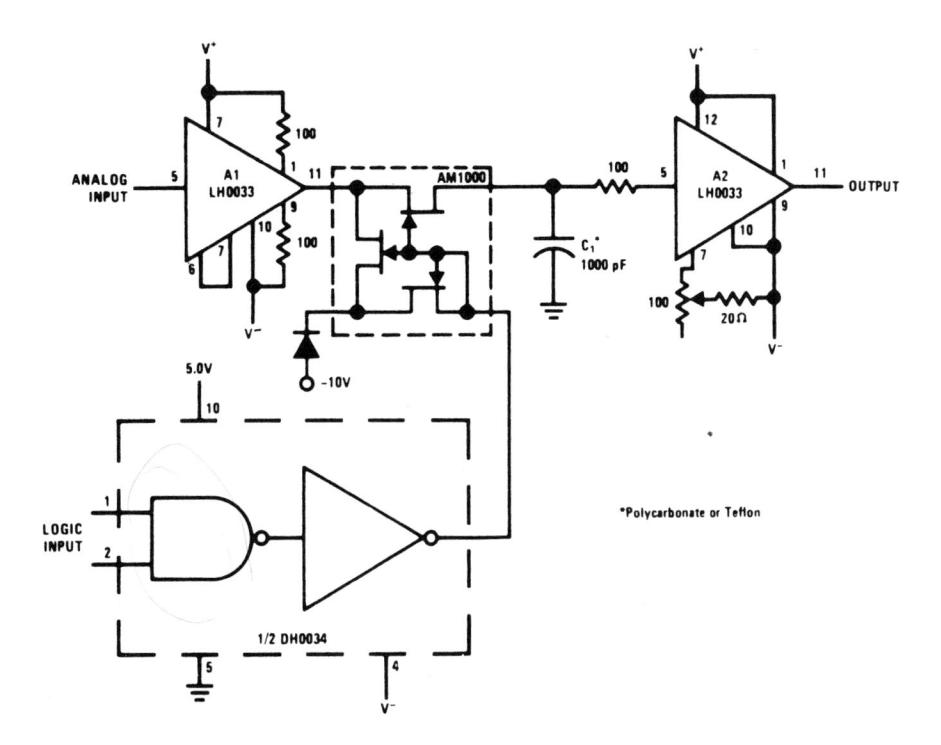

High-slew-rate sample-and-hold circuit. Amplifier A1 is used to buffer high-speed analog signals. Acquisition time is limited by the time constant of the switch *on* resistance and sampling capacitor and is typically 200 or 300 nsec.

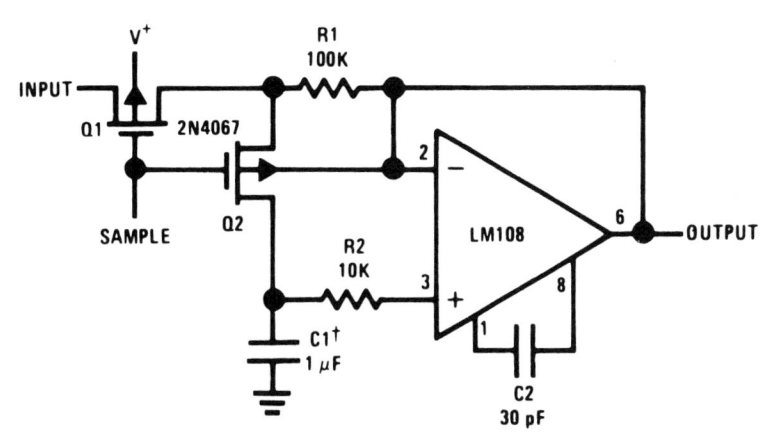

†Teflon, polyethylene or polycarbonate dielectric capacitor

Worst case drift less than 3 mV/sec

A sample-and-hold system that eliminates leakage in FET switches. When using P-channel MOS switches, the substrate must be connected to a voltage that is always more positive than the input signal. The source-to-substrate junction becomes forward-biased if this is not done. The troublesome leakage current of a MOS device occurs across the substrate-to-drain junction. Here, this current is routed to the output of the buffer amplifier through R1, so that it does not contribute to the error current.

SECTION **14** *Data Acquisition, Multiplexing, & Transmission* **451**

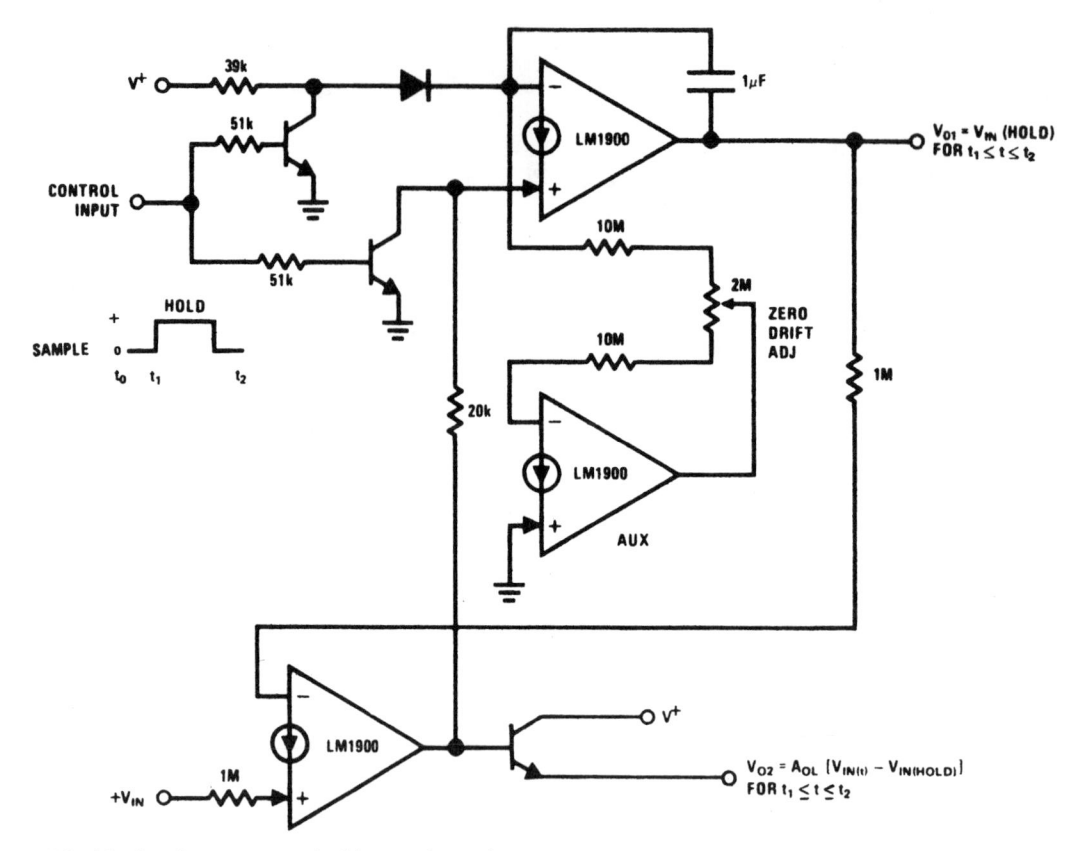

Sample-and-hold circuit compares held sample with new $+V_{IN}$.

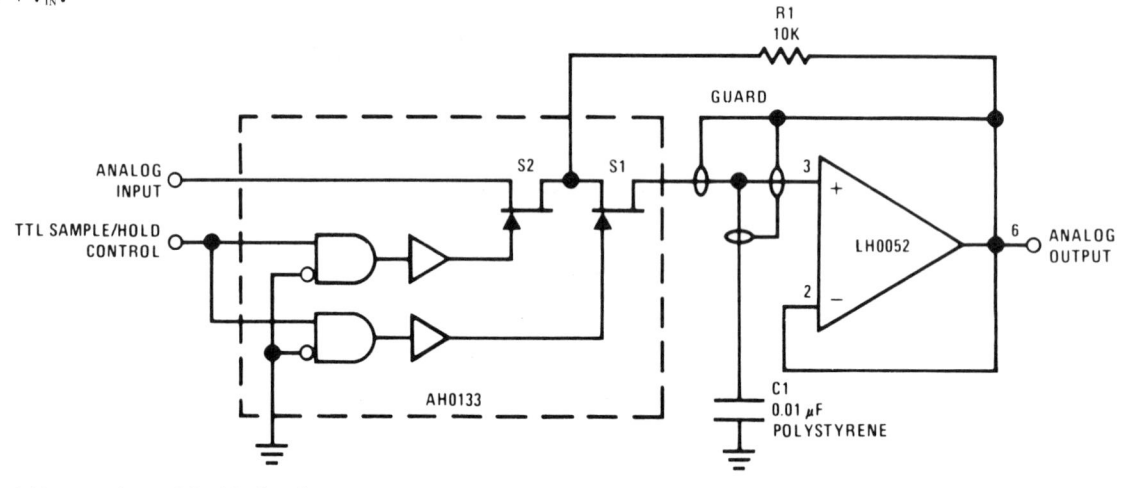

Low-drift sample-and-hold circuit.

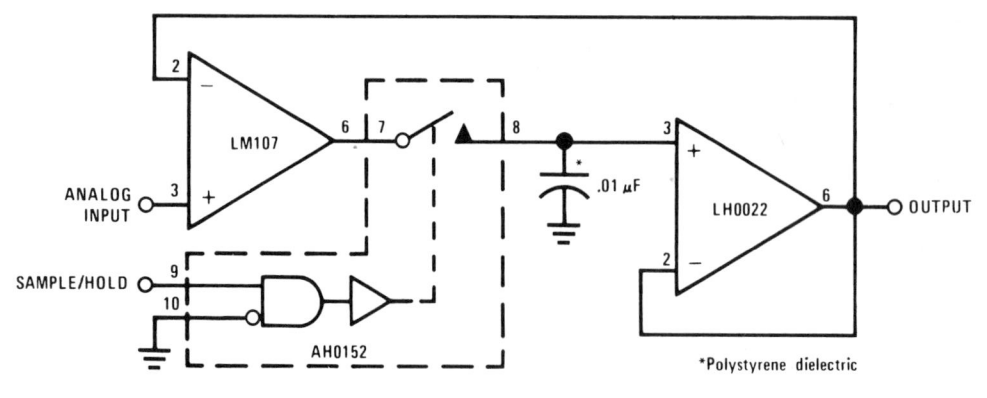

Precision sample-and-hold circuit.

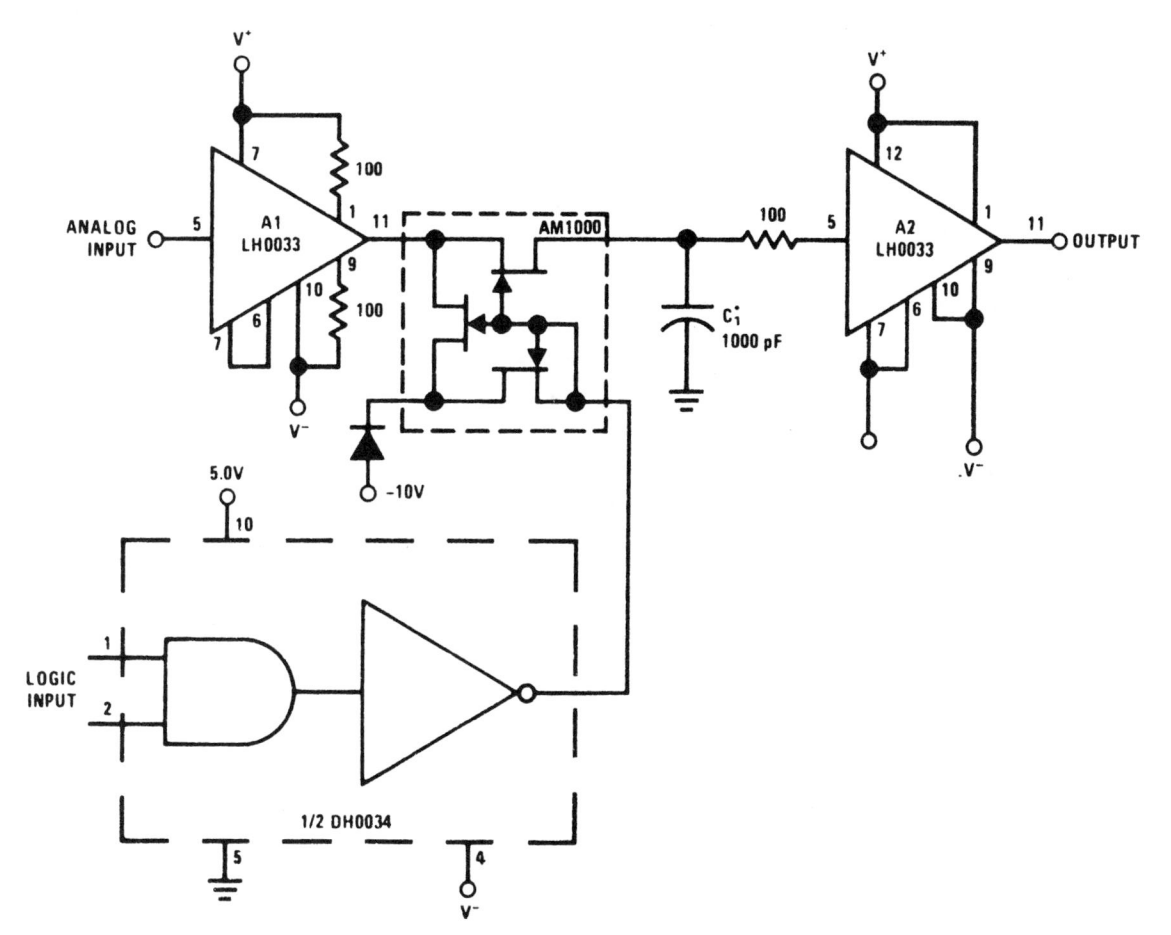

High-speed sample-and-hold circuit. The capacitor should be polycarbonate or Teflon.

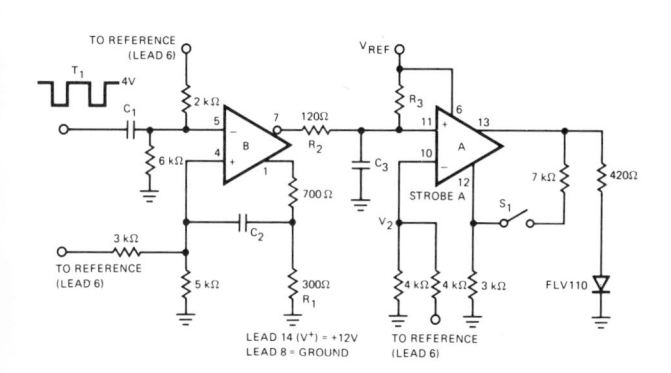

This minimum-frequency detector uses input pulse conditioning to achieve predictable operation with varying input pulse levels and widths. When power is applied, capacitor C_3 starts charging towards V_{REF}. If the clamp output of the first comparator (B) goes low (causing capacitor C_3 to discharge before the threshold voltage V_2 is reached), the output of comparator A is low and the display is off. However, if the voltage across capacitor C_3 reaches the threshold voltage (V_2) before discharging, the output of A is high. With S_1 closed, output A applies 3V to the STROBE A input and maintains drive to the display.

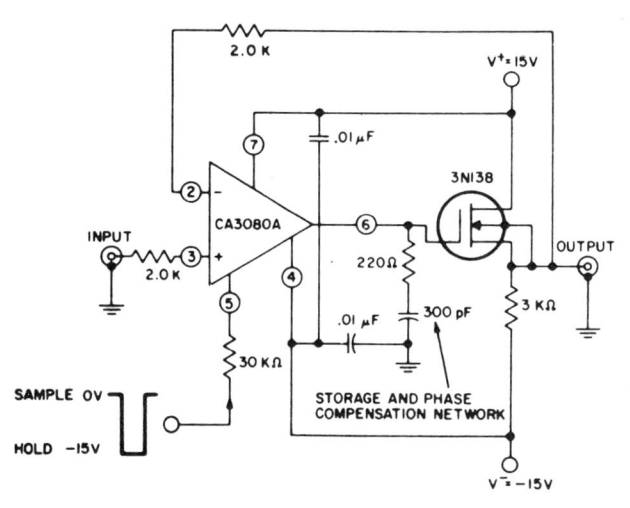

A sample-and-hold circuit using RCA CA3080A. The slew rate (in sample mode) is 1.3V/μsec, and the time required for output to settle within ±3 mV of a 4V step is 3 μsec.

SECTION **14** *Data Acquisition, Multiplexing, & Transmission* **453**

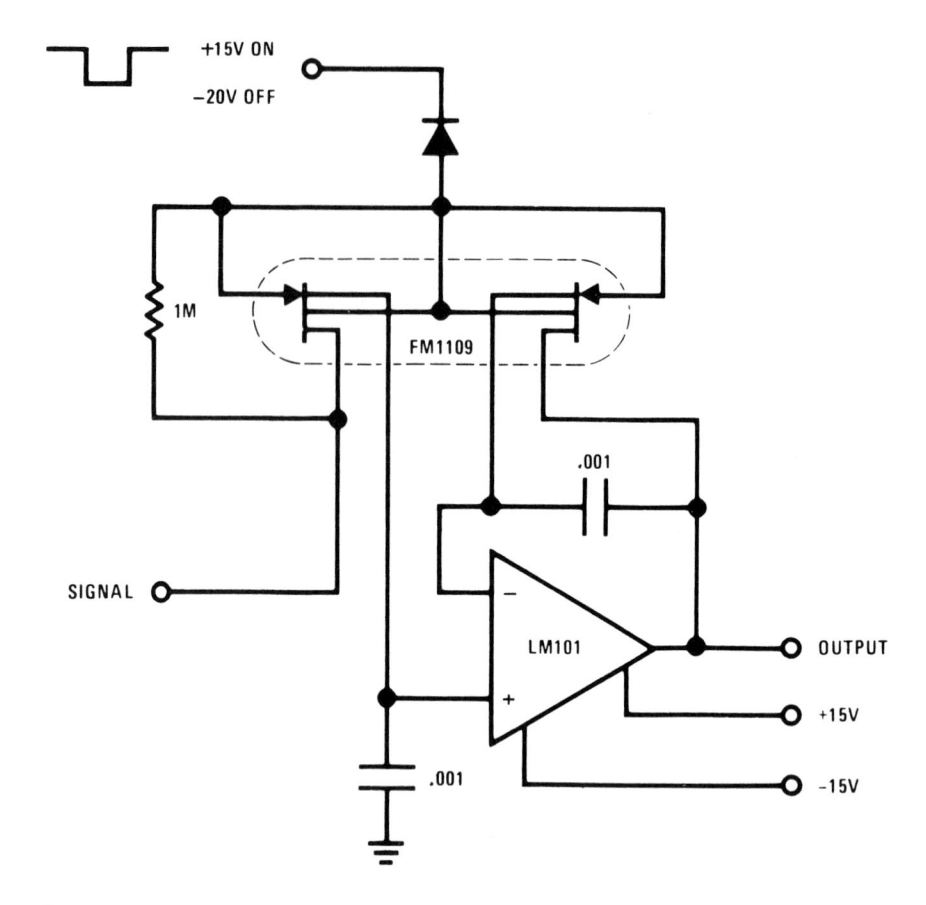

A JFET sample-and-hold circuit in which the logic voltage is applied simultaneously to the sample-and-hold JFETs. By matching input impedance and feedback resistance and capacitance, errors due to $R_{DS(ON)}$ of the JFETs are minimized. The inherent matched $R_{DS(ON)}$ and matched leakage currents of the FM1109 greatly improve circuit performance.

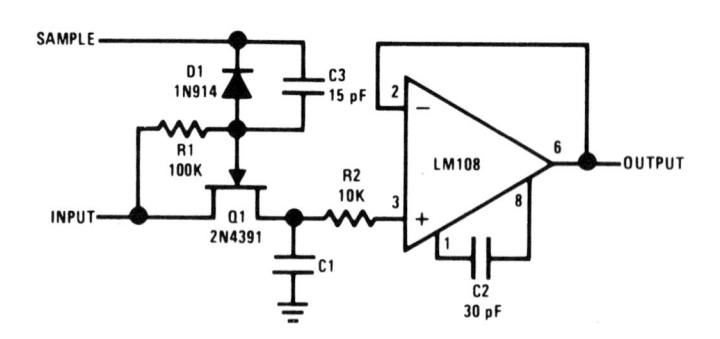

Sample-and-hold circuit. During the sample interval, Q1 is turned on, charging the hold capacitor, C1, up to the value of the input signal. When Q1 is turned off, C1 retains this voltage. The output is obtained from an op-amp that buffers the capacitor so that it is not discharged by any loading. In the *hold* mode an error is generated as the capacitor loses charge to supply circuit leakages.

454 IC SCHEMATIC SOURCEMASTER

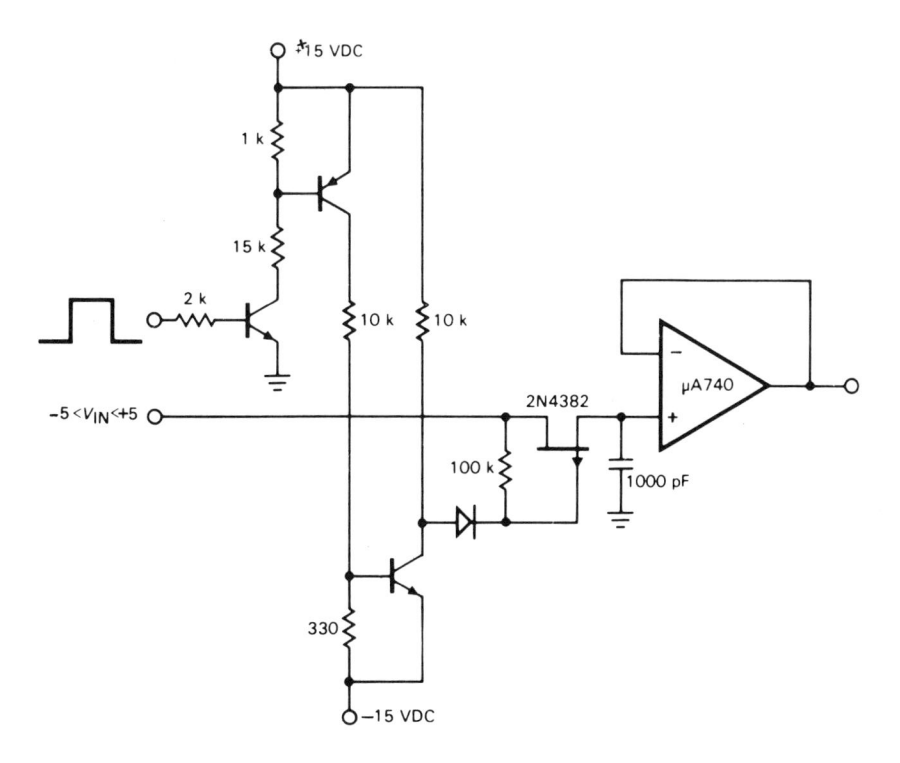

Sample-and-hold circuit. The change in capacitor voltage during the *hold* mode should be small compared with the required system accuracy. The μA740 FET operational amplifier is chosen as the buffer amplifier for its good slew rate (6V/μsec) and its low input bias current (2 nA). The FET leakage and the amplifier bias current partially compensate each other. A worst-case estimate of the change in held voltage can be made by assuming that the full amplifier bias current flows from the holding capacitor and that the FET leakage is negligible. Outboard transistors are general-purpose small-signal types: PNP = 2N4121; NPN = 2N3646.

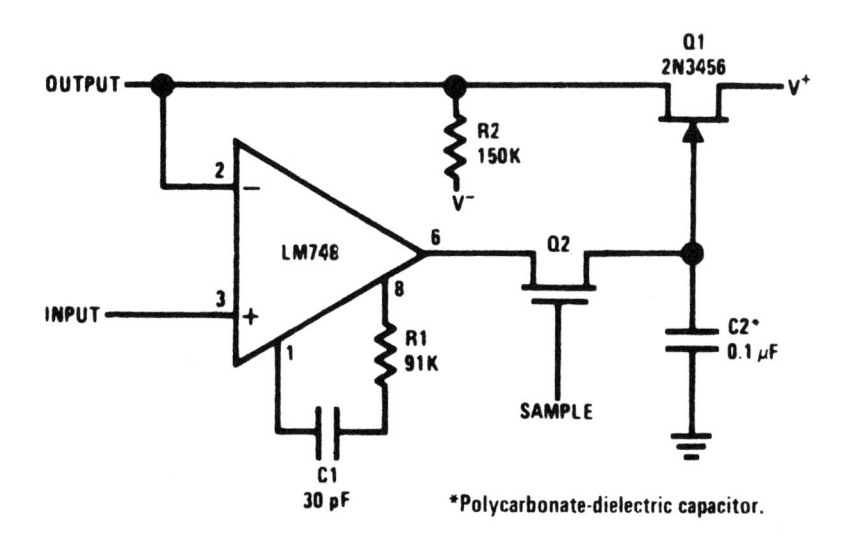

Low-drift sample-and-hold circuit.

SECTION 14 *Data Acquisition, Multiplexing, & Transmission* 455

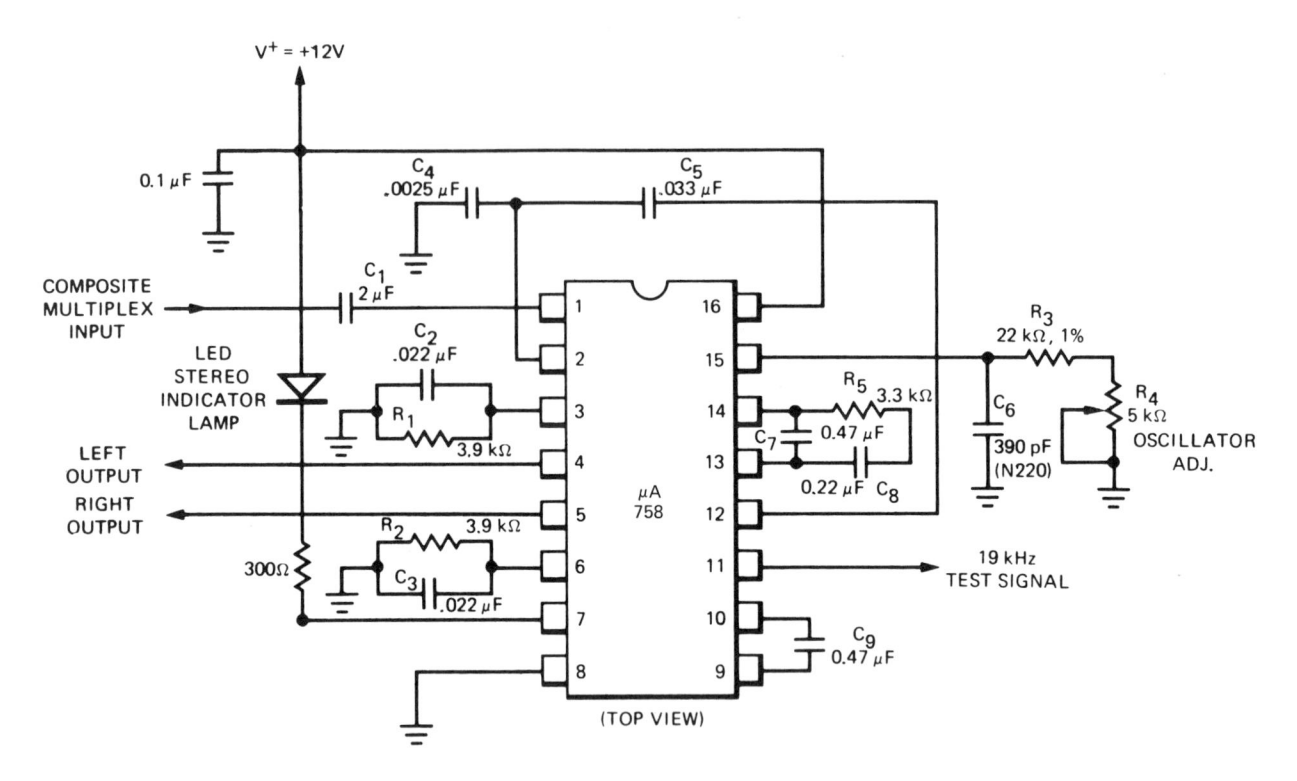

Multiplex decoder with PLL. Selection of the external components in the oscillator RC network provides a convenient means of adjusting the loop gain of the PLL by controlling the sensitivity of the VCO. The loop gain is proportional to R^2C, whereas the oscillator frequency is proportional to $1/RC$. As long as the RC product is held constant, the values of R and C can be varied to change the loop gain. Changing C6 from 390 pF to 510 pF decreases the loop gain by approximately 25% and improves both separation and distortion performance by a few decibels. However, capture range also decreases by approximately 25%, and separation becomes more sensitive to oscillator mistuning.

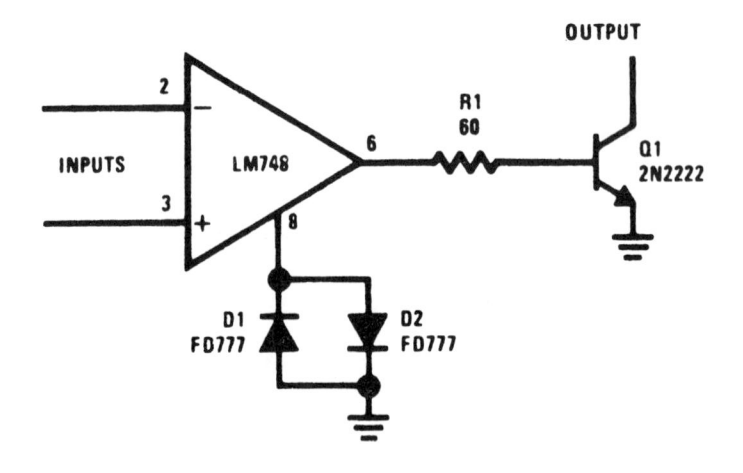

A voltage comparator for driving RTL or high-current driver.

456 IC SCHEMATIC SOURCEMASTER

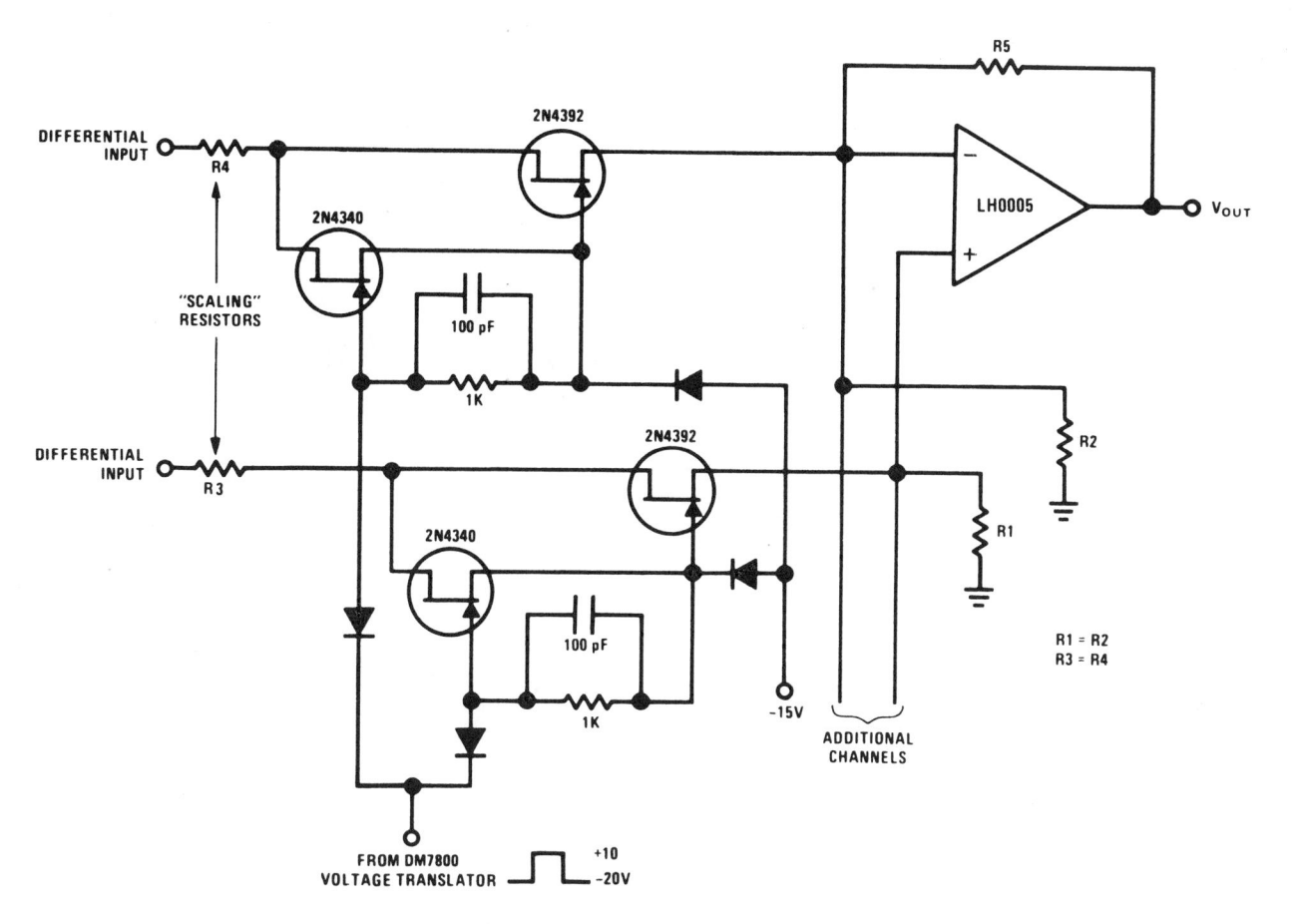

A wideband differential multiplexer provides high-frequency signal handling and high toggle rates simultaneously. Toggle rates up to 1 MHz are possible with this circuit.

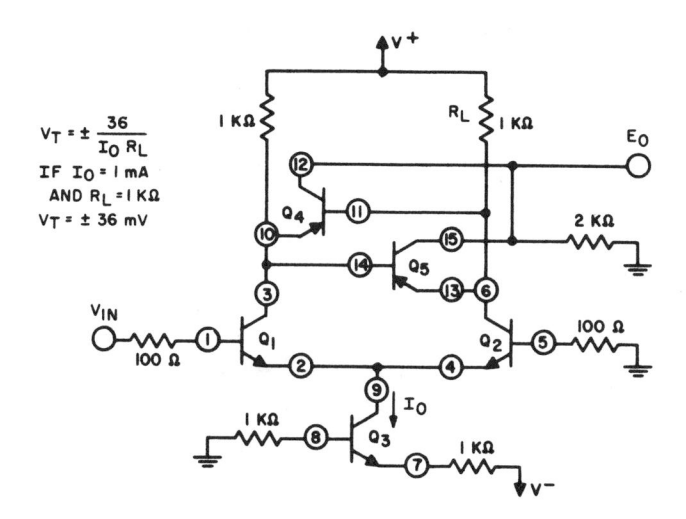

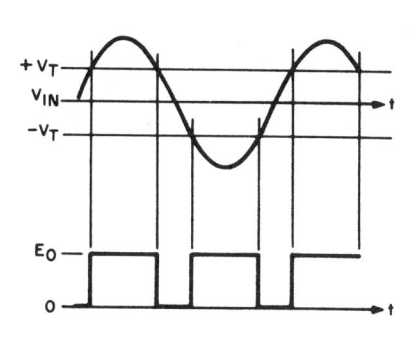

A small-signal zero-voltage detector with noise immunity. Circled numbers are IC pins. $V_T = \pm 36/I_0 R_L$; if $I_0 = 1$ mA and $R_L = 1$k, $V_T = \pm 36$ mV.

SECTION **14** Data Acquisition, Multiplexing, & Transmission 457

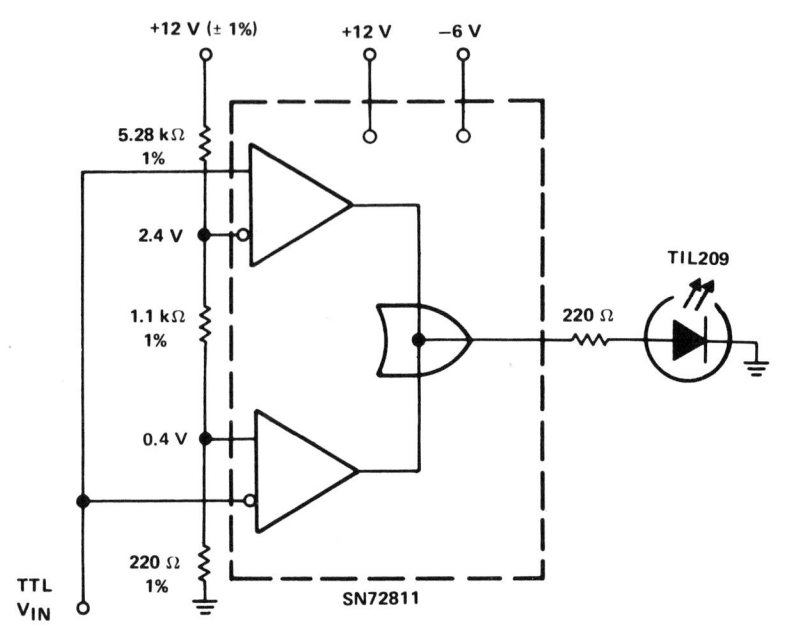

TTL VOLTAGE INPUT CONDITION	TIL209 VLED CONDITION
0 LOGIC $V_{IN} < +0.4$ V	ON
IMPROPER LEVEL $0.4V < V_{IN} < 2.4$ V	OFF
1 LOGIC $V_{IN} > 2.4$ V	ON

Basic window detector. In many testing applications it is desirable to detect parameter values below a specified minimum or above a specified maximum. With such a window detector, a dual comparator with a single output (such as the SN72811) can detect the presence or absence of a signal between two limits. In this application, a proper TTL level (high or low) at the input will result in the TIL209 visible-light-emitting diode (VLED) being turned on. An improper logic level greater than 400 mV but less than 2.4V will result in the TIL209 being turned off.

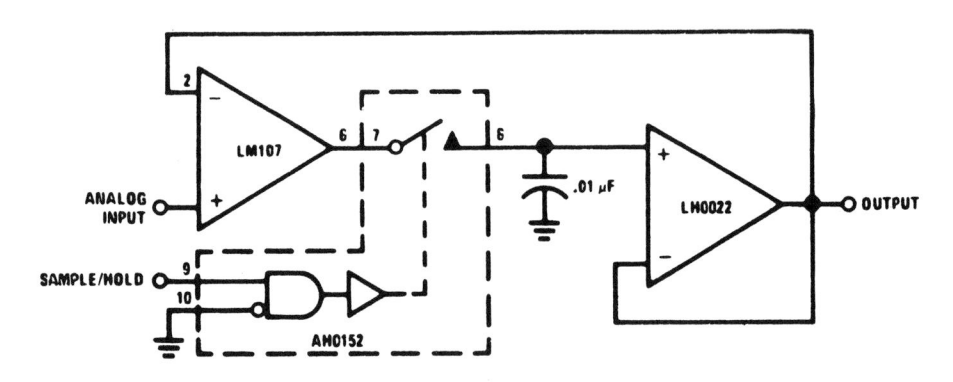

Precision sample-and-hold circuit. The capacitor should be a polystyrene dielectric.

458 IC SCHEMATIC SOURCEMASTER

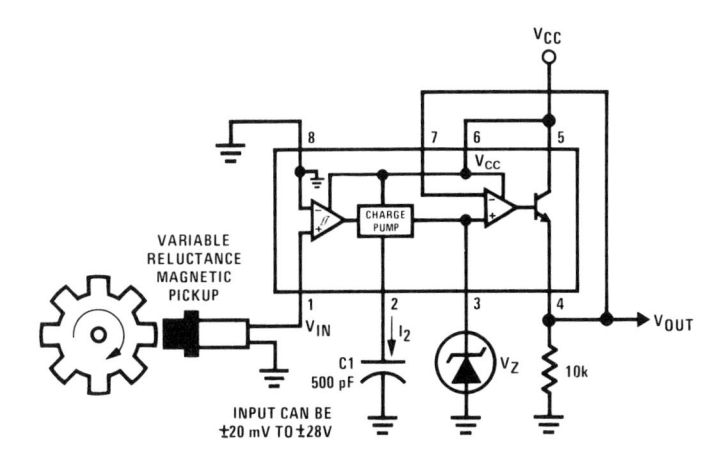

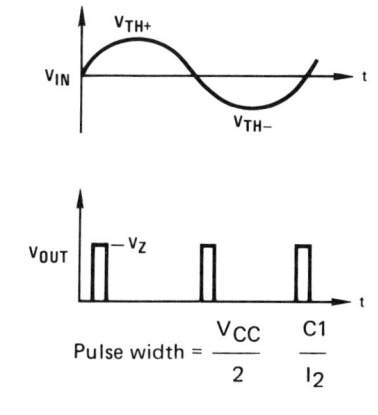

Output frequency equal twice input frequency

$$\text{Pulse width} = \frac{V_{CC}}{2} \cdot \frac{C1}{I_2}$$

Pulse height = V_{ZENER}

$$\text{Pulse width} = \frac{V_{CC}}{2} \cdot \frac{C1}{I_2}$$

"Two-shot" zero-crossing detector. At each zero crossing of the input signal, the charge pump changes the state of capacitor C1 and provides a one-shot pulse into the zener at pin 3. The width of this pulse is controlled by the internal current of pin 2 and the size of capacitor C1, as well as by the supply voltage. Since a pulse is generated by each zero crossing of the input signal, we call this a *two-shot* instead of a *one-shot* device, and this can be used for doubling the frequency that is presented to a microprocessor control system.

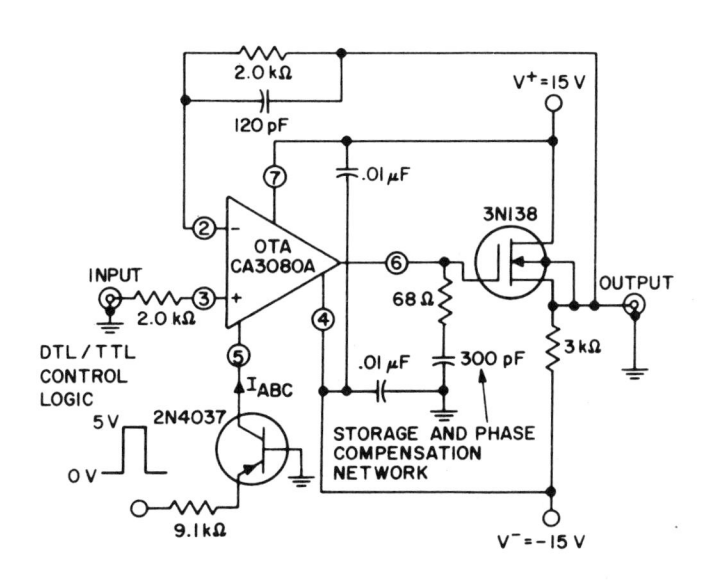

Sample-and-hold circuit for DTL/TTL control logic. The RCA 2N4037 is used to minimize capacitive feedthrough. The 9.1k resistor in series with the PNP transistor emitter establishes amplifier bias-current conditions (I_{abc}). Considerations of circuit stability and signal retention require the largest phase-compensation capacitor manageable that is compatible with applicable slew rate: choose a capacitor for maximum allowable *tilt* in the storage mode and choose a resistor so that $1/(2\pi RC) = 2$ MHz. The 120 pF capacitor in the feedback loop is used to improve transient response.

SECTION **14** Data Acquisition, Multiplexing, & Transmission **459**

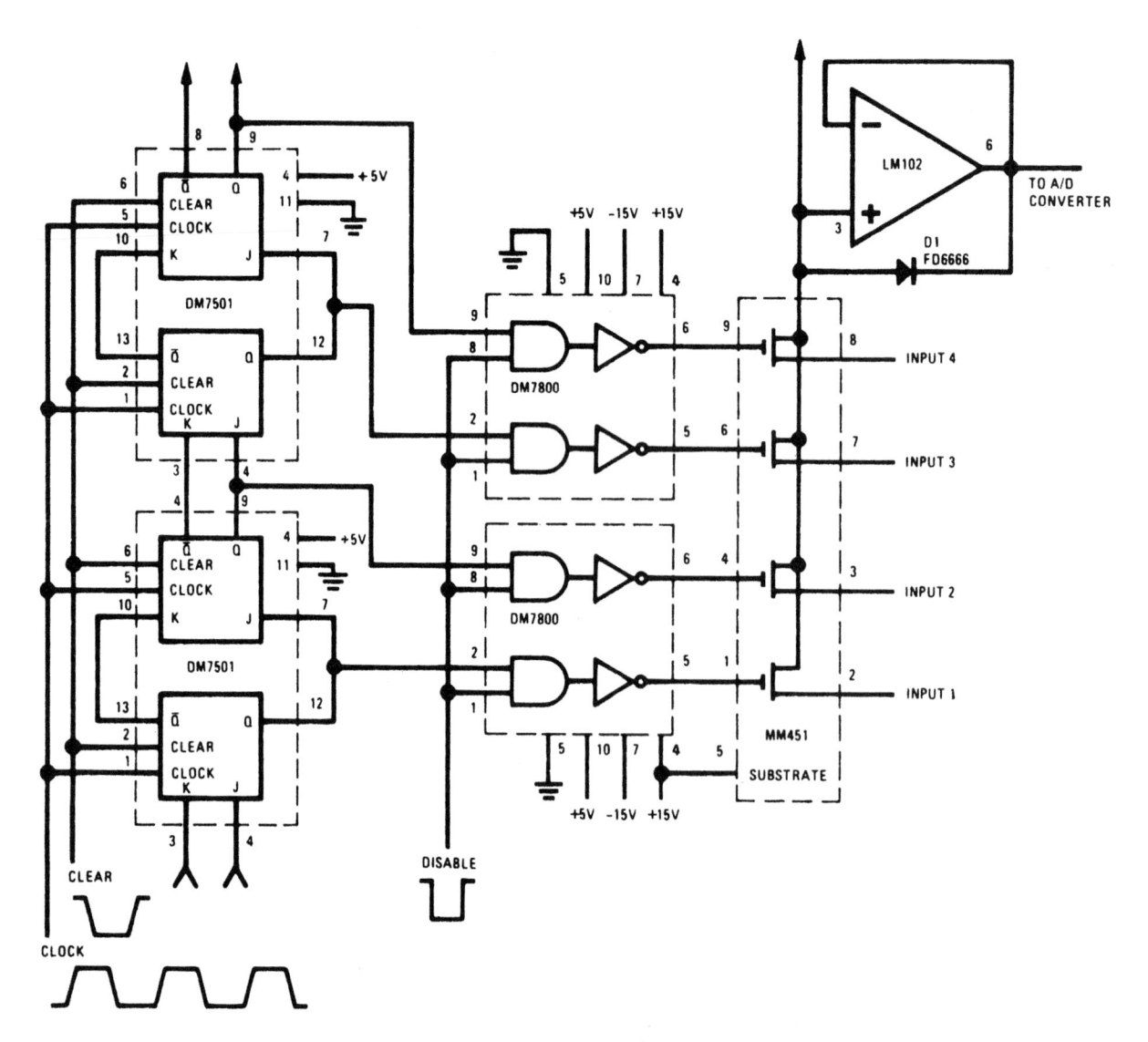

Expandable four-channel analog commutator. Two DM7501 dual flip-flops form a four-bit static shift register. The parallel outputs drive DM7800 level translators, which convert the TTL levels to voltages suitable for driving MOS devices, and this is coupled into an MM451 four-channel analog switch. An extra gate on the input of the translator can be used, as shown, to shut off all the analog switches.

In operation, a bit enters the register and cycles through at the clock frequency, turning on each analog switch in sequence. The "clear" input is used to reset the register such that all analog switches are off. The channel capacity can be expanded by connecting registers in series and hooking the output of additional analog switches to the input of the buffer amplifiers.

[Circuit diagram labels:]

f2 ○ V2
f1 ○ V1

270, 30, +6V, −6V, MC1545G, 4, 8, 5, 2, 3, 7, 6, 10, 1, +V_o, −V_o

Differential Voltage Comparator
+12 V, −6 V, 5, 11, 6, 4, MC1711CP, 10, 2, 12, 3

¼ MC700P
1, 2, 3, 4, 5, 6
Clock
11, 12, ½ MC7479P Type D, 9, 8, Q, $\overline{Q}$

+5 V

Data Input   0 = f1
             1 = f2

Slope Comparator
+12 V, −6 V
510, 0.01, ½ MC1514L, 6, 10, 14, 5, 11, 8
0.001
¼ MC3021, 2, 1, 3
510, 0.01, ½ MC1514L, 12, 13, 7, 3, 1, 0.001
−6 V, +12 V

*Capacitors Used to Filter Out Possible Oscillation During Switching

$V_1 = V_2 = 1\ V_{p-p}$
$V_o = 1\ V_{p-p}$
$A_V \approx 1$

All capacitance in µF.

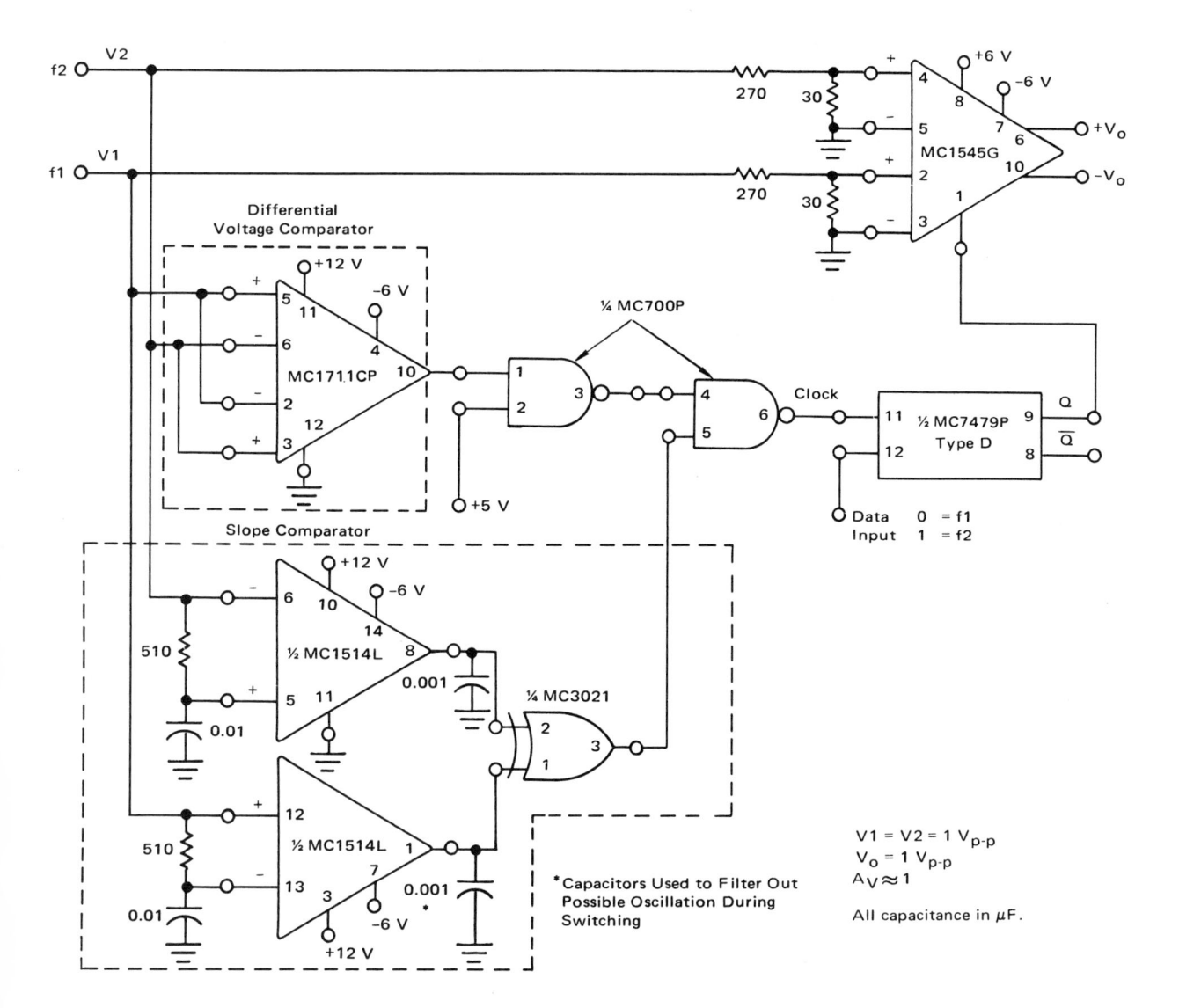

Frequency shift keyer (FSK) system. By applying a signal to each differential amplifier input pair, the output may be alternately switched between the two input signals by changing the gate voltage from one extreme to the other. With the gate level high (greater than 1.5V) the signal applied between pins 4 and 5 will be passed and the signal applied between pins 2 and 3 will be suppressed. With the gate low (less than 0.5V), the reverse situation will exist, passing the signal applied between pins 2 and 3. The output can then be coded in a binary representation: 1 = f2 transmitted; 0 = f1 transmitted. In this way a binary-to-frequency conversion is obtained, which is directly related to the binary sequence driving the gate input (pin 1).

Since the two signals, f1 and f2, are completely uncorrelated, switching from one channel to the other at random intervals could produce considerable spurious frequencies of large amplitude and generate a lot of noise in the system. The following application eliminates most of this "hash" by imposing the following constraints: the data will be allowed to change at pin 1 (the gate-logic control pin) only when (1) the sign of the rate of change of f1 and f2 are the same, and (2) the value of both f1 and f2 (sign and magnitude) are equal (within a few millivolts). That is,

$$\text{Sign}\ \frac{dV_1}{dt} = \text{Sign}\ \frac{dV_2}{dt}$$

and

$$V_1 \approx V_2$$

SECTION **14** Data Acquisition, Multiplexing, & Transmission

(a) circuit

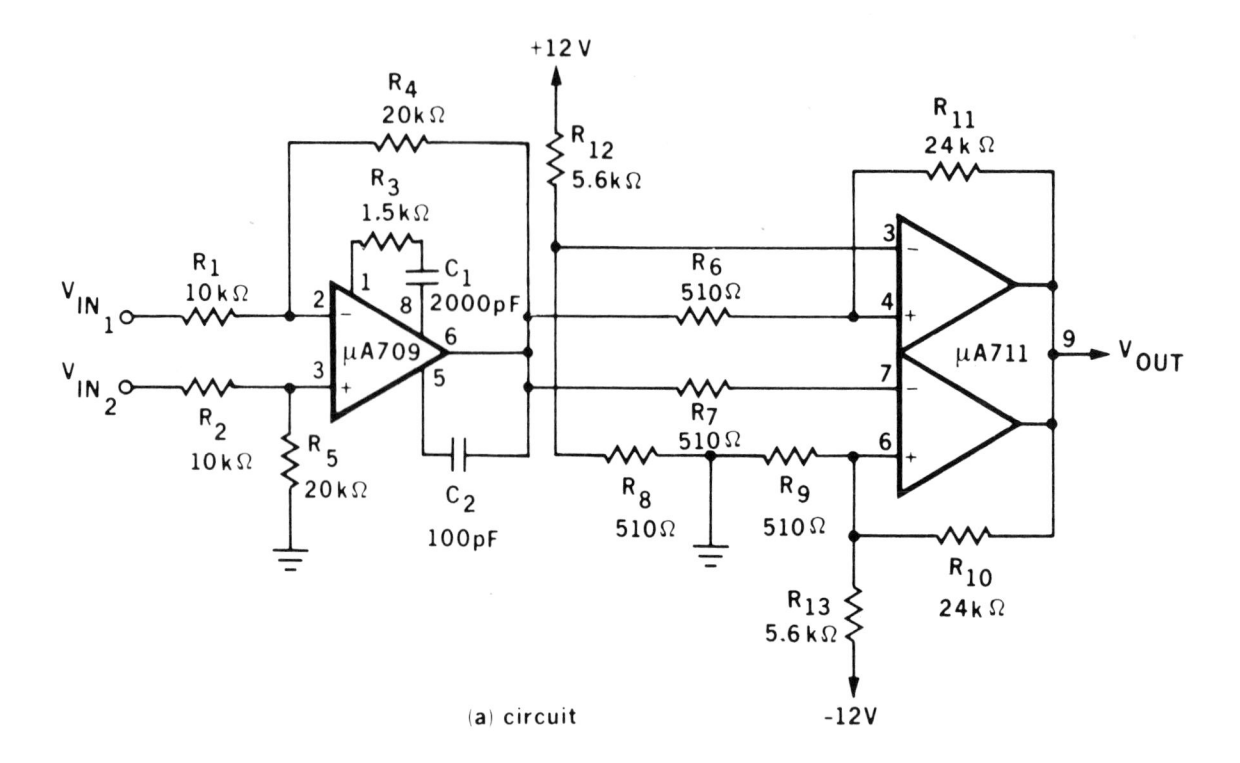

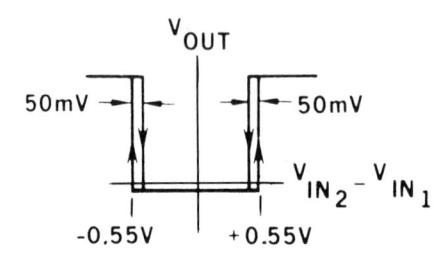

(b) transfer function

Double-ended differential threshold detector with hysteresis. In this circuit, hysteresis is provided at both the upper and lower threshold levels. A μA709 serves as both buffer and difference amplifier. With the component values shown, an output is obtained from the μA711 when the difference between the two input voltages exceeds 0.55V. A hysteresis of approximately 5 mV is obtained.

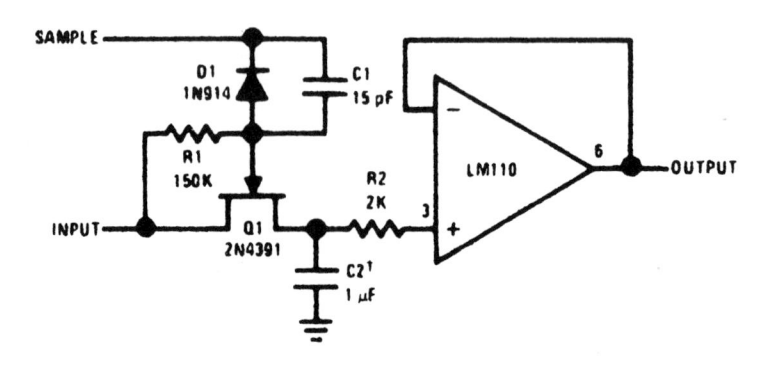

Sample-and-hold circuit. C2 should have polycarbonate, Teflon, or polyethylene dielectric.

462 IC SCHEMATIC SOURCEMASTER

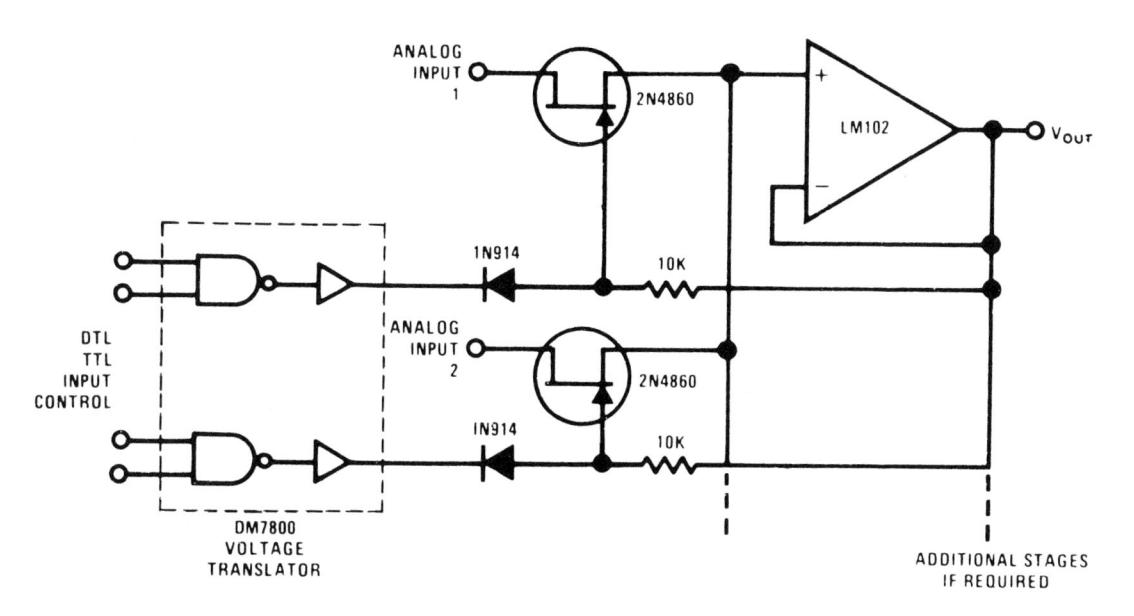

DTL/TTL-controlled buffered analog switch. This analog switch uses the 2N4860 JFET for its 25Ω R_{ON} and low leakage. The LM102 serves as a voltage buffer. This circuit can be adapted to a dual-trace oscilloscope chopper. The DM7800 monolithic IC provides adequate switch drive controlled by DTL/TTL levels.

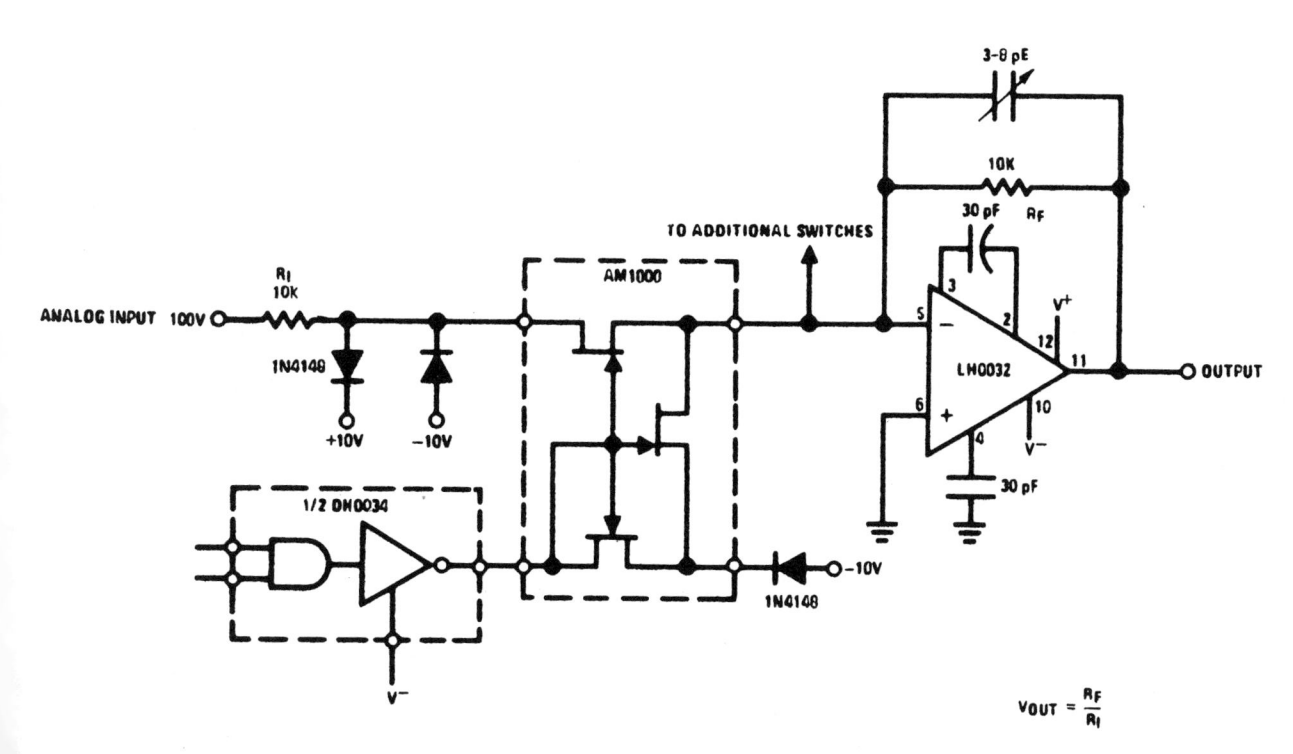

$$V_{OUT} = \frac{R_F}{R_I}$$

Current-mode multiplexer. The IC input capacitance is typically 5 pF and thus may form a significant time constant with high-value resistors. For optimum performance, the input capacitance to the inverting input should be compensated by a small capacitor across the feedback resistor. The value is strongly dependent on layout and closed-loop gain but will typically be around several picofarads.

SECTION **14** Data Acquisition, Multiplexing, & Transmission **463**

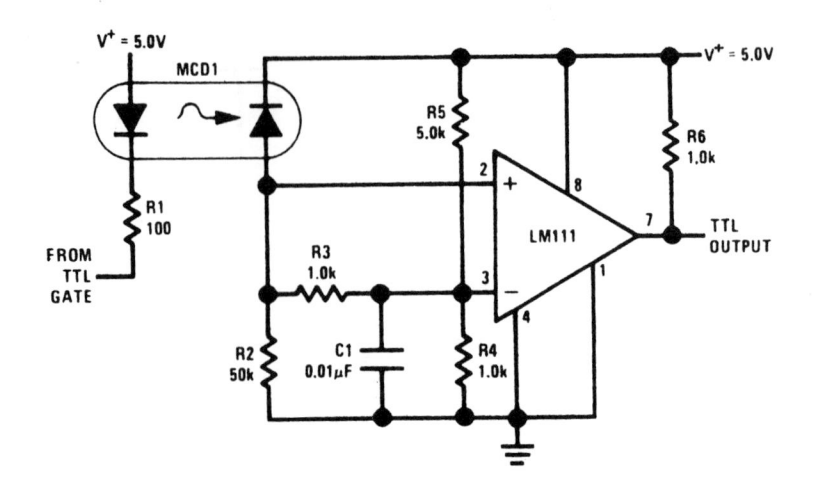

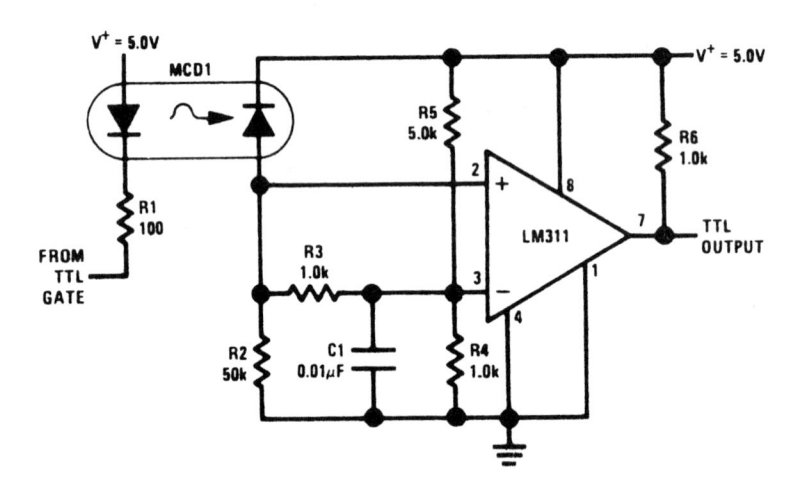

Digital transmission isolators.

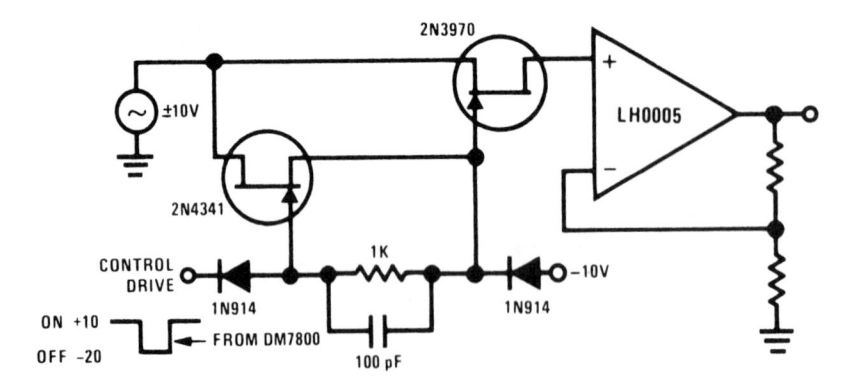

High toggle rate, high-frequency analog switch. This commutator circuit provides low-impedance gate drive to the 2N3970 analog switch for both *on* and *off* drive conditions. This circuit also approaches the ideal gate drive conditions for high-frequency signal handling by providing a low ac impedance for *off* drive and high ac impedance for *on* drive to the 2N3970. The LH0005 op-amp does the job of amplifying megahertz signals.

R1 = 10 k
C1 = 0.01 μF
R2 = 12 k
C2 = 0.01 μF
f1 = 1.6 kHz
f2 = 1.35 kHz

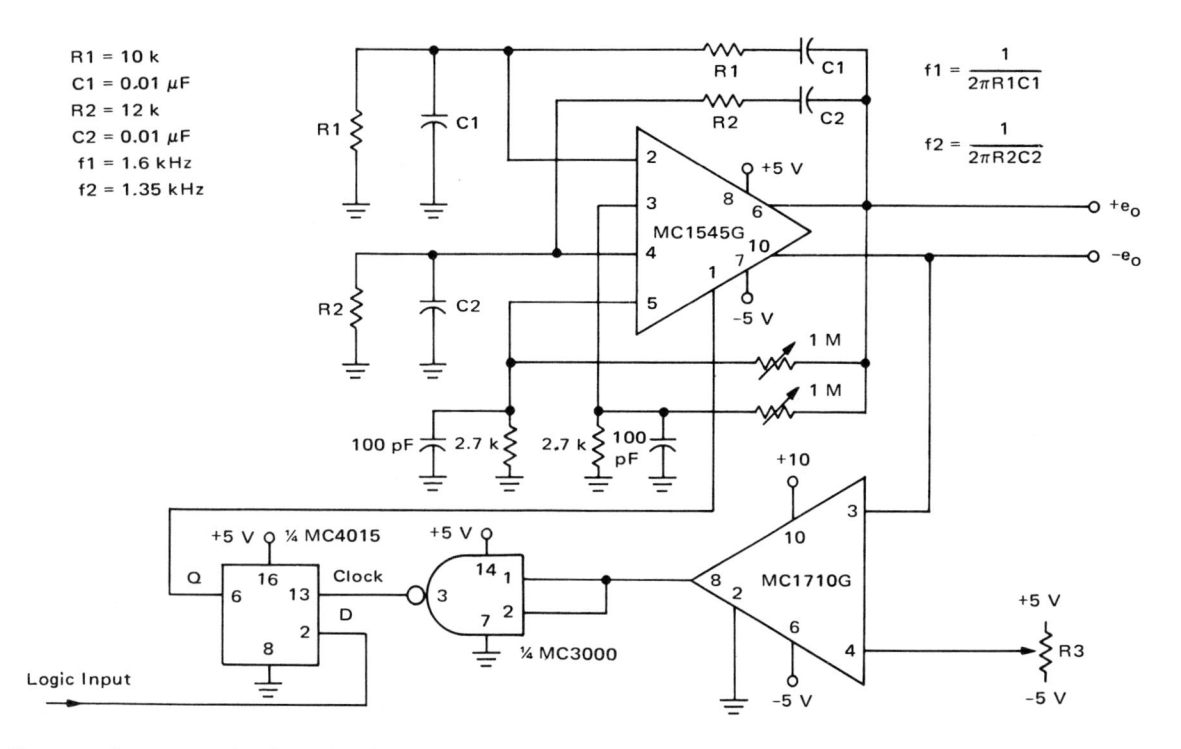

$f_1 = \dfrac{1}{2\pi R1 C1}$

$f_2 = \dfrac{1}{2\pi R2 C2}$

A self-generating FSK with adjustable level transition. The circuit is basically a dual oscillator with a logic switching network that compares the output of a reference and updates the gate input with each cycle of the output.

Each oscillator feedback network is connected to an input channel of the MC1545 and is tuned for a different frequency in the 1 to 3 kHz range. Negative feedback is applied to each channel to provide gain control.

This circuit will give smaller switching transients than would be obtainable with separate oscillators, since one oscillator will be driven at the frequency of the other oscillator while the first oscillator is off. When switching occurs, the initial conditions of the *off* oscillator are preset by the *on* oscillator. Because the RC networks do not track precisely, there will still be an initial offset that will be constant.

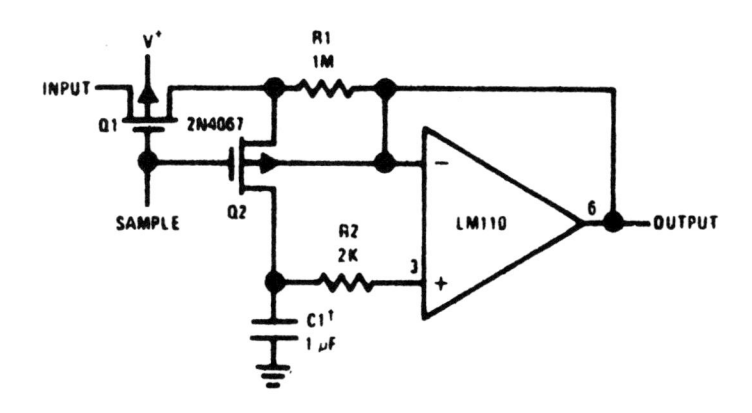

Low-drift sample-and-hold circuit. Worst-case drift is less than 3 mV/sec. C1 should have polycarbonate, Teflon, or polyethylene dielectric.

SECTION **14** *Data Acquisition, Multiplexing, & Transmission* **465**

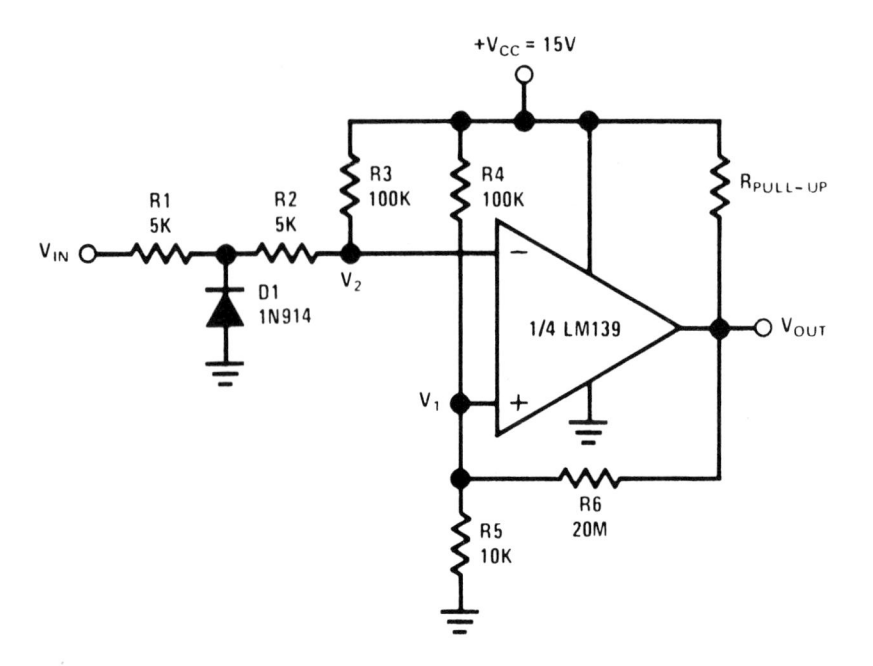

Zero-crossing detector. The LM139 can be used to square up a sine wave centered around zero by incorporating a small amount of positive feedback to improve switching times and centering the input threshold at ground. Voltage divider R4 and R5 establishes a reference voltage, V1, at the positive input. By making the series resistance of R1 plus R2 equal to R5, the switching condition, V1 = V2, will be satisfied when V_{IN} = 0. The positive-feedback resistor, R6, is made very large with respect to R5 (R6 = 2000R5). The resultant hysteresis established by this network is very small (ΔV1 < 10 mV), but it is sufficient to ensure rapid output voltage transitions. Diode D1 is used to ensure that the inverting input terminal of the comparator never goes below −100 mV. As the input terminal goes negative, D1 will clamp the node between R1 and R2 to approximately −700 mV. This sets up a voltage divider with R2 and R3, preventing V2 from going below ground. The maximum negative input overdrive is limited by the current handling ability of D1.

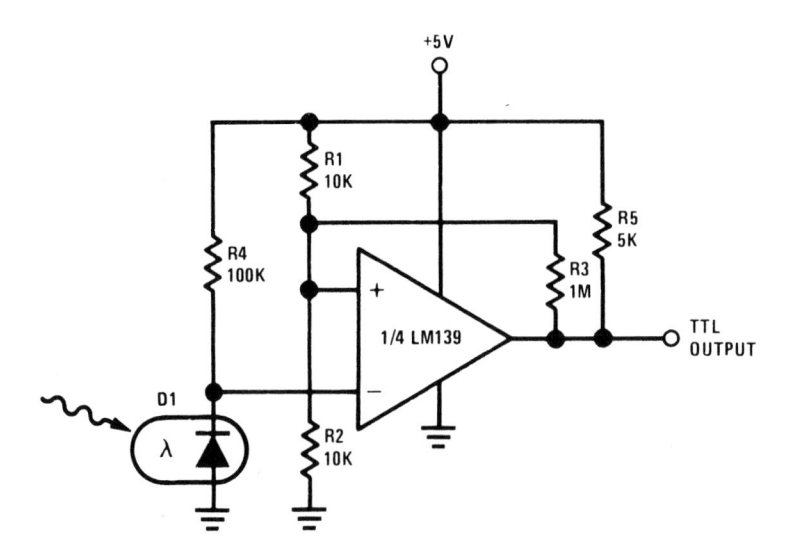

Paper tape reader (TTL output). A photodiode is used to sense the presence or absence of light passing through holes in the tape. A 1M feedback resistor gives ±5 mV of hysteresis to ensure rapid switching and noise immunity.

466 IC SCHEMATIC SOURCEMASTER

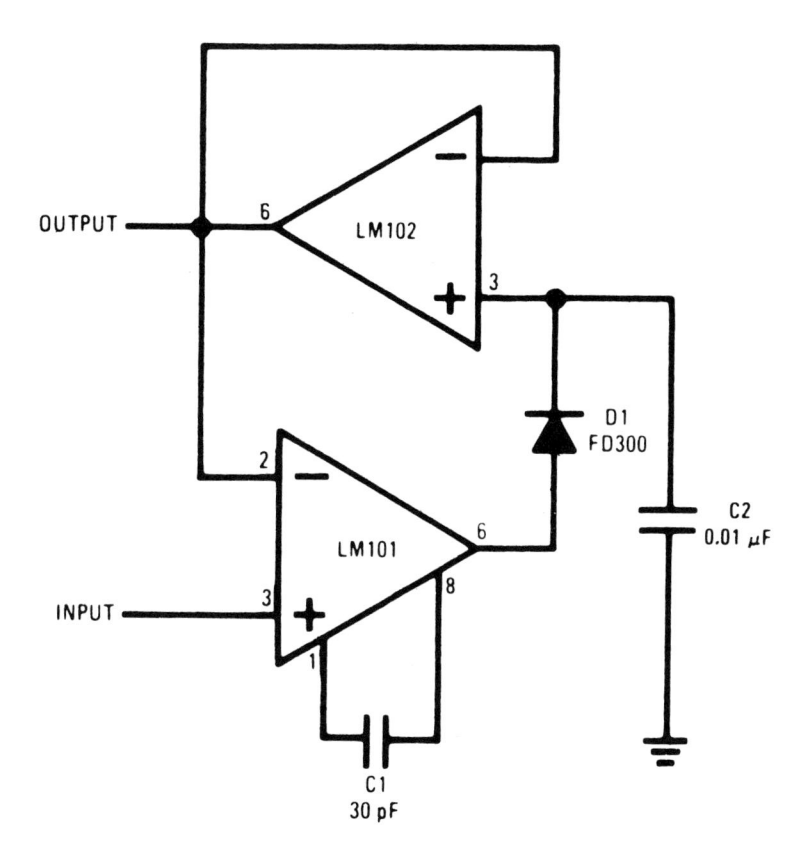

Positive peak detector with buffered output. Connected as shown, it will conduct whenever the input is greater than the output, so the output will be equal to the peak value of the input voltage. In this case, an LM102 is used as a buffer for the storage capacitor, giving low drift along with a low output resistance.

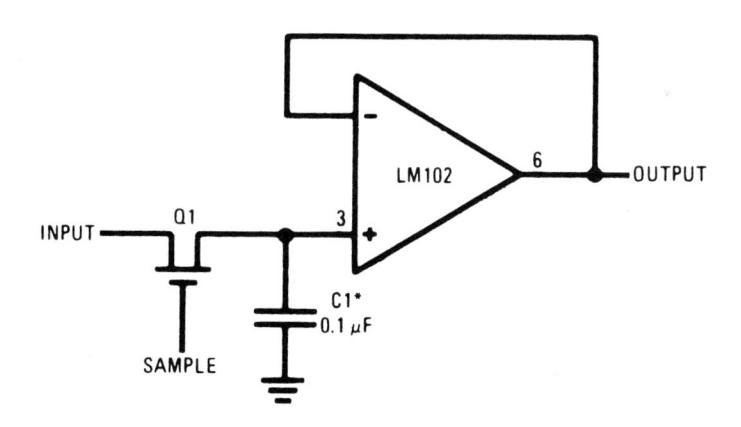

Sample-and-hold circuit. When a negative-going sample pulse is applied to the MOS switch, it will turn on hard and charge the holding capacitor to the instantaneous value of the input voltage. After the switch is turned off, the capacitor is isolated from any loading by the LM102, and it will hold the voltage impressed upon it.

The maximum input current of the LM102 is 10 nA, so with a 10 μF holding capacitor the drift rate in *hold* will be less than 1 mV/sec. If accuracies of about 1% or better are required, it is necessary to use a capacitor with polycarbonate, polyethylene, or Teflon dielectric. Most other capacitors exhibit a polarization phenomenon that causes the stored voltage to fall off after the sample interval with a time constant of several seconds. For example, if the capacitor is charged from 0 to 5V during the sample interval, the magnitude of the falloff is about 50 to 100 mV. C1 should have polycarbonate dielectric.

SECTION **14** *Data Acquisition, Multiplexing, & Transmission*

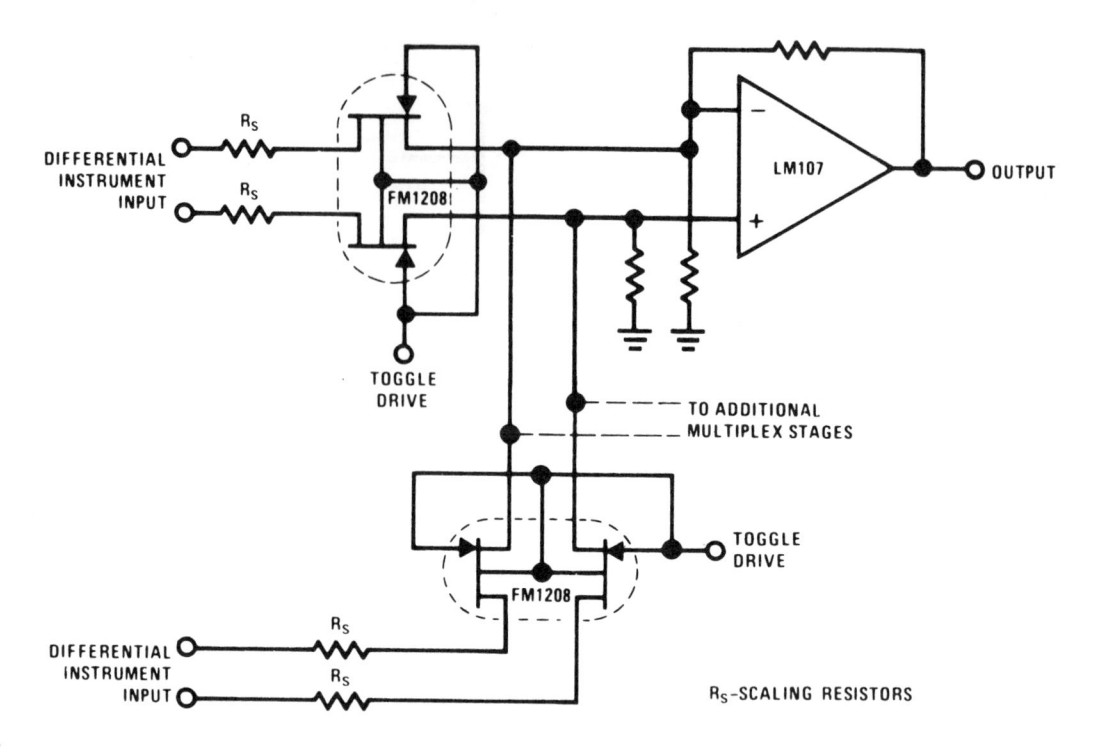

Differential analog switch. The FM1208 monolithic dual FET is used in a differential multiplexer application where $R_{DS(ON)}$ should be close-matched. $R_{DS(ON)}$ for the monolithic "dual" tracks at better than ±1% over wide temperature range (−25° to +125°C). This close tracking greatly reduces errors resulting from common-mode signals.

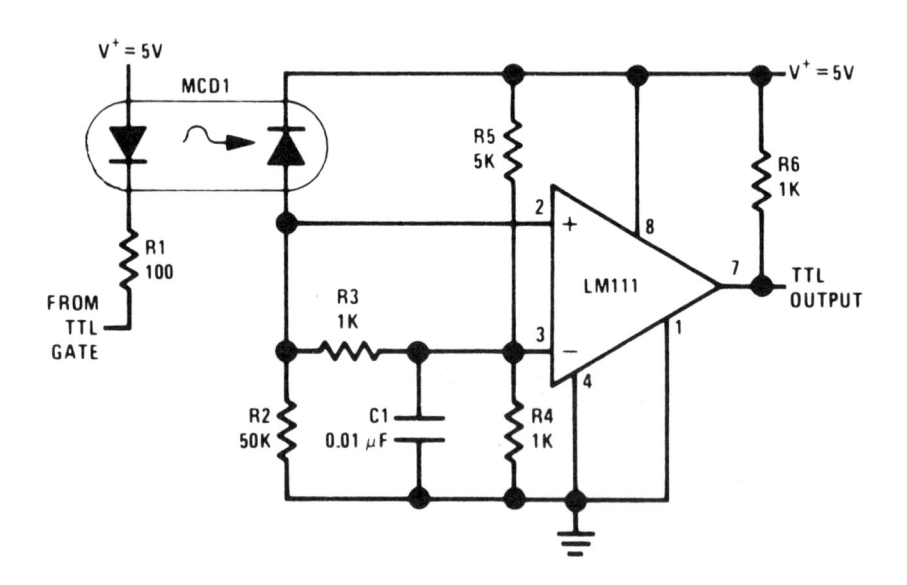

A data transmission system with near-infinite ground isolation. At the transmitting end, a TTL gate drives a gallium-arsenide LED. The light output is optically coupled to a silicon photodiode, and the comparator detects the photodiode output. The optical coupling makes possible electrical isolation of thousands of megohms at potentials of thousands of volts.

The maximum data rate of this circuit is 1 MHz. At lower rates (~200 kHz), R3 and C1 can be eliminated.

*$V_{OUT} = 0$ FOR $W < R_t C_t$
PULSE OUT $= W - R_t C_t$ FOR $W > R_t C_t$

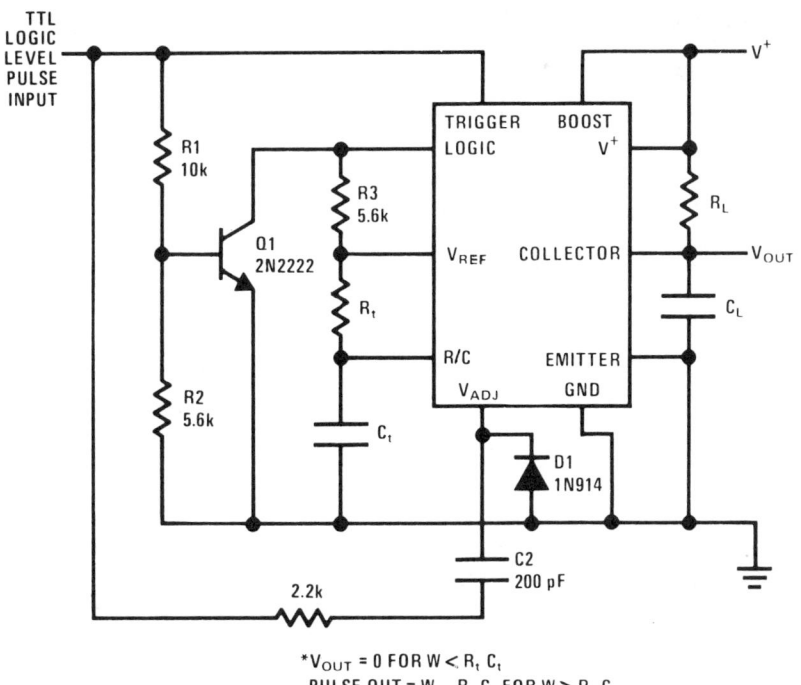

A pulse-width detector in which the logic terminal is normally held high by R3. When a trigger pulse is received, Q1 is turned on, driving the logic terminal to ground. The result of triggering the timer and reversing the logic at the same time is that the output does not change from its initial low condition. The only time the output will change states is when the trigger input stays high longer than one time period set by R_t and C_t. The output pulse width is equal to the input trigger width minus R_tC_t. C2 ensures no output pulse for short (less than RC) trigger pulses by prematurely resetting the timing capacitor when the trigger pulse drops. C_L filters the narrow spikes that would occur at the output as a result of interval delays during switching.

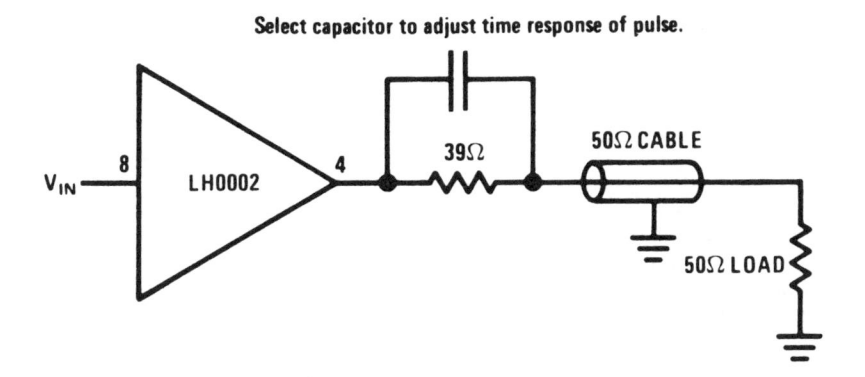

Select capacitor to adjust time response of pulse.

This line driver for long transmission lines matches the output impedance of the amplifier to the load and coaxial cable for proper line termination to minimize reflections. A capacitor can be added to adjust the time response of the waveform.

SECTION **14** Data Acquisition, Multiplexing, & Transmission

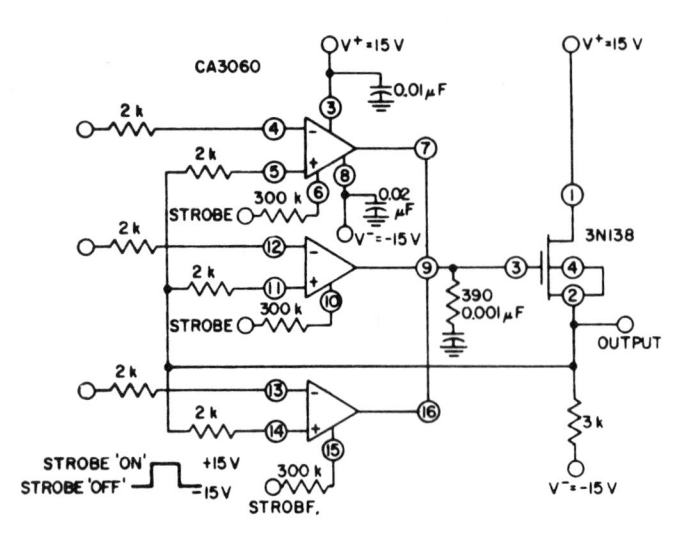

This three-channel multiplexer uses a single CA3060 and a 3N138 MOSFET as a buffer and power amplifier. When the CA3060 is connected as a high-input-impedance voltage follower and strobed on, each amplifier is activated, and the output swings to the level of the input of that amplifier. The cascade arrangement of each CA3060 amplifier with the MOSFET provides an open-loop voltage gain in excess of 100 dB, thus assuring excellent accuracy in the voltage-follower mode with 100% feedback.

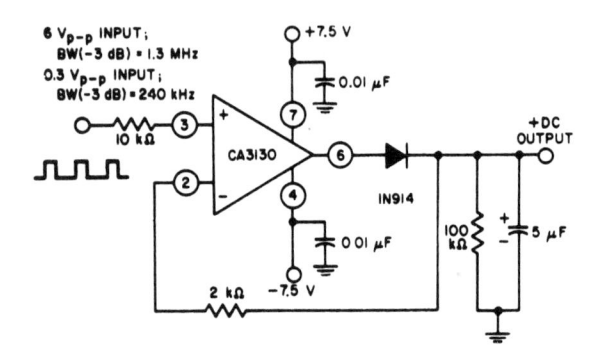

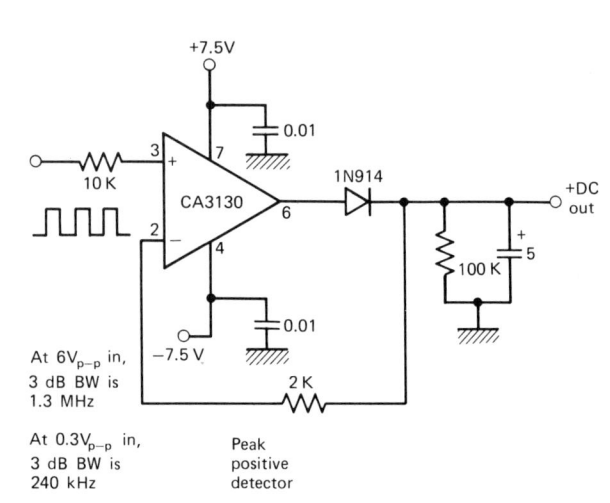

A peak detector circuit is easily implemented for both peak-positive and peak-negative circuits. For negative peak, just reverse the 1N914 diode. For large signal inputs, the bandwidth of the peak-negative circuit is much less than that of the peak-positive circuit (360 kHz at 6V in) because the second stage of the CA3130 limits it. Negative-going output signal excursion requires a positive excursion at the collector of Q11, which is loaded by the intrinsic capacitance of the circuitry in this mode. On negative excursions, the transistor functions in an active pull-down mode to discharge the intrinsic capacitance effectively.

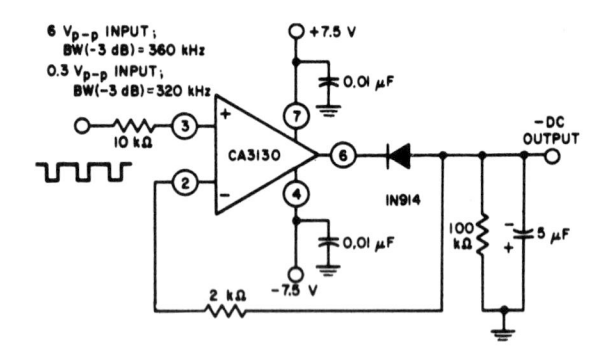

Peak-positive detector circuit

Peak-negative detector circuit

Peak detectors for both the peak-positive and the peak-negative circuits. With large signal inputs, the bandwidth of the peak-negative circuit is much less than that of the peak-positive circuit. The second stage of the CA3130 limits the bandwidth in this case. Negative-going output-signal excursion requires a positive-going signal excursion at the collector of transistor Q11, which is loaded by the intrinsic capacitance of the associated circuitry in this mode. During a negative-going signal excursion at the collector of Q11, the transistor functions in an active pull down mode, so that the intrinsic capacitance can be discharged more expeditiously.

470 IC SCHEMATIC SOURCEMASTER

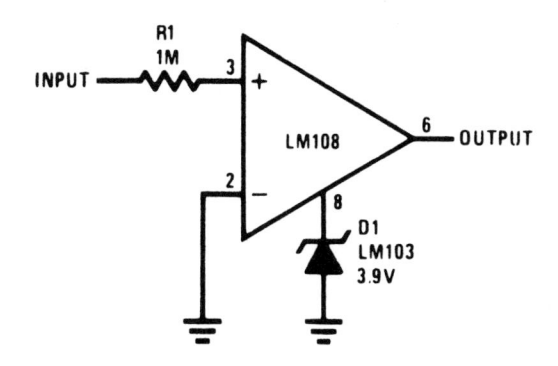

Zero-crossing detector. The output of the op-amp is clamped so that it can drive DTL or TTL directly. This is accomplished with a clamp diode on pin 8. When the output swings positive, it is clamped at the breakdown voltage of the zener. When it swings negative, it is clamped at a diode drop below ground. If the 5V logic supply is used as a positive supply for the amplifier, the zener can be replaced with an ordinary silicon diode. The maximum fan-out that can be handled by the device is one for standard DTL or TTL under worst-case conditions.

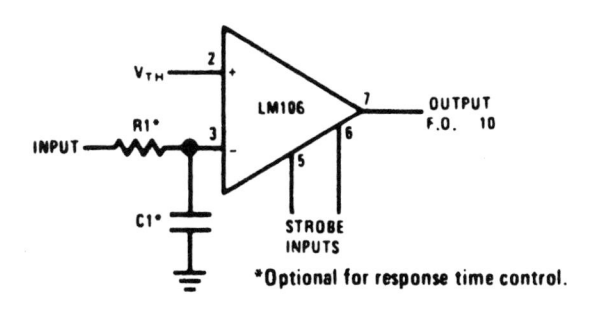

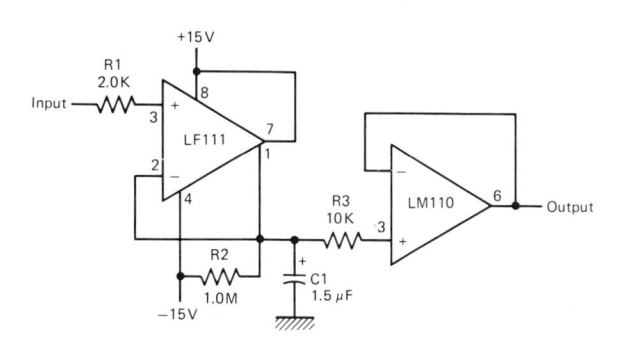

Positive-peak detector. C1 should be solid tantalum.

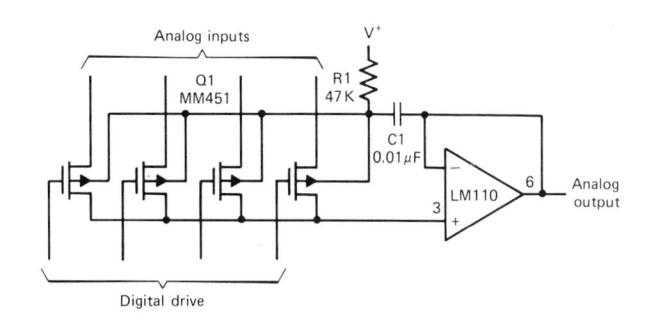

Adjustable-threshold line receiver. R1 is optional for response time control.

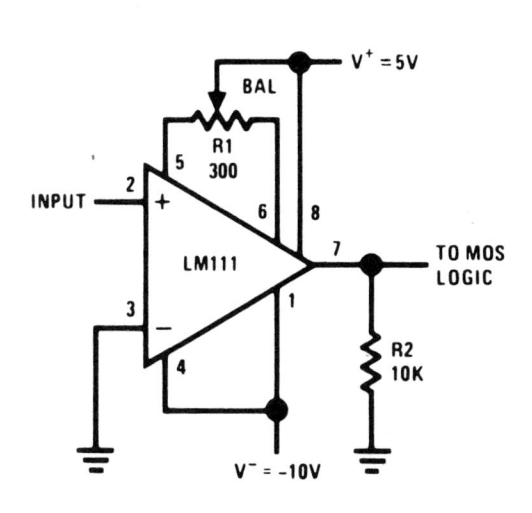

Buffer for analog switch. Switch substrates are boot strapped to reduce output capacitance of switch.

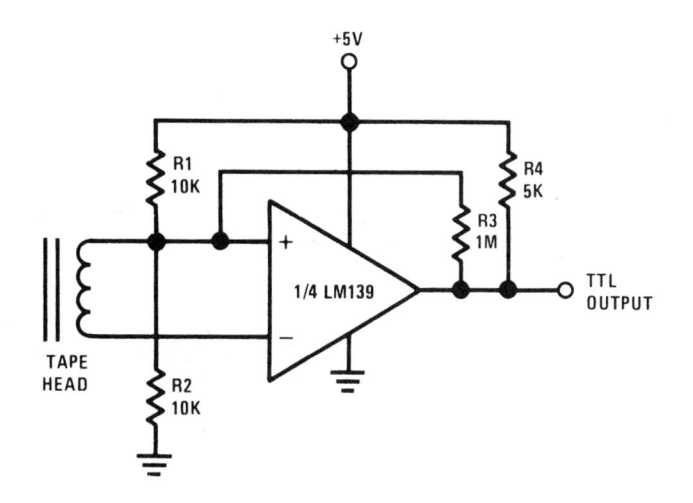

In this zero-crossing detector that can drive MOS logic, both a positive supply and the −10V supply for MOS circuits are used. Both supplies are required for the circuit to work with zero common-mode voltage.

Magnetic tape reader with TTL output. Using one LM139, four tape channels can be read simultaneously.

SECTION 14 *Data Acquisition, Multiplexing, & Transmission* **471**

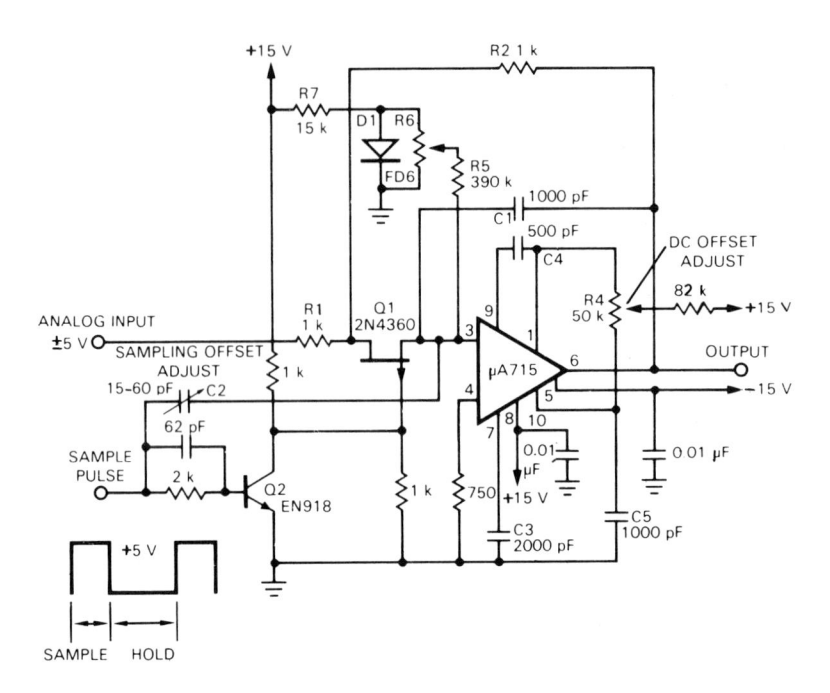

μA715 sample-and-hold circuit. During the *hold* time, the output voltage will tend to drift as the holding capacitor integrates the μA715 input bias current. This drift is compensated by R5, R6, R7, and D1. D1 acts as a temperature-dependent voltage source, causing the bias current through R5 to decrease with increasing temperature. With a 10V step input (±5V to ∓5V), the settling time to ±0.05% is 10 μsec. If C1 is decreased to 100 pF, the settling time is about 1 μsec. Temperature drift of the output in the *hold* mode is approximately 0.001%/°C for a hold time of 100 μsec.

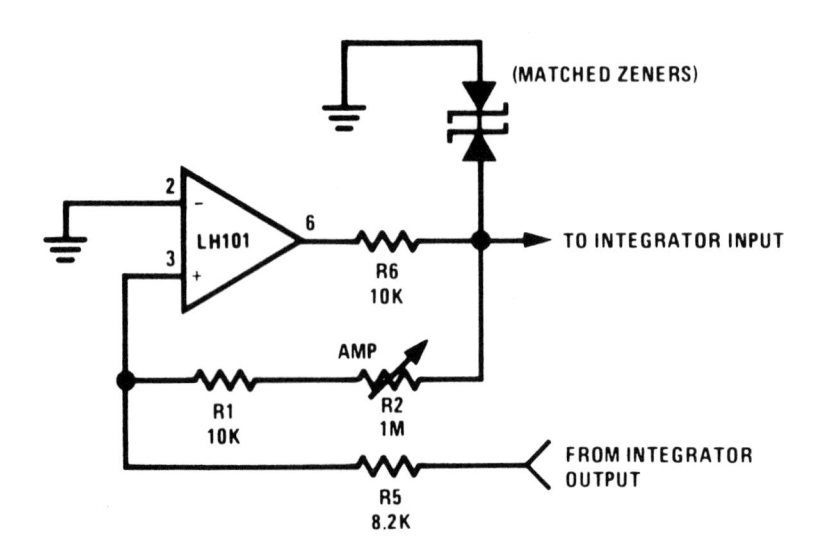

Threshold detector with regulated output. The current into the integrator should be large with respect to I_{bias} for maximum symmetry, and offset voltage should be small with respect to V_{OUT} peak.

472 IC SCHEMATIC SOURCEMASTER

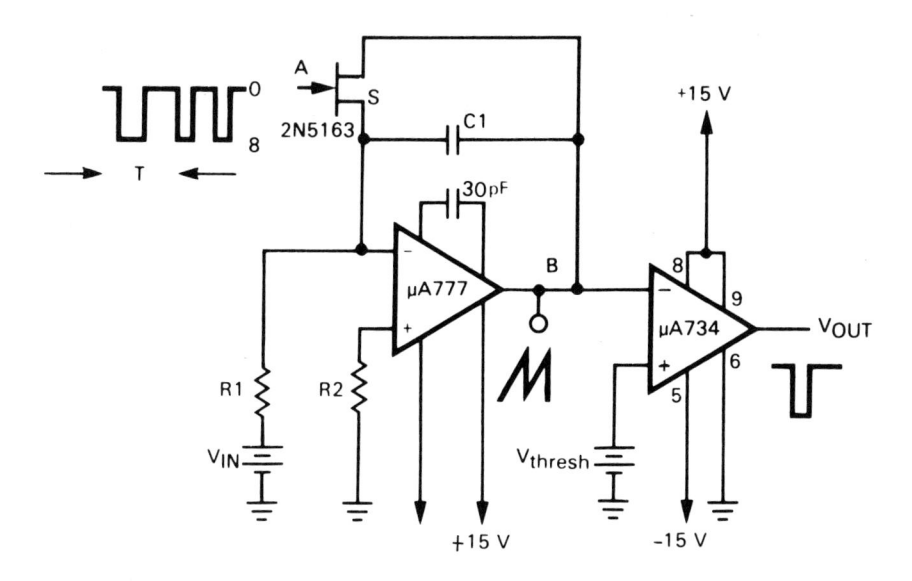

Pulse-width discriminator. A pulse appears at V_{OUT} for every negative pulse at A with width (t) longer than some desired value t_1. When the FET is turned off with a negative pulse at A, resistor R1, capacitor C1, and the μA777 form an integrator driven by the negative dc voltage V_{IN}, giving a ramp at output B. If the negative pulse at A has a width long enough for output ramp B to reach the threshold voltage of the μA734, a step change will occur at V_{OUT}. A step appears at V_{OUT} for every negative input pulse at A with width

$$T \geq T_1 = \frac{R1 C1\, V_{thresh}}{V_{IN}}$$

For R1 = R2 = 10k, C = 0.33 μF, V_{IN} = −10V, and V_A pulses with an amplitude of −8V discrimination of pulse widths from 10 μsec to 1 msec can be obtained by varying V_{thresh} from 30 mV to +3V.

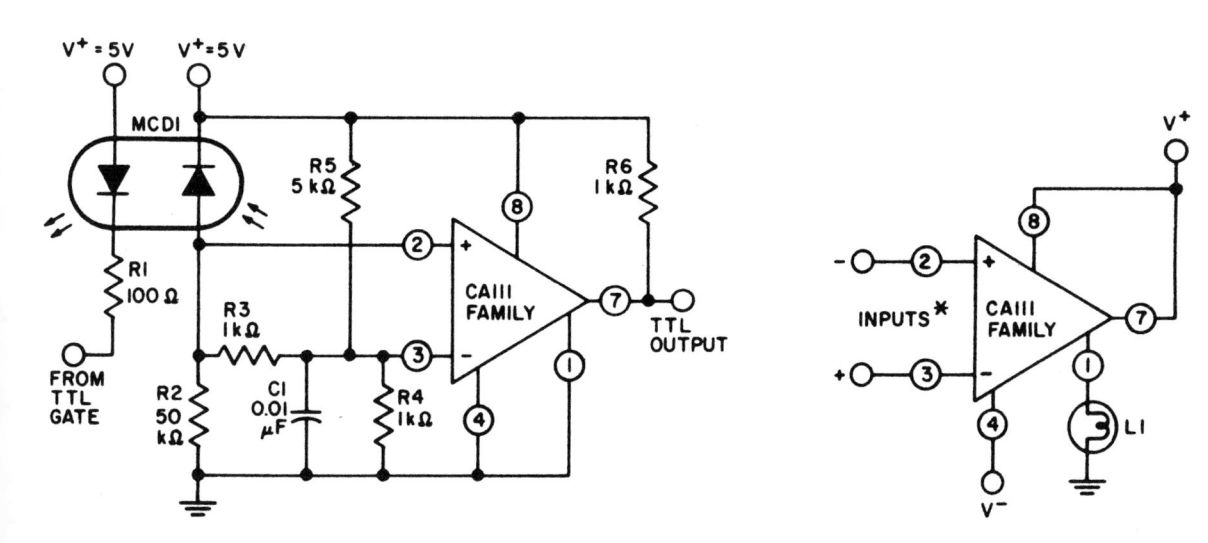

Digital transmission isolator.

Driving a ground-referred load. Input polarity is reversed when using pin 1 as output.

SECTION 14 *Data Acquisition, Multiplexing, & Transmission*

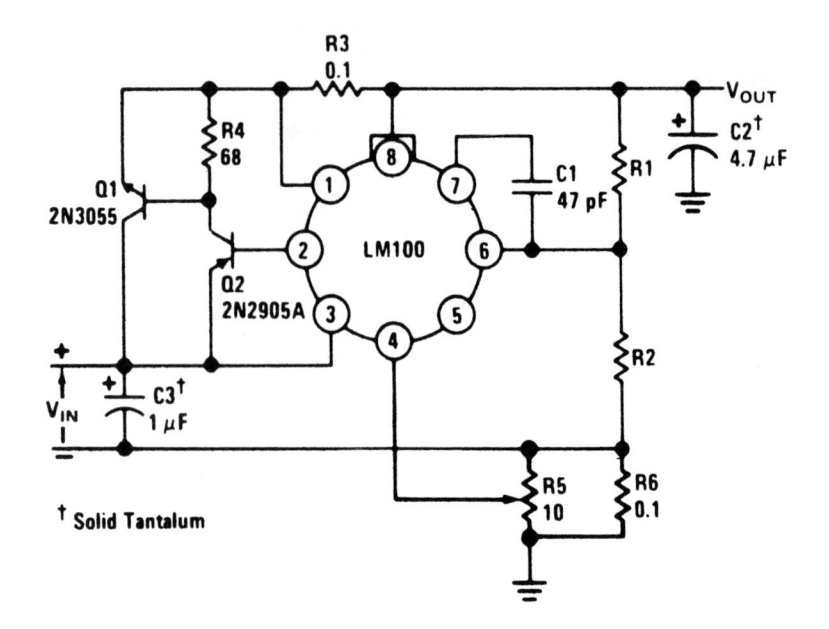

Line resistance compensator. Remote sensing of a load voltage to eliminate the effects of line resistance can be done with the LM100 by connecting the feedback resistors directly across the load rather than at the regulator output. However, it may be necessary to increase the size of the frequency compensation capacitor ordinarily used with the regulator. In certain applications, remote sensing is undesirable or the actual load is not directly accessible. An example of this is a dc motor application where it is desirable to reduce the effects of the armature resistance.

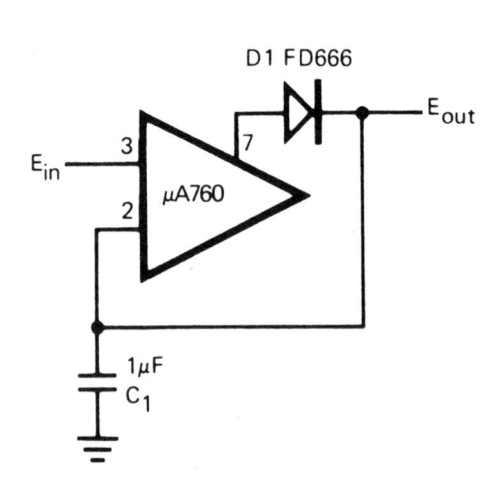

Simple fast peak detector operates with input pulse widths down to 50 nsec. The discharge current for C_1 is simply the bias current for the μA760 (typically 14 μA). If a shorter time constant is desired, a resistor can be connected between the output and V−. Without this resistor, the decay time of the voltage across C_1 is approximately 50 nsec/V.

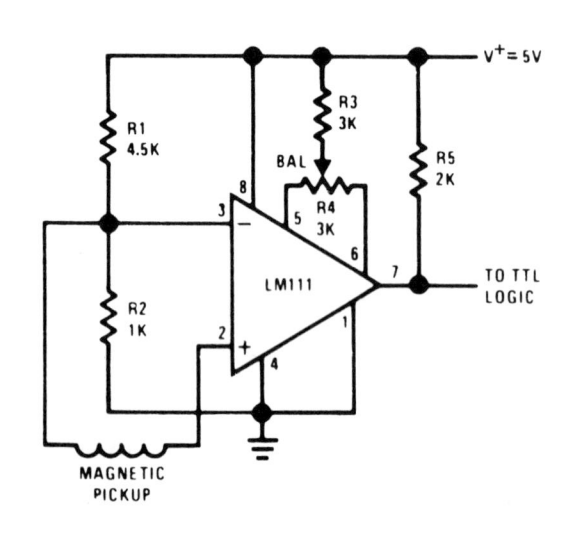

A zero-crossing detector for a magnetic pickup such as a magnetometer or shaft-position pickoff. It delivers the output signal directly to DTL or TTL circuits and operates from the 5V logic supply. The resistive divider, R1 and R2, biases the inputs 0.5V above ground within the common-mode range of the device. An optional offset balancing circuit, R3 and R4, is included.

474 IC SCHEMATIC SOURCEMASTER

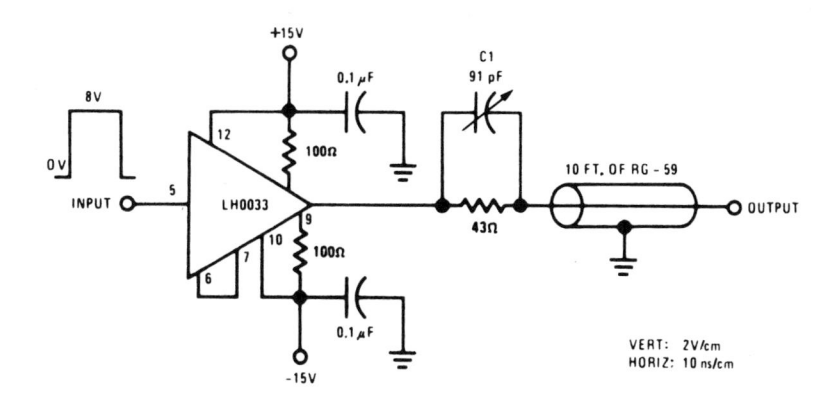

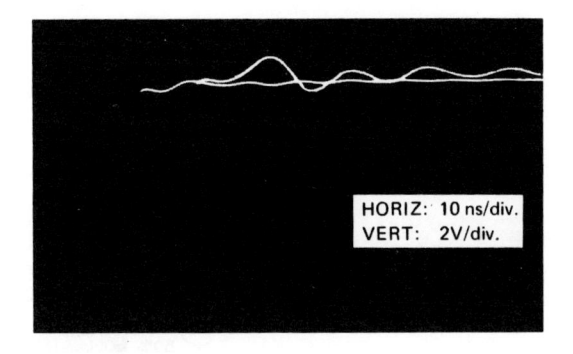

Coaxial driver. Pin 6 is shorted to pin 7, obtaining an initial offset of 5.0 mV. C1 is adjusted as a function of cable length to optimize rise and fall time. Rise time is 10 nsec, as shown in the waveform.

LH0033 pulse response into 10-foot open-ended coaxial cable.

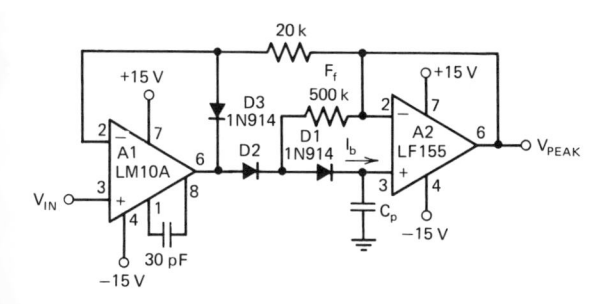

Low-drift peak detector. By adding D1 and R_f, $V_{D1} = 0$ during *hold* mode. Leakage of D2 is provided by feedback path through R_f. Leakage of circuit is essentially I_b (LF155, LF156) plus capacitor leakage of C_P. Diode D3 clamps $V_{OUT}(A1)$ to $V_{IN} - V_{D3}$ to improve speed and to limit reverse bias of D2. Maximum input frequency should be $\ll 1/(2\pi R_1 C_{D2})$, where C_{D2} is the shunt capacitance of D2.

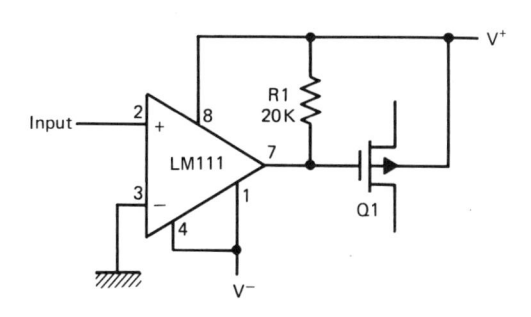

Zero-crossing detector driving an MOS analog switch. The ground terminal of the IC is connected to V^-; hence, with ±15V supplies, the signal swing delivered to the gate of Q1 is also ±15V. This type of circuit is useful where the gain or feedback configuration of an op-amp circuit must be changed at some precisely determined signal level.

SECTION **14** Data Acquisition, Multiplexing, & Transmission

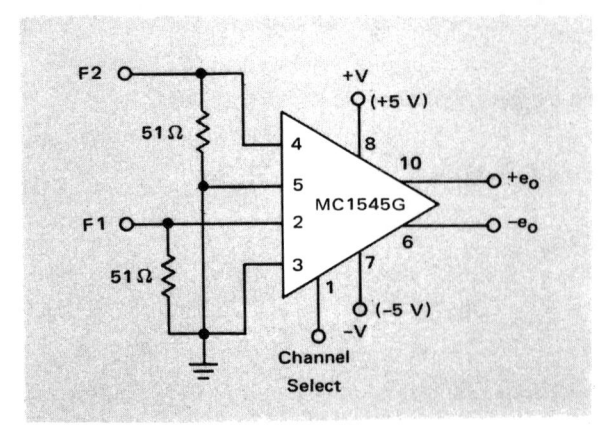

Multiplexer (FSK). Frequency F2 will be passed when the voltage at pin 1 is greater than +1.5V, and F1 will be passed when the voltage at pin is approximately 0V.

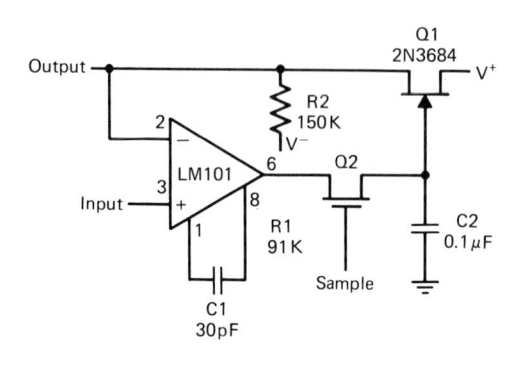

Low-drift sample-and-hold circuit. C2 should be a polycarbonate-dielectric capacitor.

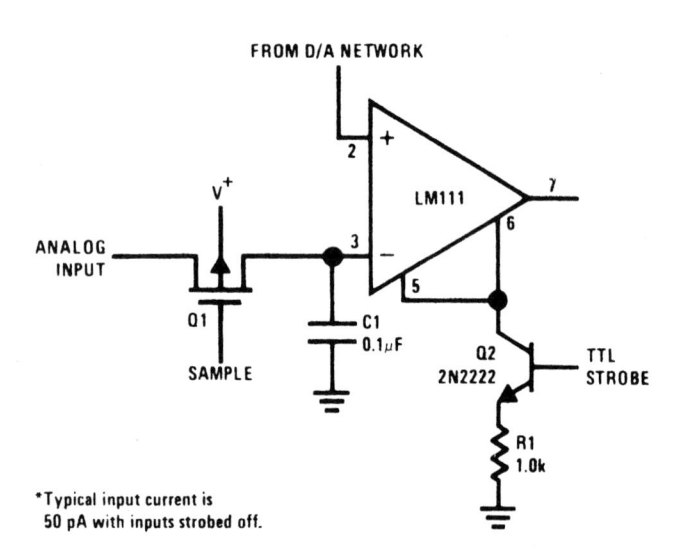

Strobing off both input and output stages. Typical input current is 50 pA with inputs strobed off.

*Typical input current is 50 pA with inputs strobed off.

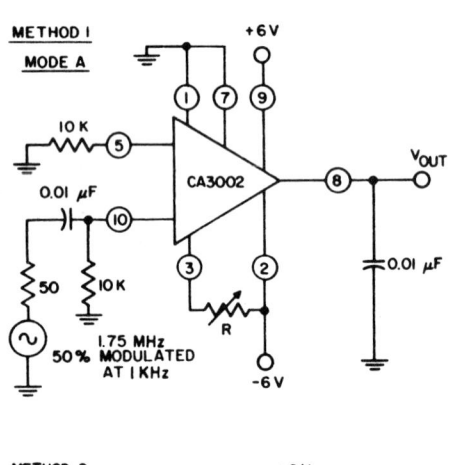

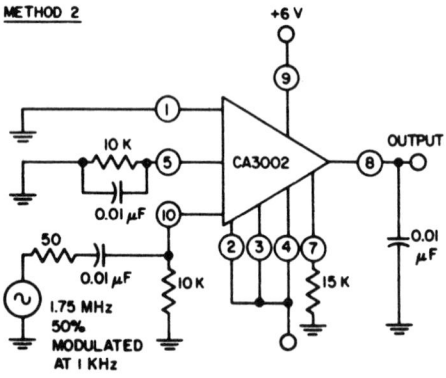

Envelope detectors using CA3002 ICs: (1) the emitter of the output transistor, Q6, can be operated at zero voltage by connection of an external resistor in the bias loop of the constant-current transistor Q3, or (2) the current in transistor Q6 can be reduced by connection of a large resistor (12,000 to 18,000Ω) in series with its emitter resistor. In the circuit for method 1, the current in the differential pair (Q2 and Q4) is increased to the point at which the common-collector output transistor Q6 is biased almost to cutoff. For this current increase, the constant-current transistor Q3 is operated with terminal 4 open, and the emitter resistor R9 is shunt-loaded by the external resistor at terminal 3. Envelope detection can be accomplished only in mode A with method 1.

476 IC SCHEMATIC SOURCEMASTER

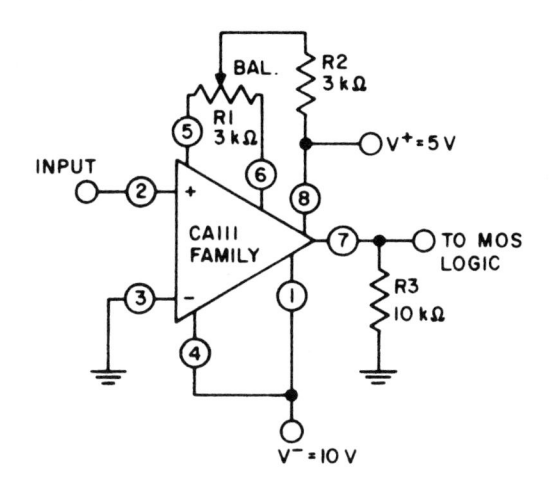

A zero-crossing detector driving MOS logic.

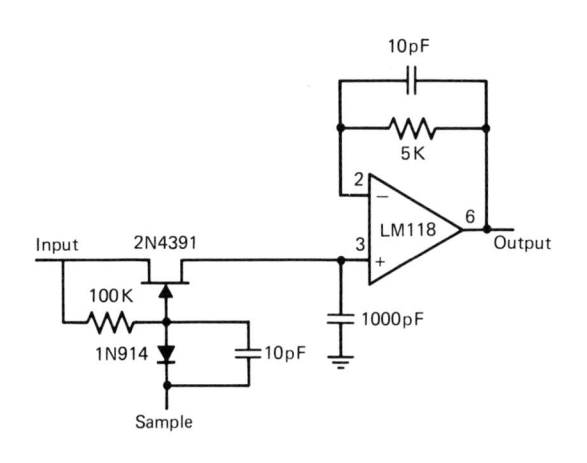

Fast sample-and-hold circuit.

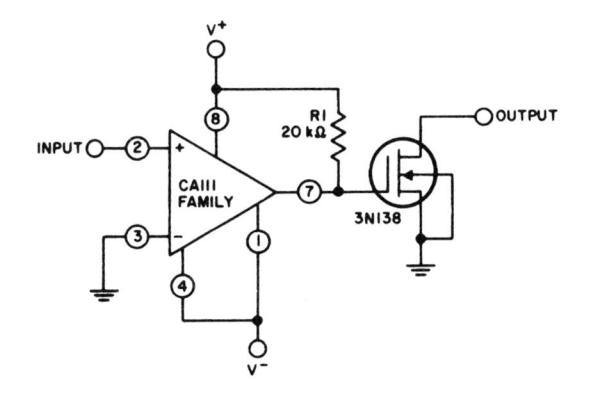

A zero-crossing detector driving N-channel MOS switch.

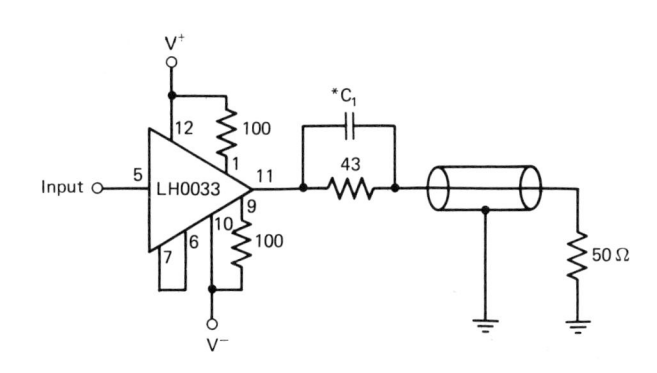

Coaxial cable driver. Select C1 for optimum pulse response.

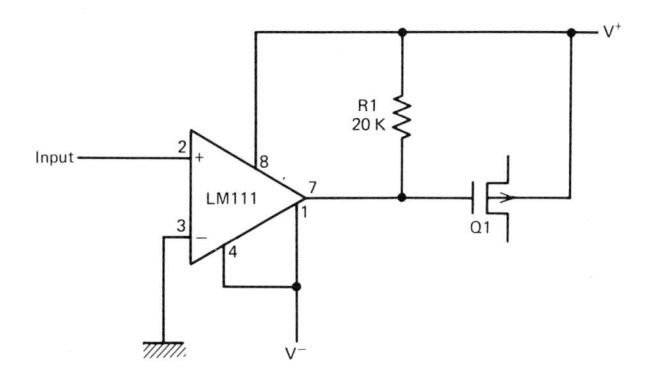

Zero-crossing detector driving P-channel MOS switch.

SECTION 14 Data Acquisition, Multiplexing, & Transmission

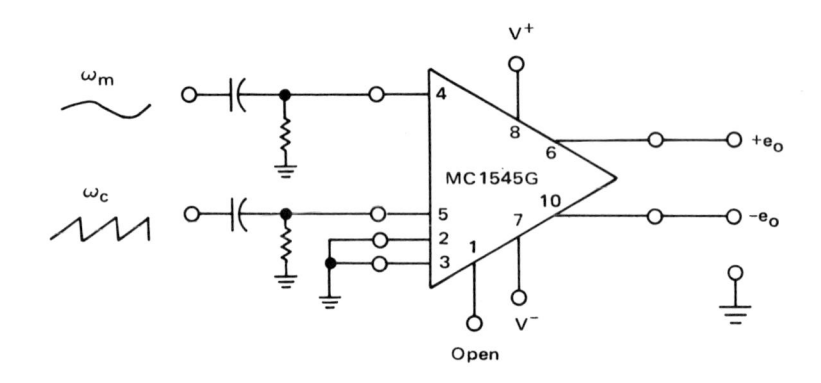

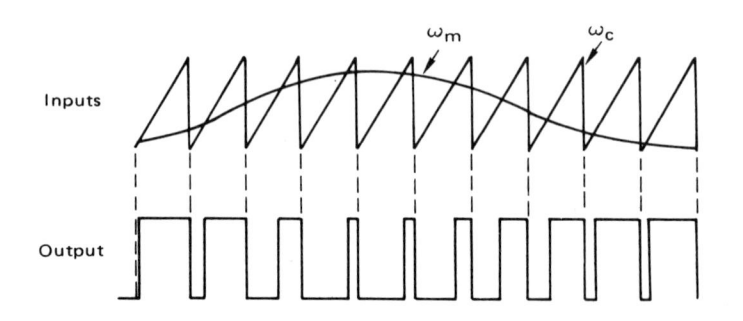

Pulse-width modulator. When the amplitude of the modulation signal is greater than that of the carrier signal, the output will be driven to one extreme if the amplifier gain is high.

To achieve a reference frequency, a sawtooth is used as the carrier signal. In this manner one edge of the output pulse will always be at the trailing edge of the sawtooth. The other edge of the output pulse is derived from the modulation signal. The inset illustrates the comparison process that varies the pulse width.

The rise time of the pulses will be approximately the time it takes the input signals to give a differential voltage large enough to drive the output to full excursion. As the voltage gain of the MC1545 has limits between 6.3V/V and 10V/V, this means the input differential voltage has a minimum of 100 mV and a maximum of 318 mV change for an output swing of $2.0V_{p-p}$. The fall time is calculated in the same manner but will be much smaller if the sawtooth has a sharp trailing edge.

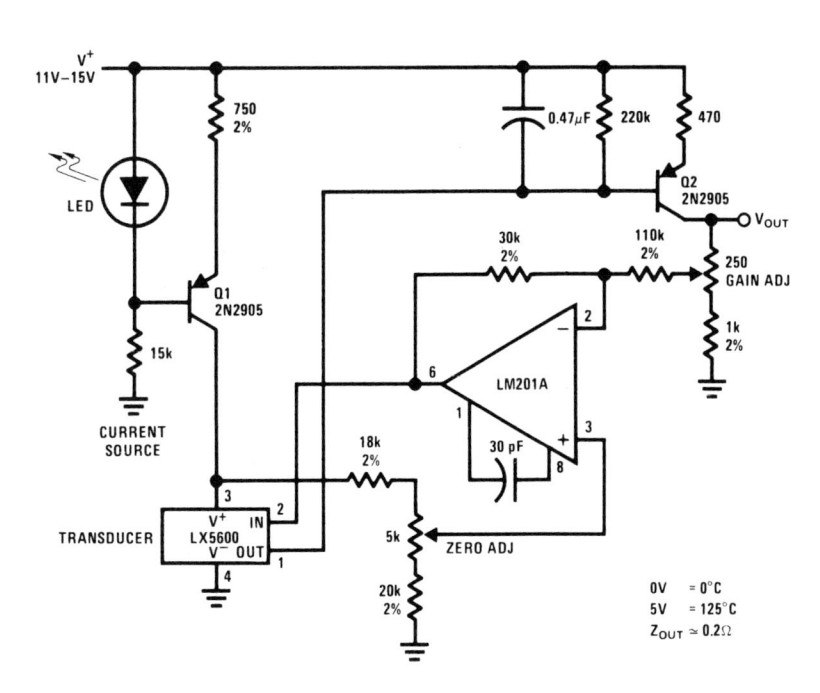

Telemetry interface. Various pieces of mobile test equipment such as instrumentation recorders or telemetry subcarrier VCOs operate best with a calibrated 0 to +5V input. In some cases a low output impedance is also helpful. The LX5600 indicates increasing temperature by a negative-going voltage with respect to the positive supply pin. Q1 supplies the thermometer with a constant current. Output transistor Q2 provides the inverting function and requires very little base drive, holding down transducer dissipation. The entire system now requires dc feedback around it to stabilize gain (and thus the full-scale calibration) and to eliminate supply and base-emitter voltage variations from the output.

The system has an open-loop gain of 5,000 to 10,000, so the attenuation of the output signal that is fed back negatively determines closed-loop gain. The reinverting op-amp actually has an attenuation of 0.27. The 250Ω pot in the output divider is thus the full-scale adjust and trims the attenuation to 0.25, or a gain of 4. Pin 3 of the transducer provides the voltage reference for the zero-adjust divider. At the calibration point of 0°C, the LM201A output must be at a nominal 2.73V below pin 3 of the transducer. Since voltages at both ends of the 30k, 110k network are known at 0°C, it is easy to predict the voltage adjustment at pin 3 of the LM201A required to fulfill this condition. Op-amp pin numbers shown are for metal can.

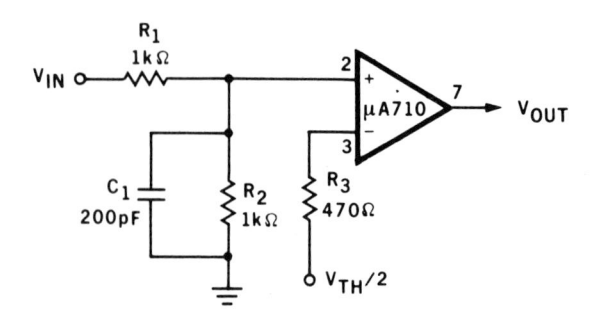

High-noise-immunity line receiver for slow, high-level logic. The resistive divider on the input of the comparator permits logic signals up to ±14V. Put a capacitor at the comparator input (C_1) to make the circuit insensitive to fast noise spikes.

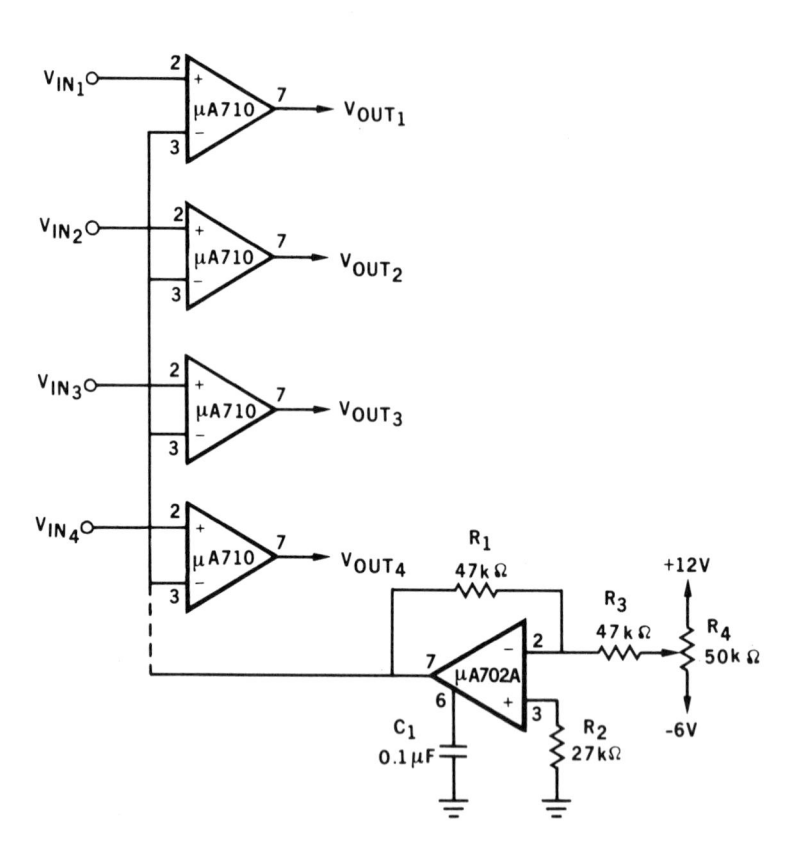

Adjustable-threshold, high-noise-immunity line receivers. It is sometimes necessary to design a piece of logic equipment to interface with any one of the many types of digital logic (RTL, DTL, CML, CTL, etc.). This circuit satisfies the requirements. The input threshold voltage can be adjusted over a ±5V range with single control. A Fairchild μA702 operational amplifier provides an adjustable reference voltage with a very low source impedance. A single amplifier can handle over 100 comparators.

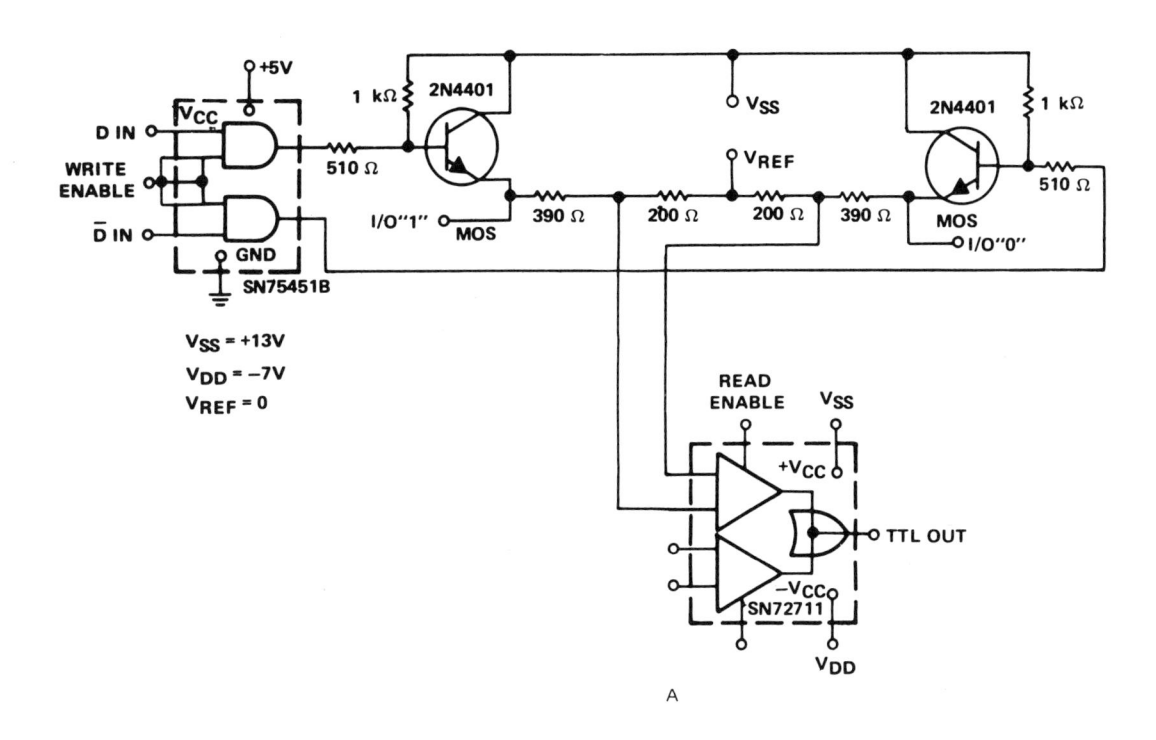

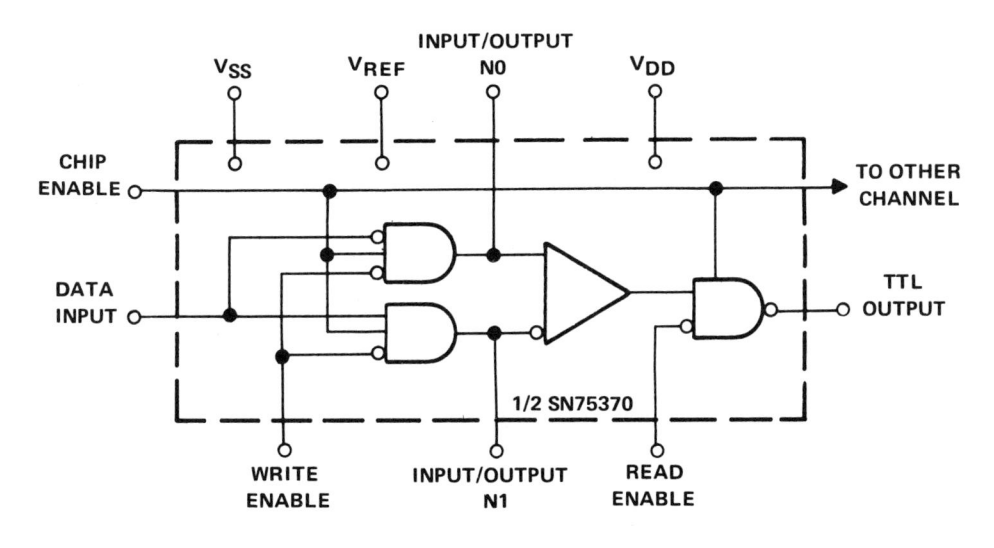

Combination write (or drive) and read (or sense) circuits. Their compactness, speed of operation, and control features make them particularly suited for more complex systems incorporating MOS circuits like the TMS4062 1024-word by 1-bit dynamic RAM. Figure A shows a circuit using a combination of discrete and integrated-circuit devices, while B provides a completely monolithic approach. The SN75370 is an entirely self-contained, dual-channel, sense-digit driver that significantly reduces component count. As a comparison, it would require 2 SN75451B packages, 1 SN72711 package, 4 2N4401 transistors, and 16 resistors to duplicate the function of 1 SN75370.

SECTION **14** *Data Acquisition, Multiplexing, & Transmission* **481**

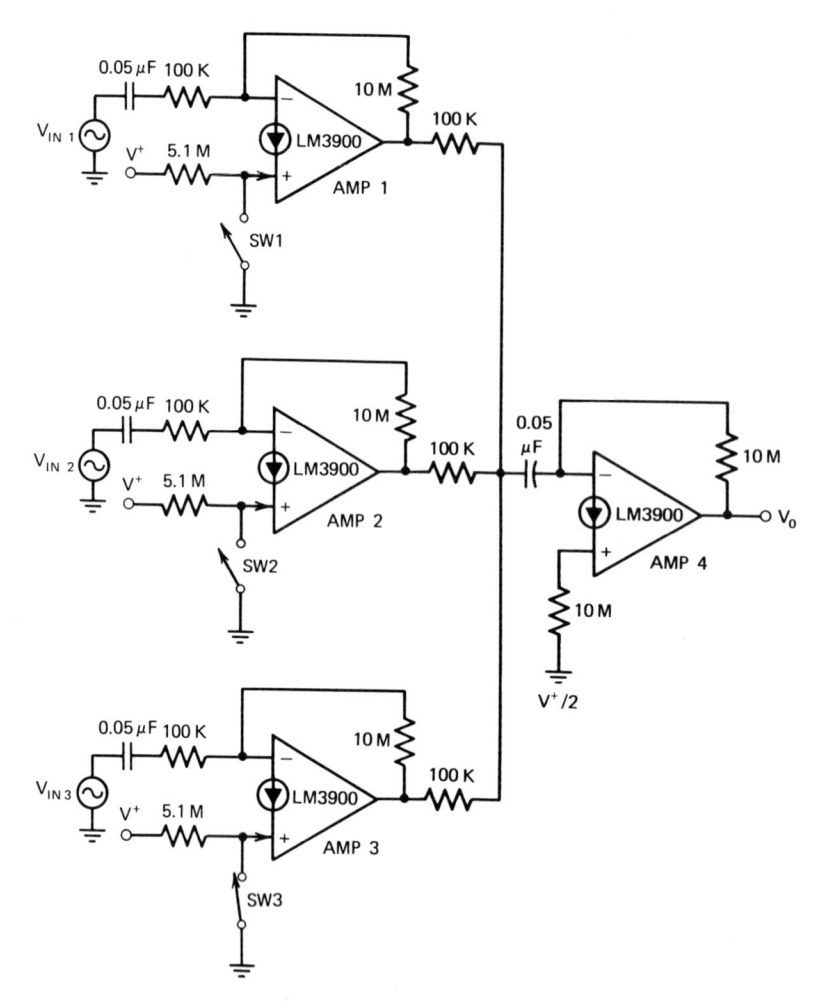

Active-channel selector. The four individual amplifiers of the LM3900 can be used for audio mixing (with many amplifiers simultaneously providing signals that are added to generate a composite output signal) or for channel selection (only one channel enabled at a time). Three amplifiers are shown being summed into a fourth. For audio mixing, all amplifiers are simultaneously active. Particular amplifiers can be gated off by making use of dc control signals applied to the (+) inputs to provide a channel-select feature.

482 IC SCHEMATIC SOURCEMASTER

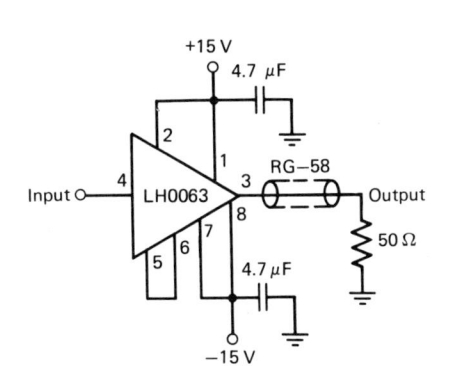

Coaxial cable driver.

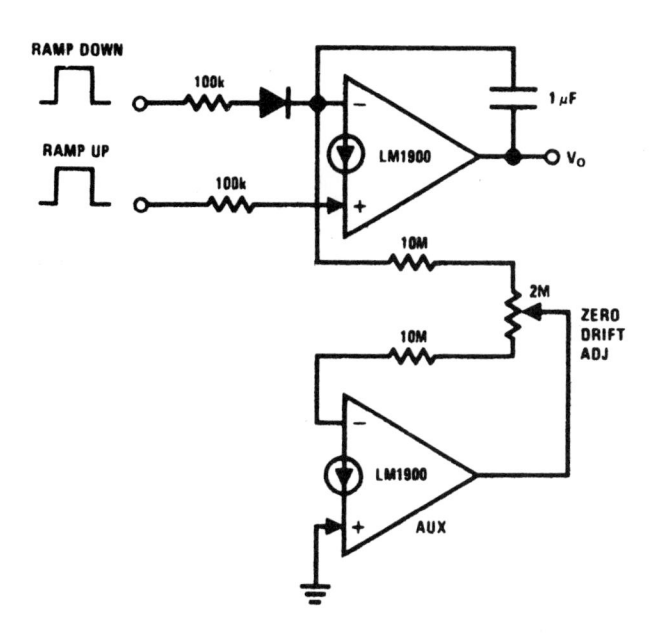

Low-drift ramp-and-hold circuit.

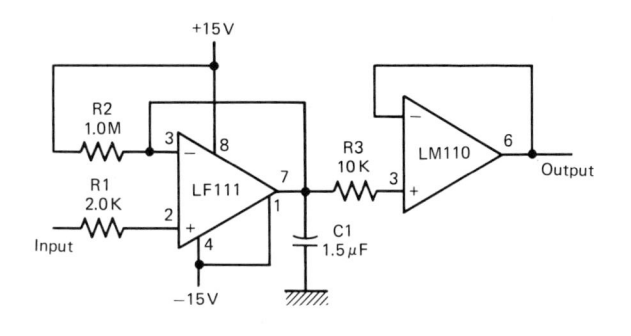

Negative peak detector. C1 should be solid tantalum.

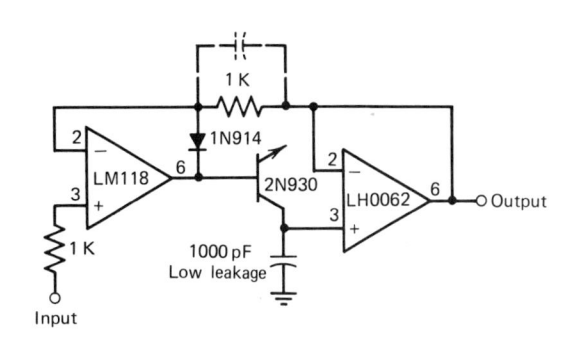

High-speed positive peak detector.

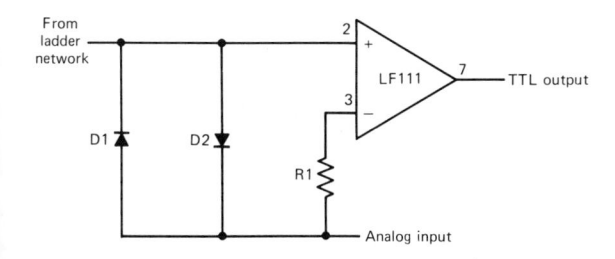

Using clamp diodes to improve response.

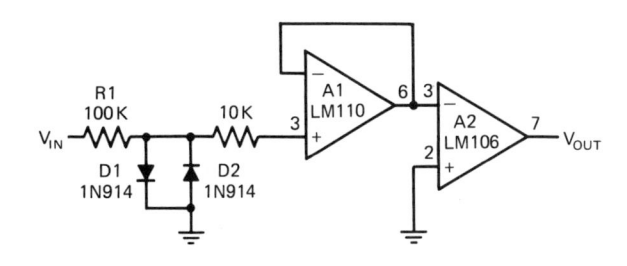

Zero-crossing detector.

SECTION **14** *Data Acquisition, Multiplexing, & Transmission* **483**

SECTION 15
LOGIC AND FUNCTIONALLY RELATED CIRCUITS

This section presents arithmetic units, logic-to-logic and logic-to-line interface circuits, a variety of gates and similar circuitry, and miscellaneous comparators.

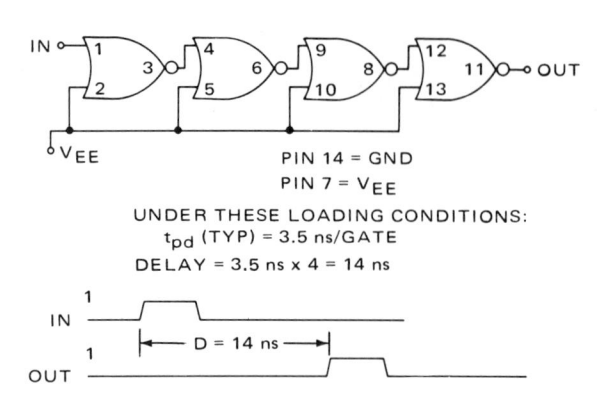

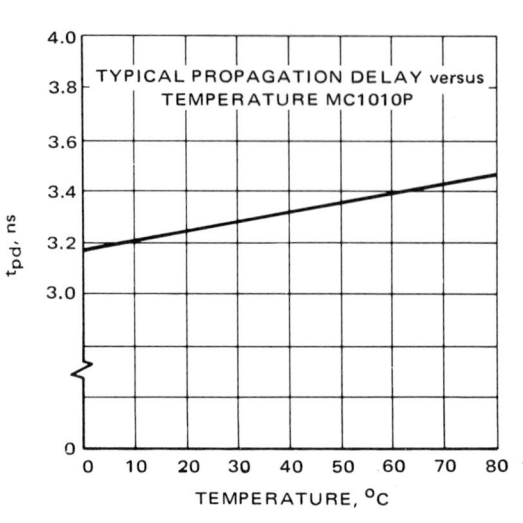

Quad-2 input NOR gate as a delay element (delay-line substitute). Since each gate has a propagation delay associated with it, the delay and output pulse width of the monostable will be equal to the propagation delay of one gate multiplied by the number of gates used. Typical delay of this circuit is 14 nsec. If the independent pulse-width monostable is used, the requirement for an even number of inversions can be avoided by connecting the input of the delay element to the NOR output, instead of to the OR output, of the first gate. This cannot be done in the fast recovery circuit, since the delay element input is connected directly to the input pulse of the monostable.

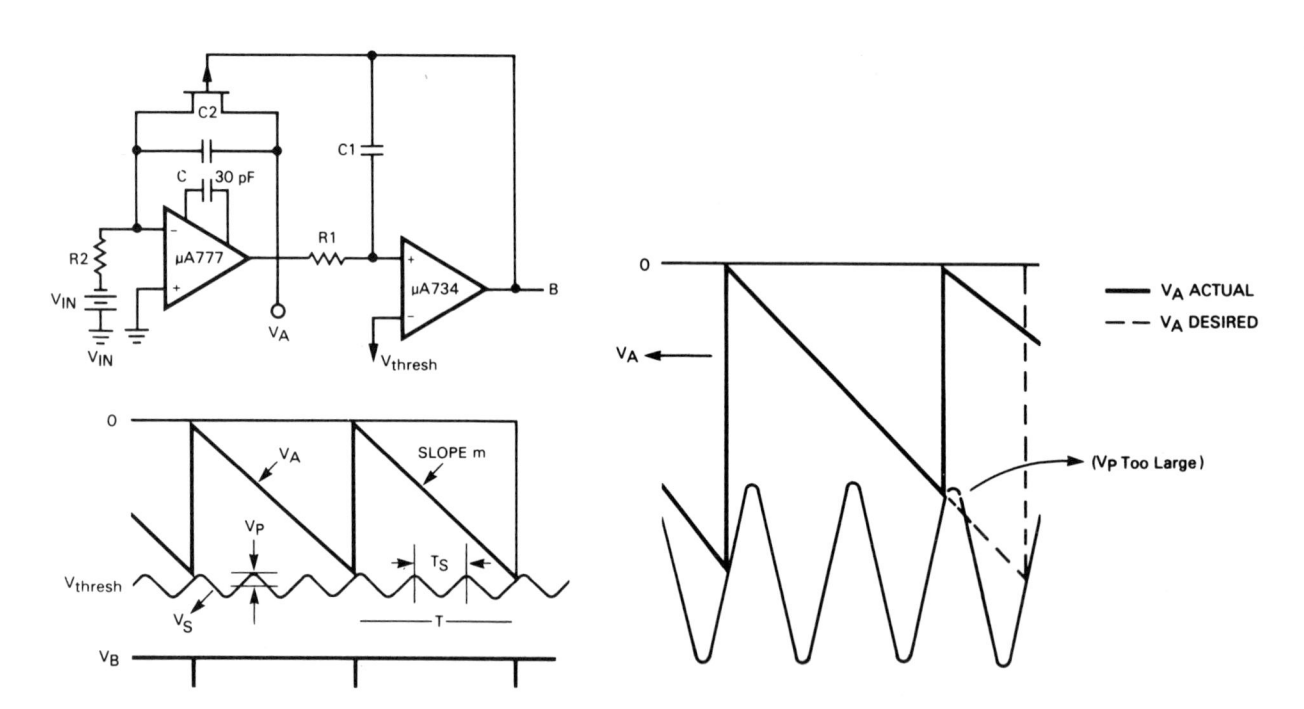

This sine-wave frequency divider has as its output a sweep voltage at A synchronized with a symmetrical signal V_S and pulses at B of frequency $1/n$. The phase of the output relative to V_S is adjusted by varying V_S dc threshold voltage. When the voltage at A is less than the threshold voltage of the µA734, output B of the µA734 is high, turning the FET off. Capacitor C2, resistor R2, and the µA777 form an integrator driven by V_{IN}, resulting in an output ramp. When the output ramp reaches the threshold voltage, output B of the µA734 goes to 0.2V and turns on the FET, which provides a discharge path for the charge across capacitor C2. When the output of the integrator exceeds V_{thresh} the µA734 again assumes the high state to pinch off the FET and complete the cycle. The signal amplitude, V_S, should be small compared to the slope of the ramp to ensure that the comparator is not false-triggered before the end of the calculated period t. The frequency division will be stable as long as the ramp voltage V_A crosses V_S on a rising portion; this permits a small variation of the signal frequency without change of the division ratio n.

486 IC SCHEMATIC SOURCEMASTER

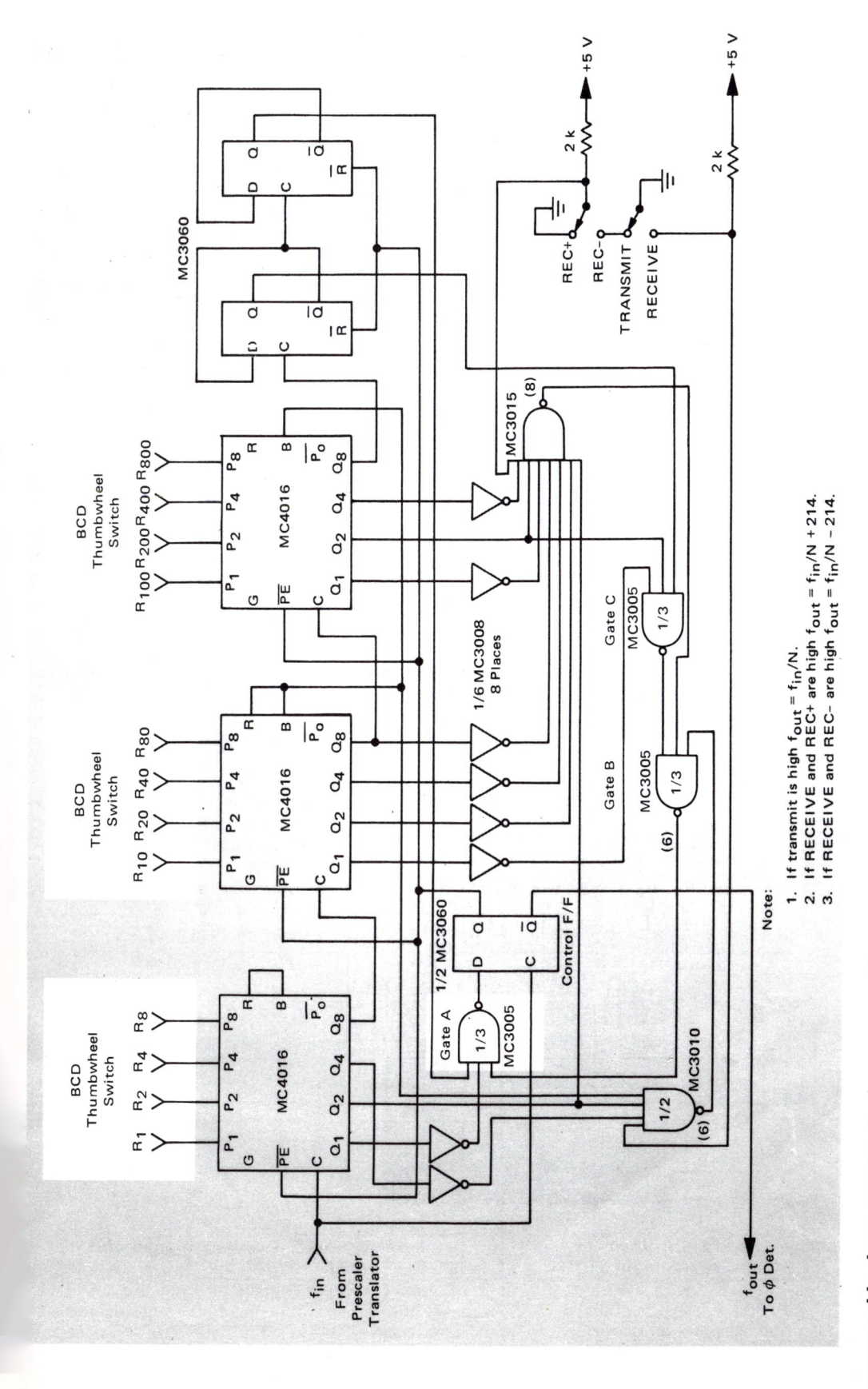

Programmable down-counter. In frequency synthesizers for the aircraft band, a programmable counter is inserted in the feedback loop to generate multiple stable output frequencies. A feature in the receive mode is the capability of automatically adding or subtracting 10.7 MHz at the output of the VCO, which reduces the design difficulties for the VCO designer.

Note:
1. If transmit is high $f_{out} = f_{in}/N$.
2. If RECEIVE and REC+ are high $f_{out} = f_{in}/N + 214$.
3. If RECEIVE and REC− are high $f_{out} = f_{in}/N - 214$.

SECTION **15** *Logic and Functionally Related Circuits* **487**

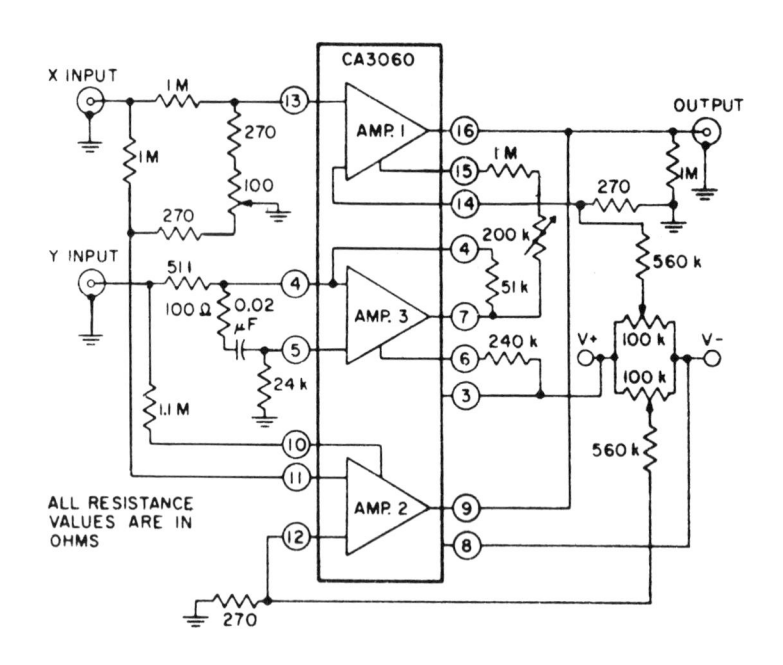

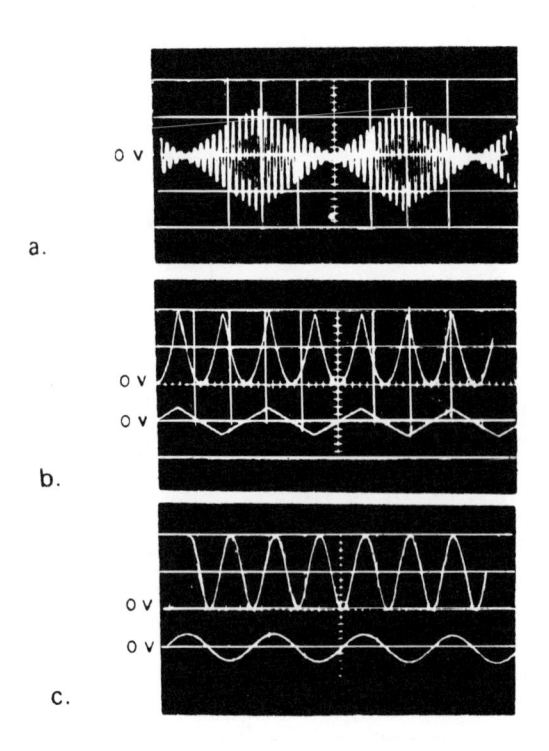

This four-quadrant multiplier consists of a single CA3060 and exhibits no level shift between input and output. Adjustment of the circuit is quite simple: with both the X and Y voltages at zero, connect terminal 10 to 8. This procedure disables amplifier 2 and permits adjusting the offset voltage of amplifier 1 to zero by means of the 100k potentiometer. Next, remove the short between terminals 10 and 8 and connect terminal 15 to 8. This step disables amplifier 1 and permits amplifier 2 to be zeroed with the other potentiometer. With ac signals on both the X and Y input, R3 and R11 are adjusted for symmetrical output signals. Inset shows circuit waveforms for suppressed carrier modulation of 1 kHz carrier with a triangular wave (a), squaring of a triangular wave (b), and squaring of a sine wave (c).

488 IC SCHEMATIC SOURCEMASTER

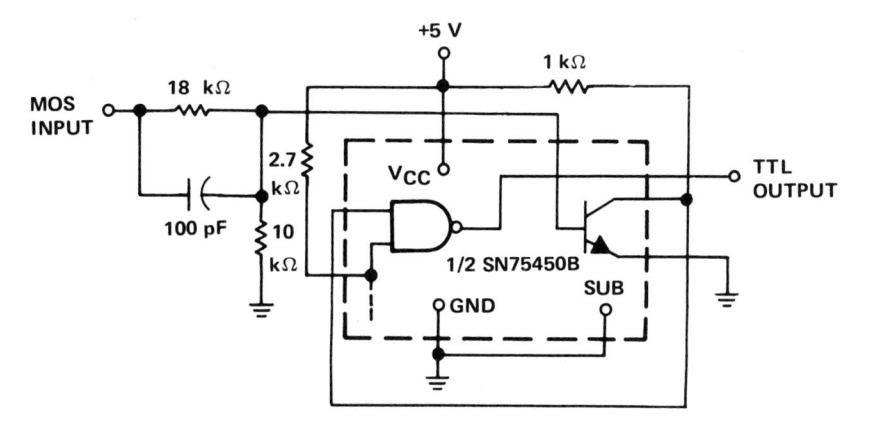

MOS-to-TTL level-shift for positive MOS levels.

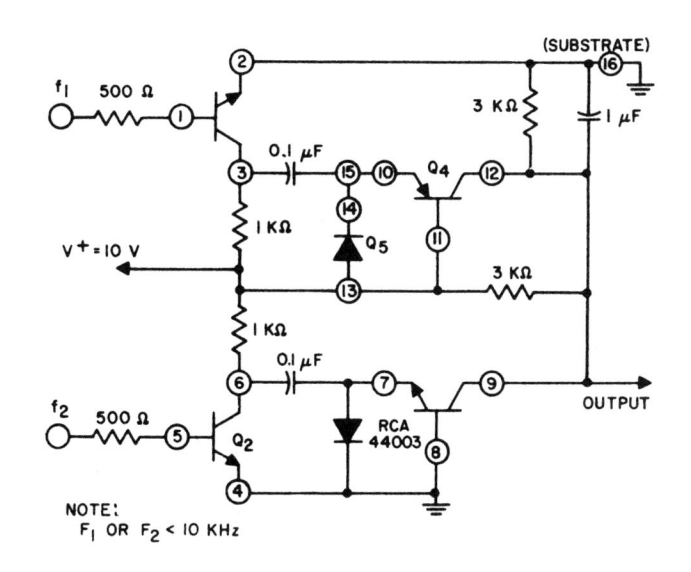

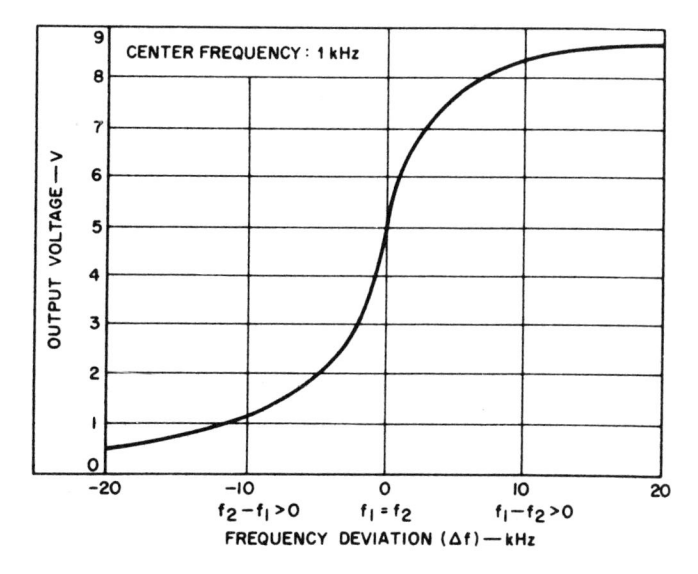

Frequency comparator using CA3096E. Chart shows comparator characteristics.

SECTION 15 *Logic and Functionally Related Circuits*

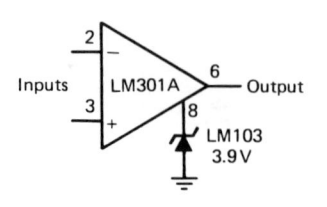

Voltage comparator for driving RTL or high-current driver.

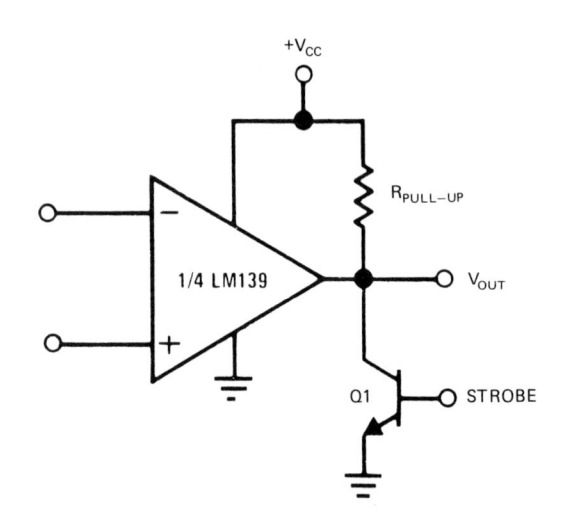

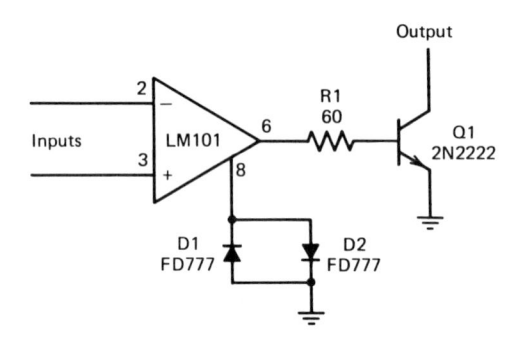

Voltage comparator for driving DTL or TTL integrated circuits.

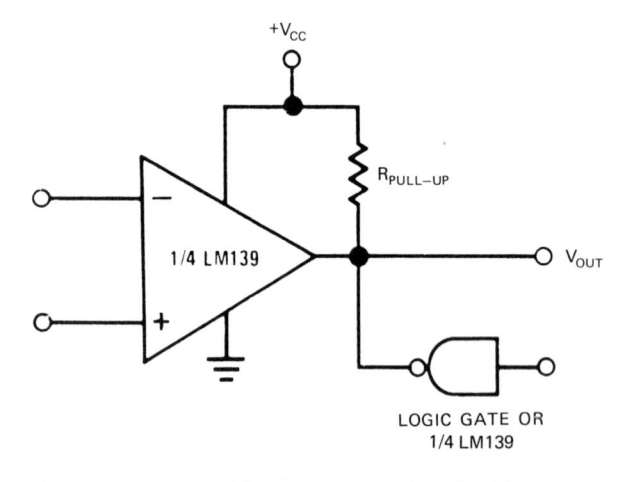

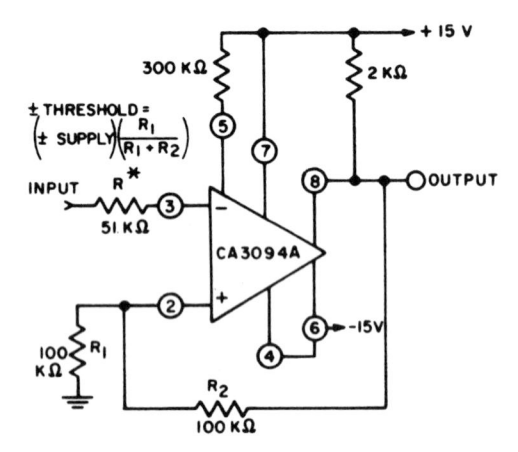

Output strobing. The output of the LM139 may be disabled by adding a clamp transistor as shown. A strobe control voltage at the base of Q1 will clamp the comparator output to ground, making it immune to any input changes.

If the LM139 is being used in a digital system, the output may be strobed using any type of gate having an uncommitted collector output (such as National's DM5401/DM7401). In addition, another comparator of the LM139 could also be used for output strobing.

Comparator and threshold combination. Asterisked resistor to be calculated; depends on values of R1 and R2: $R^* = R1R2/R1 + R2$.

490 IC SCHEMATIC SOURCEMASTER

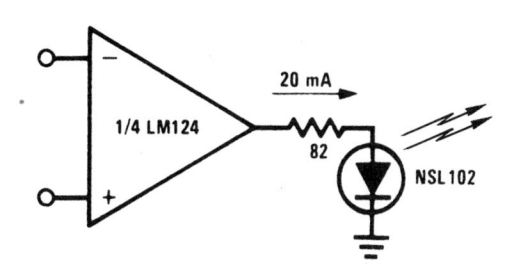

LED driver.

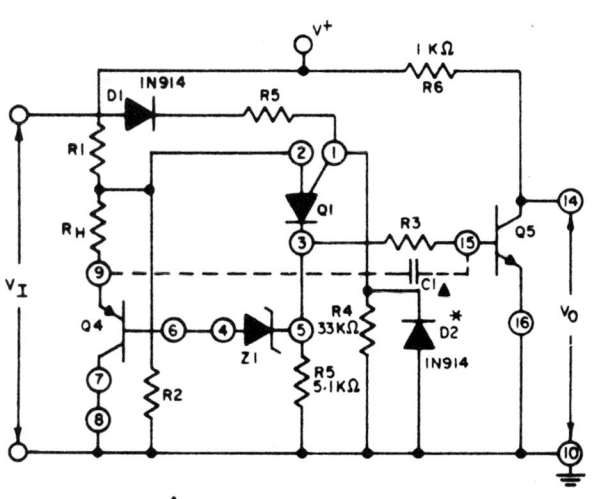

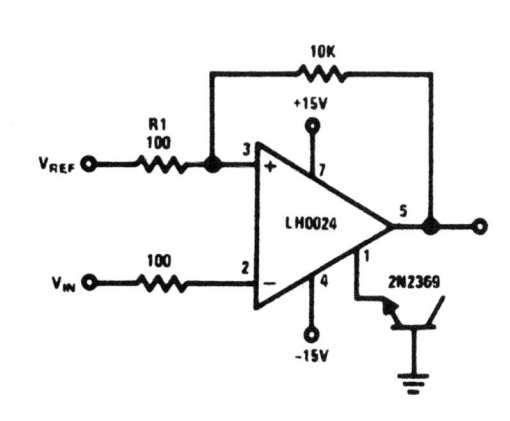

TTL compatible comparator.

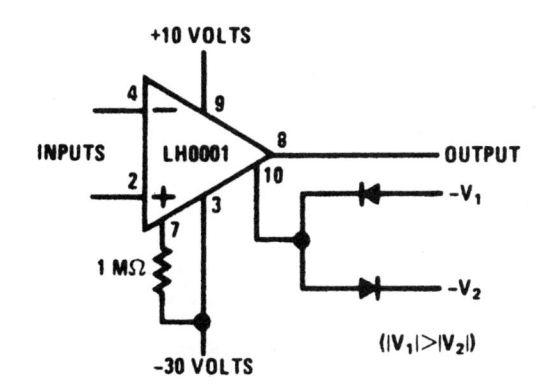

▲ OPTIONAL SPEED-UP CAPACITOR
★ REQUIRED IF V_I SWINGS BELOW GROUND

TYPICAL OPERATING CONDITIONS:
FREQUENCY IN = 0-10 KHz
SUPPLY VOLTAGE (V^+) = 15 V
R1, R2, R_H = 5.1 KΩ
R3 = 6.2 KΩ, R_S = 300 Ω
C_I = 820 pF
$V_{TH}U$ = 7.5 V, $V_{TH}L$ = 5 V
HYSTERESIS VOLTAGE = 2.5 V
UPPER THRESHOLD VOLTAGE $(V_{TH}U) \approx V^+ \frac{R2}{R1+R2}$

LOWER THRESHOLD VOLTAGE $(V_{TH}L) \approx \frac{(V^+)\left(\frac{R2\,R_H}{R2+R_H}\right)}{\frac{R2\,R_H}{R2+R_H}+R1}$

HYSTERESIS VOLTAGE = $V_{TH}U - V_{TH}L$

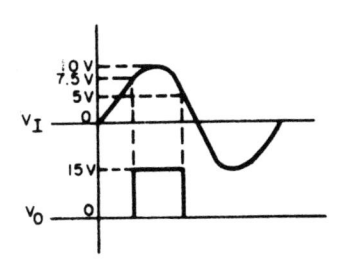

Schmitt trigger using CA3097E.

Voltage comparator for driving MOS circuits.

SECTION **15** *Logic and Functionally Related Circuits* **491**

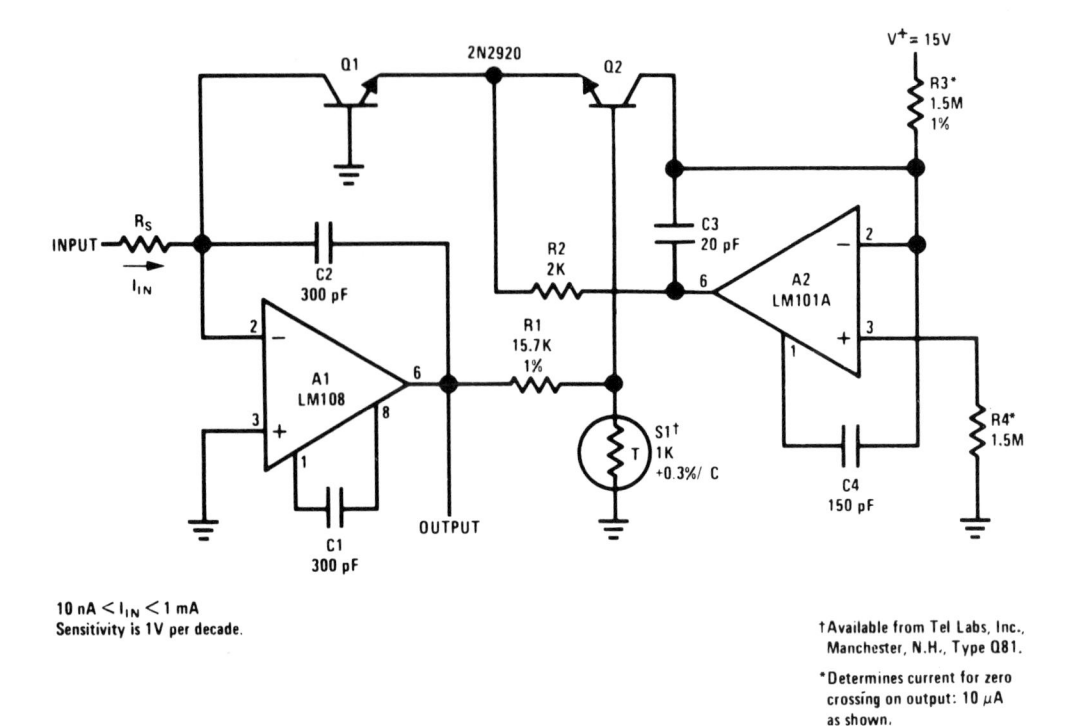

Temperature-compensated one-quadrant logarithmic converter. Q1 and Q2 should be matched devices in the same package, and S1 should be at the same temperature as these transistors. Accuracy for low input currents is determined by the error caused by the bias current of A1. At high currents, the behavior of Q1 and Q2 limits accuracy. For input currents approaching 1 mA, the 2N2920 develops logging errors in excess of 1%. If larger input currents are anticipated, bigger transistors must be used, and R2 should be reduced to ensure that A2 does not saturate.

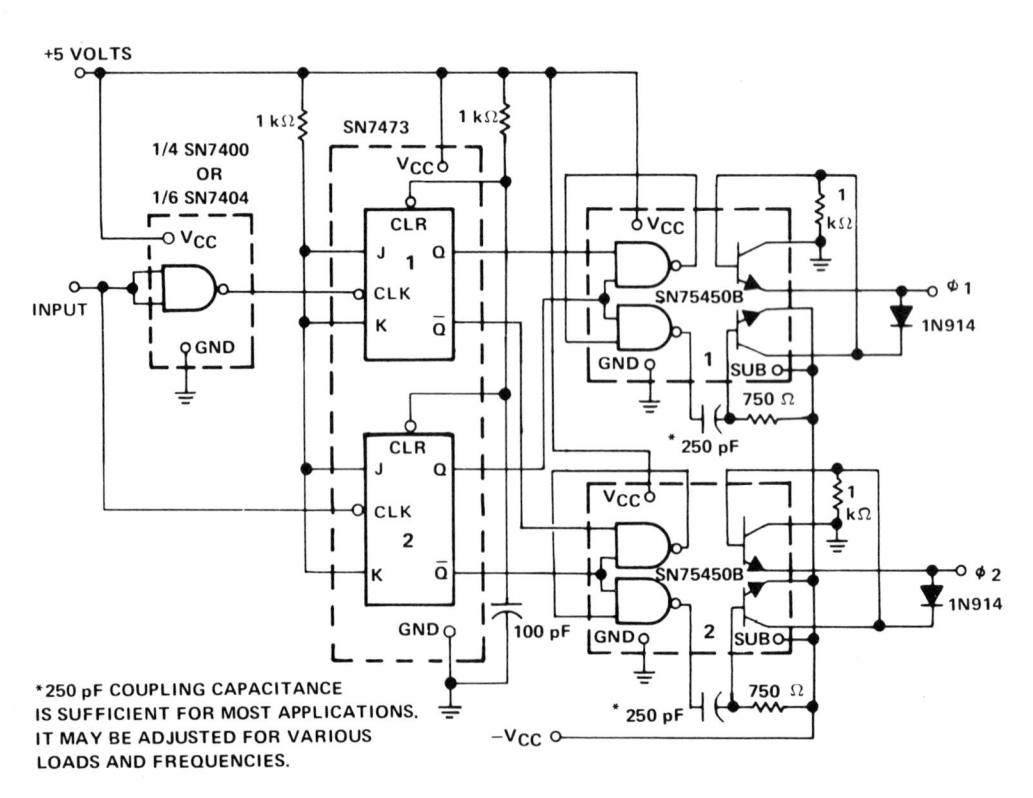

SN75450B two-phase MOS clock driver. Two 25% duty-cycle pulses are developed with 180° phase difference. Pulse-width stability over the 0° to 70°C temperature range is characteristic of this circuit.

492 IC SCHEMATIC SOURCEMASTER

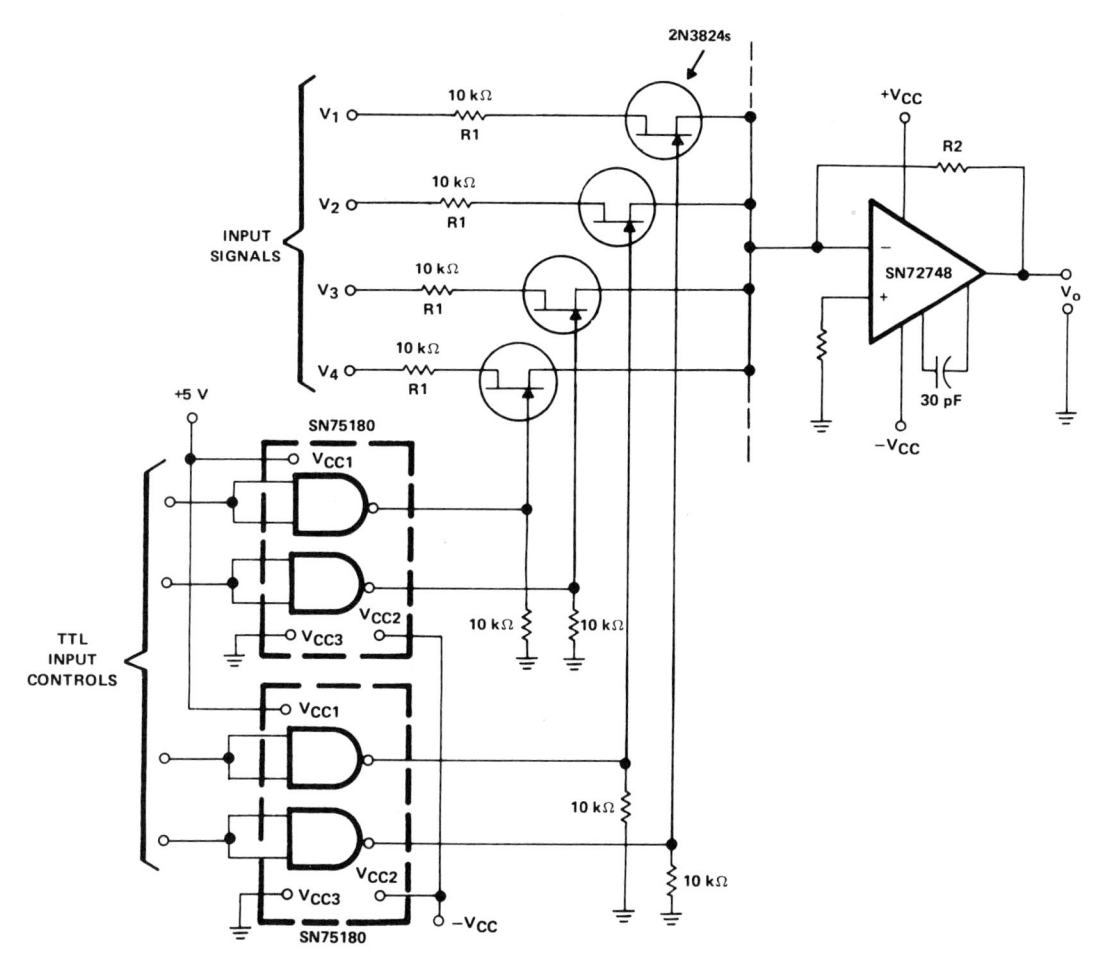

Multiple-input, logic-controlled analog amplifier. All inputs are switched on or off by TTL control signals. Actual switching is accomplished with FETs that have relatively low *on* resistances compared to signal-source and feedback resistances. The SN75180 dual-channel TTL-to-MOS converter is used to interface from standard TTL control to the negative-level gate voltages required to switch the 2N3824 off. Resulting output signals or voltages will be the product of R2 (the feedback resistor) and the sum of all input currents from the *on* sources. Isolation between inputs is greater than 40 dB at 1 kHz. With 1V input signals at 1 kHz, the signal distortion at the output is less than 2%.

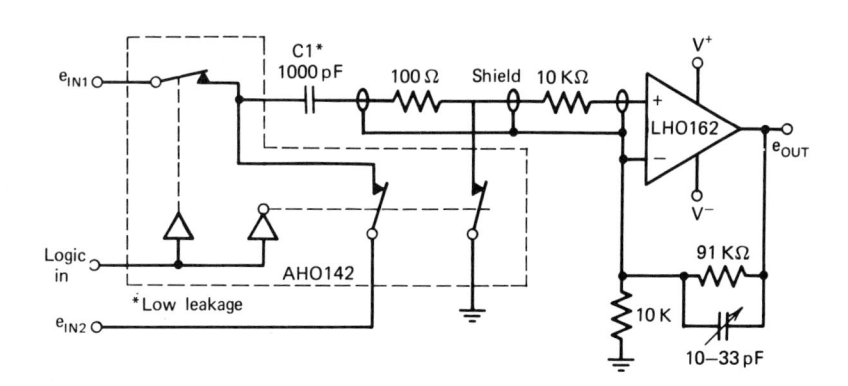

High-speed subtractor; $e_{OUT} = 10(e_{IN1} - e_{IN2})$.

SECTION **15** *Logic and Functionally Related Circuits* **493**

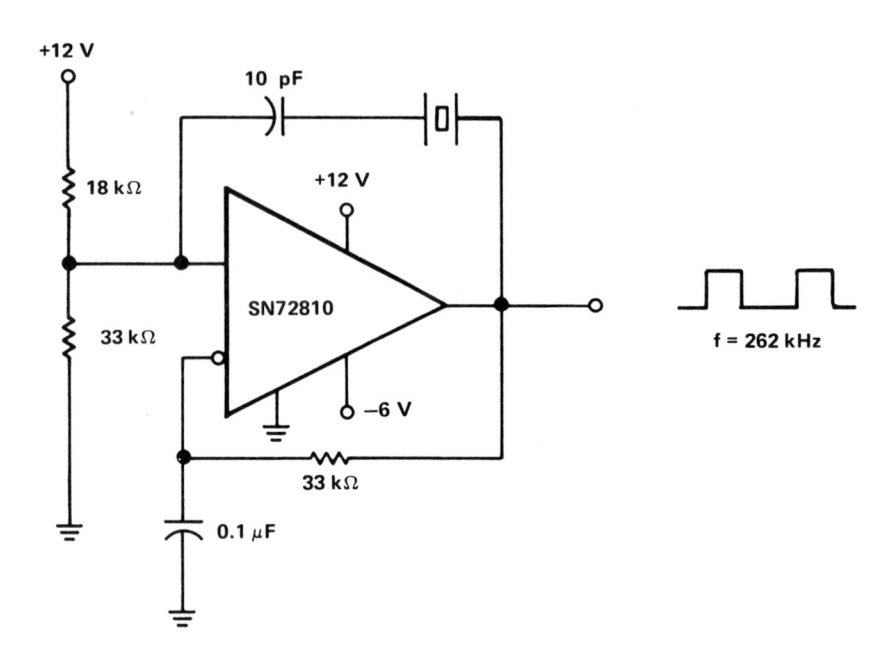

A crystal-controlled logic generator has output levels compatible with TTL. The negative-feedback RC circuit is selected to ensure that the crystal will oscillate only at its fundamental frequency. A small series capacitor trims the oscillator to the exact frequency. The duty cycle is determined by the bias level at the noninverting input. With the SN72810, the maximum operating frequency is approximately 5 MHz.

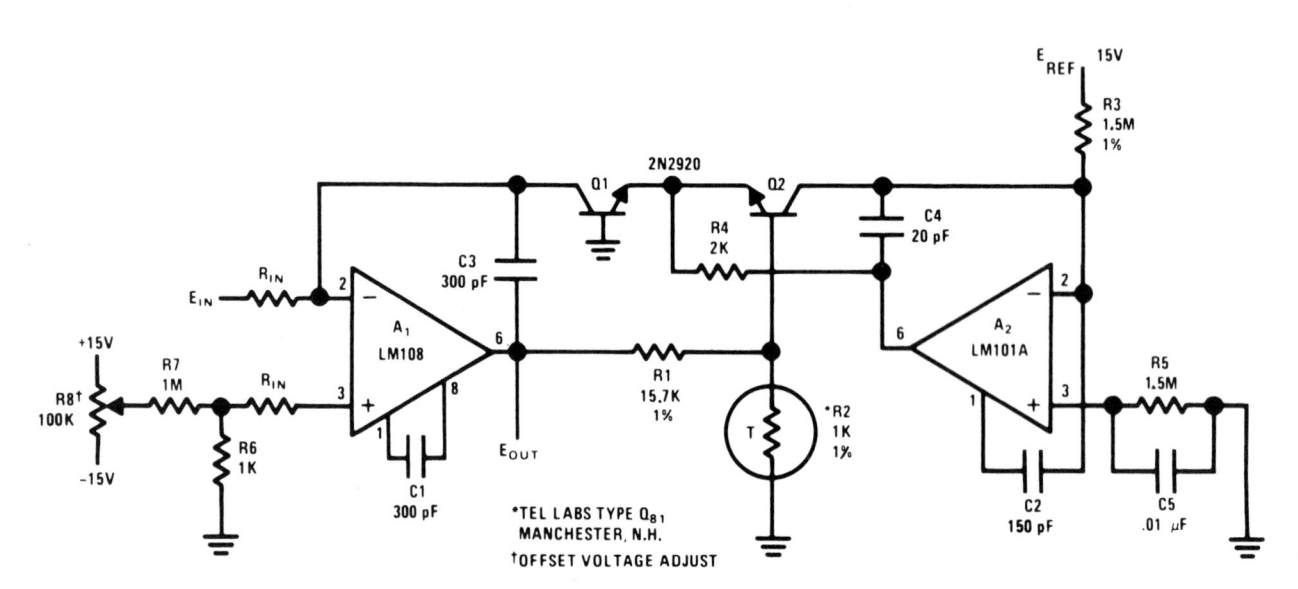

A log generator with 100 dB dynamic range. The circuit generates a logarithmic output voltage for a linear input current. Transistor Q1 is used as the nonlinear feedback element around an LM108 operational amplifier. Negative feedback is applied to the emitter of Q1 through divider R1 and R2 and the emitter–base junction of Q2. This forces the collector current of Q1 to be exactly equal to the current through the input resistor. Transistor Q2 is used as the feedback element of an LM101A operational amplifier. Negative feedback forces the collector current of Q2 to equal the current through R3. For the values shown, this current is 10 μA.

The output is proportional to the logarithm of the input voltage. The coefficient of the log term is directly proportional to absolute temperature. Without compensation, the scale factor will also vary directly with temperature. However, by making R2 directly proportional to temperature, constant gain is obtained. The temperature compensation is typically 1% over a temperature range of −25° to 100°C for the resistor specified.

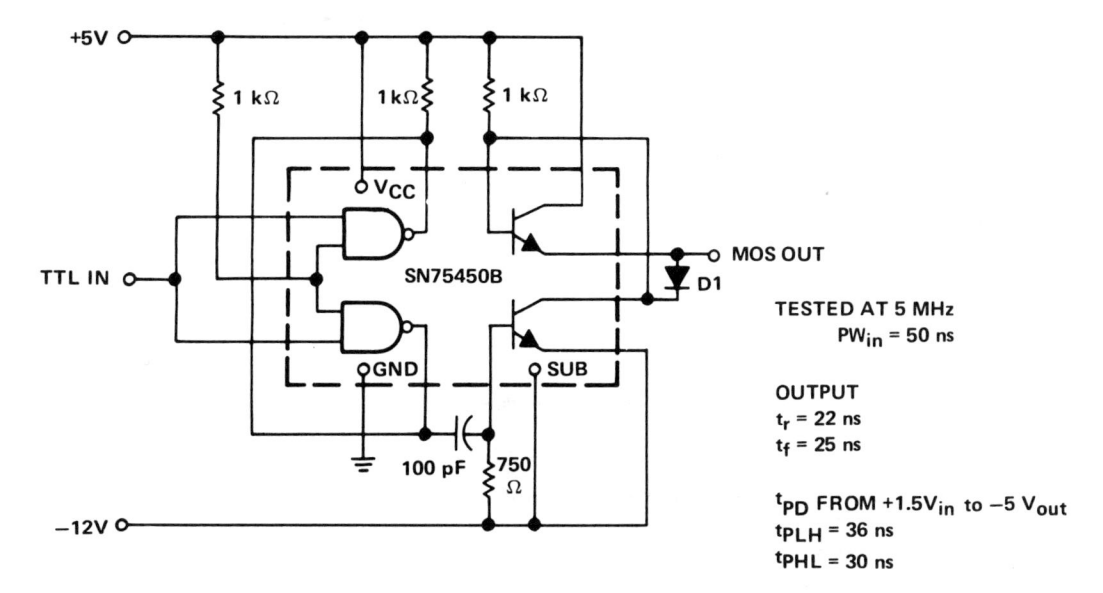

A TTL-to-MOS clock driver using the SN75450B. Level shifting from TTL levels to negative MOS logic levels is accomplished by ac (capacitive) coupling between the TTL gates and the output drive transistors of the SN75450B. Adjustment of the coupling capacitor to achieve an RC value approximately equal to twice the desired pulse width will yield proper output characteristics. For example, the 100 pF capacitor and 750Ω base termination shown have an RC time constant of 75 nsec. The ideal frequency is $(2RC)^{-1}$ or about 6.65 MHz. Good performance may be observed from 4 to 8 MHz.

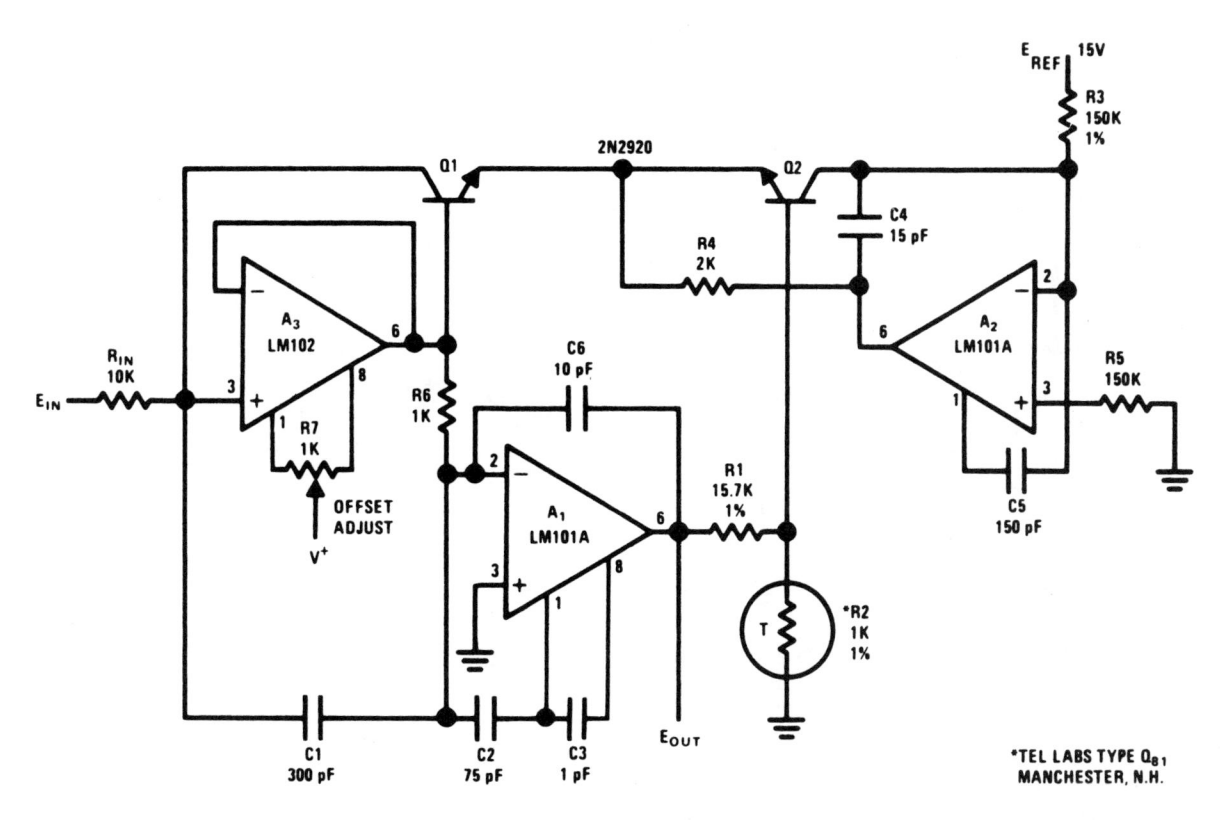

A fast log generator optimizes speed rather than dynamic range. Transistor Q1 is diode-connected to allow the use of feedforward compensation on an LM101A. This compensation extends the bandwidth to 10 MHz and increases the slew rate. To prevent errors resulting from the finite h_{FE} of Q1 and the bias current of the LM101A, an LM102 voltage follower buffers the base current and input current. Although the log circuit will operate without the LM102, accuracy will degrade at low input currents. Amplifier A2 is also compensated for maximum bandwidth.

SECTION **15** *Logic and Functionally Related Circuits* **495**

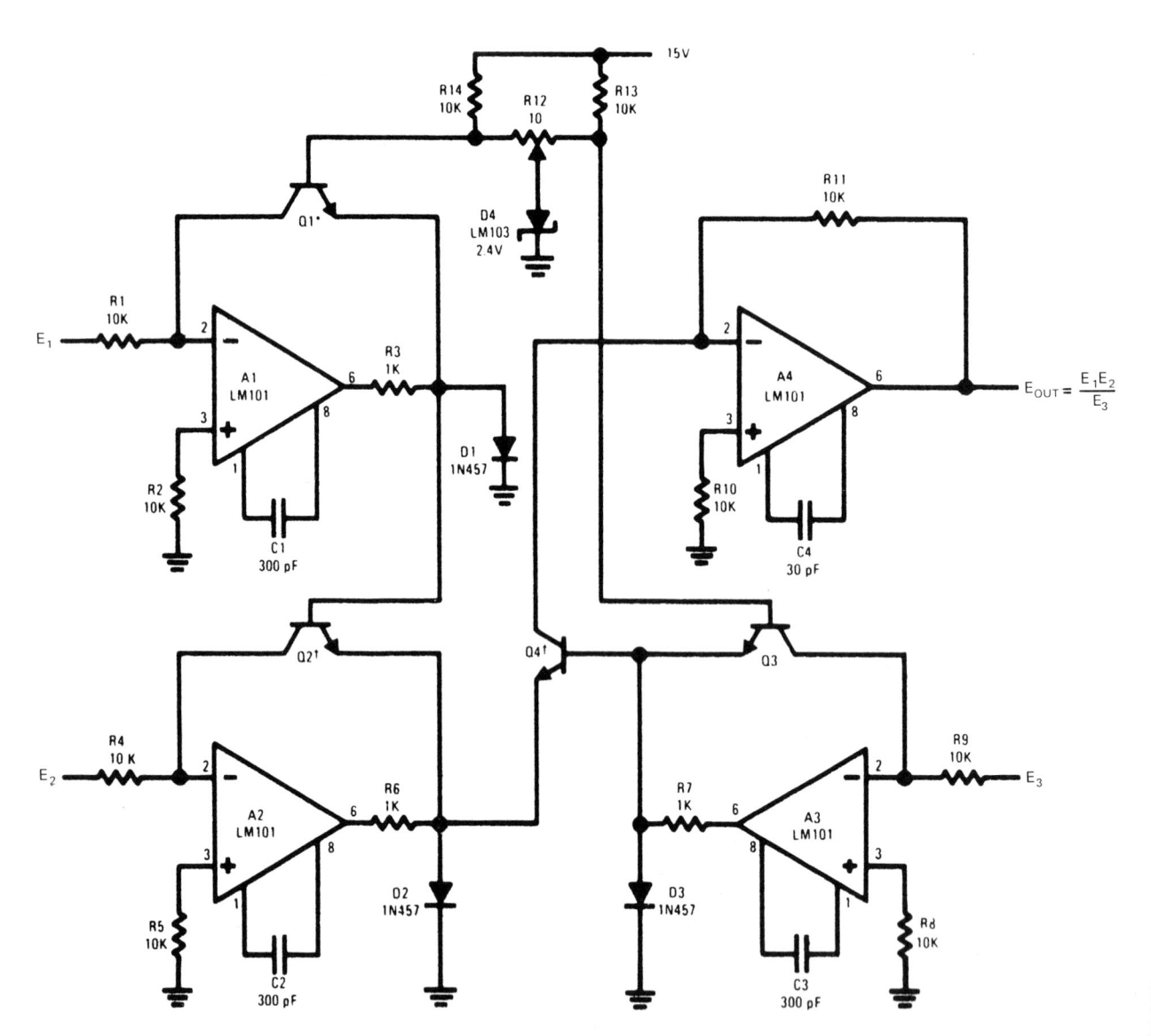

An analog multiplier/divider can give 1% accuracy for input voltages from 500 mV to 50V. To get this precision at lower input voltages, the offset of the amplifiers handling them must be individually balanced. Zener D4 increases the collector-base voltage across the logging transistors to improve high-current operation. It is not needed when these transistors are running at currents less than 0.3 mA. At currents above 0.3 mA, the lead resistances of the transistors can become important (0.25Ω is 1% at 1 mA), so the transistors should be installed with short leads and no sockets. Q1 and Q3 and Q2 and Q4, are matched 2N3728 pairs.

$$E_{OUT} = \frac{E_1 E_2}{E_3}$$

$$V_{OUT} = \frac{(X)(Y)}{(Z)} \quad ; \text{ positive inputs only.}$$

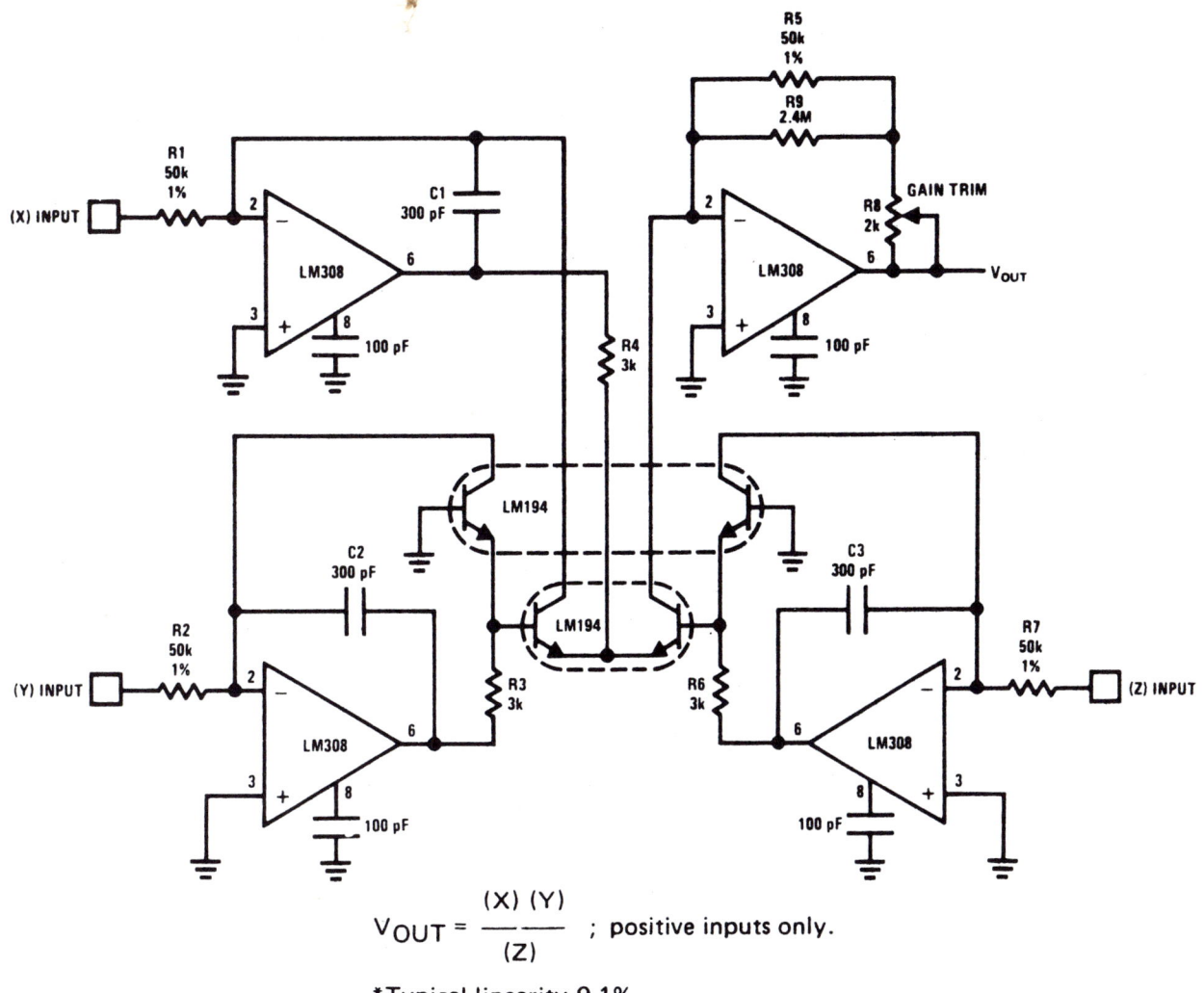

*Typical linearity 0.1%

High-accuracy one-quadrant multiplier/divider.

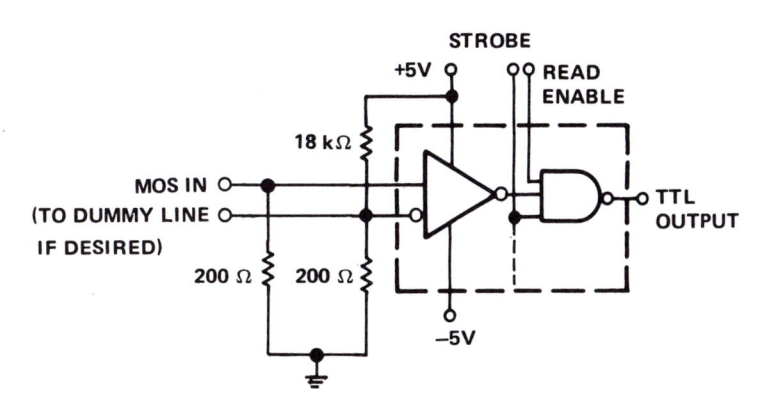

MOS-to-TTL level shift.

SECTION **15** *Logic and Functionally Related Circuits*

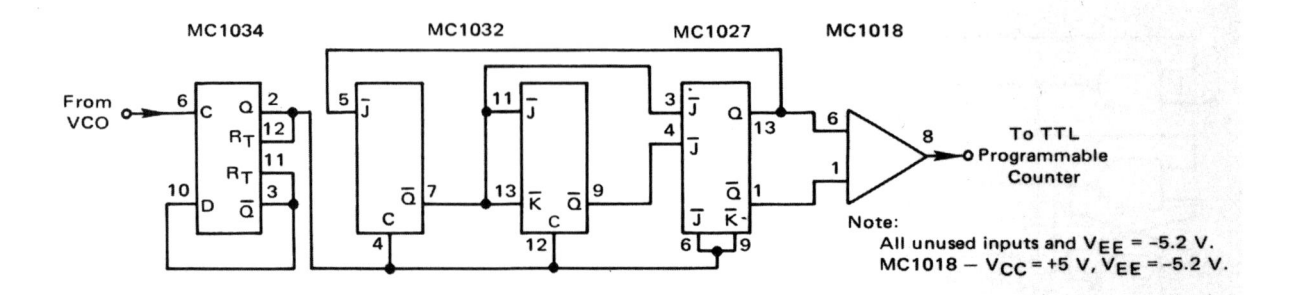

Note:
All unused inputs and V_{EE} = −5.2 V.
MC1018 − V_{CC} = +5 V, V_{EE} = −5.2 V.

A fixed divide-by-10 prescaler is used between the VCO output and the programmable counter section of avionics synthesizer. The prescaler is a high-speed divide-by-2 MC1034 flip-flop followed by a divide-by-5 section using an MC1032 and MC1027.

The output of the high-speed ECL prescaler has a maximum output frequency of 13.865 MHz. Since the negative-voltage logic swing is between −0.75 and −1.6V, translation is required to provide TTL levels to the programmable counter section. This translation is provided by means of an MC1018 driven complementary from the last ECL flip-flop.

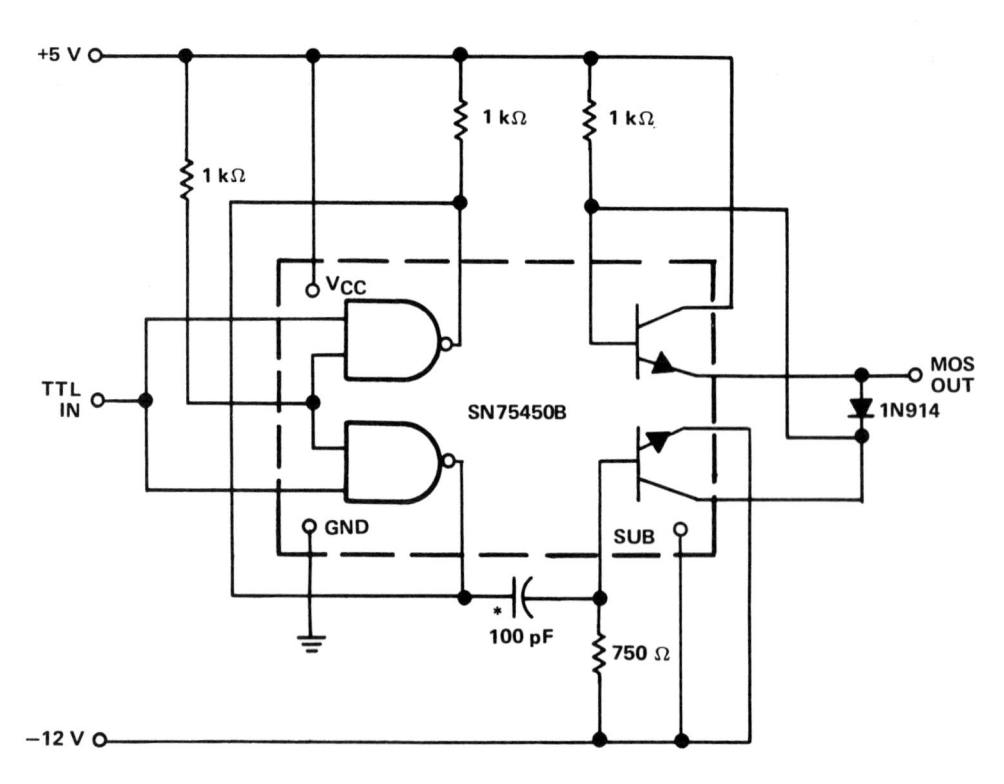

TTL-to-MOS interface. The ac coupling takes care of the level-shifting problem, and the active pullup and active pulldown allow fast charging and discharging of capacitive loads. This circuit, operating into a 220 pF load, has an output rise time and fall time of 22 and 26 nsec when tested at 5 MHz, with a TTL input having a pulse width of 50 nsec. Propagation delays are typically t_{PLH} = 36 nsec and t_{PHL} = 30 nsec. The value of the capacitor may be adjusted for different frequencies of operation.

498 IC SCHEMATIC SOURCEMASTER

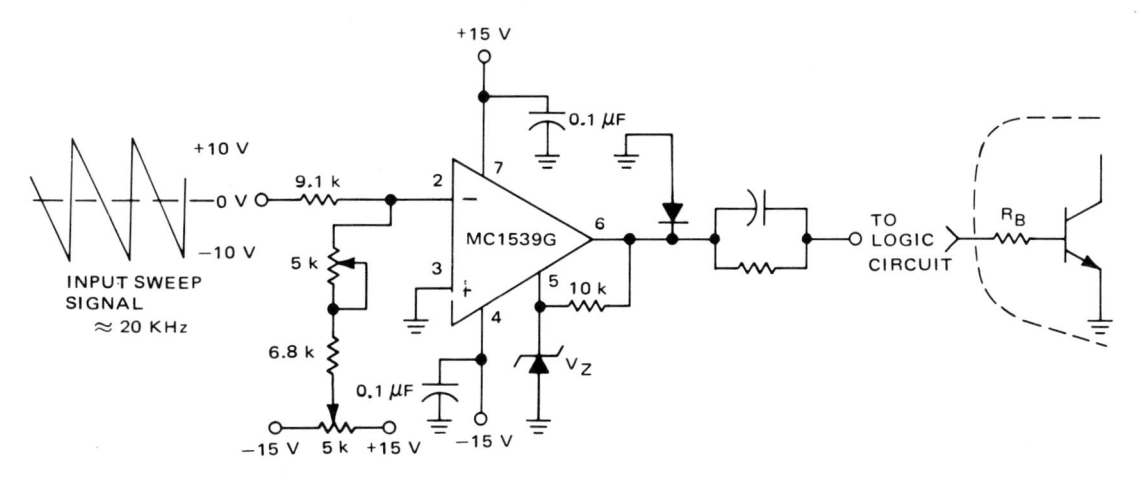

Voltage comparator. The zener diode connected to pin 5 limits the positive-going waveform at the output to about 2V below the zener voltage. The silicon diode connected to the output limits the output negative excursion to protect the logic circuit that is being driven. The output parallel RC network matches the input characteristics of the logic circuit, eliminating any reduction in response time due to RC charging time and helps minimize the output current overload problem.

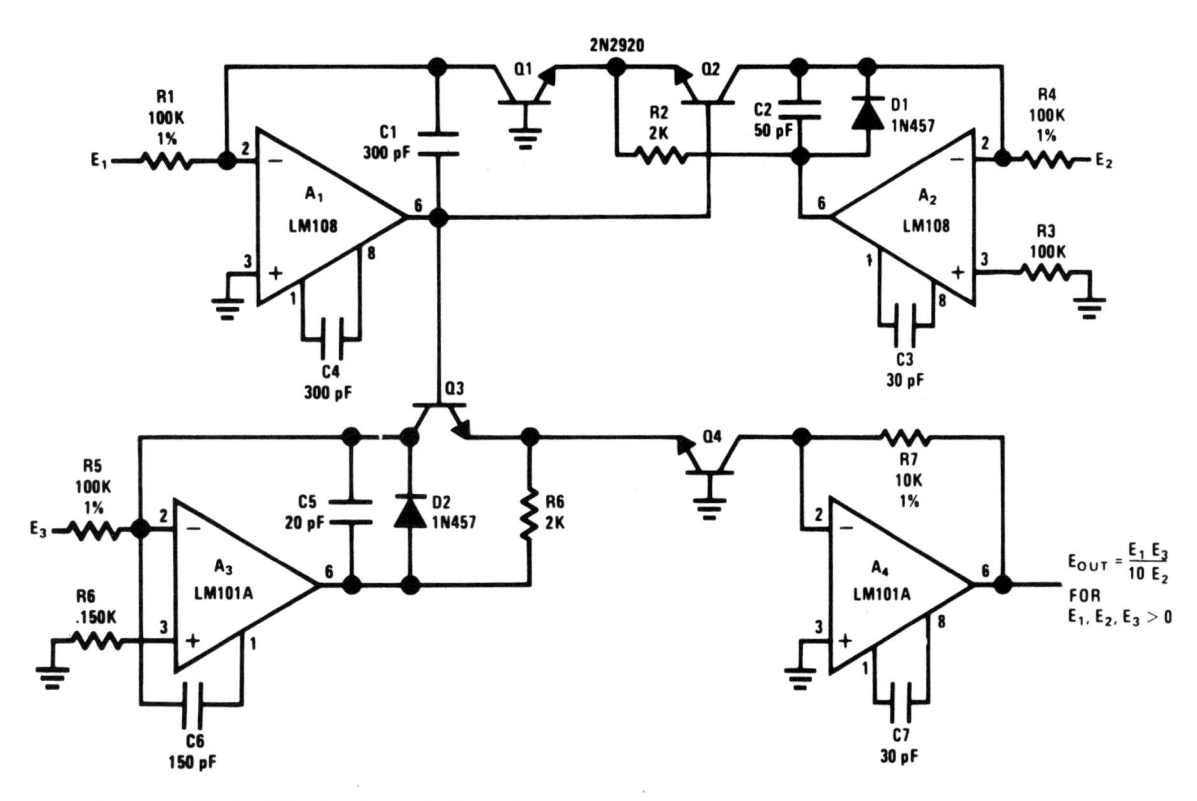

A one-quadrant multiplier/divider is basically a log generator driving an antilog generator. The log generator output from A1 drives the base of Q3 with a voltage proportional to the log of E_1/E_2. Transistor Q3 adds a voltage proportional to the log of E_3 and drives antilog transistor Q4. The collector current of Q4 is converted to an output voltage by A4 and R7, with the scale factor set by R7 at $E_1E_3/10E_2$.

SECTION **15** *Logic and Functionally Related Circuits* **499**

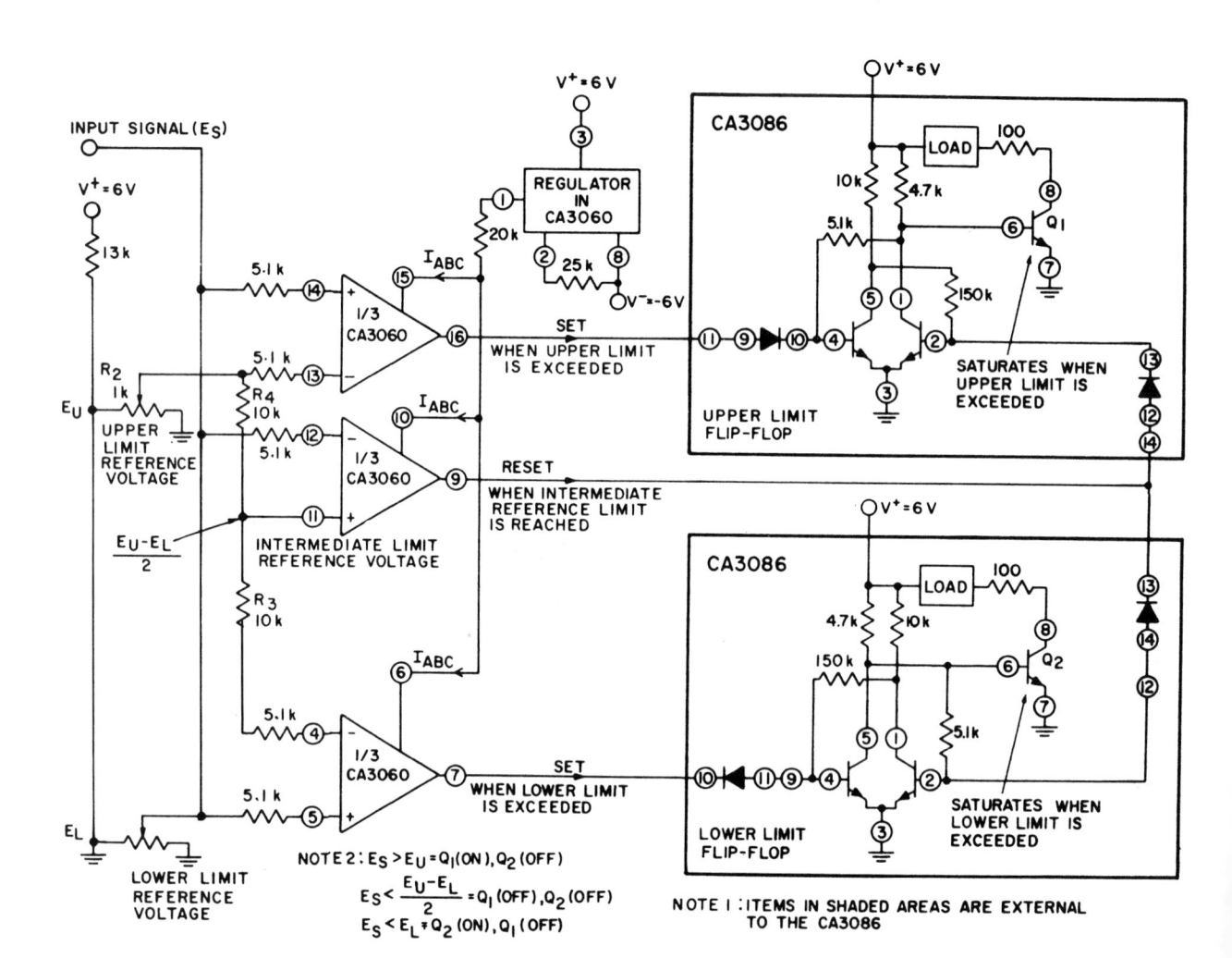

Trilevel comparator circuit. Power is provided for the CA3060 via terminals 3 and 8 by ±6V supplies, and the built-in regulator provides amplifier bias current (I_{ABC}) to the three amplifiers via terminal 1. Lower-limit and upper-limit reference voltages are selected by appropriate adjustment of potentiometers R1 and R2. When resistors R3 and R4 are equal in value, the intermediate-limit reference voltage is automatically established at a value midway between the lower-limit and upper-limit values. Appropriate variation of resistors R3 and R4 permits selection of other values of intermediate-limit voltages. Input signal E_S is applied to the three comparators via terminals 5, 12, and 14. The *set* output lines trigger the appropriate flip-flop whenever the input signal reaches a limit value. When the input signal returns to an intermediate value, the common flip-flop *reset* line is energized. The loads shown are 5V, 25 mA lamps.

500 IC SCHEMATIC SOURCEMASTER

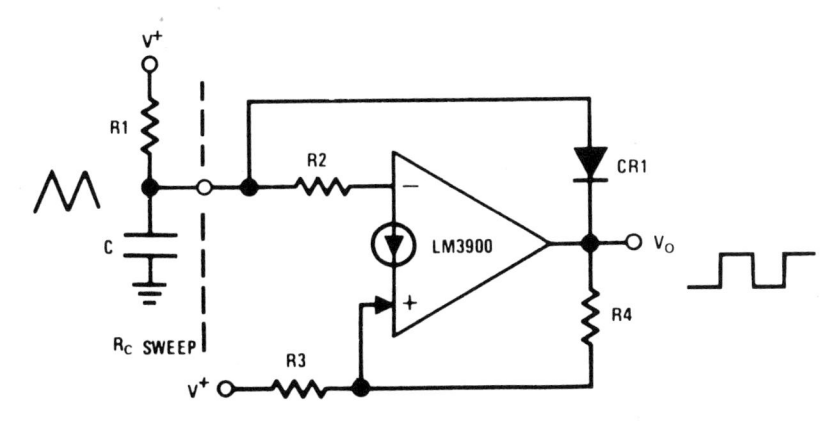

Adding a diode to the LM3900 Schmitt trigger results in a **"programmable unijunction"** function. For a low input voltage, the output voltage of the LM3900 is high and CR1 is off. When the input voltage rises to the high trip voltage, the output falls to essentially 0V and CR1 goes on, discharging input capacitor C. The low trip voltage must be larger than 1V to guarantee that the forward drop of CR1 added to the output voltage of the LM3900 will be less than the low trip voltage. The discharge current can be increased by using smaller values for R2 to provide pulldown currents larger than the 1.3 mA bias current source.

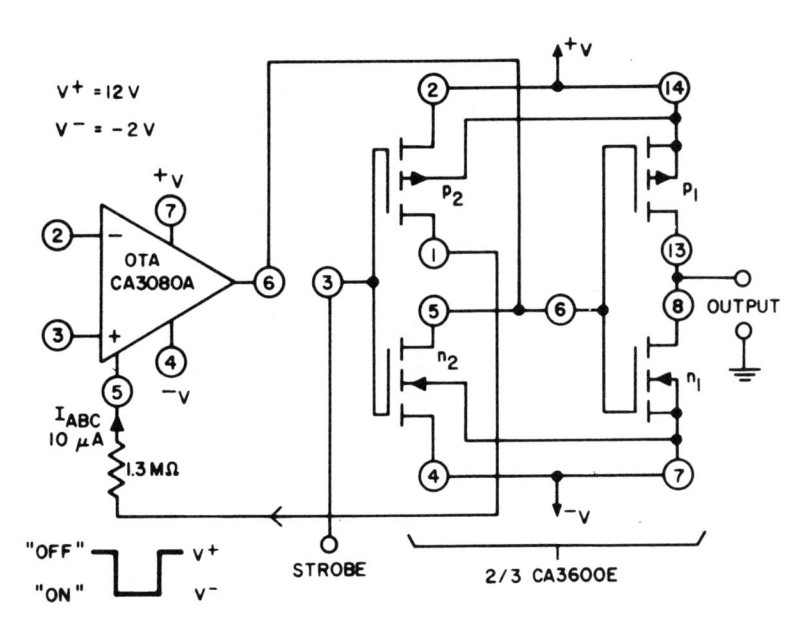

A programmable micropower comparator employs the combination of an op-amp (CA3080A) and CMOS transistor pairs in the CA3600E. Quiescent power consumption of the circuit is about 10 μW. When the comparator is strobed transistor P1 is driven into conduction and the operational transconductance amplifier (OTA) becomes active. Under these conditions, the circuit consumes 420 μW and responds to a differential input signal in about 8 μsec.

SECTION **15** *Logic and Functionally Related Circuits*

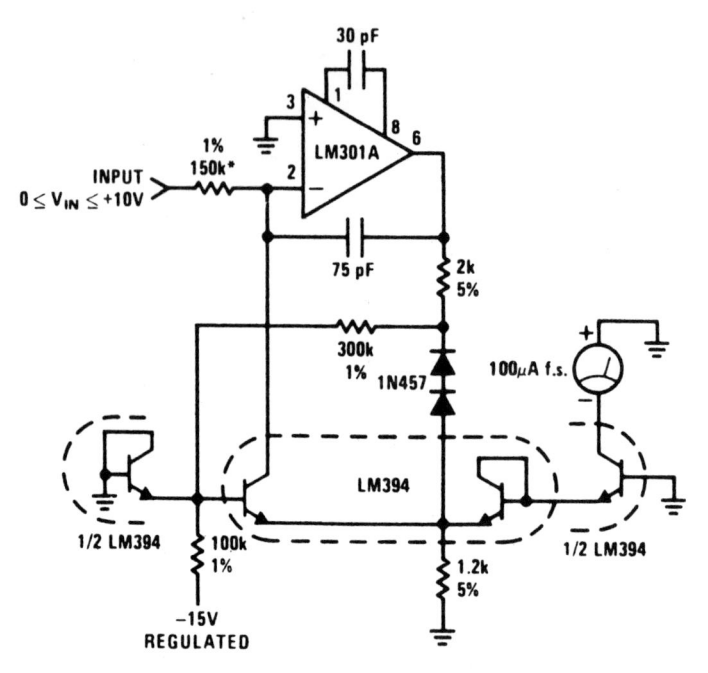

Low cost accurate square root circuit. $I_{OUT} = 10^{-5}\sqrt{V_{IN}}$. Trim 150k resistor for full-scale accuracy.

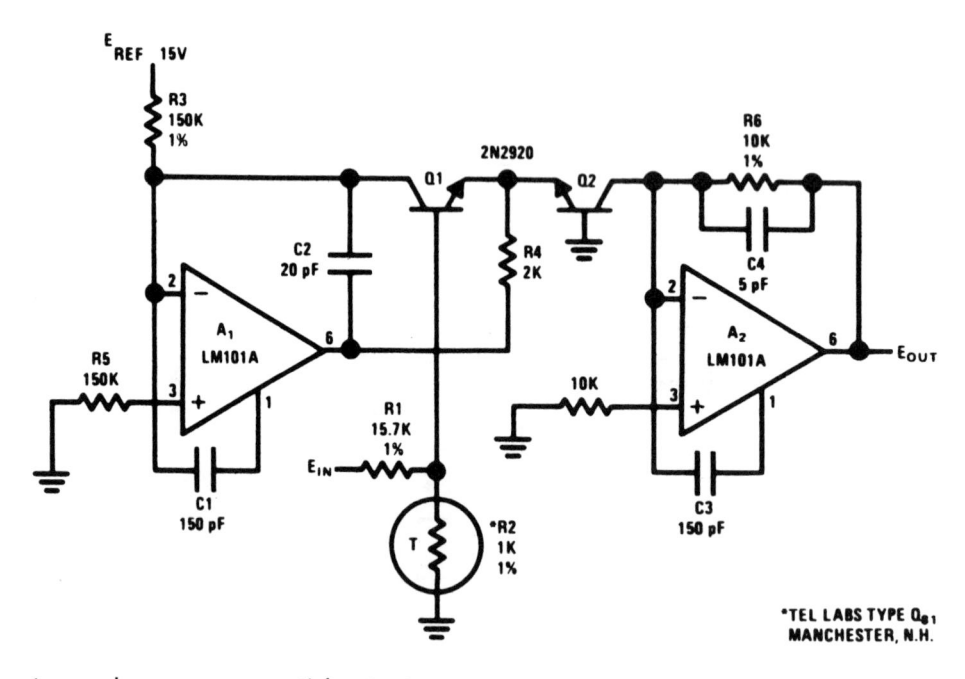

An antilog generator produces an exponential output from a linear input. Amplifier A1 in conjunction with transistor Q1 drives the emitter of Q2 in proportion to the input voltage. The collector current of Q2 varies exponentially with the emitter-base voltage. This current is converted to a voltage by amplifier A2. Many nonlinear functions such as $X^{\frac{1}{2}}$, X^2, X^3, $1/X$, XY, and X/Y are easily generated with the use of logs. Multiplication becomes addition, division becomes subtraction, and powers become gain coefficients of log terms.

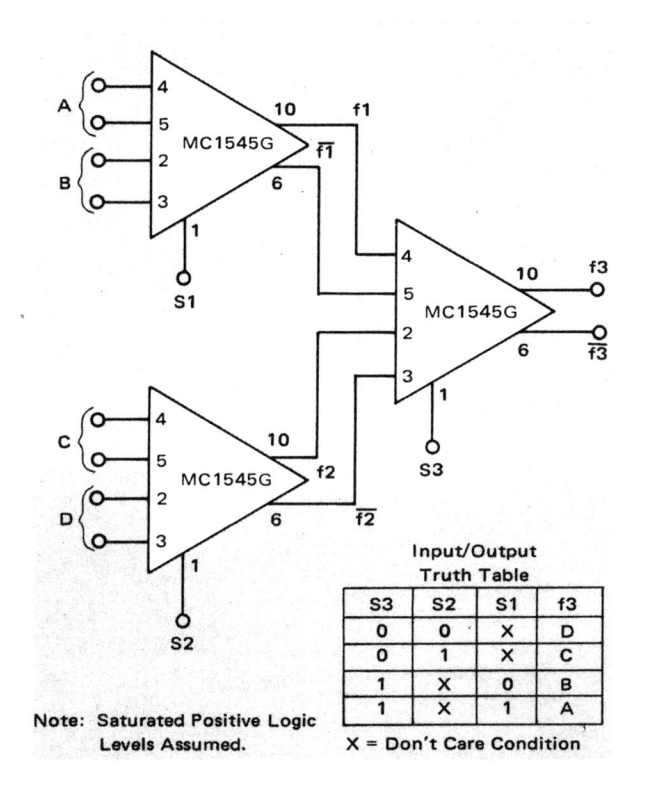

One-of-N data selector for data processing.

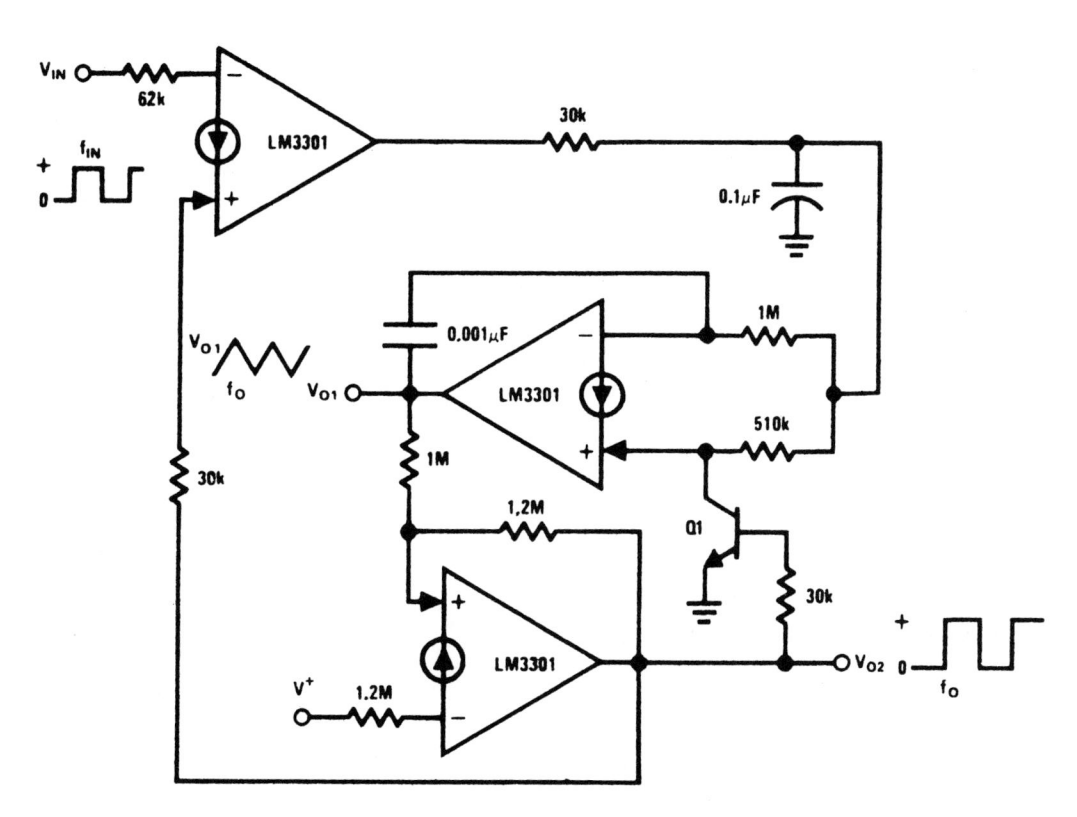

Input/Output Truth Table

S3	S2	S1	f3
0	0	X	D
0	1	X	C
1	X	0	B
1	X	1	A

X = Don't Care Condition

Note: Saturated Positive Logic Levels Assumed.

Phase-locked loop.

SECTION **15** *Logic and Functionally Related Circuits* **503**

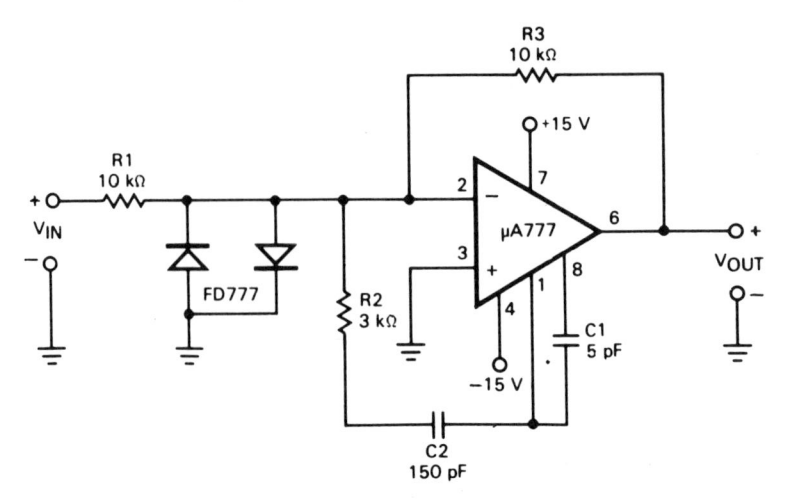

This unity-gain pulse amplifier uses both a feedforward path and a diode clamp to raise the slew rate and dampen the overshoot in the high-speed pulse response of the μA777. The feedforward path, consisting of resistor R2 and capacitor C2, bypasses the first-stage lateral PNP transistors at frequencies above about 3555 kHz and thereby raises the slew rate from the specified 0.5V/μsec to approximately 20V/μsec. The two diodes minimize the overshoot, so that it is less than 1% for any output voltage in the tested range of pulse amplitudes from 0.5 to 10.0V peak-to-peak.

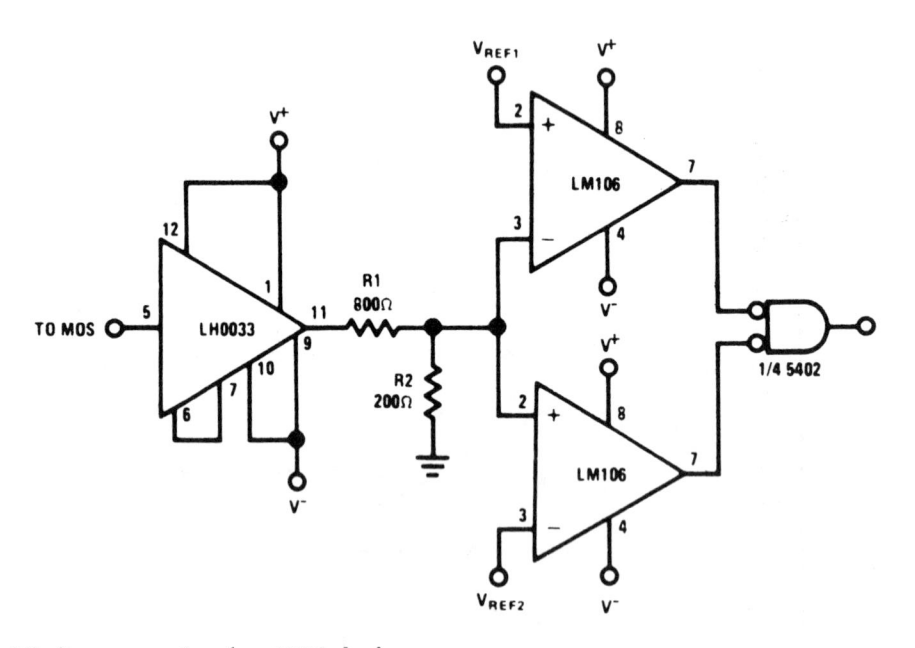

High-speed dual-limit comparator for MOS logic. The LH0033 is used as a buffer between MOS logic and a high-speed dual-limit comparator. The device's high input impedance prevents loading of the MOS logic signal. The LH0033 adds about 1.5 nsec to the total delay of the comparator. Adjustment of the voltage divider (R1, R2) allows interface to TTL, DTL, and other high-speed logic forms.

504 IC SCHEMATIC SOURCEMASTER

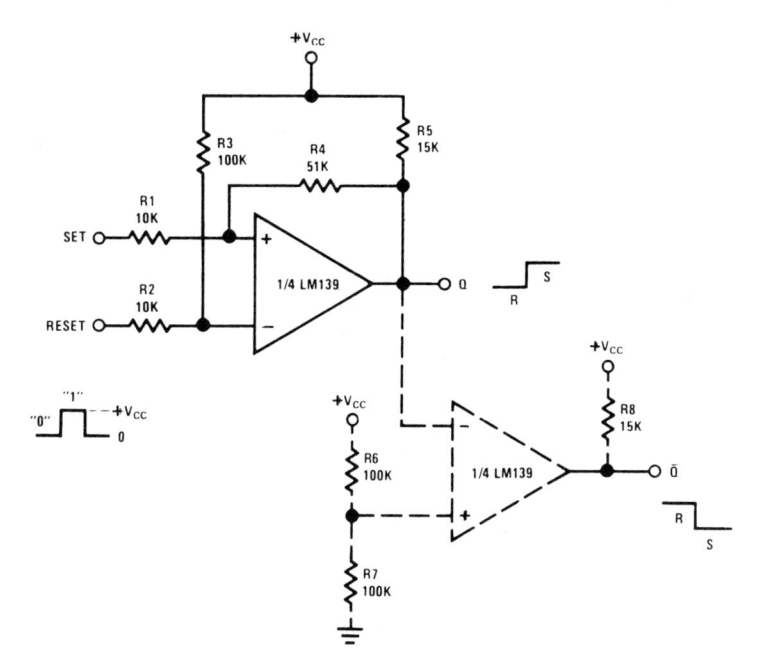

Bistable multivibrator (flip-flop). A reference voltage is provided at the inverting input by a voltage divider (R2 and R3). A pulse applied to the *set* terminal will switch the output high. Resistor divider network R1, R4, and R5 now clamps the noninverting input to a voltage greater than the reference voltage. A pulse now applied to the *reset* input will pull the output low. If both Q and Q̄ outputs are needed, another comparator can be added as shown by broken lines.

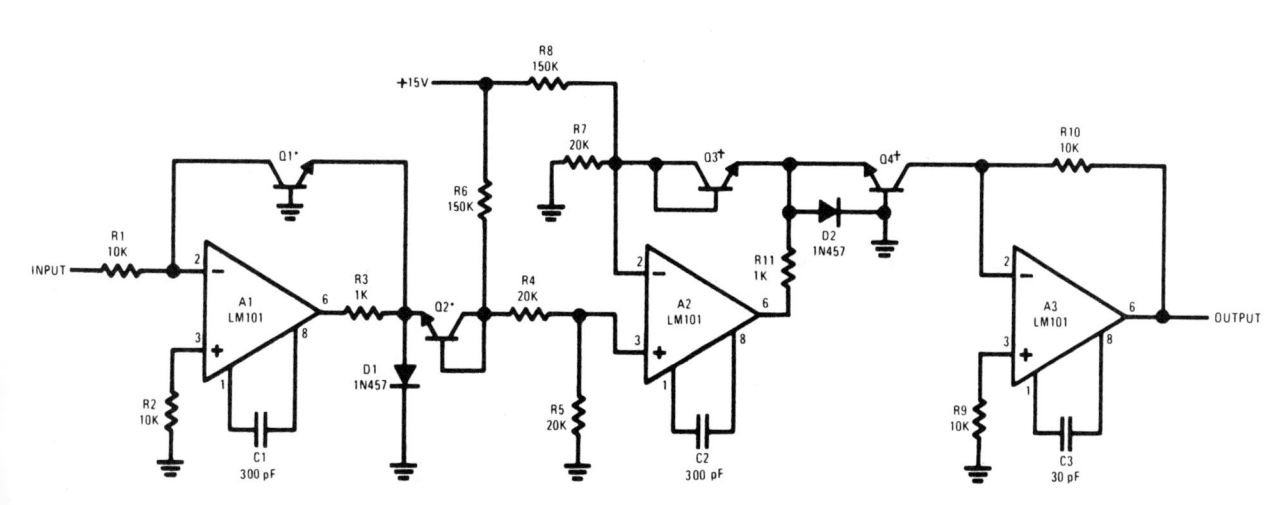

Root extractor. A1 and Q1 form the log converter for the input signal. This feeds Q2, which produces a level shift to give zero voltage into the R4–R5 divider for a 1V input. This divider reduces the log voltage by the ratio for the root desired and drives buffer amplifier A2. A2 has a second level-shifting diode, Q3, the feedback network of which gives the output voltage needed to get a 1V output from the antilog generator, consisting of A3 and Q4, with a unity input. The offset voltages of the transistors are nulled out by unbalancing R6 and R8 to give 1V output for 1V input, since any root of 1 is 1.

SECTION **15** *Logic and Functionally Related Circuits* **505**

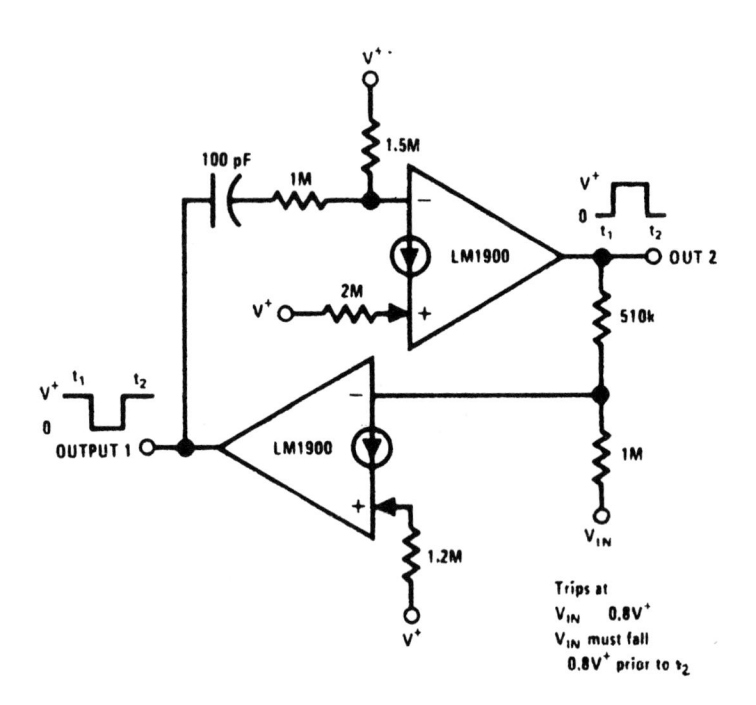

One-shot with dc input comparator.

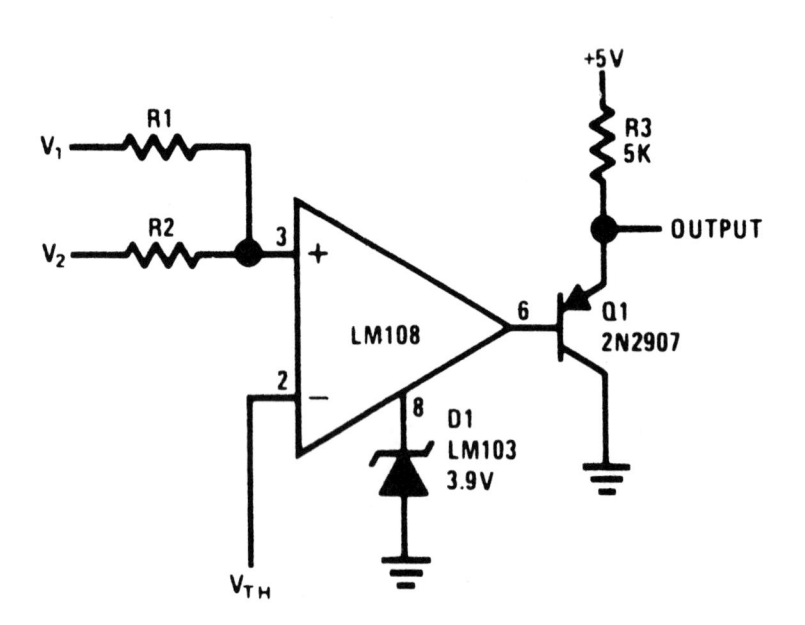

Trips at
$V_{IN} \; 0.8V^+$
V_{IN} must fall
$0.8V^+$ prior to t_2

Comparator for voltages of opposite polarity. The output changes state when the voltage on the junction of R1 and R2 is equal to V_{TH}. Mathematically, this is expressed by

$$V_{TH} = V_2 + \frac{R2(V_1 - V_2)}{R1 + R2}$$

506 IC SCHEMATIC SOURCEMASTER

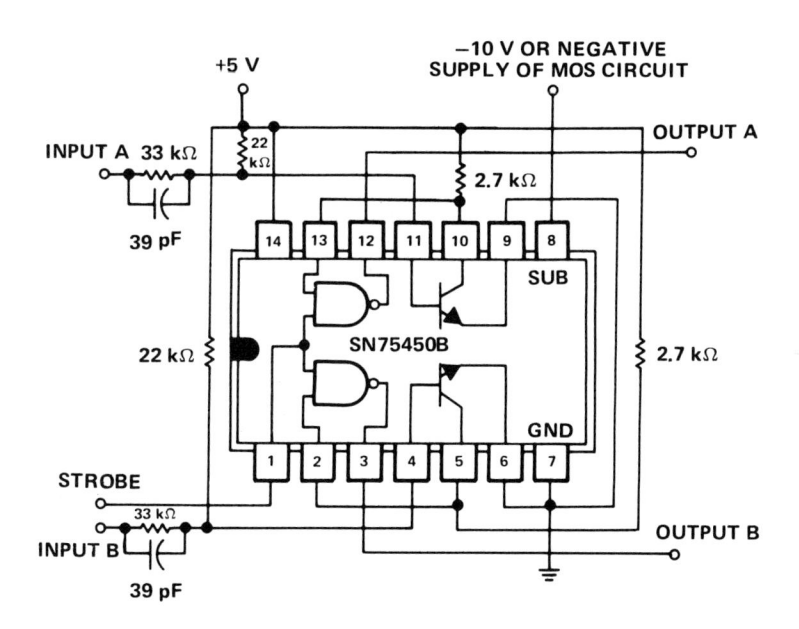

Dual MOS-to-TTL driver.

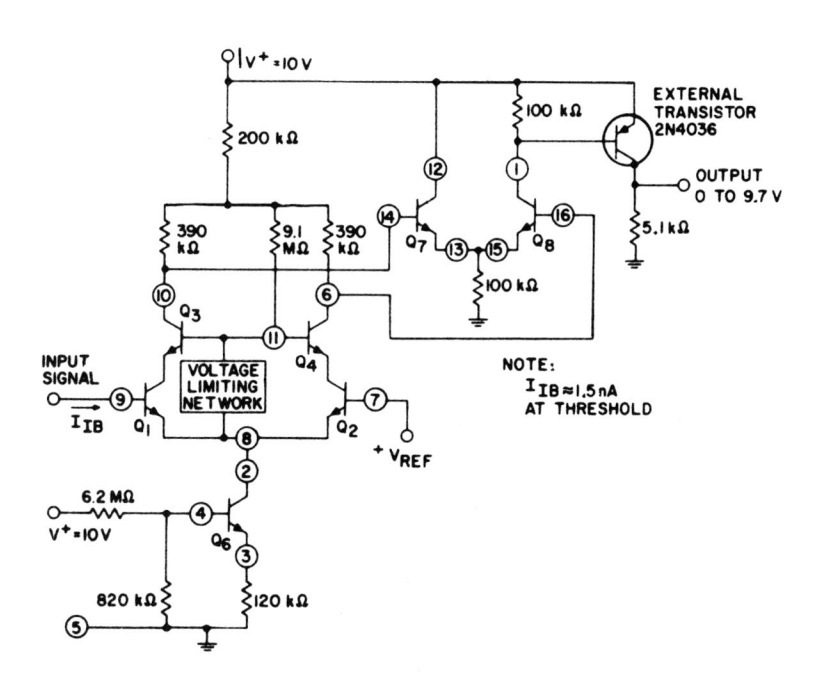

Low input-bias-current comparator circuit. Voltage-limiting network is built into superbeta IC. To use it, just connect a resistor from the positive supply of the differential amplifier to pin 11; the value should be such that current through the resistor is about half that measured at pin 8 under quiescent conditions.

SECTION 15 *Logic and Functionally Related Circuits* **507**

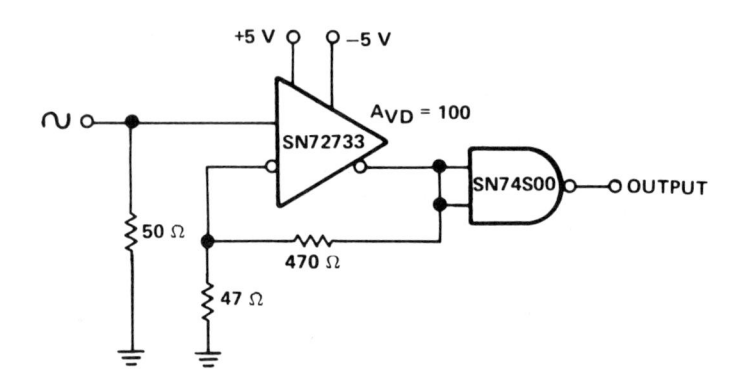

Fast Schmitt trigger. Shaping of signals of various amplitudes and shapes to meet TTL levels can be achieved by integrated-circuit Schmitt triggers at frequencies up to 22 MHz. Beyond this frequency, it is necessary to use fast triggers made from video amplifiers, as shown. The threshold voltage for the positive-going 500 mV input signal is determined by the output voltage of the SN72733 divided by the feedback ratio. For a negative-going signal, the threshold level is typically 0V.

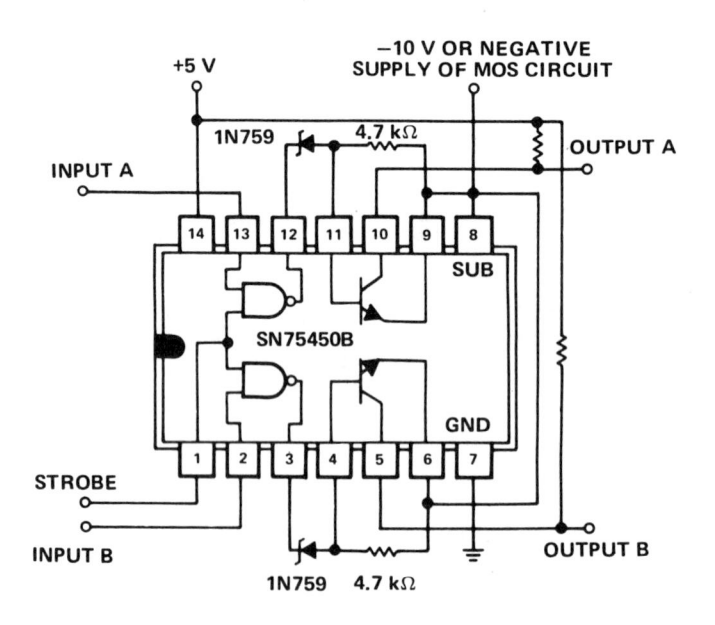

Dual TTL-to-MOS interface, with gate outputs level-shifted through zeners to proper drive levels for the output transistors. Transistor emitters, as well as the substrate, are connected to the negative MOS supply. Collectors are terminated to the +15V supply, but they could be terminated to ground.

508 IC SCHEMATIC SOURCEMASTER

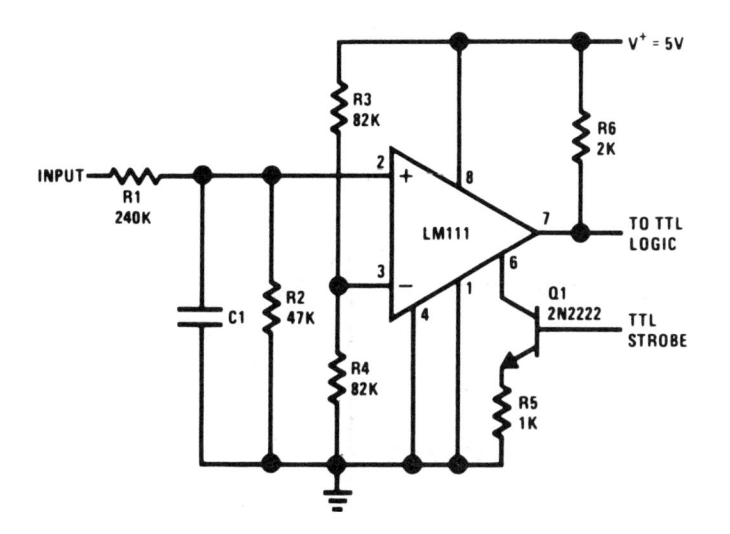

Digital interface between high-level logic and DTL or TTL. The input signal, with 0V and 30V logic states is attenuated to 0V and 5V by R1 and R2. R3 and R4 set up a 2.5V threshold level for the comparator so that it switches when the input goes through 15V. The response time of the circuit can be controlled with C1, if desired, to make it insensitive to fast noise spikes. Because of the low error currents of the LM111, it is possible to get input impedances even higher than the 300k obtained with the indicated resistor values.

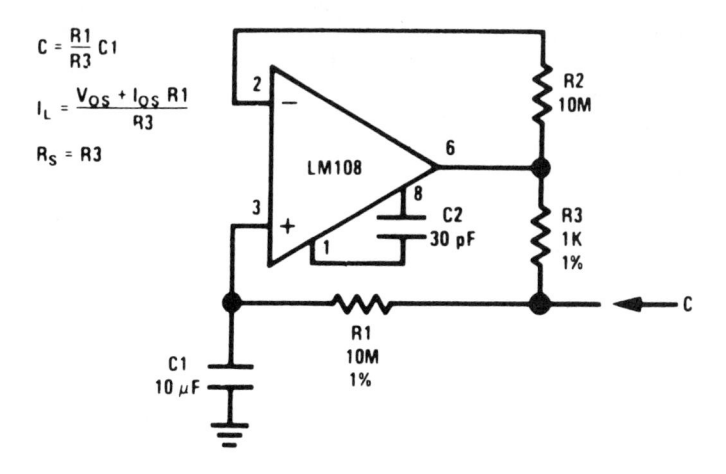

$$C = \frac{R1}{R3} C1$$

$$I_L = \frac{V_{OS} + I_{OS} R1}{R3}$$

$$R_S = R3$$

Capacitance multiplier. The performance of the circuit is described by the equations given, where C is the effective output capacitance, I_L is the leakage current of this capacitance, and R_S is the series resistance of the multiplied capacitance. The series resistance is relatively high, so high-Q capacitors cannot be realized. Hence, such applications as tuned circuits and filters are ruled out. However, the multiplier can still be used in timing circuits or in servo compensation networks where some resistance is usually connected in series with the capacitor or where the effect of the resistance can be compensated for.

The leakage current of the multiplied capacitance is not a function of the applied voltage. It persists even with no voltage on the output. Therefore, it can generate offset errors in a circuit rather than the scaling errors caused by conventional capacitors.

SECTION **15** *Logic and Functionally Related Circuits* **509**

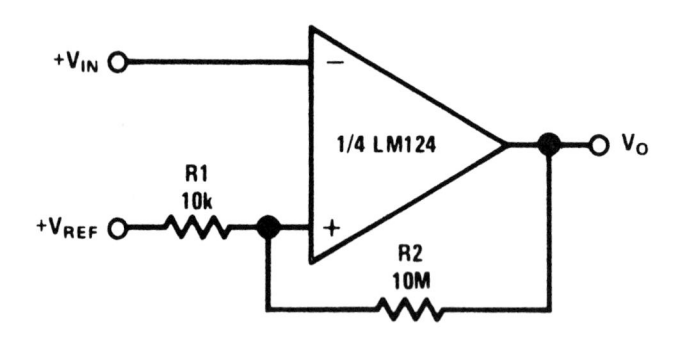

Comparator with hysteresis.

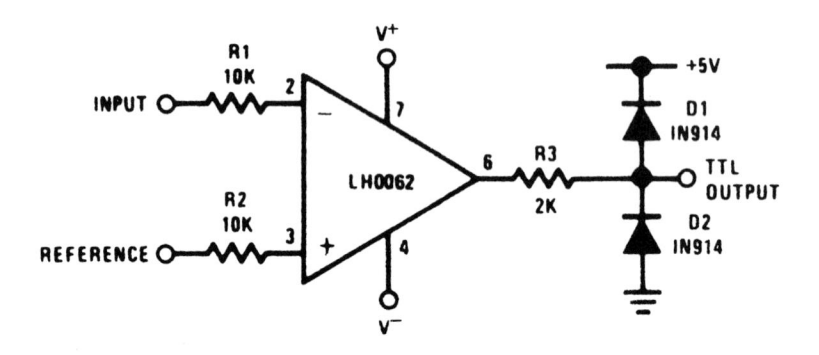

Fast precision voltage comparator.

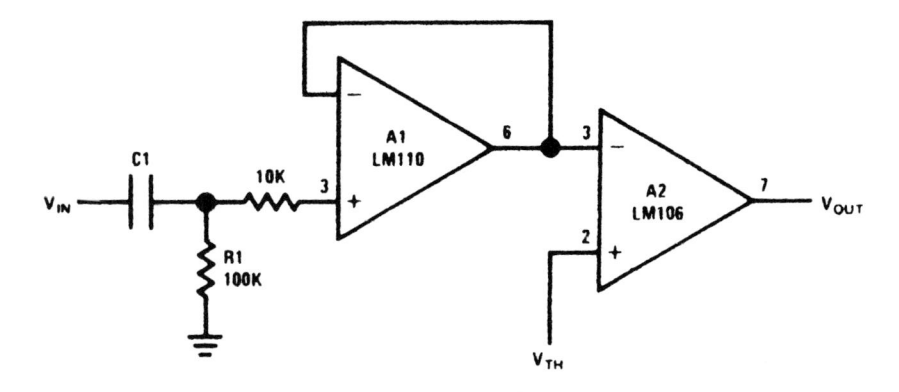

Comparator for ac-coupled signals.

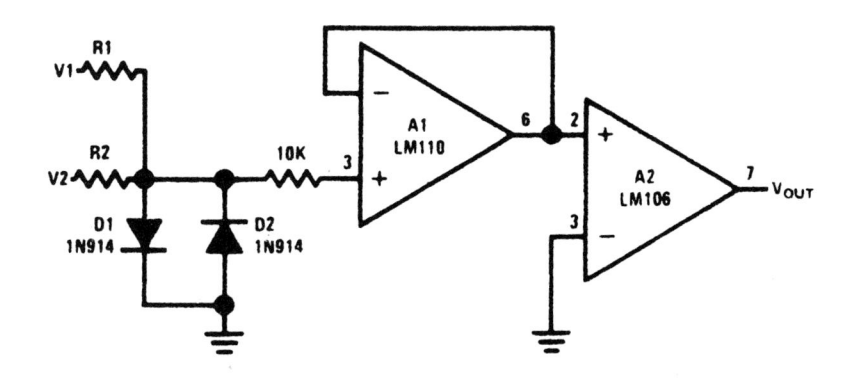

Comparator for signals of opposite polarity.

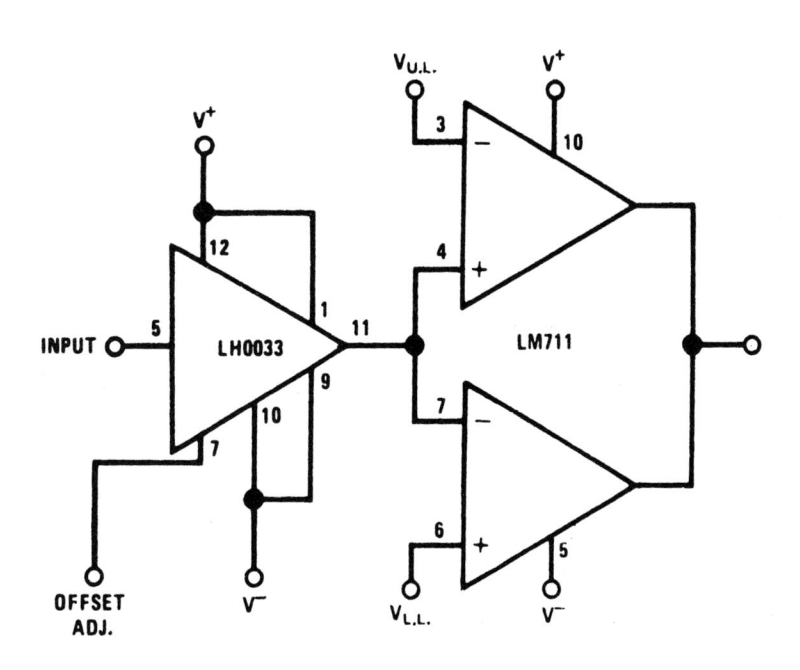

High-input-impedance comparator with offset adjust.
(No-go = logic 1; go = logic 0.)

SECTION **15** *Logic and Functionally Related Circuits* 511

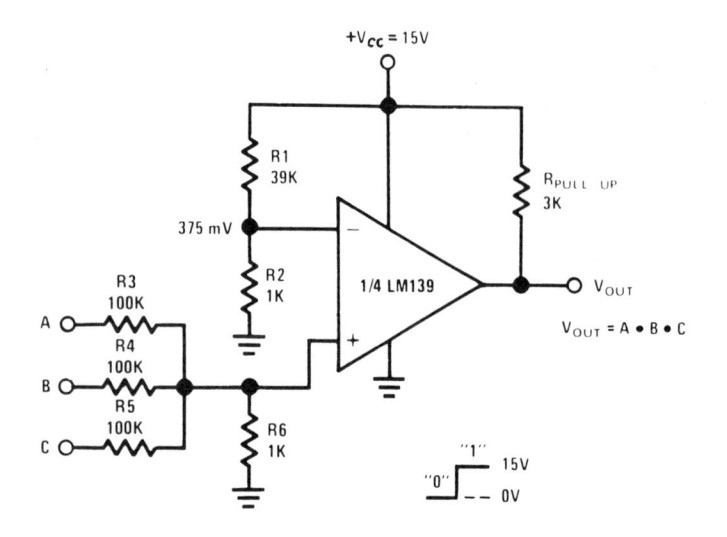

Three-input AND gate. Resistors R1 and R2 establish a reference voltage at the inverting input to the comparator. The noninverting input is the sum of the voltages at the inputs divided by the voltage dividers comprised of R3, R4, R5, and R6. The output will go high only when all three inputs are high, causing the voltage at the noninverting input to go above that at the inverting-input. The circuit values shown work for a "0" equal to ground and a "1" equal to +15V. The resistor values can be altered if different logic levels are desired. If more inputs are required, diodes are recommended to improve the voltage margin when all but one of the inputs are in the "1" state.

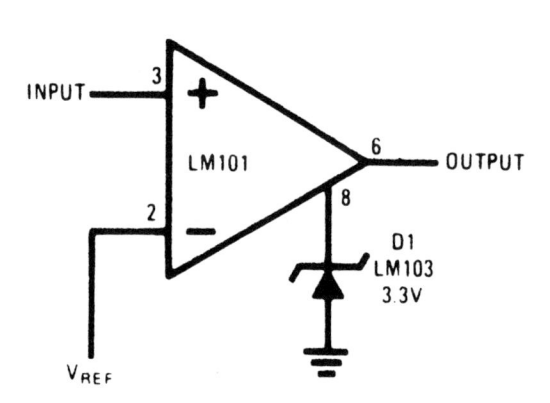

Comparator for driving DTL and TTL integrated circuits. Here, the clamping scheme makes the output signal directly compatible with DTL or TTL integrated circuits. An LM103 breakdown diode clamps the output at 0V and 4V in the low or high states, respectively. This diode has a sharp breakdown and low equivalent capacitance. When working as a comparator, the amplifier operates open-loop, so normally no frequency compensation is needed. Nonetheless, the stray capacitance between pins 5 and 6 of the amplifier should be minimized to prevent low-level oscillations when the comparator is in the active region. If this becomes a problem, a 3 pF capacitor on the normal compensation terminals will eliminate it.

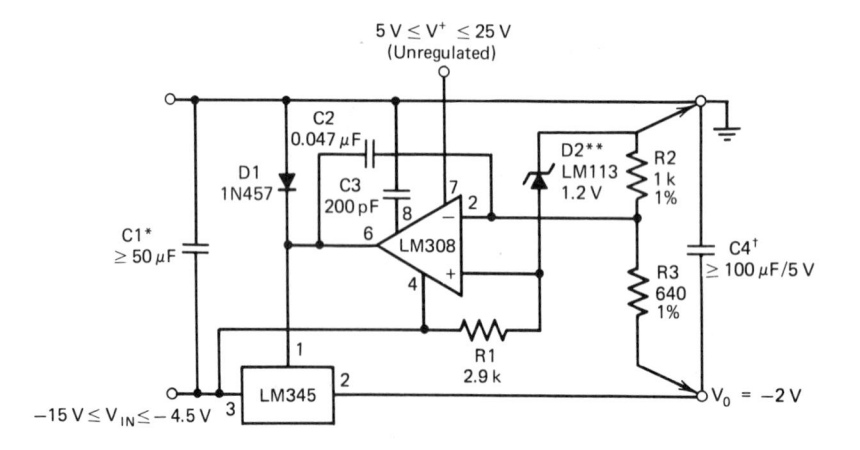

−2V ECL termination regulator. C1 is unnecessary if the power supply filter capacitor is within 3 inches of the regulator. C4 should be within 2 inches of LM345. There is no upper limit on C4; unlimited capacitance can be added at extended distances from the regulator. D2 sets initial output voltage accuracy. (LM113 is available in ±5, ±2, and ±1% tolerances.)

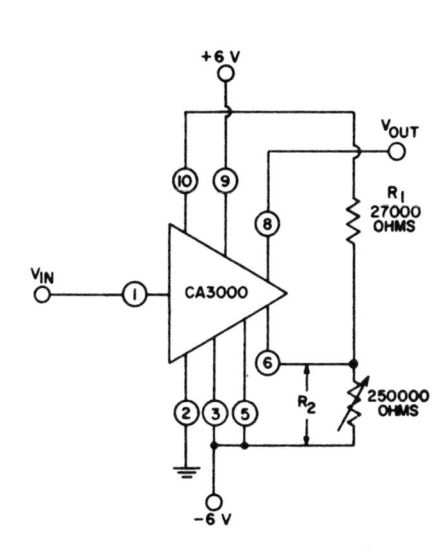

Schmitt trigger. The CA3000 can be operated as an accurate, predictable Schmitt trigger, provided saturation of either side of the differential amplifier is prevented (hysteresis is less predictable if saturation occurs). Nonsaturating operation is accomplished by operation in mode B (pins 3 and 5 shorted together). Large values are required for external resistors R1 and R2 because they receive the total collector current from pin 10. Because of the high impedances, resistor R2 is actually a parallel combination of the input impedance (100k) of the CA3000 and the 250k external resistor.

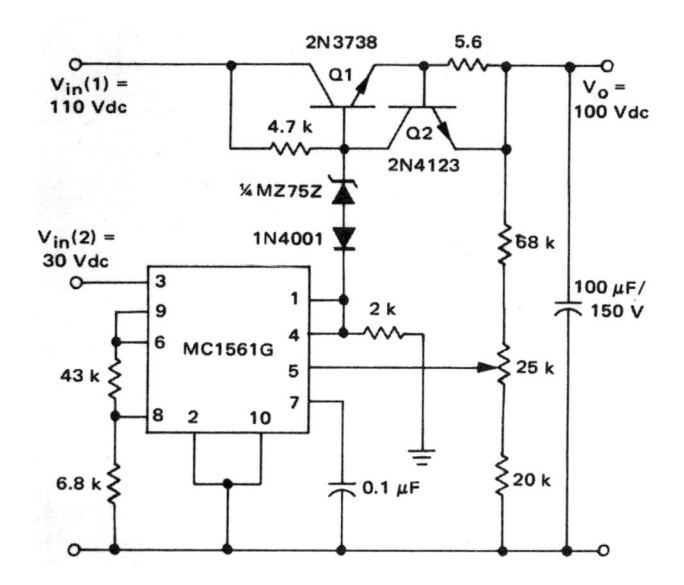

NPN voltage boost. The zener diode in the base lead of the NPN is used to shift the output voltage of the MC1561 approximately 25V to the desired high-voltage level, in this case 100V. Another voltage shift is accomplished by the resistor divider on the output to accommodate the required 25V reference to the MC1561. The 2k resistor is used to bias the zener diode so the current through the 4.7k resistor can be controlled by the MC1561. The 1N4001 diode protects the MC1561 from supplying load current under short-circuit conditions, and Q2 serves to limit base current to Q1.

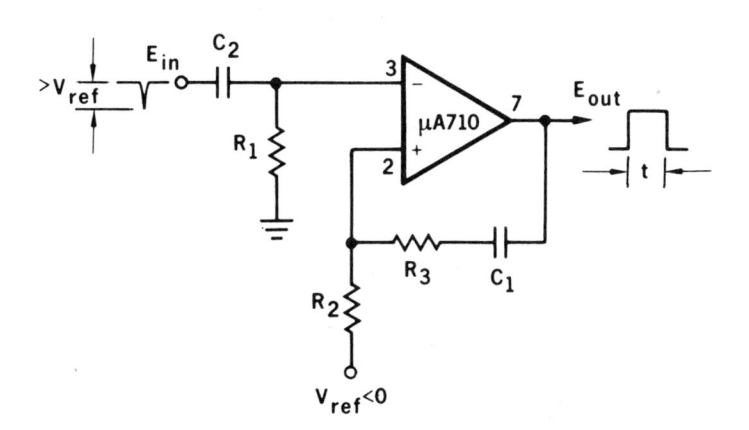

A one-shot multivibrator circuit gives a positive output pulse for a negative trigger with $V_{ref} < 0$. For $V_{ref} > 0$, a negative output pulse for a positive trigger is realized. Positive outputs for positive trigger and negative outputs for negative trigger can also be obtained by connecting the input differentiating capacitor to the noninverting input; however, the interaction between the trigger and feedback circuitry must be taken into account.

SECTION **15** *Logic and Functionally Related Circuits*

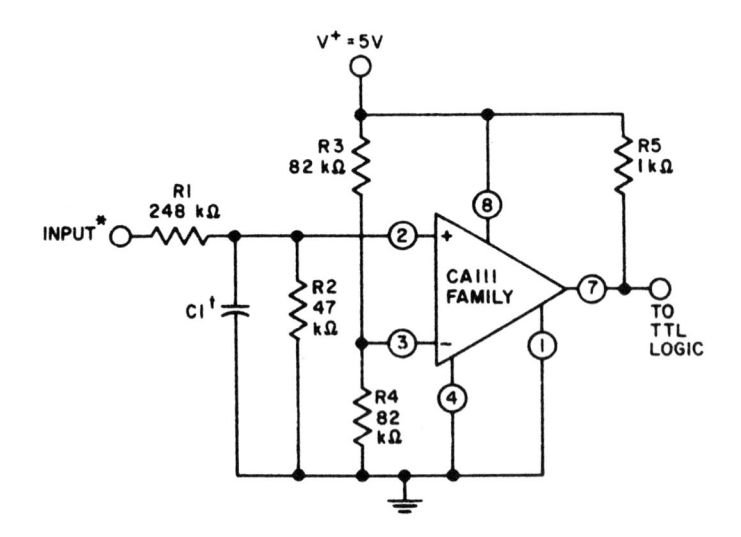

TTL interface with high-level logic. Values shown are for a 0 to 30V logic swing and a 15V threshold. The capacitor may be added to control speed and reduce noise spikes.

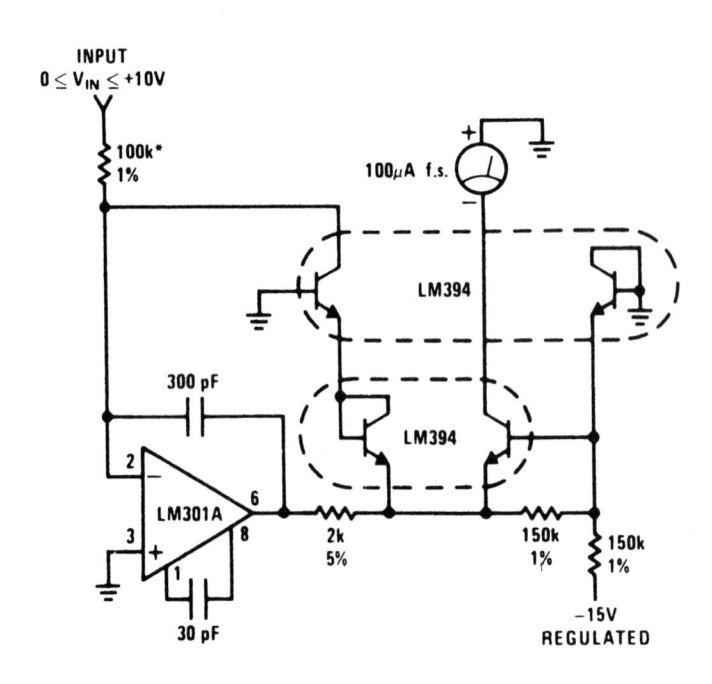

Low-cost accurate squaring circuit. $I_{OUT} = 10^{-6}(V_{IN})^2$. Trim 100k resistor for full-scale accuracy.

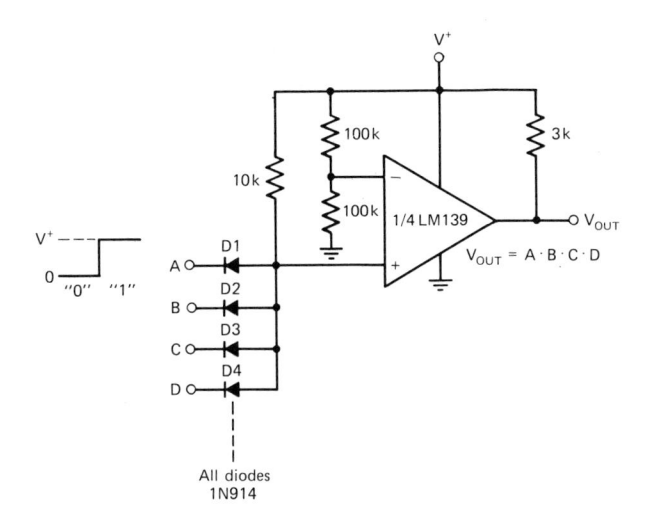

Large fan-in AND gate.

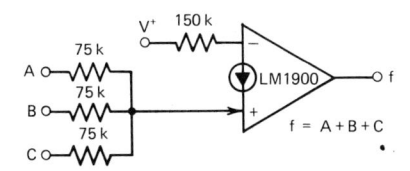

$V_{OUT} = A \cdot B \cdot C \cdot D$

All diodes 1N914

In this OR gate, $f = A + B + C$.

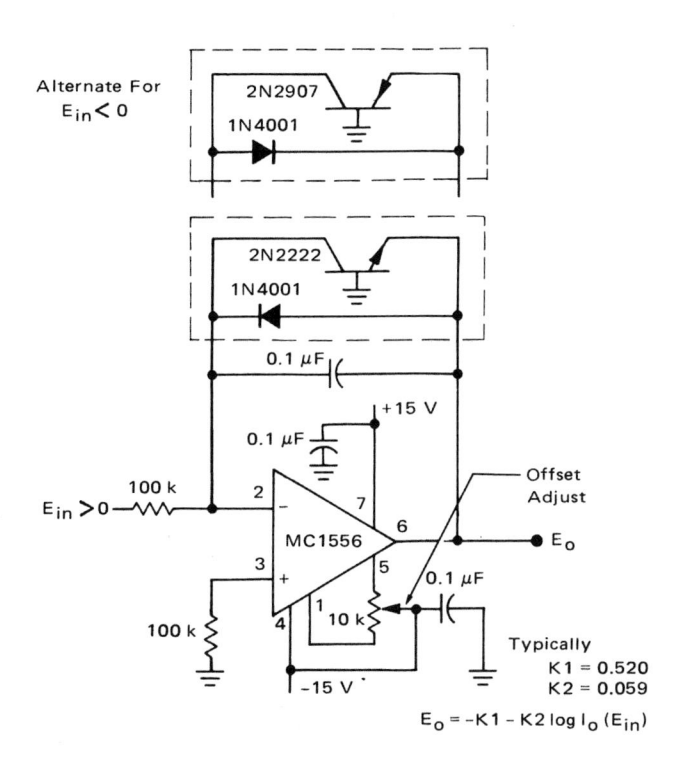

$E_o = -K1 - K2 \log_{10}(E_{in})$

Typically
K1 = 0.520
K2 = 0.059

A logarithmic amplifier relies on the exponential relationship between the base-emitter voltage and collector current of a transistor. Use of an NPN transistor implies operation with only positive input voltages. A PNP can be used just as successfully to accept negative input voltages. A diode is employed from collector to emitter to protect the base-emitter junction from degrading effects of reverse breakdown resulting from accidental reversal of the input voltage polarity. Noise pickup is eliminated by the 0.1 μF capacitor across the diode. Offset adjust is used to eliminate errors resulting from the input offset voltage.

SECTION **15** *Logic and Functionally Related Circuits* **515**

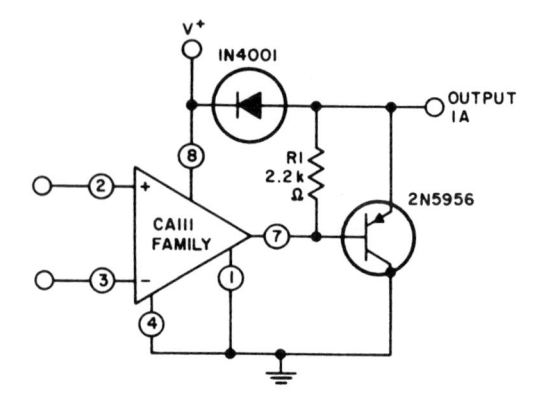

Comparator and solenoid driver.

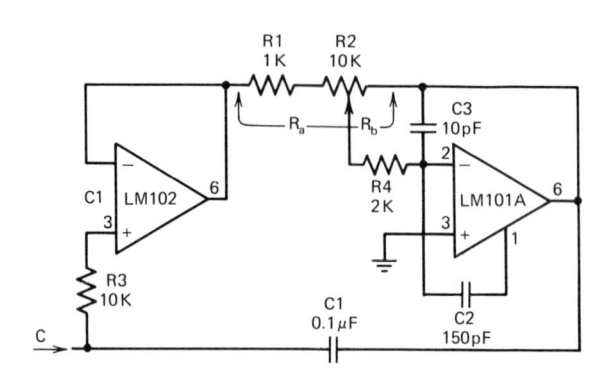

Variable capacitance multiplier. $C = 1 + R_b/R_a$.

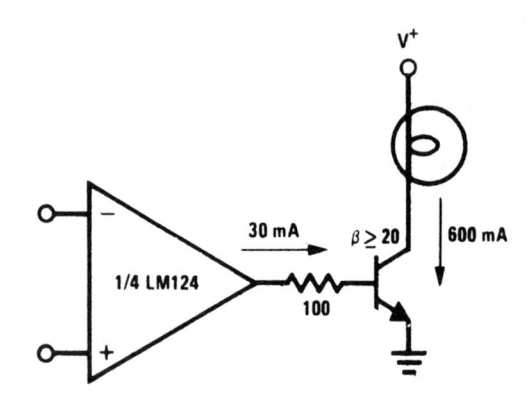

Lamp driver.

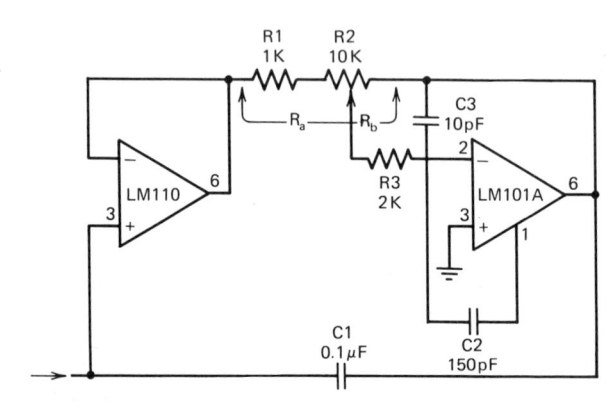

Variable capacitance multiplier. $C = \left(1 + \dfrac{R_b}{R_a}\right)C1$.

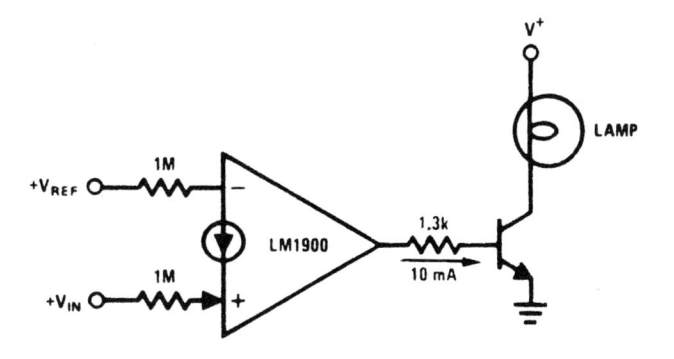

Power comparator

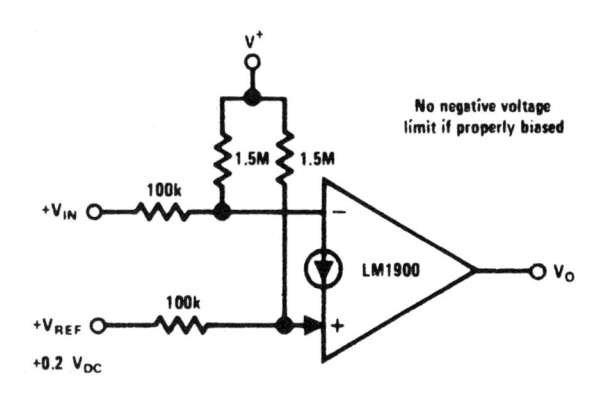

Low-voltage comparator

Comparator

Comparators. When driving either input from a low-impedance source, a limiting resistor should be placed in series with the input lead to limit the peak input current. Currents as large as 20 mA will not damage the device, but the current mirror on the noninverting input will saturate and cause a loss of mirror gain at current levels in the milliampere region—especially at high operating temperatures.

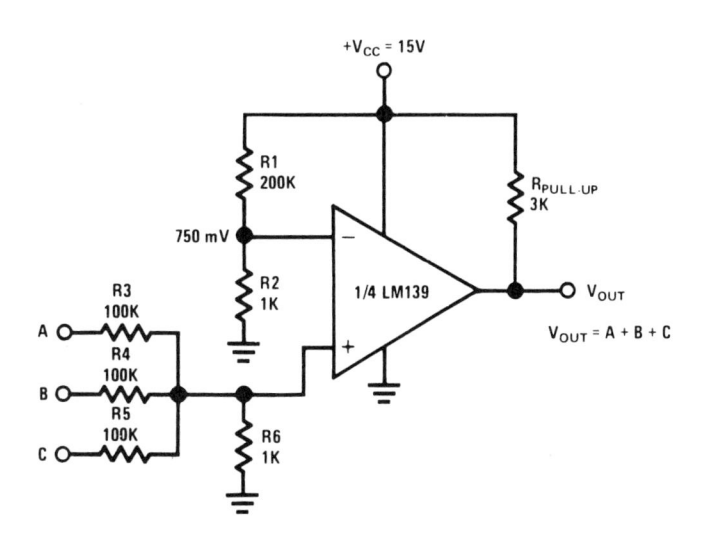

Three-input OR gate. A logic "1" at any of the inputs will produce a logic "1" at the output. A NOR gate may be implemented by simply reversing the comparator inputs. Resistor R6 may be added for the OR or NOR function at the expense of noise immunity if so desired.

SECTION **15** *Logic and Functionally Related Circuits* **517**

SECTION 16
HOBBY, FUN, AND EXPERIMENTER CIRCUITS

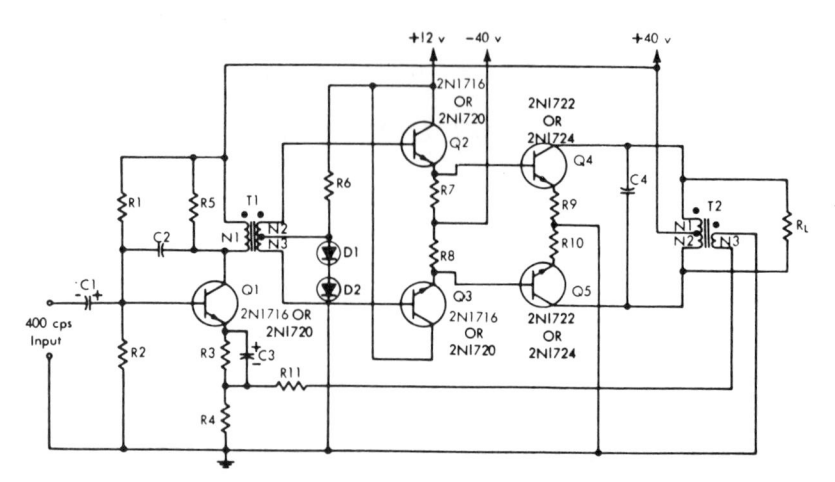

R_L – 68, 35 w	R5 – 2.21 K, 1/2 w	C1 – 40 µf, 25 v	T1: N1 = 755 T, #30 AWG; N2 = N3 = 330 T, #28 AWG Bifilar Wound. Core – Magnetic Metals 75EI SL14 or equivalent – 1 x 1 interleaved.
R1 – 4.32 K, 1/2 w	R6 – 390, 1/2 w	C2 – 500 pf, 100 v	
R2 – 3.32 K, 1/2 w	R7 & R8 – 2.00 K, 1 w	C3 – 1000 µf, 25 v	
R3 – 1.00 K, 1 w	R9 & R10 – 1.00, 2 w	C4 – 2.0 µf, 100 v	
R4 – 33.2, 1/2 w	R11 – 1.00 K, 1/2 w	D1 & D2 – TI 1N538	T2: N1 = N2 = 100 T, #20 AWG Bifilar Wound; N3 = 67 T, #28 AWG. Core – Magnetic Metals 100 EI SL14 or equivalent – Butt Joint.
		Q1, Q2, & Q3 – TI 2N1716 OR TI 2N1720	
		Q4 & Q5 – TI 2N1722 OR TI 2N1724	

NOTES:
1. All Resistance Values in ohms – 5% Tolerance
2. Resistor Wattage Ratings at 125°C Ambient
3. Capacitor Voltage Ratings at 125°C Ambient
4. Q1 on Heat Sink with $\theta_{C-HS} + \theta_{HS-A} \leq 40$ C°/w
5. Q2 and Q3 on same Heat Sink. $\theta_{C-HS} + \theta_{HS-A} \leq 40$ C°/w each. h_{FE}'s matched within 10%.
6. Q4 and Q5 on Heat Sinks with $\theta_{C-HS} + \theta_{HS-A} \leq 1.5$ C°/w. h_{FE}'s matched within 10%.

35W, 400 Hz servo amplifier. At rated power output, power gain is 45 dB and voltage gain is 37 dB. Circuit input resistance is 700 Ω. Total harmonic distortion is 5%.

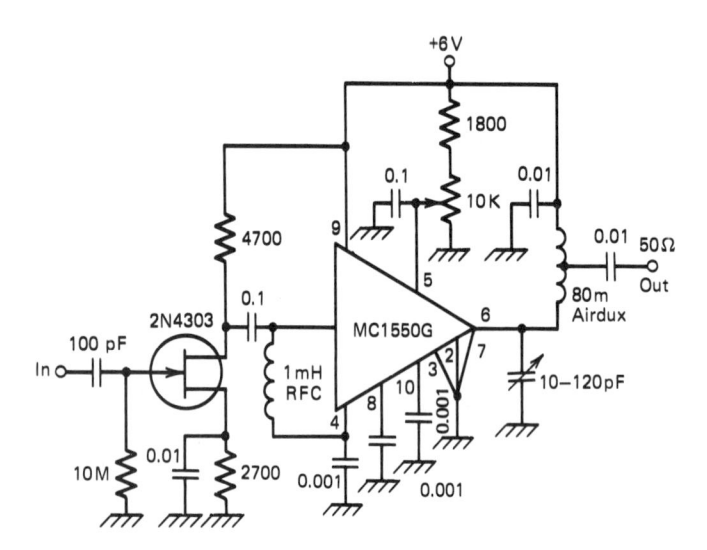

Impedance adapter/amplifier allows communications receivers to operate satisfactorily from such inadequate antennas as whips, bedsprings, window screens, and the like.

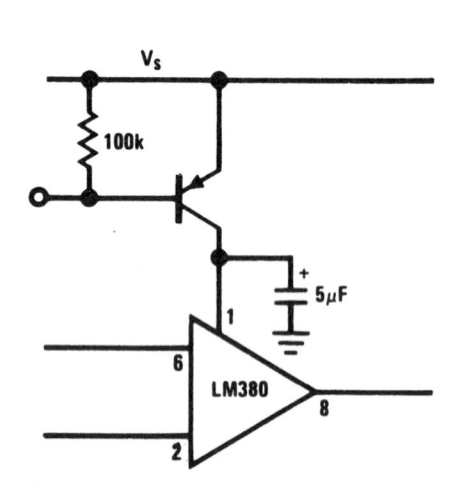

Muting (operating in a squelched mode) may be done with the LM380 by pulling the bypass pin high during the mute period. Any inexpensive general-purpose PNP transistor can be used to accomplish this function as diagrammed. During the mute cycle, the output stage will be switched off and will remain off until the PNP transistor is turned off again. Muting attack and release action is smooth and fast.

520 IC SCHEMATIC SOURCEMASTER

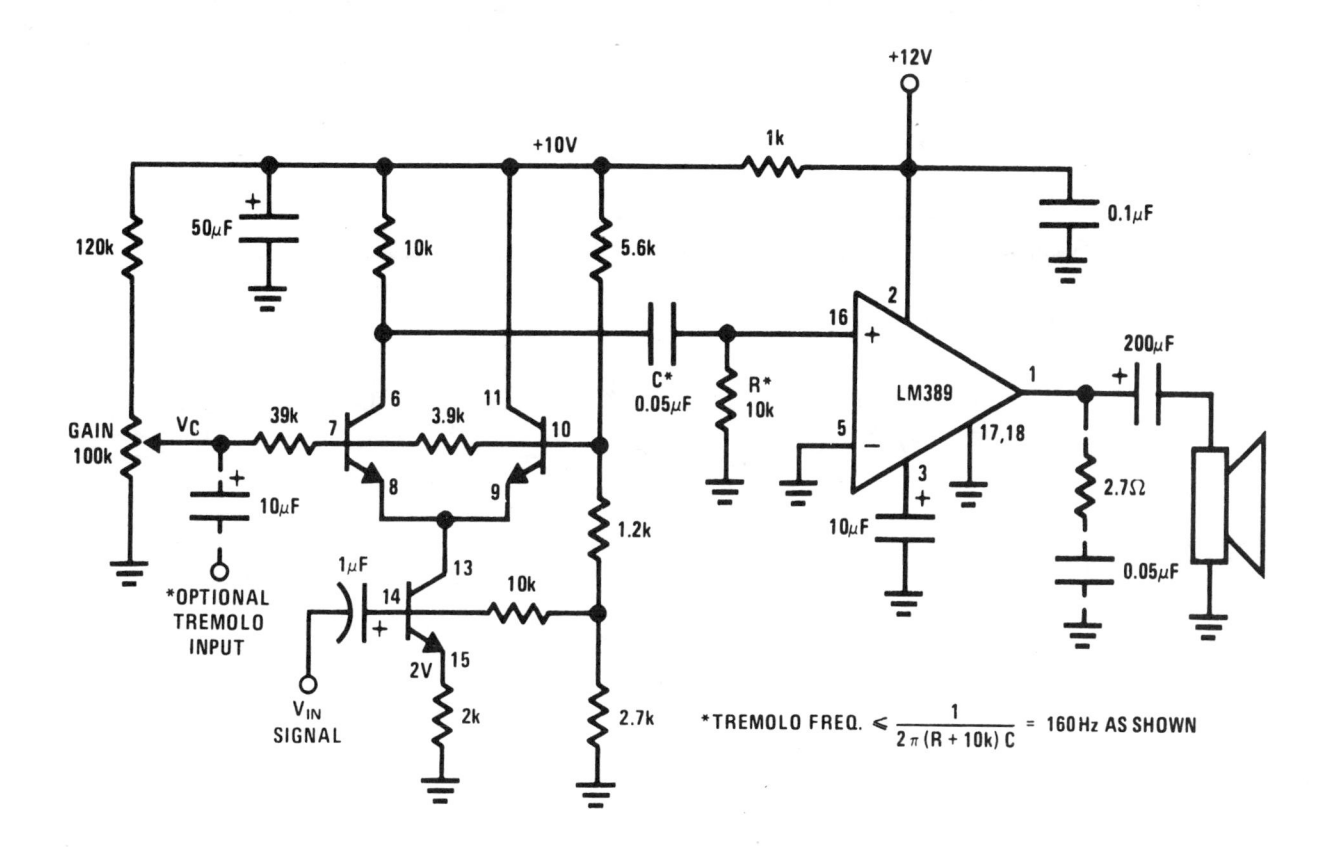

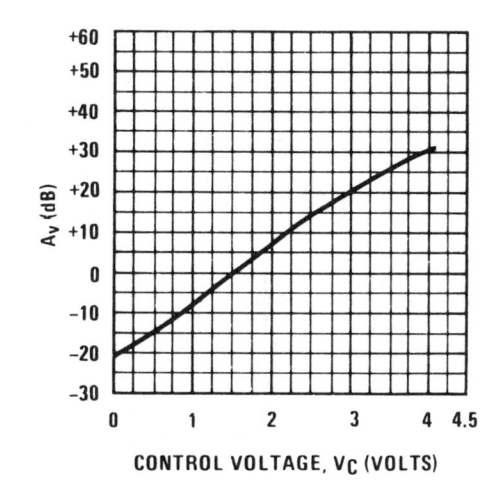

$$*\text{TREMOLO FREQ.} \leq \frac{1}{2\pi(R+10k)C} = 160\,\text{Hz AS SHOWN}$$

Voltage-controlled amplifier circuit for tremolo effect. Here, the transistors of the LM389 form a differential pair with an active current-source tail. This configuration, known technically as a variable-transconductance multiplier, has an output proportional to the product of the two input signals. Multiplication occurs as a result of the dependence of the transistor transconductance on the emitter-current bias. As shown, the emitter current is set up to a quiescent value of 1 mA by the resistive string. Gain control voltage V_C varies from 0V (minimum gain = −20 dB) to 4.5V (maximum gain = +30 dB), giving a total dynamic range of 50 dB. V_{IN} signal levels should be restricted to less than 100 mV for good distortion performance. The output of the differential gain stage is capacitively fed to the power amplifier via the RC network shown, where it is used to drive the speaker.

SECTION **16** *Circuits Hobby, Fun, and Experimenter*

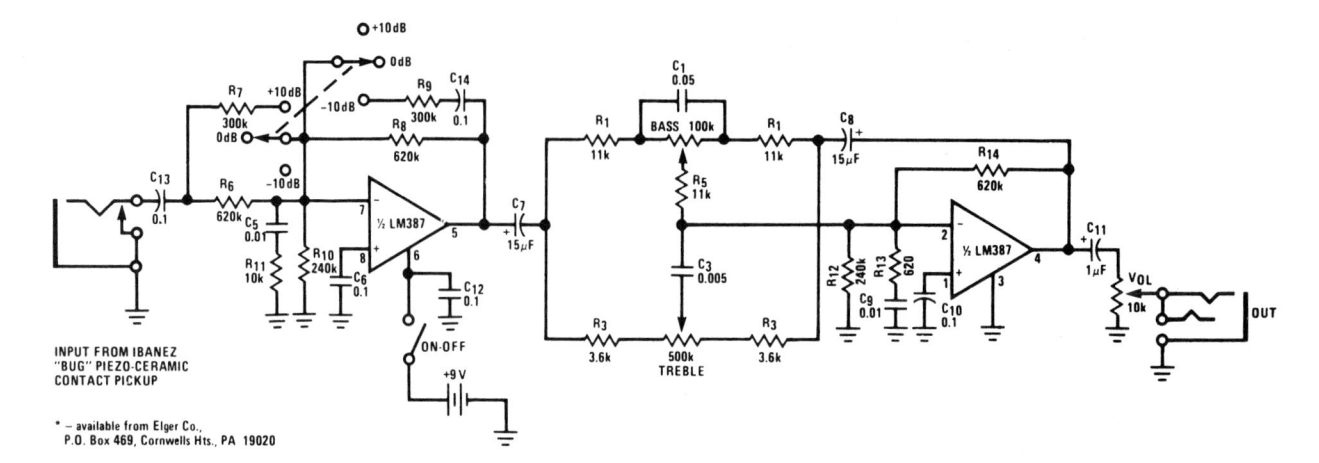

Acoustic-pickup preamp. Contact pickups designed for detection of vibrations produced by acoustic stringed musical instruments (for example, guitar, violin, dulcimer, etc.) require preamplification for optimum performance. Here the LM387 is configured as an acoustic-pickup preamp, with bass/treble tone control, volume control, and switchable ±10 dB gain select. The pickup used is the Ibanez "bug," a flat-response ceramic unit that is easy to use, inexpensive, and has excellent tone response. By using half the LM387 as the controllable gain stage and half as an active two-band tone control block, the complete circuit is made with only one 8-pin IC and requires very little space, allowing custom built-in designs where desired.

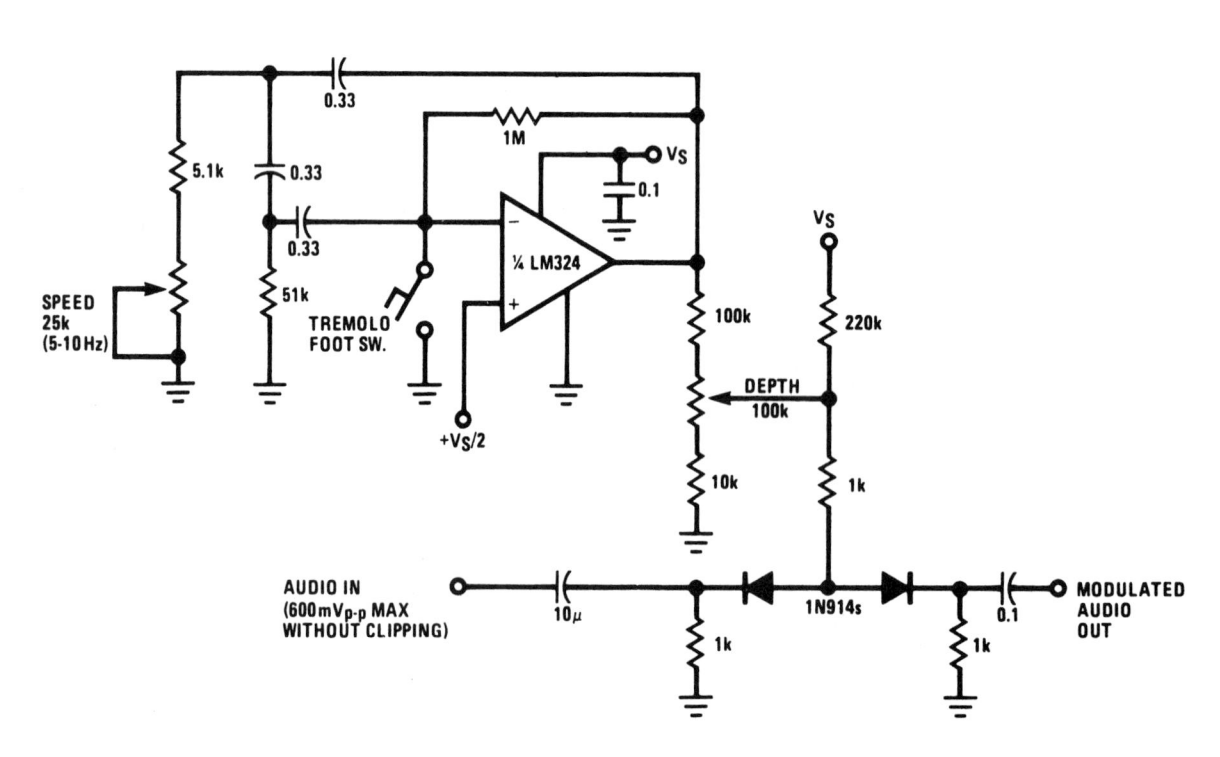

Tremolo circuit. Tremolo is amplitude modulation of the incoming signal by a low-frequency oscillator. A phase-shift oscillator using the LM324 operates at an adjustable rate (5 to 10 Hz) set by the *speed* pot. A portion of the oscillator output is taken from the *depth* pot and used to modulate the *on* resistance of two 1N914 diodes operating as voltage-controlled attenuators. Take care to restrict the incoming signal level to less than $0.6V_{p-p}$ or undesirable clipping will occur. For signals greater than 25 mV, THD will be high but is usually acceptable. Applications requiring low THD require the use of a light-detecting resistor (LDR) or a voltage-controlled gain block.

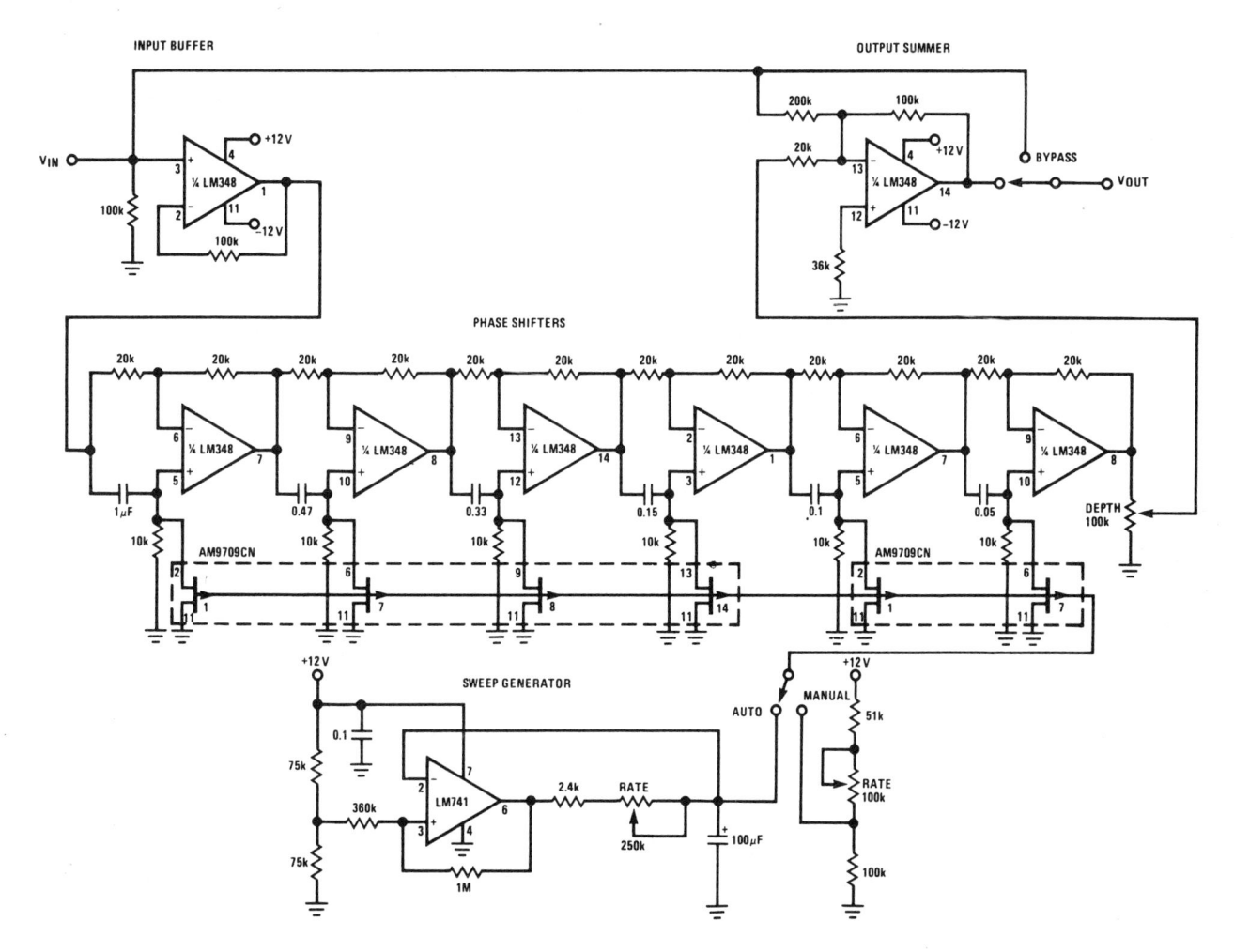

Phase shifter. A popular musical special-effect circuit called a "phase shifter" can be designed with minimum parts by using two quad op-amps, two quad JFET devices, and one LM741 op-amp. The sound effect produced is similar to a rotating-speaker or Doppler phase-shift characteristic, giving a whirling, ethereal, "inside-out" type of sound. The method used by recording studios is called *flanging*, where two tape recorders playing the same material are summed together while varying the speed of one by pressing on the tape-reel flange. The time delay introduced will cause some signals to be summed out of phase, and cancellation will occur. This phase cancellation produces the special effect and, when viewed in the frequency domain, is akin to a comb filter with variable rejection frequencies. Each stage shifts 90° at the frequency given by $1/(2\pi RC)$, where C is the positive input capacitor and R is the resistance to ground. Six phase-shift stages are used, each spaced one octave apart, distributed about the center of the audio spectrum (160 Hz to 3.2 kHz). JFETs are used to shift the frequency at which there is 90° delay by using them as voltage adjustable resistors. As shown, the resistance varies from 100Ω (FET full on) to 10k (FET full off), allowing a wide variation of frequency shift (relative to the 90° phase-shift point). The gate voltage is adjusted from 5V to 8V (optimum for the AM9709CN), either manually (via foot-operated rheostat) or automatically by the LM741 triangle-wave generator. Rate is adjustable from as slow as 0.05 Hz to a maximum of 5 Hz. The output of the phase-shift stages is proportionally summed back with the input in the output summing stage.

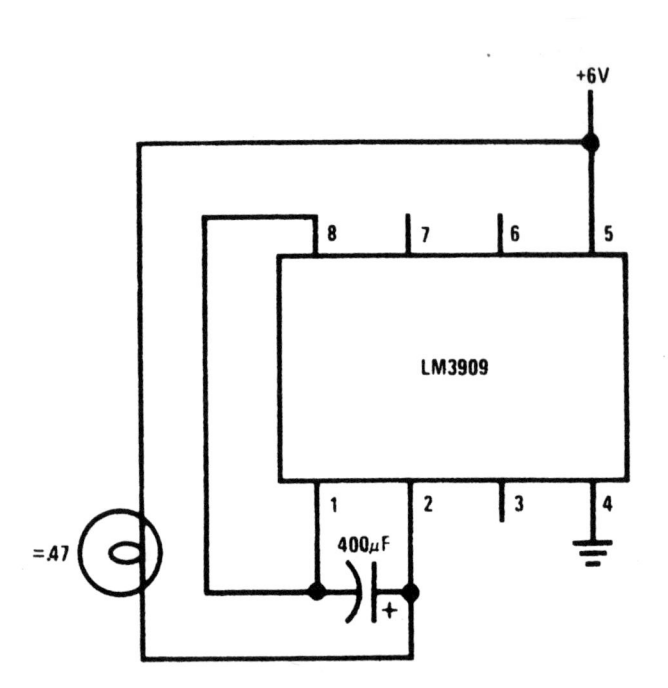

Incandescent bulb flasher blinks at rate of 1.5 Hz.

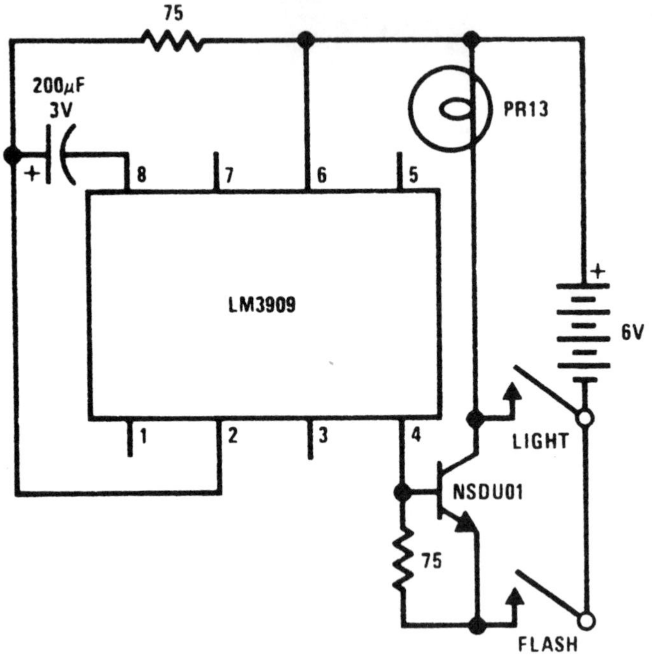

Emergency lantern/flasher blinks at 1.5 Hz rate.

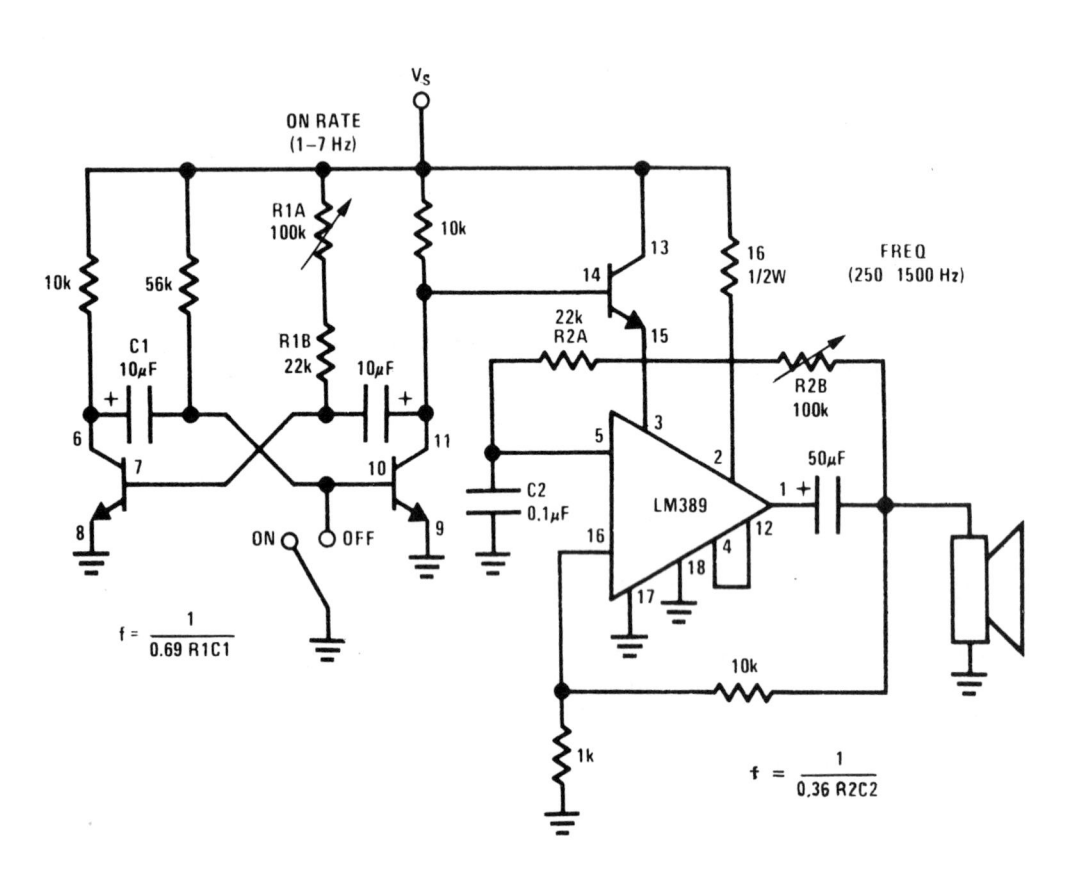

$$f = \frac{1}{0.69\, R1C1}$$

$$f = \frac{1}{0.36\, R2C2}$$

This wailing siren uses small-signal NPNs that are included as part of National IC package; $f = 1/0.36\, R2C2$.

524 IC SCHEMATIC SOURCEMASTER

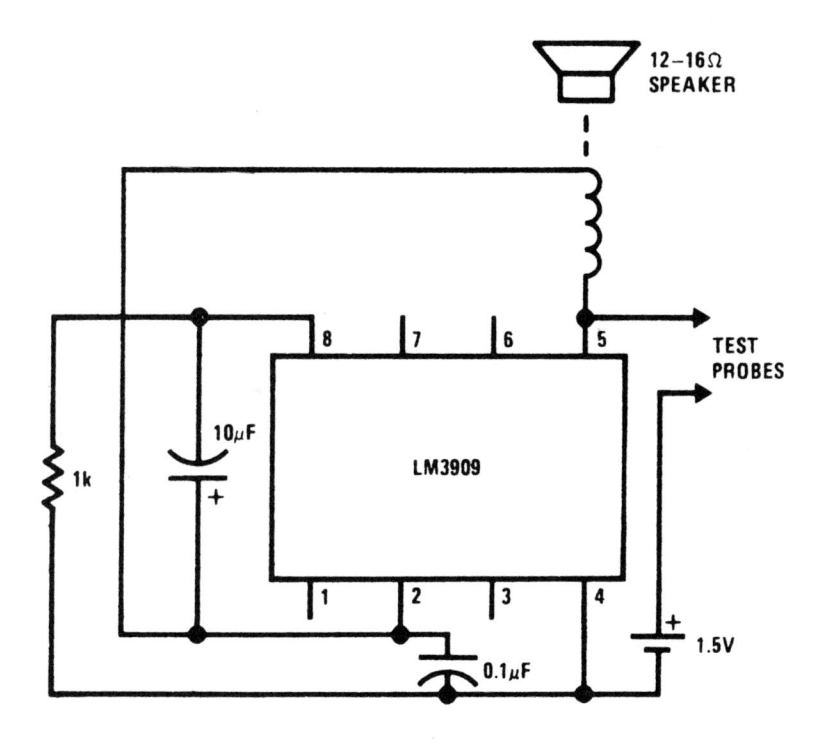

"Buzz box" continuity and coil checker. Differences between shorts, coils, and a few ohms of resistance can be heard.

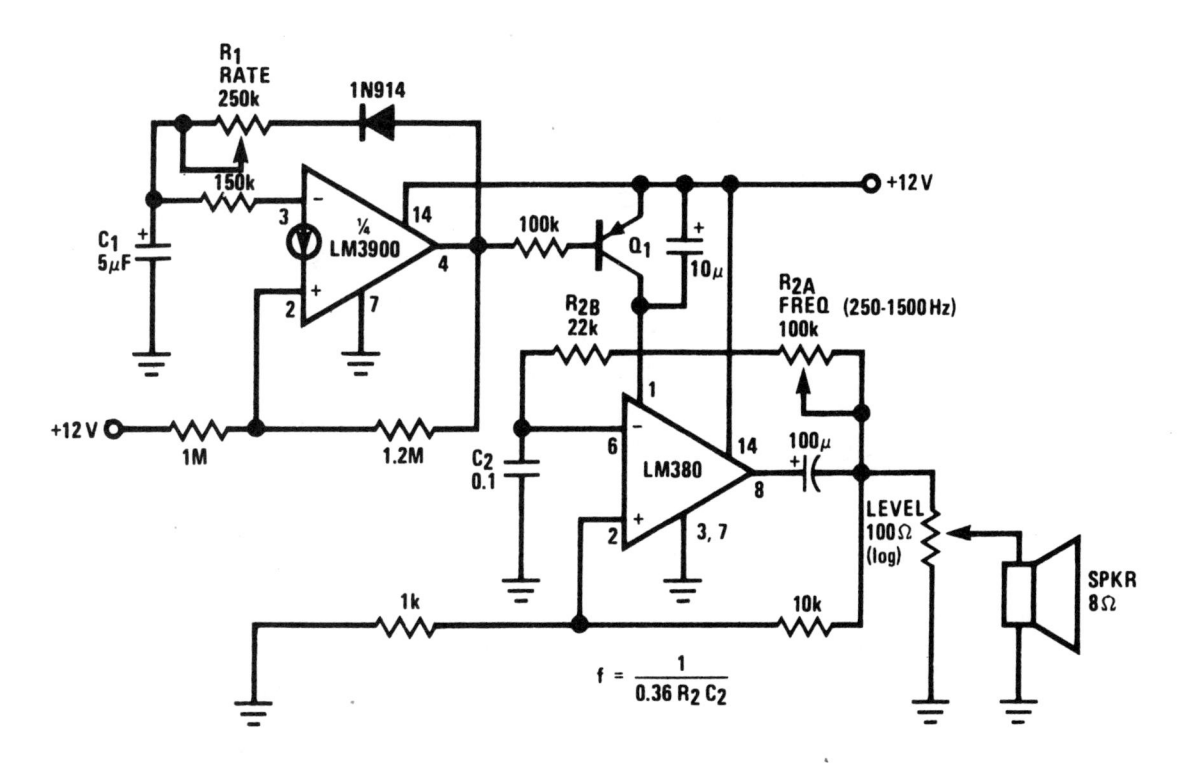

Programmable siren with frequency and rate adjustment. The LM380 operates as an astable oscillator with frequency determined by R2-C2. Adding Q1 and driving its base with the output of an LM3900 wired as a second astable oscillator acts to gate the output of the LM380 on and off at a rate fixed by R1-C1. Changing just about any component will alter the siren effect.

$$f = \frac{1}{0.36 R_2 C_2}$$

SECTION **16** Circuits Hobby, Fun, and Experimenter **525**

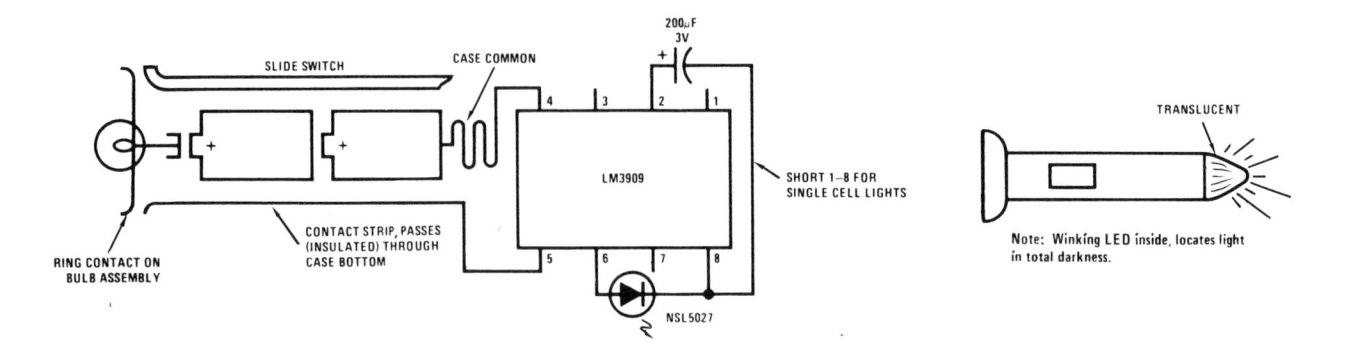

Flashlight finder. LM3909, capacitor, and LED are installed in a white translucent cap on the flashlight's back end. Only one contact strip (in addition to the case connection) is needed for flasher power. Drawing current through the bulb simplifies wiring and causes negligible loss since bulb resistance cold is typically less than 2Ω.

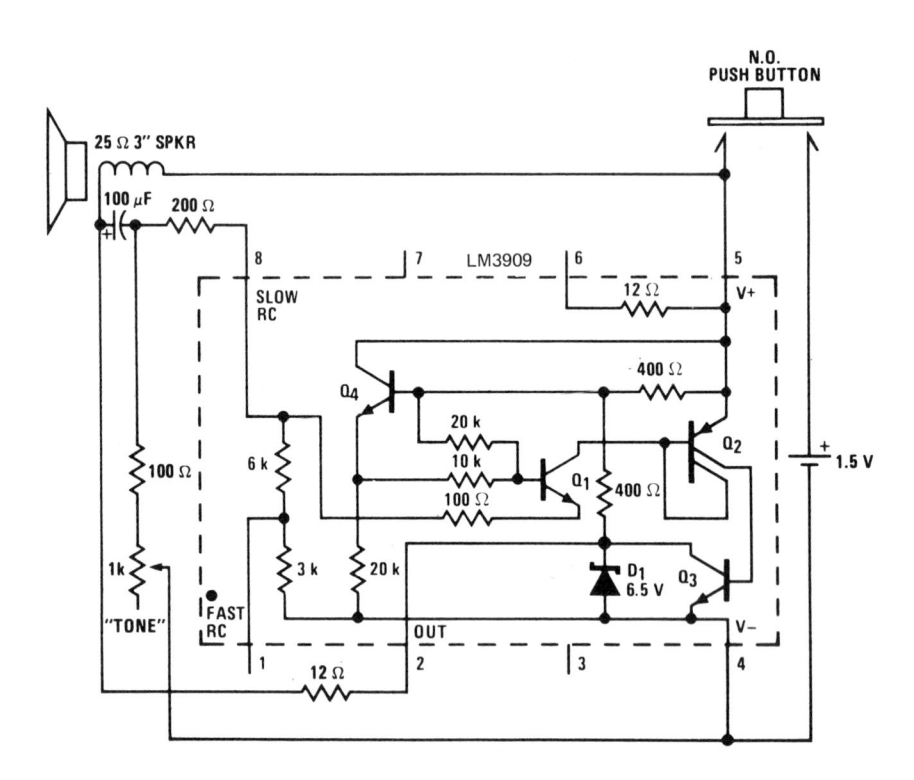

Electronic trombone. Build this in a cubical box of about 64 cubic inches, with one end able to slide in and out like a piston. Stiffen the box with thin layers of pressed wood or the like. Mount speaker, circuit, battery, etc. on the sliding end, with the speaker facing out through a 2¼-inch hole. In the prototype, a tube was provided (2½ inches long, 5/16 inches inside diameter) to bleed air in and out as the piston was moved while not affecting resonant frequency. Slide tones can be generated or a tune can be played by properly positioning the piston part and working the pushbutton. Position and direction of the piston are intuitive, so it is not difficult to play a reasonable semblance of a tune after a few tries. The 12Ω resistor in series with pin 2 (output transistor Q3's collector) and the speaker decouples voltages generated by the resonating speaker system from the low-impedance switching action of Q3. The 100 μF feedback capacitor would normally set a low or even subaudio oscillation frequency.

526 IC SCHEMATIC SOURCEMASTER

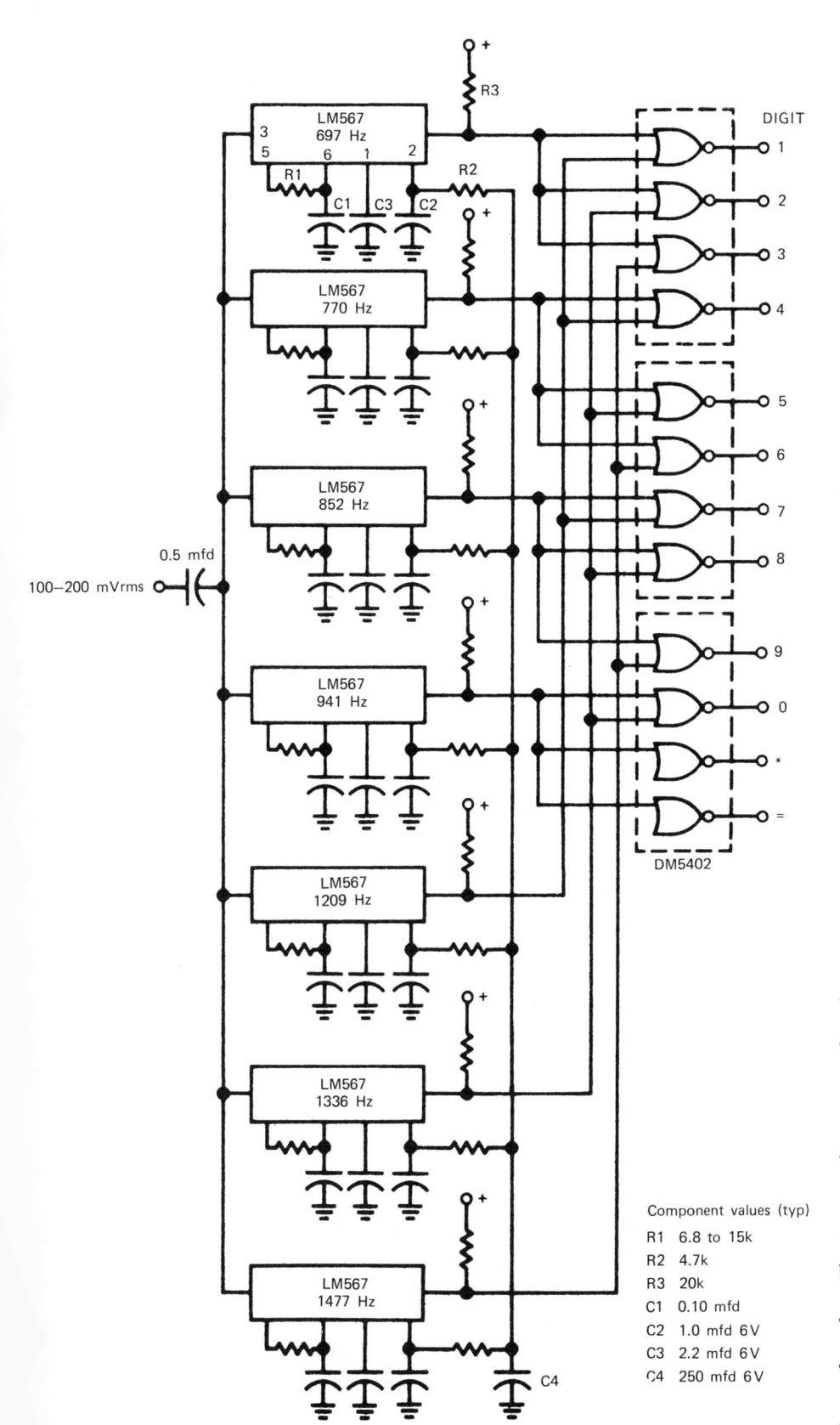

Touch-Tone decoder. The center frequency (f_0) of the tone decoder is equal to the free-running frequency of the VCO. This is given by

$$f_0 \cong \frac{1}{R1C1}$$

The bandwidth (BW) of the filter may be found from the approximation

$$BW = 1070 \sqrt{\frac{V_i}{f_0 C2}} \text{ in \% of } f_0$$

where V_i = input voltage (volts RMS), $V_i \leq 200$ mV; C2 = capacitance at pin 2 (μF).

Component values (typ)
R1 6.8 to 15k
R2 4.7k
R3 20k
C1 0.10 mfd
C2 1.0 mfd 6V
C3 2.2 mfd 6V
C4 250 mfd 6V

SECTION **16** *Circuits Hobby, Fun, and Experimenter* **527**

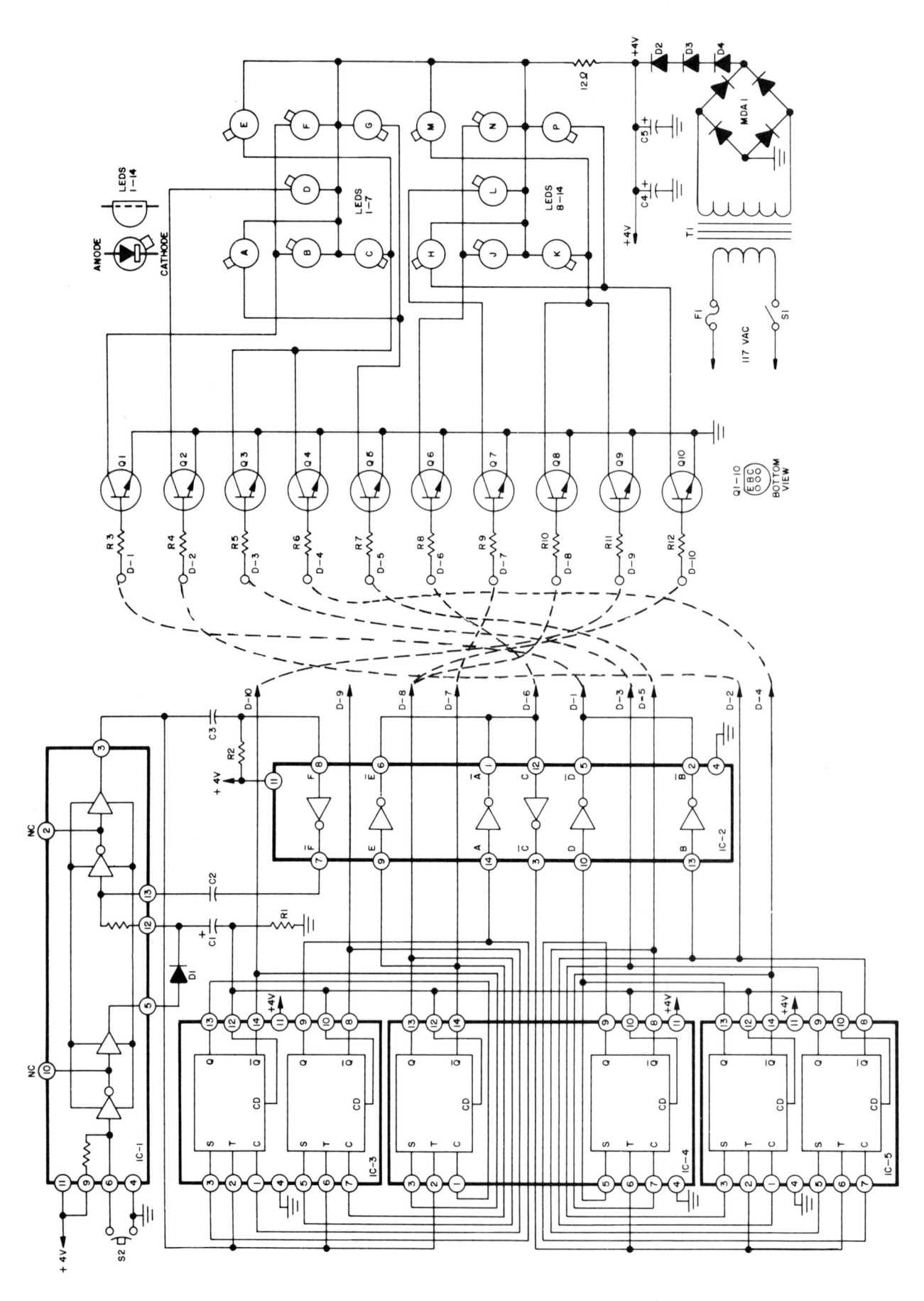

528　IC SCHEMATIC SOURCEMASTER

PARTS LIST

- 4 HEP 154 (D1-D4)
- 14 HEP P2003 (LED 1-14)
- 1 HEP 175 (MDA-1)
- 5 HEP 453
- 1 HEP 571 (1C1)
- 1 HEP 573 (1C2)
- 3 HEP 572 (1C3-1C5)
- 10 HEP 54 (Q1-Q10)
- 1 Capacitor, Electrolytic, 10 mfd, 10V, Calectro A1-104 (C1)
- 2 Capacitors, Ceramic disk, 0.1 mfd, 75V, Calectro A1-032 (C2-C3)
- 2 Capacitors, Electrolytic, 2200 mfd, 25V, Calectro A1-134 (C4-C5)
- 1 Fuse, 0.1A (F1)
- 1 Switch, SPST, Calectro E2-110 (S1)
- 1 Switch, Pushbutton, Calectro E2-142 (S2)
- 1 Resistor, 220Ω ± 10%, ½ Watt, Calectro B1-376 (R1)
- 1 Resistor, 10KΩ ± 10%, ½ Watt, Calectro B1-396 (R2)
- 10 Resistors, 470Ω ± 10%, ½ Watt, Calectro B1-380 (R3-R12)
- 1 Resistor, 12 Ω ± 10%, ½ Watt
- 1 Transformer, Filament, 117V AC to 6.3 V @ 1.2A, Calectro D1-745

This Motorola **electronic dice** circuit simulates a roll of the dice, providing a truly uncontrolled random number for each roll. Push the button and all possible combinations flash past at 2500 counts per second. Release the button and the last count remains until the button is pushed again or the power is disconnected. Only superhuman effort could manipulate these dice to obtain a particular number, since each appears for only 400 μsec in a tricky 1-3-5-6-4-2 sequence for each die face.

The circuit consists of 5 common ICs, 10 inexpensive transistors driving 14 LEDs, a small ac power supply, and a handful of resistors, capacitors, and hardware items. Wiring and lead dress are not critical, but a large number of connections (about 200) require time and care to make the unit function.

Half of a HEP 571 dual buffer and one section of a HEP 573 hex inverter are capacitively coupled as an oscillator. Pushing the button grounds the other half of the dual buffer, raising its output to a high state and biasing on the buffer and inverter oscillator, maintaining continuous oscillations so long as the button is down. Simultaneously, all flip-flops are reset to zero as capacitor C1 is charged through R1. The 2.5 kHz oscillations are coupled to two cascaded divide-by-six counters, each consisting of three J-K flip-flops (for a total of 3 HEP 572 IC dual J-K flip-flops). Each modulo-six counter consists of three flip-flops connected as a shift register, except that the outputs are cross-coupled to the inputs. After the counters have been reset to zero at the beginning of each play, each counter cycles through counts 000, 100, 110, 111, 011, 001, and repeat, with the first counter advancing through all six states in order to advance the second through one state.

Parts List

- 4 HEP 154 (D1-D4)
- 14 HEP P2003 (LED 1-14)
- 1 HEP 175 (MDA-1)
- 5 HEP 453
- 1 HEP 571 (1C1)
- 1 HEP 573 (1C2)
- 3 HEP 572 (1C3-1C5)
- 10 HEP 54 (Q1-Q10)
- 1 electrolytic, 10 μF 10V, Calectro A1-104 (Ca)
- 2 capacitors, 0.1 μF, 75V, Calectro A1-032 (C2-C3)
- 2 electrolytics, 2200 μF, 25V, Calectro A1-134 (C4-C5)
- 1 fuse, 0.1A (F1)
- 1 switch, SPST, Calectro E2-110 (S1)
- 1 pushbutton, Calectro E2-142 (S2)
- 1 resistor, 220Ω ± 10%, Calectro B1-376 (R1)
- 1 resistor, 10k ± 10%, Calectro B1-396 (R2)
- 10 resistors, 470Ω ± 10%, Calectro B1-380 (R3-R12)
- 1 resistor, 12Ω ± 10%
- 1 transformer, 117V to 6.3V at 1.2A, Calectro D1-745

SECTION **16** Circuits Hobby, Fun, and Experimenter

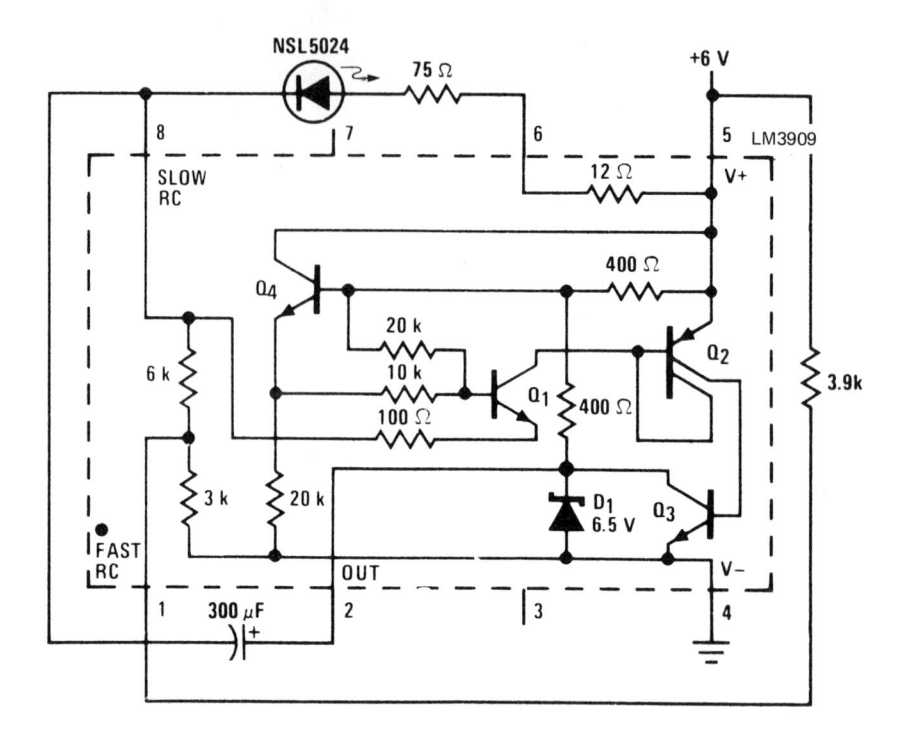

6V flasher. The 3.9k resistor connected from pin 1 to 6V supply raises voltage at the bottom of the 6k RC resistor. The charging current through that resistor is greatly reduced, decreasing the flashing rate to about 1 Hz. This biasing method also ensures starting of oscillation even under unfavorable conditions.

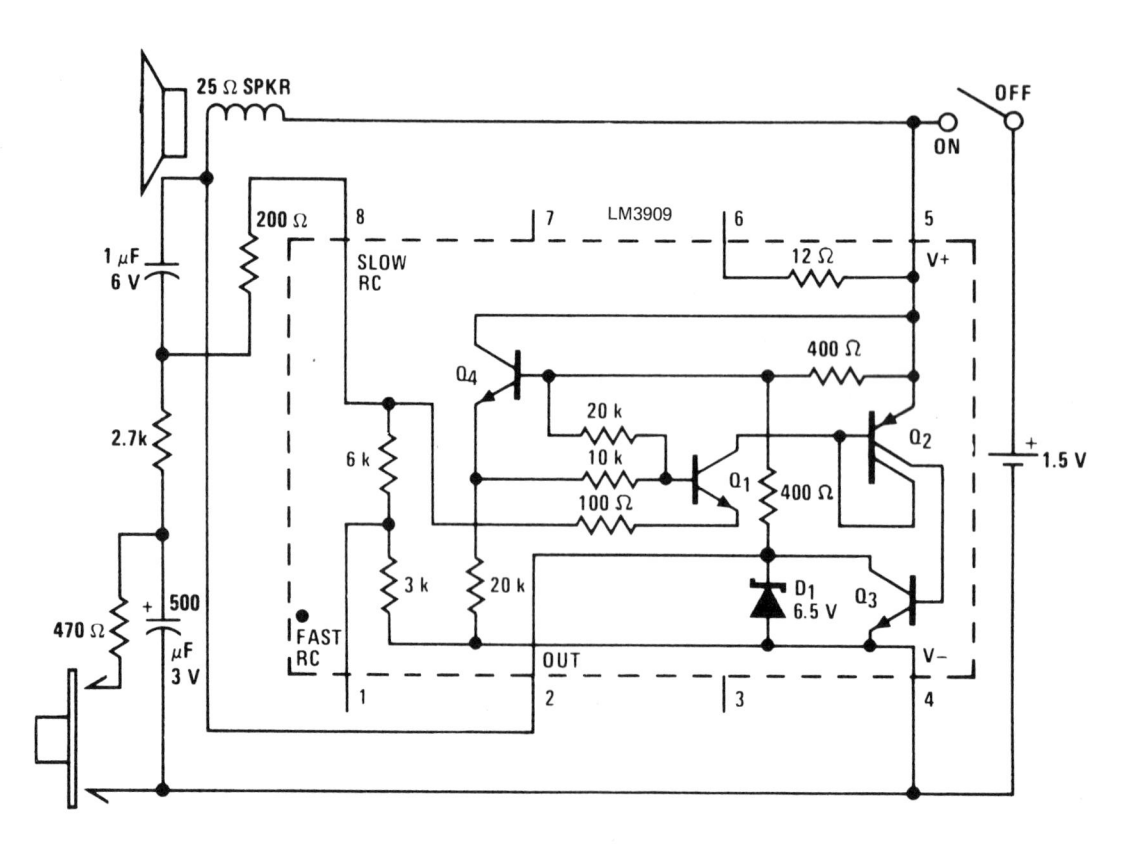

This siren produces a rapidly rising wail upon pressing the button and a slower "coasting down" upon release. If it is desirable to have the tone stop sometime after the button is released, an 18k resistor may be placed between pins 8 and 6 of the IC. The sound is then much like that of a motor-driven siren.

530 IC SCHEMATIC SOURCEMASTER

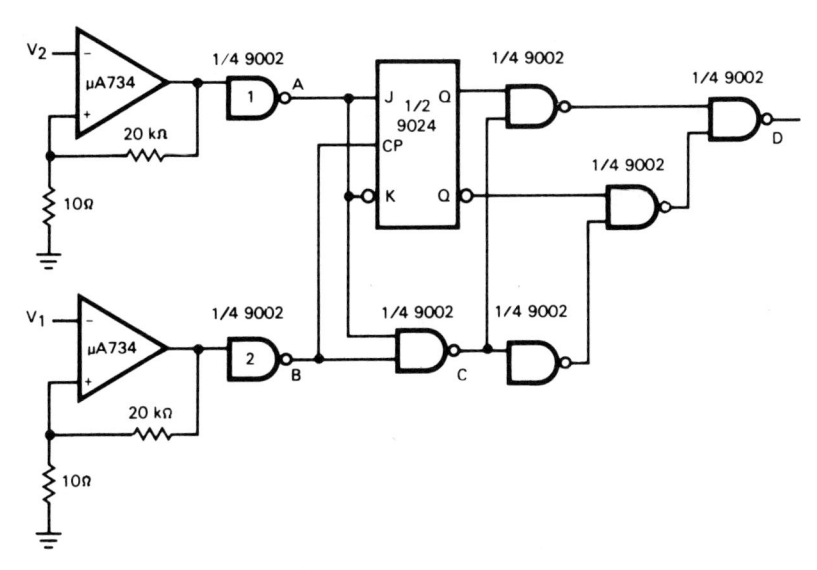

Schematic, wide range phase meter

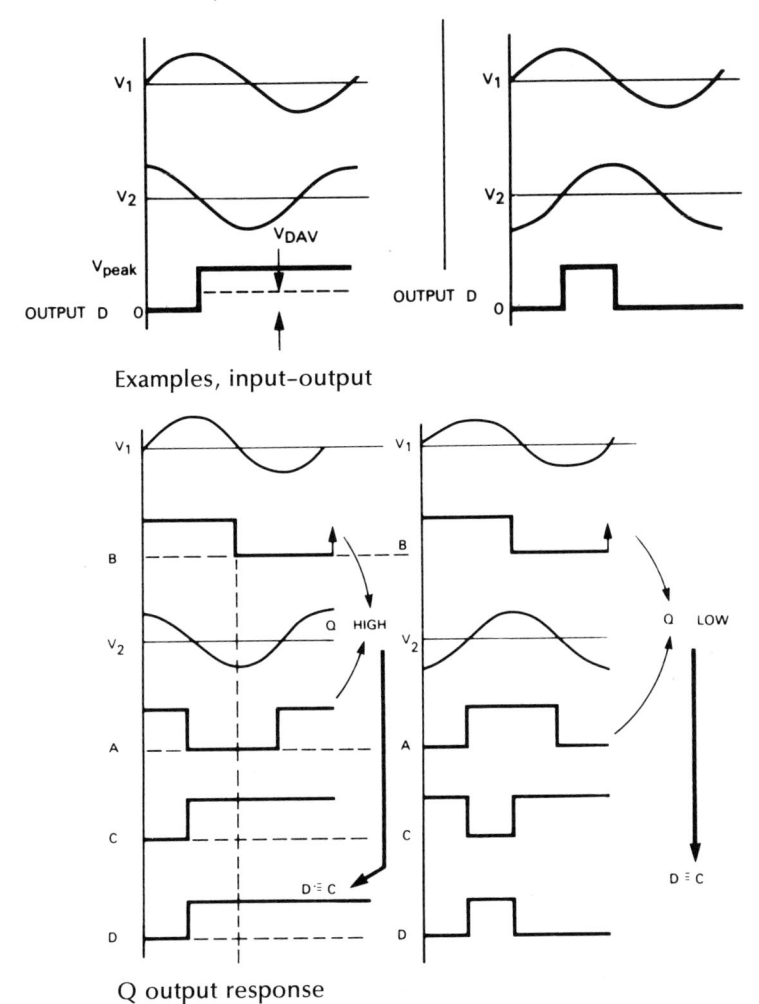

Examples, input–output

Q output response

A high-impedance wide-range phase meter for use in servo systems and related applications. For the two input sine waves (V_1 and V_2), output D has an average dc voltage V_{DAV} proportional to the magnitude and sign of the phase difference (ϕ) of V_1 and V_2.

$$\phi = \frac{2\pi V_{DAV}}{V_{peak}} - \pi$$

where V_{peak} is the amplitude of output D. Both AND gates 1 and 2 act as buffers to increase the fan-out of the Fairchild μA734s. The outputs of A and B are then NAND-gated to give output C, which has a duty cycle proportional to the magnitude of the phase difference ϕ.

SECTION **16** Circuits Hobby, Fun, and Experimenter

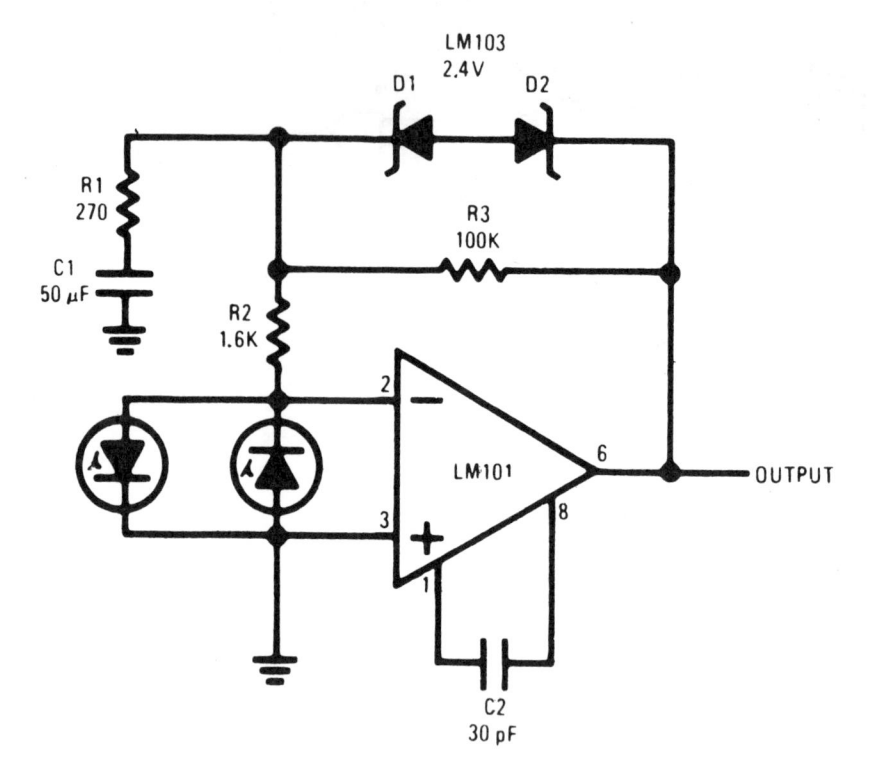

This saturating servo preamplifier with rate feedback operates in the linear mode until the output voltage reaches approximately 3V with 30 μA output current from the solar cell sensors. At this point the breakdown diodes in the feedback loop begin to conduct, drastically reducing the gain. However, a rate signal will still be developed because current is being fed back into the rate network (R1, R2, and C1) just as it would if the amplifier had remained in the linear operating region. In fact, the amplifier will not actually saturate until the error current reaches 6 mA, which would be the same as having a linear amplifier with a ±600V output swing.

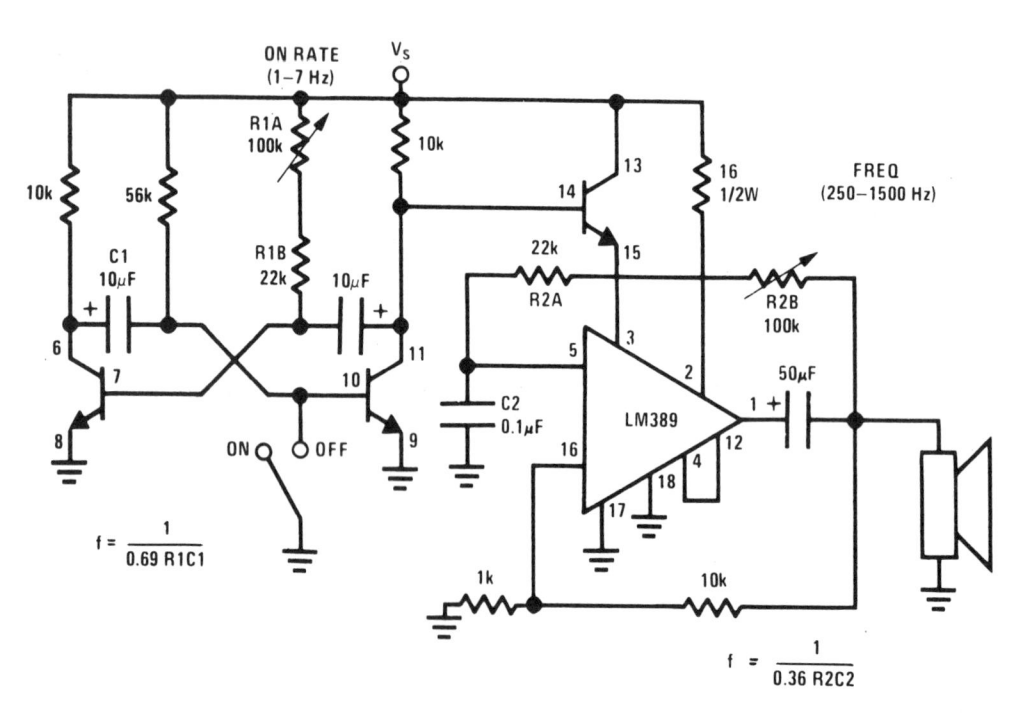

$$f = \frac{1}{0.69\ R1C1}$$

$$f = \frac{1}{0.36\ R2C2}$$

This siren circuit uses one of the LM389 transistors to gate the power amplifier on and off by applying muting. The other transistors form a cross-coupled multivibrator circuit that controls the rate of the square-wave oscillator. The power amplifier is used as the square-wave oscillator, with individual frequency adjust provided by potentiometer R2B.

532 IC SCHEMATIC SOURCEMASTER

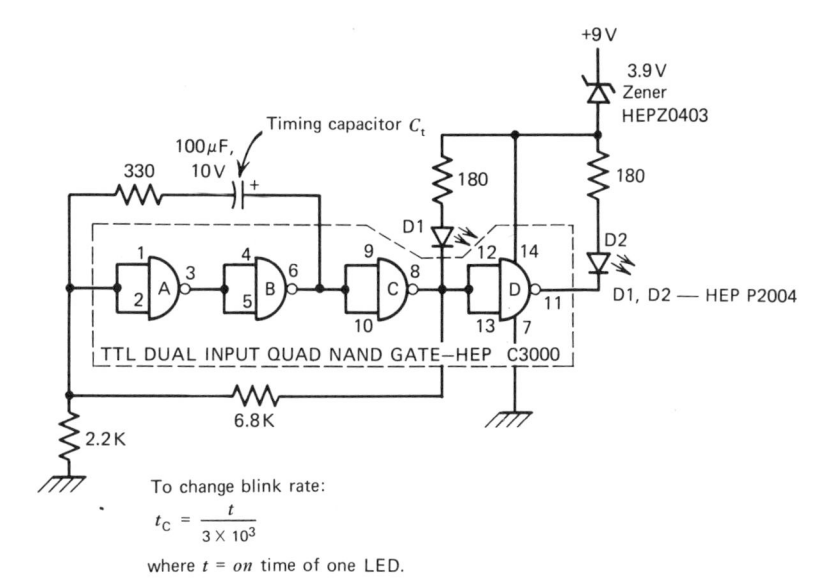

A two-LED blinker circuit for model railroad crossing signals or any other hobby application. The blink rate is controlled by the electrolytic labeled C_t. To change the timing of the blinking LEDs, divide the desired *on* time of one LED by 3000; the answer will be in microfarads.

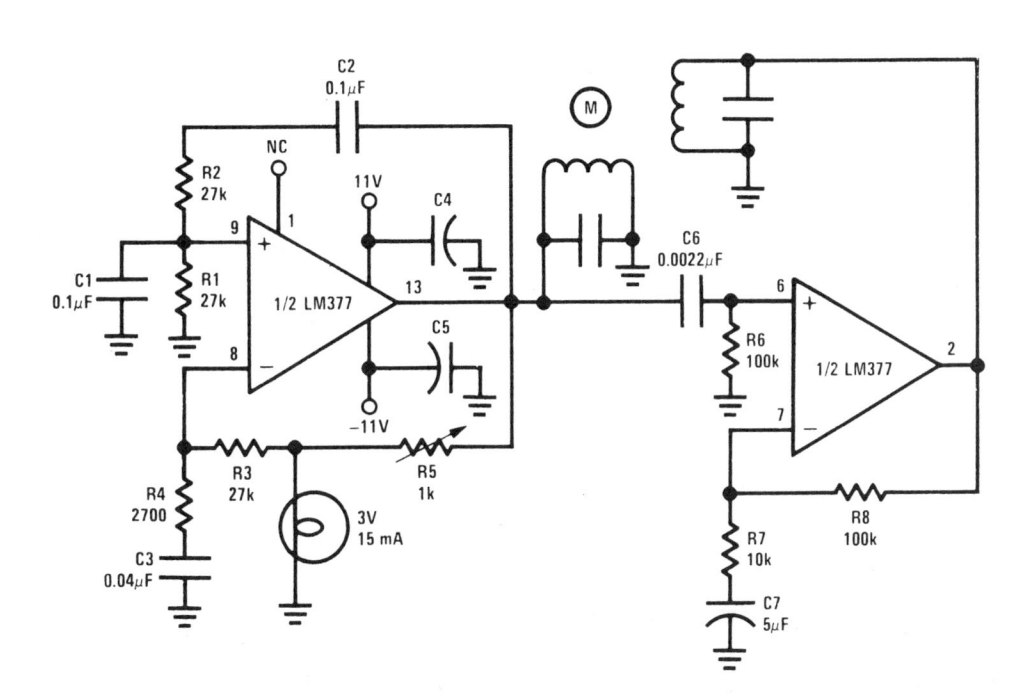

To change blink rate:
$$t_C = \frac{t}{3 \times 10^3}$$
where t = *on* time of one LED.

Two-phase motor drive. Applications such as a constant- (or selectable-) speed phonograph turntable drive are adequately met by this circuit. A split supply is used to simplify the circuit, reduce parts count, and eliminate several large bypass capacitors. An incandescent lamp is used in a simple amplitude stabilization loop. Input dc is minimized by balancing dc resistance at plus and minus amplifier inputs (R1 = R3 and R6 = R8). High-frequency stability is assured by increasing closed-loop gain from approximately 3 at 60 Hz to about 30 above 60 kHz, with the network consisting of R3, R4, and C3. The interstage coupling C6–R6 network shifts phase by 85° at 60 Hz to provide the necessary two-phase motor drive signal. The gain of the phase-shift network is purposely low, so that the buffer amplifier will operate at a gain of 10 for adequate high-frequency stability. The importance of supply bypassing, careful layout, and prevention of output ground loops is stressed. The motor windings are tuned to 60 Hz with shunt capacitors. This circuit will drive 8Ω loads to 3W each.

SECTION **16** *Circuits Hobby, Fun, and Experimenter*

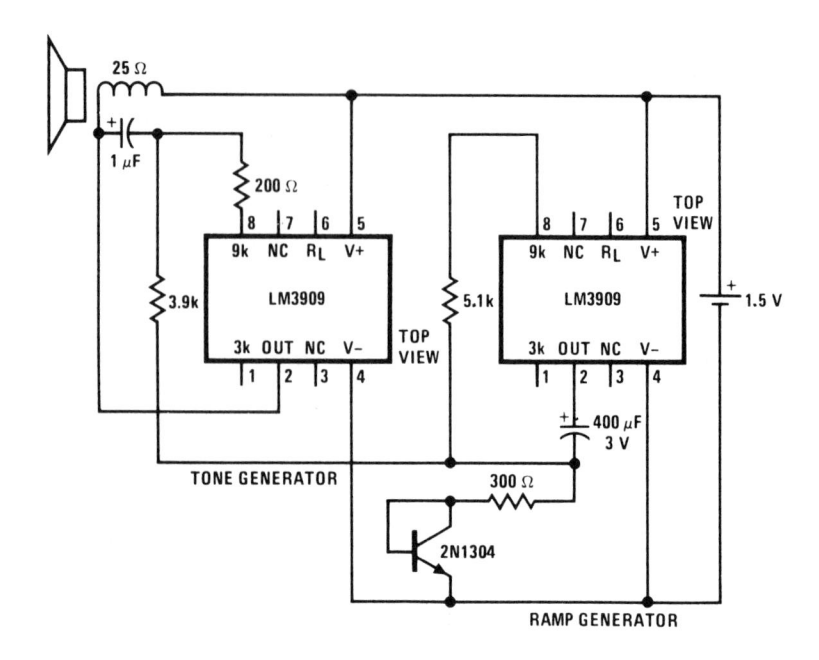

A whooper siren sounds somewhat like the electronic sirens used on city police cars, ambulances, and airport crash wagons. The rapid modulation makes the tone seem louder for the same amount of power input. Instead of a pushbutton, a rapidly rising and falling modulating voltage is generated by the second LM3909 and its associated 400 µF capacitor. The 2N1304 transistor is used as a low-voltage (germanium) diode. This transistor, along with the large feedback resistor (5.1k to pin 8), forces a ramp generator into an unusual mode of operation having longer *on* than *off* periods, which raises the average tone and makes the modulations seem more even.

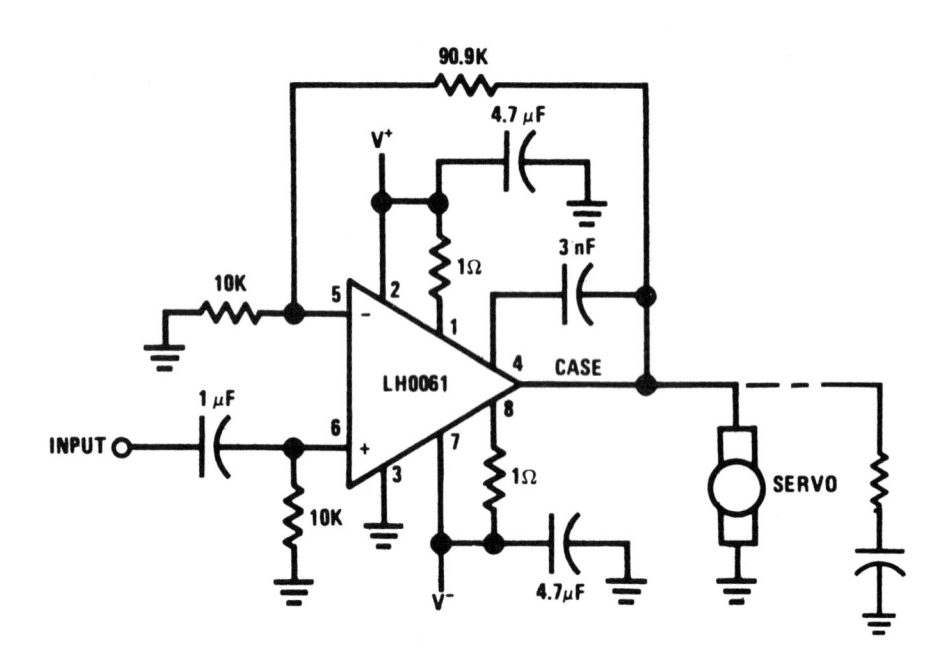

Ac servo amplifier.

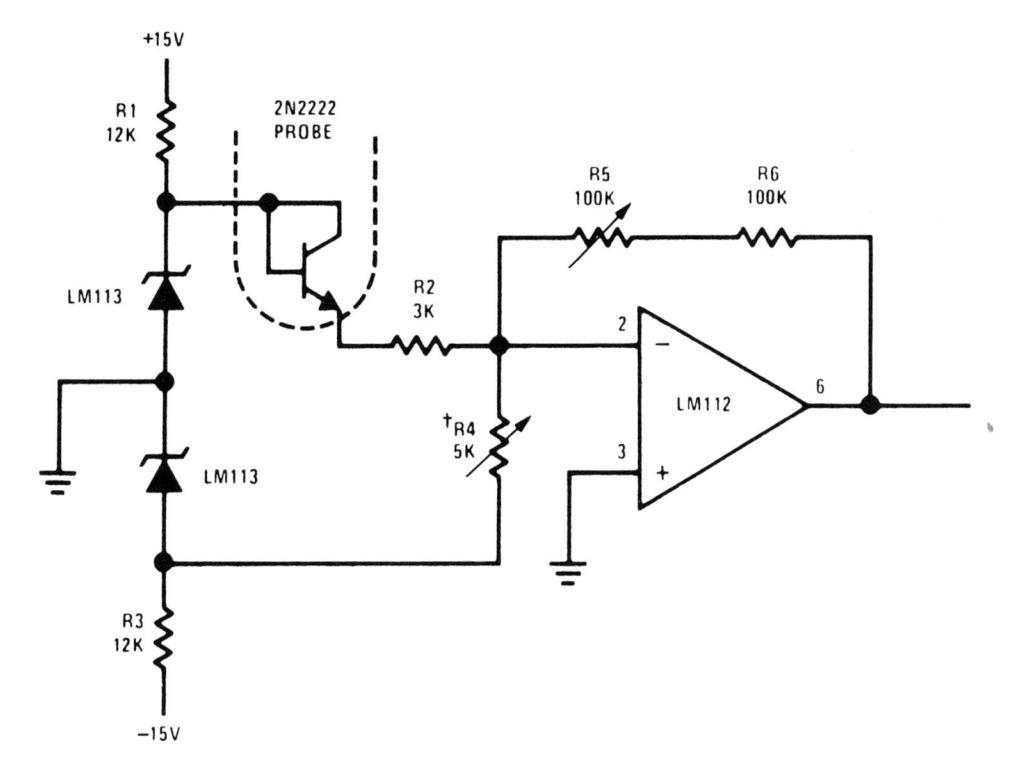

This electronic thermometer provides better than 1°C accuracy over a 100°C range. If the operating current of the sensing transistor is made proportional to absolute temperature, the nonlinearity of emitter-to-base voltage can be minimized. Over a −55 to 125°C temperature range, the nonlinearity is less than 2 mV (1°C temperature change); adjust R4 for 0V at 0°C; adjust R5 for 100 mV/°C.

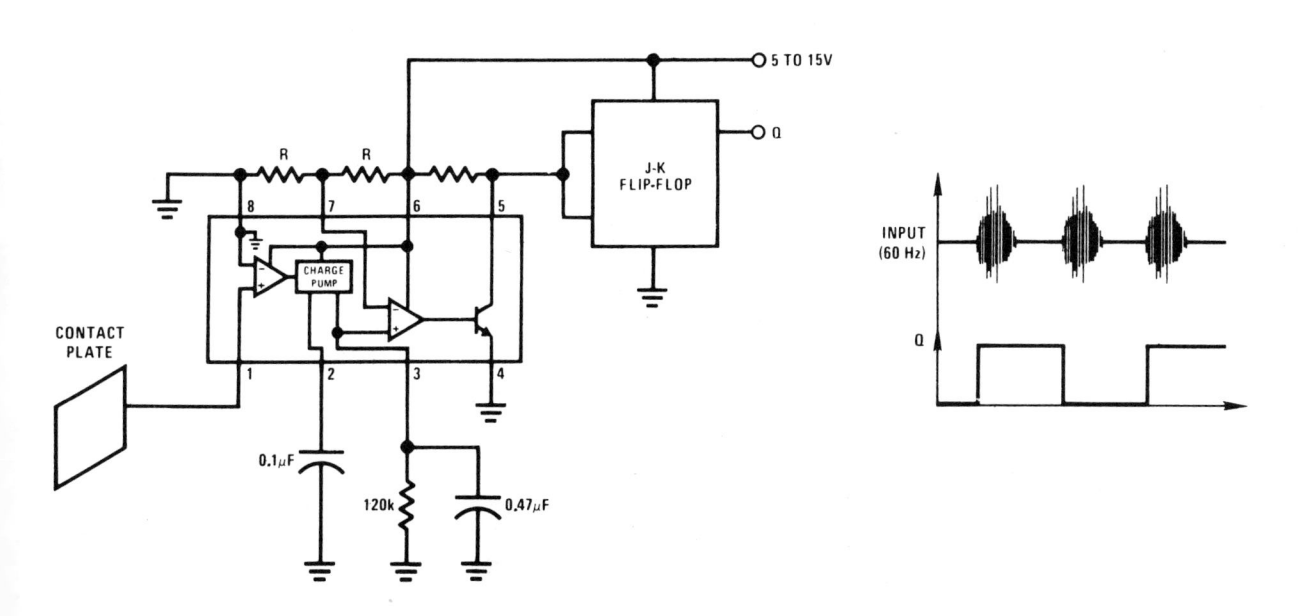

Finger-touch or contact switch.

SECTION **16** Circuits Hobby, Fun, and Experimenter **535**

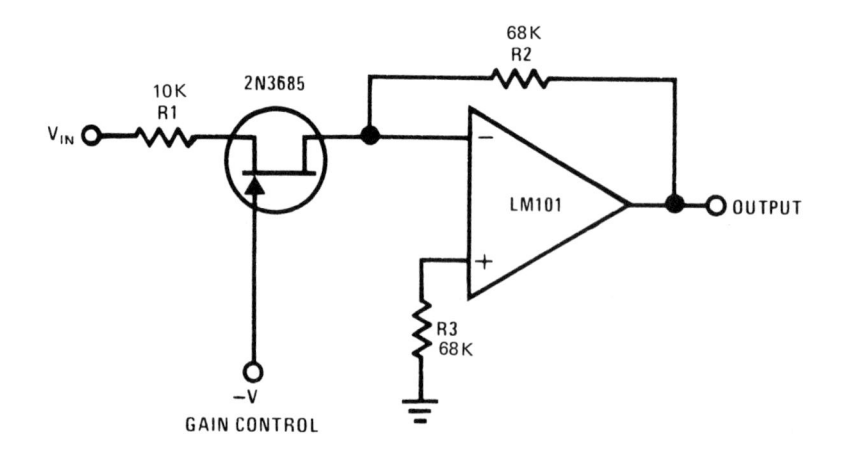

Variable attenuator. The 2N3685 JFET acts as a voltage variable resistor with an $R_{DS(ON)}$ of $800\,\Omega$ max. The JFET will have linear resistance over several decades of resistance, providing a nearly ideal electronic gain control.

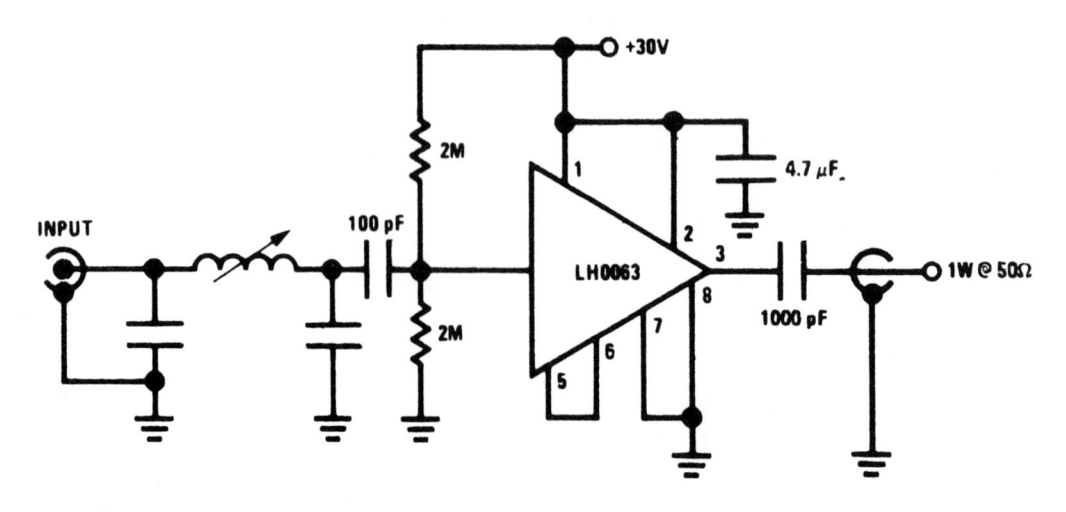

1W CW final amplifier.

10	HEP 154 (D1–D11)		
10	HEP P2003 (D12–D21)		
1	HEP 310 (Q1)		
10	HEP R1001 (SCR1–SCR10)		
11	Resistors, 1000 Ω ± 20%, ¼ Watt	10	Capacitors, 0.33 mfd, 50V
10	Resistors, 33K Ω ± 20%, ¼ Watt	10	Capacitors, Ceramic disk, 0.005 mfd, 50V, Calectro A1-125
10	Resistors, 47 Ω ± 20%, ¼ Watt	1	Capacitor, Electrolytic, 5 mfd, 50V
1	Resistor, 47K Ω ± 20%, ¼ Watt	1	Capacitor, Electrolytic, 15 mfd, 25V
1	Resistor, 100 Ω ± 20%, ¼ Watt	1	Switch, SPST, Calectro E2-110 (S1)
1	Resistor, 2200 Ω ± 20%, ¼ Watt	1	Switch, SPST, Pushbutton, Calectro E2-140 (S2)
2	Resistors, 10K Ω ± 20%, ¼ Watt	1	Battery, 9V, NEDA 1604 or equivalent

Electronic roulette wheel.

Parts list

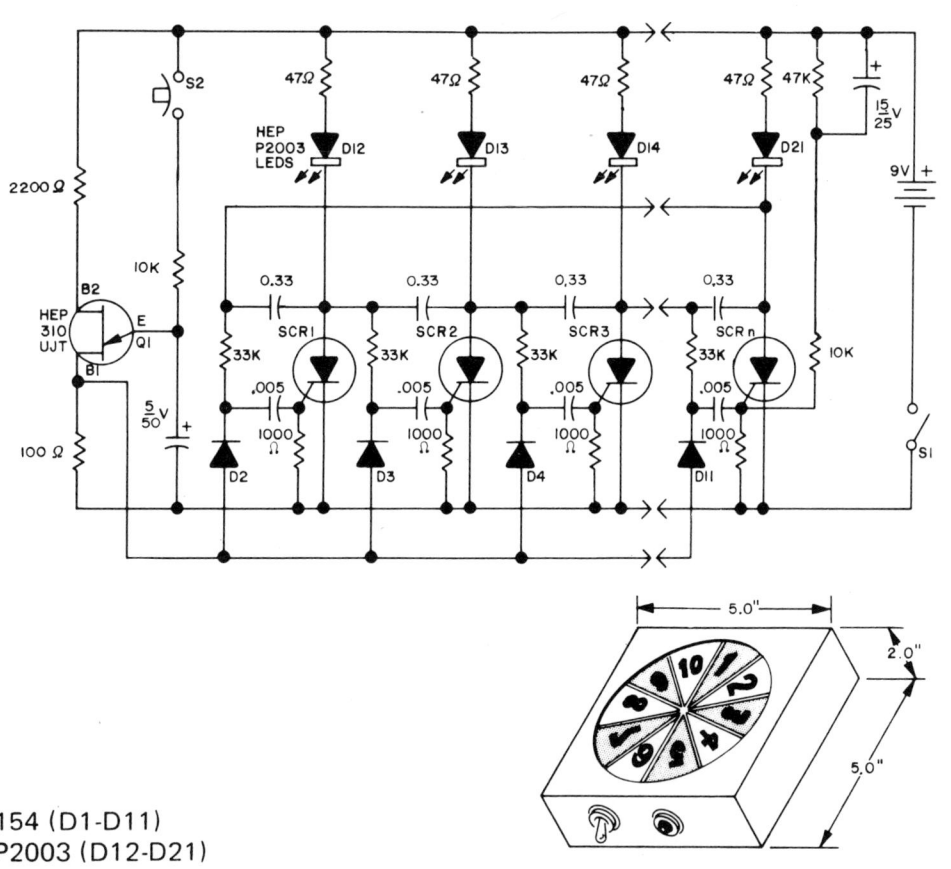

- 10 HEP 154 (D1–D11)
- 10 HEP P2003 (D12–D21)
- 1 HEP 310 (Q1)
- 10 HEP R1001 (SCR1–SCR10)
- 11 resistors, 1000 Ω
- 10 resistors, 33k
- 10 resistors, 47 Ω
- 1 resistor, 47k
- 1 resistor, 100 Ω
- 1 resistor, 2200 Ω
- 2 resistors, 10k
- 10 capacitors, 0.33 μF, 50V
- 10 capacitors, 0.005 μF, 50V, calectro A1-125
- 1 electrolytic, 5 μF, 50V
- 1 electrolytic, 15 μF, 25V
- 1 switch, Calectro E2-110 (S1)
- 1 pushbutton, Calectro E2-140 (S2)
- 1 battery, 9V

SECTION **16** *Circuits Hobby, Fun, and Experimenter*

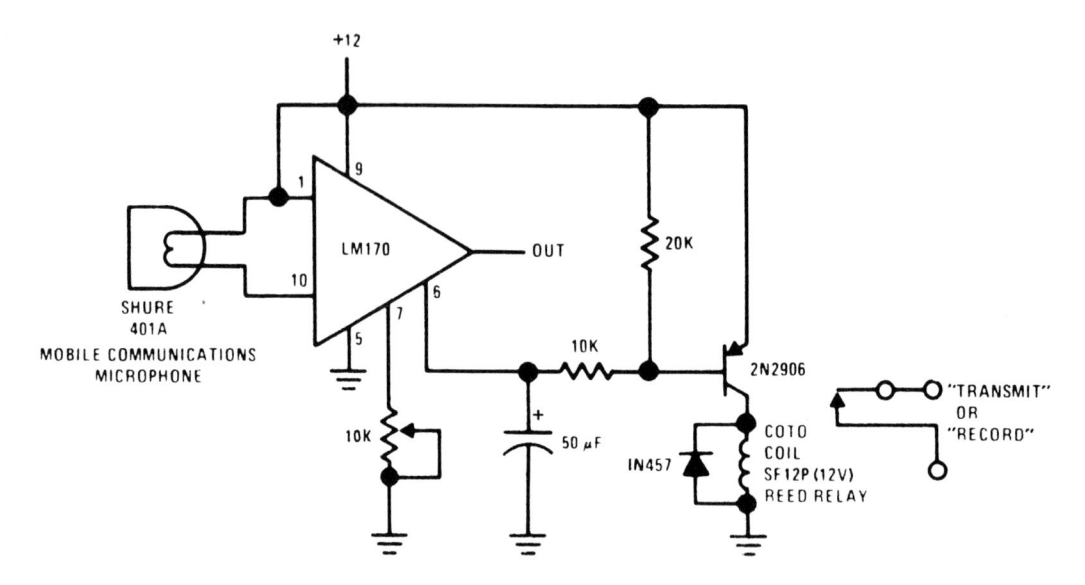

This transmitter or tape recorder voice-operated switch can switch high-powered electronic or electromechanical devices. Automatic transmit/receive operation is possible in communication systems or tape recorder motors and may be switched on at the first syllable of infrequent speech, such as in dictation. To handle large amounts of power, all that is needed is a small PNP power transistor driving a relay.

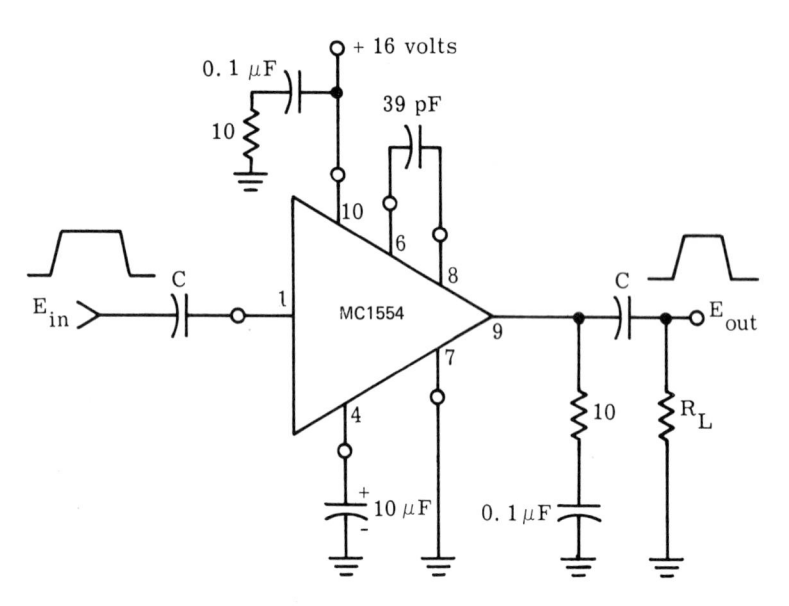

NOTE: Voltage Gain = 18 For Connection Shown

Pulse power amplifier delivers up to 3W peak. (Do not allow current to exceed 500 mA.)

538 IC SCHEMATIC SOURCEMASTER

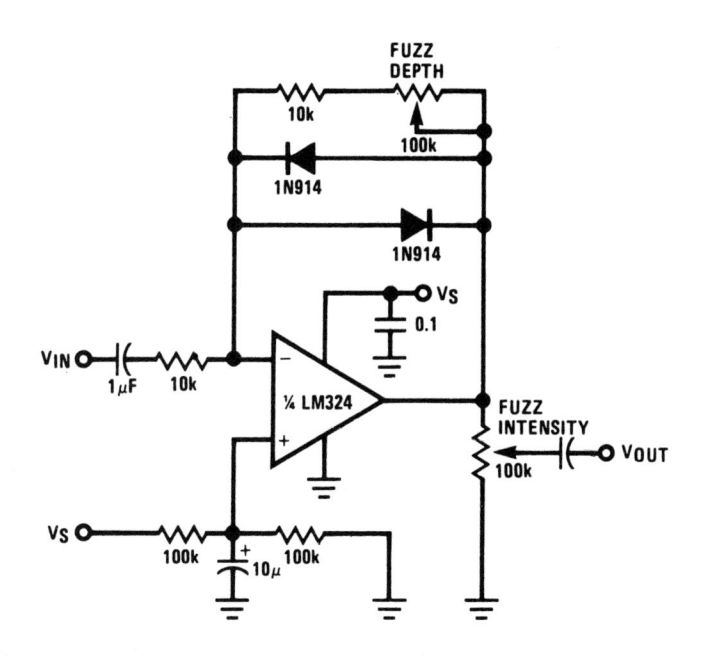

Fuzz box. Two diodes in the feedback of an LM324 create the musical instrument effect known as *fuzz*. The diodes limit the output swing to ±0.7V by clipping the output waveform. The resultant square wave contains predominantly odd-ordered harmonics and sounds similar to a clarinet. The level at which clipping begins is controlled by the *fuzz depth* pot, while the output level is determined by *fuzz intensity*.

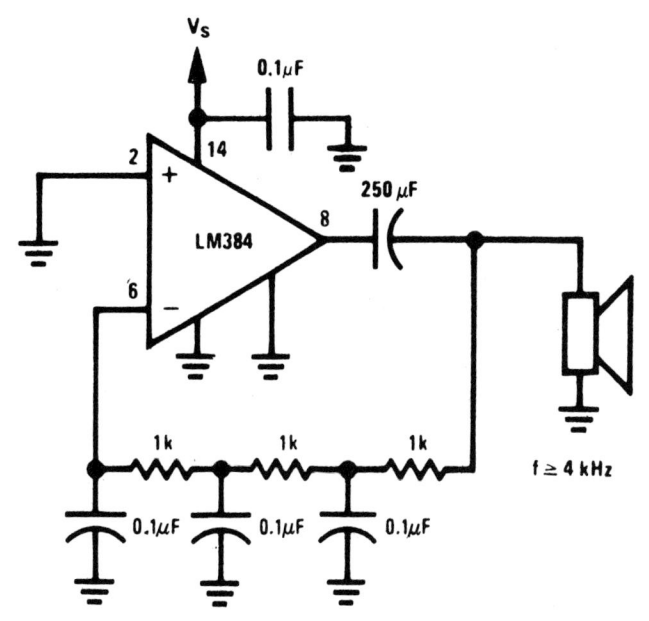

Phase-shift oscillator generates a clean sine wave at 4 kHz for code practice or single-tone-burst communications systems.

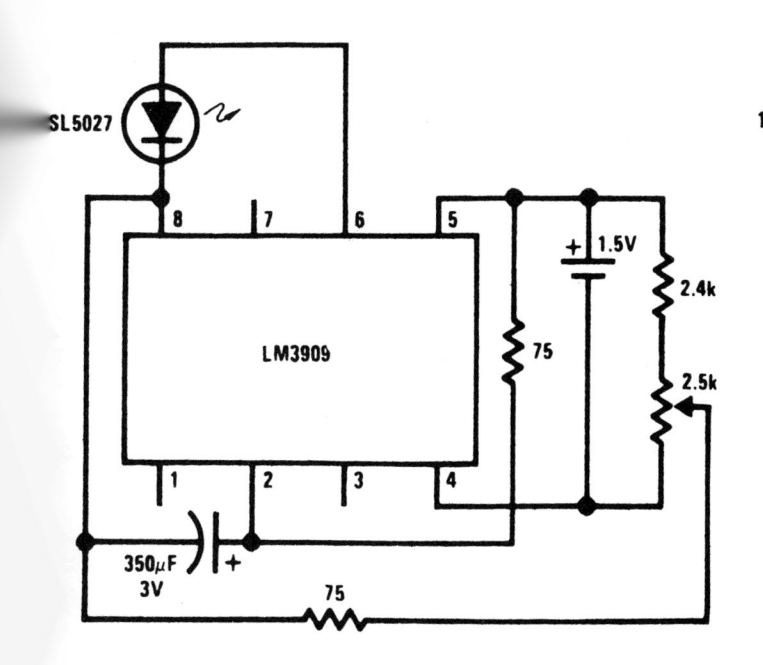

Variable flasher with adjustable 0 to 20 Hz rate.

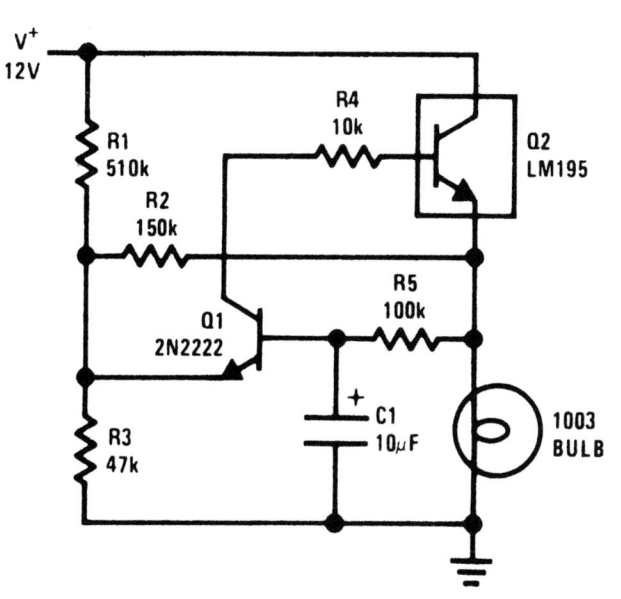

1.0A lamp flasher.

SECTION **16** *Circuits Hobby, Fun, and Experimenter* **539**

PARTS LIST FOR LED DICE:

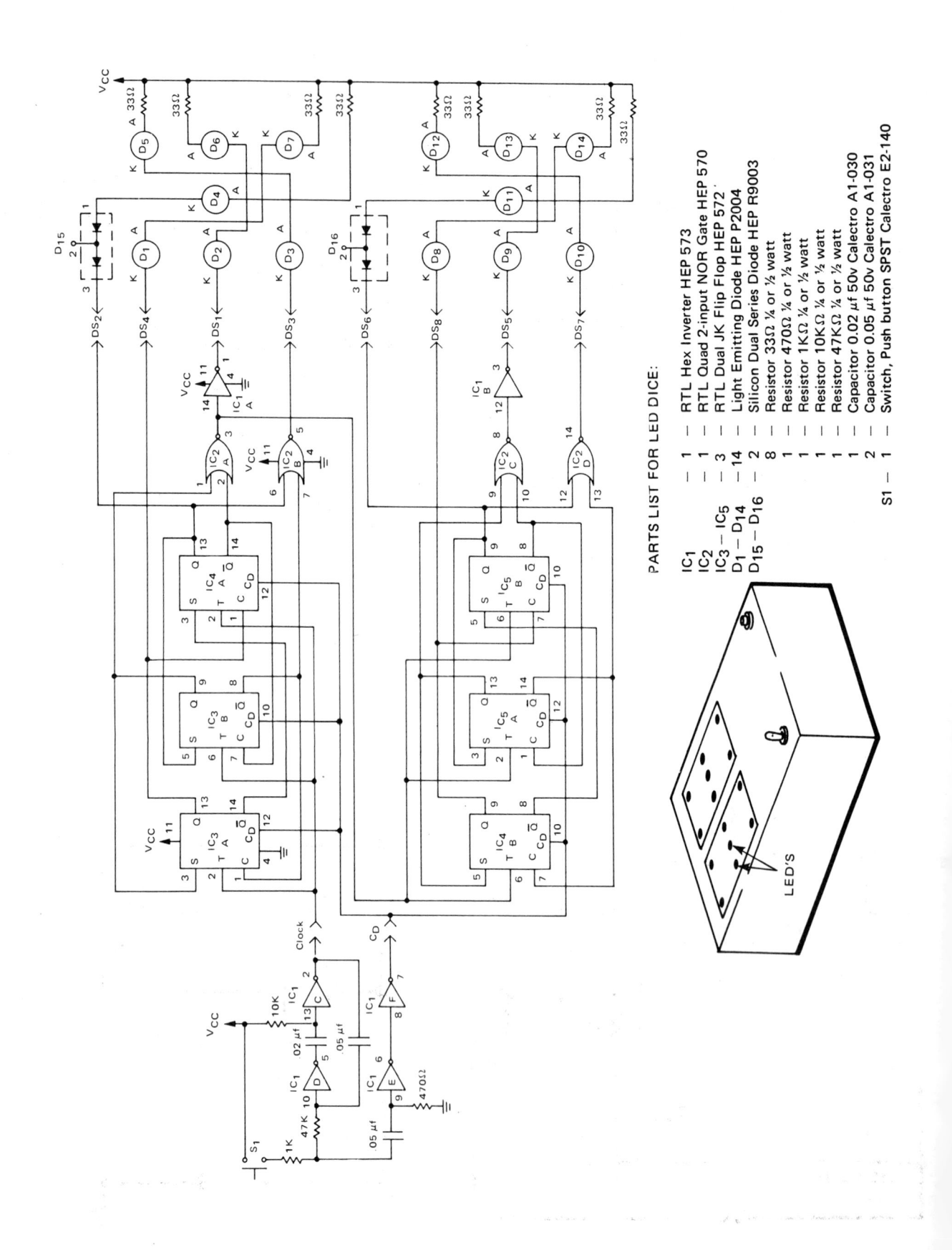

IC$_1$	—	1	— RTL Hex Inverter HEP 573
IC$_2$	—	1	— RTL Quad 2-input NOR Gate HEP 570
IC$_3$ – IC$_5$	—	3	— RTL Dual JK Flip Flop HEP 572
D$_1$ – D$_{14}$	—	14	— Light Emitting Diode HEP P2004
D$_{15}$ – D$_{16}$	—	2	— Silicon Dual Series Diode HEP R9003
		8	— Resistor 33Ω ¼ or ½ watt
		1	— Resistor 470Ω ¼ or ½ watt
		1	— Resistor 1KΩ ¼ or ½ watt
		1	— Resistor 10KΩ ¼ or ½ watt
		1	— Resistor 47KΩ ¼ or ½ watt
		1	— Capacitor 0.02 μf 50v Calectro A1-030
		2	— Capacitor 0.05 μf 50v Calectro A1-031
S$_1$	—	1	— Switch, Push button SPST Calectro E2-140

540 IC SCHEMATIC SOURCEMASTER

Electronic dice. Pressing button S1 "tosses" dice. Use 4.5V power source for V_{CC}.

IC$_1$	— 1	RTL hex inverter HEP 573
IC$_2$	— 1	RTL quad 2-input NOR gate HEP 570
IC$_3$-IC$_5$	— 3	RTL dual J-K flip-flop HEP 572
D$_1$-D$_{14}$	—14	light-emitting diode HEP P2004
D$_{15}$-D$_{16}$	— 2	silicon dual series diode HEP R9003
	8	resistor 33Ω
	1	resistor 470Ω
	1	resistor 1k
	1	resistor 10k
	1	resistor 47k
	1	capacitor 0.02 μF 50V Calectro A1-030
	2	capacitor 0.05 μF 50V Calectro A1-031
S1	— 1	pushbutton, Calectro E2-140

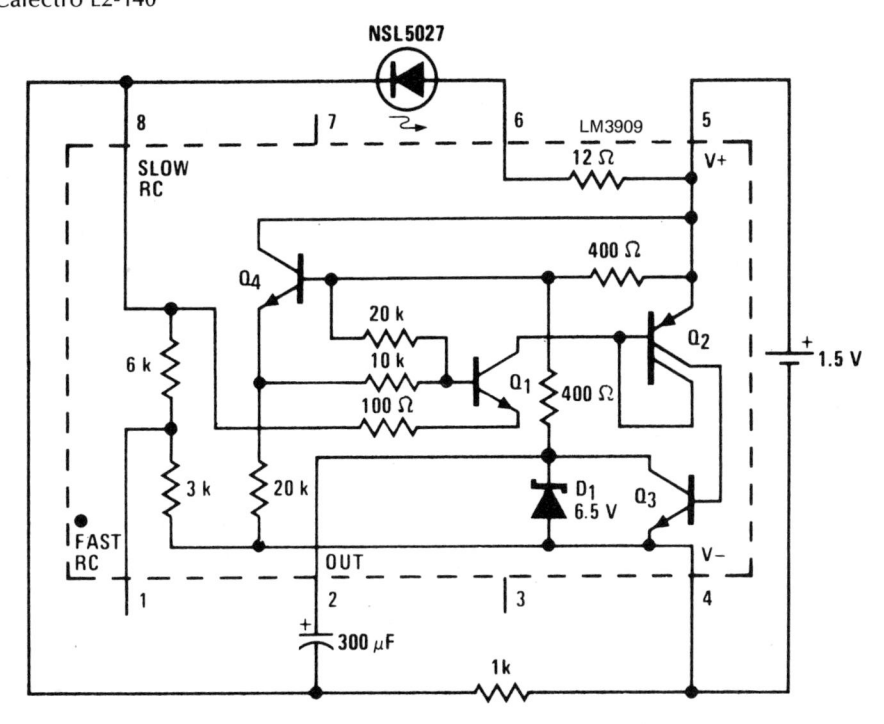

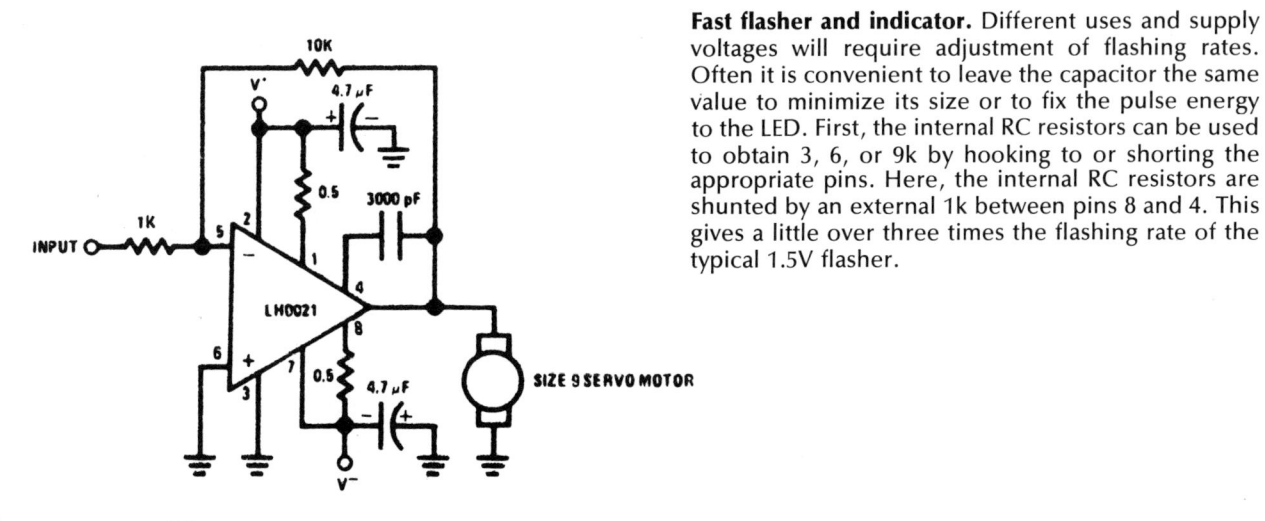

Fast flasher and indicator. Different uses and supply voltages will require adjustment of flashing rates. Often it is convenient to leave the capacitor the same value to minimize its size or to fix the pulse energy to the LED. First, the internal RC resistors can be used to obtain 3, 6, or 9k by hooking to or shorting the appropriate pins. Here, the internal RC resistors are shunted by an external 1k between pins 8 and 4. This gives a little over three times the flashing rate of the typical 1.5V flasher.

Dc servo amplifier.

SECTION **16** *Circuits Hobby, Fun, and Experimenter*

INDEX

Absolute-temperature indicator, 357
Ac amplifiers, 68, 74. See also Amplifier
Ac bridge sensor firing circuit, 345
Ac-coupled amplifier
    bootstrapped, 78
    inverting, 98
    noninverting, 100
    single-supply, 89
    see also Amplifier
Ac-coupled-signal comparator, 511
Accurate square root extractor, 502
Ac/dc biomedical amplifier, 125
Ac/dc converter, 13. See also Power supply
Ac/dc millivoltmeter, 142
Ac/dc voltmeter, 106
Acoustic-pickup preamplifier, 522
Ac servo amplifier, 534
Active-channel selector, 482
Active filters
    bandpass, 404
    bandpass (20 kHz), 405
    bass boost, 415
    high-pass (1 kHz), 417
    low-pass, 405
    see also Filter
Active tone controls, 269, 402, 410
Actuator, liquid-drain, 353
Ac voltmeter, see Voltmeter, ac
A/D converters
    CMOS, 432, 435
    comparator, 438, 439
    single-ramp, 430
    successive-approximation, 443
    with waveforms, 433
ADF oscillator, 324
Adjustable, see listings under Variable
Adjustable-duty-cycle pulse generator, 160
Adjustable-gain dc instrumentation amplifier, 123
Adjustable level translation, 465
Adjustable-Q notch filter, 403, 413, 416
Adjustable-rate
    flasher (1.5 V), 539
    siren, 525
Adjustable ripple-free regulator, 27
Adjustable threshold line receiver, 471
Adjustable voltage references, 11
AF amplifier
    floating-ground (4 W), 236
    with RF amplifier, 281
    see also Amplifier, audio
AF signal generator, 146
AF sine-wave oscillator, 113
AGC amplifier, see Amplifier, AGC
AGC/squelch, 327
AGC using built-in detector, 326
AGC'd audio amplifier, 225
AGC'd differential amplifier, 295
Air
    movement detector, 347
    velocity indicator, 396
Aircraft
    ADF oscillator, 324
    synthesizer element, 487
Airspeed indicator, 396
Alarm
    automotive overspeed, 387, 394
    burglar, 368
    fluid level, 341
    fluid seepage, 342
    intrusion, 368
    power shutoff, 428
    seepage for fluids, 342

siren, 524
temperature, 340, 371
universal, 359
ALC for cassette deck, 246
Alternating flasher, 387, 530, 541
AM
    amplifier, gain-controlled, 314
    demodulator, synchronous, 302, 330
    detector, synchronous, 330
    detector and amplifier, 321
    detector and converter, 316
    -FM IF amplifier/detector, 306
    -FM-phono-tape amplifier, 274
    -FM reverberation unit, 271
    IF strip (455 kHz), 332
    modulators, 308, 315, 323, 325, 333
    for 6 meters, 295
    radio, 297, 298, 307, 310, 320, 325
        automotive, 389
        converter, 316
        low-cost, 307
        using varactors, 297
Amateur-repeater discriminator window, 316
Ammeter
    100 nA full scale, 125
    1 mA to 10 A full scale, 10
Amperex amplifiers, 243
Amplifier, ac, 74
Amplifier, ac coupled, 78, 89
Amplifier, AGC, 299
    45 MHz, 284
    differential, 295
Amplifier, analog, logic controlled, 493
Amplifier, audio
    600 nW, 68
    500 mW, 99, 230
    1 W, 241, 276
    2 W, 231, 233, 246
    2 W TV sound IF, 254
    2 W phono, 231
    2-4 W stereo, 234
    2.5 W bridge-type, 238
    3 W power, 71
    3.5 W bridge-type, 245
    4 W, 235, 244, 279
    4 W stereo, 265
    4 W stereo (8-track), 232
    5 W, 224, 275
    5 W with PCB layout, 247, 248
    6 W hi-fi, 266
    8 W stereo, 270
    10 W, 229, 238
    12 W (8 ohms), 239
    12 W stereo, 266
    15 W hi-fi, 251
    20 W stereo, 237
    20 W (bass-boosted) stereo, 240
    30 W hi-fi, 261
    60 W stereo, 261
    90 W hi-fi (4-Ω), 250
    90 W power, 241
    with AGC, 225
    ambience (4-channel), 265
    automotive hi-fi, 393
    bass-boosted, 234, 263, 268
    battery-operated, 229
    booster, 240
    bridge-type, 69, 79, 224, 236, 262, 263, 271
    crystal-pickup, see Amplifier, phono
    gain-controlled, 255
    gain-of-20, 238, 244
    gain-of-200, 238, 239

    headphone (stereo), 226
    high-gain, 249
    line-operated, 249
    low-distortion, 81, 253
    and noise squelch, 292
    with output of 500 mV, 239
    public-address (mini), 260
    rear-channels (quad stereo), 265
    simple, 224, 274
    and squelch (FM), 296
    stereo (4 W), 232, 234, 265
    stereo (8 W), 270
    stereo (12 W), 266
    stereo (20 W), 237, 240
    stereo (60 W), 261
    stereo, with bass boost, 268
    stereo headphone, 226
    stereo (multiple-input), 274
    with supply of 36V, 262
    with tone controls, 249. See also Filter
    unity gain, 137. See also Voltage follower; Follower; and Unity-gain
    see also Hi-fi amplifier
Amplifier, balanced, 83
Amplifier, balanced-bridge, 69
Amplifier, bandpass
    10.7 MHz, 400
    see also Filter
Amplifier, biomedical (40 dB), 125
Amplifier, bootstrapped ac, 68
Amplifier, bridge
    1-3.5 W, 245
    2.5 W, 238
    4 W, 235, 244
    10 W (8 ohms), 238
    12 W, 263
    high-impedance, 79, 236
    single-supply, 132
Amplifier, bridge (current), 103, 121
Amplifier, bridge-transducer, 69
Amplifier, buffer, 75, 100
Amplifier, cascade, 71
Amplifier, cascode, 90
    10.7 MHz IF, 288
    50-100 MHz RF, 295
Amplifier, CMOS (single-supply), 86
Amplifier, compensated
    dc (drift), 73
    NAB, 239, 257, 270
    noninverting, 85
    for thermocouple, 134
Amplifier, complementary-symmetry
    12 W hi-fi, 239
    15 W hi-fi, 251
    see also Amplifier, audio
Amplifier, crystal-pickup, 110, 231, 244, 246, 251, 256, 258, 263, 266, 276
Amplifier, crystal-transducer, 110
Amplifier, dc
    drift-compensated, 73
    servo, 541
    summing, 441
Amplifier, detector, see Detector/amplifier
Amplifier, differential, 70, 87, 93, 101
    guarded, 78
    variable-gain, 93
Amplifier, driver, 245
Amplifier, FET operational, 89
Amplifier, floating-input, 121
Amplifier, gain-of-1 (inverting), 90
Amplifier, gain-of-5 (difference), 70

543

Amplifier, floating-input, 121
Amplifier, gain-of-1 (inverting), 90
Amplifier, gain-of-5 (difference), 70
Amplifier, gain-of-9, 71
Amplifier, gain-of-10, 70
Amplifier, gain-of-20, 238, 244
Amplifier, gain-of-35, 66
Amplifier, gain-of-100 (operational), 79
Amplifier, gain-of-200, 238, 239
Amplifier, general-purpose, 81
Amplifier, ground-referenced, 92
Amplifier, high-speed and low-drift, 66, 80
Amplifier, hi-fi, see Hi-fi amplifier
Amplifier, IF
  2-stage, 297
  455 kHz, 285
  10.7 MHz, 286, 290, 291, 318
    cascode, 288
  60 MHz, 297
  and audio amplifier for TV, 2 W, 254
  and converter, 316
  and detector
    AM/FM, 306
    with stereo decoder, 309
  and limiter (10.7 MHz), 317
  strip
    455 kHz, 332
    10.7 MHz, 290, 291
  with AGC, 284
  differential mode, 286
  dual-conversion, 329
  FM, 303, 333
  quadrature-detector, 306
  single-sideband, 329
Amplifier, impedance adapter and, 520
Amplifier, instrumentation, 84, 88, 116, 120, 126, 127, 131, 133, 136, 138, 139
  20 db, 121
  34 V, 110
  40 dB, 125
  x100, 108, 121
  dc, 120
  differential-type, 116, 139
  high-common-mode rejection ratio, 140, 141
  open-loop, 130
  variable-gain, 123, 141
  wide-range, 140
Amplifier, inverting 98, 357
  1W, 66
  20 dB, 69
  ac-coupled, 98
  balanced, 83
  fast, 67
  guarded, 78
  high-impedance, 103
  low-distortion, 81
  and noninverting, 82, 259
  stereo, 234
  unity-gain, 90
Amplifier, level-shifting, 87
Amplifier, level-isolation, 83
Amplifier, line-operated, 249
Amplifier, logarithmic, 515
Amplifier, low-drift, 66, 80, 93
Amplifier, meter, 108, 121, 125
Amplifier, mike, 256
Amplifier, noninverting, 83
  1W, 67
  ac-coupled, 100
  alternating-current, 100
  compensated, 85
  guarded, 77
  high-impedance, 102
  high-voltage, 82
  stereo, 234
Amplifier, offset-adjusted, 85, 112
Amplifier, operational, see Op-amp

Amplifier, pH meter, 128
Amplifier, phono, 243, 244
  crystal/ceramic, 244, 246, 251, 254, 256, 263, 266, 274, 276
  2 W, 231
  4 W, 279
  34 dB of gain, 258
  magnetic-pickup, 253, 254, 257
  500 mV output, 253
  with tone controls, 251, 263
  with tone and volume controls, 254
  see also tape
Amplifier, photovoltaic, 422
Amplifier, picoampere-level, 128
Amplifier, power, 94, 294
  1 A, 99
  30 V, 264
  36 V (audio), 262
  audio, 274
    2 W, 246
    5 W, 224, 275
    90 W, 241, 250
    for 30 V supply, 264
    for 36 V supply, 262
  differential, 71
  low-distortion, 253
  operational (op-amp), 86
  pulse, 538
  single-supply, 71
  split-supply, 71
  switching, 92, 354, 368
Amplifier, power (switching), 92, 354, 368
Amplifier, pulse, 504, 538
Amplifier, radiation-detector, 128
Amplifier, record/playback, 243, 246, 257, 270
Amplifier, recovery, 245, 271
Amplifier, rectified-line-voltage, 249
Amplifier, reset-stabilized, 89
Amplifier, RF, 281
  dc to 120 MHz, 281, 295
  15 MHz, 295, 335
  45 MHz wideband, 284
  58 MHz wideband, 309
  100 MHz wideband, 101
  with AGC, 284
  and doubler, 292
  VHF (various ranges), 318
Amplifier, sense, 95
Amplifier, servo, 534
  30 V (8 ohms), 264
  35 W, 520
  8-ohm, 30 V supply, 264
  ac, 534
  bridge-type, 244
  dc, 541
  see also Audio
Amplifier, solar-cell, 95
Amplifier, source follower, 91. See also Follower
Amplifier, squaring, 87
Amplifier, summing, 76, 89, 441
Amplifier, symmetrical, 103
Amplifier, tape-head, 243, 246, 257, 270
Amplifier, tape playback, see Tape playback, amplifier
Amplifier, thermocouple, 126, 133, 134, 137
Amplifier, transducer, 85, 110
Amplifier, tremolo, 82
Amplifier, tuned-in/tuned-out, 323
Amplifier, VHF, 318. See also Amplifier, RF
Amplifier, video (20 dB), 222. See also Wideband
Amplifier, voltage-controlled, 82
  variable, 97
  see also VCA
Amplifier, wideband, 93
  45 MHz, 284

  58 MHz, 309
  60 MHz (IF), 297
  high-impedance, 101
  RF/IF (30 dB), 281
Amplifier, zeroing, 131
Amplifier/detector
  FM/
  FM/AM, 306
  and limiter, 303
  and stereo decoder, 309
  see also Amplifier, IF
Amplifier/doubler, 292. See also Amplifier, RF
Amplifier/limiter
  500 kHz, 311
  10.7 MHz, 317
  see also Amplifier, IF
Amplifier/mixer, 20 dB audio, 252
Amplifying buffer, 69
Amplitude modulator, 295, 308, 315, 323, 325, 333
Analog, and digital circuits, 428
Analog ac voltmeter, 115, 138
Analog ammeter, 1 mA to 10 A, 110
Analog amplifier, logic-controlled, 493
Analog commutating strobe, 476
Analog commutator (4-channel), 460
Analog frequency meter, 111
Analog meter amplifier, 125
Analog multiplier, 425
Analog multiplier/divider, 496
Analog switch, see Switch, analog
Analog tachometer, 393
  buffer, 463, 471
  differential-type, 468
  four-channel, 460
  high-rate, 464
  TTL/DTL, 463
  with zero-crossing detector, 475
Analog thermometer, 116-118, 371, 535
Analog timer, 376
  for long delays, 374
Analog-to-digital, see A/D convertors
Analog voltmeter, see Voltmeter, analog
AND
  gate, 515
    large fan-in, 515
    3-input, 512
  muting, 296
Antenna impedance matcher and adapter, 520
Antilog generator, 499, 502
Antiskid, automotive, 384, 294
Approximation, successive, 443
Arithmetic square-rooter, 502
Arithmetic subtractor, high-speed, 493
Astable multivibrator, see Multivibrator, astable
Astable repeat cycle timer, 380
Attack/release muting, 520
Attenuator, variable, 536
Audible liquid-level alarm, 353
Audio, see Amplifier, audio; and Stereo
Audio alarm and fluid level sense, 341, 353
Audio amplifier, see Amplifier, audio
Audio bass boost, 415
Audio booster, 268
Audio booster amplifier, 240
Audio circuits, 223
Audio driver for car radio, 397
Audio dynamic range expander, 255
Audio equalizer, 10-band, 273
Audio mixer, see Mixer, audio
Audio output stages, 2 W, 266
Audio panner, 2-channel, 275
Audio playback amplifier, tape, 270
Audio power amplifier, see Amplifier, power, audio

Audio preamplifier
  fixed-gain, 270
  low-noise, 229
  and mixer, 276
  stereo, 260
  with tone controls, 269
Audio reverb unit, stereo, 245, 271
Audio signal generator, 146, 337
Audio system, 2-channel, 274
Audio tone controls, 402, 410. See also Stereo and Preamp
Audio oscillator (tone unit), 336
Automatic direction finding, 324
Automatic gain control, see Amplifier, AGC
Automatic level control, tape, 246
Automatic power shutoff/alarm, 428
Automatic-wail siren, 534
Automotive
  AM radio, 325, 388, 389, 397
  clutch/transmission, 392
  external-temperature readout, 386
  IF strip, 306
  flasher, 387, 395
  hi-fi amplifier (7 W), 393
  high-speed warning (overspeed), 387, 394
  overspeed latch, 391
  reverb, monaural, 371
  speed switch, 385
  stereo audio booster, 268
  tach converter, 397
  tachometer, 384
  thermometer, 386, 396
  zero-crossing detector, 446, 459
Autopolarity DVM (3.5-digit), 114
Autopolarity voltmeter, 107

Balance
  control, 224
  indicator (bridge), 142
  offset, 131
  and tone controls, 267
Balanced bridge amplifier, 69
Balanced differential amplifier, 295
Balanced line mike preamplifier, 256, 277
Balanced mixer, see Mixer, balanced
Balanced modulator, 291, 311
Balancer, offset, 93, 131
Bandpass amplifier, 10 MHz, 400
Bandpass filter, see Filter, bandpass
Band reject, see Bandstop/filter; and Notch/filter
Bandstop filter, 400, 403, 406, 412
Basic voltage regulator, 27
Basic window detector, 458
Bass, treble, midrange controls, 402
Bass boost circuit, 415
Bass-boosted audio amplifier, 234, 240, 263, 268
Bass-boosted stereo, 240, 268
Battery charger (60 Hz), 12
Battery charger regulator, 392
Battery-operated audio amplifier, 229
Battery-operated intercoms, 283
Battery-operated long-time delay, 376
Battery-operated meter amplifier, 125
Battery-operated seepage alarm, 342
Battery voltage monitor, 350
Baxandall tone control, 251
BC, see Broadcast radio
BCD analog-to-digital converter, 437, 443
BCD comparator for A/D converter, 439
Bench supply
  dual, 9
  variable (25 V), 61
Biamplifier, 5 W/5 W, 242
Bilateral current source, 49
Binary-coded decimal, see listings under BCD
Binary-controlled 4-channel multiplexer, 449
Binary ladder in D/A converter, 436
Binary multiplying D/A converter, 437
Binary-weighted network, see Network, binary-ladder
Biomedical 40 dB amplifier, 125
Bipolar output reference, 63, 133. See also Reference
Biquad filter, see Filter, biquad
Bistable multivibrator, 505
Blinker/flasher
  1.5 Hz rate, 524
  for models, 533
Boost circuit, bass, 234, 415
Boost/cut, treble, for stereo, 261
Boosted-bass audio amplifier, 234, 240, 263, 268
Booster
  audio, 268
  current, 50
  op-amp power, 89
  stereo (20 W), 240
  see also Power booster
Booster amplifier, 240
Booster follower, 69
Bootstrapped ac amplifier, 68, 78
Bounceless switch, 363
Bridge
  full-wave power supply, 2
  Wein (oscillator), 113, 147, 149, 150, 152, 158, 168, 169
Bridge amplifier, see Amplifier, bridge
Bridge balance indicator, 142
Bridge current amplifier, 103, 121
Bridge-transducer amplifier, 69
Broadband DSB modulator, 311
Broadcast radio
  AM, 297, 298, 307, 310, 320, 325
  FM, 318
  FM front end, 308
Buffer
  amplifying, 69
  analog switch, 471
  digital-to-analog-converter, 440
Buffer amplifier, 75, 100
Buffered analog switch, 463
Buffered current monitor, 48
Buffered positive-peak detector, 467
Buffered reference, 59
Buffered tone control, 249
Burglar alarm, 368
Butterworth filter, 4-pole, 407
Button-controlled siren, 530
Buzz box (continuity tester), 146, 525

Cable driver, 475, 477, 483
Calibrator
  crystal (100 kHz), 113, 293
  portable, 134, 139
Capacitance multiplier, 509, 516
Capacitive-load isolator, 93
Capacitor-input power supply, 2
Car, see Automotive
Car radio, 388, 389
  amplifier (4 W), 386
  audio driver, 397
Carrier-operated relay, 323
Carrier regenerator, 321
Carrier suppressor (DSB), 311
Cascade differential amplifiers, 70, 71
Cascode amplifier, see Amplifier, cascode
Cassette tape player
  power booster, 240
  record/play amplifier, 243, 246
CATV switching supply, 80 W, 222
Cell replacement standard, 60

Celsius thermometer, 121, 348, 535
Celsius/Fahrenheit thermometer, 122
Centigrade, see listings under Celsius
Ceramic, see listings under Crystal
Ceramic-pickup phono amplifier, 231, 244, 246, 251, 263, 266, 276
Channel selector (dc controlled), 237
Charger, battery (60 Hz), 12
Charger regulator, automotive, 392
Checker, continuity, 146, 525
Choke-input power supply, 2
Chopper for dual-trace scope, 463
Circuit interrupter, ground-fault, 124
Clamp, precision 5 mA, 132
Clamped diodes, 483
Class A audio amplifier, see Amplifier, audio
Class B audio amplifier, 10 W, 229
Clock driver
  MOS, 492
  TTL, 153
  TTL to MOS, 495
Clutch/transmission control, 392
CMOS
  analog-to-digital converter, 432, 435
  digital-to-analog converter, 434
  digital voltmeter (4.5-digit), 143
  follower amplifier, 137
  operational amplifier, 86
  phase-reversal detector, 351
  postamplifier, 130
  power source, 5
  programmable comparator, 501
  regulated supply, 4
  sample and hold, 453
Coaxial-cable driver, 475, 477, 483
Code-practice oscillator, 296, 336
Coil continuity checker, 146, 525
Cold-junction compensated amplifier, 134
Common-mode
  instrumentation amplifier, 110
  rejection op-amp, 100
  volume and tone controls, 224
Communications circuits, 281
Communications discriminator window, 316
Communications-receiver RF amplifier, 337
Commutating analog strobe, 476
Commutator, analog, 460, 464
Comparator, 423, 517
  ac-signal, 511
  A/D converter, 438, 439
  dc-input (one-shot), 506
  differential-voltage, 461
  dual-limit, 504
  frequency, 489
  frequency doubler with, 334
  go/no-go, 511
  high impedance (with offset adj), 511
  with hysteresis, 510
  lamp driver and, 353
  limit (with lamp driver), 349
  low-input-bias, 507
  low-voltage, 517
  opposite polarity, 506, 511
  power, 517
  programmable, 501
  ramp, 430
  relay driver and, 516
  RTL voltage, 490
  slope (for FSK system), 461
  with solenoid driver, 346
  temperature-sensing, 371
  threshold combination, 490
  trilevel, 500
  TTL-compatible, 491
  TTL-driving, 512
  voltage, 456, 499, 510
  voltage (DTL/TTL), 490
Compensation, frequency, 112

INDEX 545

Compensated amplifier, see Amplifier, compensated
Compensated shunt regulator, 26
Compensated summer, 89
Compensated tape preamplifier, 270
Compensated thermocouple amplifier, 134
Compensator, line-resistance, 474
Complementary-symmetry amplifier, see Amplifier, complementary-symmetry
Composite AM/FM IF amplifier/detector, 306
Conditioner
  premultiplexing, 126
  pulse, 453
Constant-current and voltage supply, 3, 29
Constant-on flashlight finder, 526
Constant-speed motor drive, 533
Constant-voltage power supply, 3
Contact switch, 6, 535
Continuity checker, 146, 525
Continuous wave, 294
Control
  antiskid, automotive, 384
  balance, 224
  bass, 268
  clutch/transmission, 392
  dc gain, noninverting, 99
  equalizer, 273
  heater, 3-phase, 366
  phase, 360
  power, zero-voltage switching, 353
  proportional-speed, 364
  repeater, 323
  speed, motor, 349, 351, 371
  synchronous heat staging, 360
  temperature, with hysteresis, 341
  thermocouple temperature, 345
  tone, 224, 249, 251, 261, 269
  volume, common-mode, 224
  zero-voltage switching, 353, 366, 367
Control center, stereo preamp and, 269, 278
Controlled-frequency multivibrator, 154
Controlled-hysteresis squelch, 323
Controller, light, 422
Controller, light-activated triac, 425
Controller, temperature, 349, 359, 370
  proportioning, 344
  sensitive, 352
Control switching with triacs, 343
Control triac, power, 428
Counter, programmable down-, 487
Counter, pulse, 162, 163
Converter
  ac-to-dc (fast), 13
  A/D and D/A, 429
  A/D (CMOS), 432
  A/D (single-ramp), 430
  D/A, 434, 440
  D/A (binary-weighted), 441
  D/A (buffered), 440
  D/A high-speed, 442
  D/A (temperature-compensated), 436
  D/A (using ladder network), 441
  high-speed 12-bit, 433
  IF amplifier and, 316
  log (1 quadrant), 492
  log (root extractor), 505
  sine-to-square, 87
  temperature-to-frequency, 386
Converter/detector, 455 kHz, 287
Converter/mixer, 100 MHz, 299
COR, 323
Crossover, biamplifier-type, 242
Crowbar regulator
  5 V, 19
  6 V, 56
  SCR, 356
Crystal calibrator, 100 kHz, 113, 293

Crystal cartridge, see Crystal-pickup amplifier; Crystal-pickup preamplifier
Crystal-controlled logic generator, 494
Crystal-controlled oscillator, 153, 157, 159
  100 kHz, 158, 169
  1 MHz, 155
  10.7 MHz, 156, 157
  and calibrator, 293
  square-wave, 160
Crystal-pickup amplifier, 110, 231, 244, 246, 251, 256, 258, 263, 266, 276
Crystal-pickup preamplifier, 244
Crystal-set radio with amplifier, 321
Crystal-transducer amplifier, 110. See also Crystal-pickup amplifier
Current amplifier, bridge, 103, 121
Current-boosted voltage follower, 72
Current booster, 50
Current-compensated follower, 93
Current-controlled oscillators, 161
Current-level detector, 423
Current-limited audio amplifier, 262
Current limiter, switchback, 47
Current limiting regulator, 21, 31
Current limiting sources, 15
Current-mode multiplexer, 463
Current monitor, 48, 346
Current-rated circuits
  100 nA full-scale meter, 121, 125
  100 uA microammeter, 110
  1 mA full-scale milliammeter, 110
  1 mA IC power supply, 58
  5 mA clamp (current sink), 132
  10 mA current regulator, 49
  15 mA switching regulator (200 V), 17
  40 mA variable-voltage regulator, 30
  50 mA variable-voltage regulator, 56
  90 mA variable-voltage power supply, 10
  100 mA current booster, 50
  100 mA current regulator, 56
  100 mA regulator (0-250 V), 37
  200 mA regulator, 43, 44
  250 mA regulator, 28
  250 mA regulator and power supply (15 V), 6
  500 mA regulator (5 V), 41
  500 mA regulator (300 V), 33
  1 A lamp flasher, 539
  1 A power amplifier, 99
  1 A reference (20 V), 63
  1 A regulator (high-stability), 45, 46
  1 A regulator (negative), 55
  1 A regulator (positive), 55
  1 A regulator and power supply, 4
  1 A regulator and power supply (65 V), 5
  1 A voltage follower, 66
  2 A current regulator, 42
  2 A shunt regulator, 44
  2 A variable regulator (20 V), 36
  3 A switching regulator, 14, 20, 44
  4 A switching regulator, 14
  5 A constant-current source, 29
  6 A variable-voltage regulator, 22
  10 A full-scale ammeter, 110
  10 A regulator (5 V), 43
  10 A regulator (15 V), 33
  10 A regulator (foldback-limited), 45
Current regulator, see Regulators, rated by current
Current, sink, see Sink, current
Current source, see Source, current
Current-switching triac, 428
Current-varying temperature indicator, 362
CW 40m QRP transmitter, 294

DAC, see D/A converter
D/A converter, 434

  4 to 10 bits, 440
  9-bit, 434
  BCD-multiplying, 437
  buffer for, 440
  high-speed, 442
  with ladder network, 441
  strobing device for, 115
  temperature-compensated, 436
Data-processing data selector, 503
Data selector, one-of-N, 503
Data transmission system, 468
dB, 269
Dc/ac milliammeter, 142
Dc amplifier, see Amplifier, dc
Dc biomedical amplifier, 125
Dc-controlled channels elector, 237
Dc-coupled low-pass filter, 418
Dc gain control, noninverting, 99
Dc input comparator and one-shot, 506
Dc instrumentation amplifier, 120. See also Amplifier, instrumentation
Dc microammeter, 126
Dc-output RMS detector, 135
Dc restoration, 330
Dc servo amplifier, 541
Dc summing amplifier, 441. See also Summer
Dc-to-120-MHz oscillator/mixer, 295
Dc-tuned table radio, 297
Dc voltmeter, analog, 123, 139
Decade multiplier, 304
Decade sine-wave oscillator, 120
Decibel meter, 269
Decoder, stereo, 305, 309, 315, 456
Decoder, Touch-Tone, 527
Decoupling, power supply, 270
Delay circuit, audio, 271
Delay circuits, time, 374
Delayed-dropout timer, 374
Delay-line substitute, 486
Demodulator
  as balanced mixer, 329
  FM, 300
  FSK, 304, 327, 330
  IRIG channel 13, 335
  satellite picture, 319
  SSB suppressed-carrier, 326
  stereo, 305, 309, 320
  see also PLL demodulator
Detector
  air motion, 347
  burglar alarm, 535
  differential threshold, 462
  envelope, 476
  ground-fault, 124
  level (logic), 422
  line-phase reversal, 351
  liquid-level, 347
  minimum-frequency, 452
  negative-peak, 483
  nuclear-particle, 364
  overtemperature, 359
  peak, 325, 467, 470, 471, 475, 483
  product, 305, 321, 326
    balanced mixer, 329
    FSK, 330
    SSB, 305, 321, 330
  pulse-width, 469
  quadrature, 306
  regulated-output threshold, 472
  synchronous AM, 330
  temperature (remote), 371
  true—RMS, 135
  window (TTL), 458
  zero-crossing, 113, 446, 448, 457, 459, 466, 471, 475
Detector/amplifier
  AGC, 326
  AM/FM IF, 306

limiter, 303
  and stereo decoder, 309
Detector/converter
  455 kHz, 287
  preamplifier, 316
Detector for magnetic transducer, 115, 128
Differential amplifier, see Amplifier, differential
Differential analog switch, 468
Differential bridge amplifier, 132
Differential comparator, 365
Differential integrator, 162, 163
Differential-input instrumentation amplifier, 84, 136, 139, 141
Differential input/output amplifier, 70
Differential multiplexer, wideband, 457
Differential thermometer, 117
Differential threshold detector, 462
Differential voltage comparator, 461
Digital and analog circuits, 429
Digital speedometer, 394
Digital thermometer (°F/°C), 122
Digital-to-analog, see D/A converter
Digital transmission isolators, 464, 473
Digital voltmeter, see Voltmeter, digital
Dimmer, lamp, 360
Discriminator, FM, 317
Discriminator, pulse-width, 473
Discriminator window, repeater, 316
Divide-by-10 prescaler, 498
Divider/multiplier
  analog, 496
  1-quadrant, 497, 499
Divider, sine-wave frequency, 486
Divider/timer chain, 377
Double-balanced mixer, 310
Double conversion, 329
Double-ended differential threshold detector, 462
Doubler
  frequency, 281, 292, 334
  frequency for microprocessor, 456
  half-wave, 2
  low-frequency, 312
  VHF (150 to 300 MHz), 313
Double sideband, see listings under DSB
Double-sideband suppressed-carrier, see listings under DSSC
Double-tuner filter, 291
Down counter, programmable, 487
Drain valve actuator, 353
Drift-compensated dc amplifier, 73
Drift-compensated sample and hold, 472
Drive and sense circuits, 481
Driver
  clock
    MOS, 492
    MOS to TTL, 507
    TTL, 153, 343
    TTL to MOS, 495
  coaxial, 127, 131, 475, 477
  gas-discharge display, 17
  ground-referred-load, 359
  ladder network, 438
  lamp, 347, 349, 353, 361, 364, 516
  LED, 346, 491
  line
    long coaxial cables, 469, 475, 477
    zero-crossing detector, 446
  meter (1 MHz), 108
  MOS clock, 492
  MOS to TTL, 507
  relay, 340, 346, 358, 361
    strobed, 347, 358
  RTL, and voltage comparator, 456
  strobed, 347, 358
  shield/line, 127, 131
  TTL clock, 153, 343

TTL to MOS, 495
Driver amplifier, reverb unit, 245
Dropout timer (relay), 374
DSB modulator, 330
  balanced, 311
  suppressed-carrier, 318
DSB regenerator (suppressed-carrier), 321
DSSC modulator, 318, 330
DSSC regenerator, 321
DTL driver zero-crossing detector, 113
DTL/TTL
  analog switch, 463
  comparator, 490, 512
  -compatible multiplexer, 449
  interface (from hi-level logic), 509
  sample and hold, 459
  voltage comparator, 490
  zero-crossing detector, 471, 474
Dual-conversion IF strip, 329
Dual-limit comparator, 504
Dual MOS-to-TTL driver, 507
Dual-output power supply, see Dual power supply
Dual-output temperature control, 361
Dual power supply
  +10 V, 4
  15 V, 4, 6
  25 V (bench-type), 9, 61
  41 V, 7
  Bench, 25 V, 61
  Bench (8 to 25 V), 9
Dual regulator, see Regulators, dual
Dual-supply
  astable-multivibrator, 158, 159
  noninverting amplifier, 67
Dual-trace scope chopper, 463
DVM, see Voltmeter, digital
Dwell/tachometer, 390
Dynamic-range expander, 255

ECL termination regulator, 512
Effective-voltage detector, 135
Electronic bounceless switch, 363
Electronic dice, 528, 540
Electronic siren, 530, 624
Electronic thermometer, 116, 118, 121, 535
Electronic thermostat, 3-wire, 356
Electronic trombone, 526
Elliptic filter (4th order, 1 kHz), 409
Emergency road flasher, 387, 395, 524
Emitter-coupled logic, 512
Emitter-follower IC, 70, 74
Encoder, tone, 146, 296, 337
Engine tachometer, 388
Envelope detector, 476
Equalized RIAA amplifier, 253
Equalized stereo tape preamplifier, 270
Equalizer, audio, 273, 410
Equalization, RIAA/NAB, 269
Error voltage generator, 431, 435
Exclusive-OR muting, 296
Expanded-scale ac voltmeter, 115, 140
Expander, dynamic-range (audio), 255
Exponential-function generator, 502
Extractor, root, 505

Fan-in AND gate, 515
Fast ac/dc converter, 13
Fast comparator, precision, 510
Fast-rate flasher, 541
Fast log generator, 495
Fast peak detector, 474
Fast summing amplifier, 441
Fast-turnon tape amplifier, 241, 275
Fast voltage follower, 74, 76
Feedback, tone compensation, 270
Feedback, transformer, in oscillator, 153
FET-input mike preamplifier, 277

FET level-isolation amplifier, 83
FET operational amplifier, 66, 76, 89
FET regulated power supply, 4
FET tone control, buffered, 249
Filter
  bandpass, 400
    1 kHz, 400, 401, 403, 404, 408, 415, 419
    20 kHz, 405
    high-Q, 405
    multiamplifier, 413
    Q of 25, 408
    three-amplifier, 413
    with waveforms, 418
  bandstop, 400, 403, 406
    and low-pass, 412
  biquad
    bandpass (1 kHz), 403
    notch (3-band), 406
    3-band, 413
  Butterworth, 4-pole (1 kHz), 407
  double-tuned, 291
  elliptic, 4-pole (1 kHz), 409
  high-pass, 419
    100 Hz cutoff, 408
    1 kHz, 417
  high-Q
    bandpass, 405
    notch, 404, 406, 412, 415
  low-pass, 405, 412, 416
    10 kHz cutoff, 405
    12 dB/octave, 314
    dc-coupled, 418
    simple, 416
    single-pole, 332
  notch, 400, 417
    60 Hz, 404, 417
    4.5 MHz, 406, 415
    adjustable-Q, 403, 413, 416
    high-Q, 404, 406, 412
    tunable, 403, 404, 414
Final amplifier, 1 W CW, 536
Find adjustment, 15 V source, 59
Finger-touch switch, 535
Fixed-voltage regulator (5 V), 50
Fixed-gain stereo preamplifier, 268
Flanger, special-effect, 523
Flanging circuit, 523
Flasher
  1 pps single-supply, 539
  automotive, 387
  electronic (6 V), 530
  fast, 541
  high-voltage, 346
  incandescent, 524
Flashlight finder, 526
Flat-response amplifier, 264
Flat-response hi-fi preamplifier, 270
Flip-flop, 505
Floating regulator
  50 V, 23
  60 V, 32
Floating-ground AF amplifier (4 W), 236
Floating-input meter amplifier, 121
Floating-load amplifier, 244
  differential-type, 70
Floating op-amp regulator, 32
Floating-speaker amplifier, 244
Fluid detector, 347
Fluid-level
  control, 350
  detector, 347
  monitor with audio warning, 341
Fluid seepage alarm, 342
FM
  audio squelch, 296
  demodulator, 300
  discriminator, 317
  discriminator window, 316

INDEX 547

front end standard broadcast, 308
generator, 313
IF, limiter, discriminator, 317
IF strip, 303, 306, 332
oscillator (52 MHz), 290
radio, 328
   limiter, 303
   reverb enhancement, 271
   subcarrier regenerator, 320
   wireless loudspeakers, 322
remote speaker system, 322
retransmitted-signal receiver, 322
retransmitter, 322
scanner noise squelch, 292
squelch, with controlled hysteresis, 323
squelched amplifier, 296
stereo receiver, complete, 328
transmitter/receiver, wireless, 322
tuner, with muting, 318
Foldback
  voltage regulator, 30
    5 V, 21
    10 A, 45
  power supply, (20 V, 60 W), 60
  power supply, current-limiting, 15
Follower
  bias-current-compensated, 93
  booster, 69
  CMOS, 137
  fast, 74
  guarded, 77
  high-current, 79
  pulse-amplifying, 504
  single-supply, 94
  source, 91
  voltage, 92
    1 A, 66
    current-boosted, 72
    in D/A converter, 434
    fast, 76
    split-supply, 67
Four-quadrant mulitplier, 118
Fourth-order elliptic filter (1 kHz), 409
Framing pulse, 365
Free-running multivibrator
  100 Hz, 161
  100 kHz, 156
Free-running staircase generator, 162, 163
Frequency
  comparator, 489
  compensation circuit, 112
  -controlled oscillator, 154
  divider, sine-wave, 486
  doubler, 281, 292, 334
    for microprocessor, 459
    less than 1 MHz, 312
    150 to 300 MHz VHF, 313
  doubling tachometer, 395
  generator, FM, 313
  meter, analog, 111
  modulation, see FM
  multiplier (x10), 304
  -rated circuits, see Frequency-valued circuits
  -shift keying, see FSK
  standard (100 kHz), 113
  synthesizer, PLL, 324
  -to-voltage converter, 397
Frequency-valued circuits
  0 to 20 Hz flasher (1.5 V), 539
  1 Hz function generator, 167
  1 pps flasher, 530
  1.5 pps flasher, 524
  10 Hz sine-wave oscillator, 113
  60 Hz battery charger, 12
  60 Hz notch filter, 403, 412, 417
    adjustable, 403
    high-Q, 406

100 Hz cutoff high-pass filter, 407
100 Hz free-running multivibrator, 161
100 Hz ac voltmeter, 138
400 Hz servo amplifier (35 W), 520
1 kHz filters
  bandpass, 400, 415, 403
  biquad, 403
  Butterworth, 407
  elliptic, 409
  high-pass, 417
1 kHz signal generator (sine-wave), 146
2 kHz FSK demodulator, 304
2.4 kHz synchronous demodulator, 302
4 kHz phase-shift oscillator, 539
10 kHz cutoff low-pass filter, 405
10 kHz phase shifter, 227
20 kHz bandpass filter, 405
20 kHz inverter, 1 kW, 16
38 kHz subcarrier generator, 320
50 kHz to 40 MHz frequency meter, 111
50 kHz square-wave generator, 115
100 kHz crystal calibrator, 113
100 kHz crystal oscillator, 158, 169, 293
100 kHz square-wave generator, 115, 156, 161
455 kHz intermediate-frequency
  amplifier, 285
  converter/detector, 287
  strip, 332
  table radio, 297
455 kHz modulated oscillator, 158
455 kHz signal generator, 110
500 kHz ac voltmeter, 138
500 kHz limiter/amplifier, 311
1 MHz bandwidth preamplifier, 139
1 MHz crystal oscillator, 155
1 MHz doubler, 312
1 MHz function generator, 167
1 MHz meter driver, 108
1 MHz multiplexer, 457
1.5 MHz bandwidth 60 MHz IF amplifier, 297
4.5 MHz notch filter, 406
  high-Q, 415
7-MHz sine-wave oscillator, 294
40m transmitter (QRP), 294
40m VFO, 294
9 MHz double-balanced mixer, 310
10 MHz bandpass amplifier, 400
10-12 MHz frequency synthesis, 324
10 MHz oscillator, 156
  sine-wave, 155
10 MHz tuned amplifier, 323
10.7 MHz cascode amplifier, 288
10.7 MHz crystal oscillator, 156, 157
10.7 MHz intermediate-frequency
  amplifier, 318
  amplifier/limiter, 317
  strips, 286, 290, 291
12 MHz gain-controlled amplifier, 314
20 MHz video amplifier, 222
45 MHz wideband amplifier, 284
50 MHz amplitude modulator, 295
50 MHz crystal oscillator, 157
50 MHz RF preamp (30 dB), 293
50-120 MHz RF amplifier, 295
6 m and 2 m cascode amplifiers, 295
52 MHz modulated oscillator, 290
58 MHz wideband amplifier, 309
60 MHz IF amplifier, 297
88-108 MHz FM front end, 308
100 MHz cascode RF amplifier, 295
100 MHz crystal oscillator, 153
100 MHz mixer, 299
150 MHz to 300 MHz doubler, 313
160 MHz video amplifier (10 dB), 222
Front end
  AM receiver, 316

FM receiver, 308
FSK
  demodulator, 327, 330
    2 kHz, 304
  multiplexer, 476
  self-generating, 465
  system, 461
Full-wave power supply, 2
Function generator, 164, 167
Fuzz box, 539

Gain control
  dc noninverting, 99
  variable, 95
Gain-controlled amplifier, 225, 255
  12 MHz, 314
Gain-of-1 inverting amplifier, 90
Gain-of-5 amplifier, 70
Gain-of-9 amplifier, 71
Gain-of-10 amplifier, 70
Gain-of-10 RF mixer, 332
Gain-of-20 amplifier, 238, 244
Gain-of-35 amplifier, 66
Gain-of-100
  audio preamplifier/mixer, 276
  bandpass filter, 403, 404
  operational amplifier, 79
Gain-of-200 amplifier (audio), 238, 239
Gain-programmable operational amplifier, 88
Gas discharge display driver, 17
Gas engine tachometer, 388
Gate
  AND, 515
  OR, 515
Gated astable multivibrator, 166
Gated oscillator, 162
Generator
  antilog, 499, 502
  error voltage (A/D), 431
  FM, 313
  function, 164, 167
  high-current, 52
  log, 494, 495, 499
  logic (TTL), 494
  marker (100 kHz), 293
  noise, 168
  pulse, 147, 150, 160, 165, 166
  random-number, 536, 540
  ramp, 294, 430
    error, 435
  RIAA inverse response, 249
  signal
    3-phase, 154
    455 kHz, 110
    audio, 146, 337
    modulated, 158
    sine-wave, 113
    tone, 337
  sine-wave, 113, 120
  square-wave, 115, 149, 150, 153, 163, 166, 167, 169
    see also Square wave oscillator/generator
  staircase, 162, 163
  time-delay, 378
  triangular-wave, 160
  waveform, see Waveform generator
Go/no-go
  battery monitor, 350
  comparator, 511
Graphic equalizer, 273
Grounded-load current source, 48
Ground-fault interrupter, 124
Ground-isolated TTL data transmission, 468
Ground referenced amplifier, 92
Ground referenced differential amplifier, 100

Ground referenced load driver, 359
Ground referenced thermometer, 120, 121, 347
Ground returned audio amplifier, 244
Guarded differential amplifier, 78
Guarded inverting amplifier, 78
Guarded noninverting amplifier, 77
Guarded voltage follower, 77
Gyrator
   operational transconductance amplifier, 416

Half-watt audio amplifier, 230
Half-wave power supply, 2
Hartley oscillator, 163
Headphone amplifier, stereo, 226
Heat staging controller, 360
Heater control, 3-phase, 366
Heater supply, negative, 62
HF oscillator, 10 MHz, 156
HF sine-wave oscillator, 155
Hi-fi amplifier
   6 W (8 ohms), 266
   10 W, 238
   12 W, 239
   15 W, 251
   30 W, 261
   90 W, 241, 250
   automotive, 393
   with performance curves, 235
Hi-fi audio booster, 268
Hi-fi FM wireless speaker system, 322
Hi-fi preamplifier
   with controls, 227, 269
   fixed-gain, 270
High-accuracy sample hold, 448
High-beta cascode amplifier, 90
High-CMRR instrumentation amplifier, 138, 141
High-compliance current sink, 51, 97
High-current
   driver and voltage comparator, 459
   generator, 52
   load switching, 343
   output buffer, 350
   regulator, 22, 24, 32
   voltage follower, 79
High-frequency
   ac voltmeter, 138
   analog switch, 464
   sine-wave oscillator, 155
High-gain audio amplifier, 249
High-gain preamplifier, 78
High-immunity instrumentation amplifier, 136
High-impedance
   amplifier, 68
      10 M input, 78
      audio, 236
      bridge-type, 79
      inverting, 103
      noninverting, 102
      wideband, 101
   comparator, 511
   dc voltmeter, 123, 139
   detector, with AGC, 326
High-level logic interface, 514
   to TTL, 509
High/low
   battery monitor, 350
   temperature-to-current transducer, 362
High-output regulator, 28
High-pass filter, see Filter, high-pass
High-power audio amplifier, 250
High-power regulator, 17
High-Q filters, see Filter, high-Q
High-speed
   amplifier, 66

   low-drift, 80
   digital-to-analog converter, 442
   sample and hold, 450, 451, 453
   subtractor, 493
   warning, automotive, 394
High-stability regulator
   1 A, 46.
   -12 V, 24
High-voltage
   flasher, 346
   noninverting amplifier, 81
   self-regulator, 42
High-Z, see High-impedance
Hysteresis
   comparator, 462, 510
   -squelch preamplifier, 323
   and temperature control, 341
   threshold detector, 462

Ice warning alarm for cars, 386
IC power supply, 58
IF amplifier, see Amplifier, IF
IF converter and detector, 455 kHz, 287
IF oscillator, 10.7 MHz, 156, 157
IF/RF amplifier, dc to 120 MHz, 295
IF signal generator, 455 kHz, 110
Impedance adapter (matcher/amplifier), 520
Impedance-rated and resistance-rated circuits
   4-ohm audio amplifiers
      1 W battery-operated, 229
      2.5 W, 231
      90 W, 250
   4-ohm audio booster, 268
   4-ohm-output car radio, 389
   8-ohm audio amplifiers
      5 W, 247, 248, 275
      10 W, 238
      70 W, 250
   8-ohm power amplifier (30 V supply), 264
   8-ohm stereo amplifier, 234
      4 W, 265
      12 W, 239, 266
   16-ohm audio amplifiers
      5 W, 247, 248
      power bridge-type, 263
      stereo (4 W per channel), 265
   45-ohm audio amplifier, 249
   50-ohm coaxial line driver, 269, 475, 477
   50-ohm 7 MHz oscillator, 294
   300-ohm stereo headphone amplifier, 226
   500 k-input amplifier, 89
   10 M-input amplifier, ac-coupled, 78
Incandescent flasher, 524
Induction-motor speed control, 369
Inductive-load driver, strobed, 358
Inductor, simulated, 400, 410
Injector, signal, 146
Input comparator and one-shot, 506
Input-lockup monostable multivibrator, 165
Input offset adjusting, 112, 125
Input/output, 115, 476
Instrumentation amplifier, see Amplifier, instrumentation
Instrumentation prescaler, divide-by-10, 498
Instrumentation-recorder interface, 479
Instrumentation shield driver, 127
Integral-cycle temperature control, 358
Integrator, 166
   differential, 162, 163
   in frequency doubler, 334
   regulator, 472
Intercom, 284, 286
   battery-operated, 283
   simple, 283, 284

   2-station, 288, 293
   using 4-ohm speakers, 283
Interface
   digital, 509
   optical-to-TTL, 424
   process-control, 342
   telemetry gear, 479
   TTL-to-MOS, 498
Intermediate frequency, see *listings under* IF
Interrupter, ground-fault, 124
Intrusion alarm, triac-controlled, 368
Inverse RIAA response generator, 249
Inverter
   high-impedance, 103
   line-operated kilowatt, 16
   ramp polarity, 430
   unity-gain, 91
Inverting amplifier, see Amplifier, inverting
I/O strobing device, 115, 476
IRIG channel 13 TV demodulator, 335
Isolated power switch, 423
Isolated-sensor zero-voltage switch, 367
Isolation amplifier, 87
Isolator
   capacitive-load, 93
   digital transmission, 473
ITOS/ESSA picture demodulator, 319

JFET current monitor, 48
JFET sample and hold, 454

Kelvin thermometer, 357

Laboratory power supply, 3
Ladder network driver, 436, 438, 439
Ladder network in A/D, 438
Lamp dimmer, 360
Lamp driver, 361, 364, 516
   and comparator, 353
   and level detector, 347
Lamp flasher, 539, 541
Large-fan-in AND gate, 515
LC oscillator (sine-wave), 155
LCD digital voltmeter, 106
LED
   driver, 346, 491
   flasher for models, 533
   thermometer (low-cost), 122
Level comparator, 500
Level detector, 423
   with lamp driver, 347
   photodiode, 422
Level isolation amplifier, 83
Level-rated circuits, see Level-valued circuits
Level sensor and alarm, 341
Level shift, MOS-to-TTL, 497
Level shifter
   analog-to-digital, 432
   MOS-to-TTL, 489
Level shifting isolation amplifier, 87
Level-switch, line-operated, 354
Level translators, 460, 465
Level-valued circuits (relative—gain or loss)
   6 dB/octave preemphasis, 314
   10 dB video amplifier (160 MHz), 222
   15 dB magnetic tape preamplifier (stereo), 228
   20 dB inverting amplifier, 69
   20 dB video amplifier, 222
   30 dB RF preamplifier (50 MHz), 293
   30 dB wideband amplifier, 281
   34 dB phono amplifier
      crystal pickup, 2 W, 246
      magnetic pickup, 253
      18 V supply, 258
   37 dB servo amplifier, 520

40 dB amplifier
  biquad bandpass filter and, 403
  hi-fi (15 W), 251
  noninverting, 125
  stereo headphones, 226
46 dB audio amplifiers,
  2 W output, 246
   low-distortion, 225
    phono (magnetic pickup), 256
50 dB wideband amplifier (58 MHz), 309
68 dB bridge-type amplifier, 262
76 dB gain-controlled 12 MHz amplifier, 314
80 dB and 86 dB IF amplifiers, 290, 297
95 dB (dynamic range) preamplifiers, 272
100 dB log generator, 494
130 dB CMOS amplifier, 74
Light-activated intrusion alarm, 368
Light-activated triac control, 425
Light-controlled amplifiers, 422
Light-controlled oscillator, 427
Light controller, 422, 424
Light-emitting diode, see LED
Light-intensity regulator, 426
Light-operated relay, 427
Light-tracking analog multiplier, 425
Limit comparator and lamp driver, 349
Limiter/amplifier/detector, 303
Limiting amplifier, 500 kHz, 311
Limiting current sources, 15
Line driver, 131
  for coaxial cables, 469, 475, 477, 483
  and zero-crossing detector, 446
Line-operated
  audio amplifier, 249
  inverter (1 kW), 16
  level switch, 354
  one-shot timer, 382
  SCR firing circuit, 345
Line phase reversal detector, 351
Line receiver
  adjustable-threshold, 471
  high-noise-immunity logic, 480
Line resistance compensator, 474
Line-voltage phase-reversal detector, 351
Linear and switching regulator, 20
Linear audio mixer, 252
Linear-to-exponential generator, 502
Liquid-crystal display, 106
Liquid-crystal DVM, 106
Liquid detector, 347
Liquid-level
  alarm, 353
  control circuit, 350
  sensor and alarm, 341
Load switching, high-current, 343
Log amplifier, 102, 515
Log converter, 505
  temperature-compensated, 492
Log-driven antilog generator, 499
Log generator, 499
  100 dB range, 494
  fast, 495
Logarithmic, see listings under Log
Logic—AND gate (3 inputs), 512, 515
Logic-controlled
  analog amplifier, 493
  analog switch, 463
  mute, 296
  -shutdown amplifier, 127
  zero-crossing detector, 471
Logic driver, TTL, 343
Logic-driving zero-crossing detector, 113
Logic gate
  AND function, 512, 515
  NOR function, 517
  OR function, 515
Logic generator, crystal-controlled, 494

Logic—NOR (3-input gate), 517
Logic-OR gate, 515
Logic-powered control, 366
Logic-triggered timer chain, 377
Logic window detector, 458
Long-delay timer, 374
Long-interval timer, 376
Loop, phase-locked, 503
Loudspeaker, see Speaker
Low-current ammeter, analog, 110
Low-current fast summer, 74
Low-distortion amplifiers, 81, 225, 253
Low-distortion Wien bridge, 168
Low-drift
  amplifier, 93
  operational amplifier, 98
  peak detector, 475
  ramp and hold, 483
  sample and hold, 450, 452, 476
  thermocouple amplifier, 137
Low-duty-cycle thermometer, 370
Low-frequency
  doubler, 312
  generator (square-wave), 150
  operational amplifier, 79, 85
  RF mixer, 332
Low-impedance headphone amplifier, 226
Low-impedance mike preamplifier, 257
Low-input-bias comparator, 507
Low-input-capacitance preamplifier, 139
Low-level power switch, 340
Low-noise
  audio amplifier, 93
  audio preamplifier, 139, 229, 278
  balanced-line mike preamplifier, 256
Low-output-impedance amplifier, 83
Low-pass filter, see Filter, low-pass
Low-power IC voltage source, 58
Low-voltage comparator, 517
Low-voltage proportional control, 364
Low-voltage short-proof regulator, 19

Machine control, solid-state, 362
Magnetic-cartridge
  amplifiers, 253, 254
  preamplifiers, 257, 267, 268
  zero-crossing detector, 474
Magnetic tape
  preamplifier, 228, 232
  reader (TTL output), 471
  stereo preamplifier, 228
Magnetic-transducer detector, 115, 128
Marker generator
  50 to 100 kHz, 115
  100 kHz, 293
Matching device for antennas, 520
Matrix amplifier, 4-channel stereo, 265
Medium-speed D/A converter buffer, 440
Meter
  ac voltage, 138
  analog voltage, 133, 135
  dc voltage, 139
  degrees Kelvin (direct-reading in kelvins), 357
  digital voltage, 106, 107, 114, 119, 143
  dwell and tach, 390
  frequency (to 40 MHz), 111
  millivolt-reading, 142
  phase (wide-range), 531
  precision digital, 143
  temperature-indicating, 346, 371
  voltage (digital), 106, 107, 114, 119, 143
  VU, 269
  wind-velocity, 396
Meter amplifier
  1 MHz driver, 108
  analog, 125
  floating-input, 121

Mic, see listings under Mike
Microammeter, 110
  dc, 126
  100 nA full-scale, 125
Microamplifier, 68
Microphone, see listings under Mike
Micropower
  comparator, 501
  instrumentation amplifier, 126
  thermometer, 129
Microprocessor frequency doubler, 459
Mike amplifier, crystal/ceramic, 256
Mike mixer, 235, 237
Mike preamplifier
  balanced-line, 256, 277
  high-gain, 249
  low-impedance, 257
  squelched, 276
  various types of, 258, 277
Millivoltmeter, ac, 142
Miniature, PA system, 260
Minimum-frequency detector, 453
Minute timer, 380
Missile telemetry demodulator, 335
Mixer, audio, 237, 242, 252
  2-channel stereo with preamplifier, 276
  3-channel, 482
  4-channel, 235, 276
  hi-fi, 242
  linear, 252
  with preamp, 235
Mixer, balanced, 303, 310, 329
  double, 310
  from product detector, 329
Mixer and oscillators, dc to 120 MHz, 295
Model-railroad crossing signals, 533
Modulated oscillator, 455 kHz, 158
Modulated-oscillator tremolo circuit, 522
Modulation, balanced, 291
Modulator
  AM, see Modulator, amplitude
  amplitude, 295, 308, 315, 323, 325, 333
  DSB, 317, 330
  high-efficiency SSB, 312
  pulse-width, 288, 294, 478
  suppressed-carrier, 283
  switching-supply, 59
Moisture detector, 347
Monaural reverb to synthesize stereo, 271
Monitor
  battery condition, 350
  current, 48, 346
Monostable multivibrator, see Multivibrator, monostable
MOS clock driver, 492, 495
MOS comparator, 423, 491
  dual-limit, 504
MOS interface
  from TTL, 498
  to TTL, 489, 497, 507
MOS sample and hold, 467
MOS-to-TTL
  driver, dual, 507
  level shifter, 489, 497
MOS zero-crossing detector, 471, 475, 477
Motor drive
  turntable, 533
  two-phase, 354
Motor speed control, 349, 351, 360, 371
  induction-type, 369
Moving-air detector, 347
Multichannel audio mixer, 482
Multicylinder tachometer for cars, 388
Multi-input
  analog amplifier, 493
  audio amplifier system, 274
  inverter, 1 kW, 16
  stereo preamplifier, 260

Multi-interval timer, 378
Multi-output supply, 9
Multiphase sine-wave generator, 154
Multiplex decoder
  with phase-locked loop, 456
  stereo, 305, 315
Multiplex signal conditioner, 126
Multiplexer
  3-channel, 470
  4-channel, 460
  16-channel, 447
  binary-controlled, 449
  current-mode, 463
  differential-mode, wideband, 457
  frequency-shift-keyed, 476
  high-frequency, 464
Multiplier
  1-quadrant, and divider, 497, 499
  4-quadrant, 118, 488
  analog, 425
  analog, and divider, 496
  capacitance, 509, 516
  D/A converter (BCD), 437
  frequency, 281, 292, 334
  x10, 304
  VHF (150 MHz), 313
Multiplier-phototube supply, 8
Multirange 3.5-digit DVM, 114
Multistation intercom, 283
Multivibrator
  astable, 158, 159
    1 pps, 160
    1 A switching (flasher), 539
    1.5 V (flasher), 541
    100 Hz, 161
    gated, 166
    timer, 376
  bistable (flip-flop), 505
  free-running (100 kHz), 156, 161
  monostable (1-shot), 164, 165, 168, 169, 352, 513
    with input lockout, 165
    reset timer, 382
    variable delay, 164
    see also One-shot
  siren-producing, 532
  voltage-controlled, 154
Music circuit
  fuzz box, 539
Music phase shifter, 523
Music tremolo circuit (VCO), 521
Mute, logic-controlled, 296
Muting circuit, 520
Muting on FM tuner, 318
Mux, see listings under Multiplex

NAB
  tape amplifiers, 239, 257, 270
  tape preamplifiers, 241, 255, 272, 274, 275, 278
Nanoammeter
  100 nA full-scale, 125
  floating-input, 121
Nanowatt amplifier, 68, 70
N-channel MOS zero-crossing detector, 477
Negative
  heater supply, 62
  -peak detector, 470, 483
  -pulse triggered timer, 381
  references, 11, 57
  regulator, 28, 31
    1 A switching, 18
    10 V at 1 A, 55
    15 V, 34, 37
    30 V, 53
    105 V, 54
  see also Regulators
Network, binary-ladder, 436

in A/D converter, 438, 441
  comparator, 439
  driver, 438
Network, electronic crossover, 242
Nicad-battery charger, 12
Nimbus picture demodulator, 319
Noise generator, 168
Noise-immune instrumentation amplifier, 136
Noise-immune zero-voltage detector, 457
Noise squelch
  FM receiver, 296
  FM scanner, 292
Noise switch, 538
No-leakage sample and hold, 451
Nonground-returned audio amplifier, 244
Noninverting amplifier, see Amplifier, noninverting
Noninverting dc gain control, 99
Nonlinear op-amp, 84
NOR gate
  3-input, 517
  delay element for, 486
Notch filter, see Filter, notch
NPN voltage boost, 513
Nuclear-particle detector, 364
Number generation circuit, random
  1 through 6, representing dice, 528
  1 through 36, representing roulette, 536

Octave equalizer, 10-band, 273
Offset-adjusting
  amplifier, 85, 112
  comparator, high-impedance, 511
Offset balancer, 93, 131
Offset-compensated sample and hold, 446
One-hour timer, 379
One-minute timers, 380. See also Timer
One-of-N data selector, 503
One-quadrant multiplier/divider, 499
One-second timer, See Timer
One-shot, 164, 168, 169, 365, 513
  control, 365
  line-operated timer, 382
  reset timer, 382
  with input comparator, 506
  with input lockout, 165
  with power capability, 352
  see also Multivibrator, monostable
Op-amp, 88
  CMOS, 86
  differential I/O, 70
  FET, 66, 76, 86, 89
  gain-of-100, 79
  low-drift, 98
  low-frequency, 79, 85
  nonlinear, 84
  power, 86
  programmable, 87, 88
  self-zeroing, 72
  superbeta, 67, 75
  temperature-compensated, 84
  various types of, 75
Op-amp astable multivibrator, 159
Op-amp bandpass filter, 1 kHz, 419
Op-amp with follower preamplifier, 75
Op-amp multivibrator, 100 Hz, 161
Op-amp postamplifier, 70, 74, 130
Op-amp power booster, 89
OP-amp square-wave generator, 169
Op-amp voltage regulator, 32
  10 A, 49
Open-loop instrumentation amplifier, 130
Operational amplifier, see Op-amp
Operational transconductance amplifier, see listings under OTA
Opposing-voltage comparator, 506, 511
Opto

isolated switch, 426
  isolator, 424, 426
  sensor/interface, 424
OR gate, 515
OR muting, 296
Oscillator
  50 kHz, square-wave, 115
  455 kHz, modulated, 158
  1 MHz, 155
  10 MHz, 156
  code-practice, 336
  crystal, 157, 163
    100 kHz, 158, 159, 169
    square-wave, 160
  current-controlled, 161
  gated, 162
  Hartley, 163
  IF (10.7 MHz), 156, 157
  LC (simple), 163
  modulated
    455 kHz, 158
    52 MHz, 290
  phase-shift, 296, 539
  pulse-output, 147
  relaxation flasher, 387
  RF
    1 MHz, 155
    10 MHz, 156
    crystal, 153
  sine-wave, 113, 120, 146, 149, 150, 155, 158, 167, 296, 539
    3-phase, 154
    7 MHz, 294
    10 MHz, 155
    audio, 146
    decade, 120
  square-wave, 149, 150, 153, 163, 166
    50 kHz, 115
    crystal, 160
    see also Square-wave oscillator/generator
  strobe-controlled, 153
  timing for A/D converters, 430
  TTL-output, 155
  variable-frequency (5-10 MHz), 294
  various types of, 159
  voltage-controlled, 148, 152, 322, 521
  Wien bridge, 147, 149, 150, 158, 168
Oscillator/calibrator, 100 kHz, 113, 293
Oscillator/doubler, 334
  150 MHz, 313
Oscillator/mixer, 100 MHz, 295, 299
  dc-120 MHz, 295
Oscillator sirens, 525, 532, 534
Oscillator/transmitter, 40m, 294
Oscilloscope chopper, dual-trace, 463
Oscilloscope diff-amp input, 116
OTA amplitude modulator, 323, 325, 333
OTA gyrator filter, 416
Output buffer, high-current, 350
Output reference, bipolar, 63, 133
Output strobing circuits, 490
Oven control, 347
Overrange-indicating DVM, 107, 114
Overspeed indicator, 387
Overtemperature detector, 359
Overvoltage protection, 356

Pan-pot, 2-channel, 275
Paper tape reader with TTL output, 466
Passive tone controls, 361, 411
PC board layout, 5 W audio amplifier, 247, 248
P-channel MOS
  sample and hold, 451
  zero-crossing detector, 477
Peak detector, 325, 470
  low-drift, 475

INDEX 551

negative, 470, 483
positive, 467, 470, 471, 483
Performance curves, AF amplifier, 235
Period timer, wide-range, 381. See also Timer
Phase-locked loop, 300. See also listings
Phase control circuit, 360 under PLL
Phase meter, wide-range, 531
Phase-reversal detector, 351
Phase-shift oscillator, 296, 539
Phase shifter
 -45 degrees, 227
 1-10 kHz, 227
 musical, 523
 variable, 226
pH meter picoampere-level amplifier, 128
Phono amplifier, See Amplifier, phono
Phono motor drive, 533
Phono preamplifier, magnetic pickup, 257, 267, 268, 272
 low-noise (46 dB), 256
 wideband, 279
Phono/radio/tape amplifier, 274
Phono/radio/tape preamplifier, 278
Photo amplifiers, 422
Photodetector/power-amplifier, 426
Photodiode-sensor amplifier, 425
Photodiode level detector, 422
Photodiode paper tape reader, 466
Photoelectric control circuit, 423
Photomultiplier dynide supply, 8
Photovoltaic, see Solar cell
Photovoltaic amplifier, 95
Picoammeter amplifier, 128
Pickup amplifier, 243. see also Amplifier, phono
Pickup preamplifier, acoustic, 522
Picture demodulator, weather satellite, 319
Piezoelectric-transducer amplifier, 110
 See also Amplifier, phono
Pink-noise generator, 168
Playback amplifier, tape (NAB), 270
Playback and record amplifier, 243
 cassette, 246
Playback preamplifier, NAB, 255, 274, 275
PLL, 503
PLL demodulator
 FM receiver (stereo), 328
 stereo, 300, 305, 315, 320, 456
 weather-satellite, 319
PLL frequency synthesizer, 324
Polarity inverter, ramp, 430
Portable calibrator, 134-139
Portable FM squelched amplifier, 296
Positioning circuit, stereo, 275
Positive current source, 62
Positive peak detector, 467, 470, 471, 483
Positive voltage references, 11
Positive regulators
 1 A, 55
 5 V, 200 mA, 43
 15 V, 34, 37
 50 V, 26
 floating, 23
 See also Regulators
Postamplifier, CMOS, 137
Postamplifier, op-amp, 70, 74, 130
Power amplifier, see Amplifier, power
Power booster
 hi-fi (10 W), 268
 op-amp, 89
 stereo (20 W), 240
 see also Booster
Power comparator, 517
Power control
 3-phase, 357
 triac, 428

zero-voltage-switching, 353
Power one-shot, 352, 365
Power op-amp, 86, 264
Power PNP, 66
Power-rated circuits, See Power-valued circuits
Power reference 20 V, 63
Power solenoid driver, 340
Power supply, 2
 10 V reference, 8
 14 V reference, 8
 20 V (60W), 60
 25 V variable, 61
 28 V regulated, 10
 40 V tracking, 7
 41 V (dual), 7
 constant-coltage, 3
 high-noise-immunity, 7
 multioutput, 9
 photomultiplier, 8
 programmable, 7
 switching (80 W), 222
 switching-regulator, 59
 see also Regulated power supply
Power switch, low-level, 340
Power switching, voice controlled, 355
Power switching shutoff/alarm, 428
Power-valued circuits
 500 nW amplifier, 70
 600 nW amplifier, 68
 10 uW instrumentation amplifier, 121
 500 mW audio hi-fi amplifier, 230
 500 mW audio power amplifier, 99
 1 W audio amplifier, 241, 276
 1 W battery-operated audio amplifier, 229
 1 W CW final amplifier, 536
 1 W inverting amplifier, 66
 1 W noninverting amplifier, 67
 2 W audio amplifier, 233
 2 W audio power amplifier, 246, 266
 2 W per channel stereo amplifier, 265
 2 W phono amplifier, 231
 2 W TV sound amplifier, 254
 2.5 W bridge-type amplifier, 238
 2.5 W class A audio amplifier, 231
 3 W differential amplifier, 71
 3 W pulse power amplifier, 538
 3.5 W bridge-type amplifier, 245
 4 W audio amplifier, 279
 4 W audio amplifier, for 8-track, 232
 4 W bridge-type amplifiers, 235, 236, 244
 4 W car radio amplifier, 386
 4 W hi-fi amplifier, 270
 4 W servo amplifier, 244
 4 W stereo amplifier, 234, 265
 5 W audio amplifier, with PC layout, 247
 5 W audio power amplifier, 224, 275
 5 W/5 W biamplifier, 242
 5 W car radio (AM), 289
 7 W hi-fi amplifier, automotive, 393
 8 W stereo amplifier, 270
 10 W audio booster, 268
 10 W class B audio amplifier, 229
 10 W hi-fi amplifier, 237
 10 W per channel stereo booster, 240
 10 W RMS audio amplifier, 238
 12 W bridge-type amplifier, 263
 12 W hi-fi amplifier, 239
 12 W stereo amplifier, 266
 15 W hi-fi amplifier, 251
 20 W audio booster, 240
 20 W bass-boosted stereo, 240
 20 W stereo amplifier, 237
 27 W-soldering-iron's thermal probe, 341
 30 W hi-fi amplifier, 261
 35 W servo amplifier (400 Hz), 520
 60 W regulated supply (20 V), 60
 60 W stereo amplifier, 261

70 W audio power amplifier, 250
80 W CATV switching supply, 222
90 W audio power amplifier, 241
90 W hi-fi amplifier, 250
1000 W line-operated inverter, 16
Preamplifier, See Preamp
Preamp
 acoustic-pickup, 521
 and control center, stereo, 278
 fixed-gain (audio), 270
 high-gain, 78
 low-noise, audio, 139, 229, 278
 magnetic-cartridge phono (RIAA), 257, 267, 268, 272
  low-noise (46 dB), 256
  wideband, 279
 magnetic-tape (stereo), 228, 272, 275
 microphone, 249
  balanced-line, 277
  balanced-line (low noise), 256
  with hysteresis, 276
  low-impedance, 257
 and mixer (20 dB), 252
 and mixer (4-channel), 235
 and mixer (stereo), 276
 RF 6m (30 dB), 293
 servo (saturating), 532
 squelch (with hysteresis), 323
 stereo, 227
  complete, 260
  fixed gain, 268
  low-distortion, 269
 tape, 228, 272, 275
 tape playback, 232, 255, 270
Precision dual 15 V regulator, 23
Precision sample and hold, 452, 458
Precision 2.15 V source, 51
Preemphasis, 6 dB/octave, 314
Premultiplex signal conditioner, 126
Prescaler, divide-by-10, 498
Presettable analog timer, 379
Printed circuit, 247, 248
Probe, temperature, 371
Probe, thermal, for solder iron, 341
Process-control interface, 342
Processor, voise, 314
Product detector, see Detector, product
Programmable
 down counter, 486
 micropower comparator, 501
 op-amp, 87, 88
 power supplies, 7
 siren, 525
 unijunction transistor, 501
Proportional-power regulator, 18
Proportional-speed controller, 363
Proportional temperature controller, 344
Protection
 overvoltage, 356
 safe-area, 241
 short-circuit, 99
Proximity detector and touch switch, 535
PTT, voise-controlled, 538
Public address, miniature, 260
Pulse amplifier
 power, 538
 unity-gain, 504
Pulse conditioner, 453
Pulse counter, 162, 163
Pulse generator, 147, 150, 160, 165, 166
 square-wave, 149
Pulse peak detector (50 ns), 474
Pulse regulator, 28
Pulse-triggered timer, 381
Pulse-width
 detector, 469
 discriminator, 473
 modulator, 288, 294, 478

Pulsing 60 Hz battery charger, 12
Pump control, on/off switching, 350
Punched-tape reader, TTL, 466
Push-to-talk, 538
PUT, 501
PWM, see Pulse-width

Q-of-25 bandpass filter, 400
Q-of-25 bandpass filter, 1 kHz, 408
Q-of-50 bandpass filter, biquad, 403
QRP final amplifier, 1 W, 536
QRP transmitter, 40 m, 294
Quadrature detector FM IF strip, 306
Quad/stereo reverb unit, 271
Quad 2-input NOR delay element, 486
Quasi-complementary power amplifier, 264
    90 W, 250
    see also Amplifier, audio
Quiescent balance control, 224

Radiation detector, 364
    amplifier, 128
Radio
    AM, see AM, radio
    FM, see FM, radio
    reverb attachment, 271
    with stereo amplifier, 274
Ramp
    comparator and gate, 430
    and error generator, 435
    generator, 160, 294, 430
    and hold, low-drift, 483
    polarity inverter, 430
Random-number generator, 536, 540
    1 through 6 (times two), for dice, 536
    1 through 36, for roulette, 536
Range expander, dynamic, audio, 255
RC filter, low-pass active, 418
RC PLL stereo decoder, 305
RC timer, pulse-triggered, 381
Reader
    magnetic-tape, 471
    paper-tape, 466
Read/write circuits, 471
Rear-channel ambience amplifier, 265
Receiver/converter IF amplifier, 316
Receiver, FM stereo, complete, 328
Receiver, line, adjustable threshold, 471
Receiver, line, high-noise-immunity, 480
Receiver for 2 MHz or less (TRF), 321
Receiving-antenna impedance matcher, 520
Recorder interface, instrumentation-type, 479
Recording/playback amplifier, 243, 257, 270
    cassette, 246
Recovery amplifier, 245, 271
Rectified-line-voltage amplifier, 249
Reference
    bipolar output, 133
    external, for temperature transducer, 117
    internal, for temperature transducer, 117
    single-supply, 58
    square-wave, 62
    voltage, 54, 57
        6.9 V, 63
        10 V, 8
        14 V, 8, 51
Referencing signals to ground, 100
Regenerator, carrier (SSB), 321
Regulated-output threshold detector, 472
Regulated power supply, 4
    5 V, 5
    5 V foldback, 15
    15 V dual, 6
    20 V, 60W, 60
    28 V, 10
    65 V tracking, 6

    see also Power supply
Regulators, descriptively rated
    basic, 27
    battery-charger, 392
    current, see Regulators, rated by current
    high-current, 22, 24, 32
    high-output, 28
    high-power, 17
    light-intensity, 426
    negative-voltage, 28, 31. See also
Regulators, rated by voltage
    pulse, 28
    self-, 50
    short-proof, 19
    slow-turnon 15 V, 27
    stepdown switching, 55
    switching, see Switching regulator
    synchronizer for, 18
    temperature-compensated, 26, 35, 41
    tracking, see Tracking regulator
    variable-voltage, 22, 50
        see also Variable regulator
    zero-voltage switching, 27
    see also Shunt regulator
Regulators, dual
    3 V, 18
    12 V tracking, 52, 53
    15 V, 39, 40
    15 V tracking, 12, 23
Regulators, floating, 23, 32
Regulators, rated by current, 42, 46
    10 mA current, 49
    100 mA current, 56
    200 mA, 44
    200 mA (5 V), 43
    250 mA, 28
    1 A, 46
    1 A (5 V), 45
    2 A current, 41, 42
    2 A and 3 A shunt, 44
    3 A switching, 14, 20
    5 A, 29
    switching, 18
Regulators, rated by voltage
    0-13 V, 30
    2 V, 31
    2-7 V, 34, 36
    3 V, 39
    3 V switching, 18
    4.5-34 V (1 A), 55
    5 V, 19
    5 V (500 mA), 21, 41, 47
    5 V (1 A), 45
    5 V aT 3 A, 20
    5 V (10 A), 43
    5 V current-limited, 21
    5 V fixed, 50
    5 V high-current, 22
    5 V pass, 35
    5 V switching/linear, 20
    5 V protected, 29
    5-7 V shunt, 28, 32
    6 V, 56
    7-23 V, 36
    7-30 V (50 mA), 56
    10 V, 24, 49
    10 V (1 A) negative, 55
    10 V (3 A), 20
    12 V, 24, 51
    12 V dual, 53
    12 V dual (tracking), 52, 53
    15 V, 27, 35, 37
    15 V (10 A), 33
    15 V dual, 23, 34, 39, 40
    15 V dual (tracking), 12
    15 V negative, 34, 37
    15 V positive, 37
    15 V slow-turnon, 27

    18 V, 24, 25
    28 V switching, 19
    30 V negative, 53
    37 V variable, 34
    50 V positive, 26
    50 V variable, 38
    105 V negative, 54
    250 V max at 100 mA, 37
    300 V at 500 mA, 33
    2 kV, 40
    current-rated, 46
        200 mA, 44
        250 mA, 28
        1 A, 46, 55
        3 A, 14
        4 A, 14
        see also Regulators, rated by current
    high-current, 24, 32
    high-power, 17
    negative, 31
    self-regulating, 50
    short-proof, 19
    stepdown, 55
    temperature-compensated, 26
    variable, 22, 50
Relaxation oscillator, 387
Relay
    carrier-operated, 323
    light-activated, 427
Relay driver, 358, 361
    comparator and, 516
    power, 340
    strobed-input, 347
    strobing, 358
Relay/solenoid driver, 358, 361
Remote load voltage sensor, 474
Remote-shutdown regulator, 29
Remote speaker system for FM, 322
Remote temperature sensor, 371
    alarm for sensor, 340
Remote 2-way discriminator window, 316
Repeat cycle timer, astable, 380
Repeater
    control module, 323
    discriminator window, 316
Reset stabilized amplifier, 89
Reset timer, monostable, 382
Resistance bridge temperature control, 370
Response curves for 4 W audio amplifier, 235
Response improvement with diode clamp, 483
Response inverse RIAA, 249
Restoration, dc, 330
Reverb, stereo, 245
Reverb enhancement system, 271
Reverberation, 245
RF amplifier, See Amplifier, RF
RF mixer, low-frequency, 332
RF oscillator, See Oscillator, RF
RF preamplifier, 50 MHz (30 dB), 293
RF signal generator, 455 kHz, 110
RIAA
    and NAB equalization, 269
    -equalized stereo preamplifier, 268
    inverse-response generator, 249
    phono amplifier, 246
        crystal and ceramic cartridges, 256
        magnetic cartridges, 254
    phono preamplifier, 272, 278
        magnetic cartridges, 267
RMS detector, 135
Root extractor, 502, 505
Root mean square, 135
Roulette wheel, electronic, 536
RTL driver and voltage comparator, 456
RTL voltage comparator, 490

INDEX   553

Safe-area protection, 241
Safety circuit, GFI, 124
Sample and hold, 446
　CMOS, 453
　compensated, 472
　DTL/TTL, 459
　high-accuracy, 448
　high-slew-rate, 451
　high-speed, 450, 453
　JFET, 454
　low-drift, 450, 452, 465, 476
　low-leakage, 451
　minimum-feedthrough, 459
　MOS, 467
　with offset adjust, 449
　precision, 452, 458
　slow, 448
　with offset adjust, 449
　with V comparison, 452
Satellite weather-picture demodulator, 319
Saturating servo preamplifier, 532
Sawtooth generator, 160
SCA decoder, 315
Scanner noise squelch (FM), 292
Schmitt trigger, 491, 513
　fast, 508
　as programmable unijunction, 501
SCR crowbar overvoltage protector, 356
Second-IF oscillator (455 kHz), 158
Seepage alarm for fluids, 342
Selector
　active-channel, 482
　channel (dc controlled), 237
　data (1 of n), 503
Self-generating FSK, 465
Self-regulator, 50
　high-voltage, 42
Self-zeroing operational amplifier, 72
Sense amplifier, 96
Sense/drive circuits, 481
Sensitive-bridge balance indicator, 142
Sensor
　liquid-level (with alarm), 341
　remote-load voltage, 474
　temperature, 359
　　analog, 116-118
　　with alarm, 340
　　remote, 371
Series voltage regulator, 15 V, 35
Servo amplifier, See Amplifier, servo
Servo driver, wideband, 264
Servomotor drive control, 355
Servo phase meter, 531
Servo preamplifier, saturating, 532
Servo proportional speed control, 363
Shaft-position-pickoff zero-crossing detector, 474
Shield/line driver, 131
　instrumentation-type, 127
Shifter
　level (MOS-to-TTL), 497
　phase
　　—45°, 227
　　flanger, recording, 523
　　variable, 226
Short-circuit protection, 99
Short-proof audio bridge amplifier, 262
Short-proof low-voltage regulator, 19
Shunt regulator
　2 A and 3 A, 44
　5 V, 28
　5-7 V, 32
　6 V, 56
　compensated, 26
Shutoff and alarm, high-power, 428
Sideband carrier suppression, 311
Signal conditioner, premultiplex, 126
Signal generator, See Generator, signal

Signal injector, 146
Simulated inductor, 400, 410
Sine-wave frequency divider, 486
Sine-wave oscillator, See Oscillator, sine-wave
Sine-wave-to-square-wave converter, 87
Single-ended class B amplifier (audio), 229
Single-IC amplifier, 224
Single-pole low-pass filter, 332
Single-sideband suppressed carrier, see SSSC
Single-supply
　ac amplifier, 89
　CMOS op-amp, 86
　diff-amp, 132
　follower, 94
　noninverting amplifier, 67
　reference, buffered, 59
　thermocouple amplifier, 133
Single-ramp analog-to-digital converter, 430
Sink, current, 49
　clamp (5 mA), 132
　and current source, 47, 57
　high-compliance, 51, 97
　precision, 49
　time delays, 375
　various types, and sources, 57
　voltage-controlled, 76
Siren, electronic, 525, 530
　alarm, 524
　automatic-wail, 534
　button-controlled, 530
　multivibrator, 532
Slope comparator, FSK, 461
Slow sample and hold, 448
Small-signal zero-voltage detector, 457
Small-system audio reverb, 271
Solar cell amplifier, 95, 422
Soldering-iron thermal probe, 341
Solenoid driver
　with comparator, 346, 516
　with strobe, 347, 358
Solenoid driver, power, 340
Solid-state relay with triac, 343
Sound-activated switch, 365
Sound system
　2-channel, 274
　TV (2 W, 24 V) with IF, 254
Source
　current, 51
　　bilateral, 49
　　with current sink, 47, 57
　　grounded-load, 48
　　limiting, 15
　　positive/negative, 62
　　and regulator, 42
　　voltage-controlled, 95
　signal (455 kHz), 110
　voltage-reference, 54
　2.15 V (precision), 51
Source follower, 91
Speaker, hi-fi wireless, for FM, 322
Speaker power booster, 268
Special-effects flanger, 523
Speed alarm, automotive, 387, 394
Speed control
　motor, 349, 351, 360
　　induction winding, 369
　　universal winding, 371
　proportional, 364
Speed switch, automotive, 385
Split-supply voltage follower, 67
Spring reverb, stereo, 245
Square-root extractor, 502, 505
Square-wave oscillator/generator, 149, 150, 153, 163, 166, 167
　100 kHz, 115, 161

crystal-controlled, 160
op-amp, 169
and triangle generator, 164
Square-wave siren, 532
Square-wave synchronizer, 18
Square-wave voltage reference, 62
Squaring amplifier, 87
Squaring circuit, low-cost, 514
Squelch
　with AGC, 327
　FM-radio noise-operated, 296
　FM scanner noise-operated, 292
Squelch preamplifier with hysteresis, 323
Squelched microphone preamplifier, 276
SSB
　intermediate-frequency (IF) strip, 329
　modulator scheme, 312
　product detector, 305, 321
　suppressed-carrier demodulator, 326
　suppressed-carrier modulator, 283
SSSC
　demodulator, 326
　modulator, 283
Staircase generator, 162, 163
Standard, 100 kHz frequency, 113
Standard cell replacement, 60
Statistical voltage standard, 61
Stepdown switching regulator, 55
Stereo
　ambience amplifier, rear-channel, 265
　amplifier
　　4 W, 234, 265
　　4 W for 8-track player, 232
　　8 W, 270
　　12 W, 8 ohms, 266
　　20 W, 237
　　20 W bass-boosted, 240
　　60 W, 261
　　with bass boost, 268
　　for headphones, 226
　　for phono, magnetic pickup, 254
　　for phono, with tone controls, 267
　　for phono/radio/tape, 274
　audio power booster, 240, 268
　decoder/demodulator, 309, 320
　　PLL-type, 300, 305
　mixer, audio, 235, 276
　panner, audio-positioning, 275
　preamplifier, 227
　　complete, 260
　　and control center, 278
　　full-performance, 269
　　for magnetic tape, 228, 270
　　and mixer, 235
　　for tuner, 278
　receiver, complete, 328
　reverb unit, 245, 271
　subcarrier regenerator, 320
　synthesis through reverberation, 271
Storecast decoder/demodulator, 315
Strobe-control
　multiplexer, 470
　oscillator, 153
　relay driver, 347, 358
Strobing
　circuits, 490
　for input and output, 476
　sense amplifier, 95
Subbroadcast-band receiver (AM), 321
Subcarrier regenerator for FM set, 320
Subcarrier-VCO interface, 479
Subtractor, high-speed, 493
Subtractor for test gear, 129
Successive approximation, 443
Summer
　250 lHz power bandwidth, 441
　3.5 MHz bandwidth, 441
　fast, 74

Summing amplifier, 74, 76, 89, 441
  dc, 441
  fast, 75
Sump-pump alarm, 353
Sump-pump on/off control, 350
Superbeta
  cascode amplifier, 90
  op-amp, 67, 75
  op-amp, with diode drive, 91
Supply, reference, 54
Suppressed-carrier modulator (DSB), 283, 318
Suppressed-carrier regenerator (DSB), 321
Suppression of spurious oscillation, 283
Switch
  analog, 447
    buffer, 463, 471
    differential-type, 468
    four-channel, 460
    high-rate, 464
    TTL/DTL, 463
    with zero-crossing detector, 475
  bounceless, 363
  level, line-operated, 354
  low-level power, 340
  optically-isolated, 426
  sound-activated, 355
  touch, 535
  zero-voltage, 27
Switch-back current limiter, 47
Switch buffer, analog, 471
Switch controller, 363
Switching power amplifier, 92, 354, 368
Switching regulator, 11, 55
  3 A, 14, 20
  5 V, 19, 20
  10 V (3A), 20
  15 V (500 mA), 21
  28 V, 19
  current, 18
  high-current, 22
  for SSB, 312
  stepdown, 55
  variable (5 to 200 V) 17
  zero-voltage, 27
Switching-regulator modulator, 312
Switching-regulator supply, 59
Switching supply, CATV (80 W), 222
Symmetrical amplifier, 102, 103
Synchronizer, square-wave, 18
Synchronous AM demodulator, 330
  2400 Hz, 302
Synchronous heat-staging control, 360
Synthesizer
  down counter for, 487
  frequency, using PLL, 324
  stereo-effect, 271

Table radio, see Radio
Tach, see Tachometer
Tach converter, 397
Tachometer
  for auto, car or boat, 384
  and dwell meter, 390
  frequency-doubling, 395
  gasoline-engine, 388
  simple, 393
  with zero-crossing detector, 446
Tape playback amplifier, 239, 257, 270
  cassette, 243
  stereo (phono/radio), 274
Tape-player power booster, 240
Tape preamplifier, NAB, 232, 241, 255, 272, 274, 275
  stereo, 228
  stereo (radio/phono), 278
Tape reader, magnetic, 471
Tape recorder voice control, 355, 538

Telemetry demodulator, 335
Telemetry interface, 479
Telephone tone-pad decoder, 527
Temperature-compensated, D/A converter, 436
Temperature-compensated D/A converter, 436
Temperature-compensated op-amp, 84
Temperature-compensated regulator, 26, 35, 41
Temperature control, 352
  with hysteresis, 341
  2-output, 361
  for oven, 348
  thermocouple, 345
Temperature controller, 348, 349, 358, 359, 370
  proportioning, 344
Temperature detection, 359
Temperature meter, analog, 116-118
Temperature probe, 371
Temperature sensor, 340, 359
  remote, 371
Temperature-to-current transducer, 362
Temperature-to-frequency converter, 386
Temperature-transducer reference, 117
Termination regulator, ECL, 512
Test equipment subtractor, 129
Tester
  coil, 146
  continuity, 146, 525
Tester, RIAA, inverse response generator, 249
Thermal probe for soldering iron, 341
Thermal sensor and comparator, 371
Thermocouple amplifier, 126, 133, 134, 137
Thermocouple temperature control, 345
Thermometer
  analog, 116-118
  automotive, 386, 396
  Celsius, 121, 348, 535
  Celsius and Fahrenheit, 122
  differential, 117
  ground-reffered, 120, 121, 347
  Kelvin, 357
  low-duty-cycle, 370
  with meter output, 346
Thermostat, 3-wire, 356
Three-band tone control, 402
Three-phase
  heater control, 366
  signal generator, 154
Three-wire thermostat, 356
Threshold detector
  and comparator combo, 490
  differential, 462
  with regulator output, 472
Thyristor firing circuit, 345
Time delay, 374, 378
  2-terminal, 381
Timer
  1 minute, 380
  1 hour, 379
  analog, 376
  chain, logic-triggered, 377
  presettable, 379
  reset one-shot, 382
  wide-range, 381
  various types, 374
Timing oscillator for A/D converter, 431
Tone generator, 296, 336
  sine-wave, 146
Tone compensation feedback, 270
Tone control, 269, 402, 410
  and amplifier, 254, 263
    phono, 251, 263, 267
  and balance control, 260
  bass, 268

  buffered, 249
  common-mode, 224
  passive, 261
Tone decoder, 527
Tone encoder, 296
Touch-tone decoder, 527
Touch switch, 535
Tracking regulator
  12 V (dual), 52, 53
  15 V, 12
  15 V (dual), 23
  40 V (dual), 7
  65 V, 1 A, 5
  precision, 23
  wide-range, 16
Transducer, temperature-to-current, 362
Transducer amplifier, 85, 110
Transformer, impedance, 94
Transformer-feedback oscillator, 153
Transformer-input mike preamp, 277
Transformerless mike preamp, 277
Transient-free power switch, 363
Translator, level, 460
Transmission and clutch control, automotive, 392
Transmission, data (TTL), 468
Transmission isolator, digital, 464, 473
Transmission-line driver, 469
Transmitter, 40 m QRP, 294
Transmitter final, 1 W, 536
Transmitter voice control, 355
Treble control, passive, 261
Tremolo, modulated-oscillator, 522
Tremolo circuit, 82
Tremolo voltage-controlled oscillator, 521
TRF receiver for less than 2 MHz, 321
Triac control, 343, 428
  intrusion alarm, 368
Triangular-wave generator, 160, 167
  and square-wave generator, 164
Trigger, Schmitt, 491, 508, 513
Tri-level comparator, 500
Trimmed-output thermometer, 116-118, 371
Tri-wave generator, 167
Trombone, electronic, 526
True instrumentation amplifier, 131
True-RMS detector, 135
TTL-compatible comparator, 491
TTL clock driver, 153
TTL data transmission, 468
TTL driver, 343, 348
TTL-driving
  divide-by-10 prescaler, 498
  magnetic-tape reader, 471
  magnetic-transducer detector, 115
  oscillator (1 MHz), 155
  paper-tape reader, 466
  zero-crossing detector, 113
TTL/DTL
  analog switch, 463
  sample and hold, 459
  voltage comparator, 490
  zero-crossing detector, 471
TTL interface, high-level logic, 514
TTL pulse-width detector, 469
TTL-to-MOS
  clock driver, 495
  interface, 498, 508
Tunable notch filter, 403, 404, 414
Tuned amplifier for VHF, See listings under VHF
Tuned-input/tuned-output amplifier, 323
Tuned radio-frequency, 321
Tuner
  AM, with amplifier, 320
  and tape preamplifier, stereo, 278
TV demodulator for Channel 13, 335

INDEX 555

TV sound amplifier (2W, 24 V) with IF, 254
Twin-tee network, 417
Two-phase
  fast-turnon tape amplifier, 241
  MOS clock driver, 492
  motor drive, 533
Two-shot
  frequency doubler, 459
  zero-crossing detector, 459

UHF crystal oscillator, 153
Ultralow-distortion amplifier, 81
Uncompensated booster follower, 69
Unity-gain
  amplifier, 70
  CMOS amplifier, 137
  inverting amplifier, 90
  pulse amplifier, 504
  voltage follower, 92
  see also Follower
Universal alarm system, 359
Universal antenna impedance adapter, 520
Universal motor speed control, 349

Varactor, 290, 297
Varactor FM generator, 313
Varactor modulator, 290
Varactor-tuned table radio, 297
Variable attenuator, 536
Varriable-capacitance multiplier, 516
Variable-delay one-shot, 164
Variable-duty-cycle pulse generator, 160
Variable-feedback crystal oscillator, 153
Variable flasher, 539
Variable-frequency oscillator, 146, 294
Variable-frequency pulse generator, 147
Variable gain control circuit, 95
Variable-gain diff-amp, 93
Variable-gain instrumentation amplifier, 141
Variable phase shifter, 226
Variable-rate siren, 525
Variable-speed control, induction motor, 369
Variable 20 V, reference, 63
Variable regulator, 50
  50 mA, 56
  6 A, 22
  5 V, 36
  5-35 V (1 A), 55
  7 V, 34
  7-23 V, 36
  7-37 V, 36
  12 V, negative, 24
  18 V, 24, 25
  25 V, 25
  37 V, 34
  50 V, 48
Variable-voltage power supply
  3.5 to 20 V, 10
  25 V, 61
  bench-type, 9
  regulated, 4
VCA, 97, 101
  tremolo, 82
  see also Amplifier, voltage-controlled
VCO, 148, 152, 161, 322, 431, 456
  multivibrator, 154
  for tremolo effect, 521
Velocity-of-air indicator, 396
Vertical-input scope amplifier, 116
VFO, 146, 294
Video amplifier, 20 dB gain, 222
VHF crystal oscillator, 153
  50-100 MHz, 156
VHF doubler, 150 to 300 MHz, 313
VHF tuned amplifier, various ranges, 318
Voice-operated control, See VOX

Voice-controlled switch, 355, 538
Voice processor, 314
Voltage boost, NPN, 513
Voltage comparator, 499
  ac-coupled, 511
  differential 461
  DTL/RTL, 490
  go/no-go, 511
  with hysteresis, 462
  low-current, 507
  MOS & TTL, 491
  precision, 510
  and relay driver, 516
  RTL, 490
  and RTL driver, 456
  tri-level, 500
  TTL, 491
Voltage-comparison sample and hold, 452
Voltage control
  amplifier, see Amplifier, voltage-controlled; VCA
  AM radio, 297
  capacitor, see VVC
  current sink, 76
  current source, 95
  microphone preamplifier, 276
  multivibrator, 154
  oscillator, see VCO
  VFO, 154
Voltage follower, 92
  1 A, 66
  compensated, 93
  current-boosted, 72
  fast, 76
  guarded, 77
  high-current, 79
  in D/A converter, 434
  single-supply version, 94
  split-supply version, 67
Voltage generator, error (A/D), 431
Voltage monitor, battery, 350
Voltage-rated circuits, see Voltage-valued circuits
Voltage reference
  6.9 V, 63
  negative and positive versions, 57
  square-wave, 62
Voltage regulator, see Regulators, rated by voltage
Voltage-regulator output sync, 18
Voltage source, 2.15 V, 51
Voltage standard, statistical, 61
Voltage-valued circuits
  10 mV to 100 V analog meter, 108
  15 mV ac voltmeter, 138
  500 mV full-scale dc voltmeter, 139
  500 mV output tape amplifier, 239
  1.5 V 20 dB amplifier, 69
  1.5 V flasher, variable-rate, 539-541
  1.5 V time delay, 10 seconds, 375
  2 V voltage regulator, 31, 34
  ECL termination, 512
  2 V source, 51
  2.15 V source, 51
  3 V regulator, 18, 39
  5 V regulated supply, 5, 15
  5 V regulator, 40, 50
    200 mA, 43
    500 mA, 41, 47
    1 A, 45
    3 A switching, 20
    10 A, 43
    current-limited, 21
    crowbar, 19
    high-current switching, 22
    NPN pass, 34
    remote-shutdown, 29
    shunt, 28, 32

    variable, 36, 50
  6 V bridge amplifier (audio, 2.5 W), 238
  6 V crystal oscillator/calibrator, 293
  6 V flasher, 530
  6 V IF amplifier for 455 kHz, 285
  6 V regulator and crowbar, 56
  6 V timers, 375
  7 V regulator, 34, 36
  9.9 V reference, 63
  9 V audio mixer, 242
    and preamplifier, 276
  9 V TRF receiver, simple, 321
  10 V calibrator, portable, 134
  10 V negative regulator, 55
  10 V op-amp voltage regulator, 49
  10 V stable regulator, 24
  10 V switching regulator, 20
  12 V flasher for cars, 395
  12 V phono preamp, 46 dB, 256
  12 V regulated power supply, 7
  12 V regulator, 24, 51-53
  12 V stereo headphone amplifier, 226
  13 V variable-voltage regulator, 30
  14 V reference supply, 8, 51
  14 V regulator/charger 392
  15 V current-limiting sources, 15
  15 V output adjustment, 59
  15 V preamplifiers for microphones, 277
  15 V regulated supply, dual, 6
  15 V regulator
    10 A, 33
    dual, 6, 39, 40
    dual tracking, 23
    negative, 34, 37
    positive, 34, 37
    series, 35
    slow-turnon, 27
    stable, 38
    tracking, 12
    variable, 36
  15 V timer, 375
  18 V regulator, variable, 24, 25
  18 V phono amplifier, 34 dB, 258
  20 V power supply, variable, 10
    regulated, 60
  20 V reference
    1 A, 63
    power, 63
  22 V audio amplifier
    5 W, 224
    20 W stereo, 237
  23 V variable regulator, 36
  24 V mike preamplifier, 277
  24 V TV sound amplifier with IF, 254
  25 V power supply
    fixed, 3
    variable, 61
  25 V regulator, variable, 25
  28 V audio mixer, 242
  28 V power supply, regulated, 10
  28 V regulator
    switching, 19
    variable (6 A), 22
  30 V amplifier
    audio power, 264
    magnetic-pickup phono, 254
  30 V regulator
    negative, 53
    variable (50 mA), 56
  34 V common-mode instrumentation amplifier, 110
  37 V variable regulator, 34
  38 V power amplifier (1 A), 99
  40 V audio mixer, 242
  40 V dual tracking power supply, 7
  41 V dual power supply, 7
  50 V regulator, 25, 26
    variable, 38

60 V regulator, floating, 32
65 V regulated power supply (1 A), 5
100 V common-mode-range instrumentation amplifier, 140
105 V regulator, negative, 54
200 V switching regulator, 17
250 V variable regulator, 100 mA, 37
300 V regulator, 500 mA, 33
2 kV voltage regulator, 40
Voltmeter, ac
  analog, 115
  and dc, 142
  expanded scale, 140
  wide-range analog, 133, 138
Voltmeter, analog, 123, 135
  10 mV to 100 V, 108
  ac, 115
  ac (wide-range), 133
  draws 20 mW, 135
  with voltage-limiting, 123
Voltmeter, digital, 119
  3.5-digit, 107, 114
  4.5-digit, 109, 143
  ac/dc (0 to 20 V), 106
  high-impedance dc, 139
  with LCD readout, 106
  multirange, 114
  with overrange blinking, 107
  wideband ac, 138
Volume control, common-mode, 224
Volume units, 269
VOX, 538
  for tape recorder, 355
VU meter hookup, 269
VVC-tuned table radio, 297
VVC modulator, 52 MHz, 290

Wailing siren, 524, 534
Water, see listings under Fluid and Liquid
Water seepage alarm, 342
Wattage-rated circuits, see Power-valued circuits
Waveform generators
  pulse, 165, 166
  sine, 149, 150, 154, 155, 158
  square, 149, 150, 153, 163, 164, 166, 169
  staircase, 162, 163
  triangle, 160
  triangle and square, 164
  triangle, square, and sine, 167
Waveforms, A/D converter, 433
Weather-satellite picture demodulator, 319
Whooper siren, 534
Wideband ac voltmeter, 138
Wideband amplifier, see Amplifier, wideband
Wideband differential multiplexer, 457
Wideband preamplifier
  and audio mixer, 276
  low-noise, 139
  magnetic-phono, 279
Wideband servo driver, 264
Wide-CMRR instrumentation amplifier, 138
Wide-common-mode-range amplifier, 101
Wide-range
  ac voltmeter, analog, 133
  differential thermometer, 117
  instrumentation amplifier, 138
  phaser meter, 531
  timer, 381
  tracking regulator, 16
  VCO, 148

Width detector, pulse, 469
Width-of-pulse,
  discriminator, 473
  modulator, 478
Wien bridge
  amplitude-stabilized, 113
  power, 169
Wien-bridge oscillator, 147, 149, 150, 152, 158, 168
Wind indicator, 396
Window detector, 364, 458
Window repeater discriminator, 316
Wireless FM transmitter and receiver, 322
Write and read circuits, 481

Xtal, see listings under Crystal

Zero-crossing
  detector, 113, 446, 448, 466, 471, 477
    magnetic, 474
    MOS, 471, 475
    two-shot, 459
  temperature controller, 344
  see also ZVS
Zeroing amplifier, 131
Zero voltage detector (small-signal), 457
Zero-voltage switching, see ZVS
ZVS, 27, 345
  lamp dimmer, 352
  on/off control, 367
    power, 353
  regulator, 27
  switch controller, 27
  transient-free control, 366